Fundamental Equations of Dynamics

KINEMATICS

Particle Rectilinear Motion

Variable a

$$a = \frac{dv}{dt}$$

$$v = \frac{ds}{dt}$$

$$v \, dv = a \, ds$$

Constant $a = a_c$

$$v = v_0 + a_c t$$

$$s = s_0 + v_0 t + \tfrac{1}{2} a_c t^2$$

$$v^2 = v_0^2 + 2a_c(s - s_0)$$

Particle Curvilinear Motion

x, y, z Coordinates

$$v_x = \dot{x} \qquad a_x = \ddot{x}$$
$$v_y = \dot{y} \qquad a_y = \ddot{y}$$
$$v_z = \dot{z} \qquad a_z = \ddot{z}$$

r, θ, z Coordinates

$$v_r = \dot{r} \qquad a_r = \ddot{r} - r\dot{\theta}^2$$
$$v_\theta = r\dot{\theta} \qquad a_\theta = r\ddot{\theta} + 2\dot{r}\dot{\theta}$$
$$v_z = \dot{z} \qquad a_z = \ddot{z}$$

n, t, Coordinates

$$v = \dot{s} \qquad a_t = \dot{v} = v\frac{dv}{ds}$$

$$a_n = \frac{v^2}{\rho} \qquad \rho = \left| \frac{[1 + (dy/dx)^2]^{3/2}}{d^2y/dx^2} \right|$$

Relative Motion

$$\mathbf{v}_B = \mathbf{v}_A + \mathbf{v}_{B/A} \qquad \mathbf{a}_B = \mathbf{a}_A + \mathbf{a}_{B/A}$$

Rigid Body Motion About a Fixed Axis

Variable α

$$\alpha = \frac{d\omega}{dt}$$

$$\omega = \frac{d\theta}{dt}$$

$$\omega \, d\omega = \alpha \, d\theta$$

Constant $\alpha = \alpha_c$

$$\omega = \omega_0 + \alpha_c t$$

$$\theta = \theta_0 + \omega_0 t + \tfrac{1}{2}\alpha_c t^2$$

$$\omega^2 = \omega_0^2 + 2\alpha_c(\theta - \theta_0)$$

For Point P

$$s = \theta r \qquad v = \omega r \qquad a_t = \alpha r \qquad a_n = \omega^2 r$$

Relative General Plane Motion-Translating Axes

$$\mathbf{v}_B = \mathbf{v}_A + \mathbf{v}_{B/A(\text{pin})} \qquad \mathbf{a}_B = \mathbf{a}_A + \mathbf{a}_{B/A(\text{pin})}$$

Relative General Plane Motion-Trans. and Rot. Axis

$$\mathbf{v}_B = \mathbf{v}_A + \boldsymbol{\Omega} \times \mathbf{r}_{B/A} + (\mathbf{v}_{B/A})_{\text{rel}}$$

$$\mathbf{a}_B = \mathbf{a}_A + \dot{\boldsymbol{\Omega}} \times \mathbf{r}_{B/A} + \boldsymbol{\Omega} \times (\boldsymbol{\Omega} \times \mathbf{r}_{B/A}) + 2\boldsymbol{\Omega} \times (\mathbf{v}_{B/A})_{\text{rel}} + (\mathbf{a}_{B/A})_{\text{rel}}$$

KINETICS

Mass Moment of Inertia $\quad I = \int r^2 \, dm$

Parallel-Axis Theorem $\quad I = I_G + md^2$

Radius of Gyration $\quad k = \sqrt{\dfrac{I}{m}}$

Equations of Motion

Particle	$\Sigma \mathbf{F} = m\mathbf{a}$
Rigid Body	$\Sigma F_x = m(a_G)_x$
(Plane Motion)	$\Sigma F_y = m(a_G)_y$
	$\Sigma M_G = I_G\alpha \, / \, \Sigma M_P = \Sigma(M_k)_P$

Principle of Work and Energy

$$T_1 + U_{1-2} = T_2$$

Kinetic Energy

Particle	$T = \tfrac{1}{2}mv^2$
Rigid Body (Plane Motion)	$T = \tfrac{1}{2}mv_G^2 + \tfrac{1}{2}I_G\omega^2$

Work

Variable force	$U_F = \int F \cos\theta \, ds$
Constant force	$U_{F_c} = (F_c \cos\theta)\,\Delta s$
Weight	$U_W = -W\,\Delta y$
Spring	$U_s = -(\tfrac{1}{2}ks_2^2 - \tfrac{1}{2}ks_1^2)$
Couple moment	$U_M = M\,\Delta\theta$

Power and Efficiency

$$P = \frac{dU}{dt} = \mathbf{F} \cdot \mathbf{v} \qquad \varepsilon = \frac{P_{\text{out}}}{P_{\text{in}}} = \frac{U_{\text{out}}}{U_{\text{in}}}$$

Conservation of Energy Theorem

$$T_1 + V_1 = T_2 + V_2$$

Potential Energy

$$V = V_g + V_e, \text{ where } V_g = \pm Wy, \; V_e = +\tfrac{1}{2}ks^2$$

Principle of Linear Impulse and Momentum

Particle	$m\mathbf{v}_1 + \Sigma \int \mathbf{F}\, dt = m\mathbf{v}_2$
Rigid Body	$m(\mathbf{v}_G)_1 + \Sigma \int \mathbf{F}\, dt = m(\mathbf{v}_G)_2$

Conservation of Linear Momentum

$$\Sigma(\text{syst. } m\mathbf{v})_1 = \Sigma(\text{syst. } m\mathbf{v})_2$$

Coefficient of Restitution $\quad e = \dfrac{(v_B)_2 - (v_A)_2}{(v_A)_1 - (v_B)_1}$

Principle of Angular Impulse and Momentum

Particle	$(\mathbf{H}_O)_1 + \Sigma \int \mathbf{M}_O \, dt = (\mathbf{H}_O)_2,$ where $H_O = (d)(mv)$
Rigid Body (Plane Motion)	$(\mathbf{H}_G)_1 + \Sigma \int \mathbf{M}_G \, dt = (\mathbf{H}_G)_2,$ where $H_G = I_G\omega$
	$(\mathbf{H}_O)_1 + \Sigma \int \mathbf{M}_O \, dt = (\mathbf{H}_O)_2,$ where $H_O = I_G\omega + (d)(mv_G)$

Conservation of Angular Momentum

$$\Sigma(\text{syst. } \mathbf{H})_1 = \Sigma(\text{syst. } \mathbf{H})_2$$

R. C. HIBBELER

FIFTH EDITION

Engineering Mechanics

STATICS AND DYNAMICS

MACMILLAN PUBLISHING COMPANY, INC.
NEW YORK
Collier Macmillan Publishers
LONDON

PRINTED IN THE UNITED STATES OF AMERICA

The two parts of this book are also published separately as *Engineering Mechanics:
Statics,* and *Engineering Mechanics: Dynamics.* Copyright © 1974, 1978, 1983, and
1986 by R. C. Hibbeler.

Macmillan Publishing Company
866 Third Avenue, New York, New York 10022

Collier Macmillan Canada, Inc.

LIBRARY OF CONGRESS CATALOGING IN PUBLICATION DATA

Hibbeler, R. C.
    Engineering mechanics—statics and mechanics / R.C. Hibbeler. —
5th ed.
        p.   cm.
    "The two parts of this book are also published separately as
Engineering mechanics—statics and Engineering mechanics—dynamics"—
T.p. verso.
    Includes index.
    ISBN 0-02-354461-9
    1. Mechanics, Applied.   I. Title.
TA350.H48 1989
620.1—dc19   88-22107
                                                                     CIP

PRINTING: 3 4 5 6 7 8   YEAR: 0 1 2 3 4 5 6 7 8

Preface

The purpose of this book is to provide the student with a clear and thorough presentation of the theory and applications of engineering mechanics. To achieve this objective the author has by no means worked alone on this task, for to a large extent this book has been shaped by the comments and suggestions of more than a hundred reviewers in the teaching profession and many of the author's students who have used its previous editions.

Several changes have been made in preparing this new edition. Overall, particular attention has been given to developing accurate definitions and explanations of the concepts. A greater emphasis has been placed on drawing free-body diagrams by adding specific sections and examples throughout the text on this topic. Also, the importance of selecting an appropriate coordinate system and associated sign convention for vector components is stressed when the equations of kinematics, the equations of equilibrium or the equations of motion are applied. Some rewriting and rearrangement of material has occurred. In Statics, the reduction of a simple distributed loading is covered at the end of Chapter 4, and the last chapters of this volume have been renumbered. In Dynamics, relative motion analysis using Cartesian vectors and scalar components has been combined into a single section. Lastly, most of the problems in this edition are new, and there are more of them than in the previous edition. In this regard, extra care has been taken in the presentation and solution of the problems, and all the problem sets have been reviewed and the solutions checked by the author's colleagues to ensure both their clarity and numerical accuracy.

Organization and Approach. The contents of each chapter are organized into well-defined sections. Selected groups of sections contain an explanation of specific topics, illustrative example problems, and a set of homework problems. The topics within each section are placed into subgroups defined by boldface titles. The purpose of this is to present a structured method for introducing each new definition or concept, and to make the book convenient for later reference and review.

A "procedure for analysis" is given at the end of many sections of the book in order to provide the student with a review or summary of the material and a logical and orderly method to follow when applying the theory. As in the previous editions, the example problems are solved using this outlined method in order to clarify its numerical application. It is to be understood, however, that once the relevant principles have been mastered and enough confidence and judgment have been obtained, the student can then develop his or her own procedures for solving problems. In most cases, it is felt that the first step in any procedure should require drawing a diagram. By doing so, the student forms the habit of tabulating the necessary data while focusing on the physical aspects of the problem and its associated geometry. If this step is correctly performed, applying the relevant equations of mechanics becomes somewhat methodical, since the data can be taken directly from the diagram. This step is particularly important when solving problems involving equilibrium or kinetics, and for this reason, drawing a free-body diagram is strongly emphasized throughout the book.

Since mathematics provides a systematic means of applying the principles of mechanics, the student is expected to have prior knowledge of algebra, geometry, trigonometry, and, for complete coverage, some calculus. Vector analysis is introduced at points where it is most applicable. Its use often provides a convenient means for presenting concise derivations of the theory, and it makes possible a simple and systematic solution of many complicated three-dimensional problems. Occasionally, the example problems are solved using more than one method of analysis so that the student develops the ability to use mathematics as a tool whereby the solution of any problem may be carried out in the most direct and effective manner.

Problems. Numerous problems in the book depict realistic situations encountered in engineering practice. It is hoped that this realism will both stimulate the student's interest in engineering mechanics and provide a means for developing the skill to reduce any such problem from its physical description to a model or symbolic representation to which the principles of mechanics may be applied. As in the previous edition, an effort has been made to include some problems which may be solved using a numerical procedure executed on either a desk-top computer or a programmable pocket calculator. Suitable numerical techniques along with associated computer programs are given in Appendix B. The intent here is to broaden the student's capacity for using other forms of mathematical analysis *without* sacrificing the time needed to focus on the application of the principles of mechanics. Problems of this type which either can or must be solved using numerical procedures are identified by a "square" symbol (■) preceding the problem number.

Throughout the text there is an approximate balance of problems using either SI or FPS units. Furthermore, in any set, an attempt has been made to arrange the problems in order of increasing difficulty.* The answers to all but every fourth problem are listed in the back of the book. To alert the user to a problem without a reported answer, an asterisk (*) is placed before the problem number.

Contents: **Statics.** The subject of statics is presented in 11 chapters, in which the principles introduced are first applied to simple situations. Most often, each principle is applied first to a particle, then to a rigid body subjected to a coplanar system of forces, and finally to the general case of three-dimensional force systems acting on a rigid body.

The text begins in Chapter 1 with an introduction to mechanics and a discussion of units. The notion of a vector and the properties of a concurrent force system are introduced in Chapter 2. This theory is then applied to the equilibrium of particles in Chapter 3. Chapter 4 contains a general discussion of both concentrated and distributed force systems and the methods used to simplify them. The principles of rigid-body equilibrium are developed in Chapter 5 and then applied to specific problems involving the equilibrium of trusses, frames, and machines in Chapter 6, and to the analysis of internal forces in beams and cables in Chapter 7. Applications to problems involving frictional forces are discussed in Chapter 8, and topics related to the center of gravity and centroid are treated in Chapter 9. If time permits, sections concerning more advanced topics, indicated by stars, (★) may be covered. Most of these topics are included in Chapter 10 (area and mass moments of inertia) and Chapter 11 (virtual work and potential energy). Note that this material also provides a suitable reference for basic principles when it is discussed in more advanced courses.

At the discretion of the instructor, some of the material may be presented in a different sequence with no loss in continuity. For example, it is possible to introduce the concept of a force and all the necessary methods of vector analysis by first covering Chapter 2 and Sec. 4.1. Then, after covering the rest of Chapter 4 (force and moment systems), the equilibrium methods in Chapters 3 and 5 can be discussed.

Contents: **Dynamics.** The subject of dynamics is presented in the last 11 chapters. In particular, the kinematics of a particle is discussed in Chapter 12, followed by a discussion of particle kinetics in Chapter 13 (equation of motion), Chapter 14 (work and energy), and Chapter 15 (impulse and momentum). The concepts of particle dynamics contained in these four chapters are then summarized in a "review" section and the student is given the chance to identify and solve a variety of different types of problems. A similar sequence

---

*Review problems, wherever they appear, are presented in random order.

of presentation is given for the planar motion of a rigid body: Chapter 16 (planar kinematics), Chapter 17 (equations of motion), Chapter 18 (work and energy), and Chapter 19 (impulse and momentum), followed by a summary and review set of problems for these chapters. If desired, it is possible to cover Chapters 12 through 19 in the following order with no loss in continuity: Chapters 12 and 16 (kinematics), Chapters 13 and 17 (equations of motion), Chapters 14 and 18 (work and energy), and Chapters 15 and 19 (impulse and momentum).

Time permitting, some of the material involving three-dimensional rigid-body motion may be included in the course. The kinematics and kinetics of this motion are discussed in Chapters 20 and 21, respectively. Chapter 22 (vibrations) may be included if the student has the necessary mathematical background. Sections of the book which are considered to be beyond the scope of the basic dynamics course are indicated by a star and may be omitted. As in *Statics,* however, this more advanced material provides a suitable reference for basic principles when it is covered in other courses.

Acknowledgments. I have endeavored to write this book so that it will appeal to both the student and instructor. Through the years many people have helped in its development and I should like to acknowledge their valued suggestions and comments. Specifically, I wish to personally thank the following individuals who have contributed to this edition, namely, Professors Robert D. Celmer, University of Hartford; Major William Conner, U.S. Military Academy, West Point; Mitsunori Denda, Rutgers University; Robert W. Fuessle, Bradley University; John E. Griffith, University of Southern Florida; William Q. Gurley, University of Tennessee, Chattanooga; Edward E. Hornsey, University of Missouri, Rolla; Jim Kauzlarich, University of Virginia; Mohammad Khosrowjerdi, Western New England College; William M. Lee, U.S. Naval Academy, Annapolis; Captain Al Lewis, U.S. Military Academy, West Point; Will Liddell, Jr., Auburn University at Montgomery; John C. McWhorter, Mississippi State University; Nasser Moshrefi, Michigan Technical University; John Peddieson, Tennessee Technical University; Robert Schmidt, University of Detroit; Robert Snell, Kansas State University; Henry L. Sundberg, Jr., Western New England College; Liang-Ning Tao, Illinois Institute of Technology; Theodore Tauchert, University of Kentucky; Donald M. Wallace, Norwich University; Han-Chin Wu, University of Iowa.

Many thanks are also extended to all my students and to those in the teaching profession who have freely taken the time to send me their suggestions and comments. Since the list is too long to mention, it is hoped that those who have given help in this manner will accept this anonymous recognition. Furthermore, I greatly appreciate the personal support and attention given to me by my editors and the staff at Macmillan. Lastly, I should like to acknowledge the encouragement and assistance of my wife, Conny, during the time it has taken to prepare the manuscript for publication.

Russell Charles Hibbeler

Contents

Contents

DYNAMICS

Contents

1

General Principles

1.1 Mechanics

Mechanics can be defined as that branch of the physical sciences concerned with the state of rest or motion of bodies that are subjected to the action of forces. A thorough understanding of this subject is required for the study of structural engineering, machine design, fluid flow, electrical devices, and even the molecular and atomic behavior of elements.

In general, mechanics is subdivided into three branches: *rigid-body mechanics, deformable-body mechanics,* and *fluid mechanics.* This book treats only rigid-body mechanics, since this subject forms a suitable basis for the design and analysis of many engineering problems, and it provides part of the necessary background for the study of the mechanics of deformable bodies and the mechanics of fluids.

Rigid-body mechanics is divided into two areas: statics and dynamics. *Statics* deals with the equilibrium of bodies, that is, those that are either at rest or move with a constant velocity; whereas *dynamics* is concerned with the accelerated motion of bodies. Although statics can be considered as a special case of dynamics, in which the acceleration is zero, statics deserves separate treatment in engineering education, since most structures or frameworks are designed with the intention that they remain in equilibrium.

Historical Development

The subject of statics developed very early in history, because the principles involved could be formulated simply from measurements of geometry and force. For example, the writings of Archimedes (287–212 B.C.) deal with the principle of the lever. Studies of the pulley, inclined plane, and wrench are also recorded in ancient writings—at times when the requirements of engineering were limited primarily to building construction.

Since the principles of dynamics depend upon an accurate measurement of time, this subject developed much later. Galileo Galilei (1564–1642) was one of the first major contributors to this field. His work consisted of experiments using pendulums and falling bodies. The most significant contributions in dynamics, however, were made by Isaac Newton (1642–1727), who is noted for his formulation of the three fundamental laws of motion and the law of universal gravitational attraction. Shortly after these laws were postulated, important techniques for their application were developed by Euler, D'Alembert, Lagrange, and others.

1.2 Fundamental Concepts

Before beginning our study of rigid-body mechanics, it is important to understand the meaning of certain fundamental concepts and principles.

Basic Quantities

The following four quantities are used throughout rigid-body mechanics.

Length. Length is needed to locate the position of a point in space and thereby describe the size of a physical system. Once a standard unit of length is defined, one can then quantitatively define distances and geometrical properties of a body as multiples of the unit length.

Time. Time is conceived as a succession of events. Although the principles of statics are time independent, this quantity does play an important role in the study of dynamics.

Mass. Mass is a property of matter by which we can compare the action of one body with that of another. This property manifests itself as a gravitational attraction between two bodies and provides a quantitative measure of the resistance of matter to a change in velocity.

Force. In general, force is considered as a ''push'' or ''pull'' exerted by one body on another. This interaction can occur when there is either direct contact between the bodies, such as a person pushing on a wall, or it can occur through a distance when the bodies are physically separated. Examples of the latter type include gravitational, electrical, and magnetic forces. In any case, a force is completely characterized by its magnitude, direction, and point of application.

Idealizations

In mechanics models or idealizations are used in order to simplify application of the theory. A few of the more important idealizations will now be defined. Others that are of importance will be discussed at points where they are needed.

Particle. A *particle* has a mass but a size that can be neglected. For example, the size of the earth is insignificant compared to the size of its orbit, and therefore the earth can be modeled as a particle when studying its orbital motion. When a body is idealized as a particle, the principles of mechanics reduce to a rather simplified form since the geometry of the body will not be involved in the analysis of the problem.

Rigid Body. A *rigid body* can be considered as a combination of a large number of particles in which all the particles remain at a fixed distance from one another both before and after applying a load. As a result, the material properties of any body that is assumed to be rigid will not have to be considered when analyzing the forces acting on the body. In most cases the actual deformations occurring in structures, machines, mechanisms, and the like are relatively small, and the rigid-body assumption is suitable for analysis.

Concentrated Force. A *concentrated force* represents the effect of a loading which is assumed to act at a *point* on a body. We can represent the effect of the loading by a concentrated force, provided the area over which the load is applied is *small* compared to the overall size of the body.

Newton's Three Laws of Motion

The entire subject of rigid-body mechanics is formulated on the basis of Newton's three laws of motion. These laws, which apply to the motion of a particle as measured from a nonaccelerating reference frame, may be briefly stated as follows:

First Law. A particle originally at rest, or moving in a straight line with constant velocity, will remain in this state provided the particle is *not* subjected to an unbalanced force.

Second Law. A particle acted upon by an *unbalanced force* **F** experiences an acceleration **a** that has the same direction as the force and a magnitude that is directly proportional to the force.* If **F** is applied to a particle of mass m, this law may be expressed mathematically as

$$\mathbf{F} = m\mathbf{a} \qquad (1-1)$$

Third Law. The mutual forces of action and reaction between two particles are equal, opposite, and collinear.

---

*Stated another way, the unbalanced force acting on the particle is proportional to the time rate of change of the particle's linear momentum.

Newton's Law of Gravitational Attraction

Shortly after formulating his three laws of motion, Newton postulated a law governing the gravitational attraction between any two particles. This law can be expressed mathematically as

$$F = G\frac{m_1 m_2}{r^2} \qquad (1\text{--}2)$$

where F = force of gravitation between the two particles
G = universal constant of gravitation; according to experimental evidence, $G = 66.73(10^{-12})$ m³/(kg · s²)
m_1, m_2 = mass of each of the two particles
r = distance between the two particles

Weight

Any two particles or bodies have a mutual attractive (gravitational) force acting between them. In the case of a particle located at or near the surface of the earth, however, the only gravitational force having any sizable magnitude is that between the earth and the particle. Consequently, this force, termed the *weight*, will be the only gravitational force considered in our study of mechanics.

From Eq. 1–2, we can develop an approximate expression for finding the weight W of a particle having a mass $m_1 = m$. If we assume the earth to be a nonrotating sphere of constant density and having a mass m_2, then if r is the distance between the earth's center and the particle, we have

$$W = G\frac{mm_2}{r^2}$$

Letting $g = Gm_2/r^2$ yields

$$\boxed{W = mg} \qquad (1\text{--}3)$$

By comparison with Eq. 1–1, we term g the acceleration due to gravity. Since it depends upon r, it can be seen that the weight of a body is *not* an absolute quantity. Instead, its magnitude depends upon where the measurement was made. For most engineering calculations, however, g is determined at sea level and at a latitude of 45°, which is considered the "standard location."

1.3 Units of Measurement

The four quantities—length, time, mass, and force—are not all independent from one another; in fact, they are *related* by Newton's second law of motion, $\mathbf{F} = m\mathbf{a}$. Hence, the *units* used to define force, mass, length, and time cannot *all* be selected arbitrarily. The equality $\mathbf{F} = m\mathbf{a}$ is maintained only if three of the four units, called *base units*, are *arbitrarily defined* and the fourth unit is *derived* from the equation.

SI Units

The International System of units, abbreviated SI after the French "Système International d'Unités," is a modern version of the metric system which has received worldwide recognition. As shown in Table 1–1, the SI system specifies length in meters (m), time in seconds (s), and mass in kilograms (kg). The unit of force, called a newton (N), is *derived* from $\mathbf{F} = m\mathbf{a}$. Thus, 1 newton is equal to a force required to give 1 kilogram of mass an acceleration of 1 m/s^2 ($N = kg \cdot m/s^2$).

If the weight of a body located at the "standard location" is to be determined in newtons, then Eq. 1–3 must be applied. Here $g = 9.806\ 65$ m/s^2; however, for calculations, the value $g = 9.81$ m/s^2 will be used. Thus,

$$W = mg \qquad (g = 9.81 \text{ m/s}^2) \qquad (1\text{--}4)$$

Therefore, a body of mass 1 kg has a weight of 9.81 N, a 2-kg body weighs 19.62 N, and so on.

U.S. Customary

In the U.S. Customary system of units (FPS) length is measured in feet (ft), time in seconds (s), and force in pounds (lb), Table 1–1. The unit of mass, called a *slug,* is *derived* from $\mathbf{F} = m\mathbf{a}.$ Hence, 1 slug is equal to the amount of matter accelerated at 1 ft/s^2 when acted upon by a force of 1 lb (slug = $lb \cdot s^2/ft$).

In order to determine the mass of a body having a weight measured in pounds, we must apply Eq. 1–3. If the measurements are made at the "standard location," then $g = 32.2$ ft/s^2 will be used for calculations. Therefore,

$$m = \frac{W}{g} \qquad (g = 32.2 \text{ ft/s}^2) \qquad (1\text{--}5)$$

so that a body weighing 32.2 lb has a mass of 1 slug, a 64.4-lb body has a mass of 2 slugs, and so on.

Table 1–1 **Systems of Units**

Name	Length	Time	Mass	Force
International System of Units (SI)	meter (m)	second (s)	kilogram (kg)	newton* (N) $\left(\dfrac{kg \cdot m}{s^2}\right)$
U.S. Customary (FPS)	foot (ft)	second (s)	slug* $\left(\dfrac{lb \cdot s^2}{ft}\right)$	pound (lb)

*Derived unit.

1.4 The International System of Units

The SI system of units is used extensively in this book since it is intended to become the worldwide standard for measurement. Consequently, the rules for its use and some of its terminology relevant to mechanics will now be presented.

Prefixes

When a numerical quantity is either very large or very small the units used to define its size may be modified by using a prefix. Some of the prefixes used in the SI system are shown in Table 1–2. Each represents a multiple or submultiple of a unit which, if applied successively, moves the decimal point of a numerical quantity to every third place.* For example, 4 000 000 N = 4 000 kN (kilo-newton) = 4 MN (mega-newton), or 0.005 m = 5 mm (milli-meter). Notice that the SI system does not include the multiple deca (10) or the submultiple centi (0.01), which form part of the old metric system. Except for some volume and area measurements, the use of these prefixes is to be avoided in science and engineering.

Table 1–2 **Prefixes**

	Exponential Form	Prefix	SI Symbol
Multiple			
1 000 000 000	10^9	giga	G
1 000 000	10^6	mega	M
1 000	10^3	kilo	k
Submultiple			
0.001	10^{-3}	milli	m
0.000 001	10^{-6}	micro	μ
0.000 000 001	10^{-9}	nano	n

Rules for Use

The following rules are given for the proper use of the various SI symbols:

1. A symbol is *never* written with a plural "s," since it may be confused with the unit for second (s).
2. Symbols are always written in lowercase letters, with two exceptions: symbols for the two largest prefixes shown in Table 1–2, giga and mega, are capitalized as G and M, respectively; and symbols named after an individual are capitalized, e.g., N.

*The kilogram is the only base unit that is defined with a prefix.

3. Quantities defined by several units which are multiples of one another are separated by a *dot* to avoid confusion with prefix notation, as illustrated by $N = kg \cdot m/s^2 = kg \cdot m \cdot s^{-2}$. Also, $m \cdot s$ (meter-second), whereas ms (milli-second).

4. The exponential power represented for a unit having a prefix refers to both the unit *and* its prefix. For example, $\mu N^2 = (\mu N)^2 = \mu N \cdot \mu N$, *not* $\mu(N \cdot N)$. Likewise, $1 \ mm^2$ represents $1 \ (mm)^2$, *not* $1m(m^2)$.

5. Physical constants or numbers having several digits on either side of the decimal point should be reported with a *space* between every three digits rather than with a comma; e.g., 73 569.213 427. In the case of four digits on either side of the decimal, the spacing is optional; e.g., 8537 or 8 537. Furthermore, always try to use decimals and avoid fractions; that is, write 15.25, *not* $15\frac{1}{4}$.

6. When performing calculations, represent the numbers in terms of their *base or derived units* by converting all prefixes to powers of 10. The final result should then be expressed using a *single prefix*. Also, after calculation, it is best to keep numerical values between 0.1 and 1000; otherwise, a suitable prefix should be chosen. For example,

$$(50 \ kN)(60 \ nm) = [50(10^3) \ N][60(10^{-9}) \ m]$$
$$= 3000(10^{-6}) \ N \cdot m = 3 \ mN \cdot m$$

7. Compound prefixes should not be used; e.g., $k\mu s$ (kilo-micro-second) should be expressed as ms (milli-second) since $1 \ k\mu s = 1(10^3)(10^{-6}) \ s = 1(10^{-3}) \ s = 1 \ ms$.

8. With the exception of the base unit the kilogram, in general avoid the use of a prefix in the denominator of composite units. For example, do not write N/mm, but rather kN/m.

9. Although not expressed in multiples of 10, the minute, hour, etc., are retained for practical purposes as multiples of the second. Furthermore, plane angular measurement is made using radians (rad). In this book, degrees will often be used, where $180° = \pi$ rad.

1.5 Numerical Calculations

Numerical work in engineering practice is most often performed by using electronic calculators and computers. It is important, however, that the answers be reported with both justifiable accuracy and appropriate significant figures. In this section we will discuss these topics together with some other important aspects involved in all engineering calculations.

Dimensional Homogeneity

The terms of any equation used to describe a physical process must be *dimensionally homogeneous*, that is, each term must be expressed in the same units. Provided this is the case, all the terms of an equation can then be combined if numerical values are substituted for the variables. Consider, for example, the equation $s = vt + \frac{1}{2}at^2$, where, in SI units, s is the position in meters (m), t is time in seconds (s), v is velocity in m/s, and a is acceleration in m/s^2. Regardless of how this equation is evaluated, it maintains its dimensional homogeneity. In the form stated each of the three terms is expressed in meters [m, (m/s̶)s̶, (m/s̶^{2})s̶^{2}], or solving for a, $a = 2s/t^2 - 2v/t$, the terms are each expressed in units of m/s^2 [m/s^2, m/s^2, (m/s)/s].

Since problems in mechanics involve the solution of dimensionally homogeneous equations, the fact that all terms of an equation are represented by a consistent set of units can be used as a partial check for algebraic manipulations of an equation.

Significant Figures

A *significant figure* is any digit from 1 to 9, or a zero that is *not used* to show the position of the decimal point for a number. For example, the numbers 5605, 34.43, 0.05892, and 500.0 each have four significant figures. The number of significant figures in numbers having zeros can be clarified by using powers of 10. There are two ways of doing this. The format for *scientific notation* specifies one digit to the left of the decimal point, with the remaining digits to the right; for example, 45,000 expressed to three significant figures would be $4.50(10^4)$. Using *engineering notation*, which is preferred here, the exponent is displayed in multiples of three in order to facilitate conversion of SI units to those having an appropriate prefix. Thus, 45,000 expressed to three significant figures would be $45.0(10^3)$.

Rounding-Off Numbers

For numerical calculations, the accuracy obtained from the solution of a problem generally can never be better than the accuracy of the problem data. This is what is to be expected, but often computers or hand calculators involve more figures in the answer than the number of significant figures used for the data. For this reason, a calculated result should always be "rounded off" to an appropriate number of significant figures.

To ensure accuracy, the following rules for rounding off a number to n significant figures apply:

1. If the $n + 1$ digit is *less than 5*, the $n + 1$ digit and others following it are dropped. For example, 2.326 and 0.451 rounded off to $n = 2$ significant figures would be 2.3 and 0.45.
2. If the $n + 1$ digit is equal to 5 with zeros following it, then round off the nth digit to an *even number*. For example, 1245 and 0.8655 rounded off to three significant figures become 1240 and 0.866.

3. If the $n + 1$ digit is *greater than 5* or equal to 5 with any nonzero digits following it, then increase the nth digit by 1 and drop the $n + 1$ digit and others following it. For example, 0.007 238 7 and 565.500 3 rounded off to three significant figures become 0.00724 and 566.

Calculations

As a general rule, to ensure accuracy of a final result when performing calculations with numbers of unequal accuracy, always retain one extra significant figure in the more accurate numbers than in the least accurate number *before* beginning the computations. Then round off the final result so that it has the same number of significant figures as the least accurate number. In particular, when adding numbers, retain one more *decimal* digit (digits to the right of the decimal point) in the more accurate numbers than is contained in the least accurate number. Also, try to work out the computations so that numbers which are approximately equal are not subtracted since accuracy is often lost from this calculation.

Most of the example problems in this book are solved with the assumption that any measured *data* are accurate to three significant figures.* Consequently, intermediate calculations are often worked out to four significant figures and the answers are generally reported to three significant figures.

Conversion of Units

In some cases it may be necessary to convert from one system of units to another. In this regard, Table 1–3 provides a set of direct conversion factors between FPS and SI units for the fundamental quantities. Also, in the FPS system, recall that 1 ft = 12 in. (inches), 5280 ft = 1 mi (mile), 1000 lb = 1 kip (kilo-pound), and 2000 lb = 1 ton.

Table 1–3   Conversion Factors

Quantity	Unit of Measurement (FPS)	Equals	Unit of Measurement (SI)
Force	lb		4.448 2 N
Mass	slug		14.593 8 kg
Length	ft		0.304 8 m

When derived units are present, a general procedure using a simple cancellation technique should be applied. The following examples illustrate this method.

---

*Of course, some numbers, such as π, e, or numbers used in derived formulas, are exact and are therefore accurate to an infinite number of significant figures.

Example 1–1

Convert 2 km/h to m/s.

SOLUTION

Since 1 km = 1000 m and 1 h = 3600 s, the factors of conversion are arranged as follows:

$$2 \text{ km/h} = \frac{2 \cancel{\text{km}}}{\cancel{\text{h}}} \left(\frac{1000 \text{ m}}{\cancel{\text{km}}} \right) \left(\frac{1 \cancel{\text{h}}}{3600 \text{ s}} \right)$$

$$= \frac{2000 \text{ m}}{3600 \text{ s}} = 0.556 \text{ m/s} \qquad\qquad \textit{Ans.}$$

Example 1–2

Convert the quantity 300 lb · s to appropriate SI units.

SOLUTION

Using Table 1–3, 1 lb = 4.448 2 N, then

$$300 \text{ lb} \cdot \text{s} = 300 \cancel{\text{lb}} \cdot \text{s} \left(\frac{4.448 \text{ 2 N}}{\cancel{\text{lb}}} \right)$$

$$= 1334.5 \text{ N} \cdot \text{s} = 1.33 \text{ kN} \cdot \text{s} \qquad\qquad \textit{Ans.}$$

Example 1–3

Evaluate each of the following and express with SI units having an appropriate prefix: (a) (50 mN)(6 GN), (b) (400 mm)(0.6 MN)^2, (c) 45 MN^3/900 Gg.

SOLUTION

First convert each number to base units, perform the indicated operations, then choose an appropriate prefix (see Rule 6 on p. 7).

Part (a)

$$(50 \text{ mN})(6 \text{ GN}) = [50(10^{-3}) \text{ N}][6(10^9) \text{ N}]$$
$$= 300(10^6) \text{ N}^2$$
$$= 300(10^6) \, \cancel{\text{N}}^{\,2} \left(\frac{1 \text{ kN}}{10^3 \, \cancel{\text{N}}}\right)\left(\frac{1 \text{ kN}}{10^3 \, \cancel{\text{N}}}\right)$$
$$= 300 \text{ kN}^2 \qquad\qquad Ans.$$

Note carefully the convention kN^2 = (kN)^2 = 10^6 N^2 (Rule 4 on p. 7).

Part (b)

$$(400 \text{ mm})(0.6 \text{ MN})^2 = [400(10^{-3}) \text{ m}][0.6(10^6) \text{ N}]^2$$
$$= [400(10^{-3}) \text{ m}][0.36(10^{12}) \text{ N}^2]$$
$$= 144(10^9) \text{ m} \cdot \text{N}^2$$
$$= 144 \text{ Gm} \cdot \text{N}^2 \qquad\qquad Ans.$$

Note that we can also write

$$144(10^9)\text{m} \cdot \text{N}^2 = 144(10^9)\text{m} \cdot \cancel{\text{N}}^{\,2}\left(\frac{1 \text{ MN}}{10^6 \, \cancel{\text{N}}}\right)\left(\frac{1 \text{ MN}}{10^6 \, \cancel{\text{N}}}\right)$$
$$= 0.144 \text{ m} \cdot \text{MN}^2$$

Part (c)

$$45 \text{ MN}^3/900 \text{ Gg} = \frac{45(10^6 \text{ N})^3}{900(10^6) \text{ kg}}$$
$$= 0.05(10^{12}) \text{ N}^3/\text{kg}$$
$$= 0.05(10^{12})\cancel{\text{N}}^{\,3}\left(\frac{1 \text{ kN}}{10^3 \, \cancel{\text{N}}}\right)^3 \frac{1}{\text{kg}}$$
$$= 0.05(10^3) \text{ kN}^3/\text{kg}$$
$$= 50 \text{ kN}^3/\text{kg} \qquad\qquad Ans.$$

Here we have used rules 4 and 8 of p. 7.

1.6 General Procedure for Analysis

The most effective way of learning the principles of engineering mechanics is to *solve problems*. To be successful at this, it is important to present the work in a *logical* and *orderly manner,* as suggested by the following sequence of steps:

1. Read the problem carefully and try to correlate the actual physical situation with the theory studied.
2. Draw any necessary diagrams and tabulate the problem data.
3. Apply the relevant principles, generally in mathematical form.
4. Solve the necessary equations algebraically as far as practical, then making sure they are dimensionally homogeneous, use a consistent set of units and complete the solution numerically. Report the answer with no more significant figures than the accuracy of the given data.
5. Study the answer with technical judgment and common sense to determine whether or not it seems reasonable.
6. Once the solution has been completed, review the problem. Try to think of other ways of obtaining the same solution.

In applying this general procedure, do the work as neatly as possible. Being neat generally stimulates clear and orderly thinking, and vice versa.

PROBLEMS

1–1. Determine the number of cubic feet contained in a cubic meter.

1–2. A beam has a weight of 5450 lb. Determine its mass in kilograms. Express the answer using an appropriate prefix.

1–3. Determine the weight in newtons of an object that has a mass of (a) 15 kg, (b) 25 mg, (c) 35 Mg. Express each answer using an appropriate prefix.

*1–4. Represent each of the following quantities using an appropriate prefix: (a) 0.000431 kg, (b) 35.3(10^3) μN, (c) 53.2(10^{-4}) km.

1–5. The specific weight (weight/volume) of steel is 420 lb/ft^3. Determine its density (mass/volume) in SI units.

1–6. Represent each of the following combinations of units in the correct SI form: (a) N/μm, (b) MN/ks^2, (c) kN/ms, (d) μMN.

1–7. Represent each of the following combinations of units in the correct SI form using an appropriate prefix:
(a) N/mm,
(b) MN/(kg · ms),
(c) Mg/ms.

*1–8. A concrete column has a diameter of 350 mm and a length of 2 m. If the density (mass/volume) of concrete is 2.45 Mg/m^3, determine the weight of the column in pounds.

1–9. Convert 568 mi/h to m/s. Express the answer using an appropriate prefix.

1–10. Evaluate each of the following and express the answers with SI units having an appropriate prefix: (a) (0.08 Mg)^2(30 ms), (b) (0.50 mm)^3(850 Mg)^2.

1–11. Evaluate each of the following and express with an appropriate prefix: (a) (515 mm)^2, (b) (43.5 N)^3, (c) [134(10^4)]$^{1/3}$ μg.

*1–12. Using Table 1–3, determine your own mass in kilograms, your weight in newtons, and your height in meters.

1–13. Convert 856 lb · in^3 to N · m^3. Express the result using an appropriate prefix.

1–14. A machine weighs 500 lb on earth. Specify (a) its mass in slugs, (b) its mass in kilograms, and (c) its weight in newtons. If the machine is on the moon, where the acceleration due to gravity is $g_m = 5.30$ ft/s^2, determine (d) its weight in pounds, and (e) its mass in kilograms.

1–15. Evaluate each of the following and express the answers with SI units having an appropriate prefix:
(a) (200 MN)(5 mg)/(45 km),
(b) (150 μg)(0.00356 MN),
(c) (750 Gm)/(52.1 Mg).

*1–16. Convert each of the following and express the answer using an appropriate prefix: (a) 175 lb/ft^3 to kN/m^3, (b) 6 ft/h to mm/s, (c) 835 lb · ft to kN · m.

1–17. Evaluate each of the following and express the answers with SI units having an appropriate prefix:
(a) (3.51 mm)(775 MN),
(b) (835 μs)/(15 mg),
(c) (0.095 Mg)(25 mm)/(515 kg).

1–18. Two particles have a mass of 500 kg and 800 kg, respectively. If they are 12 m apart, determine the force of gravitational attraction between them.

2

Force Vectors

In this chapter we will introduce the concept of a concentrated force and give the procedures for adding forces, resolving them into components, and projecting them along an axis. Since force is a vector quantity, we must use the rules of vector algebra whenever forces are considered. We will begin our study by defining scalar and vector quantities and then develop some of the basic rules of vector algebra.

2.1 Scalars and Vectors

Most of the physical quantities in mechanics can be expressed mathematically by means of scalars and vectors.

Scalar

A quantity characterized by a positive or negative number is called a *scalar*. Mass, volume, and length are scalar quantities often used in statics. In this book, scalars are indicated by letters in italic type, such as the scalar A. The mathematical operations involving scalars follow the same rules as those of elementary algebra.

Vector

A *vector* is a quantity that has both a magnitude and direction and obeys the parallelogram law of addition. This law, which will be described later, utilizes a form of construction that accounts for the combined magnitude and direction of the vector. Vector quantities commonly used in statics are position, force, and moment vectors.

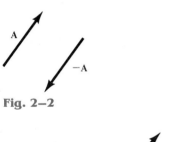

Fig. 2–1

A vector is represented graphically by an arrow, which is used to define its magnitude, direction, and sense. The *magnitude* of the vector is indicated by the *length* of the arrow, the *direction* is defined by the *angle* between a reference axis and the arrow's *line of action,* and the *sense* is indicated by the *arrowhead*. For example, the vector **A** shown in Fig. 2–1 has a *magnitude* of 4 units, a line of action which is *directed* 20° above the horizontal axis, and a *sense* which is upward and to the right. The point O is called the *tail* of the vector, the point P is the *tip* or *head*.

For handwritten work, a vector is generally represented by a letter with an arrow written over it, such as $\vec{A}$. The magnitude is designated by $|\vec{A}|$ or simply A. In this book *vectors* will be symbolized in *boldface type;* for example, **A** is used to designate the vector "A". Its *magnitude,* which is *always* a *positive quantity,* is symbolized in *italic type,* written as $|A|$ or simply A when it is understood that A is a positive scalar.

2.2 Vector Operations

Fig. 2–2

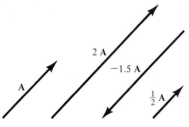

Scalar Multiplication and Division

Fig. 2–3

Multiplication and Division of a Vector by a Scalar

The product of vector **A** and scalar a, yielding a**A,** is defined as a vector having a magnitude $|aA|$. The *sense* of a**A** is the *same* as **A** provided a is *positive;* it is *opposite* to **A** if a is *negative.* Consequently, the negative of a vector is formed by multiplying the vector by the scalar (-1), Fig. 2–2. Division of a vector by a scalar can be defined using the laws of multiplication, since $\mathbf{A}/a = (1/a)\mathbf{A}$, $a \neq 0$. Graphic examples of these operations are shown in Fig. 2–3.

Vector Addition

Two vectors **A** and **B,** Fig. 2–4a, may be added to form a "resultant" vector $\mathbf{R} = \mathbf{A} + \mathbf{B}$ by using the *parallelogram law*. To do this, **A** and **B** are joined at their tails, Fig. 2–4b. Parallel dashed lines drawn from the head of each vector intersect at a common point, thereby forming the adjacent sides of a parallelogram. As shown, the resultant **R** is the diagonal of the parallelogram, which extends from the tails of **A** and **B** to the intersection of the dashed lines.

We can also add **B** to **A** using a *triangle construction*, which is a special case of the parallelogram law, whereby vector **B** is added to vector **A** in a "head-to-tail" fashion, i.e., by placing the tail of **B** at the head of **A**, Fig. 2–4c. The resultant **R** extends from the tail of **A** to the head of **B**. In a similar

Fig. 2—4   Vector Addition

manner, **R** can also be obtained by adding **A** to **B,** Fig. 2–4d. By comparison, it is seen that vector addition is commutative; in other words, the vectors can be added in either order, i.e., **R = A + B = B + A.**

If the two vectors **A** and **B** are *collinear,* i.e., both have the same line of action, the parallelogram law reduces to an *algebraic* or *scalar addition* $R = A + B$ as shown in Fig. 2–5.

Fig. 2—5   Addition of Collinear Vectors

Vector Subtraction

The resultant *difference* between two vectors **A** and **B** may be expressed as

$$\mathbf{R'} = \mathbf{A} - \mathbf{B} = \mathbf{A} + (-\mathbf{B})$$

This vector sum is shown graphically in Fig. 2–6. Subtraction is therefore defined as a special case of addition, so that the rules of vector addition also apply to vector subtraction.

Fig. 2—6

Resolution of a Vector

By using the parallelogram law it is possible to resolve a vector into ''components'' having known lines of action. For example, if **R** in Fig. 2–7 is to be resolved into two components, acting along the lines *a* and *b*, one starts at the *head* of **R** and extends a dashed line *parallel* to the *a* axis until it intersects the *b* axis. Likewise, a dashed line parallel to the *b* axis is drawn from the *head* of **R** to the point of intersection with the *a* axis. The two components **A** and **B** are then drawn such that they extend from the tail of **R** to the points of intersection, as shown in the figure.

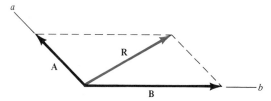

Fig. 2—7

17

2.3 Vector Addition of Forces

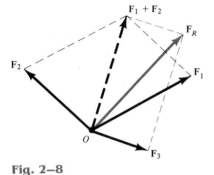

Fig. 2–8

Experimental evidence has shown that a force is a vector quantity since it has a specified magnitude and direction and it adds according to the parallelogram law. Two common problems in statics involve finding either the resultant force, knowing its components, or resolving a known force into components. As described above, both of these problems require application of the parallelogram law.

In particular, if more than two forces are to be added, successive applications of the parallelogram law can be carried out in order to obtain the resultant force. For example, if three forces $\mathbf{F}_1$, $\mathbf{F}_2$, $\mathbf{F}_3$ act at O, Fig. 2–8, the resultant of any two of the forces is found, say $\mathbf{F}_1 + \mathbf{F}_2$, and then this resultant is added to the third force, yielding the resultant of all three forces; i.e., $\mathbf{F}_R = (\mathbf{F}_1 + \mathbf{F}_2) + \mathbf{F}_3$. Using the parallelogram law to add more than two forces, as shown here, often requires extensive geometric and trigonometric calculation for the numerical value of the resultant. Instead, problems of this type are easily solved by using the "rectangular-component method," which is explained in Sec. 2.4.

PROCEDURE FOR ANALYSIS

Problems that involve the addition of two forces and contain at most *two unknowns* can be solved by using the following procedure:

Parallelogram Law. Make a sketch showing the vector addition using the parallelogram law. If possible, determine the interior angles of the parallelogram from the geometry of the problem. Recall that the sum total of these angles is 360°. Unknown angles, along with known and unknown force magnitudes, should be clearly labeled on this sketch. Redraw a half portion of the constructed parallelogram to illustrate the triangular head-to-tail addition of the components.

Trigonometry. By using trigonometry, the two unknowns can be determined from the data listed on the triangle. If the triangle does *not* contain a 90° angle, the law of sines and/or the law of cosines may be used for the solution. These formulas are given in Fig. 2–9 for the triangle shown. The following examples numerically illustrate this method for solution.

Fig. 2–9

Sine law:
$$\frac{A}{\sin a} = \frac{B}{\sin b} = \frac{C}{\sin c}$$

Cosine law:
$$C = \sqrt{A^2 + B^2 - 2AB \cos c}$$

Example 2–1

The screw eye in Fig. 2–10a is subjected to two forces, $\mathbf{F}_1$ and $\mathbf{F}_2$. Determine the magnitude and direction of the resultant force.

(a)

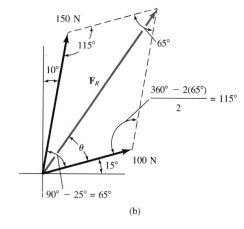

(b)

Fig. 2–10(a, b)

SOLUTION

Parallelogram Law. The parallelogram law of addition is shown in Fig. 2–10b. The two unknowns are the magnitude of $\mathbf{F}_R$ and the angle θ (theta). From Fig. 2–10b, the vector triangle, Fig. 2–10c, is constructed.

Trigonometry. F_R is determined by using the law of cosines:

$$F_R = \sqrt{(100)^2 + (150)^2 - 2(100)(150)\cos 115°}$$
$$= \sqrt{10\ 000 + 22\ 500 - 30\ 000(-0.4226)} = 212.6 \text{ N}$$
$$= 213 \text{ N} \qquad\qquad Ans.$$

The angle θ is determined by applying the law of sines, using the computed value of F_R.

$$\frac{150}{\sin \theta} = \frac{212.6}{\sin 115°}$$

$$\sin \theta = \frac{150}{212.6}(0.9063)$$

$$\theta = 39.8°$$

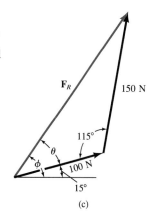

Thus, the direction ϕ (phi) of $\mathbf{F}_R$, measured from the horizontal, is

$$\phi = 39.8° + 15.0° = 54.8° \quad \angle\phi \qquad Ans. \qquad \textbf{Fig. 2–10c}$$

Example 2–2

Resolve the 200-lb force shown acting on the pin, Fig. 2–11a, into components in the (a) x and y directions, and (b) x' and y directions.

F = 200 lb

(a)

Fig. 2–11

SOLUTION

In each case the parallelogram law is used to resolve **F** into its two components, and then the triangle law is applied to determine the numerical results by trigonometry.

Part (a). The vector addition $\mathbf{F} = \mathbf{F}_x + \mathbf{F}_y$ is shown in Fig. 2–11b. In particular, note that the length of the components is scaled along the x and y axes by first constructing dashed lines parallel to the axes in accordance with the parallelogram law. From the vector triangle, Fig. 2–11c,

$$F_x = 200 \cos 40° = 153 \text{ lb} \qquad Ans.$$
$$F_y = 200 \sin 40° = 129 \text{ lb} \qquad Ans.$$

(b)

(c)

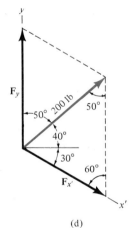

(d)

Part (b). The vector addition $\mathbf{F} = \mathbf{F}_{x'} + \mathbf{F}_y$ is shown in Fig. 2–11d. Note carefully how the parallelogram is constructed. Applying the law of sines and using the data listed on the vector triangle, Fig. 2–11e, yields

$$\frac{F_{x'}}{\sin 50°} = \frac{200}{\sin 60°}$$

$$F_{x'} = 200\left(\frac{\sin 50°}{\sin 60°}\right) = 177 \text{ lb} \qquad Ans.$$

$$\frac{F_y}{\sin 70°} = \frac{200}{\sin 60°}$$

$$F_y = 200\left(\frac{\sin 70°}{\sin 60°}\right) = 217 \text{ lb} \qquad Ans.$$

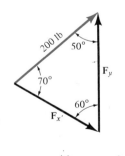

(e)

Example 2–3

The force **F** acting on the frame shown in Fig. 2–12a has a magnitude of 500 N and is to be resolved into two components acting along struts AB and AC. Determine the angle θ so that the component $\mathbf{F}_{AC}$ is directed from A toward C and has a magnitude of 400 N.

(a)   (b)   (c)

Fig. 2–12

SOLUTION

By using the parallelogram law, the vector addition of the two components yielding the resultant is shown in Fig. 2–12b. Note carefully how the resultant force is resolved into the two components $\mathbf{F}_{AB}$ and $\mathbf{F}_{AC}$, which have specified lines of action. The corresponding vector triangle is shown in Fig. 2–12c. The angle φ can be determined by using the law of sines:

(d)

$$\frac{400}{\sin\phi} = \frac{500}{\sin 60°}$$

$$\sin\phi = \left(\frac{400}{500}\right)\sin 60°$$

$$\phi = 43.9°$$

Hence,

$$\theta = 180° - 60° - 43.9° = 76.1° \quad \text{Ans.}$$

Using this value for θ, apply the law of cosines and show that $\mathbf{F}_{AB}$ has a magnitude of 560 N.

Notice that **F** can also be directed at an angle θ as shown in Fig. 2–12d, and still produce the required component $\mathbf{F}_{AC}$. Show that in this case θ = 16.1° and $F_{AB} = 161$ N.

Example 2–4

The ring shown in Fig. 2–13a is subjected to two forces, F_1 and F_2. If it is required that the resultant force have a magnitude of 1 kN and be directed vertically downward, determine (a) the magnitudes of F_1 and F_2 provided $\theta = 30°$, and (b) the magnitudes of F_1 and F_2 if F_2 is to be a minimum.

(a)

Fig. 2–13

(b)   (c)

SOLUTION

Part (a). A sketch of the vector addition according to the parallelogram law is shown in Fig. 2–13b. From the vector triangle constructed in Fig. 2–13c, the unknown magnitudes F_1 and F_2 can be determined by using the law of sines.

$$\frac{F_1}{\sin 30°} = \frac{1000}{\sin 130°}$$

$$F_1 = 653 \text{ N} \qquad Ans.$$

$$\frac{F_2}{\sin 20°} = \frac{1000}{\sin 130°}$$

$$F_2 = 446 \text{ N} \qquad Ans.$$

Part (b). By the vector triangle, F_2 may be added to F_1 in various ways to yield F_R, Fig. 2–13d. In particular, the *minimum* length or magnitude of F_2 will occur when its line of action is *perpendicular* to F_1. Any other direction, such as OA or OB, yields a larger value for F_2. Hence, when $\theta = 90° - 20° = 70°$, F_2 is minimum. From the triangle shown in Fig. 2–13e, it is seen that

$$F_1 = 1000 \sin 70° = 940 \text{ N} \qquad Ans.$$

$$F_2 = 1000 \sin 20° = 342 \text{ N} \qquad Ans.$$

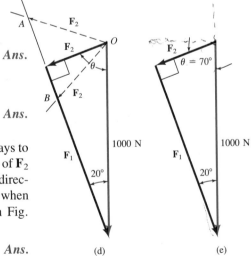

(d)   (e)

PROBLEMS

2–1. Determine the magnitude of the resultant force $\mathbf{F}_R = \mathbf{F}_1 + \mathbf{F}_2$ and its direction, measured counterclockwise from the positive x axis.

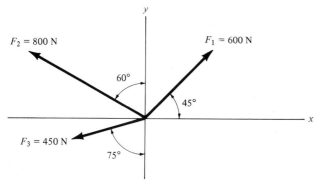

2–2. Determine the magnitude of the resultant force $\mathbf{F}_R = \mathbf{F}_1 + \mathbf{F}_3$ and its direction, measured counterclockwise from the positive x axis.

2–3. Determine the x and y components of $\mathbf{F}_1$ and $\mathbf{F}_2$.

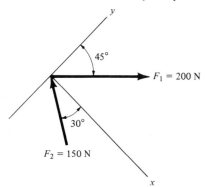

Prob. 2–3

*2–4. Determine the magnitude of the resultant force $\mathbf{F}_R = \mathbf{F}_1 + \mathbf{F}_2$ and its direction, measured counterclockwise from the positive x axis.

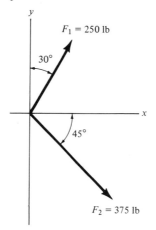

Probs. 2–4/2–5

2–5. Determine the magnitude of the resultant force $\mathbf{F}_R = \mathbf{F}_1 - \mathbf{F}_2$ and its direction, measured counterclockwise from the positive x axis.

2–6. Determine the magnitude of the resultant force $\mathbf{F}_R = \mathbf{F}_1 + \mathbf{F}_2$ and its direction, measured clockwise from the positive u axis.

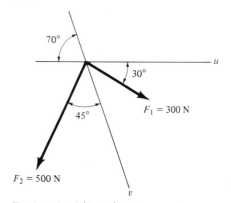

Probs. 2–6/2–7/2–8

2–7. Resolve the force $\mathbf{F}_1$ into components acting along the u and v axes and determine the magnitudes of the components.

*2–8. Resolve the force $\mathbf{F}_2$ into components acting along the u and v axes and determine the magnitudes of the components.

23

2–9. The arm is held in position by the deltoid muscle AB, which is attached to the humerus at A. If the force exerted on this bone by the deltoid is 32 N, in the direction shown, determine the force components acting along (a) the x and y axes, and (b) the u and y axes.

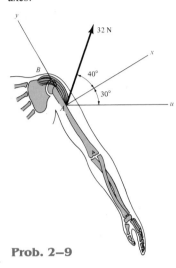

Prob. 2–9

2–10. The pin is subjected to the two forces shown. F is directed along the inclined plane and the 200-N force is vertical. Determine the magnitude of F so that the resultant force F_R is directed normal to the plane, i.e., along the positive n axis. What is the magnitude of F_R?

Prob. 2–10

2–11. The plate is subjected to the forces acting on members A and B as shown. If $\theta = 45°$, determine the magnitude of the resultant of these forces and its direction measured counterclockwise from the positive x axis.

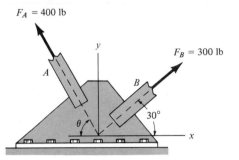

Probs. 2–11/2–12

*2–12. Determine the angle θ for connecting member A to the plate so that the resultant force of $\mathbf{F}_A$ and $\mathbf{F}_B$ is directed vertically upwards. What is the magnitude of the resultant force?

2–13. The chisel exerts a force of 20 lb on the wood dowel rod which is turning in a lathe. Resolve this force into components acting (a) along the n and t axes and (b) along the x and y axes.

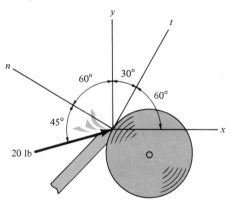

Probs. 2–13/2–14

2–14. The chisel exerts a force of 20 lb on the wood dowel rod which is turning in a lathe. Resolve this force into components acting (a) along the n and y axes and (b) along the x and t axes.

2–15. Determine the magnitude and direction of the resultant $\mathbf{F}_R = \mathbf{F}_1 + \mathbf{F}_2 + \mathbf{F}_3$ of the three forces by first finding the resultant $\mathbf{F}' = \mathbf{F}_1 + \mathbf{F}_2$ and then forming $\mathbf{F}_R = \mathbf{F}' + \mathbf{F}_3$. Specify its direction measured counterclockwise from the positive x axis.

Probs. 2–15/2–16

*2–16. Determine the magnitude and direction of the resultant $\mathbf{F}_R = \mathbf{F}_1 + \mathbf{F}_2 + \mathbf{F}_3$ of the three forces by first finding the resultant $\mathbf{F}' = \mathbf{F}_1 + \mathbf{F}_3$ and then forming $\mathbf{F}_R = \mathbf{F}' + \mathbf{F}_2$. Specify its direction measured counterclockwise from the positive x axis.

2–17. The plug is to be pulled out using two cords. Cord A is subjected to a force of 1.50 kN and is directed at 30° from the vertical. Determine the force $\mathbf{F}_B$ in cord B if the plug starts to come out when $\theta = 30°$. For this to occur, the resultant force on the plug is to be directed horizontally to the right. Also calculate the magnitude of the resultant force.

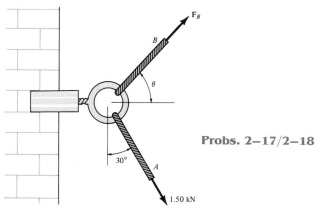

Probs. 2–17/2–18

2–18. The plug is to be pulled out using two cords. Cord A is subjected to a force of 1.50 kN and is directed at 30° from the vertical. If the resultant force acting on the plug is to be 2.50 kN, directed horizontally to the right, determine the force $\mathbf{F}_B$ in cord B and the corresponding angle θ.

2–19. Two forces $\mathbf{F}_1$ and $\mathbf{F}_2$ act on the beam. If each force acts at an angle of θ with the vertical and has a magnitude of F, determine the magnitude of the resultant force.

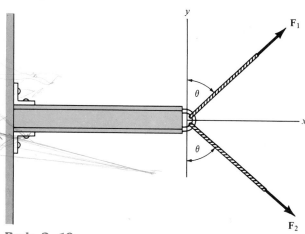

Prob. 2–19

*2–20. Determine the angle θ ($0° \leq \theta \leq 90°$) for cable AB so that the 500-N vertical force has a component of 200 N directed along cable AB from A towards B. What is the corresponding component of force acting along cable CB? Set $\phi = 60°$.

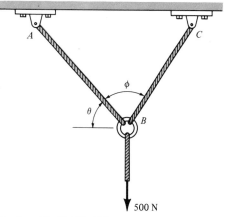

Probs. 2–20/2–21

2–21. Determine the angle ϕ ($0° \leq \phi \leq 90°$) between cables BA and BC so that the 500-N vertical force has a component of 250 N which acts along cable CB, from C to B. Set $\theta = 60°$.

2–22. Two forces, having a magnitude of 15 lb and 8 lb, act on the ring. If the resultant force has a magnitude of 20 lb, determine the angle θ between the forces.

Prob. 2–22

2–23. The truck is to be towed using two ropes. Determine the magnitudes of forces $\mathbf{F}_A$ and $\mathbf{F}_B$ acting on each rope in order to develop a resultant force of 950 N directed along the positive x axis. Set $\theta = 50°$.

Probs. 2–23/2–24

*2–24. The truck is to be towed using two ropes. If the resultant force is to be 950 N, directed along the positive x axis, determine the magnitudes of forces $\mathbf{F}_A$ and $\mathbf{F}_B$ acting on each rope and the angle θ of $\mathbf{F}_B$ so that the magnitude of $\mathbf{F}_B$ is a *minimum*. $\mathbf{F}_A$ acts at 20° from the x axis as shown.

2–25. Two forces have the same magnitude F and produce a resultant force having a magnitude of F_R when the angle between the tails of the two forces is 2ϕ. Show that the resultant force has a magnitude of $4F_R$ when the angle between the two forces is 2θ, where $\theta = \cos^{-1}(4\cos\phi)$.

2–26. Three cables pull on the tree stump such that they create a resultant force having a magnitude of 900 lb. If two of the cables are subjected to known forces, as shown in the figure, determine the direction θ of the third cable so that the magnitude of force $\mathbf{F}$ in this cable is a *minimum*. All forces lie in the x-y plane. What is the magnitude of $\mathbf{F}$? *Hint:* First find the resultant of the two known forces. Force $\mathbf{F}$ acts in this direction.

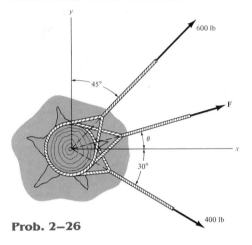

Prob. 2–26

2–27. The barge is being towed by two tugs A and B. Determine the magnitudes of the two towing forces $\mathbf{F}_A$ and $\mathbf{F}_B$ if the resultant force has a magnitude $F_R = 2$ kN and is directed along the positive x axis. Set $\theta = 25°$.

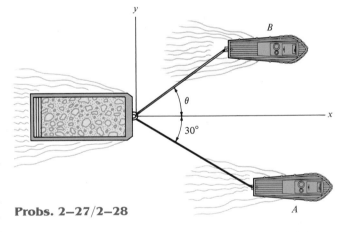

Probs. 2–27/2–28

*2–28. If the resultant $\mathbf{F}_R$ of the two forces acting on the barge is to be directed along the positive x axis and have a magnitude of 2 kN, determine the angle θ of the cable attached to the tug at B such that the force $\mathbf{F}_B$ in this cable is a *minimum*. What is the magnitude of force in each cable when this occurs?

2.4 Addition of a System of Coplanar Forces

It was shown in Sec. 2.3 that a force **F** can always be resolved into two components. In this section we will discuss the resolution of **F** into its *rectangular components* **F**$_x$ and **F**$_y$ which lie along the x and y axes, respectively, Fig. 2–14a. Although the axes are shown here to be horizontal and vertical, they may in general be directed at any inclination, as long as they remain perpendicular to one another, Fig. 2–14b. In both cases, by the parallelogram law, we require:

$$\mathbf{F} = \mathbf{F}_x + \mathbf{F}_y$$

and

$$\mathbf{F}' = \mathbf{F}'_x + \mathbf{F}'_y$$

As shown in Fig. 2–14 the sense of direction of each component is *graphically* represented by an *arrowhead*. For *analytical* work, however, we must establish a notation for representing the directional sense of the rectangular components. This can be done in one of two ways.

Scalar Notation

Since the x and y axes have designated positive and negative directions, the magnitude and directional sense of the components of a force can be expressed in terms of *algebraic scalars*. For example, the components of **F** in Fig. 2–14a can be represented by positive scalars F_x and F_y since their sense of direction is along the *positive* x and y axes, respectively. In a similar manner, the components of **F**' in Fig. 2–14b are F'_x and $-F'_y$. Here the y component is negative since **F**'$_y$ is directed along the negative y axis. It is important to keep in mind that this scalar notation is to be used only for computational purposes, not for graphical representations in figures. Throughout the text, the *head of a vector arrow* in any figure *graphically* indicates the sense of a vector; algebraic signs are not used for this purpose. Thus, the vectors in Figs. 2–14a and 2–14b are designated by using boldface (vector) notation.* Whenever italic symbols are written near vector arrows in figures, they indicate the *magnitude* of the vector, which is always a positive quantity.

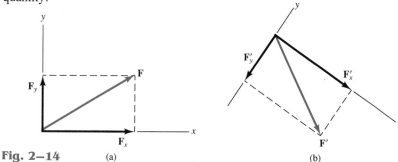

Fig. 2–14   (a)   (b)

*Negative signs are used only in figures with boldface notation when showing equal but opposite pairs of vectors as in Fig. 2–2.

Cartesian-Vector Notation

It is also possible to represent the components of a force in terms of Cartesian unit vectors. By doing this the methods of vector algebra are easier to apply, and we will see that this becomes particularly advantageous for solving problems in three dimensions. In two dimensions the Cartesian unit vectors $\mathbf{i}$ and $\mathbf{j}$ are used to designate the directions of the x and y axes, respectively, Fig. 2–15a.† These vectors have a dimensionless magnitude of unity and their sense (or arrowhead) will be described analytically by a plus or minus sign, depending on whether they are pointing along the positive or negative x or y axis.

As shown in Fig. 2–15a, the *magnitude* of each component of $\mathbf{F}$ is *always a positive quantity*, which is represented by the (positive) scalars F_x and F_y. Therefore, having established notation to represent the magnitude and the direction of each component, we can express $\mathbf{F}$ in Fig. 2–15a as a *Cartesian vector*, i.e.,

$$\mathbf{F} = F_x\mathbf{i} + F_y\mathbf{j}$$

And in the same way, $\mathbf{F}'$ in Fig. 2–15b can be expressed as

$$\mathbf{F}' = F_x'\mathbf{i} + F_y'(-\mathbf{j})$$

or simply

$$\mathbf{F}' = F_x'\mathbf{i} - F_y'\mathbf{j}$$

(a)

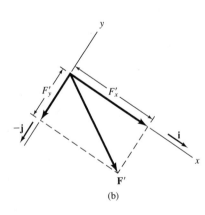

(b)

Fig. 2–15

---

†For handwritten work unit vectors are usually indicated using a circumflex, e.g., $\hat{i}$ and $\hat{j}$.

Coplanar Force Resultants

Either of the above two methods of representing the rectangular components of a force can be used to determine the resultant sum of several *coplanar forces*, i.e., forces that all lie in the same plane. To do this, each force is first resolved into its x and y components and then the respective components are added using *scalar algebra* since they are collinear. The resultant force is then formed by adding the resultants of the x and y components using the parallelogram law. For example, consider the three forces in Fig. 2–16a, which have x and y components as shown in Fig. 2–16b. To solve this problem using Cartesian vector notation, each force is first represented as a Cartesian vector, i.e.,

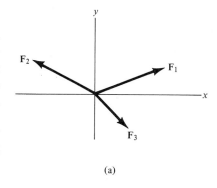

(a)

$$\mathbf{F}_1 = F_{1x}\mathbf{i} + F_{1y}\mathbf{j}$$
$$\mathbf{F}_2 = -F_{2x}\mathbf{i} + F_{2y}\mathbf{j}$$
$$\mathbf{F}_3 = F_{3x}\mathbf{i} - F_{3y}\mathbf{j}$$

The vector resultant is therefore

$$\mathbf{F}_R = \mathbf{F}_1 + \mathbf{F}_2 + \mathbf{F}_3$$
$$= F_{1x}\mathbf{i} + F_{1y}\mathbf{j} - F_{2x}\mathbf{i} + F_{2y}\mathbf{j} + F_{3x}\mathbf{i} - F_{3y}\mathbf{j}$$
$$= (F_{1x} - F_{2x} + F_{3x})\mathbf{i} + (F_{1y} + F_{2y} - F_{3y})\mathbf{j}$$
$$= (F_{Rx})\mathbf{i} + (F_{Ry})\mathbf{j}$$

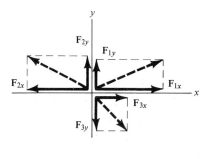

(b)

If scalar notation is used, then from Fig. 2–16b, since x is positive to the right and y is positive upward, we have

$(\xrightarrow{+})$ $F_{Rx} = F_{1x} - F_{2x} + F_{3x}$

$(+\uparrow)$ $F_{Ry} = F_{1y} + F_{2y} - F_{3y}$

These results are the *same* as the $\mathbf{i}$ and $\mathbf{j}$ components of $\mathbf{F}_R$ determined above. In either case, the resultant force is then determined *graphically* from the parallelogram law as shown in Fig. 2–16c.

In the general case, the x and y components of the resultant of any number of coplanar forces can be represented symbolically by the algebraic sum of the x and y components of all the forces, i.e.,

(c)

Fig. 2–16

$$\boxed{\begin{aligned} F_{Rx} &= \Sigma F_x \\ F_{Ry} &= \Sigma F_y \end{aligned}} \qquad (2\text{–}1)$$

Once these components are determined, the *magnitude* of $\mathbf{F}_R$ is found from the Pythagorean theorem, Fig. 2–16c, that is,

$$F_R = \sqrt{F_{Rx}^2 + F_{Ry}^2}$$

If the angle θ is used to specify the direction of $\mathbf{F}_R$, Fig. 2–16c, then by trigonometry

$$\theta = \tan^{-1}\left(\frac{F_{Ry}}{F_{Rx}}\right)$$

The above concepts are illustrated numerically in the following examples.

Example 2–5

Determine the x and y components of $\mathbf{F}_1$ and $\mathbf{F}_2$ shown in Fig. 2–17a. Express each force as a Cartesian vector.

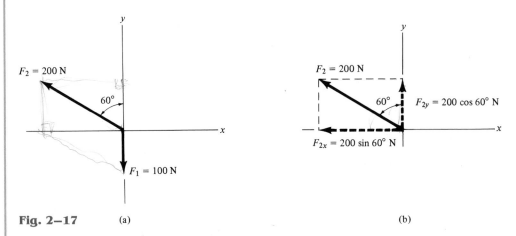

Fig. 2–17   (a)   (b)

SOLUTION

Scalar Notation. Since $\mathbf{F}_1$ acts along the negative y axis, and the magnitude of $\mathbf{F}_1$ is 100 N, then the components written in scalar form are

$$F_{1x} = 0, \qquad F_{1y} = -100 \text{ N} \qquad\qquad Ans.$$

or alternatively,

$$F_{1x} = 0, \qquad F_{1y} = 100 \text{ N} \downarrow \qquad\qquad Ans.$$

By the parallelogram law, $\mathbf{F}_2$ is resolved into x and y components, Fig. 2–17b. The magnitude of each component is determined by trigonometry. Since $\mathbf{F}_{2x}$ acts in the $-x$ direction, and $\mathbf{F}_{2y}$ acts in the $+y$ direction, we have

$$F_{2x} = -200 \sin 60° = -173 \text{ N} = 173 \text{ N} \leftarrow \qquad Ans.$$
$$F_{2y} = 200 \cos 60° = 100 \text{ N} = 100 \text{ N} \uparrow \qquad Ans.$$

Cartesian Vector Notation. Having computed the magnitudes of the components of $\mathbf{F}_2$, Fig. 2–17b, we can express each force as a Cartesian vector.

$$\mathbf{F}_1 = 0\mathbf{i} + 100\,(-\mathbf{j})$$
$$= \{-100\mathbf{j}\} \text{ N} \qquad\qquad Ans.$$

and

$$\mathbf{F}_2 = 200 \sin 60°\,(-\mathbf{i}) + 200 \cos 60°\,(\mathbf{j})$$
$$= \{-173\mathbf{i} + 100\mathbf{j}\} \text{ N} \qquad\qquad Ans.$$

Example 2–6

Determine the x and y components of the force $\mathbf{F}$ shown in Fig. 2–18a.

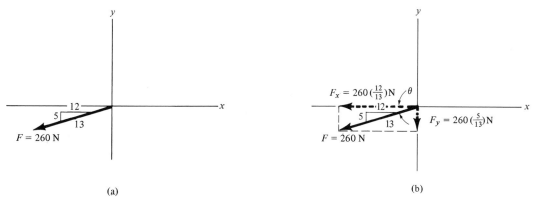

(a)   (b)

Fig. 2–18

SOLUTION

The force is resolved into its x and y components as shown in Fig. 2–18b. Here the *slope* of the line of action for the force is indicated. From this "slope triangle" we could obtain the direction angle θ, e.g., $\theta = \tan^{-1}\left(\frac{5}{12}\right)$, and then proceed to determine the magnitudes of the components in the same manner as for $\mathbf{F}_2$ in Example 2–5. An easier method, however, consists of using proportional parts of similar triangles, i.e.,

$$\frac{F_x}{260} = \frac{12}{13} \qquad F_x = 260\left(\frac{12}{13}\right) = 240 \text{ N}$$

Similarly,

$$F_y = 260\left(\frac{5}{13}\right) = 100 \text{ N}$$

Notice that the magnitude of the *horizontal component*, F_x, was obtained by multiplying the force magnitude by the ratio of the *horizontal leg* of the slope triangle divided by the hypotenuse; whereas the magnitude of the *vertical component*, F_y, was obtained by multiplying the force magnitude by the ratio of the *vertical leg* divided by the hypotenuse. Hence, using scalar notation,

$$F_x = -240 \text{ N} = 240 \text{ N} \leftarrow \qquad\qquad Ans.$$
$$F_y = -100 \text{ N} = 100 \text{ N} \downarrow \qquad\qquad Ans.$$

$\mathbf{F}$ may also be written as a Cartesian vector, i.e.,

$$\mathbf{F} = \{-240\mathbf{i} - 100\mathbf{j}\} \text{ N} \qquad\qquad Ans.$$

Example 2–7

The pin in Fig. 2–19a is subjected to two forces F_1 and F_2. Determine the magnitude and direction of the resultant force.

(a)

SOLUTION

Scalar Notation. This problem can be solved by using the parallelogram law; however, here we will resolve each force into its x and y components, Fig. 2–19b, and sum these components algebraically. Indicating the "positive" sense of the x and y force components alongside Eqs. 2–1, we have

$$\xrightarrow{+}F_{Rx} = \Sigma F_x; \qquad F_{Rx} = 600 \cos 30° - 400 \sin 45°$$
$$= 236.8 \text{ N} \rightarrow$$

$$+\uparrow F_{Ry} = \Sigma F_y; \qquad F_{Ry} = 600 \sin 30° + 400 \cos 45°$$
$$= 582.8 \text{ N} \uparrow$$

The resultant force, shown in Fig. 2–19c, has a *magnitude* of

$$F_R = \sqrt{(236.8)^2 + (582.8)^2}$$
$$= 629 \text{ N} \qquad \qquad \textit{Ans.}$$

(b)

The *direction*, defined by the angle θ in Fig. 2–19c, is

$$\theta = \tan^{-1}\left(\frac{582.8}{236.8}\right) = 67.9° \qquad \qquad \textit{Ans.}$$

Cartesian Vector Notation. From Fig. 2–19b, each force expressed as a Cartesian vector is

$$\mathbf{F}_1 = 600 \cos 30°\mathbf{i} + 600 \sin 30°\mathbf{j}$$
$$\mathbf{F}_2 = -400 \sin 45°\mathbf{i} + 400 \cos 45°\mathbf{j}$$

Thus

$$\mathbf{F}_R = \mathbf{F}_1 + \mathbf{F}_2 = (600 \cos 30° - 400 \sin 45°)\mathbf{i}$$
$$+ (600 \sin 30° + 400 \cos 45°)\mathbf{j}$$
$$= \{236.8\mathbf{i} + 582.8\mathbf{j}\} \text{ N}$$

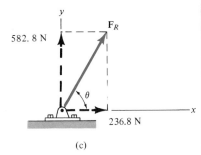

(c)

Fig. 2–19

The magnitude and direction of $\mathbf{F}_R$ are determined as shown above. Comparing the two methods of solution, it is seen that use of scalar notation is more efficient since the scalar components can be found *directly*, without first having to express each force as a Cartesian vector. Cartesian vector analysis will later be shown to be more advantageous for solving three-dimensional problems.

Example 2–8

The end of the boom O in Fig. 2–20a is subjected to three concurrent and coplanar forces. Determine the magnitude and direction of the resultant force.

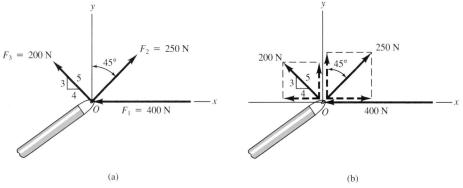

(a)

Fig. 2–20

(b)

SOLUTION

Each force is resolved into its x and y components as shown in Fig. 2–20b. Summing the x components, we have

$$\xrightarrow{+} F_{Rx} = \Sigma F_x; \qquad F_{Rx} = -400 + 250 \sin 45° - 200(\tfrac{4}{5})$$
$$= -383.2 \text{ N} = 383.2 \text{ N} \leftarrow$$

The negative sign indicates that F_{Rx} acts to the left, i.e., in the negative x direction as noted by the small arrow.

$$+\uparrow F_{Ry} = \Sigma F_y; \qquad F_{Ry} = 250 \cos 45° + 200(\tfrac{3}{5})$$
$$= 296.8 \text{ N} \uparrow$$

(c)

The resultant force, shown in Fig. 2–20c, has a *magnitude* of

$$F_R = \sqrt{(-383.2)^2 + (296.8)^2}$$
$$= 485 \text{ N} \qquad\qquad \textit{Ans.}$$

The *direction,* defined by the angle θ in Fig. 2–20c, is

$$\theta = \tan^{-1}\left(\frac{296.8}{383.2}\right) = 37.8° \qquad\qquad \textit{Ans.}$$

Note that the single force $\mathbf{F}_R$ creates the *same effect* on the boom as the three forces in Fig. 2–20a.

PROBLEMS

2–29. Express F_1 and F_2 as Cartesian vectors.

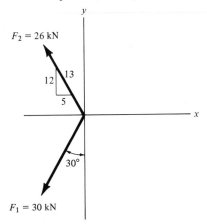

Probs. 2–29/2–30

2–30. Determine the magnitude of the resultant force and its direction measured counterclockwise from the positive x axis.

2–31. Express F_1 and F_2 as Cartesian vectors.

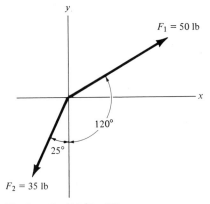

Probs. 2–31/2–32

*2–32. Determine the magnitude of the resultant force and its direction measured counterclockwise from the positive x axis.

2–33. Express F_1, F_2, and F_3 as Cartesian vectors.

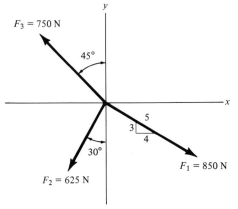

Probs. 2–33/2–34

2–34. Determine the magnitude of the resultant force and its direction measured counterclockwise from the positive x axis.

2–35. Solve Prob. 2–5 by summing the rectangular or x, y components of the forces to obtain the resultant force.

*2–36. Solve Prob. 2–4 by summing the rectangular or x, y components of the forces to obtain the resultant force.

2–37. Solve Prob. 2–16 by summing the rectangular or x, y components of the forces to obtain the resultant force.

2–38. Solve Prob. 2–23 by summing the rectangular or x, y components of the forces to obtain the resultant force.

2–39. Solve Prob. 2–27 by summing the rectangular or x, y components of the forces to obtain the resultant force.

*2–40. Four concurrent forces act on the plate. Determine the magnitude of the resultant force and its direction, measured counterclockwise from the positive x axis.

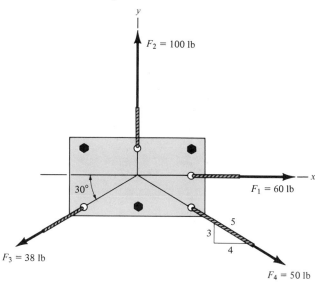

Prob. 2–40

2–41. Express each force acting on the ring in Cartesian vector form. Show that the resultant force is zero to the nearest tenth of a kilonewton.

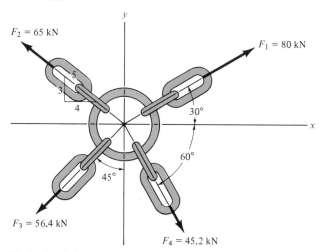

Prob. 2–41

2–42. Determine the magnitude and direction θ of $\mathbf{F}_A$ so that the resultant force is directed along the positive x axis and has a magnitude of 1250 N.

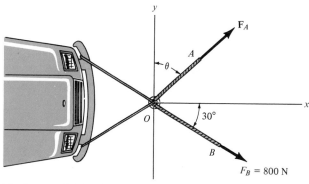

Probs. 2–42/2–43

2–43. Determine the magnitude and direction, measured counterclockwise from the positive x axis, of the resultant force acting on the ring at O, if $F_A = 750$ N and $\theta = 45°$.

*2–44. The three concurrent forces acting on the screw eye produce a resultant force $\mathbf{F}_R = \mathbf{0}$. If $F_2 = \frac{2}{3}F_1$, and $\mathbf{F}_1$ is to be 90° from $\mathbf{F}_2$ as shown, determine the required magnitude of $\mathbf{F}_3$ expressed in terms of F_1 and the angle θ.

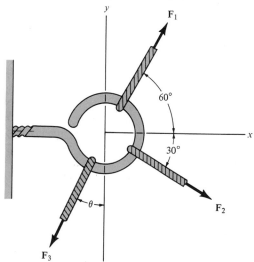

Prob. 2–44

2–45. Three forces act on the bracket. Determine the magnitude and direction θ of $\mathbf{F}_1$ so that the resultant force is directed along the positive x' axis and has a magnitude of 800 N.

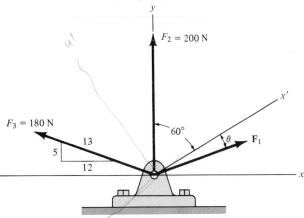

2–47. Determine the magnitude of force $\mathbf{F}$ so that the resultant $\mathbf{F}_R$ of the three forces is as small as possible. What is the magnitude of $\mathbf{F}_R$?

Prob. 2–47

Probs. 2–45/2–46

2–46. If $F_1 = 300$ N and $\theta = 10°$, determine the magnitude and direction, measured counterclockwise from the positive x' axis, of the resultant force of the three forces acting on the bracket.

2.5 Cartesian Vectors

Fig. 2–21

Right–handed coordinate system

The operations of vector algebra, when applied to solving problems in three dimensions, are greatly simplified if the vectors are first represented in Cartesian vector form. In this section we will present a general method for doing this, then in Sec. 2.6 we will apply this method to solving problems involving the addition of forces. Similar applications will be illustrated for the position and moment vectors given in later sections of the text.

Right-Handed Coordinate System

A right-handed coordinate system will be used in developing the theory of vector algebra that follows. A rectangular or Cartesian coordinate system is said to be right-handed provided the thumb of the right hand points in the direction of the positive z axis when the right-hand fingers are curled about the axis and directed from the positive x to the positive y axis, Fig. 2–21. Furthermore, according to this rule, the z axis for a two-dimensional problem as in Fig. 2–20 would be directed outward, perpendicular to the page.

Rectangular Components of a Vector

A vector **A** may have one, two, or three rectangular components along the x, y, z coordinate axes, depending upon how the vector is oriented relative to the axes. In general, though, when **A** is directed within an octant of the x, y, z frame, Fig. 2–22, then by two successive applications of the parallelogram law, we may resolve the vector into components as $\mathbf{A} = \mathbf{A}' + \mathbf{A}_z$ and $\mathbf{A}' = \mathbf{A}_x + \mathbf{A}_y$. Combining these equations, **A** is then represented by the vector sum of its *three* rectangular components,

$$\mathbf{A} = \mathbf{A}_x + \mathbf{A}_y + \mathbf{A}_z \qquad (2\text{--}2)$$

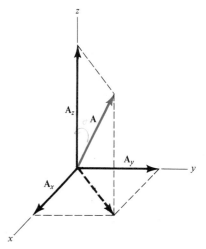

Fig. 2—22

Unit Vector

In general, a *unit vector* is a vector having a magnitude of 1. If **A** is a vector having a magnitude $A \neq 0$, then a unit vector having the *same direction* as **A** is represented by

$$\mathbf{u}_A = \frac{\mathbf{A}}{A} \qquad (2\text{--}3)$$

Rewriting this equation gives

$$\mathbf{A} = A\mathbf{u}_A \qquad (2\text{--}4)$$

Since vector **A** is of a certain kind, e.g., a force vector, it is customary to use the proper set of units for its description. The magnitude A also has this same set of units; hence, from Eq. 2–3, the *unit vector will be dimensionless* since the units will cancel out. Equation 2–4 indicates that vector **A** may therefore be expressed in terms of both its magnitude and direction *separately;* i.e., A (a positive scalar) defines the *magnitude* of **A,** and $\mathbf{u}_A$ (a dimensionless vector) defines the *direction* of **A,** Fig. 2–23.

Fig. 2—23

Cartesian Unit Vectors

In three dimensions, the set of Cartesian unit vectors, **i, j, k,** is used to designate the directions of the x, y, z axes respectively. As stated in Sec. 2–4, the *sense* (or arrowhead) of these vectors will be described analytically by a plus or minus sign, depending upon whether they are pointing along the positive or negative x, y, or z axis. Thus, the positive unit vectors are shown in Fig. 2–24.

Fig. 2—24

Fig. 2–25

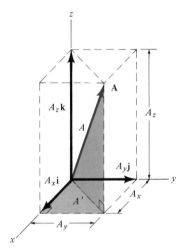

Fig. 2–26

Cartesian Vector Representation

Using the Cartesian unit vectors, the three vector components of Eq. 2–2 may be written in "Cartesian vector form." Since the components act in the positive **i, j,** and **k** directions, Fig. 2–25, we have

$$\mathbf{A} = A_x\mathbf{i} + A_y\mathbf{j} + A_z\mathbf{k} \qquad (2–5)$$

There is a distinct advantage to writing vectors in terms of their Cartesian components. In doing so, the *magnitude* and *direction* of each *component vector* are *separated,* and as stated previously, this will simplify the operations of vector algebra, particularly in three dimensions.

Magnitude of a Cartesian Vector

It is always possible to obtain the magnitude of **A** if the vector is expressed in Cartesian vector form. As shown in Fig. 2–26 from the shaded right triangle, the magnitude $A' = \sqrt{A_x^2 + A_y^2}$. Similarly, from the colored right triangle, $A = \sqrt{A'^2 + A_z^2}$. Combining these equations yields

$$A = \sqrt{A_x^2 + A_y^2 + A_z^2} \qquad (2–6)$$

*Hence, the magnitude of **A** is equal to the positive square root of the sum of the squares of its components.*

Direction of a Cartesian Vector

The *direction* of vector **A** is defined by the *coordinate direction angles* α (alpha), β (beta), and γ (gamma), measured between the *tail* of **A** and the *positive x, y, z* axes, Fig. 2–27. Note that regardless of where **A** is directed, each of these angles will be between 0° and 180°. To determine α, β, and γ, consider the projection of **A** onto the *x, y, z* axes, Fig. 2–28. Referring to the colored right triangles shown in each figure, we have

$$\cos \alpha = \frac{A_x}{A} \qquad \cos \beta = \frac{A_y}{A} \qquad \cos \gamma = \frac{A_z}{A} \qquad (2–7)$$

These numbers are known as the *direction cosines* of **A.** Once they have been obtained, the coordinate direction angles α, β, γ can then be determined from the inverse cosines.

   *An easy way of obtaining the direction cosines of **A** is to form a unit vector in the direction of **A.** Provided **A** is expressed in Cartesian vector form as $\mathbf{A} = A_x\mathbf{i} + A_y\mathbf{j} + A_z\mathbf{k}$ (Eq. 2–5), we have

$$\mathbf{u}_A = \frac{\mathbf{A}}{A} = \frac{A_x}{A}\mathbf{i} + \frac{A_y}{A}\mathbf{j} + \frac{A_z}{A}\mathbf{k} \qquad (2\text{--}8)$$

where $A = \sqrt{(A_x)^2 + (A_y)^2 + (A_z)^2}$ (Eq. 2–6). By comparison with Eqs. 2–7, it is seen that *the* **i, j,** *and* **k** *components of* $\mathbf{u}_A$ *represent the direction cosines of* **A,** i.e.,

$$\mathbf{u}_A = \cos\alpha\mathbf{i} + \cos\beta\mathbf{j} + \cos\gamma\mathbf{k} \qquad (2\text{--}9)$$

Since the magnitude of a vector is equal to the square root of the sum of the squares of the magnitudes of its components, and $\mathbf{u}_A$ has a magnitude of 1, then from Eq. 2–9 an important relation between the direction cosines can be formulated as

$$\cos^2\alpha + \cos^2\beta + \cos^2\gamma = 1 \qquad (2\text{--}10)$$

This equation is useful for determining one of the coordinate direction angles if the other two are known. (See Example 2–10.)

Finally, if the magnitude and coordinate direction angles of **A** are given, **A** may be expressed in Cartesian vector form as

$$\begin{aligned} \mathbf{A} &= A\mathbf{u}_A \\ &= A\cos\alpha\mathbf{i} + A\cos\beta\mathbf{j} + A\cos\gamma\mathbf{k} \\ &= A_x\mathbf{i} + A_y\mathbf{j} + A_z\mathbf{k} \end{aligned} \qquad (2\text{--}11)$$

Fig. 2–27

(a)

(b)

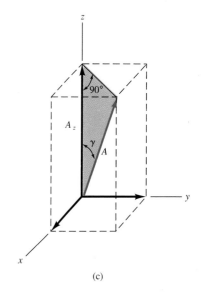
(c)

Fig. 2–28

2.6 Addition and Subtraction of Cartesian Vectors

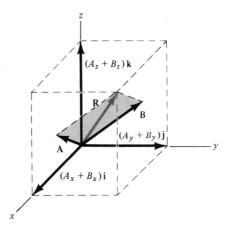

Fig. 2–29

The vector operations of addition and subtraction of two or more vectors are greatly simplified if the vectors are expressed in terms of their Cartesian components. For example, consider the two vectors $\mathbf{A}$ and $\mathbf{B}$, both of which are directed within the positive octant of the x, y, z frame, Fig. 2–29. If $\mathbf{A} = A_x\mathbf{i} + A_y\mathbf{j} + A_z\mathbf{k}$ and $\mathbf{B} = B_x\mathbf{i} + B_y\mathbf{j} + B_z\mathbf{k}$, then the resultant vector, $\mathbf{R}$, has components which represent the scalar sums of the $\mathbf{i}$, $\mathbf{j}$, and $\mathbf{k}$ components of $\mathbf{A}$ and $\mathbf{B}$, i.e.,

$$\mathbf{R} = \mathbf{A} + \mathbf{B} = (A_x + B_x)\mathbf{i} + (A_y + B_y)\mathbf{j} + (A_z + B_z)\mathbf{k}$$

Vector subtraction, being a special case of vector addition, simply requires a scalar subtraction of the respective $\mathbf{i}$, $\mathbf{j}$, and $\mathbf{k}$ components of $\mathbf{A}$ and $\mathbf{B}$. For example,

$$\mathbf{R}' = \mathbf{A} - \mathbf{B} = (A_x - B_x)\mathbf{i} + (A_y - B_y)\mathbf{j} + (A_z - B_z)\mathbf{k}$$

Concurrent Force Systems

In particular, the above concept of vector addition may be generalized and applied to a system of several concurrent forces. In this case, the force resultant of the system is the vector sum of all the forces in the system and can be written as

$$\mathbf{F}_R = \Sigma\mathbf{F} = \Sigma F_x\mathbf{i} + \Sigma F_y\mathbf{j} + \Sigma F_z\mathbf{k} \qquad (2\text{–}12)$$

Here ΣF_x, ΣF_y, and ΣF_z represent the algebraic sums of the respective x, y, z, or $\mathbf{i}$, $\mathbf{j}$, $\mathbf{k}$ components of each force in the system.

The following examples numerically illustrate the methods used to apply the above theory to the solution of problems involving force as a vector quantity.

Example 2–9

Determine the magnitude and the coordinate direction angles of the resultant force acting on the ring in Fig. 2–30a.

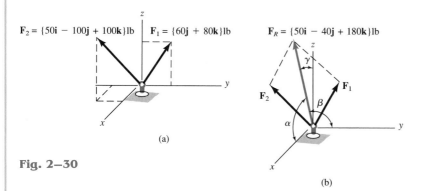

$F_2 = \{50i - 100j + 100k\}$lb $F_1 = \{60j + 80k\}$lb

$F_R = \{50i - 40j + 180k\}$lb

(a)

(b)

Fig. 2–30

SOLUTION

Since each force is represented in Cartesian vector form, the resultant force, shown in Fig. 2–30b, is

$$\mathbf{F}_R = \Sigma \mathbf{F} = \mathbf{F}_1 + \mathbf{F}_2 = (60\mathbf{j} + 80\mathbf{k}) + (50\mathbf{i} - 100\mathbf{j} + 100\mathbf{k})$$
$$= \{50\mathbf{i} - 40\mathbf{j} + 180\mathbf{k}\} \text{ lb}$$

The magnitude of $\mathbf{F}_R$ is found from Eq. 2–6, i.e.,

$$F_R = \sqrt{(50)^2 + (-40)^2 + (180)^2}$$
$$= 191.0 \text{ lb} \qquad\qquad Ans.$$

The coordinate direction angles α, β, γ are determined from the components of the unit vector acting in the direction of $\mathbf{F}_R$.

$$\mathbf{u}_{F_R} = \frac{\mathbf{F}_R}{F_R} = \frac{50}{191.0}\mathbf{i} - \frac{40}{191.0}\mathbf{j} + \frac{180}{191.0}\mathbf{k}$$
$$= 0.2617\mathbf{i} - 0.2094\mathbf{j} + 0.9422\mathbf{k}$$

so that

$$\cos \alpha = 0.2617 \qquad \alpha = 74.8° \qquad\qquad Ans.$$
$$\cos \beta = -0.2094 \qquad \beta = 102° \qquad\qquad Ans.$$
$$\cos \gamma = 0.9422 \qquad \gamma = 19.6° \qquad\qquad Ans.$$

These angles are shown in Fig. 2–30b. In particular, note that $\beta > 90°$ since the $\mathbf{j}$ component of $\mathbf{u}_{F_R}$ is negative.

Example 2–10

Express force **F** shown in Fig. 2–31 in Cartesian vector form.

Fig. 2–31

SOLUTION

Since only two coordinate direction angles are specified, Fig. 2–31, the third angle α is determined from Eq. 2–10, i.e.,

$$\cos^2 \alpha + \cos^2 \beta + \cos^2 \gamma = 1$$

$$\cos^2 \alpha + \cos^2 60° + \cos^2 45° = 1$$

$$\cos \alpha = \sqrt{1 - (0.707)^2 - (0.5)^2} = \pm 0.5$$

Hence,

$$\alpha = \cos^{-1} (0.5) = 60° \quad \text{or} \quad \alpha = \cos^{-1} (-0.5) = 120°$$

By inspection of Fig. 2–31, however, it is necessary that $\alpha = 60°$.

Using Eq. 2–11, with $F = 200$ N, we have

$$\mathbf{F} = F \cos \alpha \mathbf{i} + F \cos \beta \mathbf{j} + F \cos \gamma \mathbf{k}$$

$$= 200 \cos 60° \mathbf{i} + 200 \cos 60° \mathbf{j} + 200 \cos 45° \mathbf{k}$$

$$= \{100.0\mathbf{i} + 100.0\mathbf{j} + 141.4\mathbf{k}\} \text{ N} \qquad \textit{Ans.}$$

By applying Eq. 2–6, note that indeed the magnitude of $F = 200$ N.

$$F = \sqrt{F_x^2 + F_y^2 + F_z^2}$$

$$= \sqrt{(100.0)^2 + (100.0)^2 + (141.4)^2} = 200 \text{ N}$$

Example 2–11

Express **F** shown acting on the hook in Fig. 2–32*a* in Cartesian vector form.

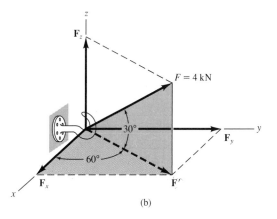

(a)   (b)

Fig. 2–32

SOLUTION

In this case the angles 60° and 30° defining the direction of **F** are *not* coordinate direction angles. Why? By two successive applications of the parallelogram law, however, **F** can be resolved into its *x*, *y*, *z* components as shown in Fig. 2–32*b*. First, from the shaded triangle,

$$F' = 4 \cos 30° = 3.46 \text{ kN}$$
$$F_z = 4 \sin 30° = 2.00 \text{ kN}$$

Next, using **F'** and the colored triangle,

$$F_x = 3.46 \cos 60° = 1.73 \text{ kN}$$
$$F_y = 3.46 \sin 60° = 3.00 \text{ kN}$$

Thus,

$$\mathbf{F} = \{1.73\mathbf{i} + 3.00\mathbf{j} + 2.00\mathbf{k}\} \text{ kN} \qquad Ans.$$

Example 2–12

Express force **F** shown in Fig. 2–33*a* in Cartesian vector form.

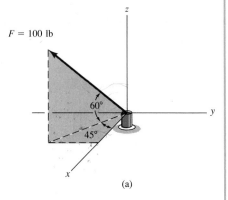

$F = 100$ lb

(a)

SOLUTION

As in Example 2–11, the angles of 60° and 45° defining the direction of **F** are *not* coordinate direction angles. The two successive applications of the parallelogram law needed to resolve **F** into its x, y, z components are shown in Fig. 2–33*b*. By trigonometry, the magnitudes of the components are

$$F_z = 100 \sin 60° = 86.6 \text{ lb}$$
$$F' = 100 \cos 60° = 50 \text{ lb}$$
$$F_x = 50 \cos 45° = 35.4 \text{ lb}$$
$$F_y = 50 \sin 45° = 35.4 \text{ lb}$$

Realizing that $\mathbf{F}_y$ has a direction defined by $-\mathbf{j}$, we have

$$\mathbf{F} = \mathbf{F}_x + \mathbf{F}_y + \mathbf{F}_z$$
$$\mathbf{F} = \{35.4\mathbf{i} - 35.4\mathbf{j} + 86.6\mathbf{k}\} \text{ lb} \qquad \textit{Ans.}$$

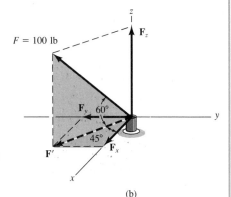

$F = 100$ lb

(b)

To show that the magnitude of this vector is indeed 100 lb, apply Eq. 2–6,

$$F = \sqrt{F_x^2 + F_y^2 + F_z^2}$$
$$= \sqrt{(35.4)^2 + (-35.4)^2 + (86.6)^2} = 100 \text{ lb}$$

If needed, the coordinate direction angles of **F** can be determined from the components of the unit vector acting in the direction of **F.** Hence,

$$\mathbf{u} = \frac{\mathbf{F}}{F} = \frac{F_x}{F}\mathbf{i} + \frac{F_y}{F}\mathbf{j} + \frac{F_z}{F}\mathbf{k}$$

$$= \frac{35.4}{100}\mathbf{i} - \frac{35.4}{100}\mathbf{j} + \frac{86.6}{100}\mathbf{k}$$

$$= 0.354\mathbf{i} - 0.354\mathbf{j} + 0.866\mathbf{k}$$

so that

$$\alpha = \cos^{-1}(0.354) = 69.3°$$
$$\beta = \cos^{-1}(-0.354) = 111°$$
$$\gamma = \cos^{-1}(0.866) = 30.0°$$

These results are shown in Fig. 2–33*c*.

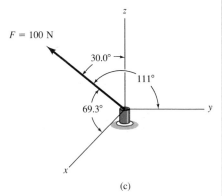

$F = 100$ N

(c)

Fig. 2–33

Example 2–13

Two forces act on the pipe shown in Fig. 2–34a. Specify the direction of $\mathbf{F}_2$ so that the resultant force $\mathbf{F}_R$ acts along the positive y axis and has a magnitude of 800 N.

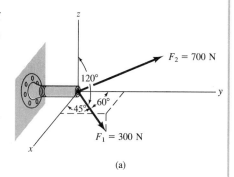

(a)

SOLUTION

To solve this problem, the resultant force and its two components, $\mathbf{F}_1$ and $\mathbf{F}_2$, will each be expressed in Cartesian vector form. Then, as shown in Fig. 2–34b, it is necessary that $\mathbf{F}_R = \mathbf{F}_1 + \mathbf{F}_2$.

Applying Eq. 2–11,

$$\mathbf{F}_1 = F_1\mathbf{u}_{F_1} = F_1\cos\alpha_1\mathbf{i} + F_1\cos\beta_1\mathbf{j} + F_1\cos\gamma_1\mathbf{k}$$
$$= 300\cos 45°\mathbf{i} + 300\cos 60°\mathbf{j} + 300\cos 120°\mathbf{k}$$
$$= \{212.1\mathbf{i} + 150\mathbf{j} - 150\mathbf{k}\}\ \text{N}$$

$$\mathbf{F}_2 = F_2\mathbf{u}_{F_2} = F_{2x}\mathbf{i} + F_{2y}\mathbf{j} + F_{2z}\mathbf{k}$$

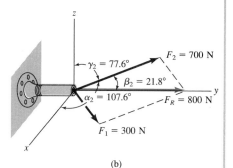

(b)

Fig. 2–34

According to the problem statement, the resultant force $\mathbf{F}_R$ has a magnitude of 800 N and acts in the $+\mathbf{j}$ direction. Hence,

$$\mathbf{F}_R = (800\ \text{N})(+\mathbf{j}) = \{800\mathbf{j}\}\ \text{N}$$

It is required that

$$\mathbf{F}_R = \mathbf{F}_1 + \mathbf{F}_2$$
$$800\mathbf{j} = 212.1\mathbf{i} + 150\mathbf{j} - 150\mathbf{k} + F_{2x}\mathbf{i} + F_{2y}\mathbf{j} + F_{2z}\mathbf{k}$$
$$800\mathbf{j} = (212.1 + F_{2x})\mathbf{i} + (150 + F_{2y})\mathbf{j} + (-150 + F_{2z})\mathbf{k}$$

To satisfy this equation, the corresponding $\mathbf{i}$, $\mathbf{j}$, and $\mathbf{k}$ components on the left and right sides must be equal. This is equivalent to stating that the x, y, z components of $\mathbf{F}_R$ be equal to the corresponding x, y, z components of $(\mathbf{F}_1 + \mathbf{F}_2)$. Hence,

$$0 = 212.1 + F_{2x} \qquad F_{2x} = -212.1\ \text{N}$$
$$800 = 150 + F_{2y} \qquad F_{2y} = 650\ \text{N}$$
$$0 = -150 + F_{2z} \qquad F_{2z} = 150\ \text{N}$$

Since the magnitudes of $\mathbf{F}_2$ and its components are known, we can use Eq. 2–11 to determine α, β, γ.

$$-212.1 = 700\cos\alpha_2 \qquad \alpha_2 = \cos^{-1}\left(\frac{-212.1}{700}\right) = 108° \quad \textit{Ans.}$$

$$650 = 700\cos\beta_2 \qquad \beta_2 = \cos^{-1}\left(\frac{650}{700}\right) = 21.8° \quad \textit{Ans.}$$

$$150 = 700\cos\gamma_2 \qquad \gamma_2 = \cos^{-1}\left(\frac{150}{700}\right) = 77.6° \quad \textit{Ans.}$$

These results are shown in Fig. 2–34b.

PROBLEMS

*2–48. Express each force as a Cartesian vector and then determine the resultant force $\mathbf{F}_R$. Find the magnitude and coordinate direction angles of the resultant force and sketch this vector on the coordinate system.

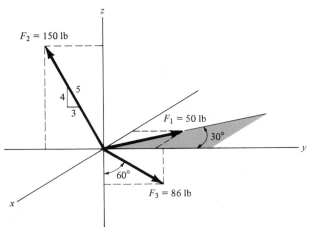

Prob. 2–48

2–50. Express each force as a Cartesian vector and then determine the resultant force $\mathbf{F}_R$. Find the magnitude and coordinate direction angles of the resultant force and sketch this vector on the coordinate system.

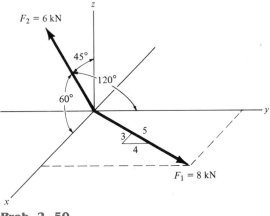

Prob. 2–50

2–49. Express each force as a Cartesian vector and then determine the resultant force $\mathbf{F}_R$. Find the magnitude and coordinate direction angles of the resultant force and sketch this vector on the coordinate system.

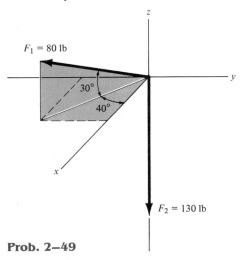

Prob. 2–49

2–51. The force $\mathbf{F}$ acting on the stake has a component of 40 N acting in the x-y plane as shown. Express $\mathbf{F}$ as a Cartesian vector.

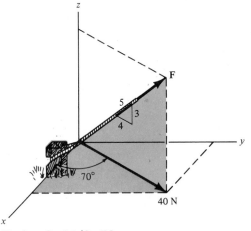

Probs. 2–51/2–52

*2–52. Determine the magnitude and coordinate direction angles of the force $\mathbf{F}$ acting on the stake.

2–53. The cable at the end of the crane boom exerts a force of 250 lb on the boom as shown. Express **F** as a Cartesian vector.

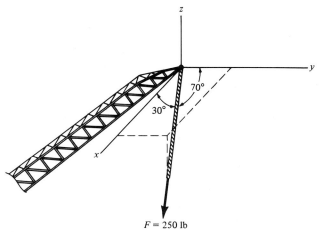

Prob. 2–53

2–54. The two forces F_1 and F_2 acting at A have a resultant force of $F_R = \{-100\mathbf{k}\}$ lb. Determine the magnitude and coordinate direction angles of F_2.

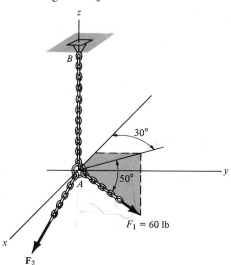

Probs. 2–54/2–55

2–55. Determine the coordinate direction angles of the force F_1 and indicate them on the figure.

*2–56. Three forces act on the hook. If the resultant force $\mathbf{F}_R$ has a magnitude and direction as shown, determine the magnitude and coordinate direction angles of force $\mathbf{F}_3$.

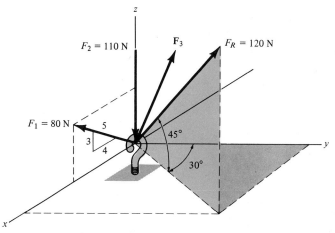

Probs. 2–56/2–57

2–57. Determine the coordinate direction angles of $\mathbf{F}_1$ and $\mathbf{F}_R$.

2–58. The pole is subjected to the force **F**, which has components acting along the x, y, z axes as shown. If the magnitude of **F** is 3 kN, and $\beta = 30°$ and $\gamma = 75°$, determine the magnitudes of its three components.

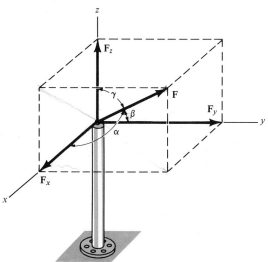

Probs. 2–58/2–59

2–59. The pole is subjected to the force **F** which has components $F_x = 1.5$ kN and $F_z = 1.25$ kN. If $\beta = 75°$, determine the magnitudes of **F** and $\mathbf{F}_y$.

*2–60. The bracket is subjected to the two forces shown. Express each force in Cartesian vector form and then determine the resultant force $\mathbf{F}_R$. Find the magnitude and coordinate direction angles of the resultant force.

2–61. Determine the magnitude and coordinate direction angles of $\mathbf{F}_3$ so that the resultant of the three forces acts along the positive y axis and has a magnitude of 600 lb.

Prob. 2–60

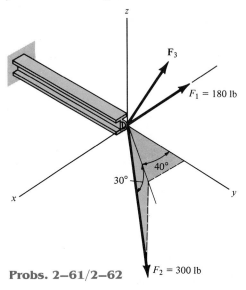

Probs. 2–61/2–62

2–62. Determine the magnitude and coordinate direction angles of $\mathbf{F}_3$ so that the resultant of the three forces is zero.

2.7 Position Vectors

In this section we will introduce the concept of a position vector. It will be shown in Sec. 2.8 that this vector is of importance in formulating a Cartesian force vector directed between any two points in space, and in Chapter 4 we will use it for finding the moment of a force.

x, y, z Coordinates

Throughout this text we will use a *right-handed* coordinate system to reference the location of points in space. Furthermore, we will use the convention followed in many technical books, and that is to require the positive z axis to be directed *upward* (the zenith direction) so that it measures the height of an object or the altitude of a point. The x, y axes then lie in the horizontal plane, Fig. 2–35. Points in space are located relative to the origin of coordinates, O, by successive measurements along the x, y, z axes. For example, in Fig. 2–35 the coordinates of point A are obtained by starting at O and measuring $x_A = 0$ along the x axis, $y_A = +2$ m along the y axis, and $z_A = 0$ along the z axis. Thus, $A(0, 2, 0)$. In a similar manner, measurements along the x, y, z axes from O to B yield the coordinates of B, i.e., $B(4, 2, -6)$. Also notice that $C(6, -1, 4)$.

Fig. 2–35

Position Vector

The *position vector* **r** is defined as a fixed vector which locates a point in space relative to another point. For example, if **r** extends from the origin of coordinates, O, to point $P(x, y, z)$, Fig. 2–36a, then **r** can be expressed in Cartesian vector form as

$$\mathbf{r} = x\mathbf{i} + y\mathbf{j} + z\mathbf{k}$$

In particular, note how the head-to-tail vector addition of the three components yields vector **r**, Fig. 2–36b. Starting at the origin O, one travels x in the $+\mathbf{i}$ direction, then y in the $+\mathbf{j}$ direction, and finally z in the $+\mathbf{k}$ direction to arrive at point P.

In the more general case, the position vector may be directed from point A to point B in space, Fig. 2–37a. As noted, this vector is designated by the symbol **r.** As a matter of convention in this text we will *sometimes* refer to this vector with *two subscripts* to indicate from and to the point where it is directed; thus, **r** can also be designated as $\mathbf{r}_{AB}$. Also, note that $\mathbf{r}_A$ and $\mathbf{r}_B$ in Fig. 2–37a are referenced with only one subscript since they extend from the origin of coordinates.

From Fig. 2–37a, by the head-to-tail vector addition, we require

$$\mathbf{r}_A + \mathbf{r} = \mathbf{r}_B$$

Solving for **r** and expressing $\mathbf{r}_A$ and $\mathbf{r}_B$ in Cartesian vector form yields

$$\mathbf{r} = \mathbf{r}_B - \mathbf{r}_A = (x_B\mathbf{i} + y_B\mathbf{j} + z_B\mathbf{k}) - (x_A\mathbf{i} + y_A\mathbf{j} + z_A\mathbf{k})$$

or

$$\boxed{\mathbf{r} = (x_B - x_A)\mathbf{i} + (y_B - y_A)\mathbf{j} + (z_B - z_A)\mathbf{k}} \qquad (2\text{–}13)$$

Thus, the **i, j, k** *components of the position vector* **r** *may be formed by taking the coordinates of the head of the vector,* $B(x_B, y_B, z_B)$, *and subtracting from them the corresponding coordinates of the tail,* $A(x_A, y_A, z_A)$. Again note how the head-to-tail addition of these three components yields **r**, i.e., going from A to B, Fig. 2–37b, one first travels $(x_B - x_A)$ in the $+\mathbf{i}$ direction, then $(y_B - y_A)$ in the $+\mathbf{j}$ direction, and finally $(z_B - z_A)$ in the $+\mathbf{k}$ direction.

(a)

(b)

Fig. 2–36

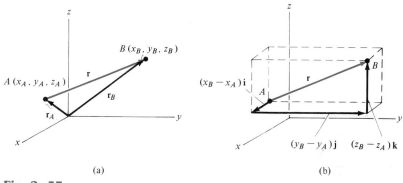

(a)   (b)

Fig. 2–37

Example 2–14

Determine the magnitude and direction of the position vector extending from *A* to *B* in Fig. 2–38*a*.

SOLUTION

The coordinates of *A* and *B* are *A*(1, 0, −3), *B*(−2, 2, 3). Hence,

$$\mathbf{r} = (-2 - 1)\mathbf{i} + (2 - 0)\mathbf{j} + (3 - (-3))\mathbf{k}$$
$$= \{-3\mathbf{i} + 2\mathbf{j} + 6\mathbf{k}\} \text{ m}$$

As shown in Fig. 2–38*b*, the three components of **r** can be obtained in a more *direct manner* by realizing that in going from *A* to *B* one must move along the *x* axis {−3**i**} m, along the *y* axis {2**j**} m, and finally along the *z* axis {6**k**} m.

The magnitude of **r** is thus

$$r = \sqrt{(-3)^2 + (2)^2 + (6)^2} = 7 \text{ m} \qquad \textit{Ans.}$$

Formulating a unit vector in the direction of **r,** we have

$$\mathbf{u} = \frac{\mathbf{r}}{r} = \frac{-3}{7}\mathbf{i} + \frac{2}{7}\mathbf{j} + \frac{6}{7}\mathbf{k}$$

The components of the unit vector yield the coordinate direction angles

$$\alpha = \cos^{-1}\left(\frac{-3}{7}\right) = 115° \qquad \textit{Ans.}$$

$$\beta = \cos^{-1}\left(\frac{2}{7}\right) = 73.4° \qquad \textit{Ans.}$$

$$\gamma = \cos^{-1}\left(\frac{6}{7}\right) = 31.0° \qquad \textit{Ans.}$$

These angles are measured from a localized coordinate system placed at the tail of **r** as shown in Fig. 2–38*c*.

(a)

(b)

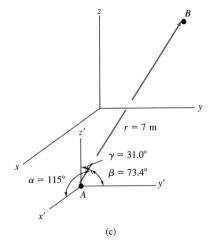
(c)

Fig. 2–38

2.8 Force Vector Directed Along a Line

Quite often in three-dimensional statics problems, the direction of a force is specified by two points through which its line of action passes. Such a situation is shown in Fig. 2–39, where the force **F** is directed along the cord AB. We can formulate **F** as a Cartesian vector by realizing that it acts in the *same direction* as the position vector **r** directed from point A to point B on the cord. This common direction is specified by the *unit vector* $\mathbf{u} = \mathbf{r}/r$. Hence,

$$\mathbf{F} = F\mathbf{u} = F\left(\frac{\mathbf{r}}{r}\right)$$

Although we have symbolically represented **F** in Fig. 2–39, note that it has units of force, and unlike **r**, which has units of length, **F** cannot be scaled along the x, y, z axes.

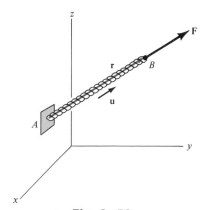

Fig. 2–39

PROCEDURE FOR ANALYSIS

When **F** is directed along a line which extends from point A to point B, then **F** can be expressed in Cartesian vector form as follows:

Position Vector. Determine the position vector **r** directed from A to B, and compute its magnitude r.

Unit Vector. Determine the unit vector $\mathbf{u} = \mathbf{r}/r$ which defines the *direction* of *both* **r** and **F**.

Force Vector. Determine **F** by combining its magnitude F and direction **u**, i.e., $\mathbf{F} = F\mathbf{u}.$

This procedure is illustrated numerically in the following example problems.

Example 2–15

The man shown in Fig. 2–40*a* pulls on the cord with a force of 70 lb. Represent this force, acting on the support *A*, as a Cartesian vector and determine its direction.

SOLUTION

Force **F** is shown in Fig. 2–40*b*. The *direction* of this vector, **u**, is determined from the position vector **r**, which extends from *A* to *B*, Fig. 2–40*b*. To formulate **F** as a Cartesian vector we use the following procedure.

Position Vector. The coordinates of the end points of the cord are *A*(0, 0, 30) and *B*(12, −8, 6). Forming the position vector by subtracting the corresponding *x*, *y*, and *z* coordinates of *A* from those of *B*, we have

$$\mathbf{r} = (12 - 0)\mathbf{i} + (-8 - 0)\mathbf{j} + (6 - 30)\mathbf{k}$$
$$= \{12\mathbf{i} - 8\mathbf{j} - 24\mathbf{k}\} \text{ ft}$$

Show on Fig. 2–40*a* how one can write **r** *directly* by going from *A* {12**i**} ft, then {−8**j**} ft, and finally {−24**k**} ft to get to *B*.

The magnitude of **r**, which represents the *length* of cord *AB*, is

$$r = \sqrt{(12)^2 + (-8)^2 + (-24)^2} = 28 \text{ ft}$$

Unit Vector. Forming the unit vector that defines the direction of both **r** and **F** yields

$$\mathbf{u} = \frac{\mathbf{r}}{r} = \frac{12}{28}\mathbf{i} - \frac{8}{28}\mathbf{j} - \frac{24}{28}\mathbf{k}$$

Force Vector. Since **F** has a *magnitude* of 70 lb and a *direction* specified by **u**, then

$$\mathbf{F} = F\mathbf{u} = 70 \text{ lb}\left(\frac{12}{28}\mathbf{i} - \frac{8}{28}\mathbf{j} - \frac{24}{28}\mathbf{k}\right)$$
$$= \{30\mathbf{i} - 20\mathbf{j} - 60\mathbf{k}\} \text{ lb} \qquad \textit{Ans.}$$

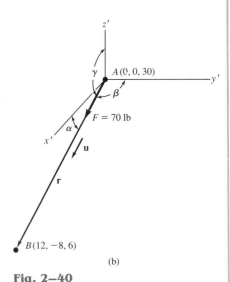

As shown in Fig. 2–40*b*, the coordinate direction angles are measured between **r** (or **F**) and the *positive axes* of a localized coordinate system with origin placed at *A*. From the components of the unit vector:

$$\alpha = \cos^{-1}\left(\frac{12}{28}\right) = 64.6° \qquad \textit{Ans.}$$

$$\beta = \cos^{-1}\left(\frac{-8}{28}\right) = 107° \qquad \textit{Ans.}$$

$$\gamma = \cos^{-1}\left(\frac{-24}{28}\right) = 149° \qquad \textit{Ans.}$$

Fig. 2–40

Example 2–16

The circular plate shown in Fig. 2–41*a* is partially supported by the cable *AB*. If the force of the cable on the hook at *A* is $F = 500$ N, express **F** as a Cartesian vector.

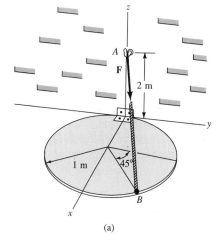

SOLUTION

As shown in Fig. 2–41*b*, **F** acts in the same direction as the position vector **r,** which extends from *A* to *B*.

Position Vector. The coordinates of the end points of the cable are $A(0, 0, 2)$ and $B(1.707, 0.707, 0)$, as indicated in the figure. Thus,

$$\mathbf{r} = (1.707 - 0)\mathbf{i} + (0.707 - 0)\mathbf{j} + (0 - 2)\mathbf{k}$$
$$= \{1.707\mathbf{i} + 0.707\mathbf{j} - 2\mathbf{k}\} \text{ m}$$

Note how one can calculate these components *directly* by going from *A*, $\{-2\mathbf{k}\}$ m along the *z* axis, then $\{1.707\mathbf{i}\}$ m along the *x* axis, and finally $\{0.707\mathbf{j}\}$ m parallel to the *y* axis to get to *B*.

The magnitude of **r** is

$$r = \sqrt{(1.707)^2 + (0.707)^2 + (-2)^2} = 2.72 \text{ m}$$

Unit Vector. Thus,

$$\mathbf{u} = \frac{\mathbf{r}}{r} = \frac{1.707}{2.72}\mathbf{i} + \frac{0.707}{2.72}\mathbf{j} - \frac{2}{2.72}\mathbf{k}$$
$$= 0.627\mathbf{i} + 0.260\mathbf{j} - 0.735\mathbf{k}$$

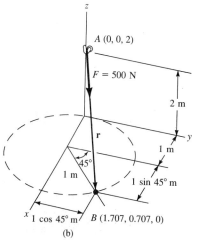

Force Vector. Since $F = 500$ N and **F** has the direction **u**, we have

$$\mathbf{F} = F\mathbf{u} = 500 \text{ N}(0.627\mathbf{i} + 0.260\mathbf{j} - 0.735\mathbf{k})$$
$$= \{314\mathbf{i} + 130\mathbf{j} - 368\mathbf{k}\} \text{ N} \qquad \textit{Ans.}$$

Using these components, notice that indeed the magnitude of **F** is 500 N, i.e.,

$$F = \sqrt{(314)^2 + (130)^2 + (-368)^2} = 500 \text{ N}$$

Fig. 2–41

Example 2–17

The cables exert forces $F_{AB} = 100$ N and $F_{AC} = 120$ N on the ring at A as shown in Fig. 2–42a. Determine the magnitude of the resultant force acting at A.

SOLUTION

The resultant force $\mathbf{F}_R$ is shown graphically in Fig. 2–42b. We can express this force as a Cartesian vector by first formulating $\mathbf{F}_{AB}$ and $\mathbf{F}_{AC}$ as Cartesian vectors and then adding their components. The directions of $\mathbf{F}_{AB}$ and $\mathbf{F}_{AC}$ are specified by forming unit vectors $\mathbf{u}_{AB}$ and $\mathbf{u}_{AC}$ along the cables. These unit vectors are obtained from the associated position vectors $\mathbf{r}_{AB}$ and $\mathbf{r}_{AC}$. With reference to Fig. 2–42b for $\mathbf{F}_{AB}$ we have

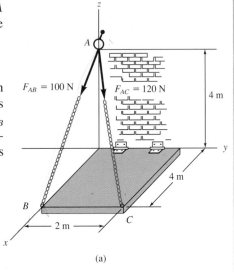

(a)

$$\mathbf{r}_{AB} = (4 - 0)\mathbf{i} + (0 - 0)\mathbf{j} + (0 - 4)\mathbf{k}$$
$$= \{4\mathbf{i} - 4\mathbf{k}\} \text{ m}$$
$$r_{AB} = \sqrt{(4)^2 + (-4)^2} = 5.66 \text{ m}$$
$$\mathbf{F}_{AB} = 100 \text{ N}\left(\frac{\mathbf{r}_{AB}}{r_{AB}}\right) = 100 \text{ N}\left(\frac{4}{5.66}\mathbf{i} - \frac{4}{5.66}\mathbf{k}\right)$$
$$\mathbf{F}_{AB} = \{70.7\mathbf{i} - 70.7\mathbf{k}\} \text{ N}$$

For $\mathbf{F}_{AC}$ we have

$$\mathbf{r}_{AC} = (4 - 0)\mathbf{i} + (2 - 0)\mathbf{j} + (0 - 4)\mathbf{k}$$
$$= \{4\mathbf{i} + 2\mathbf{j} - 4\mathbf{k}\} \text{ m}$$
$$r_{AC} = \sqrt{(4)^2 + (2)^2 + (-4)^2} = 6 \text{ m}$$
$$\mathbf{F}_{AC} = 120 \text{ N}\left(\frac{\mathbf{r}_{AC}}{r_{AC}}\right) = 120 \text{ N}\left(\frac{4}{6}\mathbf{i} + \frac{2}{6}\mathbf{j} - \frac{4}{6}\mathbf{k}\right)$$
$$= \{80\mathbf{i} + 40\mathbf{j} - 80\mathbf{k}\} \text{ N}$$

The resultant force is therefore

$$\mathbf{F}_R = \mathbf{F}_{AB} + \mathbf{F}_{AC} = (70.7\mathbf{i} - 70.7\mathbf{k}) + (80\mathbf{i} + 40\mathbf{j} - 80\mathbf{k})$$
$$= \{150.7\mathbf{i} + 40\mathbf{j} - 150.7\mathbf{k}\} \text{ N}$$

The magnitude of $\mathbf{F}_R$ is thus

$$F_R = \sqrt{(150.7)^2 + (40)^2 + (-150.7)^2}$$
$$= 217 \text{ N} \qquad \qquad Ans.$$

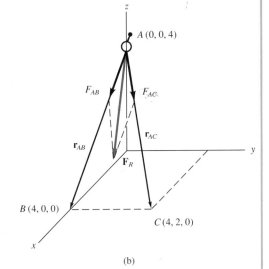

(b)

Fig. 2–42

PROBLEMS

2–63. Express the position vector **r** in Cartesian vector form; then determine its magnitude and coordinate direction angles.

Prob. 2–63

*2–64. Express the position vector **r** in Cartesian vector form; then determine its magnitude and coordinate direction angles.

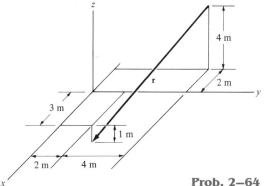

Prob. 2–64

2–65. Express the position vector **r** in Cartesian vector form; then determine its magnitude and coordinate direction angles.

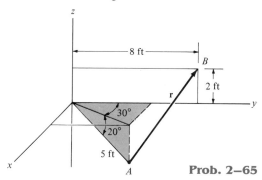

Prob. 2–65

2–66. The 5-m-long cable is attached to the beam at *A*. If $z = 3$ m, determine the coordinate location $+x$, $+y$ where the end of the cable should be staked to the ground. Choose the value such that $x = 2y$.

Probs. 2–66/2–67

2–67. The cable is attached to the beam at *A* and to the stake at *B*, which has coordinates of $x = 2$ m and $y = 1$ m. If $z = 3$ m, determine the length of the cable and the angle between the positive *z* axis and the cable.

*2–68. Determine the length of the crankshaft *AB* by first formulating a Cartesian position vector from *A* to *B* and then determining its magnitude.

Prob. 2–68

2–69. Determine the distance between the end points A and B on the wire. *Hint:* Formulate a position vector directed from A to B and then determine its magnitude.

Prob. 2–69

2–70. The positions of point A on the building and point B on the antenna have been measured relative to the electronic distance meter (EDM) at O. Determine the distance between A and B. *Hint:* Formulate a position vector directed from A to B, then determine its magnitude.

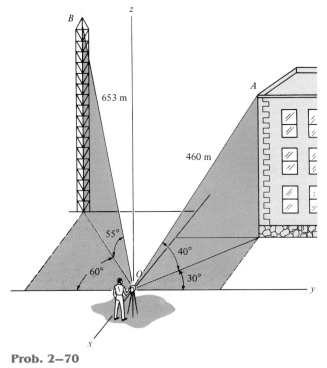

Prob. 2–70

2–71. Determine the lengths of cords ACB and CO. The knot at C is located midway between A and B.

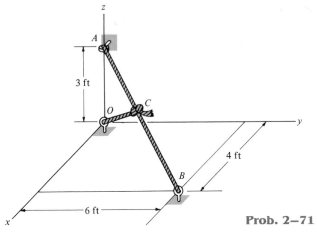

Prob. 2–71

***2–72.** Express force $\mathbf{F}$ as a Cartesian vector; then determine its coordinate direction angles.

Prob. 2–72 x

2–73. Express each of the forces in Cartesian vector form and then determine the magnitude and coordinate direction angles of the resultant force.

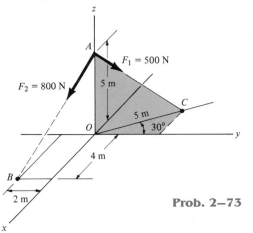

Prob. 2–73

2–74. Express each of the forces in Cartesian vector form and then determine the magnitude and coordinate direction angles of the resultant force.

Prob. 2–74

*2–76. Express each of the forces in Cartesian vector form and then determine the magnitude and coordinate direction angles of the resultant force.

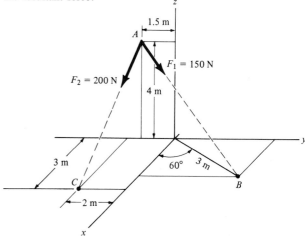

Prob. 2–76

2–75. Express each of the forces in Cartesian vector form and then determine the magnitude and coordinate direction angles of the resultant force.

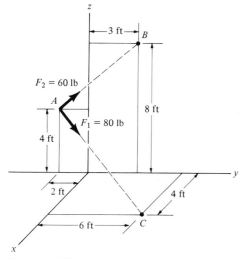

Prob. 2–75

2–77. The force which the mooring rope AB exerts on the bow of the ship is $\mathbf{F} = \{400\mathbf{i} - 300\mathbf{j} - 250\mathbf{k}\}$ N. If the length of the rope is 20 m, determine the coordinates $B(x, -y, 0)$ of the capstan. Assume rope AB is along a straight line. Also, how high, z, is A above B?

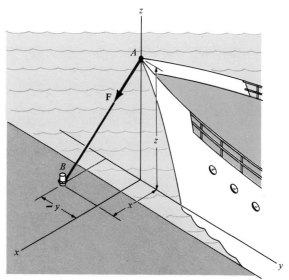

Prob. 2–77

2–78. Express force **F** as a Cartesian vector; then determine its coordinate direction angles.

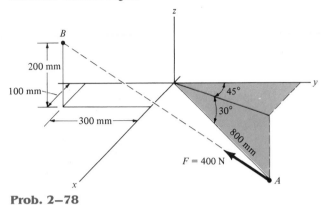

Prob. 2–78

2–79. The cord exerts a force of $\mathbf{F} = \{12\mathbf{i} + 9\mathbf{j} - 8\mathbf{k}\}$ lb on the hook. If the cord is 8 ft long, determine the location x, y of the point of attachment B, and the height z of the hook.

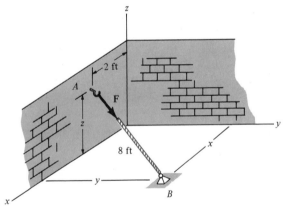

Probs. 2–79/2–80

*2–80. The cord exerts a force of $F = 30$ lb on the hook. If the cord is 8 ft long, $z = 4$ ft, and the x component of the force is $F_x = 25$ lb, determine the location x, y of the point of attachment B of the cord to the ground.

2–81. The chandelier is supported by three chains which are concurrent at point O. If the force in each chain has a magnitude of 60 lb, express each force as a Cartesian vector and determine the magnitude and coordinate direction angles of the resultant force.

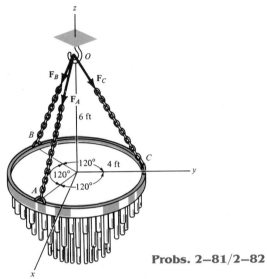

Probs. 2–81/2–82

2–82. The chandelier is supported by three chains which are concurrent at point O. If the resultant force at O has a magnitude of 130 lb and is directed along the negative z axis, determine the force in each chain assuming $F_A = F_B = F_C$.

2–83. The cable, attached to the shear-leg derrick, exerts a force on the derrick of 350 lb. Express this force as a Cartesian vector.

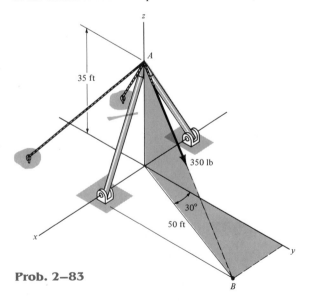

Prob. 2–83

*2–84. The pipe is supported at its ends by a cord AB. If the cord exerts a force of $F = 12$ lb on the pipe at A, express this force as a Cartesian vector.

Prob. 2–84

2–85. Two boys pull on the pole with the forces shown. Represent each force as a Cartesian vector and then determine the magnitude and coordinate direction angles of the resultant force.

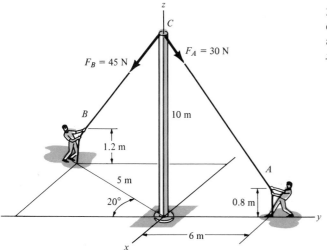

Prob. 2–85

2–86. The pole is held in place by three cables. If the force of each cable acting on the pole is shown, determine the position $(x, 0, z)$ for fixing cable DA so that the resultant force exerted on the pole is directed along its axis, from D towards O.

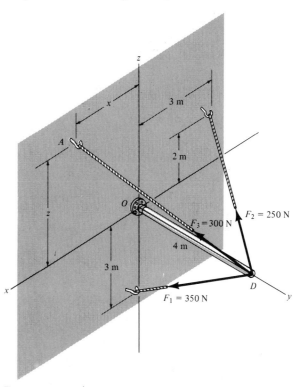

Probs. 2–86/2–87

2–87. The pole is held in place by three cables. If the force of each cable acting on the pole is shown, determine the magnitude and coordinate direction angles α, β, γ of the resultant force. Set $x = 3.50$ m, $z = 3$ m.

2.9 Dot Product

Fig. 2–43

Occasionally in statics one has a need to find the component of a force along a line or to find the angle between two lines. In two dimensions, these problems can readily be solved by trigonometry since the geometry is easy to visualize. In three dimensions, however, this is often difficult, and consequently vector methods should be employed for the solution. The dot product defines a particular method for "multiplying" two vectors and is used to solve the above-mentioned problems.

The *dot product* of vectors **A** and **B**, written **A · B**, and read "**A** dot **B**," is defined as the product of the magnitudes of **A** and **B** and the cosine of the angle θ between their tails, Fig. 2–43. Expressed in equation form,

$$\mathbf{A} \cdot \mathbf{B} = AB \cos \theta \qquad (2\text{–}14)$$

where $0° \leq \theta \leq 180°$. The dot product is often referred to as the *scalar product* of vectors, since the result is a *scalar* and not a vector.

Laws of Operation

1. Commutative law:

$$\mathbf{A} \cdot \mathbf{B} = \mathbf{B} \cdot \mathbf{A}$$

2. Multiplication by a scalar:

$$a(\mathbf{A} \cdot \mathbf{B}) = (a\mathbf{A}) \cdot \mathbf{B} = \mathbf{A} \cdot (a\mathbf{B}) = (\mathbf{A} \cdot \mathbf{B})a$$

3. Distributive law:

$$\mathbf{A} \cdot (\mathbf{B} + \mathbf{D}) = (\mathbf{A} \cdot \mathbf{B}) + (\mathbf{A} \cdot \mathbf{D})$$

It is easy to prove the first and second laws by using Eq. 2–14. The proof of the distributive law is left as an exercise (see Prob. 2–88).

Cartesian Vector Formulation

Equation 2–14 may be used to find the dot product of each of the Cartesian unit vectors. For example, $\mathbf{i} \cdot \mathbf{i} = (1)(1) \cos 0° = 1$ and $\mathbf{i} \cdot \mathbf{j} = (1)(1) \cos 90° = 0$. In a similar manner,

$$\mathbf{i} \cdot \mathbf{i} = 1 \qquad \mathbf{j} \cdot \mathbf{j} = 1 \qquad \mathbf{k} \cdot \mathbf{k} = 1$$
$$\mathbf{i} \cdot \mathbf{j} = 0 \qquad \mathbf{i} \cdot \mathbf{k} = 0 \qquad \mathbf{k} \cdot \mathbf{j} = 0$$

These results should not be memorized; rather, it should be clearly understood how each is obtained.

Consider now the dot product of two general vectors **A** and **B** which are expressed in Cartesian vector form. Then

$$\mathbf{A} \cdot \mathbf{B} = (A_x\mathbf{i} + A_y\mathbf{j} + A_z\mathbf{k}) \cdot (B_x\mathbf{i} + B_y\mathbf{j} + B_z\mathbf{k})$$
$$= A_xB_x(\mathbf{i} \cdot \mathbf{i}) + A_xB_y(\mathbf{i} \cdot \mathbf{j}) + A_xB_z(\mathbf{i} \cdot \mathbf{k})$$
$$+ A_yB_x(\mathbf{j} \cdot \mathbf{i}) + A_yB_y(\mathbf{j} \cdot \mathbf{j}) + A_yB_z(\mathbf{j} \cdot \mathbf{k})$$
$$+ A_zB_x(\mathbf{k} \cdot \mathbf{i}) + A_zB_y(\mathbf{k} \cdot \mathbf{j}) + A_zB_z(\mathbf{k} \cdot \mathbf{k})$$

Carrying out the dot-product operations, the final result becomes

$$\mathbf{A} \cdot \mathbf{B} = A_xB_x + A_yB_y + A_zB_z \qquad (2\text{--}15)$$

Thus, to determine the dot product of two Cartesian vectors, multiply their corresponding x, y, z components and sum their products algebraically. Since the result is a scalar, be careful *not* to include any unit vectors in the final result.

Applications

The dot product has two important applications in mechanics.

1. *The angle formed between two vectors or intersecting lines.* The angle θ between vectors **A** and **B** in Fig. 2–43 can be determined from Eq. 2–14 and written as

$$\theta = \cos^{-1}\left(\frac{\mathbf{A} \cdot \mathbf{B}}{AB}\right) \qquad 0° \leq \theta \leq 180°$$

Here $\mathbf{A} \cdot \mathbf{B}$ is computed from Eq. 2–15. In particular, notice that if $\mathbf{A} \cdot \mathbf{B} = 0$, $\theta = \cos^{-1} 0 = 90°$, so that **A** will be *perpendicular* to **B.**

2. *The component of a vector parallel to a line.* The component of vector **A** parallel to or collinear with the line aa' in Fig. 2–44 is defined by $\mathbf{A}_{\parallel}$, where $A_{\parallel} = A \cos \theta$. This component is sometimes referred to as the *projection* of **A** onto the line, since a right angle is formed in the construction. If the *direction* of the line is specified by the unit vector **u,** then, since $u = 1$, we can determine $A_{\parallel}$ directly from the dot product (Eq. 2–14), i.e.,

$$A_{\parallel} = A \cos \theta = \mathbf{A} \cdot \mathbf{u}$$

Hence, the scalar projection of **A** *along a line is determined from the dot product of* **A** *and the unit vector* **u** *which defines the direction of the line.* The component $\mathbf{A}_{\parallel}$ represented as a *vector* is therefore

$$\mathbf{A}_{\parallel} = A \cos \theta\,\mathbf{u} = (\mathbf{A} \cdot \mathbf{u})\mathbf{u}$$

Note that the component of **A** which is *perpendicular* to line aa' can also be obtained, Fig. 2–44. Since $\mathbf{A} = \mathbf{A}_{\parallel} + \mathbf{A}_{\perp}$, then $\mathbf{A}_{\perp} = \mathbf{A} - \mathbf{A}_{\parallel}$. If θ is first determined from the dot product, $\theta = \cos^{-1}(\mathbf{A} \cdot \mathbf{u}/A)$, then $A_{\perp} = A \sin \theta$. Alternatively, if $A_{\parallel}$ is known, then by the Pythagorean theorem we can also write $A_{\perp} = \sqrt{A^2 - A_{\parallel}^2}$.

The above two applications are illustrated numerically in the following example problems.

Fig. 2–44

Example 2–18

The frame shown in Fig. 2–45a is subjected to a horizontal force $\mathbf{F} = \{300\mathbf{j}\}$ N acting at its corner. Determine the projected component of this force along member AB.

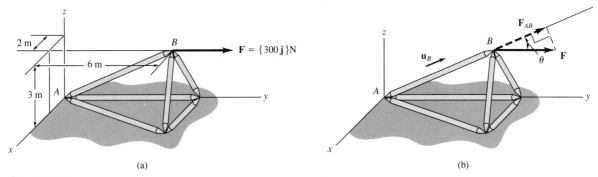

Fig. 2–45

SOLUTION

The magnitude of the component of $\mathbf{F}$ along AB is equal to the dot product of $\mathbf{F}$ and the unit vector $\mathbf{u}_B$ which defines the direction of AB, Fig. 2–45b. Since

$$\mathbf{u}_B = \frac{\mathbf{r}_B}{r_B} = \frac{2\mathbf{i} + 6\mathbf{j} + 3\mathbf{k}}{\sqrt{(2)^2 + (6)^2 + (3)^2}} = 0.286\mathbf{i} + 0.857\mathbf{j} + 0.429\mathbf{k}$$

Then

$$F_{AB} = F \cos \theta = \mathbf{F} \cdot \mathbf{u}_B = (300\mathbf{j}) \cdot (0.286\mathbf{i} + 0.857\mathbf{j} + 0.429\mathbf{k})$$
$$= (0)(0.286) + (300)(0.857) + (0)(0.429)$$
$$= 257.1 \text{ N}$$

Expressing $\mathbf{F}_{AB}$ in Cartesian vector form, we have

$$\mathbf{F}_{AB} = F_{AB}\mathbf{u}_B = 257.1 \text{ N}(0.286\mathbf{i} + 0.857\mathbf{j} + 0.429\mathbf{k})$$
$$= \{73.5\mathbf{i} + 220\mathbf{j} + 110\mathbf{k}\} \text{ N} \qquad \textit{Ans.}$$

Note that, if needed, the angle θ between $\mathbf{F}$ and $\mathbf{F}_{AB}$, Fig. 2-45b, can be determined as follows:

$$\cos \theta = \frac{\mathbf{F} \cdot \mathbf{F}_{AB}}{F \, F_{AB}} = \frac{0(73.5) + 300(220) + 0(110)}{300(257.1)}$$
$$\theta = \cos^{-1}(0.8557) = 31.2°$$

Example 2–19

The pipe in Fig. 2–46a is subjected to the force $F = 80$ lb at its end B. Determine the angle θ between $\mathbf{F}$ and the pipe segment BA, and the magnitudes of the components of $\mathbf{F}$, which are parallel and perpendicular to BA.

SOLUTION

Angle θ. Position vectors along BA and BC are first determined.

$$\mathbf{r}_{BA} = \{-2\mathbf{i} - 2\mathbf{j} + 1\mathbf{k}\} \text{ ft}$$
$$\mathbf{r}_{BC} = \{-3\mathbf{j} + 1\mathbf{k}\} \text{ ft}$$

Thus,

$$\cos\theta = \frac{\mathbf{r}_{BA} \cdot \mathbf{r}_{BC}}{r_{BA} r_{BC}} = \frac{(-2)(0) + (-2)(-3) + (1)(1)}{3\sqrt{10}}$$
$$= 0.7379$$
$$\theta = 42.5° \qquad \qquad Ans.$$

(a)

Components of F. The line BA is defined by $\mathbf{u}_{BA}$. Since $\mathbf{F} = (80 \text{ lb})\mathbf{u}_{BC}$, then from Fig. 2–46b, the component of $\mathbf{F}$ in the direction of BA is

$$F_{BA} = \mathbf{F} \cdot \mathbf{u}_{BA} = (80 \text{ lb})\mathbf{u}_{BC} \cdot \mathbf{u}_{BA}$$
$$= (80 \text{ lb})\left(\frac{\mathbf{r}_{BC}}{r_{BC}}\right) \cdot \left(\frac{\mathbf{r}_{BA}}{r_{BA}}\right)$$
$$= (80 \text{ lb})\left(\frac{(-2)(0) + (-2)(-3) + (1)(1)}{3\sqrt{10}}\right)$$
$$= 59.0 \text{ lb} \qquad \qquad Ans.$$

The magnitude of the perpendicular component is

$$F_\perp = \sqrt{F^2 - F_{BA}^2}$$
$$= \sqrt{(80)^2 - (59.0)^2}$$
$$= 54.0 \text{ lb} \qquad \qquad Ans.$$

or, since θ is known,

$$F_\perp = F \sin\theta$$
$$= 80 \sin 42.5°$$
$$= 54.0 \text{ lb} \qquad \qquad Ans.$$

(b)

Fig. 2–46

63

PROBLEMS

*2–88. Given the three vectors **A**, **B**, and **D**, show that
A · (**B** + **D**) = (**A** · **B**) + (**A** · **D**).

2–89. Determine the angle θ between the two position vectors.

Probs. 2–89/2–90

2–90. Determine the projected component of vector r_1 acting in the direction of r_2. Express the result as a Cartesian vector.

2–91. A force of **F** = {500**k**} N acts at point A. Determine the magnitudes of the components **F**$_1$ and **F**$_2$ which act along the OA axis and perpendicular to it.

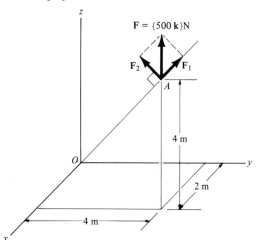

Prob. 2–91

*2–92. Determine the angle θ between the two position vectors.

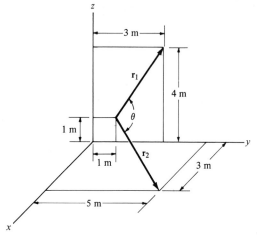

Probs. 2–92/2–93

2–93. Determine the projected component of vector r_1 acting in the direction of r_2. Express the result as a Cartesian vector.

2–94. Determine the magnitude of the projected component of the position vector **r** along the Oa axis.

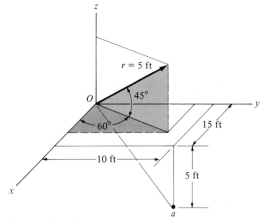

Prob. 2–94

2–95. Determine the magnitude of the projected component of force **F** along the *Oa* axis.

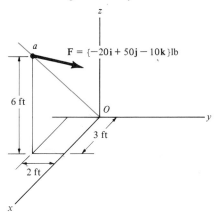

Prob. 2–95

2–98. Cable *BC* exerts a force of $F = 56$ N on the top of the flagpole. Determine the magnitude of the projected component of this force along the z axis of the pole.

Prob. 2–98

*2–96. The cables each exert a force of 400 N on the post. Determine the magnitude of the projected component of F_1 along the line of action of F_2.

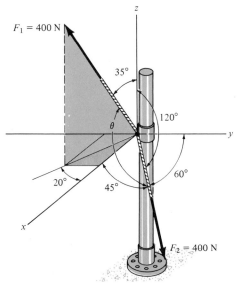

Probs. 2–96/2–97

2–97. Determine the angle θ between the two cables attached to the post.

2–99. Determine the angles θ and ϕ between the axis *OA* of the pole and each cable, *AB* and *AC*.

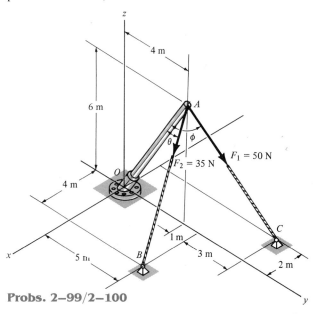

Probs. 2–99/2–100

*2–100. The two cables exert the forces shown on the pole. Determine the magnitude of the projected component of each force acting along the axis *OA* of the pole.

2–101 Determine the magnitudes of the projected components of the force $\mathbf{F} = \{60\mathbf{i} + 12\mathbf{j} - 40\mathbf{k}\}$ N in the direction of the cables AB and AC.

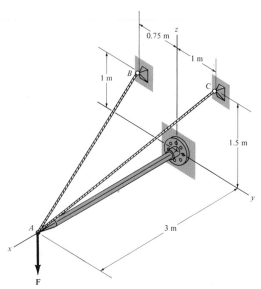

Prob. 2–101

2–102. Cable OA is used to support column OB. Determine the angle θ it makes with beam OC.

Probs. 2–102/2–103

2–103. Cable OA is used to support column OB. Determine the angle ϕ it makes with beam OD.

*2–104. The force $\mathbf{F}$ acts at the end A of the pipe assembly. Determine the magnitudes of the components $\mathbf{F}_1$ and $\mathbf{F}_2$ which act along the axis of AB and perpendicular to it.

Prob. 2–104

2–105. Determine the magnitude of the projected component of the 100-lb force acting along the axis BC of the pipe.

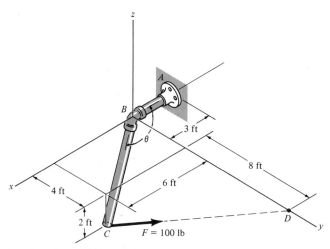

Probs. 2–105/2–106

2–106. Determine the angle θ between pipe segments BA and BC.

REVIEW PROBLEMS

2–107. The pith ball is subjected to an electrostatic repulsive force **F** which has measured components of $F_x = 40$ mN and $F_z = 20$ mN. If the angle $\beta = 135°$, determine the magnitudes of **F** and **F**$_y$.

Prob. 2–107

*2–108. Determine the angles θ and ϕ between the wire segments.

Prob. 2–108

2–109. Determine the magnitude and coordinate direction angles of **F**$_2$ so that the resultant of the two forces acts upward along the z axis of the pole and has a magnitude of 275 N.

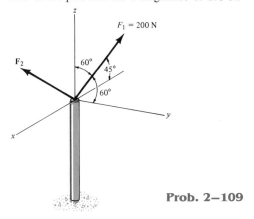

Prob. 2–109

2–110. The window is held open by cable AB. Determine the length of the cable and express the 30-N force acting at A along the cable as a Cartesian vector.

Prob. 2–110

2–111. Cable AB exerts a force of 80 N on the end of the 3-m-long boom. Determine the magnitude of the projected component of this force along AO.

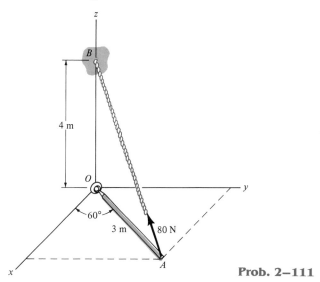

Prob. 2–111

67

*2–112. Using chains, two tractors exert forces of $\mathbf{F}_A$ and $\mathbf{F}_B$ on the trunk of a tree as shown. Determine the magnitude of the resultant force. If the same magnitudes of forces $\mathbf{F}_A$ and $\mathbf{F}_B$ are maintained, in what direction should the tractors pull on the trunk so as to cause the *maximum* resultant force? What is the magnitude of this maximum force?

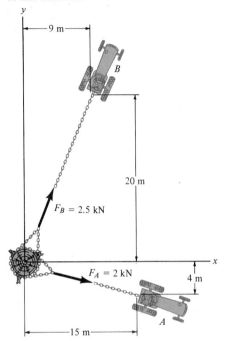

Prob. 2–112

2–113. The load at A creates a force of 60 lb in wire AB. Express this force as a Cartesian vector acting on A and directed toward B as shown.

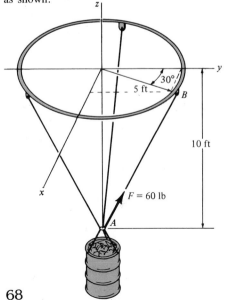

Prob. 2–113

2–114. Determine the design angle θ ($\theta < 90°$) between the two struts so that the 500-lb horizontal force has a component of 600 lb directed from A toward C. What is the component of force acting along member BA?

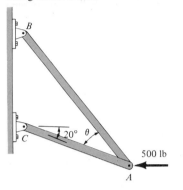

Prob. 2–114

2–115. If $\theta = 20°$ and $\phi = 35°$, determine the magnitudes of $\mathbf{F}_1$ and $\mathbf{F}_2$ so that the resultant force has a magnitude of 20 lb and is directed along the positive x axis.

Probs. 2–115/2–116

*2–116. If $F_1 = F_2 = 30$ lb, determine the angles θ and ϕ so that the resultant force is directed along the positive x axis and has a magnitude of $F_R = 20$ lb.

2–117. Determine the magnitudes of the two components of the 600-N force, one directed along the cable AC and the other along the axis of strut AB.

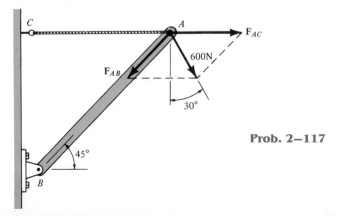

Prob. 2–117

3

Equilibrium of a Particle

In this chapter the methods of resolving a force into components and expressing a force as a Cartesian vector will be used to solve problems involving the equilibrium of a particle. Recall that the dimensions or size of a particle are assumed to be neglected, and therefore a particle can be subjected *only* to a system of *concurrent forces*. To simplify the discussion, the equilibrium of a particle subjected to a coplanar force system will be considered first. Then, in the last part of the chapter, equilibrium problems involving three-dimensional force systems will be considered.

3.1 Condition for the Equilibrium of a Particle

A particle is in *equilibrium* provided it is at rest if originally at rest, or has a constant velocity if originally in motion. Most often, however, the term "equilibrium" or more specifically "static equilibrium" is used to describe an object at rest. To maintain a state of equilibrium, it is *necessary* to satisfy Newton's first law of motion, which states that if the *resultant force* acting on a particle is *zero*, then the particle is in equilibrium. This condition may be stated mathematically as

$$\Sigma \mathbf{F} = \mathbf{0} \qquad (3-1)$$

where $\Sigma \mathbf{F}$ is the vector *sum of all the forces* acting on the particle.

Not only is Eq. 3–1 a necessary condition for equilibrium, but it is also a *sufficient* condition. This follows from Newton's second law of motion, which can be written as $\Sigma \mathbf{F} = m\mathbf{a}$. Since the force system satisfies Eq. 3–1, then $m\mathbf{a} = \mathbf{0}$, and therefore the particle's acceleration $\mathbf{a} = \mathbf{0}$, and consequently the particle indeed moves with constant velocity or remains at rest.

3.2 The Free-Body Diagram

To correctly apply the equation of equilibrium, it is necessary to account for *all* the known and unknown forces ($\Sigma \mathbf{F}$) which act *on* the particle. The best way to do this is to draw the particle's free-body diagram. This diagram is a sketch of the particle which represents it as being isolated or "free" from its surroundings. On this sketch one then shows *all* the forces which the surroundings exert *on* the particle. Provided this diagram is correctly drawn, it will then be easy to apply Eq. 3–1.

Before presenting a formal procedure as to how to draw a free-body diagram, we will first discuss two types of supports often encountered in particle equilibrium problems.

Cables and Pulleys

Throughout this book, except in Sec. 7.4, all cables are assumed to have negligible weight and they cannot be stretched. A cable can support *only* a tension or "pulling" force, and this force always acts in the direction of the cable. Later it will be shown that the tension force developed in a *continuous cable* which passes over a frictionless pulley must have a *constant* magnitude to keep the cable in equilibrium. (see Example 5–7). Hence, for any angle θ, shown in Fig. 3–1, the cable is subjected to a constant tension T throughout its length.

Fig. 3–1

Springs

If a *linear elastic spring* is used for a support, the length of the spring will change in direct proportion to the force acting on it. A characteristic that defines the "elasticity" of a spring is the *spring constant* or *stiffness k*. Specifically, the magnitude of force developed by a linear elastic spring which has a stiffness k, and is deformed (compressed or elongated) a distance s measured from its unloaded position, is

$$F = ks \qquad (3\text{–}2)$$

Note that s is determined from the difference in the spring's deformed length l and its undeformed length l_o, i.e., $s = l - l_o$. For example, the spring shown

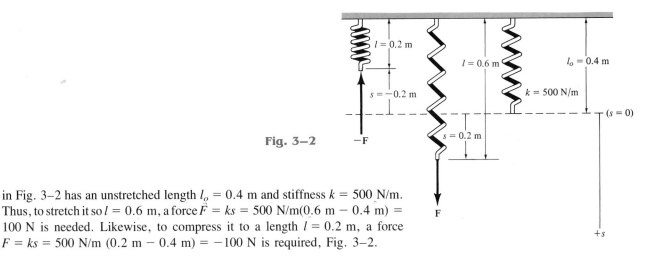

Fig. 3-2

in Fig. 3–2 has an unstretched length $l_o = 0.4$ m and stiffness $k = 500$ N/m. Thus, to stretch it so $l = 0.6$ m, a force $\vec{F} = ks = 500$ N/m$(0.6$ m $- 0.4$ m$) = 100$ N is needed. Likewise, to compress it to a length $l = 0.2$ m, a force $F = ks = 500$ N/m $(0.2$ m $- 0.4$ m$) = -100$ N is required, Fig. 3–2.

PROCEDURE FOR DRAWING A FREE-BODY DIAGRAM

The importance of drawing a free-body diagram before applying the equation of equilibrium to the solution of a problem cannot be overemphasized. To construct a free-body diagram, the following three steps are necessary.

Step 1. Imagine the particle to be *isolated* or cut "free" from its surroundings. Draw or sketch its outlined shape.

Step 2. Indicate on this sketch *all* the forces that act *on the particle.* These forces can be *active forces,* which tend to set the particle in motion, such as those caused by attached cables, weight, or magnetic and electrostatic interaction. Also, *reactive forces* will occur, such as those caused by the constraints or supports that tend to prevent motion. To account for all these forces, it may help to trace around the particle's boundary, carefully noting each force acting on it.

Step 3. The forces that are *known* should be labeled with their proper magnitudes and directions. Letters are used to represent the magnitudes and directions of forces that are unknown. In particular, if a force has a known line of action but unknown magnitude, the "arrowhead," which defines the sense of the force, can be *assumed.* The correct sense will become apparent after solving for the unknown magnitude. By definition, the *magnitude* of a force is *always positive* so that, if the solution yields a "negative" scalar, the *minus sign* indicates that the arrowhead or sense of the force is opposite to that which was originally assumed.

Application of the above steps is illustrated in the following two examples.

Example 3–1

The crate in Fig. 3–3a has a weight of 20 lb. Draw a free-body diagram of the cord *BD* and the ring at *B*.

(a)

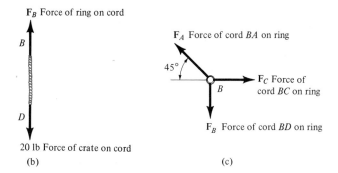

$\mathbf{F}_B$ Force of ring on cord

20 lb Force of crate on cord

(b)

$\mathbf{F}_A$ Force of cord *BA* on ring

$\mathbf{F}_C$ Force of cord *BC* on ring

$\mathbf{F}_B$ Force of cord *BD* on ring

(c)

Fig. 3–3

SOLUTION

If we imagine the cord *BD* to be isolated from its surroundings, then by inspection there are only two forces that act on it; namely, the 20-lb force at *D* caused by the weight of the crate, Fig. 3–3b, and the force $\mathbf{F}_B$ at *B* which is caused by the ring. Since these forces tend to *stretch* the cord, we can state that the cord is in *tension*. (This must be the case, since compressive forces would cause the cord to collapse.)

When the ring at *B* is isolated from its surroundings, it is realized that three forces act on it. All these forces are caused by the attached cords, Fig. 3–3c. Notice that $\mathbf{F}_B$ shown here is equal but opposite to that shown in Fig. 3–3b, a consequence of Newton's third law.

Example 3–2

The sphere in Fig. 3–4a has a mass of 6 kg and is supported as shown. Draw a free-body diagram of the knot at C.

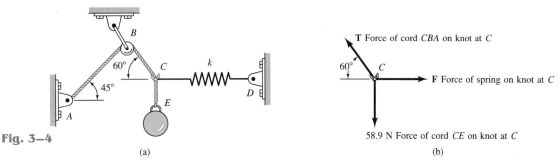

Fig. 3–4

(a)   (b)

SOLUTION

By inspection three forces act on C; they are caused by cords CBA and CE, and the spring CD. Note that the weight of the sphere, $W = (6 \text{ kg})(9.81 \text{ m/s}^2) = 58.9 \text{ N}$, is "transmitted" to point C by the supporting cord CE. That is, C is subjected to the force of the cable CE; it is not subjected *directly* to the weight of the sphere. Thus, the free-body diagram of the knot at C is shown in Fig. 3–4b.

3.3 Coplanar Force Systems

Many particle equilibrium problems involve a coplanar force system. If the forces lie in the x-y plane, they can each be resolved into their respective **i** and **j** components and Eq. 3–1 can be written as

$$\Sigma \mathbf{F} = \mathbf{0}$$
$$\Sigma F_x \mathbf{i} + \Sigma F_y \mathbf{j} = \mathbf{0}$$

For this vector equation to be satisfied, both the x and y components must equal zero, otherwise $\Sigma \mathbf{F} \neq \mathbf{0}$. Hence, we require

$$\boxed{\begin{array}{c} \Sigma F_x = 0 \\ \Sigma F_y = 0 \end{array}} \qquad (3\text{–}3)$$

These scalar equilibrium equations state that the algebraic sum of the x and y components of all the forces acting on the particle be equal to zero. As a result, Eqs. 3–3 can be solved for at most two unknowns, generally represented as angles and magnitudes of forces shown on the particle's free-body diagram.

Scalar Notation

Since each of the two equilibrium equations requires the resolution of vector components along a specified axis (x or y), we will use scalar notation to represent the components when applying these equations. By doing this, the sense of direction for each component is accounted for by an *algebraic sign* which corresponds to the arrowhead direction of the component as indicated graphically on the free-body diagram. In particular, if a force component has an unknown magnitude, then the arrowhead sense of the force on the free-body diagram can be *assumed*. Since the magnitude of a force is *always positive*, then if the *solution* yields a *negative scalar*, it indicates that the sense of the force shown on the free-body diagram is opposite to that which was assumed. For example, consider the free-body diagram of the particle subjected to two forces as shown in Fig. 3–5. For the sake of discussion, we have *assumed* the *unknown force* $\mathbf{F}$ acts to the right to maintain equilibrium. Application of the equation of equilibrium along the x axis yields

$$\xrightarrow{+}\Sigma F_x = 0; \qquad\qquad +F + 10 \text{ N} = 0$$

Both terms are ''positive'' since the forces act in the positive x direction, as indicated by the arrow placed alongside the equation. When the equation is solved, $F = -10$ N. *Here* the negative sign refers to the fact that $\mathbf{F}$ in Fig. 3–5 is shown in the opposite sense of direction. In other words, $\mathbf{F}$ must act to the left to hold the particle in equilibrium.

Fig. 3–5

PROCEDURE FOR ANALYSIS

The following procedure provides a method for solving coplanar force problems involving particle equilibrium:

Free-Body Diagram. Draw a free-body diagram of the particle. As outlined in Sec. 3.2, this requires that all the known and unknown force magnitudes and angles be labeled on the diagram. The sense of a force having an unknown magnitude can be assumed.

Equations of Equilibrium. Establish the x, y axes in *any* suitable direction and apply the two equations of equilibrium, $\Sigma F_x = 0$ and $\Sigma F_y = 0$. For application, components are positive if they are directed along the positive axes, and negative if they are directed along the negative axes. If more than two unknowns exist and the problem involves a spring, apply $F = ks$ (Eq. 3–2) to relate the spring force to the deformation s of the spring.

The following example problems numerically illustrate this solution procedure.

Example 3–3

Determine the tension in cords AB and AD for equilibrium of the 10-kg crate shown in Fig. 3–6a.

SOLUTION

The cord tensions can be obtained by investigating the equilibrium of the ring at A, because this "particle" is subjected to the force of both cords.

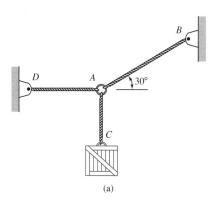

(a)

Free-Body Diagram. As shown in Fig. 3–6b, there are three concurrent forces *acting on the ring*. The cord tension forces $\mathbf{T}_B$ and $\mathbf{T}_D$ have unknown magnitudes but known directions. Cord AC exerts a downward force on A equal to the weight of the crate, $W = (10 \text{ kg})(9.81 \text{ m/s}^2) = 98.1$ N.

Equations of Equilibrium. Since the equilibrium equations require a summation of the x and y components of each force, $\mathbf{T}_B$ must be resolved into x and y components. These components, shown dashed on the free-body diagram, have magnitudes of $T_B \cos 30°$ and $T_B \sin 30°$, respectively. Applying the equations of equilibrium, we have

$$\xrightarrow{+} \Sigma F_x = 0; \qquad T_B \cos 30° - T_D = 0 \qquad (1)$$
$$+ \uparrow \Sigma F_y = 0; \qquad T_B \sin 30° - 98.1 = 0 \qquad (2)$$

Solving Eq. (2) for T_B and substituting into Eq. (1) to obtain T_D yields

$$T_B = 196 \text{ N} \qquad \qquad Ans.$$
$$T_D = 170 \text{ N} \qquad \qquad Ans.$$

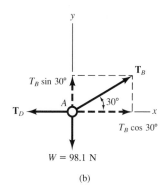

(b)

Fig. 3–6

The accuracy of these results, of course, depends on the accuracy of the data, i.e., measurements of geometry and loads. For most engineering work involving a problem such as this, the data as measured to three significant figures would be sufficient. Also, note that here we have neglected the weights of the cords, a reasonable assumption since they would be small in comparison with the weight of the crate.

Example 3–4

If the cylinder at A in Fig. 3–7a has a weight of 20 lb, determine the weight of B and the force in each cord needed to hold the system in the equilibrium position shown.

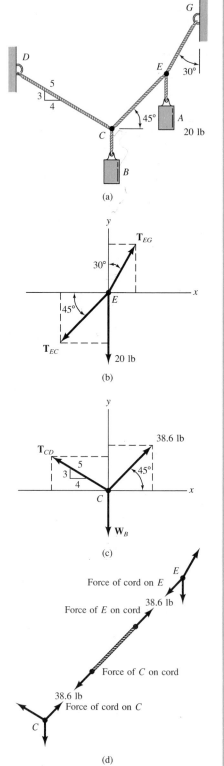

SOLUTION

Since the weight of A is known, the unknown tension in cords EG and EC can be determined by investigating the equilibrium of the knot at E. Why?

Free-Body Diagram. There are three forces acting on E, as shown in Fig. 3–7b.

Equations of Equilibrium. Resolving each force into its x and y components using trigonometry, and applying the equations of equilibrium, we have

$$\xrightarrow{+} \Sigma F_x = 0; \qquad T_{EG} \sin 30° - T_{EC} \cos 45° = 0 \qquad (1)$$

$$+ \uparrow \Sigma F_y = 0; \qquad T_{EG} \cos 30° - T_{EC} \sin 45° - 20 = 0 \qquad (2)$$

Solving Eq. (1) for T_{EG} in terms of T_{EC} and substituting the result into Eq. (2) allows a solution for T_{EC}. One then obtains T_{EG} from Eq. (1). The results are

$$T_{EC} = 38.6 \text{ lb} \qquad\qquad Ans.$$

$$T_{EG} = 54.6 \text{ lb} \qquad\qquad Ans.$$

Using the calculated result for T_{EC}, the equilibrium of the knot at C can now be investigated to determine the tension in CD and the weight of B.

Free-Body Diagram. As shown in Fig. 3–7c, $T_{EC} = 38.6$ lb "pulls" on C. The reason for this becomes clear when one draws the free-body diagram of cord CE and applies the principle of action, equal but opposite force reaction (Newton's third law), Fig. 3–7d.

Equations of Equilibrium. Noting the components of $\mathbf{T}_{CD}$ are proportional to the slope of the cord as defined by the 3–4–5 triangle, we have

$$\xrightarrow{+} \Sigma F_x = 0; \qquad 38.6 \cos 45° - (\tfrac{4}{5}) T_{CD} = 0 \qquad (3)$$

$$+ \uparrow \Sigma F_y = 0; \qquad (\tfrac{3}{5}) T_{CD} + 38.6 \sin 45° - W_B = 0 \qquad (4)$$

Solving Eq. (3) and substituting the result into Eq. (4) yields

$$T_{CD} = 34.1 \text{ lb} \qquad\qquad Ans.$$

$$W_B = 47.8 \text{ lb} \qquad\qquad Ans.$$

Fig. 3–7

Example 3–5

Determine the required length of cord AC in Fig. 3–8a so that the 8-kg lamp is suspended in the position shown. The *unstretched* length of the spring AB is $l'_{AB} = 0.4$ m, and the spring has a stiffness of $k_{AB} = 300$ N/m.

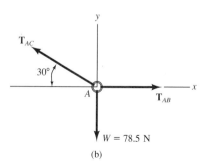

(a)

SOLUTION

If the force in spring AB is known, the stretch in the spring can be found ($F = ks$). Using the problem geometry, it is then possible to calculate the required length of AC.

Free-Body Diagram. The lamp has a weight $W = 8(9.81) = 78.5$ N. The free-body diagram of the ring at A is shown in Fig. 3–8b.

Equations of Equilibrium

$\xrightarrow{+} \Sigma F_x = 0;$ $\qquad T_{AB} - T_{AC} \cos 30° = 0$

$+ \uparrow \Sigma F_y = 0;$ $\qquad T_{AC} \sin 30° - 78.5 = 0$

Solving, we obtain

$$T_{AC} = 157.0 \text{ N}$$
$$T_{AB} = 136.0 \text{ N}$$

The stretch in spring AB is therefore

$T_{AB} = k_{AB}s_{AB};$ $\qquad 136.0 \text{ N} = 300 \text{ N/m}(s_{AB})$

$$s_{AB} = 0.453 \text{ m}$$

so that the stretched length of spring AB is

$$l_{AB} = l'_{AB} + s_{AB}$$
$$l_{AB} = 0.4 \text{ m} + 0.453 \text{ m} = 0.853 \text{ m}$$

The horizontal distance from C to B, Fig. 3–8a, requires

$$2 \text{ m} = l_{AC} \cos 30° + 0.853 \text{ m}$$
$$l_{AC} = 1.32 \text{ m} \qquad\qquad\qquad Ans.$$

Fig. 3–8

(b)

PROBLEMS

3–1. Determine the magnitudes of $\mathbf{F}_1$ and $\mathbf{F}_2$ so that particle P is in equilibrium.

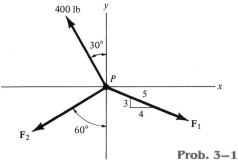

Prob. 3–1

3–2. Determine the magnitude and angle θ of $\mathbf{F}_1$ so that particle P is in equilibrium.

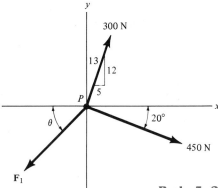

Prob. 3–2

3–3. Determine the magnitude and angle θ of $\mathbf{F}$ so that particle P is in equilibrium.

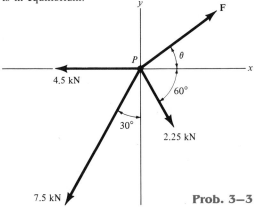

Prob. 3–3

*3–4. The members of a truss are pin-connected at joint O. Determine the magnitudes of $\mathbf{F}_1$ and $\mathbf{F}_2$ for equilibrium. Set $\theta = 60°$.

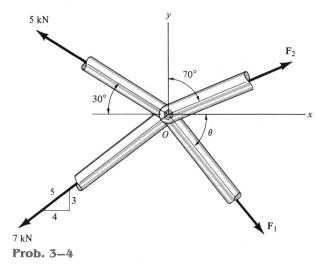

Prob. 3–4

3–5. The members of a truss are pin-connected at joint O. Determine the magnitude of $\mathbf{F}_1$ and its angle θ for equilibrium. Set $F_2 = 6$ kN.

3–6. Determine the force in cables AB and AC necessary to support the 12-kg traffic light.

Prob. 3–6

3–7. The street-lights at *A* and *B* are suspended from the two poles as shown. If each light has a weight of 50 lb, determine the tension in each of the three supporting cables and the height *h* of the pole *DE*. Cable *AB* is horizontal.

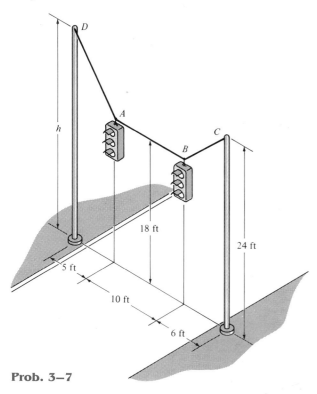

Prob. 3–7

*3–8. The sling is used to support a drum having a weight of 900 lb. Determine the force in cords *AB* and *AC* for equilibrium.

Prob. 3–8

3–9. The 200-kg engine is suspended from a vertical chain at *A*. A second chain is wrapped around the engine and held in position by the spreader bar *BC*. Determine the compressive force acting along the axis of the bar and the tension forces in segments *BA* and *CA* of the chain. *Hint:* Analyze equilibrium first at *A*, then at *B*.

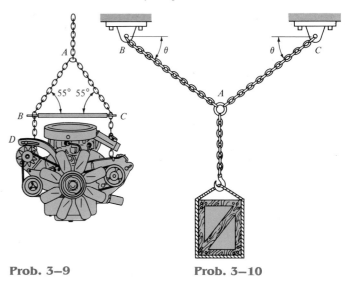

Prob. 3–9   **Prob. 3–10**

3–10. Determine the tension *T* in chains *AB* and *AC* as a function of the angle θ. The crate has a weight *W*. Plot the results as a curve representing *T* (vertical axis) versus θ (horizontal axis) for $0° \leqslant \theta \leqslant 90°$.

3–11. Determine the maximum weight of the flowerpot that can be supported without exceeding a cable tension of 50 lb in either cable *AB* or *AC*.

Prob. 3–11

*3–12. Determine the forces in cables AC and AB needed to hold the 20-kg ball D in equilibrium. Take $F = 300$ N and $d = 1$ m.

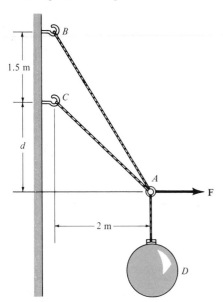

Probs. 3–12/3–13

3–13. The ball D has a mass of 20 kg. If a force of $F = 100$ N is applied horizontally to the ring at A, determine the largest dimension d so that the force in cable AC is zero.

3–14. The man attempts to pull the log at C by using the three ropes. Determine the direction θ in which he should pull on his rope with a force of 80 lb, so that he exerts a maximum force on the log. Compute the force on the log for this case. Also, determine the direction in which he should pull in order to maximize the force in the rope attached to B. What is this maximum force?

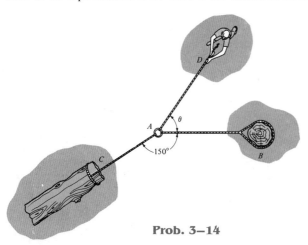

Prob. 3–14

3–15. The 30-kg pipe is supported at A by a system of five cords. Determine the force in each cord for equilibrium.

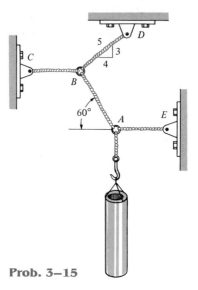

Prob. 3–15

*3–16. If the cords suspend the two buckets in the equilibrium position shown, determine the weight of bucket B. Bucket A has a weight of 80 lb.

Probs. 3–16/3–17

3–17. If the cords suspend the two buckets in the equilibrium position shown, determine the weight of bucket A. Bucket B has a weight of 20 lb.

■**3–18.** The cord *AB* of length 5 ft is attached to the end *B* of a spring having an unstretched length of 5 ft. The other end of the spring is attached to a roller *C* so that the spring remains horizontal as it stretches. If a 10-lb weight is suspended from *B*, determine the angle θ of cord *AB* for equilibrium.

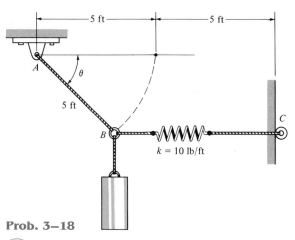

Prob. 3–18

3–19. The 2-m-long cord *AB* is attached to a spring *BC* having an unstretched length of 2 m. If the cord sags downward an amount $\theta = 30°$ as shown, determine the applied vertical force **F**. The spring has a stiffness of $k = 50$ N/m.

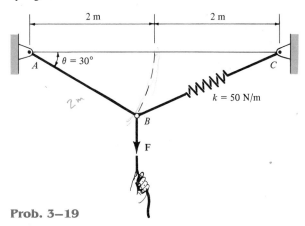

Prob. 3–19

*** 3–20.** The elastic cord (or spring) *ABC* has a stiffness of $k = 300$ N/m and an unstretched length of 400 mm. Determine the mass *D* suspended from the cord if the displacement of the pulley from the ceiling is $d = 100$ mm. Neglect the size of the pulley.

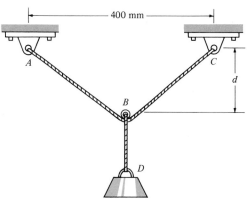

Probs. 3–20/3–21

3–21. The elastic cord (or spring) *ABC* has a stiffness of $k = 300$ N/m and an unstretched length of 400 mm. Determine the distance *d* the pulley *B* is displaced from the ceiling if the suspended mass *D* is 3 kg. Neglect the size and weight of the pulley.

■**3–22.** A horizontal force of 10 lb is applied to the ends of the cord *AB* and spring *AC*. If the spring has an unstretched length of 2 ft, determine the angle θ for equilibrium.

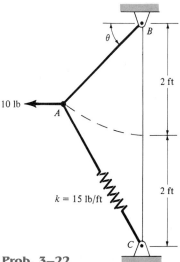

Prob. 3–22

■3–23. The springs on the rope assembly are originally un-stretched when $\theta = 0°$. Determine the tension in each rope when a horizontal force of $F = 30$ N is applied to the ring at A. Neglect the size of the pulleys at B and D.

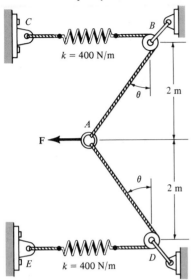

3–26. Determine the force in rope AB as a function of the angles θ and ϕ and the supported weight W. From this expression, com-pute the maximum tension in this rope and state the specific values of θ and ϕ at which this maximum value occurs.

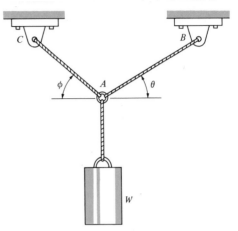

Prob. 3–26

3–27. The "scale" consists of a known weight W which is sus-pended at A from a cord of total length L. Determine the weight w at B if A is at a distance y for equilibrium. Neglect the sizes and weights of the pulleys.

Probs. 3–23/3–24

*3–24. The springs on the rope assembly are originally un-stretched when $\theta = 0°$. Determine the horizontal force F required to displace the ring at A so that $\theta = 30°$. Neglect the size of the pulleys at B and D.

3–25. The boy has a mass of 60 kg and attempts to cross the creek by using the pulley and 15-m-long rope shown. If he leaves the shore at A, determine how close s he comes to the shore at B once he reaches a state of equilibrium.

Prob. 3–27

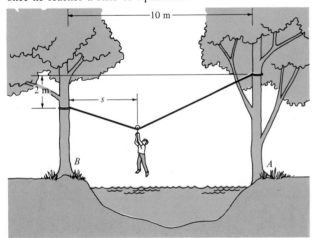

Prob. 3–25

3.4 Three-Dimensional Force Systems

It was shown in Sec. 3.1 that particle equilibrium requires

$$\Sigma \mathbf{F} = \mathbf{0} \qquad (3\text{--}4)$$

If the forces acting on the particle are resolved into their respective **i, j, k** components, we can then write

$$\Sigma F_x\mathbf{i} + \Sigma F_y\mathbf{j} + \Sigma F_z\mathbf{k} = \mathbf{0}$$

To ensure that Eq. 3–4 is satisfied, we must therefore require that the following three scalar component equations be satisfied:

$$\boxed{\begin{aligned} \Sigma F_x &= 0 \\ \Sigma F_y &= 0 \\ \Sigma F_z &= 0 \end{aligned}} \qquad (3\text{--}5)$$

These equations represent the *algebraic sums* of the x, y, z force components acting on the particle. Using them we can solve for at most three unknowns, generally represented as angles or magnitudes of forces shown on the particle's free-body diagram.

PROCEDURE FOR ANALYSIS

The following procedure provides a method for solving three-dimensional force equilibrium problems.

Free-Body Diagram. Draw a free-body diagram of the particle and label all the known and unknown forces on this diagram.

Equations of Equilibrium. Establish the x, y, z coordinate axes with origin located at the particle and apply the equations of equilibrium. Use the three scalar Eqs. 3–5 in cases where it is easy to resolve each force acting on the particle into its x, y, z components. If this appears difficult, first express each force acting on the particle in Cartesian vector form, and then substitute these vectors into Eq. 3–4. By setting the respective **i, j, k** components equal to zero, the three scalar Eqs. 3–5 can be generated. If more than three unknowns exist and the problem involves a spring, consider using $F = ks$ to relate the spring force to the deformation s of the spring.

The following example problems numerically illustrate this solution procedure.

Example 3–6

The 90-lb cylinder shown in Fig. 3–9a is supported by two cables and a spring having a stiffness $k = 500$ lb/ft. Determine the force in the cables and the stretch of the spring for equilibrium. Cable AD lies in the x-y plane and cable AC lies in the x-z plane.

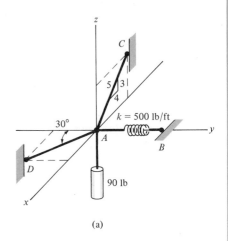

(a)

SOLUTION

The stretch of the spring can be determined once the force in AB is determined.

Free-Body Diagram. The connection at A is chosen for the analysis since the cable forces are concurrent at this point, Fig. 3–9b.

Equations of Equilibrium. By inspection, each force can easily be resolved into its x, y, z components, and therefore the three scalar equations of equilibrium can be directly applied. Considering components directed along the positive axes as "positive," we have

$$\Sigma F_x = 0; \qquad F_D \sin 30° - \tfrac{4}{5}F_C = 0 \qquad (1)$$
$$\Sigma F_y = 0; \qquad -F_D \cos 30° + F_B = 0 \qquad (2)$$
$$\Sigma F_z = 0; \qquad \tfrac{3}{5}F_C - 90 \text{ lb} = 0 \qquad (3)$$

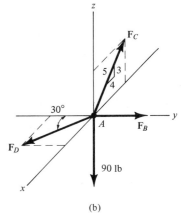

(b)

Fig. 3–9

Solving Eq. (3) for F_C, then Eq. (1) for F_D, and finally Eq. (2) for F_B, we get

$$F_C = 150 \text{ lb} \qquad\qquad Ans.$$
$$F_D = 240 \text{ lb} \qquad\qquad Ans.$$
$$F_B = 208 \text{ lb} \qquad\qquad Ans.$$

The stretch of the spring is therefore

$$F_B = ks_{AB}$$
$$208 \text{ lb} = 500 \text{ lb/ft } (s_{AB})$$
$$s_{AB} = 0.416 \text{ ft} \qquad\qquad Ans.$$

Example 3–7

Determine the magnitude and coordinate direction angles of **F** in Fig. 3–10a that are required for equilibrium of point O.

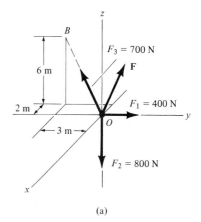

(a)

SOLUTION

Free-Body Diagram. Four forces act on point O, Fig. 3–10b.

Equations of Equilibrium. A Cartesian vector analysis will be used for the solution in order to formulate the components of $\mathbf{F}_3$. Hence,

$$\Sigma \mathbf{F} = 0; \qquad\qquad \mathbf{F}_1 + \mathbf{F}_2 + \mathbf{F}_3 + \mathbf{F} = 0 \qquad\qquad (1)$$

Expressing each of these forces in Cartesian vector form, noting that the coordinates of B are $B(-2, -3, 6)$, we have

$$\mathbf{F}_1 = \{400\mathbf{j}\} \text{ N}$$

$$\mathbf{F}_2 = \{-800\mathbf{k}\} \text{ N}$$

$$\mathbf{F}_3 = F_3 \mathbf{u}_B = F_3\left(\frac{\mathbf{r}_B}{r_B}\right) = 700 \text{ N}\left[\frac{-2\mathbf{i} - 3\mathbf{j} + 6\mathbf{k}}{\sqrt{(-2)^2 + (-3)^2 + (6)^2}}\right]$$

$$= \{-200\mathbf{i} - 300\mathbf{j} + 600\mathbf{k}\} \text{ N}$$

$$\mathbf{F} = F_x\mathbf{i} + F_y\mathbf{j} + F_z\mathbf{k}$$

Substituting into Eq. (1) yields

$$400\mathbf{j} - 800\mathbf{k} - 200\mathbf{i} - 300\mathbf{j} + 600\mathbf{k} + F_x\mathbf{i} + F_y\mathbf{j} + F_z\mathbf{k} = 0$$

Equating the respective **i**, **j**, **k** components to zero, we have

$$\Sigma F_x = 0; \qquad -200 + F_x = 0 \qquad F_x = 200 \text{ N}$$

$$\Sigma F_y = 0; \qquad 400 - 300 + F_y = 0 \qquad F_y = -100 \text{ N}$$

$$\Sigma F_z = 0; \qquad -800 + 600 + F_z = 0 \qquad F_z = 200 \text{ N}$$

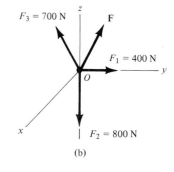

(b)

Thus,

$$\mathbf{F} = \{200\mathbf{i} - 100\mathbf{j} + 200\mathbf{k}\} \text{ N}$$

$$F = \sqrt{(200)^2 + (-100)^2 + (200)^2} = 300 \text{ N} \qquad\qquad Ans.$$

$$\mathbf{u}_F = \frac{\mathbf{F}}{F} = \frac{200}{300}\mathbf{i} - \frac{100}{300}\mathbf{j} + \frac{200}{300}\mathbf{k}$$

$$\alpha = \cos^{-1}\left(\frac{200}{300}\right) = 48.2° \qquad\qquad Ans.$$

$$\beta = \cos^{-1}\left(\frac{-100}{300}\right) = 110° \qquad\qquad Ans.$$

$$\gamma = \cos^{-1}\left(\frac{200}{300}\right) = 48.2° \qquad\qquad Ans.$$

(c)

Fig. 3–10

The magnitude and correct direction of **F** are shown in Fig. 3–10c.

Example 3–8

Determine the force developed in each cable used to support the 40-lb crate shown in Fig. 3–11a.

SOLUTION
Free-Body Diagram. As shown in Fig. 3–11b, the free-body diagram of point A is considered in order to "expose" the three unknown forces in the cables.

Equations of Equilibrium

$$\Sigma \mathbf{F} = \mathbf{0}; \qquad \mathbf{F}_B + \mathbf{F}_C + \mathbf{F}_D + \mathbf{W} = \mathbf{0} \qquad (1)$$

Since the coordinates of points B and C are $B(-3, -4, 8)$ and $C(-3, 4, 8)$, we have

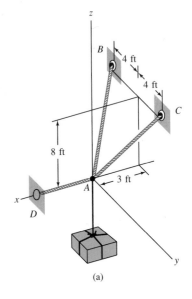

(a)

$$\mathbf{F}_B = F_B \left[\frac{-3\mathbf{i} - 4\mathbf{j} + 8\mathbf{k}}{\sqrt{(-3)^2 + (-4)^2 + (8)^2}} \right]$$

$$= -0.318F_B\mathbf{i} - 0.424F_B\mathbf{j} + 0.848F_B\mathbf{k}$$

$$\mathbf{F}_C = F_C \left[\frac{-3\mathbf{i} + 4\mathbf{j} + 8\mathbf{k}}{\sqrt{(-3)^2 + (4)^2 + (8)^2}} \right]$$

$$= -0.318F_C\mathbf{i} + 0.424F_C\mathbf{j} + 0.848F_C\mathbf{k}$$

$$\mathbf{F}_D = F_D\mathbf{i}$$

$$\mathbf{W} = -40\mathbf{k}$$

Substituting these forces into Eq. (1) gives

$$-0.318F_B\mathbf{i} - 0.424F_B\mathbf{j} + 0.848F_B\mathbf{k} - 0.318F_C\mathbf{i} + 0.424F_C\mathbf{j}$$
$$+ 0.848F_C\mathbf{k} + F_D\mathbf{i} - 40\mathbf{k} = \mathbf{0}$$

Equating the respective $\mathbf{i}$, $\mathbf{j}$, $\mathbf{k}$ components to zero yields

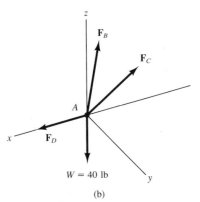

$$\Sigma F_x = 0; \qquad -0.318F_B - 0.318F_C + F_D = 0 \qquad (2)$$
$$\Sigma F_y = 0; \qquad -0.424F_B + 0.424F_C = 0 \qquad (3)$$
$$\Sigma F_z = 0; \qquad 0.848F_B + 0.848F_C - 40 = 0 \qquad (4)$$

Fig. 3–11

Equation (3) states $F_B = F_C$. Thus solving Eq. (4) for F_B and F_C and substituting the result into Eq. (2) to obtain F_D, we have

$$F_B = F_C = 23.6 \text{ lb} \qquad Ans.$$
$$F_D = 15.0 \text{ lb} \qquad Ans.$$

Example 3–9

The 100-kg cylinder shown in Fig. 3–12a is supported by three cords, one of which is connected to a spring. Determine the tension in each cord for equilibrium and the stretch in the spring.

(a)

SOLUTION

Free-Body Diagram. The force in each cord can be determined by investigating the free-body diagram of point A, Fig. 3–12b. The weight of the cylinder is $W = 100(9.81) = 981$ N.

Equations of Equilibrium

$$\Sigma F = 0; \qquad \mathbf{F}_B + \mathbf{F}_C + \mathbf{F}_D + \mathbf{W} = 0 \qquad (1)$$

Each of these vectors can be expressed in Cartesian vector form. Using Eq. 2–11 for $\mathbf{F}_C$, and noting point $D(-1, 2, 2)$ for $\mathbf{F}_D$, we have

$$\mathbf{F}_B = F_B\mathbf{i}$$

$$\mathbf{F}_C = F_C \cos 120°\mathbf{i} + F_C \cos 135°\mathbf{j} + F_C \cos 60°\mathbf{k}$$

$$= -0.5F_C\mathbf{i} - 0.707F_C\mathbf{j} + 0.5F_C\mathbf{k}$$

$$\mathbf{F}_D = F_D\left[\frac{-1\mathbf{i} + 2\mathbf{j} + 2\mathbf{k}}{\sqrt{(-1)^2 + (2)^2 + (2)^2}}\right]$$

$$= -0.333F_D\mathbf{i} + 0.667F_D\mathbf{j} + 0.667F_D\mathbf{k}$$

$$\mathbf{W} = -981\mathbf{k}$$

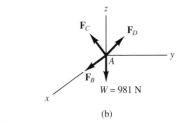

(b)

Fig. 3–12

Substituting these forces into Eq. (1) yields

$$F_B\mathbf{i} - 0.5F_C\mathbf{i} - 0.707F_C\mathbf{j} + 0.5F_C\mathbf{k} - 0.333F_D\mathbf{i} + 0.667F_D\mathbf{j}$$

$$+0.667F_D\mathbf{k} - 981\mathbf{k} = 0$$

Equating the respective **i**, **j**, **k** components to zero,

$$\Sigma F_x = 0; \qquad F_B - 0.5F_C - 0.333F_D = 0 \qquad (2)$$

$$\Sigma F_y = 0; \qquad -0.707F_C + 0.667F_D = 0 \qquad (3)$$

$$\Sigma F_z = 0; \qquad 0.5F_C + 0.667F_D - 981 = 0 \qquad (4)$$

Solving Eq. (3) for F_D in terms of F_C and substituting into Eq. (4) yields F_C. F_D is determined from Eq. (3). Substituting the results into Eq. (2) yields F_B. Hence,

$$F_C = 813 \text{ N} \qquad\qquad Ans.$$

$$F_D = 862 \text{ N} \qquad\qquad Ans.$$

$$F_B = 693.7 \text{ N} \qquad\qquad Ans.$$

The stretch in the spring is therefore

$$F = ks; \qquad\qquad 693.7 = 1500s$$

$$s = 0.462 \text{ m} \qquad\qquad Ans.$$

PROBLEMS

*3–28. Determine the magnitudes of $\mathbf{F}_1$, $\mathbf{F}_2$, and $\mathbf{F}_3$ for equilibrium of particle P.

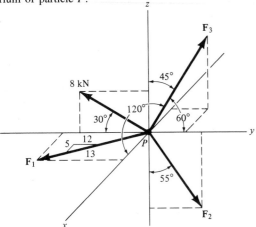

Prob. 3–28

3–29. Determine the magnitudes of $\mathbf{F}_1$, $\mathbf{F}_2$, and $\mathbf{F}_3$ for equilibrium of particle P.

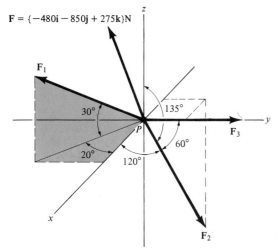

$\mathbf{F} = \{-480\mathbf{i} - 850\mathbf{j} + 275\mathbf{k}\}\text{N}$

Prob. 3–29

3–30. Determine the magnitudes of $\mathbf{F}_1$, $\mathbf{F}_2$, and $\mathbf{F}_3$ for equilibrium of particle P.

Prob. 3–30

3–31. The 50-kg pot is supported from A by the three cables. Determine the force acting in each cable for equilibrium. Take $d = 2.5$ m.

Probs. 3–31/3–32

*3–32. Determine the height d of cable AB so that the force in cables AD and AC is one-half as great as the force in cable AB. What is the force in each cable for this case? The flower pot has a mass of 50 kg.

■**3–33.** Determine the magnitudes of forces **F**₁, **F**₂, and **F**₃ necessary to hold the force **F** = {−9**i** − 8**j** − 5**k**} kN in equilibrium.

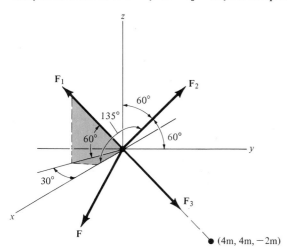

Prob. 3–33

3–34. Three cables are used to support the end of a sign which exerts a force of **F** = {250**i** + 450**j** − 150**k**} N at A. Determine the force developed in each cable.

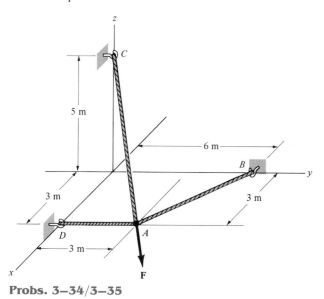

Probs. 3–34/3–35

3–35. The three cables are used to support the force **F**. Determine **F** expressed as a Cartesian vector so that it creates a tension of 300 N in cable AB, a tension of 200 N in cable AC, and a tension of 400 N in cable AD.

*3–36.** If cable AB is subjected to a tension of 500 N, determine the tension in cables AC and AD and the magnitude of the vertical force **F**$_{AO}$.

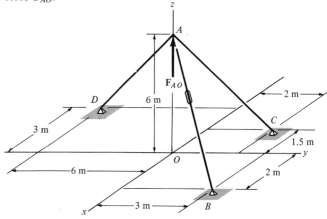

Prob. 3–36

■**3–37.** Determine the force in each cable needed to support the 3500-lb platform. Set d = 2 ft.

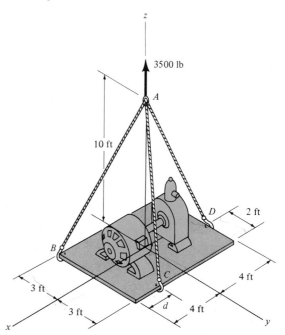

Probs. 3–37/3–38

3–38. Determine the force in each cable needed to support the 3500-lb platform. Set d = 4 ft.

3–39. Determine the force in each cable needed to support the 500-lb bucket.

Prob. 3–39

3–41. The 9500-lb crucible is supported by three cables. Determine the force in each cable for equilibrium of the hook at *A*.

Prob. 3–41

*3–40.** Determine the force in each of the three cables needed to lift the tractor-tread assembly which has a weight of 10 kN.

Prob. 3–40

■3–42.** The ring has a weight of $W = 80$ lb and is being hoisted using three springs, each having an unstretched length of $L_0 = 1.5$ ft and stiffness of $k = 50$ lb/ft. Determine the vertical distance d from the top of the ring to point *A* for equilibrium.

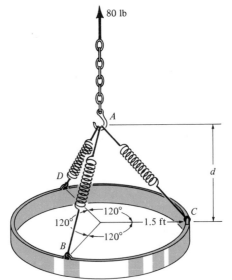

Prob. 3–42

REVIEW PROBLEMS

3–43. Blocks D and F weigh 5 lb each and block E weighs 8 lb. If the cord connecting the blocks passes over small frictionless pulleys, determine the sag s for equilibrium.

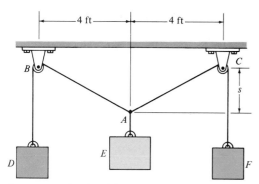

Probs. 3–43/3–44

*3–44. If blocks D and F weigh 5 lb each, determine the weight of block E if $s = 3$ ft.

■3–45. The ring of negligible size is subjected to a vertical force of 200 lb. Determine the required length l of wire AC such that the tension acting in AC is 160 lb. Also, what is the force acting in wire AB? *Hint:* Use the equilibrium condition to determine the required angle θ for attachment, then determine l using trigonometry applied to $\triangle ABC$.

Prob. 3–45

3–46. The boom supports a bucket and its contents, which have a total mass of 300 kg. Determine the forces developed in struts AD and AE and the tension in cable AB for equilibrium. The force in each strut acts along its axis.

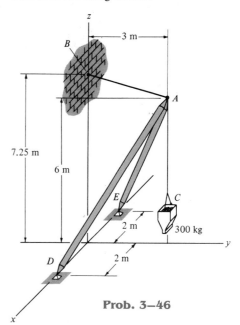

Prob. 3–46

■3–47. The lamp has a mass of 15 kg and is supported by a pole AO and cables AB and AC. If the force in the pole acts along its axis, determine the forces in AO, AB, and AC for equilibrium.

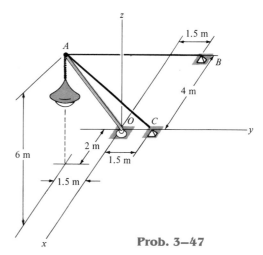

Prob. 3–47

91

*3–48. The motor at B winds up the cord attached to the 65-lb crate with a constant speed. Determine the force in rope CD supporting the pulley and the angle θ for equilibrium. Neglect the size of the pulley at C.

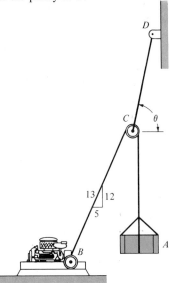

Prob. 3–48

3–49. The bulldozer is used to pull down the chimney using the cable and *small* pulley arrangement shown. If the tension force in AB is 500 lb, determine the tension in cable CAD and the angle θ which the cable makes at the pulley.

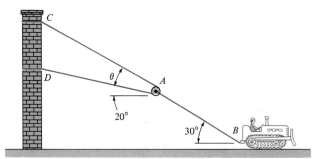

Prob. 3–49

3–50. Determine the tension in cables AB, AC, and AD, required to hold the 60-lb crate in equilibrium.

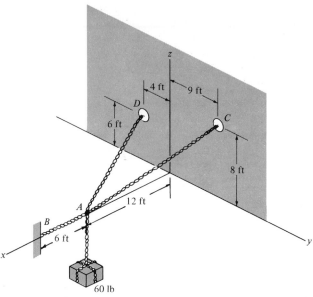

Prob. 3–50

3–51. The spider has a mass of 0.7 g and is suspended from a portion of its web as shown. Determine the force which each of the three web "strings" exerts on the twigs at C, D, and E. String CB is horizontal. *Hint:* First analyze the equilibrium at point A; then, using the result for the force in AB, analyze the equilibrium at B.

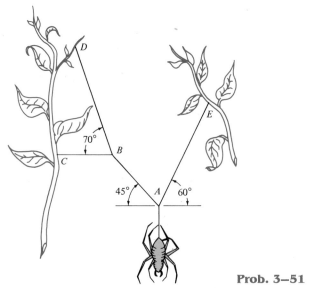

Prob. 3–51

4

Force System Resultants

In Chapter 3 it was shown that the condition for the equilibrium of a particle or a concurrent force system simply requires that the resultant of the force system be equal to zero. In Chapter 5 it will be shown that such a restriction is necessary but not sufficient for the equilibrium of a rigid body. A further restriction must be made with regard to the nonconcurrency of the applied force system, giving rise to the concept of *moment*. In this chapter, a formal definition of a moment will be developed and ways of finding the moment of a force about a point or axis will be discussed. We will also develop methods for determining the resultants of nonconcurrent force systems. This development is important since the equations for force-system simplification are similar to the equations of equilibrium for a rigid-body. Furthermore, the resultants of a force system will influence the state of equilibrium or motion of a rigid body in the same way as the force system, and therefore we can study rigid-body behavior in a simpler manner by using the resultants.

4.1 Cross Product

The moment of a force will be formulated using Cartesian vectors in the next section. Before doing this, however, it is first necessary to expand our knowledge of vector algebra and introduce the cross-product method of vector multiplication.

The cross product of two vectors **A** and **B** yields the vector **C**, which is written

$$\mathbf{C} = \mathbf{A} \times \mathbf{B}$$

and is read "**C** equals **A** cross **B**."

Magnitude

The *magnitude* of **C** is defined as the product of the magnitudes of **A** and **B** and the sine of the angle θ between their tails ($0° \leqslant \theta \leqslant 180°$). Thus, $C = AB \sin \theta$.

Direction

Vector **C** has a *direction* that is perpendicular to the plane containing **A** and **B** such that the direction of **C** is specified by the right-hand rule, i.e., curling the fingers of the right hand from vector **A** (cross) to vector **B,** the thumb then points in the direction of **C,** as shown in Fig. 4–1.

Knowing both the magnitude and direction of **C,** we can write

$$\mathbf{C} = \mathbf{A} \times \mathbf{B} = (AB \sin \theta)\mathbf{u}_C \qquad (4-1)$$

where the scalar $AB \sin \theta$ defines the *magnitude* of **C** and the unit vector $\mathbf{u}_C$ defines the *direction* of **C.** The terms of Eq. 4–1 are illustrated graphically in Fig. 4–2.

Fig. 4–1

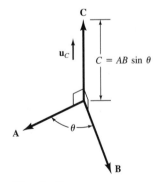

Fig. 4–2

Laws of Operation

1. The commutative law is *not* valid, i.e.,

$$\mathbf{A} \times \mathbf{B} \neq \mathbf{B} \times \mathbf{A}$$

Rather,

$$\mathbf{A} \times \mathbf{B} = -\mathbf{B} \times \mathbf{A}$$

This is shown in Fig. 4–3 by using the right-hand rule. The cross product $\mathbf{B} \times \mathbf{A}$ yields a vector that acts in the opposite direction to $\mathbf{C}$, i.e., $\mathbf{B} \times \mathbf{A} = -\mathbf{C}$.

2. Multiplication by a scalar:

$$a(\mathbf{A} \times \mathbf{B}) = (a\mathbf{A}) \times \mathbf{B} = \mathbf{A} \times (a\mathbf{B}) = (\mathbf{A} \times \mathbf{B})a$$

This property is easily shown, since the magnitude of the resultant vector ($|a|AB \sin \theta$) and its direction are the same in each case.

3. The distributive law:

$$\mathbf{A} \times (\mathbf{B} + \mathbf{D}) = (\mathbf{A} \times \mathbf{B}) + (\mathbf{A} \times \mathbf{D})$$

The proof of this identity is left as an exercise (see Prob. 4–1). It is important to note that *proper order* of the cross products must be maintained, since they are not commutative.

Fig. 4–3

Cartesian Vector Formulation

Equation 4–1 may be used to find the cross product of a pair of Cartesian unit vectors. For example, to find $\mathbf{i} \times \mathbf{j}$, the *magnitude* of the resultant vector is $(i)(j)(\sin 90°) = (1)(1)(1) = 1$, and its *direction* is determined using the right-hand rule. As shown in Fig. 4–4, the resultant vector points in the $+\mathbf{k}$ direction. Thus, $\mathbf{i} \times \mathbf{j} = (1)\mathbf{k}$. In a similar manner,

$$\mathbf{i} \times \mathbf{j} = \mathbf{k} \quad \mathbf{i} \times \mathbf{k} = -\mathbf{j} \quad \mathbf{i} \times \mathbf{i} = 0$$
$$\mathbf{j} \times \mathbf{k} = \mathbf{i} \quad \mathbf{j} \times \mathbf{i} = -\mathbf{k} \quad \mathbf{j} \times \mathbf{j} = 0$$
$$\mathbf{k} \times \mathbf{i} = \mathbf{j} \quad \mathbf{k} \times \mathbf{j} = -\mathbf{i} \quad \mathbf{k} \times \mathbf{k} = 0$$

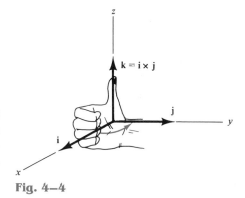

Fig. 4–4

These results should *not* be memorized; rather, it should be clearly understood how each is obtained by using the right-hand rule and the definition of the cross product. A simple scheme shown in Fig. 4–5 is helpful for obtaining the same results when the need arises. If the circle is constructed as shown, then "crossing" two unit vectors in a *counterclockwise* fashion around the circle yields the *positive* third unit vector, e.g., $\mathbf{k} \times \mathbf{i} = \mathbf{j}$. Moving *clockwise*, a *negative* unit vector is obtained, e.g., $\mathbf{i} \times \mathbf{k} = -\mathbf{j}$.

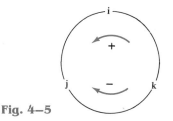

Fig. 4–5

Consider now the cross product of two general vectors **A** and **B** which are expressed in Cartesian vector form. We have

$$\mathbf{A} \times \mathbf{B} = (A_x\mathbf{i} + A_y\mathbf{j} + A_z\mathbf{k}) \times (B_x\mathbf{i} + B_y\mathbf{j} + B_z\mathbf{k})$$
$$= A_xB_x(\mathbf{i} \times \mathbf{i}) + A_xB_y(\mathbf{i} \times \mathbf{j}) + A_xB_z(\mathbf{i} \times \mathbf{k})$$
$$+ A_yB_x(\mathbf{j} \times \mathbf{i}) + A_yB_y(\mathbf{j} \times \mathbf{j}) + A_yB_z(\mathbf{j} \times \mathbf{k})$$
$$+ A_zB_x(\mathbf{k} \times \mathbf{i}) + A_zB_y(\mathbf{k} \times \mathbf{j}) + A_zB_z(\mathbf{k} \times \mathbf{k})$$

Carrying out the cross-product operations and combining terms yields

$$\mathbf{A} \times \mathbf{B} = (A_yB_z - A_zB_y)\mathbf{i} - (A_xB_z - A_zB_x)\mathbf{j} + (A_xB_y - A_yB_x)\mathbf{k} \qquad (4\text{--}2)$$

This equation may also be written in a more compact determinant form as

$$\mathbf{A} \times \mathbf{B} = \begin{vmatrix} \mathbf{i} & \mathbf{j} & \mathbf{k} \\ A_x & A_y & A_z \\ B_x & B_y & B_z \end{vmatrix} \qquad (4\text{--}3)$$

Thus, to find the cross product of any two Cartesian vectors **A** and **B,** it is necessary to expand a determinant whose first row of elements consists of the unit vectors **i, j,** and **k** and whose second and third rows represent the x, y, z components of the two vectors **A** and **B,** respectively.*

---

*A determinant having three rows and three columns can be expanded using three minors, each of which is multiplied by one of the three terms in the first row. There are four elements in each minor, e.g.,

$$\begin{vmatrix} A_{11} & A_{12} \\ A_{21} & A_{22} \end{vmatrix}$$

By *definition,* this notation represents the terms $(A_{11}A_{22} - A_{12}A_{21})$, which is simply the product of the two elements of the arrow slanting downward to the right $(A_{11}A_{22})$ *minus* the product of the two elements intersected by the arrow slanting downward to the left $(A_{12}A_{21})$. For a 3×3 determinant, such as Eq. 4–3, the three minors can be generated in accordance with the following scheme:

For element **i:** $\begin{vmatrix} \mathbf{i} & \mathbf{j} & \mathbf{k} \\ A_x & A_y & A_z \\ B_x & B_y & B_z \end{vmatrix} = \mathbf{i}(A_yB_z - A_zB_y)$

For element **j:** $\begin{vmatrix} \mathbf{i} & \mathbf{j} & \mathbf{k} \\ A_x & A_y & A_z \\ B_x & B_y & B_z \end{vmatrix} = -\mathbf{j}(A_xB_z - A_zB_x)$

For element **k:** $\begin{vmatrix} \mathbf{i} & \mathbf{j} & \mathbf{k} \\ A_x & A_y & A_z \\ B_x & B_y & B_z \end{vmatrix} = \mathbf{k}(A_xB_y - A_yB_x)$

Adding the results and noting that the **j** element *must include the minus sign* yields the expanded form of **A** × **B** given by Eq. 4–2.

4.2 Moment of a Force—Scalar Formulation

The *moment* of a force about a point or axis provides a measure of the tendency of the force to cause a body to rotate about the point or axis. For example, consider the horizontal force F_x, which acts perpendicular to the handle of the wrench and is located a distance d_y from point O, Fig. 4–6a. It is seen that this force tends to cause the pipe to turn about the z axis and so the force creates a moment $(M_O)_z$ about this axis. The larger the force or the length d_y, the greater the effect. This tendency for rotation caused by F_x is sometimes called a *torque,* but most often it is called the *moment of a force* or simply the *moment* $(M_O)_z$. In particular, note that the moment (or z) axis is perpendicular to the shaded (x-y) plane which contains both F_z and d_y and that this axis intersects the plane at point O. Now consider applying the force F_z to the wrench, Fig. 4–6b. This force will *not* rotate the pipe about the z axis. Instead, it tends to rotate it about the x axis. Keep in mind that although it may not be possible to actually ''rotate'' or turn the pipe in this manner, F_z still creates the *tendency* for rotation and so the moment $(M_O)_x$ is produced. As before, the force and distance d_y lie in the shaded plane which is perpendicular to the moment axis. Lastly, if a force F_y is applied to the wrench, Fig. 4–6c, no moment or tendency for rotation is produced about point O. This lack of turning effect results, since the force line of action passes through O and therefore no tendency for rotation is possible.

Consider now the general case of the force F and point O which lie in a shaded plane as shown in Fig. 4–7a. The moment M_O about point O, or about an axis passing through O and perpendicular to the plane, is a *vector quantity* since it has a specified magnitude and direction and adds according to the parallelogram law.

(a)

(b)

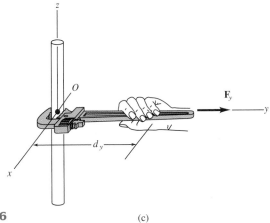

Fig. 4–6

(c)

97

Magnitude

The magnitude of M_O is

$$M_O = Fd \qquad\qquad (4\text{--}4)$$

where d is referred to as the *moment arm* or perpendicular distance from the axis at point O to the line of action of the force. Units of moment magnitude consist of force times distance, e.g., N · m or lb · ft.

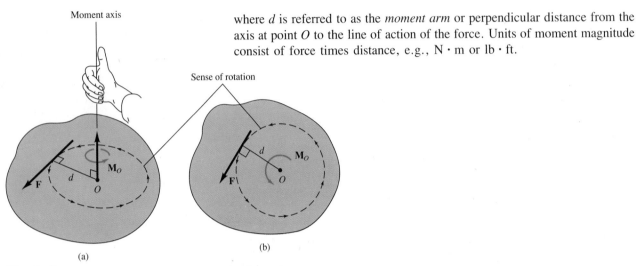

Fig. 4–7

Direction

The direction of M_O will be specified by using the "right-hand rule." To do this, the fingers of the right hand are curled such that they follow the sense of rotation, which would occur if the force could rotate about point O, Fig. 4–7a. The *thumb* then *points* along the *moment axis* so that it gives the direction and sense of the moment vector, which is *upward* and *perpendicular* to the shaded plane containing F and d. By this definition, the moment M_O can be considered as a *sliding vector* and therefore acts at any point along the moment axis.

In three dimensions, M_O is illustrated by a vector arrow with a curl on it to *distinguish* it from a force vector, Fig. 4–7a. Many problems in mechanics, however, involve coplanar force systems that may be conveniently viewed in two dimensions. For example, a two-dimensional view of Fig. 4–7a is given in Fig. 4–7b. Here M_O is simply represented by the (counterclockwise) curl, which indicates the action of F. The arrowhead on this curl is used to show the *sense of rotation* caused by F. Using the right-hand rule, however, realize that the direction and sense of the moment vector in Fig. 4–7b are indicated by the thumb, which points *out* of the page, since the fingers follow the curl. In particular, notice that *this curl or sense of rotation can always be determined by observing in which direction the force would "orbit" about point O* (counterclockwise in Fig. 4–7b). In two dimensions we will often refer to finding the moment of a force "about a point" (O). Keep in mind that the moment *actually* acts about an axis which is perpendicular to the plane of F and d, and intersects the plane at the point (O), Fig. 4–7a.

4.3 Moment of a Force—Vector Formulation

The moment of a force **F** about point O, or actually about an axis passing through O and perpendicular to the plane containing O and **F**, Fig. 4–8a, can also be expressed using the vector cross product, namely

$$\boxed{\mathbf{M}_O = \mathbf{r} \times \mathbf{F}} \qquad (4\text{–}5)$$

Here **r** represents a position vector drawn from O to *any point A* lying on the line of action of **F**. It will now be shown that the moment $\mathbf{M}_O$, when determined by this cross product, has the proper magnitude and direction.

Magnitude

By definition the angle θ for the cross product $\mathbf{r} \times \mathbf{F}$ is measured between the *tails* of **r** and **F**. Hence, **r** must be treated as a sliding vector so that θ can be properly constructed, Fig. 4–8b. As shown, the moment arm $d = r \sin \theta$ so that from Eq. 4–1 we have

$$M_O = rF \sin \theta = F (r \sin \theta) = Fd$$

which agrees with Eq. 4–4.

Direction

The direction and sense of $\mathbf{M}_O$ are determined by the right-hand rule as it applies to the cross product. Thus, extending **r** to the dashed position and curling the right-hand fingers from **r** toward **F,** "**r** cross **F,**" the thumb is directed upward or perpendicular to the plane in the *same direction* as $\mathbf{M}_O$, the moment of the force about point O, Fig. 4–8b. Note that the "curl" of the fingers like the curl around the moment vector indicates the sense of rotation caused by the force. Since the cross product is not commutative, it is important that the *proper order* of **r** and **F** be maintained in Eq. 4–5.

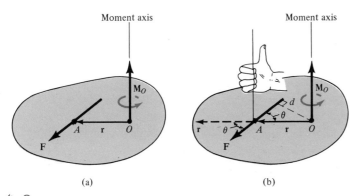

(a)   (b)

Fig. 4–8

Cartesian Vector Formulation

If the position vector **r** and force **F** are expressed in Cartesian vector form, Fig. 4–9, then from Eq. 4–5 we have

$$\mathbf{M}_O = \mathbf{r} \times \mathbf{F} = \begin{vmatrix} \mathbf{i} & \mathbf{j} & \mathbf{k} \\ r_x & r_y & r_z \\ F_x & F_y & F_z \end{vmatrix} \tag{4–6}$$

where r_x, r_y, r_z represent the Cartesian components of the position vector drawn from point O to *any point* on the line of action of the force

F_x, F_y, F_z represent the Cartesian components of the force vector

If the determinant is expanded, then like Eq. 4–2 we have

$$\mathbf{M}_O = (r_y F_z - r_z F_y)\mathbf{i} - (r_x F_z - r_z F_x)\mathbf{j} + (r_x F_y - r_y F_x)\mathbf{k} \tag{4–7}$$

The physical meaning of these three moment components becomes evident by studying Fig. 4–9a. For example, the **i** component of $\mathbf{M}_O$ is computed from the moments of forces $\mathbf{F}_z$ and $\mathbf{F}_y$ about the x axis. Since the lines of action of these forces pass through points D and E, respectively, then the magnitude of the moment of $\mathbf{F}_z$ about point A on the x axis is $r_y F_z$. By the right-hand rule this component acts in the positive **i** direction. Likewise, $\mathbf{F}_y$ contributes a moment component of $r_z F_y(-\mathbf{i})$. Force component $\mathbf{F}_x$ does *not* create a moment about the x axis since this force is *parallel* to the x axis. As an exercise, establish the **j** and **k** components of $\mathbf{M}_O$ in this manner, and show that indeed the expanded form of the determinant, Eq. 4–7, represents the moment of **F** about point O. Note that once computed, $\mathbf{M}_O$ will be *perpendicular* to the shaded plane containing vectors **r** and **F,** Fig. 4–9b.

Fig. 4–9   (a)   (b)

Example 4–1

For each case illustrated in Fig. 4–10, determine the moment of the force about point O.

SOLUTION (*SCALAR ANALYSIS*)

The line of action of each force is extended as a dashed line in order to establish the moment arm d. Also, the tendency of rotation caused by the force is shown as a colored curl. Thus,

Fig. 4–10a, $M_O = (100 \text{ N})(2 \text{ m}) = 200 \text{ N} \cdot \text{m}$ ↓   *Ans.*

Fig. 4–10b, $M_O = (50 \text{ N})(0.75 \text{ m}) = 37.5 \text{ N} \cdot \text{m}$ ↓   *Ans.*

Fig. 4–10c, $M_O = (40 \text{ lb})(4 \text{ ft} + 2 \cos 30° \text{ ft}) = 229 \text{ lb} \cdot \text{ft}$ ↓   *Ans.*

Fig. 4–10d, $M_O = (60 \text{ lb})(1 \sin 45° \text{ ft}) = 42.4 \text{ lb} \cdot \text{ft}$ ↑   *Ans.*

Fig. 4–10e, $M_O = (7 \text{ kN})(4 \text{ m} - 1 \text{ m}) = 21.0 \text{ kN} \cdot \text{m}$ ↑   *Ans.*

Fig. 4–10

Example 4–2

Determine the moment of the 800-N force acting on the frame in Fig. 4–11 about points A, B, C, and D.

SOLUTION (*SCALAR ANALYSIS*)

In general, $M = Fd$, where d is the moment arm or *perpendicular distance* from the *point* on the moment axis to the *line of action* of the force. Hence,

$M_A = 800 \text{ N}(2.5 \text{ m}) = 2000 \text{ N} \cdot \text{m} \downarrow$ *Ans.*

$M_B = 800 \text{ N}(1.5 \text{ m}) = 1200 \text{ N} \cdot \text{m} \downarrow$ *Ans.*

$M_C = 800 \text{ N}(0) = 0$ (line of action of **F** passes through C)   *Ans.*

$M_D = 800 \text{ N}(0.5 \text{ m}) = 400 \text{ N} \cdot \text{m} \uparrow$ *Ans.*   **Fig. 4–11**

Example 4–3

Determine the point of application P and the direction of a 20-lb force that lies in the plane of the square plate shown in Fig. 4–12a, so that this force creates the greatest counterclockwise moment about point O. What is this moment?

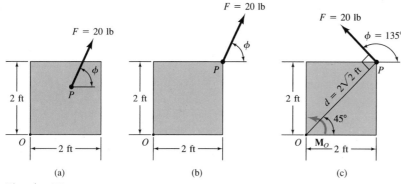

Fig. 4–12

SOLUTION (*SCALAR ANALYSIS*)

Since the maximum moment created by the force is required, the force must act on the plate at a distance *farthest* from point O. As shown in Fig. 4–12b, the point of application must therefore be at the diagonal corner. In order to produce *counterclockwise* rotation of the plate about O, **F** must act at an angle $45° < \phi < 225°$. The greatest moment is produced when the line of action of **F** is *perpendicular* to d, i.e., $\phi = 135°$, Fig. 4–12c. The maximum moment is therefore

$$M_O = Fd = (20 \text{ lb})(2\sqrt{2} \text{ ft}) = 56.6 \text{ lb} \cdot \text{ft} \uparrow \qquad Ans.$$

By the right-hand rule, $\mathbf{M}_O$ is directed out of the page.

Example 4–4

The pole in Fig. 4–13*a* is subjected to a 60-N force that is directed from C to B. Determine the magnitude of the moment created by this force about point A.

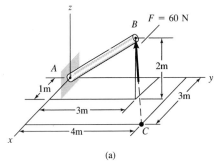

(a)

SOLUTION (*VECTOR ANALYSIS*)

As shown in Fig. 4–13*b*, either one of two position vectors can be used for the solution, since $\mathbf{M}_A = \mathbf{r}_B \times \mathbf{F}$ or $\mathbf{M}_A = \mathbf{r}_C \times \mathbf{F}$. The position vectors are represented as

$$\mathbf{r}_B = \{1\mathbf{i} + 3\mathbf{j} + 2\mathbf{k}\} \text{ m} \quad \text{and} \quad \mathbf{r}_C = \{3\mathbf{i} + 4\mathbf{j}\} \text{ m}$$

The force has a magnitude of 60 N and a direction specified by the unit vector $\mathbf{u}_F$, directed from C to B. Thus,

$$\mathbf{F} = (60 \text{ N})\mathbf{u}_F = (60) \left[\frac{(1-3)\mathbf{i} + (3-4)\mathbf{j} + (2-0)\mathbf{k}}{\sqrt{(-2)^2 + (-1)^2 + (2)^2}} \right]$$

$$= \{-40\mathbf{i} - 20\mathbf{j} + 40\mathbf{k}\} \text{ N}$$

Substituting into the determinant formulation, Eq. 4–6, and following the scheme for determinant expansion as stated in the footnote on page 96, we have

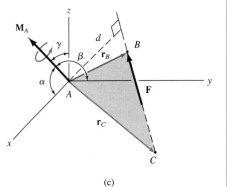

(b)

$$\mathbf{M}_A = \mathbf{r}_B \times \mathbf{F} = \begin{vmatrix} \mathbf{i} & \mathbf{j} & \mathbf{k} \\ 1 & 3 & 2 \\ -40 & -20 & 40 \end{vmatrix}$$

$$\mathbf{M}_A = [3(40) - 2(-20)]\mathbf{i} - [1(40) - 2(-40)]\mathbf{j} + [1(-20) - 3(-40)]\mathbf{k}$$

or

$$\mathbf{M}_A = \mathbf{r}_C \times \mathbf{F} = \begin{vmatrix} \mathbf{i} & \mathbf{j} & \mathbf{k} \\ 3 & 4 & 0 \\ -40 & -20 & 40 \end{vmatrix}$$

$$\mathbf{M}_A = [4(40) - 0(-20)]\mathbf{i} - [3(40) - 0(-40)]\mathbf{j} + [3(-20) - 4(-40)]\mathbf{k}$$

In both cases,

$$\mathbf{M}_A = \{160\mathbf{i} - 120\mathbf{j} + 100\mathbf{k}\} \text{ N} \cdot \text{m}$$

The *magnitude* of $\mathbf{M}_A$ is therefore

$$M_A = \sqrt{(160)^2 + (-120)^2 + (100)^2} = 224 \text{ N} \cdot \text{m} \qquad \textit{Ans.}$$

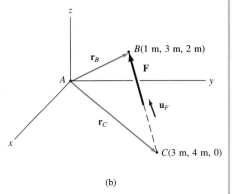

(c)

Fig. 4–13

Note that $\mathbf{M}_A$ acts perpendicular to the shaded plane containing vectors $\mathbf{F}$, $\mathbf{r}_B$, and $\mathbf{r}_C$, Fig. 4–13*c*. (How would you find its coordinate direction angles $\alpha = 44.3°$, $\beta = 122°$, $\gamma = 63.4°$?) Had this problem been worked using a scalar approach, where $M_A = Fd$, notice the difficulty that might arise in obtaining the moment arm d.

4.4 Transmissibility of a Force and the Principle of Moments

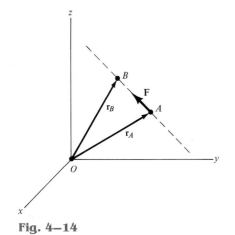

Fig. 4—14

Using the definition of a moment formalized by the cross product, we will now present two important concepts often used throughout our study of mechanics.

Transmissibility of a Force

Consider the force **F** applied at point A in Fig. 4–14. The moment created by **F** about point O is $\mathbf{M}_O = \mathbf{r}_A \times \mathbf{F}$; however, it was shown that the position vector "**r**" can extend from O to *any point* on the line of action of **F**. Consequently, **F** may be applied at point B and the same moment $\mathbf{M}_O = \mathbf{r}_B \times \mathbf{F}$ will be computed. Hence, **F** has the properties of a *sliding vector* and can therefore act at *any point along its line of action and still create the same moment about point O*. We refer to **F** in this regard as being "transmissible" and we will discuss this property further in Sec. 4.7.

Principle of Moments

A concept often used in mechanics is the *principle of moments,* which is sometimes referred to as *Varignon's theorem* since it was originally developed by the French mathematician Varignon (1654–1722). It states that *the moment of a force about a point is equal to the sum of the moments of the force's components about the point*. The proof follows directly from the distributive law of the vector cross product. To show this, consider the force **F** and two of its components, where $\mathbf{F} = \mathbf{F}_1 + \mathbf{F}_2$, Fig. 4–15. We have

$$\mathbf{M}_O = \mathbf{r} \times \mathbf{F}_1 + \mathbf{r} \times \mathbf{F}_2 = \mathbf{r} \times (\mathbf{F}_1 + \mathbf{F}_2) = \mathbf{r} \times \mathbf{F}$$

This concept has important applications to the solution of problems and proofs of theorems that follow, since it allows us to consider the moments of a force's components rather than the force itself.

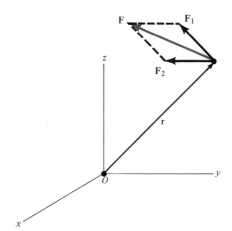

Fig. 4—15

Example 4–5

A 200-N force acts on the bracket shown in Fig. 4–16a. Determine the moment of the force about point A.

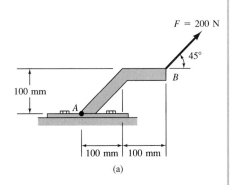

(a)

SOLUTION I

The moment arm d can be computed by trigonometry, using the construction shown in Fig. 4–16b. From triangle BCD,

$$CB = d = 100 \cos 45° = 70.71 \text{ mm} = 0.070\ 71 \text{ m}$$

Thus,

$$M_A = Fd = 200 \text{ N}(0.070\ 71 \text{ m}) = 14.1 \text{ N} \cdot \text{m} \curvearrowleft$$

According to the right-hand rule, $\mathbf{M}_A$ is directed in the $+\mathbf{k}$ direction since the force tends to rotate *counterclockwise* about point A. Hence, reporting the moment as a Cartesian vector, we have

$$\mathbf{M}_A = \{14.1\mathbf{k}\} \text{ N} \cdot \text{m} \qquad Ans.$$

(b)

SOLUTION II

The 200-N force may be resolved into x and y components, as shown in Fig. 4–16c. In accordance with the principle of moments, the moment of $\mathbf{F}$ computed about point A is equivalent to the sum of the moments produced by the two force components. Assuming counterclockwise rotation as positive, i.e., in the $+\mathbf{k}$ direction, we can add the moment of each force algebraically, since they both act along the same moment axis (perpendicular to the x-y plane, passing through A).

$$\curvearrowright +M_A = (200 \sin 45°)(0.20) - (200 \cos 45°)(0.10)$$
$$= 14.1 \text{ N} \cdot \text{m} \curvearrowleft$$

Thus

$$\mathbf{M}_A = \{14.1\mathbf{k}\} \text{ N} \cdot \text{m} \qquad Ans.$$

By comparison, it is seen that Solution II provides a more *convenient method* for analysis than Solution I since the moment arm for each component force is easier to establish.

(c)

Fig. 4–16

Example 4–6

The force **F** acts at the end of the beam shown in Fig. 4–17a. Determine the moment of the force about point O.

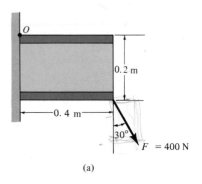

SOLUTION I (*SCALAR ANALYSIS*)

The force is resolved into its x and y components as shown in Fig. 4–17b and the moments of the components are computed about point O. Taking positive moments as counterclockwise, i.e., in the $+\mathbf{k}$ direction, we have

$$\zeta+ M_O = 400 \sin 30°(0.2) - 400 \cos 30°(0.4) \qquad (1)$$
$$= -98.6 \text{ N} \cdot \text{m} = 98.6 \text{ N} \cdot \text{m} \downarrow$$

(a)

or

$$\mathbf{M}_O = \{-98.6\mathbf{k}\} \text{ N} \cdot \text{m} \qquad \qquad \textit{Ans.}$$

SOLUTION II (*VECTOR ANALYSIS*)

Using a Cartesian vector approach, the force and position vector shown in Fig. 4–17c can be represented as

$$\mathbf{r} = \{0.4\mathbf{i} - 0.2\mathbf{j}\} \text{ m}$$
$$\mathbf{F} = \{400 \sin 30°\mathbf{i} - 400 \cos 30°\mathbf{j}\} \text{ N}$$
$$= \{200.0\mathbf{i} - 346.4\mathbf{j}\} \text{ N}$$

(b)

Hence the required moment becomes

$$\mathbf{M}_O = \mathbf{r} \times \mathbf{F} = \begin{vmatrix} \mathbf{i} & \mathbf{j} & \mathbf{k} \\ 0.4 & -0.2 & 0 \\ 200.0 & -346.4 & 0 \end{vmatrix}$$

$$\mathbf{M}_O = 0\mathbf{i} - 0\mathbf{j} + [0.4(-346.4) - (-0.2)(200.0)]\mathbf{k} \qquad (2)$$
$$\mathbf{M}_O = \{-98.6\mathbf{k}\} \text{ N} \cdot \text{m} \qquad \qquad \textit{Ans.}$$

By comparison, it is seen that the scalar analysis (Solution I) provides a more *convenient method* for analysis than Solution II since the moment arm for each component force is easy to establish. Hence, *this method is generally recommended for solving problems displayed in two dimensions.* Although the Cartesian vector analysis (Solution II) *automatically* accounts for the sign of the moment of each component force, Eq. (2), these directions can easily be established *directly* from the diagram, Fig. 4–17b, Eq. (1). Hence, there really is no need to set up and evaluate a determinant. As a result, *Cartesian vector analysis is generally recommended only for solving three-dimensional problems,* where the moment arms and force components are often more difficult to determine.

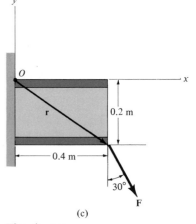

(c)

Fig. 4–17

PROBLEMS

4–1. If **A, B,** and **D** are given vectors, prove the distributive law for the vector cross product, i.e., $\mathbf{A} \times (\mathbf{B} + \mathbf{D}) = (\mathbf{A} \times \mathbf{B}) + (\mathbf{A} \times \mathbf{D})$.

4–2. Prove the triple scalar product identity $\mathbf{A} \cdot \mathbf{B} \times \mathbf{C} = \mathbf{A} \times \mathbf{B} \cdot \mathbf{C}$.

4–3. Given the three nonzero vectors **A, B,** and **C,** show that if $\mathbf{A} \cdot (\mathbf{B} \times \mathbf{C}) = 0$, the three vectors *must* lie in the same plane.

*4–4. Determine the magnitude and directional sense of the moment of the force at A about point O.

Probs. 4–4/4–5

4–5. Determine the magnitude and directional sense of the moment of the force at A about point P.

4–6. Determine the magnitude and directional sense of the moment of the force at A about point O.

Probs. 4–6/4–7

4–7. Determine the magnitude and directional sense of the moment of the force at A about point P.

*4–8. Determine the magnitude and directional sense of the moment of the force at A about point O.

Probs. 4–8/4–9

4–9. Determine the magnitude and directional sense of the moment of the force at A about point P.

4–10. Determine the moment about point A of each of the three forces acting on the beam.

Probs. 4–10/4–11

4–11. Determine the moment about point B of each of the three forces acting on the beam.

107

*4–12. Determine the moment of each force about the bolt passing through point A.

Prob. 4–12

4–13. Determine the moment of each of the three forces about point A. Solve the problem first by using each force as a whole, and then by using the principle of moments.

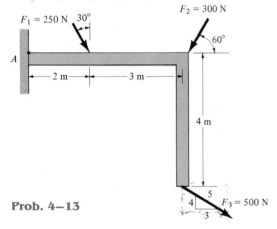

Prob. 4–13

4–14. The two boys push on the gate with forces of $F_A = 30$ lb and $F_B = 50$ lb as shown. Determine the moment of each force about C. Which way will the gate rotate, clockwise or counterclockwise? Neglect the thickness of the gate.

Probs. 4–14/4–15

4–15. Two boys push on the gate as shown. If the boy at B exerts a force of $F_B = 30$ lb, determine the magnitude of the force $\mathbf{F}_A$ the boy at A must exert in order to prevent the gate from turning, i.e., so the resultant moment about C of both forces is zero. Neglect the thickness of the gate.

*4–16. Determine the direction θ ($0° \leq \theta \leq 180°$) of the force $\mathbf{F}$ so that $\mathbf{F}$ produces (a) the maximum moment about point A and (b) the minimum moment about point A. Compute the moment in each case.

Probs. 4–16/4–17

4–17. Determine the moment of the force $\mathbf{F}$ about point A as a function of θ. Plot the results of M (ordinate) versus θ (abscissa) for $0° \leq \theta \leq 180°$.

4–18. The torque wrench ABC is used to measure the moment or torque applied to a bolt when the bolt is located at A and a force is applied to the handle at C. The mechanic reads the torque on the scale at B. If an extension AO of length d is used on the wrench, determine the required scale reading if the desired torque on the bolt at O is to be $\mathbf{M}$.

Prob. 4–18

4–19. The towline exerts a force of $P = 4$ kN at the end of the 20-m-long crane boom. If $\theta = 30°$, determine the placement x of the hook at A so that this force creates a maximum moment about point O.

Probs. 4–19/4–20

*4–20. The towline exerts a force of $P = 4$ kN at the end of the 20-m-long crane boom. If $x = 25$ m, determine the position θ of the boom so that this force creates a maximum moment about point O.

4–21. The force **F** lies in the x-y plane and acts at point A on the plate. Determine the magnitude and orientation θ of **F** if it has an x component of $\mathbf{F}_x = \{30\mathbf{i}\}$ N and the moment of **F** about point O is $\mathbf{M}_O = \{-300\mathbf{k}\}$ N · m.

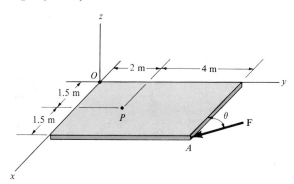

Probs. 4–21/4–22

4–22. The force **F** lies in the x-y plane and acts at point A on the plate. Determine the magnitude and orientation θ of **F** if it has a y component of $\mathbf{F}_y = \{-20\mathbf{j}\}$ N and the moment of **F** about point P is $\mathbf{M}_P = \{-180\mathbf{k}\}$ N · m.

4–23. Determine the moment of the force at A about point O. Express the result as a Cartesian vector.

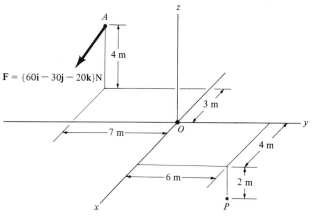

Probs. 4–23/4–24

*4–24. Determine the moment of the force at A about point P. Express the result as a Cartesian vector.

4–25. Determine the moment of the force **F** at A about point O. Express the result as a Cartesian vector.

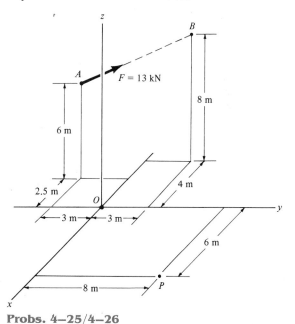

Probs. 4–25/4–26

4–26. Determine the moment of the force **F** at A about point P. Express the result as a Cartesian vector.

4–27. The man pulls on the rope with a force of 20 N. Determine the moment that this force exerts about the base of the pole at O. Solve the problem two ways, i.e., by using a position vector from O to A, then O to B.

Prob. 4–27

*4–28. The cable exerts a 200-N force on the telephone pole as shown. Determine the moment of this force at the base A of the pole. Solve the problem two ways, i.e., by using a position vector from A to C, then A to B.

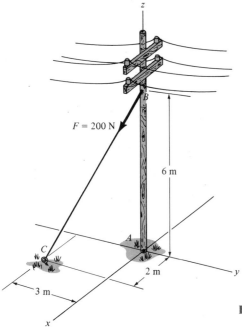

Prob. 4–28

4–29. Using Cartesian vectors, compute the moment of each of the three forces acting on the pipe assembly about point O. Add these moments and calculate the magnitude and coordinate direction angles of the resultant moment.

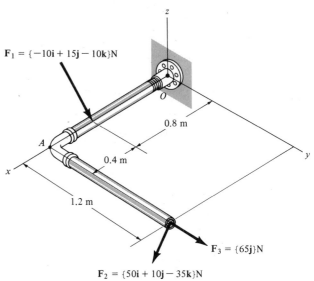

$F_1 = \{-10i + 15j - 10k\}N$

$F_3 = \{65j\}N$

$F_2 = \{50i + 10j - 35k\}N$

Probs. 4–29/4–30

4–30. Using Cartesian vectors, compute the moment of each of the three forces acting on the pipe assembly about point A. Add these moments and calculate the magnitude and coordinate direction angles of the resultant moment.

4–31. The 6-m boom lies in the y-z plane. If a force of $F = 4$ kN acts on the cable AB, determine the moment of this force about point O. Realizing that $M = Fd$, find the shortest distance d from point O to the cable.

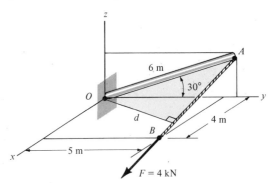

Prob. 4–31

*4–32. The curved rod lies in the *x-y* plane and has a radius of 3 m. If a force of $F = 80$ N acts at its end as shown, determine the moment of this force about point O.

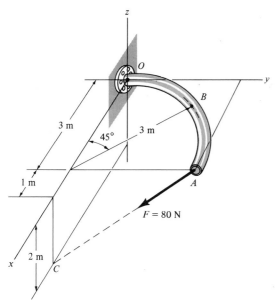

Probs. 4–32/4–33

4–33. The curved rod lies in the *x-y* plane and has a radius of 3 m. If a force of $F = 80$ N acts at its end as shown, determine the moment of this force about point B.

4–34. A force $\mathbf{F} = \{30\mathbf{i} - 50\mathbf{j} + 40\mathbf{k}\}$ N creates a moment of $\mathbf{M}_O = \{200\mathbf{i} - 80\mathbf{j} - 250\mathbf{k}\}$ N · m about the origin of coordinates O. Determine the *x* and *y* coordinates of the point where the line of action of $\mathbf{F}$ intersects the *x-y* plane.

4–35. A force of $\mathbf{F} = \{-5\mathbf{i} + 3\mathbf{j} - 4\mathbf{k}\}$ kN produces a moment of $\mathbf{M}_O = \{-17\mathbf{i} - 7\mathbf{j} + 16\mathbf{k}\}$ kN · m about point O. If the force acts at point P having a *y* coordinate of $y = 2$ m, determine the *x* and *z* coordinates.

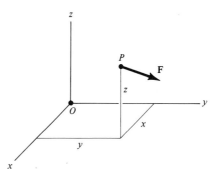

Probs. 4–35/4–36

*4–36. The force $\mathbf{F} = \{-5\mathbf{i} + 3\mathbf{j} - 4\mathbf{k}\}$ kN creates a moment about point O of $\mathbf{M}_O = \{-17\mathbf{i} - 3\mathbf{j} + 19\mathbf{k}\}$ kN · m. If the force passes through a point P having a *z* coordinate of $z = 3$ m, determine the *x* and *y* coordinates of the point. Also, realizing that $M_O = Fd$, determine the perpendicular distance d from point O to the line of action of $\mathbf{F}$.

4–37. Strut AB of the 1-m-diameter hatch door exerts a force of 450 N on point B. Determine the moment of this force about point O.

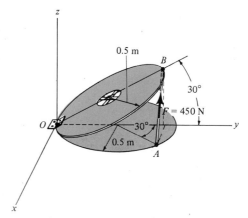

Prob. 4–37

111

4.5 Moment of a Force About a Specified Axis

Recall that when the moment of a force is computed about a point, the moment and its axis are *always* perpendicular to the plane containing the force and the moment arm. In some problems it is important to find the *component* of this moment along a specified axis that passes through the point. To solve this problem either a scalar or vector analysis can be used.

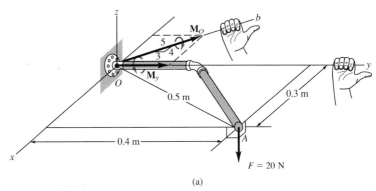

(a)

Fig. 4–18a

Scalar Analysis

As a numerical example of this problem consider the pipe asembly shown in Fig. 4–18a, which lies in the horizontal plane and is subjected to the vertical force of $F = 20$ N applied at point A. The moment of this force about point O has a *magnitude* of $M_O = (20$ N$)(0.5$ m$) = 10$ N · m and a *direction* defined by the right-hand rule, as shown in Fig. 4–18a. This moment tends to turn the pipe about the Ob axis. For practical reasons, however, it may be necessary to determine the *component* of M_O about the y axis, M_y, since this component tends to unscrew the pipe from the collar at O. From Fig. 4–18a, M_y has a magnitude of $M_y = \frac{3}{5}(10$ N · m$) = 6$ N · m and a sense of direction shown by the vector resolution. Rather than performing this *two-step* process of first finding the moment of the force about point O and then resolving the moment along the y axis, it is also possible to solve this problem *directly*. To do so, it is necessary to determine the perpendicular or moment-arm distance from the line of action of $\mathbf{F}$ to the y axis. From Fig. 4–18a this distance is 0.3 m. Thus the *magnitude* of the moment of the force about the y axis is $M_y = 0.3(20$ N$) = 6$ N · m and the *direction* is determined by the right-hand rule as shown.

In general, then, the magnitude of the moment of a force **F** about a specified axis aa' can be determined from the equation

$$M_a = F d_a \qquad\qquad (4\text{--}8)$$

where d_a is the *perpendicular or shortest distance* from the force line of action to the axis. The direction is determined from the thumb of the right hand when the fingers are curled in accordance with the direction of rotation as produced by the force. In particular, note that a *force will not contribute a moment about an axis if its line of action is parallel to the axis or its line of action passes through the axis.*

Vector Analysis

The above two-step solution to the problem in Fig. 4–18a can also be performed using a vector analysis. Here the moment about point O is first determined from $\mathbf{M}_O = \mathbf{r}_A \times \mathbf{F} = (0.3\mathbf{i} + 0.4\mathbf{j}) \times (-20\mathbf{k}) = \{-8\mathbf{i} + 6\mathbf{j}\}\text{N} \cdot \text{m}$, Fig. 4–18b. The component or projection of this moment along the y axis is then determined from the dot product (Sec. 2.9). Since the unit vector for the axis (or line) is $\mathbf{u} = \mathbf{j}$, then $M_y = \mathbf{M}_O \cdot \mathbf{u} = (-8\mathbf{i} + 6\mathbf{j}) \cdot \mathbf{j} = 6 \text{ N} \cdot \text{m}$.

A vector analysis such as this is particularly advantageous for finding the moment of a force about an axis when the force components or the appropriate moment arms are difficult to determine. For this reason, the above two-step process of first determining the moment of the force about a point (O) on the axis, and then finding the component of the moment along the axis will now

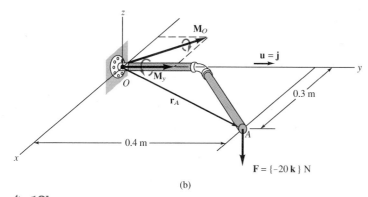

(b)

Fig. 4–18b

be generalized and applied to a body of arbitrary shape. To do so, consider the body in Fig. 4–19, which is subjected to the force **F** acting at point A. Here we wish to determine the effect of **F** in tending to rotate the body about the aa' axis. This tendency for rotation is measured by the moment component $\mathbf{M}_a$. To determine $\mathbf{M}_a$ we first compute the moment of **F** about an *arbitrary point* O that lies on the axis. In this case, $\mathbf{M}_O$ is expressed by the cross product $\mathbf{M}_O = \mathbf{r} \times \mathbf{F}$, where **r** is directed from O to A. Since $\mathbf{M}_O$ acts along the moment axis bb', which is perpendicular to the plane containing **r** and **F**, the component or projection of $\mathbf{M}_O$ onto the aa' axis is then represented by $\mathbf{M}_a$. The *magnitude* of $\mathbf{M}_a$ is determined by the dot product, $M_a = M_O \cos\theta = \mathbf{M}_O \cdot \mathbf{u}_a$, where $\mathbf{u}_a$ is a unit vector used to define the direction of the aa' axis. Combining these two steps as a general expression, we have $M_a = (\mathbf{r} \times \mathbf{F}) \cdot \mathbf{u}_a$. Since the dot product is commutative, we can also write

$$M_a = \mathbf{u}_a \cdot (\mathbf{r} \times \mathbf{F})$$

In vector algebra, this combination of dot and cross product yielding the scalar M_a is termed the *mixed triple product*. Provided the Cartesian components of each of the vectors are known, the mixed triple product may be written in determinant form as

$$M_a = (u_{a_x}\mathbf{i} + u_{a_y}\mathbf{j} + u_{a_z}\mathbf{k}) \cdot \begin{vmatrix} \mathbf{i} & \mathbf{j} & \mathbf{k} \\ r_x & r_y & r_z \\ F_x & F_y & F_z \end{vmatrix}$$

or simply

$$M_a = \mathbf{u}_a \cdot (\mathbf{r} \times \mathbf{F}) = \begin{vmatrix} u_{a_x} & u_{a_y} & u_{a_z} \\ r_x & r_y & r_z \\ F_x & F_y & F_z \end{vmatrix} \qquad (4\text{–}9)$$

where $u_{a_x}, u_{a_y}, u_{a_z}$ represent the Cartesian components of the unit vector defining the direction of the aa' axis

r_x, r_y, r_z represent the Cartesian components of the position vector drawn from any point O on the aa' axis to any point A on the line of action of the force

F_x, F_y, F_z represent the Cartesian components of the force vector.

Once the magnitude of $\mathbf{M}_a$ is determined from Eq. 4–9, we can express it in Cartesian vector form since its *direction* is defined by $\mathbf{u}_a$. We have

$$\mathbf{M}_a = M_a\mathbf{u}_a = [\mathbf{u}_a \cdot (\mathbf{r} \times \mathbf{F})]\mathbf{u}_a \qquad (4\text{–}10)$$

The following examples numerically illustrate application of the above concepts.

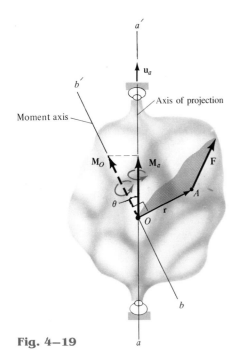

Fig. 4–19

Example 4–7

The force $\mathbf{F} = \{-40\mathbf{i} + 20\mathbf{j} + 10\mathbf{k}\}$ N acts at point A shown in Fig. 4–20a. Compute the moment of this force about the x axis and about the Oa axis.

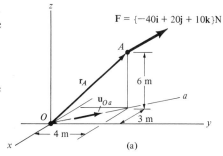

(a)

SOLUTION I *(VECTOR ANALYSIS)*

We can solve this problem by using the position vector $\mathbf{r}_A$. Why? Since $\mathbf{r}_A = \{-3\mathbf{i} + 4\mathbf{j} + 6\mathbf{k}\}$ m, and $\mathbf{u}_x = \mathbf{i}$, then applying Eq. 4–9,

$$M_x = \mathbf{i} \cdot (\mathbf{r}_A \times \mathbf{F}) = \begin{vmatrix} 1 & 0 & 0 \\ -3 & 4 & 6 \\ -40 & 20 & 10 \end{vmatrix}$$

$$M_x = 1[4(10) - 6(20)] - 0[(-3)(10) - 6(-40)] + 0[(-3)(20) - 4(-40)]$$
$$= -80 \text{ N} \cdot \text{m} \qquad\qquad Ans.$$

The negative sign indicates that the sense of $\mathbf{M}_x$ is opposite to $\mathbf{i}$. Also, since $\mathbf{u}_{Oa} = -\frac{3}{5}(\mathbf{i}) + \frac{4}{5}(\mathbf{j})$, then

$$M_{Oa} = \mathbf{u}_{Oa} \cdot (\mathbf{r}_A \times \mathbf{F}) = \begin{vmatrix} -\frac{3}{5} & \frac{4}{5} & 0 \\ -3 & 4 & 6 \\ -40 & 20 & 10 \end{vmatrix}$$

$$= -\tfrac{3}{5}[4(10) - 6(20)] - \tfrac{4}{5}[(-3)(10) - 6(-40)] + 0[(-3)(20) - 4(-40)]$$
$$= -120 \text{ N} \cdot \text{m} \qquad\qquad Ans.$$

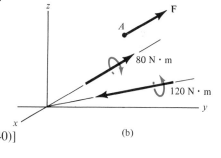

(b)

The components are shown in Fig. 4–20b.

SOLUTION II *(SCALAR ANALYSIS)*

Since the force components and moment arms are easy to determine for computing M_x a scalar analysis can be used to solve this problem. Referring to Fig. 4–20c, only the 10-N and 20-N forces contribute moments about the x axis. (The line of action of the 40-N force is *parallel* to this axis and hence its moment about x is zero.) Using the right-hand rule, we have

$$M_x = (10 \text{ N})(4 \text{ m}) - (20 \text{ N})(6 \text{ m}) = -80 \text{ N} \cdot \text{m} \qquad Ans.$$

Although not required here, note also that

$$M_y = (10 \text{ N})(3 \text{ m}) - (40 \text{ N})(6 \text{ m}) = -210 \text{ N} \cdot \text{m}$$
$$M_z = (40 \text{ N})(4 \text{ m}) - (20 \text{ N})(3 \text{ m}) = 100 \text{ N} \cdot \text{m}$$

(c)

Fig. 4–20

Notice that to determine M_{Oa} by this scalar method requires much more effort, since the force components of 40 N and 20 N are not perpendicular to the direction of Oa. The vector analysis yields a more direct solution.

Example 4–8

The rod shown in Fig. 4–21a is supported by two brackets at A and B. Determine the magnitude of the moment $\mathbf{M}_{AB}$ produced by force $\mathbf{F} = \{-600\mathbf{i} + 200\mathbf{j} - 300\mathbf{k}\}$ N, which tends to rotate the rod about the AB axis.

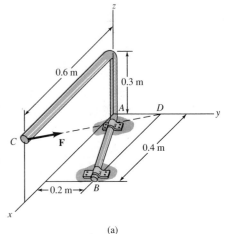

(a)

SOLUTION

A vector analysis using Eq. 4–9 will be considered for the solution since the scalar component of the moment arm is difficult to determine. In general, the magnitude of $\mathbf{M}_{AB}$ is

$$M_{AB} = \mathbf{u}_B \cdot (\mathbf{r} \times \mathbf{F}) \tag{1}$$

Each of the terms in this equation will now be identified.

Unit vector $\mathbf{u}_B$ defines the direction of the AB axis of the rod, Fig. 4–21b, where

$$\mathbf{u}_B = \frac{\mathbf{r}_B}{r_B} = \frac{0.4\mathbf{i} + 0.2\mathbf{j}}{\sqrt{(0.4)^2 + (0.2)^2}} = 0.894\mathbf{i} + 0.447\mathbf{j}$$

Vector $\mathbf{r}$ is directed from *any point* on the AB axis to *any* point on the line of action of the force. For example, position vectors $\mathbf{r}_C$ and $\mathbf{r}_D$ are suitable, Fig. 4–21b. (Although not shown, $\mathbf{r}_{BC}$ or $\mathbf{r}_{BD}$ can also be used.) For simplicity, choose $\mathbf{r}_D$, where

$$\mathbf{r}_D = \{0.2\mathbf{j}\} \text{ m}$$

The force is

$$\mathbf{F} = \{-600\mathbf{i} + 200\mathbf{j} - 300\mathbf{k}\} \text{ N}$$

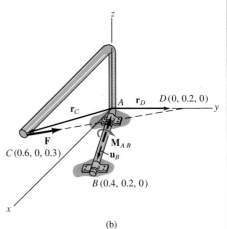

(b)

Fig. 4–21

Substituting these vectors into the determinant form of Eq. (1) and expanding, we have

$$M_{AB} = \mathbf{u}_B \cdot (\mathbf{r}_D \times \mathbf{F}) = \begin{vmatrix} 0.894 & 0.447 & 0 \\ 0 & 0.2 & 0 \\ -600 & 200 & -300 \end{vmatrix}$$

$$= 0.894[0.2(-300) - 0(200)] - 0.447[0(-300) - 0(-600)]$$
$$+ 0[0(200) - 0.2(-600)]$$

$$= -53.64 \text{ N} \cdot \text{m}$$

The negative sign indicates that the sense of $\mathbf{M}_{AB}$ is opposite to that of $\mathbf{u}_B$. Expressing $\mathbf{M}_{AB}$ as a Cartesian vector,

$$\mathbf{M}_{AB} = M_{AB}\mathbf{u}_B = (-53.64 \text{ N} \cdot \text{m})(0.894\mathbf{i} + 0.447\mathbf{j})$$
$$= \{-48.0\mathbf{i} - 24.0\mathbf{j}\} \text{ N} \cdot \text{m} \qquad \qquad \textit{Ans.}$$

The result is shown in Fig. 4–21b.

PROBLEMS

4–38. Determine the moment of the force **F** about the *Oa* axis. Express the result as a Cartesian vector.

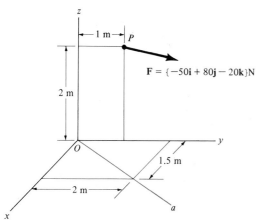

Prob. 4–38

4–39. Determine the moment of the force **F** about the *Aa* axis. Express the result as a Cartesian vector.

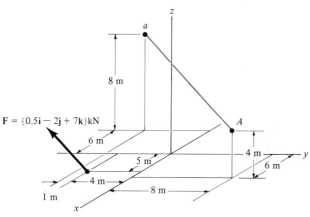

Prob. 4–39

*4–40. Determine the moment of the force **F** about the *aa* axis. Express the result as a Cartesian vector.

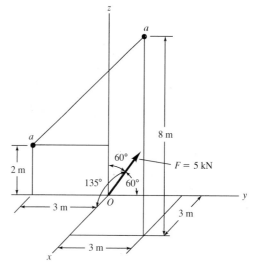

Prob. 4–40

4–41. Determine the magnitude of the moment of each of the three forces about the axis *AB*. Solve the problem (a) using a Cartesian vector approach and (b) using a scalar approach.

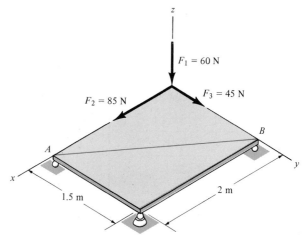

Prob. 4–41

117

4–42. The hood of the automobile is supported by the strut AB, which exerts a force of 24 lb on the hood. Determine the moment of this force about the hinged axis y. Express the result as a Cartesian vector.

Prob. 4–42

4–43. The wooden shaft is held in a lathe. The cutting tool exerts a force $\mathbf{F}$ on the shaft in the direction shown. Determine the moment of this force about the x axis of the shaft. Express the result as a Cartesian vector. The distance OA is 25 mm.

Prob. 4–43

*4–44. Determine the moment of the force $\mathbf{F}$ which is directed along ED of the block about the axis BA. Solve the problem by using two different position vectors $\mathbf{r}$. Express the result as a Cartesian vector.

Probs. 4–44/4–45

4–45. Determine the moment of the force $\mathbf{F}$ which is directed along ED of the block about the axis OC. Solve the problem by using two different position vectors $\mathbf{r}$. Express the result as a Cartesian vector.

4–46. Determine the magnitude of the moment of the force $\mathbf{F} = \{50\mathbf{i} - 20\mathbf{j} - 80\mathbf{k}\}$ N about the base line AB of the tripod.

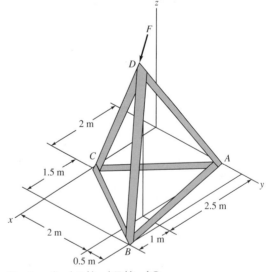

Probs. 4–46/4–47/4–48

4–47. Determine the magnitude of the moment of the force $\mathbf{F} = \{50\mathbf{i} - 20\mathbf{j} - 80\mathbf{k}\}$ N about the base line BC of the tripod.

*4–48. Determine the magnitude of the moment of the force $\mathbf{F} = \{50\mathbf{i} - 20\mathbf{j} - 80\mathbf{k}\}$ N about the base line CA of the tripod.

4-49. Determine the moments of the force **F** about the x, y, and z axes. Solve the problem (a) using a Cartesian vector approach and (b) using a scalar approach. Express each result as a Cartesian vector.

Probs. 4-49/4-50

4-50. Determine the moment of the force **F** about an axis extending between A and C. Express the result as a Cartesian vector.

4-51. Determine the moment that the force of $F = 80$ N, which is applied to the handle of the wrench, exerts about the z axis of the bolt. Solve the problem (a) using a Cartesian vector approach and (b) using a scalar approach. Express the result as a Cartesian vector.

Prob. 4-51

*4-52. The force of $F = 80$ lb acts along the edge DB of the tetrahedron. Determine the magnitude of the moment of this force about the edge AC.

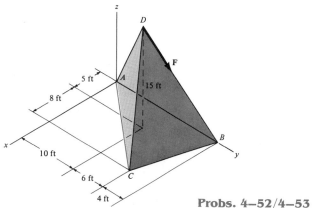

Probs. 4-52/4-53

4-53. If the moment of the force **F** about the edge AC of the tetrahedron has a magnitude of $M = 200$ lb · ft and is directed from C towards A, determine the magnitude of **F**.

4-54. A horizontal force of $\mathbf{F} = \{-50\mathbf{i}\}$ N is applied to the handle of the pipe wrench. Determine the moment that this force exerts along the axis OA (z axis) of the pipe assembly. Both the wrench and pipe assembly, $OABC$, lie in the y-z plane. *Suggestion:* Use a scalar analysis.

Probs. 4-54/4-55

4-55. Determine the magnitude of the horizontal force $\mathbf{F} = -F\mathbf{i}$ acting on the handle of the wrench so that this force produces a component of moment along the OA axis (z axis) of the pipe assembly of $\mathbf{M}_z = \{4\mathbf{k}\}$ N · m. Both the wrench and the pipe assembly, $OABC$, lie in the y-z plane. *Suggestion:* Use a scalar analysis.

119

4.6 Moment of a Couple

Fig. 4–22

A *couple* is defined as two parallel forces that have the same magnitude, opposite directions, and are separated by a perpendicular distance d, Fig. 4–22. Since the resultant force of the two forces composing the couple is zero, the only effect of a couple is to produce a rotation or tendency of rotation in a specified direction.

The moment produced by a couple, called a couple moment, is equivalent to the sum of the moments of both couple forces, computed about any arbitrary point O in space. To show this, consider position vectors $\mathbf{r}_A$ and $\mathbf{r}_B$, directed from O to points A and B lying on the line of action of $-\mathbf{F}$ and $\mathbf{F}$, Fig. 4–23. The couple moment computed about O is therefore

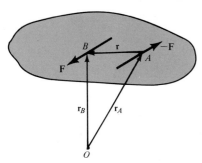

Fig. 4–23

$$\mathbf{M} = \mathbf{r}_A \times (-\mathbf{F}) + \mathbf{r}_B \times (\mathbf{F})$$
$$= (\mathbf{r}_B - \mathbf{r}_A) \times \mathbf{F}$$

By the triangle law of vector addition, $\mathbf{r}_A + \mathbf{r} = \mathbf{r}_B$ or $\mathbf{r} = \mathbf{r}_B - \mathbf{r}_A$, so that

$$\mathbf{M} = \mathbf{r} \times \mathbf{F} \qquad (4\text{–}11)$$

This result indicates that a couple moment is a *free vector*, i.e., it can act at *any point*, since $\mathbf{M}$ depends *only* upon the position vector directed *between* the forces and *not* the position vectors $\mathbf{r}_A$ or $\mathbf{r}_B$, directed from point O to the forces. Consequently, this concept is unlike the moment of a force, which requires a definite point (or axis) about which moments are computed.

Scalar Formulation

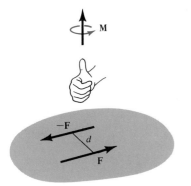

Fig. 4–24

The moment of a couple, $\mathbf{M}$, Fig. 4–24, is defined as having a *magnitude* of

$$\boxed{M = Fd} \qquad (4\text{–}12)$$

where F is the magnitude of one of the forces and d is the perpendicular distance or moment arm between the forces. The *direction* and sense of the couple moment are determined by the right-hand rule, where the thumb indicates the direction when the fingers are curled with the sense of rotation caused by the forces. In all cases, $\mathbf{M}$ acts perpendicular to the plane containing the two forces.

Vector Formulation

The moment of a couple can also be expressed by the vector cross product using Eq. 4–11,

$$\boxed{\mathbf{M} = \mathbf{r} \times \mathbf{F}} \qquad (4\text{–}13)$$

Application of this equation is easily remembered if one thinks of taking the moments of both forces about a point lying on the line of action of one of the forces. For example, if moments are taken about point A in Fig. 4–23 the moment of $-\mathbf{F}$ is zero and the moment of $\mathbf{F}$ is defined from Eq. 4–13. Notice in the formulation that $\mathbf{r}$ is crossed with the force $\mathbf{F}$ to which it is directed.

Equivalent Couples

Two couples are said to be equivalent if they produce the same moment. Since the moment produced by a couple is always perpendicular to the plane containing the couple forces, it is therefore necessary that the forces of equal couples lie either in the same plane or in planes that are *parallel* to one another. In this way, the direction of each couple moment will be the same, that is, perpendicular to the parallel planes. For example, the two couples shown in Fig. 4–25 are equivalent. One couple is produced by a pair of 100-N forces separated by a distance of $d = 0.5$ m, and the other is produced by a pair of 200-N forces separated by a distance of 0.25 m. Since the planes in which the forces act are parallel to the x-y plane, the moment produced by each of the couples may be expressed as $\mathbf{M} = \{50\mathbf{k}\}$ N · m.

Fig. 4–25

Resultant Couple

Since couple moments are free vectors, they may be applied at any point P on a body and added vectorially. For example, the two couples acting on different planes of the rigid body in Fig. 4–26a may be replaced by their corresponding moments $\mathbf{M}_1$ and $\mathbf{M}_2$, shown in Fig. 4–26b. These free vectors may then be moved to the *arbitrary point P* and added to obtain their resultant vector sum $\mathbf{M}_R = \mathbf{M}_1 + \mathbf{M}_2$, shown in Fig. 4–26c.

The following examples numerically illustrate these concepts. In general, problems displayed in two dimensions should be solved using a scalar analysis since the geometry is easy to compute.

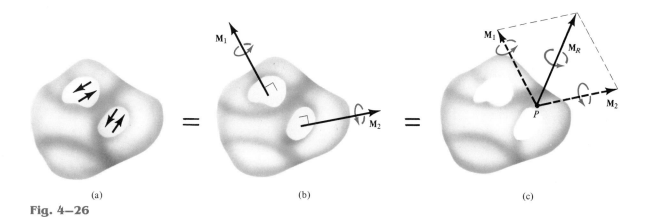

(a)   (b)   (c)

Fig. 4–26

Example 4-9

A couple acts at the end of the beam shown in Fig. 4–27a. Replace it by an equivalent couple having a pair of forces that act through (a) points A and B, and (b) points D and E.

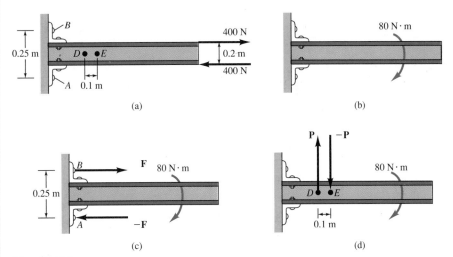

(a)

(b)

(c)

(d)

Fig. 4-27

SOLUTION (*SCALAR ANALYSIS*)

The couple has a magnitude of $M = Fd = 400(0.2) = 80$ N · m and a direction that is into the page since the forces tend to rotate clockwise. **M** is a free vector so that it can be placed at any point on the beam, Fig. 4–27b.

Part (a). To preserve the direction of **M,** *horizontal* forces acting through points A and B must be directed as shown in Fig. 4–27c. The magnitude of each force is

$$M = Fd$$
$$80 = F(0.25)$$
$$F = 320 \text{ N} \qquad\qquad Ans.$$

Part (b). To generate the required clockwise rotation, forces acting through points D and E must be *vertical* and directed as shown in Fig. 4–27d. The magnitude of each force is

$$M = Pd$$
$$80 = P(0.1)$$
$$P = 800 \text{ N} \qquad\qquad Ans.$$

Example 4–10

Determine the moment of the couple acting on the beam shown in Fig. 4–28a.

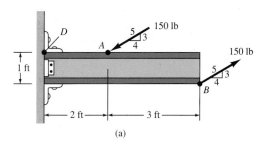

(a)

Fig. 4–28

SOLUTION (*SCALAR ANALYSIS*)

Here it is somewhat difficult to determine the perpendicular distance between the forces and compute the couple moment as $M = Fd$. Instead, we can resolve each force into its horizontal and vertical components, $F_x = \frac{4}{5}(150) = 120$ lb and $F_y = \frac{3}{5}(150) = 90$ lb, Fig. 4–28b, and then use the principle of moments. The couple moment can be computed about *any point*. For example, if point D is chosen, we have

$$\zeta+M = 120 \text{ lb}(0 \text{ ft}) - 90 \text{ lb}(2 \text{ ft}) + 90 \text{ lb}(5 \text{ ft}) + 120 \text{ lb}(1 \text{ ft})$$
$$= 390 \text{ lb} \cdot \text{ft} \,\uparrow \qquad\qquad Ans.$$

(b)

It is easier to compute the moments about point A or B in order to *eliminate* the moment of the force acting at the moment point. For point A, Fig. 4–28b, we have

$$\zeta+M = 90 \text{ lb}(3 \text{ ft}) + 120 \text{ lb}(1 \text{ ft})$$
$$= 390 \text{ lb} \cdot \text{ft} \,\uparrow \qquad\qquad Ans.$$

Show that one obtains this same result if moments are summed about point B. Notice also that the couple in Fig. 4–28a has been replaced by *two* couples in Fig. 4–28b. Using $M = Fd$, one couple has a moment of $M_1 = 90 \text{ lb}(3 \text{ ft}) = 270 \text{ lb} \cdot \text{ft}$ and the other has a moment of $M_2 = 120 \text{ lb}(1 \text{ ft}) = 120 \text{ lb} \cdot \text{ft}$. By the right-hand rule both couple moments are counterclockwise and are therefore directed out of the page. Since these couples are free vectors they can be moved to any point and added, which yields $M = 270 \text{ lb} \cdot \text{ft} + 120 \text{ lb} \cdot \text{ft} = 390 \text{ lb} \cdot \text{ft} \,\uparrow$, the same result computed above. **M** is a free vector and can therefore act at any point on the beam, Fig. 4–28c.

(c)

Example 4–11

Determine the couple moment acting on the pipe shown in Fig. 4–29a. Segment AB is directed $30°$ below the x-y plane.

(a)

Fig. 4–29

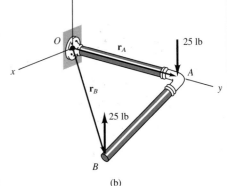

(b)

SOLUTION I (VECTOR ANALYSIS)

The moment of the two couple forces can be computed about *any point*. If point O is used, Fig. 4–29b, we have

$$\mathbf{M} = \mathbf{r}_A \times (-25\mathbf{k}) + \mathbf{r}_B \times (25\mathbf{k})$$
$$= (8\mathbf{j}) \times (-25\mathbf{k}) + (6 \cos 30°\mathbf{i} + 8\mathbf{j} - 6 \sin 30°\mathbf{k}) \times (25\mathbf{k})$$
$$= -200\mathbf{i} - 129.9\mathbf{j} + 200\mathbf{i}$$
$$= \{-130\mathbf{j}\} \text{ lb} \cdot \text{in.} \qquad \textit{Ans.}$$

It is *easier* to take moments of the couple forces about a point lying on the line of action of one of the forces, e.g., point A, Fig. 4–29c. In this case the moment of the force at A is zero, so that

$$\mathbf{M} = \mathbf{r}_{AB} \times (25\mathbf{k})$$
$$= (6 \cos 30°\mathbf{i} - 6 \sin 30°\mathbf{k}) \times (25\mathbf{k})$$
$$= \{-130\mathbf{j}\} \text{ lb} \cdot \text{in.} \qquad \textit{Ans.}$$

(c)

SOLUTION II (SCALAR ANALYSIS)

Although this problem is shown in three dimensions, the geometry is simple enough to use the scalar equation $M = Fd$. The perpendicular distance between the lines of action of the forces is $d = 6 \cos 30° = 5.20$ in., Fig. 4–29d. Hence, taking moments of the forces about either point A or B yields

$$M = Fd = 25(5.20) = 129.9 \text{ lb} \cdot \text{in.}$$

Applying the right-hand rule, $\mathbf{M}$ acts in the $-\mathbf{j}$ direction. Thus,

$$\mathbf{M} = \{-130\mathbf{j}\} \text{ lb} \cdot \text{in.} \qquad \textit{Ans.}$$

(d)

Example 4–12

Replace the two couples acting on the triangular block in Fig. 4–30a by a single resultant couple.

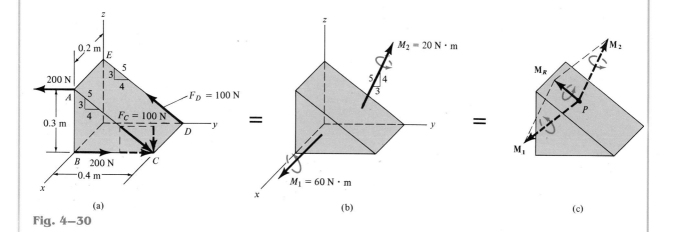

(a)   (b)   (c)

Fig. 4–30

SOLUTION *(VECTOR ANALYSIS)*

The couple M_1, caused by the forces at A and B, can easily be determined from a scalar formulation.

$$M_1 = Fd = 200(0.3) = 60 \text{ N} \cdot \text{m}$$

By the right-hand rule, M_1 acts in the $+\mathbf{i}$ direction, Fig. 4–30b. Hence,

$$M_1 = \{60\mathbf{i}\} \text{ N} \cdot \text{m}$$

Vector analysis will be used to determine M_2, caused by forces at C and D. If moments are computed about point D, Fig. 4–30a, $M_2 = \mathbf{r}_{CD} \times \mathbf{F}_C$, then,

$$\begin{aligned}
M_2 = \mathbf{r}_{CD} \times \mathbf{F}_C &= (0.2\mathbf{i}) \times [100(\tfrac{4}{5})\mathbf{j} - 100(\tfrac{3}{5})\mathbf{k}] \\
&= (0.2\mathbf{i}) \times [80\mathbf{j} - 60\mathbf{k}] = 16(\mathbf{i} \times \mathbf{j}) - 12(\mathbf{i} \times \mathbf{k}) \\
&= \{12\mathbf{j} + 16\mathbf{k}\} \text{ N} \cdot \text{m}
\end{aligned}$$

Try to establish M_2 by using a scalar formulation, Fig. 4–30b.

Since M_1 and M_2 are free vectors, they may be moved to some arbitrary point P on the block and added vectorially, Fig. 4–30c. The resultant couple moment becomes

$$M_R = M_1 + M_2 = \{60\mathbf{i} + 12\mathbf{j} + 16\mathbf{k}\} \text{ N} \cdot \text{m} \qquad \textit{Ans.}$$

PROBLEMS

*4–56.   Determine the magnitude and sense of the couple moment. Each force has a magnitude of $F = 65$ lb.

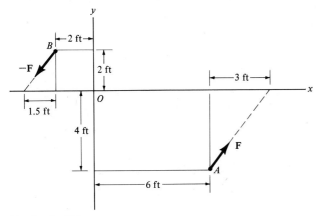

Prob. 4–56

4–57.   Determine the magnitude and sense of the couple moment. Each force has a magnitude of $F = 8$ kN.

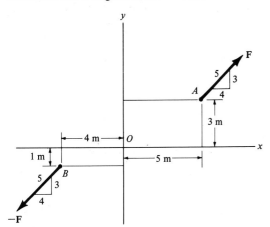

Prob. 4–57

4–58.   Determine the couple moment. Express the result as a Cartesian vector. Each force has a magnitude of $F = 120$ lb.

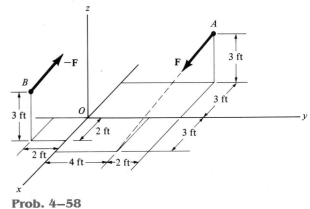

Prob. 4–58

4–59.   Determine the couple moment. Express the result as a Cartesian vector.

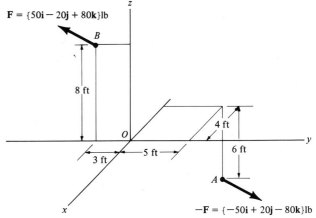

Prob. 4–59

*4–60. Two couples act on the beam as shown. Determine the magnitude of **F** so that the resultant couple moment is 300 lb · ft counterclockwise. Where on the beam does the resultant couple moment act?

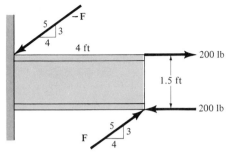

Prob. 4–60

4–61. Two couples act on the frame. If the resultant couple moment is to be zero, determine the distance d between the 80-lb couple forces.

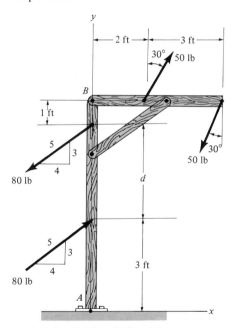

Probs. 4–61/4–62/4–63

4–62. Two couples act on the frame. If $d = 4$ ft, determine the resultant couple moment. Compute the result by resolving each force into x and y components and (a) finding the moment of each couple (Eq. 4–13) and (b) summing the moments of all the force components about point A.

4–63. Two couples act on the frame. If $d = 4$ ft, determine the resultant couple moment. Compute the result by resolving each force into x and y components and (a) finding the moment of each couple (Eq. 4–13) and (b) summing the moments of all the force components about point B.

*4–64. The ends of the triangular plate are subjected to three couples. Determine the magnitude of the force **F** so that the resultant couple moment is 600 N · m clockwise.

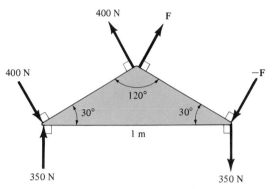

Prob. 4–64

4–65. Three couple moments act on the pipe assembly. Determine the magnitude of the resultant couple moment if $M_2 = 50$ N · m and $M_3 = 35$ N · m.

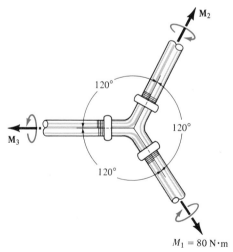

Probs. 4–65/4–66

4–66. Three couple moments act on the pipe assembly. Determine the magnitudes of **M**$_2$ and **M**$_3$ so that the resultant couple moment is zero.

4–67. Express the moment of the couple acting on the pipe assembly in Cartesian vector form. Solve the problem (a) using Eq. 4–13, and (b) summing the moments of each force about point O.

Prob. 4–67

*4–68. The gear reducer is subjected to the two couple moments shown. Express the resultant couple moment in Cartesian vector form and specify its magnitude and coordinate direction angles. The coordinate direction angles for $\mathbf{M}_1$ are $\alpha_1 = 120°$, $\beta_1 = 45°$, $\gamma_1 = 60°$.

Probs. 4–68/4–69

4–69. The gear reducer is subjected to the two couple moments shown. Determine the coordinate direction angles of $\mathbf{M}_1$ so that the resultant couple moment acts in the $+\mathbf{k}$ direction. What is the magnitude of the resultant couple moment?

4–70. Determine the resultant couple moment of the two couples that act on the pipe assembly. The distance from A to B is $d = 400$ mm. Express the result as a Cartesian vector.

Probs. 4–70/4–71

4–71. Determine the distance d between A and B so that the resultant couple moment has a magnitude of $M_R = 20$ N · m.

*4–72. Determine the couple moment that acts on the assembly. Express the result as a Cartesian vector. Member BA lies in the x-y plane.

Prob. 4–72

4.7 Movement of a Force on a Rigid Body

Many problems in statics, including the reduction of a force system to its simplest possible form, require moving a force from one point to another on a rigid body. Since a force tends to *translate* a body in the direction of the force and *rotate* it about an axis not located on the line of action of the force, it is important that these two "external" effects remain the same if the force is moved from one point to another on the body. Two cases for the location of the point O to which the force is moved will now be considered.

1. *Point O is on the line of action of the force.* Consider the rigid body shown in Fig. 4–31a, which is subjected to the force $\mathbf{F}$ applied to point A. In order to move the force to point O we will first apply equal but opposite forces $\mathbf{F}$ and $-\mathbf{F}$ at O, as shown in Fig. 4–31b. This step in no way alters the external effects on the body; however, the two forces indicated by the slash across them can be canceled, leaving the force at point O as required, Fig. 4–31c. By using this construction procedure, an *equivalent system* has been maintained between each of the diagrams, as indicated by the equal sign. Note that the force has simply been "transmitted" along its line of action, from point A, Fig. 4–31a, to point O, Fig. 4–31c. In other words, the force can be considered as a *sliding vector* since it can act at any point O along its line of action. This concept is referred to as the *principle of transmissibility*, which can be formally stated as follows: *The external effects on a rigid body remain unchanged when a force, acting at a given point on the body, is applied to another point lying on the line of action of the force.* It is important to realize that only the *external effects* caused by the force remain *unchanged* after moving it. Certainly the *internal effects* depend upon where $\mathbf{F}$ is located. For example, when $\mathbf{F}$ acts at A, the internal forces in the body have a high intensity around A; whereas if $\mathbf{F}$ acts at O, the effect of $\mathbf{F}$ on generating internal forces at A will be less.

Fig. 4–31

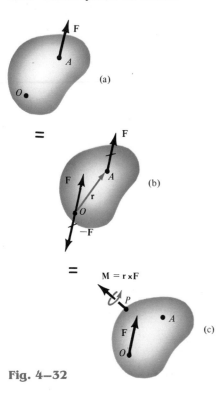

Fig. 4–32

2. *Point O is not on the line of action of the force.* This case is shown in Fig. 4–32a. Following the same procedure as above, we first apply equal but opposite forces **F** and −**F** at *O*, Fig. 4–32b. Here the two forces indicated by a slash across them form a couple which has a moment defined by the cross product **M = r × F.** Since the couple moment is a *free vector,* it may be applied at any point *P* on the body as shown in Fig. 4–32c. In addition to this couple, **F** now acts at point *O* as required. From the construction, note that the couple moment is determined by taking the moment of **F** about *O* when the force was located at its original point *A.* Furthermore, the line of action of **M** is *always* perpendicular to the plane containing **F** and **r.**

The foregoing concepts regarding the movement of a force to any point on a body may be summarized by the following two statements:

1. If the force is to be moved to a *point O located on its line of action,* by the principle of transmissibility, simply move the force to the point.
2. If the force is to be moved to a *point O that is not located on its line of action,* an equivalent system is maintained when the force is moved to point *O* and a couple moment is placed on the body. The magnitude and direction of the couple moment are determined by finding the moment about point *O* of the force when it is at its original location. In all cases the couple moment will be *perpendicular* to the force.

As a physical illustration of these two statements, consider holding the end of a stick of negligible weight. If a vertical force **F** is applied at the other end, and the stick is held in the vertical position, Fig. 4–33a, then, by the principle of transmissibility, the same force is felt at the grip, Fig. 4–33b. When the stick is held in the horizontal position, Fig. 4–33c, the force has the effect of producing *both* a downward force at the grip and a clockwise twist, Fig. 4–33d. The twist can be thought of as being caused by a *couple* that is produced when **F** is moved to the grip. The couple moment has the *same* magnitude and direction as the moment of **F** about the grip, Fig. 4–33c.

Fig. 4–33   (a)   (b)   (c)   (d)

4.8 Resultants of a Force and Couple System

When a rigid body is subjected to a system of forces and couples, it is often simpler to study their effects by using their resultants rather than the force–couple system itself. To show how to simplify a system of forces and couples to their resultants, consider the general force–couple system acting on the rigid body in Fig. 4–34a. This system can be simplified by moving the forces and couples to the arbitrary point O. In this regard, the couple moment $\mathbf{M}$ is simply moved to O since it is a free vector. Forces $\mathbf{F}_1$ and $\mathbf{F}_2$ are sliding vectors and since O does not lie on the line of action of these forces, each must be moved to O in accordance with the procedure stated in Sec. 4.7. For example, when $\mathbf{F}_1$ is applied at O, a corresponding couple moment $\mathbf{M}_1 = \mathbf{r}_1 \times \mathbf{F}_1$ must also be applied to the body, Fig. 4–34b. By vector addition, the force and couple system shown in Fig. 4–34b can now be reduced to an *equivalent* resultant force $\mathbf{F}_R = \mathbf{F}_1 + \mathbf{F}_2$ and resultant couple moment $\mathbf{M}_{R_O} = \mathbf{M} + \mathbf{M}_1 + \mathbf{M}_2$ as shown in Fig. 4–34c. Note that both the magnitude and direction of $\mathbf{F}_R$ are independent of the location of point O; however, $\mathbf{M}_{R_O}$ depends upon this location, since the moments $\mathbf{M}_1$ and $\mathbf{M}_2$ are computed using the position vectors $\mathbf{r}_1$ and $\mathbf{r}_2$. Realize also that $\mathbf{M}_{R_O}$ is a free vector and can act at *any point* on the body, although point O is generally chosen as its point of application.

(a)

(b)

Fig. 4–34

(c)

PROCEDURE FOR ANALYSIS

The above method for simplifying any force and couple system to a single resultant force acting at point O and a resultant couple moment will now be stated in general terms.

Three-Dimensional Systems

A Cartesian vector analysis is generally used to solve problems involving three-dimensional force and couple systems for which the force components and moment arms are difficult to determine.

Force Summation. The resultant force is equivalent to the vector sum of all the forces in the system, i.e.,

$$\mathbf{F}_R = \Sigma\mathbf{F} \qquad (4\text{–}14)$$

Moment Summation

The resultant couple moment is equivalent to the vector sum of all the couples in the system plus the moments about point O of all the forces in the system, i.e.,

$$\mathbf{M}_{R_O} = \Sigma\mathbf{M}_O \qquad (4\text{–}15)$$

Coplanar Force Systems

Since force components and moment arms are easy to determine in two dimensions, a scalar analysis provides the most convenient solution to problems involving coplanar force systems.

Force Summation. The *resultant force* $\mathbf{F}_R$ is equivalent to the vector sum of its two components $\mathbf{F}_{R_x}$ and $\mathbf{F}_{R_y}$. Each component is found from the scalar (algebraic) sum of the components of all the forces in the system that act in the same direction, i.e.,

$$\boxed{\begin{aligned} F_{R_x} &= \Sigma F_x \\ F_{R_y} &= \Sigma F_y \end{aligned}}$$

(4–16)

Moment Summation. The *resultant couple moment* $\mathbf{M}_{R_o}$ is perpendicular to the plane containing the forces and is equivalent to the scalar (algebraic) sum of all the couples in the system *plus* the moments about point O of all the forces in the system, i.e.,

$$\boxed{M_{R_o} = \Sigma M_O}$$

(4–17)

When computing the moments of the forces about O, it is generally advantageous to use the *principle of moments,* i.e., compute the moments of the *components* of each force rather than the moment of the force itself.

The following two examples numerically illustrate these procedures.

Example 4–13

Replace the forces acting on the pipe shown in Fig. 4–35a by an equivalent single resultant force and couple system acting at point A.

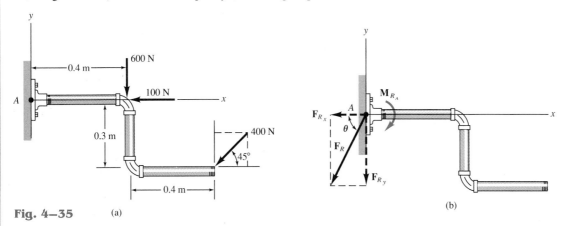

Fig. 4–35   (a)

SOLUTION (*SCALAR ANALYSIS*)

The principle of moments will be applied to the 400-N force, whereby the moments of its two rectangular components will be considered.

Force Summation.   The resultant force has x and y components of

$$\xrightarrow{+} F_{R_x} = \Sigma F_x; \; F_{R_x} = -100 - 400 \cos 45° = -382.8 \text{ N} = 382.8 \text{ N} \leftarrow$$
$$+\uparrow F_{R_y} = \Sigma F_y; \; F_{R_y} = -600 - 400 \sin 45° = -882.8 \text{ N} = 882.8 \text{ N} \downarrow$$

As shown in Fig. 4–35b, $\mathbf{F}_R$ has a magnitude of

$$F_R = \sqrt{(F_{R_x})^2 + (F_{R_y})^2} = \sqrt{(382.8)^2 + (882.8)^2} = 962 \text{ N} \quad \textit{Ans.}$$

and a direction defined by

$$\theta = \tan^{-1}\left(\frac{F_{R_y}}{F_{R_x}}\right) = \tan^{-1}\left(\frac{882.8}{382.8}\right) = 66.6° \; \theta \nearrow \qquad \textit{Ans.}$$

Moment Summation.   The resultant couple moment $\mathbf{M}_{R_A}$ is determined by summing moments about point A. Noting that the moment of each force in the system is perpendicular to the x-y plane, and assuming that positive moments act counterclockwise, i.e., in the $+\mathbf{k}$ direction, we have

$$\zeta + M_{R_A} = \Sigma M_A;$$
$$M_{R_A} = 100(0) - 600(0.4) - (400 \sin 45°)(0.8) - (400 \cos 45°)(0.3)$$
$$= -551 \text{ N} \cdot \text{m} = 551 \text{ N} \cdot \text{m} \downarrow \qquad \textit{Ans.}$$

In conclusion, when $\mathbf{M}_{R_A}$ and $\mathbf{F}_R$ act on the pipe, Fig. 4–35b, they will produce the *same* reaction at the support A (an external effect) as that produced by the force system in Fig. 4–35a.

Example 4–14

A beam is subjected to a couple moment $\mathbf{M}$ and forces $\mathbf{F}_1$ and $\mathbf{F}_2$ as shown in Fig. 4–36a. Replace this system by an equivalent single resultant force and couple moment acting at point O.

SOLUTION (*VECTOR ANALYSIS*)

Expressing the forces and couple moment as Cartesian vectors, we have

$$\mathbf{F}_1 = \{-800\mathbf{k}\} \text{ N}$$

$$\mathbf{F}_2 = (300 \text{ N})\mathbf{u}_{CB} = (300 \text{ N})\left(\frac{\mathbf{r}_{CB}}{r_{CB}}\right)$$

$$= 300\left[\frac{-0.15\mathbf{i} + 0.1\mathbf{j}}{\sqrt{(-0.15)^2 + (0.1)^2}}\right] = \{-249.6\mathbf{i} + 166.4\mathbf{j}\} \text{ N}$$

$$\mathbf{M} = -500(\tfrac{4}{5})\mathbf{j} + 500(\tfrac{3}{5})\mathbf{k} = \{-400\mathbf{j} + 300\mathbf{k}\} \text{ N} \cdot \text{m}$$

(a)

Force Summation

$$\mathbf{F}_R = \Sigma\mathbf{F}; \qquad \mathbf{F}_R = \mathbf{F}_1 + \mathbf{F}_2 = -800\mathbf{k} - 249.6\mathbf{i} + 166.4\mathbf{j}$$

$$= \{-249.6\mathbf{i} + 166.4\mathbf{j} - 800\mathbf{k}\} \text{ N} \qquad \textit{Ans.}$$

Moment Summation

$$\mathbf{M}_{R_O} = \Sigma\mathbf{M}_O;$$

$$\mathbf{M}_{R_O} = \mathbf{M} + \mathbf{r}_C \times \mathbf{F}_1 + \mathbf{r}_B \times \mathbf{F}_2$$

$$= (-400\mathbf{j} + 300\mathbf{k}) + (1\mathbf{k}) \times (-800\mathbf{k}) + \begin{vmatrix} \mathbf{i} & \mathbf{j} & \mathbf{k} \\ -0.15 & 0.1 & 1 \\ -249.6 & 166.4 & 0 \end{vmatrix}$$

$$= (-400\mathbf{j} + 300\mathbf{k}) + (\mathbf{0}) + (-166.4\mathbf{i} - 249.6\mathbf{j}) \qquad \textit{Ans.}$$

$$= \{-166.4\mathbf{i} - 649.6\mathbf{j} + 300\mathbf{k}\} \text{ N} \cdot \text{m}$$

The results are shown in Fig. 4–36b.

(b)

Fig. 4–36

4.9 Further Reduction of a Force and Couple System

Simplification to a Single Resultant Force

Consider now a special case for which the system of forces and couples acting on a rigid body, Fig. 4–37a, reduces at point O to a resultant force $\mathbf{F}_R = \Sigma\mathbf{F}$ and resultant couple $\mathbf{M}_{R_O} = \Sigma\mathbf{M}_O$, which are *perpendicular* to one another, Fig. 4–37b. Using the method of Sec. 4.7, $\mathbf{M}_{R_O}$ can always be *eliminated* by moving $\mathbf{F}_R$ to a point P located on axis bb, which is *perpendicular* to both $\mathbf{F}_R$ and $\mathbf{M}_{R_O}$, Fig. 4–37c. Point P is chosen such that the moment arm d satisfies the scalar equation $M_{R_O} = F_R d$ or $d = M_{R_O}/F_R$. Furthermore, P is located, in this case, to the *left* of O to preserve the correct direction of $\mathbf{M}_{R_O}$ when $\mathbf{F}_R$ is moved *back* to O, Fig. 4–37b. (By the right-hand rule, $\mathbf{F}_R$, acting through P in Fig. 4–37c, tends to rotate the body about the aa axis, which is in the same direction as $\mathbf{M}_{R_O}$.)

We refer to the force and couple system in Fig. 4–37a as being *equivalent or equipollent* to the single force "system" in Fig. 4–37c, because each system produces the *same* resultant force and resultant moment when replaced at point O.

If a system of forces is either concurrent, coplanar, or parallel, it can always be reduced, as in the above case, to a single resultant force $\mathbf{F}_R$ acting through a unique point P. This is because in each of these cases $\mathbf{F}_R$ and $\mathbf{M}_{R_O}$ will always be perpendicular to each other when the force system is simplified at any point O.

Fig. 4–37

Fig. 4–38

Concurrent force systems have been treated in detail in Chapter 2. Obviously, all the forces act at a point for which there is no resultant couple, so the point P is automatically specified, Fig. 4–38.

Coplanar force systems, such as shown in Fig. 4–39a, can be reduced to a single resultant force, because when each force in the system is moved to any point O in the x-y plane, it produces a couple moment that is *perpendicular* to the plane, i.e., in the $\pm\mathbf{k}$ direction. The resultant moment $\mathbf{M}_{R_O} = \Sigma\mathbf{r} \times \mathbf{F}$ is thus perpendicular to the resultant force $\mathbf{F}_R$, Fig. 4–39b.

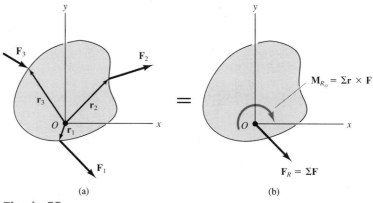

(a) (b)

Fig. 4–39

Parallel force systems, such as shown in Fig. 4–40a, can be reduced to a single resultant force, because when each (vertical) force is moved to any point O in the x-y plane, it produces a couple that has moment components only about the x and y axes. The resultant moment $\mathbf{M}_{R_O} = \Sigma\mathbf{r} \times \mathbf{F}$ is thus perpendicular to the resultant force $\mathbf{F}_R$, Fig. 4–40b.

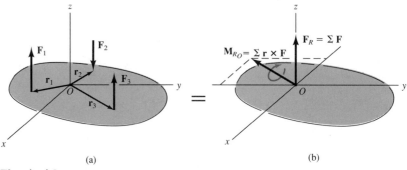

(a) (b)

Fig. 4–40

PROCEDURE FOR ANALYSIS

The technique used to reduce a coplanar or parallel force system to a single resultant force follows the general procedure outlined in the previous section. The simplification requires the following two steps:

Force Summation. The resultant force $\mathbf{F}_R$ equals the sum of all the forces of the system, i.e.,

$$\mathbf{F}_R = \Sigma \mathbf{F}$$

Moment Summation. The line of action of $\mathbf{F}_R$ is determined by equating the moment of $\mathbf{F}_R$ about an arbitrary point O, $\mathbf{M}_{R_O}$, to the sum of the moments about point O of all the forces in the system, $\Sigma \mathbf{M}_O$, i.e.,

$$\mathbf{M}_{R_O} = \Sigma \mathbf{M}_O$$

Most often a scalar analysis can be used to apply these equations, since the force components and the moment arms are easily determined for either coplanar or parallel force systems.

Reduction to a Wrench

In the general case, the force and couple system acting on a body, Fig. 4–34a, will reduce to a single resultant force $\mathbf{F}_R$ and couple $\mathbf{M}_{R_O}$ at O which are *not* perpendicular. Instead, $\mathbf{F}_R$ will act at an angle θ from $\mathbf{M}_{R_O}$. Fig. 4–34c. As shown in Fig. 4–41a, however, $\mathbf{M}_{R_O}$ may be resolved into two components: one perpendicular, $\mathbf{M}_\perp$, and the other parallel, $\mathbf{M}_\parallel$, to the line of action of $\mathbf{F}_R$. The perpendicular component $\mathbf{M}_\perp$ may be *eliminated* by moving $\mathbf{F}_R$ to point P, as shown in Fig. 4–41b. This point lies on axis bb, which is perpendicular to both $\mathbf{M}_{R_O}$ and $\mathbf{F}_R$. In order to maintain an equivalency of loading, the distance from O to P is $d = M_\perp / F_R$. Furthermore, when $\mathbf{F}_R$ is applied at P, the moment of $\mathbf{F}_R$ tending to cause rotation of the body *about* O is in the *same direction* as $\mathbf{M}_\perp$, Fig. 4–41b. Finally, since $\mathbf{M}_\parallel$ is a free vector, it may be moved to P so that it coincides with $\mathbf{F}_R$, Fig. 4–41c. This combination of a *collinear* force and couple is called a *wrench* or *screw*. The *axis of the wrench* has the same line of action as the force. Hence, the wrench tends to cause both a translation along and a rotation about this axis. Comparing Fig. 4–41a to Fig. 4–41c, it is seen that a general force and couple system acting on a body can be reduced to a wrench. The axis of the wrench and a point through which this axis passes are unique and can always be determined.

(a)

||

(b)

||

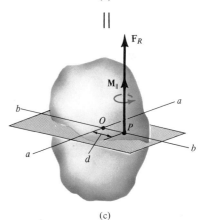

(c)

Fig. 4–41

Example 4–15

Replace the system of forces acting on the beam shown in Fig. 4–42a by an equivalent single resultant force. Specify the distance the force acts from point A.

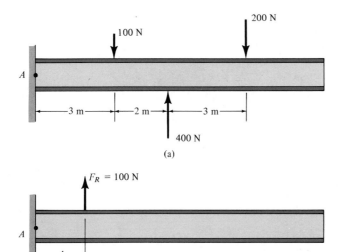

(a)

$F_R = 100$ N

(b)

Fig. 4–42

SOLUTION

Force Summation. If "positive" forces are assumed to act upward, then from Fig. 4–42a the force resultant $\mathbf{F}_R$ is

$$+\uparrow F_R = \Sigma F; \quad F_R = -100 \text{ N} + 400 \text{ N} - 200 \text{ N} = 100 \text{ N} \uparrow \qquad Ans.$$

Moment Summation. Moments will be summed about point A. Considering counterclockwise rotations as positive, i.e., positive moment vectors are directed out of the page, then from Fig. 4–42a and b we require the moment of $\mathbf{F}_R$ about A to equal the moments of the force system about A, i.e.,

$$\zeta +M_{R_A} = \Sigma M_A;$$
$$100 \text{ N}(d) = -(100 \text{ N})(3 \text{ m}) + (400 \text{ N})(5 \text{ m}) - (200 \text{ N})(8 \text{ m})$$
$$(100)d = 100$$
$$d = 1 \text{ m} \qquad Ans.$$

Using a clockwise sign convention would yield the same result. Since d is *positive*, $\mathbf{F}_R$ acts to the right of A as shown in Fig. 4–42b.

Example 4–16

The beam AE in Fig. 4–43a is subjected to a system of coplanar forces. Determine the magnitude, direction, and location on the beam of a single resultant force which is equivalent to the given system of forces.

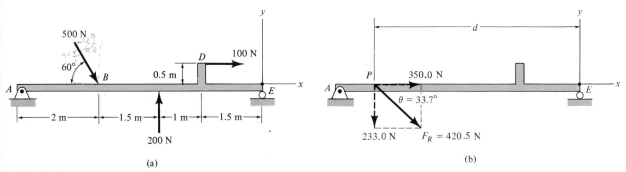

Fig. 4–43

SOLUTION

Force Summation. The origin of coordinates is arbitrarily located at point E as shown in Fig. 4–43a. Resolving the 500-N force into x and y components, and summing the force components, yields

$$\xrightarrow{+} F_{R_x} = \Sigma F_x; \quad F_{R_x} = 500 \cos 60° + 100 = 350.0 \text{ N} \rightarrow$$
$$+ \uparrow F_{R_y} = \Sigma F_y; \quad F_{R_y} = -500 \sin 60° + 200 = -233.0 \text{ N}$$
$$= 233.0 \text{ N} \downarrow$$

The magnitude and direction of the resultant force shown in Fig. 4–43b are, therefore,

$$F_R = \sqrt{(350.0)^2 + (233.0)^2} = 420.5 \text{ N} \qquad Ans.$$

$$\theta = \tan^{-1}\left(\frac{233.0}{350.0}\right) = 33.7° \qquad Ans.$$

Moment Summation. Moments will be summed about point E. Since P lies on the x axis, $\mathbf{F}_{R_x}$ (350.0 N) does not create a moment about E (principle of transmissibility), only $\mathbf{F}_{R_y}$ (233.0 N) does. Hence, from Fig. 4–43a and b, we require that the moment of $\mathbf{F}_R$ about point E equal the moments of the force system about E, i.e.,

$$\zeta + M_{R_E} = \Sigma M_E;$$
$$233.0(d) = (500 \sin 60°)(4) + (500 \cos 60°)(0) - (100)(0.5) - (200)(2.5)$$

$$d = \frac{1182.1}{233.0} = 5.07 \text{ m} \qquad Ans.$$

Example 4–17

The frame shown in Fig. 4–44a is subjected to three coplanar forces. Replace this loading by an equivalent single resultant force and specify where the resultant's line of action intersects members AB and BC.

SOLUTION

Force Summation. Resolving the 75-lb force into x and y components and summing the force components yields

$$\xleftarrow{+}F_{R_x} = \Sigma F_x; \qquad F_{R_x} = 75(\tfrac{3}{5}) + 15 = 60 \text{ lb} \leftarrow$$

$$+\downarrow F_{R_y} = \Sigma F_y; \qquad F_{R_y} = 75(\tfrac{4}{5}) + 90 = 150 \text{ lb} \downarrow$$

As shown in Fig. 4–44b,

$$F_R = \sqrt{(60)^2 + (150)^2} = 161.6 \text{ lb} \qquad \textit{Ans.}$$

$$\theta = \tan^{-1}\left(\frac{150}{60}\right) = 68.2° \quad \theta \nearrow \qquad \textit{Ans.}$$

(a)

Moment Summation. Moments will be summed about point A. If the line of action of $\mathbf{F}_R$ intersects AB, Fig. 4–44b, we require the moment of the components of $\mathbf{F}_R$ in Fig. 4–44b about A to equal the moments of the force system in Fig. 4–44a about A, i.e.,

$$\zeta + M_{R_A} = \Sigma M_A;$$
$$0(150) + y(60) = 4(15) + 0(90) + 7(75)(\tfrac{3}{5}) - 2(75)(\tfrac{4}{5})$$
$$y = 4.25 \text{ ft} \qquad \textit{Ans.}$$

By the principle of transmissibility, $\mathbf{F}_R$ can also intersect BC, Fig. 4–44b, in which case we have

$$\zeta + M_{R_A} = \Sigma M_A;$$
$$7(60) - x(150) = 4(15) + 0(90) + 7(75)(\tfrac{3}{5}) - 2(75)(\tfrac{4}{5})$$
$$x = 1.10 \text{ ft} \qquad \textit{Ans.}$$

We can also solve for these positions by assuming $\mathbf{F}_R$ acts at the arbitrary point (x, y) on its line of action, Fig. 4–44b. Then moments about A yield

$$\zeta + M_{R_A} = \Sigma M_A;$$
$$y(60) - x(150) = 4(15) + 0(90) + 7(75)(\tfrac{3}{5}) - 2(75)(\tfrac{4}{5})$$
$$60y - 150x = 255$$

which is the equation of the dashed line in Fig. 4–44b. To find the points of intersection with the frame, set $x = 0$, then $y = 4.25$ ft, and set $y = 7$ ft, then $x = 1.10$ ft.

(b)

Fig. 4–44

Example 4–18

The slab in Fig. 4–45a is subjected to four parallel forces. Determine the magnitude and direction of a single resultant force equivalent to the given force system, and locate its point of application on the slab.

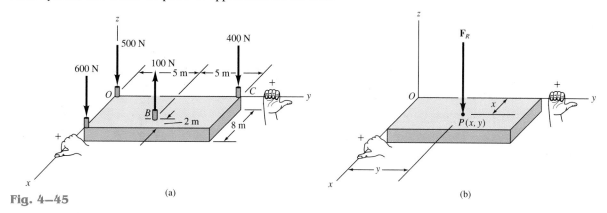

Fig. 4–45

(a)   (b)

SOLUTION (*SCALAR ANALYSIS*)
Force Summation. From Fig. 4–45a, the resultant force is

$$+\downarrow F_R = \Sigma F; \qquad F_R = 600 - 100 + 400 + 500$$
$$= 1400 \text{ N} \downarrow \qquad\qquad \textit{Ans.}$$

Moment Summation. We require the moment about the x axis of the resultant force, Fig. 4–45b, to be equal to the sum of the moments about the x axis of all the forces in the system, Fig. 4–45a. The moment arms are determined from the y coordinates since these coordinates represent the *perpendicular distances* from the x axis to the lines of action of the forces. Using the right-hand rule, where positive moments act in the $+\mathbf{i}$ direction, we have

$$M_{R_x} = \Sigma M_x;$$
$$-1400y = 600(0) + 100(5) - 400(10) + 500(0)$$
$$-1400y = -3500 \qquad y = 2.50 \text{ m} \qquad \textit{Ans.}$$

In a similar manner, assuming that positive moments act in the $+\mathbf{j}$ direction, a moment equation can be written about the y axis using moment arms defined by the x coordinates of each force.

$$M_{R_y} = \Sigma M_y;$$
$$1400x = 600(8) - 100(6) + 400(0) + 500(0)$$
$$1400x = 4200 \qquad x = 3.00 \text{ m} \qquad \textit{Ans.}$$

Hence, a force of $F_R = 1400$ N placed at point $P(3.00 \text{ m}, 2.50 \text{ m})$ on the slab, Fig. 4–45b, is equivalent to the parallel force system acting on the slab in Fig. 4–45a.

Example 4–19

Three parallel forces act on the rim of the circular plate in Fig. 4–46a. Determine the magnitude and direction of a single resultant force equivalent to the given force system and locate its point of application, P, on the plate.

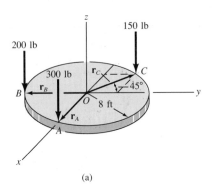

(a)

SOLUTION (*VECTOR ANALYSIS*)

Force Summation. From Fig. 4–46a, the force resultant $\mathbf{F}_R$ is

$$\mathbf{F}_R = \Sigma \mathbf{F}; \qquad \mathbf{F}_R = -300\mathbf{k} - 200\mathbf{k} - 150\mathbf{k}$$
$$= \{-650\mathbf{k}\} \text{ lb} \qquad\qquad Ans.$$

Moment Summation. Choosing point O as a reference for computing moments and assuming that $\mathbf{F}_R$ acts at a point $P(x, y)$, Fig. 4–46b, we require

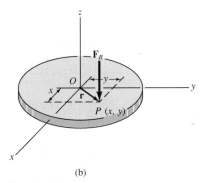

(b)

Fig. 4–46

$$\mathbf{M}_{R_O} = \Sigma \mathbf{M}_O; \ \mathbf{r} \times \mathbf{F}_R = \mathbf{r}_A \times (-300\mathbf{k}) + \mathbf{r}_B \times (-200\mathbf{k}) + \mathbf{r}_C \times (-150\mathbf{k})$$

$$(x\mathbf{i} + y\mathbf{j}) \times (-650\mathbf{k}) = \ (8\mathbf{i}) \times (-300\mathbf{k}) + (-8\mathbf{j}) \times (-200\mathbf{k})$$
$$+ (-8 \sin 45°\mathbf{i} + 8 \cos 45°\mathbf{j}) \times (-150\mathbf{k})$$

$$650x\mathbf{j} - 650y\mathbf{i} = 2400\mathbf{j} + 1600\mathbf{i} - 848.5\mathbf{j} - 848.5\mathbf{i}$$

Equating the corresponding $\mathbf{j}$ and $\mathbf{i}$ components yields

$$650x = 2400 - 848.5 \qquad\qquad (1)$$
$$-650y = 1600 - 848.5 \qquad\qquad (2)$$

Solving these equations, we obtain the coordinates of point P,

$$x = 2.39 \text{ ft} \qquad y = -1.16 \text{ ft} \qquad\qquad Ans.$$

The negative sign indicates that it was wrong to have assumed a $+y$ position for $\mathbf{F}_R$ as shown in Fig. 4–46b.

As a review, try to establish Eqs. (1) and (2) by using a scalar analysis, i.e., apply the sum of moments about the x and y axes, respectively.

PROBLEMS

4–73. Replace the force at A by an equivalent force and couple moment at point O.

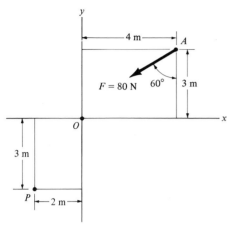

Probs. 4–73/4–74

4–74. Replace the force at A by an equivalent force and couple moment at point P.

4–75. Replace the force and couple moment system by an equivalent force and couple moment acting at point O.

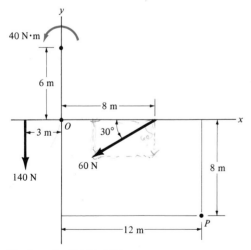

Probs. 4–75/4–76

*4–76. Replace the force and couple moment system by an equivalent force and couple moment acting at point P.

4–77. Replace the force and couple system by an equivalent force and couple moment acting at point O.

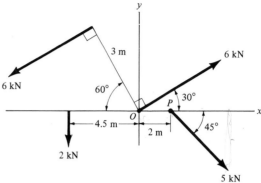

Probs. 4–77/4–78

4–78. Replace the force and couple system by an equivalent force and couple moment acting at point P.

4–79. Replace the force system by a resultant force and specify where its line of action intersects the y axis.

Prob. 4–79

143

*4–80. Replace the coplanar force system by a resultant force and specify where its line of action intersects the x axis.

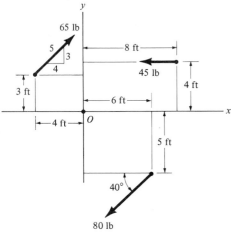

Probs. 4–80/4–81

4–81. Replace the coplanar force system by a resultant force and specify where its line of action intersects the y axis.

4–82. The force of 30 N is exerted on the handle of the wrench. Replace this force by an equivalent force and couple moment acting at point O.

Prob. 4–82

4–83. Determine the magnitude and direction θ of force $\mathbf{F}$ and its placement d on the beam so that the loading system is equivalent to a resultant force of 12 kN acting vertically downward at point A and a clockwise couple moment of 50 kN · m.

Probs. 4–83/4–84

*4–84. Determine the magnitude and direction θ of force $\mathbf{F}$ and its placement d on the beam so that the loading system is equivalent to a resultant force of 10 kN acting vertically downward at point A, and a clockwise couple moment of 45 kN · m.

4–85. Replace the force system acting on the beam by an equivalent force and couple moment at point A.

Probs. 4–85/4–86

4–86. Replace the force system acting on the beam by an equivalent force and couple moment at point B.

4–87. Replace the force system acting on the beam by an equivalent force and couple moment at point *A*.

Probs. 4–87/4–88

*4–88. Replace the force system acting on the beam by an equivalent force and couple moment at point *B*.

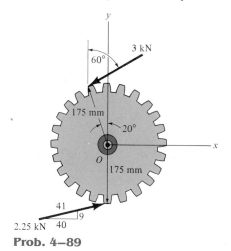

Prob. 4–89

4–89. The gear is subjected to the two forces shown. Replace these forces by an equivalent force and couple moment acting at point *O*.

4–90. The loadings of the various components of the truck are shown. Replace this system of forces by an equivalent force and couple moment acting at point *A*.

Probs. 4–90/4–91

4–91. The loadings of the various components of the truck are shown. Replace this system of forces by a resultant force and specify its location measured from point *A*.

*4–92. Determine the magnitude of the vertical force **F** acting tangent to the gear so that **F** and **M** can be replaced by a single resultant force at point *A*.

Prob. 4–92

4–93. Replace the force system acting on the frame by a single resultant force and specify where the resultant's line of action intersects member AB measured from point A.

Probs. 4–93/4–94

4–94. Replace the force system acting on the frame by an equivalent force and couple moment acting at point C.

4–95. Replace the force at A by an equivalent force and couple moment at point O. Express the results in Cartesian vector form.

Probs. 4–95/4–96

*4–96. Replace the force at A by an equivalent force and couple moment at point P. Express the results in Cartesian vector form.

4–97. Replace the force and couple moment system by an equivalent force and couple moment at point O. Express the results in Cartesian vector form.

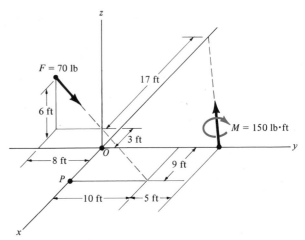

Probs. 4–97/4–98

4–98. Replace the force and couple moment system by an equivalent force and couple moment at point P. Express the results in Cartesian vector form.

4–99. Simplify the force system consisting of $\mathbf{F}_1 = \{30\mathbf{i} + 20\mathbf{j} + 15\mathbf{k}\}$ lb and $\mathbf{F}_2 = \{40\mathbf{i} - 50\mathbf{j} + 12\mathbf{k}\}$ lb to an equivalent force and couple moment acting at point P. Express the results in Cartesian vector form. Solve the problem two ways, (a) by direct application of Eqs. 4–14 and 4–15, and (b) by replacing the forces by an equivalent force and couple moment at point O, and then determining the equivalent force and couple moment at point P.

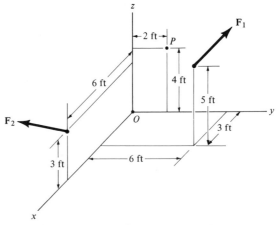

Prob. 4–99

146

*4–100. Replace the force and couple moment system by an equivalent force and couple moment at point O. Express the results in Cartesian vector form.

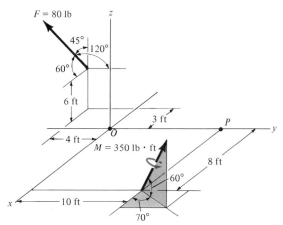

Probs. 4–100/4–101

4–101. Replace the force and couple moment system by an equivalent force and couple moment at point P. Express the results in Cartesian vector form.

4–102. A force and couple act on the pipe assembly. Replace this system by an equivalent single resultant force. Specify the point where the line of action of the resultant force intersects the x axis. The pipe lies in the x-y plane.

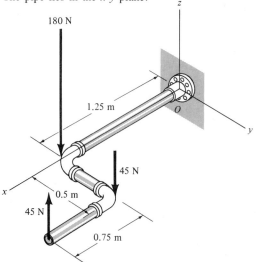

Prob. 4–102

4–103. The force $\mathbf{F} = \{-4\mathbf{i} + 2\mathbf{j} - 3\mathbf{k}\}$ kN acts on the end of the beam. Replace this force by an equivalent force and couple moment acting at point O.

Prob. 4–103

*4–104. The bracket is subjected to the force $\mathbf{F}$. Replace this force by an equivalent force and couple moment acting at point A.

Prob. 4–104

147

4–105. Replace the force **F** having a magnitude of $F = 50$ lb and acting at point A by an equivalent force and couple moment at point C.

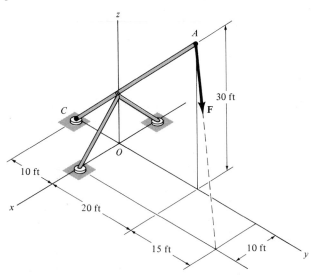

Prob. 4–105

4–106. The forces and couple moments which are exerted on the toe and heel plates of a snow ski are $\mathbf{F}_t = \{-50\mathbf{i} + 80\mathbf{j} - 158\mathbf{k}\}$ N, $\mathbf{M}_t = \{-6\mathbf{i} + 4\mathbf{j} + 2\mathbf{k}\}$ N · m, and $\mathbf{F}_h = \{-20\mathbf{i} + 60\mathbf{j} - 250\mathbf{k}\}$ N, $\mathbf{M}_h = \{-20\mathbf{i} + 8\mathbf{j} + 3\mathbf{k}\}$ N · m, respectively. Replace this system by an equivalent force and couple moment acting at point O. Express the results in Cartesian vector form.

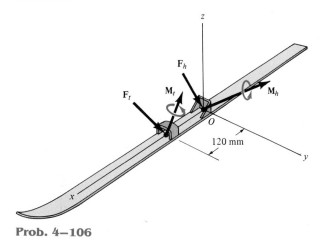

Prob. 4–106

4–107. The belt passing over the pulley is subjected to forces $\mathbf{F}_1$ and $\mathbf{F}_2$, each having a magnitude of 40 N. $\mathbf{F}_1$ acts in the $-\mathbf{k}$ direction. Replace these forces by an equivalent force and couple moment at point A. Express the result in Cartesian vector form. Set $\theta = 0°$ so that $\mathbf{F}_2$ acts in the $-\mathbf{j}$ direction.

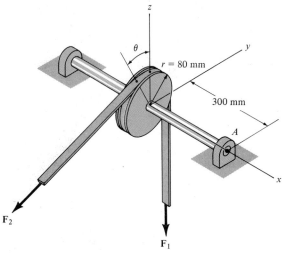

Probs. 4–107/4–108

*4–108. The belt passing over the pulley is subjected to two forces $\mathbf{F}_1$ and $\mathbf{F}_2$, each having a magnitude of 40 N. $\mathbf{F}_1$ acts in the $-\mathbf{k}$ direction. Replace these forces by an equivalent force and couple moment at point A. Express the result in Cartesian vector form. Take $\theta = 45°$.

4–109. Three parallel bolting forces act on the circular plate. Determine the resultant force, and specify its location (x, z) on the plate. $F_A = 200$ lb, $F_B = 100$ lb, and $F_C = 400$ lb.

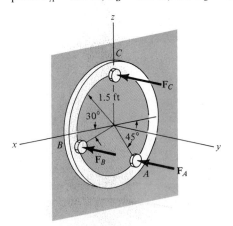

Probs. 4–109/4–110

4–110. The three parallel bolting forces act on the circular plate. If the force at A has a magnitude of $F_A = 200$ lb, determine the magnitudes of $\mathbf{F}_B$ and $\mathbf{F}_C$ so that the resultant force $\mathbf{F}_R$ of the system has a line of action that coincides with the y axis. *Hint:* This requires $\Sigma M_x = 0$ and $\Sigma M_z = 0$. Why?

4–111. A biomechanical model of the lumbar region of the human trunk is shown. The forces acting in the four muscle groups consist of $F_R = 35$ N for the rectus, $F_O = 45$ N for the oblique, $F_L = 23$ N for the lumbar latissimus dorsi, and $F_E = 32$ N for the erector spinae. These loadings are symmetric with respect to the y-z plane. Replace this system of parallel forces by an equivalent force and couple moment acting at the spine, point O. Express the result in Cartesian vector form.

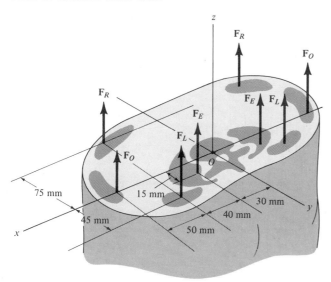

Probs. 4–111/4–112

*4–112. A biomechanical model of the lumbar region of the human trunk is shown. The forces acting in the four muscle groups consist of $F_R = 35$ N for the rectus, $F_O = 45$ N for the oblique, $F_L = 23$ N for the lumbar latissimus dorsi, and $F_E = 32$ N for the erector spinae. These loadings are symmetric with respect to the y-z plane. Determine the resultant force equivalent to the given parallel force system, and locate its point of application (x, y) on the trunk.

4–113. The pipe assembly is subjected to the action of a wrench at B and a couple at A. Simplify this system to a single resultant wrench and specify the location of the wrench along the axis of pipe CD, measured from point C.

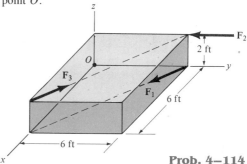

Prob. 4–113

4–114. The three forces acting on the block each have a magnitude of 10 lb. Replace this system by a wrench and specify the point where the wrench intersects the z axis measured from point O.

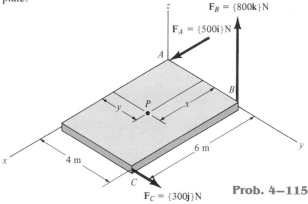

Prob. 4–114

4–115. Replace the three forces acting on the plate by a wrench. Specify the magnitude of the force and couple moment for the wrench and the point $P(x, y)$ where its line of action intersects the plate.

Prob. 4–115

149

*4–116. Replace the three forces acting on the plate by a wrench. Specify the magnitude of the force and couple moment for the wrench and the point $P(y, z)$ where its line of action intersects the plate.

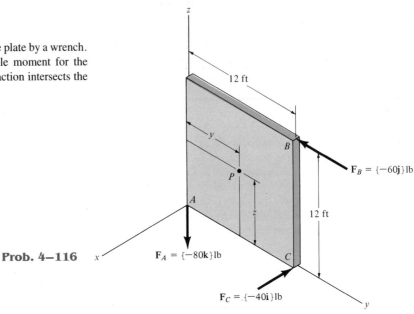

Prob. 4–116

$\mathbf{F}_B = \{-60\mathbf{j}\}\,\text{lb}$

$\mathbf{F}_A = \{-80\mathbf{k}\}\,\text{lb}$

$\mathbf{F}_C = \{-40\mathbf{i}\}\,\text{lb}$

12 ft

12 ft

4.10 Reduction of a Simple Distributed Loading

In many situations a very large surface area of a body may be subjected to *distributed loadings* such as those caused by wind, fluids, or simply the weight of material supported over the body's surface. The *intensity* of these loadings at each point is a force per unit area, defined as the *pressure p*, which can be measured in units of lb/ft^2 or pascals (Pa), where 1 Pa $= 1$ N/m^2.

In this section we will consider the most common case of a distributed pressure loading, which is *uniform* along an axis of a flat rectangular body upon which the loading is applied.* An example of such a loading is shown in Fig. 4–47*a*. The direction of the intensity of the pressure load is indicated by arrows shown on the *load-intensity diagram*. The entire loading on the plate is therefore a system of parallel forces, infinite in number and each acting on a separate differential area of the plate. Here the *loading function, $p = p(x)$* Pa, is only a function of x since the pressure is uniform along the y axis. If we multiply $p = p(x)$ by the *width a* m of the plate, we obtain $w = [p(x)\,\text{N/m}^2]a$ m $= w(x)$ N/m. This loading function, shown in Fig. 4–4*b*, is a measure of load distribution along any *line* that is parallel to the x axis. As noted, it is measured as a force per unit length, rather than a force per unit area. Consequently, the load-intensity diagram for $w = w(x)$ can be represented by a system of *coplanar* parallel forces, Fig. 4–47*b*. Using the methods of Sec. 4.9, this system of forces can be simplified to a single resultant force $\mathbf{F}_R$ and its location $\bar{x}$ specified, Fig. 4–47*c*.

---

*The more general case of a nonuniform surface loading acting on a body is considered in Sec. 9.5.

Magnitude of Resultant Force

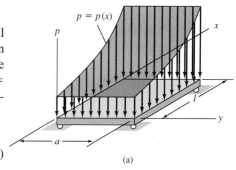

(a)

From Eq. 4–16 ($F_R = \Sigma F$) the magnitude of $\mathbf{F}_R$ is equivalent to the sum of all the forces in the system. In this case integration must be used since there is an infinite number of parallel forces $d\mathbf{F}$ acting along the plate, Fig. 4–47b. Since $d\mathbf{F}$ acts on an element of length dx, we have at any point x, $dF = w(x)dx = dA$. In other words, the magnitude of $d\mathbf{F}$ is represented by the colored differential *area* dA under the loading curve. For the entire plate length,

$$+\downarrow F_R = \Sigma F; \qquad \boxed{F_R = \int_l w(x)\,dx = \int_A dA = A} \qquad (4\text{–}18)$$

Hence, the magnitude of the resultant force is equal to the total area under the loading diagram $w = w(x)$.

Location of Resultant Force

(b)

Applying Eq. 4–17 ($M_{R_O} = \Sigma M_O$) the location $\bar{x}$ of the line of action of $\mathbf{F}_R$ can be determined by equating the moments of the force resultant and the force distribution about point O. Since $d\mathbf{F}$ produces a moment of $x\,dF = x\,dA$ about O, Fig. 4–47b, then for the entire plate, Fig. 4–47c,

$$\curvearrowleft + M_{R_O} = \Sigma M_O; \qquad \bar{x}F_R = \int_A x\,dA$$

Solving for $\bar{x}$, using Eq. 4–18, we can write

$$\boxed{\bar{x} = \dfrac{\displaystyle\int_A x\,dA}{\displaystyle\int_A dA}} \qquad (4\text{–}19)$$

(c)

This equation represents the x coordinate for the geometric center or centroid of the *area* under the distributed-loading diagram $w(x)$. *Therefore, the resultant force has a line of action which passes through the centroid C (geometric center) of the area defined by the distributed-loading diagram $w(x)$,* Fig. 4–47c.

Once $\bar{x}$ is determined, $\mathbf{F}_R$ by symmetry passes through point ($\bar{x}$, $a/2$) on the surface of the plate, Fig. 4–47d. If we consider the three-dimensional pressure loading $p(x)$, Fig. 4–47a, we can therefore conclude that *the resultant force has a magnitude equal to the volume under the distributed loading curve $p = p(x)$ and a line of action which passes through the centroid (geometric center) of this volume.* Detailed treatment of the integration techniques for computing the centroids of volumes or areas is given in Chapter 9. In many cases, however, the distributed-loading diagram is in the shape of a rectangle, triangle, or other simple geometric form. The centroids for such common shapes do not have to be determined from Eq. 4–19; rather, they can be obtained directly from the tabulation given on the inside back cover of this book.

The following examples numerically illustrate these concepts.

(d)

Fig. 4–47

Example 4–20

In each case, determine the magnitude and location of the resultant of the distributed load acting on the beams in Fig. 4–48.

Fig. 4–48

(a)   (b)   (c)   (d)

SOLUTION

Uniform Loading. As indicated $w = 400$ lb/ft, which is constant over the entire beam, Fig. 4–48a. This loading forms a rectangle, the area of which is equal to the resultant force, Fig. 4–48b, i.e.,

$$F_R = (400 \text{ lb/ft})(10 \text{ ft}) = 4000 \text{ lb} \qquad \textit{Ans.}$$

The location of $\mathbf{F}_R$ passes through the geometric center or centroid C of the area, so that

$$\bar{x} = 5 \text{ ft} \qquad \textit{Ans.}$$

Triangular Loading. Here the loading varies uniformly in intensity from 0 to 600 N/m, Fig. 4–48c. These values can be verified by substitution of $x = 0$ and $x = 6$ m into the loading function $w = 100x$ N/m. The area of this triangular loading is equal to $\mathbf{F}_R$, Fig. 4–48d. From the table on the inside back cover, $A = \frac{1}{2}bh$, so that

$$F_R = \frac{1}{2}(6 \text{ m})(600 \text{ N/m}) = 1800 \text{ N} \qquad \textit{Ans.}$$

The line of action of $\mathbf{F}_R$ passes through the centroid C of the triangle. Using the table on the inside back cover, this point lies at a distance of one third the length of the beam, measured from the right side. Hence,

$$\bar{x} = 6 \text{ m} - \frac{1}{3}(6 \text{ m}) = 4 \text{ m} \qquad \textit{Ans.}$$

Example 4–21

Determine the magnitude and location of the resultant of the distributed load acting on the beam shown in Fig. 4–49a.

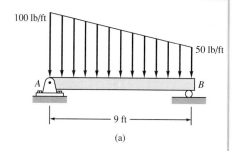

(a)

SOLUTION

The area of the loading diagram is a *trapezoid*, and therefore the solution can be obtained directly from the area and centroid formulas for a trapezoid listed on the inside back cover. Since these formulas are not easily remembered, instead we will solve this problem by using "composite" areas. In this regard, we can divide the trapezoidal loading into a rectangular and triangular loading as shown in Fig. 4–49b. The magnitude of the force represented by each of these loadings is equal to its associated *area*,

$$F_1 = \tfrac{1}{2}(9 \text{ ft})(50 \text{ lb/ft}) = 225 \text{ lb}$$
$$F_2 = (9 \text{ ft})(50 \text{ lb/ft}) = 450 \text{ lb}$$

(b)

The lines of action of these parallel forces act through the *centroid* of their associated areas and therefore intersect the beam at

$$\bar{x}_1 = \tfrac{1}{3}(9 \text{ ft}) = 3 \text{ ft}$$
$$\bar{x}_2 = \tfrac{1}{2}(9 \text{ ft}) = 4.5 \text{ ft}$$

The two parallel forces $\mathbf{F}_1$ and $\mathbf{F}_2$ can be reduced to a single resultant $\mathbf{F}_R$. The magnitude of $\mathbf{F}_R$ is

$$+\downarrow F_R = \Sigma F; \qquad F_R = 225 + 450 = 675 \text{ lb} \qquad \textit{Ans.}$$

With reference to point A, Fig. 4–49b and c, we can define the location of $\mathbf{F}_R$. We require

$$\zeta + M_{R_A} = \Sigma M_A; \qquad \bar{x}(675) = 3(225) + 4.5(450)$$
$$\bar{x} = 4 \text{ ft} \qquad \textit{Ans.}$$

(c)

Note: The trapezoidal area in Fig. 4–49a can also be divided into two triangular areas as shown in Fig. 4–49d. In this case

$$F_1 = \tfrac{1}{2}(9 \text{ ft})(100 \text{ lb/ft}) = 450 \text{ lb}$$
$$F_2 = \tfrac{1}{2}(9 \text{ ft})(50 \text{ lb/ft}) = 225 \text{ lb}$$

and

$$\bar{x}_1 = \tfrac{1}{3}(9 \text{ ft}) = 3 \text{ ft}$$
$$\bar{x}_2 = \tfrac{1}{3}(9 \text{ ft}) = 3 \text{ ft}$$

Using these results, show that $F_R = 675$ lb and $\bar{x} = 4$ ft as obtained above.

(d)

Fig. 4–49

Example 4–22

Determine the magnitude and location of the resultant force acting on the beam in Fig. 4–50a.

Fig. 4–50

SOLUTION

As shown, the colored differential area element $dA = w\,dx = 60x^2\,dx$ will be used for integration. Applying Eq. 4–18, by summing these elements from $x = 0$ to $x = 2$ m, we obtain the resultant force $\mathbf{F}_R$,

$$F_R = \Sigma F;$$

$$F_R = \int_A dA = \int_0^2 60x^2\,dx = 60\left[\frac{x^3}{3}\right]_0^2 = 60\left[\frac{2^3}{3} - \frac{0^3}{3}\right]$$

$$= 160 \text{ N} \qquad\qquad\qquad Ans.$$

Since the element of area dA is located an arbitrary distance x from O, the location $\bar{x}$ of $\mathbf{F}_R$ *measured from* O, Fig. 4–50b, is determined from Eq. 4–19.

$$\bar{x} = \frac{\displaystyle\int_A x\,dA}{\displaystyle\int_A dA} = \frac{\displaystyle\int_0^2 x(60\,x^2)\,dx}{160} = \frac{60\left[\dfrac{x^4}{4}\right]_0^2}{160} = \frac{60\left[\dfrac{2^4}{4} - \dfrac{0^4}{4}\right]}{160}$$

$$= 1.5 \text{ m} \qquad\qquad\qquad Ans.$$

These results may be checked by using the table on the inside back cover, where it is shown that for an exparabolic area of length a, height b, and shape shown in Fig. 4–50a,

$$A = \frac{ab}{3} = \frac{240(2)}{3} = 160 \text{ N} \quad \text{and} \quad \bar{x} = \frac{3}{4}a = \frac{3}{4}(2) = 1.5 \text{ m}$$

Example 4–23

A distributed loading of $p = 800x$ Pa acts over the top surface of the beam shown in Fig. 4–51a. Determine the magnitude and location of the resultant force.

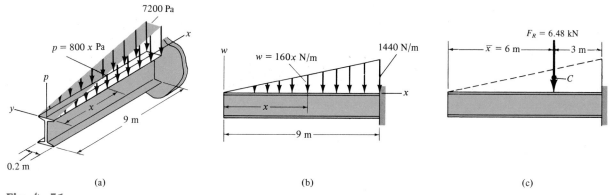

Fig. 4–51

(a)   (b)   (c)

SOLUTION

The loading function $p = 800x$ Pa indicates that the load intensity varies uniformly from $p = 0$ at $x = 0$ to $p = 7200$ Pa at $x = 9$ m. Since the intensity is uniform along the width of the beam, the loading may be viewed in two dimensions as shown in Fig. 4–51b. Here

$$w = (800 \, x \, \text{N/m}^2)(0.2 \, \text{m})$$
$$= (160x) \, \text{N/m}$$

At $x = 9$ m, note that $w = 1440$ N/m. Thus the magnitude of the resultant force is

$$F_R = \tfrac{1}{2}(9 \, \text{m})(1440 \, \text{N/m}) = 6480 \, \text{N} = 6.48 \, \text{kN} \qquad \textit{Ans.}$$

The *line of action* of $\mathbf{F}_R$ passes through the *centroid C* of the triangle. Hence,

$$\bar{x} = 9 \, \text{m} - \tfrac{1}{3}(9 \, \text{m}) = 6 \, \text{m} \qquad \textit{Ans.}$$

The results are shown in Fig. 4–51c.

We may also view the resultant $\mathbf{F}_R$ as *acting* through the *centroid* of the *volume* of the loading diagram $p = p(x)$ in Fig. 4–51a. Hence $\mathbf{F}_R$ intersects the *x-y* plane at the point (6 m, 0) on the *x* axis. Furthermore, the *magnitude* of $\mathbf{F}_R$ is equal to the *volume* under the loading diagram, i.e.,

$$F_R = V = \tfrac{1}{2}(7200 \, \text{N/m}^2)(9 \, \text{m})(0.2 \, \text{m}) = 6.48 \, \text{kN} \qquad \textit{Ans.}$$

PROBLEMS

4–117. The masonry support creates the loading distribution acting on the end of the beam. Simplify this load to a single resultant force and specify its location measured from point O.

Prob. 4–117

4–118. The electrical transmission tower supports the weights of the cables suspended from its arms and is subjected to the wind loading which acts along its height. Replace this system of loads by a single resultant force and moment acting at its base A.

Prob. 4–118

4–119. Replace the distributed loading by a single resultant force and specify its location measured from point A.

Prob. 4–119

*4–120. The loading on the book shelf is distributed as shown. Determine the magnitude of the resultant force and its location measured from point O.

Probs. 4–120/4–121

4–121. The loading on the book shelf is distributed as shown. Determine the magnitude of the resultant force and its location measured from point A.

4–122. Replace the loading by an equivalent force and couple moment acting at point O.

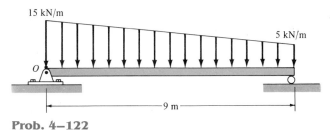

Prob. 4–122

4–123. Replace the loading by a resultant force and specify the location of the force on the beam, measured from point B.

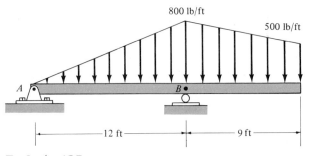

Prob. 4–123

156

*4–124. Replace the distributed loading by a resultant force, and specify the location of the force on the beam measured from the pin at *C*.

Prob. 4–124

4–125. The column is used to support the floor which exerts a force of 3000 lb on the top of the column. The effect of soil pressure along its side is distributed as shown. Replace this loading by a single resultant force and specify where it acts along the column measured from its base *A*.

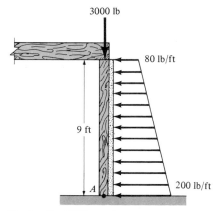

Prob. 4–125

4–126. Replace the loading by an equivalent force and couple moment acting at point *O*.

Prob. 4–126

4–127. The bricks on top of the beam and the supports at the bottom create the distributed loading shown in the second figure. Determine the required intensity *w* and dimension *d* of the right support so that the resultant force and couple moment about point *A* of the system are both zero.

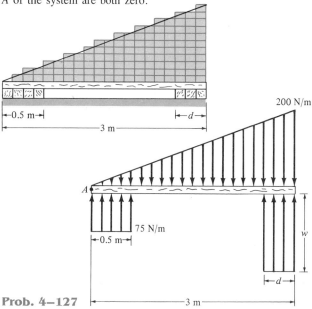

Prob. 4–127

*4–128. Replace the distributed loading by a resultant force and specify where the resultant's line of action intersects member *AB*, measured from *A*.

Probs. 4–128/4–129

4–129. Replace the distributed loading by a resultant force and specify where the resultant's line of action intersects member *BC*, measured from *C*.

157

4–130. The beam is subjected to the parabolic loading. Replace this loading by a resultant force and couple moment at point O.

Prob. 4–130

4–131. The distribution of soil loading on the bottom of a building slab is shown. The center portion of the loading is parabolic. Simplify this loading to a resultant force and specify its location measured from point O.

Prob. 4–131

*****4–132.** The form is used to cast a concrete wall having a width of 5 m. Determine the resultant force the wet concrete exerts along the form AB if the pressure distribution due to the concrete can be approximated as shown. Specify the location of the resultant force, measured from point B.

Prob. 4–132

4–133. Determine the magnitude of the resultant force and its location measured from point O.

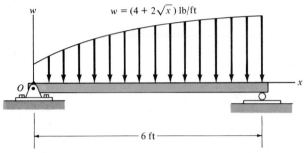

Prob. 4–133

■**4–134.** Determine the resultant of the distributed loading and its placement measured from point A. Evaluate the integrals using Simpson's rule.

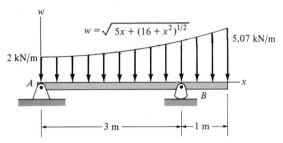

Prob. 4–134

4–135. Determine the resultant force of the distributed loading and its position measured from point A.

Prob. 4–135

REVIEW PROBLEMS

*4–136. Three parallel forces act on the circular slab. Determine the resultant force, and specify its location (x, y) on the slab.

Prob. 4–136

4–137. The pole supports a 22-lb traffic light. Using Cartesian vectors, determine the moment of the weight of the traffic light about the base of the pole at A.

Prob. 4–137

4–138. A force of 90 N acts on the handle of the paper cutter at A. Determine the moment created by this force about the hinge at O if $\theta = 60°$. At what angle θ should the force be applied so that the clockwise moment it creates about point O is a maximum? What is this maximum moment?

Prob. 4–138

4–139. The curved rod has a radius of 4 ft and is supported by the cable AB, which exerts a force of 80 lb on the rod. Determine the magnitude of the moment of this force about point C.

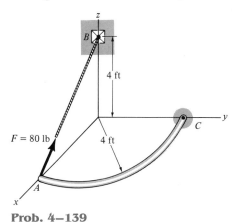

Prob. 4–139

159

*4–140. Replace the force **F**, having a magnitude of $F = 40$ lb and acting at B, by an equivalent force and couple moment at A.

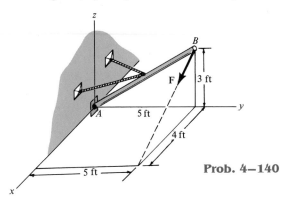

Prob. 4–140

4–141. The crane is used to lift the block that has a mass of 200 kg. Determine the clockwise moment of the block's weight about point A at the base of the boom. What magnitude of cable tension **T** is needed to create a counterclockwise moment of the same magnitude about point A?

Prob. 4–141

4–142. The system of four of the system forces acts on the roof truss. Determine the resultant force and specify its location along AB, measured from point A.

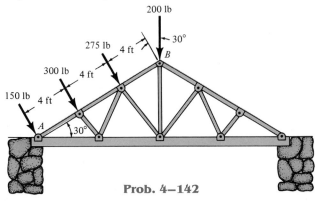

Prob. 4–142

4–143. Express the moment of the couple acting on the pipe in Cartesian vector form. What is the magnitude of the couple moment?

Prob. 4–143

*4–144. Wind has blown sand over a platform such that the intensity of load can be approximated by the function $w = (\frac{1}{2}x^3)$ N/m. Simplify this distributed loading to a resultant force and specify the magnitude and location of the force measured from A.

Prob. 4–144

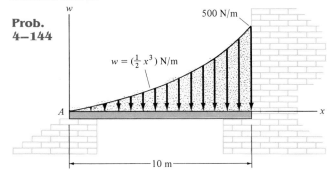

4–145. Replace the loading system acting on the beam by an equivalent force and couple system at point O.

Prob. 4–145

5

Equilibrium of a Rigid Body

In this chapter the fundamental concepts of rigid-body equilibrium will be discussed. It will be shown that equilibrium requires both a *balance of forces,* to prevent the body from translating with accelerated motion, and a *balance of moments,* to prevent the body from rotating.

Many types of engineering problems involve symmetric loadings and can be solved by projecting all the forces acting on a body onto a single plane. Hence, in the first part of this chapter, the equilibrium of a body subjected to a *coplanar* or *two-dimensional force system* will be considered. Ordinarily the geometry of such problems is not very complex, so a scalar solution is suitable for analysis. The more general discussion of rigid bodies subjected to *three-dimensional force systems* is given in the second part of this chapter. It will be seen that many of these types of problems can best be solved by using vector analysis.

5.1 Conditions for Rigid-Body Equilibrium

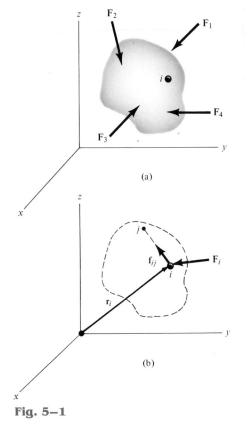

(a)

(b)

Fig. 5–1

In Chapter 3 it was stated that a particle is in equilibrium if it remains at rest or moves with constant velocity. For this to be the case, it is both necessary and sufficient to require the resultant force acting on the particle to be equal to zero. Using this fact, we will now develop the conditions required to maintain equilibrium for a rigid body. To do this, consider the rigid body in Fig. 5–1a, which is fixed in the x, y, z reference and is either at rest or moves with the reference at constant velocity. A free-body diagram of the arbitrary ith particle of the body is shown in Fig. 5–1b. There are two types of forces which act on it. The *internal forces*, represented symbolically as

$$\sum_{\substack{j=1 \\ (j \neq i)}}^{n} \mathbf{f}_{ij} = \mathbf{f}_i$$

are forces which all the other particles exert on the ith particle and produce the resultant $\mathbf{f}_i$. Although only one of these particles is shown in Fig. 5–1b, the summation extends over all n particles composing the body. For this summation note that it is meaningless for $i = j$ since the ith particle cannot exert a force on itself. The resultant *external force* $\mathbf{F}_i$ represents, for example, the effect of gravitational, electrical, magnetic, or contact forces between the ith particle and adjacent bodies or particles *not* included within the body. If the particle is in equilibrium, then applying Newton's first law we have

$$\mathbf{F}_i + \mathbf{f}_i = \mathbf{0}$$

When the equation of equilibrium is applied to each of the other particles of the body, similar equations will result. If all these equations are added together *vectorially*, we obtain

$$\Sigma \mathbf{F}_i + \Sigma \mathbf{f}_i = \mathbf{0}$$

The summation of the internal forces if carried out will equal zero since the internal forces between particles will occur in equal but opposite collinear pairs, Newton's third law. Consequently, only the sum of the *external forces* will remain; and therefore, letting $\Sigma \mathbf{F}_i = \Sigma \mathbf{F}$, the above equation can be written as

$$\Sigma \mathbf{F} = \mathbf{0}$$

Let us now consider the moments of the forces acting on the particle about the arbitrary point O, Fig. 5–1b. Using the particle equilibrium equation and the distributive law of the vector cross product yields

$$\mathbf{r}_i \times (\mathbf{F}_i + \mathbf{f}_i) = \mathbf{r}_i \times \mathbf{F}_i + \mathbf{r}_i \times \mathbf{f}_i = \mathbf{0}$$

Similar equations can be written for the other particles of the body, and adding them together vectorially, we obtain

$$\Sigma \mathbf{r}_i \times \mathbf{F}_i + \Sigma \mathbf{r}_i \times \mathbf{f}_i = \mathbf{0}$$

The second term is zero since, as stated above, the internal forces occur in equal but opposite collinear pairs, and by the transmissibility of a force, as discussed in Sec. 4.4, the moment of each pair of forces about point O is therefore zero. Hence, using the notation $\Sigma \mathbf{M}_O = \Sigma \mathbf{r}_i \times \mathbf{F}_i$, we can write the above equation as

$$\Sigma \mathbf{M}_O = \mathbf{0}$$

Hence the *equations of equilibrium* for a rigid body can be summarized as follows:

$$\boxed{\begin{array}{l} \Sigma \mathbf{F} = \mathbf{0} \\ \Sigma \mathbf{M}_O = \mathbf{0} \end{array}} \tag{5–1}$$

These equations require that a rigid body will be in equilibrium provided the sum of all the external forces acting on the body is equal to zero and the sum of the moments of the external forces about a point is equal to zero. The fact that these conditions are *necessary* for equilibrium has now been proven. They are also *sufficient* conditions. To show this, let us assume that the body is *not* in equilibrium, and yet the force system acting on it satisfies Eqs. 5–1. Suppose that an *additional force* $\mathbf{F}'$ is required to hold the body in equilibrium. As a result, the equilibrium equations become

$$\Sigma \mathbf{F} + \mathbf{F}' = \mathbf{0}$$
$$\Sigma \mathbf{M}_O + \mathbf{M}'_O = \mathbf{0}$$

where $\mathbf{M}'_O$ is the moment of $\mathbf{F}'$ about O. Since $\Sigma \mathbf{F} = \mathbf{0}$ and $\Sigma \mathbf{M}_O = \mathbf{0}$, then we require $\mathbf{F}' = \mathbf{0}$ (also $\mathbf{M}'_O = \mathbf{0}$). Consequently, the additional force $\mathbf{F}'$ is not required for holding the body, and indeed Eqs. 5–1 are also sufficient conditions.

Equilibrium in Two Dimensions

5.2 Free-Body Diagrams

Successful application of the equations of equilibrium requires a complete specification of *all* the known and unknown forces that act *on* the body. The best way to account for these forces is to draw the body's free-body diagram. This diagram is a sketch of the outlined shape of the body which represents it as being *isolated* or "free" from its surroundings. On this sketch one then shows *all* the forces (or their resultants) which the surroundings exert *on the body*. Obviously, if the free-body diagram is correctly drawn, the effects of all the applied forces and couples acting on the body can be accounted for when the equations of equilibrium are applied. For this reason, *a thorough understanding of how to draw a free-body diagram is of primary importance for solving problems in mechanics*. It is the preliminary step before applying the equations of equilibrium.

Table 5–1 **Supports for Rigid Bodies Subjected to Two-Dimensional Force Systems**

Types of Connection	Reaction	Number of Unknowns
(1) cable		One unknown. The reaction is a tension force which acts away from the member in the direction of the cable.
(2) weightless link	or	One unknown. The reaction is a force which acts along the axis of the link.
(3) roller		One unknown. The reaction is a force which acts perpendicular to the surface at the point of contact.
(4) roller or pin in confined smooth slot	or	One unknown. The reaction is a force which acts perpendicular to the slot.
(5) rocker		One unknown. The reaction is a force which acts perpendicular to the surface at the point of contact.
(6) smooth contacting surface		One unknown. The reaction is a force which acts perpendicular to the surface at the point of contact.
(7) member pin connected to collar on smooth rod	or	One unknown. The reaction is a force which acts perpendicular to the rod.

Table 5—1 (Contd.)

Types of Connection	Reaction	Number of Unknowns
(8) smooth pin or hinge	or	Two unknowns. The reactions are two components of force, or the magnitude and direction ϕ of the resultant force. Note that ϕ and θ are not necessarily equal (usually not, unless the rod shown is a link as in (2)).
(9) member fixed connected to collar on smooth rod		Two unknowns. The reactions are the couple moment and the force which acts perpendicular to the rod.
(10) fixed support	or	Three unknowns. The reactions are the couple and the two force components, or the couple and the magnitude and direction ϕ of the resultant force.

Support Reactions

Before presenting a formal procedure as to how to draw a free-body diagram, we will first consider the various types of reactions that occur at supports and points of support between bodies. For two-dimensional problems, i.e., bodies subjected to coplanar force systems, the supports most commonly encountered are given in Table 5–1. (In all cases the angle θ is assumed to be known.) Carefully study each of the symbols used to represent these supports and the types of reactions they exert on their contacting members. In this regard, if *the support prevents translation in a given direction, then a force is developed on the member in that direction. Likewise, if rotation is prevented, a couple moment is exerted on the member*. For example, a member in contact with the smooth surface (6) is prevented from translating *only* in the contact direction, perpendicular or normal to the surface. Hence, the surface exerts only a *normal force* **F** on the member at the point of contact. The magnitude of this force represents *one unknown*. Since the member is free to rotate on the surface, a couple moment cannot be developed by the surface on the member at the point of contact. The pin or hinge support (8) prevents translation of the connecting member at its point of connection. In this case, unlike that of the smooth surface, translation can be prevented in *any* direction. Hence, a force

F must be developed at the support such that it has *two unknowns,* its magnitude F and direction ϕ, or, equivalently, the magnitudes of its two components **F**$_x$ and **F**$_y$. Since the connecting member is allowed to rotate freely in the plane about the pin, a pin support does not resist a couple moment acting perpendicular to this plane. The fixed support (10), however, prevents *both* planar translation and rotation of the connecting member at the point of connection. Therefore, this type of support can exert both a force and a couple moment on the member. Note that the moment acts *perpendicular* to the plane of the page since rotation is prevented in the plane. Hence, there are *three unknowns* at a fixed support.

The concentrated forces and couple moments shown in Table 5–1 actually represent the *resultants* of *distributed surface loads* that exist between each support and its contacting member. Although it is these resultants which are determined in practice, it is generally not important to determine the actual load distribution, since the surface area over which it acts is considerably *smaller* than the *total surface area* of the contacting member.

External and Internal Forces

Since a rigid body is a composition of particles, both *external* and *internal* loadings may act on it. It is important to realize, however, that if the free-body diagram for the body is drawn, the forces that are *internal* to the body are *not represented* on the free-body diagram. As discussed in Sec. 5.1, these forces always occur in equal but opposite collinear pairs, and therefore their *net effect* on the body is zero.

In some problems, a free-body diagram for a ''system'' of connected bodies may be used for an analysis. An example would be the free-body diagram of an entire automobile (system) composed of its many parts. Obviously, the connecting forces between its parts would represent *internal forces* which would *not* be included on the free-body diagram of the automobile. To summarize then, internal forces act between particles which are located *within* a specified system which is contained within the boundary of the free-body diagram. Particles or bodies outside this boundary exert external forces on the system and these alone must be shown on the free-body diagram.

Weight and the Center of Gravity

When a body is subjected to a gravitational field, each of its particles has a specified weight as defined by Newton's law of gravitation, $F = Gm_1m_2/r^2$, Eq. 1–2. If we assume the size of the body to be ''small'' in relation to the size of the earth, then it is appropriate to consider these gravitational forces to be represented as a *system of parallel forces* acting on the particles contained within the boundary of the body. It was shown in Sec. 4.9 that such a system can be reduced to a single resultant force acting through a specified point. We refer to this force resultant as the *weight* **W** of the body, and to the location of its point of application as the *center of gravity G*. The methods used for its calculation will be developed in Chapter 9.

In the examples and problems that follow, if the weight of the body is important for the analysis, this force will then be reported in the problem statement. Also, when the body is *uniform* or made of homogeneous material, the center of gravity will be located at the body's *geometric center;* however, if the body is nonhomogeneous or has an unusual shape, then its center of gravity will be given.

PROCEDURE FOR DRAWING A FREE-BODY DIAGRAM

To construct a free-body diagram for a rigid body or group of bodies considered as a single system, the following steps should be performed:

Step 1. Imagine the body to be *isolated* or cut "free" from its constraints and connections, and draw (sketch) its outlined shape.

Step 2. Identify all the external forces and couples that act on the body. Forces and couples generally encountered are those due to (1) applied loadings, (2) reactions occurring at the supports or at points of contact with other bodies (see Table 5–1), and (3) the weight of the body. To account for all these forces, it may help to trace over the boundary, carefully noting each force or couple acting on it.

Step 3. Indicate the dimensions of the body necessary for computing the moments of forces. The forces and couples that are known should be labeled with their proper magnitudes and directions. Letters are used to represent the magnitudes and direction angles of forces and couples that are *unknown*. In particular, if a force or couple has a known line of action but unknown magnitude, the arrowhead which defines the sense of the vector can be assumed. The correctness of the assumed sense will become apparent after solving the equilibrium equations for the unknown magnitude. By definition, the *magnitude* of a vector is *always positive,* so that if the solution yields a "negative" scalar, the *minus sign* indicates that the vector's sense is *opposite* to that which was originally assumed.

Before proceeding, review this section; then carefully study the following examples. Afterward, attempt to draw the free-body diagrams for the objects in Figs. 5–2 through 5–6 without "looking" at the solutions. Further practice in drawing free-body diagrams should be gained by solving *all* the problems given at the end of this section.

Example 5–1

Draw the free-body diagram of the uniform beam shown in Fig. 5–2a. The beam has a mass of 100 kg.

Fig. 5–2a   (a)

SOLUTION

The free-body diagram of the beam is shown in Fig. 5–2b. Since the support at A is a fixed wall, there are three reactions acting *on the beam* at A, denoted as $\mathbf{A}_x$, $\mathbf{A}_y$, and $\mathbf{M}_A$. The magnitudes of these vectors are *unknown*, and their sense has been *assumed*. (How does one obtain the *correct* sense of these vectors?) The weight of the beam, $W = 100(9.81) = 981$ N, acts through the beam's center of gravity G, 3 m from A since the beam is uniform.

Fig. 5–2b

Example 5–2

Draw the free-body diagram for the bell crank *ABC* shown in Fig. 5–3*a*.

SOLUTION

The free-body diagram is shown in Fig. 5–3*b*. The pin support at *B* exerts force components **B**$_x$ and **B**$_y$ *on the crank,* each having a known line of action but unknown magnitude. The link at *C* exerts a force **F**$_C$ acting in the direction of the link and having an unknown magnitude. The dimensions of the crank are also labeled on the free-body diagram, since this information will be useful in computing the moments of the forces. As usual, the sense of the three unknown forces has been assumed. The correct sense will become apparent after solving the equilibrium equations.

Although not part of this problem, three-dimensional views of the free-body diagrams of the pin and two pin leaves at *B* are shown in Fig. 5–3*c*. Since the leaves are *fixed-connected* to the wall, there are three unknowns at this support, namely **B**$''_x$, **B**$''_y$, **M**$''_B$. These reactions are shown to be equal in magnitude and direction on each leaf due to the symmetry of the loading and geometry. Note carefully how the principle of action—equal but opposite collinear reaction is used when applying the forces **B**$'_x$ and **B**$'_y$ to each leaf and the pin. All of these unknowns can be obtained from the equations of equilibrium once **B**$_x$ and **B**$_y$ are obtained.

(a)

Fig. 5–3a

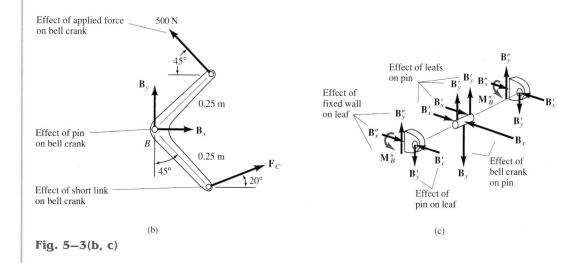

(b)

(c)

Fig. 5–3(b, c)

Example 5–3

Two smooth balls A and B, each having a mass of 2 kg, rest between the inclined planes shown in Fig. 5–4a. Draw the free-body diagrams for ball A, ball B, and balls A and B together.

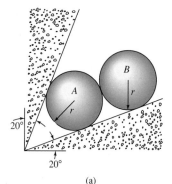

Fig. 5–4a   (a)

SOLUTION

The free-body diagram for ball A is shown in Fig. 5–4b. Note that the weight of the ball is calculated as $W = 2(9.81) = 19.62$ N. Since all contacting surfaces are *smooth,* the reactive forces $\mathbf{T}, \mathbf{F}, \mathbf{R}$ act in a direction *normal* to the tangent at their surfaces of contact.

The free-body diagram of ball B is shown in Fig. 5–4c. Can you identify each of the three forces acting *on the ball?* In particular, note that $\mathbf{R}$, representing the force of ball A on ball B, Fig. 5–4c, is equal and opposite to $\mathbf{R}$ representing the force of ball B on ball A, Fig. 5–4b. This is a consequence of Newton's third law of motion.

The free-body diagram of both balls combined ("system") is shown in Fig. 5–4d. Here the contact force $\mathbf{R}$, which acts between A and B, is considered as an *internal* force and hence is not shown on the free-body diagram. That is, it represents a pair of equal but opposite collinear forces which cancel each other.

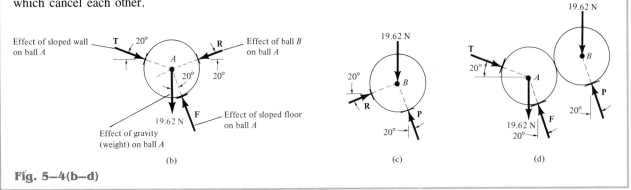

Fig. 5–4(b–d)

Example 5–4

The free-body diagram of each object in Fig. 5–5 is drawn and the forces acting on the object are identified. The weight of the objects is neglected except where indicated.

SOLUTION

Fig. 5–5

Example 5–5

The highway sign shown in Fig. 5–6a has a mass of 100 kg with a center of gravity at G. It is supported by pins at C and D and a cable AB. Draw a free-body diagram of the sign and the supporting frame. Neglect the weight of the frame.

Fig. 5–6

(a)

(b)

SOLUTION

By observation, the frame, sign, and the loading are all symmetrical about the vertical x-y plane, hence the problem may be analyzed using a system of *coplanar forces*. The free-body diagram is shown in Fig. 5–6b. Note that the force **T** that the cable exerts on the frame has a known line of action indicated by the 3-4-5 slope triangle. The force components C'_x and C'_y represent the horizontal and vertical reactions of *both* pins C and D, respectively. Consequently, after the solution for these reactions is obtained, *half* the magnitude is applied at C and half at D.

PROBLEMS

5–1. Draw the free-body diagram of the 50-kg roll of paper which has a center of mass at G and rests on the smooth blade of the paper hauler. Explain the significance of each force acting on the diagram. (See Fig. 5-2*b*.)

Prob. 5–1

5–2. Draw the free-body diagram of the beam subjected to the forces and couple moment and supported by a pin at A and short link CB. Explain the significance of each force acting on the diagram. (See Fig. 5–2*b*.)

Prob. 5–2

5–3. Draw the free-body diagram of member ABC supported by a pin at A and a horizontal short link BD. Explain the significance of each force acting on the diagram. (See Fig. 5–2*b*.)

Prob. 5–3

*5–4. Draw the free-body diagram of the crane boom AB which has a weight of 650 lb and center of gravity at G. The boom is supported by a pin at A and cable BC. The load of 1250 lb is suspended from a cable attached at B. Explain the significance of each force acting on the diagram. (See Fig. 5–2*b*.)

Prob. 5–4

5–5. Draw the free-body diagram for the tower which supports the two cables, F and H, each having a weight of 750 lb. The tower is pinned at its base A and held by guy wires BC and DE. Explain the significance of each force acting on the diagram. (See Fig. 5–2*b*.)

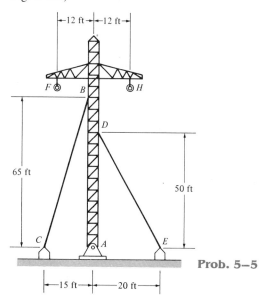

Prob. 5–5

173

5–6. Draw the free-body diagram of the dumpster D of the truck, which has a weight of 5000 lb and a center of gravity at G. It is supported by a pin at A and a pin-connected hydraulic cylinder BC (short link). Explain the significance of each force acting on the diagram. (See Fig. 5–2b.)

Prob. 5–6

5–7. Draw the free-body diagram of the oil drill rig which is supported by the hydraulic cylinder CD (short link) and the pin at A. The rig has a weight of 125 kip and center of gravity at G. Explain the significance of each force acting on the diagram. (See Fig. 5–2b.)

Prob. 5–7

*5–8. Draw the free-body diagram of member ABC which is supported by a smooth collar at A, roller at B, and short link CD. Explain the significance of each force acting on the diagram. (See Fig. 5–2b.)

Prob. 5–8

5–9. Draw the free-body diagram of member AB, which is supported by a roller at A and a pin at B. Explain the significance of each force acting on the diagram. (See Fig. 5–2b.)

Prob. 5–9

5–10. Draw the free-body diagram of the wagon and load which have a weight of 50 lb and center of gravity at G. The wagon is being held in equilibrium by the handle force **T**. The surface at A and B is smooth. Explain the significance of each force acting on the diagram. (See Fig. 5–2b.)

Prob. 5–10

5.3 Equations of Equilibrium

In Sec. 5.1 we developed the two equations which are both necessary and sufficient for the equilibrium of a rigid body, namely, $\Sigma\mathbf{F} = \mathbf{0}$ and $\Sigma\mathbf{M}_O = \mathbf{0}$. When the body is subjected to a system of forces, which all lie in the x-y plane, then the forces can be resolved into their x and y components. Consequently, the conditions for equilibrium in two dimensions are

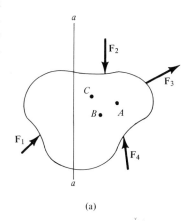

(a)

$$
\boxed{
\begin{aligned}
\Sigma F_x &= 0 \\
\Sigma F_y &= 0 \\
\Sigma M_O &= 0
\end{aligned}
}
\qquad (5\text{–}2)
$$

Here ΣF_x and ΣF_y represent, respectively, the algebraic sums of the x and y components of all the forces acting on the body, and ΣM_O represents the algebraic sum of the moments of all these force components about an axis perpendicular to the x-y plane and passing through the arbitrary point O, which may lie either on or off the body.

Alternative Sets of Equilibrium Equations

Although Eqs. 5–2 are *most often* used for solving equilibrium problems involving coplanar force systems, two *alternative* sets of three independent equilibrium equations may also be used. One such set is

$$
\begin{aligned}
\Sigma F_a &= 0 \\
\Sigma M_A &= 0 \\
\Sigma M_B &= 0
\end{aligned}
\qquad (5\text{–}3)
$$

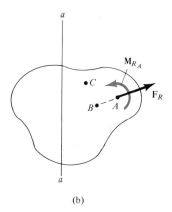

(b)

When using these equations it is required that the moment points A and B do *not* lie on a line that is *perpendicular* to the a axis. To prove that Eqs. 5–3 provide the *conditions* for equilibrium, consider the free-body diagram of an arbitrarily shaped body shown in Fig. 5–7a. Using the methods of Sec. 4.8, the system of forces may be replaced by a single resultant force $\mathbf{F}_R = \Sigma\mathbf{F}$, acting at point A, and a resultant couple $\mathbf{M}_{R_A} = \Sigma\mathbf{M}_A$, Fig. 5–7b. For equilibrium it is required that $\mathbf{F}_R = \mathbf{0}$ and $\mathbf{M}_{R_A} = \mathbf{0}$. Hence, if $\Sigma M_A = 0$, Eq. 5–3, it is necessary that $\mathbf{M}_{R_A} = \mathbf{0}$. Furthermore, in order that $\mathbf{F}_R$ satisfy $\Sigma F_a = 0$, it must have no component along the a axis, and therefore its line of action must be perpendicular to the a axis, Fig. 5–7c. Finally, if it is required that $\Sigma M_B = 0$, where B does not lie on the line of action of $\mathbf{F}_R$, then $\mathbf{F}_R = \mathbf{0}$, and indeed the body shown in Fig. 5–7a must be in equilibrium.

A second alternative set of equilibrium equations is

$$
\begin{aligned}
\Sigma M_A &= 0 \\
\Sigma M_B &= 0 \\
\Sigma M_C &= 0
\end{aligned}
\qquad (5\text{–}4)
$$

(c)

Fig. 5–7

175

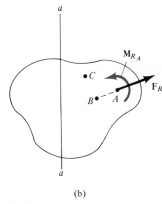

(b)

Fig. 5–7b

Here it is necessary that points A, B, and C do not lie on the same line. To prove that these equations when satisfied ensure equilibrium, consider again the free-body diagram in Fig. 5–7b. If it is required that $\Sigma M_A = 0$, then the resultant couple $\mathbf{M}_{R_A} = \mathbf{0}$. $\Sigma M_B = 0$ is satisfied if the line of action of $\mathbf{F}_R$ passes through point B as shown, and finally, if $\Sigma M_C = 0$, where C does not lie on line AB, it is necessary that $\mathbf{F}_R = \mathbf{0}$, and the body in Fig. 5–7a must then be in equilibrium.

PROCEDURE FOR ANALYSIS

The following procedure provides a method for solving coplanar force equilibrium problems:

Free-Body Diagram. Draw a free-body diagram of the body as discussed in Sec. 5.2. Briefly this requires showing all the external forces and couple moments acting *on the body*. The magnitudes of these vectors must be labeled and their directions specified. Dimensions of the body, necessary for computing the moments of forces, are also included on the free-body diagram. Identify the unknowns. The sense of a force or couple having an *unknown* magnitude but known line of action can be *assumed*.

Equations of Equilibrium. Establish the x, y, z axes and apply the equations of equilibrium: $\Sigma F_x = 0$, $\Sigma F_y = 0$, $\Sigma M_O = 0$ (or the alternative sets of Eqs. 5–3 or 5–4). To *avoid* having to solve simultaneous equations, apply the moment equation $\Sigma M_O = 0$ about a point (O) *that lies at the intersection of the lines of action of two unknown forces*. In this way, the moments of these unknowns are *zero* about O, and one can obtain a *direct solution* for the third unknown. When applying the force equations $\Sigma F_x = 0$ and $\Sigma F_y = 0$, orient the x and y axes along lines that will provide the simplest resolution of the forces into their x and y components. If the solution of the equilibrium equations yields a *negative* scalar for an unknown force or couple, it indicates that the sense is *opposite* to that which was assumed on the free-body diagram.

The following example problems numerically illustrate this procedure.

Example 5–6

Determine the horizontal and vertical components of reaction for the beam loaded as shown in Fig. 5–8a. Neglect the weight and thickness of the beam in the calculations.

Fig. 5–8

SOLUTION

Free-Body Diagram. Can you identify each of the forces shown on the free-body diagram of the beam, Fig. 5–8b? For simplicity in applying the equilibrium equations, the 600-N force is represented by its x and y components as shown. Also, note that a 200-N force acts on the beam at B, and is independent of the force components $\mathbf{B}_x$ and $\mathbf{B}_y$ which represent the effect of the pin on the beam.

Equations of Equilibrium. Applying the equilibrium equation $\Sigma F_x = 0$ to the force system on the free-body diagram yields

$$\xrightarrow{+}\Sigma F_x = 0; \qquad 600 \cos 45° \text{ N} - B_x = 0$$
$$B_x = 424 \text{ N} \qquad\qquad Ans.$$

A direct solution for $\mathbf{A}_y$ can be obtained by applying the moment equation $\Sigma M_B = 0$ about point B. For the calculation, it should be apparent that forces $600 \cos 45°$ N, $\mathbf{B}_x$, and $\mathbf{B}_y$ create zero moment about B. Assuming counterclockwise rotation about B to be positive (in the $+\mathbf{k}$ direction), Fig. 5–8b, we have

$$\zeta+\Sigma M_B = 0; \quad 100 \text{ N}(2 \text{ m}) + (600 \sin 45° \text{ N})(5 \text{ m}) - A_y(7 \text{ m}) = 0$$
$$A_y = 332 \text{ N} \qquad\qquad Ans.$$

Summing forces in the y direction, using the result $A_y = 332$ N, gives

$$+\uparrow\Sigma F_y = 0; \quad 332 \text{ N} - 600 \sin 45° \text{ N} - 100 \text{ N} - 200 \text{ N} + B_y = 0$$
$$B_y = 393 \text{ N} \qquad\qquad Ans.$$

Example 5–7

The cord shown in Fig. 5–9a supports a force of 100 lb and wraps over the frictionless pulley. Determine the tension in the cord and the horizontal and vertical components of reaction at pin A.

0.5 ft

A

B

$\theta = 30°$

100 lb

Fig. 5–9a

(a)

SOLUTION

Free-Body Diagrams. The free-body diagrams of the cord and pulley are shown in Fig. 5–9b. Note that the principle of action, equal but opposite reaction, must be carefully observed when drawing each of these diagrams: the cord exerts an unknown load distribution p along part of the pulley surface, whereas the pulley exerts an equal but opposite effect on the cord. For the solution, however, it is simpler to *combine* the free-body diagrams of the pulley and the contacting portion of the cord, so that the distributed load becomes *internal* to the system and is therefore eliminated from the analysis, Fig. 5–9c.

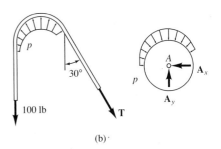

(b)

Equations of Equilibrium. Summing moments about point A to eliminate $\mathbf{A}_x$ and $\mathbf{A}_y$, Fig. 5–9c, we have

$$\zeta + \Sigma M_A = 0; \qquad 100 \text{ lb}(0.5 \text{ ft}) - T(0.5 \text{ ft}) = 0$$
$$T = 100 \text{ lb} \qquad\qquad Ans.$$

It is seen that the tension remains *constant* as the cord passes over the pulley. (This of course is true for *any angle* θ at which the cord is directed and for *any radius* r of the pulley.) Using the result for T, a force summation is applied to determine the reaction at pin A.

$$\xrightarrow{+} \Sigma F_x = 0; \qquad -A_x + 100 \sin 30° \text{ lb} = 0$$
$$A_x = 50.0 \text{ lb} \qquad\qquad Ans.$$

$$+ \uparrow \Sigma F_y = 0; \qquad A_y - 100 \text{ lb} - 100 \cos 30° \text{ lb} = 0$$
$$A_y = 187 \text{ lb} \qquad\qquad Ans.$$

$r = 0.5$ ft

A

r

A_y A_x

100 lb

$\theta = 30°$ T

(c)

Fig. 5–9(b, c)

Example 5–8

Determine the reactions at the fixed support A for the loaded frame shown in Fig. 5–10a.

SOLUTION

Free-Body Diagram. The free-body diagram for the *entire frame* is shown in Fig. 5–10b. There are three unknowns at the fixed support, represented by the magnitudes of A_x, A_y, and M_A. Note that the pin forces at B and D are *not* shown on this diagram since they are *internal*, i.e., they act *within* the boundary of the free-body diagram. Their effect on the body will cancel.

(a)

Equations of Equilibrium

$\xrightarrow{+} \Sigma F_x = 0;$ $\qquad\qquad A_x = 0$ $\qquad\qquad$ *Ans.*

$+ \uparrow \Sigma F_y = 0;$ $\qquad A_y - 400 \text{ N} - 200 \text{ N} = 0$

$\qquad\qquad\qquad A_y = 600 \text{ N}$ $\qquad\qquad$ *Ans.*

$\zeta + \Sigma M_A = 0;$ $\quad M_A - 400 \text{ N}(1 \text{ m}) - 200 \text{ N}(2 \text{ m}) = 0$

$\qquad\qquad\qquad M_A = 800 \text{ N} \cdot \text{m}$ $\qquad\qquad$ *Ans.*

Point A was chosen for summing moments since the lines of action of the *unknown* forces A_x and A_y pass through this point, and therefore these forces were not included in the moment summation. Note, however, that M_A must be *included* in the moment summation. This couple moment is a free vector and represents the effect of the fixed support on the frame.

Although only *three* independent equilibrium equations can be written for a rigid body, it is a good practice to *check* the calculations using a fourth equilibrium equation. The latter equation may be obtained from one of the other sets of equilibrium Eqs. 5–3 or 5–4. For example, the above computations may be verified by summing moments about point C:

$\zeta + \Sigma M_C = 0;$ $\quad 400 \text{ N}(1 \text{ m}) - 600 \text{ N}(2 \text{ m}) + 800 \text{ N} \cdot \text{m} \equiv 0$

$\qquad\qquad 400 \text{ N} \cdot \text{m} - 1200 \text{ N} \cdot \text{m} + 800 \text{ N} \cdot \text{m} \equiv 0$

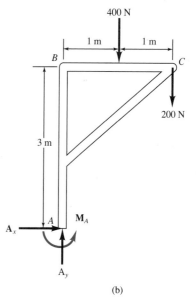

(b)

Fig. 5–10

Example 5–9

The uniform beam shown in Fig. 5–11a is subjected to a force and couple. If the beam is supported at A by a smooth wall and at B and C either at the top or bottom by smooth contacts, determine the reactions at these supports. Neglect the weight of the beam.

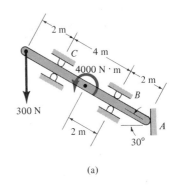

(a)

SOLUTION

Free-Body Diagram. As shown in Fig. 5–11b, all the support reactions act normal to the surface of contact since the contacting surfaces are smooth. The reactions at B and C are shown acting in the positive y' direction. This assumes that only the contacts located on the bottom of the beam are used for support.

Equations of Equilibrium. Using the x, y coordinate system in Fig. 5–11b, we have

$$\xrightarrow{+} \Sigma F_x = 0; \qquad C_{y'} \sin 30° + B_{y'} \sin 30° - A_x = 0 \qquad (1)$$

$$+\uparrow \Sigma F_y = 0; \quad -300 \text{ N} + C_{y'} \cos 30° + B_{y'} \cos 30° = 0 \qquad (2)$$

$$\downarrow+\Sigma M_A = 0; \qquad -B_{y'}(2 \text{ m}) + 4000 \text{ N} \cdot \text{m} - C_{y'}(6 \text{ m})$$
$$+(300 \cos 30° \text{ N})(8 \text{ m}) = 0 \quad (3)$$

When writing the moment equation, it should be noticed that the line of action of the force component 300 sin 30° N passes through point A, and therefore this force is not included in the moment equation.

Solving Eqs. (2) and (3) simultaneously, we obtain

$$B_{y'} = -1000.0 \text{ N} \qquad \qquad Ans.$$
$$C_{y'} = 1346.4 \text{ N} \qquad \qquad Ans.$$

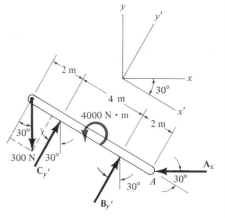

(b)

Fig. 5–11

Since $B_{y'}$ is a negative scalar, the sense of $\mathbf{B}_{y'}$ is opposite to that shown on the free-body diagram in Fig. 5–11b. Therefore, the top contact at B serves as the support rather than the bottom one. Retaining the negative sign for $B_{y'}$ (Why?) and substituting the results into Eq. (1), we obtain

$$1346.4 \sin 30° \text{ N} - 1000.0 \sin 30° \text{ N} - A_x = 0$$
$$A_x = 173.2 \text{ N} \qquad \qquad Ans.$$

Example 5–10

The beam shown in Fig. 5–12a is pin-connected at A and rests against a roller at B. Compute the horizontal and vertical components of reaction at the pin A.

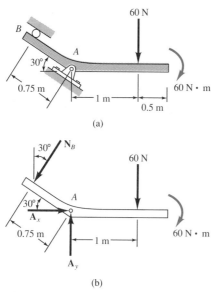

(a)

(b)

Fig. 5–12

SOLUTION

Free-Body Diagram. As shown in Fig. 5–12b, the reaction N_B is perpendicular to the beam at B, since the support is a roller. Also, horizontal and vertical components of reaction are represented at A, even though the base of the pin support is tilted.

Equations of Equilibrium. Summing moments about A, we obtain a direct solution for N_B,

$\zeta + \Sigma M_A = 0;$ $-60 \text{ N} \cdot \text{m} - 60 \text{ N}(1 \text{ m}) + N_B(0.75 \text{ m}) = 0$

$$N_B = 160 \text{ N}$$

Using this result,

$\xrightarrow{+} \Sigma F_x = 0;$ $A_x - 160 \sin 30° \text{ N} = 0$

$A_x = 80.0 \text{ N}$ *Ans.*

$+ \uparrow \Sigma F_y = 0;$ $A_y - 160 \cos 30° \text{ N} - 60 \text{ N} = 0$

$A_y = 199 \text{ N}$ *Ans.*

181

Example 5–11

A force of 150 lb acts on the end of the beam shown in Fig. 5–13a. Determine the magnitude and direction of the reaction at the pin A and the tension in the cable.

(a)

Fig. 5–13

(b)

SOLUTION

Free-Body Diagram. The forces acting on the beam are shown in Fig. 5–13b.

Equations of Equilibrium. Summing moments about point A to obtain a direct solution for the cable tension yields

$$\zeta + \Sigma M_A = 0; \quad -(\tfrac{3}{5}T)(2 \text{ ft}) - (\tfrac{4}{5}T)(3 \text{ ft}) + 150 \text{ lb}(10 \text{ ft}) = 0$$
$$-3.6T + 150 \text{ lb}(10 \text{ ft}) = 0 \quad (1)$$
$$T = 416.7 \text{ lb} \qquad Ans.$$

Note that by the principle of transmissibility it is also possible to locate **T** at C, even though this point is not on the beam, Fig. 5–13b. In this case, the vertical component of **T** creates *zero moment* about A and the moment arm of the horizontal component ($\tfrac{4}{5}T$) becomes 4.5 ft. Hence, $\Sigma M_A = 0$ yields Eq. (1) directly since $(\tfrac{4}{5}T)(4.5) = 3.6T$.

Summing forces to obtain A_x and A_y, using the result for T, we have

$$\xrightarrow{+} \Sigma F_x = 0; \qquad -A_x + (\tfrac{4}{5})(416.7 \text{ lb}) = 0$$
$$A_x = 333.3 \text{ lb}$$

$$+\uparrow \Sigma F_y = 0; \qquad (\tfrac{3}{5})416.7 \text{ lb} - 150 \text{ lb} - A_y = 0$$
$$A_y = 100 \text{ lb}$$

Thus,
$$F_A = \sqrt{(333.3 \text{ lb})^2 + (100 \text{ lb})^2}$$
$$= 348.0 \text{ lb} \qquad Ans.$$

$$\theta = \tan^{-1} \frac{100 \text{ lb}}{333.3 \text{ lb}} = 16.7° \quad \theta \nearrow \qquad Ans.$$

Example 5–12

The 100-kg uniform beam AB shown in Fig. 5–14a is supported at A by a pin and at B and C by a continuous cable which wraps around a frictionless pulley located at D. If a maximum tension force of 800 N can be developed in the cable before it breaks, determine the greatest distance d at which the 6-kN force can be placed on the beam. What are the horizontal and vertical components of reaction at A just before the cable breaks?

(a)   (b)

Fig. 5–14

SOLUTION

Free-Body Diagram. Since the cable is continuous and passes over a frictionless pulley, the entire cable is subjected to its maximum tension of 800 N when the 6-kN force acts at d. Hence, the cable exerts an 800-N force at points C and B *on the beam* in the direction of the cable, Fig. 5–14b.

Equations of Equilibrium. By summing moments of the force system about point A it is possible to obtain a direct solution for the dimension d. Why?

$\zeta + \Sigma M_A = 0;$

$$-(6000 \text{ N})(d) - 981 \text{ N}(5 \text{ m}) + (800 \sin 60° \text{ N})(8 \text{ m}) + 800 \text{ N}(10 \text{ m}) = 0$$

$$d = 1.44 \text{ m} \qquad \qquad Ans.$$

Summing forces in the x and y directions, we have

$\xrightarrow{+} \Sigma F_x = 0; \qquad \qquad -A_x + 800 \cos 60° \text{ N} = 0$

$$A_x = 400 \text{ N} \qquad \qquad Ans.$$

$+ \uparrow \Sigma F_y = 0;$

$$A_y - 6000 \text{ N} - 981 \text{ N} + 800 \sin 60° \text{ N} + 800 \text{ N} = 0$$

$$A_y = 5.49 \text{ kN} \qquad \qquad Ans.$$

5.4 Two- and Three-Force Members

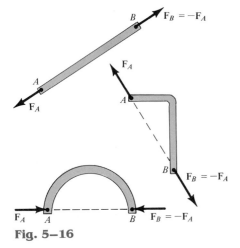

Fig. 5–15

Fig. 5–16

The solution to some equilibrium problems can be simplified if one is able to recognize members that are subjected to only two or three forces.

Two-Force Members

When a member is subjected to *no couples* and forces are applied at only two points on a member, the member is called a *two-force member*. An example of this situation is shown in Fig. 5–15a. The forces at A and B are first summed to obtain their respective *resultants* $\mathbf{F}_A$ and $\mathbf{F}_B$, Fig. 5–15b. These two forces will maintain *translational or force equilibrium* provided $\mathbf{F}_A$ is of equal magnitude and opposite direction to $\mathbf{F}_B$. Furthermore, *rotational or moment equilibrium* is satisfied if $\mathbf{F}_A$ is *collinear* with $\mathbf{F}_B$. As a result, the line of action of both forces is known, since it always passes through A and B. Hence, only the force magnitude must be determined or stated. Other examples of two-force members held in equilibrium are shown in Fig. 5–16. As will be shown throughout Chapter 6, this concept of first identifying a two-force member and then applying equal but opposite collinear forces at its ends will be of great importance to us in simplifying the force analysis of members which are pin-connected.

Three-Force Members

If a member is subjected to three coplanar forces, then it is necessary that the forces be either concurrent or parallel if the member is to be in equilibrium. To show this, consider the body in Fig. 5–17 and suppose that any two of the three forces acting on the body have lines of action that intersect at point O. To satisfy moment equilibrium about O, i.e., $\Sigma M_O = 0$, the third force must also pass through O, which then makes the force system *concurrent*. If two of the three forces are parallel, the point of concurrency, O, is considered to be at "infinity" and the third force must be parallel to the other two forces to intersect at this "point." Since the three forces are concurrent at a point, only the force equilibrium equations ($\Sigma F_x = 0$, $\Sigma F_y = 0$) must be satisfied.

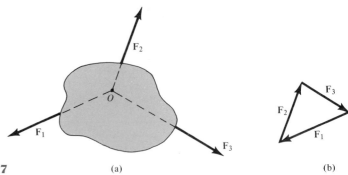

Fig. 5–17

(a)

(b)

Example 5–13

The lever *ABC* is pin-supported at *A* and connected to a short link *BD* as shown in Fig. 5–18*a*. If the weight of the members is negligible, determine the force developed on the lever at *A*.

(a)

SOLUTION

Free-Body Diagrams. As shown by the free-body diagram, Fig. 5–18*b*, the short link *BD* is a *two-force member,* so the *resultant forces* at pins *D* and *B* must be equal, opposite, and collinear. Although the magnitude of the force is unknown, the line of action is known, since it passes through *B* and *D*.

Lever *ABC* is a *three-force member,* and therefore the three nonparallel forces acting on it must be concurrent at *O*, Fig. 5–18*c*. In particular, note that the force **F** on the lever is equal but opposite to **F** acting at *B* on the link. Why? The distance *CO* must be 0.5 m, since the lines of action of **F** and the 400-N force are known.

(b)

Equations of Equilibrium. By requiring the force system to be concurrent at *O*, it is necessary that $\Sigma M_O = 0$. Hence, the angle θ which defines the line of action of $\mathbf{F}_A$ can be determined from trigonometry,

$$\theta = \tan^{-1}\left(\frac{0.7}{0.4}\right) = 60.3° \quad \measuredangle_\theta \qquad \textit{Ans.}$$

Applying the force equilibrium equations yields

$\xrightarrow{+} \Sigma F_x = 0; \qquad F_A \cos 60.3° - F \cos 45° + 400 \text{ N} = 0$

$+\uparrow \Sigma F_y = 0; \qquad F_A \sin 60.3° - F \sin 45° = 0$

Solving, we get

$$F_A = 1075 \text{ N} \qquad \textit{Ans.}$$
$$F = 1320 \text{ N}$$

We can also solve this problem by representing the force at *A* by its two components $\mathbf{A}_x$ and $\mathbf{A}_y$ and applying $\Sigma M_A = 0$, $\Sigma F_x = 0$, $\Sigma F_y = 0$. Once A_x and A_y are determined, how would you find F_A and θ?

(c)

Fig. 5–18

185

PROBLEMS

5–11. Determine the normal forces of the blade on the 50-kg roll of paper in Prob. 5–1.

*5–12. Determine the support reactions on the beam in Prob. 5–2.

5–13. Determine the reactions at the supports for member *ABC* in Prob. 5–3.

5–14. Determine the horizontal and vertical components of reaction at the pin *A* and the force in cable *CB* of the crane boom in Prob. 5–4.

5–15. Determine the reactions at the pin *A* and the force of the hydraulic cylinder on the truck dumpster in Prob. 5–6.

*5–16. Determine the reactions at the supports *A*, *B*, and *C* of member *ABC* in Prob. 5–8.

5–17. Determine the reactions at the supports *A* and *B* of member *AB* in Prob. 5–9.

5–18. Determine the handle force **T** on the wagon in Prob. 5–10, and the reactions on the wheels at *A* and *B*.

5–19. The crane is used to support a load of 5500 lb. If *B* exerts only a horizontal force on the crane, whereas *A* exerts both horizontal and vertical forces, determine these reactions for equilibrium.

Prob. 5–19

*5–20. Determine the reactions at the fixed support *A* for equilibrium of the beam.

Prob. 5–20

5–21. The link is subjected to the loadings shown. Determine the force **F**_B in the cable at *B* and the magnitude of the resultant force acting on the pin at *A*. Set $F_C = 230$ N.

Probs. 5–21/5–22

5–22. Determine the forces **F**_B and **F**_C which can be applied to the link so that the magnitude of the resultant force acting at the pin *A* is 850 N.

5-23. The man is pulling a load of 8 lb with one arm held as shown. Determine the force $\mathbf{F}_H$ this exerts on the ulna bone H, and the tension developed in the biceps muscle B. Neglect the weight of the man's arm.

8 lb

13 in.

$\mathbf{T}_B$

B

1.75 in.

H

$\mathbf{F}_H$

Prob. 5-23

*5-24. Determine the reactions at the supports A and B for equilibrium of the beam.

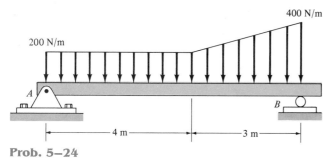

400 N/m

200 N/m

A

B

4 m

3 m

Prob. 5-24

5-25. The power pole supports the three lines, each line exerting a vertical force on the pole due to its weight as shown. Determine the reactions at the fixed support D. If it is possible for wind or ice to snap the lines, determine which line(s) when removed create(s) a condition for the greatest moment reaction at D. What is this moment? Neglect the weight of accumulated ice or wind on the members.

3 ft

400 lb

800 lb

450 lb

2 ft

4 ft

D

Prob. 5-25

5-26. The weight W and location d of the center of gravity G of the "light" airplane can be obtained by placing a scale under the wheels at A and B. If the reading at A is 560 lb and *each* of the *two* front wheels at B is 1430 lb, determine W and d.

N6002U

G

A

B

d

8.5 ft

Prob. 5-26

5–27. The bulk head AD is subjected to both water and soil-backfill pressures. Assuming AD is "pinned" to the ground at A, determine the horizontal and vertical reactions there and also the required tension in the ground anchor BC necessary for equilibrium. The bulk head has a mass of 800 kg.

Prob. 5–27

5–29. The ramp of a ship has a weight of 200 lb and a center of gravity at G. Determine the cable force in CD needed to just start lifting the ramp, i.e., so the reaction at B becomes zero. Also, determine the horizontal and vertical components of force at the hinge (pin) at A.

Prob. 5–29

*5–28. The boom supports the two vertical loads. Neglect the size of the collars at D and B and the thickness of the boom, and compute the horizontal and vertical components of force at the pin A and the force in cable CB.

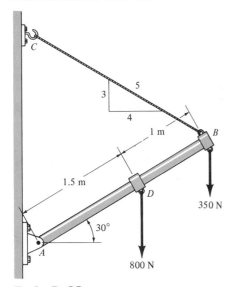

Prob. 5–28

5–30. Determine the reactions at the roller B, the rocker C, and where the beam contacts the smooth plane at A. Neglect the thickness of the beam.

Prob. 5–30

5–31. The mechanism shown was thought by its inventor to be a perpetual-motion machine. It consists of the stand A, two smooth idler wheels B and C, and in between a uniform hollow cylindrical ring D suspended in the manner shown. The ring has a weight W and it was expected to revolve in the direction indicated by the arrow. Draw a free-body diagram of the ring and using an appropriate equation of equilibrium show that it will not work.

Prob. 5–31

*5–32. The automobile is being towed at a constant velocity up the incline using the cable at C. The automobile has a mass of 5 Mg and center of mass at G. The tires are free to roll so that rolling resistance can be neglected. Determine the force $\mathbf{T}$ in the towing cable at C and the reaction on *each* of the four tires.

Prob. 5–32

5–33. Determine the magnitude of force at the pin A and in the cable BC needed to support the 500-lb load. Neglect the weight of the boom AB.

Prob. 5–33

5–34. The uniform pipe has a mass of 100 kg and a center of mass at G. Determine the reactions at the smooth supports A, B, and C.

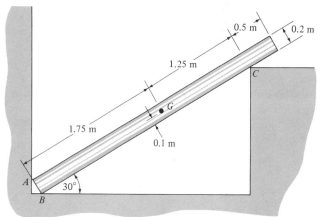

Prob. 5–34

189

5–35. The uniform bar has a weight W. Determine the weight W' which must be suspended at the end of the cord for equilibrium. Neglect the size of the pulley at C.

Prob. 5–35

*5–36. The telephone pole of negligible thickness is subjected to the cable force of 80 lb directed as shown. It is supported by the cable BCD and the pole can be assumed pinned at its base A. In order to provide clearance for a sidewalk right of way, a strut CE is attached to the pole at its midheight $h = 15$ ft and two cables, BC and CD', are attached to the strut at C, as shown by the dashed line (cable CD is removed). Determine the tension in cable CD' and the resultant force at A.

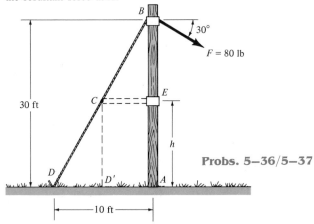

Probs. 5–36/5–37

5–37. The telephone pole of negligible thickness is subjected to the cable force of 80 lb directed as shown. It is supported by the cable BCD and can be assumed pinned at its base A. In order to provide clearance for a sidewalk right of way, a strut CE is attached to the pole and two cables, BC and CD', are attached to the strut at C, as shown by the dashed lines (cable CD is removed). If the tension in CD' is to be twice the tension in BCD, determine the height h for placement of the strut CE.

5–38. Determine the tension in the supporting cable BC and the horizontal and vertical components of reaction at A necessary to support the bucket of weight W. Express the results in terms of W and the angles θ and ϕ.

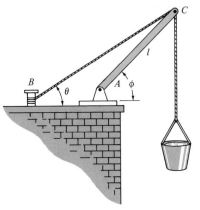

Prob. 5–38

5–39. The gantry $ABCD$ has a mass of 120 kg and center of mass at G_g, and the *uniform* adjustable cantilever beam EF has a mass of 80 kg. Determine the greatest mass m of the safe S the gantry can support without tipping over, i.e., so the wheels at D do not leave the ground. The wheels at A are fixed from rolling and can be assumed pinned.

Prob. 5–39

*5–40. It is easier to mow a lawn by pulling a lawn mower rather than pushing it. To illustrate this, consider the four-wheel mower shown in the figure having a mass of 6 kg and center of mass at G. Determine the magnitude of force **P** required to push it over the smooth obstruction at A and compare the result with pulling it over A. In both cases the force **P** is to be directed $\theta=30°$ along the handle as shown. *Hint:* It is required that the vertical reaction of the *ground* at A be equal to zero. Only the normal force at the obstruction contact needs to be considered.

Prob. 5–40

5–41. The smooth pipe rests against the wall at the points of contact A, B, and C. Determine the reactions at these points needed to support the vertical force of 45 lb. Neglect the pipe's thickness in the calculation.

Prob. 5–41

5–42. A uniform beam having a weight of 300 lb supports a vertical load of 1200 lb. If the ground pressure varies linearly as shown, determine the load intensities w_1 and w_2, measured in lb/ft, necessary for equilibrium.

Prob. 5–42

5–43. Determine the distance d for placement of the load **P** for equilibrium of the smooth bar in the position θ as shown. Neglect the weight of the bar.

Prob. 5–43

*5–44. The thin rod of length l is supported by the smooth tube. Determine the distance a needed for equilibrium if the applied load is **P**.

Prob. 5–44

5–45. The horizontal beam is supported by springs at its ends. Each spring has a stiffness of $k = 5$ kN/m and is originally unstretched so that the beam is in the horizontal position. Determine the angle of tilt of the beam if a load of 800 N is applied at point C as shown.

Probs. 5–45/5–46

5–46. The horizontal beam is supported by springs at its ends. If the stiffness of the spring at A is $k_A = 5$ kN/m, determine the required stiffness of the spring at B so that if the beam is loaded with the 800 N it remains in the horizontal position both before and after loading.

5–47. The uniform rod has a length l and weight W. It is supported at one end A by a smooth wall and the other end by a cord of length s which is attached to the wall as shown. Show that for equilibrium it is required that $h = [(s^2 - l^2)/3]^{1/2}$.

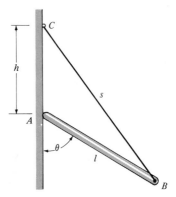

Prob. 5–47

*5–48. The rigid metal strip of negligible weight is used as part of an electromagnetic switch. If the stiffness of the springs at A and B is $k = 5$ N/m, and the strip is originally horizontal when the springs are unstretched, determine the smallest force needed to close the contact gap at C.

Probs. 5–48/5–49

5–49. The rigid metal strip of negligible weight is used as part of an electromagnetic switch. Determine the maximum stiffness k of the springs at A and B so that the contact at C closes when the vertical force developed there is 0.5 N. Originally the strip is horizontal as shown.

5–50. The rigid beam of negligible weight is supported horizontally by two springs and a pin. If the springs are uncompressed when the load is removed, determine the force in each spring when the load **P** is applied. Also, compute the vertical deflection of end *C*. Assume the spring stiffness *k* is large enough so that only small deflections occur.

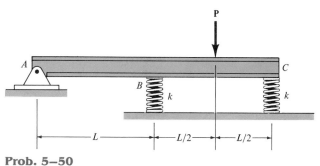

Prob. 5–50

5–51. Three uniform boxes, each having a weight *W* and length *a*, are stacked as shown. Determine the maximum distance *d* the top box can extend out from the bottom one so the stack does not topple over.

Prob. 5–51

■*5–52.** The disk has a mass of 20 kg and is supported on the smooth cylindrical surface by a spring having a stiffness of $k = 400$ N/m and unstretched length of $l_0 = 1$ m. The spring remains in the horizontal position since its end *A* is attached to the small roller guide which has negligible weight. Determine the angle θ to the nearest degree for equilibrium of the roller.

Prob. 5–52

5–53. The uniform beam has a weight *W* and length *l* and is supported by a pin at *A* and a cable *BC*. Determine the horizontal and vertical components of reaction at *A* and the tension in the cable necessary to hold the beam in the position shown.

Prob. 5–53

5–54. The rod BC is supported by two cords, each of length a, which are attached to the pin at A. If weights W and $2W$ are suspended from the ends of the rod, determine the angle θ for equilibrium, measured from the horizontal. Express the answer in terms of a and l. Neglect the weight of the rod.

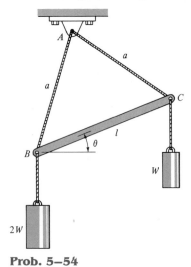

Prob. 5–54

5–55. The uniform rod has a length L and weight W. It rests on the two smooth planes at A and B. Determine the force $\mathbf{P}$ needed to hold it in the equilibrium position shown. The force is applied parallel to the plane AC.

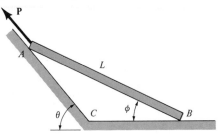

Prob. 5–55

***5–56.** The uniform rod of length L and weight W is supported on the smooth planes. Determine its position θ for equilibrium. Neglect the thickness of the rod.

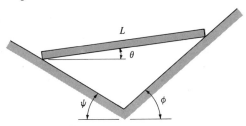

Prob. 5–56

Equilibrium in Three Dimensions

5.5 Free-Body Diagrams

The first step in solving three-dimensional equilibrium problems, as in the case of two dimensions, is to draw a free-body diagram of the body (or group of bodies considered as a system). When drawing this diagram, it is necessary to include *all the reactions* that can occur at each point of support.

Support Reactions

The reactive forces and couples acting at various types of supports and connections, when the members are viewed in three dimensions, are listed in Table 5–2. It is important to recognize the symbols used to represent each of the supports and to clearly understand how the forces and couples are developed by each support. As in the two-dimensional case, *a force is developed by a support that restricts the translation of the attached member, whereas a couple is developed when rotation of the attached member is prevented.* For example, in Table 5–2, the ball-and-socket joint (4) prevents any translation of the connecting member; therefore, a force must act on the member at the point of connection. This force has three components having unknown magnitudes, F_x, F_y, F_z. Provided these components are known, one can obtain the magnitude of force, $F = \sqrt{F_x^2 + F_y^2 + F_z^2}$, and the force's orientation defined by the coordinate direction angles α, β, γ, Eqs. 2–7.* Since the connecting member is allowed to rotate freely about *any* axis, no moment is resisted by a ball-and-socket joint.

It should be noted that the *single* bearing supports (5) and (6), the *single* pin (7), and the *single* hinge (8) can support both force and couple-moment components. Most often, however, these supports are used in conjunction with other bearings, pins, or hinges to hold the body in equilibrium. If this is the case, the *force reactions* at the supports may *alone* be adequate for supporting the body. When this occurs, the couple moments become redundant and may be neglected on the free-body diagram.† The reason for this should become clear after studying the examples which follow, but essentially the couple moments will not be developed at these supports when the rotation of the body is prevented not by the supporting couple moments, but rather by reactions developed at the other supports.

---

*The three unknowns may also be represented as an unknown force magnitude F and two unknown coordinate direction angles. The third direction angle is obtained using the identity $\cos^2 \alpha + \cos^2 \beta + \cos^2 \gamma = 1$, Eq. 2–10.

†It is also necessary that the physical body *maintain* its rigidity when loaded and the supports be *properly aligned* when connected to the body.

Table 5–2 Supports for Rigid Bodies Subjected to Three-Dimensional Force Systems

Types of Connection	Reaction	Number of Unknowns
(1) cable		One unknown. The reaction is a force which acts away from the member in the direction of the cable.
(2) smooth surface support		One unknown. The reaction is a force which acts perpendicular to the surface at the point of contact.
(3) roller on a smooth surface		One unknown. The reaction is a force which acts perpendicular to the surface at the point of contact.
(4) ball and socket		Three unknowns. The reactions are three rectangular force components.
(5) single journal bearing		Four unknowns. The reactions are two force and two couple components which act perpendicular to the shaft.
(6) single thrust bearing		Five unknowns. The reactions are three force and two couple components.

Table 5-2 (Contd.)

Types of Connection	Reaction	Number of Unknowns
(7) single smooth pin		Five unknowns. The reactions are three force and two couple components.
(8) single hinge		Five unknowns. The reactions are three force and two couple components.
(9) fixed support		Six unknowns. The reactions are three force and three couple components.

Free-Body Diagrams

The general procedure used to draw the free-body diagram of a rigid body has been outlined in Sec. 5.2. The importance of this step cannot be overemphasized when solving equilibrium problems. Essentially it requires first "isolating" the body by drawing its outlined shape. This is followed by a careful *labeling* of *all* the applied external forces and couples and the support reactions. To do this, x, y, z axes are established with the origin located at a convenient point. As a general rule, *components of reaction* having an *unknown magnitude* are shown acting on the free-body diagram in the *positive sense*. In this way, if any negative values are obtained, they will indicate that the components act in the negative coordinate direction.

Example 5–14

Several examples of objects along with their associated free-body diagrams are shown in Fig. 5–19. In all cases, the x, y, z axes are established and the unknown reaction components are indicated in the positive sense. The weight of the members is neglected.

Properly aligned journal
bearings at A, B, C

The force reactions at the bearings
are sufficient for force and moment
equilibrium since they prevent
the shaft from rotating about
each of the coordinate axes.

(a)

Pin at A,
cable BC.

Moment components are developed
by the pin on the rod in order to prevent
rotation about the x and z axes.

(b)

Fig. 5–19(a, b)

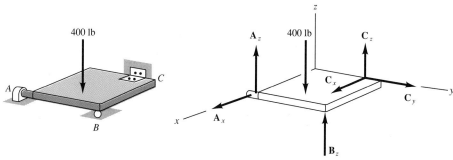

Properly aligned journal bearing at A and hinge at C. Roller at B.

Only force reactions are developed at the bearing and hinge.

(c)

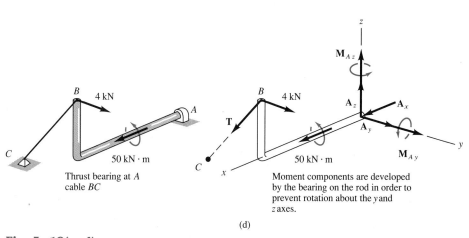

Thrust bearing at A cable BC

Moment components are developed by the bearing on the rod in order to prevent rotation about the y and z axes.

(d)

Fig. 5–19(c, d)

5.6 Equations of Equilibrium

As stated in Sec. 5.1, the conditions for equilibrium of a rigid body subjected to a three-dimensional force system require that both the *resultant* force and *resultant* couple moment acting on the body be equal to *zero*.

Vector Equations of Equilibrium

The two conditions for equilibrium of a rigid body may be expressed mathematically in vector form as

$$\Sigma \mathbf{F} = \mathbf{0}$$
$$\Sigma \mathbf{M}_O = \mathbf{0}$$

(5–5)

where $\Sigma \mathbf{F}$ is the vector sum of all the external forces acting on the body and $\Sigma \mathbf{M}_O$ is the sum of the moments of all these forces about any point O located either on or off the body.

Scalar Equations of Equilibrium

If all the applied external forces and moments are expressed in Cartesian vector form and substituted into Eqs. 5–5, we have

$$\Sigma \mathbf{F} = \Sigma F_x \mathbf{i} + \Sigma F_y \mathbf{j} + \Sigma F_z \mathbf{k} = \mathbf{0}$$
$$\Sigma \mathbf{M}_O = \Sigma M_x \mathbf{i} + \Sigma M_y \mathbf{j} + \Sigma M_z \mathbf{k} = \mathbf{0}$$

Since the $\mathbf{i}, \mathbf{j}$, and $\mathbf{k}$ components are independent from one another, the above equations are satisfied provided

$$\Sigma F_x = 0$$
$$\Sigma F_y = 0$$
$$\Sigma F_z = 0$$

(5–6a)

and

$$\Sigma M_x = 0$$
$$\Sigma M_y = 0$$
$$\Sigma M_z = 0$$

(5–6b)

These *six scalar equilibrium equations* may be used to solve for at most six unknowns shown on the free-body diagram. Equations 5–6a express the fact that the sum of the external force components acting in the x, y, and z directions must be zero, and Eqs. 5–6b require the sum of the moment components about the x, y, and z axes to be zero.

PROCEDURE FOR ANALYSIS

The following procedure provides a method for solving three-dimensional equilibrium problems.

Free-Body Diagram. Construct the free-body diagram for the body. Be sure to include *all* the forces and couples that act *on* the body. These interactions are commonly caused by the externally applied loadings, contact forces exerted by adjacent bodies, support reactions, and the weight of the body if it is significant compared to the magnitudes of the other applied forces. Establish the *x*, *y*, *z* axes with origin located at a convenient point. Orientate the axes so that they are parallel to as many of the external forces and moments as possible. Identify the unknowns, and in general show all the unknown components having a positive sense if the sense cannot be determined. Dimensions of the body, necessary for computing the moments of forces, are also included on the free-body diagram.

Equations of Equilibrium. Apply the equations of equilibrium. In many cases, problems can be solved by *direct application* of the six scalar equations $\Sigma F_x = 0$, $\Sigma F_y = 0$, $\Sigma F_z = 0$, $\Sigma M_x = 0$, $\Sigma M_y = 0$, $\Sigma M_z = 0$, Eqs. 5–6; however, if the force components or moment arms seem difficult to determine, it is recommended that the solution be obtained by using vector equations: $\Sigma \mathbf{F} = \mathbf{0}$, $\Sigma \mathbf{M}_O = \mathbf{0}$, Eqs. 5–5. In any case, it is *not necessary* that the set of axes chosen for force summation *coincide* with the set of axes chosen for moment summation. Instead, it is recommended that one *choose the direction of an axis for moment summation such that it intersects the lines of action of as many unknown forces as possible*. The moments of forces passing through points on this axis or forces which are parallel to the axis will then be zero. Furthermore, *any set of three nonorthogonal axes* may be chosen for either the force or moment summations. By the proper choice of axes, it may be possible to solve directly for an unknown quantity, or at least reduce the need for solving a large number of simultaneous equations for the unknowns.

5.7 Constraints for a Rigid Body

To ensure the equilibrium of a rigid body, it is not only necessary to satisfy the equations of equilibrium, but the body must also be properly held or constrained by its supports. Three situations may occur where the conditions for proper constraint have not been met.

Redundant Constraints

When all the reactive forces on a body can be determined from the equations of equilibrium, the problem is called *statically determinate*. Problems having more unknown forces than equilibrium equations are called *statically indeterminate*. Statical indeterminacy can arise if a body has redundant constraints, i.e., more than are necessary to maintain equilibrium. For example, the two-dimensional problem, Fig. 5–20a, and the three-dimensional problem, Fig. 5–20b, shown together with their free-body diagrams, are both statically indeterminate because of the redundancy in the number of support reactions. In the two-dimensional case, there are five unknowns, that is, M_A, A_x, A_y, B_y, and C_y, for which only three equilibrium equations can be written ($\Sigma F_x = 0$, $\Sigma F_y = 0$, and $\Sigma M_O = 0$, Eqs. 5–2). Two more equations are needed for a complete solution. Consequently, this problem is termed "indeterminate to the second degree" (five unknowns minus three equations). The three-dimensional problem has eight unknowns, for which only six equilibrium equations can be written, Eqs. 5–6. It is also "indeterminate to the second degree." The additional equations needed to solve indeterminate problems of the type shown in Fig. 5–20 are generally obtained from the deformation conditions at the points of support. These equations involve the physical properties of the body which are studied in subjects dealing with the mechanics of deformation, such as "mechanics of materials."

Fig. 5–20

(a)

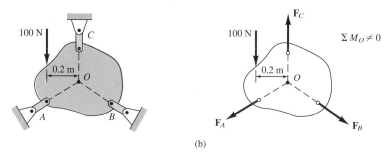

(b)

Fig. 5–21

Improper Constraints

In some cases, there may be as many unknown forces as there are equations of equilibrium; however, *instability* of the body can develop because of *improper constraining* action by the supports. In the case of three-dimensional problems, the body is improperly constrained if the support reactions *all intersect a common axis*. For two-dimensional problems, this axis is *perpendicular* to the plane of the forces and therefore appears as a point. Hence, when all the reactive forces are *concurrent* at this point, the body is improperly constrained. Examples of both cases are given in Fig. 5–21. From the free-body diagrams it is seen that the summation of moments about the x axis, Fig. 5–21a, or point O, Fig. 5–21b, will *not* be equal to zero; thus rotation about the x axis or point O will take place.* Furthermore, in both cases it becomes *impossible* to solve *completely* for all the unknowns, since one can write a moment equation that does not involve any of the unknown support reactions. (This reduces by one the number of available equilibrium equations.)

*For the three-dimensional problem, $\Sigma M_x = (400 \text{ N})(0.6 \text{ m}) \neq 0$, and for the two-dimensional problem, $\Sigma M_O = (100 \text{ N})(0.2 \text{ m}) \neq 0$.

Fig. 5–22

(a)

(b)

(a)

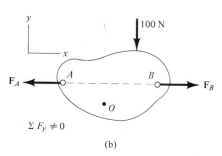

(b)

Fig. 5–23

Another way in which improper constraining leads to instability occurs when the *reactive forces* are all *parallel*. Three- and two-dimensional examples of this are shown in Fig. 5–22. In both cases, the summation of forces along the x axis will not equal zero.

In some cases, a body may have *fewer* reactive forces than equations of equilibrium that must be satisfied. The body then becomes only *partially constrained*. For example, consider the body shown in Fig. 5–23a with its corresponding free-body diagram in Fig. 5–23b. If O is a point not located on the line AB, the equations $\Sigma F_x = 0$ and $\Sigma M_O = 0$ will be satisfied by proper choice of the reactions $\mathbf{F}_A$ and $\mathbf{F}_B$. The equation $\Sigma F_y = 0$, however, will not be satisfied for the loading conditions and therefore equilibrium will not be maintained.

Proper constraining therefore requires that (1) the lines of action of the reactive forces do not intersect points on a common axis, and (2) the reactive forces must not all be parallel to one another. When the number of reactive forces needed to properly constrain the body in question is a *minimum*, the problem will be statically determinate and the equations of equilibrium can be used to determine *all* the reactive forces.

Example 5–15

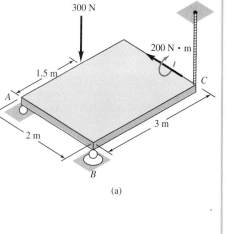

(a)

The homogeneous plate shown in Fig. 5–24a has a mass of 100 kg and is subjected to a force and couple moment along its edges. If it is supported in the horizontal plane by means of a roller at A, a ball-and-socket joint at B, and a cord at C, determine the components of reaction at the supports.

SOLUTION (SCALAR ANALYSIS)

Free-Body Diagram. Can you identify each of the forces shown on the free-body diagram of the plate, Fig. 5–24b?

Equations of Equilibrium. Since the three-dimensional geometry is rather simple, a *scalar analysis* provides a *direct solution* to this problem. A force summation yields

$\Sigma F_x = 0;$ $B_x = 0$ *Ans.*

$\Sigma F_y = 0;$ $B_y = 0$ *Ans.*

$\Sigma F_z = 0;$ $A_z + B_z + T_C - 300 \text{ N} - 981 \text{ N} = 0$ (1)

Recall that the moment of a force about an axis is equal to the product of the force magnitude and the perpendicular distance (moment arm) from the line of action of the force to the axis. The sense of the moment is determined by the right-hand rule. Hence, summing moments of the forces on the free-body diagram, with positive moments acting along the positive x or y axis, we have

(b)

Fig. 5–24

$\Sigma M_x = 0;$ $T_C(2 \text{ m}) - 981 \text{ N}(1 \text{ m}) + B_z(2 \text{ m}) = 0$ (2)

$\Sigma M_y = 0;$

$300 \text{ N}(1.5 \text{ m}) + 981 \text{ N}(1.5 \text{ m}) - B_z(3 \text{ m}) - A_z(3 \text{ m}) - 200 \text{ N} \cdot \text{m} = 0$ (3)

The components of force at B can be eliminated if the x', y', z' axes are used. We obtain

$\Sigma M_{x'} = 0;$ $981 \text{ N}(1 \text{ m}) + 300 \text{ N}(2 \text{ m}) - A_z(2 \text{ m}) = 0$ (4)

$\Sigma M_{y'} = 0;$

$-300 \text{ N}(1.5 \text{ m}) - 981 \text{ N}(1.5 \text{ m}) - 200 \text{ N} \cdot \text{m} + T_C(3 \text{ m}) = 0$ (5)

Solving Eqs. (1) to (3) or the more convenient Eqs. (1), (4), and (5) yields

     $A_z = 790 \text{ N}$ $B_z = -217 \text{ N}$ $T_C = 707 \text{ N}$ *Ans.*

The negative sign indicates that $\mathbf{B}_z$ acts downward.

Example 5-16

The windlass shown in Fig. 5–25a is supported by two smooth journal bearings A and B which are properly aligned on the shaft. Determine the magnitude of the vertical force **P** that must be applied to the handle to maintain equilibrium of the 100-kg crate. Also calculate the reactions at the bearings.

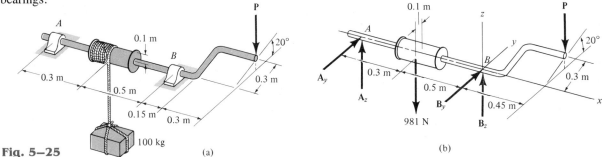

Fig. 5–25

(a) (b)

SOLUTION (SCALAR ANALYSIS)

Free-Body Diagram. Since the bearings at A and B are aligned correctly, *only* force reactions occur at these supports, Fig. 5–25b. Why are there no moment reactions?

Equations of Equilibrium. Summing moments about the x axis yields a direct solution for **P**. Why? For a scalar moment summation, it is necessary to compute the moment of each force as the product of the force magnitude and the *perpendicular distance* from the x axis to the line of action of the force. Using the right-hand rule and assuming positive moments act in the $+\mathbf{i}$ direction, we have

$$\Sigma M_x = 0; \qquad 981\ \text{N}(0.1\ \text{m}) - P(0.3\cos 20°\ \text{m}) = 0$$
$$P = 348.0\ \text{N} \qquad\qquad Ans.$$

Using this result and summing moments about the y and z axes yields

$$\Sigma M_y = 0;\ -981\ \text{N}(0.5\ \text{m}) + A_z(0.8\ \text{m}) + (348.0\ \text{N})(0.45\ \text{m}) = 0$$
$$A_z = 417.4\ \text{N} \qquad\qquad Ans.$$
$$\Sigma M_z = 0; \qquad\qquad -A_y(0.8\ \text{m}) = 0 \qquad A_y = 0 \qquad\qquad Ans.$$

The reactions at B are obtained by a force summation, using the results computed above.

$$\Sigma F_y = 0; \qquad\qquad 0 + B_y = 0 \qquad B_y = 0 \qquad\qquad Ans.$$
$$\Sigma F_z = 0; \quad 417.4 - 981 + B_z - 348.0 = 0 \qquad B_z = 911.6\ \text{N} \qquad Ans.$$

As shown on the free-body diagram, the *supports* do not provide resistance against translation in the x direction. Hence, the windlass is only partially constrained.

Example 5–17

Determine the tension in cables *BC* and *BD* and the reactions at the ball-and-socket joint *A* for the mast shown in Fig. 5–26a.

SOLUTION (*VECTOR ANALYSIS*)

Free-Body Diagram. There are five unknown force magnitudes shown on the free-body diagram, Fig. 5–26b.

Equations of Equilibrium. Expressing each force in Cartesian vector form, we have

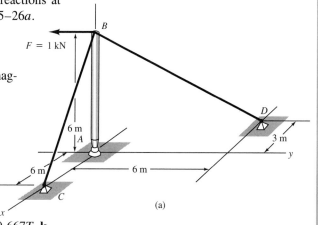

$$\mathbf{F} = \{-1000\mathbf{j}\} \text{ N}$$
$$\mathbf{F}_A = A_x\mathbf{i} + A_y\mathbf{j} + A_z\mathbf{k}$$
$$\mathbf{T}_C = 0.707T_C\mathbf{i} - 0.707T_C\mathbf{k}$$
$$\mathbf{T}_D = T_D\left(\frac{\mathbf{r}_{BD}}{r_{BD}}\right) = -0.333T_D\mathbf{i} + 0.667T_D\mathbf{j} - 0.667T_D\mathbf{k}$$

Applying the force equation of equilibrium gives

$$\Sigma\mathbf{F} = \mathbf{0}; \qquad \mathbf{F} + \mathbf{F}_A + \mathbf{T}_C + \mathbf{T}_D = \mathbf{0}$$
$$(A_x + 0.707T_C - 0.333T_D)\mathbf{i} + (-1000 + A_y + 0.667T_D)\mathbf{j}$$
$$+ (A_z - 0.707T_C - 0.667T_D)\mathbf{k} = \mathbf{0}$$

$$\Sigma F_x = 0; \qquad A_x + 0.707T_C - 0.333T_D = 0 \qquad (1)$$
$$\Sigma F_y = 0; \qquad A_y + 0.667T_D - 1000 = 0 \qquad (2)$$
$$\Sigma F_z = 0; \qquad A_z - 0.707T_C - 0.667T_D = 0 \qquad (3)$$

Summing moments about point *A*, we have

$$\Sigma\mathbf{M}_A = \mathbf{0}; \qquad \mathbf{r}_B \times (\mathbf{F} + \mathbf{T}_C + \mathbf{T}_D) = \mathbf{0}$$
$$6\mathbf{k} \times (-1000\mathbf{j} + 0.707T_C\mathbf{i} - 0.707T_C\mathbf{k}$$
$$- 0.333T_D\mathbf{i} + 0.667T_D\mathbf{j} - 0.667T_D\mathbf{k}) = \mathbf{0}$$

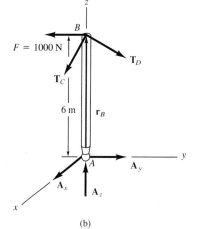

Evaluating the cross product and combining terms yields

$$(-4T_D + 6000)\mathbf{i} + (4.24T_C - 2T_D)\mathbf{j} = \mathbf{0}$$

$$\Sigma M_x = 0; \qquad -4T_D + 6000 = 0 \qquad (4)$$
$$\Sigma M_y = 0; \qquad 4.24T_C - 2T_D = 0 \qquad (5)$$

Fig. 5–26

The moment equation about the *z* axis, $\Sigma M_z = 0$, is automatically satisfied. Why? Solving Eqs. (1) to (5) we have

$$T_C = 707 \text{ N} \qquad T_D = 1500 \text{ N} \qquad\qquad Ans.$$
$$A_x = 0.0 \text{ N} \qquad A_y = 0.0 \text{ N} \qquad A_z = 1500 \text{ N} \qquad Ans.$$

Since the mast is a two-force member, note that the values of $A_x = A_y = 0.0$ could have been determined *by inspection*.

Example 5–18

Rod *AB* shown in Fig. 5–27*a* is used to support the 200-N force. Determine the reactions at the ball-and-socket joint *A* and at the smooth collar *B*.

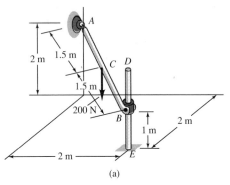

(a)

SOLUTION (*VECTOR ANALYSIS*)

Free-Body Diagram. Here we have considered the rod and attached collar together as shown in Fig. 5–27*b*.

Equations of Equilibrium. Representing each force on the free-body diagram in Cartesian vector form, we have

$$\mathbf{F}_A = A_x\mathbf{i} + A_y\mathbf{j} + A_z\mathbf{k}$$
$$\mathbf{F}_B = B_x\mathbf{i} + B_y\mathbf{j}$$
$$\mathbf{F} = \{-200\mathbf{k}\} \text{ N}$$

Applying the force equation of equilibrium,

$\Sigma\mathbf{F} = 0;$ $\mathbf{F}_A + \mathbf{F}_B + \mathbf{F} = 0$

$$(A_x + B_x)\mathbf{i} + (A_y + B_y)\mathbf{j} + (A_z - 200)\mathbf{k} = 0$$

$\Sigma F_x = 0;$ $A_x + B_x = 0$ (1)

$\Sigma F_y = 0;$ $A_y + B_y = 0$ (2)

$\Sigma F_z = 0;$ $A_z - 200 = 0$ (3)

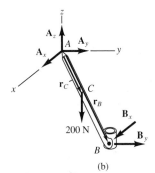

(b)

Fig. 5–27

Summing moments about point *A* yields

$\Sigma\mathbf{M}_A = 0;$ $(\mathbf{r}_C \times \mathbf{F}) + (\mathbf{r}_B \times \mathbf{F}_B) = 0$

Since $\mathbf{r}_C = \frac{1}{2}\mathbf{r}_B$, then

$$(1\mathbf{i} + 1\mathbf{j} - 0.5\mathbf{k}) \times (-200\mathbf{k}) + (2\mathbf{i} + 2\mathbf{j} - 1\mathbf{k}) \times (B_x\mathbf{i} + B_y\mathbf{j}) = 0$$

Expanding and rearranging terms gives

$$(B_y - 200)\mathbf{i} + (-B_x + 200)\mathbf{j} + (2B_y - 2B_x)\mathbf{k} = 0$$

$\Sigma M_x = 0;$ $B_y - 200 = 0$ (4)

$\Sigma M_y = 0;$ $-B_x + 200 = 0$ (5)

$\Sigma M_z = 0;$ $2B_y - 2B_x = 0$ (6)

Solving Eqs. (1) to (6), we get

$$A_x = A_y = -200 \text{ N} \qquad \textit{Ans.}$$
$$A_z = B_x = B_y = 200 \text{ N} \qquad \textit{Ans.}$$

The negative sign indicates that $\mathbf{A}_x$ and $\mathbf{A}_y$ have a sense which is opposite to that shown on the free-body diagram, Fig. 5–27*b*.

Example 5–19

The bent rod in Fig. 5–28a is supported at A by a journal bearing, at D by a ball-and-socket joint, and at B by means of cable BC. Using only *one equilibrium equation*, obtain a direct solution for the tension in cable BC. The bearing at A is capable of exerting force components only in the z and y directions, since it is properly aligned on the shaft.

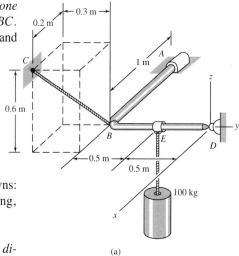

(a)

SOLUTION (*VECTOR ANALYSIS*)
Free-Body Diagram. As shown in Fig. 5–28b, there are six unknowns: the three force components at the ball-and-socket joint, two at the bearing, and the tension force in the cable.

Equations of Equilibrium. The cable tension T_B may be obtained *directly* by summing moments about an axis passing through points D and A. Why? The direction of the axis is defined by the unit vector **u**, where

$$\mathbf{u} = \frac{\mathbf{r}_{DA}}{r_{DA}} = -\frac{1}{\sqrt{2}}\mathbf{i} - \frac{1}{\sqrt{2}}\mathbf{j}$$
$$= -0.707\mathbf{i} - 0.707\mathbf{j}$$

Hence, the sum of the moments about this axis is zero provided

$$\Sigma M_{DA} = \mathbf{u} \cdot \Sigma(\mathbf{r} \times \mathbf{F}) = 0$$

Here **r** represents a position vector drawn from *any point* on the axis DA to any point on the line of action of force **F** (see Eq. 4–9). With reference to Fig. 5–28b, we can therefore write

$$\mathbf{u} \cdot (\mathbf{r}_B \times \mathbf{T}_B + \mathbf{r}_E \times \mathbf{W}) = 0$$

$$(-0.707\mathbf{i} - 0.707\mathbf{j}) \cdot [(-1\mathbf{j}) \times \left(\frac{0.2}{0.7}T_B\mathbf{i} - \frac{0.3}{0.7}T_B\mathbf{j} + \frac{0.6}{0.7}T_B\mathbf{k}\right)$$
$$+ (-0.5\mathbf{j}) \times (-981\mathbf{k})] = 0$$

$$(-0.707\mathbf{i} - 0.707\mathbf{j}) \cdot [(-0.857T_B + 490.5)\mathbf{i} + 0.286T_B\mathbf{k}] = 0$$

$$-0.707(-0.857T_B + 490.5) + 0 + 0 = 0$$

$$T_B = \frac{490.5}{0.857} = 572 \text{ N} \qquad\qquad Ans.$$

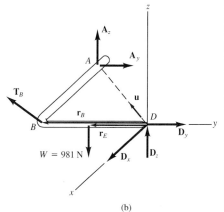

(b)

Fig. 5–28

The advantage of using Cartesian vectors for this solution should be noted. It would be especially tedious to determine the perpendicular distance from the DA axis to the line of action of $\mathbf{T}_B$ using scalar methods.

Note: In Example 5–17, a direct solution for A_z is possible by summing moments about an axis passing through the supports at C and D, Fig. 5–26a. If this is done only the moment of **F** and A_z must be considered. Try and apply the above technique and determine the result $A_z = 1500$ N.

PROBLEMS

5–57. The wrench is used to tighten the bolt at A. If the force $F = 6$ lb is applied to the handle as shown, determine the magnitudes of the resultant force and moment that the bolt head exerts on the wrench. The force $\mathbf{F}$ is in a plane parallel to the x-z plane.

Prob. 5–57

5–58. The air-conditioning unit is hoisted to the roof of a building using the three cables. If the unit has a weight of 900 lb and a center of gravity G, located at $x = 5.5$ ft, $y = 6$ ft, determine the tension in each of the cables for equilibrium.

Probs. 5–58/5–59

5–59. The air-conditioning unit is hoisted to the roof of a building using the three cables. If the tensions in the cables are $T_A = 250$ lb, $T_B = 300$ lb, and $T_C = 200$ lb, determine the weight of the unit and the location (x, y) of its center of gravity G.

*5–60. The platform has a mass of 3 Mg and center of mass located at G. If it is lifted using the three cables, determine the force in each of these cables.

Probs. 5–60/5–61

5–61. The platform has a mass of 2 Mg and center of mass located at G. If it is lifted using the three cables, determine the force in each of these cables. Solve for each force by using a single moment equation of equilibrium.

5–62. The uniform rod AB has a mass of 5 kg and is supported by a ball-and-socket joint at A, a cord BC, and a smooth wall at B. Determine the x, y, z components of reaction at the supports.

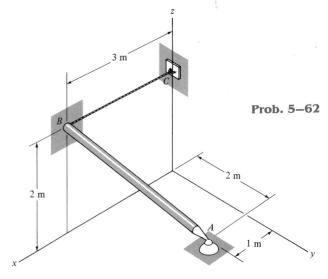

Prob. 5–62

5–63. The shaft is supported by a thrust bearing at A, a cable at B, and a journal bearing at C. Determine the x, y, z components of reaction at these supports needed to hold the shaft in equilibrium. The bearings are properly aligned on the shaft.

Prob. 5–63

■*5–64. The power line is subjected to a tension of 1700 lb. If the insulator AB weighs 50 lb, with center of gravity at G, determine the angle θ it makes with the pole.

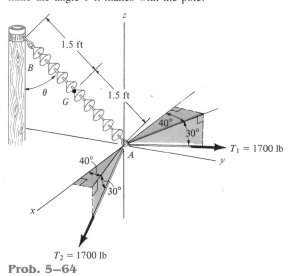

Prob. 5–64

5–65. The hatch door has a weight of 80 lb and center of gravity at G. If the force $\mathbf{F}$ applied to the handle at C has coordinate direction angles of $\alpha=60°$, $\beta=45°$, and $\gamma=60°$, determine the magnitude of $\mathbf{F}$ needed to hold the door slightly open as shown. The hinges are in proper alignment and exert only force reactions on the door. Determine the x, y, z components of these reactions if A exerts only x and z components of force and B exerts x, y, z force components.

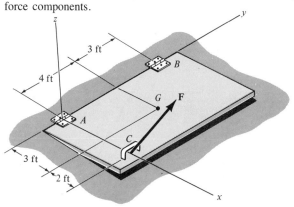

Prob. 5–65

5–66. The hatch door has a weight of 80 lb and center of gravity at G. If the force $\mathbf{F}$ applied to the handle at C has coordinate direction angles of $\alpha=60°$, $\beta=45°$, and $\gamma=60°$, determine the magnitude of $\mathbf{F}$ needed to hold the door slightly open as shown. If the hinge at A becomes loose from its attachment and is ineffective, what are the x, y, z components of reaction at hinge B?

5–67. Determine the forces in the cables and the x, y, z components of reaction acting on the smooth collar at A necessary to hold the 50-lb sign in equilibrium. The center of gravity for the sign is at G.

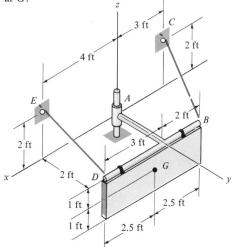

Prob. 5–67

*5–68. Member *AB* is supported by a cable *BC* and at *A* by a smooth fixed *square* rod which fits loosely through the square hole of the collar. If $\mathbf{F} = \{20\mathbf{i} - 40\mathbf{j} - 75\mathbf{k}\}$ lb, determine the *x*, *y*, *z* components of reaction at *A* and the tension in the cable.

Probs. 5–68/5–69

5–69. Member *AB* is supported by a cable *BC* and at *A* by a smooth fixed *square* rod which fits loosely through the square hole of the collar. Determine the tension in the cable *BC* if the force $\mathbf{F} = -\{45\mathbf{k}\}$ lb.

5–70. The member is supported by cable *BC* and at *A* by a smooth fixed *square* rod which fits loosely through the square hole of the collar. Determine the *x*, *y*, *z* components of reaction at *A* and the tension in the cable needed to hold the member in equilibrium.

Prob. 5–70

5–71. Determine the *x*, *y*, *z* components of reaction at the pin *A* and the tension in the cable *BC* necessary for equilibrium of the rod.

Prob. 5–71

212

*5–72. The circular door has a weight of 55 lb and a center of gravity located at G. Determine the x, y, z components of reaction at the hinge A and the force acting along the strut CB needed to hold the door in equilibrium. Set $\theta = 45°$.

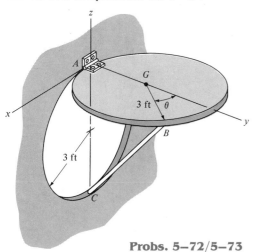

Probs. 5–72/5–73

5–73. The circular door has a weight of 55 lb and a center of gravity located at G. Determine the x, y, z components of reaction at the hinge A and the force acting along the strut CB needed to hold the door in equilibrium. Set $\theta = 90°$.

5–74. The circular plate has a weight W and center of gravity at its center. If it is supported by three vertical cords tied to its edge, determine the largest distance d from the center to where any vertical force P can be applied so as not to cause the force in any one of the cables to become zero.

Probs. 5–74/5–75

5–75. Solve Prob. 5–74 if the plate's weight W is neglected.

*5–76. The member is supported by cable BC and at A by a pin. If $\mathbf{F} = \{50\mathbf{i} + 80\mathbf{j} - 125\mathbf{k}\}$ lb and $\mathbf{M} = \{-400\mathbf{i} + 300\mathbf{j} + 500\mathbf{k}\}$ lb · ft, determine the x, y, z components of reaction at A and the tension in the cable.

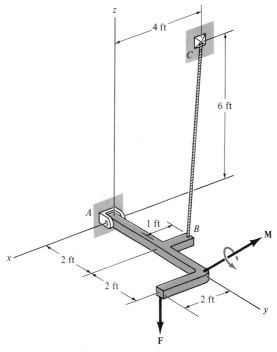

Probs. 5–76/5–77

5–77. The member is supported by cable BC and at A by a pin. If the tension in the cable is to be $T_{BC} = 250$ lb and $\mathbf{M} = \{-400\mathbf{i} + 300\mathbf{j} + 500\mathbf{k}\}$ lb · ft, determine the magnitude of $\mathbf{F}$ if its coordinate direction angles are $\alpha = 60°$, $\beta = 60°$, $\gamma = 135°$. Also, compute the x, y, z components of reaction at A.

5–78. Determine the x and z components of reaction at the journal bearing A and the tension in cords BC and BD necessary for equilibrium of the rod.

Prob. 5–78

5–79. Both pulleys are fixed to the shaft and as the shaft turns with constant angular velocity, the power of pulley A is transmitted to pulley B. Determine the horizontal tension $\mathbf{T}$ in the belt on pulley B and the x, y, z components of reaction at the journal bearing C and thrust bearing D if $\theta = 0°$. The bearings are in proper alignment and exert only force reactions on the shaft.

Probs. 5–79/5–80

*5–80. Both pulleys are fixed to the shaft and as the shaft turns with constant angular velocity, the power of pulley A is transmitted to pulley B. Determine the horizontal tension $\mathbf{T}$ in the belt on pulley B and the x, y, z components of reaction at the journal bearing C and thrust bearing D if $\theta = 45°$. The bearings are in proper alignment and exert only force reactions on the shaft.

5–81. The member is braced by two cables AB and CD. If the attachment at the foundation E can be approximated by a ball-and-socket joint, determine the x, y, z components of force at E and the tension in each of the cables when the member is subjected to the force $\mathbf{F}$.

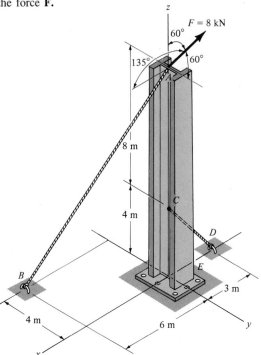

Prob. 5–81

5–82. The boom supports the force of $\mathbf{F} = \{-8\mathbf{k}\}$ kN. Determine the x, y, z components of reaction at the ball-and-socket joint A and the tension in the two cables BC and BD. Set $x = 2$ m.

*5–84. The boom supports a load having a weight of $W = 850$ lb. Determine the x, y, z components of reaction at the ball-and-socket joint A and the tension in cables BC and DE.

Probs. 5–82/5–83

Probs. 5–84/5–85

5–83. The boom supports the force of $\mathbf{F} = \{-8\mathbf{k}\}$ kN. If cable BC or BD can support a maximum tension of 4 kN before it breaks, determine the greatest distance x the force can be positioned along the boom. Also, determine the x, y, z components of reaction at the ball-and-socket joint A.

5–85. Cable BC or DE can support a maximum tension of 900 lb before it breaks. Determine the greatest weight W that can be suspended from the end of the boom. Also, determine the x, y, z components of reaction at the ball-and-socket joint A.

215

5–86. The bent rod is supported at A, B, and C by smooth journal bearings. Compute the x, y, z components of reaction at the bearings if the rod is subjected to forces $F_1 = 300$ lb and $F_2 = 250$ lb. F_1 lies in the y-z plane. The bearings are in proper alignment and exert only force reactions on the rod.

Probs. 5–86/5–87

5–87. The bent rod is supported at A, B, and C by smooth journal bearings. Determine the magnitude of $\mathbf{F}_2$ which will cause the reaction $\mathbf{C}_y$ at the bearing C to be equal to zero. The bearings are in proper alignment and exert only force reactions on the rod $F_1 = 300$ lb.

*5–88. The bent rod is supported at A, B, and C by smooth journal bearings. Compute the x, y, z components of reaction at the bearings if the rod is subjected to a force of $\mathbf{F} = \{-500\mathbf{i} + 400\mathbf{j} + 200\mathbf{k}\}$ N and couple moment of $\mathbf{M} = \{600\mathbf{j}\}$ N · m. The bearings are in proper alignment and exert only force reactions on the rod.

Prob. 5–88

5–89. Rod AB is supported by a ball-and-socket joint at A and a cable at B. Also, the rod rests against the smooth wall at B and against the thin smooth rod at C. Determine the x, y, z components of reaction at these supports if the rod is subjected to a 50 lb vertical force as shown.

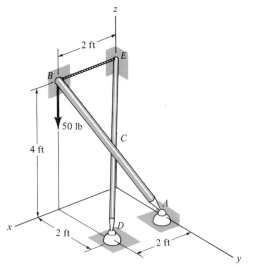

Prob. 5–89

REVIEW PROBLEMS

5–90. The lawn roller has a weight of 200 lb and is to be lifted over the 0.2-ft-high step. Determine the magnitude of the minimum force **P** required to (a) push it and (b) pull it over the step if in each case the force is directed at $\theta = 30°$ along the linkage AB as shown.

Prob. 5–90

5–91. The boom AC is supported at A by a ball-and-socket joint and by two cables BDC and CE. Cable BDC is continuous and passes over a frictionless pulley at D. Calculate the tension in the cables and the x, y, z components of reaction at A if a crate, having a weight of 80 lb, is suspended from the boom.

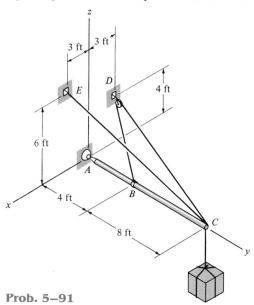

Prob. 5–91

*5–92. Determine the force reactions acting along the short links at joints A, B, D, and F of the space truss.

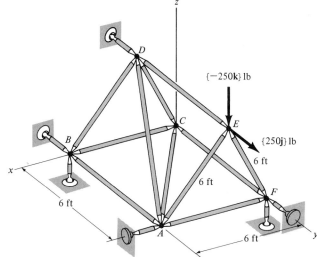

Prob. 5–92

5–93. The uniform concrete slab has a weight of 5500 lb. Determine the tension in each of the three parallel supporting cables when the slab is held in the horizontal plane as shown.

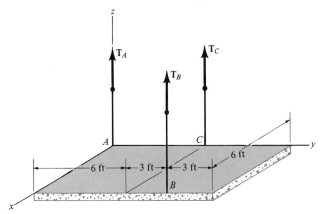

Prob. 5–93

217

5–94. The uniform ladder, having a length of 18 ft, rests along the wall of a building at A and on the roof at B. If the ladder has a weight of 25 lb and the surfaces at A and B are smooth, determine the angle θ for equilibrium.

Prob. 5–94

5–95. The pole for a power line is subjected to the two cable forces of 60 lb, each force lying in a plane parallel to the x-y plane. If the tension in the guy wire AB is 80 lb, determine the x, y, z components of reaction at the base of the pole, O.

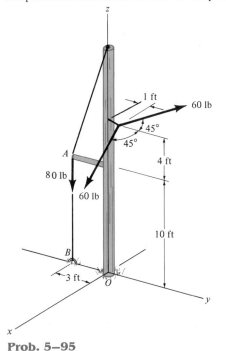

Prob. 5–95

*5–96. If the spring is unstretched when $\theta=0°$, determine the weight of the uniform bar AB if it hangs in the equilibrium position at $\theta=30°$.

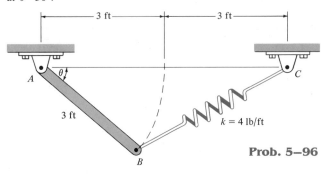

Prob. 5–96

5–97. Compute the horizontal and vertical components of force at pin B. The belt is subjected to a tension of $T = 100$ N and passes over each of the three pulleys.

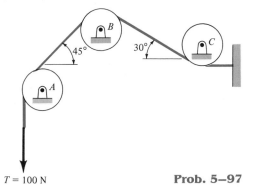

$T = 100$ N   **Prob. 5–97**

5–98. The 300-kg beam is hoisted by two cables at A and B. If the cables are directed as shown when the beam is supported horizontally, determine the location $\bar{x}$ of the center of mass G of the beam and compute the tension, T_1 and T_2, in each cable.

T_1 **Prob. 5–98** T_2

6

Structural Analysis

In this chapter we will use the equations of equilibrium to analyze structures composed of pin-connected members. The analysis is based upon the principle that if a structure is in equilibrium, then each of its members is also in equilibrium. By applying the equations of equilibrium to the various parts of a simple truss, frame, or machine, we will be able to determine all the forces acting at the connections. Later, in Chapter 7, we extend this analysis and show how to determine the internal forces in a member.

The topics in this chapter are very important since they provide practice in drawing free-body diagrams, using the principle of action, equal but opposite collinear force reaction, and applying the equations of equilibrium.

6.1 Simple Trusses

A *truss* is a structure composed of slender members joined together at their end points. The members commonly used in construction consist of wooden struts or metal bars. The joint connections are usually formed by bolting or welding the ends of the members to a common plate, called a *gusset plate*, as shown in Fig. 6–1, or by simply passing a large bolt or pin through each of the members.

Fig. 6–1

Truss joint

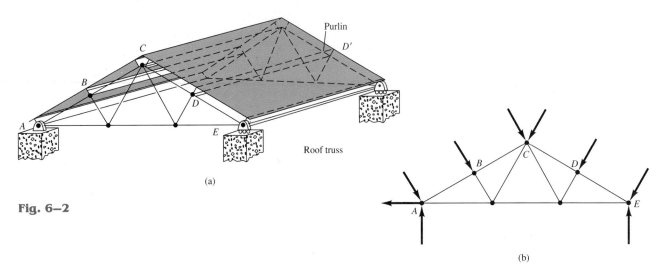

Purlin

Roof truss

(a)

Fig. 6—2

(b)

Planar Trusses

Planar trusses lie in a single plane and are often used to support roofs and bridges. The truss *ABCDE,* shown in Fig. 6–2a, is an example of a typical roof-supporting truss. In this figure, the roof load is transmitted to the truss *at the joints* by means of a series of *purlins,* such as *DD′*. Since the imposed loading acts in the same plane as the truss, Fig. 6–2b, the analysis of the forces developed in the truss members is two-dimensional.

In the case of a bridge, such as shown in Fig. 6–3a, the load on the deck is first transmitted to *stringers,* then to *floor beams,* and finally to the *joints B, C,* and *D* of the two supporting side trusses. Like the roof truss, the bridge truss loading is also coplanar, Fig. 6–3b.

When bridge or roof trusses extend over large distances, a rocker or roller is commonly used for supporting one end, joint *E* in Figs. 6–2a and 6–3a. This type of support allows freedom for expansion or contraction due to temperature or application of loads.

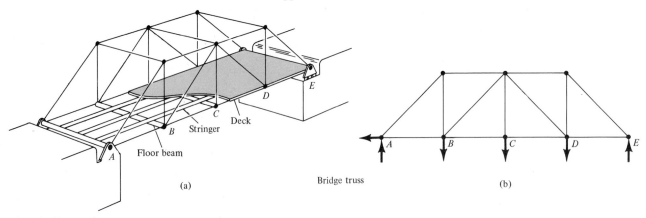

Floor beam

Stringer

Deck

Bridge truss

(a)

(b)

Fig. 6—3

Assumptions for Design

To design both the members and the connections of a truss, it is first necessary to determine the *force* developed in each member when the truss is subjected to a given loading. In this regard, two important assumptions will be made:

1. *All loadings are applied at the joints.* In most situations, such as for bridge and roof trusses, this assumption is true. Frequently in the force analysis the weights of the members are neglected, since the forces supported by the members are usually large in comparison with their weights. If the member's weight is to be included in the analysis, it is generally satisfactory to apply it as a vertical force, half of its magnitude applied at each end of the member.
2. *The members are joined together by smooth pins.* In cases where bolted or welded joint connections are used, this assumption is satisfactory provided the center lines of the joining members are *concurrent,* as in the case of point *A* in Fig. 6–1.

Because of these two assumptions, *each truss member acts as a two-force member,* and therefore the forces at the ends of the member must be directed along the axis of the member. If the force tends to *elongate* the member, it is a *tensile force* (**T**), Fig. 6–4*a*; whereas if the force tends to *shorten* the member, it is a *compressive force* (**C**), Fig. 6–4*b*. In the actual design of a truss it is important to state whether the nature of the force is tensile or compressive. Most often, compression members must be made *thicker* than tension members, because of the buckling or column effect that occurs when a member is in compression.

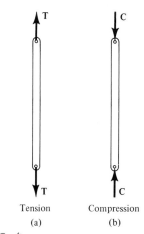

Tension
(a)

Compression
(b)

Fig. 6–4

Simple Truss

To prevent collapse, the framework of a truss must be rigid. Obviously, the four-bar frame *ABCD* in Fig. 6–5 will collapse unless a diagonal, such as *AC*, is added for support. The simplest framework which is rigid or stable is a *triangle*. Consequently, a *simple truss* is constructed by *starting* with a basic triangular element, such as *ABC* in Fig. 6–6, and connecting two members (*AD* and *BD*) to form an additional element. Thus it is seen that as each additional element of two members is placed on the truss, the number of joints is increased by one.

Fig. 6–5

Fig. 6–6

6.2 The Method of Joints

If a truss is in equilibrium, then each of its joints must also be in equilibrium. Hence, the method of joints consists of satisfying the equilibrium conditions for the forces exerted *on the pin* at each joint of the truss. Since the truss members are all straight two-force members lying in the same plane, the force system acting at each pin is *coplanar and concurrent*. Consequently, rotational or moment equilibrium is automatically satisfied at the joint (or pin), and it is only necessary to satisfy $\Sigma F_x = 0$ and $\Sigma F_y = 0$ to ensure translational or force equilibrium.

When using the method of joints, it is *first* necessary to draw the joint's free-body diagram before applying the equilibrium equations. In this regard, recall that the *line of action* of each member force acting on the joint is *specified* from the geometry of the truss, since the force in a member passes along the axis of the member. As an example, consider the pin at joint B of the truss in Fig. 6–7a. Three forces act on the pin, namely, the 500-N force and the forces exerted by members BA and BC. The free-body diagram is shown in Fig. 6–7b. As shown, $\mathbf{F}_{BA}$ is "pulling" on the pin, which means that member BA is in *tension;* whereas $\mathbf{F}_{BC}$ is "pushing" on the pin, and consequently member BC is in *compression*. These effects are clearly demonstrated by isolating the joint with small segments of the member connected to the pin, Fig. 6–7c. Notice that pushing or pulling on these small segments indicates the effect of the member being either in compression or tension.

In all cases, the analysis should start at a joint having at least one known force and at most two unknown forces, as in Fig. 6–7b. In this way, applica-

(a)

(b)

(c)

Fig. 6–7

tion of $\Sigma F_x = 0$ and $\Sigma F_y = 0$ yields two algebraic equations which can be solved for the two unknowns. When applying these equations, the correct sense of an unknown member force can be determined using one of two possible methods.

1. *Always assume* the *unknown member forces* acting on the joint's free-body diagram to be in *tension,* i.e., ''pulling'' on the pin. If this is done, then numerical solution of the equilibrium equations will yield *positive scalars for members in tension and negative scalars for members in compression.* Once an unknown member force is found, use its *correct* magnitude and sense (T or C) on subsequent joint free-body diagrams.
2. The *correct* sense of direction of an unknown member force can, in many cases, be determined ''by inspection.'' For example, $\mathbf{F}_{BC}$ in Fig. 6–7b must push on the pin (compression) since its horizontal component $F_{BC} \sin 45°$ must balance the 500-N force ($\Sigma F_x = 0$). Likewise, $\mathbf{F}_{BA}$ is a tensile force since it balances the vertical component $F_{BC} \cos 45°$ ($\Sigma F_y = 0$). In more complicated cases, the sense of an unknown member force can be *assumed;* then, after applying the equilibrium equations, the assumed sense can be verified from the numerical results. A *positive* answer indicates that the sense is *correct,* whereas a *negative* answer indicates that the sense shown on the free-body diagram must be *reversed.* This is the method we will use in the example problems which follow.

PROCEDURE FOR ANALYSIS

The following procedure provides a typical means for analyzing a truss using the method of joints.

Draw the free-body diagram of a joint having at least one known force and at most two unknown forces. (If this joint is at one of the supports, it generally will be necessary to know the external reactions at the truss support.) Use one of the two methods described above for establishing the sense of an unknown force. Orient the x and y axes such that the forces on the free-body diagram can be easily resolved into their x and y components and then apply the two force equilibrium equations $\Sigma F_x = 0$ and $\Sigma F_y = 0$. Solve for the two unknown member forces and verify their correct sense.

Continue to analyze each of the other joints, where again it is necessary to choose a joint having at most two unknowns and at least one known force. In this regard, realize that once the force in a member is found from the analysis of a joint at one of its ends, the result can be used to analyze the forces acting on the joint at its other end. Strict adherence to the principle of action, equal but opposite reaction must, of course, be observed. Remember, a member in *compression* ''pushes'' on the joint and a member in *tension* ''pulls'' on the joint.

Once the force analysis of the truss has been completed, the size of the members and their connections can be determined using the theory of mechanics of materials.

Example 6–1

Determine the force in each member of the truss shown in Fig. 6–8a and indicate whether the members are in tension or compression.

(a)

SOLUTION

If the reactions at A and C had not been computed (although on the figure they are shown acting on the truss), the joint analysis would have to begin at joint B. Why?

Joint B. The free-body diagram of the pin at B is shown in Fig. 6–8b. Three forces act on the pin: the external force of 500 N and the *two* unknown forces developed by members BA and BC. Applying the equations of joint equilibrium, we have

(b)

$$\xrightarrow{+} \Sigma F_x = 0; \quad 500 - F_{BC} \sin 45° = 0 \quad F_{BC} = 707.1 \text{ N} \quad \text{(C)} \qquad Ans.$$
$$+\uparrow \Sigma F_y = 0; \quad F_{BC} \cos 45° - F_{BA} = 0 \quad F_{BA} = 500 \text{ N} \quad \text{(T)} \qquad Ans.$$

Since the forces in members BA and BC have been determined, one can proceed to analyze the forces at joints A and C.

Joint A. The free-body diagram of the pin at A is shown in Fig. 6–8c. Here the effect of the *support* on the pin is represented by the reactions $A_x = 500$ N and $A_y = 500$ N. Since BA was found to be in tension, $\mathbf{F}_{BA}$ ($= 500$ N) pulls on the pin (refer to Fig. 6–8e). Applying the equilibrium equations yields

(c)

$$\xrightarrow{+} \Sigma F_x = 0; \quad F_{AC} - 500 = 0 \quad F_{AC} = 500 \text{ N} \quad \text{(T)} \qquad Ans.$$
$$+\uparrow \Sigma F_y = 0; \quad 500 - 500 \equiv 0 \quad \text{(check)}$$

Joint C. With the forces in all the members known, the results can be checked, in part, from the free-body diagram of the last joint C, Fig. 6–8d.

(d)

$$\xrightarrow{+} \Sigma F_x = 0; \quad -500 + 707.1 \cos 45° \equiv 0 \quad \text{(check)}$$
$$+\uparrow \Sigma F_y = 0; \quad 500 - 707.1 \sin 45° \equiv 0 \quad \text{(check)}$$

The results of the analysis are summarized in Fig. 6–8e. Note that the free-body diagram of each pin shows the effects of all the connected members and external forces applied to the pin, whereas the free-body diagram of each member shows the effects of the end pins on the member.

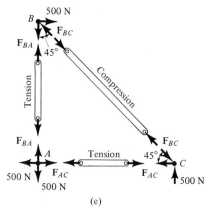

(e)

Fig. 6–8

Example 6–2

Determine the forces acting in all the members of the truss shown in Fig. 6–9a. The reactions at the supports are shown in the figure.

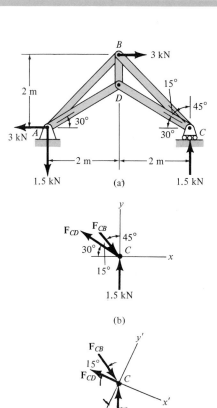

(a)

SOLUTION

Since the reactions have been computed, the analysis may begin at either joint C or A. Why?

Joint C (Fig. 6–9b)

$$\xrightarrow{+} \Sigma F_x = 0; \qquad -F_{CD} \cos 30° + F_{CB} \sin 45° = 0$$

$$+\uparrow \Sigma F_y = 0; \qquad 1.5 + F_{CD} \sin 30° - F_{CB} \cos 45° = 0$$

These two equations must be solved *simultaneously* for each of the two unknowns. Note, however, that a *direct solution* for one of the unknown forces may be obtained by applying a force summation along an axis that is *perpendicular* to the direction of the other unknown force. For example, summing forces along the y' axis, which is perpendicular to the direction of $\mathbf{F}_{CD}$, Fig. 6–9c, yields a direct solution for F_{CB}.

(b)

$$+\nearrow \Sigma F_{y'} = 0;$$

$$1.5 \cos 30° - F_{CB} \sin 15° = 0 \qquad F_{CB} = 5.02 \text{ kN} \quad (C) \qquad Ans.$$

In a similar fashion, summing forces along the y'' axis, Fig. 6–9d, yields a direct solution for F_{CD}.

$$+\nearrow \Sigma F_{y''} = 0;$$

$$1.5 \cos 45° - F_{CD} \sin 15° = 0 \qquad F_{CD} = 4.10 \text{ kN} \quad (T) \qquad Ans.$$

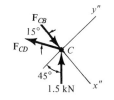

(c)

Joint D (Fig. 6–9e)

$$\xrightarrow{+} \Sigma F_x = 0; \qquad -F_{DA} \cos 30° + 4.10 \cos 30° = 0$$

$$F_{DA} = 4.10 \text{ kN} \quad (T) \qquad Ans.$$

$$+\uparrow \Sigma F_y = 0; \quad F_{DB} - 2(4.10 \sin 30°) = 0 \quad F_{DB} = 4.10 \text{ kN} \quad (T) \quad Ans.$$

Joint B (Fig. 6–9f)

$$\xrightarrow{+} \Sigma F_x = 0; \qquad F_{BA} \cos 45° - 5.02 \cos 45° + 3 = 0$$

$$F_{BA} = 0.777 \text{ kN} \quad (C) \qquad Ans.$$

$$+\uparrow \Sigma F_y = 0; \quad 0.777 \sin 45° - 4.10 + 5.02 \sin 45° \equiv 0 \quad (\text{check})$$

Check these results in part by analyzing the "last joint" A.

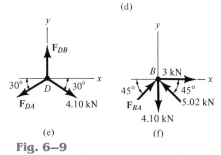

(e)   (f)

Fig. 6–9

Example 6–3

Determine the force in each member of the truss shown in Fig. 6–10a. Indicate whether the members are in tension or compression.

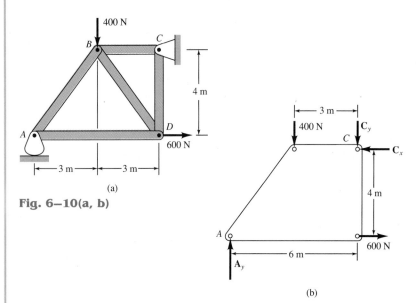

Fig. 6–10(a, b)

SOLUTION

Support Reactions. No joint can be analyzed until the support reactions are first determined. Why? A free-body diagram of the entire truss is given in Fig. 6–10b. Applying the equations of equilibrium, we have

$\xrightarrow{+} \Sigma F_x = 0$; $600 - C_x = 0$ $C_x = 600$ N

$\zeta + \Sigma M_C = 0$; $-A_y(6) + 400(3) + 600(4) = 0$ $A_y = 600$ N

$+ \uparrow \Sigma F_y = 0$; $600 - 400 - C_y = 0$ $C_y = 200$ N

The analysis can now start at either joint A or C. The choice is arbitrary, since there are one known and two unknown member forces acting on the pin at each of these joints.

Joint A (Fig. 6–10c). As shown on the free-body diagram, there are three forces that act on the pin at joint A. The inclination of $\mathbf{F}_{AB}$ is determined from the geometry of the truss. By inspection, can you see why this force is assumed to be compressive and $\mathbf{F}_{AD}$ tensile? Applying the equations of equilibrium, we have

$+ \uparrow \Sigma F_y = 0$; $600 - \frac{4}{5}F_{AB} = 0$ $F_{AB} = 750$ N   (C)   *Ans.*

$\xrightarrow{+} \Sigma F_x = 0$; $F_{AD} - \frac{3}{5}(750) = 0$ $F_{AD} = 450$ N   (T)   *Ans.*   **Fig. 6–10c**

Joint D (Fig. 6–10*d*). The pin at this joint is chosen next since, by inspection of Fig. 6–10*a*, the force in *AD* is known and the unknown forces in *DB* and *DC* can be determined. Applying the equations of equilibrium, Fig. 6–10*d*, we have

$$\xrightarrow{+} \Sigma F_x = 0; \quad -450 + \tfrac{3}{5}F_{DB} + 600 = 0 \quad F_{DB} = -250 \text{ N}$$

The negative sign indicates $\mathbf{F}_{DB}$ acts in the *opposite sense* to that shown in Fig. 6–10*d*.* Hence,

$$F_{DB} = 250 \text{ N} \quad \text{(T)} \qquad\qquad \textbf{\textit{Ans.}}$$

To determine $\mathbf{F}_{DC}$, we can either correct the sense of $\mathbf{F}_{DB}$ and then apply $\Sigma F_y = 0$, or apply this equation and retain the negative sign for F_{DB}, i.e.,

$$+\uparrow \Sigma F_y = 0; \quad -F_{DC} - \tfrac{4}{5}(-250) = 0 \quad F_{DC} = 200 \text{ N} \quad \text{(C)} \qquad \textbf{\textit{Ans.}}$$

Joint C (Fig. 6–10*e*)

$$\xrightarrow{+} \Sigma F_x = 0; \quad F_{CB} - 600 = 0 \quad F_{CB} = 600 \text{ N} \quad \text{(C)} \qquad \textbf{\textit{Ans.}}$$
$$+\uparrow \Sigma F_y = 0; \qquad\qquad 200 - 200 \equiv 0 \quad \text{(check)}$$

The analysis is summarized in Fig. 6–10*f*, which shows the correct free-body diagram for each pin and member.

*The proper sense could have been determined by inspection, prior to applying $\Sigma F_x = 0$.

(d)

(e)

Fig. 6–10(d, e)

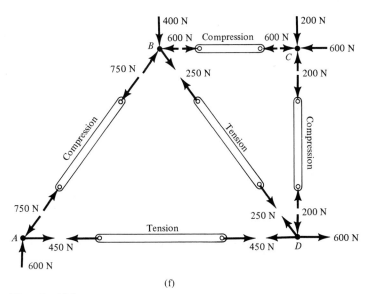

(f)

Fig. 6–10f

6.3 Zero-Force Members

Truss analysis using the method of joints is greatly simplified if one is able to first determine those members which support *no loading*. These *zero-force members* are used to increase the stability of the truss during construction and to provide support if the applied loading is changed.

The zero-force members of a truss can generally be determined *by inspection* of each of its joints. For example, consider the truss shown in Fig. 6–11a. If a free-body diagram of the pin at joint A is drawn, Fig. 6–11b, it is seen that members AB and AF are zero-force members. Note that this is also the case for members DC and DE, Fig. 6–11c. As a general rule then, *if only two members form a truss joint and no external load or support reaction is applied to the joint, then the members must be zero-force members.* The load on the truss in Fig. 6–11a is therefore supported by only five members as shown in Fig. 6–11d. Now consider the truss shown in Fig. 6–12a. The free-body diagram of the pin at joint D is shown in Fig. 6–12b. By orienting the y axis along members DC and DE and the x axis along member DA, it is

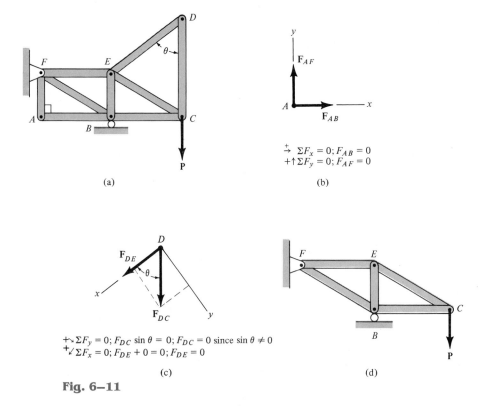

$$\xrightarrow{+} \Sigma F_x = 0; F_{AB} = 0$$
$$+\uparrow \Sigma F_y = 0; F_{AF} = 0$$

(a)   (b)

$$+\nwarrow \Sigma F_y = 0; F_{DC} \sin\theta = 0; F_{DC} = 0 \text{ since } \sin\theta \neq 0$$
$$+\swarrow \Sigma F_x = 0; F_{DE} + 0 = 0; F_{DE} = 0$$

(c)   (d)

Fig. 6–11

seen that *DA* is a zero-force member. Note that this is also the case for member *CA*, Fig. 6–12c. In general, then, *if three members form a truss joint for which two of the members are collinear, the third member is a zero-force member provided no external force or support reaction is applied to the joint.* The truss shown in Fig. 6–12d is therefore suitable for supporting the load **P.**

(a)

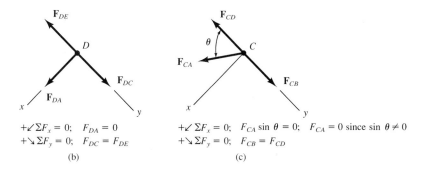

$+\swarrow \Sigma F_x = 0;$ $F_{DA} = 0$

$+\searrow \Sigma F_y = 0;$ $F_{DC} = F_{DE}$

(b)

$+\swarrow \Sigma F_x = 0;$ $F_{CA} \sin \theta = 0;$ $F_{CA} = 0$ since $\sin \theta \neq 0$

$+\searrow \Sigma F_y = 0;$ $F_{CB} = F_{CD}$

(c)

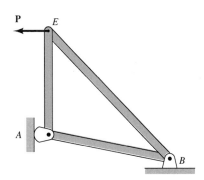

Fig. 6–12   (d)

229

Example 6–4

Using the method of joints, determine all the members of the *Fink truss* shown in Fig. 6–13*a* which are subjected to zero force.

Fig. 6–13   (a)

(b)

(c)

(d)

(e)

(f)

SOLUTION

Looking for joint situations which are similar to those outlined in Figs. 6–11 and 6–12, we have

Joint G (Fig. 6–13*b*)

$$+\uparrow \Sigma F_y = 0; \qquad\qquad F_{GC} = 0 \qquad\qquad \textbf{\textit{Ans.}}$$

Note that we could *not* conclude that GC is a zero-force member by analyzing joint C, where there are four unknowns.

Joint D (Fig. 6–13*c*)

$$+\swarrow \Sigma F_y = 0; \qquad\qquad F_{DF} = 0 \qquad\qquad \textbf{\textit{Ans.}}$$

Joint F (Fig. 6–13*d*)

$$+\uparrow \Sigma F_y = 0; \quad F_{FC} \cos \theta = 0, \text{ since } \theta \neq 0; \quad F_{FC} = 0 \qquad \textbf{\textit{Ans.}}$$

Note that if joint B is analyzed, Fig. 6–13*e*,

$$+\searrow \Sigma F_y = 0; \qquad 2 - F_{BH} = 0 \qquad F_{BH} = 2 \text{ kN} \quad (C)$$

Consequently, the numerical value of F_{HC} must satisfy $\Sigma F_y = 0$, Fig. 6–13*f*, and therefore HC is *not* a zero-force member.

PROBLEMS

6–1. Determine the force in each member of the truss and state if the members are in tension or compression.

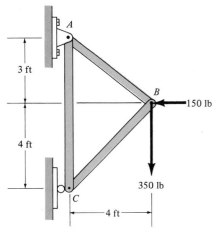

Prob. 6–1

6–2. Determine the force in each member of the truss and state if the members are in tension or compression. *Hint:* The horizontal force component at *A* must be zero. Why?

Prob. 6–2

6–3. Determine the force in each member of the truss and state if the members are in tension or compression. *Hint:* The resultant force at the pin *E* acts along member *ED*. Why?

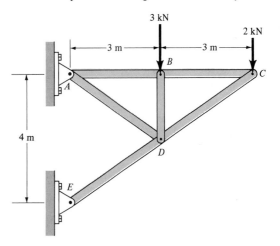

Probs. 6–3/6–4

*6–4. Each member of the truss is uniform and has a mass of 8 kg/m. Remove the external loads of 3 kN and 2 kN and determine the approximate force in each member due to the weight of the truss. Indicate whether the members are in tension or compression. Solve the problem by *assuming* the weight of each member can be represented as a vertical force, half of which is applied at each end of the member.

6–5. Determine the force in each member of the truss. State whether the members are in tension or compression.

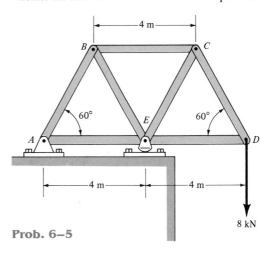

Prob. 6–5

6–6. Determine the force in each member of the truss and state if the members are in tension or compression. *Hint:* The vertical component of force at *C* must equal zero. Why?

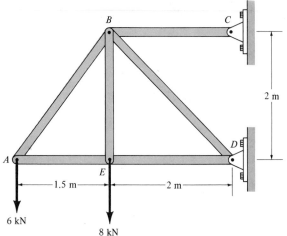

Probs. 6–6/6–7

6–9. Determine the force in each member of the truss in terms of the load *P*, and indicate whether the members are in tension or compression.

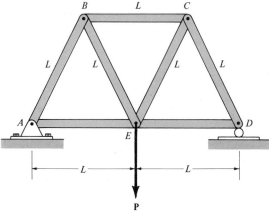

Probs. 6–9/6–10

6–7. Each member of the truss is uniform and has a mass of 8 kg/m. Remove the external loads of 6 kN and 8 kN and determine the approximate force in each member due to the weight of the truss. Indicate whether the members are in tension or compression. Solve the problem by *assuming* the weight of each member can be represented as a vertical force, half of which is applied at each end of the member.

*6–8. Determine the force in each member of the truss and state if the members are in tension or compression.

Prob. 6–8

6–10. Each member of the truss is uniform and has a weight *W*. Remove the external force **P** and determine the approximate force in each member due to the weight of the truss. Indicate whether the members are in tension or compression. Solve the problem by *assuming* the weight of each member can be represented as a vertical force, half of which is applied at each end of the member.

6–11. Determine the force in each member of the truss and indicate if the members are in tension or compression.

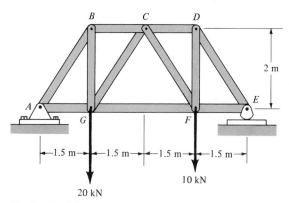

Prob. 6–11

*6–12. Determine the force in each member of the truss and state if the members are in tension or compression.

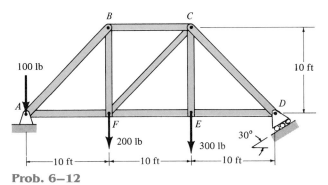

100 lb

B   C

10 ft

A   D

F   E

200 lb   300 lb   30°

—10 ft—|—10 ft—|—10 ft—

Prob. 6–12

6–13. Determine the force in each member of the truss and indicate whether the members are in tension or compression. *Hint:* Start the analysis at joint B by summing the forces along member BG to determine the force in BG. Then analyze joint G, etc.

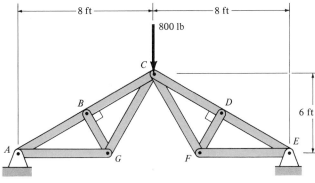

—8 ft—|—8 ft—

800 lb

C

B   D

A   E

G   F

6 ft

Prob. 6–13

6–14. Determine the force in members IC and CG of the truss and indicate whether the members are in tension or compression. Also, indicate all zero-force members.

B   C   D

2 m

I   J

2 m

A   E

H   G   F

—1.5 m—|—1.5 m—|—1.5 m—|—1.5 m—

6 kN   6 kN

Probs. 6–14/6–15

6–15. Determine the force in members JE and GF of the truss and indicate whether the members are in tension or compression. Also, indicate all zero-force members.

*6–16. Determine the force in each member of the truss. State whether the members are in tension or compression.

6 kN

B

6 kN

A   C

F

3 m

E   D

3 m

—3 m—|—3 m—|—3 m—

Prob. 6–16

6–17. Determine the force in each member of the truss in terms of the external loading, and indicate whether the members are in tension or compression.

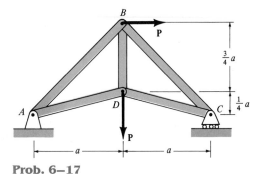

Prob. 6–17

6–19. Determine the force in each member of the truss and indicate whether the members are in tension or compression.

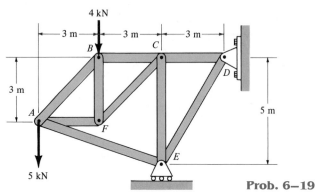

Prob. 6–19

6–18. Determine the force in each member of the truss in terms of the external loading and indicate whether the members are in tension or compression.

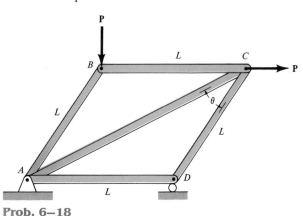

Prob. 6–18

*6–20. Determine the force in each member of the truss in terms of the load P, and indicate whether the members are in tension or compression.

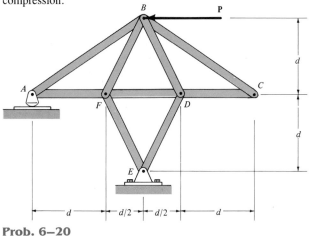

Prob. 6–20

6.4 The Method of Sections

The *method of sections* is used to determine the loadings acting within a body. It is based on the principle that if a body is in equilibrium, then any part of the body is also in equilibrium. To apply this method, one passes an *imaginary section* through the body, thus cutting it into two parts. When a free-body diagram of one of the parts is drawn, the loads acting at the section must be *included* on the free-body diagram. One then applies the equations of equilibrium to the part in order to determine the loading at the section. For example, consider the two truss members shown colored in Fig. 6–14. The internal

Fig. 6–14

loads at the section indicated by the dashed line can be obtained using one of the free-body diagrams shown on the right. Clearly, it can be seen that equilibrium requires that the member in tension be subjected to a "pull" **T** at the section, whereas the member in compression is subjected to a "push" **C.**

The method of sections can also be used to "cut" or section several members of an entire truss. If either of the two parts of the truss is isolated as a free-body diagram, we can then apply the equations of equilibrium to that part to determine the member forces at the "cut section." Since only *three* independent equilibrium equations ($\Sigma F_x = 0$, $\Sigma F_y = 0$, $\Sigma M_O = 0$) can be applied to the isolated part of the truss, one should try to select a section that, in general, passes through not more than *three* members in which the forces are unknown. For example, consider the truss in Fig. 6–15a. If the force in member GC is to be determined, section aa would be appropriate. The free-body diagrams of the two parts are shown in Fig. 6–15b and c. In particular, note that the line of action of each cut member force is specified from the *geometry* of the truss, since the force in a member passes along its axis. Also, the member forces acting on one part of the truss are equal but opposite to those acting on the other part—Newton's third law. As noted above, members assumed to be in *tension* (BC and GC) are subjected to a "pull," whereas the member in *compression* (GF) is subjected to a "push."

The three unknown member forces $\mathbf{F}_{BC}$, $\mathbf{F}_{GC}$, and $\mathbf{F}_{GF}$ can be obtained by applying the three equilibrium equations to the free-body diagram in Fig. 6–15b. If, however, the free-body diagram in Fig. 6–15c is considered, the three support reactions $\mathbf{D}_x$, $\mathbf{D}_y$, and $\mathbf{E}_x$ will have to be determined *first*. Why? (This, of course, is done in the usual manner by considering a free-body diagram of the *entire truss*.) When applying the equilibrium equations, one should consider ways of writing the equations so as to yield a *direct solution* for each of the unknowns, rather than having to solve simultaneous equations. For example, summing moments about C in Fig. 6–15b would yield a direct solution for $\mathbf{F}_{GF}$ since $\mathbf{F}_{BC}$ and $\mathbf{F}_{GC}$ create zero moment about C. Likewise,

Fig. 6–15

235

$\mathbf{F}_{BC}$ can be directly obtained by summing moments about G. Finally, $\mathbf{F}_{GC}$ can be found directly from a force summation in the vertical direction since $\mathbf{F}_{GF}$ and $\mathbf{F}_{BC}$ have no vertical components. This ability to *directly determine* the force in a particular truss member is one of the main advantages of using the method of sections.*

As in the method of joints, there are two ways in which one can determine the correct sense of an unknown member force.

1. *Always assume* that the unknown member forces at the cut section are in *tension*, i.e., "pulling" on the member. By doing this, the numerical solution of the equilibrium equations will yield *positive scalars for members in tension and negative scalars for members in compression.*

2. The correct sense of an unknown member force can in many cases be determined "by inspection." For example, $\mathbf{F}_{BC}$ is a tensile force as represented in Fig. 6–15b, since moment equilibrium about G requires that $\mathbf{F}_{BC}$ create a moment opposite to that of the 1000-N force. Also, $\mathbf{F}_{GC}$ is tensile since its vertical component must balance the 1000-N force acting downward. In more complicated cases, the sense of an unknown member force may be *assumed*. If the solution yields a *negative* scalar, it indicates that the force's sense is *opposite* to that shown on the free-body diagram. This is the method we will use in the example problems which follow.

PROCEDURE FOR ANALYSIS

The following procedure provides a means for applying the method of sections to determine the forces in the members of a truss.

Free-Body Diagram.   Make a decision as to how to "cut" or section the truss through the members where forces are to be determined. Before isolating the appropriate section, it may first be necessary to determine the truss's *external* reactions, so that the three equilibrium equations are used *only* to solve for member forces at the cut section. Draw the free-body diagram of that part of the sectioned truss which has the least number of forces acting on it. Use one of the two methods described above for establishing the sense of an unknown member force.

Equations of Equilibrium.   Try to apply the three equations of equilibrium such that simultaneous solution of equations is avoided. In this regard, moments should be summed about a point that lies at the intersection of the lines of action of two unknown forces, so that the third unknown force is determined directly from the moment equation. If two of the unknown forces are *parallel,* forces may be summed *perpendicular* to the direction of these unknowns to determine *directly* the third unknown force.

The following examples numerically illustrate these concepts.

---

*By comparison, if the method of joints were used to determine, say, the force in member GC, it would be necessary to analyze joints A, B, and G in sequence.

Example 6-5

Determine the force in members GE, GC, and BC of the bridge truss shown in Fig. 6-16a. Indicate whether the members are in tension or compression.

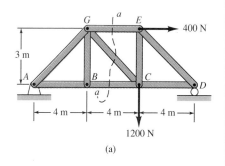

(a)

SOLUTION

Section aa in Fig. 6-16a has been chosen since it cuts through the *three* members whose forces are to be determined. In order to use the method of sections, however, it is *first* necessary to determine the external reactions at A or D. Why? A free-body diagram of the entire truss is shown in Fig. 6-16b. Applying the equations of equilibrium, we have

$\xrightarrow{+} \Sigma F_x = 0;$ $\quad\quad 400\ N - A_x = 0 \quad\quad A_x = 400\ N$

$\zeta + \Sigma M_A = 0;$ $\quad -1200\ N(8\ m) - 400\ N(3\ m) + D_y(12\ m) = 0$

$\quad\quad\quad\quad\quad\quad D_y = 900\ N$

$+\uparrow \Sigma F_y = 0;$ $\quad A_y - 1200\ N + 900\ N = 0 \quad\quad A_y = 300\ N$

(b)

Free-Body Diagrams. The free-body diagrams of the sectioned truss are shown in Fig. 6-16c and d. For the analysis the free-body diagram in Fig. 6-16c will be used since it involves the least number of forces.

Equations of Equilibrium. Summing moments about point G eliminates $\mathbf{F}_{GE}$ and $\mathbf{F}_{GC}$ and yields a direct solution for F_{BC}.

$\zeta + \Sigma M_G = 0;$ $\quad -300\ N(4\ m) - 400\ N(3\ m) + F_{BC}(3\ m) = 0$

$\quad\quad\quad\quad\quad\quad F_{BC} = 800\ N \quad (T) \quad\quad\quad\quad\quad\quad$ ***Ans.***

In the same manner, by summing moments about point C we obtain a direct solution for F_{GE}.

$\zeta + \Sigma M_C = 0;$ $\quad\quad -300\ N(8\ m) + F_{GE}(3\ m) = 0$

$\quad\quad\quad\quad\quad\quad F_{GE} = 800\ N \quad (C) \quad\quad\quad\quad\quad\quad$ ***Ans.***

Since $\mathbf{F}_{BC}$ and $\mathbf{F}_{GE}$ have no vertical components, summing forces in the y direction directly yields F_{GC}, i.e.,

$+\uparrow \Sigma F_y = 0;$ $\quad 300\ N - \frac{3}{5}F_{GC} = 0 \quad\quad F_{GC} = 500\ N \quad (T) \quad\quad$ ***Ans.***

Obtain these results by applying the equations of equilibrium to the free-body diagram shown in Fig. 6-16d.

(d)

Fig. 6-16

Example 6–6

Determine the force in member CF of the truss shown in Fig. 6–17a. Indicate whether the member is in tension or compression. The reactions at the supports are shown in the figure.

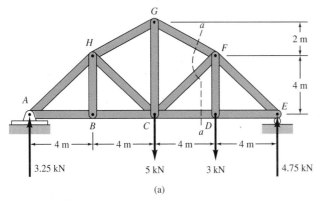

3.25 kN   5 kN   3 kN   4.75 kN

(a)

Fig. 6–17

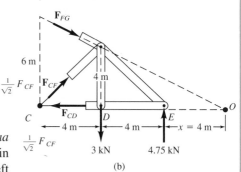

(b)

SOLUTION

Free-Body Diagram. The truss is sectioned along the dashed line aa shown in Fig. 6–17a. This section will "expose" the internal force in member CF as "external" on a free-body diagram of either the right or left portion of the truss. The free-body diagram of the right portion, which is the easiest to analyze, is shown in Fig. 6–17b. There are three unknowns, F_{FG}, F_{CF}, and F_{CD}.

Equations of Equilibrium. The most direct method for solving this problem requires application of the moment equation about a point that eliminates two of the unknown forces. Hence, to obtain $\mathbf{F}_{CF}$, we will eliminate $\mathbf{F}_{FG}$ and $\mathbf{F}_{CD}$ by summing moments about point O, Fig. 6–17b. Note that the location of point O measured from E is determined from proportional triangles, i.e., $4/(4 + x) = 6/(8 + x)$, $x = 4$ m. Or, in another manner, the slope of member GF has a drop of 2 m to a horizontal distance of $CD = 4$ m. Since FD is 4 m, Fig. 6–17a, then from D to O, Fig. 6–17b, the distance must be 8 m.

An easy way to determine the moment of $\mathbf{F}_{CF}$ about point O is to resolve $\mathbf{F}_{CF}$ into its two rectangular components and then use the principle of transmissibility to move $\mathbf{F}_{CF}$ to point C. We have

$$\zeta + \Sigma M_O = 0; \quad -\frac{1}{\sqrt{2}} F_{CF}(12 \text{ m}) + (3 \text{ kN})(8 \text{ m}) - (3.5 \text{ kN})(4 \text{ m}) = 0$$

$$F_{CF} = 0.589 \text{ kN (C)} \qquad \textit{Ans.}$$

Example 6–7

Determine the force in member *EB* of the truss shown in Fig. 6–18*a*. Indicate whether the member is in tension or compression. The reactions of the supports are shown in the figure.

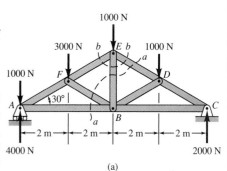

(a)

SOLUTION

Free-Body Diagrams. By the method of sections, any imaginary vertical section that cuts through *EB*, Fig. 6–18*a*, will also have to cut through three other members for which the forces are unknown. For example, section *aa* cuts through *ED*, *EB*, *FB*, and *AB*. If a free-body diagram of the left side of this section is considered, Fig. 6–18*b*, it is possible to determine $\mathbf{F}_{ED}$ by summing moments about *B* to eliminate the other three unknowns; however, $\mathbf{F}_{EB}$ cannot be determined from the remaining two equilibrium equations. One possible way of determining $\mathbf{F}_{EB}$ is to first determine $\mathbf{F}_{ED}$ from section *aa*, then use this result on section *bb*, Fig. 6–18*a*, which is shown in Fig. 6–18*c*. Here the force system is concurrent and our sectioned free-body diagram is the same as the free-body diagram for the pin at *E* (method of joints).

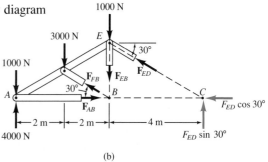

(b)

Equations of Equilibrium. In order to determine the moment of $\mathbf{F}_{ED}$ about point *B*, Fig. 6–18*b*, we will resolve it into its rectangular components and, by the principle of transmissibility, extend this force to point *C* as shown. The moments of 1000 N, F_{AB}, F_{FB}, F_{EB}, and $F_{ED}\cos 30°$ are all zero about *B*. Therefore,

(c)

Fig. 6–18

$$\curvearrowleft + \Sigma M_B = 0; \quad 1000(4) + 3000(2) - 4000(4) + F_{ED}\sin 30°(4) = 0$$
$$F_{ED} = 3000 \text{ N} \quad (\text{C})$$

Considering now the free-body diagram of section *bb*, Fig. 6–18*c*, we have

$$\xrightarrow{+}\Sigma F_x = 0; \quad F_{EF}\cos 30° - 3000\cos 30° = 0$$
$$F_{EF} = 3000 \text{ N} \quad (\text{C})$$

$$+\uparrow \Sigma F_y = 0; \quad 2(3000\sin 30°) - 1000 - F_{EB} = 0$$
$$F_{EB} = 2000 \text{ N} \quad (\text{T}) \qquad\qquad \textit{Ans.}$$

PROBLEMS

6–21. Determine the force developed in members *BC*, *BG*, *HG*, and *CG* of the bridge truss, and indicate whether the members are in tension or compression. After the truss is sectioned, use a single equation of equilibrium for the calculation of each force.

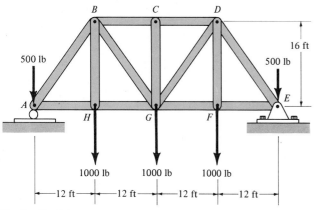

Prob. 6–21

6–22. The *Howe bridge truss* is subjected to the loading shown. Determine the force in members *CD*, *JD*, and *DI*, and indicate whether the members are in tension or compression.

Probs. 6–22/6–23

6–23. The *Howe bridge truss* is subjected to the loading shown. Determine the force in members *DE*, *EH*, and *HG*, and indicate whether the members are in tension or compression.

*6–24. Determine the force in members *BC*, *CG*, and *GF* of the *Warren truss*. Indicate if the members are in tension or compression.

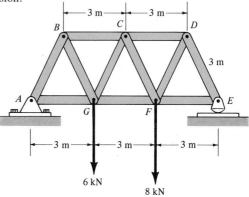

Probs. 6–24/6–25

6–25. Determine the force in members *CD*, *CF*, and *FG* of the *Warren truss*. Indicate if the members are in tension or compression.

6–26. The *Pratt roof truss* supports the vertical loading shown. Determine the force in members *BC*, *CK*, and *KJ*, and indicate whether the members are in tension or compression.

Probs. 6–26/6–27

6–27. The *Pratt roof truss* supports the vertical loading shown. Determine the force in members *DE* and *DJ*, and indicate whether the members are in tension or compression.

6–28. Determine the force in members *BC*, *HC*, and *HG*. After the truss is sectioned use a single equation of equilibrium for the calculation of each force. Indicate if the members are in tension or compression.

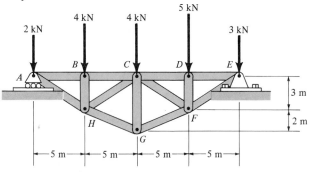

Probs. 6–28/6–29

6–29. Determine the force in members *CD*, *CF*, and *CG*. Indicate if the members are in tension or compression.

6–30. Determine the force in members *DE*, *DH*, and *IH* of the truss, and indicate whether the members are in tension or compression. After the truss is sectioned, use a single equation of equilibrium for the calculation of each force.

Prob. 6–30

6–31. Determine the force in members *BC*, *CG*, and *GF* of the truss. Indicate if the members are in tension or compression.

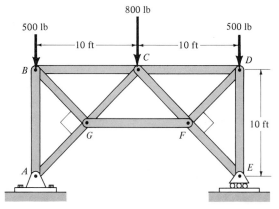

Prob. 6–31

6–32. Determine the force in members *HG*, *HC*, and *BC* of the truss and indicate whether the members are in tension or compression. After the truss is sectioned, use a single equation of equilibrium for the calculation of each force.

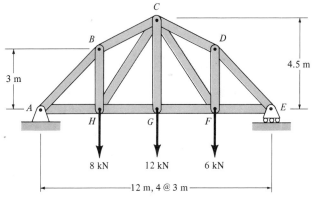

Probs. 6–32/6–33

6–33. Determine the force in members *GF*, *CF*, and *CD* of the truss and indicate whether the members are in tension or compression. After the truss is sectioned, use a single equation of equilibrium for the calculation of each force.

241

6–34. Determine the force in members *CD*, *PM*, and *ML* of the truss and indicate whether the members are in tension or compression. Also, indicate all zero-force members.

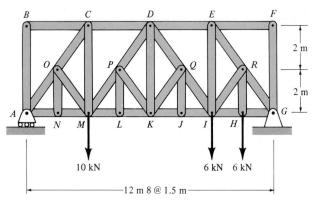

Probs. 6–34/6–35

6–35. Determine the force in members *DE*, *DQ*, and *KJ* of the truss and indicate whether the members are in tension or compression. Also, indicate all zero-force members.

*6–36. Determine the force in members *DE*, *JI*, and *DO* of the *K truss*. Indicate if the members are in tension or compression. *Hint:* Use sections *aa* and *bb*.

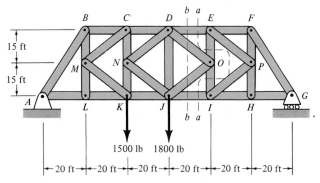

Probs. 6–36/6–37

6–37. Determine the force in members *CD* and *KJ* of the *K truss*. Indicate if the members are in tension or compression. *Hint:* Note section *aa* used to find the force in members *DE* and *JI*.

6–38. Determine the force in members *CB*, *CG*, and *GF* of the symmetrical truss. Indicate if the members are in tension or compression.

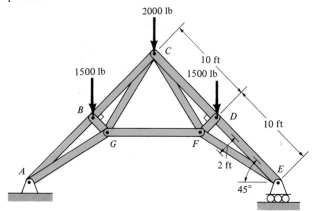

Prob. 6–38

*6.5 Space Trusses

A *space truss* consists of members joined together at their ends to form a stable three-dimensional structure. The simplest element of a space truss is a *tetrahedron*, formed by connecting six members together, as shown in Fig. 6–19. Any additional members added to this basic element would be redundant in supporting the force **P**. A *simple space truss* can be built from this basic tetrahedral element by adding three additional members and a joint forming a system of multiconnected tetrahedrons.

Assumptions for Design

The members of a space truss may be treated as two-force members provided the external loading is applied at the joints and the joints consist of ball-and-socket connections. These assumptions are justified if the welded or bolted connections of the joined members intersect at a common point and the weight of the members can be neglected. In cases where the weight of a member is to be included in the analysis, it is generally satisfactory to apply it as a vertical force, half of its magnitude applied at each end of the member.

Fig. 6–19

PROCEDURE FOR ANALYSIS

Either the method of sections or the method of joints can be used to determine the forces developed in the members of a simple space truss.

Method of Sections. If only a *few* member forces are to be determined, the method of sections may be used. When an imaginary section is passed through a truss, and the truss is separated into two parts, the force system acting on one of the parts must satisfy the *six* scalar equilibrium equations: $\Sigma F_x = 0$, $\Sigma F_y = 0$, $\Sigma F_z = 0$, $\Sigma M_x = 0$, $\Sigma M_y = 0$, $\Sigma M_z = 0$ (Eqs. 5–6). By proper choice of the section and axes for summing forces and moments, many of the unknown member forces in a space truss can be computed *directly*, using a single equilibrium equation.

Method of Joints. Generally, if the forces in *all* the members of the truss must be determined, the method of joints is most suitable for the analysis. When using the method of joints, it is necessary to solve the three scalar equilibrium equations $\Sigma F_x = 0$, $\Sigma F_y = 0$, $\Sigma F_z = 0$ at each joint. The solution of many simultaneous equations can be avoided if the force analysis begins at a joint having at least one known force and at most three unknown forces. If the three-dimensional geometry of the force system at the joint is hard to visualize, it is recommended that a Cartesian vector analysis be used in the solution. The following example numerically illustrates this procedure.

Example 6–8

Determine the forces acting in the members of the space truss shown in Fig. 6–20a. Indicate whether the members are in tension or compression.

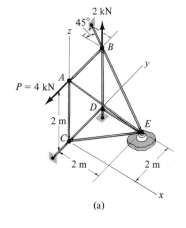

(a)

SOLUTION

Since there are one known force and three unknown forces acting at joint A, the force analysis of the truss will begin at joint A.

Joint A (Fig. 6–20b). Expressing each force that acts on the free-body diagram of joint A in vector notation yields

$$\mathbf{P} = \{-4\mathbf{j}\} \text{ kN}, \qquad \mathbf{F}_{AB} = F_{AB}\mathbf{j}, \qquad \mathbf{F}_{AC} = -F_{AC}\mathbf{k},$$

$$\mathbf{F}_{AE} = F_{AE}\left(\frac{\mathbf{r}_{AE}}{r_{AE}}\right) = F_{AE}(0.577\mathbf{i} + 0.577\mathbf{j} - 0.577\mathbf{k})$$

For equilibrium,

$$\Sigma \mathbf{F} = 0; \qquad \mathbf{P} + \mathbf{F}_{AB} + \mathbf{F}_{AC} + \mathbf{F}_{AE} = 0$$

$$-4\mathbf{j} + F_{AB}\mathbf{j} - F_{AC}\mathbf{k} + 0.577F_{AE}\mathbf{i} + 0.577F_{AE}\mathbf{j} - 0.577F_{AE}\mathbf{k} = 0$$

$$\Sigma F_x = 0; \qquad 0.577F_{AE} = 0$$

$$\Sigma F_y = 0; \qquad -4 + F_{AB} + 0.577F_{AE} = 0$$

$$\Sigma F_z = 0; \qquad -F_{AC} - 0.577F_{AE} = 0$$

$$F_{AC} = F_{AE} = 0 \qquad\qquad Ans.$$

$$F_{AB} = 4 \text{ kN} \quad (T) \qquad\qquad Ans.$$

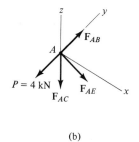

(b)

Since F_{AB} is known, joint B may be analyzed next.

Joint B (Fig. 6–20c)

$$\Sigma F_x = 0; \qquad -R_B \cos 45° + 0.707F_{BE} = 0$$

$$\Sigma F_y = 0; \qquad -4 + R_B \sin 45° = 0$$

$$\Sigma F_z = 0; \qquad 2 + F_{BD} - 0.707F_{BE} = 0$$

$$R_B = F_{BE} = 5.66 \text{ kN} \quad (T), \qquad F_{BD} = 2 \text{ kN} \quad (C) \qquad Ans.$$

The *scalar* equations of equilibrium may also be applied directly to the force systems on the free-body diagrams of joints D and C, since the force components are easily determined. Show that

$$F_{DE} = F_{DC} = F_{CE} = 0 \qquad\qquad Ans.$$

(c)

Fig. 6–20

PROBLEMS

6–39. Determine the force in each member of the space truss and indicate whether the members are in tension or compression. The truss is supported by ball-and-socket joints at A, B, and E. Set $\mathbf{F} = \{800\mathbf{j}\}$ N. *Hint:* The support reaction at E acts along member EC. Why?

6–41. Determine the force in each member of the space truss and indicate whether the members are in tension or compression. The truss is supported by ball-and-socket joints at C, D, E, and G.

Probs. 6–39/6–40

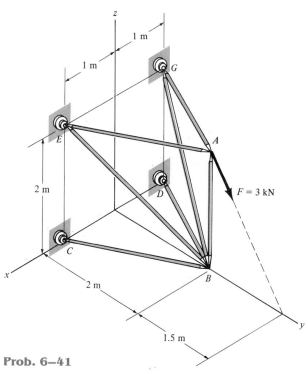

Prob. 6–41

*6–40. Determine the force in each member of the space truss and indicate whether the members are in tension or compression. The truss is supported by ball-and-socket joints at A, B, and E. Set $\mathbf{F} = \{-200\mathbf{i} + 400\mathbf{j}\}$ N. *Hint:* The support reaction at E acts along member EC. Why?

6–42. The space truss is supported by a ball-and-socket joint at D and short links at C and E. Determine the force in each member and indicate if the members are in tension or compression. Take $F_1 = \{-500\mathbf{k}\}$ lb and $F_2 = \{400\mathbf{j}\}$ lb.

*6–44.** Determine the force in each member of the space truss and indicate whether the members are in tension or compression. The truss is supported by a ball-and-socket joint at A and short links at B and C.

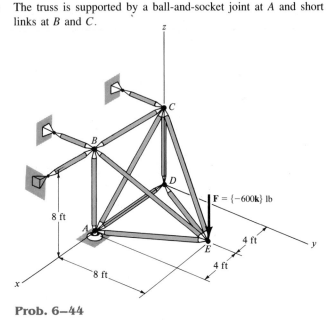

Prob. 6–44

Probs. 6–42/6–43

6–43. The space truss is supported by a ball-and-socket joint at D and short links at C and E. Determine the force in each member and indicate if the members are in tension or compression. Take $F_1 = \{200\mathbf{i} + 300\mathbf{j} - 500\mathbf{k}\}$ lb and $F_2 = \{400\mathbf{j}\}$ lb.

6.6 Frames and Machines

Frames and machines are two common types of structures which are often composed of pin-connected *multiforce members,* i.e., members that are subjected to more than two forces. *Frames* are generally stationary and are used to support loads, whereas *machines* contain moving parts and are designed to transmit and alter the effect of forces. Provided a frame or machine is properly constrained and contains no more supports or members than are necessary to prevent collapse, then the forces acting at the joints and supports can be determined by applying the equations of equilibrium to each member. Once the forces at the joints are obtained, it is then possible to *design* the size of the members, connections, and supports using the theory of mechanics of materials.

Free-Body Diagrams

In order to determine the forces acting at the joints and supports of a frame or machine, the structure must be disassembled and the free-body diagrams of its parts must be drawn. In this regard, the following important points *must* be observed:

1. Isolate each part by drawing its *outlined shape.* Then show all the forces and/or couple moments that act on the part. Make sure to *label* or *identify* each known and unknown force and couple moment and indicate any dimensions used for taking moments. Most often the equations of equilibrium are easier to apply if the forces are represented by their rectangular components. As usual, the sense of an unknown force or couple moment can be assumed.
2. Identify the two-force members in the structure, and represent their free-body diagrams as having two equal but opposite forces acting at their points of application. The line of action of the forces is defined by the line joining the two points where the forces act (see Sec. 5.4). *This step is particularly important in order to avoid solving an unnecessary number of equilibrium equations.* (See Example 6–14.)
3. Forces common to any two *contacting* members act with equal magnitudes but opposite sense on the respective members. If the two members are treated as a *"system" of connected members,* then these forces are *"internal"* and are *not shown* on the *free-body diagram of the system;* however, if the free-body diagram of *each member* is drawn, the forces are *"external"* and *must* be shown on the free-body diagrams.

The following examples graphically illustrate application of these points in drawing the free-body diagrams of a dismembered frame or machine. Each example should be thoroughly understood. In all cases, the weights of the members are neglected.

Example 6–9

For the frame shown in Fig. 6–21a, draw the free-body diagram of (a) each member, (b) the pin at B, and (c) the two members connected together.

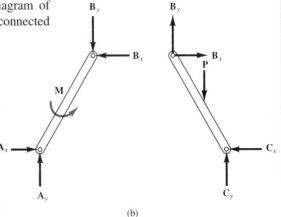

Fig. 6–21

(a)

(b)

SOLUTION

Part (a). By inspection the members are *not* two-force members. Instead, as shown on the free-body diagrams, Fig. 6–21b, BC is subjected to *three forces*, namely, the *resultants* of the two components of reaction at pins B and C and the external force **P**. Likewise, AB is subjected to the *resultant* pin-reactive forces at A and B and the external couple **M**.

Part (b). It can be seen in Fig. 6–21a that the pin at B is subjected to only *two forces*, i.e., the force of member BC on the pin and the force of member AB on the pin. For *equilibrium* these forces and therefore their respective components must be equal but opposite, Fig. 6–21c. Notice carefully how Newton's third law is applied between the pin and its contacting members, i.e., the effect of the pin on the two members, Fig. 6–21b, and the equal but opposite effect of the two members on the pin, Fig. 6–21c. Realize also that B_x and B_y shown equal but opposite in Fig. 6–21b on members AB and BC is *not* the effect of Newton's third law; instead, it results from the *equilibrium* of the pin, Fig. 6–21c.

Part (c). The free-body diagram of both members connected together, yet removed from the supporting pins at A and C, is shown in Fig. 6–21d. Note that the force components B_x and B_y are *not shown* on this diagram since they form equal but opposite collinear pairs of *internal* forces (Fig. 6–21b) and therefore cancel out.* Also note that to be consistent when later applying the equilibrium equations, the unknown force components at A and C must act in the *same sense* as those shown in Fig. 6–21b.

*This is similar to not including internal forces exerted between adjacent particles of a rigid body when drawing the free-body diagram of the entire rigid body.

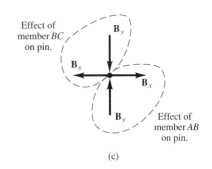

Effect of member BC on pin.

Effect of member AB on pin.

(c)

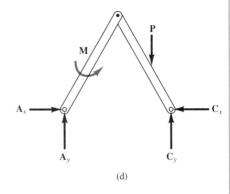

(d)

Example 6–10

For the frame shown in Fig. 6–22a, draw the free-body diagrams of (a) each of the three members, and (b) members *ABC* and *BD* together.

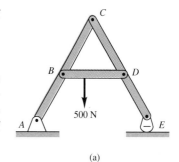

SOLUTION

Part (a). By inspection, none of the three members of the frame are two-force members. Instead, each is subjected to three forces. The free-body diagrams are shown in Fig. 6–22b. Notice that equal but opposite force reactions occur at *B*, *C*, and *D*. Draw a free-body diagram of one of the pins at *B*, *C*, or *D* and show why this is so.

(a)

Fig. 6–22a

Part (b). The free-body diagram of *ABC* and *BD* together is shown in Fig. 6–22c. Since the entire frame is in equilibrium, the force system on these two members also satisfies the equilibrium equations. Why not show the force components $\mathbf{B}_x$ and $\mathbf{B}_y$ on this diagram?

(b)

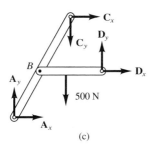

(c)

Fig. 6–22

249

Example 6–11

For the frame shown in Fig. 6–23, draw the free-body diagrams of (a) the entire frame and (b) each of its parts.

(a)

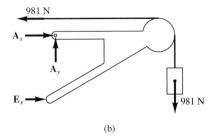

(b)

Fig. 6–23

SOLUTION

Part (a). The free-body diagram for the entire frame *including* the pulley, cord, and 100-kg mass is shown in Fig. 6–23b. Since the cord is continuous, it maintains a constant tension of 981 N throughout its length. Consequently, the 981-N force shown acting to the left represents the effect of the wall on the cord.

Part (b). By inspection, member *BD* is a two-force member, and therefore equal but opposite collinear forces act at its end points. The free-body diagrams are shown in Fig. 6–23c. Each of these diagrams should be carefully studied. In particular, note that the pulley is held in equilibrium by the two 981-N force components which the pin at *C* exerts on the pulley. The free-body diagram of the pin at *C* has been included here since *three force interactions* occur *on* this pin. The force components C_x and C_y represent the effect of member *ABC*, the force components C'_x and C'_y represent the effect of member *EDC*, and finally the two 981-N force components represent the effect of the pulley. Writing the equilibrium equations $\Sigma F_x = 0$ and $\Sigma F_y = 0$ for the pin at *C* will thus enable one to obtain two equations that relate the unknown force components C_x, C'_x and C_y, C'_y. Note that if only *two force interactions* occur at a pin connection, such as pin *D* in Fig. 6–23c, the force components acting must always be equal in magnitude but with opposite sense, and thus it is generally not necessary to consider its free-body diagram for the analysis.

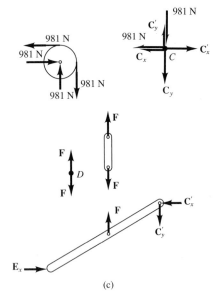

(c)

Example 6–12

Draw the free-body diagram of each part of the smooth piston and link mechanism shown in Fig. 6–24a.

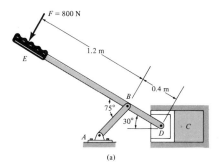

Fig. 6–24a   (a)

SOLUTION

By inspection member *AB* is a two-force member. The free-body diagrams of the parts are shown in Fig. 6–24b. Since the pins at *B* and *D* *connect only two parts together*, the forces there are shown as equal but opposite on the separate free-body diagrams of their connected members. In particular, four components of force act on the piston: $\mathbf{D}_x$ and $\mathbf{D}_y$ represent the effect of the pin (or lever *EBD*), $\mathbf{N}_w$ is the *resultant force* of the wall, and **P** is the resultant compressive force caused by the material within the cylinder *C*.

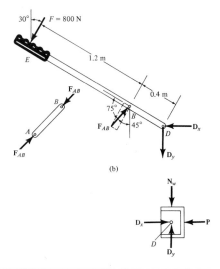

Fig. 6–24b

251

Example 6–13

For the frame shown in Fig. 6–25a draw the free-body diagrams of (a) the entire frame including the pulleys and cords, (b) the frame without the pulleys and cords, and (c) each of the pulleys.

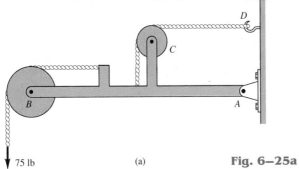

75 lb   (a)   **Fig. 6–25a**

SOLUTION

Part (a): When the entire frame including the pulleys and cords is considered, then the interactions at the points where the pulleys and cords are connected to the frame become pairs of *internal forces* which cancel each other and are therefore not shown on the free-body diagram, Fig. 6–25b.

Part (b): When the cords and pulleys are removed, their effect *on the frame* must be shown, Fig. 6–25c.

Part (c): The force components B_x, B_y, C_x, C_y of the pins on the pulleys, Fig. 6–25d, are equal but opposite to the force components exerted by the pins on the frame, Fig. 6–25c.

T_D

A_x

A_y

75 lb   (b)

Fig. 6–25(b–d)

C_y

B_y 75 lb T_B C_x

B_x A_x

(c) A_y

75 lb

B_x

75 lb B_y

T

C_x

T C_y

(d)

252

Before proceeding, it would be highly advised to cover over the solutions to the previous examples and attempt to draw the requested free-body diagrams. When doing so, make sure the work is neat and that all the forces and couple moments are properly labeled.

Equations of Equilibrium

Provided the structure (frame or machine) is properly supported and contains no more supports or members than are necessary to prevent its collapse, then the unknown forces at the supports and connections can be determined from the equations of equilibrium. If the structure lies in the *x-y* plane, then for *each* free-body diagram drawn the loading must satisfy $\Sigma F_x = 0$, $\Sigma F_y = 0$, and $\Sigma M_O = 0$. The selection of the free-body diagrams used for the analysis is *completely arbitrary*. They may represent each of the members of the structure, a portion of the structure, or its entirety. For example, consider finding the six components of the pin reactions at A, B, and C for the frame shown in Fig. 6–26*a*. If the frame is dismembered, Fig. 6–26*b*, these unknowns can be determined by applying the three equations of equilibrium to each of the two members (total of six equations). The free-body diagram of the *entire frame* can also be used for part of the analysis, Fig. 6–26*c*. Hence, if so desired, all six unknowns can be determined by applying the three equilibrium equations to the entire frame, Fig. 6–26*c*, and also to either one of its members. Furthermore, the answers can be checked in part by applying the three equations of equilibrium to the remaining "second" member. In general, then, this problem can be solved by writing *at most* six equilibrium equations using free-body diagrams of the members and/or the combination of connected members. Any more than six equations written would *not* be unique from the original six and would only serve to check the results.

(a)

(b)

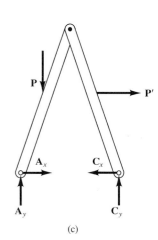

(c)

Fig. 6–26

PROCEDURE FOR ANALYSIS

The following procedure provides a method for determining the *joint reactions* of frames and machines (structures) composed of multiforce members.

Free-Body Diagrams. Draw the free-body diagram of the entire structure, a portion of the structure, or each of its members. The choice should be made so that it leads to the most direct solution to the problem.

Forces common to two members which are in contact act with equal magnitude but opposite sense on the respective free-body diagrams of the members. Recall that all *two-force members,* regardless of their shape, have equal but opposite collinear forces acting at the ends of the member. The unknown forces acting at the joints of multiforce members should be represented by their rectangular components. In many cases it is possible to tell by inspection the proper sense of the unknown forces; however, if this seems difficult, the sense can be assumed.

Equations of Equilibrium. Count the total number of unknowns to make sure that an equivalent number of equilibrium equations can be written for solution. Recall that in general three equilibrium equations can be written for each rigid body represented in two dimensions. Many times, the solution for the unknowns will be straightforward if moments are summed about a point that lies at the intersection of the lines of action of as many unknown forces as possible. If after obtaining the solution an unknown force magnitude is found to be negative, it means the sense of the force is the reverse of that shown on the free-body diagrams.

The examples that follow illustrate this procedure. All these examples should be *thoroughly understood* before proceeding to solve the problems.

Example 6–14

Determine the horizontal and vertical components of force which the pin at C exerts on member CB of the frame in Fig. 6–27a.

(a)

SOLUTION I

Free-Body Diagrams. By inspection it can be seen that AB is a two-force member. The free-body diagrams are shown in Fig. 6–27b.

Equations of Equilibrium. The *three unknowns*, C_x, C_y, and F_{AB}, can be determined by applying the three equations of equilibrium to member CB.

$\zeta + \Sigma M_C = 0$; $2000(2) - (F_{AB} \sin 60°)(4) = 0$ $F_{AB} = 1154.7 \text{ N}$

$\overset{+}{\rightarrow} \Sigma F_x = 0$; $1154.7 \cos 60° - C_x = 0$ $C_x = 577 \text{ N}$ *Ans.*

$+ \uparrow \Sigma F_y = 0$; $1154.7 \sin 60° - 2000 + C_y = 0$ $C_y = 1000 \text{ N}$ *Ans.*

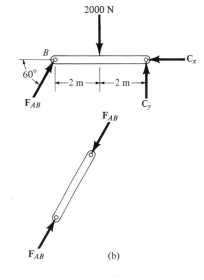

(b)

SOLUTION II

Free-Body Diagrams. If one does not recognize that AB is a two-force member, then more work is involved in solving this problem. The free-body diagrams are shown in Fig. 6–27c.

Equations of Equilibrium. The *six unknowns*, A_x, A_y, B_x, B_y, C_x, C_y, are determined by applying the three equations of equilibrium to each member.

Member AB

$\zeta + \Sigma M_A = 0$; $B_x(3 \sin 60°) - B_y(3 \cos 60°) = 0$ (1)

$\overset{+}{\rightarrow} \Sigma F_x = 0$; $A_x - B_x = 0$ (2)

$+ \uparrow \Sigma F_y = 0$; $A_y - B_y = 0$ (3)

Member BC

$\zeta + \Sigma M_C = 0$; $2000(2) - B_y(4) = 0$ (4)

$\overset{+}{\rightarrow} \Sigma F_x = 0$; $B_x - C_x = 0$ (5)

$+ \uparrow \Sigma F_y = 0$; $B_y - 2000 + C_y = 0$ (6)

The results for C_x and C_y can be determined by solving these equations in the following sequence: (4), (1), (5), then (6). The results are:

$$B_y = 1000 \text{ N}$$
$$B_x = 577 \text{ N}$$
$$C_x = 577 \text{ N} \qquad \textit{Ans.}$$
$$C_y = 1000 \text{ N} \qquad \textit{Ans.}$$

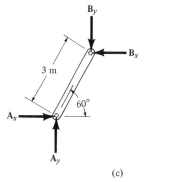

(c)

By comparison, Solution I is simpler since the requirement that $\mathbf{F}_{AB}$ in Fig. 6–27b be equal, opposite, and collinear at the ends of member AB automatically satisfies Eqs. (1), (2), and (3) above and therefore eliminates the need to write these equations. *As a result, always identify the two-force members before starting the analysis!*

Fig. 6–27

Example 6–15

The beam shown in Fig. 6–28a is pin-connected at B. Determine the reactions at its supports. Neglect its weight and thickness.

(a)

(b)

Fig. 6–28

SOLUTION

Free-Body Diagrams. The free-body diagrams of the two beam segments are shown in Fig. 6–28b.

Equations of Equilibrium

Segment BC

$\xrightarrow{+}\Sigma F_x = 0;$ $B_x = 0$

$\zeta + \Sigma M_B = 0;$ $-8 \text{ kN}(1 \text{ m}) + C_y(2 \text{ m}) = 0$

$+\uparrow \Sigma F_y = 0;$ $B_y - 8 \text{ kN} + C_y = 0$

Segment AB

$\xrightarrow{+}\Sigma F_x = 0;$ $A_x - 10 \text{ kN}(\tfrac{3}{5}) + B_x = 0$

$\zeta + \Sigma M_A = 0;$ $M_A - 10 \text{ kN}(\tfrac{4}{5})(2 \text{ m}) - B_y(4 \text{ m}) = 0$

$+\uparrow \Sigma F_y = 0;$ $A_y - 10 \text{ kN}(\tfrac{4}{5}) - B_y = 0$

Solving each of these equations successively, using previously calculated results, we obtain

$A_x = 6 \text{ kN}$ $A_y = 12 \text{ kN}$ $M_A = 32 \text{ kN} \cdot \text{m}$ *Ans.*

$B_x = 0$ $B_y = 4 \text{ kN}$

$C_y = 4 \text{ kN}$ *Ans.*

Example 6–16

Determine the horizontal and vertical components of force which the pin at C exerts on member $ABCD$ of the frame shown in Fig. 6–29a.

SOLUTION

Free-Body Diagrams. By inspection, the three components of reaction at the supports can be determined from a free-body diagram of the entire frame, Fig. 6–29b. Also, the free-body diagram of each frame member is shown in Fig. 6–29c. Notice that member BE is a two-force member. As shown by the colored dashed lines, the forces at B, C, and E have equal magnitudes but opposite directions on the separate free-body diagrams.

Equations of Equilibrium. The six unknowns A_x, A_y, T_B, C_x, C_y and D_x will be determined from the equations of equilibrium applied to the entire frame and then to member CEF. We have

Entire Frame

$$\downarrow +\Sigma M_A = 0; \quad -981 \text{ N}(2 \text{ m}) + D_x(2.8 \text{ m}) = 0 \quad D_x = 700.7 \text{ N}$$
$$\overset{+}{\rightarrow}\Sigma F_x = 0; \qquad A_x - 700.7 \text{ N} = 0 \quad A_x = 700.7 \text{ N}$$
$$+\uparrow\Sigma F_y = 0; \qquad A_y - 981 \text{ N} = 0 \quad A_y = 981 \text{ N}$$

Member CEF

$$\downarrow +\Sigma M_C = 0; \quad -981 \text{ N}(2 \text{ m}) - (T_B \sin 45°)(1.6 \text{ m}) = 0$$
$$T_B = -1734.2 \text{ N}$$
$$\overset{+}{\rightarrow}\Sigma F_x = 0; \qquad -C_x - (-1734.2 \cos 45° \text{ N}) = 0$$
$$C_x = 1230 \text{ N} \qquad\qquad Ans.$$
$$+\uparrow\Sigma F_y = 0; \quad C_y - (-1734.2 \sin 45° \text{ N}) - 981 \text{ N} = 0$$
$$C_y = -245 \text{ N} \qquad\qquad Ans.$$

Since the magnitudes of forces $\mathbf{T}_B$ and $\mathbf{C}_y$ were calculated as negative quantities, they were assumed to be acting in the wrong sense on the free-body diagrams, Fig. 6–29c. The correct sense of these forces might have been determined "by inspection" *before* applying the equations of equilibrium to member CEF. As shown in Fig. 6–29c, moment equilibrium about point E on member CEF indicates that $\mathbf{C}_y$ must actually act *downward* to counteract the moment created by the 981-N force about point E. Similarly, summing moments about point C, it is seen that the vertical component of force $\mathbf{T}_B$ must actually act *upward*. The above calculations can be checked in part by applying the three equilibrium equations to member $ABCD$, Fig. 6–29c.

Fig. 6–29

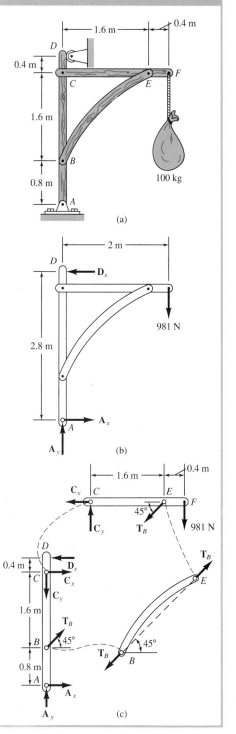

(a)

(b)

(c)

Example 6–17

The smooth disk shown in Fig. 6–30a is pinned at D and has a weight of 20 lb. Neglecting the weights of the other members, determine the horizontal and vertical components of reaction at pins B and D.

0.5 ft

3.5 ft

3 ft

Fig. 6–30 (a)

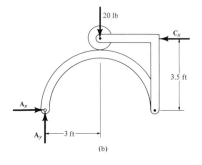

20 lb

C_x

3.5 ft

A_x

A_y ⊢—3 ft—⊣

(b)

SOLUTION

Free-Body Diagrams. By inspection, the three components of reaction at the supports can be determined from a free-body diagram of the entire frame, Fig. 6–30b. Also, free-body diagrams of the members are shown in Fig. 6–30c.

Equations of Equilibrium. The eight unknowns can of course be obtained by applying the eight equilibrium equations to each member—three to member AB, three to member BCD, and two to the disk. (Moment equilibrium is automatically satisfied for the disk.) If this is done, however, all the results can be obtained only from a simultaneous solution of some of the equations. (Try it and find out.) To avoid this situation, it is best to first determine the three support reactions on the *entire* frame; then, using these results, the remaining five equilibrium equations can be applied to two other parts in order to solve successively for the other unknowns.

⊢—3 ft—⊣

D_x C_x

D_y

3.5 ft

B_x

B_y

20 lb

D_x

D_y

N_D

   Entire Frame

$\zeta+\Sigma M_A = 0;$ $-20\ \text{lb}(3\ \text{ft}) + C_x(3.5\ \text{ft}) = 0$ $C_x = 17.1\ \text{lb}$

$\xrightarrow{+}\Sigma F_x = 0;$ $A_x - 17.1\ \text{lb} = 0$ $A_x = 17.1\ \text{lb}$

$+\uparrow\Sigma F_y = 0;$ $A_y - 20\ \text{lb} = 0$ $A_y = 20\ \text{lb}$

   Member AB

$\xrightarrow{+}\Sigma F_x = 0;$ $17.1\ \text{lb} - B_x = 0$ $B_x = 17.1\ \text{lb}$ *Ans.*

$\zeta+\Sigma M_B = 0;$ $-20\ \text{lb}(6\ \text{ft}) + N_D(3\ \text{ft}) = 0$ $N_D = 40\ \text{lb}$

$+\uparrow\Sigma F_y = 0;$ $20\ \text{lb} - 40\ \text{lb} + B_y = 0$ $B_y = 20\ \text{lb}$ *Ans.*

   Disk

$\xrightarrow{+}\Sigma F_x = 0;$ $D_x = 0$ *Ans.*

$+\uparrow\Sigma F_y = 0;$ $40\ \text{lb} - 20\ \text{lb} - D_y = 0$ $D_y = 20\ \text{lb}$ *Ans.*

N_D

20 lb

A_x B_x

A_y ⊢—3 ft—⊢—3 ft—⊣ B_y

(c)

Example 6–18

Determine the tension in the cables and also the force **P** required to support the 600-N force using the frictionless pulley system shown in Fig. 6–31a.

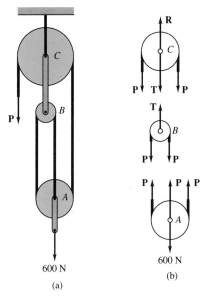

Fig. 6–31　(a)　(b)

SOLUTION

Free-Body Diagrams. A free-body diagram of each pulley *including* its pin and a portion of the contacting cable is shown in Fig. 6–31b. Since the cable is *continuous* and the pulleys are frictionless, the cable has a *constant tension* of P N acting throughout its length (see Example 5–7). The link connection between pulleys B and C is a two-force member and therefore it has an unknown tension of T N acting on it. Notice that the *principle of action, equal but opposite reaction* must be carefully observed for forces **P** and **T** when the *separate* free-body diagrams are drawn.

Equations of Equilibrium. The three unknowns are obtained as follows:

　Pulley A

$$+\uparrow \Sigma F_y = 0; \qquad 3P - 600 \text{ N} = 0 \qquad P = 200 \text{ N} \qquad\qquad Ans.$$

　Pulley B

$$+\uparrow \Sigma F_y = 0; \qquad\quad T - 2P = 0 \qquad T = 400 \text{ N} \qquad\qquad Ans.$$

　Pulley C

$$+\uparrow \Sigma F_y = 0; \qquad R - 2P - T = 0 \qquad R = 800 \text{ N} \qquad\qquad Ans.$$

Example 6–19

A man having a weight of 150 lb supports himself by means of the cable and pulley system shown in Fig. 6–32a. If the seat has a weight of 15 lb, determine the equilibrium force that he must exert on the cable at A and the force he exerts on the seat. Neglect the weight of the cables and pulleys.

(a)

SOLUTION I

Free-Body Diagrams. The free-body diagrams of the man, seat, and pulley C are shown in Fig. 6–32b. The *two* cables are subjected to tensions T_A and T_E, respectively. The man is subjected to three forces: his weight, the tension T_A of cable AC, and the reaction N_s of the seat.

Equations of Equilibrium. The three unknowns are obtained as follows:

  Man

$$+\uparrow \Sigma F_y = 0; \qquad T_A + N_s - 150 \text{ lb} = 0 \qquad (1)$$

  Seat

$$+\uparrow \Sigma F_y = 0; \qquad T_E - N_s - 15 \text{ lb} = 0 \qquad (2)$$

  Pulley C

$$+\uparrow \Sigma F_y = 0; \qquad 2T_E - T_A = 0 \qquad (3)$$

(b)

The magnitude of force T_E can be determined by adding Eqs. (1) and (2) to eliminate N_s and then using Eq. (3). The other unknowns are then obtained by resubstitution of T_E.

$$T_A = 110 \text{ lb} \qquad\qquad \textit{Ans.}$$
$$T_E = 55 \text{ lb}$$
$$N_s = 40 \text{ lb} \qquad\qquad \textit{Ans.}$$

SOLUTION II

Free-Body Diagram. By using the dashed section shown in Fig. 6–32a, the man, pulley, and seat can be considered as a *single system*, Fig. 6–32c. Here N_s and T_A are *internal* forces and hence are not included on the ''combined'' free-body diagram.

Equations of Equilibrium. Applying $\Sigma F_y = 0$ yields a *direct* solution for T_E.

$$+\uparrow \Sigma F_y = 0; \quad 3T_E - 15 \text{ lb} - 150 \text{ lb} = 0 \qquad T_E = 55 \text{ lb}$$

The other unknowns can be obtained from Eqs. (2) and (3).

(c)

Fig. 6–32

Example 6–20

  Determine the horizontal and vertical force components acting at the pin connections B and C of the frame shown in Fig. 6–33a.

Fig. 6–33

(a)

SOLUTION

Free-Body Diagrams. The free-body diagrams for members CD, BED, and the pulley are shown in Fig. 6–33b. Notice that member CD is a two-force member. Furthermore, the pulley is held in equilibrium by 981-N force components exerted on it by the pin at point E. By inspection, both force and moment equilibrium of the pulley is satisfied.

Equations of Equilibrium. The three unknowns will be determined by applying the equilibrium equations to member BED. We have

$$\zeta + \Sigma M_B = 0; \quad -981 \text{ N}(0.7 \text{ m}) + T_C \sin 45° \, (0.9 \text{ m}) = 0$$
$$T_C = 1079 \text{ N} \qquad \qquad Ans.$$

$$\xrightarrow{+} \Sigma F_x = 0; \quad B_x - 981 \text{ N} - 1079 \cos 45° \text{ N} = 0$$
$$B_x = 1740 \text{ N} \qquad \qquad Ans.$$

$$+ \uparrow \Sigma F_y = 0; \quad 1079 \sin 45° \text{ N} - 981 \text{ N} + B_y = 0$$
$$B_y = 218 \text{ N} \qquad \qquad Ans.$$

(b)

Example 6–21

The 100-kg block is held in equilibrium by means of the pulley and continuous cable system shown in Fig. 6–34a. If the cable is attached to the pin at B, compute the forces which this pin exerts on each of its connecting members.

SOLUTION

Free-Body Diagrams. A free-body diagram of each member of the frame is shown in Fig. 6–34b. By inspection, members AB and CB are two-force members. Furthermore, the cable must be subjected to a force of 490.5 N in order to hold pulley D in equilibrium. A free-body diagram of the pin at B is needed, since *four interactions* occur at this pin. These are caused by the attached cable (490.5 N), member AB ($\mathbf{F}_{AB}$), member CB ($\mathbf{F}_{CB}$), and pulley B ($\mathbf{B}_x$ and $\mathbf{B}_y$).

Equations of Equilibrium. Applying the equations of force equilibrium to pulley B, we have

$\xrightarrow{+}\Sigma F_x = 0; \quad B_x - 490.5 \cos 45° \text{ N} = 0 \qquad B_x = 346.8 \text{ N} \qquad\qquad Ans.$

$+\uparrow \Sigma F_y = 0; \qquad B_y - 490.5 \sin 45° \text{ N} - 490.5 \text{ N} = 0$

$$B_y = 837.3 \text{ N} \qquad\qquad Ans.$$

Using these results, equilibrium of the pin requires that

$+\uparrow \Sigma F_y = 0; \ \tfrac{4}{5}F_{CB} - 837.3 \text{ N} - 490.5 \text{ N} = 0 \quad F_{CB} = 1660 \text{ N} \qquad Ans.$

$\xrightarrow{+}\Sigma F_x = 0; \ F_{AB} - \tfrac{3}{5}(1660 \text{ N}) - 346.8 \text{ N} = 0 \quad F_{AB} = 1343 \text{ N} \qquad Ans.$

It may be noted that the two-force member CB is subjected to bending as caused by the force $\mathbf{F}_{CB}$. From the standpoint of design, it would be better to make this member *straight* (from C to B) so that the force $\mathbf{F}_{CB}$ would only create tension in the member.

(a)

Fig. 6–34a

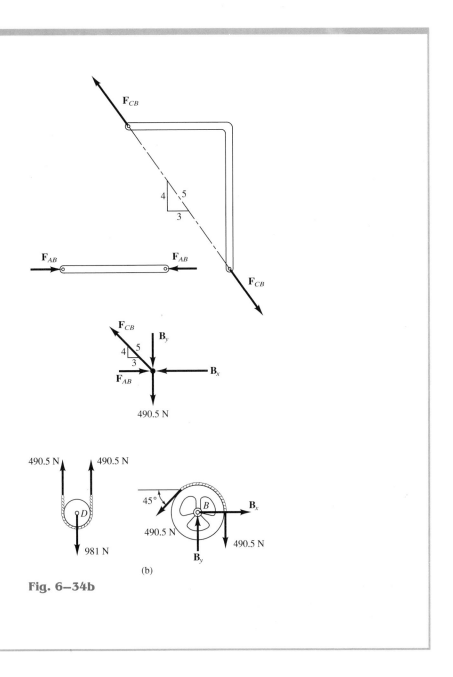

Fig. 6—34b

(b)

Before solving the following problems, it is suggested that a brief review be made of all the previous examples. This may be done by covering each solution and trying to locate the two-force members, drawing the free-body diagrams, and conceiving ways of applying the equations of equilibrium to obtain the solution.

PROBLEMS

6–45. The pipe cutter is clamped around the pipe P. If the wheel at A exerts a normal force of $F_A = 80$ N on the pipe, determine the normal forces of wheels B and C on the pipe. Also compute the pin reaction on the wheel at C. The three wheels each have a radius of 7 mm and the pipe has an outer radius of 10 mm.

Prob. 6–45

6–46. The compound beam is pin-supported at C and supported by a roller at A and B. There is a hinge (pin) at D. Determine the reactions at the supports.

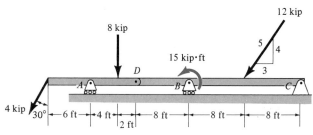

Prob. 6–46

6–47. The compound beam is fixed at A and supported by a rocker at B and C. There are hinges (pins) at D and E. Determine the reactions at the supports.

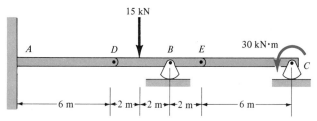

Prob. 6–47

*6–48. The compound beam is fixed supported at A and supported by rockers at B and C. If there are hinges (pins) at D and E, determine the reactions at the supports A, B, and C.

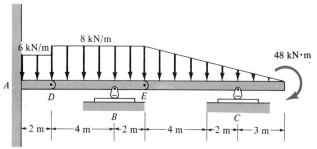

Prob. 6–48

6–49. Determine the reactions at the supports A and B.

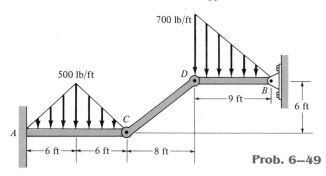

Prob. 6–49

6–50. Determine the horizontal and vertical components of force at pins A and C of the two-member frame.

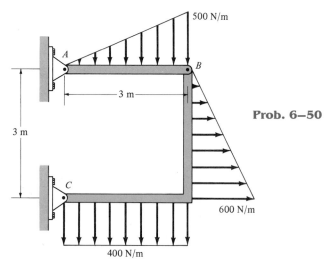

Prob. 6–50

6–51. Determine the horizontal and vertical components of force at pins A and C of the two-member frame.

Prob. 6–51

*6–52. Determine the horizontal and vertical components of force at each pin. The suspended cylinder has a weight of 80 lb.

Prob. 6–52

6–53. Determine the horizontal and vertical components of force which the pins exert on member ABC.

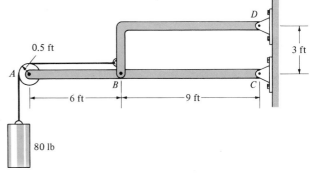

Prob. 6–53

6–54. Determine the horizontal and vertical components of force at pins A, B, and C, and the reactions at the fixed support D of the three-member frame.

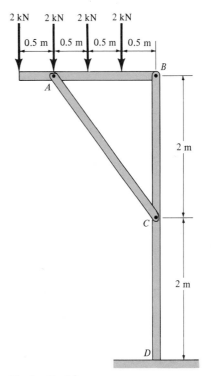

Prob. 6–54

265

6–55. Determine the horizontal and vertical force components at pins A, B, and C for equilibrium of the frame.

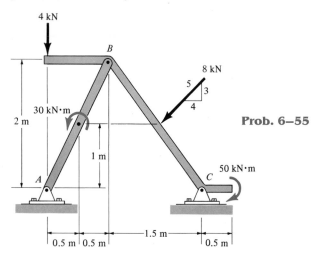

Prob. 6–55

*6–56. Determine the horizontal and vertical components of force which the pins at D and B exert on member BCD. The pin at C is attached to member BCD and passes through the smooth slot in member ACE.

Probs. 6–56/6–57

6–57. Determine the horizontal and vertical components of force which the pins at A and E exert on member ACE. The pin at C is attached to member BCD and passes through the smooth slot in member ACE.

6–58. Determine the horizontal and vertical components of force that the pins at B and C exert on member AC of the frame.

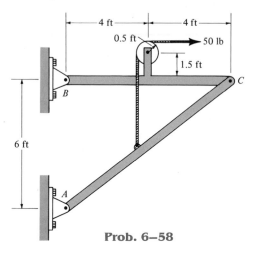

Prob. 6–58

6–59. Determine the horizontal and vertical components of force at pin D and the components of reaction at the fixed support B of member DCB. The pin at C is attached to member DCB and passes through the smooth slot in member ACE.

Probs. 6–59/6–60/6–61

*6–60. Determine the horizontal and vertical components of force which the pins at A and E exert on member ACE. The pin at C is attached to member DCB and passes through the smooth slot in member ACE.

6–61. Solve Prob. 6–59 by assuming the pin at C is attached to member ACE and a vertical slot is in member DCB.

6–62. Determine the horizontal and vertical components of force which the pins exert on member *AB* of the frame.

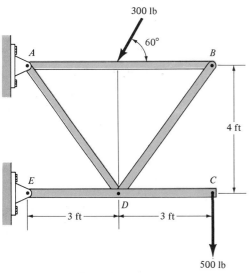

Probs. 6–62/6–63

6–63. Determine the horizontal and vertical components of force which the pins exert on member *EDC* of the frame.

*6–64. Determine the horizontal and vertical components of reaction at *A* and *B*. The pin at *C* is fixed to member *AE* and fits through a smooth slot in member *BD*.

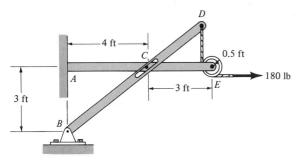

Prob. 6–64

6–65. Determine the horizontal and vertical components of force which the pins exert on member *ABC* of the frame.

Probs. 6–65/6–66

6–66. Determine the horizontal and vertical components of force which the pins exert on member *DE* of the frame.

6–67. Determine the horizontal and vertical components of force which the pins at *C* and *D* exert on member *ECD* of the frame.

Probs. 6–67/6–68

*6–68. Determine the horizontal and vertical components of force which the pins at *B* and *D* exert on member *ABD* of the frame.

6–69. The toggle clamp is subjected to a force **F** at the handle. Determine the vertical clamping force acting at *E*.

Prob. 6–69

6–70. The truck is used to tow the disabled car. If the car has a mass of 1.40 Mg and center of mass at G_C, determine the force acting along the pin-connected hydraulic cylinder at *B* (a two-force member) and the horizontal and vertical components of force developed at the pin *C*. The car exerts only vertical reactions at its front and back wheels *F* and *D*.

Probs. 6–70/6–71

6–71. If the truck in Prob. 6–70 has a mass of 2.30 Mg and center of mass at G_T, determine the vertical reaction of the ground on both its front and both its rear tires when it lifts the car in the position shown.

*6–72.** The lever mechanism for a machine press serves as a toggle which develops a large force at *E* when a small force is applied at the handle *H*. To show this is the case, compute the force at *E* if someone applies a vertical force of 80 N at *H*. The smooth head at *D* is able to slide freely downward. All members are pin-connected.

Prob. 6–72

6–73. The crab clamp is used to hold the two pieces of smooth wood together. If the compressive force developed in region *BC* of the screw is 500 N, determine the vertical clamping force on the wood. The screw is pin-connected at its end *B* and passes through the pin-connected (swivel) block at *C*.

6–74. The crab clamp is used to hold the two pieces of smooth wood together. If a vertical clamping force of 800 N is required on the wood, determine the compressive force that must be developed in region *BC* of the screw. The parts are pin-connected at all points indicated.

Prob. 6–74

6–75. The mechanism for a lift truck is shown. If the crate has a mass of 250 kg, determine the force required in the hydraulic cylinder *DF* needed to support the load in the position shown. All members are pin-connected. The pins at *A*, *E*, and *F* are fixed to the frame of the truck.

Prob. 6–75

6–77. The hydraulic, truck-mounted crane is used to lift a beam which has a mass of 1 Mg. Determine the horizontal and vertical components of force acting at each of the joints when the crane is in the position shown. Assume that all joints are pin-connected.

Prob. 6–77

*6–76. The symmetric coil tong supports the coil which has a mass of 800 kg and center of mass at *G*. Determine the horizontal and vertical components of force the linkage exerts on plate *DEIJH* at points *D* and *E*. The coil exerts only vertical reactions at *K* and *L*.

Prob. 6–76

6–78. The pumping unit is used to recover oil. When the walking beam *ABC* is horizontal, the force acting in the wireline at the well head is 250 lb. Determine the torque **M** which must be exerted by the motor in order to overcome this load. The horse-head *C* weighs 60 lb and has a center of gravity at G_c. The walking beam *ABC* has a weight of 130 lb and a center of gravity at G_B, and the counterweight has a weight of 200 lb and a center of gravity at G_w. The pitman, *AD*, is pin-connected at its ends and has negligible weight.

Prob. 6–78

6–79. The jack works on the principle of a ratchet, alternately using two pawls *AB* and *CD* for support. If the contacts at *E* and *F* are smooth, determine the force **P** that must be applied to the handle to lift a mass of 60 kg. When the handle is released, the pawl *CD* supports the load. If each pawl is a two-force member, determine the force in pawl *AB* when the load is lifted and the force in pawl *CD* when the handle is released. The jack handle is pinned to the jack frame at *G*.

Prob. 6–79

*6–80. The bucket of the backhoe and its contents have a weight of 1,500 lb and a center of gravity at *G*. Determine the forces in the hydraulic cylinder *LI* and in links *HI* and *KI* in order to hold the load in the position shown. The bucket is pinned at *J*. *Hint: LI, HI,* and *KI* are all two-force members.

Probs. 6–80/6–81

6–81. The bucket of the backhoe and its contents have a weight of 1500 lb and a center of gravity at *G*. Determine the horizontal and vertical components of reaction at the pins *F* and *A* and the forces in the hydraulic cylinders *DE* and *BC*.

6–82. The mechanism consists of identical meshed gears *A* and *B* and arms which are fixed to the gears. The spring attached to the ends of the arms has an unstretched length of 100 mm and a stiffness of *k* = 250 N/m. If a torque of *M* = 6 N · m is applied to gear *A*, determine the angle *θ* through which each arm rotates. The gears are each pinned to fixed supports at their centers.

Prob. 6–82

6–83. The flatbed trailer has a weight of 7000 lb and center of gravity at G_T. It is pin-connected to the cab at *D*. The cab has a weight of 6000 lb and center of gravity at G_C. Determine the range of values *x* for the position of the 2000-lb load *L* so that no axle is subjected to more than 5500 lb. The load has a center of gravity at G_L.

Prob. 6–83

*6–84. The toggle press is subjected to a horizontal force **P** which creates a vertical force **F** at the piston. Determine F as a function of P and θ, and plot the result of F versus θ for $0° \leq \theta \leq 90°$.

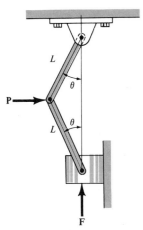

Prob. 6–84

6–85. If each of the three uniform links of the mechanism has a length L and weight W, determine the angle θ for equilibrium. The spring, which always remains vertical, is unstretched when $\theta = 0°$.

Prob. 6–85

6–86. The spring has an unstretched length of 0.3 m. Determine the angle θ for equilibrium if the uniform links each have a mass of 5 kg.

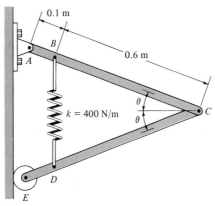

Prob. 6–86

6–87. Determine the horizontal and vertical components of reaction at pins A, B, and C of the two-member arch.

Prob. 6–87

271

*6–88. The two-bar mechanism consists of a lever arm AB and smooth link CD which has a fixed collar at its end C and a roller at the other end D. Determine the force **P** needed to hold the lever in the position θ. The spring has a stiffness k and unstretched length of 2L. The roller contacts either the top or bottom portion of the horizontal guide.

Prob. 6–88

6–89. The structure is subjected to the force of 450 lb which lies in a plane parallel to the y-z plane. Member AB is supported by a ball-and-socket joint at A and fits through a snug hole at B. Member CD is supported by a pin at C. Determine the x, y, z components of reaction at A and C.

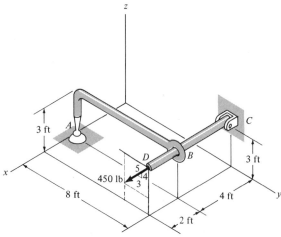

Prob. 6–89

6–90. The structure is subjected to the loading shown. Member AD is supported by a cable AB and roller at C and fits through a smooth circular hole at D. Member ED is supported by a roller at D and a pole that fits in a smooth snug circular hole at E. Determine the x, y, z components of reaction at E and the tension in cable AB.

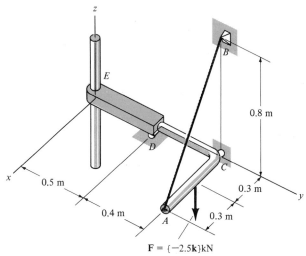

Prob. 6–90

6–91. The four-member "A" frame supports a vertical force of 600 N. If it is assumed that the supports at A and E are smooth collars, G is a pin, and all other joints are ball-and-sockets, determine the x, y, z force components which member BD exerts on members EDC and FG.

Prob. 6–91

*6–92. The three-member frame is connected at its ends using ball-and-socket joints. Determine the x, y, z components of reaction at B and the tension in member ED. The force acting at D is $\mathbf{F} = \{250\mathbf{i} - 350\mathbf{k}\}$ lb.

6–93. The three-member frame is connected at its ends using ball-and-socket joints. Determine the x, y, z components of reaction at B and the tension in member ED. The force acting at D is $\mathbf{F} = \{135\mathbf{i} + 200\mathbf{j} - 180\mathbf{k}\}$ lb.

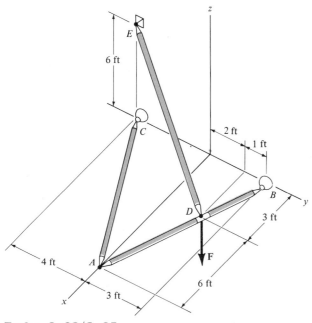

Probs. 6–92/6–93

REVIEW PROBLEMS

6–94. For the given loading, determine the force in members *CD*, *CJ*, and *KJ* of the *Pratt roof truss*. Indicate whether the members are in tension or compression.

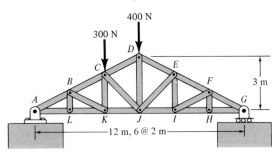

Prob. 6–94

6–95. The hoist supports the 125-kg engine. Determine the force the load creates in member *DB* and in member *FB*, which contains the hydraulic cylinder *H*.

Prob. 6–95

*6–96. Determine the force **P** needed to suspend the 80-lb weight. Also, what are the reactions at *A* and *B*?

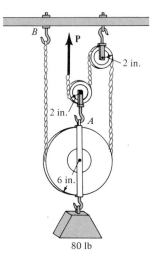

Prob. 6–96

6–97. Determine the force developed in members *GB* and *GF* of the bridge truss, and indicate whether the members are in tension or compression.

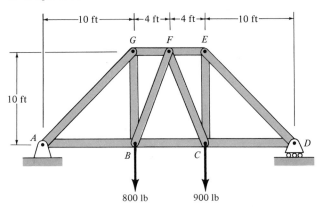

Prob. 6–97

274

6–98. Determine the force in each member of the truss and indicate whether the members are in tension or compression.

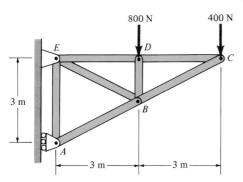

Prob. 6–98

6–99. Determine the resultant forces at pins B and C on member ABC of the four-member frame.

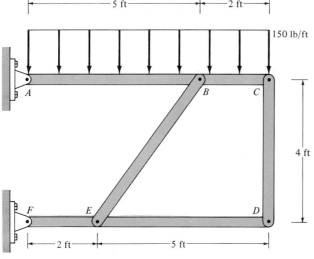

Prob. 6–99

*6–100. Compute the force in members BG, HG, and BC of the truss and indicate whether the members are in tension or compression.

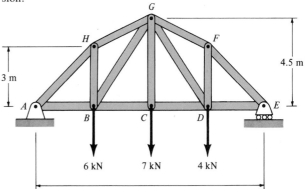

Prob. 6–100

6–101. Determine the force in members CP, PE, and KR of the truss and indicate whether the members are in tension or compression. Also, indicate all zero-force members.

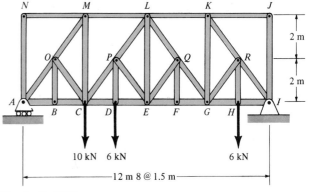

Prob. 6–101

6–102. Determine the force that the jaws J of the metal cutters exert on the smooth cable C if 100-N forces are applied to the handles. The jaws are pinned at E and A, and D and B. There is also a pin at F.

6–103. Determine the force in members BD, AD, and AF of the space truss and state whether the members are in tension or compression. The truss is supported by short links at A, B, D, and F. The external reactions are listed as the answers to Prob. 5–92.

Prob. 6–102

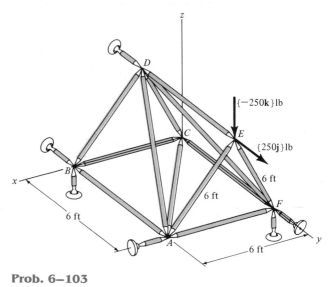

Prob. 6–103

7

Internal Forces

In this chapter we will develop a technique for finding the internal loading at specific *points* within a structural member, and then we will generalize this method to find the point-to-point variation of loading along the axis of a beam. A graph showing this variation of internal load will allow us to find the critical points where the *maximum* internal loading occurs. This information is important for the proper design of a structure using the theory of mechanics of materials. The analysis of cables will be treated in the last part of the chapter.

7.1 Internal Forces Developed in Structural Members

The design of a structural member requires an investigation of the forces acting *within* the member which are necessary to balance the forces acting external to it. The *method of sections* can be used to determine these *internal* forces. In order to illustrate the procedure, consider the "simply supported" beam shown in Fig. 7–1*a* which is subjected to the forces $\mathbf{F}_1$ and $\mathbf{F}_2$. The *support reactions* $\mathbf{A}_x$, $\mathbf{A}_y$, and $\mathbf{B}_y$ can be determined by applying the equations of equilibrium, using the free-body diagram of the *entire beam,* Fig. 7–1*b*. If the *internal loadings* at point *C* are to be determined, it is necessary to pass an imaginary section through the beam, cutting it into two segments at that point, Fig. 7–1*a*. Doing this "exposes" the internal forces as *external* on the free-body diagram of each segment, Fig. 7–1*c*. Since both segments (*AC* and *BC*) were prevented from translating *and* rotating *before* the beam was sectioned, equilibrium of each segment is maintained if rectangular force components $\mathbf{A}_C$ and $\mathbf{V}_C$ and a resultant couple moment $\mathbf{M}_C$ are developed at the cut section. Note that these loadings must be equal in magnitude and opposite in direction on each of the segments (Newton's third law). The magnitude of each unknown can be determined by applying the three equations of equilibrium to either segment *AC* or *CB*. A *direct solution* for $\mathbf{A}_C$ is obtained by applying $\Sigma F_x = 0$; $\mathbf{V}_C$ is obtained directly from $\Sigma F_y = 0$; and $\mathbf{M}_C$ is determined by summing moments about point *C*, $\Sigma M_C = 0$, in order to eliminate the moments of the unknowns $\mathbf{A}_C$ and $\mathbf{V}_C$.

(a)

(b)

Fig. 7–1

(c)

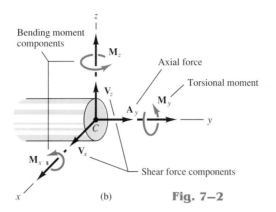

(a)

(b)

Fig. 7–2

In structural mechanics, the force components **A,** acting normal to the beam at the cut section, and **V,** acting tangent to the section, are termed the *axial force* and *shear force,* respectively. The couple moment **M** is referred to as the *bending moment,* Fig. 7–2a. In three dimensions, a general internal force and moment resultant will act at the section. The *x, y, z* components of these loadings are shown in Fig. 7–2b. Here A_y is the *axial force,* V_x and V_z are *shear force components,* M_y is a *torsional or twisting moment,* and M_x and M_z are *bending moment components.* For structural applications, these *resultant loadings* are often represented at the geometric center or centroid (*C*) of the section's cross-sectional area. In general, the magnitude for each loading will be different at various points along the axis of a structural member. In all cases, however, the method of sections can be used to determine their values.

Free-Body Diagrams

Since frames and machines are composed of *multiforce members,* each of these members will generally be subjected to axial, shear, and bending loadings. For example, consider the frame shown in Fig. 7–3a. If the dashed section is passed through the frame to determine the internal loadings at points *F, G,* and *H,* the resulting free-body diagram of the top portion of this section is shown in Fig. 7–3b. At each point where a member is sectioned there is an unknown axial force, shear force, and bending moment. As a result, we cannot apply the *three* equations of equilibrium to this section in order to obtain these *nine unknowns.** Instead, to solve this problem one has to *first dismember* the frame and determine the reactions at the connections of the members using the techniques of Sec. 6.6. Once this is done, one can section *each member* at its appropriate point and then apply the three equations of equilibrium to determine **A, V,** and **M.** For example, the free-body diagram of segment *DG,* Fig. 7–3c, can be used to determine the internal loadings at *G* provided the pin reactions D_x and D_y are known.

*Recall that this method of analysis worked well for trusses since truss members are *straight two-force members* which only support an axial load.

(a)

(b)

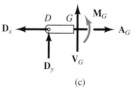

(c)

Fig. 7–3

PROCEDURE FOR ANALYSIS

The following procedure provides a means for applying the method of sections to determine the internal loadings at a specific location in a member.

Support Reactions.   Before the member is "cut" or sectioned, it may first be necessary to determine the member's support reactions, so that the equilibrium equations are used only to solve for the internal loadings when the member is sectioned. If the member is part of a frame or machine, the reactions at its connections are determined using the methods of Sec. 6.6.

Free-Body Diagram.   Keep all distributed loadings, couple moments, torques, and forces acting on the member in their *exact location,* then pass an imaginary section through the member, perpendicular to its axis at the point where the internal loading is to be determined. Draw a free-body diagram of one of the "cut" segments on either side of the section and indicate the x, y, z components of the force and moment resultants at the section. In particular, if the member is subjected to a *coplanar* system of forces, only **A, V,** and **M** act at the section. In many cases it may be possible to tell by inspection the proper sense of the unknown loadings; however, if this seems difficult, the sense can be assumed.

Equations of Equilibrium.   Apply the equations of equilibrium to obtain the unknown internal loadings. Generally, moments should be summed at the section about axes passing through the *centroid* or geometric center of the member's cross-sectional area, in order to eliminate the unknown axial and shear forces and thereby obtain direct solutions for the moment components. If the solution of the equilibrium equations yields a negative scalar, the assumed sense of the quantity is opposite to that shown on the free-body diagram.

The following examples numerically illustrate this procedure.

Example 7–1

The bar is fixed at its end and is loaded as shown in Fig. 7–4a. Determine the internal axial force at sections through points B and C.

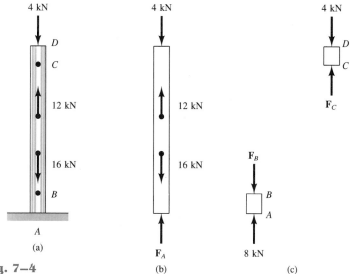

Fig. 7–4

SOLUTION

Support Reactions. A free-body diagram of the entire bar is shown in Fig. 7–4b. By inspection, only an axial force $\mathbf{F}_A$ acts at the fixed support since the loads are symmetrically applied along the bar's axis.

$$+\uparrow\Sigma F_z = 0; \quad F_A - 16\text{ kN} + 12\text{ kN} - 4\text{ kN} = 0 \qquad F_A = 8\text{ kN}$$

Free-Body Diagrams. The internal forces at B and C will be found using the free-body diagrams of the sectioned bar shown in Fig. 7–4c. In particular, segments AB and DC will be chosen here since they contain the *least* number of forces.

Equations of Equilibrium
   Segment AB

$$+\uparrow\Sigma F_z = 0; \qquad 8\text{ kN} - F_B = 0 \qquad F_B = 8\text{ kN} \qquad\qquad Ans.$$

   Segment DC

$$+\uparrow\Sigma F_z = 0; \qquad F_C - 4\text{ kN} = 0 \qquad F_C = 4\text{ kN} \qquad\qquad Ans.$$

Try working this problem in the following manner: Determine F_B from segment BD. (Note that this approach does *not require* solution for the support reaction at A.) Using the result for F_B, isolate segment BC to determine F_C.

Example 7–2

The circular shaft is subjected to three concentrated torques as shown in Fig. 7–5a. Determine the internal torques at sections through points B and C.

Fig. 7–5a   (a)

SOLUTION

Support Reactions.   Since the shaft is subjected only to collinear torques, a torque reaction occurs at the support, Fig. 7–5b. Using the right-hand rule to define the positive directions of the torques, we require

$\Sigma M_x = 0;$
$$-10 \text{ N} \cdot \text{m} + 15 \text{ N} \cdot \text{m} + 20 \text{ N} \cdot \text{m} - T_D = 0 \qquad T_D = 25 \text{ N} \cdot \text{m}$$

(b)   (c)

Fig. 7–5(b, c)

Free-Body Diagrams.   The internal torques at B and C will be found using the free-body diagrams of the shaft segments AB and CD shown in Fig. 7–5c.

Equations of Equilibrium.   Applying the equation of moment equilibrium along the shaft's axis, we have

  Segment AB

$\Sigma M_x = 0;$ $-10 \text{ N} \cdot \text{m} + 15 \text{ N} \cdot \text{m} - T_B = 0$ $T_B = 5 \text{ N} \cdot \text{m}$ *Ans.*

  Segment CD

$\Sigma M_x = 0;$ $T_C - 25 \text{ N} \cdot \text{m} = 0$ $T_C = 25 \text{ N} \cdot \text{m}$ *Ans.*

Try to solve for T_C by using segment CA. Note that this approach does not require a solution for the support reaction at D.

Example 7–3

Determine the internal axial force, shear force, and bending moment acting just to the left, point B, and just to the right, point C, of the 6-kN concentrated force shown on the beam in Fig. 7–6a.

(a)

SOLUTION

Support Reactions. The free-body diagram of the beam is shown in Fig. 7–6b. When computing the external reactions, note that the 9-kN · m couple is a free vector and can therefore act anywhere on the beam. Here we will only determine A_y since segments AB and AC will be used for the analysis.

$$\zeta + \Sigma M_D = 0; \quad 9 \text{ kN} \cdot \text{m} + (6 \text{ kN})(6 \text{ m}) - A_y(9 \text{ m}) = 0$$
$$A_y = 5 \text{ kN}$$

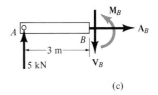

(b)

Free-Body Diagrams. The free-body diagrams of the left segments AB and AC of the beam are shown in Fig. 7–6c and d. Note that the 9-kN · m couple moment is *not included* on these diagrams. Although it is a free vector, the couple moment must be kept in its original position until *after* the section is made and the appropriate body isolated; only then will the true loading situation be represented.

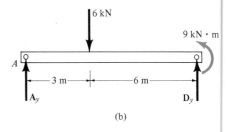

(c)

Equations of Equilibrium
  Segment AB

$$\xrightarrow{+} \Sigma F_x = 0; \quad A_B = 0 \qquad \qquad \text{Ans.}$$
$$+ \uparrow \Sigma F_y = 0; \quad 5 \text{ kN} - V_B = 0 \quad V_B = 5 \text{ kN} \qquad \text{Ans.}$$
$$\zeta + \Sigma M_B = 0; \quad -(5 \text{ kN})(3 \text{ m}) + M_B = 0 \quad M_B = 15 \text{ kN} \cdot \text{m} \qquad \text{Ans.}$$

  Segment AC

$$\xrightarrow{+} \Sigma F_x = 0; \quad A_C = 0 \qquad \qquad \text{Ans.}$$
$$+ \uparrow \Sigma F_y = 0; \quad 5 \text{ kN} - 6 \text{ kN} - V_C = 0 \quad V_C = -1 \text{ kN} \qquad \text{Ans.}$$
$$\zeta + \Sigma M_C = 0; \quad -(5 \text{ kN})(3 \text{ m}) + M_C = 0 \quad M_C = 15 \text{ kN} \cdot \text{m} \qquad \text{Ans.}$$

(d)

Note that the moment arm for the 5-kN force in both cases is approximately 3 m, since B and C are "almost" coincident.

Fig. 7–6

Example 7–4

Determine the internal axial force, shear force, and bending moment acting at a section through point B of the two-member frame shown in Fig. 7–7a.

SOLUTION

Support Reactions. A free-body diagram of each member is shown in Fig. 7–7b. Since CD is a two-force member, the equations of equilibrium have to be applied only to member AC.

$\zeta+\Sigma M_A = 0;$ $-400 \text{ lb}(4 \text{ ft}) + (\frac{3}{5})F_{DC}(8 \text{ ft}) = 0$ $F_{DC} = 333.3 \text{ lb}$

$\xrightarrow{+}\Sigma F_x = 0;$ $-A_x + (\frac{4}{5})(333.3 \text{ lb}) = 0$ $A_x = 266.7 \text{ lb}$

$+\uparrow\Sigma F_y = 0;$ $A_y - 400 \text{ lb} + \frac{3}{5}(333.3 \text{ lb}) = 0$ $A_y = 200 \text{ lb}$

Free-Body Diagrams. Passing an imaginary section perpendicular to the axis of member AC through point B yields the free-body diagrams of segments AB and BC shown in Fig. 7–7c. Notice that it is important when constructing these diagrams to keep the distributed loading exactly as it is until *after* the section is made. Only then can it be simplified to a single resultant force. Why? Also, notice that $\mathbf{A}_B$, $\mathbf{V}_B$, and $\mathbf{M}_B$ act with equal magnitude but opposite direction on each segment—Newton's third law.

Equations of Equilibrium. Applying the equations of equilibrium to segment AB, we have

$\xrightarrow{+}\Sigma F_x = 0;$ $A_B - 266.7 \text{ lb} = 0$ $A_B = 266.7 \text{ lb}$ *Ans.*

$+\uparrow\Sigma F_y = 0;$ $200 \text{ lb} - 200 \text{ lb} - V_B = 0$ $V_B = 0$ *Ans.*

$\zeta+\Sigma M_B = 0;$ $M_B - 200 \text{ lb}(4 \text{ ft}) + 200 \text{ lb}(2 \text{ ft}) = 0$

$$M_B = 400 \text{ lb} \cdot \text{ft}$$ *Ans.*

As an exercise, try to obtain the foregoing results using segment BC.

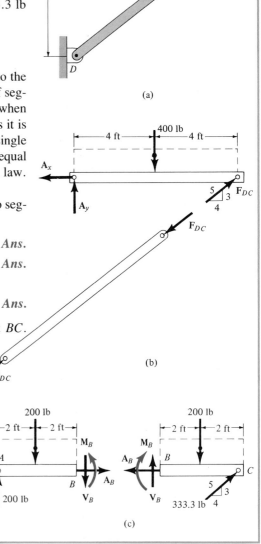

Fig. 7–7

Example 7–5

Determine the axial force, shear force, and bending moment acting at point E of the frame loaded as shown in Fig. 7–8a.

(a)

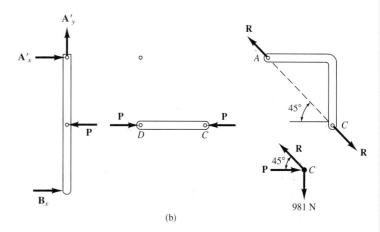

(b)

Fig. 7–8

SOLUTION

Support Reactions. The free-body diagram of each member of the frame is shown in Fig. 7–8b. Note that members AC and CD are two-force members. The pins at A and C are also included here since there are *three interactions* at each pin. At pin A A_x and A_y represent the effect of the support on the pin, A_x' and A_y' represent the effect of member ADB on the pin, and R represents the effect of member AC on the pin. Identify the three forces acting on pin C.

Applying the force equilibrium equation in the vertical direction to pin C gives

$$+\uparrow \Sigma F_y = 0; \quad R \sin 45° - 981 \text{ N} = 0 \qquad R = 1387.6 \text{ N}$$

Since we wish to determine the external loadings at E, it is not necessary to determine **P**.

(c)

Free-Body Diagram. The free-body diagram of segment CE is shown in Fig. 7–8c.

Equations of Equilibrium

$$\xrightarrow{+}\Sigma F_x = 0; \quad 1387.6 \cos 45° \text{ N} - V_E = 0 \qquad V_E = 981 \text{ N} \qquad Ans.$$

$$+\uparrow \Sigma F_y = 0; \quad -1387.6 \sin 45° \text{ N} + A_E = 0 \qquad A_E = 981 \text{ N} \qquad Ans.$$

$$\zeta+\Sigma M_E = 0; \quad 1387.6 \cos 45° \text{ N}(0.5 \text{ m}) - M_E = 0 \qquad M_E = 490 \text{ N} \cdot \text{m} \qquad Ans.$$

285

Example 7–6

A force of $\mathbf{F} = \{-3\mathbf{i} + 7\mathbf{j} - 4\mathbf{k}\}$ kN acts at the corner of a beam extended from a fixed wall as shown in Fig. 7–9a. Determine the internal loadings at a section passing through point A.

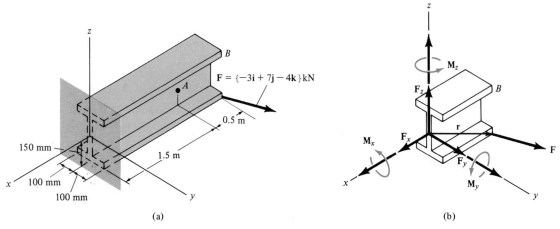

Fig. 7–9

SOLUTION

This problem can be solved by considering a segment of the beam that will *not* involve the support reactions.

Free-Body Diagram. The free-body diagram of segment AB is shown in Fig. 7–9b. The components of the resultant force $\mathbf{F}_A$ and moment $\mathbf{M}_A$ pass through the *centroid* or geometric center of the cross-sectional area at A.

Equations of Equilibrium

$$\Sigma\mathbf{F} = 0; \qquad \mathbf{F}_A - 3\mathbf{i} + 7\mathbf{j} - 4\mathbf{k} = 0$$

$$\mathbf{F}_A = \{3\mathbf{i} - 7\mathbf{j} + 4\mathbf{k}\} \text{ kN} \qquad \textit{Ans.}$$

$$\Sigma\mathbf{M}_A = 0; \quad \mathbf{M}_A + \mathbf{r}\times\mathbf{F} = \mathbf{M}_A + \begin{vmatrix} \mathbf{i} & \mathbf{j} & \mathbf{k} \\ -0.5 & 0.1 & -0.15 \\ -3 & 7 & -4 \end{vmatrix} = 0$$

$$\mathbf{M}_A = \{-0.650\mathbf{i} + 1.55\mathbf{j} + 3.20\mathbf{k}\} \text{ kN} \cdot \text{m} \qquad \textit{Ans.}$$

Note that $\mathbf{F}_x = \{3\mathbf{i}\}$ kN represents the axial force A, whereas $\mathbf{F}_y = \{-7\mathbf{j}\}$ kN and $\mathbf{F}_z = \{4\mathbf{k}\}$ kN are components of the shear force $V = \sqrt{F_y^2 + F_z^2}$. Also, the torsional moment is $\mathbf{M}_x = \{-0.65\mathbf{i}\}$ kN · m, and the bending moment is determined from its components $\mathbf{M}_y = \{1.55\mathbf{j}\}$ kN · m and $\mathbf{M}_z = \{3.20\mathbf{k}\}$ kN · m; i.e., $M_b = \sqrt{(M_y)^2 + (M_z)^2}$.

PROBLEMS

7–1. Three torques act on the shaft. Determine the torques at sections passing through points A, B, C, and D.

Prob. 7–1

7–2. The shaft is supported by smooth bearings at A and B and subjected to the torques shown. Determine the torques at sections passing through points C, D, and E.

Prob. 7–2

7–3. Three torques act on the shaft. Determine the torques at sections passing through points A, B, C, and D.

Prob. 7–3

*7–4. Two beams are attached to the column such that structural connections transmit the loads shown. Determine the axial force, shear force, and bending moment acting in the column at a section passing through point A.

Prob. 7–4

7–5. Determine the force **T** needed to hold the bell crank in equilibrium, then draw a free-body diagram of the pin at A and calculate the shearing force in the pin between the bell crank and the pin leaf at B.

Prob. 7–5

287

7–6. Determine the shear force and moment at the beam sections passing through points C and D.

Prob. 7–6

7–7. Determine the shear and moment acting at a section passing through point C in the beam.

Prob. 7–7

*7–8. Determine the shear, axial load, and moment at a section passing through point C.

Prob. 7–8

7–9. Determine the axial force, shear force, and moment at a section passing through (a) point C, which is just to the right of the roller at A, and (b) point D, which is just to the left of the 3000-lb concentrated force.

Prob. 7–9

7–10. Determine the axial force, shear force, and moment at a section passing through point C. Assume the support at A can be approximated by a pin and B acts as a roller.

Prob. 7–10

7–11. The pliers are used to grip the tube. If a force of 20 lb is applied to the handles, determine the shear force and moment at a section passing through point E. Assume the jaws of the pliers exert only normal forces on the tube.

Prob. 7–11

*7–12. Determine the axial force, shear force, and moment at sections passing through points C and D of the beam.

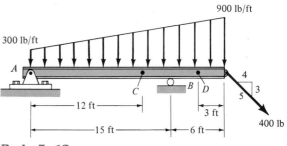

Prob. 7–12

■**7–13.** Determine the shear force and moment acting at a section passing through point C of the beam. Use Simpson's rule to evaluate the integrals used for the calculation.

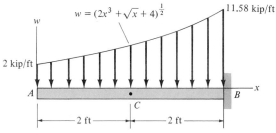

Prob. 7–13

7–14. Determine the axial force, shear force, and moment acting at a section passing through point C.

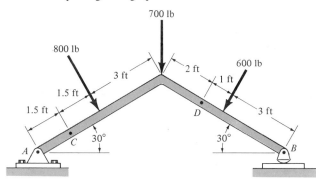

Probs. 7–14/7–15

7–15. Determine the axial force, shear force, and moment acting at a section passing through point D.

***7–16.** Determine the ratio of a/b for which the shear force will be zero at the midpoint C of the beam.

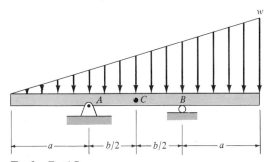

Prob. 7–16

7–17. The 9-kip force is supported by the floor panel DE, which in turn is simply supported at its ends by floor beams. These beams transmit their loads to the girder AB. Determine the shear and moment acting at a section passing through point C in the girder.

Prob. 7–17

7–18. Determine the axial force, shear force, and moment in the beam at sections passing through points D and E. Point E is just to the right of the 4-kip load.

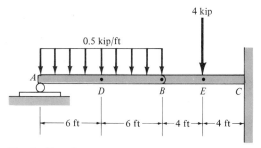

Prob. 7–18

7–19. Determine the axial force, shear force, and moment at sections passing through points E and F. Member BC is pinned at B and there is a smooth slot in it at C. The pin at C is fixed to member CD.

Prob. 7–19

289

*7–20. Determine the axial force, shear force, and moment at a section passing through point C.

Prob. 7–20

7–21. Determine the axial force, shear force, and moment at a section passing through point D of the two-member frame.

Prob. 7–21

7–22. Determine the axial force, shear force, and moment at a section passing through point D of the two-member frame.

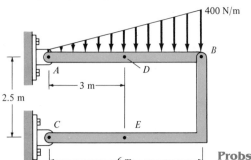

Probs. 7–22/7–23

7–23. Determine the axial force, shear force, and moment at a section passing through point E of the two-member frame.

*7–24. Determine the axial force, shear force, and moment acting at sections passing through points B and C on the curved member.

Prob. 7–24

7–25. Two forces act at the corners of the concrete block. If the block has a specific weight of 150 lb/ft^3, determine the x, y, z components of force and moment acting at a horizontal section taken through point O.

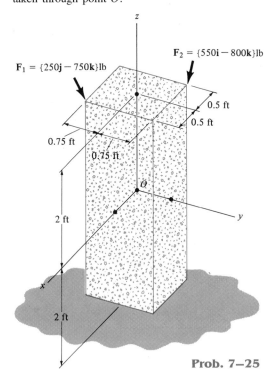

Prob. 7–25

7–26. A vertical force of 80 N is applied to the handle of the pipe wrench as shown. Determine the x, y, z components of internal force and moment at point A.

Prob. 7–26

7–27. Two forces act on the beam. Determine the x, y, z components of force and moment acting at a vertical section passing through point A.

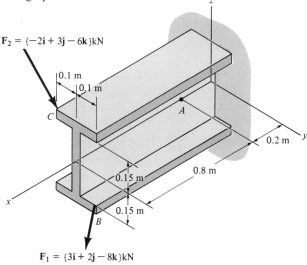

Prob. 7–27

*7–28. Determine the x, y, z components of force and moment at a section passing through point C in the pipe assembly. Neglect the weight of the pipe. Take $\mathbf{F}_1 = \{350\mathbf{j} - 400\mathbf{k}\}$ lb and $\mathbf{F}_2 = \{150\mathbf{i} - 300\mathbf{k}\}$ lb.

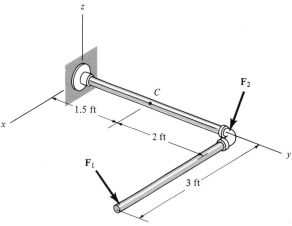

Probs. 7–28/7–29

7–29. Determine the x, y, z components of force and moment at a section passing through point C in the pipe assembly. Neglect the weight of the pipe. Take $\mathbf{F}_1 = \{-80\mathbf{i} + 200\mathbf{j} - 300\mathbf{k}\}$ lb and $\mathbf{F}_2 = \{250\mathbf{i} - 150\mathbf{j} - 200\mathbf{k}\}$ lb.

7–30. Determine the x, y, z components of force and moment at a section passing through point D of the rod. Take $\mathbf{F} = \{-7\mathbf{i} + 12\mathbf{j} - 5\mathbf{k}\}$ kN. The supports at A, B, and C are journal bearings.

Prob. 7–30

291

*7.2 Shear and Moment Equations and Diagrams

Beams are structural members which are designed to support loadings applied perpendicular to their axis. In general, beams are long, straight bars, having a constant cross-sectional area. Often beams are classified as to how they are supported. For example, a *simply supported beam* is pinned at one end and roller supported at the other, Fig. 7–10, whereas a *cantilevered beam* is fixed at one end and free at the other. The actual design of a beam requires a detailed knowledge of the *variation* of the internal shear force *V* and bending moment *M* acting at *each point* along the axis of the beam. After the force and bending-moment analysis is complete, one can then use the theory of mechanics of materials to determine the beam's required cross-sectional area.

The *variations* of *V* and *M* as a function of the position *x* of an *arbitrary point* along the beam's axis can be obtained by using the method of sections discussed in Sec. 7.1. Here, however, it is necessary to locate the imaginary section at an arbitrary distance *x* from the end of the beam rather than at a specified point. If the results are plotted, the graphical variations of *V* and *M* as a function of *x* are termed the *shear diagram* and *bending-moment diagram,* respectively.

In general, the internal shear and bending-moment variations will be discontinuous, or their slope will be discontinuous at points where a distributed load changes or where concentrated forces or couples are applied. Because of this, shear and bending-moment functions must be determined for *each segment* of the beam located between any two discontinuities of loading. For example, sections located at x_1, x_2, and x_3 will have to be used to describe the variation of *V* and *M* throughout the length of the beam in Fig. 7–10. These functions will be valid *only* within regions from *O* to *a* for x_1, from *a* to *b* for x_2, and from *b* to *l* for x_3.

The internal axial force will not be considered in the following discussion for two reasons. In most cases, the loads applied to a beam act perpendicular to the beam's axis and hence produce only an internal shear force and bending moment. For design purposes, the beam's resistance to shear, and particularly to bending, is more important than its ability to resist axial force.

Fig. 7–10

Sign Convention

Before presenting a method for determining the shear and bending moment as functions of *x* and later plotting these functions (shear and bending-moment diagrams), it is first necessary to establish a *sign convention* so as to define "positive" and "negative" shear force and bending moment acting in the beam. [This is analogous to assigning coordinate directions *x* positive to the right and *y* positive upward, when plotting a function $y = f(x)$.] Although the choice of a sign convention is arbitrary, here we will choose the one used in the majority of books on mechanics. It is illustrated in Fig. 7–11*a*. On the *left-hand face* (L.H.F.) of a beam segment, the internal shear force **V** acts downward and the internal moment **M** acts counterclockwise. In accordance with Newton's third law, an equal and opposite shear force and bending moment must act on the *right-hand face* (R.H.F.) of a segment. Perhaps an easy way to remember this sign convention is to isolate a small beam segment and note that *positive shear tends to rotate the segment clockwise*, Fig. 7–11*b*, and a *positive bending moment tends to bend the segment, if it were elastic, concave upward*, Fig. 7–11*c*.

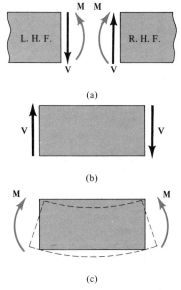

Fig. 7–11

PROCEDURE FOR ANALYSIS

The following procedure provides a method for constructing the shear and bending-moment diagrams for a beam.

Support Reactions. Determine all the reactive couples and forces acting on the beam and resolve the forces into components acting perpendicular and parallel to the beam's axis.

Shear and Moment Functions. Specify separate coordinates *x* having an origin at the beam's *left end* and extending to regions of the beam *between* concentrated forces and/or couples and where there is no discontinuity of distributed loading. Section the beam perpendicular to its axis at each distance *x* and from the free-body diagram of one of the segments, determine the unknowns *V* and *M* at the cut section as functions of *x*. On the free-body diagram, **V** and **M** should be shown acting in their *positive sense,* in accordance with the sign conventions given in Fig. 7–11. *V* is obtained from $\Sigma F_y = 0$ and *M* is obtained by summing moments about point *S* located at the cut section, $\Sigma M_S = 0$.

Shear and Moment Diagrams. Plot the shear diagram (*V* versus *x*) and the moment diagram (*M* versus *x*). If computed values of the functions describing *V* and *M* are *positive,* the values are plotted above the *x* axis, whereas *negative* values are plotted below the *x* axis. Generally, it is convenient to plot the shear and bending-moment diagrams directly below the free-body diagram of the beam.

The following examples numerically illustrate this procedure.

Example 7–7

Draw the shear and bending-moment diagrams for the beam shown in Fig. 7–12a.

(a)

SOLUTION

Support Reactions. The support reactions have been computed, Fig. 7–12a.

Shear and Moment Functions. The beam is sectioned at an arbitrary distance x from point A, extending within the region AB, and the free-body diagram of the left segment is shown in Fig. 7–12b. The unknowns **V** and **M** are indicated acting in the *positive sense* on the right-hand face of the segment according to the established sign convention. Why? Applying the equilibrium equations yields

(b)

$$+ \uparrow \Sigma F_y = 0; \qquad V = 25 \text{ kN} \qquad (1)$$
$$\zeta + \Sigma M_S = 0; \qquad M = 25x \text{ kN} \cdot \text{m} \qquad (2)$$

A free-body diagram for a left segment of the beam extending a distance x within the region BC is shown in Fig. 7–12c. As always, **V** and **M** are shown acting in the positive sense. Hence,

(c)

$$+ \uparrow \Sigma F_y = 0; \qquad 25 - 50 - V = 0$$
$$V = -25 \text{ kN} \qquad (3)$$
$$\zeta + \Sigma M_S = 0; \qquad M + 50(x - 2) - 25(x) = 0$$
$$M = (100 - 25x) \text{ kN} \cdot \text{m} \qquad (4)$$

Shear and Moment Diagrams. When Eqs. (1) to (4) are plotted within the regions in which they are valid, the shear and bending-moment diagrams shown in Fig. 7–12d are obtained. The shear diagram indicates that the internal shear force is always 25 kN (positive) within beam segment AB. Just to the right of point B, the shear force changes sign and remains at a constant value of -25 kN for segment BC. The moment diagram starts at zero, increases linearly to point C at $x = 2$ m, where $M_{max} = 25$ kN(2 m) $= 50$ kN · m, and thereafter decreases back to zero.

It is seen in Fig. 7–12d that the graph of the shear and moment diagrams is discontinuous at points of concentrated force, i.e., points A, B, and C. For this reason, as stated earlier, it is necessary to express both the shear and bending-moment functions separately for regions between concentrated loads. It should be realized, however, that all loading discontinuities are mathematical, arising from the *idealization of a concentrated force and couple moment*. Physically, all loads are applied over a finite area, and if this load variation could be accounted for, the shear and bending-moment diagrams would actually be continuous over the beam's entire length.

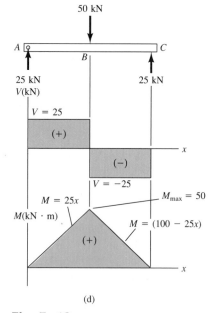

(d)

Fig. 7–12

Example 7–8

Draw the shear and bending-moment diagrams for the beam shown in Fig. 7–13a.

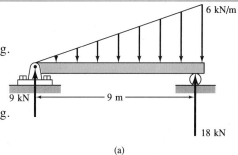

(a)

SOLUTION

Support Reactions. The support reactions have been computed, Fig. 7–13a.

Shear and Moment Functions. A free-body diagram for a left segment of the beam having a length x is shown in Fig. 7–13b. The distributed loading acting on this segment has an intensity of $\frac{2}{3}x$ at its end and is replaced by a resultant force *after* the segment is isolated as a free-body diagram. The *magnitude* of the resultant force is equal to $\frac{1}{2}(x)(\frac{2}{3}x) = \frac{1}{3}x^2$. This force *acts through the centroid* of the distributed loading area, a distance $\frac{1}{3}x$ from the right end. Applying the two equations of equilibrium yields*

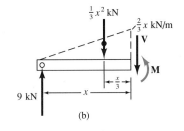

(b)

$$+\uparrow \Sigma F_y = 0; \qquad 9 - \frac{1}{3}x^2 - V = 0$$

$$V = \left(9 - \frac{x^2}{3}\right) \text{ kN} \qquad (1)$$

$$\downarrow + \Sigma M_S = 0; \qquad M + \frac{1}{3}x^2\left(\frac{x}{3}\right) - 9x = 0$$

$$M = \left(9x - \frac{x^3}{9}\right) \text{ kN} \cdot \text{m} \qquad (2)$$

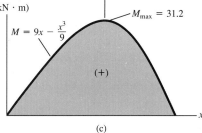

Shear and Moment Diagrams. The shear and bending-moment diagrams shown in Fig. 7–13c are obtained by plotting Eqs. (1) and (2).

The point of *zero shear* can be found using Eq. (1):

$$V = 9 - \frac{x^2}{3} = 0$$

$$x = 5.20 \text{ m}$$

This value of x happens to represent the point on the beam where the *maximum moment* occurs (see Sec. 7.3). Using Eq. (2), we have

$$M_{max} = \left(9(5.20) - \frac{(5.20)^3}{9}\right) \text{ kN} \cdot \text{m}$$

$$= 31.2 \text{ kN} \cdot \text{m}$$

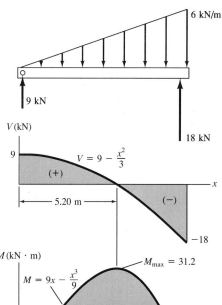

(c)

Fig. 7–13

*Note from Eqs. (2) and (1) that $dM/dx = V$ and $dV/dx = -w$. This is a consequence of Eqs. 7–1 and 7–2, which will be presented in the next section.

PROBLEMS

For each of the following problems, establish the x axis with the origin at the right side of the beam, and obtain the internal shear and moment as a function of x. Use these results to plot the shear and moment diagrams.

7–31. Draw the shear and moment diagrams for the beam (a) in terms of the parameters shown; (b) set $P = 800$ lb, $a = 5$ ft, $L = 12$ ft.

Prob. 7–31

*7–32. Draw the shear and moment diagrams for the beam (a) in terms of the parameters shown; (b) set $P = 9$ kN, $a = 2$ m, $L = 6$ m.

Prob. 7–32

7–33. Draw the shear and moment diagrams for the beam (a) in terms of the parameters shown; (b) set $w = 500$ lb/ft, $L = 10$ ft.

Prob. 7–33

7–34. Draw the shear and moment diagrams for the beam.

Prob. 7–34

7–35. Draw the shear and moment diagrams for the horizontal portion ABC of the frame. Note that B is a rigid joint. Neglect the thickness of the members.

Prob. 7–35

*7–36. Draw the shear and moment diagrams for the beam (a) in terms of the parameters shown; (b) set $M = 500$ N · m, $L = 8$ m.

Prob. 7–36

7–37. Draw the shear and moment diagrams for the beam.

Prob. 7–37

7–38. Draw the shear and moment diagrams for the beam (a) in terms of the parameters shown; (b) set $w = 250$ lb/ft, $L = 12$ ft.

Prob. 7–38

7–39. Draw the shear and moment diagrams for beam ABC. Note that there is a pin at B. Solve the problem (a) in terms of the parameters shown; (b) set $w = 5$ kN/m, $L = 12$ m.

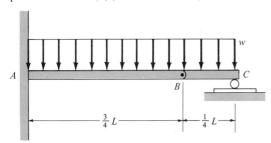

Prob. 7–39

*7–40. Draw the shear and moment diagrams for the beam.

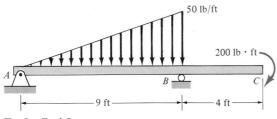

Prob. 7–40

7–41. Draw the shear and moment diagrams for the compound beam. The beam is pin-connected at E and F.

Prob. 7–41

7–42. Draw the shear and moment diagrams for the beam.

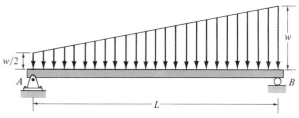

Prob. 7–42

7–43. Draw the shear and moment diagrams for the beam.

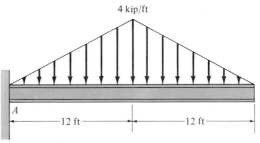

Prob. 7–43

297

*7–44. Draw the shear and moment diagrams for the beam.

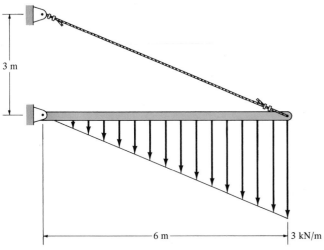

Prob. 7–44

7–45. Draw the shear and moment diagrams for the beam.

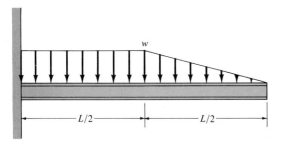

Prob. 7–45

7–46. Draw the shear and moment diagrams for the beam (a) in terms of the parameters shown; (b) set $w = 500$ N/m, $L = 3$ m.

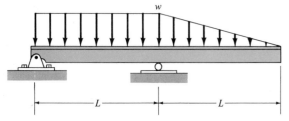

Prob. 7–46

7–47. Determine the internal axial force, shear force, and moment in the curved member as a function of θ.

Prob. 7–47

*7–48. Determine the internal axial force, shear force, and moment in the curved member as a function of θ.

Prob. 7–48

7–49. The cantilevered beam is made of material having a specific weight (weight per unit volume) of γ. Determine the shear force and moment in the beam as a function of x.

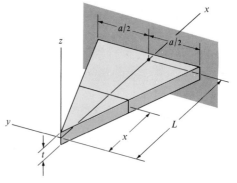

Prob. 7–49

7–50. The beam will fail when the maximum moment developed in it is M_{max}. Determine the position x of the concentrated force **P** and its smallest magnitude that will cause the beam to fail.

Prob. 7–50

7–51. Determine the distance a as a fraction of the beam's length L for locating the roller support so that the moment in the beam at B is zero.

Prob. 7–51

*7–52. Determine the distance a between the supports in terms of the beam's length L so that the moment in the *symmetric* beam is zero at the beam's center.

Prob. 7–52

7–53. Express the internal x, y, z components of force and moment in the rod as a function of y, where $0 \leqslant y \leqslant 4$ ft.

Prob. 7–53

* 7.3 Relations Between Distributed Load, Shear, and Moment

In cases where a beam is subjected to several concentrated forces, couples, and distributed loads, the method of constructing the shear and bending-moment diagrams discussed in Sec. 7.2 may become quite tedious. In this section a simpler method for constructing these diagrams is discussed—a method based upon differential relations that exist between the load, shear, and bending moment.

Fig. 7–14

(b)

Consider the beam *AD* shown in Fig. 7–14*a*, which is subjected to an arbitrary distributed loading $w = w(x)$ and a series of concentrated forces and couple moments. In the following discussion, the *distributed load* will be considered *positive* when the *loading acts downward* as shown. The free-body diagram for a small segment of the beam having a length Δx is shown in Fig. 7–14*b*. Since this segment has been chosen at a point *x* along the beam which is *not* subjected to a concentrated force or couple moment, any results obtained will not apply at points of concentrated loading. The internal shear force and bending moment shown on the free-body diagram are assumed to act in the *positive sense* according to the established sign convention, Fig. 7–14. Note that both the shear force and moment acting on the right-hand face must be increased by a small, finite amount in order to keep the segment in equilibrium. The distributed loading has been replaced by a resultant force $\Delta F = w(x)\,\Delta x$ that acts at a fractional distance $\varepsilon(\Delta x)$ from the right end, where $0 < \varepsilon < 1$ [for example, if $w(x)$ is *uniform*, $\varepsilon = \frac{1}{2}$]. Applying the equations of equilibrium, we have

$$+\uparrow \Sigma F_y = 0; \qquad V - w(x)\,\Delta x - (V + \Delta V) = 0$$
$$\Delta V = -w(x)\,\Delta x$$

$$\zeta + \Sigma M_O = 0; \quad -V\,\Delta x - M + w(x)\,\Delta x[\varepsilon(\Delta x)] + (M + \Delta M) = 0$$
$$\Delta M = V\,\Delta x - w(x)\varepsilon(\Delta x)^2$$

Dividing by Δx and taking the limit as $\Delta x \to 0$ the above equations become

$$\frac{dV}{dx} = -w(x) \tag{7-1}$$

$$\frac{\text{Slope of}}{\text{shear diagram}} = \frac{-\text{Distributed}}{\text{load intensity}}$$

and

$$\frac{dM}{dx} = V \tag{7-2}$$

$$\frac{\text{Slope of}}{\text{moment diagram}} = \text{Shear}$$

These two equations provide a convenient means for plotting the shear and moment diagrams for a beam. At a specific point in a beam, Eq. 7–1 states that the *slope* of the shear *diagram* is equal to the intensity of the *distributed loading,* while Eq. 7–2 states that the slope of the moment diagram is equal to the shear force at the point. In particular, if the shear is equal to zero, $dM/dx = 0$, and therefore *a point of zero shear corresponds to a point of maximum (or possibly minimum) moment.*

Equations 7–1 and 7–2 may be rewritten in the form $dV = -w(x)\, dx$ and $dM = V\, dx$. Noting that $w(x)\, dx$ and $V\, dx$ represent the differential areas under the distributed-loading and shear diagrams, one can integrate these areas between two points B and C along the beam, Fig. 7–14a, and write

$$\Delta V_{BC} = -\int w(x)\, dx \tag{7-3}$$

$$\frac{\text{Change}}{\text{in shear}} = \frac{-\text{Area under}}{\text{loading curve}}$$

and

$$\Delta M_{BC} = \int V\, dx \tag{7-4}$$

$$\frac{\text{Change}}{\text{in moment}} = \frac{\text{Area under}}{\text{shear diagram}}$$

Equation 7–3 states that the change in shear between points B and C is equal to the negative of the *area* under the distributed-loading curve between these points, Fig. 7–14a. Similarly, from Eq. 7–4, the change in moment between points B and C is equal to the area under the shear diagram for region BC. Because two integrations are involved, first to determine the moment, Eq.

7–4, we can state that if the loading curve $w = w(x)$ is a polynomial of degree n, then $V = V(x)$ will be a curve of degree $n + 1$, and $M = M(x)$ will be a curve of degree $n + 2$.*

From the derivations it should be clear that the above equations do not apply at points where a *concentrated* force or couple moment acts. These two special cases create *discontinuities* in the shear and moment diagrams, respectively. For example, consider the beam shown in Fig. 7–14a. The free-body diagrams for the colored elements loaded by the force and couple moment are shown in Fig. 7–15a and b, respectively. From Fig. 7–15a it is seen that force equilibrium requires the change in shear to be

$$+ \uparrow \Sigma F_y = 0; \qquad \qquad \Delta V = -F \qquad \qquad (7\text{–}5)$$

Thus, when **F** acts *downward* on the beam, ΔV is negative so that the shear "jumps" *downward*. Likewise, if **F** acts *upward*, the jump (ΔV) is *upward*. From Fig. 7–15b, letting $\Delta x \to 0$, moment equilibrium requires the change in moment to be

$$\zeta + \Sigma M_O = 0; \qquad \qquad \Delta M = M' \qquad \qquad (7\text{–}6)$$

In this case, if an external moment **M'** is applied *clockwise*, ΔM is positive so that the moment diagram shows a jump *upward*, and when **M'** acts *counterclockwise*, the jump (ΔM) must be *downward*.

The following examples illustrate application of the above equations for the construction of the shear and moment diagrams. After working through these examples, it is recommended that Examples 7–7 and 7–8 be solved using this method.

(a)

Fig. 7–15
(b)

---

*Refer back to Example 7–8, Eqs. (1) and (2). Also, see the footnote for Example 7–8.

Example 7–9

Draw the shear and bending-moment diagrams for the beam shown in Fig. 7–16a.

50 lb/ft

8 ft

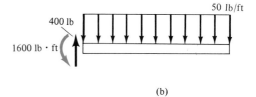

400 lb

1600 lb · ft

50 lb/ft

(b)

Fig. 7–16   (a).

SOLUTION

Support Reactions (Fig. 7–16b). The reactions at the fixed support have been calculated and are shown on the free-body diagram of the beam.

Shear Diagram (Fig. 7–16c). The shear at the end points is plotted first. From the sign convention, Fig. 7–11, $V = +400$ at $x = 0$ (R.H.F.) and $V = 0$ at $x = 8$. Since $dV/dx = -w = -50$, a straight *negative* sloping line, shown colored, connects the end points.

Moment Diagram (Fig. 7–16d). From our sign convention, Fig. 7–11, the moments at the beam's end points, $M = -1600$ at $x = 0$ (R.H.F.) and $M = 0$ at $x = 8$, are plotted first. Successive values of shear on the shear diagram indicate the slope $dM/dx = V$, shown colored in Fig. 7–16d, is always positive yet *linearly decreasing* from $dM/dx = 400$ at $x = 0$ to $dM/dx = 0$ at $x = 8$. Thus, due to the integrations, w a constant becomes V a sloping line (first-degree curve) and M a parabola (second-degree curve).

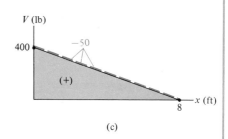

V (lb)

400

−50

(+)

x (ft)

8

(c)

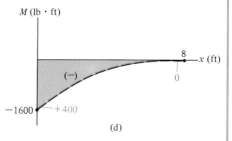

M (lb · ft)

8

x (ft)

0

(−)

−1600   +400

(d)

Example 7–10

Draw the shear and bending-moment diagrams for the beam shown in Fig. 7–17a.

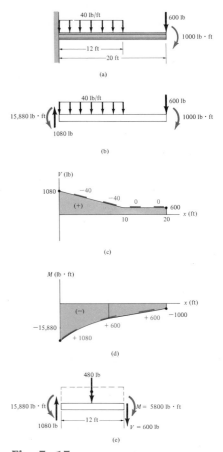

Fig. 7–17

SOLUTION

Support Reactions (Fig. 7–17b). The reactions at the fixed support have been calculated and are shown on the free-body diagram of the beam.

Shear Diagram (Fig. 7–17c). Using the established sign convention, Fig. 7–11, the shear at the ends of the beam is plotted first, i.e., $x = 0$, $V = +1080$; $x = 20$, $V = +600$.

Since the uniform distributed load is downward and *constant*, then the slope of the shear diagram is $dV/dx = -w = -40$ for $0 \leq x < 12$ as indicated.

The magnitude of shear at $x = 12$ is $+600$. This can be determined by first finding the area under the load diagram between $x = 0$ and $x = 12$, i.e., $\Delta V = -\int w(x)\,dx = -40(12) = -480$. Thus $V|_{x=12} = V|_{x=0} + (-480) = 1080 - 480 = 600$. Also, we can obtain this value by using the method of sections, Fig. 7–17e, where for equilibrium $V = +600$.

Since the load between $12 < x \leq 20$ is $w = 0$, then $dV/dx = 0$ as indicated. This brings the shear to the required value of $V = 600$ at $x = 20$.

Moment Diagram (Fig. 7–17d). Again, using the established sign convention, Fig. 7–11, the moments at the ends of the beam are plotted first, i.e., $x = 0$, $M = -15,880$; $x = 20$, $M = -1000$.

Each value of shear gives the slope of the moment diagram since $dM/dx = V$. As indicated, at $x = 0$, $dM/dx = +1080$ and at $x = 12$, $dM/dx = +600$. For $0 < x < 12$ specific values of the shear diagram are positive but linearly decreasing. Hence, the moment diagram is parabolic with a linear decreasing slope.

The magnitude of moment at $x = 12$ ft is -5800. This can be found from the trapezoidal area under the shear diagram, i.e., $\Delta M = \int V\,dx = 600(12) + \frac{1}{2}(1080 - 600)(12) = +10,080$, so that $M|_{x=12} = M|_{x=0} + 10,080 = -15,880 + 10,080 = -5800$. The more "basic" method of sections can also be used, where equilibrium at $x = 12$ requires $M = -5800$, Fig. 7–17e.

The moment diagram has a constant slope for $12 < x \leq 20$ since $dM/dx = V = +600$. This brings the value of $M = -1000$ at $x = 20$, as required.

Example 7–11

Draw the shear and moment diagrams for the beam in Fig. 7–18a.

Fig. 7–18　　　　　　　　(a)

(b)

SOLUTION

Support Reactions (Fig. 7–18b).　The reactions at the supports are shown on the free-body diagram.

Shear Diagram (Fig. 7–18c).　The end points $x = 0$, $V = +35$ and $x = 8$, $V = -35$ are plotted first.

Since there is no distributed load on the beam, the slope of the shear diagram throughout the beam's length is zero, i.e., $dV/dx = 0$. There is a discontinuity or "jump" of the shear diagram, however, at each concentrated force. From Eq. 7–5, $\Delta V = -F$, the change in shear is negative when the load acts downward and positive when the load acts upward. Stated another way, the "jump" follows the force, i.e., a downward force causes a downward jump, and vice versa. Thus, the 20-kN force at $x = 2$ m changes the shear force from 35 kN to 15 kN; the 30-kN force at $x = 4$ m changes the shear from 15 kN to -15 kN, etc. We can *also* obtain numerical values for the shear at a specified point in the beam by using the method of sections, as for example, $x = 2^+$ m, $V = 15$ kN in Fig. 7–18e.

(c)

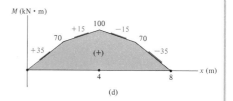

(d)

Moment Diagram (Fig. 7–18d).　The end points $x = 0$, $M = 0$ and $x = 8$, $M = 0$ are plotted first.

Since the shear is constant in each region of the beam, the moment diagram has a corresponding constant positive or negative slope as indicated on the diagram. Numerical values of the moment at any point can be computed from the *area* under the shear diagram. For example, at $x = 2$ m, $\Delta M = \int V \, dx = 35(2) = 70$, so $M|_{x=2} = M|_{x=0} + 70 = 0 + 70 = 70$. Also, by the method of sections, we can determine the moment at a specified point, as, for example, $x = 2^+$ m, $M = 70$ kN · m, Fig. 7–18e.

(e)

Example 7–12

Sketch the shear and bending-moment diagrams for the beam shown in Fig. 7–19a.

SOLUTION

Support Reactions. The reactions are calculated and indicated on the free-body diagram, Fig. 7–19b.

Shear Diagram. As in Example 7–11, the shear diagram can be constructed by "following the load" on the free-body diagram. In this regard, beginning at A, $V_A = +100$ lb. No load acts between A and C, so the shear remains constant, i.e., $dV/dx = w(x) = 0$. At C the 600-lb force is down, so the shear jumps down 600 lb, from 100 lb to -500 lb. Again the shear is constant (no load) and ends at -500 lb, point B. Notice that no jump or discontinuity in shear occurs at D, the point where the 4000-lb · ft moment is applied, Fig. 7–19a. Use the method of sections to show this.

Moment Diagram. The moment at each end of the beam is zero. These two points are plotted first, Fig. 7–19d. The slope of the moment diagram from A to C is constant since $dM/dx = V = +100$. The value of the moment at C can be determined by the method of sections or by computing the rectangular area under the shear diagram between A and C, i.e., $\Delta M_{AC} = M_C - M_A = (100 \text{ lb})(10 \text{ ft}) = 1000$ lb · ft. Since $M_A = 0$, then $M_C = 0 + 1000$ lb · ft $= 1000$ lb · ft. From C to D the slope is $dM/dx = V = -500$, Fig. 7–19c. The area under the shear diagram between points C and D is $\Delta M_{CD} = M_D - M_C = (-500 \text{ lb})(5 \text{ ft}) = -2500$ lb · ft. Thus $M_D = 1000 - 2500 = -1500$ lb · ft. A jump occurs at point D due to the concentrated couple moment of 4000 lb · ft. Using Eq. 7–6, the jump is *positive* since the couple moment is *clockwise*. Thus, at $x = 15^+$ ft, the moment is $M_D = -1500 + 4000 = 2500$ lb · ft. This value can *also* be determined by the method of sections. From point D the slope of $dM/dx = -500$ is maintained until the diagram closes to zero at B, Fig. 7–19d.

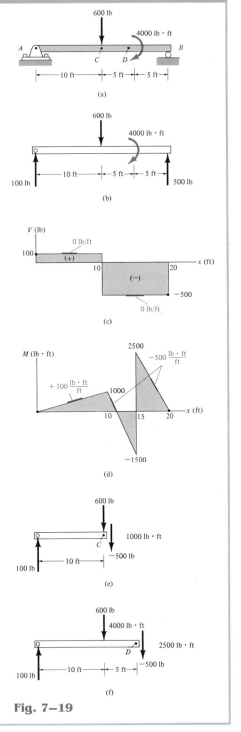

Fig. 7–19

PROBLEMS

7–54. Draw the shear and moment diagrams for the beam in Prob. 7–34.

7–55. Draw the shear and moment diagrams for the beam in Prob. 7–37.

*7–56. Draw the shear and moment diagrams for the beam in Prob. 7–40.

7–57. Draw the shear and moment diagrams for the beam in Prob. 7–42.

7–58. Draw the shear and moment diagrams for the beam in Prob. 7–43.

7–59. Draw the shear and moment diagrams for the beam in Prob. 7–45.

*7–60. Draw the shear and moment diagrams for the beam. There is a pin at *C*.

Prob. 7–60

7–61. Draw the shear and moment diagrams for the beam.

Prob. 7–61

7–62. Draw the shear and moment diagrams for the beam.

Prob. 7–62

7–63. Draw the shear and moment diagrams for the beam.

Prob. 7–63

*7–64. Draw the shear and moment diagrams for the beam.

Prob. 7–64

7–65. Draw the shear and moment diagrams for the beam.

Prob. 7–65

7–66. Draw the shear and moment diagrams for the beam.

Prob. 7–66

7–67. Draw the shear and moment diagrams for the beam.

Prob. 7–67

7–71. Draw the shear and moment diagrams for the beam.

Prob. 7–71

***7–68.** Draw the shear and moment diagrams for the beam.

Prob. 7–68

***7–72.** Draw the shear and moment diagrams for the beam.

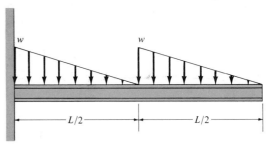

Prob. 7–72

7–69. Draw the shear and moment diagrams for the beam.

Prob. 7–69

7–73. The beam consists of two segments pin-connected at *B*. Draw the shear and moment diagrams for the beam.

Prob. 7–73

7–70. Draw the shear and moment diagrams for the beam.

Prob. 7–70

7–74. The two segments of the girder are pin-connected together by a short vertical link *BC*. Draw the shear and moment diagrams for the girder.

Prob. 7–74

*7.4 Cables

Flexible cables and chains are often used in engineering structures for support and to transport loads from one member to another. When used for supporting suspension bridges and trolley wheels, cables form the main load-carrying element of the structure. In the force analysis of such systems, the weight of the cable itself may be neglected; however, when cables are used for transmission lines and guys for radio antennas and derricks, the cable weight may become important and must be included in the structural analysis. Three cases will be considered in the analysis that follows: (1) a cable subjected to concentrated loads, (2) a cable subjected to a distributed load, and (3) a cable subjected to its own weight. Regardless of which loading conditions are present, provided the loading is coplanar with the cable, the requirements for equilibrium are formulated in an identical manner:

When deriving the necessary relations between the tensile force, sag, length, and span of a cable, we will make the assumption that the cable is *perfectly flexible* and *inextensible*. Due to flexibility, the cable offers no resistance to bending, and therefore, the tensile force acting in the cable is always tangent to the cable at points along its length. Being inextensible, the cable has a constant length both before and after the load is applied. As a result, once the load is applied, the geometry of the cable is fixed, and the cable or a segment of it can be treated as a rigid body.

Cable Subjected to Concentrated Loads

When a cable of negligible weight supports several concentrated loads, the cable takes the form of several straight-line segments, each of which is subjected to a constant tensile force. Consider, for example, the cable shown in Fig. 7–20, where the distances h, L_1, L_2, and L_3 and the loads P_1 and P_2 are known. The problem here is to determine the *nine unknowns* consisting of the tensions in each of the *three* segments, the four components of reaction at A and B, and the sags y_C and y_D at the *two* points C and D. To do this, we can write *two* equations of force equilibrium at each of points A, B, C, and D. This results in a total of *eight equations*.* To complete the solution, it will be necessary to know something about the geometry of the cable in order to obtain the necessary ninth equation. For example, if the cable's total *length L* is specified, then the Pythagorean theorem can be used to relate each of the three segmental lengths, written in terms of h, y_C, y_D, L_1, L_2, and L_3, to the total length L. Unfortunately, this type of problem cannot be easily solved by hand. Another possibility, however, is to specify one of the sags instead of the cable length. By doing this, the equilibrium equations are then sufficient for obtaining the force components and the remaining sag. Once the sag is obtained, the length of the cable can be determined by trigonometry. The following example illustrates a procedure for performing the equilibrium analysis for this type of problem.

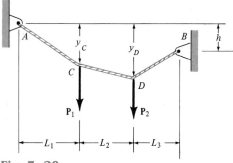

Fig. 7–20

*As will be shown in the following example, the eight equilibrium equations can *also* be written for the entire cable, or any part thereof. But *no more* than *eight* equations are available.

Example 7–13

Determine the tension in each segment of the cable shown in Fig. 7–21a.

(a)

Fig. 7–21(a–c)

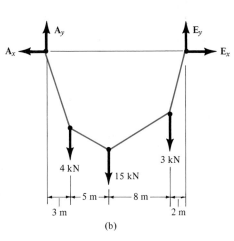

(b)

SOLUTION

By inspection, there are four unknown external reactions (A_x, A_y, E_x, and E_y) and four unknown cable tensions, one in each cable segment. These eight unknowns along with the two unknown sags y_B and y_D can be determined from *ten* available equilibrium equations. One method is to apply these equations as force equilibrium ($\Sigma F_x = 0$, $\Sigma F_y = 0$) at each of the five points A through E. Here we will take a more direct approach. Consider the free-body diagram for the entire cable, Fig. 7–21b. Thus,

$$\xrightarrow{+}\Sigma F_x = 0; \qquad\qquad -A_x + E_x = 0$$

$$\zeta+\Sigma M_E = 0; \quad -A_y(18) + 4(15) + 15(10) + 3(2) = 0 \qquad A_y = 12 \text{ kN}$$

$$+\uparrow\Sigma F_y = 0; \qquad\quad 12 - 4 - 15 - 3 + E_y = 0 \qquad E_y = 10 \text{ kN}$$

Since the sag $y_C = 12$ m is known, we will now consider the leftmost section, which cuts cable BC, Fig. 7–21c.

$$\zeta+\Sigma M_C = 0; \quad A_x(12) - 12(8) + 4(5) = 0 \qquad A_x = E_x = 6.33 \text{ kN}$$

$$\xrightarrow{+}\Sigma F_x = 0; \qquad\qquad T_{BC} \cos\theta_{BC} - 6.33 = 0$$

$$+\uparrow\Sigma F_y = 0; \qquad\quad 12 - 4 - T_{BC} \sin\theta_{BC} = 0$$

$$\theta_{BC} = 51.6°$$

Thus,

$$T_{BC} = 10.2 \text{ kN} \qquad\qquad\qquad Ans.$$

(c)

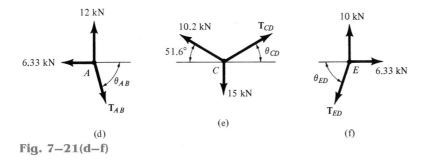

(d)   (e)   (f)

Fig. 7–21(d–f)

Proceeding now to analyze the equilibrium of points A, C, and E in sequence, we have

Point A (Fig. 7–21d)

$$\xrightarrow{+}\Sigma F_x = 0; \qquad T_{AB} \cos \theta_{AB} - 6.33 = 0$$

$$+ \uparrow \Sigma F_y = 0; \qquad -T_{AB} \sin \theta_{AB} + 12 = 0$$

$$\theta_{AB} = 62.2°$$

$$T_{AB} = 13.6 \text{ kN} \qquad\qquad Ans.$$

Point C (Fig. 7–21e)

$$\xrightarrow{+}\Sigma F_x = 0; \qquad T_{CD} \cos \theta_{CD} - 10.2 \cos 51.6° = 0$$

$$+ \uparrow \Sigma F_y = 0; \qquad T_{CD} \sin \theta_{CD} + 10.2 \sin 51.6° - 15 = 0$$

$$\theta_{CD} = 47.9°$$

$$T_{CD} = 9.44 \text{ kN} \qquad\qquad Ans.$$

Point E (Fig. 7–21f)

$$\xrightarrow{+}\Sigma F_x = 0; \qquad 6.33 - T_{ED} \cos \theta_{ED} = 0$$

$$+ \uparrow \Sigma F_y = 0; \qquad 10 - T_{ED} \sin \theta_{ED} = 0$$

$$\theta_{ED} = 57.7°$$

$$T_{ED} = 11.8 \text{ kN} \qquad\qquad Ans.$$

Note that the maximum cable tension is in segment AB since this segment has the greatest slope (θ) and it is required that for any segment the horizontal component $T \cos \theta = A_x$ (a constant). Also, since the angles that the cable segments make with the horizontal have now been determined, one can determine the sags y_B and y_D, Fig. 7–21a, using trigonometry.

Cable Subjected to a Distributed Load

Consider the weightless cable shown in Fig. 7–22a, which is subjected to a loading function $w = w(x)$ *as measured in the x direction*. The free-body diagram of a small segment of the cable having a length Δs is shown in Fig. 7–22b. Since the tensile force in the cable changes continuously in both magnitude and direction along the length of the cable, this change is denoted on the free-body diagram by $\Delta \mathbf{T}$. The distributed load is represented by its resultant force $w(x)(\Delta x)$, which acts at a fractional distance $\varepsilon(\Delta x)$ from point O, where $0 < \varepsilon < 1$. Applying the equations of equilibrium yields

$$\xrightarrow{+} \Sigma F_x = 0; \qquad -T \cos \theta + (T + \Delta T) \cos (\theta + \Delta\theta) = 0$$

$$+ \uparrow \Sigma F_y = 0; \qquad -T \sin \theta - w(x)(\Delta x) + (T + \Delta T) \sin (\theta + \Delta\theta) = 0$$

$$\underset{+}{\downarrow} \Sigma M_O = 0; \qquad w(x)(\Delta x)\varepsilon(\Delta x) - T \cos \theta \, \Delta y + T \sin \theta \, \Delta x = 0$$

Dividing each of the preceding equations by Δx and taking the limit as $\Delta x \to 0$, and hence $\Delta y \to 0$, $\Delta \theta \to 0$, and $\Delta T \to 0$, we obtain

$$\frac{d(T \cos \theta)}{dx} = 0 \tag{7-7}$$

$$\frac{d(T \sin \theta)}{dx} - w(x) = 0 \tag{7-8}$$

$$\frac{dy}{dx} = \tan \theta \tag{7-9}$$

Integrating Eq. 7–7, we have

$$T \cos \theta = \text{constant} = F_H \tag{7-10}$$

where F_H represents the horizontal component of tensile force at any point along the cable.

Integrating Eq. 7–8 gives

$$T \sin \theta = \int w(x) \, dx \tag{7-11}$$

Dividing Eq. 7–11 by Eq. 7–10 eliminates T. Then, using Eq. 7–9, we obtain

$$\tan \theta = \frac{dy}{dx} = \frac{1}{F_H} \int w(x) \, dx$$

Performing a second integration yields

$$\boxed{y = \frac{1}{F_H} \int \left(\int w(x) \, dx \right) dx} \tag{7-12}$$

This equation is used to determine the curve for the cable, $y = f(x)$. The force F_H and the two constants, say C_1 and C_2, resulting from the integration of Eq. 7–12 are determined by applying the boundary conditions for the cable.

(a)

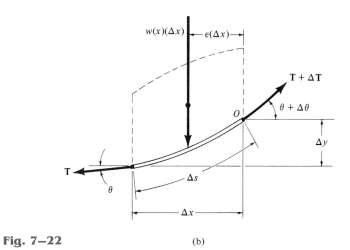

Fig. 7—22

(b)

Example 7–14

The cable of a suspension bridge supports half of the uniform road surface between the two columns at A and B, as shown in Fig. 7–23a. If this loading is w_o N/m, determine the maximum force developed in the cable and the required length of the cable. The span length L and sag h are known.

Fig. 7–23a

(a)

SOLUTION

For reasons of symmetry, the origin of coordinates has been placed at the cable's center. The curve of the cable can be found by using Eq. 7–12. Noting that $w(x) = w_o$, we have

$$y = \frac{1}{F_H} \int \left(\int w_o \, dx \right) dx$$

Integrating this equation twice gives

$$y = \frac{1}{F_H} \left(\frac{w_o x^2}{2} + C_1 x + C_2 \right) \tag{1}$$

The constants of integration may be determined by using the boundary conditions $y = 0$ at $x = 0$ and $dy/dx = 0$ at $x = 0$. Substituting into Eq. (1) yields $C_1 = C_2 = 0$. The curve then becomes

$$y = \frac{w_o}{2F_H} x^2 \tag{2}$$

This is the equation of a *parabola*. The constant F_H may be obtained by using the boundary condition $y = h$ at $x = L/2$. Thus,

$$F_H = \frac{w_o L^2}{8h} \tag{3}$$

Therefore, Eq. (2) becomes

$$y = \frac{4h}{L^2}x^2 \qquad (4)$$

The maximum tension in the cable may be determined using Eq. 7–10. Since $T = F_H/\cos\theta$, for $0 \le \theta < \pi/2$, T_{max} will occur when θ is *maximum*, i.e., at point B, Fig. 7–23a. From Eq. (2) this occurs when

$$\left.\frac{dy}{dx}\right|_{max} = \tan\theta_{max} = \left.\frac{w_o}{F_H}x\right|_{x=L/2}$$

or

$$\theta_{max} = \tan^{-1}\left(\frac{w_o L}{2F_H}\right) \qquad (5)$$

Therefore,

$$T_{max} = \frac{F_H}{\cos(\theta_{max})} \qquad (6)$$

Using the triangular relationship shown in Fig. 7–23b, which is based on Eq. (5), Eq. (6) may be written as

$$T_{max} = \frac{\sqrt{4F_H^2 + w_o^2 L^2}}{2}$$

(b)

Fig. 7–23b

Substituting Eq. (3) into the above equation yields

$$T_{max} = \frac{w_o L}{2}\sqrt{1 + \left(\frac{L}{4h}\right)^2} \qquad Ans.$$

For a differential segment of cable length,

$$ds = \sqrt{(dx)^2 + (dy)^2} = \sqrt{1 + \left(\frac{dy}{dx}\right)^2}\, dx$$

Hence, the total length of the cable, $\mathscr{L}$, can be determined by integration, i.e.,

$$\mathscr{L} = \int ds = 2\int_0^{L/2}\sqrt{1 + \left(\frac{w_o x}{F_H}\right)^2}\, dx \qquad (7)$$

Integrating and substituting the limits yields

$$\mathscr{L} = \frac{L}{2}\sqrt{1 + \left(\frac{w_o L}{2F_H}\right)^2} + \frac{F_H}{w_o}\sinh^{-1}\left(\frac{w_o L}{2F_H}\right)$$

Using Eq. (3) and rearranging terms, the final result can be written as

$$\mathscr{L} = \frac{L}{2}\left[\sqrt{1 + \left(\frac{4h}{L}\right)^2} + \frac{L}{4h}\sinh^{-1}\left(\frac{4h}{L}\right)\right] \qquad Ans.$$

Cable Subjected to Its Own Weight

When the weight of the cable becomes important in the force analysis, the loading function along the cable becomes a function of the arc length s rather than the projected length x. A generalized loading function $w = w(s)$ acting along the cable is shown in Fig. 7–24a. The free-body diagram for a segment of the cable is shown in Fig. 7–24b. Applying the equilibrium equations to the force system on this diagram, one obtains relationships identical to those given by Eqs. 7–7 to 7–9, but with ds replacing dx. Therefore, it may be shown that

$$T \cos \theta = F_H$$

$$T \sin \theta = \int w(s) \, ds$$

$$\frac{dy}{dx} = \frac{1}{F_H} \int w(s) \, ds \tag{7--13}$$

To perform a direct integration of Eq. 7–13, it is necessary to replace dy/dx by ds/dx. Since

$$\frac{dy}{dx} = \sqrt{\left(\frac{ds}{dx}\right)^2 - 1}$$

we obtain

$$\frac{ds}{dx} = \left\{1 + \frac{1}{F_H^2}\left(\int w(s) \, ds\right)^2\right\}^{1/2}$$

Separating the variables and integrating yields

$$x = \int \frac{ds}{\left\{1 + \dfrac{1}{F_H^2}\left(\displaystyle\int w(s) \, ds\right)^2\right\}^{1/2}} \tag{7--14}$$

The two constants of integration, say C_1 and C_2, are found using the boundary conditions for the cable.

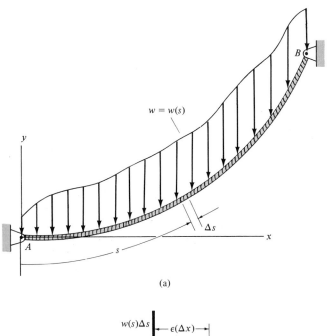

$w = w(s)$

Δs

A

s

(a)

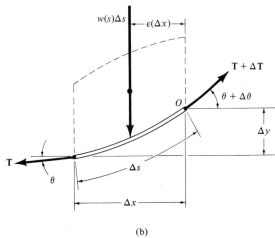

$w(s)\Delta s$

$\epsilon(\Delta x)$

$T + \Delta T$

$\theta + \Delta\theta$

O

Δy

T

θ

Δs

Δx

Fig. 7–24

(b)

Example 7–15

Determine the deflection curve, the length, and the maximum tension in the uniform cable shown in Fig. 7–25. The cable weighs $w_o = 5$ N/m.

Fig. 7–25

SOLUTION

For reasons of symmetry, the origin of coordinates is located at the center of the cable. Applying Eq. 7–14, where $w(s) = w_o$, yields

$$x = \int \frac{ds}{\left\{ 1 + \dfrac{1}{F_H^2} \left(\int w_o\, ds \right)^2 \right\}^{1/2}}$$

Integrating the term under the integral sign in the denominator, we have

$$x = \int \frac{ds}{\left\{ 1 + \dfrac{1}{F_H^2} \left(w_o s + C_1 \right)^2 \right\}^{1/2}}$$

Substituting $u = (1/F_H)(w_o s + C_1)$ so that $du = (w_o/F_H)\, ds$, a second integration yields

$$x = \frac{F_H}{w_o}(\sinh^{-1} u + C_2)$$

or

$$x = \frac{F_H}{w_o}\left\{ \sinh^{-1}\left[\frac{1}{F_H}(w_o s + C_1) \right] + C_2 \right\} \tag{1}$$

From Eq. 7–13,

$$\frac{dy}{dx} = \frac{1}{F_H} \int w_o\, ds$$

or

$$\frac{dy}{dx} = \frac{1}{F_H}(w_o s + C_1)$$

Since $dy/dx = 0$ at $s = 0$, $C_1 = 0$ and

$$\frac{dy}{dx} = \tan\theta = \frac{w_o s}{F_H} \tag{2}$$

where θ is the angle of the cable at any point. The constant C_2 may be evaluated by using the condition $s = 0$ at $x = 0$ in Eq. (1), in which case $C_2 = 0$. To obtain the deflection curve, $y = f(x)$, solve for s in Eq. (1), which yields

$$s = \frac{F_H}{w_o} \sinh\left(\frac{w_o}{F_H} x \right) \tag{3}$$

Now, substitute Eq. (3) into Eq. (2), in which case

$$\frac{dy}{dx} = \sinh\left(\frac{w_o}{F_H} x \right)$$

Hence

$$y = \frac{F_H}{w_o} \cosh\left(\frac{w_o}{F_H}x\right) + C_3 \qquad (4)$$

If the boundary condition $y = 0$ at $x = 0$ is applied, the constant $C_3 = -F_H/w_o$, and therefore the deflection curve becomes

$$y = \frac{F_H}{w_o}\left[\cosh\left(\frac{w_o}{F_H}x\right) - 1\right]$$

This equation defines the shape of a *catenary curve*. The constant F_H is obtained by using the boundary condition that $y = h$ at $x = L/2$, in which case

$$h = \frac{F_H}{w_o}\left[\cosh\left(\frac{w_o L}{2F_H}\right) - 1\right] \qquad (5)$$

Since $w_o = 5$ N/m, $h = 6$ m, and $L = 20$ m, Eqs. (4) and (5) become

$$y = \frac{F_H}{5 \text{ N/m}}\left[\cosh\left(\frac{5 \text{ N/m}}{F_H}x\right) - 1\right] \qquad (6)$$

$$6 \text{ m} = \frac{F_H}{5 \text{ N/m}}\left[\cosh\left(\frac{50 \text{ N}}{F_H}\right) - 1\right] \qquad (7)$$

Equation (7) can be solved for F_H by using a trial-and-error procedure. The result is

$$F_H = 45.8 \text{ N}$$

and therefore Eq. (6) becomes

$$y = 9.16[\cosh (0.109x) - 1] \text{ m} \qquad \textit{Ans.}$$

Using Eq. (3), with $x = 10$ m, the half-length of the cable is

$$\frac{\mathscr{L}}{2} = \frac{45.8 \text{ N}}{5 \text{ N/m}} \sinh\left[\frac{5 \text{ N/m}}{45.8 \text{ N}}(10 \text{ m})\right] = 12.1 \text{ m}$$

Hence,

$$\mathscr{L} = 24.2 \text{ m} \qquad \textit{Ans.}$$

Since $T = F_H/\cos \theta$, the maximum tension occurs when θ is maximum, i.e., at $s = \mathscr{L}/2 = 12.1$ m. Using Eq. (2) yields

$$\left.\frac{dy}{dx}\right|_{\max} = \tan \theta_{\max} = \frac{5 \text{ N/m}(12.1 \text{ m})}{45.8 \text{ N}} = 1.32$$

$$\theta_{\max} = 52.9°$$

Thus,

$$T_{\max} = \frac{45.8 \text{ N}}{\cos 52.9°} = 75.9 \text{ N} \qquad \textit{Ans.}$$

PROBLEMS

7–75. Determine the tension in each segment of the cable and the cable's total length.

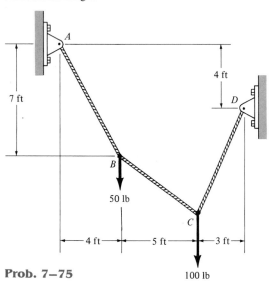

Prob. 7–75

*7–76.** Cable *ABCD* supports the 4-kg flowerpot *E* and 6-kg flowerpot *F*. Determine the maximum tension in the cable and the sag of point *B*.

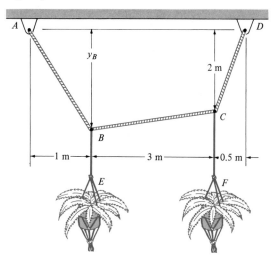

Prob. 7–76

7–77. The cable supports the loading shown. Determine the distance x_B the force at point *B* acts from *A*. Set $P = 40$ lb.

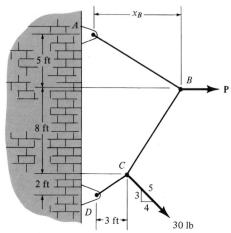

Probs. 7–77/7–78

7–78. The cable supports the loading shown. Determine the magnitude of the horizontal force **P** so that $x_B = 6$ ft.

7–79. The cable supports the three loads shown. Determine the sags y_B and y_D of points *B* and *D*. Take $P_1 = 400$ lb, $P_2 = 250$ lb.

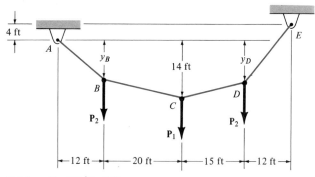

Probs. 7–79/7–80

*7–80.** The cable supports the three loads shown. Determine the magnitude of **P**$_1$ if $P_2 = 300$ lb and $y_B = 8$ ft. Also find the sag y_D.

320

7–81. Determine the force **P** needed to hold the cable in the position shown, i.e., so segment *BC* remains horizontal. Also, compute the sag y_B and the maximum tension in the cable.

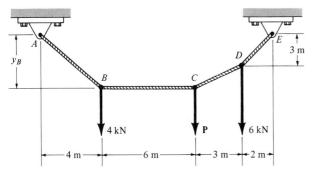

Prob. 7–81

7–82. If the side loading on the suspension bridge in Example 7–14 is $w_o = 0.75$ kN/m and $L = 190$ m, $h = 20$ m, determine the maximum tension in the cable.

7–83. The cable will break when the maximum tension reaches $T_{max} = 12$ kN. Determine the uniform distributed load w required to develop this maximum tension. Take $h = 6$ m.

Prob. 7–83

*7–84. The cable is subjected to the triangular loading. If the slope of the cable at point *O* is zero, determine the equation of the curve $y = f(x)$ which defines the cable shape *OB*, and the maximum tension developed in the cable.

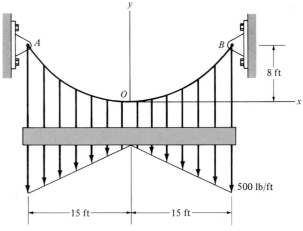

Prob. 7–84

7–85. The cable is subjected to the parabolic loading $w = 150(1 - (x/50)^2)$ lb/ft, where x is in ft. Determine the equation $y = f(x)$ which defines the cable shape *AB* and the maximum tension in the cable.

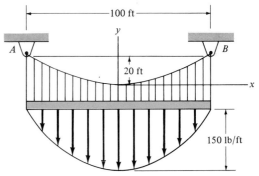

Prob. 7–85

7–86. The cable *AB* is subjected to a uniform loading of 200 N/m. If the weight of the cable is neglected and the slope angles at points *A* and *B* are 30° and 60°, respectively, determine the deflection curve of the cable and the maximum tension developed in the cable.

Prob. 7–86

■7–87. A 50-ft cable is suspended between two points a distance of 15 ft apart and at the same elevation. If the minimum tension in the cable is 350 lb, determine the total weight of the cable and the maximum tension in the cable.

*7–88. The cable supports a girder which weighs 850 lb/ft. Determine the tension in the cable at points *A*, *B*, and *C*.

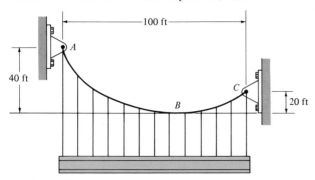

Prob. 7–88

7–89. A 200-lb cable is attached between two points at a distance of 75 ft apart having equal elevations. If the maximum tension developed in the cable is 120 lb, determine the length of the cable and the sag.

7–90. Determine the maximum uniform loading *w* lb/ft that the cable can support if it is capable of sustaining a maximum tension of 3000 lb before it will break.

Prob. 7–90

■7–91. A 50-m-long chain has a total mass of 100 kg and is suspended between two points 15 m apart at the same elevation. Determine the maximum tension and the sag in the chain.

■*7–92. A chain has a mass of 3 kg/m and is supported at points which are 3 m apart and at the same elevation. If the sag in the chain is 1 m, determine the maximum tension in the chain.

■7–93. The telephone wire has a mass of 500 g/m. If the cable has a length of 32 m between the poles, determine the maximum tension in the cable and its sag.

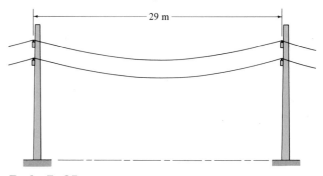

Prob. 7–93

7–94. The uniform beam weighs 800 lb/ft and is held in the horizontal position by means of the cable AB, which has a weight of 10 lb/ft. If the slope of the cable at A is 15°, determine the length of the cable.

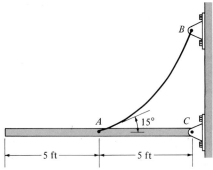

Prob. 7–94

■**7–95.** A telephone line (cable) stretches between two points which are 150 ft apart and at the same elevation. The line sags 5 ft and the cable has a weight of 0.3 lb/ft. Determine the length of the cable and the maximum tension in the cable.

■*7–96.** The cable has a mass of 0.5 kg/m and is 25 m long. Determine the vertical and horizontal components of force it exerts on the top of the tower.

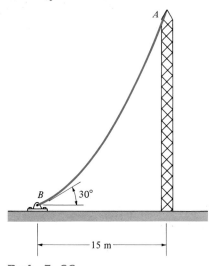

Prob. 7–96

■**7–97.** A cable has a weight of 5 lb/ft. If it can span 300 ft and has a sag of 15 ft, determine the length of the cable. The ends of the cable are supported at the same elevation.

■**7–98.** The power line is supported at A by the tower. If the cable weighs 0.75 lb/ft, and the sag $s = 3$ ft, determine the resultant horizontal force the cable exerts at A.

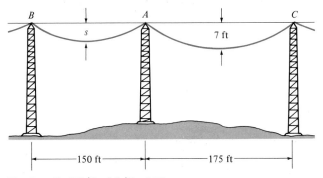

Probs. 7–98/7–99/7–100

7–99. The power line is supported at A by the tower. If the cable weighs 0.75 lb/ft, determine the total length of the cable, BAC. Set $s = 3$ ft.

*7–100.** The power line is supported at A by the tower. If the cable weighs 0.75 lb/ft, determine the required sag s so that the resultant horizontal force the cable exerts at A is zero.

7–101. Show that the deflection curve of the cable discussed in Example 7–15 reduces to Eq. (4) in Example 7–14 when the *hyperbolic cosine function* is expanded in terms of a series and only the first two terms are retained. (The answer indicates that the *catenary* may be replaced by a *parabola* in the analysis of problems in which the sag is small. In this case, the cable weight is assumed to be uniformly distributed along the horizontal.)

■**7–102.** A cable has a weight of 3 lb/ft and is supported at points that are 500 ft apart and at the same elevation. If it has a length of 600 ft, determine the sag.

REVIEW PROBLEMS

7–103. Draw the shear and moment diagrams for the beam.

2 kN/m

A

5 kN · m

B

5 m

Prob. 7–103

*7–104. A uniform cord is suspended between two points having the same elevation. Determine the sag-to-span ratio so that the maximum tension in the cord equals the cord's total weight.

7–105. Determine the axial force, shear force, and moment at a section passing through point D.

150 N/m

A

B

D

4 m

4 m

3 m

C

4 m

Prob. 7–105

7–106. Determine the axial force, shear force, and the moment as a function of $0° \leq \theta \leq 180°$ and $0 \leq x \leq 2$ ft for the beam loaded as shown.

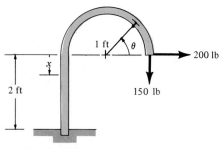

1 ft

θ

200 lb

x

2 ft

150 lb

Prob. 7–106

7–107. The buoyant (or vertical) component of force acting on the weather balloon is 600 N. If cable OB is 70 m in length and has a mass of 500 g/m, determine the altitude h of the balloon.

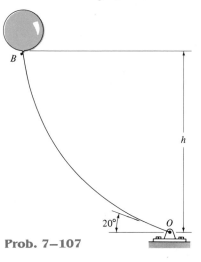

B

h

20°

O

Prob. 7–107

*7–108. Determine the axial force, shear force, and moment at a section passing through point C.

0.5 ft

B

2 ft

A

81 lb · ft

C

3 ft

8 ft

4 ft

Prob. 7–108

7–109. Draw the shear and moment diagrams for the beam.

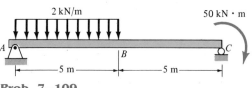

2 kN/m

50 kN · m

A

B

C

5 m

5 m

Prob. 7–109

7–110. Draw the shear and moment diagrams for the beam.

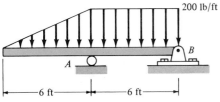

200 lb/ft

B

A

6 ft

6 ft

Prob. 7–110

7–111. Determine the axial force, shear force, and moment at a section passing through point D of the two-member frame.

400 N/m

200 N/m

B

A

D

3 m

2.5 m

C

6 m

Prob. 7–111

*7–112. Draw the shear and moment diagrams for the beam.

8 kN

8 kN

15 kN/m

A

B

C

D

1 m

1.5 m

1 m

1 m

Prob. 7–112

8

Friction

In the previous chapters the surfaces of contact between two bodies were considered to be perfectly *smooth*. Because of this, the force of interaction between the bodies always acts *normal* to the surface at points of contact. In reality, however, all surfaces are *rough,* and depending upon the nature of the problem, the ability of a body to support a *tangential* as well as a *normal* force at its contacting surface must be considered. The tangential force is caused by friction, and in this chapter we will show how to analyze problems involving frictional forces. Specific application will include frictional forces on screws, bearings, disks, and belts. The analysis of rolling resistance is given in the last part of the chapter.

8.1 Characteristics of Dry Friction

Friction may be defined as a force of resistance acting on a body which prevents or retards slipping of the body relative to a second body with which it is in contact. This force always acts *tangent* to the surface at points of contact with other bodies and is directed so as to oppose the possible or existing motion of the body at these points.

(a)

(b)

(c)

Resultant Normal
N and Frictional Forces

(d)

Fig. 8–1(a–d)

In general, two types of friction can occur between surfaces. *Fluid friction* exists when the contacting surfaces are separated by a film of fluid (gas or liquid). The nature of fluid friction is studied in fluid mechanics since it depends upon knowledge of the velocity of the fluid and the fluid's ability to resist shearing force. In this book only the effects of *dry friction* will be presented. This type of friction is often called *Coulomb friction,* since its characteristics were extensively studied by C. A. Coulomb in 1781. Specifically, dry friction occurs between the contacting surfaces of bodies in the absence of a lubricating fluid.

Theory of Dry Friction

The theory of dry friction can best be explained by considering what effects are caused by pulling horizontally on a block of uniform weight W which is resting on a rough horizontal surface, Fig. 8–1a. To properly develop a full understanding of the nature of friction, it is necessary to consider the surfaces of contact to be *nonrigid or deformable.* The other portion of the block, however, will be considered rigid. As shown on the free-body diagram of the block, Fig. 8–1b, the floor exerts a *distribution* of both *normal force* ΔN_n and *frictional force* ΔF_n along the contacting surface. For equilibrium, the normal forces must act *upward* to balance the block's weight W and the frictional forces act to the left to prevent the applied force P from moving the block to the right. Close examination of the contacting surfaces between the floor and block reveals how these frictional and normal forces develop, Fig. 8–1c. It can be seen that many microscopic irregularities exist between the two surfaces and, as a result, reactive forces ΔR_n are developed at each of the protuberances.* These forces act at all points of contact and, as shown, each reactive force contributes both a frictional component ΔF_n and a normal component ΔN_n.

Equilibrium. For simplicity in the following analysis, the effect of the distributed normal and frictional loadings will be indicated by their *resultants* N and F, which are represented on the free-body diagram of the block as shown in Fig. 8–1d. Clearly, the distribution of ΔF_n in Fig. 8–1b indicates that F always acts *tangent to the contacting surface, opposite* to the direction of P. The normal force N is determined from the distribution of ΔN_n in Fig. 8–1b and is directed upward to balance the block's weight W. Notice that it acts a distance x to the right of the line of action of W, Fig. 8–1d. This location, which coincides with the centroid of the loading diagram in Fig. 8–1b, is necessary in order to balance the "tipping effect" caused by P. For example, if P is applied at a height h from the surface, Fig. 8–1d, then moment equilibrium about point O is satisfied if $Wx = Ph$ or $x = Ph/W$. In particular, note that the block will be on the verge of *tipping* if $x = a/2$.

---

*Besides mechanical interactions as explained here, a detailed treatment of the nature of frictional forces must also include the effects of temperature, density, cleanliness, and atomic or molecular attraction between the contacting surfaces. See D. Tabor, *Journal of Lubrication Technology,* 103, 169, 1981.

Impending Motion. In cases where h is small or the surfaces of contact are rather "slippery," the frictional force **F** may *not* be great enough to balance the magnitude of **P**, and consequently the block will tend to slip *before* it can tip. In other words, as the magnitude of **P** is slowly increased, the magnitude of **F** correspondingly increases until it attains a certain *maximum value* F_s, called the *limiting static frictional force*, Fig. 8–1e. When this value is reached, the block is in *unstable equilibrium*, since any further increase in **P** will cause deformations and fractures at the points of surface contact and consequent motion of the block. Experimentally, it has been determined that the magnitude of the limiting static frictional force **F**$_s$ is *directly proportional* to the magnitude of the resultant normal force **N**. This may be expressed mathematically as

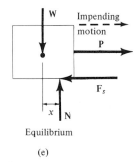

Equilibrium

(e)

Fig. 8–1e

$$\boxed{F_s = \mu_s N} \tag{8–1}$$

where the constant of proportionality, μ_s (mu "sub" s), is called the *coefficient of static friction*.

Typical values for μ_s, found in many engineering handbooks, are given in Table 8–1. Here it may be seen that, as in the case of aluminum on aluminum, it is possible for μ_s to be greater than 1. Furthermore, it should be noted that μ_s is dimensionless and depends only upon the characteristics of the two surfaces in contact. A wide range of values is given for each value of μ_s, since experimental testing was done under variable conditions of roughness and cleanliness of the contacting surfaces. For applications, therefore, it is important that both caution and judgment be exercised when selecting a coefficient of friction for a given set of conditions. When an exact calculation of F_s is required, the coefficient of friction should be determined directly by an experiment that involves the two materials to be used.

Motion. If the magnitude of **P** is increased so that it becomes greater than F_s, the frictional force at the contacting surfaces drops slightly to a smaller value F_k, called the *kinetic frictional force*. The block will *not* be held in equilibrium ($P > F_k$); instead, it begins to slide with increasing speed, Fig. 8–1f. The drop made in the frictional force magnitude, from F_s (static) to F_k (kinetic), can be explained by again examining the surfaces of contact, Fig. 8–1g. Here it is seen that when $P > F_s$, then P has the capacity to shear off the peaks at the contact surfaces and cause the block to "lift" somewhat out of its settled position and "ride" on top of the peaks. Once the block begins to slide, high local temperatures at the points of contact cause momentary adhesion (welding) of these points. The continued shearing of these welds is the dominant mechanism creating friction. Since the resultant contact forces $\Delta \mathbf{R}_n$ are aligned slightly more in the vertical direction than before, Fig. 8–1c, they thereby contribute *smaller* frictional components, $\Delta \mathbf{F}_n$, as when the irregularities are meshed.

Experiments with sliding blocks indicate that the magnitude of the resultant frictional force **F**$_k$ is directly proportional to the magnitude of the resultant normal force **N**. This may be expressed mathematically as

Table 8–1 Typical Values for μ_s

Contact Materials	Coefficient of Static Friction (μ_s)
Metal on ice	0.03–0.05
Wood on wood	0.30–0.70
Leather on wood	0.20–0.50
Leather on metal	0.30–0.60
Aluminum on aluminum	1.10–1.70

(f)

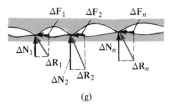

(g)

Fig. 8–1(f, g)

$$F_k = \mu_k N \qquad\qquad (8\text{--}2)$$

where the constant of proportionality, μ_k, is called the *coefficient of kinetic friction*. Typical values for μ_k are approximately 25 per cent *smaller* than those listed in Table 8–1 for μ_s.

The above effects can be summarized by reference to the graph in Fig. 8–2, which shows the variation of the frictional force **F** versus the applied load **P.** Here the frictional force is categorized in three different ways, namely, **F** is a *static-frictional force* if equilibrium is maintained, **F** is a *limiting static-frictional force* $\mathbf{F}_s$ when its magnitude reaches a maximum value needed to maintain equilibrium, and finally **F** is termed a *kinetic-frictional force* $\mathbf{F}_k$ when sliding occurs at the contacting surface. Notice also from the graph that for very large values of **P** or for high speeds, due to aerodynamic effects, $\mathbf{F}_k$ and likewise μ_k decrease somewhat.

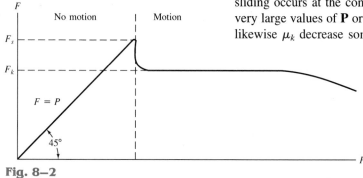

Fig. 8–2

Characteristics of Dry Friction

As a result of *experiments* that pertain to the foregoing discussion, the following rules which apply to bodies subjected to dry friction may be stated.

1. The frictional force acts *tangent* to the contacting surfaces in a direction *opposed* to the *relative motion* or tendency for motion of one surface against another.
2. The magnitude of the maximum static frictional force $\mathbf{F}_s$ that can be developed is independent of the area of contact provided the normal pressure is not very low nor great enough to severely deform or crush the contacting surfaces of the bodies.
3. The magnitude of the maximum static frictional force is generally greater than the magnitude of the kinetic frictional force for any two surfaces of contact. However, if one of the bodies is moving with a *very low velocity* over the surface of another, F_k becomes approximately equal to F_s, i.e., $\mu_s \approx \mu_k$.
4. When *slipping* at the point of contact is *about to occur,* the magnitude of the maximum static frictional force is proportional to the magnitude of the normal force at the point of contact, such that $F_s = \mu_s N$, Eq. 8–1.
5. When *slipping* at the point of contact is *occurring,* the magnitude of the kinetic frictional force is proportional to the magnitude of the normal force at the point of contact, such that $F_k = \mu_k N$, Eq. 8–2.

Angle of Friction

It should be observed that Eqs. 8–1 and 8–2 have a specific, yet *limited,* use in the solution of friction problems. In particular, the frictional force acting at a contacting surface is determined from $F_k = \mu_k N$ *only* if *relative motion* is occurring between the two surfaces. Furthermore, if two bodies are *stationary,* the magnitude of frictional force, F, *does not necessarily* equal $\mu_s N$; instead, F must satisfy the inequality $F \leq \mu_s N$. Only when *impending motion* occurs does F reach its upper limit, $F = F_s = \mu_s N$. This situation may be better understood by considering the block shown in Fig. 8–3a, which is acted upon by a force **P.** In this case consider $P = F_s$, so that the block is on the *verge of sliding.* For equilibrium, the normal force **N** and frictional force $\mathbf{F}_s$ combine to create a resultant $\mathbf{R}_s$. The angle ϕ_s that $\mathbf{R}_s$ makes with **N** is called the *angle of static friction.* From the figure,

Impending motion
(a)

$$\phi_s = \tan^{-1}\left(\frac{F_s}{N}\right) = \tan^{-1}\left(\frac{\mu_s N}{N}\right) = \tan^{-1}\mu_s$$

Provided the block is *not in motion,* any horizontal force $P < F_s$ causes a resultant **R** which has a line of action directed at an angle ϕ from the vertical such that $\phi \leq \phi_s$. If **P** creates uniform *motion* of the block, then $P = F_k$. In this case, the resultant $\mathbf{R}_k$ has a line of action defined by ϕ_k, Fig. 8–3b. This angle is referred to as the *angle of kinetic friction,* where

Motion
(b)

Fig. 8–3

$$\phi_k = \tan^{-1}\left(\frac{F_k}{N}\right) = \tan^{-1}\left(\frac{\mu_k N}{N}\right) = \tan^{-1}\mu_k$$

By comparison, $\phi_s > \phi_k$.

Angle of Repose

An experimental method which can be used to measure the coefficient of friction between two contacting surfaces consists of placing a block of one material having a weight W on a plane made of another material. The plane is inclined to the angle θ_s, at which point the block is on the *verge of sliding* and therefore $F_s = \mu_s N$. The free-body diagram of the block at this instant is shown in Fig. 8–4. Applying the force equations of equilibrium yields $N = W \cos \theta_s$ and $F_s = \mu_s N = W \sin \theta_s$ so that

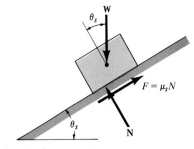

$$\theta_s = \tan^{-1}\left(\frac{F_s}{N}\right) = \tan^{-1}\left(\frac{\mu_s N}{N}\right) = \tan^{-1}\mu_s$$

Fig. 8–4

The angle θ_s is referred to as the *angle of repose,* and by comparison it is equal to the angle of static friction ϕ_s. Once it is measured, the coefficient of static friction is obtained from $\mu_s = \tan \theta_s$. Note that this calculation is independent of the weight of the block.

8.2 Problems Involving Dry Friction

(a)

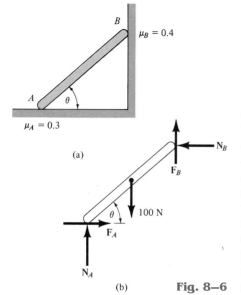

(b)

Fig. 8–5

If a rigid body is in equilibrium when it is subjected to a system of forces that includes the effect of friction, the force system must not only satisfy the equations of equilibrium but *also* the laws that govern the frictional forces.

Types of Frictional Problems

In general, there are three types of mechanics problems involving dry friction. They can easily be classified once the free-body diagrams are drawn and the total number of unknowns are identified and compared with the total number of available equilibrium equations. Each type of problem will now be explained and illustrated graphically by examples. In all cases the geometry and dimensions for the problem are assumed to be known.

Equilibrium. Problems in this category are strictly equilibrium problems which require *the total number of unknowns to be equal to the total number of available equilibrium equations*. Once the frictional forces are determined from the solution, however, their numerical values must be checked to be sure they satisfy the inequality $F \leqslant \mu_s N$; otherwise, slipping will occur and the body will not remain in equilibrium. A problem of this type is shown in Fig. 8–5a. Here we must determine the frictional forces at A and C to check if the equilibrium position of the bars can be maintained. If the bars are uniform and have known weights of 100 N each, then the free-body diagrams are shown in Fig. 8–5b. There are six unknown force components which can be determined *strictly* from the six equilibrium equations (three for each member). Once $\mathbf{F}_A$, N_A, $\mathbf{F}_C$, and N_C are determined, then the bars will remain in equilibrium provided $F_A \leqslant 0.3 N_A$ and $F_C \leqslant 0.5 N_C$ are satisfied.

Impending Motion at All Points. In this case *the total number of unknowns will equal the total number of available equilibrium equations plus the total number of available frictional equations, $F = \mu N$.* In particular, if *motion is impending* at the points of contact, then $F_s = \mu_s N$; whereas if the body is *slipping*, then $F_k = \mu_k N$. For example, consider the problem of finding the smallest angle θ at which the 100-N bar in Fig. 8–6a can be placed against the wall without slipping. The free-body diagram is shown in Fig. 8–6b. Here there are *five* unknowns: F_A, N_A, F_B, N_B, θ. For the solution there are *three* equilibrium equations and *two* static frictional equations which apply at *both* points of contact, so that $F_A = 0.3 N_A$ and $F_B = 0.4 N_B$. (It should also be noted that the bar will not be in a state where motion impends *unless* the bar slips at *both* points A and B simultaneously.)

Fig. 8–6

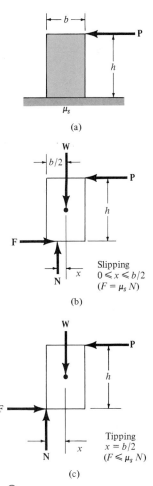

Fig. 8–7

Tipping or Impending Motion at Some Points. Here *the total number of unknowns will be less than the number of available equilibrium equations plus the total number of frictional equations or conditional equations for tipping.* As a result of this, several possibilities for motion or impending motion will exist and the problem will involve a determination of the kind of motion which actually occurs. For example, consider the two-member frame shown in Fig. 8–7a. In this problem we wish to determine the horizontal force **P** needed to cause movement of the frame. If each member has a weight of 100 N, then the free-body diagrams are as shown in Fig. 8–7b. There are *seven* unknowns: N_A, F_A, N_C, F_C, B_x, B_y, P. For a unique solution we must satisfy the *six* equilibrium equations (three for each member) and only *one* of two possible static frictional equations. This means that as **P** increases its magnitude it will either cause slipping at A and no slipping at C, so that $F_A = 0.3N_A$ and $F_C \leq 0.5N_C$, or slipping occurs at C and no slipping at A, in which case $F_C = 0.5N_C$ and $F_A \leq 0.3N_A$. The actual situation can be determined by calculating P for each case and then choosing the case for which P is *smallest*. If in both cases the *same value* for P is calculated, then slipping at both points occurs simultaneously; i.e., the *seven unknowns* will satisfy *eight equations*. As a second example, consider a block having a width b, height h, and weight W which is resting on a rough surface, Fig. 8–8a. The force **P** needed to cause motion is to be determined. Inspection of the free-body diagram, Fig. 8–8b, indicates there are *four unknowns*, namely, P, F, N, and x. For a unique solution, however, we must satisfy the *three* equilibrium equations and either *one* static friction equation or *one* conditional equation which requires the block not to tip. Hence two possibilities of motion exist. Either the block will *slip*, Fig. 8–8b, in which case $F = \mu_s N$ and the value obtained for x must satisfy $0 \leq x \leq b/2$, or the block will *tip*, Fig. 8–8c, in which case $x = b/2$ and the frictional force satisfies the inequality $F \leq \mu_s N$. The solution yielding the *smallest* value of P will define the type of motion the block undergoes. If it happens that the same value of P is calculated for both cases, then slipping and tipping will occur simultaneously, i.e., the *four unknowns* will satisfy *five equations*.

Fig. 8–8

Equilibrium Versus Frictional Equations

It was stated earlier that the frictional force *always* acts so as to either oppose the relative motion or impede the motion of a body over its contacting surface. Realize, however, that we can *assume* the sense of the frictional force in problems which require F to be an "equilibrium force" and satisfy the inequality $F < \mu_s N$. The correct sense is made known *after* solving the equations of equilibrium for F. For example, if F is a negative scalar, the sense of **F** is the reverse of that which was assumed. This convenience of *assuming* the sense of **F** is possible because the equilibrium equations equate to zero the *components of vectors* acting in the *same direction*. In cases where the frictional equation $F = \mu N$ is used in the solution of a problem, however, the convenience of *assuming* the sense of **F** is *lost,* since the frictional equation relates only the *magnitudes* of two perpendicular vectors. Consequently, **F** *must always* be shown acting with its *correct sense* on the free-body diagram whenever the frictional equation is used for the solution of a problem.

PROCEDURE FOR ANALYSIS

The following procedure provides a method for solving equilibrium problems involving dry friction.

Free-Body Diagrams. Draw the necessary free-body diagrams and determine the number of unknowns or equations required for a complete solution. Unless stated in the problem, *always* show the frictional forces as *unknowns,* i.e., *do not assume that* $F = \mu N$. Recall that only three equations of coplanar equilibrium can be written for each body. Consequently, if there are more unknowns than equations of equilibrium, then it will be necessary to apply the frictional equation at some, if not all, points of contact to obtain the extra number of equations needed for a complete solution.

Equations of Friction and Equilibrium. Apply the equations of equilibrium and the necessary frictional equations (or conditional equations if tipping is involved) and solve for the unknowns. If the problem involves a three-dimensional force system such that it becomes difficult to obtain the force components or the necessary moment arms, apply the equations of equilibrium using Cartesian vectors.

The following example problems numerically illustrate this procedure.

Example 8–1

The uniform crate shown in Fig. 8–9a has a mass of 20 kg. If a force $P = 80$ N is applied to the crate, determine if the crate remains in equilibrium. The coefficient of static friction is $\mu_s = 0.3$.

(a)

Fig. 8–9

SOLUTION

Free-Body Diagram. As shown in Fig. 8–9b, the *resultant* normal force $\mathbf{N}_C$ must act a distance x from the crate's center line in order to counteract the tipping effect caused by **P**. There are *three unknowns:* F, N_C, and x, which can be determined strictly from the *three* equations of equilibrium.

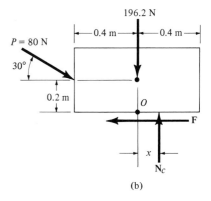

(b)

Equations of Equilibrium

$\xrightarrow{+} \Sigma F_x = 0;$ $80 \cos 30° - F = 0$

$+\uparrow \Sigma F_y = 0;$ $-80 \sin 30° + N_C - 196.2 = 0$

$\zeta+\Sigma M_O = 0;$ $80 \sin 30°(0.4) - 80 \cos 30°(0.2) + N_C(x) = 0$

Solving,

$$F = 69.3 \text{ N}$$
$$N_C = 236 \text{ N}$$
$$x = -0.00908 \text{ m} = -9.08 \text{ mm}$$

Since x is negative it indicates the *resultant* normal force acts (slightly) to the *left* of the crate's center line. No tipping will occur since $x \leqslant 0.4$ m. Also, the *maximum* frictional force which can be developed at the surface of contact is $F_{max} = \mu_s N_C = 0.3(236 \text{ N}) = 70.8$ N. Since $F = 69.3$ N < 70.8 N, the crate will *not slip,* although it is very close to doing so.

Example 8–2

The pipe shown in Fig. 8–10*a* is gripped between two levers that are pinned together at *C*. If the coefficient of static friction between the levers and the pipe is $\mu = 0.3$, determine the maximum angle θ at which the pipe can be gripped without slipping. Neglect the weight of the pipe.

SOLUTION

Free-Body Diagram. As shown in Fig. 8–10*b*, there are five unknowns: N_A, F_A, N_B, F_B, and θ. The *three* equations of equilibrium and *two* frictional equations at *A* and *B* apply. The frictional forces act toward *C* to prevent upward motion of the pipe.

Equations of Friction and Equilibrium. The frictional equations are

$$F_s = \mu_s N; \qquad\qquad F_A = \mu N_A$$
$$F_B = \mu N_B$$

Using these results, and applying the equations of equilibrium, yields

$$\xrightarrow{+}\Sigma F_x = 0; \quad N_A \cos (\theta/2) + \mu N_A \sin (\theta/2) - N_B \cos (\theta/2) - \mu N_B \sin (\theta/2) = 0 \quad (1)$$

$$\overset{\curvearrowright}{+}\Sigma M_O = 0; \qquad\qquad \mu N_B(r) - \mu N_A(r) = 0 \qquad\qquad\qquad (2)$$

$$+\uparrow\Sigma F_y = 0; \quad N_A \sin (\theta/2) - \mu N_A \cos (\theta/2) + N_B \sin (\theta/2) - \mu N_B \cos (\theta/2) = 0 \quad (3)$$

From either Eq. (1) or (2) it is seen that $N_A = N_B$. This could also have been determined directly from the symmetry of *both* geometry and loading. Substituting the result into Eq. (3), we obtain

$$\sin (\theta/2) - \mu \cos (\theta/2) = 0$$

so that

$$\tan (\theta/2) = \frac{\sin (\theta/2)}{\cos (\theta/2)} = \mu = 0.3$$

$$\theta = 2 \tan^{-1} 0.3 = 33.4° \qquad \textit{Ans.}$$

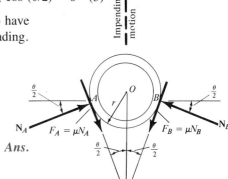

(a)

(b)

Fig. 8–10

Example 8–3

The uniform plank having a weight W and length l is supported at its ends A and B, where the coefficient of static friction is μ, Fig. 8–11a. Determine the greatest angle θ so the plank does not slip. Neglect the thickness of the plank for the calculation.

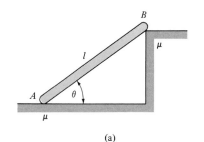

(a)

SOLUTION

Free-Body Diagram. As shown in Fig. 8–11b, there are *five* unknowns: F_A, N_A, F_B, N_B, and θ. These can be determined from the *three* equilibrium equations and *two* frictional equations applied at points A and B. The frictional forces must be drawn with their correct sense so that they oppose the tendency for motion of the plank. Why?

Equations of Friction and Equilibrium. Writing the frictional equations,

$$F = \mu_s N; \qquad\qquad F_A = \mu N_A$$
$$F_B = \mu N_B$$

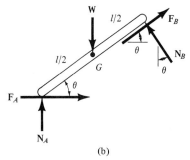

(b)

Fig. 8–11

Using these results and applying the equations of equilibrium yields

$$\xrightarrow{+}\Sigma F_x = 0; \quad \mu N_A + \mu N_B \cos\theta - N_B \sin\theta = 0 \qquad (1)$$

$$+\uparrow\Sigma F_y = 0; \quad N_A - W + N_B \cos\theta + \mu N_B \sin\theta = 0 \qquad (2)$$

$$\downarrow+\Sigma M_G = 0; \quad -N_A\left(\frac{l}{2}\cos\theta\right) + \mu N_A\left(\frac{l}{2}\sin\theta\right) + N_B\left(\frac{l}{2}\right) = 0 \qquad (3)$$

Notice that moments were summed about the center of the plank G in order to eliminate **W.** We can now solve Eqs. (1) and (3), which reduce to

$$\mu N_A = N_B(\sin\theta - \mu\cos\theta)$$
$$N_B = N_A(\cos\theta - \mu\sin\theta)$$

Thus,

$$\mu N_A = N_A(\cos\theta - \mu\sin\theta)(\sin\theta - \mu\cos\theta)$$
$$\mu = \sin\theta\cos\theta - \mu\cos^2\theta - \mu\sin^2\theta + \mu^2\sin\theta\cos\theta$$
$$\mu = (1 + \mu^2)\sin\theta\cos\theta - \mu(\sin^2\theta + \cos^2\theta)$$

Since $\sin^2\theta + \cos^2\theta = 1$ and $\sin 2\theta = 2\sin\theta\cos\theta$, then

$$2\mu = \left(\frac{1 + \mu^2}{2}\right)\sin 2\theta$$

or

$$\theta = \frac{1}{2}\sin^{-1}\left(\frac{4\mu}{1 + \mu^2}\right) \qquad\qquad Ans.$$

Example 8–4

The homogeneous block shown in Fig. 8–12a has a weight of 20 lb and rests on the incline for which $\mu_s = 0.55$. Determine the largest angle of tilt, θ, of the plane before the block moves.

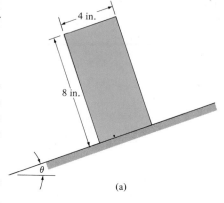

(a)

SOLUTION

Free-Body Diagram. As shown in Fig. 8–12b, the dimension x is used to locate the position of the resultant normal force N_B under the block. There are *four* unknowns, θ, N_B, F_B, and x. *Three* equations of equilibrium are available. The *fourth* equation is obtained by investigating the conditions for tipping or sliding of the block.

Equations of Equilibrium. Applying the equations of equilibrium yields

$$+\swarrow \Sigma F_x = 0; \qquad 20 \sin \theta - F_B = 0 \qquad\qquad (1)$$

$$+\nwarrow \Sigma F_y = 0; \qquad N_B - 20 \cos \theta = 0 \qquad\qquad (2)$$

$$\curvearrowright + \Sigma M_O = 0; \qquad -20 \sin \theta(4) + 20 \cos \theta(x) = 0 \qquad (3)$$

(Impending Motion of Block.) This requires use of the frictional equation

$$F_s = \mu_s N; \qquad\qquad F_B = 0.55 N_B \qquad\qquad (4)$$

Solving Eqs. (1) to (4) yields

$$N_B = 17.5 \text{ lb} \qquad F_B = 9.64 \text{ lb} \qquad \theta = 28.8° \qquad x = 2.2 \text{ in.}$$

Since $x = 2.2$ in. > 2 in., the block will tip *before* sliding.

(b)

Fig. 8–12

(Tipping of Block.) This requires

$$x = 2 \text{ in.} \qquad\qquad (5)$$

Solving Eqs. (1) to (3) using Eq. (5) yields

$$N_B = 17.9 \text{ lb} \qquad F_B = 8.94 \text{ lb}$$

$$\theta = 26.6° \qquad\qquad\qquad \textit{Ans.}$$

Note: If we *first* assumed that the block tips, then the results for F_B would have to be checked with the maximum *possible* static frictional force, i.e.,

$$F_B = 8.94 \overset{?}{<} (0.55)(17.9 \text{ lb}) = 9.84 \text{ lb}$$

Since the inequality holds, indeed the block will tip before it slips.

Example 8-5

Beam *AB* is subjected to a uniform load of 200 N/m and is supported at *B* by column *BC*, Fig. 8–13a. If the coefficients of static friction at *B* and *C* are $\mu_B = 0.2$ and $\mu_C = 0.5$, determine the force **P** needed to pull the column out from under the beam. Neglect the weight of the members.

(a)

SOLUTION

Free-Body Diagrams. The free-body diagram of beam *AB* is shown in Fig. 8–13b. Applying $\Sigma M_A = 0$, we obtain $N_B = 400$ N. This result is shown on the free-body diagram of the column, Fig. 8–13c. Referring to this member, the *four* unknowns F_B, P, F_C, and N_C are determined from the *three* equations of equilibrium and *one* frictional equation applied either at *B* or *C*.

(b)

Equations of Equilibrium and Friction.

$$\xrightarrow{+}\Sigma F_x = 0; \qquad P - F_B - F_C = 0 \qquad (1)$$

$$+\uparrow \Sigma F_y = 0; \qquad N_C - 400 = 0 \qquad (2)$$

$$\curvearrowright+\Sigma M_C = 0; \qquad P(0.25) - F_B(1) = 0 \qquad (3)$$

(Column Slips Only at B.) This requires $F_C \leq \mu_C N_C$ and

$$F_B = \mu_B N_B; \qquad F_B = 0.2(400) = 80 \text{ N}$$

Using this result and solving Eqs. (1) to (3), we obtain

$$P = 320 \text{ N}$$
$$F_C = 240 \text{ N}$$
$$N_C = 400 \text{ N}$$

Since $F_C = 240 \text{ N} > \mu_C N_C = 0.5(400 \text{ N}) = 200 \text{ N}$, the other case of movement must be investigated.

(Column Slips Only at C.) Here $F_B \leq \mu_B N_B$ and

$$F_C = \mu_C N_C; \qquad F_C = 0.5 N_C \qquad (4)$$

Solving Eqs. (1) to (4) yields

$$P = 267 \text{ N} \qquad\qquad Ans.$$
$$N_C = 400 \text{ N}$$
$$F_C = 200 \text{ N}$$
$$F_B = 66.7 \text{ N}$$

Obviously, this case of movement occurs first since it requires a *smaller* value for *P*.

Fig. 8–13

(c)

Example 8–6

Determine the normal force that must be exerted on the 100-kg spool shown in Fig. 8–14a to push it up the 20° incline at constant velocity. The coefficients of static and kinetic friction at the points of contact are $(\mu_s)_A = 0.18$, $(\mu_k)_A = 0.15$ and $(\mu_s)_B = 0.45$, $(\mu_k)_B = 0.4$.

(a)

SOLUTION

Free-Body Diagram. As shown in Fig. 8–14b, there are four unknowns N_A, F_A, N_B, and F_B acting on the spool. These can be determined from the *three* equations of equilibrium and *one* frictional equation, which applies either at A or B. If slipping only occurs at B, the spool *rolls* up the incline; whereas if slipping only occurs at A, the spool will *slide* up the incline. The problem requires determination of N_B.

Equations of Equilibrium and Friction

$$+\nearrow\Sigma F_x = 0; \qquad -F_A + N_B - 981\sin 20° = 0 \qquad (1)$$

$$+\nwarrow\Sigma F_y = 0; \qquad N_A - F_B - 981\cos 20° = 0 \qquad (2)$$

$$\curvearrowright+\Sigma M_O = 0; \qquad -F_B(400 \text{ mm}) + F_A(400 \text{ mm}) = 0 \qquad (3)$$

(b)

Fig. 8–14

(Spool Rolls up Incline.) In this case $F_A \leq 0.18 N_A$ and

$$(F_k)_B = (\mu_k)_B N_B; \qquad F_B = 0.40 N_B \qquad (4)$$

The direction of the frictional force at B must be correctly specified. Why? Since the spool is being forced up the plane, $\mathbf{F}_B$ acts downward to prevent the clockwise rolling motion of the spool, Fig. 8–14b. Solving Eqs. (1) to (4), we have

$$N_A = 1146 \text{ N} \qquad F_A = 224 \text{ N} \qquad N_B = 559 \text{ N} \qquad F_B = 224 \text{ N}$$

The assumption regarding no slipping at A should be checked.

$$F_A \leq (\mu_s)_A N_A; \qquad 224 \text{ N} \overset{?}{\leq} 0.18(1146) = 206 \text{ N}$$

The inequality does *not apply,* and therefore slipping occurs at A and not at B. Hence, the other case of motion must be investigated.

(Spool Slides up Incline.) In this case, $F_B \leq 0.45 N_B$ and

$$(F_k)_A = (\mu_k)_A N_A; \qquad F_A = 0.15 N_A \qquad (5)$$

Solving Eqs. (1) to (3) and (5) yields

$$N_A = 1084 \text{ N} \qquad F_A = 163 \text{ N} \qquad N_B = 498 \text{ N} \qquad F_B = 163 \text{ N}$$

The validity of the solution ($N_B = 498$ N) can be checked by testing the assumption that indeed no slipping occurs at B.

$$F_B \leq (\mu_s)_B N_B; \qquad 163 \text{ N} < 0.45(498 \text{ N}) = 224 \text{ N} \quad \text{(check)}$$

PROBLEMS

8–1. Determine the horizontal force P needed to just start pushing the 300-lb crate up the plane. Take $\mu_s = 0.3$.

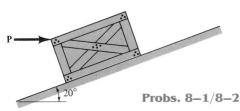

Probs. 8–1/8–2

8–2. Determine the range of values for which the horizontal force P will prevent the crate from slipping down or up the inclined plane. Take $\mu_s = 0.1$.

8–3. Determine the magnitude of force **P** needed to start towing the 40-kg crate. Also determine the location of the resultant normal force acting on the crate, measured from point A. Take $\mu_s = 0.3$.

Prob. 8–3

*8–4. The tractor exerts a towing force $T = 400$ lb. Determine the normal reactions at each of the two front and two rear tires and the tractive frictional force **F** on each rear tire needed to pull the load forward at constant velocity. The tractor has a weight of 7500 lb and a center of gravity located at G_T. An additional weight of 600 lb is added to its front having a center of gravity at G_A. Take $\mu_s = 0.4$.

Prob. 8–4

8–5. The car has a mass of 5 Mg and is coasting up the slope. When its speed becomes zero, the driver applies the brakes to the rear wheels. Determine if the frictional force acting on the rear wheels will hold the car in equilibrium. The coefficient of static friction at A is $\mu_A = 0.3$. The front wheels are free to roll.

Prob. 8–5

8–6. The board can be adjusted vertically by tilting it up and sliding the smooth pin A along the vertical guide G. When placed horizontally, the bottom C bears along the edge of the guide, where $\mu_s = 0.4$. Determine the largest dimension d which will support any applied force **F** without causing the board to slip downward.

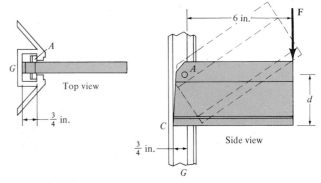

Prob. 8–6

341

8–7. The ring has a mass of 0.5 kg and is resting on the surface of the table. In an effort to move the ring a normal force **P** from the finger is exerted on it. If this force is directed towards the ring's center O as shown, determine its magnitude when the ring is on the verge of slipping at A. The coefficient of static friction at A is $\mu_A = 0.2$ and at B, $\mu_B = 0.3$.

Prob. 8–7

■*8–8. The spool has a mass of 100 kg and a center of mass at G. If the coefficient of static friction between it and the inclined plane is $\mu_A = 0.4$, determine the greatest angle θ of the inclined plane without causing the spool to slip.

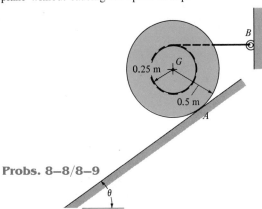

Probs. 8–8/8–9

8–9. The spool has a mass of 100 kg and a center of mass at G. If the coefficient of static friction between it and the inclined plane is $\mu_A = 0.4$, determine the tension in the cable at B and the friction force at A if $\theta = 20°$.

8–10. An ant falls into the bowl which is in the shape of a spherical cup. Determine if it is possible for the ant to crawl out if the coefficient of static friction is $\mu_s = 0.25$.

Prob. 8–10

8–11. Determine how far d the man can walk up the plank without causing the plank to slip. The coefficient of static friction at A and B is $\mu_s = 0.3$. The man has a weight of 200 lb and a center of gravity at G. Neglect the thickness and weight of the plank.

Prob. 8–11

*8–12. The door stop of negligible weight is pin-connected at A and the coefficient of static friction at B is $\mu_s = 0.3$. Determine the required distance s from A to the floor so that the stop will resist opening of the door for any force P applied to the handle. Neglect the weight of the door stop.

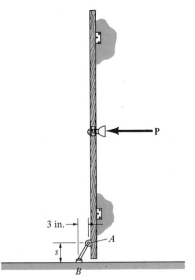

Prob. 8–12

8–13. The 5-kg cylinder is suspended from two equal-length cords. The end of each cord is attached to a ring of negligible mass, which passes along a horizontal shaft. If the coefficient of static friction between each ring and the shaft is $\mu_s = 0.5$, determine the greatest distance d at which the rings can be separated and still support the cylinder.

Prob. 8–13

8–14. The spool has a mass of 200 kg and rests against the wall and on the beam. If the coefficient of static friction at A and B is $\mu_A = 0.4$ and $\mu_B = 0.5$, respectively, determine the smallest vertical force P that can be applied to the cable that will cause the spool to turn.

Probs. 8–14/8–15

8–15. The spool has a mass of 200 kg and rests against the wall and on the beam. If the coefficient of static friction at B is $\mu_B = 0.3$, and the wall is smooth, determine the friction force developed at B when the vertical force applied to the cable is $P = 800$ N.

*8–16. The semicircular hoop of weight W is suspended by the small peg at A. A horizontal force P is slowly applied at B. If the hoop begins to slip at A when $\theta = 30°$, determine the coefficient of static friction between the hoop and the peg.

Prob. 8–16

8–17. The friction tongs are used to drag the 100-kg pallet and load along the floor. Determine the tension force in the chain and the minimum coefficient of static friction at the shoe A of the tongs so slipping of the tongs does not occur. The coefficient of static friction between the pallet and the floor is $\mu_s = 0.4$.

Probs. 8–17/8–18

8–18. The coefficient of static friction between the shoes at A and B of the tongs and the pallet is $\mu_s' = 0.5$ and between the pallet and the floor $\mu_s = 0.4$. If a horizontal towing force of $P = 300$ N is applied to the tongs, determine the largest mass that can be towed.

8–19. The friction hook is made from a fixed frame which is shown colored and a cylinder of negligible weight. A piece of paper is placed between the wall and the cylinder. If $\mu_s = 0.3$ at all points of contact, determine the design angle θ so that any weight W of paper p can be held.

8–23. The uniform beam has a mass of 500 kg/m. It is supported by a pipe at A and a block at B. If the coefficients of static friction at B and C are $\mu_B = 0.4$ and $\mu_C = 0.6$, respectively, determine the horizontal force P required to move the beam. Neglect the weight of the block and assume it will not tip. Also neglect the thickness of the beam.

Prob. 8–23

*8–24.** The crate has a weight of 200 lb and a center of gravity at G. Determine the horizontal force P required to tow it. Also, determine the location of the resultant normal force measured from A. Take $h = 4$ ft and $\mu_s = 0.4$.

Probs. 8–19/8–20

*8–20.** The friction hook is made from a fixed frame which is shown colored and a cylinder of negligible weight. A piece of paper is placed between the smooth wall and the cylinder. If $\theta = 20°$, determine the smallest coefficient of static friction μ_s at all points of contact so that any weight W of paper p can be held.

8–21. The semicylinder of mass m and radius r lies on the rough inclined plane for which $\phi = 10°$ and the coefficient of static friction is $\mu_s = 0.3$. Determine if the semicylinder slides down the plane, and if not, compute the angle of tip of its base AB. The center of mass G is at $4r/3\pi$. See table inside back cover.

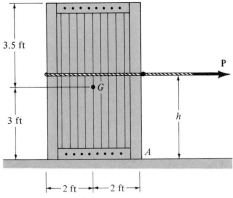

Probs. 8–24/8–25

8–25. The crate has a weight of 200 lb and a center of gravity at G. Determine the height h of the tow rope so that the crate slips and tips at the same time. What horizontal force P is required to do this? Take $\mu_s = 0.4$.

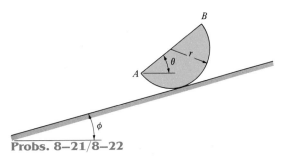

Probs. 8–21/8–22

8–22. The semicylinder of mass m and radius r lies on the rough inclined plane. If the inclination $\phi = 15°$, determine the smallest coefficient of static friction which will prevent the semicylinder from slipping. The center of mass G is at $4r/3\pi$. See table inside back cover.

8–26. The refrigerator has a weight of 180 lb and rests on a tile floor for which $\mu_s = 0.25$. If the man pushes horizontally on it in the direction shown, determine the smallest magnitude of force needed to move the refrigerator. Also, if the man has a weight of 150 lb, determine the smallest coefficient of friction between his shoes and the floor so that he does not slip.

Probs. 8–26/8–27

8–27. The refrigerator has a weight of 180 lb and rests on a tile floor for which $\mu_s = 0.25$. Also, the man has a weight of 150 lb and the coefficient of static friction between the floor and his shoes is $\mu'_s = 0.6$. If he pushes horizontally on the refrigerator, determine if he can move it. If so does the refrigerator slip or tip?

*8–28. Two blocks A and B have a weight of 10 lb and 6 lb, respectively. They are resting on the incline for which the coefficients of static friction are $\mu_A = 0.15$ and $\mu_B = 0.25$. Determine the incline angle θ for which both blocks begin to slide. Also find the required stretch or compression in the connecting spring for this to occur. The spring has a stiffness of $k = 2$ lb/ft.

Probs. 8–28/8–29

8–29. Two blocks A and B have a weight of 10 lb and 6 lb, respectively. They are resting on the incline for which the coefficients of static friction are $\mu_A = 0.15$ and $\mu_B = 0.25$. Determine the angle θ which will cause motion of one of the blocks. What is the friction force under each of the blocks when this occurs? The spring has a stiffness of $k = 2$ lb/ft and is originally unstretched.

8–30. A man pushes against the stack of four uniform boxes each weighing 20 lb. If the coefficient of static friction between each box is $\mu_s = 0.6$ and between the floor and a box $\mu'_s = 0.4$, determine the greatest horizontal force P that can be applied without causing any slipping or tipping.

Prob. 8–30

8–31. The bed has a weight of 300 lb and a center of gravity at G. If the coefficient of static friction at A and B is $\mu_s = 0.2$, determine if the 200-lb man can push the bed to the left. The coefficient of static friction between his shoes and the floor is $\mu'_s = 0.4$. Assume the man only exerts a horizontal force on the bed.

Probs. 8–31/8–32

*8–32. The bed has a weight of 300 lb and a center of gravity at G. If the coefficient of static friction at A and B is $\mu_s = 0.2$, determine if the 200-lb man can pull the bed to the right. The coefficient of static friction between his shoes and the floor is $\mu'_s = 0.35$. Assume the man only exerts a horizontal force on the bed.

8–33. The crate has a weight of 350 lb and is being towed by the man who has a weight of 180 lb. If the coefficient of static friction between the crate and floor is $\mu_s = 0.3$ and between the man and the floor $\mu'_s = 0.5$, determine if the man can move the crate. What is the friction force between the man and the floor when the crate or the man begin to slip?

Prob. 8–33

8–34. The three bars have a weight of $W_A = 20$ lb, $W_B = 40$ lb, and $W_C = 60$ lb, respectively. If the coefficients of static friction at the surfaces of contact are as shown, determine the smallest horizontal force P needed to move block A.

Prob. 8–34

8–35. The beam AB has a negligible mass and thickness and is subjected to a triangular distributed loading. It is supported at one end by a pin and at the other end by a post having a mass of 50 kg and negligible thickness. Determine the minimum force P needed to move the post. The coefficients of static friction at B and C are $\mu_B = 0.4$ and $\mu_C = 0.2$, respectively.

Prob. 8–35

*8–36. The coefficients of static and kinetic friction between the drum and brake bar are $\mu_s = 0.4$ and $\mu_k = 0.3$, respectively. If $M = 50$ N · m and $P = 85$ N, determine the horizontal and vertical components of reaction at the pin O. Neglect the weight and thickness of the brake. The drum has a mass of 25 kg.

Probs. 8–36/8–37

8–37. The coefficient of static friction between the drum and brake bar is $\mu_s = 0.4$. If the moment $M = 35$ N · m, determine the smallest force P that can be applied to the brake bar in order to prevent the drum from rotating. Also determine the horizontal and vertical components of reaction at the pin O. Neglect the weight and thickness of the brake. The drum has a mass of 25 kg.

8–38. The paper-towel dispenser carries two rolls of paper. The one in use is called the stub roll A and the other is the fresh roll B. They weigh 2 lb and 5 lb, respectively. If the coefficients of static friction at the points of contact C and D are $\mu_C = 0.2$ and $\mu_D = 0.5$, determine the initial vertical force P that must be applied to the paper on the stub roll in order to pull down a sheet. The stub roll is pinned at its center, whereas the fresh roll is not. Neglect friction at the pin.

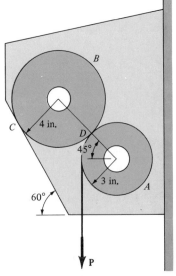

Prob. 8–38

8–39. Two blocks A and B, each having a mass of 6 kg, are connected by the linkage shown. If the coefficients of static friction at the contacting surfaces are $\mu_A = 0.2$ and $\mu_B = 0.8$, determine the largest vertical force P that may be applied to pin C without causing the blocks to slip. Neglect the weight of the links.

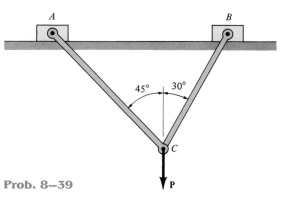

Prob. 8–39

*8–40. The symmetrical crab hook is used to lift packages by means of friction developed between the shoes A and B and a package. Determine the smallest coefficient of static friction at the shoes so that the package of weight W can be lifted.

Prob. 8–40

8–41. The 45-kg disk rests on the surface for which the coefficient of static friction is $\mu_A = 0.2$. Determine the largest couple moment M that can be applied to the bar without causing motion.

Probs. 8–41/8–42

8–42. The 45-kg disk rests on the surface for which the coefficient of static friction is $\mu_A = 0.15$. If $M = 50$ N · m, determine the friction force at A.

8–43. Each of the two cylinders has a mass of 50 kg. If the coefficients of static friction at the points of contact are $\mu_A = 0.5$, $\mu_B = 0.5$, $\mu_C = 0.5$, and $\mu_D = 0.6$, determine the couple moment M needed to rotate cylinder E.

Prob. 8–43

*8–44. The roll of paper of weight W rests between the two supports at A and B. If the coefficient of static friction at these points is $\mu_s = 0.3$, determine the horizontal force P needed to cause motion.

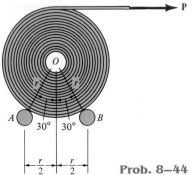

Prob. 8–44

8–45. Specify the angle θ and the magnitude P of the smallest force needed to tow the 120-kg sled up the incline at constant velocity. The coefficient of kinetic friction between the ground and the sled is $\mu_k = 0.2$. *Hint:* Using the force equilibrium equations, relate $P = f(\theta)$, then set $dP/d\theta = 0$ to determine θ.

Prob. 8–45

8–46. The block of weight W is being pulled up the inclined plane of slope α using a force P. If P acts at the angle ϕ as shown, show that for slipping to occur, $P = W \sin(\alpha + \theta)/\cos(\phi - \theta)$, where θ is the angle of friction; $\theta = \tan^{-1} \mu$.

Probs. 8–46/8–47

8–47. Determine the angle ϕ at which P should act on the block so that the magnitude of P is as small as possible. What is the corresponding value of P? The block weighs W and the slope α is known.

*8–48. The 200-mm uniform rod has a weight W and is placed in a glass as shown. If the smallest angle at which it can lean against the glass without slipping is $\theta = 60°$, determine the coefficient of static friction between the glass and rod.

Prob. 8–48

8–49. The uniform pole has a weight W and length L. Its end B is tied to a supporting cord, and end A is placed against the wall for which the coefficient of static friction is μ. Determine the largest angle θ at which the pole can be placed without slipping.

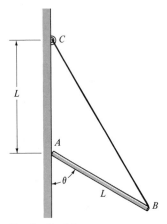

Prob. 8–49

8–50. The uniform beam has a weight W and length $4a$. It rests on the fixed rails at A and B. If the coefficient of static friction at the rails is μ, determine the horizontal force P, applied perpendicular to the face of the beam, which will cause the beam to move.

Prob. 8–50

8–51. The disk has a weight W and lies on the plane which has a coefficient of static friction μ. Determine the maximum height h to which the plane can be lifted without causing the disk to slip.

Prob. 8–51

8.3 Wedges

A *wedge* is a simple machine which is often used to transform an applied force into much larger forces, directed at approximately right angles to the applied force. Also, wedges can be used to give small displacements or adjustments to heavy loads.

Consider, for example, the wedge shown in Fig. 8–15a, which is used to *lift* a block of weight W by applying a force P to the wedge. Free-body diagrams of the wedge and block are shown in Fig. 8–15b. Here we have excluded the weight of the wedge since it is usually *small* compared to the weight of the block. Also, note that the frictional forces F_1 and F_2 must oppose the motion of the wedge. Likewise, the frictional force F_3 of the wall on the block must act downward so as to oppose the block's upward motion. The locations of the resultant normal forces are not important in the force analysis since neither the block nor wedge will "tip." Hence the moment equilibrium equations will not be considered. There are seven unknowns consisting of the applied force P, needed to cause motion of the wedge, and six normal and frictional forces. The seven available equations consist of two force equilibrium equations ($\Sigma F_x = 0$, $\Sigma F_y = 0$) applied to the wedge and block (four equations total) and the frictional equation $F = \mu N$ applied at each surface of contact (three equations total).

Realize that if the block is to be *lowered,* the frictional forces will all act in a sense opposite to that shown in Fig. 8–15b. The applied force P will act to the right as shown if the coefficient of friction is small or the wedge angle θ is large. Otherwise, P would have to *pull* on the wedge in order to remove it. If P is *removed,* and friction forces hold the block in place, then the wedge is referred to as *self-locking.*

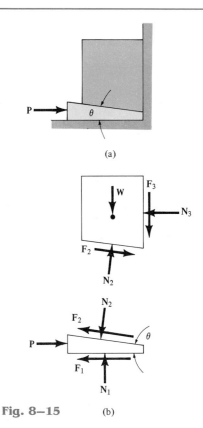

Fig. 8–15 (b)

Example 8–7

The uniform stone has a mass of 500 kg and is held in the horizontal position using a wedge as shown in Fig. 8–16a. If the coefficient of static friction is $\mu_s = 0.3$ at the surface in contact with the wedge, determine the force **P** needed to remove the wedge. Is the wedge self-locking? Assume that the stone does not slip at A.

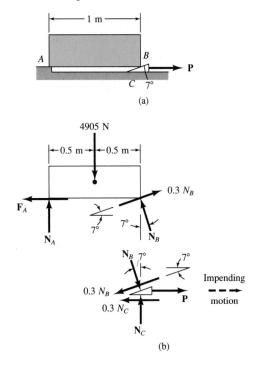

Fig. 8–16

(b)

SOLUTION

Since the wedge is to be removed, slipping is about to occur at the surfaces of contact. Thus, $F = \mu_s N$, and the free-body diagrams are shown in Fig. 8–16b. Note that on the stone at A, $F_A < \mu_s N_A$, since slipping does not occur there. From the free-body diagram of the stone,

$$\zeta + \Sigma M_A = 0; \quad -4905(0.5) + (N_B \cos 7°)(1) + (0.3N_B \sin 7°)(1) = 0$$
$$N_B = 2383.1 \text{ N}$$

Using this result for the wedge, we have

$$\xrightarrow{+} \Sigma F_x = 0; \quad 2383.1 \sin 7° - 0.3(2383.1) \cos 7° + P - 0.3N_C = 0$$
$$+ \uparrow \Sigma F_y = 0; \quad N_C - 2383.1 \cos 7° - 0.3(2383.1) \sin 7° = 0$$
$$N_C = 2452.5 \text{ N}$$
$$P = 1154.9 \text{ N} = 1.15 \text{ kN} \qquad \qquad \textit{Ans.}$$

Since P is positive, indeed the wedge must be pulled out. Obviously, if P is zero, the wedge will remain in place (self-locking) and the frictional forces $\mathbf{F}_C$ and $\mathbf{F}_B$ developed at the points of contact will satisfy $F_B < \mu_s N_B$ and $F_C < \mu_s N_C$.

*8.4 Frictional Forces on Screws

In most cases screws are used as fasteners; however, in many types of machines, screws are incorporated to transmit power or motion from one part of the machine to another. A *square-threaded screw* is most commonly used for the latter purpose, especially when large forces are applied along the axis of the screw. This type of screw will be considered in the following analysis. The analysis of other types of screws, such as the V-thread, is based on the same principles.

A *screw* may be thought of simply as an inclined plane or wedge wrapped around a cylinder. A nut initially at position A on the screw shown in Fig. 8–17 will move up to B when rotated 360° around the screw. This rotation is equivalent to translating the nut up an inclined plane of height L and length $l = 2\pi r$, where r is the mean radius of the thread. The rise L for a single revolution is referred to as the *lead* of the screw, where the *lead angle* is given by $\theta = \tan^{-1}(L/2\pi r)$.

Fig. 8–17

Frictional Analysis

When a screw is subjected to large axial loads, the frictional forces developed at the thread become important in the analysis to determine the force needed to turn the screw. Consider, for example, the square-threaded jack screw shown in Fig. 8–18, which supports the vertical load **W** and twisting moment **M**.* If moments are summed about the axis of the screw, **M** can be thought of as being equivalent to the moment of a horizontal force **S** acting at the mean radius r of the thread, so that $M = Sr$. The reactive forces of the jack to loads S and W are actually distributed over the circumference of the screw thread in contact with the screw hole in the jack, that is, within region h shown in Fig. 8–18. For simplicity, this portion of thread can be imagined as being un-

*For applications, **M** is developed by applying a horizontal force **P** at a right angle to the end of a lever that would be fixed to the screw.

Fig. 8–18

wound from the screw and represented as a simple block resting on an inclined plane having the screw's lead angle θ, Fig. 8–19a. The inclined plane represents the inside *supporting thread* of the jack base. The block is subjected to the total axial load **W** acting on the jack, the horizontal force **S** (which is related to the applied moment **M**), and the *resultant force* **R** exerted by the plane. As shown, force **R** has components acting normal, **N,** and tangent, **F,** to the contacting surfaces.

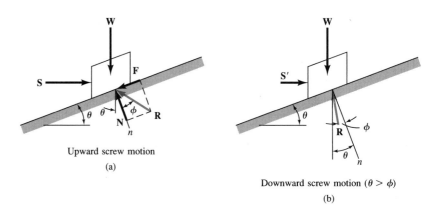

Upward screw motion

(a)

Downward screw motion $(\theta > \phi)$

(b)

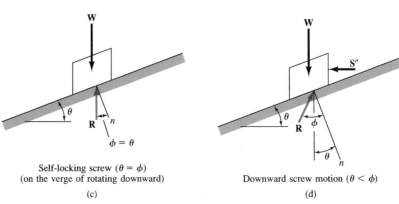

Self-locking screw $(\theta = \phi)$
(on the verge of rotating downward)

(c)

Downward screw motion $(\theta < \phi)$

(d)

Fig. 8–19

Upward Screw Motion. Provided M is great enough, the screw (and hence the block) can either be brought to the verge of upward impending motion or motion can be occurring. Under these conditions, **R** acts at an angle $(\theta + \phi)$ from the vertical as shown in Fig. 8–19a, where $\phi = \tan^{-1}(F/N) = \tan^{-1}(\mu N/N) = \tan^{-1}\mu$. Applying the two force equations of equilibrium to the block, we obtain

$$\xrightarrow{+}\Sigma F_x = 0; \qquad\qquad S - R\sin(\theta + \phi) = 0$$
$$+\uparrow\Sigma F_y = 0; \qquad\qquad R\cos(\theta + \phi) - W = 0$$

Eliminating R and solving for S, then substituting this value into the equation $M = Sr$, yields

$$\boxed{M = Wr\tan(\theta + \phi)} \qquad\qquad (8\text{–}3)$$

This equation gives the required value M necessary to cause upward impending motion of the screw when $\phi = \phi_s = \tan^{-1}\mu_s$ (the angle of static friction). If ϕ is replaced by $\phi_k = \tan^{-1}\mu_k$ (the angle of kinetic friction), Eq. 8–3 would give a smaller value M necessary to maintain uniform upward motion of the screw.

Downward Screw Motion ($\theta > \phi$). If the surface of the screw is very *slippery*, it may be possible for the screw to rotate downward if the magnitude, *not the direction,* of the moment is reduced to say $M' < M$. As shown in Fig. 8–19b, this causes the effect of **M'** to become **S'** and it requires the angle ϕ (ϕ_s or ϕ_k) to lie on the opposite side of the normal n to the plane supporting the block, such that $\theta > \phi$. For this case, Eq. 8–3 becomes

$$\boxed{M' = Wr\tan(\theta - \phi)} \qquad\qquad (8\text{–}4)$$

Self-Locking Screw. If the moment **M** (or **S**) *is removed,* the screw will remain *self-locking,* i.e., it will support the load **W** by *friction forces alone* provided $\phi \geqslant \theta$. To show this, consider the necessary limiting case when $\phi = \theta$, Fig. 8–19c. Here vertical equilibrium is maintained since **R** is vertical and thus balances **W**.

Downward Screw Motion ($\theta < \phi$). When the surface of the screw is *very rough,* the screw will not rotate downward as stated above. Instead, the direction of the applied moment must be *reversed* in order to cause the motion. The free-body diagram shown in Fig. 8–19d is representative of this case. Here **S''** is caused by the applied (reverse) moment **M''**. Hence Eq. 8–3 becomes

$$\boxed{M'' = Wr\tan(\phi - \theta)} \qquad\qquad (8\text{–}5)$$

Each of the above cases should be thoroughly understood before proceeding to solve the problems.

Example 8–8

The turnbuckle shown in Fig. 8–20 has a square thread with a mean radius of 5 mm and a lead of 2 mm. If the coefficient of static friction between the screw and the turnbuckle is $\mu_s = 0.25$, determine the moment **M** that must be applied to draw the end screws closer together. Is the turnbuckle self-locking?

2 kN

M

2 kN

Fig. 8–20

SOLUTION

The moment may be obtained by using Eq. 8–3. Why? Since friction at two screws must be overcome, this requires

$$M = 2[Wr \tan (\theta + \phi)] \tag{1}$$

Here, $W = 2000$ N, $r = 5$ mm, $\phi_s = \tan^{-1} \mu_s = \tan^{-1} (0.25) = 14.04°$, and $\theta = \tan^{-1} (L/2\pi r) = \tan^{-1} (2 \text{ mm}/[2\pi(5 \text{ mm})]) = 3.64°$. Substituting these values into Eq. (1) and solving gives

$$M = 2[(2000 \text{ N})(5 \text{ mm}) \tan (14.04° + 3.64°)]$$
$$M = 6375.1 \text{ N} \cdot \text{mm} = 6.38 \text{ N} \cdot \text{m} \qquad \textit{Ans.}$$

When the moment is *removed*, the turnbuckle will be self-locking; i.e., it will not unscrew, since $\phi_s > \theta$.

*8.5 Frictional Forces on Flat Belts

In the design of belt drives or band brakes it is necessary to determine the frictional forces developed between a belt and its contacting surface. Consider the flat belt passing over a fixed drum of radius r, as shown in Fig. 8–21a. The *total* angle of belt contact in radians is β and the coefficient of friction between the two surfaces is μ. If it is known that the tension acting in the belt on the right of the drum is $\mathbf{T}_1$, let it be required to find the tension $\mathbf{T}_2$ needed to pull the belt counterclockwise over the surface of the drum. *Obviously, T_2 must be greater than T_1 since the belt must overcome T_1 and the resistance of friction at the surface of contact.*

Frictional Analysis

A free-body diagram of the belt segment in contact with the drum is shown in Fig. 8–21b. Here the normal force $\mathbf{N}$ and the frictional force $\mathbf{F}$, acting at different points along the contacting surface of the belt, will vary both in magnitude and direction. Due to this unknown force distribution, the analysis of the problem will proceed on the basis of initially studying the forces acting on a differential element of the belt.

A free-body diagram of an element having a length $ds = r\,d\theta$ is shown in Fig. 8–21c. Assuming either impending motion or motion of the belt, the

Fig. 8–21

Fig. 8–21c

magnitude of frictional force $dF = \mu\, dN$. This force opposes the sliding motion of the belt and thereby increases the magnitude of tensile force acting in the belt by dT. Applying the two force equations of equilibrium, we have

$$\xrightarrow{+}\Sigma F_x = 0; \quad T \cos (d\theta/2) + \mu\, dN - (T + dT) \cos (d\theta/2) = 0$$

$$+\uparrow \Sigma F_y = 0; \quad dN - (T + dT) \sin (d\theta/2) - T \sin (d\theta/2) = 0$$

Since $d\theta$ is of *infinitesimal size,* $\sin (d\theta/2)$ and $\cos (d\theta/2)$ can be replaced by $d\theta/2$ and 1, respectively. Also, the *product* of two infinitesimals dT and $d\theta/2$ may be neglected when compared to infinitesimals of the first order. The above two equations therefore reduce to

$$\mu\, dN = dT$$

and

$$dN = T\, d\theta$$

Eliminating dN yields

$$\frac{dT}{T} = \mu\, d\theta$$

Integrating this equation between all the points of contact that the belt makes with the drum, and noting that $T = T_1$ at $\theta = 0$ and $T = T_2$ at $\theta = \beta$, yields

$$\int_{T_1}^{T_2} \frac{dT}{T} = \mu \int_0^\beta d\theta$$

or

$$\ln \frac{T_2}{T_1} = \mu\beta$$

Solving for T_2, we obtain

$$\boxed{T_2 = T_1 e^{\mu\beta}} \tag{8–6}$$

where T_2, T_1 = belt tensions; $\mathbf{T}_1$ opposes the direction of motion (or impending motion) of the belt measured relative to the surface, while $\mathbf{T}_2$ acts in the direction of the belt motion (or impending motion); because of friction, $T_2 > T_1$

μ = coefficient of static or kinetic friction between the belt and the surface of contact

β = angle of belt to surface contact, measured in radians

$e = 2.718. \ldots$, base of the natural logarithm

Note that Eq. 8–6 is *independent* of the *radius* of the drum and instead depends upon the angle of belt to surface contact, β. Furthermore, as indicated by the integration, this equation is valid for flat belts placed on *any shape* of contacting surface. For application, keep in mind that Eq. 8–6 is valid only when *impending motion* occurs.

Example 8–9

The maximum tension that can be developed in the belt shown in Fig. 8–22*a* is 500 N. If the pulley at *A* is free to rotate and the coefficient of static friction at the fixed drums *B* and *C* is $\mu_s = 0.25$, determine the largest mass of the cylinder that can be lifted by the belt. Assume that the force **T** applied at the end of the belt is directed vertically downward, as shown.

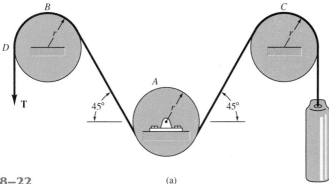

Fig. 8–22   (a)

SOLUTION

Lifting the cylinder, which has a weight $W = mg$, causes the belt to move counterclockwise over the drums at *B* and *C*; hence, the maximum tension T_2 in the belt occurs at *D*. Thus, $T_2 = 500$ N. A section of the belt passing over the drum at *B* is shown in Fig. 8–22*b*. Since $180° = \pi$ rad, the angle of contact between the drum and the belt is $\beta = (135°/180°)\pi = 3\pi/4$ rad. Using Eq. 8–6, we have

$$T_2 = T_1 e^{\mu_s \beta}; \qquad 500 \text{ N} = T_1 e^{0.25[(3/4)\pi]}$$

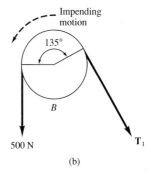

(b)

Hence,

$$T_1 = \frac{500 \text{ N}}{e^{0.25[(3/4)\pi]}} = \frac{500}{1.80} = 277.4 \text{ N}$$

Since the pulley at *A* is free to rotate, equilibrium requires that the tension in the belt remains the *same* on both sides of the pulley.

The section of the belt passing over the drum at *C* is shown in Fig. 8–22*c*. The load $W < 277.4$ N. Why? Applying Eq. 8–6, we obtain

$$T_2 = T_1 e^{\mu_s \beta}; \qquad 277.4 = W e^{0.25[(3/4)\pi]}$$

$$W = 153.9 \text{ N}$$

so that

$$m = \frac{W}{g} = \frac{153.9}{9.81}$$

$$= 15.7 \text{ kg} \qquad \qquad \textit{Ans.}$$

(c)

PROBLEMS

*8–52. Determine the smallest force P needed to lift the 3000-lb load. The coefficient of static friction between A and C and B and D is $\mu_s = 0.3$, and between A and B $\mu_s' = 0.4$. Neglect the weight of each wedge.

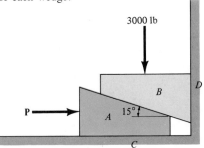

Probs. 8–52/8–53

8–53. Determine the reversed horizontal force $-P$ needed pull out wedge A. The coefficient of static friction between A and C and B and D is $\mu_s = 0.2$, and between A and B $\mu_s' = 0.1$. Neglect the weight of each wedge.

8–54. The concrete blocks each have a weight of 50 lb. If the coefficient of static friction at A is $\mu_s = 0.2$ and between each block $\mu_s' = 0.4$, determine how many blocks can be stacked as shown before they begin to topple.

Prob. 8–54

8–55. Each block has a weight of 400 lb. Determine how far the force P can compress the spring until block B slips on block A. What is the magnitude of P for this to occur?

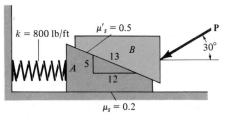

Prob. 8–55

*8–56. Column D is subjected to a vertical load of 8000 lb. It is supported on two identical wedges A and B for which the coefficient of static friction at the contacting surfaces AB and BC is $\mu_s = 0.4$. Determine the force P needed to raise wedge B and the equilibrium force P' needed to hold wedge A stationary. The contacting surfaces AD are smooth.

Probs. 8–56/8–57

8–57. Column D is subjected to a vertical load of 8000 lb. It is supported on two identical wedges A and B for which the coefficient of static friction at the contacting surfaces AB and BC is $\mu_s = 0.4$. If the forces P and P' are removed are the wedges self-locking? The contacting surfaces AD are smooth.

8–58. If the spring is compressed 60 mm and the coefficient of static friction between the tapered stub S and the slider A is $\mu_{SA} = 0.5$, determine the horizontal force P needed to move the slider forward. The stub is free to move without friction within the fixed collar C. The coefficient of static friction between A and the surface B is $\mu_{AB} = 0.4$. Neglect the weights of the slider and stub.

Prob. 8–58

8–59. If the beam AD is loaded as shown, determine the horizontal force P which must be applied to the wedge in order to remove it from under the beam. The coefficients of static friction at the wedge's top and bottom surfaces are $\mu_{CA} = 0.25$ and $\mu_{CB} = 0.35$, respectively. If $P = 0$, is the wedge self-locking?

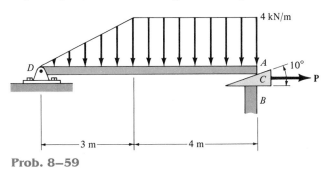

Prob. 8–59

*8–60. The wedge is used to level the member. For the loading shown, determine the horizontal force P that must be applied to move the wedge to the right. The coefficient of static friction between the wedge and the two surfaces of contact is $\mu_s = 0.25$. Neglect the size and weight of the wedge.

Probs. 8–60/8–61

8–61. The wedge is used to level the member. For the loading shown, determine the reversed horizontal force $-\mathbf{P}$ that must be applied to pull the wedge out to the left. The coefficient of static friction between the wedge and the two surfaces of contact is $\mu_s = 0.15$. Neglect the weight of the wedge.

8–62. The coefficient of static friction between wedges B and C is $\mu_s = 0.6$ and between the surfaces of contact B and A and C and D $\mu'_s = 0.4$. If the spring is compressed 200 mm when in the position shown, determine the smallest force P needed to move wedge C to the left. Neglect the weight of the wedges.

Probs. 8–62/8–63

8–63. The coefficient of static friction between the wedges B and C is $\mu_s = 0.6$ and between the surfaces of contact B and A and C and D $\mu'_s = 0.4$. If $P = 50$ N, determine the largest allowable compression of the spring without causing wedge C to move to the left. Neglect the weight of the wedges.

*8–64. The wedge blocks are used to hold the specimen in a tension testing machine. Determine the design angle θ of the wedges so that the specimen will not slip regardless of the applied load. The coefficients of static friction are $\mu_A = 0.1$ at A and $\mu_B = 0.6$ at B. Neglect the weight of the blocks.

Prob. 8–64

8–65. The cylinder has a mass of 50 kg. The coefficient of static friction between it and the wedge is $\mu_{AB} = 0.3$, between the wedge and the vertical surface $\mu_{BC} = 0.2$, and between the cylinder and inclined surface $\mu_{AD} = 0.5$. Determine the force P needed to push the wedge downward.

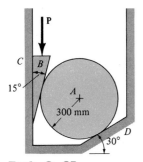

Prob. 8–65

8–66. Determine the clam... force on the board A if the screw of the "C" clamp is tighte..u with a twist of $M = 8\ \text{N} \cdot \text{m}$. The single square-threaded screw has a mean radius of 10 mm, a lead of 3 mm, and the coefficient of static friction is $\mu_s = 0.35$.

Probs. 8–66/8–67

8–67. If the required clamping force at the board A is to be 50 N, determine the torque M that must be applied to the handle of the "C" clamp to tighten it down. The single square-threaded screw has a mean radius of 10 mm, a lead of 3 mm, and the coefficient of static friction is $\mu_s = 0.35$.

*8–68. The shaft has a square-threaded screw with a lead of 9 mm and a mean radius of 15 mm. If it is in contact with a plate gear having a mean radius of 20 mm, determine the resisting torque M on the plate gear which can be overcome if a torque of $7\ \text{N} \cdot \text{m}$ is applied to the shaft. The coefficient of static friction at the screw is $\mu_s = 0.2$. Neglect friction of the bearings located at A and B.

Prob. 8–68

8–69. The device is used to pull the battery cable terminal C from the post of a battery. If the required pulling force is 85 lb, determine the torque M that must be applied to the handle on the screw to tighten it. The screw has square threads, a mean diameter of 0.2 in., a lead of 0.08 in., and a coefficient of static friction of $\mu_s = 0.5$.

Prob. 8–69

8–70. Determine the horizontal force P applied perpendicular to the handle of the jack screw necessary to start lifting the 3-kN load. The square-threaded screw has a lead of 5 mm and a mean diameter of 60 mm. The coefficient of static friction for the screw is $\mu_s = 0.2$.

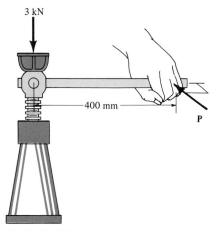

Prob. 8–70

8–71. The square-threaded screw has a mean diameter of 20 mm and a lead of 4 mm. If the weight of the plate A is 5 lb, determine the smallest coefficient of static friction between the screw and the plate so that the plate does not travel down the screw when the plate is suspended as shown.

Prob. 8–71

*8–72. The lolly column is used to support the upper floor. If the force applied perpendicular to the handle is F = 80 N to tighten the screw, determine the compressive force in the column. The square threaded screw on the jack has a coefficient of static friction of $\mu_s = 0.4$, mean diameter of 25 mm, and a lead of 3 mm.

Probs. 8–72/8–73

8–73. If the force **F** is removed from the handle of the jack in Prob. 8–72, determine if the screw is self-locking.

8–74. The hand clamp is constructed using a square-threaded screw having a mean diameter of 36 mm, a lead of 4 mm, and a coefficient of static friction at the screw of $\mu_s = 0.3$. To tighten the screw, a force of $F = 20$ N is applied perpendicular to the handle. Determine the clamping force in the board AB.

50 mm

A B

F

Probs. 8–74/8–75

8–75. The hand clamp is constructed using a square-threaded screw having a mean diameter of 36 mm, a lead of 4 mm, and a coefficient of static friction at the screw of $\mu_s = 0.3$. If the clamping force in the board AB is 300 N, determine the reversed force $-\mathbf{F}$ that must be applied perpendicular to the handle in order to loosen the screw.

*8–76. The fixture clamp consists of a square-threaded screw having a coefficient of static friction of $\mu_s = 0.3$, mean diameter of 3 mm, and a lead of 1 mm. The five points indicated are pin connections. Determine the clamping force at the smooth blocks D and E when a torque of $M = 0.08$ N · m is applied to tighten the screw.

30 mm

45°

E B

D

30 mm

40 mm

C A

40 mm   40 mm

Prob. 8–76

$M = 0.08$ N · m

8–77. The braking mechanism consists of two pinned arms and a square-threaded screw with left and right-hand threads. Thus, when turned, the screw draws the two arms together. If the lead of the screw is 4 mm, the mean diameter 12 mm, and the coefficient of static friction is $\mu_s = 0.35$, determine the tension in the screw when a torque of 5 N · m is applied to tighten the screw. If the coefficient of static friction between the brake pads A and B and the circular shaft is $\mu_s' = 0.5$, determine the maximum torque M the brake can resist.

5 N · m

300 mm

M

A B

200 mm

300 mm

C D

Prob. 8–77

8–78. A turnbuckle, similar to that shown in Fig. 8–20, is used to tension member BC of the truss. The coefficient of static friction between the square-threaded screws and the turnbuckle is $\mu_s = 0.5$. The screws have a mean radius of 6 mm and a lead of 3 mm. If a torque of $M = 10$ N · m is applied to the turnbuckle, to draw the screws closer together, determine the force in each member of the truss. No external forces act on the truss.

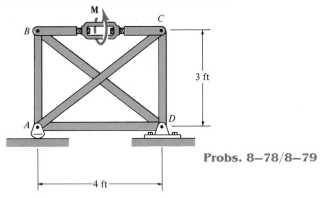

M

B C

3 ft

A D

4 ft

Probs. 8–78/8–79

8–79. A turnbuckle, similar to that shown in Fig. 8–20, is used to tension member BC of the truss. The coefficient of static friction between the square-threaded screws and the turnbuckle is $\mu_s = 0.5$. The screws have a mean radius of 6 mm and a lead of 3 mm. Determine the torque M which must be applied to the turnbuckle to draw the screws closer together, so that a tension force of 500 N is developed in member BC.

*8–80. A cylinder having a mass of 250 kg is to be supported by the cord which wraps over the pipe. Determine the smallest vertical force F needed to support the load if the cord passes (a) once over the pipe, $\beta = 180°$, and (b) two times over the pipe, $\beta = 540°$. Take $\mu_s = 0.2$.

F

Probs. 8–80/8–81

8–81. A cylinder having a mass of 250 kg is to be supported by the cord which wraps over the pipe. Determine the largest vertical force F that can be applied to the cord without moving the cylinder. The cord passes (a) once over the pipe, $\beta = 180°$, and (b) two times over the pipe, $\beta = 540°$. Take $\mu_s = 0.2$.

8–82. Determine the *minimum* tension in the rope at points A and B that is necessary to maintain equilibrium. Take $\mu_s = 0.3$ between the rope and the fixed post D. The rope is wrapped only once around the post.

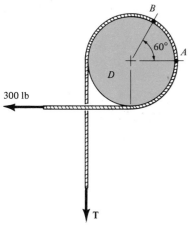

B

60°
A

D

300 lb

T

Probs. 8–82/8–83

8–83. Determine the *maximum* tension in the rope at points A and B that is necessary to maintain equilibrium. Take $\mu_s = 0.3$ between the rope and the fixed post D. The rope is wrapped only once around the post.

*8–84. The rope brake is used as a means of measuring an engine's horsepower. This is done by connecting the engine to the drum and then winding a rope once around the drum and suspending from it a counterweight A of 400 lb. If the spring balance B measures a tension in the other end of the rope of 25 lb, determine the torque M developed by the engine needed to turn the drum at a constant rate. Also, what is the coefficient of kinetic friction between the rope and the drum?

B

M

8 in.

A

Prob. 8–84

8–85. A cord having a weight of 0.5 lb/ft and a total length of 10 ft is suspended over a peg P as shown. If the coefficient of static friction between the peg and cord is $\mu_s = 0.5$, determine the longest length h which one side of the suspended cord can have without causing motion. Neglect the size of the peg and the length of cord draped over it.

P

h

Prob. 8–85

8–86. The cord supporting the 6-kg block passes around three pegs, A, B, C, where $\mu_s = 0.2$. Determine the range of values of P for which the block will not move up or down.

Prob. 8–86

8–87. The beam is being hoisted using a rope which is wrapped around the beam and passes through a ring at A as shown. If the end of the rope is subjected to a tension **T** and the coefficient of static friction between the rope and ring is $\mu_s = 0.3$, determine the angle θ for equilibrium.

Prob. 8–87

*8–88. Show that the frictional relationship between the belt tensions, the coefficient of friction μ, and the angular contacts α and β for the V-belt is $T_2 = T_1 e^{\mu\beta/\sin(\alpha/2)}$.

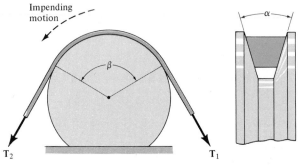

Prob. 8–88

8–89. A conveyor belt is used to transport granular material. If the frictional resistance on the top of the belt is $F = 500$ N, determine the smallest stretch of the spring attached to the movable axle of the idle pulley B so that the belt does not slip at the driving pulley A when the torque **M** is applied. What minimum torque M is required to keep the belt moving? The coefficient of static friction between the belt and the wheel at A is $\mu_A = 0.2$.

Prob. 8–89

8–90. The uniform beam has a mass of 80 kg and is supported by a cord which is wrapped over two pegs for which $\mu_s = 0.5$. Determine the minimum distance x for the placement of a 1-kN force without causing the beam to move.

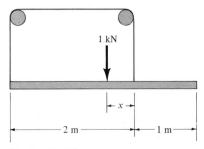

Prob. 8–90

8–91. Blocks A and B have a mass of 100 kg and 150 kg, respectively. If the coefficient of static friction between A and B and between B and C is $\mu_s = 0.25$, and between the ropes and the pegs D and E $\mu_s' = 0.5$, determine the smallest force F needed to cause motion of block B if P = 30 N.

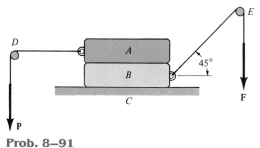

Prob. 8–91

*8–92. Block A has a mass of 50 kg and rests on surface B for which $\mu_s = 0.25$. If the coefficient of static friction between the cord and the fixed peg at C is $\mu_s' = 0.3$, determine the greatest mass of the suspended cylinder D without causing motion.

Probs. 8–92/8–93

8–93. Block A has a mass of 50 kg and rests on surface B for which $\mu_s = 0.25$. If the mass of the suspended cylinder D is 4 kg, determine the frictional force acting on A. The coefficient of static friction between the cord and the fixed peg at C is $\mu_s' = 0.3$.

8–94. Determine the lever force P needed to prevent the wheel from rotating. The coefficient of static friction between the belt and the wheel is $\mu_s = 0.3$. The wheel is pin-connected at its center, B. Set M = 250 N · m.

Probs. 8–94/8–95

8–95. Determine the torque M that can be resisted by the band brake if a force of P = 30 N is applied to the handle of the lever. The coefficient of static friction between the belt and the wheel is $\mu_s = 0.3$. The wheel is pin-connected at its center, B.

*8–96. Blocks A and B weigh 50 lb and 30 lb, respectively. Using the coefficients of static friction indicated, determine the greatest weight of block D without causing motion.

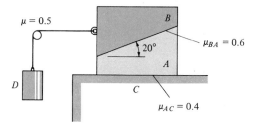

Probs. 8–96/8–97

8–97. Blocks A, B, and D weigh 50 lb, 30 lb, and 12 lb, respectively. Using the coefficients of static friction indicated, determine the frictional force between blocks A and B and between block A and the floor C.

8–98. The cylinder A has a mass of 75 kg. Determine the smallest force P applied to the handle of the lever required for equilibrium. The coefficient of static friction between the belt and the wheel is $\mu_s = 0.3$. The drum is pin-connected at its center, B.

8–99. Determine the largest mass of cylinder A that can be supported from the drum if a force $P = 20$ N is applied to the handle of the lever. The coefficient of static friction between the belt and the wheel is $\mu_s = 0.3$. The drum is pin-supported at its center, B.

Probs. 8–98/8–99

*8.6 Frictional Forces on Collar Bearings, Pivot Bearings, and Disks

Pivot and *collar bearings* are commonly used to support the *axial load* on a rotating shaft. These two types of support are shown in Fig. 8–23. Provided the bearings are not lubricated, or only partially lubricated, the laws of dry friction may be applied to determine the moment **M** needed to turn the shaft when the shaft supports an axial force **P**.

Fig. 8–23

Frictional Analysis

Consider, for example, the end of a collar-bearing shaft shown in Fig. 8–24, which is subjected to an axial force **P** and has a total bearing or contact area $\pi(R_2^2 - R_1^2)$. In the following analysis, the normal pressure p is considered to be *uniformly distributed* over this area—a reasonable assumption provided the bearing is new and evenly supported. Since $\Sigma F_z = 0$, p, measured as a force per unit area, is $p = P/\pi(R_2^2 - R_1^2)$.

The moment needed to cause impending rotation of the shaft can be determined from moment equilibrium of the frictional forces $d\mathbf{F}$ developed at the bearing surface by applying $\Sigma M_z = 0$. A small area element $dA = (r\, d\theta)(dr)$, shown in Fig. 8–24, is subjected to both a normal force $dN = p\, dA$ and an associated frictional force

$$dF = \mu_s\, dN = \mu_s\, p\, dA = \frac{\mu_s P}{\pi(R_2^2 - R_1^2)}\, dA$$

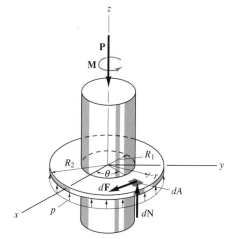

Fig. 8–24

The normal force does not create a moment about the z axis of the shaft; however, the frictional force does, namely $dM = r\, dF$. Integration is needed to compute the total moment created by all the frictional forces acting on differential areas dA. Therefore, for impending rotational motion,

$$\Sigma M_z = 0; \qquad\qquad M - \int_A r\, dF = 0$$

Using $dA = (r\, d\theta)(dr)$ and integrating over the entire bearing area yields

$$M = \int_{R_1}^{R_2} \int_0^{2\pi} r\left[\frac{\mu_s P}{\pi(R_2^2 - R_1^2)}\right] (r\, d\theta\, dr) = \frac{\mu_s P}{\pi(R_2^2 - R_1^2)} \int_{R_1}^{R_2} r^2\, dr \int_0^{2\pi} d\theta$$

or

$$M = \tfrac{2}{3}\mu_s P\left(\frac{R_2^3 - R_1^3}{R_2^2 - R_1^2}\right) \tag{8-7}$$

This equation gives the magnitude of moment required for impending rotation of the shaft. The frictional moment developed at the end of the shaft, when it is *rotating* at constant speed, can be found by substituting μ_k for μ_s in Eq. 8–7.

When $R_2 = R$ and $R_1 = 0$, as in the case of a pivot bearing, Fig. 8–23a, Eq. 8–7 reduces to

$$M = \tfrac{2}{3}\mu_s PR \tag{8-8}$$

Recall from the initial assumption that both Eqs. 8–7 and 8–8 apply only for bearing surfaces subjected to *constant pressure*. If the pressure is not uniform, a variation of the pressure as a function of the bearing area must be determined before integrating to obtain the moment. The following example illustrates this concept.

Example 8–10

The uniform bar shown in Fig. 8–25a has a total mass m. If it is assumed that the normal pressure acting at the contacting surface varies linearly along the length of the bar as shown, determine the couple **M** required to rotate the bar. Assume that the bar's width a is negligible in comparison to its length l. The coefficient of static friction is equal to μ.

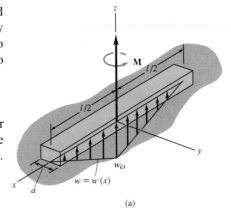

(a)

SOLUTION

A free-body diagram of the bar is shown in Fig. 8–25b. Since the bar has a total weight of $W = mg$, the intensity w_o of the distributed load at the center $(x = 0)$ is determined from vertical force equilibrium, Fig. 8–25a.

$$+\uparrow \Sigma F_z = 0; \quad -mg + 2\left[\frac{1}{2}\left(\frac{l}{2}\right)w_o\right] = 0 \qquad w_o = \frac{2mg}{l}$$

Since $w = 0$ at $x = l/2$, the distributed load expressed as a function of x is

$$w = w_o\left(1 - \frac{2x}{l}\right) = \frac{2mg}{l}\left(1 - \frac{2x}{l}\right)$$

The magnitude of the normal force acting on a segment of area having a length dx is therefore

$$dN = w\, dx = \frac{2mg}{l}\left(1 - \frac{2x}{l}\right)dx$$

The magnitude of the frictional force acting on the same element of area is

$$dF = \mu\, dN = \frac{2\mu mg}{l}\left(1 - \frac{2x}{l}\right)dx$$

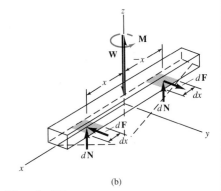

(b)

Fig. 8–25

Hence, the moment created by this force about the z axis is

$$dM = x\, dF = \frac{2\mu mg}{l}x\left(1 - \frac{2x}{l}\right)dx$$

The summation of moments about the z axis of the bar is determined by integration, which yields

$$\Sigma M_z = 0; \qquad M - 2\int_0^{l/2}\frac{2\mu mg}{l}x\left(1 - \frac{2x}{l}\right)dx = 0$$

$$M = \frac{4\mu mg}{l}\left(\frac{x^2}{2} - \frac{2x^3}{3l}\right)\Big|_0^{l/2}$$

$$M = \frac{\mu mgl}{6} \qquad\qquad Ans.$$

*8.7 Frictional Forces on Journal Bearings

When a shaft or axle is subjected to lateral loads, a *journal bearing* is commonly used for support. Well-lubricated journal bearings are subjected to the laws of fluid mechanics, in which the viscosity of the lubricant, the speed of rotation, and the amount of clearance between the shaft and bearing are needed to determine the frictional resistance of the bearing. When the bearing is not lubricated or only partially lubricated, however, a reasonable analysis of the frictional resistance can be based on the laws of dry friction.

Frictional Analysis

A typical journal-bearing support is shown in Fig. 8–26a. As the shaft rotates in the direction shown in the figure, it rolls up against the wall of the bearing to some point A where slipping occurs. If the lateral load acting at the end of the shaft is $\mathbf{W}$, it is necessary that the bearing reactive force $\mathbf{R}$ acting at A be equal and opposite to $\mathbf{W}$, Fig. 8–26b. The moment needed to maintain constant rotation of the shaft can be found by summing moments about the z axis of the shaft; i.e.,

$$\Sigma M_z = 0; \qquad -M + (R \sin \phi_k)r = 0$$

or

$$M = Rr \sin \phi_k \qquad (8\text{–}9)$$

where ϕ_k is the angle of kinetic friction defined by $\tan \phi_k = F/N = \mu_k N/N = \mu_k$. In Fig. 8–26c, it is seen that $r \sin \phi_k = r_f$. The dashed circle with radius r_f is called the *friction circle*, and as the shaft rotates, the reaction $\mathbf{R}$ will always be tangent to it. If the bearing is partially lubricated, μ_k is small, and therefore $\mu_k = \tan \phi_k \approx \sin \phi_k \approx \phi_k$. Under these conditions, a reasonable approximation to the moment needed to overcome the frictional resistance becomes

$$M \approx Rr\mu_k \qquad (8\text{–}10)$$

The following example illustrates a common application of this analysis.

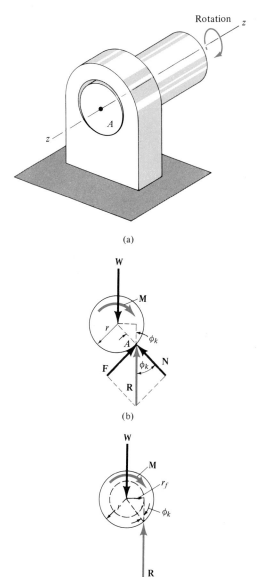

Fig. 8–26

(a)

(b)

(c)

369

Example 8–11

The 100-mm-diameter pulley shown in Fig. 8–27a fits loosely on a 10-mm-diameter shaft for which the coefficient of static friction is $\mu_s = 0.4$. Determine the minimum tension T in the belt to (a) raise the 100-kg block and (b) lower the block. Assume no slipping occurs between the belt and pulley and neglect the weight of the pulley.

50 mm

$r = 5$ mm

100 kg

T

(a)

Fig. 8–27

(b)

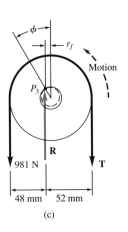

(c)

SOLUTION

Part (a). A free-body diagram of the pulley is shown in Fig. 8–27b. When the pulley is subjected to belt tensions of 981 N each, the pulley makes contact with the shaft at point P_1. As the tension T is *increased*, the pulley will roll around the shaft to point P_2 before motion impends. From the figure, the friction circle has a radius $r_f = r \sin \phi$. Using the simplification $\sin \phi \approx \phi$, $r_f \approx r\mu_s = (5 \text{ mm})(0.4) = 2$ mm, so that summing moments about P_2 gives

$$\zeta+\Sigma M_{P_2} = 0; \qquad 981 \text{ N}(52 \text{ mm}) - T(48 \text{ mm}) = 0$$
$$T = 1063 \text{ N} = 1.06 \text{ kN} \qquad\qquad \textit{Ans.}$$

If a more exact analysis is used, then $\phi = \tan^{-1} 0.4 = 21.8°$. Thus, the radius of the friction circle would be $r_f = r \sin \phi = 5 \sin 21.8° = 1.86$ mm. Therefore,

$$\zeta+\Sigma M_{P_2} = 0; \quad 981 \text{ N}(50 \text{ mm} + 1.86 \text{ mm}) - T(50 \text{ mm} - 1.86 \text{ mm}) = 0$$
$$T = 1057 \text{ N} = 1.06 \text{ kN} \qquad\qquad \textit{Ans.}$$

Part (b). When the block is lowered, the resultant force **R** acting on the shaft passes through point P_3, as shown in Fig. 8–27c. Summing moments about this point yields

$$\zeta+\Sigma M_{P_3} = 0; \qquad 981 \text{ N}(48 \text{ mm}) - T(52 \text{ mm}) = 0$$
$$T = 906 \text{ N} \qquad\qquad \textit{Ans.}$$

*8.8 Rolling Resistance

If a *rigid* cylinder of weight **W** rolls at constant velocity along a *rigid* surface, the normal force exerted by the surface on the cylinder acts at the tangent point of contact, as shown in Fig. 8–28a. Under these conditions, provided the cylinder does not encounter frictional resistance from the air, motion will continue indefinitely. Actually, however, no materials are perfectly rigid, and therefore the reaction of the surface on the cylinder consists of a distribution of normal pressure. For example, consider the cylinder to be made of a very hard material, and the surface upon which it rolls to be soft. Due to its weight, the cylinder compresses the surface underneath it, Fig. 8–28b. As the cylinder rolls, the surface material in front of the cylinder *retards* the motion since it is being *deformed*, whereas the material in the rear is *restored* from the deformed state and therefore tends to *push* the cylinder forward. The normal pressures acting on the cylinder in this manner are represented in Fig. 8–28b by their resultant forces $\mathbf{N}_d$ and $\mathbf{N}_r$. Unfortunately, the magnitude of the force of *deformation* and its horizontal component is *always greater* than that of *restoration*, and consequently a horizontal driving force **P** must be applied to the cylinder to maintain the motion, Fig. 8–28b.*

Rolling resistance is caused primarily by this effect, although it is also, to a smaller degree, the result of surface adhesion and relative micro-sliding between the surfaces of contact. Because the actual force **P** needed to overcome these effects is difficult to determine, a simplified method will be developed here to explain one way engineers have analyzed this phenomenon. To do this, we will consider the resultant of the entire normal pressure, $\mathbf{N} = \mathbf{N}_d + \mathbf{N}_r$, acting on the cylinder, Fig. 8–28c. As shown in Fig. 8–28d, this force acts at an angle θ with the vertical. To keep the cylinder in equilibrium, i.e., rolling at a constant rate, it is necessary that **N** be *concurrent* with the driving force **P** and the weight **W**. Summing moments about point A gives $Wa = P(r\cos\theta)$. Since the deformations are generally very small in relation to the cylinder's radius, $\cos\theta \approx 1$; hence,

$$Wa \approx Pr$$

or

$$\boxed{P \approx \frac{Wa}{r}} \qquad (8\text{–}11)$$

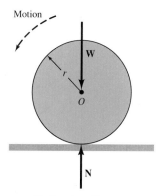

Rigid surface of contact

(a)

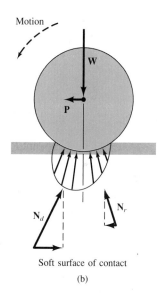

Soft surface of contact

(b)

(c)

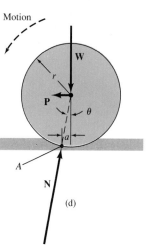

(d)

Fig. 8–28

---

* Actually, the deformation force $\mathbf{N}_d$ causes *energy* to be stored in the material as its magnitude is increased, whereas the restoration force $\mathbf{N}_r$, as its magnitude is decreased, allows some of this energy to be released. The remaining energy is *lost* since it is used to heat up the surface, and if the cylinder's weight is very large, it accounts for permanent deformation of the surface. Work must be done by the horizontal force **P** to make up for this loss.

The distance a is termed the *coefficient of rolling resistance,* which has the dimension of length. For instance, $a \approx 0.5$ mm for a mild steel wheel rolling on a steel rail. For hardened steel ball bearings on steel $a \approx 0.1$ mm. Experimentally, though, this factor is difficult to measure, since it depends on such parameters as the rate of rotation of the cylinder, the elastic properties of the contacting surfaces, and the surface finish. For this reason, little reliance is placed on the data for determining a. The analysis presented here does, however, indicate why a heavy load (W) offers greater resistance to motion (P) than a light load under the same conditions. Furthermore, since the force needed to *roll* the cylinder over the surface will be much less than that needed to *slide* the cylinder across the surface, the analysis indicates why roller or ball bearings are often used to minimize the frictional resistance between moving parts.

Example 8–12

A 10-kg steel wheel shown in Fig. 8–29a has a radius of 100 mm and rests on an inclined plane made of wood. If θ is increased so that the wheel begins to roll down the incline with constant velocity when $\theta = 1.2°$, determine the coefficient of rolling resistance.

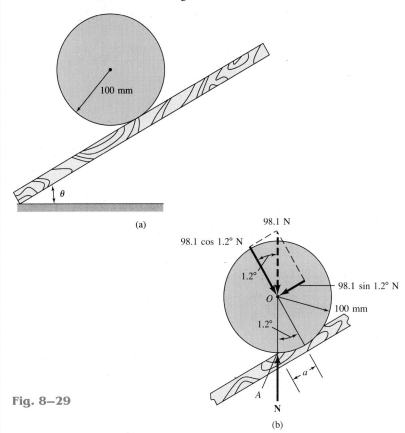

(a)

(b)

Fig. 8–29

SOLUTION

As shown on the free-body diagram, Fig. 8–29b, when the wheel has impending motion, the normal reaction **N** acts at point A defined by the dimension a. Resolving the weight into rectangular components parallel and perpendicular to the incline, and summing moments about point A, yields (approximately)

$$\zeta + \Sigma M_A = 0; \quad 98.1 \cos 1.2°(a) - 98.1 \sin 1.2°(100) = 0$$

Solving, we obtain

$$a = 2.1 \text{ mm} \qquad\qquad Ans.$$

PROBLEMS

*8–100. Assuming the pressure of the sanding disk D on the surface to be uniform, determine the vertical couple forces developed at the handles which are necessary to hold it in equilibrium. The disk is pushed against the surface with a total horizontal force of 18 lb. Take $\mu_k = 0.59$.

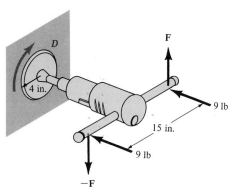

Prob. 8–100

8–101. The collar bearing uniformly supports an axial force of $P = 500$ lb. If the coefficient of static friction is $\mu_s = 0.3$, determine the torque M required to overcome friction.

Probs. 8–101/8–102

8–102. The collar bearing uniformly supports an axial force of $P = 500$ lb. If a torque of $M = 3$ lb · ft is applied to the shaft and causes it to rotate at constant velocity, determine the coefficient of kinetic friction at the surface of contact.

8–103. The annular ring bearing is subjected to a thrust of 800 lb. If $\mu_s = 0.35$, determine the torque M that must be applied to overcome friction.

Prob. 8–103

*8–104. The plate clutch consists of a plate A that slides over the rotating shaft S. The shaft is fixed to the driving plate gear B. If the gear C, which is in mesh with B, is subjected to a torque of $M = 0.6$ N · m, determine the force P, which must be applied via the control arm, to stop the rotation. The coefficient of kinetic friction between the plates A and D is $\mu_k = 0.4$. Assume the bearing pressure between A and D to be uniform.

Prob. 8–104

8–105. Because of wearing at the edges, the pivot bearing is subjected to a conical pressure distribution at its surface of contact. Determine the torque M required to overcome friction and turn the shaft which supports an axial force **P**. The coefficient of static friction is μ. For the solution, it is necessary to determine the peak pressure p_o in terms of P and the bearing radius R.

Prob. 8–105

8–106. The pivot bearing is subjected to a parabolic pressure distribution at its surface of contact. If the coefficient of static friction is μ, determine the torque M required to overcome friction and turn the shaft if it supports an axial force **P**.

$$p = p_0 \left(1 - \frac{r^2}{R^2}\right)$$

Prob. 8–106

8–107. Assuming that the variation of pressure at the bottom of the pivot bearing is defined as $p = p_o(R_2/r)$, determine the torque M needed to overcome friction if the shaft is subjected to an axial force **P**. The coefficient of static friction is μ. For the solution, it is necessary to determine p_o in terms of P and the bearing dimensions R_1 and R_2.

Prob. 8–107

$$p = p_o \frac{R_2}{r}$$

*8–108. The pivot bearing is subjected to a pressure distribution at its surface of contact which varies as shown. If the coefficient of static friction is μ, determine the torque M required to overcome friction if the shaft supports an axial force **P**.

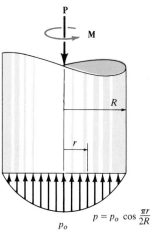

$$p = p_o \cos \frac{\pi r}{2R}$$

Prob. 8–108

8–109. The conical bearing is subjected to a constant pressure distribution at its surface of contact. If the coefficient of static friction is μ, determine the torque M required to overcome friction if it supports an axial force **P**.

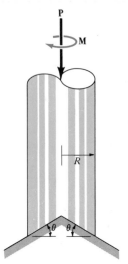

Prob. 8–109

8–110. The *Morse taper shank* of angle θ is used on a lathe. If the contact pressure between the socket B and the shank is uniform, and the coefficient of static friction is μ, determine the torque M needed to overcome friction if the shank is subjected to an axial force **P**.

Prob. 8–110

8–111. A beam having a uniform weight W rests on the rough horizontal surface having a coefficient of static friction μ. If the horizontal force **P** is applied perpendicular to the beam's length, determine the location d of the point O about which the beam begins to rotate.

Prob. 8–111

*8–112. A pulley having a diameter of 80 mm and mass of 1.25 kg is supported loosely on a shaft having a diameter of 20 mm. Determine the torque M that must be applied to the pulley to cause it to rotate with constant motion. The coefficient of kinetic friction between the shaft and pulley is $\mu_k = 0.4$. Also calculate the angle θ which the normal force at the point of contact makes with the horizontal. The shaft itself cannot rotate.

Prob. 8–112

8–113. The axle of the pulley fits loosely in a 50-mm-diameter pin hole. If $\mu_k = 0.30$, determine the minimum tension T required to turn the pulley clockwise at constant velocity if $F = 80$ N. Neglect the weight of the pulley and assume that the cord does not slip on the pulley.

Prob. 8–113

8–114. The axle of the pulley fits loosely in a 50-mm-diameter pin hole. If $\mu_k = 0.30$, determine the minimum tension T required to turn the pulley counterclockwise at constant velocity if $F = 60$ N. Neglect the weight of the pulley and assume that the cord does not slip on the pulley.

8–115. A disk having an outer diameter of 8 in. fits loosely over a fixed shaft having a diameter of 3 in. If the coefficient of static friction between the disk and the shaft is $\mu_s = 0.15$, determine the smallest vertical force F, acting on the rim, which must be applied to the disk to cause it to slip over the shaft. The disk weighs 10 lb.

Prob. 8–115

*8–116. The collar fits *loosely* around a fixed shaft that has a radius of 2 in. If the coefficient of kinetic friction between the shaft and the collar is $\mu_k = 0.3$, determine the tension T in the horizontal segment of the belt so that the collar rotates counterclockwise with a constant angular speed. Assume that the belt does not slip on the collar; rather, the collar slips on the shaft. Neglect the weight and thickness of the belt and collar. The radius, measured from the center of the collar to the mean thickness of the belt, is 2.25 in.

Probs. 8–116/8–117

8–117. The collar fits *loosely* around a fixed shaft that has a radius of 2 in. If the coefficient of kinetic friction between the shaft and the collar is $\mu_s = 0.3$, determine the tension T in the horizontal segment of the belt so that the collar rotates clockwise with a constant angular speed. Assume that the belt does not slip on the collar; rather, the collar slips on the shaft. Neglect the weight and thickness of the belt and collar. The radius, measured from the center of the collar to the mean thickness of the belt, is 2.25 in.

8–118. The pulley has a radius of 3 in. and fits loosely on the 0.5-in.-diameter shaft. If the loadings acting on the belt cause the pulley to rotate uniformly, determine the frictional force between the shaft and the pulley and compute the coefficient of kinetic friction. The pulley weighs 18 lb.

Probs. 8–118/8–119

8–119. The pulley has a radius of 3 in. and fits loosely on the 0.5-in.-diameter shaft. If the loadings acting on the belt cause the pulley to rotate uniformly, determine the frictional force between the shaft and the pulley and compute the coefficient of kinetic friction. Neglect the weight of the pulley.

*8–120. The 5-kg pulley has a diameter of 240 mm and the axle has a diameter of 40 mm. If the coefficient of kinetic friction between the axle and the pulley is $\mu_k = 0.15$, determine the vertical tension T in the rope required to lift the 80-kg block.

Probs. 8–120/8–121

8–121. Solve Prob. 8–120 if the tension T is applied horizontally to the right.

8–122. The trailer has a total weight of 850 lb and center of gravity at G which is directly over its axle. If the axle has a diameter of 1 in., the radius of the wheel is $r = 1.5$ ft, and the coefficient of kinetic friction at the bearing is $\mu_k = 0.08$, determine the horizontal force P needed to pull the trailer.

Prob. 8–122

8–123. A wheel on a freight car carries a load of 20 kip. If the axle of the car has a diameter of 2 in., determine the horizontal force P that must be applied to the axle to rotate the wheel. The coefficient of kinetic friction is $\mu_k = 0.05$.

Prob. 8–123

*8–124. Experimentally it is found that a disk having a diameter of 120 mm rolls with a constant speed down an inclined plane having a slope of 12 mm/m. Determine the coefficient of rolling resistance for the disk.

8–125. A steel railroad wheel has a diameter of 2 ft. If the rolling resistance of the wheel on horizontal steel rails is 3 lb per thousand pounds weight, determine the coefficient of rolling resistance.

8–126. The carriage is used to support a crane load of $F = 1.25(10^6)$ lb. The nine axles shown in the arrangement serve to equalize the load on each of the eight wheels. (By comparison, note that a single supporting beam would *deform* and the wheels would share an unequal amount of load.) Determine the horizontal force that must be applied to the carriage in order to overcome the resistance to rolling. The coefficient of rolling resistance is 0.007 in. and each wheel has a diameter of 36 in.

Prob. 8–126

8–127. A wagon has a weight of 600 lb and is supported by four wheels, each of which has a diameter of 20 in. If the coefficient of rolling resistance is 0.023 in. between the road and each wheel, determine the magnitude of horizontal force required to overcome the rolling resistance of the wheels.

*8–128. A car has a mass of 1.5 Mg. Determine the horizontal force *P* that must be applied to overcome the rolling resistance of the tires if the coefficient of rolling resistance is 0.05 in. The tires have a diameter of 2.75 ft.

8–129. The hand cart has wheels with a diameter of 6 in. If a crate having a weight of 1500 lb is placed on the cart, determine the force *P* that must be applied to the handle to overcome the rolling resistance. The coefficient of rolling resistance is 0.04 in. Neglect the weight of the cart.

Prob. 8–129

8–130. The cylinder is subjected to a load that has a weight *W*. If the coefficients of rolling resistance for the cylinder's top and bottom surfaces are a_A and a_B, respectively, show that a force having a magnitude of $P = [W(a_A + a_B)]/2r$ is required to move the load and thereby roll the cylinder forward. Neglect the weight of the cylinder.

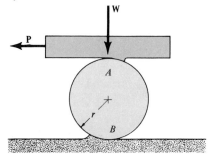

Prob. 8–130

8–131. The 1.4-Mg machine is to be moved over a level surface using a series of rollers for which the coefficient of rolling resistance is 0.5 mm at the ground and 0.2 mm at the bottom surface of the machine. Determine the appropriate diameter of the rollers so that the machine can be pushed forward with a horizontal force of $P = 250$ N. *Hint:* Use the result of Prob. 8–130.

Prob. 8–131

*8–132. A large stone having a mass of 500 kg is moved along the incline using a series of 150-mm-diameter rollers for which the coefficient of rolling resistance is 3 mm at the ground and 4 mm at the bottom surface of the stone. Determine the magnitude of force **T** needed to tow the stone up the plane at a constant speed. *Hint:* Use the result of Prob. 8–130.

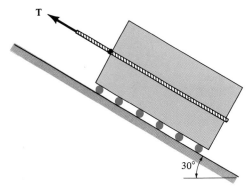

Probs. 8–132/8–133

8–133. A large stone having a mass of 500 kg is moved along the incline using a series of 150-mm-diameter rollers for which the coefficient of rolling resistance is 3 mm at the ground and 4 mm at the bottom surface of the stone. Determine the magnitude of force **T** needed to allow the stone to descend the plane at a constant speed. *Hint:* Use the result of Prob. 8–130.

REVIEW PROBLEMS

8–134. The lawn roller has a mass of 80 kg. If the arm *BA* is held at an angle of 30° from the horizontal and the coefficient of rolling resistance for the roller is 25 mm, determine the force *F* needed to push the roller at constant speed. Neglect friction developed at the axle, *A*, and assume that the force **F** acting on the handle is applied along arm *BA*.

Prob. 8–134

8–135. A cable is attached to a 60-lb plate *B*, passes over a fixed disk at *C*, and is attached to the block at *A*. Using the coefficients of static friction shown in the figure, determine the smallest weight of block *A* that will prevent sliding motion of *B* down the plane.

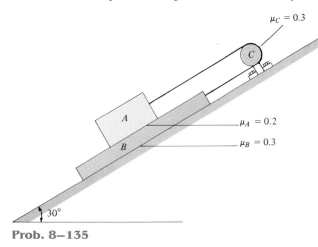

Prob. 8–135

*8–136. A man attempts to lift the uniform 40-lb ladder to an upright position by applying a force **F** perpendicular to the ladder at rung *R*. Determine the coefficient of static friction between the ladder and the ground at *A* if the ladder begins to slip on the ground when his hands reach a height of 6 ft.

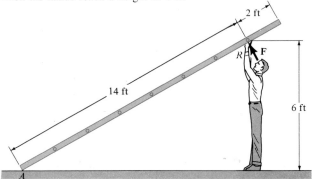

Prob. 8–136

8–137. A man attempts to support a stack of books horizontally by applying a compressive force of $F = 120$ N to the ends of the stack with his hands. If each book has a mass of 0.95 kg, determine the greatest number of books that can be supported in the stack. The coefficient of static friction between the man's hands and a book is $\mu_h = 0.6$ and between any two books, $\mu_b = 0.4$.

Prob. 8–137

8–138. A cable is wrapped around the inner core of a spool. Determine the magnitude of the vertical force **P** that must be applied to the end of the cable to cause impending motion of the spool. The coefficient of static friction between the spool and the contacting surfaces at *A* and *B* is $\mu_s = 0.6$. The spool has a mass of 100 kg.

Prob. 8–138

8–139. The jeep and driver have a mass of 1.2 Mg and mass center at G. Determine the steepest slope θ which the jeep can climb at constant speed if (a) driving power is applied only to the rear wheels B, while the front wheels A are free to roll, and (b) driving power is applied to all four wheels. The coefficient of static friction between the wheels and the ground is $\mu_s = 0.6$.

Prob. 8–139

*8–140. A 200-mm-diameter post is driven 3 m into sand for which $\mu_s = 0.2$. If the normal pressure acting *completely around the post* varies linearly with depth as shown, determine the frictional torque M that must be overcome to rotate the post.

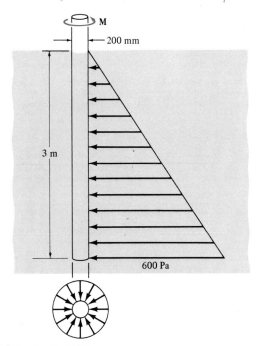

Prob. 8–140

8–141. The collar fits *loosely* around a fixed shaft that has a radius of 2 in. If the coefficient of kinetic friction between the shaft and the collar is $\mu_k = 0.3$, determine the tension T in the horizontal segment of the belt so that the belt can be lowered in the direction of the 20-lb force with a constant speed. Assume that the belt does not slip on the collar; rather, the collar slips on the shaft. Neglect the weight and thickness of the belt and collar. The radius measured from the center of the collar to the mean thickness of the belt is 2.25 in.

Prob. 8–141

8–142. The square-threaded screw of the clamp has a mean diameter of 14 mm and a lead of 6 mm. If $\mu_s = 0.2$ for the threads, and the torque applied to the handle is 1.5 N · m, determine the compressive force F on the block after the screw is tightened.

Prob. 8–142

8–143. Block C has a mass of 50 kg and is confined between two walls by smooth rollers. If the block rests on top of a spool that has a mass of 40 kg, determine the minimum cable force P needed to move the spool. The cable is wrapped around the spool's inner core. The coefficients of static friction at A and B are $\mu_A = 0.3$ and $\mu_B = 0.6$.

Prob. 8–143

*8–144.** A pipe has a total weight of 200 lb, length $l = 8$ ft, and radius $R = 0.75$ ft. If it rests in sand for which the coefficient of static friction is $\mu_s = 0.23$, determine the torque M needed to turn it. Assume that the pressure distribution along the length of the pipe is defined by $p = p_o \sin \theta$. For the solution it is necessary to determine p_o, the peak pressure, in terms of the weight and pipe dimensions.

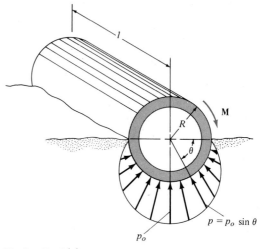

Prob. 8–144

9

Center of Gravity and Centroid

In this chapter we will discuss the method used to determine the location of the center of gravity and center of mass for a system of discrete particles, and then we will expand its application to include a body of arbitrary shape. The same method of analysis will also be used to determine the geometric center, or centroid, of lines, areas, and volumes. Once the centroid has been located, we will then show how to obtain the area and volume of a surface of revolution and determine the resultants of distributed loadings.

9.1 Center of Gravity and Center of Mass for a System of Particles

Center of Gravity

Consider the system of n particles fixed within a region of space as shown in Fig. 9–1. The weights of the particles comprise a system of parallel forces which can be replaced by a single (equivalent) resultant weight and a defined point of application. This point is called the *center of gravity G*. To find its $\bar{x}$, $\bar{y}$, $\bar{z}$ coordinates, one must use the principles outlined in Sec. 4.9. This requires that the resultant weight be equal to the total weight of all n particles; i.e., $W_R = \Sigma W$. The sum of the moments of the weights of all the particles is

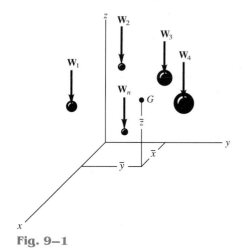

Fig. 9–1

then equal to the moment of the resultant weight about the respective x, y, and z axes.* For example, if moments are summed about the y axis, we have

$$\bar{x}W = \tilde{x}_1 W_1 + \tilde{x}_2 W_2 + \cdots + \tilde{x}_n W_n$$

where $\tilde{x}_1$, etc. represent the x coordinate positions of the particles. This calculation and the two others like it can be written symbolically as

$$\bar{x} = \frac{\Sigma \tilde{x} W}{\Sigma W} \qquad \bar{y} = \frac{\Sigma \tilde{y} W}{\Sigma W} \qquad \bar{z} = \frac{\Sigma \tilde{z} W}{\Sigma W} \qquad (9\text{--}1)$$

Here $\tilde{x}$, $\tilde{y}$, $\tilde{z}$ represent the coordinate locations of each particle in the system.

Center of Mass

To study problems concerning the motion of *matter* under the influence of force, i.e., dynamics, it is necessary to locate a point called the *center of mass*. Provided the acceleration of gravity g for every particle is constant, then $W = mg$. Substituting into Eqs. 9–1 and canceling g from both the numerator and denominator yields

$$\bar{x} = \frac{\Sigma \tilde{x} m}{\Sigma m} \qquad \bar{y} = \frac{\Sigma \tilde{y} m}{\Sigma m} \qquad \bar{z} = \frac{\Sigma \tilde{z} m}{\Sigma m} \qquad (9\text{--}2)$$

By comparison then, the location of the center of gravity *coincides* with that of the center of mass.[†] Recall, however, that particles have "weight" only when under the influence of a gravitational attraction, whereas the center of mass is independent of gravity. For example, it would be meaningless to define the center of gravity of a system of particles representing the planets of our solar system, while the center of mass of this system is important.

---

*This is not true in the exact sense since the weights are not parallel to each other, rather they are all *concurrent* at the earth's center. Furthermore, the acceleration of gravity g is actually different for each particle since it depends upon the distance from the earth's center to the particle. For all practical purposes, however, both of these effects can be neglected.

[†]If the weights of the particles are all parallel to the z axis, as shown in Fig. 9–1, then they will create no moment about this axis. However, to determine $\bar{z}$ it may be helpful to imagine the coordinate system and the particles fixed in it as being rotated 90° about either the x or y axis. This way one can "visualize" how the z-axis moments are developed.

9.2 Center of Gravity, Center of Mass, and Centroid for a Body

Center of Gravity

When the principles used to determine Eqs. 9–1 are applied to a system of particles composing a rigid body, one obtains the same form as these equations except each particle located at $(\tilde{x}, \tilde{y}, \tilde{z})$ is thought to have a *differential weight dW*. Consequently, *integration* is required rather than a discrete summation of the terms. The resulting equations are

$$\bar{x} = \frac{\int \tilde{x}\, dW}{\int dW} \qquad \bar{y} = \frac{\int \tilde{y}\, dW}{\int dW} \qquad \bar{z} = \frac{\int \tilde{z}\, dW}{\int dW} \qquad (9\text{–}3)$$

In order to properly use these equations, the differential weight dW must be expressed in terms of its associated volume dV. If γ represents the *specific weight* of the body, measured as a weight per unit volume, then $dW = \gamma\, dV$ and therefore

$$\bar{x} = \frac{\int_V \tilde{x}\gamma\, dV}{\int_V \gamma\, dV} \qquad \bar{y} = \frac{\int_V \tilde{y}\gamma\, dV}{\int_V \gamma\, dV} \qquad \bar{z} = \frac{\int_V \tilde{z}\gamma\, dV}{\int_V \gamma\, dV} \qquad (9\text{–}4)$$

Here integration must be performed throughout the entire volume of the body.

Center of Mass

The *density ρ*, or mass per unit volume, is related to γ by the equation $\gamma = \rho g$, where g is the acceleration of gravity. Substituting this relationship into Eqs. 9–4 and canceling g from both the numerators and denominators yields similar equations (with ρ replacing γ) that can be used to determine the body's *center of mass*.

Centroid

The *centroid* is a point which defines the *geometrical center* of an object. Its location can be determined from formulas similar to those used to determine the body's center of gravity or center of mass. For example, if the material composing a body is uniform or *homogeneous*, the *density or specific weight* will be *constant* throughout the body, and therefore this term will factor out of the integrals and *cancel* from both the numerators and denominators of Eqs. 9–4. The resulting formulas define the centroid of the body since they are independent of the body's weight and instead depend only on the body's volume.

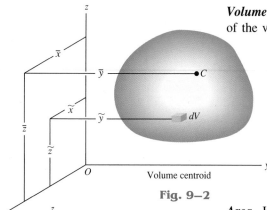

Volume. Using a differential element dV, the coordinates for the centroid C of the volume of space occupied by an object, Fig. 9–2, are

$$\bar{x} = \frac{\int_V \tilde{x} \, dV}{\int_V dV} \qquad \bar{y} = \frac{\int_V \tilde{y} \, dV}{\int_V dV} \qquad \bar{z} = \frac{\int_V \tilde{z} \, dV}{\int_V dV} \qquad (9\text{–}5)$$

Volume centroid

Fig. 9–2

Area. In a similar manner, the centroid for the surface area of an object, such as a plate or shell, Fig. 9–3, can be found by subdividing the area into differential elements dA and computing the "moments" of these area elements about the coordinate axes. Doing this gives the coordinates for the centroid of the area, namely,

$$\bar{x} = \frac{\int_A \tilde{x} \, dA}{\int_A dA} \qquad \bar{y} = \frac{\int_A \tilde{y} \, dA}{\int_A dA} \qquad \bar{z} = \frac{\int_A \tilde{z} \, dA}{\int_A dA} \qquad (9\text{–}6)$$

Area centroid

Fig. 9–3

Line. If the geometry of the object, such as a thin rod or wire, takes the form of a line, Fig. 9–4, the manner of finding its centroid is identical to the procedure outlined above. The results are

$$\bar{x} = \frac{\int_L \tilde{x} \, dL}{\int_L dL} \qquad \bar{y} = \frac{\int_L \tilde{y} \, dL}{\int_L dL} \qquad \bar{z} = \frac{\int_L \tilde{z} \, dL}{\int_L dL} \qquad (9\text{–}7)$$

Line centroid   **Fig. 9–4**

Fig. 9–5

Notice that in all of the above cases the location of C does not necessarily have to be within the object; rather, it can be located off the object in space. Also, the *centroids* of some shapes may be partially or completely specified by using *symmetry conditions*. In cases where the shape has an axis of symmetry, the centroid of the shape will lie along that axis. For example, the centroid C for the line shown in Fig. 9–5 must lie along the y axis, since for every elemental length dL at a distance $+\widetilde{x}$ to the right of the y axis, there is an identical element at a distance $-\widetilde{x}$ to the left. The total moment for all the elements about the axis of symmetry will therefore cancel; i.e., $\int \widetilde{x}\, dL = 0$ (Eq. 9–7) so that $\bar{x} = 0$. In cases where a shape has two or three axes of symmetry, it follows that the centroid lies at the intersection of these axes, Fig. 9–6.

It is important to remember that the terms $\widetilde{x}, \widetilde{y}, \widetilde{z}$ in Eqs. 9–4 to 9–7 refer to the "moment arms" or perpendicular distances from the coordinate planes to the *center of gravity or centroid of the differential element* used in the equations. If possible, this differential element should be chosen such that it is of differential size or thickness in only *one direction*. When this is done only a single integration is required to cover the entire region.

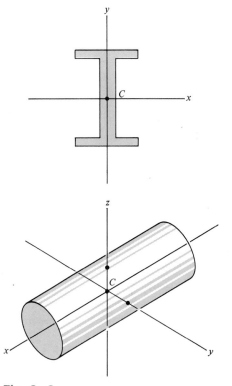

Fig. 9–6

PROCEDURE FOR ANALYSIS

The following procedure provides a method for determining the center of gravity or centroid of an object using a single integration.

Differential Element. Specify the coordinate axes and choose an appropriate differential element for integration. For lines this element dL is represented as a differential line segment; for areas the element dA is generally a rectangle, having a finite height and differential width; and for volumes the element dV is either a circular disk, having a finite radius and differential thickness, or a shell, having a finite length and radius and differential thickness. Locate the element so that it intersects the boundary of the line, area, or volume at an *arbitrary point* (x, y, z).

Size and Moment Arms. Express the length dL, area dA, or volume dV of the element in terms of the coordinates used to define the boundary of the object. Determine the perpendicular distances (moment arms) from the coordinate axes to the centroid or center of gravity of the element. For an x, y, z coordinate system, these dimensions are represented by the coordinates $\widetilde{x}, \widetilde{y}, \widetilde{z}$.

Integrations. Substitute the data computed above into the appropriate equations (Eqs. 9–4 to 9–7) and perform the integrations.* Note that integration can be accomplished when the function in the integrand is expressed in terms of the *same variable as the differential thickness of the element*. The limits of the integral are then defined from the two extreme locations of the element's differential thickness, so that when the elements are "summed" or the integration performed, the entire region is covered.

The following examples numerically illustrate this procedure.

*Formulas for integration are given in Appendix A.

Example 9–1

Locate the centroid of the rod bent into the shape of a parabolic arc, shown in Fig. 9–7.

Fig. 9–7

SOLUTION

Differential Element. The differential element is shown in Fig. 9–7. It intersects the curve at the arbitrary point (x, y).

Length and Moment Arms. The length of the element is determined using the Pythagorean theorem.

$$dL = \sqrt{(dx)^2 + (dy)^2} = \sqrt{\left(\frac{dx}{dy}\right)^2 + 1}\, dy$$

Since $x = y^2$, then $dx/dy = 2y$. Therefore,

$$dL = \sqrt{(2y)^2 + 1}\, dy$$

The centroid is located at $\widetilde{x} = x$, $\widetilde{y} = y$.

Integrations. Using Eqs. 9–7 and integrating with respect to y yields

$$\bar{x} = \frac{\displaystyle\int_L \widetilde{x}\, dL}{\displaystyle\int_L dL} = \frac{\displaystyle\int_0^1 x\sqrt{4y^2 + 1}\, dy}{\displaystyle\int_0^1 \sqrt{4y^2 + 1}\, dy} = \frac{\displaystyle\int_0^1 y^2\sqrt{4y^2 + 1}\, dy}{\displaystyle\int_0^1 \sqrt{4y^2 + 1}\, dy}$$

$$= \frac{0.6059}{1.479} = 0.410 \text{ m} \qquad\qquad Ans.$$

$$\bar{y} = \frac{\displaystyle\int_L \widetilde{y}\, dL}{\displaystyle\int_L dL} = \frac{\displaystyle\int_0^1 y\sqrt{4y^2 + 1}\, dy}{\displaystyle\int_0^1 \sqrt{4y^2 + 1}\, dy} = \frac{0.848}{1.479} = 0.573 \text{ m} \qquad Ans.$$

Example 9–2

Locate the centroid of the triangular area shown in Fig. 9–8a.

Differential Element. A rectangular differential element of thickness dx is shown in Fig. 9–8a. The element intersects the curve at the arbitrary point (x, y).

Area and Moment Arms. The area of the element is $dA = y\, dx$, and the centroid is located at $\tilde{x} = x$, $\tilde{y} = \tfrac{1}{2}y$.

Integrations. Using Eqs. 9–6 and integrating with respect to x yields

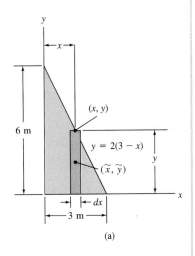

(a)

$$\bar{x} = \frac{\displaystyle\int_A \tilde{x}\, dA}{\displaystyle\int_A dA} = \frac{\displaystyle\int_0^3 xy\, dx}{\displaystyle\int_0^3 y\, dx} = \frac{\displaystyle\int_0^3 x\, 2(3-x)\, dx}{\displaystyle\int_0^3 2(3-x)\, dx} = \frac{\left.\tfrac{6}{2}x^2 - \tfrac{2}{3}x^3\right|_0^3}{\left.6x - \tfrac{2}{2}x^2\right|_0^3} = 1 \text{ m} \quad Ans.$$

$$\bar{y} = \frac{\displaystyle\int_A \tilde{y}\, dA}{\displaystyle\int_A dA} = \frac{\displaystyle\int_0^3 (\tfrac{1}{2}y)\, y\, dx}{\displaystyle\int_0^3 y\, dx} = \frac{\displaystyle\int_0^3 \tfrac{4}{2}(3-x)^2\, dx}{\displaystyle\int_0^3 2(3-x)\, dx} = \frac{\left.2(9x - \tfrac{6}{2}x^2 + \tfrac{1}{3}x^3)\right|_0^3}{\left.6x - \tfrac{2}{2}x^2\right|_0^3} = 2 \text{ m} \quad Ans.$$

Differential Element. A rectangular differential element of thickness dy is shown in Fig. 9–8b. As before, the element intersects the curve at the arbitrary point (x, y).

Area and Moment Arms. The area of the element is $dA = x\, dy$, and the centroid is located at $\tilde{x} = \tfrac{1}{2}x$, $\tilde{y} = y$.

Integrations. Using Eqs. 9–6, and integrating with respect to y yields

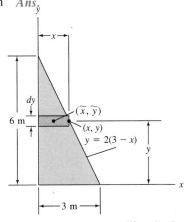

(b) **Fig. 9–8**

$$\bar{x} = \frac{\displaystyle\int_A \tilde{x}\, dA}{\displaystyle\int_A dA} = \frac{\displaystyle\int_0^6 (\tfrac{1}{2}x)x\, dy}{\displaystyle\int_0^6 x\, dy} = \frac{\displaystyle\int_0^6 \tfrac{1}{2}(3 - \tfrac{1}{2}y)^2\, dy}{\displaystyle\int_0^6 (3 - \tfrac{1}{2}y)\, dy} = \frac{\left.\tfrac{1}{2}(9y - \tfrac{3}{2}y^2 + \tfrac{1}{12}y^3)\right|_0^6}{\left.3y - \tfrac{1}{4}y^2\right|_0^6} = 1 \text{ m} \quad Ans.$$

$$\bar{y} = \frac{\displaystyle\int_A \tilde{y}\, dA}{\displaystyle\int_A dA} = \frac{\displaystyle\int_0^6 yx\, dy}{\displaystyle\int_0^6 x\, dy} = \frac{\displaystyle\int_0^6 y(3 - \tfrac{1}{2}y)\, dy}{\displaystyle\int_0^6 (3 - \tfrac{1}{2}y)\, dy} = \frac{\left.\tfrac{3}{2}y^2 - \tfrac{1}{6}y^3\right|_0^6}{\left.3y - \tfrac{1}{4}y^2\right|_0^6} = 2 \text{ m} \quad Ans.$$

Example 9–3

Locate the centroid of the area shown in Fig. 9–9a.

SOLUTION I

Differential Element. A differential element of thickness dx is shown in Fig. 9–9a. The element intersects the curve at the arbitrary point (x, y).

Area and Moment Arms. The area of the element is $dA = y\, dx$, and the centroid is located at $\widetilde{x} = x$, $\widetilde{y} = y/2$.

Integrations. Using Eqs. 9–6 and integrating with respect to x yields

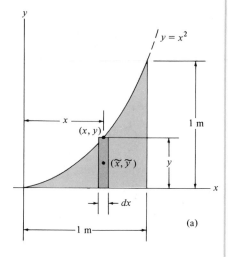

(a)

$$\bar{x} = \frac{\displaystyle\int_A \widetilde{x}\, dA}{\displaystyle\int_A dA} = \frac{\displaystyle\int_0^1 xy\, dx}{\displaystyle\int_0^1 y\, dx} = \frac{\displaystyle\int_0^1 x^3\, dx}{\displaystyle\int_0^1 x^2\, dx} = \frac{0.250}{0.333} = 0.75 \text{ m} \qquad Ans.$$

$$\bar{y} = \frac{\displaystyle\int_A \widetilde{y}\, dA}{\displaystyle\int_A dA} = \frac{\displaystyle\int_0^1 \left(\frac{y}{2}\right) y\, dx}{\displaystyle\int_0^1 y\, dx} = \frac{\displaystyle\int_0^1 \left(\frac{x^2}{2}\right) x^2\, dx}{\displaystyle\int_0^1 x^2\, dx} = \frac{0.100}{0.333} = 0.3 \text{ m} \quad Ans.$$

SOLUTION II

Differential Element. The differential element of thickness dy is shown in Fig. 9–9b. Again the element intersects the curve at the arbitrary point (x, y).

Area and Moment Arms. The area of the element is $dA = (1 - x)\, dy$, and the centroid is located at

$$\widetilde{x} = x + \left(\frac{1 - x}{2}\right) = \frac{1 + x}{2}, \qquad \widetilde{y} = y$$

Integrations. Using Eqs. 9–6, and integrating with respect to y, we obtain

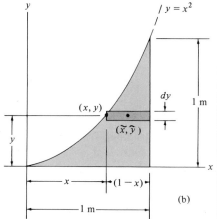

(b)

Fig. 9–9

$$\bar{x} = \frac{\displaystyle\int_A \widetilde{x}\, dA}{\displaystyle\int_A dA} = \frac{\displaystyle\int_0^1 \left(\frac{1 + x}{2}\right)(1 - x)\, dy}{\displaystyle\int_0^1 (1 - x)\, dy} = \frac{\dfrac{1}{2}\displaystyle\int_0^1 (1 - y)\, dy}{\displaystyle\int_0^1 (1 - \sqrt{y})\, dy} = \frac{0.250}{0.333} = 0.75 \text{ m} \quad Ans.$$

$$\bar{y} = \frac{\displaystyle\int_A \widetilde{y}\, dA}{\displaystyle\int_A dA} = \frac{\displaystyle\int_0^1 y(1 - x)\, dy}{\displaystyle\int_0^1 (1 - x)\, dy} = \frac{\displaystyle\int_0^1 (y - y^{3/2})\, dy}{\displaystyle\int_0^1 (1 - \sqrt{y})\, dy} = \frac{0.100}{0.333} = 0.3 \text{ m} \quad Ans.$$

Example 9–4

Locate the $\bar{x}$ centroid of the shaded area bounded by the two curves $y = x$ and $y = x^2$, Fig. 9–10.

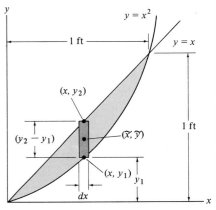

(a)

SOLUTION I

Differential Element. A differential element of thickness dx is shown in Fig. 9–10a. The element intersects the curves at *arbitrary* points (x, y_1) and (x, y_2).

Area and Moment Arm. The area of the element is $dA = (y_2 - y_1)\, dx$, and the centroid is located at $\widetilde{x} = x$.

Integrations. Using Eq. 9–6, we have

$$\bar{x} = \frac{\displaystyle\int_A \widetilde{x}\, dA}{\displaystyle\int_A dA} = \frac{\displaystyle\int_0^1 x(y_2 - y_1)\, dx}{\displaystyle\int_0^1 (y_2 - y_1)\, dx} = \frac{\displaystyle\int_0^1 x(x - x^2)\, dx}{\displaystyle\int_0^1 (x - x^2)\, dx} = \frac{\dfrac{1}{12}}{\dfrac{1}{6}} = 0.5 \text{ ft.} \quad Ans.$$

SOLUTION II

Differential Element. A differential element having a thickness dy and intersecting the curves at the arbitrary points (x_2, y) and (x_1, y) is shown in Fig. 9–10b.

Area and Moment Arm. The area of the element is $dA = (x_1 - x_2)\, dy$, and the centroid is located at

$$\widetilde{x} = x_2 + \frac{x_1 - x_2}{2} = \frac{x_1 + x_2}{2}$$

Integrations. Using Eq. 9–6, we have

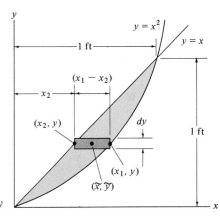

(b)

Fig. 9–10

$$\bar{x} = \frac{\displaystyle\int_A \widetilde{x}\, dA}{\displaystyle\int_A dA} = \frac{\displaystyle\int_0^1 \left(\frac{x_1 + x_2}{2}\right)(x_1 - x_2)\, dy}{\displaystyle\int_0^1 (x_1 - x_2)\, dy} = \frac{\displaystyle\int_0^1 \left(\frac{\sqrt{y} + y}{2}\right)(\sqrt{y} - y)\, dy}{\displaystyle\int_0^1 (\sqrt{y} - y)\, dy}$$

$$= \frac{\dfrac{1}{2}\displaystyle\int_0^1 (y - y^2)\, dy}{\displaystyle\int_0^1 (\sqrt{y} - y)\, dy} = \frac{\dfrac{1}{12}}{\dfrac{1}{6}} = 0.5 \text{ ft} \qquad Ans.$$

Example 9–5

Locate the $\bar{y}$ centroid for the paraboloid of revolution, which is generated by revolving the shaded area shown in Fig. 9–11a about the y axis.

SOLUTION I

Differential Element. An element having the shape of a *thin disk* is chosen, Fig. 9–11a. This element has a radius of $r = z$ and a thickness of dy. In this "disk" method of analysis, the element of planar area, dA, is always taken *perpendicular* to the axis of revolution. Here the element intersects the generating curve at the arbitrary point $(0, y, z)$.

Volume and Moment Arm. The volume of the element is $dV = (\pi z^2)\, dy$, and the centroid is located at $\tilde{y} = y$.

Integrations. Using the second of Eqs. 9–5 and integrating with respect to y yields

$$\bar{y} = \frac{\displaystyle\int_V \tilde{y}\, dV}{\displaystyle\int_V dV} = \frac{\displaystyle\int_0^{100} y(\pi z^2)\, dy}{\displaystyle\int_0^{100} (\pi z^2)\, dy} = \frac{100\pi \displaystyle\int_0^{100} y^2\, dy}{100\pi \displaystyle\int_0^{100} y\, dy} = 66.7 \text{ mm} \qquad Ans.$$

Fig. 9–11a

(a)

Fig. 9–11b

(b)

SOLUTION II

Differential Element. As shown in Fig. 9–11b, the volume element can be chosen in the form of a *thin cylindrical shell,* where the shell radius is $r = z$ and the thickness is dz. In this "shell" method of analysis, the element of planar area, dA, is always taken *parallel* to the axis of revolution. Here the element intersects the generating curve at point $(0, y, z)$.

Volume and Moment Arm. The volume of the element is $dV = 2\pi r\, dA = 2\pi z(100 - y)\, dz$, and the centroid is located at $\widetilde{y} = y + (100 - y)/2 = (100 + y)/2$.

Integrations. Using the second of Eqs. 9–5 and integrating with respect to z yields

$$
\bar{y} = \frac{\displaystyle\int_V \widetilde{y}\, dV}{\displaystyle\int_V dV} = \frac{\displaystyle\int_0^{100} \left(\frac{100 + y}{2}\right) 2\pi z(100 - y)\, dz}{\displaystyle\int_0^{100} 2\pi z(100 - y)\, dz}
$$

$$
= \frac{\pi \displaystyle\int_0^{100} z(10^4 - 10^{-4}z^4)\, dz}{2\pi \displaystyle\int_0^{100} z(100 - 10^{-2}z^2)\, dz} = 66.7 \text{ mm} \qquad \textit{Ans.}
$$

Example 9–6

Determine the location of the center of mass of the right circular cylinder shown in Fig. 9–12a. The density of the cylinder varies directly with its distance from the base, such that $\rho = 200z$ kg/m³.

SOLUTION

For reasons of material symmetry

$$\bar{x} = \bar{y} = 0 \qquad Ans.$$

Differential Element. A disk element of thickness dz is chosen for integration, Fig. 9–12a, since the *density of the entire element is constant* for a given value of z. The element is located along the z axis at the arbitrary point $(0, 0, z)$.

Volume and Moment Arm. The volume of the element is $dV = \pi(0.5)^2\, dz$. The centroid is located at $\tilde{z} = z$.

Integrations. Using an equation similar to the third of Eqs. 9–4 and integrating with respect to z, noting that $\rho = 200z$, we have

$$\bar{z} = \frac{\displaystyle\int_V \tilde{z}\rho\, dV}{\displaystyle\int_V \rho\, dV} = \frac{\displaystyle\int_0^1 z(200z)\pi(0.5)^2\, dz}{\displaystyle\int_0^1 (200z)\pi(0.5)^2\, dz}$$

$$= \frac{\displaystyle\int_0^1 z^2\, dz}{\displaystyle\int_0^1 z\, dz} = 0.667 \text{ m} \qquad Ans.$$

Note: It is not possible to use a shell element for integration such as shown in Fig. 9–12b, since the density of the material composing the shell would *vary* along the shell's height, and hence the location of $\tilde{z}$ for the element would not be easily determined.

(a)

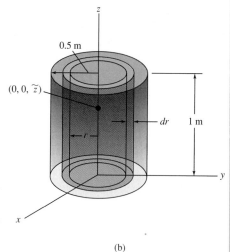

(b)

Fig. 9–12

PROBLEMS

9–1. Locate the center of gravity $\bar{x}$ of the homogeneous rod bent in the form of a semicircular arc. If the rod has a weight per unit length of 0.5 lb/ft, determine the horizontal reaction at the smooth support B and the x and y components of reaction at A.

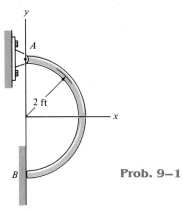

Prob. 9–1

9–2. Locate the center of mass of the homogeneous rod bent into the shape of a circular arc.

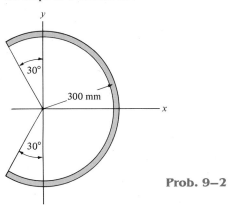

Prob. 9–2

9–3. Locate the centroid $\bar{x}$ of the circular rod. Express the answer in terms of the radius r and semiarc angle α.

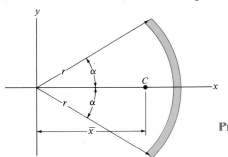

Prob. 9–3

■*9–4. Locate the center of gravity $\bar{x}, \bar{y}$ of the homogeneous rod bent into the shape of the arch shown. What is its weight if it has a mass of 3 kg/m? Solve the problem by evaluating the integrals using Simpson's rule.

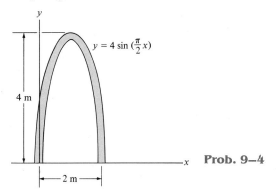

$y = 4 \sin \left(\frac{\pi}{2} x\right)$

Prob. 9–4

9–5. Determine the distance $\bar{x}$ to the center of gravity of the homogeneous rod bent into the parabolic shape. If the rod has a weight per unit length of 0.5 lb/ft, determine the reactions at the fixed support O.

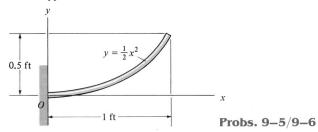

$y = \frac{1}{2} x^2$

Probs. 9–5/9–6

9–6. Determine the distance $\bar{y}$ to the center of gravity of the homogeneous rod bent into the parabolic shape.

9–7. Locate the centroid of the shaded area.

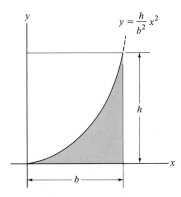

$y = \frac{h}{b^2} x^2$

Prob. 9–7

*9–8. Locate the centroid of the shaded parabolic area.

Prob. 9–8

9–9. Locate the centroid of the shaded area.

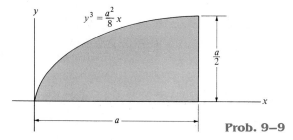

Prob. 9–9

9–10. Determine the location of the centroid of the shaded area.

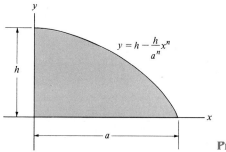

Prob. 9–10

9–11. Determine the location of the centroid of the shaded area.

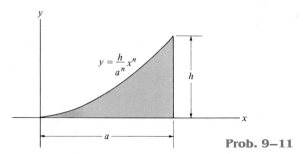

Prob. 9–11

*9–12. Locate the centroid of the shaded area.

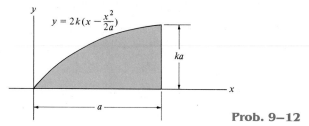

Prob. 9–12

9–13. Locate the centroid of the quarter circle.

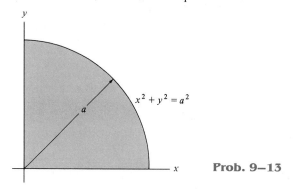

Prob. 9–13

9–14. Locate the centroid of the quarter elliptical area.

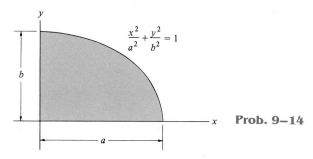

Prob. 9–14

9–15. Locate the centroid of the shaded area bounded by the parabola and the line $y = a$.

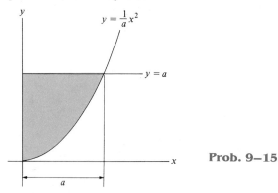

Prob. 9–15

*9–16. Locate the centroid of the shaded area.

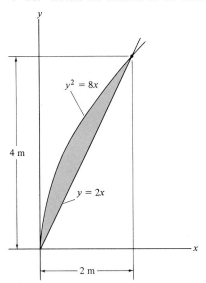

Prob. 9–16

9–17. Locate the centroid of the shaded area.

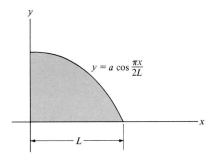

Prob. 9–17

9–18. Locate the centroid for the shaded area.

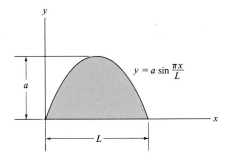

Prob. 9–18

9–19. Locate the centroid of the shaded area. *Hint:* Choose elements of thickness dy and length $[(2 - y) - y^2]$.

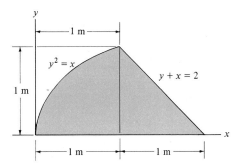

Prob. 9–19

*9–20. The plate has a thickness of 0.25 ft and a specific weight of $\gamma = 180$ lb/ft^3. Determine the location of its center of gravity. Also, compute the tension in each of the cords used to support it.

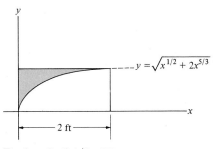

Prob. 9–20

■9–21. Locate the centroid $\bar{x}$ of the shaded area. Solve the problem by evaluating the integrals using Simpson's rule.

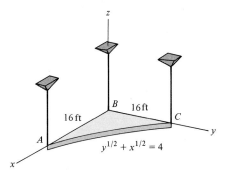

Probs. 9–21/9–22

■9–22. Locate the centroid $\bar{y}$ of the shaded area. Solve the problem by evaluating the integrals using Simpson's rule.

■**9–23.** Locate the centroid $\bar{x}$ of the shaded area. Solve the problem by evaluating the integrals using Simpson's rule.

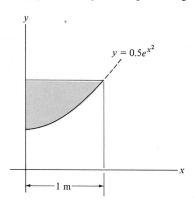

Probs. 9–23/9–24

■*9–24.** Locate the centroid $\bar{y}$ of the shaded area. Solve the problem by evaluating the integrals using Simpson's rule.

9–25. The steel plate is 0.3 m thick and has a density of 7850 kg/m^3. Determine the location of its center of gravity. Also compute the horizontal and vertical reactions at the pin and the reaction at the roller support. Assume the support reactions act at the edges of the plate. *Hint:* The normal force at B is perpendicular to the tangent at B which is found from $\tan \theta = dy/dx$.

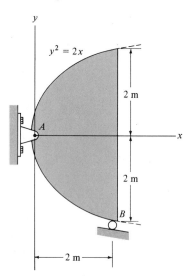

Prob. 9–25

9–26. Locate the centroid of the solid.

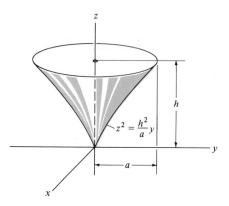

Prob. 9–26

9–27. Locate the centroid of the ellipsoid of revolution.

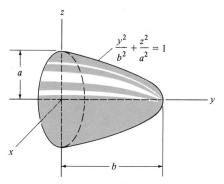

Prob. 9–27

*9–28.** Locate the centroid of the solid.

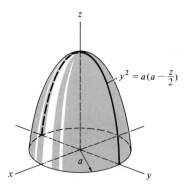

Prob. 9–28

9–29. Locate the centroid of the quarter-cone.

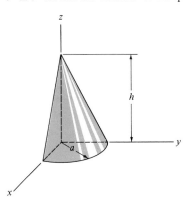

Prob. 9–29

9–30. The king's chamber of the Great Pyramid of Gîza is located at its centroid. Assuming the pyramid to be a solid, prove that this point is at $\bar{z} = \frac{1}{4}h$. *Suggestion:* Use a rectangular differential plate element having a thickness dz and area $(2x)(2y)$.

Prob. 9–30

9–31. Determine the location $\bar{z}$ of the centroid for the tetrahedron. *Suggestion:* Use a triangular "plate" element parallel to the x-y plane and of thickness dz.

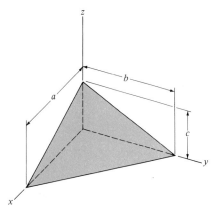

Prob. 9–31

9.3 Composite Bodies

A *composite body* consists of a series of connected "simpler" shaped bodies, which may be rectangular, triangular, semicircular, etc. Such a body can often be sectioned or divided into its composite parts, and provided the weight and location of the center of gravity of each of these "composite parts" are known, one can eliminate the need for integration to determine the center of gravity for the entire body. Following the procedure outlined in Sec. 9.1 yields formulas analogous to Eqs. 9–1, since there is a finite number of weights to be accounted for. Thus, the necessary formulas for finding the center of gravity for a three-dimensional body become

$$\bar{x} = \frac{\Sigma \widetilde{x} W}{\Sigma W} \qquad \bar{y} = \frac{\Sigma \widetilde{y} W}{\Sigma W} \qquad \bar{z} = \frac{\Sigma \widetilde{z} W}{\Sigma W} \qquad\qquad (9\text{–}8)$$

Here $\widetilde{x}$, $\widetilde{y}$, and $\widetilde{z}$ represent the *algebraic distances* from the center of gravity of each composite part to the origin of the coordinates, and ΣW represents the sum of the weights of the composite parts or simply the *total weight* of the body.

When the body has a *constant density or specific weight,* the center of gravity *coincides* with the centroid of the body. The centroid for composite lines, areas, and volumes can be found using relations analogous to Eqs. 9–8; however, the W's are replaced by L's, A's, and V's, respectively. Centroids for common shapes of lines, areas, shells, and volumes are given in the table on the inside back cover of this book.

PROCEDURE FOR ANALYSIS

The following procedure provides a method for determining the center of gravity of a body or the centroid of a composite geometrical object represented by a line, area, or volume.

Composite Parts. Using a sketch, divide the body or object into a finite number of composite parts. If a composite part has a *hole,* or geometric region having no material, then consider the composite part without the hole, and the hole as an *additional* composite part having *negative* weight or size.

Moment Arms. Establish the coordinate axes on the sketch and determine the location $\widetilde{x}$, $\widetilde{y}$, $\widetilde{z}$ of the center of gravity or centroid of each part.

Summations. Determine $\bar{x}$, $\bar{y}$, $\bar{z}$ by applying the center of gravity equations, Eqs. 9–8, or the analogous centroid equations. If an object is *symmetrical* about an axis, recall that the centroid of the object lies on the axis.

If desired, the calculations can be arranged in tabular form, as indicated in the following three examples.

Example 9–7

Locate the centroid of the wire shown in Fig. 9–13a.

SOLUTION

Composite Parts. The wire is divided into three segments as shown in Fig. 9–13b.

Moment Arms. The location of the centroid for each piece is determined and indicated on the sketch. In particular, the centroid of segment ① is determined either by integration or using the table on the inside back cover.

Summations. The calculations are tabulated as follows:

Segment	L (mm)	$\widetilde{x}$ (mm)	$\widetilde{y}$ (mm)	$\widetilde{z}$ (mm)	$\widetilde{x}L$ (mm^2)	$\widetilde{y}L$ (mm^2)	$\widetilde{z}L$ (mm^2)
1	$\pi(60) = 188.5$	60	−38.2	0	11 310	−7200	0
2	40	0	20	0	0	800	0
3	20	0	40	−10	0	800	−200
	$\Sigma L = 248.5$				$\Sigma\widetilde{x}L = 11\ 310$	$\Sigma\widetilde{y}L = -5600$	$\Sigma\widetilde{z}L = -200$

Thus,

$$\bar{x} = \frac{\Sigma\widetilde{x}L}{\Sigma L} = \frac{11\ 310}{248.5} = 45.5 \text{ mm} \qquad Ans.$$

$$\bar{y} = \frac{\Sigma\widetilde{y}L}{\Sigma L} = \frac{-5600}{248.5} = -22.5 \text{ mm} \qquad Ans.$$

$$\bar{z} = \frac{\Sigma\widetilde{z}L}{\Sigma L} = \frac{-200}{248.5} = -0.805 \text{ mm} \qquad Ans.$$

(a) (b)

Fig. 9–13

Example 9–8

Locate the centroid of the plate area shown in Fig. 9–14a.

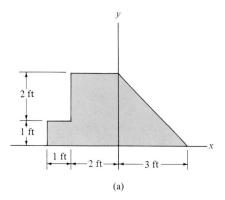

Fig. 9–14   (a)

SOLUTION

Composite Parts.   The plate is divided into three segments as shown in Fig. 9–14b. Here the area of the small rectangle ③ is considered "negative" since it must be subtracted from the larger one ②.

Moment Arms.   The centroid of each segment is located as indicated in the figure. Note that the $\tilde{x}$ coordinates of ② and ③ are *negative*.

Summations.   Taking the data from Fig. 9–14b, the calculations are tabulated as follows:

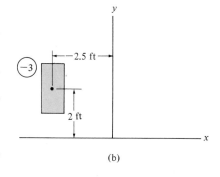

Segment	A (ft²)	$\tilde{x}$ (ft)	$\tilde{y}$ (ft)	$\tilde{x}A$ (ft³)	$\tilde{y}A$ (ft³)
1	$\frac{1}{2}(3)(3) = 4.5$	1	1	4.5	4.5
2	$(3)(3) = 9$	−1.5	1.5	−13.5	13.5
3	$-(2)(1) = -2$	−2.5	2	5	−4
	$\Sigma A = 11.5$			$\Sigma \tilde{x}A = -4$	$\Sigma \tilde{y}A = 14$

(b)

Thus,

$$\bar{x} = \frac{\Sigma \tilde{x}A}{\Sigma A} = \frac{-4}{11.5} = -0.348 \text{ ft} \qquad \textit{Ans.}$$

$$\bar{y} = \frac{\Sigma \tilde{y}A}{\Sigma A} = \frac{14}{11.5} = 1.22 \text{ ft} \qquad \textit{Ans.}$$

Example 9–9

Locate the center of mass of the composite assembly shown in Fig. 9–15a. The conical frustum has a density of $\rho_c = 8$ Mg/m³ and the hemisphere has a density of $\rho_h = 4$ Mg/m³.

(a)

Fig. 9–15a

SOLUTION

Composite Parts. The assembly can be thought of as consisting of four segments as shown in Fig. 9–15b. For the calculations, segments ③ and ④ must be considered as "negative" volumes in order that the four pieces, when added together, yield the total composite shape shown in Fig. 9–15a.

Moment Arm. Using the table on the inside back cover, the computations for the centroid $\tilde{z}$ of each piece are shown in the figure.

Summations. Because of *symmetry*, note that

$$\bar{x} = \bar{y} = 0 \qquad\qquad Ans.$$

Since $W = mg$ and g is constant, the third of Eqs. 9–8 becomes $\bar{z} = \Sigma \tilde{z}m/\Sigma m$. The mass of each piece can be computed from $m = \rho V$ and used for the calculations. Also, 1 Mg/m³ $= 10^{-6}$ kg/mm³, so that

Segment	m (kg)	$\tilde{z}$ (mm)	$\tilde{z}m$ (kg · mm)
1	$8(10^{-6})(\frac{1}{3})\pi(50)^2(200) = 4.189$	50	209.440
2	$4(10^{-6})(\frac{2}{3})\pi(50)^3 = 1.047$	-18.75	-19.635
3	$-8(10^{-6})(\frac{1}{3})\pi(25)^2(100) = -0.524$	$100 + 25 = 125$	-65.450
4	$-8(10^{-6})\pi(25)^2(100) = -1.571$	50	-78.540
	$\Sigma m = 3.141$		$\Sigma \tilde{z}m = 45.815$

Thus,

$$\bar{z} = \frac{\Sigma \tilde{z}m}{\Sigma m} = \frac{45.815}{3.141} = 14.6 \text{ mm} \qquad\qquad Ans.$$

Fig. 9–15b

(b)

PROBLEMS

*9–32. Determine the location $(\bar{x}, \bar{y})$ of the centroid of the rod. Neglect the rod's thickness.

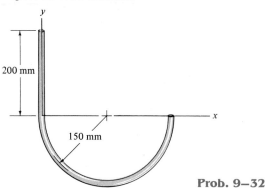

Prob. 9–32

9–33. Locate the centroid for the cold-formed metal strut having the cross section shown. Neglect the thickness of the material and slight bends at the corners.

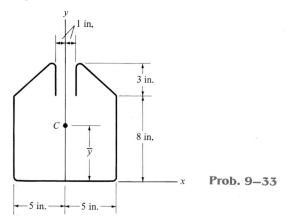

Prob. 9–33

9–34. The "roll-formed" strut has a cross section as shown. Determine the location $\bar{y}$ of its centroid C. Neglect the thickness of the material.

Prob. 9–34

9–35. Determine the location $(\bar{x}, \bar{y})$ of the centroid of the sheet-metal cross section. Neglect the thickness of the material and slight bends at the corners.

Prob. 9–35

*9–36. Determine the location $(\bar{x}, \bar{y})$ of the centroid of the sheet-metal cross section. Neglect the thickness of the material and slight bends at the corners.

Prob. 9–36

9–37. Each of the three members of the frame has a mass per unit length of 6 kg/m. Locate the position ($\bar{x}$, $\bar{y}$) of the center of gravity. Neglect the size of the pins at the joints and the thickness of the members. Also, calculate the reactions at the pin A and roller E.

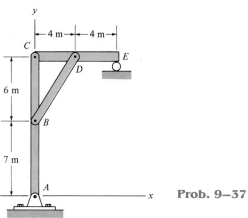

Prob. 9–37

9–38. Locate the center of gravity of the homogeneous rod. The rod has a weight of 2 lb/ft. Also, compute the x, y, z components of reaction at the fixed support A.

Prob. 9–38

9–39. Determine the distance $\bar{y}$ to the centroid C of the channel's cross-sectional area.

Prob. 9–39

*9–40. Determine the location ($\bar{x}$, $\bar{y}$) of the centroid C for the angle's cross-sectional area.

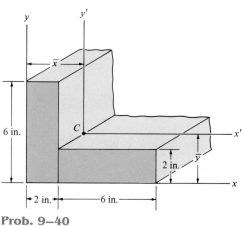

Prob. 9–40

9–41. Determine the distance $\bar{y}$ to the centroid of the beam's cross-sectional area.

Prob. 9–41

9–42. Determine the distance $\bar{y}$ to the centroid C of the cross-sectional area of the beam constructed from a channel and a plate. Assume all corners are square and neglect the size of the weld at A.

Prob. 9–42

9–43. Determine the distance $\bar{y}$ to the centroid of the cross-sectional area of the beam.

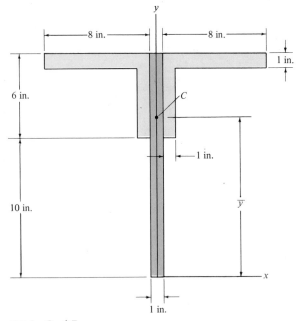

Prob. 9–43

*9–44. Determine the location $(\bar{x}, \bar{y})$ of the centroid C of the area.

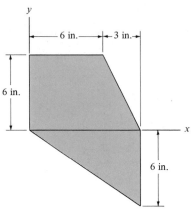

Prob. 9–44

9–45. Determine the distance $\bar{y}$ to the centroid for the beam's cross-sectional area.

Prob. 9–45

9–46. Determine the location $(\bar{x}, \bar{y})$ of the centroid of the member's cross-sectional area.

Prob. 9–46

9–47. Determine the location ($\bar{x}$, $\bar{y}$) of the centroid of the cross-sectional area.

300 mm

300 mm   400 mm

600 mm

300 mm

300 mm

Prob. 9–47

*9–48. The gravity wall is made of concrete. Determine the location ($\bar{x}$, $\bar{y}$) of the center of gravity G for the wall.

1.2 m

$\bar{x}$

G

3 m

$\bar{y}$

0.4 m

2.4 m

Prob. 9–48   0.6 m   0.6 m

9–49. Determine the distance $\bar{y}$ to the centroid of the cross-sectional area.

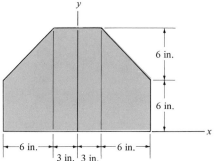

6 in.

6 in.

6 in.   6 in.

3 in. 3 in.

Prob. 9–49

9–50. Determine the distance $\bar{y}$ to the centroid of the member's cross-sectional area.

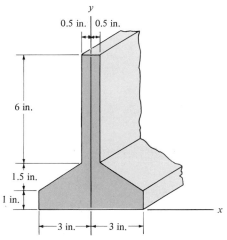

0.5 in. 0.5 in.

6 in.

1.5 in.

1 in.

3 in.   3 in.

Prob. 9–50

9–51. Determine the distance $\bar{y}$ to the centroid of the trapezoidal area in terms of the dimensions shown.

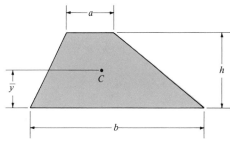

a

C

h

$\bar{y}$

b

Prob. 9–51

*9–52. Determine the distance $\bar{y}$ to the centroid of the cross-sectional area.

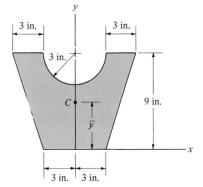

3 in.   3 in.

3 in.

C

9 in.

$\bar{y}$

3 in. 3 in.

Prob. 9–52

9–53. Determine the distance $\bar{y}$ to the centroid of the shaded area of the annulus segment.

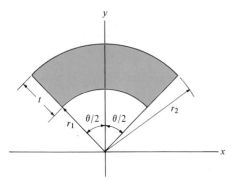

Prob. 9–53

9–54. Locate the center of gravity of the sheet-metal bracket if the material is homogeneous and has a constant thickness. If the bracket is resting on the horizontal x-y plane as shown, determine the maximum angle of tilt θ which it can have before it falls over, i.e., begins to rotate about the y axis.

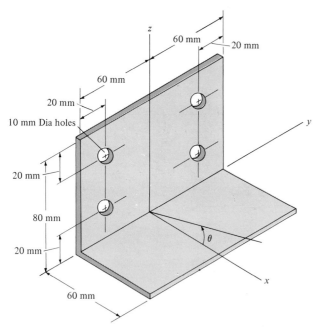

Prob. 9–54

9–55. A triangular plate made of homogeneous material has a constant thickness which is very small. If it is folded over as shown, determine the location $(0, \bar{y}, \bar{z})$ of the plate's center of gravity G.

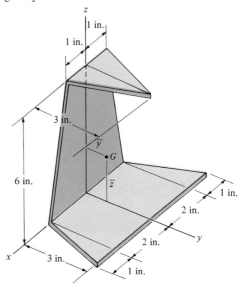

Prob. 9–55

***9–56.** Determine the location $(\bar{x}, \bar{y})$ of the center of gravity of the helicopter. The locations of the centers of gravity of the various components and their weights are indicated and tabulated in the figure.

1. Passenger compartment	300 lb
2. Engine and housing	1600 lb
3. Main roter	120 lb
4. Tail structure	230 lb
5. Tail roter and housing	60 lb

Prob. 9–56

9–57. Determine the location of the center of mass for the compressor assembly. The locations of the centers of mass of the various components and their masses are indicated and tabulated in the figure. What are the vertical reactions at blocks A and B needed to support the platform?

① Instrument panel	230 kg	
② Filter system	183 kg	
③ Piping assembly	120 kg	
④ Liquid storage	85 kg	
⑤ Structural framework	468 kg	**Prob. 9–57**

9–58. Major floor loadings in a shop are caused by the weights of the objects shown. Each force acts through its respective center of gravity G. Determine the resultant force and specify its location $\bar{x}$, $\bar{y}$ on the floor.

Prob. 9–58

9–59. A wooden buoy is made from two cones having a common radius of 1.5 ft. Find the centroid distance $\bar{z}$.

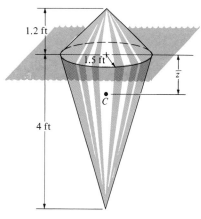

Prob. 9–59

*9–60. Determine the location $\bar{z}$ of the center of mass. The material has a density of $\rho = 3$ Mg/m³. There is a hole bored through the center.

Prob. 9–60

409

9–61. The assembly consists of a 20-in. wooden dowel rod and a tight-fitting steel collar. Determine the distance $\bar{x}$ to its center of gravity if the specific weights of the materials are $\gamma_w = 150$ lb/ft^3 and $\gamma_{st} = 490$ lb/ft^3.

Prob. 9–61

9–62. Determine the location $\bar{z}$ of the centroid of the top made from a hemisphere and a cone.

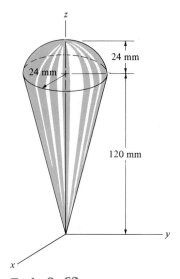

Prob. 9–62

9–63. The assembly is made from a steel hemisphere, $\rho_{st} = 7.80$ Mg/m^3, and an aluminum cylinder, $\rho_{al} = 2.70$ Mg/m^3. Determine the height h of the cylinder so that the mass center G of the assembly is located at $\bar{z} = 160$ mm.

Prob. 9–63

*9–64. Determine the distance h to which a 100-mm-diameter hole must be bored into the cone so that the center of mass G of the resulting shape is located at $\bar{z} = 115$ mm. The material has a density of 8 Mg/m^3.

Prob. 9–64

*9.4 Theorems of Pappus and Guldinus

The two *theorems of Pappus and Guldinus*, which were first developed by Pappus of Alexandria during the third century A.D. and then restated at a later time by the Swiss mathematician Paul Guldin or Guldinus (1577–1643), are used to find the surface area and volume of any object of revolution.

A *surface area of revolution* is generated by revolving a *plane curve* about a nonintersecting fixed axis in the plane of the curve; whereas a *volume of revolution* is generated by revolving a *plane area* about a nonintersecting fixed axis in the plane of the area. For example, if the *line AB* shown in Fig. 9–16 is rotated about a fixed axis, it generates the *surface area* of a cone; if the *triangular area ABC* shown in Fig. 9–17 is rotated about the axis, it generates the *volume* of a cone.

The statements and proofs of the theorems of Pappus and Guldinus follow. The proofs require that the generating curves and areas do *not* cross the axis about which they are rotated; otherwise, two sections on either side of the axis would generate areas or volumes having opposite signs and hence cancel each other.

Fig. 9–16

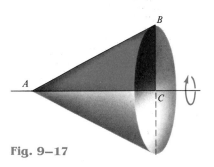

Fig. 9–17

Surface Area

The area of a surface of revolution equals the product of the length of the generating curve and the distance traveled by the centroid of the curve in generating the surface area.

Proof. When a differential length dL of the curve shown in Fig. 9–18 is revolved about an axis through a distance $2\pi r$, it generates a ring having a surface area $dA = 2\pi r\, dL$. The entire surface area, generated by revolving the entire curve about the axis, is therefore $A = 2\pi \int_L r\, dL$. Since $\int_L r\, dL = \bar{r}L$, where $\bar{r}$ locates the centroid C of the generating curve L, the area becomes $A = 2\pi\bar{r}L$. In general, though,

$$A = \theta\bar{r}L \qquad (9\text{–}9)$$

where A = surface area of revolution
θ = angle of revolution measured in radians, $\theta \leqslant 2\pi$
$\bar{r}$ = perpendicular distance from the axis of revolution to the centroid of the generating curve
L = length of the generating curve

Fig. 9–18

411

Volume

The volume of a body of revolution equals the product of the generating area and the distance traveled by the centroid of the area in generating the volume.

Proof. When the differential area dA shown in Fig. 9–19 is revolved about an axis through a distance $2\pi r$, it generates a ring having a volume $dV = 2\pi r\, dA$. The entire volume, generated by revolving A about the axis, is therefore $V = 2\pi\int_A r\, dA$. Since $\int_A r\, dA = \bar{r}A$, where $\bar{r}$ locates the centroid C of the generating area A, the volume becomes $V = 2\pi\bar{r}A$. In general, though,

$$\boxed{V = \theta\bar{r}A} \tag{9–10}$$

where V = volume of revolution

        θ = angle of revolution measured in radians, $\theta \leqslant 2\pi$

        $\bar{r}$ = perpendicular distance from the axis of revolution to the centroid of the generating area

        A = generating area

When applying the above two theorems, the generating line or area may also be considerd as a series of composite parts, in which case, $A = \theta\, \Sigma\tilde{r}_i L_i$ and $V = \theta\, \Sigma\tilde{r}_i A_i$. Here $\tilde{r}_i$ is the radial distance for each of the L_i's and A_i's that make up L and A, respectively.

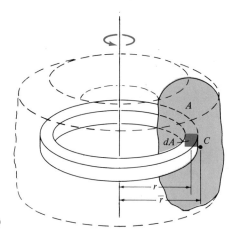

Fig. 9–19

Example 9–10

Show that the surface area of a sphere is $A = 4\pi R^2$ and its volume is $V = \frac{4}{3}\pi R^3$.

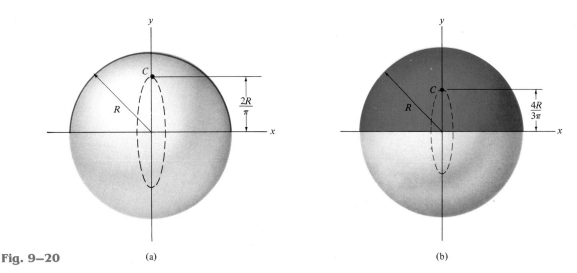

Fig. 9–20

(a)

(b)

SOLUTION

The sphere in Fig. 9–20a is generated by rotating a semicircular *arc* about the x axis. Using the table on the inside back cover, it is seen that the centroid of this arc is located at a distance $\bar{y} = 2R/\pi$ from the x axis of rotation. Since the centroid moves through an angle of $\theta = 2\pi$ rad in generating the sphere, applying Eq. 9–9 gives the surface area of the sphere as:

$$A = \theta \bar{y}L; \qquad A = 2\pi\left(\frac{2R}{\pi}\right)\pi R = 4\pi R^2 \qquad \textit{Ans.}$$

In a similar manner, the volume of the sphere is generated by rotating the semicircular *area* in Fig. 9–20b about the x axis. Using the table on the inside back cover to locate the centroid, and applying Eq. 9–10, we have

$$V = \theta \bar{r}A; \qquad V = 2\pi\left(\frac{4R}{3\pi}\right)\left(\frac{1}{2}\pi R^2\right) = \frac{4}{3}\pi R^3 \qquad \textit{Ans.}$$

Example 9–11

Determine the mass of concrete needed to construct the arched beam shown in Fig. 9–21. The density of concrete is $\rho_c = 2.1$ Mg/m^3. Also, determine the surface area of the beam.

SOLUTION

The mass of the arch may be determined by using the density ρ_c provided the volume of concrete is known. The cross-sectional area of the beam is composed of two rectangles having centroids at points C_1 and C_2 as shown in Fig. 9–21b. The volume generated by each of these areas is equal to the product of the cross-sectional area and the distance traveled by its centroid. In generating the arch, the centroids move through an angle of $\theta = \pi(90°/180°) = 1.57$ radians. Point C_1 acts at a distance of $d_1 = 4.5 - 0.1 = 4.4$ m from the center of rotation, point O; whereas C_2 acts at $d_2 = 4 + 0.15 = 4.15$ m from O. Applying Eq. 9–10, we obtain the total volume:

$V = \theta \Sigma \bar{x} A$

$\quad = 1.57$ rad$[(4.4$ m$)(1$ m$)(0.2$ m$) + (4.15$ m$)(0.15$ m$)(0.3$ m$)]$

$\quad = 1.676$ m^3

The required mass of concrete is then

$$m = \rho_c V = (2.1 \text{ Mg/m}^3)(1.676 \text{ m}^3) = 3.52 \text{ Mg} \qquad Ans.$$

The surface area of the arch is determined by dividing the perimeter into eight line segments, Fig. 9–21c, and then applying Eq. 9–9. Thus,

$A = \theta \Sigma \bar{x} L = 1.57$ rad$[4$ m$(0.15$ m$) + 2(4.15$ m$)(0.3$ m$) +$

$\qquad (4.3$ m$)(1$ m $- 0.15$ m$) + 2(4.4$ m$)(0.2$ m$) + (4.5$ m$)(1$ m$)]$

$\quad = 20.4$ m^2 $\qquad\qquad Ans.$

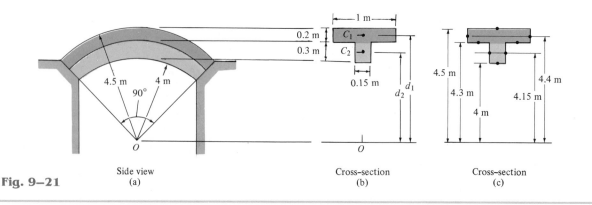

Fig. 9–21

Side view
(a)

Cross-section
(b)

Cross-section
(c)

PROBLEMS

9–65. Using integration, compute both the area and the centroidal distance $\bar{x}$ of the shaded area. Then, using the second theorem of Pappus-Guldinus, compute the volume of the solid generated by revolving the area about the y axis.

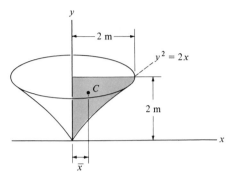

Prob. 9–65

■**9–66.** Using Simpson's rule, determine the area and the centroidal distance $\bar{y}$ of the shaded area. Then using the second theorem of Pappus-Guldinus, determine the volume formed by revolving the area about the x axis.

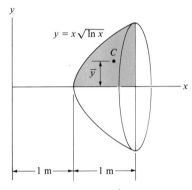

Prob. 9–66

9–67. Using integration, determine the area and the centroidal distance $\bar{y}$ of the shaded area. Then, using the second theorem of Pappus-Guldinus, determine the volume of a solid formed by revolving the area about the x axis.

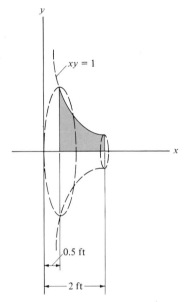

Prob. 9–67

***9–68.** Determine the volume of steel needed to produce the tapered part. The cross section is shown, although the part is 360° around. Also, compute the outside surface area of the part, excluding the ends.

Prob. 9–68

415

9–69. Determine the volume of steel needed to make the brake piston. It consists of a full circular part. Its cross section is shown in the figure.

40 mm

60 mm

80 mm

20 mm

40 mm 30 mm 20 mm   **Probs. 9–69/9–70/9–71**

9–70. Determine the volume of oil that can be contained in the brake piston. It consists of a full circular part. Its cross section is shown in the figure.

9–71. Determine the interior surface area of the brake piston. It consists of a full circular part. Its cross section is shown in the figure.

*9–72. The full circular aluminum housing is used in an automotive brake system. Half the cross section is shown in the figure. Determine its weight if aluminum has a specific weight of 169 lb/ft^3.

2.00 in.

0.25 in.

0.15 in.

3.25 in.

4.00 in.

0.25 in.

1.80 in.   **Prob. 9–72**

9–73. The starter for an electric motor has the cross-sectional area shown. If copper wiring has a density of $\rho_{cu} = 8.90$ Mg/m^3, and the steel frame has a density of $\rho_{st} = 7.80$ Mg/m^3, estimate the total mass of the starter. Neglect the size of the copper wire.

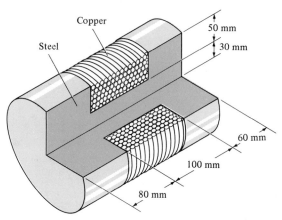

Copper

Steel

50 mm

30 mm

60 mm

100 mm

80 mm

Prob. 9–73

9–74. The process tank is used to store liquids during manufacturing. Estimate both the volume of the tank and its surface area so that the amount of steel plate needed to construct it can be determined. The tank has a flat top.

3 m   3 m

6 m

4 m

Prob. 9–74

9–75. Determine the approximate amount of aluminum necessary to make the funnel. It consists of a full circular part having a thickness of 2 mm. Its cross section is shown in the figure.

50 mm
80 mm
30 mm
60 mm
30 mm

Probs. 9–75/9–76

*9–76. Determine the approximate surface area of the funnel. It consists of a full circular part of negligible thickness.

9–77. Determine the weight of water contained in a spherical vessel having an inner diameter of 8 ft if the depth of the water is 4 ft. The specific weight of water is $\gamma = 62.4$ lb/ft^3.

9–78. Determine the surface area of the tank which consists of a cylinder and hemispherical cap. Neglect the thickness of the plates.

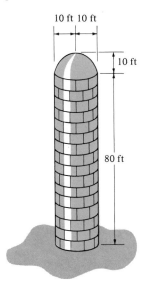

10 ft 10 ft

10 ft

80 ft

Probs. 9–78/9–79

9–79. Determine the volume of the tank which consists of a cylinder and hemispherical cap. Neglect the thickness of the plates.

*9–80. The suspension bunker is made from plates which are curved to the natural shape which a completely flexible membrane would take if subjected to a full load of coal. This curve may be approximated by a parabola, $y = 0.2x^2$. Determine the weight of coal which the bunker may contain when completely filled. Coal has a specific weight of $\gamma = 50$ lb/ft^3 and assume there is a 20% loss in volume due to air voids. Solve the problem by integration to determine the cross-sectional area of ABC, then use the second theorem of Pappus-Guldinus to find the volume.

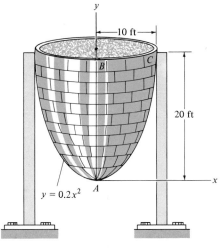

y
10 ft
B C
20 ft
x
$y = 0.2x^2$ A

Prob. 9–80

9–81. Determine the height h to which liquid should be poured into the cup so that it contacts half the surface area on the inside of the cup. Neglect the cup's thickness for the calculation.

30 mm
50 mm
h
10 mm

Prob. 9–81

417

★9.5 Resultant of a General Distributed Force System

In Sec. 4.10 we discussed the method used to simplify a distributed loading that is uniform along an axis of a rectangular surface. In this section we will generalize this method to include surfaces that have an arbitrary shape and are subjected to a variable pressure distribution. As a specific application we will find the resultant loading acting on the surface of a body that is submerged in a fluid.

Pressure Distribution over a Surface

(a)

Consider the flat plate shown in Fig. 9–22a which is subjected to the loading function $p = p(x, y)$ Pa, where a pascal Pa $= 1$ N/m². Knowing this function, we can determine the *magnitude* of the infinitesimal force dF acting on the differential area dA m² of the plate, located at the arbitrary point (x, y). This force magnitude is simply $dF = [p(x, y) \text{ N/m}^2](dA \text{ m}^2) = [p(x, y) \, dA]$ N. The entire loading on the plate is therefore represented as a system of *parallel forces* infinite in number and each acting on separate differential areas dA. This system of parallel forces will now be simplified to a single resultant force $\mathbf{F}_R$ acting through a unique point $(\bar{x}, \bar{y})$ on the plate, Fig. 9–22b.

Magnitude of Resultant Force. To determine the *magnitude* of $\mathbf{F}_R$, it is necessary to sum each of the differential forces $d\mathbf{F}$ acting over the plate's *entire surface area A*. This sum may be expressed mathematically as an integral.

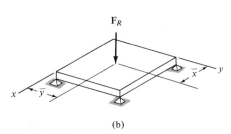

(b)

Fig. 9–22

$$F_R = \Sigma F; \qquad \boxed{F_R = \int_A p(x, y) \, dA = \int_V dV} \qquad (9\text{–}11)$$

Note that $p(x, y) \, dA = dV$, the colored differential *volume element* shown in Fig. 9–22a. Therefore, the result indicates that the *magnitude of the resultant force is equal to the total volume under the distributed-loading diagram.*

Location of Resultant Force. The location $(\bar{x}, \bar{y})$ of $\mathbf{F}_R$ is determined by setting the moments of $\mathbf{F}_R$ equal to the moments of all the forces $d\mathbf{F}$ about the respective x and y axes. From Fig. 9–22a and 9–22b, we get

$$\boxed{\bar{x} = \frac{\int_A x p(x, y) \, dA}{\int_A p(x, y) \, dA} \qquad \bar{y} = \frac{\int_A y p(x, y) \, dA}{\int_A p(x, y) \, dA}} \qquad (9\text{–}12)$$

Setting $dV = p(x, y) \, dA$, we can also write

$$\bar{x} = \frac{\int_V x \, dV}{\int_V dV} \qquad \bar{y} = \frac{\int_V y \, dV}{\int_V dV} \qquad (9\text{–}13)$$

Hence, it can be seen that the *line of action of the resultant force passes through the geometric center or centroid of the volume under the distributed loading diagram.*

*9.6 Fluid Pressure

According to Pascal's law, a fluid at rest creates a pressure p at a point that is the *same* in *all* directions. The magnitude of p, measured as a force per unit area, depends upon the specific weight γ or mass density ρ of the fluid and the depth z of the point from the fluid surface. The relationship can be expressed mathematically as

$$p = \gamma z = \rho g z \qquad\qquad (9\text{–}14)$$

where g is the acceleration of gravity. Equation 9–14 is only valid for fluids that are assumed *incompressible,* as in the case of most liquids.* Gases are compressible fluids, and since their density changes significantly with both pressure and temperature, Eq. 9–14 cannot be used.

Consider now what effect the pressure of a liquid has at points A, B, and C, located on the top surface of the submerged plate shown in Fig. 9–23. Since points A and B are both at depth z_2 from the liquid surface, the *pressure* at these points has a magnitude of $p_2 = \gamma z_2$. Likewise, point C is at depth z_1; hence, $p_1 = \gamma z_1$. In all cases, the pressure acts *normal* to the surface area dA located at the point, Fig. 9–23.

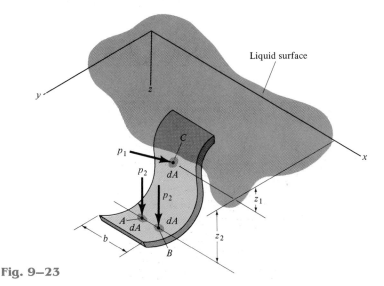

Fig. 9–23

*Note that for water, $\gamma = 62.4$ lb/ft^3, or $\gamma = 9810$ N/m^3 since $\rho = 1000$ kg/m^3.

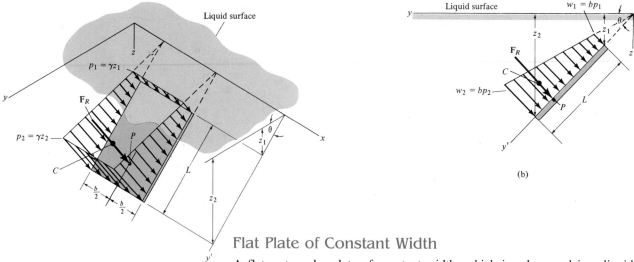

Fig. 9–24   (a)

Flat Plate of Constant Width

A flat rectangular plate of constant width, which is submerged in a liquid having a specific weight γ, is shown in Fig. 9–24a. The plane of the plate makes an angle θ with the horizontal, such that its top edge is located at a depth z_1 from the liquid surface and its bottom edge is located at a depth z_2. Since pressure varies linearly with depth, Eq. 9–14, the distribution of pressure over the plate's surface is represented by a trapezoidal volume of loading. As noted in Sec. 9.5 the magnitude of the *resultant force* $\mathbf{F}_R$ is equal to the *volume* of this loading diagram and $\mathbf{F}_R$ has a *line of action* that passes through the volume's centroid C. Note that $\mathbf{F}_R$ does *not* act at the center of the plate; rather, it acts at point P, called the *center of pressure*.

Since the plate has a *uniform width*, the loading distribution may also be viewed in two dimensions, Fig. 9–24b. Here the loading intensity is measured as force/length and varies linearly from $w_1 = bp_1 = b\gamma z_1$ to $w_2 = bp_2 = b\gamma z_2$. The magnitude of $\mathbf{F}_R$ in this case equals the trapezoidal *area* and $\mathbf{F}_R$ has a *line of action* that passes through the area's *centroid C*. For numerical applications, the area and location of the centroid for a trapezoid are tabulated on the inside back cover.

Curved Plate of Constant Width

When the submerged plate is curved, the pressure acting normal to the plate continually changes direction, and therefore calculation of the magnitude of $\mathbf{F}_R$ and its location P is more difficult than for a flat plate. Three- and two-dimensional views of the loading distribution are shown in Fig. 9–25a and b, respectively. Here integration can be used to determine both F_R and the location of the centroid C or center of pressure P.

A simpler method exists, however, for calculating the magnitude of $\mathbf{F}_R$ and its location along a curved (or flat) plate having a *constant width*. This method requires separate calculations for the horizontal and vertical *components* of $\mathbf{F}_R$. For example, the distributed loading acting on the top surface of the

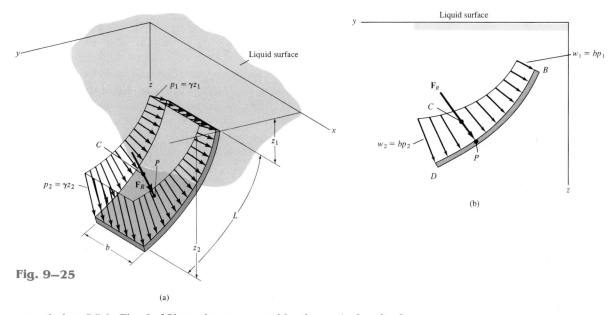

Fig. 9–25

(a)

(b)

curved plate DB in Fig. 9–25b can be represented by the *equivalent loading* shown in Fig. 9–26. Here it is seen that DB supports a horizontal force along its *vertical projection AD*. This force, $\mathbf{F}_{AD}$, has a magnitude that equals the area under the trapezoid and acts through the centroid C_{AD} of this area. The distributed loading along the *horizontal projection AB* is constant, since all points lying in this plane are at the same depth from the surface of the liquid. The magnitude of $\mathbf{F}_{AB}$ is simply the area of the rectangle. This force acts through the area's centroid C_{AB} or the midpoint of AB. In addition to $\mathbf{F}_{AB}$, the curved surface DB must also support the downward *weight of fluid* $\mathbf{W}_f$ contained within the block BDA. This force has a magnitude of $W_f = (\gamma b)(Area_{BDA})$ and acts through the centroid of BDA. Summing the three coplanar forces shown in Fig. 9–26 yields $\mathbf{F}_R = \Sigma\mathbf{F} = \mathbf{F}_{AD} + \mathbf{F}_{AB} + \mathbf{W}_f$. The center of pressure P is determined by applying the principle of moments to this force system and its resultant about a convenient reference point. The final results for $\mathbf{F}_R$ and its location P will be equivalent to those shown in Fig. 9–25.

Fig. 9–26

Flat Plate of Variable Width

The pressure distribution acting across the top surface of a submerged plate having a variable width is shown in Fig. 9–27. The resultant force of this loading equals the volume described by the plate area as its base and linear varying pressure distribution as its altitude. The shaded element shown in Fig. 9–27 may be used if integration is chosen to determine this volume. The element consists of a strip of area $dA = x\,dy'$ located at a depth z below the liquid surface. Since a uniform pressure $p = \gamma z$ (force/area) acts on dA, the magnitude of the differential force $d\mathbf{F}$ is equal to $dF = dV = p\,dA = \gamma z(x\,dy')$. Integrating over the entire volume yields Eq. 9–11, i.e.,

$$F_R = \int_V dV = V$$

From Eq. 9–13, the centroid of V defines the point through which $\mathbf{F}_R$ acts. The center of pressure, which lies on the surface of the plate just below C, has coordinates $P(\bar{x}, \bar{y}')$ defined by the equations

$$\bar{x} = \frac{\int_V \tilde{x}\,dV}{\int_V dV} \qquad \bar{y}' = \frac{\int_V \tilde{y}'\,dV}{\int_V dV}$$

This point should *not* be mistaken for the centroid of the plate *area*.

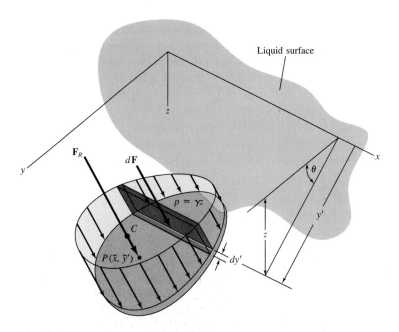

Fig. 9–27

Example 9–12

Determine the magnitude and location of the resultant hydrostatic force acting on the submerged rectangular plate AB shown in Fig. 9–28a. The plate has a width of 1.5 m; $\rho_w = 1000$ kg/m³.

(a)

SOLUTION

Since the plate has a constant width, the distributed loading can be viewed in two dimensions as shown in Fig. 9–28b. The intensity of the load at A and B is computed as

$$w_A = b\rho_w g z_A = (1.5 \text{ m})(1000 \text{ kg/m}^3)(9.81 \text{ m/s}^2)(2 \text{ m})$$
$$= 29.4 \text{ kN/m}$$
$$w_B = b\rho_w g z_B = (1.5 \text{ m})(1000 \text{ kg/m}^3)(9.81 \text{ m/s}^2)(5 \text{ m})$$
$$= 73.6 \text{ kN/m}$$

From the table on the inside back cover, the magnitude of the resultant force $\mathbf{F}_R$ created by the distributed load is

$$F_R = \text{area of trapezoid}$$
$$= \tfrac{1}{2}(3)(29.4 + 73.6) = 154.5 \text{ kN} \qquad Ans.$$

This force acts through the centroid of the area,

$$h = \frac{1}{3}\left(\frac{2(29.4) + 73.6}{29.4 + 73.6}\right)(3) = 1.29 \text{ m} \qquad Ans.$$

(b)

measured upward from B, Fig. 9–28b.

The same results can be obtained by considering two components of $\mathbf{F}_R$ defined by the triangle and rectangle shown in Fig. 9–28c. Each force acts through its associated centroid and has a magnitude of

$$F_{Re} = (29.4 \text{ kN/m})(3 \text{ m}) = 88.2 \text{ kN}$$
$$F_t = \tfrac{1}{2}(44.2 \text{ kN/m})(3 \text{ m}) = 66.3 \text{ kN}$$

Hence,

$$F_R = F_{Re} + F_t = 88.2 + 66.3 = 154.5 \text{ kN} \qquad Ans.$$

The location of $\mathbf{F}_R$ is determined by summing moments about B, Fig. 9–28b and c, i.e.,

$$\zeta + (M_R)_B = \Sigma M_B; \quad (154.5)h = 88.2(1.5) + 66.3(1)$$
$$h = 1.29 \text{ m} \qquad Ans.$$

(c)

Fig. 9–28

423

Example 9–13

Determine the magnitude of the resultant hydrostatic force acting on the surface of a seawall shaped in the form of a parabola as shown in Fig. 9–29a. The wall is 5 m long; $\rho_w = 1020 \text{ kg/m}^3$.

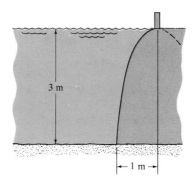

Fig. 9–29

SOLUTION

The horizontal and vertical components of the resultant force will be calculated, Fig. 9–29b. Since

$$w_B = b\rho_w g z_B = 5 \text{ m}(1020 \text{ kg/m}^3)(9.81 \text{ m/s}^2)(3 \text{ m}) = 150.0 \text{ kN/m}$$

then

$$F_x = \tfrac{1}{2}(3 \text{ m})(150.0 \text{ kN/m}) = 225.0 \text{ kN}$$

The area of the parabolic sector ABC can be determined using the table on the inside back cover. Hence, the weight of water within this region is

$$F_y = (\rho_w g b)(Area_{ABC})$$
$$= (1020 \text{ kg/m}^3)(9.81 \text{ m/s}^2)(5 \text{ m})[\tfrac{1}{3}(1 \text{ m})(3 \text{ m})] = 50.0 \text{ kN}$$

The resultant force is therefore

$$F_R = \sqrt{F_x^2 + F_y^2} = \sqrt{(225.0)^2 + (50.0)^2}$$
$$= 230 \text{ kN} \qquad\qquad Ans.$$

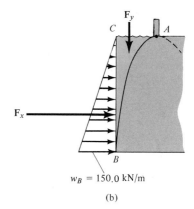

(b)

Example 9–14

Determine the magnitude and location of the resultant force acting on the triangular end plates of the water trough shown in Fig. 9–30a; $\rho_w = 1000 \text{ kg/m}^3$.

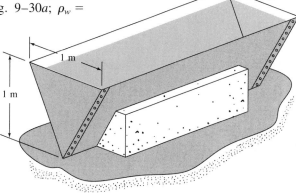

(a)

SOLUTION

The pressure loading acting on the end plate E is shown in Fig. 9–30b. The magnitude of the resultant force **F** is equal to the volume of this loading diagram. Choosing the differential volume element shown in the figure, we get

$$dF = dV = p\, dA = \rho_w g z (2x\, dz) = 19\,620zx\, dz$$

The equation of line AB is

$$x = 0.5(1 - z)$$

Hence, substituting and integrating with respect to z from $z = 0$ to $z = 1$ m yields

$$F = V = \int_V dV = \int_0^1 (19\,620)z[0.5(1 - z)]\, dz$$

$$= 9810 \int_0^1 (z - z^2)\, dz = 1635 \text{ N}$$

or

$$F = 1.64 \text{ kN} \qquad \textit{Ans.}$$

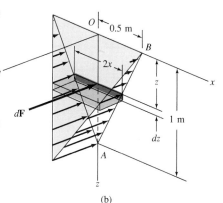

(b)

Fig. 9–30

This resultant passes through the *centroid of the volume*. Because of symmetry,

$$\bar{x} = 0 \qquad \textit{Ans.}$$

Since $\widetilde{z} = z$ for the volume element, then

$$\bar{z} = \frac{\int_V \widetilde{z}'\, dV}{\int_V dV} = \frac{\int_0^1 z(19\,620)z[0.5(1 - z)]\, dz}{1635} = \frac{9810 \int_0^1 (z^2 - z^3)\, dz}{1635}$$

$$= 0.5 \text{ m} \qquad \textit{Ans.}$$

PROBLEMS

9–82. Determine the magnitude of the resultant hydrostatic force acting on the dam and its location measured from the top surface of the water. The width of the dam is 8 m; $\rho_w = 1.0$ Mg/m^3.

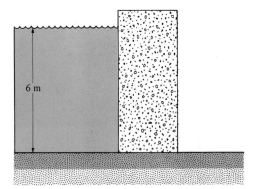

Prob. 9–82

9–83. The tank is filled with water to a depth of $d = 4$ m. Determine the resultant force the water exerts on side A and side B of the tank. If oil instead of water is placed in the tank, to what depth d should it reach so that it creates the same resultant forces? $\rho_o = 900$ kg/m^3 and $\rho_w = 1000$ kg/m^3.

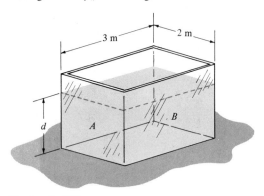

Prob. 9–83

***9–84.** The structure shown is used for temporary storage of oil at sea for later loading into ships. When it is empty the water level is at A (sea level). As oil is loaded, the water is displaced through exit ports at D. If the riser EC is filled with oil, i.e., to level C, determine the height h of the oil level above sea level. $\rho_o = 900$ kg/m^3 and $\rho_w = 1020$ kg/m^3.

Prob. 9–84

9–85. The concrete "gravity" dam is held in place by its own weight. If the density of concrete is $\rho_c = 2.5$ Mg/m^3, and water has a density of $\rho_w = 1.0$ Mg/m^3, determine the smallest width d that will prevent the dam from overturning about its end A. Water pressure does not act underneath the dam.

Prob. 9–85

9–86. Determine the magnitude of the resultant hydrostatic force acting per meter of length on the sea wall; $\rho_w = 1.0 \text{ Mg/m}^3$.

Prob. 9–86

9–89. The storage tank contains oil having a specific weight of $\gamma_o = 56 \text{ lb/ft}^3$. If the tank is 6 ft wide, calculate the resultant force acting on the inclined side BC of the tank, caused by the oil, and specify its location along BC, measured from B. Also compute the total resultant force acting on the bottom of the tank.

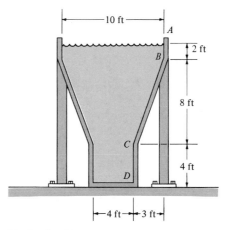

Prob. 9–89

9–87. Determine the magnitude and location of the resultant hydrostatic force acting on each of the cover plates A and B. The density of water is $\rho_w = 1.0 \text{ Mg/m}^3$.

Prob. 9–87

9–90. The gate AB is 8 m wide. Determine the horizontal and vertical components of force acting on the pin at B and the vertical reaction at the smooth support A. $\rho_w = 1.0 \text{ Mg/m}^3$.

Prob. 9–90

*9–88. Determine the magnitude of the resultant hydrostatic force acting per meter of length on the sea wall; $\gamma_w = 62.4 \text{ lb/ft}^3$.

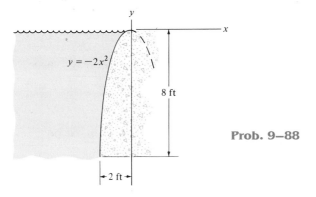

Prob. 9–88

9–91. The symmetric concrete ''gravity'' dam is held in place by its own weight. If the density of concrete is $\rho_c = 2.5$ Mg/m^3; and water has a density of $\rho_w = 1.0$ Mg/m^3, determine the smallest width d at its base that will prevent the dam from overturning about its end A. The dam has a width of 8 m.

Prob. 9–91

*9–92. The tank is used to store a liquid having a specific weight of $\gamma = 80$ lb/ft^3. If it is filled to the top, determine the magnitude of force the liquid exerts on side $ABCD$.

Probs. 9–92/9–93

9–93. The tank is used to store a liquid having a specific weight of $\gamma = 80$ lb/ft^3. If it is filled to the top, determine the magnitude of force the liquid exerts on side $ABEF$.

9–94. The drum is filled to its top ($y = 1.5$ ft) with oil having a density of $\rho_o = 55$ lb/ft^3. Determine the resultant force of the oil pressure acting on the flat end plate A of the drum, and specify its location measured from the top of the drum.

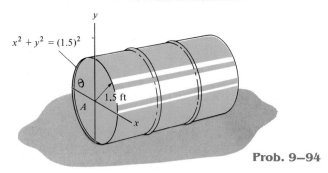

Prob. 9–94

9–95. The ends of the tank are parabolic. If it is filled to the top with water, determine the resultant force and its location $\bar{y}$ measured from the bottom of the tank. $\gamma_w = 62.4$ lb/ft^3.

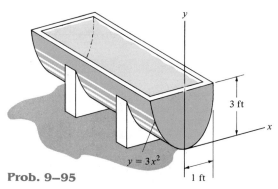

Prob. 9–95

*9–96. The end plates of the trough are in the form of a semiellipse. If it is filled to the top with water, determine the resultant force the water exerts on these plates and the location of the center of pressure, measured from the top of the trough. $\gamma_w = 62.4$ lb/ft^3.

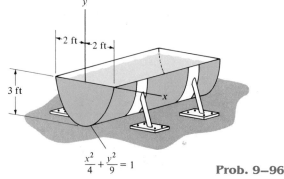

Prob. 9–96

9–97. A wind loading creates a positive pressure on one side of the chimney and a negative (suction) pressure on the other side, as shown. If this pressure loading acts uniformly along the chimney's length, determine the magnitude of the resultant force created by the wind.

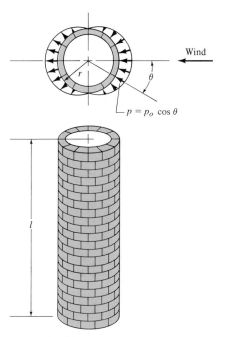

9–99. The pressure loading on the plate is described by the function $p = [-240/(x + 1) + 340]$ Pa. Determine the magnitude of the resultant force and coordinates of the point where the line of action of the force intersects the plate.

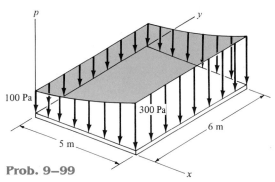

Prob. 9–99

Prob. 9–97

9–98. The support exerts the distributed loading on the pipe as shown. If the pipe weighs 50 lb and has a length of 3 ft, determine the magnitude of p_o for equilibrium. $p = 0.5 \, p_o \, (1 + \cos \theta)$.

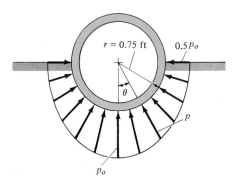

Prob. 9–98

***9–100.** The loading acting on a square plate is represented by a parabolic pressure distribution. Determine the magnitude and location of the resultant force. Also, what are the reactions at the rollers B and C and the ball-and-socket joint A? Neglect the weight of the plate.

Prob. 9–100

9–101. The load over the plate varies linearly along the sides of the plate such that $p = \frac{2}{3}[x(4 - y)]$ kPa. Determine the resultant force and its position $(\bar{x}, \bar{y})$ on the plate.

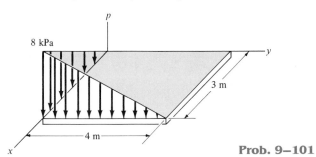

Prob. 9–101

REVIEW PROBLEMS

9–102. Determine the distance $\bar{y}$ to the centroid of the beam's cross-sectional area. Neglect the size of the corner welds at A and B for the calculation.

Prob. 9–102

9–103. Locate the center of gravity of the volume generated by revolving the shaded area about the x axis. The material is homogeneous.

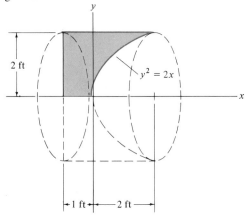

Probs. 9–103/9–104

*9–104. Using integration, compute the area and the centroidal distance $\bar{y}$ for the shaded area. Then using the second theorem of Pappus-Guldinus, compute the volume of the solid generated by revolving the shaded area about the x axis.

9–105. The rectangular plate is subjected to a distributed load over its *entire surface*. If the load is defined by the expression $p = p_o \sin(\pi x/a) \sin(\pi y/b)$, where p_o represents the pressure acting at the center of the plate, determine the magnitude and location of the resultant force acting on the plate.

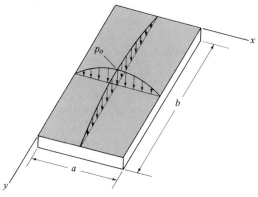

Prob. 9–105

9–106. Determine the volume of concrete needed to construct the circular curb.

Prob. 9–106

9–107. Locate the center of gravity of the homogeneous wire.

Prob. 9–107

*9–108. Determine the surface area and the volume of material required to make the casting.

Side View

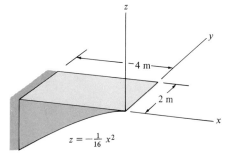

Front View

Prob. 9–108

9–109. Locate the centroid of the cross-sectional area of the beam.

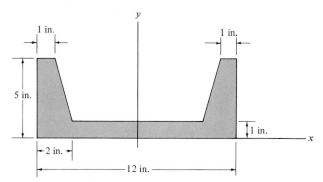

Prob. 9–109

9–110. Locate the center of gravity of the homogeneous cantilever beam.

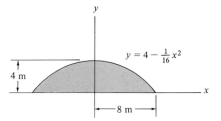

$z = -\frac{1}{16} x^2$

Prob. 9–110

9–111. Locate the centroid of the shaded area.

$y = 4 - \frac{1}{16} x^2$

4 m

8 m

Prob. 9–111

10

Moments of Inertia

In Chapter 9 we determined the *centroid* for an area by considering the *first moment* of the area about an axis, i.e., for the computation it is necessary to evaluate an integral of the form $\int x \, dA$. There are many important topics in engineering practice, however, which require evaluation of an integral of the *second moment* of an area, i.e., $\int x^2 \, dA$. This integral is referred to as the moment of inertia for an area. The terminology ''moment of inertia'' as used here is actually a misnomer; however, it has been adopted because of the similarity with integrals of the same form related to mass.

Methods used to determine both the area and mass moments of inertia will be discussed in this chapter. In particular, a body's mass moment of inertia is frequently used to solve problems in dynamics.

10.1 Definition of Moments of Inertia for Areas

The moment of inertia of an area originates whenever one computes the moment of a distributed load that varies linearly from the moment axis. A common example of this type of loading occurs in the study of fluid mechanics. It was pointed out in Sec. 9.6 that the pressure, or force per unit area, exerted at a point located a distance z below the surface of a liquid is $p = \gamma z$, Eq. 9–14,

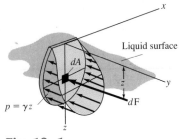

Fig. 10–1

where γ is the specific weight of the liquid. Thus, the magnitude of force exerted by a liquid on the area dA of the submerged plate shown in Fig. 10–1 is $dF = p\,dA = \gamma z\,dA$. The moment of this force about the x axis of the plate is $dM = dFz = \gamma z^2\,dA$ and, therefore, the moment created by the *entire* distributed (pressure) loading is $M = \gamma \int z^2\,dA$. Here the integral represents the moment of inertia of the area of the plate about the x axis. Since integrals of this form often arise in formulas used in fluid mechanics, mechanics of materials, structural mechanics, and machine design, the engineer should become familiar with the methods used for their computation.

Moment of Inertia

Consider the area A, shown in Fig. 10–2, which lies in the x-y plane. By definition, the moments of inertia of the differential planar area dA about the x and y axes are $dI_x = y^2\,dA$ and $dI_y = x^2\,dA$, respectively. For the entire area the *moments of inertia* are determined by integration, i.e.,

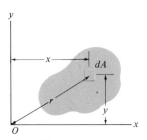

Fig. 10–2

$$\boxed{\begin{aligned} I_x &= \int_A y^2\,dA \\ I_y &= \int_A x^2\,dA \end{aligned}}$$

(10–1)

We can also formulate the second moment of the differential area dA about the pole O or z axis, Fig. 10–2. This is referred to as the polar moment of inertia, $dJ_O = r^2\,dA$. Here r is the perpendicular distance from the pole (z axis) to the element dA. For the entire area the *polar moment of inertia* is

$$\boxed{J_O = \int_A r^2\,dA = I_x + I_y}$$

(10–2)

The relationship between J_O and I_x, I_y is possible since $r^2 = x^2 + y^2$, Fig. 10–2.

From the above formulations it is seen that I_x, I_y, and J_O will *always* be *positive,* since they involve the product of distance squared and area. Furthermore, the units for moment of inertia involve length raised to the fourth power, e.g., m^4, mm^4, or ft^4, in^4.

10.2 Parallel-Axis Theorem for an Area

If the moment of inertia for an area is known about an axis passing through its centroid, we can determine the moment of inertia of the area about a corresponding parallel axis using the *parallel-axis theorem*. To derive this theorem, consider finding the moment of inertia of the shaded area shown in Fig.

10–3 about the x axis. In this case, a differential element dA is located at an arbitrary distance y' from the centroidal x' axis, whereas the *fixed distance* between the parallel x and x' axes is defined as d_y. Since the moment of inertia of dA about the x axis is $dI_x = (y' + d_y)^2\, dA$, then for the entire area,

$$I_x = \int_A (y' + d_y)^2\, dA$$

$$= \int_A y'^2\, dA + 2d_y \int_A y'\, dA + d_y^2 \int_A dA$$

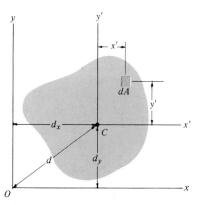

Fig. 10–3

The first term on the right represents the moment of inertia of the area about the x' axis, $\bar{I}_{x'}$. The second term is zero since the x' axis passes through the area's centroid C, i.e., $\int y'\, dA = \bar{y} \int dA = 0$ since $\bar{y} = 0$. The final result is therefore

$$\boxed{I_x = \bar{I}_{x'} + Ad_y^2} \tag{10-3}$$

A similar expression can be written for I_y, i.e.,

$$\boxed{I_y = \bar{I}_{y'} + Ad_x^2} \tag{10-4}$$

And finally, for the polar moment of inertia about an axis perpendicular to the x-y plane and passing through the pole O (z axis), Fig. 10–3, we have

$$\boxed{J_O = \bar{J}_C + Ad^2} \tag{10-5}$$

The form of each of these equations states that *the moment of inertia of an area about an axis is equal to the area's moment of inertia about a parallel axis passing through the centroid plus the product of the area and the square of the perpendicular distance between the axes.*

10.3 Radius of Gyration of an Area

The *radius of gyration* of a planar area has units of length and is a quantity often used for the design of columns in structural mechanics. Provided the areas and moments of inertia are *known*, the radii of gyration are determined from the formulas

$$\boxed{k_x = \sqrt{\frac{I_x}{A}} \qquad k_y = \sqrt{\frac{I_y}{A}} \qquad k_O = \sqrt{\frac{J_O}{A}}} \tag{10-6}$$

Note that the form of these equations is easily remembered since it is similar to that for finding the moment of inertia of a differential area about an axis. For example, $I_x = k_x^2 A$; whereas for a differential area, $dI_x = y^2\, dA$.

10.4 Moments of Inertia for an Area by Integration

When the boundaries for a planar area can be expressed by mathematical functions, Eqs. 10–1 may be integrated to determine the moments of inertia for the area. If the element of area chosen for integration has a differential size in two directions as shown in Fig. 10–2, a double integration must be performed to evaluate the moment of inertia. Most often, however, it is easier to choose an element having a differential size or thickness in only one direction, because then the evaluation requires only a single integration.

PROCEDURE FOR ANALYSIS

If a single integration is performed to determine the moment of inertia of an area about an axis, it will first be necessary to determine the most effective way to *orient* the element with respect to the axis, so that the process of integration is made as simple as possible. In this regard, two possibilities exist:

Case 1. The length of the element can be *parallel* to the axis. This situation occurs when using the rectangular element shown in Fig. 10–4 for computing I_y of the area. *Direct* application of Eq. 10–1, $I_y = \int x^2 \, dA$, can be made in this case since *all parts* of the element lie at the *same* moment-arm distance x from the y axis.*

Case 2. The length of the element can be *perpendicular* to the axis. Here Eq. 10–1 *does not apply* since all parts of the element *do not* lie at the same moment-arm distance from the axis. Instead, the *parallel-axis theorem* must be used. Hence, if the rectangular element in Fig. 10–4 is used for computing I_x for the area, it will first be necessary to compute the moment of inertia of the *element* about a horizontal axis passing through its centroid and then determine the moment of inertia of the *element* about the x axis by using the parallel-axis theorem. Integration of this result will yield I_x.

The following examples illustrate the above procedure for each case.

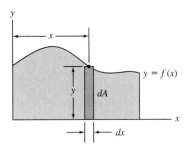

Fig. 10–4

---

* In the case of the element $dA = dx \, dy$, Fig. 10–2, the moment arms y and x are appropriate for the formulation of I_x and I_y (Eq. 10–1) since the *entire* element, because of its "small-ness," lies at the specified y and x perpendicular distances from the x and y axes.

Example 10–1

Determine the moment of inertia for the rectangular area shown in Fig. 10–5 with respect to (a) the centroidal x' axis, (b) the axis x_b passing through the base of the rectangle, and (c) the pole or z' axis perpendicular to the x'-y' plane and passing through the centroid C.

Fig. 10–5

SOLUTION (CASE 1)

Part (a). The differential element shown in Fig. 10–5 is chosen for integration. Because of its location and orientation, the *entire element* is at a distance y' from the x' axis. Here it is necessary to integrate from $y' = -h/2$ to $y' = h/2$. Since $dA = b\, dy'$, then

$$\bar{I}_{x'} = \int_A y'^2\, dA = \int_{-h/2}^{h/2} y'^2(b\, dy') = b\int_{-h/2}^{h/2} y'^2\, dy'$$

$$= \frac{1}{12}bh^3 \qquad\qquad Ans.$$

Part (b). The moment of inertia about an axis passing through the base of the rectangle can be obtained by using the result of part (a) and applying the parallel-axis theorem, Eq. 10–3.

$$I_{x_b} = \bar{I}_{x'} + Ad_y^2$$

$$= \frac{1}{12}bh^3 + bh\left(\frac{h}{2}\right)^2 = \frac{1}{3}bh^3 \qquad\qquad Ans.$$

Part (c). The moment of inertia $\bar{I}_{y'}$ may be found by interchanging the dimensions b and h in the result of part (a), in which case

$$\bar{I}_{y'} = \frac{1}{12}hb^3$$

Using Eq. 10–2, the polar moment of inertia about C is

$$\bar{J}_C = \bar{I}_{x'} + \bar{I}_{y'} = \frac{1}{12}bh(h^2 + b^2) \qquad\qquad Ans.$$

Example 10–2

Compute the moment of inertia of the shaded area shown in Fig. 10–6a about the x axis.

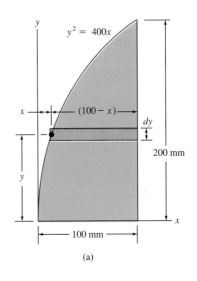

(a)

SOLUTION I (CASE **1**)

A differential element of area that is *parallel* to the x axis, as shown in Fig. 10–6a, is chosen for integration. Since the element has a thickness dy and intersects the curve at the arbitrary point (x, y), the area is $dA = (100 - x) \, dy$. Furthermore, all parts of the element lie at the same distance y from the x axis. Hence, integrating with respect to y, from $y = 0$ to $y = 200$ mm, yields

$$I_x = \int_A y^2 \, dA = \int_A y^2 (100 - x) \, dy$$

$$= \int_0^{200} y^2 \left(100 - \frac{y^2}{400} \right) dy = 100 \int_0^{200} y^2 \, dy - \frac{1}{400} \int_0^{200} y^4 \, dy$$

$$= 107(10^6) \text{ mm}^4 \qquad\qquad\qquad Ans.$$

SOLUTION II (CASE **2**)

A differential element *parallel* to the y axis, as shown in Fig. 10–6b, is chosen for integration. In this case, however, all parts of the element do *not* lie at the same distance from the x axis, and therefore the parallel-axis theorem must be used to determine the *moment of inertia of the element* with respect to the axis. For a rectangle having a base b and height h, the moment of inertia about its centroidal axis has been computed in part (a) of Example 10–1. There it was found that $\bar{I}_{x'} = \frac{1}{12}bh^3$. For the differential element shown in Fig. 10–6b, $b = dx$ and $h = y$, and thus $d\bar{I}_{x'} = \frac{1}{12} dx \, y^3$. Since the centroid of the element is at $\widetilde{y} = y/2$ from the x axis, the moment of inertia of the element about the x axis is

(b)

Fig. 10–6

$$dI_x = d\bar{I}_{x'} + dA \, \widetilde{y}^2 = \frac{1}{12} dx \, y^3 + y \, dx \left(\frac{y}{2} \right)^2 = \frac{1}{3} y^3 \, dx$$

(This result can also be concluded from part (b) of Example 10–1). Integrating with respect to x, from $x = 0$ to $x = 100$ mm, yields

$$I_x = \int dI_x = \int_A \frac{1}{3} y^3 \, dx = \int_0^{100} \frac{1}{3} (400x)^{3/2} \, dx$$

$$= 107(10^6) \text{ mm}^4 \qquad\qquad\qquad Ans.$$

Example 10–3

Determine the moment of inertia with respect to the x axis of the circular area shown in Fig. 10–7a.

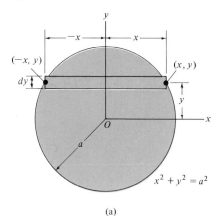

(a)

Fig. 10–7

SOLUTION I (*CASE 1*)

Using the differential element shown in Fig. 10–7a, since $dA = 2x\,dy$, we have

$$I_x = \int_A y^2\,dA = \int_A y^2(2x)\,dy$$

$$= \int_{-a}^{a} y^2(2\sqrt{a^2 - y^2})\,dy = \frac{\pi a^4}{4} \qquad Ans.$$

SOLUTION II (*CASE 2*)

When the differential element is chosen as shown in Fig. 10–7b, the parallel-axis theorem must be used. Why? The centroid for the element lies on the x axis. Hence, $\widetilde{y} = 0$. Applying Eq. 10–3, noting that $dA = 2y\,dx$, we have

$$dI_x = d\overline{I}_{x'} + dA\,\widetilde{y}^2$$

$$= \frac{1}{12}\,dx\,(2y)^3 + 2y\,dx\,(0)$$

$$= \frac{2}{3}y^3\,dx$$

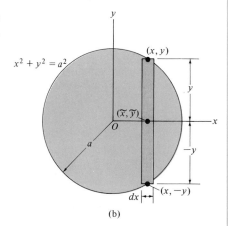

(b)

Therefore, integrating with respect to x yields

$$I_x = \int_{-a}^{a} \frac{2}{3}(a^2 - x^2)^{3/2}\,dx = \frac{\pi a^4}{4} \qquad Ans.$$

Example 10–4

Determine the moment of inertia of the shaded area shown in Fig. 10–8a about the x axis.

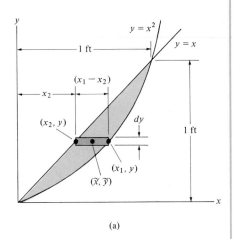

(a)

SOLUTION I (*CASE 1*)

The differential element of area parallel to the x axis is chosen for integration, Fig. 10–8a. The element intersects the curve at the arbitrary points (x_2, y) and (x_1, y). Consequently, its area is $dA = (x_1 - x_2)\, dy$. Since all parts of the element lie at the same distance y from the x axis, we have

$$I_x = \int_A y^2\, dA = \int_0^1 y^2(x_1 - x_2)\, dy = \int_0^1 y^2(\sqrt{y} - y)\, dy$$

$$I_x = \frac{2}{7}y^{7/2} - \frac{1}{4}y^4 \Big|_0^1 = 0.0357 \text{ ft}^4 \qquad\qquad Ans.$$

SOLUTION II (*CASE 2*)

The differential element of area parallel to the y axis is shown in Fig. 10–8b. It intersects the curves at the arbitrary points (x, y_2) and (x, y_1). Since all parts of its entirety do *not* lie at the same distance from the x axis, we must first use the parallel-axis theorem to find the element's moment of inertia about the x axis, then integrate this result to determine I_x. Thus,

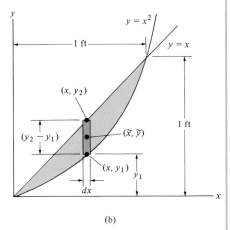

(b)

Fig. 10–8

$$dI_x = d\bar{I}_{x'} + dA\, \tilde{y}^2 = \frac{1}{12}\, dx\, (y_2 - y_1)^3 + (y_2 - y_1)\, dx \left(y_1 + \frac{y_2 - y_1}{2}\right)^2$$

$$= \frac{1}{3}(y_2^3 - y_1^3)\, dx = \frac{1}{3}(x^3 - x^6)\, dx$$

$$I_x = \frac{1}{3}\int_0^1 (x^3 - x^6)\, dx = \frac{1}{12}x^4 - \frac{1}{21}x^7 \Big|_0^1 = 0.0357 \text{ ft}^4 \qquad Ans.$$

By comparison, Solution I requires much less computation. Realize, however, that if an integral using a particular element appears difficult to evaluate, one should try solving the problem using an element oriented in the other direction.

PROBLEMS

10–1. The area has a moment of inertia about the xx axis of 3325 in^4. If the area is 28 in^2, determine its moment of inertia about the $x'x'$ axis. The $\overline{xx}$ axis passes through the centroid C of the area.

Prob. 10–1

10–2. The polar moment of inertia of the area is $J'_C = 548(10^6)$ mm^4, computed about the z' axis passing through the centroid C. The moment of inertia about the y' axis is $383(10^6)$ mm^4 and the moment of inertia about the x axis is $856(10^6)$ mm^4. Determine the area A.

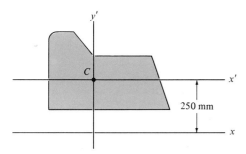

Prob. 10–2

10–3. Determine the moment of inertia of the shaded area about the x axis.

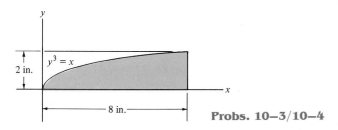

Probs. 10–3/10–4

*10–4. Determine the moment of inertia of the shaded area about the y axis.

10–5. Determine the moment of inertia of the shaded area about the x axis.

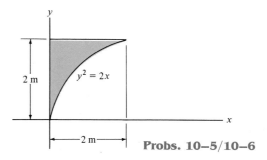

Probs. 10–5/10–6

10–6. Determine the moment of inertia of the shaded area about the y axis.

10–7. Determine the moment of inertia of the shaded area about the x axis.

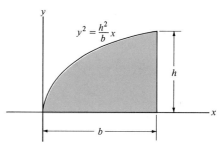

Probs. 10–7/10–8

*10–8. Determine the moment of inertia of the shaded area about the y axis.

10–9. Determine the moment of inertia of the shaded area about the y axis.

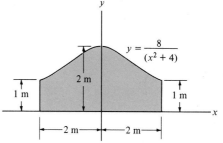

Prob. 10–9

441

10–10. Determine the moment of inertia of the shaded area about the x axis.

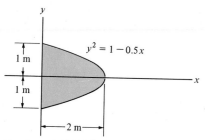

Probs. 10–10/10–11

10–11. Determine the moment of inertia of the shaded area about the y axis.

*10–12. Determine the moment of inertia of the shaded area about the x axis.

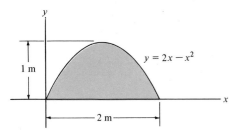

Probs. 10–12/10–13

10–13. Determine the moment of inertia of the shaded area about the y axis.

10–14. Determine the moment of inertia of the shaded area about the x axis.

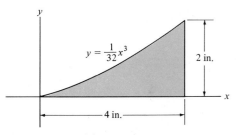

Probs. 10–14/10–15

10–15. Determine the moment of inertia of the shaded area about the y axis.

*10–16. Determine the moments of inertia I_x and I_y of the shaded area.

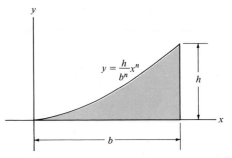

Prob. 10–16

10–17. Determine the moments of inertia I_x and I_y of the shaded area.

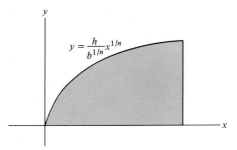

Prob. 10–17

10–18. Determine the moment of inertia of the triangular area about (a) the x axis, and (b) the centroidal x' axis.

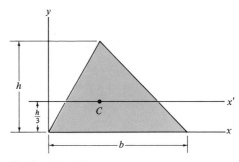

Prob. 10–18

10–19. Determine the moment of inertia of the shaded area about the x axis.

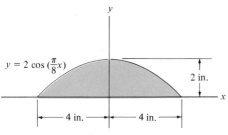

$y = 2 \cos\left(\frac{\pi}{8}x\right)$

2 in.

4 in. ——— 4 in.

Probs. 10–19/10–20

*10–20. Determine the moment of inertia of the shaded area about the y axis.

10–21. Determine the moment of inertia of the area about the x axis. Solve the problem in two ways, using rectangular differential elements: (a) having a thickness dx, and (b) having a thickness dy.

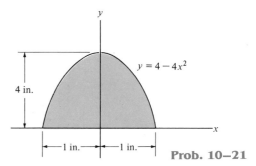

$y = 4 - 4x^2$

4 in.

1 in. ——— 1 in.

Prob. 10–21

10.5 Moments of Inertia for Composite Areas

A composite area consists of a series of connected "simpler" area shapes, such as semicircles, rectangles, and triangles. Provided the moment of inertia of each of these shapes is known or can be computed about a common axis, then the moment of inertia of the composite area equals the *algebraic sum* of the moments of inertia of all its composite parts.

PROCEDURE FOR ANALYSIS

The following procedure provides a method for determining the moment of inertia of a composite area about a reference axis.

Composite Parts. Using a sketch, divide the area into its composite parts and indicate the perpendicular distance from the *centroid* of each part to the reference axis.

Parallel-Axis Theorem. The moment of inertia of each part should be computed about its centroidal axis, which is parallel to the reference axis. For the calculation use the table given on the inside back cover of this book. If the centroidal axis does not coincide with the reference axis, the parallel-axis theorem, $I = \bar{I} + Ad^2$, should be used to determine the moment of inertia of the part about the reference axis.

Summation. The moment of inertia of the entire area about the reference axis is determined by summing the results of its composite parts. In particular, if a composite part has a "hole," the moment of inertia for the composite is found by "subtracting" the moment of inertia for the hole from the moment of inertia of the entire area including the hole.

443

Example 10–5

Compute the moment of inertia of the composite area shown in Fig. 10–9a about the x axis.

(a)   (b)

Fig. 10–9

SOLUTION

Composite Parts. The composite area is obtained by *subtracting* the circle from the rectangle as shown in Fig. 10–9b. The centroid of each area is located in the figure.

Parallel-Axis Theorem. The moments of inertia about the x axis are computed using the parallel-axis theorem and the data contained in the table on the inside back cover.

Circle

$$I_x = \bar{I}_{x'} + A d_y^2$$

$$= \frac{1}{4}\pi(25)^4 + \pi(25)^2(75)^2 = 11.4(10^6) \text{ mm}^4$$

Rectangle

$$I_x = \bar{I}_{x'} + A d_y^2$$

$$= \frac{1}{12}(100)(150)^3 + (100)(150)(75)^2 = 112.5(10^6) \text{ mm}^4$$

Summation. The moment of inertia for the composite area is thus

$$I_x = -11.4(10^6) + 112.5(10^6)$$

$$= 101(10^6) \text{ mm}^4 \qquad\qquad Ans.$$

Example 10–6

Compute the moments of inertia of the beam's cross-sectional area shown in Fig. 10–10a about the x and y centroidal axes.

(a)

SOLUTION

Composite Parts. The cross section can be considered as three composite rectangular areas A, B, and D shown in Fig. 10–10b. For the calculation, the centroid of each of these rectangles is located in the figure.

Parallel-Axis Theorem. From the table on the inside back cover, or Example 10–1, the moment of inertia of a rectangle about its centroidal axis is $\bar{I} = \frac{1}{12}bh^3$. Hence, using the parallel-axis theorem for rectangles A and D, the computations are as follows:

Rectangle A

$$I_x = \bar{I}_{x'} + Ad_y^2 = \frac{1}{12}(100)(300)^3 + (100)(300)(200)^2$$

$$= 1.425(10^9) \text{ mm}^4$$

$$I_y = \bar{I}_{y'} + Ad_x^2 = \frac{1}{12}(300)(100)^3 + (100)(300)(250)^2$$

$$= 1.90(10^9) \text{ mm}^4$$

Rectangle B

$$I_x = \frac{1}{12}(600)(100)^3 = 0.05(10^9) \text{ mm}^4$$

$$I_y = \frac{1}{12}(100)(600)^3 = 1.80(10^9) \text{ mm}^4$$

(b)

Fig. 10–10

Rectangle D

$$I_x = \bar{I}_{x'} + Ad_y^2 = \frac{1}{12}(100)(300)^3 + (100)(300)(200)^2$$

$$= 1.425(10^9) \text{ mm}^4$$

$$I_y = \bar{I}_{y'} + Ad_x^2 = \frac{1}{12}(300)(100)^3 + (100)(300)(250)^2$$

$$= 1.90(10^9) \text{ mm}^4$$

Summation. The moments of inertia for the entire cross section are thus

$$I_x = 1.425(10^9) + 0.05(10^9) + 1.425(10^9)$$

$$= 2.90(10^9) \text{ mm}^4 \qquad \textit{Ans.}$$

$$I_y = 1.90(10^9) + 1.80(10^9) + 1.90(10^9)$$

$$= 5.60(10^9) \text{ mm}^4 \qquad \textit{Ans.}$$

PROBLEMS

10–22. Determine the moment of inertia of the beam's cross-sectional area about the x axis.

Probs. 10–22/10–23

10–23. Determine the moment of inertia of the beam's cross-sectional area about the y axis.

*10–24. The beam is constructed from the two channels and two cover plates. If each channel has a cross-sectional area of $A_c = 11.8 \text{ in}^2$ and a moment of inertia about a horizontal axis passing through its own centroid, C_c, of $(\bar{I}_{\bar{x}})_{C_c} = 349 \text{ in}^4$, determine the moment of inertia of the beam about the x axis.

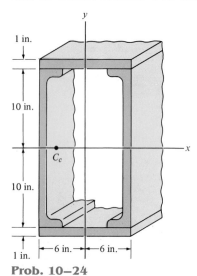

Prob. 10–24

10–25. Determine the radius of gyration k_x of the column's cross-sectional area.

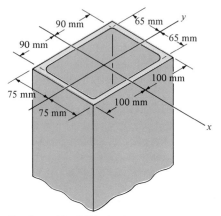

Probs. 10–25/10–26

10–26. Determine the radius of gyration k_y of the column's cross-sectional area.

10–27. Determine the distance $\bar{x}$ to the centroid for the beam's cross-sectional area, then find $I_{\bar{y}}$.

Probs. 10–27/10–28

*10–28. Determine the moment of inertia I_x.

10–29. Determine the moments of inertia I_x and I_y of the Z-section. The origin of coordinates is at the centroid C.

Prob. 10–29

10–30. The composite beam consists of a wide-flange beam and a cover plate welded together as shown. Determine the moment of inertia of the cross-sectional area with respect to the $\bar{x}\bar{x}$ centroidal axis. $\bar{y} = 77.8$ mm.

Prob. 10–30

10–31. Determine the distance $\bar{x}$ to the centroid for the beam's cross-sectional area, then find $\bar{I}_{y'}$.

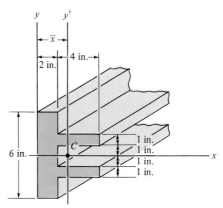

Probs. 10–31/10–32

*10–32. Determine the moment of inertia I_x.

10–33. Determine the distance $\bar{y}$ to the centroid for the beam's cross-sectional area; then determine the moment of inertia about the x' axis.

Probs. 10–33/10–34

10–34. Determine the moment of inertia of the beam's cross-sectional area about the y axis.

10–35. Determine the moment of inertia of the parallelogram about the x axis.

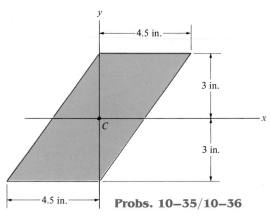

Probs. 10–35/10–36

*10–36. Determine the moment of inertia of the parallelogram about the y axis.

10–37. Determine $\bar{y}$, which locates the centroid C of the *wing channel*, and then determine the moment of inertia $\bar{I}_{x'}$ about the centroidal x' axis. Neglect the effect of rounded corners. The material has a uniform thickness of 0.5 in.

Prob. 10–37

10–38. Compute the moments of inertia $\bar{I}_x$ and $\bar{I}_y$ for the shaded area about the x and y axes.

Probs. 10–38/10–39/10–40

10–39. Determine the distance $\bar{y}$ to the centroid C of the beam's cross-sectional area and then compute the moment of inertia $\bar{I}_{x'}$ about the x' axis.

*10–40. Determine the distance $\bar{x}$ to the centroid C of the beam's cross-sectional area and then compute the moment of inertia $\bar{I}_{y'}$ about the y' axis.

10–41. Compute the polar radius of gyration, k_O, for the ring and show that for small thicknesses, $t = r_2 - r_1$, k_O is equal to the mean radius r_m of the ring.

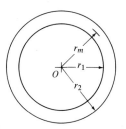

Prob. 10–41

10–42. The assembly consists of two pipes and a plate. Determine the inner radii r of the pipes so that the moment of inertia of the assembly about the x axis is $I_x = 527$ in^4.

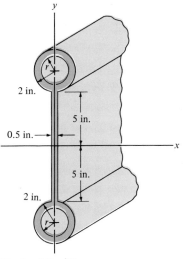

Prob. 10–42

10–43. Determine the moment of inertia of the beam's cross-sectional area about the x axis.

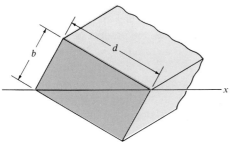

Prob. 10–43

*10–44. Determine the moment of inertia of the beam's cross-sectional area about the x axis which passes through the centroid C.

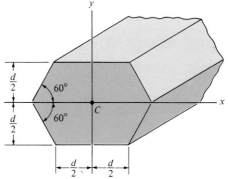

Probs. 10–44/10–45

10–45. Determine the moment of inertia of the beam's cross-sectional area about the y axis which passes through the centroid C.

*10.6 Product of Inertia for an Area

In general, the moment of inertia for an area is different for every axis about which it is computed. In some applications of structural design it is necessary to know the orientation of those axes which give, respectively, the maximum and minimum moments of inertia for the area. The method for determining this is discussed in Sec. 10.7. To use this method, however, one must first compute the product of inertia for the area as well as its moments of inertia for the given x, y axes.

The product of inertia for an element of area located at point (x, y) as shown in Fig. 10–11 is defined as $dI_{xy} = xy\, dA$. Thus, for the entire area A, the *product of inertia* is

$$I_{xy} = \int_A xy\, dA \qquad (10\text{–}7)$$

If the element of area chosen has a differential size in two directions, as shown in Fig. 10–11, a double integration must be performed to evaluate I_{xy}. Most often, however, it is easier to choose an element having a differential size or thickness in only one direction, in which case the evaluation requires only a single integration (see Example 10–7).

Fig. 10–11

449

Fig. 10–12

Fig. 10–13

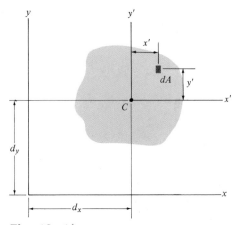

Fig. 10–14

Like the moment of inertia, the product of inertia has units of length raised to the fourth power, e.g., m^4, mm^4 or ft^4, in^4. However, since x or y may be a negative quantity, while the element of area is always positive, the product of inertia may be positive, negative, or zero, depending upon the location and orientation of the coordinate axes. For example, the product of inertia I_{xy} for an area will be *zero* if either the x or y axis is an axis of *symmetry* for the area. To show this, consider the shaded area in Fig. 10–12, where for every element dA located at point (x, y) there is a corresponding element dA located at $(x, -y)$. Since the products of inertia for these elements are respectively $xy\ dA$ and $-xy\ dA$, then the algebraic sum or integration of all the elements that are chosen in this way will cancel each other. Consequently, the product of inertia for the total area becomes zero. It also follows from the definition of I_{xy} that the "sign" of this quantity depends upon the quadrant where the area is located, Fig. 10–13.

Parallel-Axis Theorem

Consider the shaded area shown in Fig. 10–14, where x' and y' represent a set of axes passing through the *centroid* of the area, and x and y represent a corresponding set of parallel axes. Since the product of inertia of dA with respect to the x and y axes is $dI_{xy} = (x' + d_x)(y' + d_y)\ dA$, then for the entire area,

$$I_{xy} = \int_A (x' + d_x)(y' + d_y)\ dA$$

$$= \int_A x'y'\ dA + d_x \int_A y'\ dA + d_y \int_A x'\ dA + d_x d_y \int_A dA$$

The first term on the right represents the product of inertia of the area with respect to the centroidal axis, $\bar{I}_{x'y'}$. The integrals in the second and third terms are zero since the moments of the area are taken about the centroidal axis. Realizing that the fourth integral represents the total area A, we therefore have as the final result,

$$\boxed{I_{xy} = \bar{I}_{x'y'} + Ad_x\, d_y} \qquad (10\text{–}8)$$

The similarity between this equation and the parallel-axis theorem for moments of inertia should be noted. In particular, it is important that the *algebraic signs* for d_x and d_y be maintained when applying Eq. 10–8. As illustrated in Example 10–8, the parallel-axis theorem finds important application in determining the product of inertia of a *composite area* with respect to a set of x, y axes.

Example 10–7

Determine the product of inertia I_{xy} of the triangle shown in Fig. 10–15a.

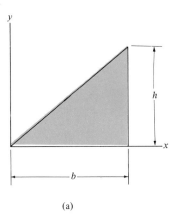

(a)

SOLUTION I

Consider the differential element that has a thickness dx and area $dA = y\,dx$, Fig. 10–15b. The product of inertia of the element about the x, y axes is determined using the parallel-axis theorem.

$$dI_{xy} = d\bar{I}_{x'y'} + dA\,\tilde{x}\,\tilde{y}$$

where $(\tilde{x}, \tilde{y})$ locates the *centroid* of the element or the origin of the x', y' axes. Since $dI_{x'y'} = 0$, due to symmetry, and $\tilde{x} = x$, $\tilde{y} = y/2$, then

$$dI_{xy} = 0 + (y\,dx)x\left(\frac{y}{2}\right) = \left(\frac{h}{b}x\,dx\right)x\left(\frac{h}{2b}x\right)$$

$$= \frac{h^2}{2b^2}x^3\,dx$$

Integrating with respect to x from $x = 0$ to $x = b$ yields

$$I_{xy} = \frac{h^2}{2b^2}\int_0^b x^3\,dx = \frac{b^2h^2}{8} \qquad \textit{Ans.}$$

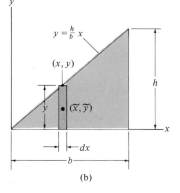

(b)

SOLUTION II

Consider a differential element that has a thickness dy and area $dA = (b - x)\,dy$, as shown in Fig. 10–15c. The *centroid* is located at point $\tilde{x} = x + (b - x)/2 = (b + x)/2$, $\tilde{y} = y$, so that the product of inertia of the element becomes

$$dI_{xy} = d\bar{I}_{x'y'} + dA\,\tilde{x}\,\tilde{y}$$

$$= 0 + (b - x)\,dy\left(\frac{b + x}{2}\right)y$$

$$= \left(b - \frac{b}{h}y\right)dy\left(\frac{b + \frac{b}{h}y}{2}\right)y = \frac{1}{2}y\left(b^2 - \frac{b^2}{h^2}y^2\right)dy$$

Integrating with respect to y from $y = 0$ to $y = h$ yields

$$I_{xy} = \frac{1}{2}\int_0^h y\left(b^2 - \frac{b^2}{h^2}y^2\right)dy = \frac{b^2h^2}{8} \qquad \textit{Ans.}$$

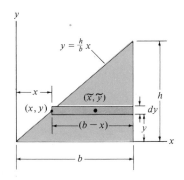

Fig. 10–15

(c)

Example 10–8

Compute the product of inertia of the beam's cross-sectional area, shown in Fig. 10–16a, about the x and y centroidal axes.

Fig. 10–16

SOLUTION

As in Example 10–6, the cross section can be considered as three composite rectangular areas A, B, and D, Fig. 10–16b. The coordinates for the centroid of each of these rectangles are shown in the figure. Due to symmetry, the product of inertia of *each rectangle* is *zero* about a set of x', y' axes that pass through the rectangle's centroid. Hence, application of the parallel-axis theorem to each of the rectangles yields

Rectangle A

$$I_{xy} = \bar{I}_{x'y'} + A d_x d_y$$
$$= 0 + (300)(100)(-250)(200)$$
$$= -1.50(10^9) \text{ mm}^4$$

Rectangle B

$$I_{xy} = \bar{I}_{x'y'} + A d_x d_y$$
$$= 0 + 0$$
$$= 0$$

Rectangle D

$$I_{xy} = \bar{I}_{x'y'} + A d_x d_y$$
$$= 0 + (300)(100)(250)(-200)$$
$$= -1.50(10^9) \text{ mm}^4$$

The product of inertia for the entire cross section is thus

$$I_{xy} = -1.50(10^9) + 0 - 1.50(10^9) = -3.00(10^9) \text{ mm}^4 \qquad Ans.$$

*10.7 Moments of Inertia for an Area About Inclined Axes

In structural mechanics, it is sometimes necessary to calculate the moments and product of inertia I_u, I_v, and I_{uv} for an area with respect to a set of inclined u and v axes when the values for θ, I_x, I_y, and I_{xy} are *known*. From Fig. 10–17 the x, y and u, v coordinates for the area element dA may be related by using the *transformation equations*

$$u = x \cos \theta + y \sin \theta$$
$$v = y \cos \theta - x \sin \theta$$

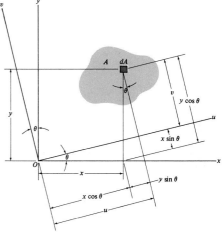

Using these equations, the moments and product of inertia of dA about the u and v axes become

$$dI_u = v^2\, dA = (y \cos \theta - x \sin \theta)^2\, dA$$
$$dI_v = u^2\, dA = (x \cos \theta + y \sin \theta)^2\, dA$$
$$dI_{uv} = uv\, dA = (x \cos \theta + y \sin \theta)(y \cos \theta - x \sin \theta)\, dA$$

Expanding each expression and integrating, realizing that $I_x = \int y^2\, dA$, $I_y = \int x^2\, dA$, and $I_{xy} = \int xy\, dA$, we obtain

Fig. 10–17

$$I_u = I_x \cos^2 \theta + I_y \sin^2 \theta - 2I_{xy} \sin \theta \cos \theta$$
$$I_v = I_x \sin^2 \theta + I_y \cos^2 \theta + 2I_{xy} \sin \theta \cos \theta$$
$$I_{uv} = I_x \sin \theta \cos \theta - I_y \sin \theta \cos \theta + I_{xy}(\cos^2 \theta - \sin^2 \theta)$$

These equations may be simplified by using the trigonometric identities $\sin 2\theta = 2 \sin \theta \cos \theta$ and $\cos 2\theta = \cos^2 \theta - \sin^2 \theta$, in which case

$$I_u = \frac{I_x + I_y}{2} + \frac{I_x - I_y}{2} \cos 2\theta - I_{xy} \sin 2\theta$$

$$I_v = \frac{I_x + I_y}{2} - \frac{I_x - I_y}{2} \cos 2\theta + I_{xy} \sin 2\theta \qquad (10\text{–}9)$$

$$I_{uv} = \frac{I_x - I_y}{2} \sin 2\theta + I_{xy} \cos 2\theta$$

Note that if the first and second equations are added together, the polar moment of inertia about the z axis passing through point O is *independent* of the orientation of the u and v axes, i.e.,

$$J_O = I_u + I_v = I_x + I_y$$

Principal Moments of Inertia

From Eqs. 10–9, it may be seen that I_u, I_v, and I_{uv} depend upon the angle of inclination, θ, of the u, v axes. We will now determine the orientation of the u, v axes about which the moments of inertia for the area, I_u and I_v, are maximum and minimum. This particular set of axes is called the *principal axes* of the area, and the corresponding moments of inertia with respect to

453

these axes are called the *principal moments of inertia*. In general, there is a set of principal axes for every chosen origin O, although in structural mechanics the area's centroid is an important location for O.

The angle $\theta = \theta_p$, which defines the orientation of the principal axes for the area, may be found by differentiating the first of Eqs. 10–9 with respect to θ and setting the result equal to zero. Thus,

$$\frac{dI_u}{d\theta} = -2\left(\frac{I_x - I_y}{2}\right)\sin 2\theta - 2I_{xy}\cos 2\theta = 0$$

Therefore, at $\theta = \theta_p$,

$$\tan 2\theta_p = \frac{-I_{xy}}{\dfrac{I_x - I_y}{2}} \tag{10-10}$$

This equation has two roots, θ_{p_1} and θ_{p_2}, which specify the inclination of the principal axes. Because of the nature of the tangent, the values of $2\theta_{p_1}$ and $2\theta_{p_2}$ are 180° apart, so that θ_{p_1} and θ_{p_2} are 90° apart. The sine and cosine of $2\theta_{p_1}$ and $2\theta_{p_2}$ can be obtained from the triangles shown in Fig. 10–18, which are based upon Eq. 10–10.

For θ_{p_1}

$$\sin 2\theta_{p_1} = -I_{xy}\bigg/\sqrt{\left(\frac{I_x - I_y}{2}\right)^2 + I_{xy}^2}$$

$$\cos 2\theta_{p_1} = \left(\frac{I_x - I_y}{2}\right)\bigg/\sqrt{\left(\frac{I_x - I_y}{2}\right)^2 + I_{xy}^2}$$

For θ_{p_2}

$$\sin 2\theta_{p_2} = I_{xy}\bigg/\sqrt{\left(\frac{I_x - I_y}{2}\right)^2 + I_{xy}^2}$$

$$\cos 2\theta_{p_2} = -\left(\frac{I_x - I_y}{2}\right)\bigg/\sqrt{\left(\frac{I_x - I_y}{2}\right)^2 + I_{xy}^2}$$

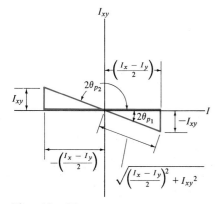

Fig. 10–18

If these two sets of trigonometric relations are substituted into the first or second of Eqs. 10–9 and simplified, the result is

$$I_{\substack{max \\ min}} = \frac{I_x + I_y}{2} \pm \sqrt{\left(\frac{I_x - I_y}{2}\right)^2 + I_{xy}^2} \tag{10-11}$$

Depending upon the sign chosen, this result gives the maximum or minimum moment of inertia for the area. Furthermore, if the above trigonometric relations for θ_{p_1} and θ_{p_2} are substituted into the third of Eqs. 10–9, it may be seen that $I_{uv} = 0$; that is, the *product of inertia with respect to the principal axes is zero*. Since it was indicated in Sec. 10.6 that the product of inertia is zero with respect to any symmetrical axis, it therefore follows that *any symmetrical axis represents a principal axis of inertia for the area*.

Example 10–9

Determine the principal moments of inertia for the beam's cross-sectional area shown in Fig. 10–19a with respect to an axis passing through the centroid.

(a)

SOLUTION

The moments and product of inertia of the cross section with respect to the x, y axes have been computed in Examples 10–6 and 10–8. The results are

$$I_x = 2.90(10^9) \text{ mm}^4 \qquad I_y = 5.60(10^9) \text{ mm}^4 \qquad I_{xy} = -3.00(10^9) \text{ mm}^4$$

Using Eq. 10–10, the angles of inclination of the principal axes u and v are

$$\tan 2\theta_p = \frac{-I_{xy}}{\dfrac{I_x - I_y}{2}} = \frac{3.00(10^9)}{\dfrac{2.90(10^9) - 5.60(10^9)}{2}} = -2.22$$

$$2\theta_{p_1} = -65.8° \qquad \text{and} \qquad 2\theta_{p_2} = 114.2°$$

Thus, as shown in Fig. 10–19b,

$$\theta_{p_1} = -32.9° \qquad \text{and} \qquad \theta_{p_2} = 57.1°$$

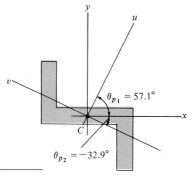

(b)

Fig. 10–19

The principal moments of inertia with respect to the u and v axes are determined by using Eq. 10–11. Hence,

$$I_{\substack{\text{max} \\ \text{min}}} = \frac{I_x + I_y}{2} \pm \sqrt{\left(\frac{I_x - I_y}{2}\right)^2 + I_{xy}^2}$$

$$= \frac{2.90(10^9) + 5.60(10^9)}{2} \pm \sqrt{\left[\frac{2.90(10^9) - 5.60(10^9)}{2}\right]^2 + [-3.00(10^9)]^2}$$

$$= 4.25(10^9) \pm 3.29(10^9)$$

or

$$(I_u)_{\text{max}} = 7.54(10^9) \text{ mm}^4 \qquad (I_v)_{\text{min}} = 0.960(10^9) \text{ mm}^4 \qquad \textit{Ans.}$$

Specifically, the maximum moment of inertia, $(I_u)_{\text{max}} = 7.54(10^9) \text{ mm}^4$, occurs with respect to the selected u axis, since *by inspection* most of the cross-sectional area is farthest away from this axis. Or stated in a general manner, $(I_u)_{\text{max}}$ occurs about the u axis since it is located within $\pm45°$ of the axis (x or y) which has the largest I (I_x or I_y). Also, this may be concluded mathematically by substituting the data with $\theta = 57.1°$ into the first of Eqs. 10–9.

*10.8 Mohr's Circle for Moments of Inertia

Equations 10–9 to 10–11 have a graphical solution that is convenient to use and generally easy to remember. Squaring the first and third of Eqs. 10–9 and adding, it is found that

$$\left(I_u - \frac{I_x + I_y}{2}\right)^2 + I_{uv}^2 = \left(\frac{I_x - I_y}{2}\right)^2 + I_{xy}^2 \qquad (10\text{–}12)$$

In a given problem, I_u and I_{uv} are *variables,* and I_x, I_y, and I_{xy} are *known constants*. Thus, Eq. 10–12 may be written in compact form as

$$(I_u - a)^2 + I_{uv}^2 = R^2$$

When this equation is plotted, the resulting graph represents a *circle* of radius

$$R = \sqrt{\left(\frac{I_x - I_y}{2}\right)^2 + I_{xy}^2}$$

having its center located at point $(a, 0)$, where $a = (I_x + I_y)/2$. The circle so constructed is called *Mohr's circle,* named after the German engineer Otto Mohr (1835–1918).

PROCEDURE FOR ANALYSIS

There are several methods for plotting Mohr's circle as defined by Eq. 10–12. The main purpose in using the circle here is to have a convenient means for transforming I_x, I_y, and I_{xy} into the principal moments of inertia. The following procedure provides a method for doing this.

Compute I_x, I_y, I_{xy}. Establish the x, y axes for the area, with the origin located at the point P of interest, and determine I_x, I_y, and I_{xy}, Fig. 10–20a.

Construct the Circle. Construct a rectangular coordinate system such that the abscissa represents the moment of inertia I, and the ordinate represents the product of inertia I_{xy}, Fig. 10–20b. Determine the center of the circle, O, which is located at a distance $(I_x + I_y)/2$ from the origin, and plot the "controlling point" A having coordinates (I_x, I_{xy}). By definition, I_x is always positive, whereas I_{xy} will be either positive or negative. Connect the controlling point A with the center of the circle, and determine the distance OA by trigonometry. This distance represents the radius of the circle, Fig. 10–20b. Finally, draw the circle.

Principal Moments of Inertia. The points where the circle intersects the abscissa give the values of the principal moments of inertia $(I_v)_{min}$ and $(I_u)_{max}$. Notice that the *product of inertia will be zero at these points*, Fig. 10–20b.

Principal Axes. To find the direction of the major principal axis, determine by trigonometry the angle 2α, *measured from the radius OA to the positive I axis*, Fig. 10–20b. This angle represents twice the angle from the x axis of the area in question to the axis of maximum moment of inertia $(I_u)_{max}$, Fig. 10–20a. Both the angle on the circle, 2α, and the angle on the area, α, *must be measured in the same sense,* as shown in Fig. 10–20. The axis for minimum moment of inertia $(I_v)_{min}$ is perpendicular to the axis for $(I_u)_{max}$.

Using trigonometry, the above procedure may be verified to be in accordance with the equations developed in Sec. 10.7.

(a)

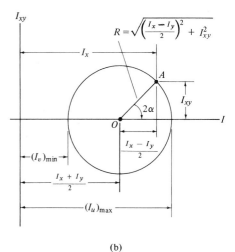

(b)

Fig. 10–20

Example 10-10

Using Mohr's circle, determine the principal moments of inertia for the beam's cross-sectional area, shown in Fig. 10–21a, with respect to an axis passing through the centroid.

Fig. 10-21

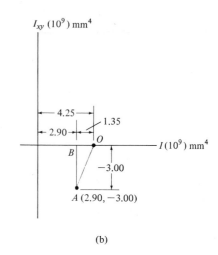

(a)

(b)

SOLUTION

Compute I_x, I_y, I_{xy}. The moments of inertia and the product of inertia have been determined in Examples 10–6 and 10–8 with respect to the x, y axes shown in Fig. 10–21a. The results are $I_x = 2.90(10^9)$ mm^4, $I_y = 5.60(10^9)$ mm^4, and $I_{xy} = -3.00(10^9)$ mm^4.

Construct the Circle. The I and I_{xy} axes are shown in Fig. 10–21b. The center of the circle, O, lies at a distance $(I_x + I_y)/2 = (2.90 + 5.60)/2 = 4.25$ from the origin. When the controlling point $A(2.90, -3.00)$ is connected to point O, the radius OA is determined from the triangle OBA using the Pythagorean theorem.

$$OA = \sqrt{(1.35)^2 + (-3.00)^2} = 3.29$$

The circle is constructed in Fig. 10–21c.

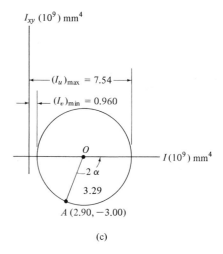

(c)

Principal Moments of Inertia. The circle intersects the I axis at points (7.54, 0) and (0.960, 0). Hence,

$$(I_u)_{max} = 7.54(10^9) \text{ mm}^4 \qquad \textit{Ans.}$$
$$(I_v)_{min} = 0.960(10^9) \text{ mm}^4 \qquad \textit{Ans.}$$

Principal Axes. As shown in Fig. 10–21c, the angle 2α is determined from the circle by measuring counterclockwise from OA to the direction of the *positive I* axis. Hence,

$$2\alpha = 180° - \sin^{-1}\left(\frac{|BA|}{|OA|}\right) = 180° - \sin^{-1}\left(\frac{3.00}{3.29}\right) = 114.2°$$

The principal axis for $(I_u)_{max} = 7.54(10^9)$ mm^4 is therefore oriented at an angle $\alpha = 57.1°$, measured *counterclockwise,* from the *positive x* axis to the *positive u* axis. The v axis is perpendicular to this axis. The results are shown in Fig. 10–21d.

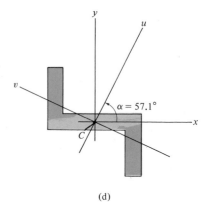

(d)

PROBLEMS

10–46. Determine the product of inertia of the shaded area with respect to the x and y axes.

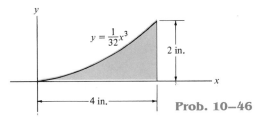

$y = \frac{1}{32}x^3$

2 in.

4 in.

Prob. 10–46

10–47. Determine the product of inertia of the shaded area with respect to the x and y axes.

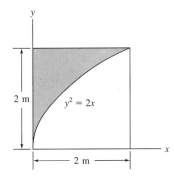

2 m

$y^2 = 2x$

2 m

Prob. 10–47

*10–48. Determine the product of inertia of the parabolic area with respect to the x and y axes.

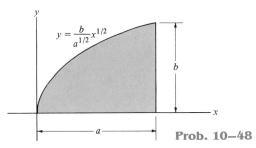

$y = \frac{b}{a^{1/2}}x^{1/2}$

b

a

Prob. 10–48

10–49. Determine the product of inertia of the shaded area with respect to the x and y axes.

2 in.

$y^3 = x$

8 in.

Prob. 10–49

10–50. Determine the product of inertia of the shaded area with respect to the x and y axes.

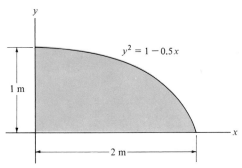

$y^2 = 1 - 0.5x$

1 m

2 m

Prob. 10–50

■10–51. Determine the product of inertia of the area with respect to the x and y axes. Use Simpson's rule to evaluate the integral.

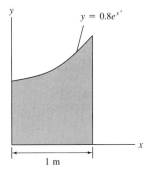

$y = 0.8e^{x^2}$

1 m

Prob. 10–51

*10–52. Determine the product of inertia of the shaded area with respect to the x and y axes.

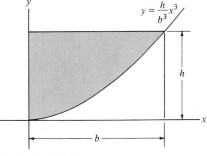

$y = \frac{h}{b^3}x^3$

h

b

Prob. 10–52

10–53. Determine the product of inertia of the quarter elliptical area with respect to the x and y axes.

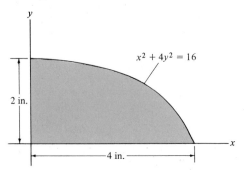

Prob. 10–53

10–54. Determine the product of inertia of the shaded area with respect to the x and y axes.

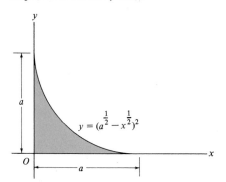

Prob. 10–54

10–55. Determine the product of inertia of the quarter circular area with respect to the x and y axes. Then use the parallel-axes theorem to determine the product of inertia about the centroidal x' and y' axes.

Prob. 10–55

*10–56. Determine the product of inertia of the area with respect to the x and y axes.

Prob. 10–56

10–57. Locate the position $\bar{x}$, $\bar{y}$ for the centroid C of the cross-sectional area and then compute the product of inertia with respect to the x' and y' axes.

Prob. 10–57

10–58. Determine the product of inertia of the cross-sectional area with respect to the x and y axes that have their origin located at the centroid C.

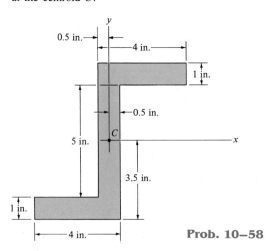

Prob. 10–58

10–59. If $\bar{x} = 61.59$ mm and $\bar{y} = 80.68$ mm, compute the product of inertia of the shaded area with respect to the x' and y' axes.

Prob. 10–59

*10–60. Determine the product of inertia of the cross-sectional area with respect to the x and y axes that have their origin located at the centroid C.

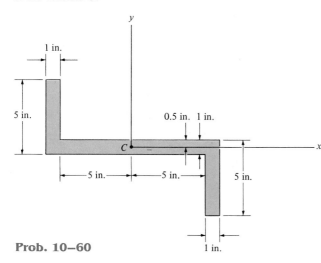

Prob. 10–60

10–61. Determine the moments of inertia I_u and I_v and the product of inertia I_{uv} for the rectangular area. The u and v axes pass through the centroid C.

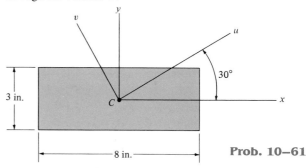

Prob. 10–61

10–62. Determine the distance $\bar{y}$ to the centroid of the area and then calculate the moments of inertia I_u and I_v of the channel section. The u and v axes have their origin at the centroid C. For the calculation, assume all corners to be square.

Prob. 10–62

10–63. Determine the moments of inertia I_u, I_v and I_{uv} of the shaded area.

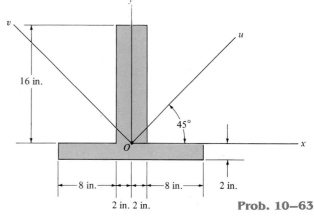

Prob. 10–63

*10–64. Determine the moments of inertia I_u and I_v of the shaded area.

Prob. 10–64

10–65. Determine the moments of inertia I_u and I_v and the product of inertia I_{uv} for the semicircular area.

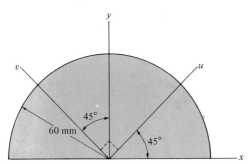

Prob. 10–65

10–66. Determine the directions of the principal axes with origin located at point O, and the principal moments of inertia for the triangular area about these axes.

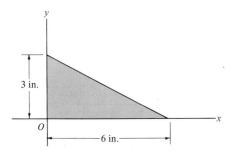

Prob. 10–66

10–67. Determine the principal moments of inertia of the cross-sectional area about the principal axes that have their origin located at the centroid C. Use the equations developed in Sec. 10.7. For the calculation, assume all corners to be square.

Prob. 10–67

*10–68. Determine the principal moments of inertia for the angle's cross-sectional area with respect to a set of principal axes that have their origin located at the centroid C. Use the equations developed in Sec. 10.7. For the calculation, assume all corners to be square.

Prob. 10–68

462

10–69. Determine the distance $\bar{y}$ to the centroid, and the orientation of the principal centroidal axes for the composite area. What are the moments of inertia with respect to these axes?

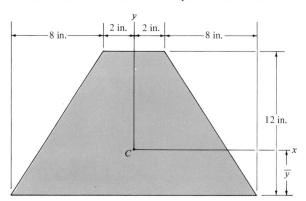

Prob. 10–69

10–70. Solve Prob. 10–66 using Mohr's circle.

10–71. Solve Prob. 10–67 using Mohr's circle.

*10–72. Solve Prob. 10–68 using Mohr's circle.

10–73. Solve Prob. 10–69 using Mohr's circle.

10.9 Mass Moment of Inertia

The mass moment of inertia of a body is a property that measures the resistance of the body to angular acceleration. Since it is used in dynamics to study rotational motion, methods for its calculation will now be discussed.

We define the *mass moment of inertia* as the integral of the "second moment" about an axis of all the differential-size elements of mass dm which compose the body.* For example, consider the rigid body shown in Fig. 10–22. The body's moment of inertia about the z axis is

$$I = \int_m r^2 \, dm \qquad (10\text{–}13)$$

Here the "moment arm" r is the perpendicular distance from the axis to the arbitrary element dm. Since the formulation involves r, the value of I is *unique* for each axis z about which it is computed. The axis which is generally chosen for analysis, however, passes through the body's mass center G. The moment of inertia computed about this axis will be defined as I_G. Note that because r is squared in Eq. 10–13, the mass moment of inertia is always a *positive quantity*. Common units used for its measurement are kg · m² or slug · ft².

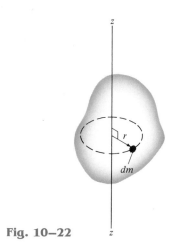

Fig. 10–22

---

* Another property of the body which measures the symmetry of the body's mass with respect to a coordinate system is the mass product of inertia. This property most often applies to the three-dimensional motion of a body and is discussed in *Engineering Mechanics: Dynamics* (Chapter 21).

(a)

(b)

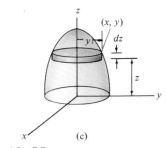

(c)

Fig. 10–23

PROCEDURE FOR ANALYSIS

For integration, we will only consider symmetric bodies having surfaces which are generated by revolving a curve about an axis. An example of such a body which is generated about the z axis is shown in Fig. 10–23.

If the body consists of material having a variable density, $\rho = \rho(x, y, z)$, the elemental mass dm of the body may be expressed in terms of its density and volume as $dm = \rho \, dV$. Substituting dm in Eq. 10–13, the body's moment of inertia is then computed using *volume elements* for integration, i.e.,

$$I = \int_V r^2 \rho \, dV \qquad (10\text{–}14)$$

In the special case of ρ being a *constant,* this term may be factored out of the integral and the integration is then purely a function of geometry,

$$I = \rho \int_V r^2 \, dV \qquad (10\text{–}15)$$

When the elemental volume chosen for integration has differential sizes in all three directions, e.g., $dV = dx \, dy \, dz$, Fig. 10–23a, the moment of inertia of the body must be computed using "triple integration." The integration process can, however, be simplified to a *single integration* provided the chosen elemental volume has a differential size or thickness in only *one direction*. Shell or disk elements are often used for this purpose.

Shell Element. If a *shell element* having a height z, radius $r = y$, and thickness dy is chosen for integration, Fig. 10–23b, then the volume is $dV = (2\pi y)(z) \, dy$. This element may be used in Eq. 10–14 or 10–15 for computing the moment of inertia I_z of the body about the z axis, since the *entire element*, due to its "thinness," lies at the *same* perpendicular distance $r = y$ from the z axis (see Example 10–11).

Disk Element. If a disk element having a radius $r = y$ and a thickness dz is chosen for integration, Fig. 10–23c, then the volume is $dV = (\pi y^2) \, dz$. In this case, however, the element is *finite* in the radial direction, and consequently parts of it *do not* all lie at the *same radial distance r* from the z axis. As a result, Eq. 10–14 or 10–15 *cannot* be used to determine I_z. Instead, to perform the integration using this element, it is first necessary to determine the moment of inertia *of the element* about the z axis and then integrate this result (see Example 10–12).

Example 10–11

Determine the moment of inertia of the right circular cylinder shown in Fig. 10–24a about the z axis. The density ρ of the material is constant.

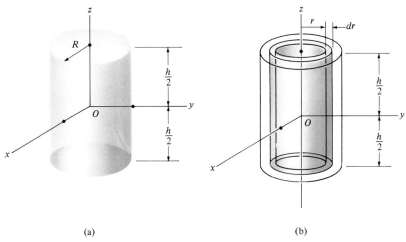

Fig. 10–24

SOLUTION

Shell Element. This problem may be solved using the *shell element* in Fig. 10–24b and single integration. The volume of the element is $dV = (2\pi r)(h)\ dr$, so that the mass is $dm = \rho\ dV = \rho(2\pi hr\ dr)$. Since the *entire element* lies at the same distance r from the z axis, the moment of inertia *of the element* is

$$dI_z = r^2\ dm = \rho 2\pi hr^3\ dr$$

Integrating over the entire region of the cylinder yields

$$I_z = \int_m r^2\ dm = \rho 2\pi h \int_0^R r^3\ dr = \frac{\rho\pi}{2}R^4 h$$

The mass of the cylinder is

$$m = \int_m dm = \rho 2\pi h \int_0^R r\ dr = \rho\pi hR^2$$

so that

$$I_z = \frac{1}{2}mR^2 \qquad\qquad Ans.$$

Example 10–12

A solid is formed by revolving the shaded area shown in Fig. 10–25a about the y axis. If the density of the material is 5 slug/ft^3, determine the moment of inertia about the y axis.

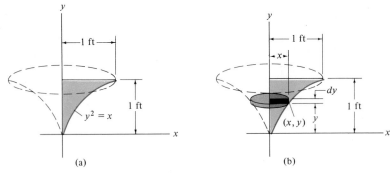

Fig. 10–25

SOLUTION

Disk Element. The moment of inertia will be computed using a *disk element*, as shown in Fig. 10–25b. Here the element intersects the curve at the arbitrary point (x, y) and has a mass

$$dm = \rho \, dV = \rho \, (\pi x^2) \, dy$$

Although all portions of the element are *not* located at the same distance from the y axis, it is still possible to determine the moment of inertia dI_y *of the element* about the y axis. In Example 10–11 it was shown that the moment of inertia of a cylinder about its longitudinal axis is $I = \frac{1}{2}mR^2$, where m and R are the mass and radius of the cylinder. Since the height of the cylinder is not involved in this formula, the moment of inertia of the disk element in Fig. 10–25b is

$$dI_y = \tfrac{1}{2}(dm)x^2 = \tfrac{1}{2}[\rho(\pi x^2) \, dy]x^2$$

Substituting $x = y^2$, $\rho = 5$ slug/ft^3, and integrating with respect to y, from $y = 0$ to $y = 1$ ft, yields the moment of inertia for the entire solid.

$$I_y = \frac{5\pi}{2} \int_0^1 x^4 \, dy = \frac{5\pi}{2} \int_0^1 y^8 \, dy = 0.873 \text{ slug} \cdot \text{ft}^2 \qquad \textit{Ans.}$$

Parallel-Axis Theorem

If the moment of inertia of the body about an axis passing through the body's mass center is known, then the body's moment of inertia may be determined about any other *parallel axis* by using the *parallel-axis theorem*. This theorem can be derived by considering the body shown in Fig. 10–26. The z' axis passes through the body's mass center G, whereas the corresponding *parallel axis z* lies at a constant distance d away. Selecting the differential element of mass dm which is located at point (x', y') and using the Pythagorean theorem, $r^2 = (d + x')^2 + y'^2$, we can express the moment of inertia of the body about the z axis as

$$I = \int_m r^2 \, dm = \int_m [(d + x')^2 + y'^2] \, dm$$

$$= \int_m (x'^2 + y'^2) \, dm + 2d \int_m x' \, dm + d^2 \int_m dm$$

Since $r'^2 = x'^2 + y'^2$, the first integral represents I_G. The second integral equals *zero,* since the z' axis passes through the body's mass center, i.e., $\int x' \, dm = \bar{x} \int dm = 0$ since $\bar{x} = 0$. Finally, the third integral represents the total mass m of the body. Hence, the moment of inertia about the z axis can be written as

$$\boxed{I = I_G + md^2} \tag{10–16}$$

where I_G = moment of inertia about the z' axis passing through the mass center G

m = mass of the body

d = perpendicular distance between the parallel axes

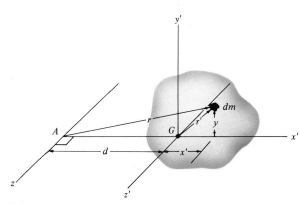

Fig. 10–26

Radius of Gyration

Occasionally, the moment of inertia of a body about a specified axis is reported in handbooks using the *radius of gyration, k*. This value depends only on the geometry of the body. And, when it and the body's mass m are known, the moment of inertia is determined from the equation

$$I = mk^2 \quad \text{or} \quad k = \sqrt{\frac{I}{m}} \tag{10–17}$$

Note the *similarity* between the definition of k in this formula and r in the equation $dI = r^2 \, dm$, which defines the moment of inertia of an elemental mass dm of the body about an axis.

Composite Bodies

If a body is constructed from a number of simple shapes such as disks, spheres, and rods, the moment of inertia of the body about any axis z can be determined by adding algebraically the moments of inertia of all the composite shapes computed about the z axis. Algebraic addition is necessary since a composite part must be considered as a negative quantity if it has already been included within another part—for example, a ''hole'' subtracted from a solid plate. The parallel-axis theorem is needed for the calculations if the center of mass of each composite part does not lie on the z axis. For the calculation, then, $I = \Sigma(I_G + md^2)$, where I_G for each of the composite parts is computed by integration or can be determined from a table, such as the one given on the inside back cover of this book.

Example 10–13

If the plate shown in Fig. 10–27a has a density of 8000 kg/m³ and a thickness of 10 mm, compute its moment of inertia about an axis directed perpendicular to the page and passing through point O.

Fig. 10–27

SOLUTION

The plate consists of two composite parts, the 250-mm-radius disk *minus* a 125-mm-radius disk, Fig. 10–27b. The moment of inertia about O can be determined by computing the moment of inertia of each of these parts about O and then *algebraically* adding the results. The computations are performed by using the parallel-axis theorem in conjunction with the data listed in the table on the inside back cover.

Disk. The moment of inertia of a thin disk about an axis perpendicular to the plane of the disk is $I_G = \frac{1}{2}mr^2$. The mass center of the disk is located at a distance of 0.25 m from point O. Thus,

$$m_d = \rho_d V_d = 8000 \text{ kg/m}^3[\pi(0.25 \text{ m})^2(0.01 \text{ m})] = 15.71 \text{ kg}$$

$$(I_O)_d = \frac{1}{2}m_d r_d^2 + m_d d^2$$

$$= \frac{1}{2}(15.71 \text{ kg})(0.25 \text{ m})^2 + (15.71 \text{ kg})(0.25 \text{ m})^2$$

$$= 1.473 \text{ kg} \cdot \text{m}^2$$

Hole. For the 125-mm-radius disk (hole), we have

$$m_h = \rho_h V_h = 8000 \text{ kg/m}^3[\pi(0.125 \text{ m})^2(0.01 \text{ m})] = 3.93 \text{ kg}$$

$$(I_O)_h = \frac{1}{2}m_h r_h^2 + m_h d^2$$

$$= \frac{1}{2}(3.93 \text{ kg})(0.125 \text{ m})^2 + (3.93 \text{ kg})(0.25 \text{ m})^2$$

$$= 0.276 \text{ kg} \cdot \text{m}^2$$

The moment of inertia of the plate about point O is therefore

$$I_O = (I_O)_d - (I_O)_h$$

$$= 1.473 \text{ kg} \cdot \text{m}^2 - 0.276 \text{ kg} \cdot \text{m}^2$$

$$= 1.20 \text{ kg} \cdot \text{m}^2 \qquad\qquad\qquad \textit{Ans.}$$

Example 10–14

The pendulum consists of two thin rods each having a weight of 10 lb and suspended from point O as shown in Fig. 10–28. Compute the pendulum's moment of inertia about an axis passing through (a) the pin at O, and (b) the mass center G of the pendulum.

SOLUTION

Part (a). Using the table on the inside back cover, the moment of inertia of rod OA about an axis perpendicular to the page and passing through the end point O of the rod is $I_O = \frac{1}{3}ml^2$. Hence,

$$(I_O)_{OA} = \frac{1}{3}ml^2 = \frac{1}{3}\left(\frac{10}{32.2}\right)(2)^2 = 0.414 \text{ slug} \cdot \text{ft}^2$$

Fig. 10–28

The same value may be computed using $I_G = \frac{1}{12}ml^2$ and the parallel-axis theorem; i.e.,

$$(I_O)_{OA} = \frac{1}{12}ml^2 + md^2 = \frac{1}{12}\left(\frac{10}{32.2}\right)(2)^2 + \frac{10}{32.2}(1)^2$$

$$= 0.414 \text{ slug} \cdot \text{ft}^2$$

For rod BC we have

$$(I_O)_{BC} = \frac{1}{12}ml^2 + md^2 = \frac{1}{12}\left(\frac{10}{32.2}\right)(2)^2 + \frac{10}{32.2}(2)^2$$

$$= 1.346 \text{ slug} \cdot \text{ft}^2$$

The moment of inertia of the pendulum is therefore

$$I_O = 0.414 + 1.346 = 1.76 \text{ slug} \cdot \text{ft}^2 \qquad \textit{Ans.}$$

Part (b). The mass center G will be located relative to the pin at O. Assuming this distance to be $\bar{y}$, Fig. 10–28, and using the formula for determining the mass center, we have

$$\bar{y} = \frac{\Sigma \tilde{y}m}{\Sigma m} = \frac{1\left(\dfrac{10}{32.2}\right) + 2\left(\dfrac{10}{32.2}\right)}{\left(\dfrac{10}{32.2}\right) + \left(\dfrac{10}{32.2}\right)} = 1.50 \text{ ft}$$

The moment of inertia I_G may be computed in the same manner as I_O, which requires successive applications of the parallel-axis theorem in order to transfer the moments of inertia of rods OA and BC to G. A more direct solution, however, involves applying the parallel-axis theorem using the result for I_O, i.e.,

$$I_O = I_G + md^2; \qquad 1.76 = I_G + \left(\frac{20}{32.2}\right)(1.50)^2$$

$$I_G = 0.362 \text{ slug} \cdot \text{ft}^2 \qquad \textit{Ans.}$$

PROBLEMS

10–74. Determine the moment of inertia I_y for the slender rod. The rod's density ρ and cross-sectional area A are constant. Express the result in terms of the rod's total mass m.

Prob. 10–74

10–75. Determine the moment of inertia of the thin ring about the z axis. The ring has a mass m.

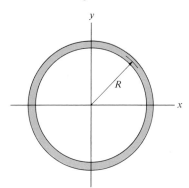

Prob. 10–75

*10–76. The right circular cone is formed by revolving the shaded area around the x axis. Determine the moment of inertia I_x and express the result in terms of the total mass m of the cone. The cone has a constant density ρ.

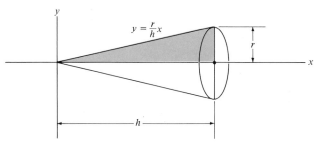

Prob. 10–76

10–77. Determine the moment of inertia I_x and express the result in terms of the total mass m of the sphere. The sphere has a constant density ρ.

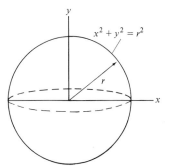

Prob. 10–77

10–78. The paraboloid is formed by revolving the shaded area around the x axis. Determine the radius of gyration k_x. The material has a constant density ρ.

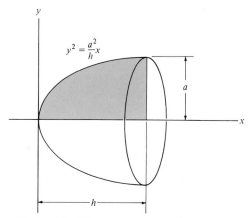

Prob. 10–78

471

10–79. A semiellipsoid is formed by rotating the shaded area about the x axis. Determine the moment of inertia of this solid with respect to the x axis and express the result in terms of the mass m of the solid. The material has a constant density ρ.

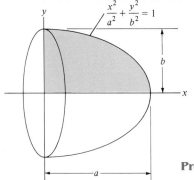

$$\frac{x^2}{a^2} + \frac{y^2}{b^2} = 1$$

Prob. 10–79

10–82. Determine the moment of inertia of the homogeneous triangular prism with respect to the y axis. Express the result in terms of the mass m of the prism. *Hint:* For integration, use thin plate elements parallel to the x-y plane and having a thickness of dz.

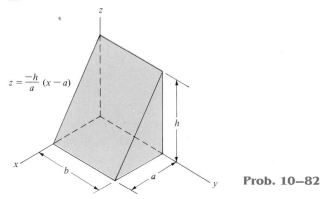

$$z = \frac{-h}{a}(x - a)$$

Prob. 10–82

*10–80. The solid is formed by revolving the shaded area around the x axis. Determine the radius of gyration k_x. The specific weight of the material is $\gamma = 380$ lb/ft^3.

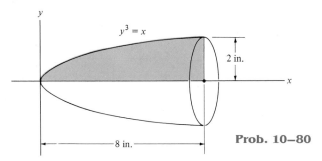

$y^3 = x$

2 in.

8 in.

Prob. 10–80

10–83. The concrete solid is formed by rotating the shaded area about the y axis. Determine the moment of inertia I_y. The specific weight of concrete is $\gamma = 150$ lb/ft^3.

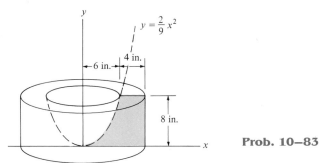

$$y = \frac{2}{9}x^2$$

6 in.   4 in.

8 in.

Prob. 10–83

10–81. The solid is formed by revolving the shaded area around the x axis. Determine the radius of gyration k_x. The density of the material is $\rho = 5$ Mg/m^3.

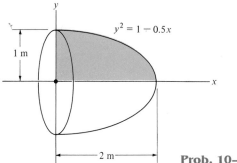

$y^2 = 1 - 0.5x$

1 m

2 m

Prob. 10–81

*10–84. The slender rods have a weight of 3 lb/ft. Determine the moment of inertia of the assembly about an axis perpendicular to the page and passing through point A.

A

2 ft

1 ft

1.5 ft   1.5 ft

Prob. 10–84

10–85. The wheel consists of a thin ring having a mass of 10 kg and four spokes made from slender rods and each having a mass of 2 kg. Determine the wheel's moment of inertia about an axis perpendicular to the page and passing through point A.

Prob. 10–85

10–86. Determine the moment of inertia of the solid steel assembly about the xx axis. Steel has a specific weight of $\gamma_{st} = 490$ lb/ft³.

Prob. 10–86

10–87. Determine the moment of inertia I_z of the frustum of the cone which has a conical depression. The material has a density of 200 kg/m³.

Prob. 10–87

***10–88.** Determine the moment of inertia of the wheel about an axis which is perpendicular to the page and passes through the center of mass G. The material has a specific weight of $\gamma = 90$ lb/ft³.

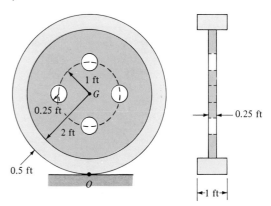

Probs. 10–88/10–89

10–89. Determine the moment of inertia of the wheel about an axis which is perpendicular to the page and passes through point O. The material has a specific weight of $\gamma = 90$ lb/ft³.

10–90. The pendulum consists of two slender rods AB and OC which have a mass of 3 kg/m. The thin plate has a mass of 12 kg/m². Determine the location $\bar{y}$ of the center of mass G of the pendulum, then calculate the moment of inertia of the pendulum about an axis perpendicular to the page and passing through G.

Prob. 10–90

473

10–91. Determine the moment of inertia of the wire triangle about an axis perpendicular to the page and passing through point O. Also, locate the mass center G and determine the moment of inertia about an axis perpendicular to the page and passing through point G. The wire has a mass of 0.3 kg/m. Neglect the size of the ring at O.

Prob. 10–91

*10–92.** The pendulum consists of a disk having a mass of 6 kg and slender rods AB and DC which have a mass of 2 kg/m. Determine the length L of DC so that the center of mass is at the bearing O. What is the moment of inertia of the assembly about an axis perpendicular to the page and passing through point O?

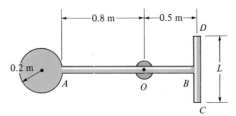

Prob. 10–92

REVIEW PROBLEMS

10–93. Determine the moment of inertia of the beam's cross-sectional area about the x axis.

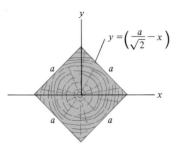

$$y = \left(\frac{a}{\sqrt{2}} - x\right)$$

Prob. 10–93

10–94. Determine the product of inertia for the angle's cross-sectional area with respect to the x' and y' axes having their origin located at the centroid C. Assume all corners to be square.

Prob. 10–94

10–95. Determine the moments of inertia I_x and I_y of the shaded area.

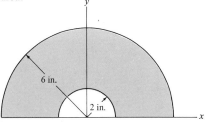

Prob. 10–95

*10–96. Determine the moment of inertia of the shaded area about the x axis.

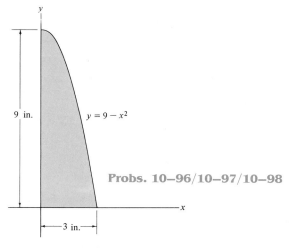

$y = 9 - x^2$

9 in.

3 in.

Probs. 10–96/10–97/10–98

10–97. Determine the moment of inertia of the shaded area about the y axis.

10–98. Determine the product of inertia of the shaded area with respect to the x and y axes.

10–99. Determine the moment of inertia of the beam's cross-sectional area with respect to the x' axis passing through the centroid C of the cross section. Neglect the size of the corner welds at A and B for the calculation. $\bar{y} = 104.3$ mm.

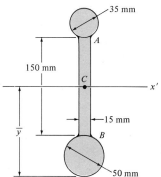

35 mm
150 mm
15 mm
50 mm

Prob. 10–99

*10–100. Determine the product of inertia of the shaded area with respect to the x and y axes.

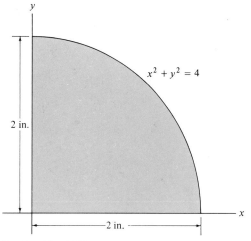

$x^2 + y^2 = 4$

2 in.

2 in.

Prob. 10–100

10–101. Determine the moment of inertia of the shaded area about the x axis.

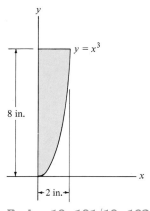

$y = x^3$

8 in.

2 in.

Probs. 10–101/10–102

10–102. Determine the moment of inertia of the shaded area about the y axis.

11

Virtual Work

In this chapter we will use the principle of virtual work and the method of potential energy to determine the equilibrium configuration or position of a series of connected rigid bodies. Although application requires more mathematical sophistication than using the equations of equilibrium, it will be shown that, once the equation of virtual work or the potential-energy function is established, the solution may be obtained *directly*, without having to dismember the system to obtain relationships between forces occurring at the connections. Furthermore, by using the potential-energy method, we will be able to investigate the "type" of equilibrium or the stability of the configuration.

11.1 Definition of Work and Virtual Work

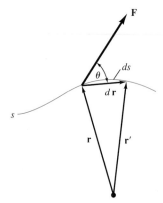

Fig. 11–1

Work of a Force

In mechanics a force **F** does work only when it undergoes a displacement in the direction of the force. For example, consider the force **F** in Fig. 11–1, which has a location on the path s specified by the position vector **r**. If the force moves along the path to a new position $\mathbf{r}' = \mathbf{r} + d\mathbf{r}$, the displacement is then $d\mathbf{r}$ and therefore the work dU is a *scalar quantity*, defined by the dot product

$$dU = \mathbf{F} \cdot d\mathbf{r}$$

Because $d\mathbf{r}$ is infinitesimal the magnitude of $d\mathbf{r}$ can be represented by ds, the differential arc segment along the path. If the angle between the tails of $d\mathbf{r}$ and **F** is θ, Fig. 11–1, then by definition of the dot product, the above equation may also be written as

$$dU = F \, ds \, \cos \theta$$

Work as expressed by this equation may be interpreted in one of two ways: either as the product of **F** and the component of displacement in the direction of the force, i.e., $ds \cos \theta$, or as the product of ds and the component of force in the direction of displacement, i.e., $F \cos \theta$. Note that if $0° \leq \theta < 90°$, then the force component and the displacement have the *same sense,* so that the work is *positive;* whereas if $90° < \theta \leq 180°$, these vectors have an *opposite sense,* and therefore the work is *negative.* Also, $dU = 0$ if the force is *perpendicular* to displacement, since $\cos 90° = 0$, or if the force is applied at a *fixed point,* in which case the displacement $ds = 0$.

The basic unit for work combines the units of force and displacement. In the SI system a *joule* (J) is equivalent to the work done by a force of 1 newton which moves 1 meter in the direction of the force (1 J = 1 N · m). In the FPS system work is defined in units of ft · lb. The moment of a force has the same combination of units; however, the concepts of moment and work are in no way related. A moment is a vector quantity, whereas work is a scalar.

Work of a Couple

The two forces of a couple do work when the couple *rotates* about an axis perpendicular to the plane of the couple. To show this, consider the body in Fig. 11–2a, which is subjected to a couple having a magnitude of $M = Fr$. Any general differential displacement of the body can be considered as a combination of a translation and rotation. When the body *translates* such that the *component of displacement* along the line of action of each force is ds_t, clearly the "positive" work of one force (Fds_t) *cancels* the "negative" work of the other ($-Fds_t$), Fig. 11–2b. Consider now a differential *rotation* $d\theta$ of the body about an axis perpendicular to the plane of the couple, which intersects the plane at point O, Fig. 11–2c. (For the derivation, any other point in the plane may also be considered.) As shown, each force undergoes a dis-

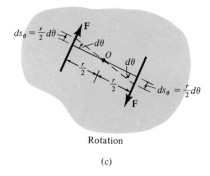

Translation

Rotation

(a) (b) (c)

Fig. 11–2

placement $ds_\theta = (r/2)\, d\theta$ in the direction of the force; hence, the total work done is

$$dU = F\left(\frac{r}{2}\, d\theta\right) + F\left(\frac{r}{2}\, d\theta\right) = (Fr)\, d\theta$$

or

$$dU = M\, d\theta$$

The resultant work is *positive* when the sense of **M** is the *same* as that of *d**θ***, and negative when they have an opposite sense. As in the case of the moment vector, the *direction and sense* of *d**θ*** are defined by the right-hand rule, where the fingers of the right hand follow the rotation or "curl" and the thumb indicates the direction of *d**θ***. Hence, the line of action of *d**θ*** must be *parallel* to the line of action of **M**. This is always the case if movement of the body occurs in the *same plane*. If the body rotates in space, however, the *component* of *d**θ*** in the direction of **M** is required; i.e., the work done is defined by the dot product, $dU = \mathbf{M} \cdot d\boldsymbol{\theta}$.

Virtual Work

The definitions of the work of a force and a couple have been presented in terms of *actual movements* expressed by differential displacements having magnitudes of *ds* and *dθ*. Consider now an *imaginary* or *virtual movement,* which indicates a displacement or rotation that is *assumed* and *does not actually exist*. These movements are first-order differential quantities and will be denoted by the symbols δ*s* and δ*θ* (delta *s* and delta *θ*), respectively. The *virtual work* done by a force undergoing a virtual displacement is

$$\delta U = F \cos \theta\, \delta s \qquad (11{-}1)$$

Similarly, when a couple undergoes a virtual rotation in the plane of the couple forces, the *virtual work* is

$$\delta U = M\, \delta \theta \qquad (11{-}2)$$

11.2 Principle of Virtual Work for a Particle and a Rigid Body

If a particle is in equilibrium, the resultant of a force system acting on it must be equal to zero. Hence, if the particle undergoes an imaginary or virtual displacement in the x, y, or z direction, the virtual work (δU) done by the force system must be equal to zero since the components $\Sigma F_x = 0$, $\Sigma F_y = 0$, $\Sigma F_z = 0$. Alternatively, this may be expressed as

$$\delta U = 0$$

For example, if the particle in Fig. 11–3 is given a virtual displacement $\delta \mathbf{x}$, only the x components of the forces acting on the particle do work. (No work is done by the y and z components since they are perpendicular to the displacement.) The virtual work equation is therefore

$$\delta U = 0; \qquad F_{1x}\,\delta x + F_{2x}\,\delta x + F_{3x}\,\delta x = 0$$

Factoring out δx, which is common to every term, yields

$$(F_{1x} + F_{2x} + F_{3x})\,\delta x = 0$$

Since $\delta x \neq 0$, this equation is satisfied only if the sum of the force components in the x direction is equal to zero, i.e., $\Sigma F_x = 0$. Two other virtual work equations can be written by assuming virtual displacements $\delta \mathbf{y}$ and $\delta \mathbf{z}$ in the y and z directions, respectively. Doing this, however, amounts to satisfying the equilibrium equations $\Sigma F_y = 0$ and $\Sigma F_z = 0$ for the particle.

In a similar manner, a rigid body that is subjected to a coplanar force system will be in equilibrium provided $\Sigma F_x = 0$, $\Sigma F_y = 0$, and $\Sigma M_O = 0$. We can also write a set of three virtual work equations for the body, each of which requires $\delta U = 0$. If these equations involve separate virtual translations in the x and y directions and a virtual rotation about an axis perpendicular to the x-y plane and passing through point O, then it can be shown that they will correspond to the above mentioned three equilibrium equations. When writing these equations, it is *not necessary* to include the work done by the *internal forces* acting on the body since a rigid body *does not deform* when subjected to an external loading, and furthermore, when the body moves through a virtual displacement, the internal forces occur in equal but opposite collinear pairs, so that the corresponding work done by each pair of forces *cancels*.

As in the case of a particle, however, no added advantage would be gained by solving rigid-body equilibrium problems using the principle of virtual work. This is because for each application of the virtual-work equation the virtual displacement, common to every term, factors out, leaving an equation that could have been obtained in a more *direct manner* by applying the equations of equilibrium.

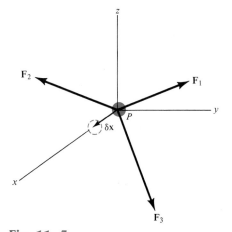

Fig. 11–3

11.3 Principle of Virtual Work for a System of Connected Rigid Bodies

The method of virtual work is most suitable for solving equilibrium problems that involve a system of several *connected* rigid bodies such as shown in Fig. 11–4. For each of these systems, the number of *independent virtual-work equations* that can be written depends upon the number of *independent virtual displacements* that can be made by the system.

Degrees of Freedom

In general, *for a system of connected bodies, the number of independent virtual displacements equals the minimum number of independent coordinates q needed to specify completely the location of all members of the system.* Each independent coordinate gives the system a *degree of freedom* that must be consistent with the constraining action of the supports. Thus, an *n*-degree-of-freedom system requires *n* independent coordinates q_n to specify the location of all its members with respect to a fixed reference point. For example, the link and sliding-block arrangement shown in Fig. 11–4a is an example of a one-degree-of-freedom system. The independent coordinate $q = \theta$ may be used to specify the location of the two connecting links and the block. The coordinate x could also be used as the independent coordinate. However, since the block is constrained to move within the slot, x is not independent of θ; rather it can be related to θ using the cosine law, $b^2 = a^2 + x^2 - 2ax \cos \theta$. The double-link arrangement, shown in Fig. 11–4b, is an example of a two-degree-of-freedom system. To specify the location of each link, the coordinate angles θ_1 and θ_2 must be known, since a rotation of one link is independent of a rotation of the other.

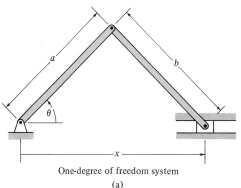

One-degree of freedom system

(a)

Two-degree of freedom system

(b)

Principle of Virtual Work

The principle of virtual work for a system of rigid bodies whose connections are *frictionless* may be stated as follows: *A system of connected rigid bodies is in equilibrium provided the virtual work done by all the external forces and couples acting on the system is zero for each independent virtual displacement of the system.* Mathematically, this may be expessed as

$$\delta U = 0 \qquad (11\text{–}3)$$

where δU represents the virtual work of all the external forces (and couples) acting on the system during any independent virtual displacement.

It has already been pointed out that if a system has *n* degrees of freedom it takes *n* independent coordinates q_n to completely specify the location of the system. Hence, for the system it is possible to write *n* independent virtual-work equations; one for each independent coordinate, while the remaining $n - 1$ coordinates are held *fixed*.

PROCEDURE FOR ANALYSIS

The following procedure provides a method for applying the equation of virtual work to solve problems involving a system of frictionless connected rigid bodies having a single degree of freedom.

Free-Body Diagram.   Draw the free-body diagram of the entire system of connected bodies and define the *independent coordinate q*. Sketch the "deflected position" of the system on the free-body diagram when the system undergoes a *positive* virtual displacement δq. From this, specify the "active" forces and couples, that is, those that do work.

Virtual Displacements.   Indicate *position coordinates s_i*, measured from a *fixed point* on the free-body diagram to each of the *i* number of "active" forces and couples. Each coordinate axis should be in the *same direction* as the line of action of the "active" force to which it is directed.

Relate each of the position coordinates s_i to the independent coordinate *q*; then *differentiate* these expressions in order to express the virtual displacements δs_i in terms of δq.

Virtual-Work Equation.   Write the *virtual-work equation* for the system assuming that, whether possible or not, all the position coordinates s_i undergo *positive* virtual displacements δs_i. Using the relations for δs_i, express the work of *each* "active" force and couple in the equation in terms of the single independent virtual displacement δq. By factoring out this common displacement, one is left with an equation that generally can be solved for an unknown force, couple, or equilibrium position.

If the system contains *n* degrees of freedom, *n* independent coordinates q_n must be specified. In this case, follow the above procedure and let *only one* of the independent coordinates undergo a virtual displacement, while the remaining $n - 1$ coordinates are held fixed. In this way, *n* virtual-work equations can be written, one for each independent coordinate.

The following examples should help to clarify application of this procedure.

Example 11–1

Determine the angle θ for equilibrium of the two-member linkage shown in Fig. 11–5a. Each member has a mass of 10 kg.

(a)

(b)

Fig. 11–5

SOLUTION

Free-Body Diagram. The system has only one degree of freedom, since the location of both links may be specified by the single independent coordinate $(q =)\ \theta$. As shown on the free-body diagram in Fig. 11–5b, when θ undergoes a *positive* (clockwise) virtual rotation $\delta\theta$, only the active forces, **F** and the two 98.1-N weights, do work. (The reactive forces $\mathbf{D}_x$ and $\mathbf{D}_y$ are fixed, and $\mathbf{B}_y$ does not move along its line of action.)

Virtual Displacements. If the origin of coordinates is established at the *fixed* pin support D, the location of **F** and **W** may be specified by the *position coordinates* x_B and y_w, as shown in the figure. In order to compute the work, note that these coordinates are in the *same direction* as the lines of action of their associated forces.

Expressing the position coordinates in terms of the independent coordinate θ and taking the derivatives yields

$$x_B = 2(1\ \cos\ \theta)\ \text{m} \qquad \delta x_B = -2\ \sin\ \theta\ \delta\theta\ \text{m} \tag{1}$$

$$y_w = \tfrac{1}{2}(1\ \sin\ \theta)\ \text{m} \qquad \delta y_w = 0.5\ \cos\ \theta\ \delta\theta\ \text{m} \tag{2}$$

It is seen by the *signs* of these equations, and indicated in Fig. 11–5b, that an *increase* in θ (i.e., $\delta\theta$) causes a *decrease* in x_B and an *increase* in y_w.

Virtual-Work Equation. If the virtual displacements δx_B and δy_w were *both positive*, then the forces **W** and **F** would do positive work since the forces and their corresponding displacements would have the same sense. Hence, the virtual-work equation for the displacement $\delta\theta$ is

$$\delta U = 0; \qquad W\ \delta y_w + W\ \delta y_w + F\ \delta x_B = 0 \tag{3}$$

Substituting Eqs. (1) and (2) into Eq. (3) in order to relate the virtual displacements to the common virtual displacement $\delta\theta$ yields

$$98.1(0.5\ \cos\ \theta\ \delta\theta) + 98.1(0.5\ \cos\ \theta\ \delta\theta) + 25(-2\ \sin\ \theta\ \delta\theta) = 0$$

Notice that the "negative work" done by **F** (force in the opposite sense to displacement) has been *accounted for* in the above equation by the "negative sign" of Eq. (1). Factoring out the *common displacement* $\delta\theta$ and solving for θ, noting that $\delta\theta \neq 0$, yields

$$(98.1\ \cos\ \theta - 50\ \sin\ \theta)\ \delta\theta = 0$$

$$\theta = \tan^{-1}\ \frac{98.1}{50} = 63.0° \qquad\qquad Ans.$$

If this problem had been solved using the equations of equilibrium, it would have been necessary to dismember the links and apply three scalar equations to *each* link. The principle of virtual work, by means of calculus, has eliminated this task so that the answer is obtained in a very direct manner.

Example 11–2

Using the principle of virtual work, determine the angle θ required to maintain equilibrium of the mechanism shown in Fig. 11–6a. Neglect the weight of the links. The spring is unstretched when $\theta = 0°$ and it maintains a horizontal position due to the roller.

SOLUTION

Free-Body Diagram. The mechanism has one degree of freedom, and therefore the location of each member may be specified using the independent coordinate θ. When θ undergoes a *positive* virtual displacement $\delta\theta$, as shown on the free-body diagram in Fig. 11–6b, links AB and EC rotate by the same amount since they have the same length, and link BC only translates. Since a couple does work *only* when it rotates, the work done by $\mathbf{M}_2$ is zero. The reactive forces at A and E do no work. Why?

Virtual Displacements. The position coordinates x_B and x_D are *parallel* to the lines of action of $\mathbf{P}$ and $\mathbf{F}$ and locate these forces with respect to the *fixed points* A and E. From Fig. 11–6b,

$$x_B = 0.4 \sin \theta \text{ m}$$
$$x_D = 0.2 \sin \theta \text{ m}$$

Thus,

$$\delta x_B = 0.4 \cos \theta \; \delta\theta \text{ m}$$
$$\delta x_D = 0.2 \cos \theta \; \delta\theta \text{ m}$$

Fig. 11–6

Virtual-Work Equation. Applying the equation of virtual work, noting that $\mathbf{F}_s$ is opposite to *positive* δx_D displacement and hence does negative work, we obtain

$$\delta U = 0; \qquad M_1 \, \delta\theta + P \, \delta x_B - F_s \, \delta x_D = 0$$

Relating each of the virtual displacements to the *common* virtual displacement $\delta\theta$ yields

$$0.5 \, \delta\theta + 2(0.4 \cos \theta \; \delta\theta) - F_s(0.2 \cos \theta \; \delta\theta) = 0$$
$$(0.5 + 0.8 \cos \theta - 0.2F_s \cos \theta) \, \delta\theta = 0 \qquad (1)$$

For the arbitrary angle θ, the spring is stretched a distance of $x_D = (0.2 \sin \theta)$ m; and therefore, $F_s = 60$ N/m$(0.2 \sin \theta)$ m $= (12 \sin \theta)$ N. Substituting into Eq. (1) and noting that $\delta\theta \neq 0$, we have

$$0.5 + 0.8 \cos \theta - 0.2(12 \sin \theta) \cos \theta = 0$$

Since $\sin 2\theta = 2 \sin \theta \cos \theta$, then

$$1 = 2.4 \sin 2\theta - 1.6 \cos \theta$$

Solving for θ by trial and error yields

$$\theta = 36.3° \qquad\qquad \textit{Ans.}$$

Example 11–3

Using the principle of virtual work, determine the force that the pin at C must exert in order to hold the mechanism shown in Fig. 11–7a in equilibrium when $\theta = 45°$. Neglect the weight of the members.

(a)

SOLUTION

Free-Body Diagram. The reaction $\mathbf{C}_x$ can be obtained by *releasing* the pin constraint at C in the x direction and allowing the frame to be displaced in this direction. The system then has only one degree of freedom, defined by the independent coordinate θ, Fig. 11–7b. When θ undergoes a *positive* virtual displacement $\delta\theta$, only $\mathbf{C}_x$ and the 200-N force do work.

Virtual Displacements. Forces $\mathbf{C}_x$ and 200 N are located from the fixed origin A using position coordinates y_B and x_C. From Fig. 11–7b, x_C can be related to θ by the "law of cosines." Hence,

(b)

Fig. 11–7

$$(0.7)^2 = (0.6)^2 + x_C^2 - 2(0.6)x_C \cos\theta \qquad (1)$$

$$0 = 0 + 2x_C\,\delta x_C - 1.2\,\delta x_C \cos\theta + 1.2x_C \sin\theta\,\delta\theta$$

$$\delta x_C = \frac{1.2x_C \sin\theta}{1.2\cos\theta - 2x_C}\,\delta\theta \qquad (2)$$

Also,

$$y_B = 0.6\sin\theta$$

$$\delta y_B = 0.6\cos\theta\,\delta\theta \qquad (3)$$

Virtual-Work Equation. When y_B and x_C undergo *positive* virtual displacements δy_B and δx_C, $\mathbf{C}_x$ and 200 N do *negative work,* since they both act in the opposite sense to $\delta\mathbf{y}_B$ and $\delta\mathbf{x}_C$. Hence,

$$\delta U = 0; \qquad -200\,\delta y_B - C_x\,\delta x_C = 0$$

Substituting Eqs. (2) and (3) into this equation, factoring out $\delta\theta$, and solving for C_x yields

$$-200(0.6\cos\theta\,\delta\theta) - C_x\frac{1.2x_C \sin\theta}{1.2\cos\theta - 2x_C}\,\delta\theta = 0$$

$$C_x = \frac{-120\cos\theta(1.2\cos\theta - 2x_C)}{1.2x_C \sin\theta} \qquad (4)$$

At the required equilibrium position $\theta = 45°$, the corresponding value of x_C can be found by using Eq. (1), in which case

$$x_C^2 - 1.2\cos 45° \, x_C - 0.13 = 0$$

Solving for the positive root yields

$$x_C = 0.981 \text{ m}$$

Thus, from Eq. (4),

$$C_x = 114 \text{ N} \qquad\qquad \textit{Ans.}$$

Example 11–4

Using the principle of virtual work, determine the equilibrium position of the two-bar linkage shown in Fig. 11–8a. Neglect the weight of the links.

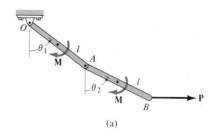

(a)

SOLUTION

The system has two degrees of freedom, since the *independent coordinates* θ_1 and θ_2 must be known to locate the position of both links. The position coordinate x_B, measured from the fixed point O, is used to specify the location of **P,** Fig. 11–8b and c.

If θ_1 is held *fixed* and θ_2 varies by an amount $\delta\theta_2$, as shown in Fig. 11–8b, the virtual-work equation becomes

$$[\delta U = 0]_{\theta_2}; \qquad P(\delta x_B)_{\theta_2} - M\,\delta\theta_2 = 0 \qquad (1)$$

where P and M represent the magnitudes of the applied force and couple acting on link AB.

When θ_2 is held *fixed* and θ_1 varies by an amount $\delta\theta_1$, as shown in Fig. 11–8c, the virtual-work equation becomes

$$[\delta U = 0]_{\theta_1}; \qquad P(\delta x_B)_{\theta_1} - M\,\delta\theta_1 - M\,\delta\theta_1 = 0 \qquad (2)$$

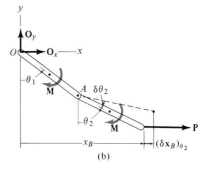

(b)

The *position coordinate* x_B may be related to the independent coordinates θ_1 and θ_2 by the equation

$$x_B = l \sin\theta_1 + l \sin\theta_2 \qquad (3)$$

To obtain the variation of δx_B in terms of $\delta\theta_2$, it is necessary to take the *partial derivative* of x_B with respect to θ_2 since x_B is a function of both θ_1 and θ_2. Hence,

$$\frac{\partial x_B}{\partial \theta_2} = l \cos\theta_2 \qquad (\delta x_B)_{\theta_2} = l \cos\theta_2\,\delta\theta_2$$

Substituting into Eq. (1), we have

$$(Pl \cos\theta_2 - M)\,\delta\theta_2 = 0$$

Since $\delta\theta_2 \neq 0$, then

$$\theta_2 = \cos^{-1}\left(\frac{M}{Pl}\right) \qquad Ans.$$

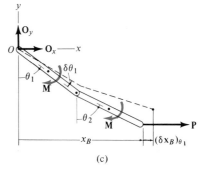

(c)

Fig. 11–8

Using Eq. (3) to obtain the variation of x_B with θ_1 yields

$$\frac{\partial x_B}{\partial \theta_1} = l \cos\theta_1 \qquad (\delta x_B)_{\theta_1} = l \cos\theta_1\,\delta\theta_1$$

Substituting into Eq. (2), we have

$$(Pl \cos\theta_1 - 2M)\,\delta\theta_1 = 0$$

Since $\delta\theta_1 \neq 0$, then

$$\theta_1 = \cos^{-1}\left(\frac{2M}{Pl}\right) \qquad Ans.$$

PROBLEMS

11–1. The thin rod of weight W rests against the smooth wall and floor. Determine the magnitude of the couple moment **M** needed to hold it in equilibrium for a given angle θ.

Prob. 11–1

■11–2. Determine the angle θ for equilibrium of the uniform 7-kg bar AC. Due to the roller guide at B, the spring remains vertical and is unstretched when $\theta = 0°$.

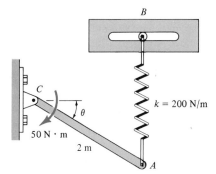

Prob. 11–2

11–3. Determine the force **F** needed to lift the block having a weight of 100 lb. *Hint:* Note that the coordinates s_A and s_B are related to the *constant* vertical length l of the cord.

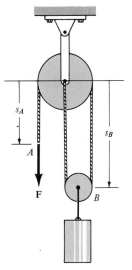

Prob. 11–3

■*11–4. The two-bar linkage is subjected to a couple moment $M = 100$ N $\cdot$ m and vertical force $P = 500$ N. Determine the angle θ for equilibrium. The spring is unstretched when $\theta = 45°$. Neglect the mass of each bar.

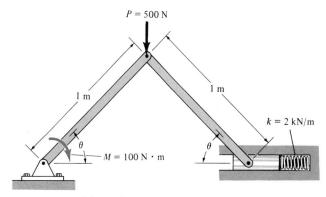

Probs. 11–4/11–5

■11–5. The two-bar linkage is subjected to a couple moment $M = 100$ N $\cdot$ m and vertical force $P = 500$ N. Also each bar is uniform and has a mass of 10 kg. Determine the angle θ for equilibrium. The spring is unstretched when $\theta = 45°$.

487

■**11–6.** Each member of the pin-connected mechanism has a mass of 8 kg. If the spring is unstretched when $\theta = 0°$, determine the angle θ for equilibrium.

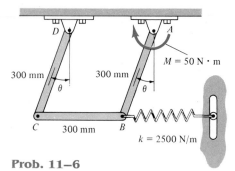

Prob. 11–6

11–7. If each of the three links of the mechanism has a weight of 20 lb, determine the angle θ for equilibrium. The spring, which always remains vertical, is unstretched when $\theta = 0°$. Set **P = 0.**

Probs. 11–7/11–8

*****11–8.** The spring is unstretched when $\theta = 0°$. If $P = 8$ lb, determine the angle θ for equilibrium. Neglect the weight of the links.

11–9. The pin-connected mechanism is constrained at A by a pin and at B by a smooth sliding block. If $P = 10$ lb, determine the angle θ for equilibrium. The spring is unstretched when $\theta = 45°$. Neglect the weights of the members.

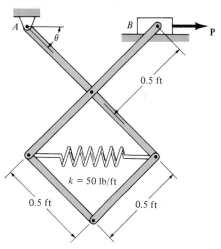

Prob. 11–9

11–10. The spring has an unstretched length of 0.3 m. Determine the angle θ for equilibrium if the uniform links each have a mass of 5 kg. Set **P = 0.**

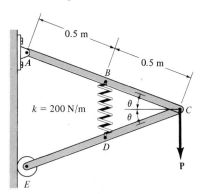

Probs. 11–10/11–11

11–11. The spring has an unstretched length of 0.3 m. Determine the angle θ for equilibrium if the vertical load **P** has a magnitude of $P = 50$ N. Neglect the weight of the links.

*11–12. The disk has a weight of 10 lb and is subjected to a horizontal force $F = 8$ lb and a couple moment $M = 8$ lb · ft. Determine the disk's rotation θ if the end of the spring wraps around the periphery of the disk as the disk turns. The spring is originally unstretched.

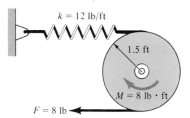

$k = 12$ lb/ft

1.5 ft

$M = 8$ lb · ft

$F = 8$ lb

Prob. 11–12

11–13. The crankshaft is subjected to a torque of $M = 50$ N · m. Determine the vertical compressive force F applied to the piston for equilibrium when $\theta = 60°$.

F

5 in.

3 in.

M

θ

A

Prob. 11–13

■11–14. The scissors jack supports a load **P.** Determine the axial force in the screw necessary for equilibrium when the jack is in the position θ. Each of the four links has a length L and is pin-connected at its center. Points B and D can move horizontally.

P

C

D

A

θ

B

Prob. 11–14

11–15. The assembly is used for exercise. It consists of four pin-connected bars, each of length L, and a spring of stiffness k and unstretched length a ($<2L$). If horizontal forces **P** and **−P** are applied to the handles so that θ is slowly decreased, determine the angle θ at which the magnitude of **P** becomes a maximum.

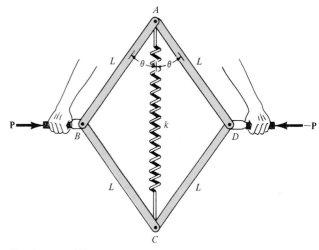

A

L

θ—θ

L

P

B

k

D

−P

L

L

C

Prob. 11–15

■*11–16. The 4-ft members of the mechanism are pin-connected at their centers. If vertical forces act at C and E as shown, determine the angle θ for equilibrium. The spring is unstretched when $\theta = 45°$. Neglect the weight of the members.

40 lb

20 lb

E

$k = 200$ lb/ft

C

2 ft

2 ft

2 ft

B

2 ft

A

θ

D

Prob. 11–16

11–17. If a force $P = 40$ N is applied perpendicular to the handle of the toggle press, determine the vertical compressive force developed at C; $\theta = 30°$.

Prob. 11–17

11–19. Rods AB and BC have a center of gravity located at their midpoints. If all contacting surfaces are smooth and BC has a mass of 100 kg, determine the appropriate mass of AB required for equilibrium.

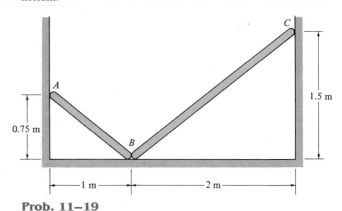

Prob. 11–19

11–18. Determine the horizontal force F required to maintain equilibrium of the slider mechanism when $\theta = 60°$.

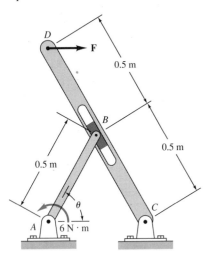

Prob. 11–18

*11–20. A horizontal force acts on the end of the link as shown. Determine the angles θ_1 and θ_2 for equilibrium of the two links. Each link is uniform and has a mass m.

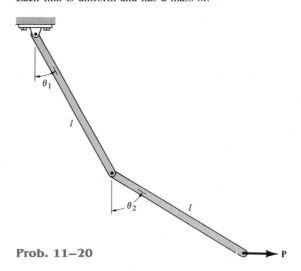

Prob. 11–20

■11–21. A 3-lb disk is attached to the end of rod *ABC*. If the rod is supported by a smooth slider block at *C* and rod *BD*, determine the angle θ for equilibrium. Neglect the weight of the rods and the slider.

11–22. Determine the mass of *A* and *B* required to hold the 400-g desk lamp in balance for any angle θ or ϕ. Neglect the weight of the mechanism and the size of the lamp.

Prob. 11–22

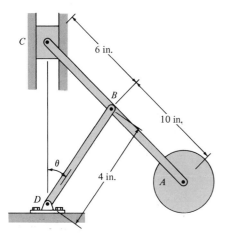

Prob. 11–21

*11.4 Conservative Forces

The work done by a force when it undergoes a *differential displacement* has been defined as $dU = F \cos \theta \, ds$. If the force is displaced over a path that has a *finite length s*, the work is determined by integrating over the path, i.e.,

$$U = \int_s F \cos \theta \, ds$$

To evaluate the integral, it is necessary to obtain a relationship between *F* and the component of displacement $ds \cos \theta$. In some instances, however, the work done by a force will be *independent* of its path and, instead, will depend only upon the initial and final locations of the force along the path. A force that has this property is called a *conservative force*.

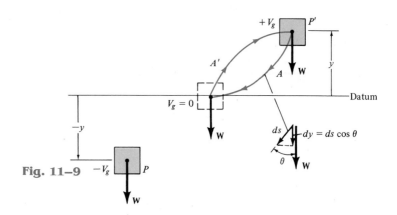

Fig. 11–9

Weight

Consider the body in Fig. 11–9, which is initially at P'. If the body is moved *down* along the *arbitrary path* A to the dashed position, then, for a given displacement ds along the path, the displacement component in the direction of $\mathbf{W}$ has a magnitude of $dy = ds\cos\theta$, as shown. Since both the force and displacement are in the same direction, the work is positive; hence,

$$U = \int_s W\cos\theta\,ds = \int_0^y W\,dy$$

or

$$U = Wy$$

In a similar manner, the work done by the weight when the body moves up a distance y back to P', along the arbitrary path A', is

$$U = -Wy$$

Why is the work negative?

The weight of a body is therefore a conservative force, since the work done by the weight depends *only* on the body's *vertical displacement* and is independent of the path along which the body moves.

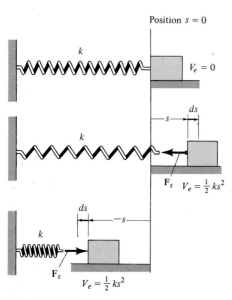

Fig. 11–10

Elastic Spring

The force developed by an elastic spring ($F_s = ks$) is also a conservative force. If the spring is attached to a body and the body is displaced along *any path*, such that it causes the spring to elongate or compress from a position s_1 to a further position s_2, the work will be negative, since the spring exerts a force $\mathbf{F}_s$ *on the body* that is opposite to the body's displacement ds, Fig. 11–10. For either extension or compression, the work is independent of the path and is simply

$$U = \int_{s_1}^{s_2} F_s\,ds = \int_{s_1}^{s_2} (-ks)\,ds$$
$$= -(\tfrac{1}{2}ks_2^2 - \tfrac{1}{2}ks_1^2)$$

Friction

In contrast to a conservative force, consider the force of *friction* exerted on a moving body by a fixed surface. The work done by the frictional force depends upon the path; the longer the path, the greater the work. Consequently, frictional forces are *nonconservative,* and the work done is dissipated from the body in the form of heat.

*11.5 Potential Energy

When a conservative force acts on a body, it gives the body the capability to do work. This capacity, measured as *potential energy,* depends upon the location of the body.

Gravitational Potential Energy

If a body is located a distance y *above* a fixed horizontal reference or datum, Fig. 11–9, the weight of the body has *positive* gravitational potential energy V_g since **W** has the capacity of doing positive work when the body is moved back down to the datum. Likewise, if the body is located a distance y *below* the datum, V_g is *negative* since the weight does negative work when the body is moved back up to the datum. At the datum $V_g = 0$.

Measuring y as *positive upward,* the gravitational potential energy of the body's weight **W** is thus

$$V_g = Wy \tag{11–4}$$

Elastic Potential Energy

The elastic potential energy V_e that a spring produces on an attached body when the spring is elongated or compressed from an undeformed position ($s = 0$) to a final position s is

$$V_e = \tfrac{1}{2}ks^2 \tag{11–5}$$

Here V_e is *always positive,* since in the deformed position the spring has the capacity of doing *positive work* in *returning* the body back to the spring's undeformed position, Fig. 11–10.

Potential Function

In the general case, if a body is subjected to *both* gravitational and elastic forces, the *potential energy or potential function V* of the body can be expressed as the algebraic sum

$$V = V_g + V_e \tag{11–6}$$

where measurement of V depends upon the location of the body with respect to a selected datum in accordance with Eqs. 11–4 and 11–5.

In general, if a system of frictionless connected rigid bodies has a *single degree of freedom* such that its position is defined by the independent coordinate q, then the potential energy for the system can be expressed as $V = V(q)$, which is defined by Eq. 11–6. The work done by all the conservative forces acting on the system in moving the system from q_1 to q_2 is measured by the *difference* in V, i.e.,

$$U_{1-2} = V(q_1) - V(q_2) \qquad (11–7)$$

For example, the potential function for a system consisting of a block of weight **W** supported by a spring, Fig. 11–11a, can be expressed in terms of its independent coordinate ($q =$) y, measured from a fixed datum located at the unstretched length of the spring, i.e.,

$$\begin{aligned} V &= V_g + V_e \\ &= -Wy + \tfrac{1}{2}ky^2 \end{aligned} \qquad (11–8)$$

If the block moves from y_1 to a farther downward position y_2, then the work of **W** and $\mathbf{F}_s$ is

$$U_{1-2} = V(y_1) - V(y_2) = -W[y_1 - y_2] + \tfrac{1}{2}ky_1^2 - \tfrac{1}{2}ky_2^2$$

*11.6 Potential-Energy Criterion for Equilibrium

System Having One Degree of Freedom

When the displacement of a frictionless connected system is *infinitesimal*, i.e., from q to $q + dq$, Eq. 11–7 becomes

$$dU = V(q) - V(q + dq)$$

or

$$dU = -dV$$

Furthermore, if the independent coordinate undergoes a *virtual displacement* δq, rather than an actual displacement dq, then $\delta U = -\delta V$. For equilibrium, the principle of virtual work requires $\delta U = 0$ and therefore, provided the potential function for the system is known, this also requires that $\delta V = 0$. We can also express this requirement as

$$\boxed{\dfrac{dV}{dq} = 0} \qquad (11–9)$$

Hence, *when a system of connected rigid bodies is in equilibrium, the first variation or change in V is zero*. This change is computed by taking the *first derivative* of the potential function and setting it equal to zero. For example, using Eq. 11–8 to determine the equilibrium position for the spring and block in Fig. 11–11a, we have

Datum

$F_s = ky_{eq}.$

(a)   (b)

Fig. 11–11

$$\frac{dV}{dy} = W - ky = 0$$

Hence, the equilibrium position $y = y_{eq}$ is

$$y_{eq} = \frac{W}{k}$$

Of course, the *same result* is obtained by applying $\Sigma F_y = 0$ to the forces acting on the free-body diagram of the block, Fig. 11–11b.

System Having n Degrees of Freedom

When the system of connected bodies has n degrees of freedom, the total potential energy stored in the system will be a function of n independent coordinates q_n; i.e., $V = V(q_1, q_2, \cdots, q_n)$. In order to apply the equilibrium criterion $\delta V = 0$, it is necessary to compute the change in potential energy δV by using the "chain rule" of differential calculus; i.e., the total variation or total differential of V must be equal to zero. Thus,

$$\delta V = \frac{\partial V}{\partial q_1} \delta q_1 + \frac{\partial V}{\partial q_2} \delta q_2 + \cdots + \frac{\partial V}{\partial q_n} \delta q_n = 0$$

Since the virtual displacements $\delta q_1, \delta q_2, \cdots, \delta q_n$ are independent of one another, it is necessary that

$$\frac{\partial V}{\partial q_1} = 0, \quad \frac{\partial V}{\partial q_2} = 0, \quad \cdots, \quad \frac{\partial V}{\partial q_n} = 0$$

As for the case of virtual work, *it is possible to write n independent potential functions for a system having n degrees of freedom.*

*11.7 Stability of Equilibrium

Once the equilibrium configuration for a body or system of connected bodies is defined, it is sometimes important to investigate the "type" of equilibrium or the stability of the configuration. For example, consider the position of a ball resting at a point on each of the three paths shown in Fig. 11–12. Each situation represents an equilibrium state for the ball. When the ball is at A, it is said to be in *stable equilibrium* because if it is given a small displacement up the hill, it will always *return* to its original, lowest, position. At A, its total potential energy is a *minimum*. When the ball is at B, it is in *neutral equilibrium*. A small displacement either to the left or right of B will not alter this condition. The ball *remains* in equilibrium in the displaced position, and therefore its potential energy is *constant*. When the ball is at C, it is in *unstable equilibrium*. Here a small displacement will cause the ball's potential energy to be *decreased,* and so it will roll farther *away* from its original, highest position. At C, the potential energy of the ball is a *maximum*.

Stable equilibrium   Neutral equilibrium

Unstable equilibrium

Fig. 11–12

Type of Equilibrium

The above example illustrates that one of three types of equilibrium positions can be specified for a body or system of connected bodies.

1. *Stable equilibrium* occurs when a small displacement of the system causes the system to return to its original position. In this case the original potential energy of the system is a minimum.
2. *Neutral equilibrium* occurs when a small displacement of the system causes the system to remain in its displaced state. In this case the potential energy of the system remains constant.
3. *Unstable equilibrium* occurs when a small displacement of the system causes the system to move farther away from its original position. In this case, the original potential energy of the system is a maximum.

System Having One Degree of Freedom

For *equilibrium* of a system having a single degree of freedom, defined by the independent coordinate q, it has been shown that the first derivative of the potential function for the system must be equal to zero; i.e., $dV/dq = 0$. If the potential function $V = V(q)$ is plotted, Fig. 11–13, the first derivative (equilibrium position) is represented as the slope dV/dq, which is zero when the function is maximum, minimum, or an inflection point.

If the *stability* of a body at the equilibrium position is to be investigated, it is necessary to compute the *second derivative* of V and evaluate it at the equilibrium position $q = q_{eq}$. As shown in Fig. 11–13a, if $V = V(q)$ is a *minimum,* then

$$\frac{dV}{dq} = 0, \quad \frac{d^2V}{dq^2} > 0 \qquad \text{stable equilibrium} \qquad (11\text{–}10)$$

$\dfrac{d^2V}{dq^2} > 0$

$\dfrac{dV}{dq} = 0$

$q_{eq.}$

Stable Equilibrium

(a)

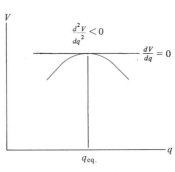

$\dfrac{d^2V}{dq^2} < 0$

$\dfrac{dV}{dq} = 0$

$q_{eq.}$

Unstable Equilibrium

(b)

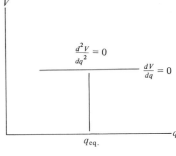

$\dfrac{d^2V}{dq^2} = 0$

$\dfrac{dV}{dq} = 0$

$q_{eq.}$

Neutral Equilibrium

(c)

Fig. 11–13

If $V = V(q)$ is a *maximum*, Fig. 11–13b, then

$$\frac{dV}{dq} = 0, \quad \frac{d^2V}{dq^2} < 0 \qquad \text{unstable equilibrium} \qquad (11\text{–}11)$$

If the second derivative is zero, it will be necessary to investigate *higher-order* derivatives to determine the stability. In particular, stable equilibrium will occur if the order of the lowest remaining nonzero derivative is *even* and the sign of this nonzero derivative is positive when it is evaluated at $q = q_{eq}$; otherwise, it is unstable.

If the system is in neutral equilibrium, Fig. 11–13c, it is required that

$$\frac{dV}{dq} = \frac{d^2V}{dq^2} = \frac{d^3V}{dq^3} = \cdots = 0 \qquad \text{neutral equilibrium} \qquad (11\text{–}12)$$

since then V must be constant at and around the "neighborhood" of q_{eq}.

System Having Two Degrees of Freedom

A criterion for investigating stability becomes increasingly complex as the number of degrees of freedom for the system increases. For a system having two degrees of freedom, with independent coordinates (q_1, q_2), it may be verified (using the calculus of functions of two variables) that equilibrium and stability occur at a point (q_{1eq}, q_{2eq}) when

$$\frac{\partial V}{\partial q_1} = \frac{\partial V}{\partial q_2} = 0$$

$$\left[\left(\frac{\partial^2 V}{\partial q_1 \, \partial q_2} \right)^2 - \left(\frac{\partial^2 V}{\partial q_1^2} \right) \left(\frac{\partial^2 V}{\partial q_2^2} \right) \right] < 0$$

$$\left(\frac{\partial^2 V}{\partial q_1^2} + \frac{\partial^2 V}{\partial q_2^2} \right) > 0$$

Both equilibrium and instability occur when

$$\frac{\partial V}{\partial q_1} = \frac{\partial V}{\partial q_2} = 0$$

$$\left[\left(\frac{\partial^2 V}{\partial q_1 \, \partial q_2} \right)^2 - \left(\frac{\partial^2 V}{\partial q_1^2} \right) \left(\frac{\partial^2 V}{\partial q_2^2} \right) \right] < 0$$

$$\left(\frac{\partial^2 V}{\partial q_1^2} + \frac{\partial^2 V}{\partial q_2^2} \right) < 0$$

PROCEDURE FOR ANALYSIS

Using potential-energy methods, the equilibrium positions and the stability of a body or a system of connected bodies having a single degree of freedom can be obtained by applying the following procedure.

Potential Function.   Formulate the potential function $V = V_g + V_e$ for the system. To do this, sketch the system so that it is located at some *arbitrary position* specified by the independent coordinate q. A horizontal *datum* is established through a *fixed point,** and the *gravitational potential energy* V_g is expressed in terms of the weight W of each member and its vertical distance y from the datum, $V_g = Wy$, Eq. 11–4. The elastic potential energy V_e of the system is expressed in terms of the stretch or compression, s, of any connecting spring and the spring's stiffness k, $V_e = \frac{1}{2}ks^2$, Eq. 11–5. Once V has been established, express the *position coordinates* y and s in terms of the independent coordinate q.

Equilibrium Position.   The equilibrium position is determined by taking the first derivative of V and setting it equal to zero, $\delta V = 0$, Eq. 11–9.

Stability.   Stability at the equilibrium position is determined by evaluating the second or higher-order derivatives of V as indicated by Eqs. 11–10 to 11–12.

The following examples numerically illustrate this procedure.

*The location of the datum is *arbitrary* since only the *changes* or differentials of V are required for investigation of the equilibrium position and its stability.

Example 11–5

The uniform link shown in Fig. 11–14a has a mass of 10 kg. The spring is unstretched when $\theta = 0°$. Determine the angle θ for equilibrium and investigate the stability at the equilibrium position.

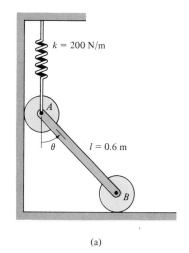

SOLUTION

Potential Function. The datum is established at the top of the link when the *spring is unstretched*, Fig. 11–14b. When the link is located at the arbitrary position θ, the spring increases its potential energy by stretching and the weight decreases its potential energy. Hence,

$$V = V_e + V_g = \frac{1}{2}ks^2 - W\left(s + \frac{l}{2}\cos\theta - \frac{l}{2}\right)$$

Since $l = s + l\cos\theta$ or $s = l(1 - \cos\theta)$, then

$$V = \frac{1}{2}kl^2(1 - \cos\theta)^2 - \frac{Wl}{2}(1 - \cos\theta)$$

Equilibrium Position. The derivative of V gives

$$\frac{dV}{d\theta} = kl^2(1 - \cos\theta)\sin\theta - \frac{Wl}{2}\sin\theta = 0$$

or

$$l\left[kl(1 - \cos\theta) - \frac{W}{2}\right]\sin\theta = 0$$

This equation is satisfied provided

$$\sin\theta = 0 \qquad \theta = 0° \qquad\qquad Ans.$$

$$\theta = \cos^{-1}\left(1 - \frac{W}{2kl}\right) = \cos^{-1}\left[1 - \frac{10(9.81)}{2(200)(0.6)}\right] = 53.8° \quad Ans.$$

Stability. Computing the second derivative of V gives

$$\frac{d^2V}{d\theta^2} = kl^2(1 - \cos\theta)\cos\theta + kl^2\sin\theta\sin\theta - \frac{Wl}{2}\cos\theta$$

$$= kl^2(\cos\theta - \cos 2\theta) - \frac{Wl}{2}\cos\theta$$

Fig. 11–14

Substituting values for the constants, with $\theta = 0°$ and $\theta = 53.8°$, yields

$$\frac{d^2V}{d\theta^2}\bigg|_{\theta=0°} = 200(0.6)^2(\cos 0° - \cos 0°) - \frac{10(9.81)(0.6)}{2}\cos 0°$$

$$= -29.4 < 0 \qquad \text{(unstable equilibrium at } \theta = 0°) \quad Ans.$$

$$\frac{d^2V}{d\theta^2}\bigg|_{\theta=53.8°} = 200(0.6)^2(\cos 53.8° - \cos 107.6°) - \frac{10(9.81)(0.6)}{2}\cos 53.8°$$

$$= 46.9 > 0 \qquad \text{(stable equilibrium at } \theta = 53.8°) \qquad Ans.$$

499

Example 11–6

Determine the mass m of the block required for equilibrium of the uniform 10-kg rod shown in Fig. 11–15a when $\theta = 20°$. Investigate the stability at the equilibrium position.

(a)

SOLUTION

Potential Function. The datum is established through point A, Fig. 11–15b. When $\theta = 0°$, the block is assumed to be suspended $(y_W)_1$ below the datum. Hence,

$$V = V_e + V_g = 98.1\left(\frac{1.5 \sin \theta}{2}\right) - m(9.81)(\Delta y) \qquad (1)$$

The distance Δy may be related to the independent coordinate θ by measuring the difference in cord lengths $B'C$ and BC. Since

$$B'C = \sqrt{(1.5)^2 + (1.2)^2} = 1.92$$
$$BC = \sqrt{(1.5 \cos \theta)^2 + (1.2 - 1.5 \sin \theta)^2} = \sqrt{3.69 - 3.60 \sin \theta}$$

then

$$\Delta y = B'C - BC = 1.92 - \sqrt{3.69 - 3.60 \sin \theta}$$

Substituting the above result into Eq. (1) yields

$$V = 98.1\left(\frac{1.5 \sin \theta}{2}\right) - m(98.1)(1.92 - \sqrt{3.69 - 3.60 \sin \theta}) \qquad (2)$$

(b)

Fig. 11–15

Equilibrium Position

$$\frac{dV}{d\theta} = 73.6 \cos \theta - \left[\frac{m(98.1)}{2}\right]\left(\frac{3.60 \cos \theta}{\sqrt{3.69 - 3.60 \sin \theta}}\right) = 0$$

$$\left.\frac{dV}{d\theta}\right|_{\theta=20°} = 69.16 - 10.58m = 0$$

$$m = \frac{69.16}{10.58} = 6.54 \text{ kg} \qquad \qquad Ans.$$

Stability. Taking the second derivative of Eq. (2), we obtain

$$\frac{d^2V}{d\theta^2} = -73.6 \sin \theta - \left[\frac{m(9.81)}{2}\right]\left(\frac{-1}{2}\right)\frac{(3.60 \cos \theta)^2}{(3.69 - 3.60 \sin \theta)^{3/2}}$$
$$- \frac{m(9.81)}{2}\left(\frac{-3.60 \sin \theta}{\sqrt{3.69 - 3.60 \sin \theta}}\right)$$

For the equilibrium position $\theta = 20°$, $m = 6.54$ kg, hence

$$\frac{d^2V}{d\theta^2} = 47.6 > 0 \qquad \text{(stable equilibrium at } \theta = 20°) \qquad Ans.$$

Example 11–7

A homogeneous block having a mass m rests on the top surface of a cylinder, Fig. 11–16a. Show that this is a condition of unstable equilibrium if $h > 2R$.

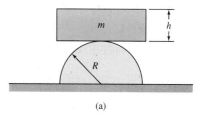

(a)

SOLUTION

Potential Function. The datum is established at the base of the cylinder, Fig. 11–16b. If the block is displaced by an amount θ from the equilibrium position, the potential function may be written in the form

$$V = V_e + V_g$$
$$= 0 + mgy$$

From Fig. 11–16b,

$$y = \left(R + \frac{h}{2}\right) \cos \theta + R\theta \sin \theta$$

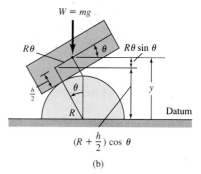

Thus,

$$V = mg\left[\left(R + \frac{h}{2}\right) \cos \theta + R\theta \sin \theta\right]$$

Fig. 11–16

Equilibrium Position

$$\frac{dV}{d\theta} = mg\left[-\left(R + \frac{h}{2}\right) \sin \theta + R \sin \theta + R\theta \cos \theta\right] = 0$$

$$= mg\left(-\frac{h}{2} \sin \theta + R\theta \cos \theta\right) = 0$$

Obviously $\theta = 0°$ is the equilibrium position that satisfies this equation.

Stability. Taking the second derivative of V yields

$$\frac{d^2V}{d\theta^2} = mg\left(-\frac{h}{2} \cos \theta + R \cos \theta - R\theta \sin \theta\right)$$

At $\theta = 0°$,

$$\left.\frac{d^2V}{d\theta^2}\right|_{\theta=0°} = -mg\left(\frac{h}{2} - R\right)$$

Since all the constants are positive, the block is in unstable equilibrium if $h > 2R$, for then $d^2V/d\theta^2 < 0$.

PROBLEMS

11–23. Solve Prob. 11–7 using the principle of potential energy, and investigate the stability at the equilibrium position.

*11–24. Solve Prob. 11–10 using the principle of potential energy, and investigate the stability at the equilibrium position.

11–25. Solve Prob. 11–19 using the principle of potential energy.

11–26. If the potential function for a conservative one-degree-of-freedom system is $V = (8x^3 - 2x^2 - 10)$ J, where x is given in meters, determine the positions for equilibrium and investigate the stability at each of these positions.

11–27. If the potential function for a conservative one-degree-of-freedom system is $V = (12 \sin 2\theta + 15 \cos \theta)$ J, where $0° < \theta < 180°$, determine the positions for equilibrium and investigate the stability at each of these positions.

*11–28. If the potential function for a conservative two-degree-of-freedom system is $V = (9y^2 + 18x^2)$ J, where x and y are given in meters, determine the equilibrium position and investigate the stability at this position.

11–29. The uniform beam has a weight W. If the contacting surfaces are smooth, determine the angle θ for equilibrium. The spring is uncompressed when $\theta = 90°$.

Prob. 11–29

■11–30. The two bars each have a weight of 8 lb. Determine the angle θ for equilibrium and investigate the stability at the equilibrium position. The spring has an unstretched length of 1 ft.

$k = 30$ lb/ft

Prob. 11–30

11–31. The bar supports a weight of $W = 500$ lb at its end. If the springs are originally unstretched when the bar is vertical, and $k_1 = 300$ lb/ft, $k_2 = 500$ lb/ft, investigate the stability of the bar when it is in the vertical position.

Prob. 11–31

*11–32. The spring of the scale has an unstretched length of a. Determine the angle θ for equilibrium when a weight W is supported on the platform. Neglect the weights of the members. What value of W would be required to keep the scale in neutral equilibrium when $\theta = 0°$?

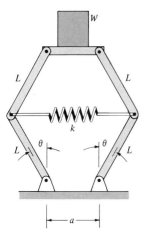

Prob. 11–32

11–33. The bent rod has a weight of 5 lb/ft. A pivot of negligible size is attached at its center A and it is balanced as shown. Determine the length L of its vertical segments so that it remains in neutral equilibrium. Neglect the thickness of the rod.

Prob. 11–33

11–34. The *Roberval balance* is in equilibrium when no weights are placed on the pans A and B. If two masses m_A and m_B are placed at *any* location a and b on the pans, show that neutral equilibrium is maintained if $m_A d_A = m_B d_B$.

Prob. 11–34

11–35. The rod supports a disk B having a weight of 800 N. It is held in equilibrium using four springs for which $k = 400$ N/m. Determine the minimum distance d between the springs so that the rod remains in stable equilibrium when it is vertical. Neglect the weight of the rod.

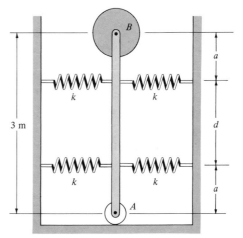

Prob. 11–35

*11–36. The truck is weighed on the highway inspection scale. If a known mass m is placed a distance s from the fulcrum B of the scale, determine the mass of the truck m_t for equilibrium if its center of gravity is located at a distance d from point C. When the scale is empty, the weight of the lever ABC balances the scale CDE.

Prob. 11–36

11–37. The cup has a hemispherical bottom and a mass m. Determine the position h of the center of mass G so that the cup is in neutral equilibrium.

Prob. 11–37

11–38. The assembly shown consists of a semicylinder and a rectangular block. If the block weighs 8 lb and the semicylinder weighs 2 lb, investigate the stability when the assembly is resting in the equilibrium position. Set $h = 4$ in.

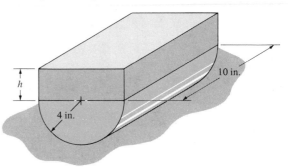

Probs. 11–38/11–39

11–39. The 2-lb semicylinder supports the block which has a specific weight of $\gamma = 80$ lb/ft^3. Determine the height h of the block which will produce neutral equilibrium in the position shown.

*11–40. The homogeneous cylinder has a conical cavity cut into it as shown. Determine the depth d of the cavity so that the cylinder balances on the pivot and remains in neutral equilibrium.

Prob. 11–40

11–41. The uniform right circular cone having a mass m is suspended from the cord as shown. Determine the angle θ at which it hangs from the wall for equilibrium. Is the cone in stable equilibrium?

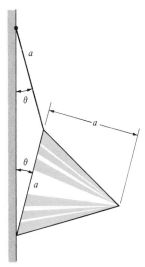

Prob. 11–41

11–42. The triangular block of weight W rests on the smooth corners which are a distance a apart. If the block has three equal sides of length d, determine the angle θ for equilibrium.

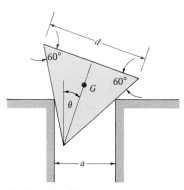

Prob. 11–42

11–43. Two uniform bars, each having a weight W, are pin-connected at their ends. If they are placed over a smooth cylindrical surface, show that the angle θ for equilibrium must satisfy the equation $\cos \theta / \sin^3 \theta = a/2r$.

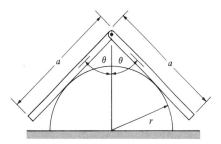

Prob. 11–43

*11–44. The homogeneous block has a mass of 10 kg and rests on the smooth corners of two ledges. Determine the angle θ for placement that will cause the block to be stable.

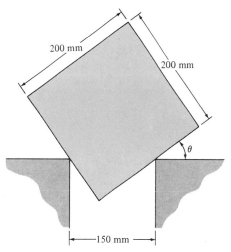

Prob. 11–44

REVIEW PROBLEMS

11–45. The three-bar mechanism of negligible weight is subjected to a couple moment $M_A = 8$ lb · ft. Determine the magnitude of the couple moment $\mathbf{M}_D$ needed to maintain the equilibrium position $\theta = 30°$, $\phi = 90°$.

Prob. 11–45

11–46. Two smooth bars of length 500 mm are pin-connected. Determine the angle θ for equilibrium and investigate the stability at this position. The bars each have a mass of 3 kg and the suspended block D has a mass of 7 kg. Cord DC has a total length of 1 m.

Prob. 11–46

11–47. Determine the force P that must be applied to the cord wrapped around the shaft at C, which is necessary to lift the mass m. As the mass is lifted, the pulley rolls on a cord that winds up on shaft B and unwinds from shaft A.

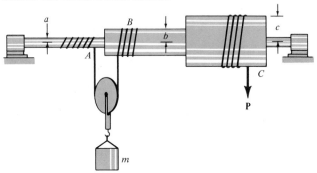

Prob. 11–47

*11–48. The uniform rod AB has a weight of 10 lb. If the spring DC is unstretched when $\theta = 90°$, determine the angle θ for equilibrium using the principle of virtual work. The spring always acts in the horizontal position because of the roller guide at D.

Probs. 11–48/11–49

11–49. Solve Prob. 11–48 using the principle of potential energy. Investigate the stability of the rod when it is in the equilibrium position.

506

■**11–50.** The spring has an unstretched length of 0.3 m. Determine the angle θ for equilibrium if the uniform links each have a mass of 5 kg.

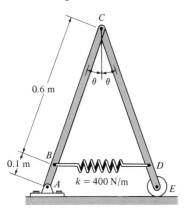

Prob. 11–50

11–51. The uniform bar AB weighs 10 lb. If the attached spring is unstretched when $\theta = 90°$, use the method of virtual work and determine the angle θ for equilibrium. Note that the spring always remains in the vertical position since the top end is attached to a roller which moves freely along the horizontal guide.

Probs. 11–51/11–52

*****11–52.** Solve Prob. 11–51 using the principle of potential energy. Investigate the stability of the bar when it is in the equilibrium position.

11–53. Compute the force developed in the spring required to keep the 6-kg rod in equilibrium when $\theta = 30°$.

Prob. 11–53

11–54. The uniform link AB has a weight of 15 lb and is pin-connected at both of its end points. The rod BD, having negligible weight, passes through a swivel block at C. If the spring has a stiffness of $k = 5$ lb/ft and is unstretched when $\theta = 0°$, determine the angle θ for equilibrium of the link. Investigate the stability at the equilibrium position. Neglect the size of the swivel block.

Prob. 11–54

Mathematical Expressions

Quadratic Formula

If $ax^2 + bx + c = 0$, then $x = \dfrac{-b \pm \sqrt{b^2 - 4ac}}{2a}$

Hyperbolic Functions

$\sinh x = \dfrac{e^x - e^{-x}}{2}$, $\cosh x = \dfrac{e^x + e^{-x}}{2}$, $\tanh x = \dfrac{\sinh x}{\cosh x}$

Trigonometric Identities

$\sin^2 \theta + \cos^2 \theta = 1$

$\sin(\theta \pm \phi) = \sin \theta \cos \phi \pm \cos \theta \sin \phi$

$\sin 2\theta = 2 \sin \theta \cos \theta$

$\cos(\theta \pm \phi) = \cos \theta \cos \phi \mp \sin \theta \sin \phi$

$\cos 2\theta = \cos^2 \theta - \sin^2 \theta$

$\cos \theta = \pm \sqrt{\dfrac{1 + \cos 2\theta}{2}}$, $\sin \theta = \pm \sqrt{\dfrac{1 - \cos 2\theta}{2}}$

$\tan \theta = \dfrac{\sin \theta}{\cos \theta}$

$1 + \tan^2 \theta = \sec^2 \theta$ $\qquad$ $1 + \text{ctn}^2 \theta = \csc^2 \theta$

Power–Series Expansions

$\sin x = x - \dfrac{x^3}{3!} + \dfrac{x^5}{5!} - \dfrac{x^7}{7!} + \cdots$

$\cos x = 1 - \dfrac{x^2}{2!} + \dfrac{x^4}{4!} - \dfrac{x^6}{6!} + \cdots$

$\sinh x = x + \dfrac{x^3}{3!} + \dfrac{x^5}{5!} + \cdots$

$\cosh x = 1 + \dfrac{x^2}{2!} + \dfrac{x^4}{4!} + \cdots$

Derivatives

$\dfrac{d}{dx}(u^n) = nu^{n-1}\dfrac{du}{dx}$

$\dfrac{d}{dx}(uv) = u\dfrac{dv}{dx} + v\dfrac{du}{dx}$

$\dfrac{d}{dx}\left(\dfrac{u}{v}\right) = \dfrac{v\dfrac{du}{dx} - u\dfrac{dv}{dx}}{v^2}$

$\dfrac{d}{dx}(\sin u) = \cos u\dfrac{du}{dx}$

$\dfrac{d}{dx}(\cos u) = -\sin u\dfrac{du}{dx}$

$\dfrac{d}{dx}(\tan u) = \sec^2 u\dfrac{du}{dx}$

$\dfrac{d}{dx}(\cot u) = -\csc^2 u\dfrac{du}{dx}$

$\dfrac{d}{dx}(\sec u) = \tan u \sec u\dfrac{du}{dx}$

$\dfrac{d}{dx}(\csc u) = -\csc u \cot u\dfrac{du}{dx}$

$\dfrac{d}{dx}(\sinh u) = \cosh u\dfrac{du}{dx}$

$\dfrac{d}{dx}(\cosh u) = \sinh u\dfrac{du}{dx}$

Integrals

$$\int x^n \, dx = \frac{x^{n+1}}{n+1} + C, \; n \neq -1$$

$$\int \frac{dx}{a+bx} = \frac{1}{b} \ln(a+bx) + C$$

$$\int \frac{dx}{a+bx^2} = \frac{1}{2\sqrt{-ba}} \ln \left[\frac{\sqrt{a}+2\sqrt{-b}}{\sqrt{a}-x\sqrt{-b}} \right] + C, \; a>0, \, b<0$$

$$\int \frac{x \, dx}{a+bx^2} = \frac{1}{2b} \ln(bx^2+a) + C$$

$$\int \frac{x^2 \, dx}{a+bx^2} = \frac{x}{b} - \frac{a}{b\sqrt{ab}} \tan^{-1} \frac{x\sqrt{ab}}{a} + C$$

$$\int \frac{dx}{a^2-x^2} = \frac{1}{2a} \ln \left[\frac{a+x}{a-x} \right] + C, \; a^2 > x^2$$

$$\int \sqrt{a+bx} \, dx = \frac{2}{3b} \sqrt{(a+bx)^3} + C$$

$$\int x\sqrt{a+bx} \, dx = \frac{-2(2a-3bx)\sqrt{(a+bx)^3}}{15b^2} + C$$

$$\int x^2\sqrt{a+bx} \, dx = \frac{2(8a^2-12abx+15b^2x^2)\sqrt{(a+bx)^3}}{105b^3} + C$$

$$\int \sqrt{a^2-x^2} \, dx = \frac{1}{2} \left[x\sqrt{a^2-x^2} + a^2 \sin^{-1} \frac{x}{a} \right] + C, \; a>0$$

$$\int x\sqrt{a^2-x^2} \, dx = -\frac{1}{3}\sqrt{(a^2-x^2)^3} + C$$

$$\int x^2\sqrt{a^2-x^2} \, dx = -\frac{x}{4}\sqrt{(a^2-x^2)^3}$$
$$+ \frac{a^2}{8}\left(x\sqrt{a^2-x^2} + a^2 \sin^{-1} \frac{x}{a} \right)$$
$$+ C, \; a>0$$

$$\int \sqrt{x^2 \pm a^2} \, dx = \frac{1}{2}\left[x\sqrt{x^2+a^2} \pm a^2 \ln(x+\sqrt{x^2 \pm a^2}) \right] + C$$

$$\int x\sqrt{x^2 \pm a^2} \, dx = \frac{1}{3}\sqrt{(x^2 \pm a^2)^3} + C$$

$$\int x^2\sqrt{x^2 \pm a^2} \, dx = \frac{x}{4}\sqrt{(x^2 \pm a^2)^3} \pm \frac{a^2}{8} x\sqrt{x^2 \pm a^2}$$
$$- \frac{a^4}{8} \ln(x+\sqrt{x^2 \pm a^2}) + C$$

$$\int \frac{dx}{\sqrt{a+bx}} = \frac{2\sqrt{a+bx}}{b} + C$$

$$\int \frac{x \, dx}{\sqrt{x^2 \pm a^2}} = \sqrt{x^2 \pm a^2} + C$$

$$\int \frac{dx}{\sqrt{a+bx+cx^2}} = \frac{1}{\sqrt{c}} \ln \left[\sqrt{a+bx+cx^2} + x\sqrt{c} + \frac{b}{2\sqrt{c}} \right]$$
$$+ C, \; c>0$$
$$= \frac{1}{\sqrt{-c}} \sin^{-1} \left(\frac{-2cx-b}{\sqrt{b^2-4ac}} \right) + C, \; c<0$$

$$\int \sin x \, dx = -\cos x + C$$

$$\int \cos x \, dx = \sin x + C$$

$$\int x \cos(ax) \, dx = \frac{1}{a^2} \cos(ax) + \frac{x}{a} \sin(ax) + C$$

$$\int x^2 \cos(ax) \, dx = \frac{2x}{a^2} \cos(ax) + \frac{a^2x^2-2}{a^3} \sin(ax) + C$$

$$\int e^{ax} \, dx = \frac{1}{a}e^{ax} + C$$

$$\int x \, e^{ax} \, dx = \frac{e^{ax}}{a^2}(ax-1) + C$$

$$\int \sinh x \, dx = \cosh x + C$$

$$\int \cosh x \, dx = \sinh x + C$$

509

B

Numerical and Computer Analysis

Occasionally the application of the laws of mechanics will lead to a system of equations for which a closed-form solution is difficult or impossible to obtain. When confronted with this situation, engineers will often use a numerical method which in most cases can be programmed on a microcomputer or "programmable" pocket calculator. Here we will briefly present a computer program for solving a set of linear algebraic equations and three numerical methods which can be used to solve an algebraic or transcendental equation, evaluate a definite integral, and solve an ordinary differential equation. Application of each method will be explained by example, and an associated computer program written in Microsoft BASIC, which is designed to run on most personal computers, is provided.* A text on numerical analysis should be consulted for further discussion regarding a check of the accuracy of each method and the inherent errors that can develop.

Linear Algebraic Equations

Application of the equations of static equilibrium or the equations of motion sometimes requires solving a set of linear algebraic equations. The computer program listed in Fig. B–1 can be used for this purpose. It is based on the

*Similar types of programs can be written or purchased for programmable pocket calculators.

```
1 PRINT"Linear system of equations":PRINT
2 DIM A(10,11)
3 INPUT"Input number of equations : ",N
4 PRINT
5 PRINT"A  coefficients"
6 FOR I = 1 TO N
7 FOR J = 1 TO N
8 PRINT "A(";I;",";J;
9 INPUT")=",A(I,J)
10 NEXT J
11 NEXT I
12 PRINT
13 PRINT"B  coefficients"
14 FOR I = 1 TO N
15 PRINT "B(";I;
16 INPUT")=",A(I,N+1)
17 NEXT I
18 GOSUB 25
19 PRINT

20 PRINT"Unknowns"
21 FOR I = 1 TO N
22 PRINT "X(";I;")=";A(I,N+1)
23 NEXT I
24 END
25 REM Subroutine Guassian
26 FOR M=1 TO N
27 NP=M
28 BG=ABS(A(M,M))
29 FOR I = M TO N
30 IF ABS(A(I,M))<=BG THEN 33
31 BG=ABS(A(I,M))
32 NP=I
33 NEXT I
34 IF NP=M THEN 40
35 FOR I = M TO N+1
36 TE=A(M,I)
37 A(M,I)=A(NP,I)
38 A(NP,I)=TE

39 NEXT I
40 FOR I = M+1 TO N
41 FC=A(I,M)/A(M,M)
42 FOR J = M+1 TO N+1
43 A(I,J)=A(I,J)-FC*A(M,J)
44 NEXT J
45 NEXT I
46 NEXT M
47 A(N,N+1)=A(N,N+1)/A(N,N)
48 FOR I = N-1 TO 1 STEP -1
49 SM=0
50 FOR J=I+1 TO N
51 SM=SM+A(I,J)*A(J,N+1)
52 NEXT J
53 A(I,N+1)=(A(I,N+1)-SM)/A(I,I)
54 NEXT I
55 RETURN
```

Fig. B–1

method of a Gauss elimination and can solve at most ten equations with ten unknowns. To do so, the equations should first be written in the following general format:

$$A_{11}x_1 + A_{12}x_2 + \cdots + A_{1n}x_n = B_1$$
$$A_{21}x_1 + A_{22}x_2 + \cdots + A_{2n}x_n = B_2$$
$$\cdot$$
$$\cdot$$
$$\cdot$$
$$A_{n1}x_1 + A_{n2}x_2 + \cdots + A_{nn}x_n = B_n$$

The "A" and "B" coefficients are "called" for when running the program. The output presents the unknowns $x_1, \ldots, x_n$.

Example B–1

Solve the two equations
$$3x_1 + x_2 = 4$$
$$2x_1 - x_2 = 10$$

SOLUTION

When the program begins to run, it first calls for the number of equations (2), then the A coefficients in the following sequence: $A_{11} = 3$, $A_{12} = 1$, $A_{21} = 2$, $A_{22} = -1$ and finally the B coefficients $B_1 = 4$, $B_2 = 10$. The output appears as:

Unknowns		
$X(1) = 2.8$		*Ans.*
$X(2) = -4.4$		*Ans.*

Simpson's Rule

This numerical method can be used to determine the area under a curve given as a graph or as an explicit function $y = f(x)$. Likewise it can be used to compute the value of a definite integral which involves the function $y = f(x)$. To do so, the area must be subdivided into an *even number* of strips or intervals having a width h. The curve between three consecutive ordinates is approximated by a parabola, and the entire area or definite integral is then determined from the formula

$$\int_{x_0}^{x_n} f(x)\, dx \simeq \frac{h}{3}[y_0 + 4(y_1 + y_3 + \cdots + y_{n-1})$$
$$+ 2(y_2 + y_4 + \cdots + y_{n-2}) + y_n] \quad \text{(B–1)}$$

The computer program for this equation is given in Fig. B–2. For its use, one must first enter the function (line 6), then the upper and lower limits of the integral (lines 7 and 8), and finally the number of intervals (line 9). The value of the integral is then given as the output.

```
1 PRINT"Simpson's rule":PRINT
2 PRINT" To execute this program :":PRINT
3 PRINT"   1- Modify right-hand side of the equation given below,
4 PRINT"      then press RETURN key"
5 PRINT"   2- Type  RUN 6":PRINT:EDIT 6
6 DEF FNF(X)=LOG(X)
7 PRINT:INPUT" Enter Lower Limit = ",A
8 INPUT" Enter Upper Limit = ",B
9 INPUT" Enter Number (even) of Intervals = ",N%
10 H=(B-A)/N%:AR=FNF(A):X=A+H
11 FOR J%=2 TO N%
12 K=2*(2-J%+2*INT(J%/2))
13 AR=AR+K*FNF(X)
14 X=X+H:NEXT J%
15 AR=H*(AR+FNF(B))/3
16 PRINT" Integral = ",AR
17 END
```

Fig. B–2

Example B–2

Evaluate the definite integral $\int_2^5 \ln x \, dx$

SOLUTION

The interval $x_0 = 2$ to $x_6 = 5$ will be divided into six equal parts ($n = 6$), each having a width $h = (5 - 2)/6 = 0.5$. We then compute $y = f(x) = \ln x$ at each point of subdivision.

n	x_n	y_n
0	2	0.693
1	2.5	0.916
2	3	1.099
3	3.5	1.253
4	4	1.386
5	4.5	1.504
6	5	1.609

Thus, Eq. B–1 becomes

$$\int_2^5 \ln x \, dx \simeq \frac{0.5}{3}[0.693 + 4(0.916 + 1.253 + 1.504)$$
$$+ 2(1.099 + 1.386) + 1.609]$$
$$\simeq 3.66 \qquad\qquad Ans.$$

This answer is equivalent to the exact answer to three significant figures. Obviously, accuracy to a larger number of significant figures can be improved by selecting a smaller interval h (or larger n).

Using the computer program, we enter the function $\ln x$, line 6 in Fig. B–2, the upper and lower limits 2 and 5, and the number of intervals $n = 6$. The output appears as

$$\text{Integral} = 3.66082 \qquad\qquad Ans.$$

The Secant Method

This numerical method is used to find the real roots of an algebraic or transcendental equation $f(x) = 0$. The method derives its name from the fact that the formula used is established from the slope of the secant line to the graph $y = f(x)$. This slope is $(f(x_n) - f(x_{n-1}))/(x_n - x_{n-1})$ and the secant formula is

$$x_{n+1} = x_n - f(x_n) \left(\frac{x_n - x_{n-1}}{f(x_n) - f(x_{n-1})} \right) \tag{B-2}$$

For application it is necessary to provide two initial guesses, x_0 and x_1, and thereby evaluate x_2 from Eq. B-2 ($n = 1$). One then proceeds to reapply Eq. B-2 with x_1 and the calculated value of x_2 and obtain x_3 ($n = 2$), etc., until the value $x_{n+1} \simeq x_n$. One can see this will occur if x_n is approaching the root of the function $f(x) = 0$ since the correction term on the right of Eq. B-2 will tend toward zero. In particular, the larger the slope the smaller the correction to x_n and the faster the root will be found. Likewise, if the slope is very small in the neighborhood of the root, the method leads to large corrections for x_n and convergence to the root is slow and may even lead to a failure to find it. In such cases other numerical techniques must be used for solution.

A computer program based on Eq. B-2 is listed in Fig. B-3. The user must supply the function (line 7) along with two initial guesses x_0 and x_1 used to approximate the solution. The output specifies the value of the root. If it cannot be determined it is so stated.

```
1 PRINT"Secant method":PRINT
2 PRINT" To execute this program :":PRINT
3 PRINT"    1) Modify right hand side of the equation given below,"
4 PRINT"       then press RETURN key."
5 PRINT"    2) Type  RUN 7"
6 PRINT:EDIT 7
7 DEF FNF(X)=.5*SIN(X)-2*COS(X)+1.3
8 INPUT"Enter point #1 =",X
9 INPUT"Enter point #2 =",X1
10 IF X=X1 THEN 14
11 EP=.00001:TL=2E-20
12 FP=(FNF(X1)-FNF(X))/(X1-X)
13 IF ABS(FP)>TL THEN 15
14 PRINT"Root can not be found.":END
15 DX=FNF(X1)/FP
16 IF ABS(DX)>EP THEN 19
17 PRINT "Root = ";X1;"     Function evaluated at this root = ";FNF(X1)
18 END
19 X=X1:X1=X1-DX
20 GOTO 12
```

Fig. B-3

Example B–3

Determine the root of the equation

$$f(x) = 0.5 \sin x - 2 \cos x + 1.30 = 0$$

SOLUTION

Guesses of the initial roots will be $x_0 = 45°$ and $x_1 = 30°$. Applying Eq. B–2,

$$x_2 = 30° - (-0.1821)\frac{(30° - 45°)}{(-0.1821 - 0.2393)} = 36.48°$$

Using this value in Eq. B–2, along with $x_1 = 30°$, we have

$$x_3 = 36.48° - (-0.0108)\frac{36.48° - 30°}{(-0.0108 + 0.1821)} = 36.89°$$

Repeating the process with this value and $x_2 = 36.48°$ yields

$$x_4 = 36.89° - (0.0005)\left(\frac{36.89° - 36.48°}{(0.0005 + 0.0108)}\right) = 36.87°$$

Thus $x = 36.9°$ is appropriate to three significant figures.

If the problem is solved using the computer program, we enter the function, line 7 in Fig. B–3. The first and second guesses must be in radians. Choosing these to be 0.8 rad, and 0.5 rad, the result appears as

Root = 0.6435022.
Function evaluated at this root = 1.66893E–06.

This result converted from radians to degrees is therefore

$$x = 36.9° \hspace{4cm} Ans.$$

Runge-Kutta Method

This numerical method is used to solve an ordinary differential equation. It consists of applying a set of formulas which are used to find specific values of y for corresponding incremental values h in x. The formulas given in general form are as follows:

First-Order Equation. To integrate $\dot{x} = f(t, x)$ step-by-step use

$$x_{i+1} = x_i + \frac{1}{6}(k_1 + 2k_2 + 2k_3 + k_4) \qquad (B-3)$$

where

$$k_1 = hf(t_i, x_i)$$
$$k_2 = hf\left(t_i + \frac{h}{2}, x_i + \frac{k_1}{2}\right)$$
$$k_3 = hf\left(t_i + \frac{h}{2}, x_i + \frac{k_2}{2}\right) \qquad (B-4)$$
$$k_4 = hf(t_i + h, x_i + k_3)$$

Second-Order Equation. To integrate $\ddot{x} = f(t, x, \dot{x})$ use

$$x_{i+1} = x_i + h\left[\dot{x}_i + \frac{1}{6}(k_1 + k_2 + k_3)\right]$$
$$\dot{x}_{i+1} = \dot{x}_i + \frac{1}{6}(k_1 + 2k_2 + 2k_3 + k_4) \qquad (B-5)$$

where

$$k_1 = hf(t_i, x_i, \dot{x}_i)$$
$$k_2 = hf\left(t_i + \frac{h}{2}, x_i + \frac{h}{2}\dot{x}_i, \dot{x}_i + \frac{k_1}{2}\right)$$
$$k_3 = hf\left(t_i + \frac{h}{2}, x_i + \frac{h}{2}\dot{x}_i + \frac{h}{4}k_1, \dot{x}_i + \frac{k_2}{2}\right) \qquad (B-6)$$
$$k_4 = hf\left(t_i + h, x_i + h\dot{x}_i + \frac{h}{2}k_2, \dot{x}_i + k_3\right)$$

To apply these equations, one starts with initial values $t_i = t_0$, $x_i = x_0$ and $\dot{x}_i = \dot{x}_0$ (for the second-order equation). Choosing an increment h for t_0 the four constants k are computed and these results are substituted into Eq. B–3 or B–5 in order to compute $x_{i+1} = x_1$, $\dot{x}_{i+1} = x_1$, corresponding to $t_{i+1} = t_1 = t_0 + h$. Repeating this process using t_1, x_1, $\dot{x}_1$ and h, values for x_2, $\dot{x}_2$ and $t_2 = t_1 + h$ are then computed, etc.

Computer programs which solve first and second-order differential equations by this method are listed in Figs. B–4 and B–5, respectively. In order to use these programs, the operator specifies the function $\dot{x} = f(t, x)$ or $\ddot{x} = f(t, x, \dot{x})$ (line 7), the initial values t_0, x_0, $\dot{x}_0$ (for second-order equation), the final time t_n, and the step size h. The output gives the values of t, x, and $\dot{x}$ for each time increment until t_n is reached.

```
1 PRINT"Runge-Kutta Method for 1-st order Differential Equation":PRINT
2 PRINT" To execute this program :":PRINT
3 PRINT"    1) Modify right hand side of the equation given below,"
4 PRINT"       then Press RETURN key"
5 PRINT"    2) Type  RUN 7"
6 PRINT:EDIT 7
7 DEF FNF(T,X)=5*T+X
8 CLS:PRINT" Initial Conditions":PRINT
9 INPUT"Input  t   = ",T
10 INPUT"       x   = ",X
11 INPUT"Final  t   = ",T1
12 INPUT"step size  = ",H:PRINT
13 PRINT"         t              x"
14 IF T>=T1+H THEN 23
15 PRINT USING"######.#####";T;X
16 K1=H*FNF(T,X)
17 K2=H*FNF(T+.5*H,X+.5*K1)
18 K3=H*FNF(T+.5*H,X+.5*K2)
19 K4=H*FNF(T+H,X+K3)
20 T=T+H
21 X=X+(K1+K2+K2+K3+K3+K4)/6
22 GOTO 14
23 END
```

Fig. B–4

```
1 PRINT"Runge-Kutta Method for 2-nd order Differential Equation":PRINT
2 PRINT" To execute this program :":PRINT
3 PRINT"    1) Modify right hand side of the equation given below,"
4 PRINT"       then Press RETURN key"
5 PRINT"    2) Type  RUN 7"
6 PRINT:EDIT 7
7 DEF FNF(T,X,XD)=
8 INPUT"Input  t   = ",T
9 INPUT"       x   = ",X
10 INPUT"    dx/dt = ",XD
11 INPUT"Final  t   = ",T1
12 INPUT"step size  = ",H:PRINT
13 PRINT"         t         x         dx/dt"
14 IF T>=T1+H THEN 24
15 PRINT USING"######.#####";T;X;XD
16 K1=H*FNF(T,X,XD)
17 K2=H*FNF(T+.5*H,X+.5*H*XD,XD+.5*K1)
18 K3=H*FNF(T+.5*H,X+(.5*H)*(XD+.5*K1),XD+.5*K2)
19 K4=H*FNF(T+H,X+H*XD+.5*H*K2,XD+K3)
20 T=T+H
21 X=X+H*XD+H*(K1+K2+K3)/6
22 XD=XD+(K1+K2+K2+K3+K3+K4)/6
23 GOTO 14
24 END
```

Fig. B–5

Example B–4

Solve the differential equation $\dot{x} = 5t + x$. Obtain the results for two steps using time increments of $h = 0.02$ s. At $t_0 = 0$, $x_0 = 0$.

SOLUTION

This is a first-order equation, so Eqs. B–3 and B–4 apply. Thus, for $t_0 = 0$, $x_0 = 0$, $h = 0.02$, we have

$$k_1 = 0.02[0 + 0] = 0$$
$$k_2 = 0.02[5(0.1) + 0] = 0.001$$
$$k_3 = 0.02[5(0.01) + 0.0005] = 0.00101$$
$$k_4 = 0.02[5(0.02) + 0.00101] = 0.00202$$

$$x_1 = 0 + \frac{1}{6}[0 + 2(0.001) + 2(0.00101) + 0.00202] = 0.00101$$

Using the values $t_1 = 0 + 0.02 = 0.02$ and $x_1 = 0.00101$ with $h = 0.02$, the value for x_2 is now computed from Eqs. B–3 and B–4.

$$k_1 = 0.02[5(0.02) + 0.00101] = 0.00202$$
$$k_2 = 0.02[5(0.03) + 0.00202] = 0.00304$$
$$k_3 = 0.02[5(0.03) + 0.00253] = 0.00305$$
$$k_4 = 0.02[5(0.04) + 0.00406] = 0.00408$$

$$x_2 = 0.001 + \frac{1}{6}[0.00202 + 2(0.00304) + 2(0.00305) + 0.00408]$$

$$x_2 = 0.00405 \qquad\qquad\qquad\qquad\qquad \textit{Ans.}$$

To solve this problem using the computer program in Fig. B–4, the function is first entered on line 7, then the data $t_0 = 0$, $x_0 = 0$, $t_n = 0.04$, and $h = 0.02$ is specified. The results appear as

t	x
0.00000	0.00000
0.02000	0.00101
0.04000	0.00405

$\textit{Ans.}$

Answers

2–46. 389 N, 42.7°.

2–47. $F = 2.03$ kN, $F_R = 7.87$ kN.

2–49. $\mathbf{F}_1 = \{53.1\mathbf{i} - 44.5\mathbf{j} + 40\mathbf{k}\}$ lb, $\mathbf{F}_2 = \{-130\mathbf{k}\}$ lb,
$\mathbf{F}_R = \{53.1\mathbf{i} - 44.5\mathbf{j} - 90\mathbf{k}\}$ lb, $F_R = 114$ lb, $\alpha = 62.1°$,
$\beta = 113°$, $\gamma = 142°$.

2–50. $\mathbf{F}_1 = \{4.80\mathbf{i} + 6.40\mathbf{j}\}$ kN, $\mathbf{F}_2 = \{3.00\mathbf{i} - 3.00\mathbf{j} + 4.24\mathbf{k}\}$ kN,
$\mathbf{F}_R = \{7.80\mathbf{i} + 3.40\mathbf{j} + 4.24\mathbf{k}\}$ kN, $F_R = 9.51$ kN, $\alpha = 34.9°$,
$\beta = 69.0°$, $\gamma = 63.5°$.

2–51. $\mathbf{F} = \{13.7\mathbf{i} + 37.6\mathbf{j} + 30.0\mathbf{k}\}$ N.

2–53. $\mathbf{F} = \{217\mathbf{i} + 85.5\mathbf{j} - 91.2\mathbf{k}\}$ lb.

2–54. 66.4 lb, $\alpha = 59.8°$, $\beta = 107°$, $\gamma = 144°$.

2–55. $\alpha = 124°$, $\beta = 71.3°$, $\gamma = 140°$.

2–57. F_1: $\alpha = 36.9°$, $\beta = 90°$, $\gamma = 53.1°$, F_R: $\alpha = 69.3°$,
$\beta = 52.2°$, $\gamma = 45°$.

2–58. $F_x = 1.28$ kN, $F_y = 2.60$ kN, $F_z = 0.776$ kN.

2–59. $F = 2.02$ kN, $F_y = 0.523$ kN.

2–61. $F_3 = 428$ lb, $\alpha = 88.3°$, $\beta = 20.6°$, $\gamma = 69.5°$.

2–62. $F_3 = 250$ lb, $\alpha = 87.0°$, $\beta = 143°$, $\gamma = 53.1°$.

2–63. $r = 7$ ft, $\alpha = 31.0°$, $\beta = 107°$, $\gamma = 115°$.

2–65. $\mathbf{r} = \{-2.35\mathbf{i} + 3.93\mathbf{j} + 3.71\mathbf{k}\}$ ft, $r = 5.89$ ft, $\alpha = 113°$,
$\beta = 48.2°$, $\gamma = 51.0°$.

2–66. $x = 3.58$ m, $y = 1.79$ m.

2–67. 3.74 m, 143°.

2–69. 9.06 in.

2–70. 567 m.

2–71. $r_{ACB} = 7.81$ ft, $r_{OC} = 3.91$ ft.

2–73. $\mathbf{F}_1 = \{-177\mathbf{i} + 306\mathbf{j} - 354\mathbf{k}\}$ N, $\mathbf{F}_2 = \{477\mathbf{i} - 239\mathbf{j} - 596\mathbf{k}\}$ N,
$F_R = 998$ N, $\alpha = 72.5°$, $\beta = 86.1°$, $\gamma = 162°$.

2–74. $\mathbf{F}_1 = \{-551\mathbf{i} + 184\mathbf{j} - 551\mathbf{k}\}$ N,
$\mathbf{F}_2 = \{-486\mathbf{i} - 135\mathbf{j} + 324\mathbf{k}\}$ N, $F_R = 1.06$ kN, $\alpha = 167°$,
$\beta = 87.4°$, $\gamma = 102°$.

2–75. $\mathbf{F}_1 = \{32.7\mathbf{i} + 65.3\mathbf{j} - 32.7\mathbf{k}\}$ lb, $\mathbf{F}_2 = \{46.9\mathbf{j} + 37.5\mathbf{k}\}$ lb,
$F_R = 117$ lb, $\alpha = 73.8°$, $\beta = 16.4°$, $\gamma = 87.6°$

2–77. $x = 14.3$ m, $y = -10.7$ m, $z = 8.94$ m.

2–78. $\mathbf{F} = \{-146\mathbf{i} - 296\mathbf{j} + 225\mathbf{k}\}$ N, $\alpha = 111°$, $\beta = 138°$,
$\gamma = 55.7°$.

2–79. $x = 7.65$ ft, $y = 4.24$ ft, $z = 3.76$ ft.

2–81. $\mathbf{F}_A = \{28.8\mathbf{i} - 16.6\mathbf{j} - 49.9\mathbf{k}\}$ lb, $\mathbf{F}_B = \{-28.8\mathbf{i} - 16.6\mathbf{j} - 49.9\mathbf{k}\}$ lb, $\mathbf{F}_C = \{33.3\mathbf{j} - 49.9\mathbf{k}\}$ lb, $F_R = 150$ lb, $\alpha = 90°$,
$\beta = 90°$, $\gamma = 180°$.

2–82. 52.1 N.

2–83. $\mathbf{F} = \{143\mathbf{i} + 248\mathbf{j} - 201\mathbf{k}\}$ lb.

2–85. $\mathbf{F}_A = \{16.4\mathbf{j} - 25.1\mathbf{k}\}$ N, $\mathbf{F}_B = \{-7.60\mathbf{i} - 20.9\mathbf{j} - 39.1\mathbf{k}\}$ N,
$F_R = 64.9$ N, $\alpha = 96.7°$, $\beta = 94.0°$, $\gamma = 172°$.

2–86. $x = 2.34$ m, $z = 1.96$ m.

2–87. $F_R = 664$ N, $\alpha = 87.2°$, $\beta = 176°$, $\gamma = 87.4°$.

2–89. 82.9°.

2–90. $\mathbf{r}_P = \{-1.29\mathbf{i} + 0.429\mathbf{j} + 0.857\mathbf{k}\}$ ft.

2–91. $F_1 = 333$ N, $F_2 = 373$ N.

2–93. $\mathbf{r}_P = \{0.577\mathbf{i} + 0.769\mathbf{j} - 0.192\mathbf{k}\}$ m.

2–94. 2.11 ft.

2–95. 31.4 lb.

2–97. 97.3°.

2–98. 48.0 N.

2–99. $\theta = 52.4°$, $\phi = 68.2°$.

2–101. $F_{PAB} = 70.5$ N, $F_{PAC} = 65.1$ N.

2–102. 70.5°.

2–103. 65.8°.

2–105. 10.5 lb.

2–106. 143°.

2–107. $F = 63.2$ mN, $F_y = 44.7$ mN.

2–109. $F_2 = 246$ N, $\alpha = 54.9°$, $\beta = 114°$, $\gamma = 44.7°$.

2–110. $r_{AB} = 492$ mm, $\mathbf{F} = \{-13.2\mathbf{i} - 18.3\mathbf{j} + 19.8\mathbf{k}\}$ N.

2–111. 48.0 N.

2–113. $\mathbf{F} = \{13.4\mathbf{i} + 23.2\mathbf{j} + 53.7\mathbf{k}\}$ lb.

2–114. 215 lb, 52.7°.

2–115. $F_1 = 14.0$ lb, $F_2 = 8.35$ lb.

CHAPTER 3

3–1. $F_1 = 435$ lb, $F_2 = 171$ lb.

3–2. 552 N, 12.9°.

3–3. 11.0 kN, 49.8°.

3–5. 4.31 kN, 4.69°.

3–6. $T_{AB} = 239$ N, $T_{AC} = 243$ N.

3–7. $F_{AB} = 50$ lb, $F_{BC} = 70.7$ lb, $F_{AD} = 70.7$ lb, $h = 23$ ft.

3–9. $F_{CA} = F_{BA} = 1.20$ kN, $F_{BC} = 687$ N.

3–10. $T = \dfrac{W}{2 \sin \theta}$.

3–11. 76.6 lb.

3–13. 2.42 m.

3–14. $F_{AC} = 160$ lb, $\theta = 90°$, $F_{AB} = 160$ lb, $\theta = 120°$.

3–15. $T_{AB} = 340$ N, $T_{AE} = 170$ N, $T_{BD} = 490$ N, $T_{BC} = 562$ N.

3–17. 14.8 lb.

3–18. 40.2°.

3–19. 22.3 N.

3–21. 185 mm.

3–22. 55.0°.

3–23. 46.1 N.

3–25. 4.11 m.

3–26. $F_{AB} = \dfrac{W}{(\sin \theta + \cos \theta \tan \phi)}$, $\theta = \phi = 0°$, $F_{AB} \to \infty$.

3–27. $w = \dfrac{2W}{(L - y)} \sqrt{(L - y)^2 - d^2}$.

3–29. $F_1 = 1.47$ kN, $F_2 = 1.43$ kN, $F_3 = 0.571$ kN.

3–30. $F_1 = 800$ N, $F_2 = 147$ N, $F_3 = 564$ N.

3–31. $F_{AB} = 580$ N, $F_{AD} = 312$ N, $F_{AC} = 312$ N.

3–33. $F_1 = 8.26$ kN, $F_2 = 3.84$ kN, $F_3 = 12.2$ kN.

3–34. $F_{AB} = 226$ N, $F_{AC} = 197$ N, $F_{AD} = 520$ N.

3–35. $\mathbf{F} = \{304\mathbf{i} + 279\mathbf{j} - 152\mathbf{k}\}$ N.

3–37. $F_{AD} = 1.70$ kip, $F_{AC} = 0.744$ kip, $F_{AB} = 1.37$ kip.

3–38. $F_{AD} = 1.42$ kip, $F_{AC} = 0.913$ kip, $F_{AB} = 1.47$ kip.

3–39. $F_{AB} = 250$ lb, $F_{AC} = 389$ lb, $F_{AD} = 194$ lb.

3–41. $F_{AB} = 3.01$ kip, $F_{AC} = 3.01$ kip, $F_{AD} = 4.26$ kip.

3–42. 1.64 ft.

3–43. 5.33 ft.

3–45. 2.34 ft, 175 lb.

3–46. $F_{AB} = 1.32$ kN, $F_{DA} = 1.42$ kN, $F_{EA} = 1.42$ kN.

3–47. $F_{OA} = 319$ N, $F_{AC} = 85.8$ N, $F_{AB} = 110$ N.

3–49. 254 lb, 20°.

3–50. $T_{AB} = 109$ lb, $T_{AC} = 47.4$ lb, $T_{AD} = 87.9$ lb.

3–51. $F_{BD} = 2.68$ mN, $F_{BC} = 1.60$ mN, $F_{AE} = 5.03$ mN.

CHAPTER 4

4–5. 8.00 kN · m ↓.

4–6. 2.13 kN · m ↾.

4–7. 2.37 kN · m ↾.

4–9. 3.15 kN · m ↾.

4–10. $M_{A1} = 1.44$ kip-ft ↓, $M_{A2} = 7.20$ kip-ft ↓, $M_{A3} = 5.20$ kip-ft ↓.

4–11. $M_{B1} = 3.96$ kip-ft ↾, $M_{B2} = 4.80$ kip-ft ↾, $M_{B3} = 1.30$ kip-ft ↾.

4–13. $M_{A1} = 433$ N · m ↓, $M_{A2} = 1.30$ kN · m ↓, $M_{A3} = 800$ N · m ↓.

4–14. $M_{CA} = 162$ lb-ft ↓, $M_{CB} = 260$ lb-ft ↾, counterclockwise.

4–15. 28.9 lb.

4–17. $M_A = 1200 \sin \theta + 800 \cos \theta$.

4–18. $m = \left(\dfrac{l}{d + l}\right)M$.

4–19. $x = 24.0$ m.

4–21. 50 N, 53.1°.

4–22. 42.5 N, 28.1°.

4–23. $\mathbf{M}_O = \{260\mathbf{i} + 180\mathbf{j} + 510\mathbf{k}\}$ N · m.

4–25. $\mathbf{M}_O = \{-84\mathbf{i} - 8\mathbf{j} - 39\mathbf{k}\}$ kN · m.

4–26. $\mathbf{M}_P = \{-116\mathbf{i} + 16\mathbf{j} - 135\mathbf{k}\}$ kN · m.

4–27. $\mathbf{M}_O = \{61.2\mathbf{i} + 81.6\mathbf{j}\}$ N · m.

4–29. $\mathbf{M}_{O1} = \{8\mathbf{j} + 12\mathbf{k}\}$ N · m, $\mathbf{M}_{O2} = \{-42\mathbf{i} + 42\mathbf{j} - 48\mathbf{k}\}$ N · m, $\mathbf{M}_{O3} = \{78\mathbf{k}\}$ N · m, $M_O = 77.6$ N · m, $\alpha = 123°$, $\beta = 49.9°$, $\gamma = 57.3°$.

4–30. $\mathbf{M}_{A1} = \{-4\mathbf{j} - 6\mathbf{k}\}$ N · m, $\mathbf{M}_{A2} = \{-42\mathbf{i} - 60\mathbf{k}\}$ N · m, $\mathbf{M}_{A3} = \mathbf{0}$, $\mathbf{M}_A = \{-42\mathbf{i} - 4\mathbf{j} - 66\mathbf{k}\}$ N · m, $M_A = 78.3$ N, $\alpha = 122°$, $\beta = 92.9°$, $\gamma = 147°$.

4–31. $\mathbf{M}_O = \{-12.0\mathbf{i} + 9.59\mathbf{j} - 16.6\mathbf{k}\}$ kN · m, $d = 5.66$ m.

4–33. $\mathbf{M}_B = \{-37.6\mathbf{i} + 90.7\mathbf{j} - 155\mathbf{k}\}$ N · m.

4–34. $x = 2$ m, $y = 5$ m.

4–35. $x = 2$ m, $z = 3$ m.

4–37. $\mathbf{M}_O = \{373\mathbf{i} - 99.9\mathbf{j} + 173\mathbf{k}\}$ N · m.

4–38. $\mathbf{M}_{oa} = \{-113\mathbf{i} - 150\mathbf{j}\}$ N · m.

4–39. $M_{aa} = \{-39.1\mathbf{i} - 42.3\mathbf{j} + 13.0\mathbf{k}\}$ kN · m

4–41. $M_{AB1} = 72$ N · m, $M_{AB2} = M_{AB3} = 0$.

4–42. $\mathbf{M}_y = \{-78.4\mathbf{j}\}$ lb-ft.

4–43. $\mathbf{M}_x = \{0.211\mathbf{i}\}$ N · m.

4–45. $\mathbf{M}_{OC} = \{-4.61\mathbf{i} - 3.46\mathbf{j}\}$ N · m.

4–46. 136 N · m.

4–47. 165 N · m.

4–49. $\mathbf{M}_x = \{15\mathbf{i}\}$ lb-ft, $\mathbf{M}_y = \{4\mathbf{j}\}$ lb-ft, $\mathbf{M}_z = \{36\mathbf{k}\}$ lb-ft.

4–50. $\mathbf{M}_{AC} = \{11.5\mathbf{i} + 8.64\mathbf{j}\}$ lb-ft.

4–51. $\mathbf{M}_z = \{-42.7\mathbf{k}\}$ N · m.

4–53. 19.8 lb.

4–54. $\mathbf{M}_z = \{35.4\mathbf{k}\}$ N · m.

4–55. 5.66 N.

4–57. 17.6 kN · m ↾.

4–58. $\mathbf{M}_C = \{-411\mathbf{i} - 257\mathbf{j} - 651\mathbf{k}\}$ lb-ft.

4–59. $\mathbf{M}_C = \{-360\mathbf{i} + 380\mathbf{j} + 320\mathbf{k}\}$ lb-ft.

4–61. 2.03 ft.

4–62. 126 lb-ft ↾.

4–63. 126 lb-ft ↾.

4–65. 39.7 N · m.

4–66. $M_2 = 80$ N · m, $M_3 = 80$ N · m.

4–67. $\mathbf{M}_C = \{-5\mathbf{i} + 8.75\mathbf{j}\}$ N · m.

4–69. $M_R = 30.0$ N · m, $\alpha_1 = 143°$, $\beta_1 = 90°$, $\gamma_1 = 53.1°$.

4–70. $\mathbf{M}_C = \{-12.1\mathbf{i} - 10\mathbf{j} - 17.3\mathbf{k}\}$ N · m.

4–71. 342 mm.

4–73. 80 N ⦨ 60°, 47.8 N · m ↾.

4–74. 80 N ⦨ 60°, 176 N · m ↾.

4–75. 178 N ↘ 73.0°, 220 N · m ↾.

4–77. 6.57 kN ↙ 57.4°, 19.9 kN · m ↾.

4–78. 6.57 kN ↙ 57.4°, 31.0 kN · m ↾.

4–79. $F_R = 610$ N →, $y = -2.57$ m.

4–81. 55.7 lb 12.9° ↘, -13.8 ft.

4–82. 30 N 60° ↘, 14.3 N · m ↓.

4–83. 4.43 kN ↙ 71.6°, 3.52 m.

4–85. 5.93 kN, 77.8° ↘, 34.8 kN · m ↓.

4–86. 5.93 kN, 77.8° ↘, 11.6 kN · m ↾.

4–87. 180 lb, 32.0° ↘, 4.79 kip-in.

4–89. 1.08 kN, 68.2° ↘, 0.901 kN · m ↾.

4–90. 9.86 kip, 81.6 kip-ft ↓.

4–91. 9.86 kip, 8.28 ft.

4–93. 129 lb, 60.1° ↘, 4.64 ft.

4–94. 129 lb, 60.1° ↘, 621 lb-ft ↾.

4–95. $\mathbf{F}_R = \{50\mathbf{i} - 30\mathbf{j} + 80\mathbf{k}\}$ N, $\mathbf{M}_{RO} = \{410\mathbf{i} + 230\mathbf{j} - 170\mathbf{k}\}$ N · m.

4–97. $\mathbf{F}_R = \{37.4\mathbf{i} + 56.1\mathbf{j} - 18.7\mathbf{k}\}$ lb, $\mathbf{M}_{RO} = \{-307\mathbf{i} + 78.4\mathbf{j} + 131\mathbf{k}\}$ lb-ft.

4–98. $\mathbf{F}_R = \{37.4\mathbf{i} + 56.1\mathbf{j} - 18.7\mathbf{k}\}$ lb, $\mathbf{M}_{RP} = \{-307\mathbf{i} - 90\mathbf{j} - 374\mathbf{k}\}$ lb-ft.

4–99. $\mathbf{F}_R = \{70\mathbf{i} - 30\mathbf{j} + 27\mathbf{k}\}$ lb, $\mathbf{M}_{RP} = \{-34\mathbf{i} - 127\mathbf{j} - 280\mathbf{k}\}$ lb-ft.

4–101. $\mathbf{F}_R = \{40\mathbf{i} - 40\mathbf{j} + 56.6\mathbf{k}\}$ lb, $\mathbf{M}_{RP} = \{-492\mathbf{i} + 574\mathbf{j} + 983\mathbf{k}\}$ lb-ft.

4–102. $\mathbf{F}_R = \{-180\mathbf{k}\}$ N, $x = 1.06$ m.

4–103. $\mathbf{F}_R = \{-4\mathbf{i} + 2\mathbf{j} - 3\mathbf{k}\}$ kN, $\mathbf{M}_R = \{-0.05\mathbf{i} + 11\mathbf{j} + 7.40\mathbf{k}\}$ kN · m.

4–105. $\mathbf{F}_R = \{14.3\mathbf{i} + 21.4\mathbf{j} - 42.9\mathbf{k}\}$ lb, $\mathbf{M}_{RC} = \{-1.93\mathbf{i} + 0.429\mathbf{j} - 0.429\mathbf{k}\}$ kip-ft.

4–106. $\mathbf{F}_R = \{-70\mathbf{i} + 140\mathbf{j} - 408\mathbf{k}\}$ N, $\mathbf{M}_{RO} = \{-26\mathbf{i} + 31.0\mathbf{j} + 14.6\mathbf{k}\}$ N · m.

4–107. $\mathbf{F}_R = \{-40\mathbf{j} - 40\mathbf{k}\}$ N, $\mathbf{M}_{RA} = \{-12\mathbf{j} + 12\mathbf{k}\}$ N · m.

4–109. $\mathbf{F}_R = \{-700\mathbf{j}\}$ lb, $z = 0.447$ ft, $x = -0.117$ ft.

4–110. $F_B = 163$ lb, $F_C = 223$ lb.

4–111. $\mathbf{F}_R = \{270\mathbf{k}\}$ N, $\mathbf{M}_{RO} = \{-2.22\mathbf{i}\}$ N · m.

4–113. $\mathbf{F}_{Rw} = \{-40\mathbf{i}\}$ N, $\mathbf{M}_{Rw} = \{-30\mathbf{i}\}$ N · m, $y = 0.20$ m.

4–114. $\mathbf{F}_w = \{-10\mathbf{j}\}$ lb, $\mathbf{M}_w = \{-14.1\mathbf{j}\}$ lb · ft, $z = 0.586$ ft.

4–115. $F_R = 990$ N, $M_R = 3.07$ kN · m, $x = 1.16$ m, $y = 2.06$ m.

4–117. 0.525 kN, 0.171 m.

4–118. 9.69 kip ↙ 68.2°, 225 kip-ft ↓.

4–119. 3.10 kN, 2.06 m.

4–121. 13.3 lb ↓, 3.09 ft.

4–122. 90 kN ↓ , 338 kN · m ↓.

4–123. 10.7 kip ↓ , 0.479 ft.

4–125. 3.25 kip, 67.2° ⤢ , 3.86 ft.

4–126. $F_R = 0$, $M_{RO} = 1.35$ kip-ft ↑.

4–127. $d = 1.50$ m, $w = 175$ N/m.

4–129. 1.35 kN, 42.0° ⤢ , 0.556 m.

4–130. 3.60 kN ↓ , 16.2 kN · m ↓.

4–131. 22.4 kip, 7.5 ft.

4–133. 43.6 lb, 3.27 ft.

4–134. 14.9 kN, 2.27 m.

4–135. $F_R = \dfrac{w_0}{a}(e^{aL} - 1)$, $\bar{x} = \dfrac{(e^{aL}aL - e^{aL} + 1)}{a(e^{aL} - 1)}$.

4–137. $\mathbf{M}_A = \{-229\mathbf{i} + 132\mathbf{j}\}$ lb-ft.

4–138. $M_O = 18$ N · m ↓, $\theta = 120°$, $(M_O)_{max} = 36$ N · m ↓.

4–139. 392 N · m.

4–141. 30.1 kN · m ↓, 8.66 kN.

4–142. 904 lb, ⤡ 66.4°, 6.10 ft.

4–143. $\mathbf{M}_C = \{37.5\mathbf{i} - 25\mathbf{j}\}$ N · m, $M_C = 45.1$ N · m.

4–145. 294 N, ⤢ 40.1°, 39.6 N · m ↓.

CHAPTER 5

5–11. $R_A = 425$ N, $R_B = 245$ N.

5–13. $F_{BD} = 2.55$ kN, $A_x = 2.55$ kN, $A_y = 0.491$ kN.

5–14. $T_{BC} = 11.1$ kip, $A_x = 10.2$ kip, $A_y = 6.15$ kip.

5–15. $F_{BC} = 3.76$ kip, $A_x = 3.26$ kip, $A_y = 3.12$ kip.

5–17. $R_A = 105$ lb, $B_x = 97.4$ lb, $B_y = 269$ lb.

5–18. $R_A = 33.3$ lb, $T = 10.1$ lb, $R_B = 14.5$ lb.

5–19. $B_x = 2.14$ kip, $A_x = 2.14$ kip, $A_y = 5.50$ kip.

5–21. $F_B = 697$ N, $F_A = 82.0$ N.

5–22. $F_B = 1.99$ kN, $F_C = 1.78$ kN.

5–23. $F_H = 59.4$ lb, $T_B = 67.4$ lb.

5–25. $D_x = 0$, $D_y = 1.65$ kip, $M_D = 1.40$ kip-ft, $(M_D)_{max} = 3.00$ kip-ft.

5–26. $W = 3.42$ kips, $d = 7.11$ ft.

5–27. $F = 311$ kN, $A_x = 460$ kN, $A_y = 7.85$ kN.

5–29. $F_{CD} = 195$ lb, $A_x = 97.5$ lb, $A_y = 31.2$ lb.

5–30. $R_A = 667$ N, $R_B = 220$ N, $R_C = 440$ N.

5–33. $F_{BC} = 1.82$ kip, $F_A = 2.06$ kip.

5–34. $R_C = 493$ N, $R_A = 247$ N, $R_B = 554$ N.

5–35. $W' = \dfrac{W}{\sqrt{2}}$.

5–37. 15.8 ft.

5–38. $F_{BC} = \dfrac{W \cos \theta}{\sin (\phi - \theta)}$, $A_x = \dfrac{W \cos \phi \cos \theta}{\sin (\phi - \theta)}$,

$A_y = W\left(1 + \dfrac{\sin \theta \cos \phi}{\sin (\phi - \theta)}\right)$.

5–39. 253 kg.

5–41. $R_C = 63.9$ lb, $R_B = 11.9$ lb, $R_A = 26.0$ lb.

5–42. $W_1 = 33.3$ lb/ft, $W_2 = 300$ lb/ft.

5–43. $d = \dfrac{a}{\cos^3 \theta}$.

5–45. 1.02°.

5–46. 2.50 kN/m.

5–47. $h = \sqrt{\dfrac{s^2 - l^2}{3}}$.

5–49. 250 N/m.

5–50. $F_B = 0.3\,P$, $F_C = 0.6\,P$, $x_C = \dfrac{0.6\,P}{k}$.

5–51. $d = \tfrac{3}{4}a$.

5–53. $F_{BC} = \dfrac{W \cos \theta}{2 \sin (\phi - \theta)}$, $A_x = \dfrac{W \cos \theta \cos \phi}{2 \sin (\phi - \theta)}$,

$A_y = W\left(1 - \dfrac{\cos \theta \sin \phi}{2 \sin (\phi - \theta)}\right)$.

5–54. $\theta = \tan^{-1}\left(\dfrac{l}{6\sqrt{a^2 - \left(\dfrac{l}{2}\right)^2}}\right)$.

5–55. $P = \dfrac{W}{2(\sin \theta + \cos \theta \cot \theta)}$.

5–57. 6 lb, 7.05 lb-ft.

5–58. $T_A = 235$ lb, $T_B = 293$ lb, $T_C = 372$ lb.

5–59. $W = 750$ lb, $x = 5.20$ ft, $y = 5.27$ ft.

5–61. $F_{AC} = F_{BC} = 4.09$ kN, $F_{DE} = 13.1$ kN.

5–62. $A_x = 24.5$ N, $A_y = -24.5$ N, $A_z = 49.1$ N, $B_y = 24.5$ N $T_{BC} = 24.5$ N.

5–63. $C_x = -175$ N, $C_z = 0$, $A_x = -525$ N, $A_y = 0$, $A_z = 750$ N, $T_B = 250$ N.

5–65. $F = 96$ lb, $B_x = 48.5$ lb, $B_y = -67.9$ lb, $B_z = 45.7$ lb, $A_x = -96.5$ lb, $A_z = -13.7$ lb.

5–66. $F = 96$ lb, $B_x = -48$ lb, $B_y = -67.9$ lb, $B_z = 32$ lb, $M_{Bx} = 96$ lb-ft, $M_{Bz} = -675$ lb-ft.

5–67. $T_{ED} = 32.1$ lb, $T_{BC} = 42.9$ lb, $M_{Ax} = 0$, $M_{Ay} = -17.9$ lb, $A_x = 3.57$ lb, $A_y = 50$ lb.

5–69. 105 lb.

5–70. $F_{BC} = 2.60$ kN, $A_y = 2.40$ kN, $A_z = 0$, $M_{Ax} = -480$ N · m, $M_{Ay} = 240$ N · m, $M_{Az} = 120$ N · m.

5–71. $F_{BC} = 101$ lb, $A_x = -233$ lb, $A_y = -140$ lb, $A_z = 77.8$ lb, $M_{Ay} = -389$ lb-ft, $M_{Az} = 93.3$ lb-ft.

5–73. $F_{CB} = 67.4$ lb, $A_x = -27.5$ lb, $A_y = -27.5$ lb, $A_z = 0$, $M_{Ay} = 165$ lb-ft, $M_{Az} = 0$.

5–74. $d = \dfrac{r}{2}\left(1 + \dfrac{W}{P}\right)$.

5–75. $d = \dfrac{r}{2}$.

5–77. $F = 10.1$ lb, $A_x = 102$ lb, $A_y = 66.4$ lb, $A_z = -207$ lb. $M_{Ay} = -529$ lb-ft, $M_{Az} = -776$ lb-ft.

5–78. $F_{BD} = 208$ N, $F_{BC} = 792$ N, $A_x = 0$, $A_z = 0$, $M_{Ax} = 0$, $M_{Az} = 700$ N · m.

5–79. $T = 58$ N, $C_y = -28.8$ N, $C_z = 87$ N, $D_x = 0$, $D_y = -79.2$ N, $D_z = 58$ N.

5–81. $T_{BA} = 14.0$ kN, $T_{DC} = 1.72$ kN, $E_x = 0.686$ N, $E_y = 0$, $E_z = 9.37$ kN.

5–82. $F_{BC} = 8.16$ kN, $F_{BD} = 6.59$ kN, $A_x = 13.2$ kN, $A_y = 0$, $A_z = 2.67$ kN.

5–83. $x = 0.981$ m, $A_x = 6.46$ kN, $A_y = 0$, $A_z = 5.38$ kN.

5–85. $W = 354$ lb, $A_x = -129$ lb, $A_y = 643$ lb, $A_z = -504$ lb.

5–86. $A_x = 633$ lb, $A_y = -141$ lb, $B_x = -721$ lb, $B_z = 895$ lb, $C_y = 200$ lb, $C_z = -506$ lb.

5–87. 674 lb.

5–89. $T_B = 25$ lb, $A_x = 25$ lb, $A_y = -25$ lb, $A_z = 50$ lb, $B_y = 25$ lb, $F_C = 0$.

5–90. (a) 305 lb, (b) 121 lb.

5–91. $F_{CE} = 110$ lb, $F_{BDC} = 62.0$ lb, $A_x = 19.4$ lb, $A_y = 192$ lb, $A_z = -25.8$ lb.

5–93. $T_A = 1.38$ kip, $T_B = 2.75$ kip, $T_C = 1.38$ kip.

5–94. 30.8°.

5–95. $O_x = 0$, $O_y = -84.9$ lb, $O_z = 80$ lb, $M_{Ox} = 948$ lb-ft, $M_{Oy} = 0$, $M_{Oz} = 0$.

5–97. $B_x = 15.9$ N, $B_y = 121$ N.

5–98. $T_1 = 1.89$ kN, $T_2 = 2.25$ kN, $\bar{x} = 2.25$ m.

CHAPTER 6

6–1. $F_{BC} = 374$ lb (C), $F_{AB} = 143$ lb (T), $F_{AC} = 264$ lb (T).

6–2. $F_{BC} = 400$ lb (C), $F_{CD} = 693$ lb (C), $F_{BD} = 667$ lb (T), $F_{AB} = 1.13$ kip (C).

6–3. $F_{CD} = 3.61$ kN (C), $F_{BC} = 3$ kN (T), $F_{BD} = 3$ kN (C), $F_{AB} = 3$ kN (T), $F_{AD} = 2.70$ kN (T), $F_{DE} = 6.31$ kN (C).

6–5. $F_{CD} = 9.24$ kN (T), $F_{DE} = 4.62$ kN (C), $F_{CE} = 9.24$ kN (C), $F_{BC} = 9.24$ kN (T), $F_{AB} = 9.24$ kN (T), $F_{BE} = 9.24$ kN (C), $F_{AE} = 4.62$ kN (C).

6–6. $F_{AB} = 7.5$ kN (T), $F_{AE} = 4.5$ kN (C), $F_{BE} = 8$ kN (T), $F_{DE} = 4.5$ kN (C), $F_{BD} = 19.8$ kN (T), $F_{BC} = 18.5$ kN (T).

6–7. $F_{AB} = 196$ N (T), $F_{AE} = 118$ N (C), $F_{BE} = 216$ N (T), $F_{DE} = 118$ N (C), $F_{BD} = 1.04$ kN (T), $F_{BC} = 857$ N (T).

6–9. $F_{CD} = 0.577\ P$ (C), $F_{DE} = 0.289\ P$ (T), $F_{CE} = 0.577\ P$ (T), $F_{BC} = 0.577\ P$ (C), $F_{BE} = 0.577\ P$ (T), $F_{AB} = 0.577\ P$ (C), $F_{AE} = 0.289\ P$ (T).

6–10. $F_{CD} = F_{AB} = 2.89\ W$ (C), $F_{BE} = F_{CE} = 1.15\ W$ (T), $F_{AE} = F_{DE} = 1.44\ W$ (T), $F_{BC} = 2.02\ W$ (C).

6–11. $F_{AB} = 21.9$ kN (C), $F_{AG} = 13.1$ kN (T), $F_{BC} = 13.1$ kN (C), $F_{BG} = 17.5$ kN (T), $F_{CG} = 3.13$ kN (T), $F_{FG} = 11.3$ kN (T), $F_{CF} = 3.13$ kN (C), $F_{CD} = 9.38$ kN (C), $F_{DE} = 15.6$ kN (C), $F_{DF} = 12.5$ kN (T), $F_{EF} = 9.38$ kN (T).

6–13. $F_{BC} = F_{CD} = F_{AB} = F_{DE} = 667$ lb (C), $F_{BG} = F_{AG} = F_{CG} = F_{DF} = F_{CF} = F_{EF} = 0$.

6–14. *AB, BC, CD, DE, HI, GI* are zero force members, $F_{CI} = 5.63$ kN (C), $F_{CG} = 9.00$ kN (T).

6–15. *AB, BC, CD, DE, HI, GI* are zero force members, $F_{EJ} = 9.38$ kN (C), $F_{FG} = 5.63$ kN (T).

6–17. $F_{BC} = 1.89\ P$ (C), $F_{CD} = 1.37\ P$ (T), $F_{AB} = 0.471\ P$ (C), $F_{BD} = 1.67\ P$ (T), $F_{AD} = 1.37\ P$ (T).

6–18. $F_{AB} = P \csc 2\theta$ (C), $F_{BC} = P \cot 2\theta$ (C), $F_{AC} = (\cot\theta \cos\theta - \sin\theta + 2\cos\theta)\ P$ (T), $F_{CD} = (\cot 2\theta + 1)\ P$ (C), $F_{AD} = P(\cot 2\theta + 1)\cos 2\theta$ (C).

6–19. $F_{DE} = 16.3$ kN (C), $F_{CD} = 8.40$ kN (T), $F_{AE} = 8.85$ kN (C), $F_{CE} = 6.20$ kN (C), $F_{AB} = 3.11$ kN (T), $F_{AF} = 6.20$ kN (T), $F_{BC} = 2.20$ kN (T), $F_{BF} = 6.20$ kN (T), $F_{CF} = 8.77$ kN (T).

6–21. $F_{GH} = 1.13$ kip (T), $F_{BC} = 1.50$ kip (C), $F_{BG} = 0.625$ kip (T), $F_{CG} = 0$.

6–22. $F_{CD} = 35$ kN (C), $F_{DJ} = 21.2$ kN (C), $F_{DI} = 0$.

6–23. $F_{DE} = 45$ kN (C), $F_{EH} = 5$ kN (T), $F_{HG} = 45$ kN (T).

6–25. $F_{CD} = 8.47$ kN (C), $F_{FG} = 8.08$ kN (T), $F_{CF} = 0.770$ kN (T).

6–26. $F_{JK} = 13.3$ kN (T), $F_{BC} = 14.9$ kN (C), $F_{CK} = 0$.

6–27. $F_{DE} = 11.9$ kN (C), $F_{DJ} = 2.67$ kN (T).

6–29. $F_{CD} = 11.3$ kN (C), $F_{CF} = 3.21$ kN (T), $F_{CG} = 6.80$ kN (C).

6–30. $F_{HI} = 2.31$ kN (T), $F_{DH} = 4.62$ kN (T), $F_{DE} = 5.33$ kN (C).

6–31. $F_{GF} = 800$ lb (T), $F_{BC} = 400$ lb (C), $F_{CG} = 566$ lb (C).

6–33. $F_{CD} = 14.0$ kN (C), $F_{GF} = 12.7$ kN (T), $F_{CF} = 0.300$ kN (C).

6–34. $F_{CD} = 7.31$ kN (C), $F_{ML} = 7.13$ kN (T), $F_{PM} = 0.313$ kN (T), *DK, BC, BA, EF, FG, QJ, ON, PL, MO, PK, QK* are zero-force members.

6–35. $F_{KJ} = 7.13$ kN (T), $F_{DE} = 6.94$ kN (C), $F_{DQ} = 0.313$ kN (C), *DK, BC, BA, EF, FG, QJ, ON, PK, MO, PL, QK* are zero-force members.

6–37. $F_{CD} = 2.53$ kip (C), $F_{JK} = 2.53$ kip (T).

6–38. $F_{GF} = 2.92$ kip (T), $F_{BC} = 9.55$ kip (C), $F_{CG} = 6.91$ kip (T).

6–39. $F_{AD} = 686$ N (T), $F_{BD} = 0$, $F_{CD} = 615$ N (C), $F_{BC} = 229$ N (T), $F_{EC} = 457$ N (C), $F_{AC} = 343$ N (T).

6–41. $F_{AB} = 2.40$ kN (C), $F_{AG} = F_{AE} = 1.01$ kN (T), $F_{BE} = F_{BG} = 1.80$ kN (T), $F_{BC} = F_{BD} = 1.34$ kN (C).

6–42. $F_{BF} = 0$, $F_{BC} = 0$, $F_{BE} = 500$ lb (T), $F_{AB} = 300$ lb (C), $F_{AC} = 583$ lb (T), $F_{AD} = 333$ lb (T), $F_{AE} = 667$ lb (C), $F_{EF} = 300$ lb (C), $F_{CD} = 300$ lb (C), $F_{CF} = 300$ lb (C), $F_{DF} = 424$ lb (T), $F_{DE} = 0$.

6–43. $F_{BF} = 0$, $F_{BC} = 0$, $F_{BE} = 500$ lb (T), $F_{AB} = 300$ lb (C), $F_{AC} = 972$ lb (T), $F_{AD} = 0$, $F_{AE} = 367$ lb (T), $F_{EF} = 300$ lb (C), $F_{CD} = 500$ lb (C), $F_{CF} = 300$ lb (C), $F_{DF} = 424$ lb (T).

6–45. $R_B = R_C = 49.5$ N, $F_C = 49.5$ N.

6–46. $A_y = 9.59$ kip, $C_x = 9.20$ kip, $C_y = 2.93$ kip, $B_y = 8.54$ kip.

6–47. $C_y = 5$ kN, $B_y = 15$ kN, $A_x = 0$, $A_y = 5$ kN, $M_A = 30$ kN · m.

6–49. $B_x = 2.80$ kip, $B_y = 1.05$ kip, $A_x = 2.80$ kip, $A_y = 5.10$ kip, $M_A = 43.2$ kip · ft.

6–50. $A_x = 1.40$ kN, $A_y = 250$ kN, $C_x = 500$ kN, $C_y = 1.70$ kN.

6–51. $A_x = 16.6$ kip, $A_y = 0.109$ kip, $C_x = 3.25$ kip, $C_y = 10.4$ kip.

6–53. $A_x = 80$ lb, $A_y = 80$ lb, $B_x = 333$ lb, $B_y = 133$ lb, $C_x = 413$ lb, $C_y = 53.3$ lb.

6–54. $A_x = C_x = 5$ kN, $A_y = C_y = 6.67$ kN, $B_x = 5$ kN, $B_y = 1.33$ kN, $D_x = 0$, $D_y = 8$ kN, $M_D = 10$ kN · m.

6–55. $B_y = A_y = 0$, $B_x = A_x = 15$ kN, $C_x = 21.4$ kN, $C_y = 8.80$ kN.

6–57. $A_x = 0.306$ kN, $A_y = 0.750$ kN, $E_y = 0.500$ kN, $E_x = 1.36$ kN.

6–58. $C_x = 16.7$ lb, $C_y = 34.4$ lb, $B_x = 66.7$ lb, $B_y = 15.6$ lb.

6–59. $D_x = 125$ lb, $D_y = 217$ lb, $B_x = 45$ lb, $B_y = 157$ lb, $M_B = 680$ lb-ft.

6–61. $D_x = 125$ lb, $D_y = 217$ lb, $B_x = 0$, $B_y = 217$ lb, $M_B = 500$ lb-ft.

6–62. $B_x = 97.4$ lb, $B_y = 130$ lb, $A_x = 52.6$ lb, $A_y = 130$ lb.

6–63. $D_x = 945$ lb, $D_y = 1000$ lb, $E_x = 945$ lb, $E_y = 500$ lb.

6–65. $A_y = 657$ N, $C_x = 0$, $C_y = 229$ N, $B_x = 0$, $B_y = 429$ N.

6–66. $E_x = 300$ N, $D_x = 300$ N, $D_y = E_y = 42.9$ N.

6–67. $C_x = 80$ lb, $C_y = 160$ lb, $D_x = 80$ lb, $D_y = 80$ lb.

6–69. $3.64\ F$.

6–70. $F_B = 29.0$ kN, $C_x = 14.5$ kN, $C_y = 16.5$ kN.

6–71. $E_y = 23.7$ kN, $A_y = 7.44$ kN.

6–73. $A_y = E_y = 417$ N.

6–74. 960 N.

6–75. 14.6 kN.

6–77. $F_{FH} = 78.5$ kN, $G_x = 78.5$ kN, $G_y = 9.81$ kN, $F_{EC} = $ 180 kN, $D_x = 0$, $D_y = 170$ kN, $F_{BA} = 170$ kN, $C_x = 0$, $C_y = 0$.

6–78. 314 lb-ft.

6–79. $F_{AB} = 916$ N, $P = 103$ N, $F_{CD} = 768$ N.

6–81. $F_{DE} = 1.16$ kip, $F_x = 1.07$ kip, $F_y = 1.06$ kip, $F_{BC} = $ 14.0 kip, $A_x = 6.25$ kip, $A_y = 11.0$ kip.

6–82. 16.1°.

6–83. 1.75 ft $\leq x \leq$ 17.4 ft.

6–85. $\theta = \sin^{-1}\left(\dfrac{8W}{kL}\right)$.

6–86. 21.7°.

6–87. $A_x = B_x = 0.25\, P$, $A_y = B_y = 0.144\, P$, $C_x = 0.250\, P$, $C_y = 1.01\, P$.

6–89. $A_x = 0$, $A_y = 1.44$ kip, $A_z = 0.540$ kip, $C_x = 0$, $C_y = -1.17$ kip, $C_z = -0.180$ kip, $(M_C)_x = 0$, $(M_C)_z = -4.14$ kip · ft.

6–90. $E_x = 0$, $E_y = 0$, $M_{Ex} = 0.500$ kN · m, $M_{Ey} = 0$, $F_{AB} = 1.56$ kN.

6–91. $F_{GFx} = 0$, $F_{GFy} = 1.20$ kN, $D_y = 0.600$ kN, $D_z = 0.600$ kN, $D_x = 0.150$ kN.

6–93. $F_{DE} = 270$ lb, $B_x = -30$ lb, $B_y = -13.3$ lb, $B_z = 0$.

6–94. $F_{CD} = 671$ N (C), $F_{CJ} = 283$ N (C), $F_{JK} = 800$ N (T).

6–95. $F_{BF} = 1.94$ kN, $F_{BD} = 2.60$ kN.

6–97. $F_{FG} = 836$ lb (C), $F_{BG} = 836$ lb (T).

6–98. $F_{BC} = 894$ N (C), $F_{CD} = 800$ N (T), $F_{BD} = 800$ N (C), $F_{DE} = 800$ N (T), $F_{AB} = 1.79$ kN (C), $F_{BE} = 894$ N (T), $F_{AE} = 800$ N (T).

6–99. $F_{BE} = 1.53$ kip, $F_{CD} = 0.350$ kip.

6–101. $F_{CP} = 2.50$ kN (C), $F_{EP} = 3.75$ kN (C), $F_{KR} = 8.75$ kN (C), MN, AN, JK, IJ, BO, CO, FQ, EQ are zero-force members.

6–102. 19.6 kN.

6–103. $F_{BD} = 144$ lb (C), $F_{AD} = 204$ lb (T), $F_{AF} = 72.2$ lb (T).

CHAPTER 7

7–1. $T_A = 680$ N · m, $T_B = 350$ N · m, $T_C = 800$ N · m, $T_D = 0$.

7–2. $T_C = 0$, $T_D = 400$ N · m, $T_E = 950$ N · m.

7–3. $T_A = 550$ N · m, $T_B = 400$ N · m, $T_C = 600$ N · m, $T_D = 0$.

7–5. $T = 709$ N, $V = 229$ N.

7–6. $M_C = 857$ lb-ft, $V_C = 386$ lb, $M_D = 600$ lb-ft, $V_D = 300$ lb.

7–7. $M_C = 48$ kip-ft, $V_C = 6$ kip.

7–9. $A_C = 0$, $M_C = 15$ kip-ft, $V_C = 2.01$ kip, $M_D = 3.77$ kip-ft, $V_D = 1.11$ kip.

7–10. $A_C = 0$, $V_C = 0.5$ kip, $M_C = 3.6$ kip-ft.

7–11. $A_E = 0$, $V_E = 133$ lb, $M_E = 133$ lb.-in.

7–13. $V_C = 5.79$ kip, $M_C = 4.95$ kip-ft.

7–14. $A_C = 406$ lb, $V_C = 903$ lb, $M_C = 1.35$ kip-ft.

7–15. $A_D = 464$ lb, $V_D = 203$ lb, $M_D = 2.61$ kip-ft.

7–17. $V_C = 3.75$ kip, $M_C = 30$ kip-ft.

7–18. $A_D = 0$, $V_D = 0$, $M_D = 9$ kip-ft, $A_E = 0$, $V_E = 7$ kip, $M_E = 12$ kip-ft.

7–19. $A_E = 250$ lb, $V_E = 245$ lb, $M_E = 490$ lb-ft, $A_F = 0$, $V_F = 308$ lb, $M_F = 1.23$ kip-ft.

7–21. $A_D = 1.20$ kN, $V_D = 0.400$ kN, $M_D = 0.600$ kN · m.

7–22. $A_D = 1.92$ kN, $V_D = 100$ N, $M_D = 900$ N · m.

7–23. $A_E = 1.92$ kN, $V_E = 800$ N, $M_E = 2.40$ kN · m.

7–25. $F_{Ox} = 550$ lb, $F_{Oy} = 250$ lb, $F_{Oz} = 2.00$ kip, $M_{Ox} = 538$ lb-ft, $M_{Oy} = 1.08$ lb-ft, $M_{Oz} = 288$ lb-ft.

7–26. $A_{Ax} = 0$, $V_{Ax} = 0$, $V_{Az} = 80$ N, $M_{Ax} = 24$ N · m, $M_{Ay} = 20$ N · m, $M_{Az} = 0$.

7–27. $A_x = 1$ kN, $A_y = 5$ kN, $A_z = 14$ kN, $(M_A)_x = $ 0.35 kN · m, $M_{Ay} = 10.4$ kN · m, $M_{Az} = 3.50$ kN · m.

7–29. $C_x = 170$ lb, $C_y = 50$ lb, $C_z = 500$ lb, $M_{Cx} = 1$ kip-ft, $M_{Cy} = 900$ lb-ft, $M_{Cz} = 260$ lb-ft.

7–30. $D_x = 116$ kN, $D_y = 65.6$ kN, $D_z = 0$, $M_{Dx} = $ 49.2 kN · m, $M_{Dy} = 87.0$ kN · m, $M_{Dz} = 26.2$ kN · m.

7–31. $0 \leq x < 5$, $V = 800$, $M = 800x$, $5 < x < 7$, $V = 0$, $M = $ 4000, $7 < x \leq 12$, $V = -800$, $M = 9600 - 800x$.

7–33. $V = 2500 - 500x$, $M = 2500x - 250x^2$.

7–34. $0 \leq x < L/2$, $V = wL/8$, $M = wLx/8$, $L/2 < x \leq L$, $V = w(5L - 8x)/8$, $M = w(-4x^2 + 5Lx - L^2)/8$.

7–35. $0 \leq x < L$, $V = P$, $M = Px$, $L < x \leq 2L$, $V = P$, $M = -P(2L - x)$.

7–37. $V = 250(10 - x)$, $M = 25(100x - 5x^2 - 6)$.

7–38. $V = -10.4x^2$, $M = -3.47x^3$.

7–39. $V = 5(10.5 - x)$, $M = -2.5(x^2 - 21x + 108)$.

7–41. $x = L^-$, $V = -11wL/18$, $M = -72wL^2/648$, $x = 3L/2$, $V = 0$, $M = wL^2/72$.

7–42. $V = w(4L^2 - 6xL - 3x^2)/12L$, $M = w(4xL^2 - 3x^2L - x^3)/12L$.

7–43. $0 \leq x < 12$, $V = 48 - x^2/6$, $M = -576 + 48x - x^3/18$, $12 < x \leq 24$, $V = 96 - 8x + x^2/6$, $M = -(24 - x)^3/18$.

7–45. $0 \leq x < L/2$, $V = w(3L - 4x)/4$, $M = -w(12x^2 - 18Lx + 7L^2)/24$, $L/2 < x \leq L$, $V = w(L - x)^2/L$, $M = -w(L - x)^3/3L$.

7–46. $0 \leq x < 3$, $V = 500(1 - x)$, $M = 250(2x - x^2)$, $3 < x \leq 6$, $V = 83.3(6 - x)^2$, $M = -27.8(6 - x)^3$.

7–47. $A = P(4\cos\theta + 3\sin\theta)/5$, $V = P(4\sin\theta - 3\cos\theta)/5$, $M = Pr(4 - 4\cos\theta - 3\sin\theta)/5$.

7–49. $V = -\dfrac{ax^2\, t\, \gamma}{2L}$, $M = -\dfrac{ax^3\, t\, \gamma}{6L}$.

7–50. $x = L/2$, $P = 4M_{max}/L$.

7–51. $a = L/3$.

7–53. $V_x = 1.50$ kip, $V_y = 0$, $V_z = 800(4 - y)$ lb, $M_x = 400(4 - y)^2$ lb-ft, $M_y = -3.00$ kip-ft, $M_z = -1.50(4 - y)$ kip-ft.

7–61. $x = 9^+$, $V = 36$, $M = -108$, $x = 27^+$, $V = 36$, $M = -108$.

7–62. $x = 4$, $V = 0$, $M = 96$, $x = 5^+$, $V = -12$, $M = 60$.

7–63. $x = 6^-$, $V = -15$, $M = -45$, $x = 11$, $V = 0$, $M = 17.5$.

7–65. $x = 2^+$, $V = -3.25$, $M = 7.5$, $x = 4^+$, $V = -3.25$, $M = 13$.

7–66. $x = 15^-$, $V = -20$, $M = -300$, $x = 30$, $V = 0$, $M = 150$.

7–67. $x = 4^+$, $V = 1050$, $M = -1600$, $x = 16$, $V = -150$, $M = 3800$.

7–69. $x = 0$, $V = 6$, $M = -18$, $x = 3$, $V = 3$, $M = -6$.

7–70. $x = 3.34$, $V = 0$, $M = 224$, $x = 8^+$, $V = 20$, $M = -210$.

7–71. $x = 3^-$, $V = -14.3$, $M = -22.9$, $x = 5^+$, $V = 45$, $M = -67.5$.

7–73. $x = 8^-$, $V = 1100$, $M = -1600$, $x = 12$, $V = 400$, $M = 0$.

7–74. $x = 3^+$, $V = -5.89$, $M = 27.3$, $x = 9$, $V = -5.89$, $M = 0$.

7–75. $T_{BC} = 46.7$ lb, $T_{AB} = 83.0$ lb, $T_{CD} = 88.1$ lb, $L = 20.2$ ft.

7–77. 4.36 ft.

7–78. 71.4 lb.

7–79. $y_B = 8.67$ ft, $y_D = 7.04$ ft.

7–81. $y_B = 3.53$ m, $P = 0.8$ kN, $T_{max} = T_{DE} = 8.17$ kN.

7–82. 184 kN.

7–83. 1.36 kN/m.

7–85. $y = x^2(75 - x^2/200)/7813$, $T_{max} = 9.28$ kip.

7–86. $y = (38.5x^2 + 577x)10^{-3}$, $T_{max} = 5.20$ kN.

7–87. 6.99 kip, $T_{max} = 3.51$ kip.

7–89. $\mathcal{L} = 94.3$ ft, $h = 25.3$ ft.

7–90. 51.9 lb/ft.

7–91. 22.6 m, 493 N.

7–93. 121 N, 5.91 m.

7–94. 5.18 ft.

7–95. 150 ft, 170 lb.

7–97. 302 ft.

7–98. 292 lb.

7–99. 326 ft.

7–102. 146 ft.

7–103. $x = 0$, $V = 10$, $M = -30$, $x = 5$, $V = 0$, $M = -5$.

7–105. $A_D = -800$ N, $V_D = 0$, $M_D = 1.20$ kN·m.

7–106. $0° \leq \theta \leq 180°$, $A = 150 \cos \theta + 200 \sin \theta$, $V = 150 \sin \theta - 200 \cos \theta$, $M = 150 \cos \theta + 200 \sin \theta - 150$, $0 \leq x \leq 2$, $A = -150$, $V = 200$, $M = -300 - 200x$.

7–107. 35.8 m.

7–109. $x = 1.25$, $V = 0$, $M = 1.56$, $x = 5$, $V = -7.5$, $M = -12.5$.

7–110. $x = 6^-$, $V = -600$, $M = -1200$, $x = 10$, $V = 0$, $M = 400$.

7–111. $A_D = 2.40$ kN, $V_D = 50$ N, $M_D = 1.35$ kN·m.

CHAPTER 8

8–1. $P = 224$ lb.

8–2. 76.4 lb $\leq P \leq 144$ lb.

8–3. $P = 140$ N, $x = 500$ mm.

8–5. The car will slide back.

8–6. $d = 2.70$ in.

8–7. $P = 14.4$ N, $(F_B)_{max} = 4.33$ N.

8–9. $F_A = 117$ N, $T = 233$ N.

8–10. It is not possible.

8–11. 10.2 ft.

8–13. 537 mm.

8–14. 1.37 kN.

8–15. 200 N.

8–17. 227 N, $\mu = 0.662$.

8–18. 54.9 kg.

8–19. 33.4°.

8–21. 24.2°.

8–22. 0.268.

8–23. $P = 7.85$ kN, $(F_C)_{max} = 11.8$ kN.

8–25. $P = 80$ lb, $h = 5$ ft.

8–26. $P = 45$ lb, $\mu = 0.3$.

8–27. Refrigerator slips ($P = 45$ lb), Man can push refrigerator ($F_{max} = 90$ lb).

8–29. $F_A = 1.48$ lb, $F_B = 0.890$ lb, $\theta = 8.53°$, $(F_B)_{max} = 1.48$ lb.

8–30. 12 lb.

8–31. The man can push the bed.

8–33. 42.4 lb, The man can move the crate.

8–34. 63.5 lb.

8–35. 355 N.

8–37. $O_x = 280$ N, $O_y = 945$ N, $P = 350$ N.

8–38. 0.990 lb.

8–39. 40.2 N.

8–41. 77.3 N·m.

8–42. 71.4 N.

8–43. 90.6 N·m.

8–45. 522 N, 11.3°.

8–47. $\phi = \theta$, $P = W \sin(\alpha + \phi)$.

8–49. $\theta = 2 \tan^{-1} \mu$.

8–50. $P = \dfrac{\mu w}{2}$.

8–51. $\mu = \dfrac{\sqrt{5}}{2} \dfrac{h}{a}$.

8–53. 106 lb.

8–54. Any number of blocks can be stacked.

8–55. $P = 1.61$ kip, $x = 1.34$ ft.

8–57. The wedges are self-locking.

8–58. 34.5 N.

8–59. $P = 5.53$ kN, The wedge is self-locking.

8–61. 4.21 kN.

8–62. 304 N.

8–63. 32.9 mm.

8–65. 914 N.

8–66. 1.98 kN.

8–67. 0.202 N·m.

8–69. 5.69 lb-in.

8–70. 51.2 N.

8–71. 0.0637.

8–73. The screw is self-locking.

8–74. 164 N.

8–75. 28.3 N.

8–77. $P = 880$ N, $T = 352$ N.

8–78. $F_{BC} = F_{AD} = 1.38$ kN (T), $F_{BD} = F_{AC} = 1.73$ kN (C), $F_{AB} = F_{CD} = 1.04$ kN (T).

8–79. 3.62 N·m.

8–81. (a) $F = 4.60$ kN, (b) $F = 16.2$ kN.

8–82. $T_A = 187$ lb, $T_B = 137$ lb.

8–83. $T_A = 481$ lb, $T_B = 658$ lb.

8–85. $h = 8.28$ ft.

8–86. 22.9 N $\leq P \leq 151$ N.

8–87. 99.2°.

8–89. 50 N·m, 0.286 m.

8–90. 0.222 m.

8–91. 2.49 kN.

8–93. 16.2 N.

8–94. 42.3 N.

8–95. 177 N · m.

8–97. F_{AB} = 15.4 lb, F_{AC} = 5.47 lb.

8–98. 9.63 N.

8–99. 156 kg.

8–101. 190 lb-in.

8–102. 0.0568.

8–103. 36.3 lb-ft.

8–105. $M = \frac{1}{2}\mu PR$.

8–106. $M = \dfrac{8\mu PR}{15}$.

8–107. $M = \frac{1}{2}\mu P(R_2 + R_1)$.

8–109. $M = \dfrac{2\mu PR}{3\cos\theta}$.

8–110. $M = \dfrac{2}{3}\dfrac{\mu P}{\sin(\theta/2)}\dfrac{(r_2{}^3 - r_1{}^3)}{(r_2{}^2 - r_1{}^2)}$.

8–111. $d = 0.140\,L$.

8–113. 89.0 N (approx).

8–114. T = 54.2 N, T = 53.9 N (approx).

8–115. F = 0.589 lb, F = 0.596 lb (approx).

8–117. T = 29.0 lb, T = 29.5 lb (approx).

8–118. μ = 0.215, μ = 0.211 (approx), F = 6 lb.

8–119. μ = 0.696, μ = 0.571 (approx).

8–121. 814 N.

8–122. P = 1.88 lb, P = 1.89 lb (approx).

8–123. P = 62.4 lb, P = 62.5 lb (approx).

8–125. 0.036 in.

8–126. 486 lb.

8–127. 1.38 lb.

8–129. 25.3 lb.

8–131. 38.5 mm.

8–133. 2.25 kN.

8–134. F = 96.7 N, F = 96.2 N (approx).

8–135. 6.67 lb.

8–137. 12 books.

8–138. 530 N.

8–139. (a) θ = 10.5°, (b) θ = 31.0°.

8–141. T = 13.8 lb, T = 13.6 lb (approx).

8–142. 620 N.

8–143. 589 N.

CHAPTER 9

9–1. B_x = 1 lb, A_x = 1 lb, A_y = 3.14 lb, $\bar{x} = \dfrac{4}{\pi}$.

9–2. $\bar{x}$ = 124 mm, $\bar{y}$ = 0.

9–3. $\bar{x} = \dfrac{r\sin\alpha}{\alpha}$.

9–5. O_x = 0, O_y = 0.574 lb, M_O = 0.305 lb-ft, $\bar{x}$ = 0.531 ft.

9–6. $\bar{y}$ = 0.183 ft.

9–7. $\bar{x} = \frac{3}{4}b$, $\bar{y} = \frac{3}{10}h$.

9–9. $\bar{x} = \frac{4}{7}a$, $\bar{y} = \frac{1}{5}a$.

9–10. $\bar{x} = \dfrac{n+1}{2(n+2)}a$, $\bar{y} = \dfrac{n}{2n+1}h$.

9–11. $\bar{x} = \dfrac{n+1}{n+2}a$, $\bar{y} = \dfrac{n+1}{2(2n+1)}h$.

9–13. $\bar{x} = \dfrac{4a}{3\pi}$, $\bar{y} = \dfrac{4a}{3\pi}$.

9–14. $\bar{x} = \dfrac{4a}{3\pi}$, $\bar{y} = \dfrac{4b}{3\pi}$.

9–15. $\bar{x} = \dfrac{3}{8}a$, $\bar{y} = \dfrac{3}{5}a$.

9–17. $\bar{x} = \dfrac{L(\pi - 2)}{\pi}$, $\bar{y} = \dfrac{a\pi}{8}$.

9–18. $\bar{x} = \dfrac{L}{2}$, $\bar{y} = \dfrac{a\pi}{8}$.

9–19. $\bar{x}$ = 0.914 m, $\bar{y}$ = 0.357 m.

9–21. 0.642 ft.

9–22. 2.04 ft.

9–23. 0.398 m.

9–25. $\bar{x}$ = 1.2 m, $\bar{y}$ = 0, R_B = 55.1 kN, A_x = 24.6 kN, A_y = 73.9 kN.

9–26. $\bar{x} = \bar{y}$ = 0, $\bar{z} = \frac{3}{8}h$.

9–27. $\bar{y} = \frac{3}{8}b$, $\bar{x} = \bar{z}$ = 0.

9–29. $\bar{z} = \dfrac{h}{4}$, $\bar{x} = \bar{y} = \dfrac{a}{\pi}$.

9–30. $\bar{z} = \dfrac{h}{4}$.

9–31. $\bar{z} = \frac{1}{4}c$.

9–33. 5.14 m.

9–34. 49.5 mm.

9–35. $\bar{x}$ = 2.85 in., $\bar{y}$ = 5.33 in.

9–37. $\bar{x}$ = 1.65 m, $\bar{y}$ = 9.24 m, E_y = 0.342 kN, A_x = 0, A_y = 1.32 kN.

9–38. $\bar{x}$ = 0.75 ft, $\bar{y}$ = 2.54 ft, $\bar{z}$ = 0.571 ft, M_{Ax} = 71 lb-ft, M_{Ay} = −21 lb-ft, M_{Az} = 0, A_x = 0, A_y = 0, A_z = 28 lb.

9–39. 2 in.

9–41. 210 mm.

9–42. 272 mm.

9–43. 11.9 in.

9–45. 37.5 mm.

9–46. $\bar{x}$ = 77.2 mm, $\bar{y}$ = 31.7 mm.

9–47. $\bar{x}$ = 339 mm, $\bar{y}$ = 432 mm.

9–49. 5.2 in.

9–50. 2.57 in.

9–51. $\bar{y} = \dfrac{1}{3}\left(\dfrac{2a+b}{a+b}\right)h$.

9–53. $\bar{y} = \dfrac{4}{3}\left(\dfrac{r_2^3 - r_1^3}{r_2^2 - r_1^2}\right)\dfrac{\sin\dfrac{\theta}{2}}{\theta}$.

9–54. $\bar{x}$ = 13.1 mm, $\bar{y}$ = 0, $\bar{z}$ = 22.5 mm, θ = 30.2°.

9–55. $\bar{y}$ = 0.75 in., $\bar{z}$ = 1.83 in.

9–57. $\bar{x}$ = 4.56 m, $\bar{y}$ = 3.07 m, R_A = 5.99 kN, R_B = 4.66 kN.

9–58. $\bar{x}$ = 19.0 ft, $\bar{y}$ = 11.0 ft, F_R = 2.83 kip.

9–59. 0.7 ft.

9–61. 8.22 in.

9–62. 101 mm.

9–63. 385 mm.

9–65. $A = 1.33$ m^2, $\bar{x} = 0.60$ m, $V = 5.03$ m^3.

9–66. $A = 0.941$ m^2, $\bar{y} = 0.569$ m, $V = 3.36$ m^3.

9–67. $A = 1.39$ ft^2, $\bar{y} = 0.541$ ft, $V = 4.71$ ft^3.

9–69. $6.79 \ (10^6)$ mm^3.

9–70. $3.83 \ (10^6)$ mm^3.

9–71. $119 \ (10^3)$ mm^2.

9–73. 39.5 kg.

9–74. $V = 207$ m^3, $A = 188$ m^2.

9–75. $143 \ (10^3)$ mm^3.

9–77. 8.36 kip.

9–78. $5.65(10^3)$ ft^2.

9–79. $27.2 \ (10^3)$ ft^3.

9–81. 29.9 mm.

9–82. 1.41 MN, 4 m.

9–83. $F_{RA} = 157$ kN, $F_{RB} = 235$ kN, $d = 4.22$ m.

9–85. 2.68 m.

9–86. 331 kN/m.

9–87. $F_{R1} = 2.45$ kN $\rightarrow$, $d' = 1.02$ m, $F_{R2} = 9.75$ kN $\downarrow$, located at the center of plate B.

9–89. $F_R = 17.2$ kip, $\searrow_{20.6°}$ $d = 5.22$ ft, F_R (bottom) $= 18.8$ kip $\downarrow$.

9–90. $B_x = 2.20$ MN, $B_y = 0.859$ MN, $A_y = 2.51$ MN.

9–91. 3.65 m.

9–93. 4 kip.

9–94. 583 lb, 1.88 ft.

9–95. 300 lb, 1.29 ft.

9–97. $\pi p_0 r d$

9–98. 12.4 lb/ft^2.

9–99. $F_R = 7.62$ kN, $\bar{x} = 2.74$ m, $\bar{y} = 3$ m.

9–101. $F_R = 24$ kN, $\bar{x} = 2$ m, $\bar{y} = 1.33$ m.

9–102. 154 mm.

9–103. $\bar{x} = 0.0833$ ft, $\bar{y} = \bar{z} = 0$.

9–105. $F = \dfrac{4}{\pi^2} p_0 ab$, Acting at $(a/2, b/2, 0)$.

9–106. 1.08 m^3.

9–107. $\bar{x} = 112$ mm, $\bar{y} = 112$ mm, $\bar{z} = 136$ mm.

9–109. $\bar{x} = 0$, $\bar{y} = 1.64$ in.

9–110. $\bar{x} = -3$ m, $\bar{y} = 1$ m, $\bar{z} = -0.3$ m.

9–111. $\bar{x} = 0$, $\bar{y} = 1.6$ ft.

CHAPTER 10

10–1. 973 in^4.

10–2. $11.1 \ (10^3)$ mm^2.

10–3. 10.7 in^4.

10–5. 3.20 m^4.

10–6. 0.762 m^4.

10–7. $I_x = \frac{2}{15}bh^3$.

10–9. 6.87 m^4.

10–10. 0.533 m^4.

10–11. 2.44 m^4.

10–13. 1.60 m^4.

10–14. 1.07 in^4.

10–15. 21.3 in^4.

10–17. $I_x = \dfrac{n}{3(3+n)}bh^3$, $I_y = \dfrac{n}{1+3n}b^3h$.

10–18. (a) $I_x = \dfrac{bh^3}{12}$, (b) $I_{\bar{x}'} = \dfrac{bh^3}{36}$.

10–19. 9.05 in^4.

10–21. 19.5 in^4.

10–22. $2.26 \ (10^3)$ in^4.

10–23. 341 in^4.

10–25. 74.7 mm.

10–26. 59.4 mm.

10–27. $\bar{x} = 170$ mm, $I_{\bar{y}'} = 722 \ (10^6)$ mm^4.

10–29. $I_x = 319 \ (10^6)$ mm^4, $I_y = 18.0 \ (10^6)$ mm^4.

10–30. $76.3 \ (10^6)$ mm^4.

10–31. $\bar{x} = 2.20$ in, $I_y = 57.9$ in^4.

10–33. $\bar{y} = 22.5$ mm, $I_{\bar{x}'} = 34.4 \ (10^6)$ mm^4.

10–34. $122 \ (10^6)$ mm^4.

10–35. 81 in^4.

10–37. $\bar{y} = 1.29$ in, $I_{\bar{x}'} = 6.74$ in^4.

10–38. $I_x = 154 \ (10^6)$ mm^4, $I_y = 91.3 \ (10^6)$ mm^4.

10–39. $\bar{y} = 80.7$ mm, $I_{\bar{x}'} = 67.6 \ (10^6)$ mm^4.

10–41. $k_O = \sqrt{\frac{1}{2}(r_2^2 + r_1^2)}$.

10–42. 1.57 in.

10–43. $I_x = \dfrac{1}{6}\left(\dfrac{b^3 d^3}{b^2 + d^2}\right)$.

10–45. $I_y = 0.187 \ d^4$.

10–46. $I_{xy} = 4$ in^4.

10–47. $I_{xy} = 1.33$ m^4.

10–49. $I_{xy} = 48$ in^4.

10–50. $I_{xy} = 0.333$ m^4.

10–51. $I_{xy} = 3.30$ m^4.

10–53. $I_{xy} = 8$ in^4.

10–54. $I_{xy} = \frac{1}{280}a^4$.

10–55. $I_{xy} = -0.0165a^4$.

10–57. $\bar{x} = 85.0$ mm, $\bar{y} = 35.0$ mm, $I_{\bar{x}\bar{y}} = -7.50 \ (10^6)$ mm^4.

10–58. 36 in^4.

10–59. $-22.5 \ (10^6)$ mm^4.

10–61. $I_u = 45.5$ in^4, $I_v = 100$ in^4, $I_{uv} = -47.6$ in^4.

10–62. $\bar{y} = 12.5$ mm, $I_u = 5.89 \ (10^6)$ mm^4, $I_v = 38.5 \ (10^6)$ mm^4.

10–63. $I_u = 3.47 \ (10^3)$ in^4, $I_v = 3.47 \ (10^3)$ in^4, $I_{uv} = 2.05 \ (10^3)$ in^4.

10–65. $I_u = 5.09 \ (10^6)$ mm^4, $I_v = 5.09 \ (10^6)$ mm^4, $I_{uv} = 0$.

10–66. $\theta_p = 16.8°$, $I_{max} = 58.1$ in^4, $I_{min} = 9.41$ in^4.

10–67. $I_{max} = 64.1$ in^4, $I_{min} = 5.33$ in^4.

10–69. x and y are principal axes, $\bar{y} = 4.67$ in, $I_x = 1.47 \ (10^3)$ in^4, $I_y = 2.50 \ (10^3)$ in^4.

10–74. $I_y = \frac{1}{3}ml^2$.

10–75. $I = mR^2$.

10–77. $I_y = \frac{2}{5}mr^2$.

10–78. $k_x = \dfrac{a}{\sqrt{3}}$.

10–79. $I_x = \frac{2}{5}mb^2$.

10–81. $k_x = 0.577$ m.

10–82. $I_y = \dfrac{m}{6}(a^2 + h^2)$.

10–83. $I_y = 2.25$ slug-ft^2.

10–85. $7.67 \text{ kg} \cdot \text{m}^2$.

10–86. 5.64 slug-ft^2.

10–87. $1.53 \text{ kg} \cdot \text{m}^2$.

10–89. 293 slug-ft^2.

10–90. $\bar{y} = 0.888 \text{ m}, I = 5.61 \text{ kg} \cdot \text{m}^2$.

10–91. $\bar{y} = 0.0577 \text{ m}, I_O = 0.450 \,(10^{-3}) \text{ kg} \cdot \text{m}^2, I_G = 0.150 \,(10^{-3}) \text{ kg} \cdot \text{m}^2$.

10–93. $I_x = \dfrac{a^4}{12}$.

10–94. $17.0 \,(10^6) \text{ mm}^4$.

10–95. $I_x = I_y = 503 \text{ in}^4$.

10–97. 32.4 in^4.

10–98. 60.8 in^4.

10–99. $30.2 \,(10^6) \text{ mm}^4$.

10–101. 307 in^4.

10–102. 10.7 in^4.

CHAPTER 11

11–1. $m = \frac{1}{2}Wl \cos \theta$.

11–2. $8.57°$.

11–3. 50 lb.

11–5. $31.4°$.

11–6. $10.7°$.

11–7. $53.1°, 90°$.

11–9. $24.9°$.

11–10. $52.2°$.

11–11. $33.4°$.

11–13. 21.6 N.

11–14. $F = 2P \cot \theta$.

11–15. $\theta = \cos^{-1}\left(\dfrac{a}{2L}\right)^{1/3}$.

11–17. 427 N.

11–18. 13.9 N.

11–19. 100 kg.

11–21. $\theta = 0°$ or $33.1°$.

11–22. $m_A = 1.6 \text{ kg}, m_B = 2 \text{ kg}$.

11–23. $\theta = 53.1°$, stable, $\theta = 90°$, unstable.

11–25. 100 kg.

11–26. at $x = 0$, unstable, at $x = 0.167 \text{ m}$, stable.

11–27. $\theta = 34.6°$, unstable.

11–29. $\theta = \sin^{-1}\left(\dfrac{W}{2k\ell}\right)$.

11–30. $\theta = 2.55°$, unstable.

11–31. $\theta = 0°$, stable.

11–33. 8 in.

11–35. 2.45 m.

11–37. $h = 0$.

11–38. At $\theta = 0°$, unstable.

11–39. 1.35 in.

11–41. $\theta = 9.46°$, stable.

11–42. $\theta = 0°, \theta = \cos^{-1}\left(\dfrac{d}{4a}\right)$.

11–43. $\dfrac{\cos \theta}{\sin^3 \theta} = \dfrac{a}{2r}$.

11–45. 4.62 lb-ft.

11–46. $\theta = 12.1°$, unstable.

11–47. $P = \left(\dfrac{b - a}{2c}\right) mg$.

11–49. $\theta = 90°$ (stable), $\theta = 17.5°$ (unstable).

11–50. $15.5°$.

11–51. $\theta = 30°, \theta = 90°$.

11–53. 211 N.

11–54. $\theta = 14.0°$, stable.

Index

529

Engineering
Mechanics
DYNAMICS

R. C. HIBBELER

FIFTH EDITION

Engineering Mechanics

DYNAMICS

MACMILLAN PUBLISHING COMPANY, INC.
NEW YORK
Collier Macmillan Publishers
LONDON

PRINTED IN THE UNITED STATES OF AMERICA

Macmillan Publishing Company
866 Third Avenue, New York, New York 10022
Collier Macmillan Canada, Inc.

LIBRARY OF CONGRESS CATALOGING IN PUBLICATION DATA

Hibbeler, R. C.
    Engineering mechanics—dynamics / R. C. Hibbeler. — 5th ed.
        p.   cm.
    Companion volume to Engineering mechanics—statics. These two
works are also published together as Engineering mechanics—statics
and dynamics.
    Includes index.
    ISBN 0-02-354661-1
    1. Dynamics. I. Title.
TA352.H5 1989
620.1′04—dc19   88-22105
                                                              CIP

PRINTING: 3 4 5 6 7 8   YEAR: 0 1 2 3 4 5 6 7 8

Preface

The purpose of this book is to provide the student with a clear and thorough presentation of the theory and applications of engineering mechanics. To achieve this objective the author has by no means worked alone on this task, for to a large extent this book has been shaped by the comments and suggestions of more than a hundred reviewers in the teaching profession and many of the author's students who have used its previous editions.

Several changes have been made in preparing this new edition. Overall, particular attention has been given to developing accurate definitions and explanations of the concepts. A greater emphasis has been placed on drawing free-body diagrams, and the importance of selecting an appropriate coordinate system and associated sign convention for vector components is stressed when the equations of kinematics or the equations of motion are applied. Some rewriting of material has occurred; and in particular, relative motion analysis using Cartesian vectors and scalar components has been combined into a single section. Lastly, most of the problems in this edition are new, and there are more of them than in the previous edition. In this regard, extra care has been taken in the presentation and solution of the problems, and all the problem sets have been reviewed and the solutions checked by the author's colleagues to ensure both their clarity and numerical accuracy.

Organization and Approach. The contents of each chapter are organized into well-defined sections. Selected groups of sections contain an explanation of specific topics, illustrative example problems, and a set of homework problems. The topics within each section are placed into subgroups defined by boldface titles. The purpose of this is to present a structured method for introducing each new definition or concept, and to make the book convenient for later reference and review.

A ''procedure for analysis'' is given at the end of many sections of the book in order to provide the student with a review or summary of the material and a logical and orderly method to follow when applying the theory. As in the previous editions, the example problems are solved using this outlined method in order to clarify its numerical application. It is to be understood, however, that once the relevant principles have been mastered and enough confidence and judgment have been obtained, the student can then develop his or her own procedures for solving problems. In most cases it is felt that the first step in any procedure should require drawing a diagram. By doing so, the student forms the habit of tabulating the necessary data while focusing on the physical aspects of the problem and its associated geometry. If this step is correctly performed, applying the relevant equations of mechanics becomes somewhat methodical, since the data can be taken directly from the diagram. This step is particularly important when solving problems involving kinetics, and for this reason, application of the free-body diagram technique is strongly emphasized throughout the book.

Since mathematics provides a systematic means of applying the principles of mechanics, the student is expected to have prior knowledge of algebra, geometry, trigonometry, and, for complete coverage, some calculus. Vector analysis is introduced at points where it is most applicable. Its use often provides a convenient means for presenting concise derivations of the theory, and it makes possible a simple and systematic solution of many complicated three-dimensional problems. Occasionally, the example problems are solved using more than one method of analysis so that the student develops the ability to use mathematics as a tool whereby the solution of any problem may be carried out in the most direct and effective manner.

Problems. Numerous problems in the book depict realistic situations encountered in engineering practice. It is hoped that this realism will both stimulate the student's interest in engineering mechanics and provide a means for developing the skill to reduce any such problem from its physical description to a model or symbolic representation to which the principles of mechanics may be applied. As in the previous editions, an effort has been made to include some problems which may be solved using a numerical procedure executed on either a desk-top computer or programmable pocket calculator. Suitable numerical techniques along with associated computer programs are given in Appendix B. The intent here is to broaden the student's capacity for using other forms of mathematical analysis *without* sacrificing the time needed to focus on the application of the principles of mechanics. Problems of this type which either can or must be solved using numerical procedures are identified by a "square" symbol (■) preceding the problem number.

Throughout the text there is an approximate balance of problems using either SI or FPS units. Furthermore, in any set, an attempt has been made to arrange the problems in order of increasing difficulty. The answers to all but every fourth problem are listed in the back of the book. To alert the user to a problem without a reported answer, an asterisk (*) is placed before the problem number.

Contents. The book is divided into 11 chapters.* In particular, the kinematics of a particle is discussed in Chapter 12, followed by a discussion of particle kinetics in Chapter 13 (equation of motion), Chapter 14 (work and energy), and Chapter 15 (impulse and momentum). The concepts of particle dynamics contained in these four chapters are then summarized in a "review" section and the student is then given the chance to identify and solve a variety of different types of problems. A similar sequence of presentation is given for the planar motion of a rigid body: Chapter 16 (planar kinematics), Chapter 17 (equations of motion), Chapter 18 (work and energy), and Chapter 19 (impulse and momentum), followed by a summary and set of review problems for these chapters. If desired, it is possible to cover Chapters 12 through 19 in the following order with no loss in continuity: Chapters 12 and 16 (kinematics), Chapters 13 and 17 (equations of motion), Chapters 14 and 18 (work and energy), and Chapters 15 and 19 (impulse and momentum).

Time permitting, some of the material involving three-dimensional rigid-body motion may be included in the course. The kinematics and kinetics of this motion are discussed in Chapters 20 and 21, respectively. Chapter 22 (vibrations) may be included if the student has the necessary mathematical background. Sections of the book which are considered to be beyond the scope of the basic dynamics course are indicated by a star (★) and may be omitted. Note, however, that this more advanced material provides a suitable reference for basic principles when it is covered in other courses.

Acknowledgments. I have endeavored to write this book so that it will appeal to both the student and instructor. Through the years many people have helped in its development and I should like to acknowledge their valued suggestions and comments. Specifically, I wish to personally thank the following individuals who have contributed to this revision, namely, Professors Reginald Amory, Northeastern University; Robert D. Celmer, University of Hartford; C. H. Chang, University of Alabama; Major William Conner, U.S. Military Academy, West Point; Mitsunori Denda, Rutgers University; Frank D'Souza, Illinois Institute of Technology; Robert W. Fuessle, Bradley University; Sherman Goplen, North Dakota State University; John E. Griffith, University of Southern Florida; William Q. Gurley, University of Tennessee, Chattanooga; Edward E. Hornsey, University of Missouri, Rolla; Mohammad Khos-

---

*The first eleven chapters of this sequence form the contents of *Engineering Mechanics: Statics.*

rowjerdi, Western New England College; William M. Lee, U.S. Naval Academy, Annapolis; Captain Al Lewis, U.S. Military Academy, West Point; Will Liddell, Jr., Auburn University at Montgomery; John C. McWhorter, Mississippi State University; Nasser Moshrefi, Michigan Technical University; Larry Oline, University of Southern Florida; David Oglesby, University of Missouri, Rolla; John Peddieson, Tennessee Technical University; James R. Ritter, Old Dominion University; Robert Schmidt, University of Detroit; Jack Smith, U.S. Naval Academy, Annapolis; Robert Snell, Kansas State University; Henry L. Sundberg, Jr., Western New England College; Liang-Ning Tao, Illinois Institute of Technology; Theodore Tauchert, University of Kentucky; V. L. Vorsteveld, University of Vermont; Donald M. Wallace, Norwich University; William H. Walston, Jr., University of Maryland; Han-Chin Wu, University of Iowa.

Many thanks are also extended to all my students and to those in the teaching profession who have freely taken the time to send me their suggestions and comments. Since the list is too long to mention, it is hoped that those who have given help in this manner will accept this anonymous recognition. Furthermore, I greatly appreciate the personal support and attention given to me by my former editor Mr. Peter Gordon and currently, Mr. David Johnstone, along with the staff at Macmillan. Lastly, I should like to acknowledge the encouragement and assistance of my wife, Conny, during the time it has taken to prepare the manuscript for publication.

Russell Charles Hibbeler

Contents

12

Kinematics of a Particle   1

Contents

Review 1: Kinematics and Kinetics of a Particle

16

Planar Kinematics of a Rigid Body

17

Planar Kinetics of a Rigid Body: Force and Acceleration

18

Planar Kinetics of a Rigid Body: Work and Energy

22

Vibrations   485

APPENDIXES

A

Mathematical Expressions   521

B

Numerical and Computer Analysis   524

C

Vector Analysis   533

Answers   537

Index   547

12

Kinematics of a Particle

Engineering mechanics involves the study of both statics and dynamics. *Statics* is concerned with the equilibrium of bodies at rest or moving with constant velocity, whereas *dynamics* is concerned with bodies having accelerated motion. In this text, the subject of dynamics will be presented in two parts: *kinematics,* which treats only the geometrical aspects of motion, and *kinetics,* which is the analysis of the forces causing the motion. For simplicity particle dynamics will be discussed first, followed by topics in rigid-body dynamics presented in two and then three dimensions.

We will begin our study of dynamics by discussing particle kinematics. Recall that a *particle* has a mass but negligible size and shape. Therefore we must limit application to those objects that have dimensions that are of no consequence in the analysis of the motion. In most problems encountered, one is interested in bodies of finite size, such as rockets, projectiles, or vehicles. Such objects may be considered as particles, provided motion of the body is characterized by motion of its mass center and any rotation of the body is neglected.

This chapter begins with a study of the absolute motion of a particle, which is motion measured with respect to a fixed coordinate system. In this regard, motion along a straight line will be studied before the more general motion along a curved path. Afterward, the relative motion between two particles will be considered, using a translating coordinate system.

12.1 Rectilinear Kinematics

Rectilinear motion occurs when a particle moves along a straight-line path. The kinematics of the motion is characterized by specifying at any given instant the particle's position, velocity, and acceleration.

Position

The straight-line path of the particle can be defined using a single coordinate axis s, Fig. 12–1a. The origin O on the path is a fixed point, and from this point the *position vector* $\mathbf{r}$ is used to specify the position of the particle P at any given instant. For *rectilinear motion,* however, the *direction* of $\mathbf{r}$ is *always* along the s axis, and so it never changes. What will change is its magnitude and its sense or arrowhead direction. For analytical work it is therefore convenient to represent $\mathbf{r}$ by an *algebraic scalar s,* representing the *position coordinate* of the particle, Fig. 12–1a. The magnitude of s (and $\mathbf{r}$) is the distance from O to P, usually measured in meters (m) or feet (ft) and the sense (or arrowhead direction of $\mathbf{r}$) is defined by the algebraic sign of s. Although the choice is arbitrary, in this case s is positive since the coordinate axis is positive to the right of the origin. Likewise it is negative if the particle is located to the left of O.

Displacement

The *displacement* of the particle is defined as the *change* in its *position*. For example, if the particle moves from P to P', Fig. 12–1b, the displacement is $\Delta \mathbf{r} = \mathbf{r}' - \mathbf{r}$. Using algebraic scalars to represent $\Delta \mathbf{r}$, we also have $\Delta s = s' - s$. Here Δs is *positive* since the particle's final position is to the *right* of its initial position, i.e., $s' > s$. Likewise, if the final position is to the *left*, Δs is *negative*.

Since the displacement of a particle is a vector quantity, it should be distinguished from the distance the particle travels. Specifically, the *distance traveled* is a *positive scalar* which represents the total length of path traversed by the particle.

Velocity

If the particle moves through a displacement $\Delta \mathbf{r}$ from P to P' during the time interval Δt, Fig. 12–1b, the *average velocity* of the particle during this time interval is

$$\mathbf{v}_{\text{avg}} = \frac{\Delta \mathbf{r}}{\Delta t}$$

If we take smaller and smaller values of Δt, the magnitude of $\Delta \mathbf{r}$ becomes smaller and smaller. Consequently, the *instantaneous velocity* is defined as $\mathbf{v} = \lim\limits_{\Delta t \to 0} (\Delta \mathbf{r}/\Delta t)$ or

Position

(a)

Displacement

(b)

Velocity

(c)

Fig. 12–1(a–c)

$$\mathbf{v} = \frac{d\mathbf{r}}{dt}$$

Representing **v** as an algebraic scalar, Fig. 12–1c, we can also write

$(\overset{+}{\to})$

$$v = \frac{ds}{dt}$$

(12–1)

Since Δt or dt is always positive, the sign used to define the *sense* of the velocity is the same as that of Δs (or ds). For example, if the particle is moving to the *right*, Fig. 12–1c, the velocity is *positive*; whereas if it is moving to the *left*, the velocity is *negative*. (This fact is emphasized here by the arrow written at the left of the above equation.) The *magnitude* of the velocity is known as the *speed* and is generally expressed in units of m/s or ft/s.

Occasionally, the term ''average speed'' is used. The *average speed* is *always* a positive scalar and is defined as the total distance traveled by a particle, s_T, divided by the elapsed time Δt, i.e.,

$$v_{sp} = \frac{s_T}{\Delta t}$$

3

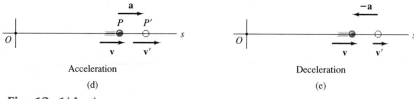

Acceleration   Deceleration

(d)   (e)

Fig. 12–1(d, e)

Acceleration

Provided the velocity of the particle is known at the two points P and P', the *average acceleration* of the particle during the time interval Δt is defined as

$$\mathbf{a}_{avg} = \frac{\Delta \mathbf{v}}{\Delta t}$$

Here $\Delta \mathbf{v}$ represents the difference in the velocity during the time interval Δt, i.e., $\Delta \mathbf{v} = \mathbf{v}' - \mathbf{v}$, Fig. 12–1d.

The *instantaneous acceleration* at time t is found by taking smaller and smaller values of Δt and corresponding smaller and smaller values of $\Delta \mathbf{v}$, so that $\mathbf{a} = \lim_{\Delta t \to 0} (\Delta \mathbf{v}/\Delta t)$ or, using algebraic scalars,

$(\xrightarrow{+})$
$$\boxed{a = \frac{dv}{dt}} \tag{12–2}$$

Taking the second time derivative of Eq. 12–1, we can also write

$(\xrightarrow{+})$
$$a = \frac{d^2 s}{dt^2}$$

Both the average and instantaneous acceleration can be either positive or negative. In particular, when the particle is *slowing down*, or its speed is decreasing, it is said to be *decelerating*. In this case, v' in Fig. 12–1e is *less* than v and so $\Delta v = v' - v$ will be negative. Consequently, a will also be negative and therefore it will act to the *left*, in the opposite *sense* to v. Also, note that when the *velocity* is *constant*, the *acceleration is zero* since $\Delta v = v - v = 0$. Units commonly used to express the magnitude of acceleration are m/s^2 or ft/s^2.

A differential relation involving the displacement, velocity, and acceleration along the path may be obtained by eliminating the time differential dt between Eqs. 12–1 and 12–2. Show that this yields

$(\xrightarrow{+})$
$$\boxed{a \, ds = v \, dv} \tag{12–3}$$

$v = \frac{ds}{dt}$

Constant Acceleration, $a = a_c$

When the acceleration is constant, each of the three kinematic equations $a_c = dv/dt$, $v = ds/dt$, and $a_c\ ds = v\ dv$ may be integrated to obtain formulas that relate a_c, v, s, and t.

Velocity as a Function of Time. Integrate $a_c = dv/dt$, assuming that initially $v = v_0$ when $t = 0$.

$$\int_{v_0}^{v} dv = \int_{0}^{t} a_c\ dt$$

$$v - v_0 = a_c(t - 0)$$

$(\xrightarrow{+})$

$$\boxed{\begin{array}{c} v = v_0 + a_c t \\ \text{Constant Acceleration} \end{array}}$$

(12–4)

Position as a Function of Time. Integrate $v = ds/dt = v_0 + a_c t$, assuming that initially $s = s_0$ when $t = 0$.

$$\int_{s_0}^{s} ds = \int_{0}^{t} (v_0 + a_c t)\ dt$$

$$s - s_0 = v_0(t - 0) + a_c(\tfrac{1}{2}t^2 - 0)$$

$(\xrightarrow{+})$

$$\boxed{\begin{array}{c} s = s_0 + v_0 t + \tfrac{1}{2}a_c t^2 \\ \text{Constant Acceleration} \end{array}}$$

(12–5)

Velocity as a Function of Position. Either solve for t in Eq. 12–4 and substitute into Eq. 12–5, or integrate $v\ dv = a_c\ ds$, assuming that initially $v = v_0$ at $s = s_0$.

$$\int_{v_0}^{v} v\ dv = \int_{s_0}^{s} a_c\ ds$$

$$\tfrac{1}{2}v^2 - \tfrac{1}{2}v_0^2 = a_c(s - s_0)$$

$(\xrightarrow{+})$

$$\boxed{\begin{array}{c} v^2 = v_0^2 + 2a_c(s - s_0) \\ \text{Constant Acceleration} \end{array}}$$

(12–6)

The magnitudes and signs of s_0, v_0, and a_c, used in these equations, are determined from the chosen origin and positive direction of the s axis. As indicated by the arrows written at the left of these equations, here we have assumed positive quantities act to the right, in accordance with the coordinate axis s shown in Fig. 12–1.

It is important to remember that the above equations are useful *only when the acceleration is constant and when* $t = 0$, $s = s_0$, $v = v_0$. A common example of this motion occurs when a body falls freely toward the earth. If air resistance is neglected and the distance of fall is short, then the constant *downward* acceleration of the body when it is close to the earth is approximately 9.81 m/s^2 or 32.2 ft/s^2.*

PROCEDURE FOR ANALYSIS

Coordinate System. Whenever the kinematic equations are applied, it is *very important* first to establish a position coordinate s along the path, and to specify its *fixed origin* and positive direction. Because the path is *rectilinear,* the lines of direction of the particle's position, velocity, and acceleration *never change*. Therefore these quantities can be represented as algebraic scalars. For analytical work the sense of s, v, and a can then be determined from their *algebraic signs*. In the examples that follow, the positive sense will be indicated by an arrow alongside each kinematic equation as it is applied.

Kinematic Equations. Often a mathematical relationship between *any two* of the four variables a, v, s, and t can be established by observation or experiment. When this is the case, the relationships between the remaining variables can be obtained by either differentiation or integration, using the kinematic equations $a = dv/dt$, $v = ds/dt$, or $a\,ds = v\,dv$.* Since each of these equations relates *three* variables, then, if a variable is *known* as a function of another variable, a third variable can be determined by *choosing the kinematic equation which relates all three*. For example, suppose that the *acceleration* is known as a function of *position*, $a = f(s)$. The *velocity* can be determined from $a\,ds = v\,dv$ since $f(s)$ can be substituted for a to yield $f(s)\,ds = v\,dv$. Solution for v requires integration. Note that the velocity *cannot* be obtained by using $a = dv/dt$, since $f(s)\,dt = dv$ contains two variables, s and t, on the left side and so it *cannot* be integrated. Whenever integration is performed, it is important that the position and velocity be known at a given instant in order to evaluate either the constant of integration if an indefinite integral is used, or the limits of integration if a definite integral is used. Lastly, keep in mind that Eqs. 12–4 through 12–6 have only a limited use. *Never apply these equations unless it is absolutely certain that the acceleration is constant.*

*Some standard differentiation and integration formulas are given in Appendix A.

*The proof is given in Example 13–2.

Example 12–1

The car in Fig. 12–2 moves in a straight line such that for a short time its velocity is defined by $v = (9t^2 + 2t)$ ft/s, where t is in seconds. Determine its position and acceleration when $t = 3$ s.

Fig. 12–2

SOLUTION

Coordinate System. The position coordinate extends from the fixed origin O to the car, positive to the right.

Position. The car's velocity is given as a function of time so that its position can be determined from $v = ds/dt$, since this equation relates v, s, and t. Noting that $s = 0$ when $t = 0$, we have*

$(\xrightarrow{+})$
$$v = \frac{ds}{dt} = (9t^2 + 2t)$$

$$\int_0^s ds = \int_0^t (9t^2 + 2t)\, dt$$

$$s\Big|_0^s = 3t^3 + t^2 \Big|_0^t$$

$$s = 3t^3 + t^2$$

When $t = 3$ s,

$$s = 3(3)^3 + (3)^2 = 90 \text{ ft} \qquad\qquad Ans.$$

Acceleration. Knowing the velocity as a function of time, the acceleration is determined from $a = dv/dt$, since this equation relates a, v, and t.

$(\xrightarrow{+})$
$$a = \frac{dv}{dt} = \frac{d}{dt}(9t^2 + 2t)$$

$$= 18t + 2$$

When $t = 3$ s,

$$a = 18(3) + 2 = 56 \text{ ft/s}^2 \rightarrow \qquad\qquad Ans.$$

Note that the formulas for constant acceleration *cannot* be used to solve this problem.

---

*The *same result* can be obtained by evaluating a constant of integration C rather than using definite limits on the integral. For example, integrating $ds = (9t^2 + 2t)\, dt$ yields $s = 3t^3 + t^2 + C$. Using the condition that at $t = 0$, $s = 0$, then $C = 0$.

Example 12–2

A small projectile is fired vertically *downward* into a fluid medium with an initial velocity of 60 m/s. If the projectile experiences a deceleration which is equal to $a = (-0.4v^3)$ m/s^2, where v is measured in m/s, determine the velocity v and position s four seconds after the projectile is fired.

SOLUTION

Coordinate System.   Since the motion is downward, the position coordinate is positive downward, with origin located at O, Fig. 12–3.

Velocity.   The acceleration is given as a function of velocity so that the velocity can be obtained from $a = dv/dt$, since this equation relates v, a, and t. (Why not use $v = v_0 + a_c t$?) Separating the variables and integrating, with $v_0 = 60$ m/s when $t = 0$, yields

$(+\downarrow)$
$$a = \frac{dv}{dt} = -0.4v^3$$

Fig. 12–3

$$\int_{60}^{v} \frac{dv}{-0.4v^3} = \int_0^t dt$$

$$\frac{1}{-0.4}\left(\frac{1}{-2}\right)\frac{1}{v^2}\bigg|_{60}^{v} = t - 0$$

$$\frac{1}{0.8}\left[\frac{1}{v^2} - \frac{1}{(60)^2}\right] = t$$

$$v = \left\{\left[\frac{1}{(60)^2} + 0.8t\right]^{-1/2}\right\} \text{ m/s}$$

Here the positive root is taken, since the projectile is moving downward. When $t = 4$ s,

$$v = 0.559 \text{ m/s} \downarrow \qquad\qquad Ans.$$

Position.   Knowing the velocity as a function of time, we obtain the position from $v = ds/dt$, since this equation relates s, v, and t. Using the initial condition $s = 0$, when $t = 0$, we have

$(+\downarrow)$
$$v = \frac{ds}{dt} = \left[\frac{1}{(60)^2} + 0.8t\right]^{-1/2}$$

$$\int_0^s ds = \int_0^t \left[\frac{1}{(60)^2} + 0.8t\right]^{-1/2} dt$$

$$s = \frac{2}{0.8}\left[\frac{1}{(60)^2} + 0.8t\right]^{1/2}\bigg|_0^t$$

$$s = \frac{1}{0.4}\left\{\left[\frac{1}{(60)^2} + 0.8t\right]^{1/2} - \frac{1}{60}\right\} \text{ m}$$

When $t = 4$ s,

$$s = 4.43 \text{ m} \qquad\qquad Ans.$$

Example 12–3

A boy tosses a ball in the vertical direction off the side of a cliff, as shown in Fig. 12–4. If the initial velocity of the ball is 15 m/s upward, and the ball is released 40 m from the bottom of the cliff, determine the maximum height s_B reached by the ball and the speed of the ball just before it hits the ground. During the entire time the ball is in motion, it is subjected to a constant downward acceleration of 9.81 m/s^2 due to gravity. Neglect the effect of air resistance.

SOLUTION

Coordinate System. The origin O for the position coordinate s is taken at ground level with positive upward, Fig. 12–4.

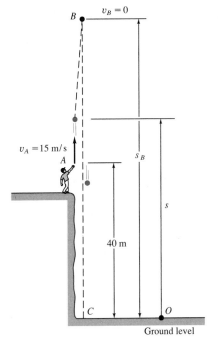

Fig. 12–4

Maximum Height. At the maximum height $s = s_B$ the velocity $v_B = 0$. Since the ball is thrown *upward* when $t = 0$, it is subjected to a velocity $v_A = +15$ m/s (positive since it is in the same sense as positive displacement). For the entire motion, the acceleration is *constant* such that $a_c = -9.81$ m/s^2 (negative since it acts in the *opposite* sense to positive velocity or positive displacement). Since a_c is *constant* throughout the entire motion, the position may be related to the velocity at the two points A and B on the path as follows:

$$(+\uparrow) \qquad v_B^2 = v_A^2 + 2a_c(s_B - s_A)$$
$$0 = (15)^2 + 2(-9.81)(s_B - 40)$$
$$s_B = 51.5 \text{ m} \qquad\qquad Ans.$$

Velocity. To obtain the velocity of the ball just before it hits the ground, Eq. 12–6 can be applied between points B and C, Fig. 12–4,

$$(+\uparrow) \qquad v_C^2 = v_B^2 + 2a_c(s_C - s_B)$$
$$= 0 + 2(-9.81)(0 - 51.5)$$
$$v_C = -31.8 \text{ m/s} = 31.8 \text{ m/s} \downarrow \qquad\qquad Ans.$$

The negative root was chosen since the ball is moving *downward*.

Similarly, Eq. 12–6 may also be applied between points A and C, i.e.,

$$(+\uparrow) \qquad v_C^2 = v_A^2 + 2a_c(s_C - s_A)$$
$$= 15^2 + 2(-9.81)(0 - 40)$$
$$v_C = -31.8 \text{ m/s} = 31.8 \text{ m/s} \downarrow \qquad\qquad Ans.$$

Note: It should be realized that the ball is subjected to a *deceleration* from A to B of 9.81 m/s^2, and then from B to C it is *accelerated* at this rate. Furthermore, even though the ball momentarily comes to *rest* at B ($v_B = 0$) the acceleration at B is 9.81 m/s^2 downward!

Example 12–4

A metallic particle is subjected to the influence of a magnetic field such that it travels vertically downward through a fluid that extends from plate A to plate B, Fig. 12–5. If the particle is released from rest at C, $s = 100\text{mm}$, and the acceleration is measured as $a = (4s)$ m/s^2, where s is in meters, determine the velocity of the particle when it reaches plate B, $s = 200$ mm, and the time it needs to travel from C to plate B.

SOLUTION

Coordinate System. As shown in Fig. 12–5, s is taken positive downward, measured from plate A.

Velocity. Knowing the acceleration as a function of position, the velocity as a function of position can be obtained by using $v\,dv = a\,ds$. Why not use the formulas for constant acceleration? Realizing that $v = 0$ at $s = 100$ mm $= 0.1$ m, we have

Fig. 12–5

$(+\downarrow)$
$$v\,dv = a\,ds$$

$$\int_0^v v\,dv = \int_{0.1}^s 4s\,ds$$

$$\tfrac{1}{2}v^2 \Big|_0^v = \frac{4}{2}s^2 \Big|_{0.1}^s$$

$$v = 2(s^2 - 0.01)^{1/2} \qquad (1)$$

At $s = 200$ mm $= 0.2$ m,

$$v_B = 0.346 \text{ m/s} = 346 \text{ mm/s} \downarrow \qquad\qquad Ans.$$

The positive root is chosen since the particle is traveling downward, i.e., in the $+s$ direction.

Time. The time for the particle to travel from C to plate B can be obtained using $v = ds/dt$ and Eq. (1), where $s = 0.1$ m when $t = 0$.

$(+\downarrow)$
$$ds = v\,dt$$
$$= 2(s^2 - 0.01)^{1/2}\,dt$$

$$\int_{0.1}^s \frac{ds}{(s^2 - 0.01)^{1/2}} = \int_0^t 2\,dt$$

$$\ln(s + \sqrt{s^2 - 0.01})\Big|_{0.1}^s = 2t \Big|_0^t$$

$$\ln(s + \sqrt{s^2 - 0.01}) + 2.30 = 2t$$

At $s = 200$ mm $= 0.2$ m,

$$t = \frac{\ln(0.2 + \sqrt{(0.2)^2 - 0.01}) + 2.30}{2} = 0.657 \text{ s} \qquad Ans.$$

Example 12–5

A particle moves along a horizontal straight line such that its velocity is given by $v = (3t^2 - 6t)$ m/s, where t is the time in seconds. If it is initially located at the origin O, determine the distance traveled during the time interval $t = 0$ to $t = 3.5$ s, the average velocity, and the average speed of the particle during this time interval.

SOLUTION

Coordinate System. Here we will assume positive motion to the right, measured from a fixed origin O.

Distance Traveled. Since the velocity is related to time, the position related to time may be found by integrating $v = ds/dt$ with $t = 0$, $s = 0$.

$(\xrightarrow{+})$

$$ds = v\, dt$$
$$= (3t^2 - 6t)\, dt$$
$$\int_0^s ds = 3\int_0^t t^2\, dt - 6\int_0^t t\, dt$$
$$s = (t^3 - 3t^2) \text{ m} \qquad (1)$$

In order to determine the distance traveled in 3.5 s, it is necessary to investigate the path of motion. The graph of the velocity function, Fig. 12–6a, reveals that for $0 < t < 2$ s the velocity is *negative*, which means the particle is traveling to the *left*, and for $t > 2$ s the velocity is *positive*, and hence the particle is traveling to the *right*. Also, $v = 0$ at $t = 2$ s. The particle's position when $t = 0$, $t = 2$ s, and $t = 3.5$ s can be computed from Eq. (1). This yields

$$s|_{t=0} = 0, \qquad s|_{t=2\,s} = -4 \text{ m}, \qquad s|_{t=3.5\,s} = 6.12 \text{ m}$$

The path is shown in Fig. 12–6b. Hence, the distance traveled in 3.5 s is

$$s_T = 4 + 4 + 6.12 = 14.1 \text{ m} \qquad \textit{Ans.}$$

Velocity. The *displacement* from $t = 0$ to $t = 3.5$ s is

$$\Delta s = s|_{t=3.5\,s} - s|_{t=0} = 6.12 - 0 = 6.12 \text{ m}$$

so that the average velocity is

$$v_{\text{avg}} = \frac{\Delta s}{\Delta t} = \frac{6.12}{3.5 - 0} = 1.75 \text{ m/s} \rightarrow \qquad \textit{Ans.}$$

The average speed is defined in terms of the *distance traveled* s_T. Hence,

$$(v_{\text{sp}})_{\text{avg}} = \frac{s_T}{\Delta t} = \frac{14.12}{3.5 - 0} = 4.03 \text{ m/s} \qquad \textit{Ans.}$$

(a)

(b)

Fig. 12–6

PROBLEMS

12–1. If a ball is thrown upwards 8 m into the air before it starts to fall, determine its initial velocity and the total time of flight.

12–2. A train is traveling along a straight track at 56 ft/s when its brakes are applied, subjecting it to a constant deceleration. If it comes to a stop in 800 ft, determine its deceleration and the time needed to make the stop.

12–3. A particle travels along a straight-line path such that in 4 s it moves from an initial position $s_A = -8$ m to a position $s_B = +3$ m. Then in another 5 s it moves from s_B to $s_C = -6$ m. Determine the particle's average velocity and average speed during the 9-s time interval.

*12–4. A truck, traveling along a straight road at 20 km/h, increases its speed to 120 km/h in 15 s. If its acceleration is constant, determine the distance traveled.

12–5. A balloon is traveling upwards at a constant speed of 6 m/s. When a bag of sand falls from it, it strikes the ground in 3 s. Determine the height of the balloon at the moment the bag is released. Also determine the highest point reached by the bag of sand.

12–6. A particle is moving along a straight line such that its position is given by $s = (4t - t^2)$ ft, where t is in seconds. Determine the distance traveled from $t = 0$ to $t = 5$ s, the average velocity, and the average speed of the particle during this time interval.

12–7. A sphere is fired downwards into a medium with an initial speed of 27 m/s. If it experiences a deceleration of $a = (-6t)$ m/s,^{2} where t is in seconds, determine the distance traveled before it stops.

*12–8. A particle travels along a straight line with a constant acceleration. When $s = 4$ ft, $v = 3$ ft/s and when $s = 10$ ft, $v = 8$ ft/s. Determine the velocity as a function of position.

12–9. A particle travels to the right along a straight path with a velocity $v = [10/(2 + s^2)]$ m/s, where s is in meters. Determine its position when $t = 4$ s if $s = 3$ m when $t = 0$.

12–10. A particle travels in a straight line such that for a short time 4 s $\leqslant t \leqslant$ 8 s its motion is described by $v = (4/a)$ ft/s, where a is in ft/s^2. If $v = 10$ ft/s when $t = 4$ s, determine the particle's acceleration when $t = 6$ s.

12–11. Tests reveal that a normal driver can *react* to a situation in 0.75 s before beginning to avoid a collision. It takes about 3 s for a driver having 0.1% alcohol in his system to do the same. If two such drivers are traveling on a straight road at 30 mph (44 ft/s) and their cars can decelerate at 2 ft/s^2, determine the shortest stopping distance d for each from the moment they see the pedestrians at A. *Moral:* If you must drink, please don't drive!

Prob. 12–11

■*12–12. A particle moves along a straight-line path with an acceleration of $a = 5/(3s^{1/3} + s^{5/2})$, where s is in meters. Determine the particle's velocity when $s = 2$ m, if it starts from rest when $s = 1$ m. Use Simpson's rule to evaluate the integral.

■12–13. A projectile, initially at the origin, moves vertically downward along a straight-line path through a fluid medium such that its velocity is defined as $v = 3(8e^{-t} + t)^{1/2}$ m/s, where t is in seconds. Plot the position s of the projectile during the first 2 s. Use the Runge-Kutta method to evaluate s with incremental values of $h = 0.25$ s.

12–14. The acceleration of a rocket traveling upwards is given by $a = (6 + 0.02s)$ m/s^2, where s is in meters. Determine the rocket's velocity when $s = 2$ km and the time needed to reach this elevation. Initially, $v = 0$ and $s = 0$ when $t = 0$.

Prob. 12–14

12–15. The position of a particle traveling along a straight path is defined by $s = (t^3 - 9t^2 + 15t)$ ft, where t is in seconds. Determine the position and the total distance traveled in $t = 6$ s. *Hint:* To determine the total distance traveled, plot the path and calculate the displacements between the times when the particle stops.

*12–16. A motorcycle starts from rest and travels on a straight road with a constant acceleration of 10 ft/s² for 8 s, after which it maintains a constant speed for 2 s. Finally it decelerates at 14 ft/s² until it stops. Determine the total distance traveled and the average speed.

12–17. The position of a particle along a straight line is given by $s = (1.5t^3 - 13.5t^2 + 22.5t)$ ft, where t is in seconds. Determine the position of the particle when $t = 6$ s and the total distance it travels during the 6-s time interval. *Hint:* Plot the path to determine the total distance traveled.

12–18. A particle moving along a straight line is subjected to a deceleration $a = (-2v^3)$ m/s², where v is in m/s. If it has a velocity $v = 8$ m/s and a position $s = 10$ m when $t = 0$, determine its velocity and position when $t = 4$ s.

12–19. Plane B is traveling a distance d ahead of plane A. Both planes are traveling at 550 ft/s when the pilot of B suddenly causes his plane to decelerate at 12 ft/s². It takes the pilot of plane A 0.75 s to react and then he decelerates at 15 ft/s². Determine the minimum distance d between the planes so as to avoid a collision. Neglect the size of the planes.

Prob. 12–19

*12–20. The acceleration of a particle along a straight line is defined by $a = (2t - 9)$ m/s², where t is in seconds. At $t = 0$, $s = 1$ m and $v = 10$ m/s. When $t = 9$ s, determine (a) the particle's position, (b) the total distance traveled, and (c) the velocity. Assume the positive direction is to the right.

12–21. At $t = 0$ bullet A is fired vertically with an initial (muzzle) velocity of 450 m/s. When $t = 3$ s, bullet B is fired upwards with a muzzle velocity of 600 m/s. Determine the time t, after A is fired, as to when bullet B passes bullet A. At what elevation does this occur?

12–22. A stone is observed to fall past a 1.25-m-high window in 0.2 s. Determine (a) the average speed of the stone while it is in view, (b) the velocity of the stone when it reaches the bottom ledge, and (c) the height above the top of the window from which it fell from rest.

12–23. A particle is moving along a straight line with an initial velocity of 6 m/s when it is subjected to a deceleration of $a = (-1.5v^{1/2})$ m/s², where v is in m/s. Determine how far it travels before it stops. How much time does this take?

*12–24. Car A starts from rest at $t = 0$ and travels along a straight road with a constant acceleration of 6 ft/s² until it reaches a speed of 80 ft/s. Afterwards it maintains this speed. Also, when $t = 0$, car B located 6000 ft down the road is traveling towards A at a constant speed of 60 ft/s. Determine the time and distance traveled by car A when they pass each other.

12–25. A bicycle travels along a straight line with an acceleration of $a = (2 - v^2/500)$ ft/s², where the velocity v is in ft/s. If it starts from rest, determine its maximum speed. How far must it travel to reach a speed of 8 ft/s?

12–26. Two particles A and B start from rest at the origin $s = 0$ and move along a straight line such that $a_A = (6t - 3)$ ft/s² and $a_B = (12t^2 - 8)$ ft/s², where t is in seconds. Determine the distance between them when $t = 4$ s and the total distance each has traveled in $t = 4$ s.

12–27. As a body is projected to a high altitude above the earth's *surface,* the variation of the acceleration of gravity with respect to altitude y must be taken into account. Neglecting air resistance, this acceleration is determined from the formula $a = -g_o[R^2/(R + y)^2]$, where g_o is the constant gravitational acceleration at sea level, R is the radius of the earth, and the positive direction is measured upward. If $g_o = 9.81$ m/s² and $R = 6356$ km, determine the minimum initial velocity (escape velocity) at which a projectile should be shot vertically from the earth's surface so that it does not fall back to the earth. *Hint:* This requires that $v = 0$ as $y \to \infty$.

*12–28. Accounting for the variation of gravitational acceleration a with respect to altitude y (see Prob. 12–27), derive an equation that relates the velocity of a freely falling particle to its altitude. Assume that the particle is released from rest at an altitude y_o from the earth's surface. With what velocity does the particle strike the earth if it is released from rest at an altitude $y_o = 500$ km? Use the numerical data in Prob. 12–27.

12.2 Graphical Solutions

When a particle's motion during a time period is erratic, it may be difficult to obtain a continuous mathematical function to describe its position, velocity, or acceleration. Instead, the motion may best be described graphically using a series of curves. If this graph describes the behavior of any two of the variables, a, v, s, t, a graph describing the behavior of the other variables can be established by using the kinematic equations $a = dv/dt$, $v = ds/dt$, $a\ ds = v\ dv$. The following situations frequently occur.

Given the s-t Graph, Construct the v-t Graph

If the position of a particle can be *experimentally determined* during a time period t, the s-t graph for the particle can be plotted, Fig. 12–7a. Since *$v = ds/dt$, the v-t graph can be established by measuring the slope (ds/dt) of the s-t graph at various times and plotting the results.* (In a graphical sense, the slope is measured with a ruler and protractor.) For example, measurement of the slopes v_0, v_1, v_2, and v_3 at the intermediate points (t_0, s_0), (t_1, s_1), (t_2, s_2), and (t_3, s_3) on the s-t graph, Fig. 12–7a, gives corresponding points on the v-t graph shown in Fig. 12–7b.

It may also be possible to establish the v-t graph *mathematically,* provided the curves of the s-t graph can be expressed in the form of an equation $s = f(t)$. Corresponding equations describing curves of the v-t graph are then determined by *differentiation* since $v = ds/dt = d\,(f(t))/dt$.

(a)

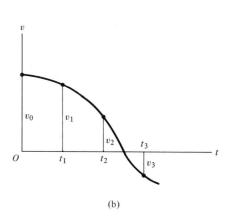

(b)

Fig. 12–7

14

Given the *v-t* Graph, Construct the *a-t* Graph

When the particle's *v-t* graph is known, as in Fig. 12–8*a*, the acceleration can be determined at any instant using $a = dv/dt$. Hence, *the a-t graph can be established by measuring the slope (dv/dt) of the v-t graph at various times and plotting the results*. For example, measurement of the slopes a_0, a_1, a_2, and a_3 at the intermediate points (t_0, v_0), (t_1, v_1), (t_2, v_2), and (t_3, v_3) on the *v-t* graph, Fig. 12–8*a*, yields corresponding points on the *a-t* graph shown in Fig. 12–8*b*.

Any segmented curves of the *a-t* graph can also be determined *mathematically* provided the equations of the corresponding curves of the *v-t* graph are known, $v = g(t)$. This is done by simply taking the *derivative* since $a = dv/dt = d(g(t))/dt$.

Since differentiation reduces a polynomial of degree *n* to that of degree *n-1*, then from the above explanation, if the *s-t* graph is parabolic (second-degree curve), the *v-t* graph will be a sloping line (first-degree curve), and the *a-t* graph will be a constant or a horizontal line (zero-degree curve).

(a)

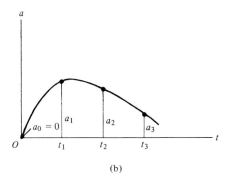

(b)

Fig. 12–8

15

Example 12–6

A car moves along a straight road such that its position is described by the graph shown in Fig. 12–9a. Construct the *v-t* and *a-t* graphs for the time period $0 \leqslant t \leqslant 30$ s.

SOLUTION

v-t Graph. Since $v = ds/dt$, the *v-t* graph can be determined by differentiating the equations defining the *s-t* graph, Fig. 12–9a. We have

$$0 \leqslant t < 10 \text{ s}; \qquad s = t^2 \qquad v = \frac{ds}{dt} = 2t$$

$$10 \text{ s} < t \leqslant 30 \text{ s}; \quad s = 20t - 100 \qquad v = \frac{ds}{dt} = 20$$

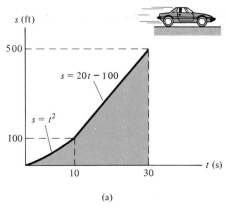

The results are plotted in Fig. 12–9b. We can also obtain specific values of *v* by measuring the *slope* of the *s-t* graph at a given instant. For example, at $t = 20$ s, the slope of the *s-t* graph is determined from the straight line from 10 s to 30 s, i.e.,

$$t = 20 \text{ s}; \qquad v = \frac{\Delta s}{\Delta t} = \frac{500 - 100}{30 - 10} = 20 \text{ ft/s}$$

a-t Graph. Since $a = dv/dt$, the *a-t* graph can be determined by differentiating the equations defining the lines of the *v-t* graph. This yields

$$0 \leqslant t < 10 \text{ s}; \qquad v = 2t \qquad a = \frac{dv}{dt} = 2$$

$$10 < t \leqslant 30 \text{ s}; \qquad v = 20 \qquad a = \frac{dv}{dt} = 0$$

The results are plotted in Fig. 12–9c. Show how to determine the specific value of the car's acceleration when $t = 5$ s by measuring the slope of the *v-t* graph.

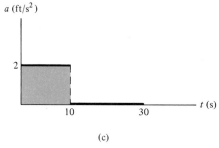

Fig. 12–9

Given the a-t Graph, Construct the v-t Graph

If the *a-t* graph is given, Fig. 12–10*a*, the *v-t* graph may be constructed using the equation $a = dv/dt$, written in integrated form as $\Delta v = \int a\, dt$. In this case, *the change in the particle's speed during a period of time is equal to the area under the a-t graph during the same time period*, Fig. 12–10*b*. (In a graphical sense, any *small* area may be approximated as a trapezoid or rectangle.) Using this method, one begins with knowing the particle's initial velocity v_0 and adding (algebraically) to this small area increments Δv determined from the *a-t* graph. In this manner, one determines successive points, $v_1 = v_0 + \Delta v$, etc., for the *v-t* graph. Notice that an algebraic addition of area is necessary, since areas lying above the *t* axis correspond to an increase in *v* ("positive" area), whereas those lying below the *t* axis indicate a decrease in *v* ("negative" area).

If the curves of the *a-t* graph can be described by a series of equations, then each of these equations may be *integrated* to yield equations describing the corresponding curves of the *v-t* graph. Hence, if the *a-t* graph is linear (first-degree curve), the integration will yield a *v-t* graph that is parabolic (second-degree curve), etc.

Given the v-t Graph, Construct the s-t Graph

When the *v-t* graph is given, Fig. 12–11*a*, it is possible to determine the *s-t* graph using $v = ds/dt$, written in integrated form as $\Delta s = \int v\, dt$. In this case *the particle's displacement during a period of time is equal to the area under the v-t graph during the same time period*, Fig. 12–11*b*. In the same manner as stated above, one begins by knowing the particle's initial position s_0 and adding (algebraically) to this small area increments Δs determined from the *v-t* graph.

If it is possible to describe the segments of the *v-t* graph by a series of equations, then each of these equations may be *integrated* to yield equations describing the *s-t* graph.

(a)

(b)

Fig. 12–10

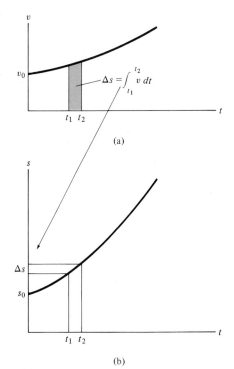

(a)

(b)

Fig. 12–11

Example 12–7

The rocket sled in Fig. 12–12a starts from rest and travels along a straight track such that it accelerates at a constant rate for 10 s and then decelerates at a constant rate. Draw the v-t and s-t graphs and determine the time t' needed to stop the sled. How far has the sled traveled?

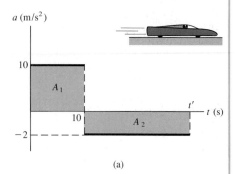

(a)

SOLUTION

v-t Graph. Since $dv = a\,dt$, the v-t graph is determined by integrating the straight-line segments of the a-t graph. Using the initial condition $v = 0$ when $t = 0$, we have

$$0 \leqslant t < 10 \text{ s}; \quad a = 10; \quad \int_0^v dv = \int_0^t 10\,dt$$

$$v = 10t$$

When $t = 10$ s, $v = 10(10) = 100$ m/s. Using this as the initial condition for the next time period, we have

$$10 \text{ s} < t \leqslant t'; \quad a = -2; \quad \int_{100}^v dv = \int_{10}^t -2\,dt$$

$$v = -2t + 120$$

When $t = t'$ we require $v = 0$. This yields

$$t' = 60 \text{ s} \qquad \textit{Ans.}$$

The results are shown in Fig. 12–12b.

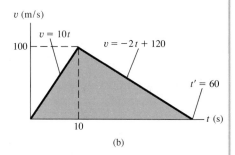

(b)

s-t Graph. Since $ds = v\,dt$, integrating the equations of the v-t graph yields the corresponding equations of the s-t graph. Using the initial condition $s = 0$ when $t = 0$, we have

$$0 \leqslant t \leqslant 10 \text{ s}; \quad v = 10t; \quad \int_0^s ds = \int_0^t 10t\,dt$$

$$s = 5t^2$$

When $t = 10$ s, $s = 5(10)^2 = 500$ m. Using this initial condition,

$$10 \leqslant t \leqslant 60 \text{ s}; \quad v = -2t + 120; \quad \int_{500}^s ds = \int_{10}^t (-2t + 120)\,dt$$

$$s - 500 = -t^2 + 120t - (-(10)^2 + 120(10))$$

$$s = -t^2 + 120t - 600$$

When $t' = 60$ s, the position is

$$s = -(60)^2 + 120(60) - 600 = 3000 \text{ m} \qquad \textit{Ans.}$$

The s-t graph is shown in Fig. 12–12c. Note that a direct solution for s when $t' = 60$ s is possible, since the *triangular area* under the v-t graph would yield the displacement $\Delta s = s - 0$ from $t = 0$ to $t' = 60$ s. Hence,

$$\Delta s = \tfrac{1}{2}(60)(100) = 3000 \text{ m} \qquad \textit{Ans.}$$

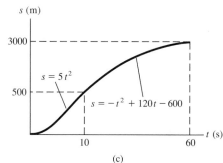

(c)

Fig. 12–12

Given the a-s Graph, Construct the v-s Graph

In some cases an *a-s* graph is known, so that points on the *v-s* graph can be determined by using $v \, dv = a \, ds$. Integrating this equation between the limits $v = v_1$ at $s = s_1$ and $v = v_2$ at $s = s_2$, we have $\frac{1}{2}(v_2^2 - v_1^2) = \int_{s_1}^{s_2} a \, ds$. Thus, small segments of area under the *a-s* graph, $\int_{s_1}^{s_2} a \, ds$, shown colored in Fig. 12–13a, equal one-half the difference in the squares of the speed, $\frac{1}{2}(v_2^2 - v_1^2)$. By approximation of the area, $\int_{s_1}^{s_2} a \, ds$, it is possible to compute the value of v_2 at s_2 if an initial value of v_1 at s_1 is known, i.e., $v_2 = (2 \int_{s_1}^{s_2} a \, ds + v_1^2)^{1/2}$, Fig. 12–13b. The *v-s* graph can be constructed in this manner starting from the initial velocity v_0.

Another way to construct the *v-s* graph is to first determine the equations which define the various curves of the *a-s* graph. Then the corresponding equations defining the curves of the *v-s* graph can be obtained directly from integration, using $v \, dv = a \, ds$.

Given the v-s Graph, Construct the a-s Graph

If the *v-s* graph is known, the acceleration *a* at any position *s* can be determined using the following graphical procedure. At any point $P(s, v)$, Fig. 12–14a, the slope dv/ds of the *v-s* graph is determined. Since $a \, ds = v \, dv$, then $a = v(dv/ds)$, Fig. 12–14b. Of course, one must work with a consistent set of units when using this procedure; for example, if v is measured in m/s and s in meters, then a will be calculated in m/s^2.

We can also determine the curves describing the *a-s* graph analytically, provided the equations of the corresponding curves of the *v-s* graph are known. As above, this requires application of $a \, ds = v \, dv$.

(a)

(b)

Fig. 12–13

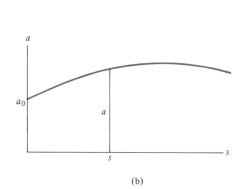

Fig. 12–14   (a)   (b)

Example 12–8

The v-s graph describing the motion of a motorcycle is shown in Fig. 12–15a. Construct the a-s graph of the motion and determine the time needed for the motorcycle to reach the position $s = 400$ ft.

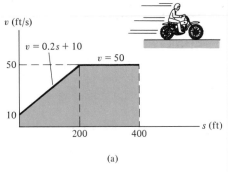

(a)

SOLUTION

a-s Graph. Since the equations of the v-s graph are given, the a-s graph can be determined using the equation $a\,ds = v\,dv$, which yields

$0 \leqslant s < 200$ ft; $v = 0.2s + 10$

$$a = v\frac{dv}{ds} = (0.2s + 10)\frac{d}{ds}(0.2s + 10)$$

$$= 0.04s + 2$$

200 ft $< s \leqslant 400$ ft; $v = 50$; $a = v\frac{dv}{ds} = (50)\frac{d}{ds}(50)$

$$= 0$$

The results are plotted in Fig. 12–15b.

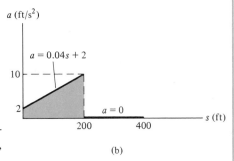

(b)

Fig. 12–15

Time. The time can be obtained using the v-s graph and $v = ds/dt$, because this equation relates v, s, and t. For the two segments of motion,

$0 \leqslant s < 200$ ft; $v = 0.2s + 10$; $dt = \dfrac{ds}{v} = \dfrac{ds}{0.2s + 10}$

$$\int_0^t dt = \int_0^s \frac{ds}{0.2s + 10}$$

$$t = 5\ln(0.2s + 10) - 5\ln 10$$

At $s = 200$ ft, $t = 5\ln(0.2(200) + 10) - 5\ln 10 = 8.05$ s.

200 ft $< s \leqslant 400$ ft; $v = 50$; $dt = \dfrac{ds}{v} = \dfrac{ds}{50}$

$$\int_{8.05}^t dt = \int_{200}^s \frac{ds}{50}$$

$$t - 8.05 = \frac{s}{50} - 4$$

$$t = \frac{s}{50} + 4.05$$

Therefore, at $s = 400$ ft,

$$t = \frac{400}{50} + 4.05 = 12.0 \text{ s} \qquad\qquad \textit{Ans.}$$

PROBLEMS

12–29. The bobsled moves down along the straight course such that its v-t graph is as shown. Construct the s-t and a-t graphs for the same 50-s time interval. When $t = 0$, $s = 0$.

12–30. A rocket car starting from rest travels along a straight road and for 10 s has an acceleration as shown. Construct the v-t graph that describes the motion and find the distance traveled in 10 s.

Prob. 12–30

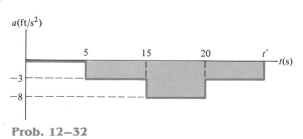

Prob. 12–29

12–31. A motorcycle starts from rest at $s = 0$ and travels along a straight road with the speed shown by the v-t graph. Determine the total distance the motorcycle travels until it stops when $t = 15$ s. Also plot the a-t and s-t graphs.

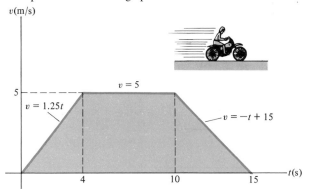

Prob. 12–31

*12–32. An airplane lands on the straight runway, originally traveling at 110 ft/s when $s = 0$. If it is subjected to the decelerations shown, determine the time t' needed to stop the plane and construct the s-t graph for the motion.

Prob. 12–32

12–33. The s-t graph for a train has been experimentally determined. From the data, construct the v-t and a-t graphs for the motion; $0 \leqslant t \leqslant 40$ s. For $0 \leqslant t \leqslant 30$ s, the curve is $s = (0.4t^2)$ m, where t is in seconds.

Prob. 12–33

12–34. A train starts from rest when $t = 0$ and $s = 0$. It travels along a straight track with a constant acceleration of 0.75 ft/s². When $t = 15$ s it maintains a constant speed until it has traveled a total distance of 500 ft. Determine its maximum speed and draw the s-t graph for the motion.

21

12–35. A two-stage rocket is fired vertically from rest at $s = 0$ with an acceleration as shown. After 30 s the first stage A burns out and the second stage B ignites. Plot the v-t and s-t graphs which describe the motion of the second stage for $0 \leqslant t \leqslant 60$ s.

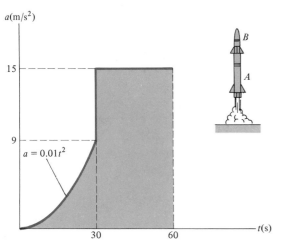

$a(\text{m/s}^2)$

15

9

$a = 0.01t^2$

30   60 $t(\text{s})$

Prob. 12–35

*12–36.** The v-t graph for the motion of a car as it moves along a straight road is shown. Draw the a-t graph and determine the maximum acceleration during the 30-s time interval. The car starts from rest at $s = 0$.

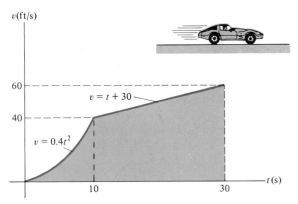

$v(\text{ft/s})$

60

40

$v = t + 30$

$v = 0.4t^2$

10   30 $t(\text{s})$

Probs. 12–36/12–37

12–37. The v-t graph for the motion of a car as it moves along a straight road is shown. Draw the s-t graph and determine the average speed and the distance traveled for the 30-s time interval. The car starts from rest at $s = 0$.

12–38. The race car travels from rest along a straight road and reaches a speed of 26 m/s in 8 s as shown on the v-t graph. The flat part of the graph is caused by shifting gears. Draw the a-t graph and determine the maximum acceleration of the car.

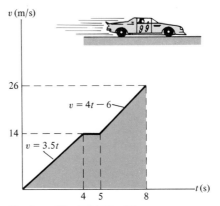

$v \, (\text{m/s})$

26

$v = 4t - 6$

14

$v = 3.5t$

4 5   8 $t(\text{s})$

Probs. 12–38/12–39

12–39. The race car travels from rest at $s = 0$ along a straight road and reaches a speed of 26 m/s in 8 s as shown on the v-t graph. The flat part of the graph is caused by shifting gears. Draw the s-t graph and determine the total distance traveled.

*12–40.** A particle is moving at $v = (4s)$ ft/s, where s is given in ft. Draw the v-t and a-t graphs for the particle for $0 \leqslant t \leqslant 5$ s. Initially, $s_0 = 10$ ft at $t = 0$.

12–41. Starting from rest at $s = 0$, a boat travels in a straight line with an acceleration as shown by the a-s graph. Determine the boat's speed when $s = 40$, 90, and 200 ft.

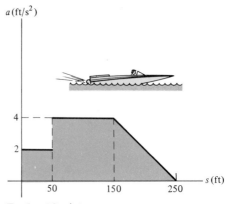

$a \, (\text{ft/s}^2)$

4

2

50   150   250 $s \, (\text{ft})$

Prob. 12–41

12–42. The car is originally traveling at 15 ft/s when it is subjected to the motion shown by the *a-t* graph. Determine the car's maximum speed and the time t' when it stops.

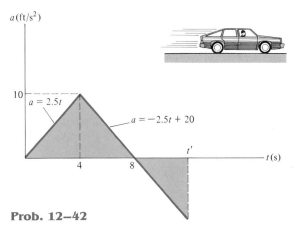

Prob. 12–42

12–43. The *v-s* graph for an airplane traveling on a straight runway is shown. Determine the acceleration of the plane at $s = 100$ m and $s = 150$ m. Draw the *a-s* graph.

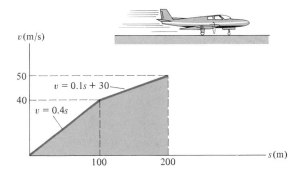

Prob. 12–43

*****12–44.** Two rockets start from rest at the same ground elevation. Rocket A accelerates vertically at 20 m/s² for 12 s and then maintains a constant speed. Rocket B accelerates at 15 m/s² until reaching a constant speed of 150 m/s and then maintains this speed. Construct the *a-t*, *v-t*, and *s-t* graphs for each rocket until $t = 20$ s. What is the distance between the rockets when $t = 20$ s?

12–45. The car starts from rest at $s = 0$ and is subjected to an acceleration as shown by the *a-s* graph. Draw the *v-s* graph and determine the time needed to travel 200 ft.

Prob. 12–45

12–46. A freight train starts from rest when $t = 0$ and travels on a straight track with a constant acceleration of 0.75 ft/s². When $t = t'$ it maintains a constant speed so that when $t = 160$ s the train has traveled 1500 ft. Determine the time t' and draw the *v-t* graph for the motion.

12.3 General Curvilinear Motion

When a particle moves along a curved path, the motion is called *curvilinear motion*. Because the path is often represented in three dimensions, *vector analysis* will be used to formulate the particle's position, velocity, and acceleration.* In this section the general aspects of curvilinear motion are discussed and in subsequent sections three types of coordinate systems often used to describe this motion will be introduced.

*A summary of some of the important concepts of vector analysis is given in Appendix C.

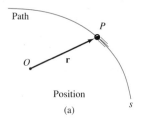

Path

P

O

r

Position

(a)

Position

Consider a particle located at point *P* on a space curve defined by the path function *s*, Fig. 12–16*a*. The position of the particle, measured from a fixed point *O*, will be designated by the *position vector* **r** = **r**(*t*). This vector is a function of time since, in general, both its magnitude and direction change as the particle moves along the curve.

Displacement

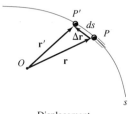

P'

ds

r'

Δ**r**

P

O

r

Displacement

(b)

Suppose that during a small time interval Δ*t* the particle moves a distance Δ*s* along the curve to a new position *P'*, defined by **r'** = **r** + Δ**r**, Fig. 12–16*b*. The *displacement* Δ**r** represents the change in the particle's position and is determined by vector subtraction; i.e., Δ**r** = **r'** − **r.**

Velocity

During the time Δ*t*, the *average velocity* of the particle is defined as

$$\mathbf{v}_{avg} = \frac{\Delta \mathbf{r}}{\Delta t}$$

The *instantaneous velocity* is determined from this equation by letting Δ*t* → 0 and consequently the direction of Δ**r** approaches the *tangent* to the curve at point *P*. Hence, **v** = $\lim\limits_{\Delta t \to 0}$ (Δ**r**/Δ*t*) or

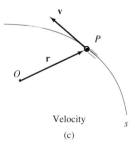

v

P

r

O

Velocity

(c)

Fig. 12–16(a–c)

$$\boxed{\mathbf{v} = \frac{d\mathbf{r}}{dt}} \qquad (12\text{–}7)$$

Since *d***r** will be tangent to the curve at *P*, the *direction* of **v** is *also tangent to the curve*, Fig. 12–16*c*. The *magnitude* of **v**, which is called the *speed*, may be obtained by noting that the magnitude of the displacement Δ**r** is the length of the straight line segment from *P* to *P'*, Fig. 12–16*b*. Realizing that this length, Δ*r*, approaches the arc length Δ*s* as Δ*t* → 0, we have *v* = $\lim\limits_{\Delta t \to 0}$ (Δ*r*/Δ*t*) = $\lim\limits_{\Delta t \to 0}$ (Δ*s*/Δ*t*), or

$$\boxed{v = \frac{ds}{dt}} \qquad (12\text{–}8)$$

Thus, the *speed* may be obtained by differentiating the path function *s* with respect to time.

(d)   s   (e)   (f)

Fig. 12–16(d–g)

Acceleration

If the particle has a velocity **v** at time t and a velocity $\mathbf{v}' = \mathbf{v} + \Delta\mathbf{v}$ at $t + \Delta t$, Fig. 12–16d, then the *average acceleration* of the particle during the time interval Δt is

$$\mathbf{a}_{\text{avg}} = \frac{\Delta\mathbf{v}}{\Delta t}$$

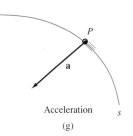

Acceleration

(g)

where $\Delta\mathbf{v} = \mathbf{v}' - \mathbf{v}$. To study this time rate of change, the two velocity vectors in Fig. 12–16d are plotted in Fig. 12–16e such that their tails are located at the fixed point O' and their heads reach points on the dashed curve. This curve is called a *hodograph*, and when constructed, it describes the locus of points for the head of the velocity vector in the same manner as the *path s* describes the locus of points for the head of the position vector, Fig. 12–16a.

To obtain the *instantaneous acceleration*, let $\Delta t \to 0$ so that in the limit $\Delta\mathbf{v}$ becomes *tangent to the hodograph* and we have $\mathbf{a} = \lim\limits_{\Delta t \to 0} (\Delta\mathbf{v}/\Delta t)$, or

$$\mathbf{a} = \frac{d\mathbf{v}}{dt} \qquad\qquad (12\text{–}9)$$

Using Eq. 12–7, we can also write

$$\mathbf{a} = \frac{d^2\mathbf{r}}{dt^2}$$

By definition of the derivative, **a** acts *tangent to the hodograph*, Fig. 12–16f, and therefore, *in general,* **a** *is not tangent to the path of motion*, Fig. 12–16g. To clarify this point, realize that $\Delta\mathbf{v}$ and consequently **a** must account for the change made in *both* the magnitude *and* direction of the velocity **v** as the particle moves from P to P', Fig. 12–16d. Just a magnitude change increases (or decreases) the ''length'' of **v,** and this in itself would allow **a** to remain tangent to the path. However, in order for the particle to follow the curve, the directional change always swings the velocity vector toward the ''inside'' or ''concave side'' of the curve, and therefore **a** cannot remain tangent to the path.

25

12.4 Curvilinear Motion: Rectangular Components

Occasionally the motion of a particle is expressed in terms of its three rectangular components, which are measured from the origin of a fixed x, y, z frame of reference.

Position

If at a given instant the particle P is at point $(x,\ y,\ z)$ on the curved path s, Fig. 12–17a, its location is then defined by the *position vector*

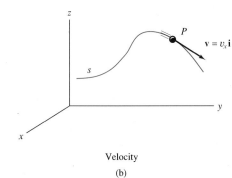

Position

(a)

$$\boxed{\mathbf{r} = x\mathbf{i} + y\mathbf{j} + z\mathbf{k}} \tag{12–10}$$

Because of the particle motion and the shape of the path, the x, y, z components of $\mathbf{r}$ are generally all functions of time; i.e., $x = x(t)$, $y = y(t)$, and $z = z(t)$, so that $\mathbf{r} = \mathbf{r}(t)$.

In accordance with the discussion in Appendix C, the *magnitude* of $\mathbf{r}$ is *always positive* and defined from Eq. C–3 as

$$r = \sqrt{x^2 + y^2 + z^2}$$

The *direction* is specified by the components of the unit vector $\mathbf{u}_r = \mathbf{r}/r$.

Velocity

(b)

Fig. 12–17(a, b)

Velocity

The first time derivative of $\mathbf{r}$ yields the velocity $\mathbf{v}$ of the particle. Hence,

$$\mathbf{v} = \frac{d\mathbf{r}}{dt} = \frac{d}{dt}(x\mathbf{i}) + \frac{d}{dt}(y\mathbf{j}) + \frac{d}{dt}(z\mathbf{k})$$

When taking the derivative, it is necessary to account for changes in *both* the magnitude and direction of the vector's components. From Eq. C–23, the derivative of the $\mathbf{i}$ component of $\mathbf{v}$ is therefore

$$\frac{d}{dt}(x\mathbf{i}) = \frac{dx}{dt}\mathbf{i} + x\frac{d\mathbf{i}}{dt}$$

The second term on the right side is zero, since the x, y, z reference frame is *fixed*, and therefore the *direction* (and the *magnitude*) of $\mathbf{i}$ does not change with time. Differentiation of the $\mathbf{j}$ and $\mathbf{k}$ components may be carried out in a similar manner, which yields the final result,

$$\boxed{\mathbf{v} = \frac{d\mathbf{r}}{dt} = v_x\mathbf{i} + v_y\mathbf{j} + v_z\mathbf{k}} \tag{12–11}$$

where

$$v_x = \dot{x}$$
$$v_y = \dot{y}$$
$$v_z = \dot{z}$$

(12–12)

The "dot" notation $\dot{x}$, $\dot{y}$, $\dot{z}$ represents the first time derivatives of the parametric equations $x = x(t)$, $y = y(t)$, and $z = z(t)$, respectively.

The velocity has a *magnitude* defined as the *positive* value of

$$v = \sqrt{v_x^2 + v_y^2 + v_z^2}$$

and a *direction* that is specified by the components of the unit vector $\mathbf{u}_v = \mathbf{v}/v$. This direction is *always tangent* to the path, as shown in Fig. 12–17b.

Acceleration

The acceleration of the particle is obtained by taking the first time derivative of Eq. 12–11 (or the second time derivative of Eq. 12–10). Using dots to represent the derivatives of the components, we have

$$\mathbf{a} = \frac{d\mathbf{v}}{dt} = a_x\mathbf{i} + a_y\mathbf{j} + a_z\mathbf{k}$$

(12–13)

where

$$a_x = \dot{v}_x = \ddot{x}$$
$$a_y = \dot{v}_y = \ddot{y}$$
$$a_z = \dot{v}_z = \ddot{z}$$

(12–14)

Hence, a_x, a_y, and a_z represent, respectively, the first time derivatives of the time functions v_x, v_y, and v_z, or the second time derivatives of the time functions x, y, and z.

The acceleration has a *magnitude* defined as the *positive* value of

$$a = \sqrt{a_x^2 + a_y^2 + a_z^2}$$

and a *direction* specified by the components of the unit vector $\mathbf{u}_a = \mathbf{a}/a$. Since $\mathbf{a}$ represents the time rate of *change* in velocity, in general $\mathbf{a}$ will *not* be tangent to the path traveled by the particle, Fig. 12–17c.

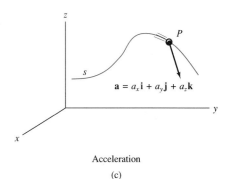

$$\mathbf{a} = a_x\mathbf{i} + a_y\mathbf{j} + a_z\mathbf{k}$$

Acceleration

(c)

Fig. 12–17c

PROCEDURE FOR ANALYSIS

Coordinate System. A rectangular coordinate system is often used to solve problems when the motion can be expressed in terms of its x, y, z components.

Kinematic Quantities. Since *rectilinear motion* occurs along each axis, a description of the motion of each component can be determined using $v = ds/dt$, $a = dv/dt$, and $a\ ds = v\ dv$ as outlined in Sec. 12.1 and formalized above. Once the components have been specified, the resultant motion is determined from the Pythagorean Theorem.

Example 12–9

At any instant the position of the kite in Fig. 12–18a is defined by the coordinates $x = (30t)$ ft and $y = (9t^2)$ ft, where t is given in seconds. Determine (a) the equation which describes the path and the distance of the kite from the boy when $t = 2$ s, (b) the magnitude and direction of the velocity when $t = 2$ s, and (c) the magnitude and direction of the acceleration when $t = 2$ s.

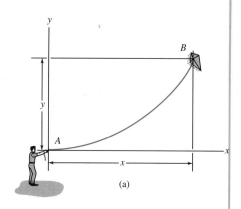

(a)

SOLUTION

Position. The equation of the path is determined by eliminating t from the expressions for x and y, i.e., $t = x/30$ so that $y = 9(x/30)^2$ or

$$y = x^2/100 \qquad\qquad Ans.$$

This is the equation of a parabola, Fig. 12–18a. When $t = 2$ s,

$$x = 30(2) = 60 \text{ ft}, \qquad y = 9(2)^2 = 36 \text{ ft}$$

The distance from A to B is therefore

$$r = \sqrt{(60)^2 + (36)^2} = 70.0 \text{ ft} \qquad\qquad Ans.$$

(b)

Velocity. Using Eqs. 12–12, the components of velocity when $t = 2$ s are

$$v_x = \frac{d}{dt}(30t) = 30 \text{ ft/s} \rightarrow$$

$$v_y = \frac{d}{dt}(9t^2) = 18t \Big|_{t=2s} = 36 \text{ ft/s} \uparrow$$

When $t = 2$ s, the magnitude of velocity is

$$v = \sqrt{(30)^2 + (36)^2} = 46.9 \text{ ft/s} \qquad\qquad Ans.$$

The direction is tangent to the path, Fig. 12–18b, where

$$\theta_v = \tan^{-1} \frac{36}{30} = 50.2° \qquad\qquad Ans.$$

(c)

Fig. 12–18

Acceleration. The components of acceleration are determined from Eqs. 12–14,

$$a_x = \frac{d}{dt}(30) = 0$$

$$a_y = \frac{d}{dt}(18t) = 18 \text{ ft/s}^2 \uparrow$$

Here the acceleration is constant for all time and thus

$$a = \sqrt{(0)^2 + (18)^2} = 18 \text{ ft/s}^2 \qquad\qquad Ans.$$

The direction, as shown in Fig. 12–18c, is

$$\theta_a = \tan^{-1} \frac{18}{0} = 90° \qquad\qquad Ans.$$

Example 12–10

The motion of a bead B sliding down along the spiral path shown in Fig. 12–19 is defined by the position vector $\mathbf{r} = \{0.5 \sin{(2t)}\mathbf{i} + 0.5 \cos{(2t)}\mathbf{j} - 0.2t\mathbf{k}\}$ m, where t is given in seconds and the arguments for sine and cosine are given in radians (π rad $= 180°$). Determine the location of the bead when $t = 0.75$ s and the magnitudes of the bead's velocity and acceleration at this instant.

SOLUTION

Position. Evaluating $\mathbf{r}$ when $t = 0.75$ s yields

$$\mathbf{r}|_{t=0.75\,\text{s}} = \{0.5 \sin{(1.5 \text{ rad})}\mathbf{i} + 0.5 \cos{(1.5 \text{ rad})}\mathbf{j} - 0.2(0.75)\mathbf{k}\} \text{ m}$$
$$= \{0.499\mathbf{i} + 0.035\mathbf{j} - 0.150\mathbf{k}\} \text{ m} \qquad Ans.$$

The magnitude of this vector represents the distance of the bead from the origin O, which is

$$r = \sqrt{(0.499)^2 + (0.035)^2 + (-0.150)^2} = 0.522 \text{ m} \qquad Ans.$$

The direction of $\mathbf{r}$ is obtained from the components of the unit vector

$$\mathbf{u}_r = \frac{\mathbf{r}}{r} = \frac{0.499}{0.522}\mathbf{i} + \frac{0.035}{0.522}\mathbf{j} - \frac{0.150}{0.522}\mathbf{k}$$
$$= 0.956\mathbf{i} + 0.067\mathbf{j} - 0.287\mathbf{k}$$

Hence, the coordinate direction angles α, β, and γ, Fig. 12–19, are

$$\alpha = \cos^{-1}{(0.956)} = 17.1° \qquad Ans.$$
$$\beta = \cos^{-1}{(0.067)} = 86.2° \qquad Ans.$$
$$\gamma = \cos^{-1}{(-0.287)} = 106° \qquad Ans.$$

Velocity. The velocity is defined by

$$\mathbf{v} = \frac{d\mathbf{r}}{dt} = \{1 \cos{(2t)}\mathbf{i} - 1 \sin{(2t)}\mathbf{j} - 0.2\mathbf{k}\} \text{ m/s}$$

Hence, when $t = 0.75$ s the magnitude of velocity, or speed, is

$$v = \sqrt{v_x^2 + v_y^2 + v_z^2}$$
$$= \sqrt{(1 \cos{(1.5 \text{ rad})})^2 + (-1 \sin{(1.5 \text{ rad})})^2 + (-0.2)^2}$$
$$= 1.02 \text{ m/s} \qquad Ans.$$

The velocity is tangent to the path as shown in Fig. 12–19.

Acceleration. The bead's acceleration $\mathbf{a}$, which is shown in Fig. 12–19, is *not* tangent to the path. Show that

$$\mathbf{a} = \frac{d\mathbf{v}}{dt} = \{-2 \sin{(2t)}\mathbf{i} - 2 \cos{(2t)}\mathbf{j}\} \text{ m/s}^2$$

and when $t = 0.75$ s the magnitude is $a = 2$ m/s^2.

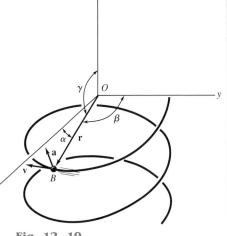

Fig. 12–19

12.5 Motion of a Projectile

Free-flight motion of a projectile is often studied in terms of its rectangular components since the projectile's acceleration *always* acts in the vertical direction. To illustrate the concepts involved in the kinematic analysis, consider a projectile launched at point (x_0, y_0), as shown in Fig. 12–20. The path is defined in the *x-y* plane such that the initial velocity is $\mathbf{v}_0$, having components $(\mathbf{v}_x)_0$ and $(\mathbf{v}_y)_0$. When air resistance is neglected, the only force acting on the projectile is its weight, which creates a *constant downward acceleration* of approximately $a_c = g = 9.81$ m/s^2 or $g = 32.2$ ft/s^2.* Hence, $a_y = -g$ and $a_x = 0$.

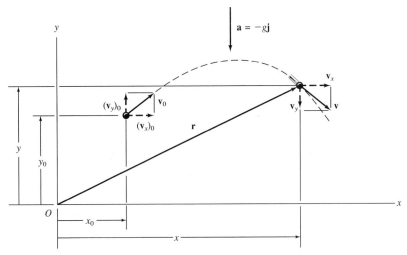

Fig. 12–20

Horizontal Motion

Since the component of acceleration in the *x* direction is $a_x = 0$, application of the constant acceleration equations, 12–4 to 12–6, yields

$$(\xrightarrow{+})\, v = v_0 + a_c t; \qquad\qquad v_x = (v_x)_0$$
$$(\xrightarrow{+})\, x = x_0 + v_0 t + \tfrac{1}{2} a_c t^2; \qquad x = x_0 + (v_x)_0 t$$
$$(\xrightarrow{+})\, v^2 = v_0^2 + 2a_c(s - s_0); \qquad v_x = (v_x)_0$$

The first and last equations indicate that *the horizontal component of velocity remains constant throughout the motion.*

Vertical Motion

Since the positive *y* axis is directed upward, the component of acceleration in the *y* direction is $a_y = -g$. Applying Eqs. 12–4 to 12–6, we get

*This assumes that the earth's gravitational field does not vary with altitude (see Example 13–2).

$$(+\uparrow)\, v = v_0 + a_c t; \qquad\qquad v_y = (v_y)_0 - gt$$
$$(+\uparrow)\, y = y_0 + v_0 t + \tfrac{1}{2} a_c t^2; \qquad y = y_0 + (v_y)_0 t - \tfrac{1}{2} g t^2$$
$$(+\uparrow)\, v^2 = v_0^2 + 2a_c(y - y_0); \qquad v_y^2 = (v_y)_0^2 - 2g(y - y_0)$$

Recall that the last equation can be formulated on the basis of eliminating the time t between the first two equations, and therefore *only two of the above three equations are independent of one another*.

To summarize, problems involving the motion of a projectile can have at most three unknowns since only three independent equations can be written. These equations consist of one equation in the horizontal direction and two in the vertical direction. Scalar analysis can be used here because the component motions along the x and y axes are *rectilinear*. Once $\mathbf{v}_x$ and $\mathbf{v}_y$ are obtained, realize that the resultant velocity $\mathbf{v}$ is defined by the *vector sum* as shown in Fig. 12–20.

PROCEDURE FOR ANALYSIS

Using the above results, the following procedure provides a method for solving problems concerning free-flight projectile motion.

Coordinate System. Establish the fixed x, y coordinate axes and sketch the trajectory of the particle. Between any *two points* on the path specify the given problem data and the *three unknowns*. In all cases the acceleration of gravity acts downward. The particle's initial and final velocities should be represented in terms of their x and y components. Remember that positive and negative position, velocity, and acceleration components always act in accordance with the associated coordinate directions.

Kinematic Equations. Depending upon the known data and what is to be determined, a choice should be made as to which three of the following four equations should be applied between the two points on the path to obtain the most direct solution to the problem.

Horizontal Motion. The *velocity* in the horizontal or x direction is *constant*, i.e., $(v_x) = (v_x)_0$, and

$$x = x_0 + (v_x)_0 t$$

Vertical Motion. In the vertical or y direction *only two* of the following three equations can be used for solution.

$$v_y = (v_y)_0 + a_c t$$
$$y = y_0 + (v_y)_0 t + \tfrac{1}{2} a_c t^2$$
$$v_y^2 = (v_y)_0^2 + 2a_c(y - y_0)$$

The following examples numerically illustrate application of this procedure.

Example 12–11

A cannon ball is fired from point A with a horizontal muzzle velocity of 120 m/s, as shown in Fig. 12–21. If the cannon is located at an elevation of 60 m above the ground, determine the time for the cannon ball to strike the ground and the range R.

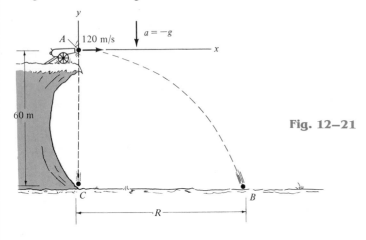

Fig. 12–21

SOLUTION

Coordinate System. The origin of coordinates is established at the initial point A, Fig. 12–21. The initial velocity of the cannon ball has components $(v_A)_x = 120$ m/s and $(v_A)_y = 0$. Also, between points A and B the acceleration is $a_y = -9.81$ m/s^2. Since $(v_B)_x = (v_A)_x = 120$ m/s, the three unknowns are $(v_B)_y$, R, and the time of flight t.

Vertical Motion. The vertical distance from A to B is known, and therefore we can obtain a direct solution for t by using the equation

$(+\uparrow)$
$$y = y_0 + (v_y)_0 t + \tfrac{1}{2}a_c t^2$$
$$-60 = 0 + 0 + \tfrac{1}{2}(-9.81)t^2$$
$$t = 3.50 \text{ s} \qquad \textit{Ans.}$$

This calculation also indicates that if an object is released from rest at A at the instant the cannon is fired (horizontally), it will strike the ground at C the same instant the cannon ball strikes the ground at B.

Horizontal Motion. Since t has been calculated, R is determined as follows:

$(\xrightarrow{+})$
$$x = x_0 + (v_x)_0 t$$
$$R = 0 + 120(3.50)$$
$$R = 420 \text{ m} \qquad \textit{Ans.}$$

Example 12–12

A ball is thrown from a position 5 ft above the ground to the roof of a 40-ft-high building, as shown in Fig. 12–22. If the initial velocity of the ball is 70 ft/s, inclined at an angle of 60° from the horizontal, determine the range or horizontal distance R from the point where the ball is thrown to where it strikes the roof.

Fig. 12–22

SOLUTION

Coordinate System. When the motion is analyzed between points O and A, the three unknowns are represented as the horizontal distance R, time of flight t_{OA}, and vertical component of velocity $(v_y)_A$. With the origin of coordinates at O, Fig. 12–22, the ball's initial velocity has components of

$$(v_x)_O = 70 \cos 60° = 35 \text{ ft/s} \rightarrow$$
$$(v_y)_O = 70 \sin 60° = 60.62 \text{ ft/s} \uparrow$$

Also, $(v_x)_A = (v_x)_O = 35$ ft/s and $a_y = -32.2$ ft/s².

Horizontal Motion

$$(\overset{+}{\rightarrow}) \qquad\qquad x_A = x_O + (v_x)_O t_{OA}$$
$$R = 0 + 35 t_{OA} \qquad\qquad (1)$$

Vertical Motion. Relating t_{OA} to the initial and final elevations of the ball,

$$(+\uparrow) \qquad\qquad y_A = y_O + (v_y)_O t_{OA} + \tfrac{1}{2} a_c t_{OA}^2$$
$$(40 - 5) = 0 + 60.62 t_{OA} + \tfrac{1}{2}(-32.2) t_{OA}^2$$

Solving for the two roots, using the quadratic formula, we have $t_{OB} = 0.712$ s and $t_{OA} = 3.05$ s. The first root (shortest time) is designated as t_{OB}, since it represents the time needed for the ball to reach point B, which has the same elevation as point A, Fig. 12–22. Substituting t_{OA} into Eq. (1) and solving for R yields

$$R = 35(3.05) = 107 \text{ ft} \qquad\qquad \textit{Ans.}$$

Using only one equation, can you show that the maximum height reached by the ball, Fig. 12–22, is $h = 62.1$ ft?

Example 12–13

When a ball is kicked from A as shown in Fig. 12–23, it just clears the top of a wall at B as it reaches its maximum height. Knowing that the distance from A to the wall is 20 m and the wall is 4 m high, determine the initial speed at which the ball was kicked. Neglect the size of the ball.

Fig. 12–23

SOLUTION

Coordinate System. The three unknowns are represented by the initial speed v_A, angle of inclination θ, and the time t_{AB} to travel from A to B, Fig. 12–23. At the highest point, B, the velocity $(v_y)_B = 0$, and $(v_x)_B = (v_x)_A = v_A \cos \theta$.

Horizontal Motion

$(\xrightarrow{+})$

$$x_B = x_A + (v_x)_A t_{AB}$$
$$20 = 0 + (v_A \cos \theta) t_{AB} \tag{1}$$

Vertical Motion

$(+\uparrow)$

$$(v_y)_B = (v_y)_A + a_c t_{AB}$$
$$0 = v_A \sin \theta - 9.81 t_{AB} \tag{2}$$

$(+\uparrow)$

$$(v_y)_B^2 = (v_y)_A^2 + 2a_c[y_B - y_A]$$
$$0 = v_A^2 \sin^2 \theta + 2(-9.81)(4 - 0) \tag{3}$$

To obtain v_A, eliminate t_{AB} from Eqs. (1) and (2), which yields

$$v_A^2 \sin \theta \cos \theta = 196.2 \tag{4}$$

Solve for v_A^2 in Eq. (4) and substitute into Eq. (3), so that

$$\frac{\sin \theta}{\cos \theta} = \tan \theta = \frac{2(9.81)(4)}{196.2} = 0.4$$

$$\theta = \tan^{-1}(0.4) = 21.8°$$

Then using Eq. (4), the required initial speed is

$$v_A = \sqrt{\frac{196.2}{(\sin 21.8°)(\cos 21.8°)}} = 23.9 \text{ m/s} \qquad \textit{Ans.}$$

PROBLEMS

12–47. A car travels east 2 km for 5 minutes, then north 3 km for 8 minutes, and then west 4 km for 10 minutes. Determine the total distance traveled and the displacement of the car. Also, what is the magnitude of the average velocity and the average speed?

*12–48. If the velocity of a particle is defined as $\mathbf{v}(t) = \{1.5t^2\mathbf{i} + 1.8t\mathbf{j} + t^3\mathbf{k}\}$ m/s, where t is in seconds, determine the displacement of the particle from $t = 1$ s to $t = 3$ s.

12–49. If the velocity of a particle is defined as $\mathbf{v}(t) = \{1.5t^2\mathbf{i} + 1.8t\mathbf{j} + t^3\mathbf{k}\}$ m/s, determine the magnitude and coordinate direction angles α, β, γ of the particle's acceleration when $t = 2$ s.

12–50. A particle is subjected to an acceleration $\mathbf{a} = \{9t^2\mathbf{i} + 12t^3\mathbf{j} - 6t\mathbf{k}\}$ ft/s². Determine the particle's position (x, y, z) when $t = 2$ s. When $t = 0$, the particle is located at point $\mathbf{r}_0 = \{2\mathbf{i} - 1\mathbf{j} + 2\mathbf{k}\}$ ft and has a velocity of $\mathbf{v}_0 = \{2\mathbf{i} - 6\mathbf{j} + 5\mathbf{k}\}$ ft/s.

12–51. A particle travels along the path $y = x^2$, where x and y are in meters. If the particle's component of velocity in the y direction is always $v_y = 3$ m/s, determine the magnitude of the particle's velocity when $t = 2$ s. When $t = 0$ the particle is at point $(1 \text{ m}, 1 \text{ m})$. How far from the origin is the particle when $t = 2$ s?

*12–52. A particle travels with a constant speed v around a helix defined by the parametric equations $x = (5 \cos 2t)$ m, $y = (5 \sin 2t)$ m, $z = (3t)$ m, where t is in seconds. Determine the particle's acceleration as a function of time.

12–53. The position of a particle is defined by $\mathbf{r} = \{5 \cos 2t\mathbf{i} + 4 \sin 2t\mathbf{j}\}$ m, where t is in seconds and the arguments for the sine and cosine are given in radians. Determine the magnitudes of the velocity and acceleration of the particle when $t = 1$ s. Also, prove that the path of the particle is elliptical.

12–54. The two particles A and B start at the origin O and travel in opposite directions along the circular path. If A and B travel at a constant speed of $v_A = 0.7$ m/s and $v_B = 1.5$ m/s, respectively, determine in $t = 2$ s, (a) the displacement of each particle, (b) the position vector to each particle, and (c) the distance between the particles.

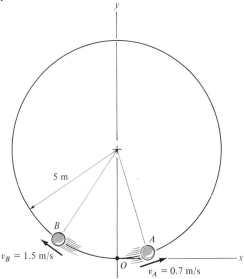

Prob. 12–54

12–55. The curvilinear motion of a particle is defined by $x = 5 \sin (4t)$, $y = 8 \cos (3t) + 2$, and $z = 14t^3$, where the x, y, z position is given in millimeters, the time in seconds, and the arguments for the sine and cosine are given in radians. Determine the magnitudes and coordinate direction angles α, β, γ of the particle's velocity and acceleration when $t = 2$ s.

*12–56. A particle travels along a path such that its position is $\mathbf{r} = \{(5 \sin t + 3)\mathbf{i} + (5 \cos t)\mathbf{j}\}$ ft, where t is in seconds and the arguments for the sine and cosine are in radians. Find the equation $y = f(x)$ which describes the path, and show that the magnitudes of the particle's velocity and acceleration are constant. What are these magnitudes, and what is the magnitude of the particle's displacement from $t = 1$ s to $t = 3$ s?

12–57. The position of particles A and B is described by the vectors $\mathbf{r}_A = \{4.5t\mathbf{i} + 13.5t(2 - t)\mathbf{j}\}$ m and $\mathbf{r}_B = \{4.5(t^2 - 2t + 2)\mathbf{i} + (4.5t - 9)\mathbf{j}\}$ m, respectively, where t is in seconds. Determine the point where the particles collide and their speeds just before the collision. How long does it take before the collision occurs?

12–58. The motion of particles A and B is described by the position vectors $\mathbf{r}_A = \{2t\mathbf{i} + (t^2 - 1)\mathbf{j}\}$ ft and $\mathbf{r}_B = \{(t + 2)\mathbf{i} + (2t^2 - 5)\mathbf{j}\}$ ft, respectively, where t is in seconds. Determine the point where the particles collide and their speeds just before the collision.

12–59. A particle moves along the path $\mathbf{r} = \{4t^4\mathbf{i} + (3t + 8t^3)\mathbf{j}\}$ in., where t is in seconds. Determine the magnitudes of the particle's velocity and acceleration when $t = 2$ s. Also determine the equation $y = f(x)$ of the path.

*12–60. The position of a particle is defined by $\mathbf{r} = \{\theta^3\mathbf{i} + \sin 2\theta\mathbf{j} + \cos^2 \theta\mathbf{k}\}$ ft, where θ is in radians. If $\theta = (2t^2)$ rad, where t is in seconds, determine the velocity $\mathbf{v}$ and acceleration $\mathbf{a}$ when $t = 1$ s. Express $\mathbf{v}$ and $\mathbf{a}$ as Cartesian vectors.

12–61. A particle is moving along the curve $y = x - (x^2/400)$, where x and y are in ft. If the velocity component in the x direction is $v_x = 2$ ft/s and remains *constant,* determine the magnitudes of the velocity and acceleration when $x = 20$ ft.

12–62. A particle moves along a hyperbolic path $x^2/16 - y^2 = 28$. If the x component of velocity is $v_x = 4$ m/s and remains constant, determine the magnitude of the particle's velocity and acceleration when it is at point $(32 \text{ m}, 6 \text{ m})$.

12–63. A particle moves along the curve $y = e^{2x}$ such that its velocity has a constant magnitude of $v = 4$ ft/s. Determine the x and y components of velocity when the particle is at $y = 5$ ft.

*12–64. The particle travels along the path defined by the parabola $y = 0.5x^2$. If the component of velocity along the x axis is $v_x = (5t)$ ft/s, where t is in seconds, determine the particle's distance from the origin O and the magnitude of acceleration when $t = 1$ s. When $t = 0$, $x = 0$, $y = 0$.

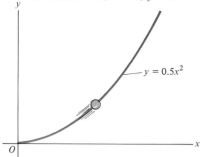

$y = 0.5x^2$

Prob. 12–64

■12–65. A particle is traveling with a velocity of $\mathbf{v} = \{3\sqrt{t}e^{-0.2t}\mathbf{i} + 4e^{-0.8t^2}\mathbf{j}\}$ m/s, where t is in seconds. Determine the magnitude of the particle's displacement from $t = 0$ to $t = 3$ s. Use Simpson's rule to evaluate the integrals. What is the magnitude of the particle's acceleration when $t = 2$ s?

12–66. It is observed that the skier leaves the ramp A and then strikes the ground at B in $t_{AB} = 5$ s. Determine his initial speed v_A and the launch angle θ_A.

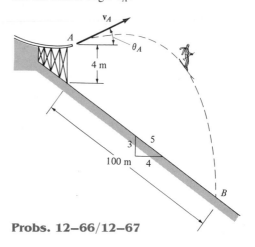

Probs. 12–66/12–67

12–67. It is observed that the skier leaves the ramp A at an angle $\theta_A = 25°$ with the horizontal. If he strikes the ground at point B, determine his initial speed v_A and the time of flight t_{AB}.

*12–68. The boy at A attempts to throw a ball over the roof of a barn with an initial speed of $v_A = 15$ m/s. Determine the angle θ_A at which the ball must be thrown so that it reaches its highest altitude at C. Also, find the distance d where he should stand to make the throw.

Probs. 12–68/12–69

12–69. The boy at A attempts to throw a ball over the roof of a barn such that it is launched at an angle $\theta_A = 40°$. Determine the minimum speed v_A at which he must throw the ball so it reaches its highest altitude at C. Also, find the distance d where the boy must stand so that he can make the throw.

12–70. The pitcher throws the baseball horizontally with a speed of 150 ft/s from a height of 5 ft. If the batter is 60 ft away, determine the time needed for the ball to arrive to the batter and the height h at which it passes the batter.

Prob. 12–70

12–71. A golf ball is struck with a velocity of 80 ft/s as shown. Determine the distance d to where it will land.

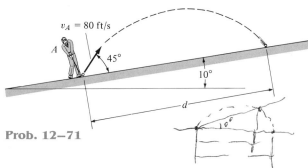

Prob. 12–71

*__12–72.__ A ball is thrown from the ground and in 3 s it is caught by someone on top of a building, at a point which has coordinates $x = 100$ ft, $y = 40$ ft, measured from the point where the ball is thrown. Determine the velocity and angle θ at which it is thrown.

12–73. Measurements of a shot which was recorded on a videotape during a basketball game are shown. The ball passed through the hoop even though it barely cleared the hands of player B who attempted to block it. Neglecting the size of the ball, determine the magnitude v_A of its initial velocity and the height h of the ball when it passes over player B.

Prob. 12–73

12–74. The box is traveling at 3 m/s when it leaves the end of the chute at A. Determine the time of flight from A to B and the range R of the trajectory.

Prob. 12–74

12–75. The ball at A is kicked with a speed of $v_A = 80$ ft/s and an angle of $\theta_A = 30°$. Determine the point $(x, -y)$ where it strikes the ground. Assume the ground has the shape of a parabola as shown.

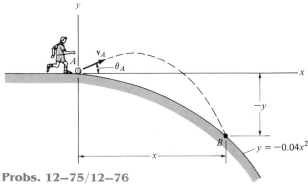

Probs. 12–75/12–76

*__12–76.__ The ball at A is kicked such that $\theta_A = 30°$. If it strikes the ground at B having coordinates $x = 15$ ft, $y = -9$ ft, determine the speed at which it is kicked and the speed at which it strikes the ground.

12–77. The man at A wishes to throw two darts at the target at B so that they arrive at the *same time*. If each dart is thrown with a speed of 10 m/s, determine the angles θ_C and θ_D at which they should be thrown and the time between each throw. Note that the first dart must be thrown at $\theta_C(>\theta_D)$, then the second dart is thrown at θ_D.

Prob. 12–77

12–78. The girl at A can throw a ball at $v_A = 10$ m/s. Calculate the maximum possible range R_{max} and the associated angle θ at which it should be thrown. Assume the ball is caught at B at the same elevation from which it is thrown.

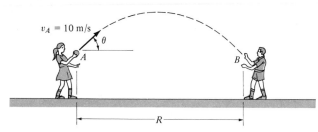

Probs. 12–78/12–79

12–79. Show that the girl at A can throw the ball to the boy at B by launching it at equal angles measured up or down from a $45°$ inclination. If $v_A = 10$ m/s, determine the range R if this value is $15°$, i.e., $\theta_1 = 45° - 15° = 30°$ and $\theta_2 = 45° + 15° = 60°$. Assume the ball is caught at the same elevation from which it is thrown.

*12–80. The baseball player A hits the baseball at $v_A = 40$ ft/s and $\theta_A = 60°$ from the horizontal. When the ball is directly overhead of player B he begins to run under it. Determine the constant speed at which B must run and the distance d in order to make the catch at the same elevation at which the ball was hit.

Prob. 12–80

12–81. The football player at A throws the ball in the y-z plane with a speed of $v_A = 50$ ft/s and an angle of $\theta_A = 60°$ with the horizontal. At the instant the ball is thrown, the player at B is running with constant speed along the line BC in order to catch it. Determine this speed, v_B, so that he makes the catch at the same elevation from which the ball was thrown.

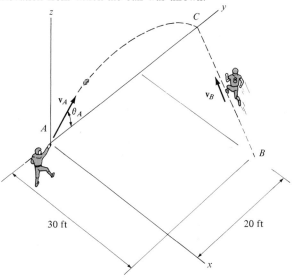

Probs. 12–81/12–82

12–82. The football player at A throws the ball in the y-z plane with a speed of $v_A = 50$ ft/s and an angle of $\theta_A = 60°$ with the horizontal. The player at B reacts within 0.1 s after seeing the ball thrown. If he can run at a maximum speed of $v_B = 23$ ft/s along the line BC, determine if he can arrive at C, running at constant speed, in order to catch the ball.

12–83. A boy at O throws a ball in the air with a speed v_0 at an angle θ_1. If he then throws another ball at the same speed v_0 at an angle $\theta_2 < \theta_1$, determine the time between the throws so the balls collide in mid air at B.

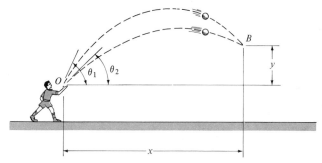

Prob. 12–83

12.6 Curvilinear Motion: Normal and Tangential Components

When the path along which a particle is moving is *known*, it is often convenient to describe the motion using n and t coordinates which act normal and tangent to the path, respectively, and at the instant considered have their *origin located at the particle*.

Planar Motion

Consider the particle P shown in Fig. 12–24a, which is moving in a plane such that at a given instant it is located along a fixed curve at position s, measured from point O. We will now consider a coordinate system that has its origin at a *fixed point* on the curve, and at the instant considered this origin happens to *coincide* with the location of the particle. The t axis is *tangent* to the curve at P and is positive in the direction of *increasing s*. We will designate this positive direction with the unit vector $\mathbf{u}_t$. A unique choice for the *normal axis* can be made by considering the fact that geometrically the curve consists of a series of differential arc segments ds. As shown in Fig. 12–24b, each segment ds is constructed from the arc of an associated circle having a *radius of curvature ρ* (rho) and *center of curvature O'*. The normal axis n which will be chosen is perpendicular to the t axis and is directed from P *toward* the center of curvature O', Fig. 12–24a. This positive direction, which is *always* on the concave side of the curve, will be designated by the unit vector $\mathbf{u}_n$. The plane which contains the n and t axes is referred to as the *osculating plane*, and in this case it is fixed in the plane of motion.*

Velocity

Since the particle is moving, s is a function of time. As indicated in Sec. 12.3, the particle's velocity $\mathbf{v}$ has a *direction* that is *always tangent to the curve*, Fig. 12–24c, and a *magnitude* that is determined by taking the time derivative of the path function $s = s(t)$, i.e., $v = ds/dt$ (Eq. 12–8). Hence

$$\mathbf{v} = v\mathbf{u}_t \qquad (12\text{–}15)$$

where

$$v = \dot{s} \qquad (12\text{–}16)$$

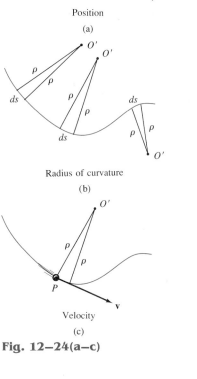

Position

(a)

Radius of curvature

(b)

Velocity

(c)

Fig. 12–24(a–c)

*The osculating plane may also be defined as that plane which has the greatest contact with the curve at a point. It is the limiting position of a plane contacting both the point and the arc segment ds. As noted above, the osculating plane is always coincident with a plane curve; however, each point on a three-dimensional curve has a unique osculating plane.

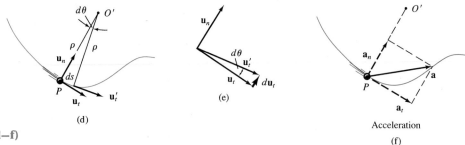

Fig. 12–24(d–f)

Acceleration

The acceleration of the particle is the time rate of change of the velocity. Thus,

$$\mathbf{a} = \dot{\mathbf{v}} = \dot{v}\mathbf{u}_t + v\dot{\mathbf{u}}_t \tag{12–17}$$

The first term, $\dot{v}\mathbf{u}_t$, represents the change in the magnitude of velocity, i.e., $\dot{v}$, whereas the second term, $v\dot{\mathbf{u}}_t$, represents the change in the velocity's direction. In order to compute the time derivative $\dot{\mathbf{u}}_t$, note that as the particle moves along the arc ds in time dt, $\mathbf{u}_t$ preserves its magnitude of unity; however, it changes its *direction,* so that it becomes $\mathbf{u}_t'$, Fig. 12–24d. The change $d\mathbf{u}_t$ in $\mathbf{u}_t$ is shown in Fig. 12–24e. Here $d\mathbf{u}_t$ stretches between the heads of $\mathbf{u}_t$ and $\mathbf{u}_t'$, which lie on an infinitesimal arc of radius $u_t = 1$. Hence, $d\mathbf{u}_t$ has a *magnitude* of $du_t = (1)\, d\theta$ and its *direction* is defined by $\mathbf{u}_n$. Consequently, $d\mathbf{u}_t = d\theta\mathbf{u}_n$ so that the time derivative becomes $\dot{\mathbf{u}}_t = \dot{\theta}\mathbf{u}_n$. Since $ds = \rho\, d\theta$, Fig. 12–24d, then $\dot{\theta} = \dot{s}/\rho$, and consequently

$$\dot{\mathbf{u}}_t = \frac{\dot{s}}{\rho}\mathbf{u}_n = \frac{v}{\rho}\mathbf{u}_n$$

Substituting into Eq. 12–17, $\mathbf{a}$ can be written as the sum of its two components,

$$\mathbf{a} = a_t\mathbf{u}_t + a_n\mathbf{u}_n \tag{12–18}$$

where

$$a_t = \dot{v} \qquad \text{or} \qquad a_t = v\frac{dv}{ds} \tag{12–19}$$

and

$$a_n = \frac{v^2}{\rho} \tag{12–20}$$

These two mutually perpendicular components are shown in Fig. 12–24f, in which case the *magnitude* of acceleration is the positive value of

$$a = \sqrt{a_t^2 + a_n^2} \tag{12–21}$$

Fig. 12–25

To summarize the above concepts consider the following two special cases of motion.

1. If the particle moves along a straight line, then $\rho \rightarrow \infty$ and from Eq. 12–20, $a_n = 0$. Thus $a = a_t = \dot{v}$, and we can conclude that the *tangential component of acceleration represents the time rate of change in the magnitude of the velocity*.
2. If the particle moves along a curve with a constant speed, then $a_t = \dot{v} = 0$ and $a = a_n = v^2/\rho$. Therefore, the *normal component of acceleration represents the time rate of change in the direction of the velocity*. Since it *always* acts toward the center of curvature, this component is sometimes referred to as the *centripetal acceleration*.

As a result of these interpretations, a particle moving along the curved path shown in Fig. 12–25 will have accelerations directed as shown in the figure.

Three-Dimensional Motion

If the particle is moving along a space curve, Fig. 12–26, then at a given instant the t axis is uniquely specified; however, an infinite number of straight lines can be constructed normal to the tangent axis at P. As in the case of planar motion, we will choose the positive n axis directed from P toward the path's center of curvature O'. This axis is referred to as the *principal normal* to the curve at P. With the n and t axes so defined, Eqs. 12–15 to 12–21 can be used to determine **v** and **a**. Since $\mathbf{u}_t$ and $\mathbf{u}_n$ are always perpendicular to one another, and lie in the osculating plane, for spatial motion a third unit vector, $\mathbf{u}_b$, defines a *binormal axis b* which is perpendicular to $\mathbf{u}_t$ and $\mathbf{u}_n$, Fig. 12–26. Using the vector cross product, these three vectors are related by the equation $\mathbf{u}_b = \mathbf{u}_t \times \mathbf{u}_n$.

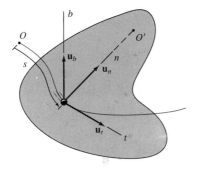

Fig. 12–26

PROCEDURE FOR ANALYSIS

Coordinate System. Provided the *path* of the particle is *known*, we can establish a set of *n* and *t* coordinates having a *fixed origin* which is coincident with the particle at the instant considered. The positive tangent axis acts in the direction of motion and the positive normal axis is directed toward the path's center of curvature. This set of axes is particularly advantageous for studying the velocity and acceleration of the particle, because the *t* and *n* components of **a** have been readily formulated.

Velocity. The particle's *velocity* is tangent to the path. Its magnitude is found from the time derivative of the path function.

$$v = \dot{s}$$

Note that v can be positive or negative depending upon whether it acts in the positive or negative s direction.

Tangential Acceleration. The *tangential component of acceleration is the result of the time rate of change in the magnitude of velocity.* This component acts in the positive s direction if the particle's speed is increasing or in the opposite direction if the speed is decreasing. The magnitudes of **v** and $\mathbf{a}_t$ are related to the path function s and time t by the equations of rectilinear motion, namely,

$$a_t = \dot{v} \qquad a_t \, ds = v \, dv$$

If a_t is *constant*, $a_t = (a_t)_c$, the above equations, when integrated, yield

$$s = s_0 + v_0 t + \tfrac{1}{2}(a_t)_c t^2$$
$$v = v_0 + (a_t)_c t$$
$$v^2 = v_0^2 + 2(a_t)_c (s - s_0)$$

Normal Acceleration. The *normal component of acceleration is the result of the time rate of change in the direction of the particle's velocity.* This component is *always* directed toward the center of curvature of the path, i.e., along the positive n axis. Its magnitude is determined from

$$a_n = \frac{v^2}{\rho}$$

In particular, if the path is expressed as $y = f(x)$, the radius of curvature ρ is computed from the equation*

$$\rho = \left| \frac{[1 + (dy/dx)^2]^{3/2}}{d^2y/dx^2} \right|$$

The following examples numerically illustrate application of these equations.

*The derivation of this result is given in any standard calculus text.

Example 12–14

A skier travels with a constant speed of 6 m/s along the parabolic path $y = \frac{1}{20}x^2$ shown in Fig. 12–27. Determine his velocity and acceleration at the instant he arrives at A. Neglect the size of the skier in the calculation.

SOLUTION

Coordinate System. Although the path has been expressed in terms of its x and y coordinates, we can still establish the origin of the n, t axes at the fixed point A and determine the components of $\mathbf{v}$ and $\mathbf{a}$ along these axes, Fig. 12–27a.

Velocity. By definition, the velocity is always directed tangent to the path. Since $y = \frac{1}{20}x^2$, $dy/dx = \frac{1}{10}x$, then $dy/dx|_{x=10} = 1$. Hence, at A, $\mathbf{v}$ makes an angle of $\theta = \tan^{-1} 1 = 45°$ with the x axis, Fig. 12–27. Therefore,

$$v_A = 6 \text{ m/s} \qquad \text{\small 45°} \nearrow \text{\small v}_A \qquad\qquad \textit{Ans.}$$

Acceleration. The acceleration is computed by using $\mathbf{a} = \dot{v}\mathbf{u}_t + (v^2/\rho)\mathbf{u}_n$. However, it is first necessary to determine the radius of curvature of the path at A (10 m, 5 m). Since $d^2y/dx^2 = \frac{1}{10}$, then

$$\rho = \left| \frac{[1 + (dy/dx)^2]^{3/2}}{d^2y/dx^2} \right| = \left| \frac{[1 + (\frac{1}{10}x)^2]^{3/2}}{\frac{1}{10}} \right|_{x=10\text{m}} = 28.3 \text{ m}$$

The acceleration becomes

$$\mathbf{a}_A = \dot{v}\mathbf{u}_t + \frac{v^2}{\rho}\mathbf{u}_n$$

$$= 0\mathbf{u}_t + \frac{(6)^2}{28.3}\mathbf{u}_n$$

$$= \{1.27\mathbf{u}_n\} \text{ m/s}^2$$

Since $\mathbf{a}_A$ acts in the direction of the positive n axis, it makes an angle of $\theta + 90° = 135°$ with the positive x axis. Hence,

$$a_A = 1.27 \text{ m/s}^2 \qquad \text{\small a}_A \text{\small 135°} \qquad\qquad \textit{Ans.}$$

Note: By using n, t coordinates, we were able to readily solve this problem since the n and t components account for the *separate* changes in the magnitude and direction of $\mathbf{v}$, and each of these changes have been directly formulated as Eqs. 12–19 and 12–20, respectively.

Fig. 12–27

43

Example 12–15

A race car C travels around the horizontal circular track that has a radius of 300 ft, Fig. 12–28. If the car increases its speed at a constant rate of 7 ft/s^2 starting from rest, determine the time needed for it to reach an acceleration of 8 ft/s^2. What is its speed at this instant?

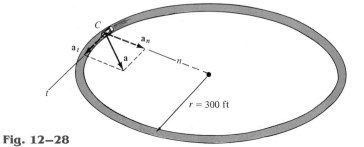

Fig. 12–28

SOLUTION

Coordinate System. The origin of the n and t axes is coincident with the car at the instant considered. The t axis is in the direction of motion, and the positive n axis is directed toward the center of the circle. This coordinate system is selected since the path is known.

Acceleration. The magnitude of acceleration can be related to its components using $a = \sqrt{a_t^2 + a_n^2}$. Here $a_t = 7$ ft/s^2. Also $a_n = v^2/\rho$, where $\rho = 300$ ft and

$$v = v_0 + (a_t)_c t$$
$$v = 0 + 7t$$

Thus

$$a_n = \frac{v^2}{\rho} = \frac{(7t)^2}{300} = 0.163t^2 \text{ ft/s}^2$$

The time needed for the acceleration to reach 8 ft/s^2 is therefore

$$a = \sqrt{a_t^2 + a_n^2}$$
$$8 = \sqrt{(7)^2 + (0.163t^2)^2}$$

Solving for the positive value of t yields

$$0.163t^2 = \sqrt{(8)^2 - (7)^2}$$
$$t = 4.87 \text{ s} \qquad\qquad Ans.$$

Velocity. The speed at time $t = 4.87$ s is

$$v = 7t = 7(4.87) = 34.1 \text{ ft/s} \qquad\qquad Ans.$$

Example 12–16

A car starts from rest at point A and travels along the horizontal track shown in Fig. 12–29a. During the motion, the increase in speed is $a_t = (0.2t)$ m/s^2, where t is in seconds. Determine the magnitude of the car's acceleration when it arrives at point B.

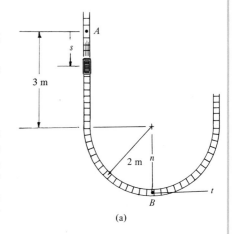

(a)

SOLUTION

Coordinate System. The position of the car at any instant is defined from the fixed point A using the position or path coordinate s, Fig. 12–29a. The car's acceleration is to be determined at B, so the origin of the n, t axes is at this point. Why are n and t coordinates selected to solve this problem?

The acceleration is computed from $\mathbf{a} = \dot{v}\mathbf{u}_t + (v^2/\rho)\mathbf{u}_n$. To apply this equation, however, it is first necessary to formulate v and $\dot{v}$ so that they may be evaluated at B. Since $v_A = 0$ when $t = 0$, then

$$a_t = \dot{v} = 0.2t \qquad (1)$$

$$\int_0^v dv = \int_0^t 0.2t \, dt$$

$$v = 0.1t^2 \qquad (2)$$

The time needed for the car to reach point B can be determined by realizing that the position of B is $s_B = 3 + 2\pi(2)/4 = 6.14$ m, Fig. 12–29a, and since $s_A = 0$ when $t = 0$ we have

$$v = \frac{ds}{dt} = 0.1t^2$$

$$\int_0^{6.14} ds = \int_0^{t_B} 0.1t^2 \, dt$$

$$6.14 = 0.0333t_B^3$$

$$t_B = 5.69 \text{ s}$$

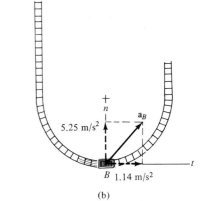

(b)

Fig. 12–29

Substituting into Eqs. (1) and (2) yields

$$(a_t)_B = 0.2(5.69) = 1.14 \text{ m/s}^2$$

$$v_B = 0.1(5.69)^2 = 3.24 \text{ m/s}$$

At B, $\rho_B = 2$ m, so that

$$(a_n)_B = \frac{v_B^2}{\rho_B} = \frac{(3.24)^2}{2} = 5.25 \text{ m/s}^2$$

The magnitude of $\mathbf{a}_B$, Fig. 12–29b, is therefore

$$a_B = \sqrt{(1.14)^2 + (5.25)^2} = 5.37 \text{ m/s}^2 \qquad \textit{Ans.}$$

PROBLEMS

*12–84. The car travels along the curve having a radius of 300 m. If its speed is uniformly increased from 15 m/s to 27 m/s in 3 s, determine the magnitude of its acceleration at the instant its speed is 20 m/s.

20 m/s

300 m

Prob. 12–84

12–85. At a given instant the jet plane has a speed of 400 ft/s and an acceleration of 70 ft/s^2 acting in the direction shown. Determine the rate of increase in the plane's speed and the radius of curvature ρ of the path.

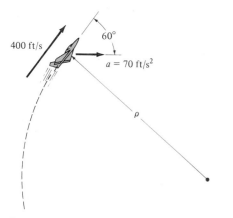

400 ft/s

60°

$a = 70$ ft/s^2

ρ

Prob. 12–85

12–86. A particle travels along the path $y = a + bx + cx^2$, where a, b, c are constants. If the speed of the particle is constant, $v = v_0$, determine the x and y components of velocity and the normal component of acceleration when $x = 0$.

12–87. A particle P moves along the curve $y = (x^2 - 9)$ ft with a constant speed of 15 ft/s. Determine the point on the curve where the maximum magnitude of acceleration occurs and compute its value.

*12–88. The truck travels in a circular path having a radius of 50 m at a speed of 4 m/s. For a short distance from $s = 0$, its speed is increased by $\dot{v} = (0.05s)$ m/s^2, where s is in meters. Determine its speed and the magnitude of its acceleration when it has moved $s = 10$ m.

$\dot{v} = (0.05s)$ m/s^2
$v = 4$ m/s

50 m

Prob. 12–88

12–89. A rocket follows a path such that its acceleration is defined by $\mathbf{a} = \{16\mathbf{i} + 4t\mathbf{j}\}$ ft/s^2. If it starts from rest at $\mathbf{r} = \mathbf{0}$, determine the speed of the rocket and the radius of curvature of its path when $t = 10$ s.

12–90. A particle moves along the curve $y = 180/x^2$ such that when it is at $x = 5$ ft it has a speed of 8 ft/s, which is increasing at 12 ft/s^2. Determine the magnitude of the particle's acceleration at this instant.

12–91. A car moves along a circular track of radius 250 ft such that its speed for a short period of time $0 \leqslant t \leqslant 4$ s is $v = 3(t + t^2)$ ft/s, where t is in seconds. Determine the magnitude of its acceleration when $t = 3$ s. How far has it traveled in $t = 3$ s?

*12–92. The plane travels along the vertical parabolic path at a constant speed of 200 m/s. Determine the magnitude of acceleration of the plane when it is at point A.

$y = 0.4x^2$

A

10 km

5 km

x

Probs. 12–92/12–93

12–93. The jet plane travels along the vertical parabolic path. When it is at point A it has a speed of 200 m/s, which is increasing at the rate of 0.8 m/s². Determine the magnitude of acceleration of the plane when it is at point A.

12–94. The automobile is originally at rest at $s = 0$. If it then starts to increase its speed according to the equation $\dot{v} = (0.05t^2)$ ft/s², where t is in seconds, determine the magnitudes of its velocity and acceleration when $t = 18$ s.

300 ft

s

240 ft

Probs. 12–94/12–95

12–95. The automobile is originally at rest at $s = 0$. If it then starts to increase its speed at $\dot{v} = (0.05t^2)$ ft/s², where t is in seconds, determine the magnitudes of its velocity and acceleration at $s = 550$ ft.

*12–96. The car B turns such that its speed is increased by $\dot{v}_B = (0.5e^t)$ m/s², where t is in seconds. If the car starts from rest when $\theta = 0°$, determine the magnitudes of the velocity and acceleration of the car when the arm AB rotates $\theta = 30°$. Neglect the size of the car.

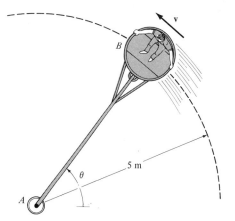

B

5 m

θ

A

Prob. 12–96

12–97. Starting from rest a bicyclist travels around a horizontal circular path, $\rho = 10$ m, at a speed $v = (0.09t^2 + 0.1t)$ m/s, where t is in seconds. Determine the magnitudes of his velocity and acceleration when he has traveled $s = 3$ m.

12–98. A boat has an initial speed of 16 ft/s. If it increases its speed along a circular path of radius $\rho = 80$ ft at the rate of $\dot{v} = (1.5s)$ ft/s, where s is in feet measured along the path, determine the normal and tangential components of the boat's acceleration at $s = 16$ ft.

12–99. A boat has an initial speed of 16 ft/s. If it increases its speed along a circular path of radius $\rho = 80$ ft at the rate of $\dot{v} = (1.5s)$ ft/s, where s is in feet measured along the path, determine the time needed for the boat to travel 50 ft.

*12–100. The ball is kicked with an initial speed of $v_A = 8$ m/s at an angle of $\theta_A = 40°$ with the horizontal. Find the equation of the path, $y = f(x)$, and then determine the ball's velocity and the normal and tangential components of acceleration when $t = 0.25$ s.

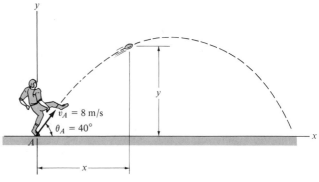

Prob. 12–100

12.7 Curvilinear Motion: Cylindrical Components

In some engineering problems it is often convenient to express the path of motion of a particle in terms of cylindrical coordinates, r, θ, and z. If motion is restricted to the plane, the polar coordinates r and θ are used.

Polar Coordinates

We can specify the location of particle P shown in Fig. 12–30a using both the *radial coordinate r*, which extends outward from the fixed origin O to the particle, and a *transverse coordinate θ*, which is the counterclockwise angle between a fixed reference line and the r axis. The angle is generally measured in degrees or radians, where 1 rad = $180°/\pi$. The positive directions of the r and θ coordinates are defined by the unit vectors $\mathbf{u}_r$ and $\mathbf{u}_\theta$, respectively. These directions are *perpendicular* to one another. Here $\mathbf{u}_r$ or the radial direction $+r$ extends from P along increasing r, when θ is held fixed, and $\mathbf{u}_\theta$ or $+\theta$ extends from P in a direction that occurs when r is held fixed and θ is increased.

Position

At any instant the position of the particle, Fig. 12–30a, is defined by the position vector

$$\mathbf{r} = r\mathbf{u}_r \qquad (12\text{–}22)$$

Position
(a)

Velocity

The instantaneous velocity **v** is obtained by taking the time derivative of **r**. Using a dot to represent time differentiation, we have

$$\mathbf{v} = \dot{\mathbf{r}} = \dot{r}\mathbf{u}_r + r\dot{\mathbf{u}}_r$$

In order to evaluate $\dot{\mathbf{u}}_r$, notice that $\mathbf{u}_r$ changes only its direction with respect to time, since by definition the magnitude of this vector is always unity. Hence, when θ changes by an increasing amount $\Delta\theta$ during the time Δt, $\mathbf{u}_r$ becomes unit vector $\mathbf{u}'_r$, Fig. 12–30b. The time change in $\mathbf{u}_r$ is then $\Delta\mathbf{u}_r$. For small angles $\Delta\theta$ this vector has a magnitude $\Delta u_r \approx 1(\Delta\theta)$ and acts in the $\mathbf{u}_\theta$ direction, i.e., $\Delta\mathbf{u}_r = \Delta\theta\mathbf{u}_\theta$. Thus,

$$\dot{\mathbf{u}}_r = \lim_{\Delta t \to 0} \frac{\Delta\mathbf{u}_r}{\Delta t} = \left(\lim_{\Delta t \to 0} \frac{\Delta\theta}{\Delta t}\right)\mathbf{u}_\theta$$

$$\dot{\mathbf{u}}_r = \dot{\theta}\mathbf{u}_\theta \qquad (12\text{–}23)$$

(b)

Substituting into the above equation, the velocity can be written in component form as

$$\boxed{\mathbf{v} = v_r\mathbf{u}_r + v_\theta\mathbf{u}_\theta} \qquad (12\text{–}24)$$

where

$$\boxed{\begin{aligned} v_r &= \dot{r} \\ v_\theta &= r\dot{\theta} \end{aligned}} \qquad (12\text{–}25)$$

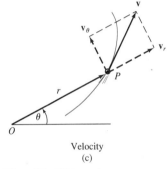

Velocity
(c)

Fig. 12–30(a–c)

These components are shown graphically in Fig. 12–30c. Note that the *radial component* of velocity $\mathbf{v}_r$ is a measure of the rate of increase or decrease in the length of the radial coordinate, i.e., $\dot{r}$, whereas the *transverse component* of velocity, $\mathbf{v}_\theta$, can be interpreted as the rate of motion along the circumference of a circle having a radius r. In particular, the term $\dot{\theta} = d\theta/dt$ is called the *angular velocity* since it provides a measure of the time rate of change of the angle θ. Common units used for this measurement are rad/s.

Since the components are mutually perpendicular, the *magnitude* of velocity is simply the positive value of

$$v = \sqrt{(\dot{r})^2 + (r\dot{\theta})^2} \qquad (12\text{–}26)$$

and the *direction* of **v** is, of course, tangent to the path at P, Fig. 12–30c.

(d)

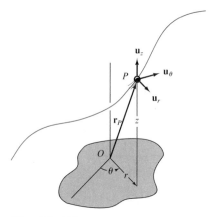

Acceleration

(e)

Fig. 12-30(d, e)

Acceleration

Taking the time derivatives of Eq. 12–24, using Eqs. 12–25, we obtain the particle's instantaneous acceleration,

$$\mathbf{a} = \dot{\mathbf{v}} = \ddot{r}\mathbf{u}_r + \dot{r}\dot{\mathbf{u}}_r + \dot{r}\dot{\theta}\mathbf{u}_\theta + r\ddot{\theta}\mathbf{u}_\theta + r\dot{\theta}\dot{\mathbf{u}}_\theta$$

To evaluate the term involving $\dot{\mathbf{u}}_\theta$, it is necessary to find the change made in the direction of $\mathbf{u}_\theta$. As shown in Fig. 12–30d, when θ changes by an increasing amount $\Delta\theta$, the new or changed position of $\mathbf{u}_\theta$ is designated as $\mathbf{u}_\theta'$. The time change in $\mathbf{u}_\theta$ is thus $\Delta\mathbf{u}_\theta$. For small angles this vector has a magnitude $\Delta u_\theta \approx 1(\Delta\theta)$ and acts in the $-\mathbf{u}_r$ direction, i.e., $\Delta\mathbf{u}_\theta = -\Delta\theta\mathbf{u}_r$. Thus,

$$\dot{\mathbf{u}}_\theta = \lim_{\Delta t \to 0} \frac{\Delta\mathbf{u}_\theta}{\Delta t} = -\left(\lim_{\Delta t \to 0} \frac{\Delta\theta}{\Delta t}\right)\mathbf{u}_r$$

$$\dot{\mathbf{u}}_\theta = -\dot{\theta}\mathbf{u}_r \qquad (12\text{--}27)$$

Substituting this result and Eq. 12–23 into the above equation for $\mathbf{a}$, we can write the acceleration in component form as

$$\boxed{\mathbf{a} = a_r\mathbf{u}_r + a_\theta\mathbf{u}_\theta} \qquad (12\text{--}28)$$

where

$$\boxed{\begin{aligned} a_r &= \ddot{r} - r\dot{\theta}^2 \\ a_\theta &= r\ddot{\theta} + 2\dot{r}\dot{\theta} \end{aligned}} \qquad (12\text{--}29)$$

The term $\ddot{\theta} = d^2\theta/dt^2 = d/dt(d\theta/dt)$ is called the *angular acceleration* since it measures the change made in the rate of change of θ during an instant of time. Common units for this measurement are rad/s².

Since $\mathbf{a}_r$ and $\mathbf{a}_\theta$ are always perpendicular, the *magnitude* of acceleration is simply the positive value of

$$a = \sqrt{(\ddot{r} - r\dot{\theta}^2)^2 + (r\ddot{\theta} + 2\dot{r}\dot{\theta})^2} \qquad (12\text{--}30)$$

The *direction* is determined from the vector addition of its two components. In general, $\mathbf{a}$ will *not* be tangent to the path, Fig. 12–30e.

Cylindrical Coordinates

If the particle P moves along a space curve as shown in Fig. 12–31, then its location may be specified by the three *cylindrical coordinates, r, θ, z*. The z coordinate is identical to that used for rectangular coordinates. Since the unit vector defining its direction, $\mathbf{u}_z$, is constant, the time derivatives of this vector are zero and therefore the position, velocity, and acceleration of the particle can be written in terms of its cylindrical coordinates as follows:

$$\mathbf{r}_P = r\mathbf{u}_r + z\mathbf{u}_z$$

$$\mathbf{v} = \dot{r}\mathbf{u}_r + r\dot{\theta}\mathbf{u}_\theta + \dot{z}\mathbf{u}_z \qquad (12\text{--}31)$$

$$\mathbf{a} = (\ddot{r} - r\dot{\theta}^2)\mathbf{u}_r + (r\ddot{\theta} + 2\dot{r}\dot{\theta})\mathbf{u}_\theta + \ddot{z}\mathbf{u}_z \qquad (12\text{--}32)$$

Fig. 12-31

PROCEDURE FOR ANALYSIS

Coordinate System. Polar coordinates are a suitable choice for solving problems for which data regarding the angular motion of the radial coordinate r is given to describe the particle's motion. Also, some paths of motion can conveniently be described in terms of these coordinates. To use cylindrical coordinates, the origin is established at a fixed point, and the radial line r is directed to the particle. Also, the transverse coordinate θ is established, measured counterclockwise from a fixed reference line to the radial line. If necessary, the equation of the path can then be determined by relating r to θ, $r = f(\theta)$.

Time derivatives. Two types of problems generally occur:

1. If the coordinates are specified as time-parametric equations, $r = r(t)$ and $\theta = \theta(t)$, the derivatives are simply $\dot{r} = dr/dt$, $\ddot{r} = d^2r/dt^2$, $\dot{\theta} = d\theta/dt$, and $\ddot{\theta} = d^2\theta/dt^2$ (see Examples 12–17 and 12–18).
2. If the time-parametric equations of motion are not given, it will be necessary to specify the path $r = f(\theta)$ and compute the relationship between the time derivatives using the chain rule of calculus. In this case,

$$\dot{r} = \frac{df(\theta)}{d\theta}\dot{\theta} \qquad \text{and} \qquad \ddot{r} = \left(\frac{d^2f(\theta)}{d\theta^2}\dot{\theta}\right)\dot{\theta} + \frac{df(\theta)}{d\theta}\ddot{\theta}$$

Thus, if two of the *four* time derivatives $\dot{r}$, $\ddot{r}$, $\dot{\theta}$, and $\ddot{\theta}$ are *known,* the other two can be obtained from these equations (see Example 12–19). In some problems, however, two of these time derivatives may *not* be known; instead the magnitude of the particle's velocity or acceleration may be specified. If this is the case, $v^2 = \dot{r}^2 + (r\dot{\theta})^2$ and $a^2 = (\ddot{r} - r\dot{\theta}^2)^2 + (r\ddot{\theta} + 2\dot{r}\dot{\theta})^2$ may be used to obtain the necessary relationships involving $\dot{r}$, $\ddot{r}$, $\dot{\theta}$, and $\ddot{\theta}$ (see Example 12–20).

Velocity and Acceleration. Once r and the time derivatives $\dot{r}$, $\ddot{r}$, $\dot{\theta}$, and $\ddot{\theta}$ have been evaluated at the instant considered, their values can be substituted into Eqs. 12–25 and 12–29 to obtain the radial and transverse components of **v** and **a.**

Motion in three dimensions requires a simple extension of the above procedure to include $\dot{z}$ and $\ddot{z}$.

Besides the examples which follow, further examples involving the calculation of a_r and a_θ can be found in the ''kinematics'' sections of Examples 13–10 to 13–12.

Example 12–17

The ball B shown in Fig. 12–32a is rotating in a horizontal circular path of radius r such that the attached cord has an angular velocity $\dot{\theta}$ and angular acceleration $\ddot{\theta}$. Express the velocity and acceleration of the ball in terms of its polar coordinates.

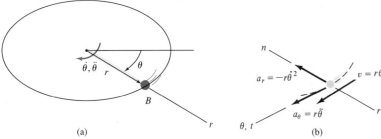

(a)

(b)

Fig. 12–32

SOLUTION

Coordinate System. Since the angular motion of the cord is reported, polar coordinates are chosen for the solution, Fig. 12–32a. Here θ is not related to r, since the radius is constant for all θ.

Time Derivatives. Equations 12–25 and 12–29 will be used for the solution, and so it is first necessary to specify the first and second time derivatives of r and θ. Since r is *constant*, we have

$$r = r, \qquad \dot{r} = 0, \qquad \ddot{r} = 0$$

Velocity
$$v_r = \dot{r} = 0 \qquad\qquad Ans.$$
$$v_\theta = r\dot{\theta} \qquad\qquad Ans.$$

Acceleration
$$a_r = \ddot{r} - r\dot{\theta}^2 = -r\dot{\theta}^2 \qquad\qquad Ans.$$
$$a_\theta = r\ddot{\theta} + 2\dot{r}\dot{\theta} = r\ddot{\theta} \qquad\qquad Ans.$$

These results are shown in Fig. 12–32b. Also shown are the n, t directions, which in this special case of circular motion happen to be *collinear* with the r and θ directions, respectively. In particular note that $v = v_r = v_t = r\dot{\theta}$. Also,

$$-a_r = a_n = \frac{v^2}{\rho} = \frac{(r\dot{\theta})^2}{r} = r\dot{\theta}^2$$

$$a_\theta = a_t = \frac{dv}{dt} = \frac{d}{dt}(r\dot{\theta}) = \frac{dr}{dt}\dot{\theta} + r\frac{d\dot{\theta}}{dt} = 0 + r\ddot{\theta}$$

Example 12–18

The rod OA, shown in Fig. 12–33a, is rotating in the horizontal x-y plane such that at any instant $\theta = (t^3)$ rad. At the same time, the collar B is sliding outward along OA so that $r = (100t^2)$ mm. If in both cases t is in seconds, determine the velocity and acceleration of the collar when $t = 1$ s.

(a)

SOLUTION

Coordinate System. Since time-parametric equations of the path are given, it is not necessary to relate r to θ.

Time Derivatives. Computing the time derivatives, and evaluating when $t = 1$ s, we have

$$r = 100t^2 \Big|_{t=1\,s} = 100 \text{ mm} \qquad \theta = t^3 \Big|_{t=1\,s} = 1 \text{ rad} = 57.3°$$

$$\dot{r} = 200t \Big|_{t=1\,s} = 200 \text{ mm/s} \qquad \dot{\theta} = 3t^2 \Big|_{t=1\,s} = 3 \text{ rad/s}$$

$$\ddot{r} = 200 \Big|_{t=1\,s} = 200 \text{ mm/s}^2 \qquad \ddot{\theta} = 6t \Big|_{t=1\,s} = 6 \text{ rad/s}^2$$

Velocity. Fig. 12–33b

$$\mathbf{v} = \dot{r}\mathbf{u}_r + r\dot{\theta}\mathbf{u}_\theta$$
$$= 200\mathbf{u}_r + 100(3)\mathbf{u}_\theta$$
$$= \{200\mathbf{u}_r + 300\mathbf{u}_\theta\} \text{ mm/s}$$

(b)

The magnitude of $\mathbf{v}$ is

$$v = \sqrt{(200)^2 + (300)^2} = 361 \text{ mm/s} \qquad \textit{Ans.}$$

From Fig. 12–33b,

$$\delta = \tan^{-1}\left(\frac{300}{200}\right) = 56.3° \qquad \delta + 57.3° = 114° \quad \textit{Ans.}$$

Acceleration. Fig. 12–33c

$$\mathbf{a} = (\ddot{r} - r\dot{\theta}^2)\mathbf{u}_r + (r\ddot{\theta} + 2\dot{r}\dot{\theta})\mathbf{u}_\theta$$
$$= [200 - 100(3)^2]\mathbf{u}_r + [100(6) + 2(200)3]\mathbf{u}_\theta$$
$$= \{-700\mathbf{u}_r + 1800\mathbf{u}_\theta\} \text{ mm/s}^2$$

The magnitude of $\mathbf{a}$ is

$$a = \sqrt{(700)^2 + (1800)^2} = 1930 \text{ mm/s}^2 \qquad \textit{Ans.}$$

From Fig. 12–33c,

$$\phi = \tan^{-1}\left(\frac{1800}{700}\right) = 68.7° \qquad \phi - 57.3° = 11.4° \qquad \textit{Ans.} \quad \textbf{Fig. 12–33}$$

(c)

Example 12–19

The searchlight shown in Fig. 12–34a casts a spot of light along the face of a wall that is located 100 m from the searchlight. Determine the magnitudes of the velocity and acceleration at which the spot appears to travel across the wall at the instant $\theta = 45°$. The searchlight is rotating about the z axis at a constant rate $\dot{\theta} = 4$ rad/s.

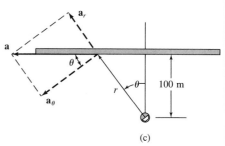

(a)

SOLUTION

Coordinate System. Why should polar coordinates be used to solve this problem? To compute the necessary time derivatives it is first necessary to relate r to θ. From Fig. 12–34a, this relation is

$$r = \frac{100}{\cos \theta} = 100 \sec \theta$$

Time Derivatives. Using the chain rule of calculus, noting that $d (\sec \theta) = \sec \theta \tan \theta \, d\theta$, and $d (\tan \theta) = \sec^2 \theta \, d\theta$, we have

$$\dot{r} = 100(\sec \theta \tan \theta)\dot{\theta}$$

$$\ddot{r} = 100(\sec \theta \tan \theta)\dot{\theta}(\tan \theta)\dot{\theta} + 100 \sec \theta(\sec^2 \theta)\dot{\theta}(\dot{\theta}) + 100 \sec \theta \tan \theta(\ddot{\theta})$$

$$= 100 \sec \theta \tan^2 \theta(\dot{\theta})^2 + 100 \sec^3 \theta(\dot{\theta})^2 + 100(\sec \theta \tan \theta)\ddot{\theta}$$

Since $\dot{\theta} = 4$ rad/s = constant, then $\ddot{\theta} = 0$, and the above equations, when $\theta = 45°$, become

$$r = 100 \sec 45° = 141.4$$

$$\dot{r} = 400 \sec 45° \tan 45° = 565.7$$

$$\ddot{r} = 1600(\sec 45° \tan^2 45° + \sec^3 45°) = 6788.2$$

Velocity Fig. 12–34b

$$\mathbf{v} = \dot{r}\mathbf{u}_r + r\dot{\theta}\mathbf{u}_\theta$$

$$= 565.7\mathbf{u}_r + 141.4(4)\mathbf{u}_\theta$$

$$= 565.7\mathbf{u}_r + 565.7\mathbf{u}_\theta$$

$$v = \sqrt{v_r^2 + v_\theta^2} = \sqrt{(565.7)^2 + (565.7)^2}$$

$$= 800 \text{ m/s} \qquad\qquad Ans.$$

Acceleration. Fig. 12–34c

$$\mathbf{a} = (\ddot{r} - r\dot{\theta}^2)\mathbf{u}_r + (r\ddot{\theta} + 2\dot{r}\dot{\theta})\mathbf{u}_\theta$$

$$= [6788.2 - 141.4(4)^2]\mathbf{u}_r + [141.4(0) + 2(565.7)4]\mathbf{u}_\theta$$

$$= 4525.8\mathbf{u}_r + 4525.8\mathbf{u}_\theta$$

$$a = \sqrt{a_r^2 + a_\theta^2} = \sqrt{(4525.8)^2 + (4525.8)^2}$$

$$= 6400 \text{ m/s}^2 \qquad\qquad Ans.$$

(b)

(c)

Fig. 12–34

Example 12–20

Due to the rotation of the forked rod, the cylindrical peg A in Fig. 12–35a travels around the slotted path, a portion of which is in the shape of a cardioid, $r = 0.5(1 - \cos \theta)$ ft, where θ is in radians. If the peg's velocity is $v = 4$ ft/s and its acceleration is $a = 30$ ft/s^2 at the instant $\theta = 180°$, determine the angular velocity $\dot{\theta}$ and angular acceleration $\ddot{\theta}$ of the fork.

$r = 0.5 (1 - \cos\theta)$ ft

(a)

SOLUTION

Coordinate System. This path is most unusual, and mathematically it is best expressed using polar coordinates rather than rectangular coordinates. Also, $\dot{\theta}$ and $\ddot{\theta}$ must be determined so r, θ coordinates are an obvious choice.

Time Derivatives. Computing the time derivatives of r using the chain rule of calculus yields

$$r = 0.5(1 - \cos \theta)$$
$$\dot{r} = 0.5(\sin \theta)\dot{\theta}$$
$$\ddot{r} = 0.5(\cos \theta)\dot{\theta}(\dot{\theta}) + 0.5(\sin \theta)\ddot{\theta}$$

Evaluating these results at $\theta = 180°$, we have

$$r = 1 \text{ ft} \qquad \dot{r} = 0 \qquad \ddot{r} = -0.5\dot{\theta}^2$$

Velocity. Since $v = 4$ ft/s, using Eq. 12–26 to determine $\dot{\theta}$ yields

$$v = \sqrt{(\dot{r})^2 + (r\dot{\theta})^2}$$
$$4 = \sqrt{(0)^2 + (1\dot{\theta})^2}$$
$$\dot{\theta} = 4 \text{ rad/s} \qquad\qquad Ans.$$

$v = 4$ ft/s $a = 30$ ft/s^2

(b)

Fig. 12–35

Acceleration. In a similar manner, $\ddot{\theta}$ can be found using Eq. 12–30.

$$a = \sqrt{(\ddot{r} - r\dot{\theta}^2)^2 + (r\ddot{\theta} + 2\dot{r}\dot{\theta})^2}$$
$$30 = \sqrt{[-0.5(4)^2 - 1(4)^2]^2 + [1\ddot{\theta} + 2(0)(4)]^2}$$
$$(30)^2 = (-24)^2 + \ddot{\theta}^2$$
$$\ddot{\theta} = 18 \text{ rad/s}^2 \qquad\qquad Ans.$$

Vectors **a** and **v** are shown in Fig. 12–35b.

PROBLEMS

12–101. A particle is moving along a circular path having a radius of $r = 0.4$ m. Its angular position as a function of time is given by $\theta = (2t^2)$ rad, where t is in seconds. Determine the magnitude of the particle's acceleration when $\theta = 30°$. The particle starts from rest when $\theta = 0°$.

12–102. A car is traveling along the circular curve having a radius $r = 400$ ft. At the instant shown, its angular rate of rotation is $\dot\theta = 0.25$ rad/s, which is decreasing at the rate $\ddot\theta = -0.008$ rad/s². Determine the radial and transverse components of the car's velocity and acceleration at this instant and sketch these components on the curve.

Probs. 12–102/12–103

12–103. A car is traveling along the circular curve of radius $r = 400$ ft with a constant speed $v = 30$ ft/s. Determine the angular rate of rotation $\dot\theta$ of the radial line r and the magnitude of the car's acceleration.

*12–104. The arm of the robot has a fixed length so that $r = 3$ ft and its grip A moves along the path $z = (3 \sin 4\theta)$ ft, where θ is in radians. If $\theta = (0.5t)$ rad, where t is in seconds, determine the magnitudes of the grip's velocity and acceleration when $t = 3$ s.

$V = 5.95 \ ft/s$
$a = 3.44 \ ft/s^2$

Probs. 12–104/12–105

12–105. For a short time the arm of the robot is extending at a constant rate such that $\dot r = 1.5$ ft/s when $r = 3$ ft, $z = (4t^2)$ ft, and $\theta = 0.5t$ rad, where t is in seconds. Determine the magnitudes of the velocity and acceleration of the grip A when $t = 3$ s.

12–106. At the instant shown, the watersprinkler is rotating with an angular speed $\dot\theta = 2$ rad/s and an angular acceleration $\ddot\theta = 3$ rad/s². If the nozzle lies in the vertical plane and water is flowing through it at a constant rate of 3 m/s, determine the magnitudes of the velocity and acceleration of a water particle as it exits the open end, $r = 0.2$ m.

$r = 0.2$ m

Prob. 12–106

12–107. A particle moves along a spiral path $r = (10/\theta)$ ft, where θ is in radians. If it maintains a constant speed of $v = 20$ ft/s, determine the magnitudes of its components v_r and v_θ as functions of θ and evaluate each at $\theta = 1$ rad. Sketch the path and show the velocity components at $\theta = 1$ rad.

*12–108. The telescopic rod rotates at $\dot\theta = 0.8$ rad/s which is decreasing at $\ddot\theta = -0.2$ rad/s², and at the same time the rod is moving in and out such that its end A is following the path $r = (0.75\cos\theta + 2)$ ft, where θ is in radians. Determine the radial and transverse components of the velocity and acceleration of A at the instant $\theta = \pi/8$.

Probs. 12–108/12–109

12–109. The telescopic rod rotates at $\dot\theta = 0.8$ rad/s which is decreasing at $\ddot\theta = -0.2$ rad/s², and at the same time the rod is moving in and out such that its end A is following the path $r = (1 + 0.75 \cos\theta)$ ft, where θ is in radians. Determine the radial and transverse components of the velocity and acceleration of A at the instant $\theta = \pi/8$.

12–110. A particle moves along the spiral path $r = (0.4e^\theta)$ ft such that $\dot\theta = (0.2t)$ rad/s, where t is in seconds. Determine the transverse and radial components of the particle's velocity and acceleration when $t = 1$ s. When $t = 0$, $\theta = \pi/2$ rad.

12–111. A horse on the merry-go-round moves according to the equations $r = 8$ ft, $\theta = (0.6t)$ rad, and $z = (1.5 \sin \theta)$ ft, where t is in seconds. Determine the cylindrical components of the velocity and acceleration of the horse when $t = 4$ s.

Prob. 12–111

*12–112. A particle moves along an Archimedean spiral $r = (8\theta)$ ft, where θ is given in radians. If $\dot\theta = 4$ rad/s (constant), compute the radial and transverse components of the particle's velocity and acceleration at the instant $\theta = \pi/2$ rad. Sketch the curve and show the components on the curve.

12–113. Solve Prob. 12–112 if the particle has an angular acceleration $\ddot\theta = 5$ rad/s² when $\dot\theta = 4$ rad/s at $\theta = \pi/2$ rad.

12–114. For a short time the roller coaster travels along the track defined by the equation $r = (200 \cos 2\theta)$ ft, where θ is in radians. If the angular velocity of the radial coordinate line is $\dot\theta = (0.008t^2)$ rad/s, where t is in seconds, determine the radial and transverse components of the car's velocity and acceleration at the instant $\theta = 30°$. When $t = 0$, $\theta = 0$.

Prob. 12–114

12–115. The car travels around the circular curve with a constant speed of 20 m/s. Determine the car's radial and transverse components of velocity and acceleration at the instant $\theta = \pi/4$ rad.

$r = (400 \cos \theta)$m

Prob. 12–115

*12–116. A particle starts from the origin at $t = 0$ and moves along a curve such that it has a velocity of $\mathbf{v} = \{8\mathbf{u}_r + 4t\mathbf{u}_\theta + 5t^2\mathbf{u}_z\}$ ft/s, where t is in seconds. Determine the magnitudes of the velocity and acceleration of the particle when $t = 2$ s. At $t = 2$ s, $r = 3$ ft.

12–117. The slotted arm AB drives the pin C through the spiral groove described by the equation $r = a\theta$. If the angular rate of rotation is constant at $\dot{\theta}$, determine the radial and transverse components of velocity and acceleration of the pin.

Probs. 12–117/12–118/12–119

12–118. Solve Prob. 12–117 if the spiral path is logarithmic, i.e., $r = a\,e^{k\theta}$.

12–119. The slotted arm AB drives the pin C through the spiral groove described by the equation $r = (1.5\theta)$ ft, where θ is in radians. If the arm starts from rest when $\theta = 60°$ and is driven at an angular rate of $\dot{\theta} = (4t)$ rad/s, where t is in seconds, determine the radial and transverse components of velocity and acceleration of the pin C when $t = 1$ s.

*12–120. The roller coaster is traveling down along the spiral ramp with a constant speed $v = 6$ m/s. If the track descends a distance of 10 m for every full revolution, $\theta = 2\pi$ rad, determine the magnitude of the roller coaster's acceleration as it moves along the track, $r = 5$ m. *Hint:* For part of the solution, note that the tangent to the ramp at any point is at an angle $\phi = \tan^{-1}[10/2\pi(5)] = 17.66°$ from the horizontal. Use this to determine the velocity components v_θ and v_z, which in turn are used to determine $\dot{\theta}$ and $\dot{z}$.

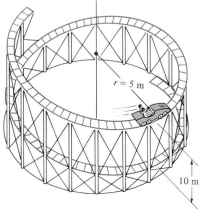

$r = 5$ m

10 m

Prob. 12–120

12–121. The rocket is traveling upwards such that when $\theta = 30°$ the radar tracking device is turning at $\dot{\theta} = 0.6$ rad/s and $\ddot{\theta} = 0.2$ rad/s^2. Determine the magnitude of velocity and acceleration of the rocket at this instant.

$\dot{\theta}, \ddot{\theta}$

r

θ

800 m

Probs. 12–121/12–122

12–122. As the rocket travels upwards, the radar tracking device has an angular position $\theta = (0.03t^2)$ rad, where t is in seconds. Determine the magnitude of velocity and acceleration of the rocket at the instant $t = 3$ s.

12–123. The slotted link is pinned at O and as a result of the constant angular velocity $\dot{\theta} = 3$ rad/s it drives the peg P along the path defined by $r = (0.6 \sin \theta + 0.3)$ ft, where θ is in radians. Compute the radial and transverse components of the velocity and acceleration of P at the instant $\theta = \pi/3$ rad.

O

r

θ

P

Probs. 12–123/12–124

*12–124. Solve Prob. 12–123 if the slotted link has an angular acceleration of $\ddot{\theta} = 4$ rad/s^2 and $\dot{\theta} = 3$ rad/s at $\theta = \pi/3$ rad.

12–125. A particle travels along the "four-leaf rose" defined by the equation $r = (5 \cos 2\theta)$ m. If the angular velocity of the radial coordinate line is $\dot{\theta} = (3t^2)$ rad/s, where t is in seconds, determine the radial and transverse components of the particle's velocity and acceleration at the instant $\theta = 30°$. When $t = 0$, $\theta = 0$.

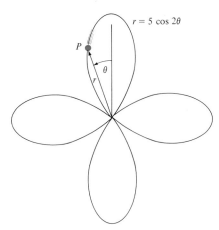

Prob. 12–125

12–126. The time rate of change of acceleration is referred to as the *jerk*, which is often used as a means of measuring passenger discomfort. Calculate this vector, $\dot{\mathbf{a}}$, in terms of its cylindrical components, using Eq. 12–32.

■**12–127.** The double collar C is pin-connected together such that one collar slides over a fixed rod and the other slides over a rotating rod AB. If the angular velocity of AB is given as $\dot{\theta} = (e^{0.5t^2})$ rad/s, where t is in seconds, and the path defined by the fixed rod is $r = |(0.4 \sin \theta + 0.2)|$ m, determine the radial and transverse components of the collar's velocity and acceleration when $t = 1$ s. When $t = 0$, $\theta = 0$. Use Simpson's rule to determine θ at $t = 1$ s.

Prob. 12–127

*12–128.** A particle travels along a path such that its position vector is defined by $\mathbf{r} = \{2 \cos (0.1t)\mathbf{i} + 1.5 \sin(0.1t)\mathbf{j} + (2t)\mathbf{k}\}$ m, where t is in seconds and the arguments for the sine and cosine are given in radians. When $t = 10$ s, determine the coordinate direction angles α, β, and γ, which the binormal axis to the osculating plane makes with the x, y, and z axes. Sketch the path in three dimensions. *Hint:* Solve for the velocity $\mathbf{v}$ and acceleration $\mathbf{a}$ of the particle in terms of their $\mathbf{i}$, $\mathbf{j}$, $\mathbf{k}$ components. The binormal is collinear to $\mathbf{v} \times \mathbf{a}$. Why?

12–129. The motion of a particle along a fixed path is defined by the parametric equations $r = 1.5$ m, $\theta = (2t)$ rad, and $z = (t^2)$ m, where t is in seconds. Determine the unit vector that specifies the direction of the binormal axis to the osculating plane with respect to a set of fixed x, y, z coordinate axes when $t = 0.25$ s. *Hint:* Formulate the particle's velocity $\mathbf{v}$ and acceleration $\mathbf{a}$ in terms of their $\mathbf{i}$, $\mathbf{j}$, $\mathbf{k}$ components. Note that $x = r \cos \theta$ and $y = r \sin \theta$. The binormal is collinear to $\mathbf{v} \times \mathbf{a}$. Why?

12.8 Absolute Dependent Motion Analysis of Two Particles

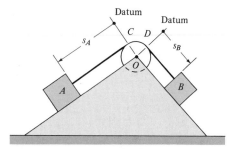

Fig. 12—36

In some types of problems the motion of one particle will *depend* upon the corresponding motion of another particle. This dependency commonly occurs if the particles are interconnected by inextensible cords which are wrapped around pulleys. For example, the movement of block *A* downward along the inclined plane in Fig. 12–36 will cause a corresponding movement of block *B* up the other incline. We can show this mathematically by first specifying the positions of the blocks using the path coordinates s_A and s_B. Note that these coordinates are (1) referenced from a *fixed* point (*O*) or datum, (2) measured along the plane in the direction of motion, and (3) have a positive sense from *C* to *A* and *D* to *B*. Since the total cord length l_T is constant, the coordinates are related by the equation

$$s_A + s_B = l_T - l_{CD} = l'$$

where l_{CD} is the length of cord passing over arc *CD*.* Taking the time derivative of this expression yields a relation between the velocities of the blocks, i.e.,

$$\frac{ds_B}{dt} + \frac{ds_A}{dt} = 0$$

or

$$v_B = -v_A$$

The negative sign indicates that when block *A* has a velocity downward, i.e., in the direction of positive s_A, it causes a corresponding upward velocity of block *B*; i.e., *B* moves in the negative s_B direction.

In a similar manner, time differentiation of the velocities yields the relation between the accelerations, i.e.,

$$a_B = -a_A$$

A more complicated example involving dependent motion of two blocks is shown in Fig. 12–37*a*. In this case, the position of block *A* is specified by s_A, and the position of the *end* of the cord from which block *B* is suspended is defined by s_B. Here we have chosen coordinates which are (1) referenced from fixed points or datums, (2) measured in the direction of motion of each block, and (3) positive to the right (s_A) and downward (s_B). The *constant* cord length *l*, which represents the total length of cord *minus* the constant cord segments which are colored, can be related to the position coordinates by the equation

---

*Note that when the blocks move, the position coordinates s_A and s_B measure the lengths of the *changing segments* of the cord. The cord length l_{CD} always remains *constant*.

$$2s_B + h + s_A = l$$

Since l and h are constants, the two time derivatives yield

$$2v_B = -v_A \qquad 2a_B = -a_A$$

Hence, when B moves downward $(+s_B)$, A moves to the left $(-s_A)$ at two times the motion.

This example can also be worked by defining the position of block B from the center of the bottom pulley (a fixed point), Fig. 12–37b. In this case

$$2(h - s_B) + h + s_A = l$$

Time differentiation yields

$$2v_B = v_A \qquad 2a_B = a_A$$

Here the signs are the same. Why?

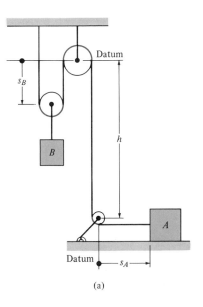

(a)

PROCEDURE FOR ANALYSIS

The above method of relating the dependent motion of one particle to that of another can be performed using algebraic scalars or position coordinates provided each particle moves along a rectilinear path. When this is the case, only the magnitudes of the velocity and acceleration of each particle will change, not their line of direction. The following procedure is required.

Position-Coordinate Equation. Establish position coordinates which have their origin referenced to a *fixed* point or datum and extend from this point to each of the particles. It is *not necessary* that the *origin* be the *same* for both of these coordinates; however, it is *important* that each coordinate axis selected be directed along the *path of motion* of the particle.

Using geometry or trigonometry, relate the coordinates to the total length of the cord, l_T, or to that portion of cord, l, which *excludes* the segments that do not change length as the particles move—such as the arc segments wrapped over pulleys.

Time Derivatives. The two successive time derivatives of the position-coordinate equation yield the required velocity and acceleration equations which relate the motions of the two particles. The signs of the terms in these equations will be consistent with those that specify the positive and negative sense of the position coordinates.

If a problem involves a *system* of two or more cords wrapped around pulleys, then the motion of a point on one cord must be related to the motion of a point on another cord using the above procedure. Separate equations are written for each cord of the system and the positions of the two particles are then related by these equations (see Examples 12–22 and 12–23).

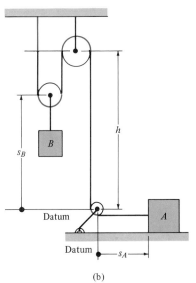

(b)

Fig. 12–37

Example 12–21

Determine the speed of block A in Fig. 12–38. Block B has an upward speed of 6 ft/s.

Fig. 12–38

SOLUTION

Position-Coordinate Equation. There is *one* cord in this system having segments which are changing length. Position coordinates s_A and s_B will be used since each is measured from a fixed point (C or D) and extends along the block's *path of motion*. In particular, s_B is directed to point E since motion of B and E is the *same*.

The colored segments of the cord in Fig. 12–38 remain at a constant length and do not have to be considered as the blocks move. The remaining length of cord, l, is also constant and is related to the changing position coordinates s_A and s_B by the equation

$$s_A + 3s_B = l$$

Time Derivative. Taking the time derivative yields

$$v_A + 3v_B = 0$$

so that when $v_B = -6$ ft/s (upward), then

$$v_A = 18 \text{ ft/s} \downarrow \qquad\qquad Ans.$$

Example 12–22

Determine the speed of block A in Fig. 12–39. Block B has an upward speed of 6 ft/s.

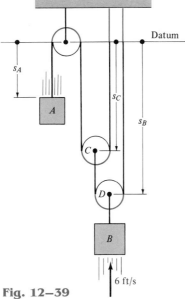

Fig. 12–39

SOLUTION

Position-Coordinate Equation. This system has two cords which have segments that change length. Motion of the end of the cord attached to A can be related to the motion of the end of the cord attached to the pulley at C using coordinates s_A and s_C. Likewise, the motion of the end of the cord at C can be related to the motion of the end of the cord at D using position coordinates s_C and s_B.

The colored segments of the cords in Fig. 12–39 do not have to be considered in the analysis. Why? For the remaining cord lengths, say l_1 and l_2, we have

$$s_A + 2s_C = l_1 \qquad s_B + (s_B - s_C) = l_2$$

Eliminating s_C yields an equation defining the positions of both blocks, i.e.,

$$s_A + 4s_B = 2l_2 + l_1$$

Time Derivative. The time derivative gives

$$v_A + 4v_B = 0$$

so that when $v_B = -6$ ft/s (upward), then

$$v_A = 24 \text{ ft/s} \downarrow \qquad\qquad \textit{Ans.}$$

Example 12–23

Determine the speed with which block B rises in Fig. 12–40 if the end of the cord at A is pulled down with a speed of 2 m/s.

Fig. 12–40

SOLUTION

Position-Coordinate Equation. By inspection this system has two cords. Three position coordinates have been selected. These are all measured from the horizontal datum passing through the fixed pin at pulley D. The position of the end of the cord at A is defined by s_A, the position of the end of the cord at C is defined by s_C, and the position of the end of the cord at E (block B) is defined by s_E.

Excluding the colored segments of the cords in Fig. 12–40, the remaining constant cord lengths l_1 and l_2 (with the hook dimensions) can be expressed in terms of the three position coordinates. We have

$$(s_A - s_C) + (s_E - s_C) + s_E = l_1$$
$$s_C + s_E = l_2$$

Eliminating s_C yields

$$s_A + 4s_E = l_1 + 2l_2$$

This equation relates the position s_E of block B to the position s_A of point A.

Time Derivative. The time derivative gives

$$v_A + 4v_E = 0$$

so that when $v_A = 2$ m/s (downward), then

$$v_E = -0.5 \text{ m/s} = 0.5 \text{ m/s} \uparrow \qquad\qquad \textit{Ans.}$$

Example 12–24

A man at A is hoisting a safe S as shown in Fig. 12–41 by walking to the right with a constant velocity $v_A = 0.5$ m/s. Determine the velocity and acceleration of the safe when it reaches the window elevation at E. The rope is 30 m long and passes over a small pulley at D.

Fig. 12–41

SOLUTION

Position-Coordinate Equation. This problem is unlike the previous examples, since rope segment DA changes both direction and magnitude. However, the ends of the rope, which define the positions of S and A, are specified by means of the x and y coordinates measured from fixed points and *directed along the paths of motion* of the ends of the rope.

The x and y coordinates may be related since the rope has a fixed length $l = 30$ m, which at all times is equal to the length of segment DA plus CD. Using the Pythagorean theorem to determine l_{DA}, we have $l_{DA} = \sqrt{(15)^2 + x^2}$; also, $l_{CD} = 15 - y$. Hence,

$$l = l_{DA} + l_{CD}$$
$$30 = \sqrt{(15)^2 + x^2} + (15 - y)$$
$$y = \sqrt{225 + x^2} - 15 \tag{1}$$

Time Derivatives. Taking the time derivative, where $v_S = dy/dt$ and $v_A = dx/dt$, yields

$$v_S = \frac{dy}{dt} = \frac{1}{2}\frac{2x}{\sqrt{225 + x^2}}\frac{dx}{dt}$$

$$= \frac{x}{\sqrt{225 + x^2}}v_A \tag{2}$$

At $y = 10$ m, x is determined from Eq. (1), i.e., $x = 20$ m. Hence, from Eq. (2) with $v_A = 0.5$ m/s,

$$v_S = \frac{20}{\sqrt{225 + (20)^2}}(0.5) = 0.4 \text{ m/s} = 400 \text{ mm/s} \uparrow \qquad Ans.$$

The acceleration is determined by taking the time derivative of Eq. (2). Since v_A is constant, then $a_A = dv_A/dt = 0$, and we have

$$a_S = \frac{d^2y}{dt^2} = \left[\frac{-x^2}{(225 + x^2)^{3/2}}\frac{dx}{dt} + \frac{1}{(225 + x^2)^{1/2}}\frac{dx}{dt}\right]v_A$$

$$= \frac{225v_A^2}{(225 + x^2)^{3/2}}$$

At $x = 20$ m, with $v_A = 0.5$ m/s, the acceleration becomes

$$a_S = \frac{225(0.5)^2}{(225 + (20)^2)^{3/2}} = 0.00360 \text{ m/s}^2 = 3.60 \text{ mm/s}^2 \uparrow \qquad Ans.$$

12.9 Relative-Motion Analysis of Two Particles Using Translating Axes

Throughout this chapter the absolute motion of a particle has been determined using a single fixed reference frame for measurement. There are many cases, however, where the path of motion for a particle is complicated, so that it may be feasible to analyze the motion in parts by using two or more frames of reference. For example, the motion of a particle located at the tip of an airplane propeller, while the plane is in flight, is more easily described if one observes first the motion of the airplane from a fixed reference and then superimposes (vectorially) the circular motion of the particle measured from a reference attached to the airplane. Any type of coordinates—rectangular, cylindrical, etc.—may be chosen to describe these two different motions.

In this section only *translating frames of reference* will be considered for the analysis. Relative-motion analysis of particles using rotating frames of reference will be treated in Secs. 16.8 and 20.4, since such an analysis depends upon prior knowledge of the kinematics of line segments.

Position

Consider particles A and B, which move along the arbitrary paths aa and bb, respectively, as shown in Fig. 12–42a. The *absolute position* of each particle, $\mathbf{r}_A$ and $\mathbf{r}_B$, is measured from the common origin O of the *fixed x, y, z* reference frame. The origin of a second frame of reference x', y', z' is attached to and moves with particle A. The axes of this frame do not rotate; rather they are *only permitted to translate* relative to the fixed frame. The *relative position* of "B with respect to A" is observed from this moving frame and is designated by $\mathbf{r}_{B/A}$, called a *relative-position vector*. Using vector addition, the three vectors shown in Fig. 12–42a can be related by the equation*

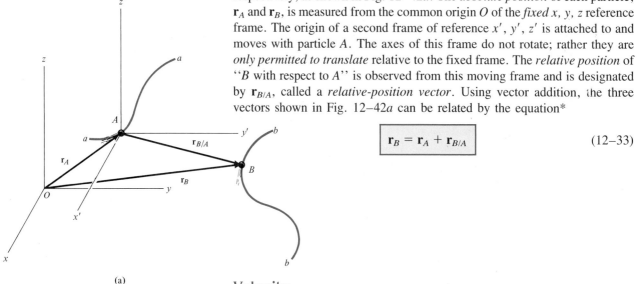

$$\boxed{\mathbf{r}_B = \mathbf{r}_A + \mathbf{r}_{B/A}} \tag{12–33}$$

(a)

Fig. 12—42a

Velocity

An equation that relates the velocities of the particles can be determined by taking the time derivative of Eq. 12–33, i.e.,

*An easy way to remember the setup of this equation, and others like it, is to note the "cancellation" of the subscript A between the two terms, i.e., $\mathbf{r}_B = \mathbf{r}_{\cancel{A}} + \mathbf{r}_{B/\cancel{A}}$.

$$\boxed{\mathbf{v}_B = \mathbf{v}_A + \mathbf{v}_{B/A}} \qquad\qquad (12\text{–}34)$$

(b)

Here $\mathbf{v}_B = d\mathbf{r}_B/dt$ and $\mathbf{v}_A = d\mathbf{r}_A/dt$ refer to *absolute velocities*, since they are measured from the fixed frame of reference; whereas the *relative velocity* $\mathbf{v}_{B/A} = d\mathbf{r}_{B/A}/dt$ is observed from the translating reference frame. It is important to note that since the x', y', z' axes translate, the *components* of $\mathbf{r}_{B/A}$ will *not* change direction and therefore the time derivative of this vector will only have to account for changes in the vector's magnitude. The above equation therefore states that the velocity of B is equal to the velocity of A plus (vectorially) the relative velocity of "B with respect to A," as measured by a *translating observer* fixed in the x', y', z' reference, Fig. 12–42b.

(c)

Fig. 12–42(b, c)

Acceleration

The time derivative of Eq. 12–34 yields a similar vector relationship between the *absolute* and *relative accelerations* of particles A and B,

$$\boxed{\mathbf{a}_B = \mathbf{a}_A + \mathbf{a}_{B/A}} \qquad\qquad (12\text{–}35)$$

Here $\mathbf{a}_{B/A}$ is the acceleration of B as seen by an observer located at A and translating with the x', y', z' reference frame. The vector addition is shown in Fig. 12–42c.

PROCEDURE FOR ANALYSIS

When applying the relative-position equation, $\mathbf{r}_B = \mathbf{r}_A + \mathbf{r}_{B/A}$, it is first necessary to specify the locations of the fixed x, y, z and translating x', y', z' axes. Usually, the origin A of the translating axes is located at a point having a *known position*, $\mathbf{r}_A$, Fig. 12–42a. A graphical representation of the vector addition $\mathbf{r}_B = \mathbf{r}_A + \mathbf{r}_{B/A}$ can be shown, and both the known and unknown quantities labeled on this sketch. Since vector addition forms a triangle there can be at most *two unknowns*, represented by the magnitudes and/or directions of the vector quantities. These unknowns can be solved for either graphically, using trigonometry (law of sines, law of cosines), or by resolving each of the three vectors $\mathbf{r}_B$, $\mathbf{r}_A$, and $\mathbf{r}_{B/A}$ into rectangular or Cartesian components, thereby generating two scalar equations. The latter method is illustrated in the example problems which follow.

The relative-motion equations $\mathbf{v}_B = \mathbf{v}_A + \mathbf{v}_{B/A}$ and $\mathbf{a}_B = \mathbf{a}_A + \mathbf{a}_{B/A}$ are applied in the same manner as explained above, except in this case the origin O of the fixed x, y, z axes does not have to be specified, Figs. 12–42b and 12–42c.

Example 12–25

Water drips from a faucet at the rate of five drops per second as shown in Fig. 12–43. Determine the vertical separation between two consecutive drops after the lower drop has attained a velocity of 3 m/s.

Datum

Fig. 12–43

SOLUTION

If the first and second drops of water are denoted as B and A, respectively, Fig. 12–43, the separation $s_{B/A}$ (position of B with respect to A) can be determined by the equation

$(+\downarrow)$ $$s_{B/A} = s_B - s_A \qquad (1)$$

Here the origin of the fixed frame of reference is located at the head of the faucet (datum) with positive s downward. The origin of the moving frame s' is at drop A. Since each drop is subjected to rectilinear motion, having a *constant* downward acceleration $a_c = g = 9.81$ m/s², Eq. 12–4 may be applied with the initial condition $v_0 = 0$ to determine the time needed for drop B to attain a velocity of 3 m/s.

$(+\downarrow)$ $$v = v_0 + a_c t$$
$$3 = 0 + 9.81 t_B$$
$$t_B = 0.306 \text{ s}$$

Since one drop falls every fifth of a second (0.2 s), drop A falls for a time

$$t_A = 0.306 \text{ s} - 0.2 \text{ s} = 0.106 \text{ s}$$

before drop B attains a velocity of 3 m/s.

The position of each drop can be found by means of Eq. 12–5, with the condition $s_0 = 0$, $v_0 = 0$.

$(+\downarrow)$ $$s = s_0 + v_0 t + \tfrac{1}{2}a_c t^2$$
$$s_B = 0 + 0 + \tfrac{1}{2}(9.81)(0.306)^2$$
$$= 0.459 \text{ m}$$
$$s_A = 0 + 0 + \tfrac{1}{2}(9.81)(0.106)^2$$
$$= 0.0549 \text{ m}$$

Applying Eq. (1) yields

$$s_{B/A} = s_B - s_A = 0.459 - 0.0549 = 0.404 \text{ m}$$
$$= 404 \text{ mm} \downarrow \qquad\qquad \textit{Ans.}$$

Example 12–26

A train, traveling at a constant speed of 60 mi/h, crosses over a road as shown in Fig. 12–44a. If an automobile A is traveling at 45 mi/h along the road, determine the relative velocity of the train with respect to the automobile.

$v_T = 60$ mi/h

SOLUTION I

Vector Analysis. The relative velocity $\mathbf{v}_{T/A}$ is measured from the translating x', y' axes attached to the automobile, Fig. 12–44a. It is determined from $\mathbf{v}_T = \mathbf{v}_A + \mathbf{v}_{T/A}$. Since $\mathbf{v}_T$ and $\mathbf{v}_A$ are known in *both* magnitude and direction, the unknowns become the components of $\mathbf{v}_{T/A}$. Using the x, y axes in Fig. 12–44a and a Cartesian vector analysis, we have

$$\mathbf{v}_T = \mathbf{v}_A + \mathbf{v}_{T/A}$$
$$60\mathbf{i} = (45 \cos 45° \, \mathbf{i} + 45 \sin 45° \, \mathbf{j}) + \mathbf{v}_{T/A}$$
$$\mathbf{v}_{T/A} = \{28.2\mathbf{i} - 31.8\mathbf{j}\} \text{ mi/h} \qquad \textit{Ans.}$$

The magnitude of $\mathbf{v}_{T/A}$ is thus

$$v_{T/A} = \sqrt{(28.2)^2 + (-31.8)^2} = 42.5 \text{ mi/h} \qquad \textit{Ans.}$$

From the direction of each component, the direction of $\mathbf{v}_{T/A}$ defined from the x axis, is

$$\tan \theta = \frac{(v_{T/A})_y}{(v_{T/A})_x} = \frac{31.8}{28.2}$$
$$\theta = 48.4° \qquad \textit{Ans.}$$

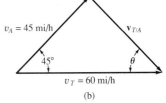

(b)

Fig. 12–44

Note that the vector addition shown in Fig. 12–44b indicates the correct sense for $\mathbf{v}_{T/A}$. This figure anticipates the answer and can be used to check it.

SOLUTION II

Scalar Analysis. The components of $\mathbf{v}_{T/A}$ can also be determined by applying a scalar analysis. We will assume these components act in the *positive* x and y directions. Thus,

$$\mathbf{v}_T = \mathbf{v}_A + \mathbf{v}_{T/A}$$
$$\begin{bmatrix} 60 \text{ mi/h} \\ \rightarrow \end{bmatrix} = \begin{bmatrix} 45 \text{ mi/h} \\ \angle 45° \end{bmatrix} + \begin{bmatrix} (v_{T/A})_x \\ \rightarrow \end{bmatrix} + \begin{bmatrix} (v_{T/A})_y \\ \uparrow \end{bmatrix}$$

Resolving each vector into its horizontal and vertical components yields

$$(\xrightarrow{+}) \qquad 60 = 45 \cos 45° + (v_{T/A})_x + 0$$
$$(+\uparrow) \qquad 0 = 45 \sin 45° + 0 + (v_{T/A})_y$$

Solving, we obtain the previous results,

$$(v_{T/A})_x = 28.2 \text{ mi/h} = 28.2 \text{ mi/h} \rightarrow$$
$$(v_{T/A})_y = -31.8 \text{ mi/h} = 31.8 \text{ mi/h} \downarrow$$

Example 12–27

Two jet planes are flying horizontally at the same elevation, as shown in Fig. 12–45a. Plane A is flying along a straight-line path, and at the instant shown it has a speed of 700 km/h and an acceleration of 50 km/h^2. Plane B is flying along a circular path having a radius of curvature of $\rho_B = 400$ km. Its speed is 600 km/h, which is decreasing at the rate of 100 km/h^2. Determine the velocity and acceleration of B as measured by the pilot in A.

SOLUTION

Velocity. The fixed x, y axes are located at an arbitrary point. Plane A is traveling with rectilinear motion and a *translating frame of reference* x', y' will be attached to it. Applying the relative-velocity equation in scalar form since the velocity vectors of both planes are parallel at the instant shown, we have

$$(+ \uparrow) \qquad v_B = v_A + v_{B/A}$$

$$600 = 700 + v_{B/A}$$

$$v_{B/A} = -100 \text{ km/h} = 100 \text{ km/h} \downarrow \qquad\qquad \textit{Ans.}$$

The vector addition is shown in Fig. 12–45b.

Acceleration. Plane B has both tangential and normal components of acceleration, since it is flying along a *curved path*. From Eq. 12–20, the magnitude of the normal component is

$$(a_B)_n = \frac{v_B^2}{\rho} = \frac{(600 \text{ km/h})^2}{400 \text{ km}} = 900 \text{ km/h}^2$$

Applying the relative-acceleration equation, we have

$$\mathbf{a}_B = \mathbf{a}_A + \mathbf{a}_{B/A}$$

$$(900\mathbf{i} - 100\mathbf{j}) = 50\mathbf{j} + \mathbf{a}_{B/A}$$

Thus,

$$\mathbf{a}_{B/A} = \{900\mathbf{i} - 150\mathbf{j}\} \text{ km/h}^2$$

The magnitude and direction of $\mathbf{a}_{B/A}$ are therefore

$$a_{B/A} = 912 \text{ km/h}^2 \qquad \theta = \tan^{-1}\frac{150}{900} = 9.46° \quad \diagdown\theta \qquad \textit{Ans.}$$

The vector addition is shown in Fig. 12–45c.

Notice that the solution to this problem is possible using a translating frame of reference, since the pilot in plane A is "translating." Observation of plane A with respect to plane B, however, must be obtained using a *rotating* set of axes attached to plane B. (This assumes, of course, that the pilot of B does not turn his eyes to follow the motion of A.) The analysis for this case is given in Example 16–19.

Fig. 12–45

PROBLEMS

12–130. The cable and pulley system is used to lift the load A. Determine the displacement of the load if the end C of the cable is pulled 4 ft upward.

Prob. 12–130

12–131. Determine the constant speed at which the cable at A must be drawn in by the motor in order to hoist the load at B 15 ft in 5 s.

Probs. 12–131/12–132

*12–132. If the motor draws in the cable such that point A is subjected to an acceleration of $a_A = (3t^2)$ ft/s^2, where t is in seconds, determine how high the load at B travels in $t = 1.5$ s starting from rest.

12–133. The pulley system supports the three blocks. If A is moving downward with a speed of 5 ft/s while C is moving upward at 2 ft/s, determine the speed of block B.

Probs. 12–133/12–134

12–134. If block A of the pulley system is moving downward at 11 ft/s while block C is moving downward at 18 ft/s, determine the relative velocity of block B with respect to C.

71

12–135. The pipe P is being lifted using the cable and pulley system shown. If point A on the cable is being drawn toward the drum with a speed of 2 m/s, determine the speed of the pipe.

Probs. 12–135/12–136

*12–136. The pipe P is being lifted using the cable and pulley system shown. Starting from rest, point A on the cable is being drawn in with an acceleration $a_A = (5t^2)$ m/s^2, where t is in seconds. If $s = 0$ when $t = 0$, determine the height s of the pipe when $t = 2$ s.

12–137. The motor draws in the cable at C with a constant velocity of $v_C = 4$ m/s. The motor draws in the cable at D with a constant acceleration of $a_D = 8$ m/s^2. If $v_D = 0$ and $h = 3$ m when $t = 0$, determine (a) the time needed for $h = 0$, and (b) the relative velocity of block A with respect to block B when this occurs.

Prob. 12–137

12–138. The hoist is used to lift the load at D. If the end A of the chain is traveling downward at $v_A = 5$ ft/s and the end B is traveling upward at $v_B = 2$ ft/s, determine the velocity of the load at D.

Prob. 12–138

12–139. If the point A on the cable is moving upward at $v_A = 14$ m/s, determine the speed of block B.

Prob. 12–139

*12–140. The crate C is being lifted by moving the roller at A downward with a constant speed of $v_A = 2$ m/s along the guide. Determine the velocity and acceleration of the crate at the instant $s = 1$ m. When the roller is at B, the crate rests on the ground. Neglect the size of the pulley in the calculation. *Hint:* Relate the coordinates x_C and x_A using the problem geometry, then take the first and second time derivatives.

Prob. 12–140

12–141. The cable at A is drawn in at a constant rate $v_A = 6$ m/s. Determine the speed at which the load at B rises as a function of its vertical position y.

Probs. 12–141/12–142

12–142. If the cable at A is drawn in with an acceleration $a_A = 2$ m/s² when it has a speed $v_A = 6$ m/s, determine the velocity and acceleration of the load at B at this instant, which occurs when $y = 3$ m.

12–143. The roller at A is moving upward with a velocity of $v_A = 3$ ft/s and has an acceleration of $a_A = 4$ ft/s², when $s_A = 4$ ft. Determine the velocity and acceleration of block B at this instant.

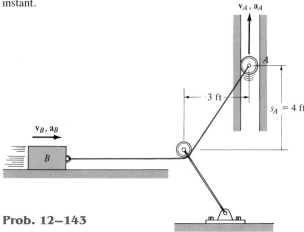

Prob. 12–143

*12–144. The truck at B is used to hoist the cabinet up to the fourth floor of a building using the rope and pulley arrangement shown. If the truck is moving forward at a constant speed of $v_B = 2$ ft/s, determine the speed at which the cabinet rises at the instant $s_A = 40$ ft. Neglect the size of the pulley. When $s_B = 0$, $s_A = 0$, so that points A and B are coincident, i.e., the rope is 100 ft long.

Probs. 12–144/12–145

12–145. Determine the acceleration of the cabinet when $s_A = 40$ ft in Prob. 12–144 if the truck is accelerating at $a_B = 0.6$ ft/s² and it has a speed of $v_B = 2$ ft/s when $s_A = 40$ ft.

73

12–146. Planes A and B are flying at the same altitude. If their velocities are $v_A = 300$ mi/h and $v_B = 250$ mi/h when the angle between their straight-line courses is 35° as shown, determine the velocity of plane A with respect to plane B.

Probs. 12–146/12–147

12–147. At a given instant the planes A and B are located as shown. If they maintain their straight-line courses and speeds, determine the distance between them in $t = 5$ min. Plane A is traveling at $v_A = 300$ mi/h and plane B is traveling at $v_B = 250$ mi/h.

12–148. Two planes, A and B, are flying at the same altitude. If their velocities are $v_A = 600$ km/h and $v_B = 500$ km/h and the angle between their straight-line courses is $\theta = 75°$, determine the velocity of plane B with respect to plane A.

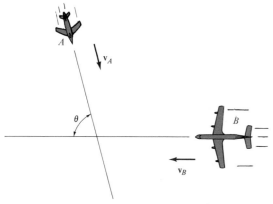

Prob. 12–148

12–149. Two boats leave the shore at the same time and travel in the directions shown. If $v_A = 20$ ft/s and $v_B = 15$ ft/s, determine the speed of boat A with respect to boat B. How long after leaving the shore will the boats be 800 ft apart?

Probs. 12–149/12–150

12–150. Each boat begins from rest at point O and heads in the direction shown at the same instant. If A has an acceleration of $a_A = 2$ ft/s^2, and B has an acceleration of $a_B = 3$ ft/s^2, determine the speed of boat A with respect to boat B at the instant they become 800 ft apart. How long will this take?

12–151. At the instant shown, the bicyclist at A is traveling at 7 m/s around the curve on the race track while increasing his speed at 0.5 m/s^2. The bicyclist at B is traveling at 8.5 m/s along the straight-a-way and increasing his speed at 0.7 m/s^2. Determine the relative velocity and relative acceleration of A with respect to B at this instant.

Prob. 12–151

*12–152. A passenger in an automobile observes that raindrops make an angle of 30° with the horizontal as the auto travels forward with a speed of 60 km/h. Compute the terminal (constant) velocity $\mathbf{v}_r$ of the rain if it is assumed to fall vertically.

Prob. 12–152

12–153. The boat can travel with a speed of 16 km/h in still water. The point of destination is located along the dashed line. If the water is moving at 4 km/h, determine the bearing angle θ at which the boat must travel to stay on course.

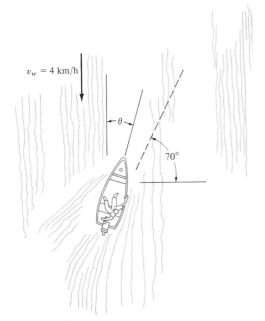

Prob. 12–153

12–154. Cars A and B are traveling around the circular race track. At the instant shown, A has a speed of 90 ft/s and is increasing its speed at the rate of 15 ft/s², whereas B has a speed of 105 ft/s and is decreasing its speed at 25 ft/s². Determine the relative velocity and relative acceleration of car A with respect to car B at this instant.

Prob. 12–154

12–155. At a given instant the football player at A throws a football C with a velocity of 20 m/s in the direction shown. Determine the constant speed at which the player at B must run so that he can catch the football at the same elevation at which it was thrown. Also calculate the relative velocity and relative acceleration of the football with respect to B at the instant the catch is made. Player B is 15 m away from A when A starts to throw the football.

Prob. 12–155

13

Kinetics of a Particle: Force and Acceleration

In Chapter 12 we developed the methods needed to formulate the acceleration of a particle in terms of its velocity and position. In this chapter we will use these concepts when applying Newton's second law of motion, $\mathbf{F} = m\mathbf{a}$, to study the effects caused by an unbalanced force acting on a particle. Depending upon the geometry of the path, the analysis of problems will be performed using rectangular, normal and tangential, or cylindrical coordinates. In the last part of the chapter, Newton's second law of motion will be used to study problems in space mechanics.

13.1 Newton's Laws of Motion

Many of the earlier notions about dynamics were dispelled after 1590 when Galileo performed experiments to study the motions of pendulums and falling bodies. The conclusions drawn from these experiments gave some insight as to the effects of forces acting on bodies in motion. The general laws of motion of a body subjected to forces had not been known, however, until 1687, when Isaac Newton first presented three basic laws governing the motion of a particle. In a slightly reworded form, Newton's three laws of motion can be stated as follows:

First Law. A particle originally at rest, or moving in a straight line with a constant velocity, will remain in this state provided the particle is not subjected to an unbalanced force.

Second Law. A particle acted upon by an unbalanced force **F** experiences an acceleration **a** that has the same direction as the force and a magnitude that is directly proportional to the force.*

Third Law. The mutual forces of action and reaction between two particles are equal, opposite, and collinear.

The first and third laws were used extensively in developing the concepts of statics. Although these laws are also considered in dynamics, Newton's second law of motion forms the basis for most of this study, since this law relates the accelerated motion of a particle to the forces that act on it. It should be noted that statics is a special case of dynamics, since Newton's second law yields the results of his first law when the resultant force is equal to zero; namely, no acceleration occurs, and therefore the particle's velocity is constant.

Measurements of force and acceleration can be recorded in a laboratory so that in accordance with the second law, if a known unbalanced force F_1 is applied to a particle, the acceleration a_1 of the particle may be measured. Since the force and acceleration are directly proportional, the constant of proportionality, m, may be determined from the ratio $m = F_1/a_1$. Provided the units of measurement are consistent, a different unbalanced force F_2 applied to the particle will create an acceleration a_2, such that $F_2/a_2 = m$.† In both cases the ratio will be the same and the acceleration and the force, both being vector quantities, will have the same direction. The scalar m is called the *mass* of the particle. Being constant during any acceleration, m provides a quantitative measure of the resistance of the particle to a change in its velocity.

If the mass of the particle is m, Newton's second law of motion may be written in mathematical form as

$$\mathbf{F} = m\mathbf{a}$$

This equation, which is referred to as the *equation of motion,* is one of the most important formulations in mechanics.‡ As stated above, its validity has been based solely upon *experimental evidence*. In 1905, however, Albert

*Stated another way, the unbalanced force acting on the particle is proportional to the time rate of change of the particle's linear momentum. See footnote ‡ below.

†Throughout this discussion, the units of force, mass, length, and time used to measure F, m, and a must be chosen such that the units for one of these quantities are defined in terms of all the others. Such is the case for the SI and FPS units, where it may be recalled that N = kg · m/s² and slug = lb · s²/ft (see Sec. 1.3 of *Statics*). Note, however, that if units of force, mass, length, and time are *all* selected arbitrarily, then $F = kma$ and k (a dimensionless constant) would have to be experimentally determined in order to preserve the equality.

‡Since m is constant, we can also write $\mathbf{F} = \dfrac{d}{dt}(m\mathbf{v})$, where $m\mathbf{v}$ is the particle's linear momentum.

Einstein developed the theory of relativity and placed limitations on the use of Newton's second law for describing general particle motion. Through experiments it was proven that *time* is not an absolute quantity as assumed by Newton; and as a result, the equation of motion fails to predict the exact behavior of a particle, especially when the particle's speed approaches the speed of light [0.3 Gm/s]. Developments of the theory of quantum mechanics by Schrödinger and others indicate further that conclusions drawn from using this equation are also invalid when particles move within an atomic distance of one another. For the most part, however, these requirements regarding particle speed and size are not encountered in engineering problems, so that their effects will not be considered in this book.

Newton's Law of Gravitational Attraction

Shortly after formulating his three laws of motion, Newton postulated a law governing the mutual attraction between any two particles. In mathematical form this law can be expressed as

$$F = G\frac{m_1 m_2}{r^2} \tag{13-1}$$

where F = force of attraction between the two particles

   G = universal constant of gravitation; according to experimental evidence $G = 66.73(10^{-12})$ m^3/(kg · s^2)

m_1, m_2 = mass of each of the two particles

   r = distance between the centers of the two particles

Any two particles or bodies have a mutually attractive gravitational force acting between them. In the case of a particle located at or near the surface of the earth, however, the only gravitational force having any sizable magnitude is that between the earth and the particle. This force is termed the "weight" and, for our purpose, it will be the only gravitational force considered.

Mass and Weight

Mass is a property of matter by which we can compare the action of one body with that of another. As indicated above, this property manifests itself as a gravitational attraction between two bodies and provides a quantitative measure of the resistance of matter to a change in velocity. It is an *absolute* quantity since the measurement of mass can be made at any location. The weight of a body, however, is *not absolute* since it is measured in a gravitational field, and hence its magnitude depends upon where the measurement is made. From Eq. 13–1, we can develop a general expression for finding the weight W of a particle having a mass $m = m_1$. Let m_2 be the mass of the earth and r the distance between the earth's center and the particle. Then if $g = Gm_2/r^2$, we have

$$W = mg$$

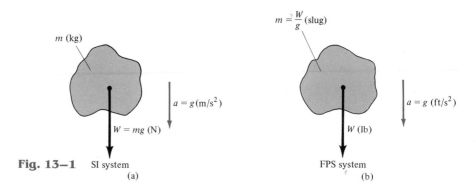

Fig. 13–1 SI system
(a)

FPS system
(b)

By comparison with $F = ma$, we term g the acceleration due to gravity. For most engineering calculations we measure g at a point on the surface of the earth at sea level, and at a latitude of 45°, considered the "standard location."

The mass and weight of a body are measured differently in the SI and FPS systems of units, and the method of defining these units should be thoroughly understood.

SI System of Units. In the SI system the mass of the body is specified in kilograms and the weight must be calculated using the equation of motion, $F = ma$. Hence, if a body has a mass m (kg) and is located at a point where the acceleration due to gravity is g (m/s²), then the weight is expressed in *newtons* as $W = mg$ (N), Fig. 13–1a. In particular, if the body is located at the "standard location," the acceleration due to gravity is $g = 9.806\ 65$ m/s². For calculations, the value $g = 9.81$ m/s² will be used, so that

$$\boxed{W = mg\ \text{(N)}\quad (g = 9.81\ \text{m/s}^2)}\qquad (13\text{–}2)$$

Therefore, a body of mass 1 kg has a weight of 9.81 N; a 2-kg body weighs 19.62 N; and so on.

FPS System of Units. In the FPS system the weight of the body is specified in pounds and the mass must be calculated from $F = ma$. Hence, if a body has a weight W (lb) and is located at a point where the acceleration due to gravity is g (ft/s²), then the mass is expressed in *slugs* as $m = W/g$ (slug), Fig. 13–1b. Since the acceleration of gravity at the standard location is approximately 32.2 ft/s²(= 9.81 m/s²), the mass of the body measured in slugs is

$$\boxed{m = \frac{W}{g}\text{(slug)}\quad (g = 32.2\ \text{ft/s}^2)}\qquad (13\text{–}3)$$

Therefore, a body weighing 32.2 lb has a mass of 1 slug; a 64.4-lb body has a mass of 2 slugs; and so on.

13.2 The Equation of Motion

When more than one force acts on a particle, the resultant force is determined by a vector summation of all the forces; i.e., $\mathbf{F}_R = \Sigma\mathbf{F}$. For this more general case, the equation of motion may be written as

$$\boxed{\Sigma\mathbf{F} = m\mathbf{a}} \qquad (13\text{–}4)$$

To illustrate application of this equation, consider the particle P shown in Fig. 13–2a, which has a mass m and is subjected to the action of two forces, $\mathbf{F}_1$ and $\mathbf{F}_2$. We can graphically account for the magnitude and direction of each force acting on the particle by drawing the particle's *free-body diagram*, Fig. 13–2b. Since the *resultant* of these forces *produces* the vector $m\mathbf{a}$, its magnitude and direction can be represented graphically on the *kinetic diagram*, shown in Fig. 13–2c. The equal sign written between the diagrams symbolizes the *graphical* equivalency between the free-body diagram and the kinetic diagram; i.e., $\Sigma\mathbf{F} = m\mathbf{a}$.* In particular, note that if $\mathbf{F}_R = \Sigma\mathbf{F} = \mathbf{0}$, then the acceleration is also zero, so that the particle will either remain at *rest* or move along a straight-line path with *constant velocity*. Such are the conditions of *static equilibrium*, Newton's first law of motion.

Inertial Frame of Reference

Whenever the equation of motion is applied, it is required that measurements of the acceleration be made from a *Newtonian* or *inertial frame of reference*. Such a *coordinate system does not rotate and is either fixed or translates in a given direction with a constant velocity (zero acceleration)*. This definition ensures that the particle's *acceleration* measured by observers in two different inertial frames of reference will always be the *same*. For example, consider the particle P moving with an acceleration $\mathbf{a}_P$ along a straight path as shown in

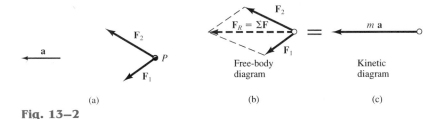

(a)   (b)   (c)

Fig. 13–2

---

*The equation of motion can also be rewritten in the form $\Sigma\mathbf{F} - m\mathbf{a} = \mathbf{0}$. The vector $-m\mathbf{a}$ is referred to as the *inertia force vector*. If it is treated in the same way as a "force vector," then the state of "equilibrium" created is referred to as *dynamic equilibrium*. This method for application of the equation of motion is often referred to as the *D'Alembert principle*, named after the French mathematician Jean le Rond d'Alembert (1717–1783).

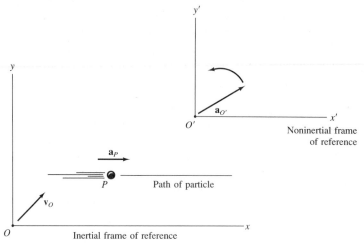

Fig. 13–3

Fig. 13–3. If the observer is *fixed* in the inertial x, y frame of reference, this acceleration, $\mathbf{a}_P$, will be measured by the observer regardless of the direction and magnitude of the velocity $\mathbf{v}_O$ of the frame of reference. On the other hand, if the observer is *fixed* in the noninertial x', y' frame of reference, Fig. 13–3, the observer will not measure the particle's acceleration as $\mathbf{a}_P$. Instead, if the frame is *accelerating* at $\mathbf{a}_{O'}$ the particle will appear to have an acceleration of $\mathbf{a}_{P/O'} = \mathbf{a}_P - \mathbf{a}_{O'}$. Also, if the frame is *rotating*, as indicated by the curl, then the particle will appear to move along a *curved path*, in which case it must have other components of acceleration (see Sec. 16–8). In any case, the measured acceleration cannot be used in Newton's law of motion to determine the forces acting on the particle.

When studying the motions of rockets and satellites it is justifiable to consider the inertial reference frame as fixed to the stars, whereas dynamics problems concerned with motions on or near the surface of the earth may be solved by using an inertial frame which is assumed fixed to the earth. Even though the earth rotates about its own axis and revolves about the sun, the acceleration created by these rotations can be neglected in most computations.

13.3 Equation of Motion for a System of Particles

The equation of motion will now be extended to include a system of n particles isolated within an enclosed region in space, as shown in Fig. 13–4a. In particular, there is no restriction in the way the particles are connected, and as a result the following analysis will apply equally well to the motion of a solid, liquid, or gas system. At the instant considered, the arbitrary ith particle, having a mass m_i, is subjected to a set of internal forces and a resultant external force. The *resultant internal force,* represented symbolically as $\mathbf{f}_i = $

$\displaystyle\sum_{j=1(j\neq i)}^{n} \mathbf{f}_{ij}$, is determined from the forces which the other particles exert on the

ith particle. Although only one of these particles is shown in Fig. 13–4a, the summation extends over all n particles within the dashed boundary. Note that it is meaningless for $j = i$ since the ith particle cannot exert a force on itself. The *resultant external force* $\mathbf{F}_i$ represents, for example, the effect of gravitational, electrical, magnetic, or contact forces between adjacent bodies or particles *not* included within the system.

The free-body and kinetic diagrams for the ith particle are shown in Fig. 13–4b. Applying the equation of motion yields

$$\Sigma\mathbf{F} = m\mathbf{a}; \qquad \mathbf{F}_i + \mathbf{f}_i = m_i\mathbf{a}_i$$

When the equation of motion is applied to each of the other particles of the system, similar equations will result. If all these equations are added together *vectorially,* we obtain

$$\Sigma\mathbf{F}_i + \Sigma\mathbf{f}_i = \Sigma m_i\mathbf{a}_i$$

The summation of the internal forces, if carried out, will equal zero since the internal forces between particles all occur in equal, but opposite collinear pairs. Consequently, only the sum of the external forces will remain and therefore the equation of motion, written for the system of particles, becomes

$$\Sigma\mathbf{F}_i = \Sigma m_i\mathbf{a}_i \qquad (13\text{–}5)$$

If $\mathbf{r}_G$ is a position vector which locates the *center of mass G* of the particles, Fig. 13–4a, then by definition of the center of mass $m\mathbf{r}_G = \Sigma m_i\mathbf{r}_i$, where $m = \Sigma m_i$ is the total mass of all the particles. Differentiating this equation twice with respect to time, assuming no mass is entering or leaving the system,* yields

$$m\mathbf{a}_G = \Sigma m_i\mathbf{a}_i$$

Substituting this result into the above equation, we obtain

$$\boxed{\Sigma\mathbf{F} = m\mathbf{a}_G} \qquad (13\text{–}6)$$

This equation states that the sum of the external forces acting on the system of particles is equal to the mass $m = \Sigma m_i$ of a single "fictitious" particle times its acceleration. This fictitious particle is located at the center of mass G of all the particles.

Since, in reality, all particles must have finite size to possess mass, Eq. 13–6 justifies application of the equation of motion to a *body* that is represented as a single particle.

*A case in which m is a function of time (variable mass) is discussed in Sec. 15.9.

Inertial coordinate system

(a)

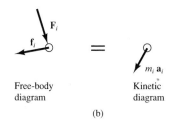

Free-body diagram

Kinetic diagram

(b)

Fig. 13–4

13.4 Equations of Motion: Rectangular Coordinates

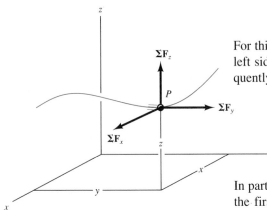

Fig. 13–5

When a particle is moving relative to an inertial x, y, z frame of reference, the forces acting on the particle, as well as its acceleration, may be expressed in terms of their **i, j, k** components, Fig. 13–5. Applying the equation of motion, we have

$$\Sigma \mathbf{F} = m\mathbf{a}$$

$$\Sigma F_x \mathbf{i} + \Sigma F_y \mathbf{j} + \Sigma F_z \mathbf{k} = m(a_x \mathbf{i} + a_y \mathbf{j} + a_z \mathbf{k})$$

For this equation to be satisfied, the respective **i, j,** and **k** components on the left side must equal the corresponding components on the right side. Consequently, we may write the following three scalar equations:

$$\boxed{\begin{aligned} \Sigma F_x &= ma_x \\ \Sigma F_y &= ma_y \\ \Sigma F_z &= ma_z \end{aligned}} \qquad (13\text{–}7)$$

In particular, if the particle is constrained to move in the x-y plane, then only the first two of these equations are used to specify the motion.

PROCEDURE FOR ANALYSIS

The equations of motion are used to solve problems which require a relationship between the forces and the accelerated motion they cause. Whenever they are applied, the unknown force and acceleration components should be identified and an equivalent number of equations should be written. If further equations are necessary for the solution, kinematics may be considered. Recall that these equations only relate the geometric properties of the motion, and therefore they become useful if the particle's position or velocity is to be related to its acceleration. Whatever the situation, the following procedure provides a general method for solving problems in kinetics.

Free-Body Diagram. Select the inertial coordinate system. Most often, rectangular or x, y, z coordinates are chosen to analyze problems for which the particle has *rectilinear motion*. If this occurs, one of the axes should extend in the direction of motion. Once the coordinates are established, draw the particle's free-body diagram. Drawing the free-body diagram is *very important* since it provides a graphical representation of *all the forces* ($\Sigma \mathbf{F}$) which act on the particle and thereby makes it possible to resolve these forces into their x, y, z components. The direction and sense of the particle's acceleration **a** should also be established. If the sense of its components is unknown, for mathematical convenience assume that they are in the same *direction* as the *positive* inertial coordinate axes. The acceleration may be sketched on the x, y, z coordinate system, *but not on* the free-body diagram, or it may be represented as the $m\mathbf{a}$ vector on the kinetic diagram.* Once the free-body diagram has been constructed, identify the unknowns under consideration.

Equations of Motion. If the forces can be resolved directly from the free-body diagram, apply the equations of motion in their scalar component form. If the geometry of the problem appears complicated, which often occurs in three dimensions, Cartesian vector analysis can be used for the solution.

Friction If the particle contacts a rough surface, it may be necessary to use the *frictional equation,* which relates the coefficient of kinetic friction μ_k to the magnitudes of the frictional and normal forces $\mathbf{F}_f$ and $\mathbf{N}$ acting at the surfaces of contact, i.e., $F_f = \mu_k N$.

Spring If the particle is connected to an *elastic spring* having negligible mass, the spring force F_s can be related to the deformation of the spring by the equation $F_s = ks$. Here k is the spring's stiffness measured as a force per unit length, and s is the stretch or compression defined as the difference between the deformed length l and the undeformed length l_0, i.e., $s = l - l_0$.

Kinematics. As stated above, if a complete solution cannot be obtained strictly from the equation of motion, the equations of kinematics may be considered. For example, if the velocity or position of the particle is to be found, it will be necessary to apply the proper kinematic equations once the particle's acceleration is determined from $\Sigma\mathbf{F} = m\mathbf{a}.$

If the *acceleration is a function of time,* use $a = dv/dt$ and $v = ds/dt$ which, when integrated, yield the particle's velocity and position.

If the *acceleration is a function of displacement,* integrate $a\, ds = v\, dv$ to obtain the velocity as a function of position.

If the *acceleration is constant,* use

$$v = v_0 + a_c t \qquad s = s_0 + v_0 t + \tfrac{1}{2}a_c t^2 \qquad v^2 = v_0^2 + 2a_c(s - s_0)$$

to determine the position or velocity of the particle.

---

*It is a convention in this text to always use the kinetic diagram as a graphical aid when developing the proofs and theory.

If the problem involves the dependent motion of several particles, use the method outlined in Sec. 12.8 to relate their accelerations.

In all cases, make sure the positive coordinate directions used for writing the kinematic equations are the same as those used for writing the equations of motion; otherwise, simultaneous solution of the equations will result in errors. Also, if the solution for an unknown vector component yields a negative scalar, it indicates that the component acts in the direction opposite to that which was assumed.

The following examples numerically illustrate application of this procedure.

Example 13–1

The 50-kg crate shown in Fig. 13–6a rests on a horizontal plane for which the coefficient of kinetic friction is $\mu_k = 0.3$. If the crate is subjected to a 400-N towing force as shown, determine the velocity of the crate in 5 s starting from rest.

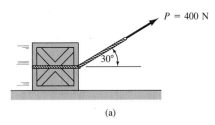

$P = 400$ N

30°

(a)

SOLUTION

This problem requires application of the equations of motion, since the crate's acceleration can be related to the force causing the motion. The crate's velocity can then be determined using kinematics.

Free-Body Diagram. The weight of the crate is $W = mg = $ 50 kg (9.81 m/s^2) = 490.5 N. As shown in Fig. 13–6b, the frictional force has a magnitude $F = \mu_k N_C$ and acts to the left, since it opposes the motion of the crate. The acceleration **a** is assumed to act horizontal, in the positive x direction. There are two unknowns, namely N_C and a.

Equations of Motion. Using the data shown on the free-body diagram, we have

$$\xrightarrow{+}\Sigma F_x = ma_x; \qquad 400 \cos 30° - 0.3\,N_C = 50a \qquad (1)$$

$$+\uparrow \Sigma F_y = ma_y; \quad N_C - 490.5 + 400 \sin 30° = 0 \qquad (2)$$

Solving Eq. (2) for N_C, substituting the result into Eq. (1), and solving for a yields

$$N_C = 290.5 \text{ N}$$

$$a = 5.19 \text{ m/s}^2$$

Kinematics. Since the acceleration is *constant*, and the initial velocity is zero, the velocity of the crate in 5 s is

$(\xrightarrow{+})$

$$v = v_0 + a_c t$$

$$= 0 + 5.19(5)$$

$$= 26.0 \text{ m/s} \rightarrow \qquad\qquad\qquad Ans.$$

Note the alternative procedure of drawing the crate's free-body *and* kinetic diagrams, Fig. 13–6c, prior to applying the equations of motion.

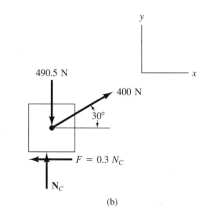

y

x

490.5 N

400 N

30°

$F = 0.3\,N_C$

N_C

(b)

490.5 N

400 N

30°

$F = 0.3\,N_C$

N_C

$=$

50**a**

(c)

Fig. 13–6

Example 13–2

A 10-kg projectile is fired vertically upward from the ground, with an initial velocity of 50 m/s, Fig. 13–7a. Determine the maximum height to which it will travel if (a) atmospheric resistance is neglected; and (b) atmospheric resistance is measured as $F_D = (0.01v^2)$ N, where v is the speed at any instant, measured in m/s.

(a)

SOLUTION

In both cases the known force on the projectile can be related to its acceleration using the equation of motion. Kinematics can then be used to relate the projectile's acceleration to its position.

Part (a) Free-Body Diagram. As shown in Fig. 13–7b, the projectile's weight is $W = mg = 10(9.81) = 98.1$ N. We assume the unknown acceleration **a** acts upward in the *positive z* direction.

Equation of Motion

$+\uparrow \Sigma F_z = ma_z;$ $\qquad -98.1 = 10a,$ $a = -9.81$ m/s^2

The result indicates that the projectile, like every object having free-flight motion near the earth's surface, is subjected to a constant downward acceleration of 9.81 m/s^2.

98.1 N

(b)

Kinematics. Initially, $z_0 = 0$ and $v_0 = 50$ m/s, and at the maximum height $z = h$, $v = 0$. Since the acceleration is *constant*, then

$(+\uparrow)$ $\qquad\qquad v^2 = v_0^2 + 2a_c(z - z_0)$

$\qquad\qquad\qquad 0 = (50)^2 + 2(-9.81)(h - 0)$

$\qquad\qquad\qquad h = 127$ m $\qquad\qquad\qquad$ *Ans.*

Part (b) Free-Body Diagram. Since the force $F_D = (0.01v^2)$ N tends to retard the upward motion of the projectile, it acts downward as shown on the free-body diagram, Fig. 13–7c.

*Equation of Motion**

$+\uparrow \Sigma F_z = ma_z;$ $\qquad -0.01v^2 - 98.1 = 10a$

$\qquad\qquad\qquad\qquad a = -0.001v^2 - 9.81$

F_D

98.1 N

(c)

Fig. 13–7

Kinematics. Here the acceleration is *not constant;* however, it can be related to the velocity and displacement by using $a\,dz = v\,dv$.

$(+\uparrow)\ a\,dz = v\,dv;$ $\quad (-0.001v^2 - 9.81)dz = v\,dv$

Separating the variables and integrating, realizing that initially $z_0 = 0$, $v_0 = 50$ m/s (positive upward), and at $z = h$, $v = 0$, we have

$$\int_0^h dz = -\int_{50}^0 \frac{v\,dv}{0.001v^2 + 9.81}$$

$$h = -500 \ln (v^2 + 9810)\Big|_{50}^0 = 113 \text{ m} \qquad Ans.$$

The answer indicates a lower elevation than that obtained in Part (a). Why?

*Note that if the projectile were fired downward, with z positive downward, the equation of motion would then be $-0.01v^2 + 98.1 = 10a$.

Example 13–3

The crate shown in Fig. 13–8a has a weight of 50 lb and is acted upon by a force having a variable magnitude $P = 20t$, where P is in pounds and t is in seconds. Compute the crate's velocity 2 s after **P** has been applied. The crate's initial velocity is $v_0 = 3$ ft/s down the plane, and the coefficient of kinetic friction between the crate and the plane is $\mu_k = 0.3$.

(a)

SOLUTION

Before reading further, do you see *why* it is necessary to apply both the equations of motion *and* kinematics to obtain a solution of this problem?

Free-Body Diagram. As shown in Fig. 13–8b, the frictional force is directed opposite to the crate's sliding motion and has a magnitude $F = \mu_k N_C = 0.3 N_C$. The mass is $m = W/g = 50/32.2 = 1.55$ slug. At this point, there are two unknowns, namely N_C and a.

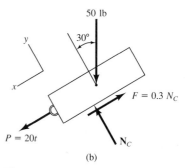

(b)

Fig. 13–8

Equations of Motion. Using the data shown on the free-body diagram, we have

$$+\swarrow \Sigma F_x = ma_x; \quad 20t - 0.3N_C + 50 \sin 30° = 1.55a \quad (1)$$
$$+\nwarrow \Sigma F_y = ma_y; \quad N_C - 50 \cos 30° = 0 \quad (2)$$

Solving for N_C in Eq. (2) ($N_C = 43.3$ lb), substituting into Eq. (1), and simplifying yields

$$a = 12.88t + 7.73 \quad (3)$$

Kinematics. Since the acceleration is a function of time, the velocity of the crate is obtained by using $a = dv/dt$ with the initial condition that $v_0 = 3$ ft/s at $t = 0$. We have

$(+\swarrow)$
$$dv = a\,dt$$
$$\int_3^v dv = \int_0^t (12.88t + 7.73)dt$$
$$v = 6.44t^2 + 7.73t + 3$$

When $t = 2$ s,

$$v = 44.2 \text{ ft/s} \swarrow \qquad\qquad Ans.$$

Example 13–4

A smooth 2-kg collar C, shown in Fig. 13–9a, is attached to a spring having a stiffness $k = 3$ N/m and an unstretched length of 0.75 m. If the collar is released from rest at A, determine its acceleration and the normal force of the rod on the collar at the instant $y = 1$ m.

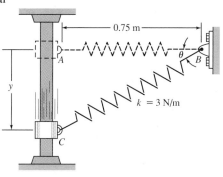

(a)

SOLUTION

Free-Body Diagram. The free-body diagram of the collar when it is located at the arbitrary position y is shown in Fig. 13–9b. Note that the weight is $W = 2(9.81) = 19.62$ N. Furthermore, the collar is *assumed* to be accelerating so that '**a**' acts downward in the *positive y* direction. There are four unknowns, namely N_C, F_s, a, and θ.

(b)

Fig. 13–9

Equations of Motion. Using the data in Fig. 13–9b,

$$\xrightarrow{+} \Sigma F_x = ma_x; \qquad -N_C + F_s \cos \theta = 0 \qquad (1)$$
$$+ \downarrow \Sigma F_y = ma_y; \qquad 19.62 - F_s \sin \theta = 2a \qquad (2)$$

From Eq. (2) it is seen that the acceleration is not constant; rather, it depends upon the magnitude and direction of the spring force. Solution is possible once F_s and θ are known.

The magnitude of the spring force is a function of the stretch s of the spring; i.e., $F_s = ks$. Here the unstretched length is $AB = 0.75$ m, Fig. 13–9a; therefore, $s = CB - AB = \sqrt{y^2 + (0.75)^2} - 0.75$. Since $k = 3$ N/m, then

$$F_s = ks = 3(\sqrt{y^2 + (0.75)^2} - 0.75) \qquad (3)$$

From Fig. 13–9a, the angle θ is related to y by trigonometry.

$$\sin \theta = \frac{y}{\sqrt{y^2 + (0.75)^2}} \qquad (4)$$

Substituting $y = 1$ m into Eqs. (3) and (4) yields $F_s = 1.50$ N and $\theta = 53.1°$. Substituting these results into Eqs. (1) and (2), we obtain

$$N_C = 0.900 \text{ N} \qquad\qquad Ans.$$
$$a = 9.21 \text{ m/s}^2 \downarrow \qquad\qquad Ans.$$

Example 13–5

The 100-kg block A shown in Fig. 13–10a is released from rest. If the mass of the pulleys and the cord is neglected, determine the speed of the 20-kg block B in 2 s.

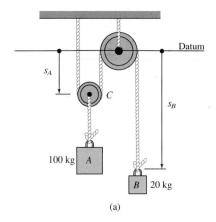

(a)

SOLUTION

Motion of blocks A and B will be analyzed separately.

Free-Body Diagram. Since the mass of the pulleys is neglected, the equilibrium condition for pulley C is shown in Fig. 13–10b. The free-body diagrams for blocks A and B are shown in Fig. 13–10c and d, respectively. Here we *assume* both blocks accelerate downward, in the direction of $+s_A$ and $+s_B$. What are the three unknowns?

Equations of Motion

Block A (Fig. 13–10c):

$$+\downarrow \Sigma F_y = ma_y; \qquad 981 - 2T = 100a_A \qquad (1)$$

Block B (Fig. 13–10d):

$$+\downarrow \Sigma F_y = ma_y; \qquad 196.2 - T = 20a_B \qquad (2)$$

(b)

Kinematics. The necessary third equation is obtained by studying the kinematics of the pulley arrangement in order to relate a_A to a_B. Using the technique developed in Sec. 12.8, the coordinates s_A and s_B measure the positions of A and B from the fixed datum, Fig. 13–10a. It is seen that

$$2s_A + s_B = l$$

where l is constant and represents the total vertical length of cord. Differentiating this expression twice with respect to time yields

$$2a_A = -a_B \qquad (3)$$

Hence when block A accelerates *downward*, block B accelerates *upward* at twice the amount. Notice, however, that in writing Eqs. (1) to (3), the *positive direction was always assumed downward*. It is important to be *consistent* in this assumption since we are seeking a simultaneous solution of equations. The solution yields

$$T = 327.0 \text{ N}$$
$$a_A = 3.27 \text{ m/s}^2$$
$$a_B = -6.54 \text{ m/s}^2$$

(c)

Since a_B is constant, the velocity of block B in 2 s is thus

$$(+\downarrow) \qquad v = v_0 + a_B t$$
$$= 0 + (-6.54)(2)$$
$$= -13.1 \text{ m/s} \qquad \textit{Ans.}$$

Fig. 13–10

The negative sign indicates that block B is moving upward. Why?

(d)

PROBLEMS

13–1. Venus and Earth have diameters of 12,400 km and 12,760 km, respectively. The mass of Venus is 0.815 times that of the Earth. If a body has a mass of 35 kg on the surface of the Earth, what would its weight be on Venus? Also, what is the acceleration of gravity on Venus?

13–2. The moon has a mass of $73.5(10^{21})$ kg and the earth's mass is $5.98(10^{24})$ kg. If their centers are 384 Mm apart, determine the gravitational attractive force between the two bodies.

13–3. Determine the gravitational attraction between the earth, which has a mass of $5.98(10^{24})$ kg, and the sun, which has a mass of $1.99(10^{30})$ kg. The distance between their mass centers is $150(10^6)$ km.

*13–4. The baggage truck A has a mass of 800 kg and is used to pull each of the 300-kg cars. Determine the tension force in the couplings at B and C if the tractive force $\mathbf{F}$ on the truck is $F = 480$ N. What is the speed of the truck when $t = 2$ s, starting from rest? The car wheels are free to roll. Neglect the mass of the wheels.

Probs. 13–4/13–5

13–5. The baggage truck A has a mass of 800 kg and is used to pull each of the 300-kg cars. If the tractive force $\mathbf{F}$ on the truck is $F = 480$ N, determine the initial acceleration of the truck. What is the acceleration of the truck if the coupling at C suddenly fails? The car wheels are free to roll. Neglect the mass of the wheels.

13–6. Two men stand on ice and hold the ends of a rope. Man A weighs 150 lb and man B weighs 200 lb. If A pulls horizontally on the rope at 30 lb, determine the force which the rope exerts on man B and the acceleration of each man. Neglect friction.

13–7. The elevator E has a mass of 500 kg and the counterweight at A has a mass of 150 kg. If the motor supplies a constant force of 5 kN on the cable at B, determine the speed of the elevator in $t = 3$ s starting from rest. Neglect the mass of the pulleys and cable.

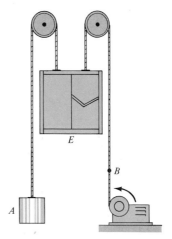

Probs. 13–7/13–8

*13–8. The elevator E has a mass of 500 kg and the counterweight at A has a mass of 150 kg. If the elevator attains a speed of 10 m/s after it rises 40 m, determine the force developed in the cable at B. Neglect the mass of the pulleys and cable.

13–9. Block B has a mass m and is released from rest when it is on top of cart A, which has a mass of $3\ m$. Determine the tension in cord CD needed to hold the cart from moving while B is sliding down A. Neglect friction.

Probs. 13–9/13–10

13–10. Block B has a mass m and is released from rest when it is on top of cart A, which has a mass of $3\ m$. Determine the tension in cord CD needed to hold the cart from moving while B is sliding down A. The coefficient of kinetic friction between A and B is μ.

13–11. At a given instant the 5-lb block A is moving downward with a speed of 4 ft/s. Determine its speed 3 s later. Block B has a weight of 6 lb, and the coefficient of kinetic friction between it and the horizontal plane is $\mu_k = 0.3$. Neglect the mass of the pulleys and cord.

Prob. 13–11

*13–12. The 200-kg crate is suspended from the cable of a crane. Determine the force in the cable in $t = 2$ s, if the crate is moving upward with (a) a constant velocity of 2 m/s, and (b) a speed of $v = (0.2t^2 + 2)$ m/s, where t is in seconds.

Prob. 13–12

13–13. The man weighs 180 lb and supports the barbells which have a weight of 100 lb. If he lifts them 2 ft in the air in 1.5 s starting from rest, determine the reaction of *both* of his feet on the ground during the lift. Assume the motion is with uniform acceleration.

Prob. 13–13

13–14. The 400-lb cylinder at A is hoisted using the motor and the pulley system shown. If the speed of point B on the cable is increased at a constant rate from zero to $v_B = 10$ ft/s in $t = 5$ s, determine the tension in the cable at B to cause the motion.

Prob. 13–14

13–15. The speed of the 3500-lb sports car is plotted over the 30-s time period. Plot the variation of the traction force $\mathbf{F}$ needed to cause the motion.

Prob. 13–15

*13–16. Block A has a mass of 20 kg and block B has a mass of 50 kg. If the spring is stretched 0.2 m at the instant shown, determine the acceleration of block B at this instant if it is (a) originally at rest, (b) moving downward with a speed of 3 m/s. Neglect friction.

$k = 30$ N/m

Prob. 13–16

13–17. A force of $F = 15$ lb is applied to the cord. Determine how high the 30-lb block A rises in 2 s starting from rest. Neglect the weight of the pulleys and cord.

Probs. 13–17/13–18

13–18. Determine the constant force $\mathbf{F}$ which must be applied to the cord in order to cause the 30-lb block A to have a speed of 12 ft/s when it has been displaced 3 ft upward starting from rest. Neglect the weight of the pulleys and cord.

13–19. A freight elevator, including its load, has a mass of 500 kg. It is prevented from rotating by using the track and wheels mounted along its sides. If the motor M develops a constant tension $T = 1.50$ kN in its attached cable, determine the velocity of the elevator when it has moved upward 3 m starting from rest. Neglect the mass of the pulleys and cables.

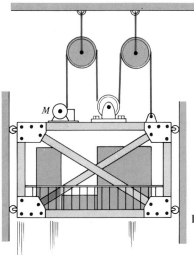

Probs. 13–19/13–20

*13–20. A freight elevator, including its load, has a mass of 500 kg. It is prevented from rotating by using the track and wheels mounted along its sides. Starting from rest, in $t = 2$ s, the motor M draws in the cable with a speed of 6 m/s, *measured relative to the elevator*. Determine the constant acceleration of the elevator and the tension in the cable. Neglect the mass of the pulleys and cables.

13–21. Determine the acceleration of the 5-kg cylinder A. Neglect the mass of the pulleys and cords. The block at B has a mass of 10 kg. Assume the surface at B is smooth.

Probs. 13–21/13–22

13–22. Determine the acceleration of the 5-kg cylinder A. Neglect the mass of the pulleys and cords. The block at B has a mass of 10 kg. The coefficient of kinetic friction between block B and the surface is $\mu_k = 0.1$.

13–23. The 40-lb suitcase slides from rest 20 ft down the smooth ramp. Determine the point where it strikes the ground at C. How long does it take to go from A to C?

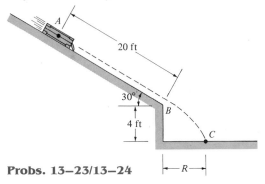

Probs. 13–23/13–24

*13–24. Solve Prob. 13–23 if the suitcase has an initial velocity down the ramp of $v_A = 10$ ft/s and the coefficient of kinetic friction along AB is $\mu_k = 0.2$.

13–25. Block *A* has a weight of 8 lb and block *B* has a weight of 6 lb. They rest on a surface for which the coefficient of kinetic friction is $\mu_k = 0.2$. If the spring has a stiffness of $k = 20$ lb/ft, and it is compressed 0.2 ft, determine the acceleration of each block just after they are released.

Prob. 13–25

13–26. The 100-kg crate is hoisted up the incline using the cable and motor *M*. For a short time, the force in the cable is $F = 800t^2$ N, where *t* is in seconds. If the crate has an initial velocity $v_1 = 2$ m/s when $t = 0$, determine its velocity when $t = 2$ s. The coefficient of kinetic friction between the crate and the incline is $\mu_k = 0.3$.

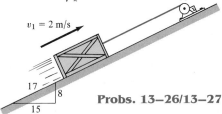

Probs. 13–26/13–27

13–27. The 100-kg crate is hoisted up the incline using the cable and motor *M*. For a short time, the force in the cable is $F = 800t^2$ N, where *t* is in seconds. If the crate has an initial velocity $v_1 = 2$ m/s at $s = 0$ and $t = 0$, determine the distance the crate moves up the plane when $t = 2$ s. The coefficient of kinetic friction between the crate and the incline is $\mu_k = 0.3$.

13–28. The crate has a mass of 80 kg and is being towed by a chain which is always directed at 20° from the horizontal as shown. If the magnitude of **T** is increased until the crate begins to slide, determine the crate's initial acceleration if the coefficient of static friction is $\mu_s = 0.5$ and the coefficient of kinetic friction is $\mu_k = 0.3$.

Probs. 13–28/13–29

13–29. The crate has a mass of 80 kg and is being towed by a chain which is always directed at 20° from the horizontal as shown. Determine the crate's acceleration in $t = 2$ s if the coefficient of static friction is $\mu_s = 0.4$ and the coefficient of kinetic friction is $\mu_k = 0.3$ and the towing force is $T = (90t^2)$ N, where *t* is in seconds.

13–30. A parachutist having a mass *m* is falling at v_0 when he opens his parachute at a very high altitude. If the atmospheric drag resistance is $F_D = kv^2$, where *k* is a constant, determine his velocity when he has fallen a distance *h*. What is his velocity when he lands on the ground? This velocity is referred to as the *terminal velocity*, which is found by letting the distance of fall $y \rightarrow \infty$.

Prob. 13–31

13–31. A parachutist having a mass *m* opens his parachute from an at-rest position at a very high altitude. If the atmospheric drag resistance is $F_D = kv^2$, where *k* is a constant, determine his velocity when he has fallen for a time *t*. What is his velocity when he lands on the ground? This velocity is referred to as the *terminal velocity*, which is found by letting the time of fall $t \rightarrow \infty$.

13–32. A car of mass *m*, is traveling at a slow velocity of v_0. If it is subjected to the drag resistance of the wind, which is proportional to its velocity, i.e., $F_D = kv$, determine the distance and the time the car will travel before its velocity becomes $0.5\,v_0$. Assume no other frictional forces act on the car.

Prob. 13–32

13–33. The device shown is used as a crash sensor in an automobile to activate the inflation of an air bag in the event of a sudden deceleration. During operation, the 0.8-kg smooth "sensing" ball B moves forward and pushes on the rod at A. As the rod rotates about C from the vertical position, the firing pin D is triggered and stabs a primer P, which then inflates the bag. If the spring on the firing pin is compressed 20 mm, determine the required compression of the lever bias spring E so that the mechanism is triggered when the car decelerates at 30 m/s². Neglect friction and the mass of the lever and the small deformation of the springs due to the slight rotation of the lever before firing.

Prob. 13–33

13–34. Each of the two blocks has a mass m. The coefficient of kinetic friction at all surfaces of contact is μ. If a horizontal force P is applied to the bottom block in each case, determine the acceleration of the bottom block.

(a)

(b)

Prob. 13–34

13–35. Each of the three plates has a mass of 10 kg. If the coefficients of static and kinetic friction at each surface of contact are $\mu_s = 0.3$ and $\mu_k = 0.2$, respectively, determine the acceleration of each plate at the instant the three horizontal forces are applied.

Prob. 13–35

*13–36. Blocks A and B each have a mass m. Determine the largest horizontal force **P** which can be applied to B so that A will not move relative to B. All surfaces are smooth.

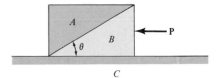

Prob. 13–36

13–37. Blocks A and B each have a mass m. Determine the largest horizontal force **P** which can be applied to B so that A will not slip up B. The coefficient of static friction between A and B is μ. Neglect any friction between B and C.

■13–38. The 10-kg crate rests on the cart for which the coefficient of static friction is $\mu_s = 0.3$ between the crate and cart. Determine the largest angle θ of the plane so that the crate does not slide on the cart when the cart is given an acceleration of $a = 6$ m/s².

Prob. 13–38

95

13–39. The conveyor belt C is moving downward at 2 m/s. If the coefficient of static friction between the conveyor and the 10-kg box B is $\mu_s = 0.8$, determine the shortest time the conveyor can stop so that the box does not shift or move on the belt.

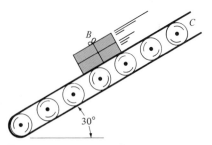

Probs. 13–39/13–40

***13–40.** The coefficients of static and kinetic friction between the conveyor and the 10-kg box B are $\mu_s = 0.8$ and $\mu_k = 0.7$, respectively. Starting from rest a point on the conveyor belt moves downward 2 m in 3 s. Determine the acceleration of the box and the conveyor.

13–41. Cylinder B has a mass m and is hoisted using the cord and pulley system shown. Determine the magnitude of force $\mathbf{F}$ as a function of the block's vertical position y so that when $\mathbf{F}$ is applied to the end of the cord the block rises with a constant acceleration $\mathbf{a}_B$. Neglect the mass of the cord and pulleys.

Prob. 13–41

13–42. The spring mechanism is used as a shock absorber for railroad cars. Determine the maximum compression of spring HI if the fixed bumper R of a 5-Mg railroad car, rolling freely at 1.5 m/s, strikes the plate P. Bar AB slides along the guide paths CE and DF. The ends of all springs are attached to their respective members and the springs are originally unstretched. Assume that P and R remain in contact during collision.

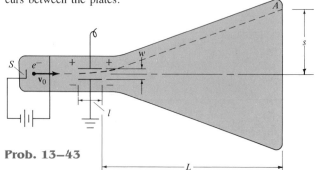

Prob. 13–42

13–43. In the cathode-ray tube, electrons having a mass m are emitted from a source point S and begin to travel horizontally with an initial velocity $\mathbf{v}_0$. While passing between the grid plates a distance l, they are subjected to a vertical force having a magnitude eV/w, where e is the charge of an electron, V the applied voltage acting across the plates, and w the distance between the plates. After passing clear of the plates, the electrons then travel in straight lines and strike the screen at A. Determine the deflection s of the electrons in terms of the dimensions of the voltage plate and tube. Neglect gravity and the slight vertical deflection which occurs between the plates.

Prob. 13–43

*13–44. Block *A* has a mass of 15 kg and block *B* has a mass of 20 kg. If the horizontal forces of 100 N and 300 N are applied, determine the acceleration of each block. Assume the surfaces of contact are smooth.

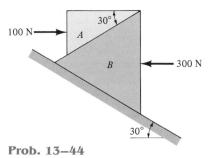

Prob. 13–44

13–45. The smooth block *B* of negligible size has a mass *m* and rests on the horizontal plane. If the board *AC* pushes on the block at an angle θ with a constant acceleration of $\mathbf{a}_0$, determine the velocity of the block along the board and the distance *s* the block moves along the board as a function of time *t*. The block starts from rest when $s = 0$, $t = 0$.

Prob. 13–45

13.5 Equations of Motion: Normal and Tangential Coordinates

When a particle moves over a curved path which is known, the equation of motion for the particle may be written in the normal and tangential directions. We have

$$\Sigma \mathbf{F} = m\mathbf{a}$$
$$\Sigma F_t \mathbf{u}_t + \Sigma F_n \mathbf{u}_n + \Sigma F_b \mathbf{u}_b = m\mathbf{a}_t + m\mathbf{a}_n$$

Here ΣF_n, ΣF_t, and ΣF_b represent the sums of all the force components acting on the particle in the normal, tangential, and binormal directions, respectively, Fig. 13–11. Note that there is no motion of the particle in the binormal direction, since the particle is constrained to move along the path. The above equation is satisfied provided

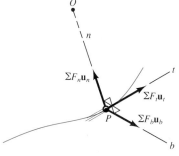

Fig. 13–11

$$\Sigma F_t = ma_t$$
$$\Sigma F_n = ma_n \qquad\qquad (13\text{--}8)$$
$$\Sigma F_b = 0$$

Recall that $a_t (= dv/dt)$ represents the time rate of change in the magnitude of velocity. Consequently, if $\Sigma \mathbf{F}_t$ acts in the direction of motion, the particle's speed will increase, whereas if it acts in the opposite direction, the particle will slow down. Likewise, $a_n (= v^2/\rho)$ represents the time rate of change in the velocity's direction. Since this vector *always* acts in the positive n direction, i.e., towards the path's center of curvature, then $\Sigma \mathbf{F}_n$, which causes $\mathbf{a}_n$, also acts in this direction. In particular, when the particle is constrained to travel in a circular path with a constant speed, there is a normal force exerted on the particle by the constraint. Since this force is always directed toward the center of the path, it is often referred to as the *centripetal force*.

PROCEDURE FOR ANALYSIS

When a problem involves the motion of a particle along a *known curved path,* normal and tangential coordinates should be considered for the analysis since the acceleration components can be readily formulated. The method for applying the equations of motion, which relate the forces to the acceleration, has been outlined in the procedure given in Sec. 13.4. Specifically, for n, t, b coordinates it may be stated as follows:

Free-Body Diagram. Establish the inertial n, t, b coordinate system at the particle and draw the particle's free-body diagram. The particle's normal acceleration $\mathbf{a}_n$ *always* acts in the positive n direction. If the tangential acceleration $\mathbf{a}_t$ is unkown, assume it acts in the positive t direction. From the diagram identify the unknowns in the problem.

Equations of Motion. Apply the equations of motion, Eqs. 13–8.

Kinematics. Formulate the tangential and normal components of acceleration; i.e., $a_t = dv/dt$ or $a_t = v\, dv/ds$ and $a_n = v^2/\rho$. If the path is defined as $y = f(x)$, the radius of curvature can be obtained from $\rho = |[1 + (dy/dx)^2]^{3/2}/(d^2y/dx^2)|$.

The following examples numerically illustrate application of this procedure.

Example 13–6

Determine the banking angle θ of the circular track so that the wheels of the sports car shown in Fig. 13–12a will not have to depend upon friction to prevent the car from sliding either up or down the curve. The car travels at a constant speed of 100 ft/s. The radius of the track is 600 ft.

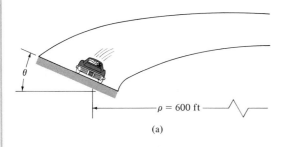

$\rho = 600$ ft

(a)

(b)

Fig. 13–12

SOLUTION

Before looking at the solution to this problem, give some thought as to why it should be solved using n, t, b coordinates.

Free-Body Diagram. As shown in Fig. 13–12b, the car is assumed to have a mass m. As stated in the problem, no frictional force acts on the car. Here $\mathbf{N}_C$ represents the *resultant* of all the wheels on the ground. Since a_n can be calculated, the unknowns are N_C and θ.

Equations of Motion. Using the n, b axes shown,

$$+\uparrow \Sigma F_b = 0; \qquad N_C \cos \theta - mg = 0 \qquad (1)$$

$$\xrightarrow{+} \Sigma F_n = ma_n; \qquad N_C \sin \theta = m\frac{v^2}{\rho} \qquad (2)$$

Eliminating N_C and m from these equations by dividing Eq. (2) by Eq. (1), we obtain

$$\tan \theta = \frac{v^2}{g\rho} = \frac{(100)^2}{32.2(600)}$$

$$\theta = \tan^{-1}(0.518)$$

$$= 27.4° \qquad\qquad Ans.$$

A force summation in the tangential direction of motion is of no consequence to the solution. If it were considered, note that $a_t = dv/dt = 0$, since the car moves with *constant speed*.

Example 13–7

The 3-kg disk D is attached to the end of a cord as shown in Fig. 13–13a. The other end of the cord is attached to a ball-and-socket joint located at the center of a platform. If the platform is rotating rapidly, and the disk is placed on it and released from rest as shown, determine the time it takes for the disk to reach a speed great enough to break the cord. The maximum tension the cord can sustain is 100 N, and the coefficient of kinetic friction between the disk and the platform is $\mu_k = 0.1$.

(a)

(b)

Fig. 13–13

SOLUTION

Free-Body Diagram. As shown in Fig. 13–13b, the disk has *both* normal and tangential components of acceleration as a result of the unbalanced forces **T** and **F**. Since sliding occurs, the frictional force has a magnitude $F = \mu_k N_D = 0.1 N_D$ and a direction that opposes the *relative motion* of the disk with respect to the platform. The weight of the disk is $W = 3(9.81) = 29.43$ N. Since a_n can be related to v, the unknowns are N_D, a_t and v.

Equations of Motion

$$\Sigma F_b = 0; \qquad\qquad N_D - 29.43 = 0 \qquad\qquad (1)$$

$$\Sigma F_t = ma_t; \qquad\qquad 0.1 N_D = 3a_t \qquad\qquad (2)$$

$$\Sigma F_n = ma_n; \qquad\qquad T = 3\left(\frac{v^2}{1}\right) \qquad\qquad (3)$$

Since the maximum tension sustained by the cord is $T = 100$ N, Eq. (3) can be solved for the critical speed v_{cr} of the disk needed to break the cord. Solving all the equations, we obtain

$$N_D = 29.43 \text{ N}$$
$$a_t = 0.981 \text{ m/s}^2$$
$$v_{cr} = 5.77 \text{ m/s}$$

Kinematics. Since a_t is *constant*, the time needed to break the cord is

$$v_{cr} = v_0 + a_t t$$
$$5.77 = 0 + (0.981)t$$
$$t = 5.89 \text{ s} \qquad\qquad Ans.$$

Example 13–8

The skier in Fig. 13–14*a* descends the smooth slope, which may be approximated by a parabola. If she has a weight of 120 lb, determine the normal force she exerts on the ground at the instant she arrives at point A, where her velocity is 30 ft/s. Also compute her acceleration at A.

$$y = \tfrac{1}{60}x^2 - 15$$

15 ft

(a)

SOLUTION

Why consider using n, t coordinates to solve this problem?

Free-Body Diagram. The free-body diagram for the skier when she is at A is shown in Fig. 13–14*b*. Since the path is *curved*, there are two components of acceleration, $\mathbf{a}_n$ and $\mathbf{a}_t$. Since a_n can be calculated, the unknowns are a_t and N_A.

120 lb

N_A

(b)

Fig. 13–14

Equations of Motion

$$+\uparrow \Sigma F_n = ma_n; \qquad N_A - 120 = \frac{120}{32.2}\left(\frac{(30)^2}{\rho}\right) \qquad (1)$$

$$\xleftarrow{+}\Sigma F_t = ma_t; \qquad 0 = \frac{120}{32.2}a_t \qquad (2)$$

The radius of curvature ρ for the path must be computed at point $A(0, -15 \text{ ft})$. Since $y = \tfrac{1}{60}x^2 - 15$, $dy/dx = \tfrac{1}{30}x$, $d^2y/dx^2 = \tfrac{1}{30}$, then at $x = 0$,

$$\rho = \left|\frac{[1 + (dy/dx)^2]^{3/2}}{d^2y/dx^2}\right|_{x=0} = \left|\frac{[1 + (0)^2]^{3/2}}{\tfrac{1}{30}}\right| = 30 \text{ ft}$$

Substituting into Eq. (1) and solving for N_A, we have

$$N_A = 232 \text{ lb} \qquad\qquad Ans.$$

Kinematics. From Eq. (2),

$$a_t = 0$$

Thus,

$$a_n = \frac{v^2}{\rho} = \frac{(30)^2}{30} = 30 \text{ ft/s}^2$$

$$a_A = a_n = 30 \text{ ft/s}^2 \uparrow \qquad\qquad Ans.$$

Example 13–9

A block having a mass of 2 kg is given an initial velocity of 1 m/s when it is at the top surface of the smooth cylinder shown in Fig. 13–15a. If the block slides along a path of radius 0.5 m, determine the angle $\theta = \theta_{\max}$ at which it begins to leave the cylinder's surface.

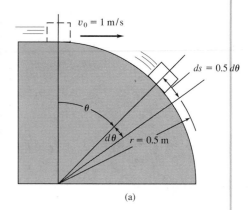

(a)

SOLUTION

Free-Body Diagram. The free-body diagram for the block, when the block is located at the *general position* θ, is shown in Fig. 13–15b. The block must have a tangential acceleration $\mathbf{a}_t$, since its *speed* is always *increasing* as it slides downward. The weight is $W = 2(9.81) = 19.62$ N. Specify the three unknowns.

Equations of Motion

$$+\swarrow \Sigma F_n = ma_n; \qquad -N_B + 19.62 \cos \theta = 2\frac{v^2}{0.5} \qquad (1)$$

$$+\searrow \Sigma F_t = ma_t; \qquad 19.62 \sin \theta = 2a_t \qquad (2)$$

These two equations contain four unknowns, N_B, v, a_t, and θ. At the instant $\theta = \theta_{\max}$, however, the block leaves the surface of the cylinder so that $N_B = 0$.

(b)

Fig. 13–15

Kinematics.
A third equation for the solution may be obtained by noting that the magnitude of tangential acceleration a_t may be related to the speed of the block v and the angle θ. Since $a_t\, ds = v\, dv$ and $ds = r\, d\theta = 0.5\, d\theta$, Fig. 13–15a, we have

$$a_t = \frac{v\, dv}{0.5\, d\theta} \qquad (3)$$

Substituting Eq. (3) into Eq. (2) and separating the variables, we have

$$v\, dv = 4.905 \sin \theta\, d\theta$$

Integrating both sides, realizing that when $\theta = 0°$, $v_0 = 1$ m/s, yields

$$\int_1^v v\, dv = 4.905 \int_{0°}^{\theta} \sin \theta\, d\theta$$

$$\left.\frac{v^2}{2}\right|_1^v = -4.905 \cos \theta \Big|_{0°}^{\theta}$$

$$v^2 = 9.81(1 - \cos \theta) + 1$$

Substituting into Eq. (1) with $N_B = 0$ and solving for $\cos \theta_{\max}$ gives

$$19.62 \cos \theta_{\max} = \frac{2}{0.5}[9.81(1 - \cos \theta_{\max}) + 1]$$

$$\cos \theta_{\max} = \frac{43.24}{58.86}; \qquad \theta_{\max} = 42.7° \qquad \textit{Ans.}$$

PROBLEMS

13-46. Compute the mass of the sun, knowing that the distance from the earth to the sun is $149.6(10^6)$ km. *Hint:* Use Eq. 13–1 to represent the force of gravity acting on the earth.

13-47. If the crest of the hill has a radius of curvature of $\rho = 200$ ft, determine the maximum constant speed at which the car can travel over it without leaving the surface of the road. Neglect the size of the car in the calculation. The car has a weight of 3500 lb.

$\rho = 200$ ft

Prob. 13-47

*13-48. Determine the constant speed of the planes on the amusement park ride if it is observed that the supporting cables are directed at $\theta = 30°$ from the vertical. Each plane including its passenger has a mass of 225 kg. Also, determine the normal force that a 50-kg passenger exerts on the seat of the plane during the motion. Neglect friction and the size of the passenger and plane.

8 m

$\theta = 30°$ $\theta = 30°$ 4 m

Prob. 13-48

13-49. The 150-lb man lies against the cushion for which the coefficient of static friction is $\mu_s = 0.5$. Determine the resultant normal and frictional forces the cushion exerts on him if, due to rotation about the z axis, he has a constant speed $v = 20$ ft/s. Neglect the size of the man.

z

8 ft

G

60°

Probs. 13-49/13-50

13-50. The 150-lb man lies against the cushion for which the coefficient of static friction is $\mu_s = 0.3$. Determine the smallest speed v he can have, due to a rotation about the z axis, so that he does not slide downward. Neglect the size of the man.

13-51. Prove that if the block is released from rest at the top point B of a smooth path of *arbitrary shape*, the speed it attains when it reaches point A is equal to the speed it attains when it falls freely through a distance h; i.e., $v = \sqrt{2gh}$.

B

h

A

Prob. 13-51

*13–52. A skier starts from rest at $A(10 \text{ m}, 0)$ and descends the smooth slope, which may be approximated by a parabola. If she has a mass of 52 kg, determine the normal force she exerts on the ground at the instant she arrives at point B. Neglect the size of the skier. *Hint:* Use the result of Prob. 13–51.

$$y = \frac{1}{20}x^2 - 5$$

Probs. 13–52/13–53

13–53. A skier starts from rest at $A(10 \text{ m}, 0)$ and descends the smooth slope, which may be approximated by a parabola. If she has a mass of 52 kg, determine the normal force she exerts on the ground at the instant she arrives at point C. Neglect the size of the skier. *Hint:* Use the result of Prob. 13–51.

13–54. The pendulum bob has a mass m and is released from rest when $\theta = 0°$. Determine the tension in the cord as a function of the angle of descent θ. Neglect the size of the bob.

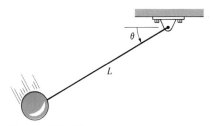

Prob. 13–54

13–55. At the instant $\theta = 60°$ the boy's mass center G has a speed of $v_G = 15$ ft/s. Determine the rate of increase in his speed and the tension in each of the two supporting cords at this instant. The boy has a weight of 60 lb. Neglect his size and the mass of the seat and cords.

Probs. 13–55/13–56

*13–56. At the instant $\theta = 60°$ the boy's mass center G is at rest. Determine the rate of increase in his speed and the tension in each of the two supporting cords when $\theta = 90°$. The boy has a weight of 60 lb. Neglect his size and the mass of the seat and cords.

13–57. A girl having a mass of 25 kg sits at the edge of the merry-go-round so her center of mass G is at a distance of 1.5 m from the axis of rotation. If the angular motion of the platform is *slowly* increased so that the girl's tangential component of acceleration can be neglected, determine the maximum speed which she can have before she begins to slip off the merry-go-round. The coefficient of static friction between the girl and the merry-go-round is $\mu_s = 0.3$.

Probs. 13–57/13–58

13–58. Solve Prob. 13–57 assuming the platform starts rotating from rest so that the girl's speed is increased uniformly at $\dot{v} = 0.5$ m/s^2.

13–59. The cylindrical plug has a weight of 2 lb and it is free to move within the confines of the smooth pipe. The spring has a stiffness of $k = 14$ lb/ft and when no motion occurs the distance $d = 0.5$ ft. Determine the force of the spring on the plug when the plug is at rest with respect to the pipe. The plug is traveling with a constant speed of 15 ft/s which is caused by the rotation of the pipe about the vertical axis.

Prob. 13–59

*13–60. If the ball has a mass of 30 kg and a speed $v = 4$ m/s at the instant it is at its lowest point, $\theta = 0°$, determine the tension in the cord at this instant. Also, determine the angle θ to which the ball swings at the instant it stops. Neglect the mass of the cord.

Probs. 13–60/13–61

13–61. The ball has a mass of 30 kg and a speed $v = 4$ m/s at the instant it is at its lowest point, $\theta = 0°$. Determine the tension in the cord and the rate at which the ball's speed is decreasing when $\theta = 20°$. Neglect the mass of the cord.

13–62. The 200-kg snowmobile with passenger is traveling down the hill at a constant speed of 6 m/s. Determine the resultant normal force and the resultant frictional force exerted on the tracks at the instant it reaches point A. Neglect the size of the snowmobile.

Probs. 13–62/13–63

13–63. The 200-kg snowmobile with passenger is traveling down the hill such that when it is at point A, it is traveling at 4 m/s and increasing its speed at 2 m/s². Determine the resultant normal force and the resultant frictional force exerted on the tracks at this instant. Neglect the size of the snowmobile.

*13–64. The block has a weight of 5 lb. The attached spring has an unstretched length of 3 ft. If at the instant $\theta = 30°$ the block has a speed $v = 4$ ft/s, determine the normal force on the block and the magnitude of the block's acceleration. Neglect friction.

Probs. 13–64/13–65

13–65. Solve Prob. 13–64 if the coefficient of kinetic friction between the block and the surface is $\mu_k = 0.1$.

13–66. The sled is freely coasting down the hill defined by the curve $y = 3(1 - e^{-0.5x})$. If it has a speed of 4 m/s when $x = 4$ m, determine the normal reaction acting on it and the rate of increase in its speed at this instant. The sled and rider have a total mass of 80 kg. Neglect friction and the size of the sled.

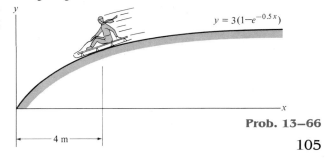

Prob. 13–66

105

13–67. The 2-kg spool S fits loosely on the inclined rod for which the coefficient of static friction is $\mu_s = 0.2$. If the spool is located 0.25 m from A, determine the minimum constant speed the spool can have so that it does not slip down the rod.

Probs. 13–67/13–68

*13–68. The 2-kg spool S fits loosely on the inclined rod for which the coefficient of static friction is $\mu_s = 0.2$. If the spool is located 0.25 m from A, determine the maximum constant speed the spool can have so that it does not slip up the rod.

13–69. If the bicycle and rider have a total weight of 180 lb, determine the resultant normal force acting on the bicycle when it is at point A while it is freely coasting at $v = 6$ ft/s. Also, compute the increase in the bicyclist's speed at this point. Neglect the resistance due to the wind.

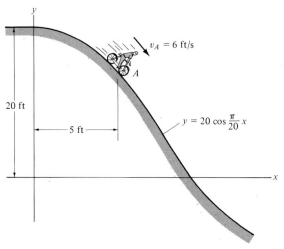

Prob. 13–69

13–70. The 2-kg block B is given a velocity of $v_A = 2$ m/s when it reaches point A. Determine the speed v of the block and the normal force N_B of the plane on the block as a function of θ. Plot these results as v vs. θ and N_B vs. θ and specify the angle at which the normal force is maximum. Neglect friction and the size of the block in the calculation.

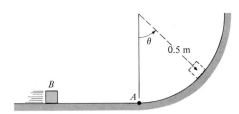

Prob. 13–70

13.6 Equations of Motion: Cylindrical Coordinates

When all the forces acting on a particle are resolved into components along the unit-vector directions $\mathbf{u}_r$, $\mathbf{u}_\theta$, and $\mathbf{u}_z$, Fig. 13–16, the equation of motion may be expressed as

$$\Sigma \mathbf{F} = m\mathbf{a}$$

$$\Sigma F_r \mathbf{u}_r + \Sigma F_\theta \mathbf{u}_\theta + \Sigma F_z \mathbf{u}_z = ma_r \mathbf{u}_r + ma_\theta \mathbf{u}_\theta + ma_z \mathbf{u}_z$$

To satisfy this equation, the respective $\mathbf{u}_r$, $\mathbf{u}_\theta$, and $\mathbf{u}_z$ components on the left side must equal the corresponding components on the right side. Consequently, we may write the following three scalar equations:

$$
\begin{array}{|c|}
\hline
\Sigma F_r = ma_r \\
\Sigma F_\theta = ma_\theta \\
\Sigma F_z = ma_z \\
\hline
\end{array}
\qquad (13\text{–}9)
$$

If the particle is constrained to move only in the r-θ plane, then only the first two of Eqs. 13–9 are used to specify the motion.

Tangential and Normal Forces

The most straightforward type of problem involving cylindrical coordinates requires the determination of the resultant force components ΣF_r, ΣF_θ, and ΣF_z causing a particle to move with a *known* acceleration. If, however, the particle's accelerated motion is not completely specified at the given instant, then some information regarding the directions or magnitudes of the forces acting on the particle must be known or computed in order to solve Eqs.

Fig. 13—16

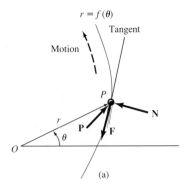

(a)

$r = f(\theta)$

Tangent

Motion

P

N

F

r

θ

O

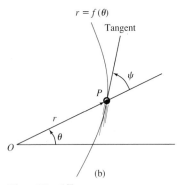

(b)

$r = f(\theta)$

Tangent

ψ

P

r

θ

O

Fig. 13–17

13–9. For example, the force **P** causes the particle in Fig. 13–17a to move along a path defined in polar coordinates; $r = f(\theta)$. The *normal force* **N** which the path exerts on the particle is always *perpendicular to the tangent of the path,* whereas the frictional force **F** always acts along the tangent in the opposite direction of motion. The *directions* of **N** and **F** can be specified relative to the radial coordinate by computing the angle ψ (psi), Fig. 13–17b, which is defined between the *extended* radial line $r = OP$ and the tangent to the curve. This angle is determined from the equation*

$$\tan \psi = \frac{r}{dr/d\theta} \qquad (13\text{–}10)$$

If it is calculated as a positive quantity, it is measured from the extended radial line to the tangent in a counterclockwise sense or in the positive direction of θ. If it is negative, it is measured in the opposite direction to positive θ. Application is illustrated numerically in Example 13–12.

PROCEDURE FOR ANALYSIS

Cylindrical or polar coordinates are a suitable choice for the analysis of a problem for which data regarding the angular motion of the radial line r is given, or in cases where the path can be conveniently expressed in terms of these coordinates. Once these coordinates have been established, the equations of motion can be applied in order to relate the forces acting on the particle to its acceleration components. The method for doing this has been outlined in the procedure for analysis given in Sec. 13.4. The following is a summary of this procedure.

Free-Body Diagram. Establish the r, θ, z inertial coordinate system and draw the particle's free-body diagram. Assume that $\mathbf{a}_r$, $\mathbf{a}_\theta$, and $\mathbf{a}_z$ act in the *positive directions* of r, θ, and z if they are unknown. Before applying the equations of motion, identify all the unknowns.

Equations of Motion. Apply the equations of motion, Eqs. 13–9.

Kinematics. Use the methods of Sec. 12.7 to determine r and the time derivatives $\dot{r}$, $\ddot{r}$, $\dot{\theta}$, $\ddot{\theta}$, and $\ddot{z}$, and evaluate the acceleration components $a_r = \ddot{r} - r\dot{\theta}^2$, $a_\theta = r\ddot{\theta} + 2\dot{r}\dot{\theta}$, and $a_z = \ddot{z}$. If any of these three components are computed as negative quantities, it indicates that they act in their negative coordinate directions.

The following examples numerically illustrate application of this procedure. Further application of Eqs. 13–9 is given in Sec. 13.7, where problems involving rockets, satellites, and planetary bodies subjected to gravitational forces are considered.

*The derivation is given in any standard calculus text.

Example 13–10

The 2-lb block in Fig. 13–18a moves on the smooth horizontal plane, such that its path is specified in polar coordinates by the parametric equations $r = (10t^2)$ ft and $\theta = (0.5t)$ rad, where t is in seconds. Determine the magnitude of the *resultant* force **F** causing the motion at the instant $t = 1$ s.

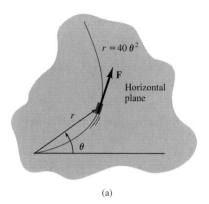

$r = 40\,\theta^2$

F

Horizontal plane

(a)

SOLUTION

Free-Body Diagram. Note that if the parameter t is eliminated between r and θ, the equation of the path becomes $r = 40\theta^2$. The block's free-body diagram when it is located at an arbitrary point along the path is shown in Fig. 13–18b. Here the r and θ components of the resultant force are represented. They, and $\mathbf{a}_r$ and $\mathbf{a}_\theta$, are assumed to act in the *positive* coordinate directions. The four unknowns are F_r, F_θ, a_r, and a_θ.

Equations of Motion

$$\Sigma F_r = ma_r; \qquad F_r = \frac{2}{32.2}(\ddot{r} - r\dot{\theta}^2) \qquad (1)$$

$$\Sigma F_\theta = ma_\theta; \qquad F_\theta = \frac{2}{32.2}(r\ddot{\theta} + 2\dot{r}\dot{\theta}) \qquad (2)$$

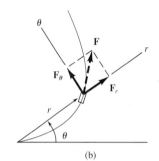

(b)

Fig. 13–18

Kinematics. Since the motion is specified, the coordinates and the required time derivatives can be computed and evaluated at $t = 1$ s.

$$r = 10t^2 \Big|_{t=1\text{s}} = 10 \text{ ft} \qquad \theta = 0.5t \Big|_{t=1\text{s}} = 0.5 \text{ rad}$$

$$\dot{r} = 20t \Big|_{t=1\text{s}} = 20 \text{ ft/s} \qquad \dot{\theta} = 0.5 \text{ rad/s}$$

$$\ddot{r} = 20 \text{ ft/s}^2 \qquad \ddot{\theta} = 0$$

Substituting these results into Eqs. (1) and (2) yields

$$F_r = \frac{2}{32.2}(20 - 10(0.5)^2) = 1.087 \text{ lb}$$

$$F_\theta = \frac{2}{32.2}(0 + 2(20)(0.5)) = 1.242 \text{ lb}$$

Hence, the resultant force has a magnitude of

$$F = \sqrt{(1.087)^2 + (1.242)^2}$$
$$= 1.65 \text{ lb} \qquad\qquad \textit{Ans.}$$

Example 13–11

The smooth 2-kg cylinder C in Fig. 13–19a has a peg P through its center which passes through the slot in arm OA. If the arm rotates in the *vertical plane* at a constant rate $\dot{\theta} = 0.5$ rad/s, determine the force that the arm exerts on the peg at the instant $\theta = 60°$.

(a)

SOLUTION

Why is it a good idea to use polar coordinates to solve this problem?

Free-Body Diagram. The free-body diagram for the cylinder is shown in Fig. 13–19b. The force of the peg, $\mathbf{F}_P$, acts perpendicular to the slot in the arm. As usual, $\mathbf{a}_r$ and $\mathbf{a}_\theta$ are assumed to act in the directions of *positive r* and θ, respectively. Identify the four unknowns.

Equations of Motion. Using the data in Fig. 13–19b, we have

$$+\swarrow\Sigma F_r = ma_r; \qquad 19.62 \sin\theta - N_C \sin\theta = 2a_r \qquad (1)$$

$$+\searrow\Sigma F_\theta = ma_\theta; \quad 19.62 \cos\theta + F_P - N_C \cos\theta = 2a_\theta \qquad (2)$$

(b)

Fig. 13–19

Kinematics. From Fig. 13–19a, r can be related to θ by the equation

$$r = \frac{0.4}{\sin\theta} = 0.4 \csc\theta$$

Since $d(\csc\theta) = -(\csc\theta \cot\theta)d\theta$ and $d(\cot\theta) = -(\csc^2\theta)\,d\theta$, then r and the necessary time derivatives become

$\dot{\theta} = 0.5 \qquad r = 0.4 \csc\theta$

$\ddot{\theta} = 0 \qquad \dot{r} = -0.4(\csc\theta \cot\theta)\dot{\theta}$

$\qquad\qquad = -0.2 \csc\theta \cot\theta$

$\qquad\quad \ddot{r} = -0.2(-\csc\theta \cot\theta)(\dot{\theta}) \cot\theta - 0.2 \csc\theta(-\csc^2\theta)\dot{\theta}$

$\qquad\qquad = 0.1 \csc\theta(\cot^2\theta + \csc^2\theta)$

Evaluating these formulas at $\theta = 60°$, we get

$\dot{\theta} = 0.5 \qquad r = 0.462$

$\ddot{\theta} = 0 \qquad \dot{r} = -0.133$

$\qquad\qquad \ddot{r} = 0.192$

$\qquad a_r = \ddot{r} - r\dot{\theta}^2 = 0.192 - 0.462(0.5)^2 = 0.0765$

$\qquad a_\theta = r\ddot{\theta} + 2\dot{r}\dot{\theta} = 0 + 2(-0.133)(0.5) = -0.133$

Substituting these results into Eqs. (1) and (2) with $\theta = 60°$ and solving yields

$$N_C = 19.4 \text{ N} \qquad F_P = -0.355 \text{ N} \qquad\qquad Ans.$$

The negative sign indicates that $\mathbf{F}_P$ acts opposite to that shown in Fig. 13–19b.

Example 13–12

A can C, having a mass of 0.5 kg, moves along a grooved horizontal slot shown in Fig. 13–20a. The slot is in the form of a spiral, which is defined by the equation $r = (0.1\theta)$ m, where θ is in radians. If the arm OA is rotating at a constant rate $\dot{\theta} = 4$ rad/s in the horizontal plane, determine the force it exerts on the can at the instant $\theta = \pi$ rad. Neglect friction and the size of the can.

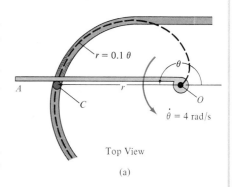

Top View

(a)

SOLUTION

Free-Body Diagram. The driving force $\mathbf{F}_C$ acts perpendicular to the arm OA, whereas the normal force of the wall of the slot on the can, $\mathbf{N}_C$, acts perpendicular to the tangent to the curve at $\theta = \pi$ rad, Fig. 13–20b. As usual, $\mathbf{a}_r$ and $\mathbf{a}_\theta$ act in the *positive directions* of r and θ, respectively. Since the path is specified, the angle ψ which the extended radial line r makes with the tangent, Fig. 13–20c, can be determined from Eq. 13–10. We have $r = 0.1\theta$, so that $dr/d\theta = 0.1$, and therefore

$$\tan\psi = \frac{r}{dr/d\theta} = \frac{0.1\theta}{0.1} = \theta$$

When $\theta = \pi$, $\psi = \tan^{-1}\pi = 72.3°$, so that $\phi = 90° - \psi = 17.7°$, as shown in Fig. 13–20$c$. Identify the four unknowns in Fig. 13–20b.

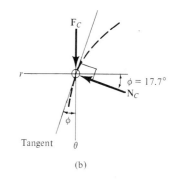

(b)

Equations of Motion. Using the data shown in Fig. 13–20b, we have

$$\xleftarrow{+}\Sigma F_r = ma_r; \qquad N_C \cos 17.7° = 0.5a_r \qquad (1)$$
$$+\downarrow\Sigma F_\theta = ma_\theta; \qquad F_C - N_C \sin 17.7° = 0.5a_\theta \qquad (2)$$

Kinematics. Computing the time derivatives of r and θ gives

$$\dot{\theta} = 4 \text{ rad/s} \qquad r = 0.1\theta$$
$$\ddot{\theta} = 0 \qquad \dot{r} = 0.1\dot{\theta} = 0.1(4) = 0.4 \text{ m/s}$$
$$\ddot{r} = 0.1\ddot{\theta} = 0$$

(c)

Fig. 13–20

At the instant $\theta = \pi$ rad,

$$a_r = \ddot{r} - r\dot{\theta}^2 = 0 - 0.1(\pi)(4)^2 = -5.03 \text{ m/s}^2$$
$$a_\theta = r\ddot{\theta} + 2\dot{r}\dot{\theta} = 0 + 2(0.4)(4) = 3.20 \text{ m/s}^2$$

Substituting these results into Eqs. (1) and (2) and solving yields

$$N_C = -2.64 \text{ N}$$
$$F_C = 0.797 \text{ N} \qquad\qquad Ans.$$

What does the negative sign for N_C indicate?

PROBLEMS

13–71. A particle, having a mass of 2 kg, moves along a three-dimensional path defined by the equations $r = (10t^2 + 3t)$ m, $\theta = (0.1t^3)$ rad, and $z = (4t^2 + 15t - 6)$ m, where t is in seconds. Determine the r, θ, and z components of force which the path exerts on the particle when $t = 2$ s.

*13–72. The girl has a mass of 50 kg. She is seated on the horse of the merry-go-round which undergoes constant rotational motion $\dot{\theta} = 1.5$ rad/s. If the path of the horse is defined by $r = 4$ m, $z = (0.5 \sin \theta)$ m, determine the maximum and minimum force F_z the horse exerts on her during the motion.

Prob. 13–72

13–73. For a short time the 250-kg roller coaster car is traveling along the spiral track at a constant speed of $v = 8$ m/s. If the track descends $d = 12$ m for every full revolution, $\theta = 2\pi$ rad, determine the magnitudes of the components of force $\mathbf{F}_r$, $\mathbf{F}_\theta$, and $\mathbf{F}_z$ which the track exerts on the car. Neglect the size of the car.

Probs. 13–73/13–74

13–74. For a short time the 250-kg roller coaster car is traveling along the spiral track at a constant speed such that its position measured from the top of the track has components $r = 10$ m, $\theta = (0.2t)$ rad, and $z = (-0.3t)$ m, where t is in seconds. Determine the magnitudes of the components of force $\mathbf{F}_r$, $\mathbf{F}_\theta$, and $\mathbf{F}_z$ which the track exerts on the car at the instant $t = 2$ s. Neglect the size of the car.

13–75. The collar, which has a weight of 3 lb, slides along the smooth rod lying in the *horizontal plane* and having the shape of a parabola $r = 4/(1 - \cos \theta)$, where θ is in radians and r is in feet. If the collar's angular rate is constant and equals $\dot{\theta} = 4$ rad/s, determine the tangential retarding force P needed to cause the motion and the normal force that the collar exerts on the rod at the instant $\theta = 90°$.

Probs. 13–75/13–76

*13–76. Solve Prob. 13–75 if the parabolic path (rod) lies in the *vertical plane*.

13–77. The 0.5-lb particle is guided along the vertical circular path using the arm OA. If the arm has an angular velocity $\dot{\theta} = 0.4$ rad/s and an angular acceleration $\ddot{\theta} = 0.8$ rad/s^2 at the instant $\theta = 30°$, determine the force of the arm on the particle. Set $r_c = 0.4$ ft.

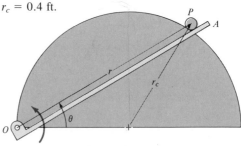

Probs. 13–77/13–78

13–78. The particle of mass m is guided along the vertical circular path of radius r_c using the arm OA. If the arm has a constant angular velocity $\dot{\theta}_0$, determine the angle θ at which the particle starts to leave the surface of the semicylinder.

13–79. Using air pressure, the 0.5-kg ball is forced to move through the smooth tube lying in the *horizontal plane* and having the shape of a logarithmic spiral. If the tangential force exerted on the ball due to the air is 6 N, determine the magnitude and direction of the rate of increase in the ball's speed at the instant $\theta = \pi/2$. Neglect the size of the ball.

$r = 0.2e^{0.1\theta}$

r

θ

$F = 6$ N

Prob. 13–79

*13–80. Using a forked rod, a smooth cylinder C having a mass of 0.5 kg is forced to move along the *horizontal slotted* path $r = (0.5\theta)$ m, where θ is in radians. If the angular position of the arm is $\theta = (0.5t^2)$ rad, where t is in seconds, determine the force of the forked rod on the cylinder and the normal force of the slot on the cylinder at the instant $t = 2$ s. The cylinder is in contact with only *one edge* of the rod and slot at any instant.

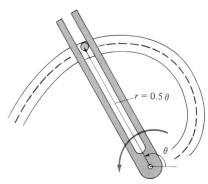

$r = 0.5\,\theta$

θ

Probs. 13–80/13–81

13–81. Using a forked rod, a smooth cylinder C having a mass of 0.5 kg is forced to move along the *vertical slotted* path $r = (0.5\theta)$ m, where θ is in radians. If the angular position of the arm is $\theta = (0.5t^2)$ rad, where t is in seconds, determine the force of the forked rod on the cylinder and the normal force of the slot on the cylinder at the instant $t = 2$ s. The cylinder is in contact with only *one edge* of the rod and slot at any instant.

13–82. Rod OA rotates counterclockwise in the horizontal plane with a constant angular rate $\dot{\theta} = 4$ rad/s. The double collar B is pin-connected together such that one collar slides over the rotating rod and the other collar slides over the circular rod described by the equation $r = (1.6 \cos\theta)$ m. If *both* collars have a total mass of 0.5 kg, determine the force which the circular rod exerts on one of the collars and the force that OA exerts on the other collar at the instant $\theta = 45°$.

$r = (1.6 \cos\theta)$ m

$\dot{\theta} = 4$ rad/s

r B

A

θ

O

0.8 m

Probs. 13–82/13–83

13–83. Solve Prob. 13–82 if the path is *vertical*.

*13–84. The forked rod is used to move the smooth 2-lb cylinder around the horizontal path in the shape of a limaçon, $r = (2 + \cos\theta)$ ft. If at all times $\dot{\theta} = 0.5$ rad/s, determine the force which the rod exerts on the cylinder at the instant $\theta = 90°$. The fork and path contact the cylinder on only one side.

2 ft

$\dot{\theta}$ θ

1 ft   3 ft

Prob. 13–84/13–85/13–86

13–85. Solve Prob. 13–84 at the instant $\theta = 60°$.

13–86. The forked rod is used to move the smooth 2-lb cylinder around the horizontal path in the shape of a limaçon, $r = (2 + \cos\theta)$ ft. If $\theta = (0.5t^2)$ rad, where t is in seconds, determine the force which the rod exerts on the cylinder at the instant $t = 1$ s. The fork and path contact the cylinder on only one side.

13–87. A car of a roller coaster travels along a track which for a short distance is defined by a conical spiral, $r = 0.75z$, $\theta = -1.5z$, where r and z are in meters and θ is in radians. If the angular motion $\dot{\theta} = 1$ rad/s is always maintained, determine the r, θ, z components of reaction exerted on the car by the tracks at the instant $z = 6$ m. The car and passenger have a total mass of 300 kg. Neglect the mass of the wheels and the size of the car.

Prob. 13–87

*13–88. The ball has a mass of 2 kg and a negligible size. It is originally traveling around the horizontal circular path of radius $r_0 = 0.5$ m such that the angular rate of rotation is $\dot{\theta}_0 = 1$ rad/s. If the attached cord ABC is drawn down through the hole at a constant speed of 0.2 m/s, determine the tension the cord exerts on the ball at the instant $r = 0.25$ m. Also, compute the angular velocity of the ball at this instant. Neglect the effects of friction between the ball and horizontal plane. *Hint:* First show that the equation of motion in the θ direction yields $a_\theta = r\ddot{\theta} + 2\dot{r}\dot{\theta} = \frac{1}{r}\frac{d(r^2\dot{\theta})}{dt} = 0$. When integrated, $r^2\dot{\theta} = C$, where the constant C is determined from the problem data.

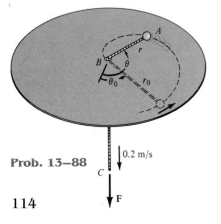

Prob. 13–88

13–89. Determine the normal and frictional driving forces that the spiral track exerts on the 200-kg motorcycle at the instant $\theta = \frac{5}{3}\pi$ rad, $\dot{\theta} = 0.4$ rad/s, and $\ddot{\theta} = 0.8$ rad/s².

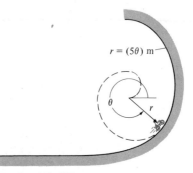

$r = (5\theta)$ m

Prob. 13–89

13–90. The pilot of an airplane executes a vertical loop which in part follows the path of a cardioid, $r = 600(1 + \cos\theta)$ ft, where θ is in radians. If his speed at A ($\theta = 0$) is a constant $v_P = 80$ ft/s, determine the vertical force the belt of his seat must exert on him to hold him to his seat when the plane is upside down at A. He weighs 150 lb. *Hint:* To determine the time derivatives necessary to compute the plane's acceleration components a_r and a_θ, take the first and second time derivatives of $r = 600(1 + \cos\theta)$ ft. Then, for further information, use Eq. 12–26 to determine $\dot{r}$ and $\dot{\theta}$. Also, take the time derivative of Eq. 12–26, noting that $\dot{v}_P = 0$, to determine $\ddot{r}$ and $\ddot{\theta}$.

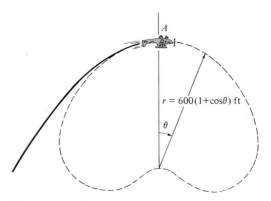

$r = 600(1 + \cos\theta)$ ft

Prob. 13–90

13–91. A truck T has a weight of 8000 lb and is traveling along a portion of a road defined by the lemniscate $r^2 = 0.2(10^6) \cos 2\theta$, where r is measured in feet and θ is in radians. If the truck maintains a constant speed of $v_T = 4$ ft/s, determine the magnitude of the resultant frictional force which must be exerted by all the wheels to maintain the motion when $\theta = 0°$. *Hint:* To determine the time derivatives necessary to compute the truck's acceleration components a_r and a_θ, take the first and second time derivatives of $r^2 = 0.2(10^6) \cos 2\theta$. Then, for further information, use Eq. 12–19 to determine $\dot{r}$ and $\dot{\theta}$. Also take the time derivative of Eq. 12–19, noting that $\dot{v}_T = 0$, to determine $\ddot{r}$ and $\ddot{\theta}$.

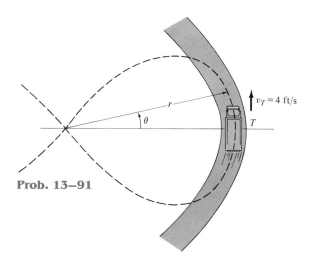

Prob. 13–91

*13.7 Central-Force Motion and Space Mechanics

If a particle is moving only under the influence of a force having a line of action which is always directed toward a fixed point, the motion is called *central-force motion*. This type of motion is commonly caused by electrostatic and gravitational forces.

In Fig. 13–21a, the particle P having a mass m is acted upon only by the central force $\mathbf{F}$. The free-body diagram for the particle is shown in Fig. 13–21b. Using polar coordinates (r, θ), the equations of motion, Eqs. 13–9, become

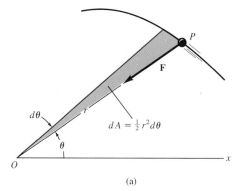

$$-F = m\left[\frac{d^2 r}{dt^2} - r\left(\frac{d\theta}{dt}\right)^2\right]$$

$$0 = m\left(r\frac{d^2\theta}{dt^2} + 2\frac{dr}{dt}\frac{d\theta}{dt}\right) \qquad (13\text{–}11)$$

(a)

The second of these equations may be written in the form

$$\frac{1}{r}\left[\frac{d}{dt}\left(r^2\frac{d\theta}{dt}\right)\right] = 0$$

so that integrating yields

$$r^2\frac{d\theta}{dt} = h \qquad (13\text{–}12)$$

(b)

Fig. 13–21

where h is a constant of integration. From Fig. 13–21a it can be seen that the shaded area described by the radius r, as r moves through an angle $d\theta$, is $dA = \frac{1}{2}r^2\, d\theta$. If the *areal velocity* is defined as

$$\frac{dA}{dt} = \frac{1}{2}r^2\frac{d\theta}{dt} = \frac{h}{2} \tag{13–13}$$

then, by comparison with Eq. 13–12, it is seen that the areal velocity for a particle subjected to central-force motion is *constant*. In other words, the particle will sweep out equal segments of area per unit time as it travels along the path. To obtain the *path of motion, $r = f(\theta)$*, the independent variable t must be eliminated from Eqs. 13–11. Using the chain rule of calculus and Eq. 13–12, the time derivatives of Eqs. 13–11 may be replaced by

$$\frac{dr}{dt} = \frac{dr}{d\theta}\frac{d\theta}{dt} = \frac{h}{r^2}\frac{dr}{d\theta}$$

$$\frac{d^2r}{dt^2} = \frac{d}{dt}\left(\frac{h}{r^2}\frac{dr}{d\theta}\right) = \frac{d}{d\theta}\left(\frac{h}{r^2}\frac{dr}{d\theta}\right)\frac{d\theta}{dt} = \left[\frac{d}{d\theta}\left(\frac{h}{r^2}\frac{dr}{d\theta}\right)\right]\frac{h}{r^2}$$

Substituting a new dependent variable (xi) $\xi = 1/r$ into the second equation, it is seen that

$$\frac{d^2r}{dt^2} = -h^2\xi^2\frac{d^2\xi}{d\theta^2}$$

Also, the square of Eq. 13–12 becomes

$$\left(\frac{d\theta}{dt}\right)^2 = h^2\xi^4$$

Substituting these last two equations into the first of Eqs. 13–11, we have

$$-h^2\xi^2\frac{d^2\xi}{d\theta^2} - h^2\xi^3 = -\frac{F}{m}$$

or

$$\frac{d^2\xi}{d\theta^2} + \xi = \frac{F}{mh^2\xi^2} \tag{13–14}$$

This differential equation defines the path over which the particle travels when it is subjected to the central force* **F.**

For application, the force of gravitational attraction will be considered. Some common examples of central-force systems which depend upon gravitation include the motion of the moon and artificial satellites about the earth, and the motion of the planets about the sun. As a typical problem in space mechanics, consider the trajectory of a space satellite or space vehicle launched into orbit with an initial velocity v_0, Fig. 13–22. It will be assumed that this velocity is initially *parallel* to the tangent at the surface of the earth, as shown in the figure.† Just after the satellite is released into free flight the

---

*In the derivation, **F** is considered positive when it is directed toward point O. If **F** is oppositely directed, the right side of Eq. 13–14 should be negative.

†The case where v_0 acts at some initial angle θ to the tangent is best described using the conservation of angular momentum (see Prob. 15–73).

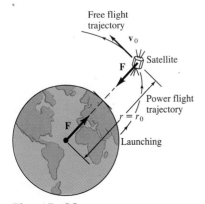

Free flight trajectory
v_0
F
Satellite
Power flight trajectory
$r = r_0$
F
Launching

Fig. 13–22

only force acting upon it is the gravitational force of the earth. (Gravitational attractions involving other bodies such as the moon or sun will be neglected, since for orbits close to the earth their effect is small in comparison with the gravitational force of the earth.) According to Newton's law of gravitation, force **F** will always act between the mass centers of the earth and the satellite, Fig. 13–22. From Eq. 13–1, this force of attraction has a magnitude of

$$F = G \frac{M_e m}{r^2}$$

where M_e and m represent the mass of the earth and the satellite, respectively, G is the gravitational constant, and r is the distance between the mass centers. Setting $\xi = 1/r$ in the above equation and substituting the result into Eq. 13–14, we obtain

$$\frac{d^2\xi}{d\theta^2} + \xi = \frac{GM_e}{h^2} \tag{13–15}$$

This second-order ordinary differential equation has constant coefficients and is nonhomogeneous. The solution is represented as the sum of the complementary and particular solutions. The complementary solution is obtained when the term on the right is equal to zero. It is,

$$\xi_c = C \cos (\theta - \phi)$$

where C and ϕ are constants of integration. The particular solution is

$$\xi_p = \frac{GM_e}{h^2}$$

Thus, the complete solution to Eq. 13–15 is

$$\xi = \xi_c + \xi_p$$
$$= \frac{1}{r} = C \cos (\theta - \phi) + \frac{GM_e}{h^2} \tag{13–16}$$

The validity of this result may be checked by substitution into Eq. 13–15.

Equation 13–16 represents the *free-flight trajectory* of the satellite. It is the equation of a conic section expressed in terms of polar coordinates. As shown in Fig. 13–23, a *conic section* is defined as the locus of point P which moves in a plane in such a way that the ratio of its distance from a fixed point F to its distance from a fixed line is constant. The fixed point is called the *focus*, and the fixed line DD is called the *directrix*. The constant ratio is called the *eccentricity* of the conic section and is denoted by e. Thus,

$$e = \frac{FP}{PA}$$

which may be written in the form

$$FP = r = e(PA) = e[p - r \cos (\theta - \phi)]$$

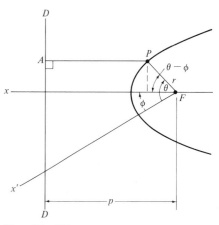

Fig. 13–23

or

$$\frac{1}{r} = \frac{1}{p} \cos (\theta - \phi) + \frac{1}{ep}$$

Comparing this equation with Eq. 13–16, it is seen that the eccentricity of the conic section for the trajectory is

$$e = \frac{Ch^2}{GM_e} \qquad (13\text{–}17)$$

and the fixed distance from the focus to the directrix is

$$p = \frac{1}{C} \qquad (13\text{–}18)$$

Provided the polar angle θ is measured from the x axis (an axis of symmetry since it is perpendicular to the directrix DD), the angle ϕ is zero, Fig. 13–23, and therefore Eq. 13–16 reduces to

$$\frac{1}{r} = C \cos \theta + \frac{GM_e}{h^2} \qquad (13\text{–}19)$$

The constants h and C are determined from the data obtained for the position and velocity of the satellite at the end of the *power-flight trajectory*. For example, if the initial height or distance to the space vehicle is r_0 (measured from the center of the earth) and its initial speed is v_0 at the beginning of its free flight, Fig. 13–24, then the constant h may be obtained from Eq. 13–12.

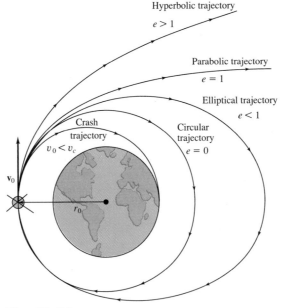

Fig. 13–24

When $\theta = \phi = 0°$, the velocity $\mathbf{v}_0$ has no radial component; therefore, from Eq. 12–25, $v_0 = r_0(d\theta/dt)$, so that

$$h = r_0^2 \frac{d\theta}{dt}$$

or

$$\boxed{h = r_0 v_0} \tag{13–20}$$

To determine C, use Eq. 13–19 with $\theta = 0°$, $r = r_0$, and substitute Eq. 13–20 for h:

$$\boxed{C = \frac{1}{r_0}\left(1 - \frac{GM_e}{r_0 v_0^2}\right)} \tag{13–21}$$

The equation for the free-flight trajectory therefore becomes

$$\boxed{\frac{1}{r} = \frac{1}{r_0}\left(1 - \frac{GM_e}{r_0 v_0^2}\right)\cos\theta + \frac{GM_e}{r_0^2 v_0^2}} \tag{13–22}$$

The type of path taken by the satellite is determined from the value of the eccentricity of the conic section as given by Eq. 13–17. If

$$\boxed{\begin{array}{ll} e = 0 & \text{free-flight trajectory is a circle} \\ e = 1 & \text{free-flight trajectory is a parabola} \\ e < 1 & \text{free-flight trajectory is an ellipse} \\ e > 1 & \text{free-flight trajectory is a hyperbola} \end{array}} \tag{13–23}$$

Each of these trajectories is shown in Fig. 13–24. From the curves it is seen that when the satellite follows a parabolic path, it is "on the border" of not returning to its initial starting point. The initial velocity at launching, $\mathbf{v}_0$, required for the satellite to follow a parabolic path is called the *escape velocity*. The speed, v_e, can be determined by using the second of Eqs. 13–23 with Eqs. 13–17, 13–20, and 13–21. It is left as an exercise to show that

$$v_e = \sqrt{\frac{2GM_e}{r_0}} \tag{13–24}$$

The speed v_c required to launch a satellite into a *circular orbit* can be found using the first of Eqs. 13–23. Since e is related to h and C, Eq. 13–17, C must be zero to satisfy this equation (from Eq. 13–20, h cannot be zero); and therefore, using Eq. 13–21, we have

$$v_c = \sqrt{\frac{GM_e}{r_0}} \tag{13–25}$$

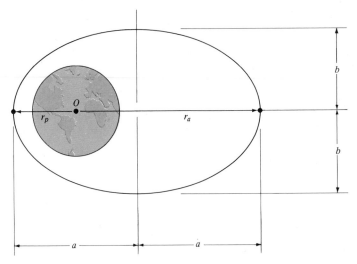

Fig. 13–25

Provided r_0 represents a minimum height for launching, in which frictional resistance from the atmosphere is neglected, speeds at launch which are less than v_c will cause the satellite to reenter the earth's atmosphere and either burn up or crash, Fig. 13–24.

All the trajectories attained by planets and most satellites are elliptical, Fig. 13–25. For an orbit about the earth, the *minimum distance* from the orbit to the center of the earth O (which is located at one of the foci of the ellipse) is r_p and can be found using Eq. 13–22 with $\theta = 0°$. Therefore,

$$\boxed{r_p = r_0} \tag{13–26}$$

This minimum distance is called the *perigee* of the orbit. The *apogee* or maximum distance r_a can be found using Eq. 13–22 with $\theta = 180°$.* Thus,

$$\boxed{r_a = \dfrac{r_0}{\dfrac{2GM_e}{r_0 v_0^2} - 1}} \tag{13–27}$$

With reference to Fig. 13–25, the semimajor axis a of the ellipse is

$$a = \frac{r_p + r_a}{2} \tag{13–28}$$

---

*Actually, the terminology perigee and apogee pertains only to orbits about the *earth*. If any other body is located at the focus of an elliptical orbit, the minimum and maximum distances are referred to respectively as the *periapsis* and *apoapsis* of the orbit.

Using analytical geometry it can be shown that the minor axis b is determined from the equation

$$b = \sqrt{r_p r_a} \tag{13-29}$$

Furthermore, by direct integration, the area of an ellipse is

$$A = \pi ab = \frac{\pi}{2}(r_p + r_a)\sqrt{r_p r_a} \tag{13-30}$$

The areal velocity has been defined by Eq. 13–13, $dA/dt = h/2$. Integrating yields $A = hT/2$, where T is the *period* of time required to make one orbital revolution. From Eq. 13–30, the period is

$$\boxed{T = \frac{\pi}{h}(r_p + r_a)\sqrt{r_p r_a}} \tag{13-31}$$

In addition to predicting the orbital trajectory of earth satellites, the theory developed in this section is valid, to a surprisingly close approximation, at predicting the actual motion of the planets traveling around the sun. In this case the mass of the sun, M_s, should be substituted for M_e when the appropriate formulas are used.

The fact that the planets do indeed follow elliptic orbits about the sun was discovered by the German astronomer Johannes Kepler (1571–1630). His discovery was made *before* Newton had developed the laws of motion and the law of gravitation. Kepler's laws, stipulated after 20 years of planetary observation, are summarized by the following three statements:

1. Every planet moves in its orbit such that the line joining it to the sun sweeps over equal areas in equal intervals of time, whatever the line's length.
2. The orbit of every planet is an ellipse with the sun placed at one of its foci.
3. The square of the period of any planet is directly proportional to the cube of the minor axis of its orbit.

A mathematical statement of the first and second laws is given by Eqs. 13–13 and 13–22. The third law can be shown from Eq. 13–31 (see Prob. 13–103).

Example 13–13

A satellite is launched 600 km from the surface of the earth, with an initial velocity of 30 Mm/h acting parallel to the tangent at the surface of the earth, Fig. 13–26. Assuming that the radius of the earth is 6378 km and that its mass is $5.976(10^{24})$ kg, determine (a) the eccentricity of the orbital path, and (b) the velocity of the satellite at apogee.

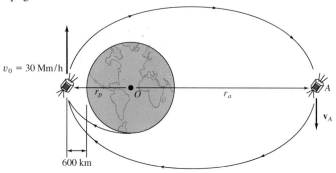

Fig. 13–26

SOLUTION

Part (a). The eccentricity of the orbit is obtained using Eq. 13–17. The constants h and C are first determined from Eqs. 13–20 and 13–21. Since

$$r_p = 6378 \text{ km} + 600 \text{ km} = 6.978(10^6) \text{ m}$$
$$v_0 = 30 \text{ Mm/h} = 8333.3 \text{ m/s}$$

then

$$h = r_p v_0 = 6.978(10^6)(8333.3) = 5.815(10^{10}) \text{ m}^2/\text{s}$$

$$C = \frac{1}{r_p}\left(1 - \frac{GM_e}{r_p v_0^2}\right)$$

$$= \frac{1}{6.978(10^6)}\left(1 - \frac{6.673(10^{-11})[5.976(10^{24})]}{6.978(10^6)(8333.3)^2}\right) = 2.54(10^{-8}) \text{ m}^{-1}$$

Hence,

$$e = \frac{Ch^2}{GM_e} = \frac{2.54(10^{-8})[5.815(10^{10})]^2}{6.673(10^{-11})[5.976(10^{24})]} = 0.215 < 1 \qquad Ans.$$

From Eq. 13–23, observe that the orbit is an *ellipse*.

Part (b). If the satellite were launched at the apogee A shown in Fig. 13–26, with a velocity $\mathbf{v}_A$, the same orbit would be maintained provided

$$h = r_p v_0 = r_a v_A = 5.815(10^{10}) \text{ m}^2/\text{s}$$

Using Eq. 13–27, we have

$$r_a = \frac{r_p}{\dfrac{2GM_e}{r_p v_0^2} - 1} = \frac{6.978(10^6)}{\left(\dfrac{2[6.673(10^{-11})][5.976(10^{24})]}{6.978(10^6)(8333.3)^2} - 1\right)} = 10.804(10^6)$$

Thus,

$$v_A = \frac{5.815(10^{10})}{10.804(10^6)} = 5382.3 \text{ m/s} = 19.4 \text{ Mm/h} \qquad Ans.$$

PROBLEMS

In the following problems, assume that the radius of the earth is 6378 km, the earth's mass is $5.976(10^{24})$ kg, the mass of the sun is $1.99(10^{30})$ kg, and the gravitational constant is $G = 66.73(10^{-12})$ m³/(kg · s²).

*13–92. A satellite is placed into orbit at a velocity of 6 km/s, parallel to the surface of the earth. Determine the proper altitude of the satellite above the earth's surface such that its orbit remains circular. What will happen to the satellite if its initial velocity is 5 km/s when placed tangentially into orbit at the calculated altitude?

13–93. A communications satellite is to be placed into an equatorial circular orbit around the earth so that it always remains directly over a point on the earth's surface. If this requires the period to be 24 hours (approximately), determine the radius of the orbit and the satellite's velocity.

13–94. A satellite is launched with an initial velocity of $v_0 = 2500$ mi/h, parallel to the surface of the earth. Determine the required altitude (or range of altitudes) above the earth's surface for launching if the free-flight trajectory is to be (a) circular, (b) parabolic, (c) elliptical, and (d) hyperbolic. Take $G = 34.4(10^{-9})$(lb · ft²)/slug², $M_e = 409(10^{21})$ slug, the earth's radius $r_e = 3960$ mi, and 1 mi = 5280 ft.

13–95. The rocket is in a free-flight elliptical orbit about the earth such that the eccentricity of its orbit is e and its perigee is r_0. Determine the minimum increment of speed it should have in order to escape the earth's gravitational field. Where along the path should this occur?

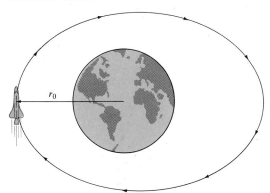

Prob. 13–95

*13–96. The rocket shown is originally in a circular orbit 6 Mm above the surface of the earth. It is required that it travel in another circular orbit having an altitude of 14 Mm. To do this, the rocket is given a short pulse of power at A so that it travels in free flight along the dashed elliptical path from the first orbit to the second orbit. Determine the necessary speed it must have at A just after the power pulse, and the time required to get to the outer orbit along the path AA'. What adjustment in speed must be made at A' to maintain the second circular orbit?

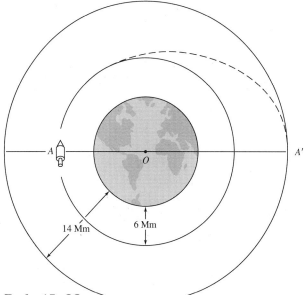

Prob. 13–96

13–97. A rocket is in circular orbit about the earth at an altitude of $h = 4$ Mm. Determine the minimum increment in speed it must have in order to escape the earth's gravitational field.

13–98. A satellite is in an elliptic orbit around the earth such that $e = 0.156$. If its perigee is 5 Mm, determine its velocity at this point and also the distance OB when it is at point B, located 135° away as shown.

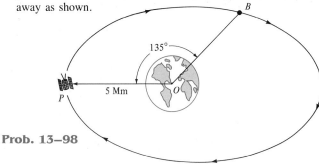

Prob. 13–98

13–99. A satellite is to be placed into an elliptical orbit about the earth such that at the perigee of its orbit it has an *altitude* of 800 km and at apogee its *altitude* is 2400 km. Determine its required launch velocity tangent to the earth's surface at perigee and the period of its orbit.

*13–100. An asteroid is in an elliptical orbit about the sun such that its periapsis is $9.30(10^9)$ km. If the eccentricity of the orbit is $e = 0.073$, determine the apoapsis of the orbit.

13–101. The rocket is traveling in a free-flight elliptical orbit about the earth such that $e = 0.76$ and its perigee is 9 Mm as shown. Determine its velocity when it is at point B. Also determine the sudden decrease in speed the rocket must experience at A in order to travel in a circular orbit about the earth.

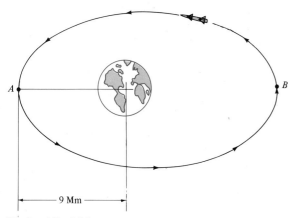

Prob. 13–101

13–102. The planet Jupiter travels around the sun in an elliptical orbit such that the eccentricity is $e = 0.048$. If the periapsis between Jupiter and the sun is $r_0 = 440(10^6)$ mi, determine (a) Jupiter's speed at periapsis and (b) the apoapsis of the orbit. Take $G = 34.4(10^{-9})$ lb · ft^2/slug^2, $M_s = 197(10^{27})$ slug, and 1 mi = 5280 ft.

13–103. Prove Kepler's third law of motion. *Hint:* Use Eqs. 13–19, 13–28, 13–29, and 13–31.

*13–104. Show that the velocity of a satellite launched into a circular orbit about the earth is given by Eq. 13–25. Determine the velocity of a satellite launched parallel to the surface of the earth so that it travels in a circular orbit 800 km from the earth's surface.

13–105. The elliptical path of a satellite has an eccentricity $e = 0.130$. If it has a speed of 15 Mm/h when it is at perigee, P, determine its speed when it arrives at apogee, A. Also, how far is it from the earth's surface when it is at A?

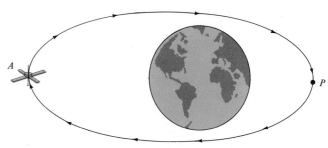

Prob. 13–105

13–106. A satellite S travels in a circular orbit around the earth. A rocket is located at the apogee of its elliptical orbit for which $e = 0.58$. Determine the sudden change in velocity that must occur at A so that the rocket can enter the satellite's orbit while in free flight along the dashed elliptical trajectory. When it arrives at B, determine the sudden adjustment in speed that must be given to the rocket in order to maintain the circular orbit.

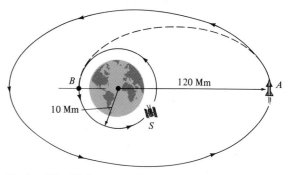

Prob. 13–106

14

Kinetics of a Particle: Work and Energy

In this chapter we will integrate the equation of motion with respect to displacement and thereby obtain the principle of work and energy. This principle is useful for solving problems which involve force, velocity, and displacement. Later in the chapter the concept of power is discussed and a method is presented for solving kinetic problems using the theorem of conservation of energy. Before presenting these topics, however, it is first necessary to define the work done by various types of forces.

14.1 The Work of a Force

In mechanics a force $\mathbf{F}$ does *work* only when it undergoes a *displacement in the direction of the force*. For example, consider the force $\mathbf{F}$ having a location on the path s which is specified by the position vector $\mathbf{r}$, Fig. 14–1. If the force moves along the path to a new position $\mathbf{r}'$, the displacement is then $d\mathbf{r} = \mathbf{r}' - \mathbf{r}$. The magnitude of $d\mathbf{r}$ is represented by ds, the differential segment along the path. If the angle between the tails of $d\mathbf{r}$ and $\mathbf{F}$ is θ, Fig. 14–1, then the work dU which is done by $\mathbf{F}$ is a *scalar quantity*, defined by

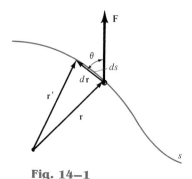

Fig. 14–1

$$dU = F \, ds \cos \theta$$

By definition of the dot product (see Eq. C–14) the above equation may also be written as

$$dU = \mathbf{F} \cdot d\mathbf{r}$$

Work as expressed by the above equation may be interpreted in one of two ways: either as the product of F and the component of displacement in the direction of the force, i.e., $ds \cos \theta$, or as the product of ds and the component of force in the direction of displacement, i.e., $F \cos \theta$. Note that if $0° \leq \theta \leq 90°$, then the force component and the displacement have the *same sense* so that the work is *positive;* whereas if $90° < \theta \leq 180°$, these vectors have an *opposite sense* and therefore the work is *negative*. Also, $dU = 0$ if the force is *perpendicular* to displacement, since $\cos 90° = 0$, or if the force is applied at a *fixed point,* in which case the displacement is zero.

The basic unit for work in the SI system is called a joule (J). This unit combines the units of force and displacement. Specifically, 1 *joule* of work is done when a force of 1 newton moves 1 meter along its line of action (1 J = 1 N · m). The moment of a force has this same combination of units (N · m); however, the concepts of moment and work are in no way related. A moment is a vector quantity, whereas work is a scalar. In the FPS system work is generally defined by writing the units as ft · lb, which is distinguished from the units for a moment written as lb · ft.

Work of a Variable Force

If a force undergoes a finite displacement along its path from $\mathbf{r}_1$ to $\mathbf{r}_2$ or s_1 to s_2, Fig. 14–2a, the work is determined by integration. If $\mathbf{F}$ is expressed as a function of position, $F = F(s)$, we have

$$U_{1-2} = \int_{\mathbf{r}_1}^{\mathbf{r}_2} \mathbf{F} \cdot d\mathbf{r} = \int_{s_1}^{s_2} F \cos \theta \, ds \qquad (14\text{–}1)$$

If the working component of the force, $F \cos \theta$, is plotted versus s, Fig. 14–2b, the integral represented in this equation can be interpreted as the *area under the curve* from position s_1 to position s_2.

(a)

Fig. 14–2

(b)

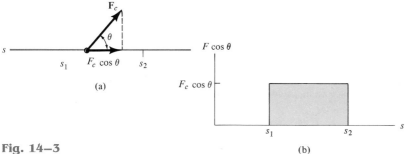

Fig. 14–3

Work of a Constant Force Moving Along a Straight Line

If the force $\mathbf{F}_c$ has a constant magnitude and acts at a constant angle θ from its straight-line path, Fig. 14–3a, then the component of $\mathbf{F}_c$ in the direction of displacement is $F_c \cos \theta$. The work done by $\mathbf{F}_c$ when it is displaced from s_1 to s_2 is determined by Eq. 14–1, in which case

$$U_{1-2} = F_c \cos \theta \int_{s_1}^{s_2} ds$$

or

$$\boxed{U_{1-2} = F_c \cos \theta \, (s_2 - s_1)} \qquad (14\text{–}2)$$

Here the work of $\mathbf{F}_c$ represents the *area under the rectangle* in Fig. 14–3b.

Work of a Weight

Consider a particle which moves up along the path s shown in Fig. 14–4 from position s_1 to position s_2. At an intermediate point, the displacement $d\mathbf{r} = dx\,\mathbf{i} + dy\,\mathbf{j} + dz\,\mathbf{k}$. Since $\mathbf{W} = -W\mathbf{j}$, applying Eq. 14–1 yields

$$U_{1-2} = \int \mathbf{F} \cdot d\mathbf{r} = \int_{\mathbf{r}_1}^{\mathbf{r}_2} (-W\mathbf{j}) \cdot (dx\,\mathbf{i} + dy\,\mathbf{j} + dz\,\mathbf{k})$$

$$= \int_{y_1}^{y_2} -W\,dy = -W(y_2 - y_1)$$

or

$$\boxed{U_{1-2} = -W\Delta y} \qquad (14\text{–}3)$$

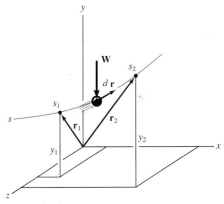

Fig. 14–4

Thus, the work done is equal to the magnitude of the particle's weight times its vertical displacement. In the case shown in Fig. 14–4 the work is *negative,* since W is downward and Δy is upward. Note, however, that if the particle is displaced *downward* $(-\Delta y)$ the work of the weight is *positive*. Why?

Work of a Spring Force

The magnitude of force developed in a linear elastic spring when the spring is displaced a distance s from its unstretched position is $F_s = ks$, where k is the spring stiffness. If the spring is elongated or compressed from a position s_1 to a further position s_2, Fig. 14–5a, the work done *on the spring* by $\mathbf{F}_s$ is *positive*, since in each case the force and displacement are in the *same direction*. We require

$$U_{1-2} = \int_{s_1}^{s_2} F_s \, ds = \int_{s_1}^{s_2} ks \, ds$$
$$= \tfrac{1}{2}ks_2^2 - \tfrac{1}{2}ks_1^2$$

This equation represents the trapezoidal area under the line $F_s = ks$ versus s, Fig. 14–5b.

If a particle (or body) is attached to a spring, then the force $\mathbf{F}_s$ exerted on the particle is *opposite* to that exerted on the spring, Fig. 14–5c. Consequently, the force will do *negative work* on the particle when the particle is moving so as to further elongate (or compress) the spring. Hence, the above equation becomes

$$\boxed{U_{1-2} = -(\tfrac{1}{2}ks_2^2 - \tfrac{1}{2}ks_1^2)} \tag{14–4}$$

When this equation is used, a mistake in sign can be eliminated if one simply notes the direction of the spring force acting on the particle and compares it with the direction of displacement of the particle—if both are in the *same direction, positive work* results; if they are *opposite* to one another, the *work is negative*.

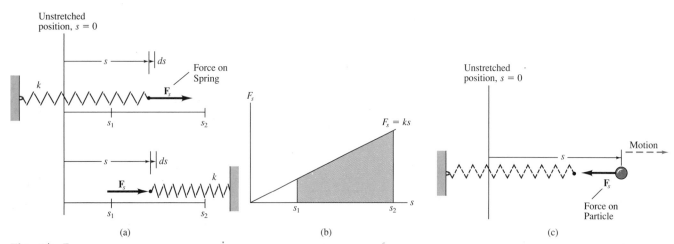

Fig. 14–5

Example 14–1

The 10-kg block shown in Fig. 14–6a rests on the smooth incline. If the spring is originally unstretched, determine the total work done by all the forces acting on the block when a horizontal force $P = 400$ N pushes the block up the plane $s = 2$ m.

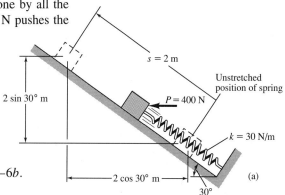

(a)

SOLUTION

The free-body diagram of the block is shown in Fig. 14–6b.

(b)

Fig. 14–6

Horizontal Force **P.** Since the force is *constant,* the work is computed using Eq. 14–2. The result can be calculated as the force times the component of displacement in the direction of the force, i.e.,

$$U_P = 400(2 \cos 30°) = 692.8 \text{ J}$$

or, the displacement times the component of force in the direction of displacement, i.e.,

$$U_P = 400 \cos 30°(2) = 692.8 \text{ J}$$

Spring Force **F$_s$.** Since the spring is originally unstretched and in the final position is stretched 2 m, the work of **F$_s$** is

$$U_s = -\tfrac{1}{2}(30)(2)^2 = -60 \text{ J}$$

Why is the work negative?

Weight **W.** Since the weight acts in the opposite direction to its vertical displacement the work is negative; i.e.,

$$U_W = -98.1(2 \sin 30°) = -98.1 \text{ J}$$

Note that it is also possible to consider the component of weight in the direction of displacement; i.e.,

$$U_W = -(98.1 \sin 30°)2 = -98.1 \text{ J}$$

Normal Force **N$_B$.** This force does *no work* since it is *always* perpendicular to the displacement.

The work of all the forces when the block is displaced 2 m is thus

$$U_T = 692.8 - 60 - 98.1 = 535 \text{ J} \qquad \textit{Ans.}$$

14.2 Principle of Work and Energy

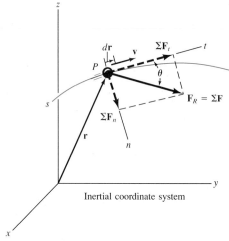

Inertial coordinate system

Fig. 14–7

Consider a particle P in Fig. 14–7, which at the instant considered is located at position $\mathbf{r}$ on the path as measured from an inertial coordinate system. If the particle has a mass m and is subjected to a system of external forces represented by the resultant $\mathbf{F}_R = \Sigma\mathbf{F}$, then the equation of motion for the particle is $\Sigma\mathbf{F} = m\mathbf{a}$. When the particle undergoes a displacement $d\mathbf{r}$ along the path, the work done by the forces is

$$\Sigma\mathbf{F} \cdot d\mathbf{r} = m\mathbf{a} \cdot d\mathbf{r}$$

If we establish the n and t axes at the particle, then $\Sigma\mathbf{F}$ can be resolved into its normal and tangential components, Fig. 14–7. Realizing that the magnitude of $d\mathbf{r}$ is ds, it can be seen that $\Sigma\mathbf{F} \cdot d\mathbf{r} = \Sigma F \, ds \cos\theta = \Sigma F_t \, ds$. In other words, the work of $\Sigma\mathbf{F}$ is computed *only from its tangential components*. The normal components $\Sigma F_n = \Sigma F \sin\theta$ do *no work* since they cannot move in the normal direction. Since $\mathbf{a} \cdot d\mathbf{r} = a_t \, ds$, the above equation can also be written as

$$\Sigma\mathbf{F} \cdot d\mathbf{r} = \Sigma F_t \, ds = ma_t \, ds$$

Using the kinematic equation $a_t \, ds = v \, dv$, and integrating both sides, assuming initially that the particle has a position $\mathbf{r} = \mathbf{r}_1$ and a speed $v = v_1$, and later $\mathbf{r} = \mathbf{r}_2$, $v = v_2$, yields

$$\Sigma \int_{\mathbf{r}_1}^{\mathbf{r}_2} \mathbf{F} \cdot d\mathbf{r} = \int_{v_1}^{v_2} mv \, dv$$

or

$$\Sigma \int_{\mathbf{r}_1}^{\mathbf{r}_2} \mathbf{F} \cdot d\mathbf{r} = \tfrac{1}{2}mv_2^2 - \tfrac{1}{2}mv_1^2 \tag{14–5}$$

Using Eq. 14–1, the final result may be written as

$$\Sigma U_{1-2} = \tfrac{1}{2}mv_2^2 - \tfrac{1}{2}mv_1^2 \tag{14–6}$$

This equation represents the *principle of work and energy* for the particle. The term on the left is the sum of the work done by *all* the forces acting on the particle as the particle moves from point 1 to point 2. The two terms on the right side, which are of the form $T = \tfrac{1}{2}mv^2$, define the particle's final and initial *kinetic energy,* respectively. These terms are *positive* scalar quantities since they do not depend on the direction of the particle's velocity. Furthermore, Eq. 14–6 is dimensionally homogeneous so that the kinetic energy has the same units as work, e.g., joules (J) or ft · lb.

When Eq. 14–6 is applied, it is often symbolized in the form

$$\boxed{T_1 + \Sigma U_{1-2} = T_2} \tag{14–7}$$

which states that the particle's initial kinetic energy plus the work done by all the forces acting on the particle as it moves from its initial to its final position is equal to the particle's final kinetic energy.

As noted from the derivation, the principle of work and energy represents an integrated form of $\Sigma F_t = ma_t$, acquired by using the kinematic equation $a_t = v\,dv/ds$. As a result, this principle will provide a convenient substitution for $\Sigma F_t = ma_t$ when solving those types of kinetic problems which involve force, velocity, and displacement, since these variables are involved in the terms of Eq. 14–7. For example, if a particle's initial speed is known, and the work of all the forces acting on the particle can be computed, then Eq. 14–7 provides a *direct means* of obtaining the final speed v_2 of a particle after it undergoes a specified displacement. If instead v_2 is determined by means of the equation of motion, a two-step process is necessary; i.e., apply $\Sigma F_t = ma_t$ to obtain a_t, then integrate $a_t = v\,dv/ds$ to obtain v_2. Note that the principle of work and energy cannot be used, for example, to determine forces directed *normal* to the path of motion, since these forces do no work on the particle. Instead $\Sigma F_n = ma_n$ must be applied. For curved paths, however, the magnitude of the normal force is a function of velocity. Hence, it is generally easier to obtain the velocity using the principle of work and energy, and then substitute this quantity into the equation of motion $\Sigma F_n = mv^2/\rho$ to obtain the normal force.

PROCEDURE FOR ANALYSIS

As stated above, the principle of work and energy is used to solve kinetic problems that involve *velocity, force,* and *displacement,* since these terms are involved in the equation. For application it is suggested that the following procedure be used.

Work (Free-Body Diagram). Establish the inertial coordinate system and draw a free-body diagram of the particle in order to account for all the forces that do work on the particle as it moves along its path.

Principle of Work and Energy. Apply the principle of work and energy, $T_1 + \Sigma U_{1-2} = T_2$. The kinetic energy at the initial and final points is always positive, since it involves the speed squared ($T = \frac{1}{2}mv^2$). The work done by each force shown on the free-body diagram is computed by using the appropriate equations developed in Sec. 14.1. Since *algebraic addition* of the work terms is required, it is important that the proper sign of each term be specified. Specifically, work is positive when the force component is in the same direction as its displacement, otherwise it is negative.

Numerical application of this procedure is illustrated in the examples following Sec. 14.3.

14.3 Principle of Work and Energy for a System of Particles

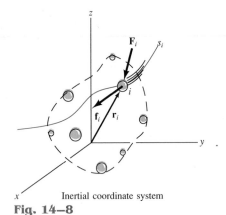

Fig. 14–8

Inertial coordinate system

The principle of work and energy can be extended to include a system of n particles isolated within an enclosed region of space as shown in Fig. 14–8. Here the arbitrary ith particle, having a mass m_i, is subjected to a resultant external force $\mathbf{F}_i$ and a resultant internal force $\mathbf{f}_i = \sum\limits_{j=1(j \neq i)}^{n} \mathbf{f}_{ij}$ which each of the other particles exerts on the ith particle. Using Eq. 14–5, the principle of work and energy written for the ith particle is thus

$$\tfrac{1}{2}m_i v_{i1}^2 + \int_{\mathbf{r}_{i1}}^{\mathbf{r}_{i2}} \mathbf{F}_i \cdot d\mathbf{r}_i + \int_{\mathbf{r}_{i1}}^{\mathbf{r}_{i2}} \mathbf{f}_i \cdot d\mathbf{r}_i = \tfrac{1}{2}m_i v_{i2}^2$$

Similar equations result if the principle of work and energy is applied to each of the other particles of the system. Since both work and kinetic energy are scalars, the results may be added together algebraically, so that

$$\Sigma\tfrac{1}{2}m_i v_{i1}^2 + \Sigma\int_{\mathbf{r}_{i1}}^{\mathbf{r}_{i2}} \mathbf{F}_i \cdot d\mathbf{r}_i + \Sigma\int_{\mathbf{r}_{i1}}^{\mathbf{r}_{i2}} \mathbf{f}_i \cdot d\mathbf{r}_i = \Sigma\tfrac{1}{2}m_i v_{i2}^2$$

We can write this equation symbolically as

$$\Sigma T_1 + \Sigma U_{1-2} = \Sigma T_2 \tag{14–8}$$

Hence, the system's initial kinetic energy (ΣT_1) plus the work done by all the external and internal forces acting on the particles of the system (ΣU_{1-2}) is equal to the system's final kinetic energy (ΣT_2). To maintain this balance of energy, strict accountability of the work done by all the forces must be made. In this regard, note that although the internal forces on adjacent particles occur in equal but opposite collinear pairs, the total work done by each of these forces will, in general, *not cancel* out since the paths over which corresponding particles travel will be *different*. There are, however, two important exceptions to this rule which often occur in practice. If the particles are contained within the boundary of a *translating rigid body*, the internal forces all undergo displacement by an equal amount, and therefore the internal work will be zero. Also, particles connected by inextensible cables make up a system which has internal forces which are displaced by an equal amount. In this case, adjacent particles exert equal but opposite internal forces which have components that undergo the same displacement, and therefore the work of these forces cancels. On the other hand, note that if the body is assumed to be *nonrigid*, the particles of the body are displaced along *different paths*, and some of the energy due to force interactions would be given off and lost as heat or stored in the body if permanent deformations occur. We will briefly discuss these effects at the end of this section and in Sec. 15–4. Throughout this text, however, the principle of work and energy will be applied to the solution of problems only when direct accountability of these energy losses does not have to be made.

The procedure for analysis outlined in Sec. 14.2 provides a method for applying Eq. 14–8; however, only one equation applies for the entire system. If the particles are connected by cords, other equations can generally be obtained by using the kinematic principles outlined in Sec. 12.8 in order to relate the particles' speeds. When doing this, assume that the particles are displaced in the *positive coordinate directions* when writing *both* the kinematic equation and the principle of work and energy (see Example 14–6).

Work of Friction Caused by Sliding

A special class of problems will now be investigated which requires a careful application of Eq. 14–8. These problems all involve cases where a body is sliding over the surface of another body in the presence of friction. Consider, for example, a block which is translating a distance s over a rough surface as shown in Fig. 14–9. If the applied force **P** just balances the *resultant* frictional force $\mu_k N$, then due to equilibrium a constant velocity **v** is maintained and one would expect Eq. 14–8 to be applied as follows:

$$\tfrac{1}{2}mv^2 + Ps - \mu_k N s = \tfrac{1}{2}mv^2$$

Fig. 14–9

Indeed this equation is satisfied if $P = \mu_k N$; however, as one realizes from experience, the sliding motion will *generate heat,* a form of energy which seems not to be accounted for in the work-energy equation. In order to explain this paradox and thereby more accurately represent the nature of friction, we should actually consider a model of the surfaces of contact which are *deformable* (nonrigid).* Recall that the rough portions at the bottom of the block act as "teeth," and when the block slides these teeth *deform slightly* and either break off or vibrate due to interlocking effects and pulling away from "teeth" at the contacting surface. As a result, frictional forces that act at these points are diminished; however, they are replaced by other frictional forces as other points of contact are made. At any instant, the *resultant* of the frictional forces remains essentially constant, i.e., $\mu_k N$; however, due to the localized deformations, the actual displacement s' of $\mu_k N$ is *not* the same displacement s as the applied force **P.** Instead, s' will be *less* than s ($s' < s$), and therefore the *external work* done by the resultant frictional force will be $\mu_k N s'$. The remaining amount of work, $\mu_k N(s - s')$, manifests itself as an increase in *internal energy,* which in fact causes the block's temperature to rise.

In summary then, Eq. 14–8 can be applied to problems involving sliding friction; however, *it should be fully realized that the work of the resultant frictional force is not represented by $\mu_k N s$; instead, this term represents both the external work of friction ($\mu_k N s'$) and internal work [$\mu_k N(s - s')$] which is converted into various forms of internal energy.*†

*See Chapter 9 of *Engineering Mechanics: Statics.*

†See B. A. Sherwood and W. H. Bernard, "Work and Heat Transfer in the Presence of Sliding Friction," *Am. J. Phys.* 52, 1001 (1984).

Example 14–2

The 3500-lb automobile shown in Fig. 14–10a is traveling down the 10° inclined road at a speed of 20 ft/s. If the driver wishes to stop his car, determine how far s his tires skid on the road if he jams on the brakes, causing his wheels to lock. The coefficient of kinetic friction between the wheels and the road is $\mu_k = 0.5$.

(a)

SOLUTION

This problem can be solved using the principle of work and energy since it involves force, velocity, and displacement.

(b)

Fig. 14–10

Work (Free-Body Diagram). As shown in Fig. 14–10b, the normal force N_A does no work since it never undergoes displacement along its line of action. The weight, 3500 lb, is displaced $s \sin 10°$ and does positive work, whereas the frictional force F_A does both external and internal work when it is *thought* to undergo a displacement s. Applying the equation of equilibrium normal to the road, we have

$$+\nwarrow\Sigma F_n = 0; \quad N_A - 3500 \cos 10° = 0 \qquad N_A = 3446.8 \text{ lb}$$

Thus,
$$F_A = 0.5 N_A = 1723.4 \text{ lb}$$

Principle of Work and Energy

$$\{T_1\} + \{\Sigma U_{1-2}\} = \{T_2\}$$

$$\left\{\frac{1}{2}\left(\frac{3500}{32.2}\right)(20)^2\right\} + \{3500(s \sin 10°) - 1723.4s\} = \{0\}$$

Solving for s yields
$$s = 19.5 \text{ ft} \qquad\qquad \textit{Ans.}$$

Note that if this problem is solved by using the equation of motion, *two steps* are involved. First, from the free-body diagram, Fig. 14–10b, the equation of motion is applied along the incline. This yields

$$+\swarrow\Sigma F_s = ma_s; \quad 3500 \sin 10° - 1723.4 = \frac{3500}{32.2}a$$

$$a = -10.3 \text{ ft/s}^2$$

Then, using the integrated form of $a \, ds = v \, dv$ (kinematics) since a is constant, we have

$$(+\swarrow) \quad v^2 = v_0^2 + 2a_c(s - s_0); \qquad (0)^2 = (-20)^2 + 2(-10.3)(s - 0)$$

$$s = 19.5 \text{ ft} \qquad \textit{Ans.}$$

Example 14–3

A 10-kg block rests on the horizontal surface shown in Fig. 14–11a. The spring, which is not attached to the block, has a stiffness $k = 500$ N/m and is initially compressed 0.2 m from C to A. After the block is released from rest at A, determine its velocity when it passes point D. The coefficient of kinetic friction between the block and the plane is $\mu_k = 0.2$.

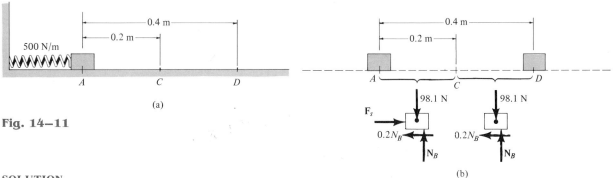

(a)

Fig. 14–11

(b)

SOLUTION

Why can this problem be solved using the principle of work and energy?

Work (Free-Body Diagrams). Two free-body diagrams for the block are shown in Fig. 14–11b. The block moves under the influence of the spring force $\mathbf{F}_s$ along the 0.2-m-long path AC, after which it continues to slide along the plane to point D. With reference to either free-body diagram, $\Sigma F_y = 0$; hence, $N_B = 98.1$ N. Only the spring and friction forces do work during the displacement—the spring force does positive work from A to C, whereas the frictional force does negative work. Why?

Principle of Work and Energy

$$\{T_A\} + \{\Sigma U_{A-D}\} = \{T_D\}$$

$$\{\tfrac{1}{2}m(v_A)^2\} + \{\tfrac{1}{2}ks_{AC}^2 - (0.2N_B)s_{AD}\} = \{\tfrac{1}{2}m(v_D)^2\}$$

$$\{0\} + \left\{\frac{1}{2}(500)(0.2)^2 - 0.2(98.1)(0.4)\right\} = \left\{\frac{1}{2}(10)(v_D)^2\right\}$$

Solving for v_D, we get

$$v_D = 0.656 \text{ m/s} \rightarrow \qquad\qquad \textit{Ans.}$$

Example 14–4

The platform P, shown in Fig. 14–12a, has negligible mass and is tied down so that the 0.4-m-long cords keep the spring compressed 0.6 m when *nothing* is on the platform. If a 2-kg block is placed on the platform and released from rest after the platform is pushed down 0.1 m, determine the maximum height h the block rises in the air, measured from the ground.

Original height of platform

0.1 m

h

P

0.4 m

$k = 200$ N/m

(a)

SOLUTION

Work (Free-Body Diagram). Since the block is released from rest and later reaches its maximum height, the initial and final velocities are zero. The free-body diagram of the block when it is still in contact with the platform is shown in Fig. 14–12b. Note that the weight does negative work and the spring force does positive work. Why? In particular, the *initial compression* in the spring is $s_1 = 0.6$ m $+ 0.1$ m $= 0.7$ m. Due to the cords, the spring's *final compression* is $s_2 = 0.6$ m (after the block leaves the platform). The bottom of the block rises from a height of (0.4 m $-$ 0.1 m) $= 0.3$ m to a final height h.

19.62 N

$\mathbf{F}_s$

(b)

Fig. 14–12

Principle of Work and Energy

$$\{T_1\} + \{\Sigma U_{1-2}\} = \{T_2\}$$
$$\{\tfrac{1}{2}mv_1^2\} + \{-(\tfrac{1}{2}ks_2^2 - \tfrac{1}{2}ks_1^2) - W\Delta y\} = \{\tfrac{1}{2}mv_2^2\}$$

Note that here $s_2 < s_1$, so that the work of the spring will indeed be positive. Thus,

$$\{0\} + \left\{-\left(\frac{1}{2}(200)(0.6)^2 - \frac{1}{2}(200)(0.7)^2\right) - (19.62)[h - (0.3)]\right\} = \{0\}$$

Solving yields

$$h = 0.963 \text{ m} \qquad\qquad Ans.$$

Example 14–5

A block having a mass of 2 kg is given an initial velocity of $v_0 = 1$ m/s when it is at the top surface of the smooth cylinder shown in Fig. 14–13a. If the block travels along a path of 0.5-m radius, determine the angle $\theta = \theta_{max}$ at which it begins to leave the cylinder's surface.

(a)

SOLUTION

Work (Free-Body Diagram). Inspection of Fig. 14–13b reveals that only the weight $W = 2(9.81) = 19.62$ N does work during the displacement. If the block is assumed to leave the surface when $\theta = \theta_{max}$ then the weight moves through a vertical displacement of $0.5(1 - \cos \theta_{max})$ m, as shown in the figure.

Principle of Work and Energy

$$\{T_1\} + \{\Sigma U_{1-2}\} = \{T_2\}$$

$$\left\{\frac{1}{2}(2)(1)^2\right\} + \{19.62(0.5)(1 - \cos \theta_{max})\} = \left\{\frac{1}{2}(2)v_2^2\right\}$$

$$v_2^2 = 9.81(1 - \cos \theta_{max}) + 1 \qquad (1)$$

(b)

Fig. 14–13

Equation of Motion. There are two unknowns in Eq. (1), θ_{max} and v_2. A second equation relating these two variables may be obtained by applying the equation of motion in the *normal direction* to the forces on the free-body diagram. Thus,

$$+\swarrow \Sigma F_n = ma_n; \qquad -N_B + 19.62 \cos \theta = 2\left(\frac{v^2}{0.5}\right)$$

When the block leaves the surface of the cylinder at $\theta = \theta_{max}$, $N_B = 0$ and $v = v_2$; hence,

$$\cos \theta_{max} = \frac{v_2^2}{4.905} \qquad (2)$$

Eliminating the unknown v_2^2 between Eqs. (1) and (2) gives

$$4.905 \cos \theta_{max} = 9.81(1 - \cos \theta_{max}) + 1$$

Solving, we have

$$\cos \theta_{max} = 0.735$$

$$\theta_{max} = 42.7° \qquad \qquad Ans.$$

This problem has also been solved in Example 13–9. If the two methods of solution are compared it will be apparent that a work-energy approach yields a more direct solution.

Example 14–6

The blocks A and B shown in Fig. 14–14a have a mass of 10 kg and 100 kg, respectively. Determine the distance B travels from the point where it is released from rest to the point where its speed becomes 2 m/s.

SOLUTION

This problem may be solved by considering the blocks separately and applying the principle of work and energy to each block. However, the work of the (unknown) cable tension can be eliminated from the analysis by considering blocks A and B together as a *system*. The solution will require simultaneous solution of the equations of work and energy *and* kinematics. To be consistent with our sign convention, we will assume both blocks move in the positive *downward* direction.

(a)

Work (Free-Body Diagram). As shown on the free-body diagram of the system, Fig. 14–14b, the cable force **T** and reactions **R**$_1$ and **R**$_2$ do *no work,* since these forces represent the reactions at the supports and consequently do not move while the blocks are being displaced.

Principle of Work and Energy. Realizing the blocks are released from rest, and as stated previously, assuming *both* move *downward,* we have

$$\{\Sigma T_1\} + \{\Sigma U_{1-2}\} = \{\Sigma T_2\}$$

$$\{\tfrac{1}{2}m_A(v_A)_1^2 + \tfrac{1}{2}m_B(v_B)_1^2\} + \{W_B\Delta s_B + W_A\Delta s_A\} = \{\tfrac{1}{2}m_A(v_A)_2^2 + \tfrac{1}{2}m_B(v_B)_2^2\}$$

$$\{0 + 0\} + \{981(\Delta s_B) + 98.1(\Delta s_A)\} = \left\{\frac{1}{2}(10)(v_A)_2^2 + \frac{1}{2}(100)(2)^2\right\} \quad (1)$$

(b)

Fig. 14–14

Kinematics. Using the methods of kinematics discussed in Sec. 12.8, it may be seen from Fig. 14–14a that at any given instant the total length l of all the vertical segments of cable may be expressed in terms of the position coordinates s_A and s_B as

$$s_A + 4s_B = l$$

Hence, a change in position yields the displacement equation

$$\Delta s_A + 4\,\Delta s_B = 0$$
$$\Delta s_A = -4\,\Delta s_B \quad (2)$$

Taking the time derivative yields

$$v_A = -4v_B = -4(2) = -8 \text{ m/s}$$

Retaining the negative sign in Eq. (2) and substituting into Eq. (1) yields

$$\Delta s_B = 0.883 \text{ m} \downarrow \qquad\qquad Ans.$$

PROBLEMS

14-1. A woman having a mass of 70 kg stands in an elevator which has a downward acceleration of 4 m/s^2 starting from rest. Determine the work done by her weight and the work of the normal force which the floor exerts on her when the elevator descends 6 m. Explain why the work of these forces is different.

14-2. The car having a mass of 2 Mg is towed along an inclined road. If the car starts from rest and attains a speed of 5 m/s after traveling a distance of 150 m, determine the constant towing force F applied to the car. Neglect friction and the mass of the wheels.

Prob. 14-2

14-3. The "air spring" A is used to protect the support structure B and prevent damage to the conveyor belt tensioning weight C in the event of a belt failure D. The force developed by the spring as a function of its deflection is shown by the graph. Determine the suspended height d of the 50-lb counterweight C so that if the conveyor belt fails the counterweight will deform the spring $s_{max} = 0.2$ ft. Neglect the mass of the pulley and belt.

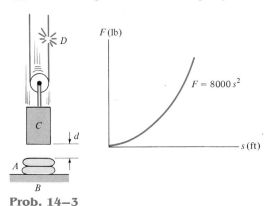

Prob. 14-3

*14-4. Marbles having a mass of 5 g fall from rest at A through the glass tube and accumulate in the can at C. Determine the placement R of the can from the end of the tube and the speed at which the marbles fall into the can. Neglect the size of the can.

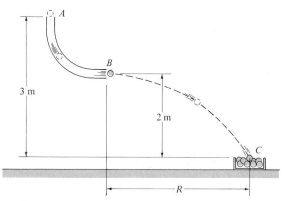

Prob. 14-4

14-5. Packages having a weight of 15 lb are transferred horizontally from one conveyor to the next using a ramp for which the coefficient of kinetic friction is $\mu_k = 0.15$. The top conveyor is moving at 6 ft/s and the packages are spaced 3 ft apart. Determine the required speed of the bottom conveyor so no sliding occurs when the packages come horizontally in contact with it. What is the spacing s between the packages on the bottom conveyor?

Prob. 14-5

14–6. The motion of a 6500-lb boat is arrested using a bumper which provides a resistance as shown in the graph. Determine the distance the boat dents the bumper if it is coasting at 3 ft/s when it approaches it.

$v = 3$ ft/s

$F = 3(10^3)s^3$

F(lb)

s(ft)

Prob. 14–6

14–7. The conveyor belt delivers each 12-kg crate to the ramp at A such that the crate's velocity is $v_A = 2.5$ m/s, directed down *along* the ramp. If the coefficient of kinetic friction between each crate and the ramp is $\mu_k = 0.3$, determine the speed at which each crate slides off the ramp at B. Assume that no tipping occurs.

$v_A = 2.5$ m/s

Prob. 14–7

3 m

$30°$

A

B

*14–8. Cylinder A has a mass of $m_A = 3$ kg and cylinder B has a mass of $m_B = 8$ kg. Determine the speed of A after it moves upward 2 m starting from rest. Neglect the mass of the cord and pulleys.

A

Prob. 14–8

B

14–9. The smooth plug has a weight of 20 lb and is pushed against a series of Belleville spring washers so that the compression in the spring is $s = 0.05$ ft. If the force of the spring on the plug is $F = (3s^{1/3})$ lb, where s is given in feet, determine the speed of the plug after it moves away from the spring. Neglect friction.

Prob. 14–9

14–10. Block A has a weight of $W_A = 60$ lb and block B has a weight of $W_B = 10$ lb. Determine the speed of A just after it moves downward 3 ft starting from rest. Neglect the mass of the cord and pulleys.

B

A

Prob. 14–10

14–11. The 50-kg block has a speed of $v_A = 8$ m/s when it reaches point A. Determine the normal force it exerts on the incline when it reaches point B. Neglect friction and the block's size.

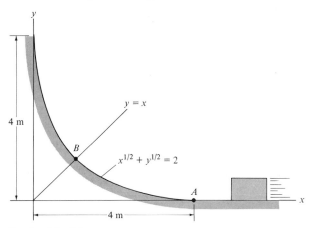

Prob. 14–11

*14–12. The spring has a stiffness $k = 50$ lb/ft and an *unstretched length* of 2 ft. As shown, it is confined by the plate and wall using cables so that its length is 1.5 ft. Determine the speed v_A of the 4-lb block when it is at A, so that it slides along the horizontal plane and strikes the plate and pushes it forward 0.25 ft before stopping. The coefficient of kinetic friction between the block and plane is $\mu_k = 0.2$. Neglect the mass of the plate and spring and the energy loss between the plate and block during the collision.

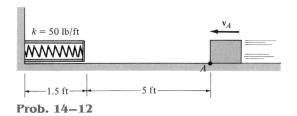

Prob. 14–12

14–13. The 120-lb man acts as a human cannon ball by being "fired" from the spring-loaded cannon shown. If the greatest acceleration he can experience is $a = 10g = 322$ ft/s^2, determine the required stiffness of the spring if it is compressed 2 ft at the moment of firing. With what velocity will he exit the cannon barrel when the cannon is fired? When the spring is compressed $s = 2$ ft then $d = 8$ ft. Neglect friction and assume the man holds himself in a rigid position throughout the motion.

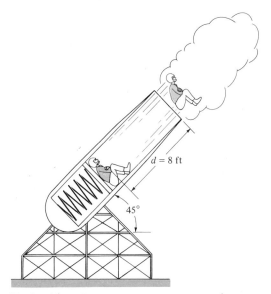

Prob. 14–13

14–14. Packages having a weight of 50 lb are delivered to the chute at $v_A = 3$ ft/s using a conveyor belt. Determine their speed when they reach points B, C, and D. Also calculate the normal force of the chute on the packages at B and C. Neglect friction and the size of the packages.

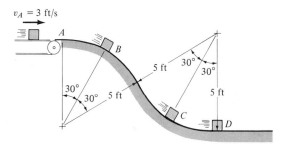

Prob. 14–14

141

14–15. The 0.5-kg ball of negligible size is fired up the vertical circular track using the spring plunger. The plunger keeps the spring compressed 0.08 m when $s = 0$. Determine how far s it must be pulled back and released so that the ball will begin to leave the track when $\theta = 135°$.

14–18. Block A has a weight of $W_A = 3$ lb and rests on a surface for which the coefficient of kinetic friction is $\mu_k = 0.3$. Block B has a weight of $W_B = 8$ lb. Determine the distance B must descend so that A has a speed of $v_A = 5$ ft/s starting from rest.

Prob. 14–18

Probs. 14–15/14–16

*14–16. The 0.5-kg ball of negligible size is fired up the vertical circular track using the spring plunger. The plunger keeps the spring compressed 0.08 m when $s = 0$. Determine how far s it must be pulled back and released so that the ball will just make it around the loop and land on the platform at B.

14–17. The smooth block has a mass m and is attached to a spring having an unstretched length when $y = d$. Determine the work done by the horizontal force $\mathbf{P}$ needed to raise the block from an at-rest position at $y = d$ to an at-rest position when the elevation is y.

14–19. The 200-kg roller coaster car is accelerated uniformly from rest at A until it reaches a maximum speed at B in $t = 3.5$ s, at which point it begins to travel freely along the spiral loop. Determine this maximum speed at B so that the car just makes the loop without falling off the track. Also, compute the constant horizontal force F needed to give the car the necessary acceleration from A to B. The radius of curvature at C is $\rho_C = 25$ m.

Prob. 14–17

Probs. 14–19/14–20

*14–20. Determine the height h to the top of the incline D to which the 200-kg roller coaster car will reach, if it is launched at B with a speed just sufficient for it to round the top of the loop at C without leaving the tracks. The radius of curvature at C is $\rho_C = 25$ m.

14–21. The collar has a mass of 5 kg and is moving at 8 m/s when $x = 0$. A force of $F = 60$ N is applied to it as shown. The direction θ of this force varies such that $\theta = 10x$, where x is in meters and θ is clockwise, measured in degrees. Determine the speed of the collar when $x = 3$ m. The coefficient of kinetic friction between the collar and the rod is $\mu_k = 0.3$.

$F = 60$ N

Prob. 14–21

■**14–22.** The force **F** acting in a constant direction on the 20-kg block has a magnitude which varies with the position s of the block. Determine how far the block slides before its velocity becomes 5 m/s. When $s = 0$ the block is moving to the right at 2 m/s. The coefficient of kinetic friction between the block and surface is $\mu_k = 0.3$.

F(N)

$F = 50s^2$

s(m)

Prob. 14–22

14–23. As indicated by the derivation, the principle of work and energy is valid for observers in *any* inertial reference frame. Show that this is so, by considering the 10-kg block which rests on the smooth surface and is subjected to a horizontal force of 6 N. If observer A is in a *fixed* frame x, determine the final speed of the block if it has an initial speed of 5 m/s and travels 10 m, both directed to the right and measured from the fixed frame. Compare the result with that obtained by an observer B, attached to the x' axis and moving at a constant velocity of 2 m/s relative to A. *Hint:* The distance the block travels will first have to be computed for observer B before applying the principle of work and energy.

Prob. 14–23

*****14–24.** The 2-kg block is subjected to a force having a constant direction and a magnitude $F = (300/(1 + s))$ N, where s is measured in meters. When $s = 4$ m, the block is moving to the left with a speed of 8 m/s. Determine its speed when $s = 12$ m. The coefficient of kinetic friction between the block and the ground is $\mu_k = 0.25$.

v   **F**

30°

s

Prob. 14–24

14–25. The steel ingot has a mass of 1800 kg. It travels along the conveyor at a speed of $v = 0.5$ m/s when it collides with the "nested" spring assembly. Determine the maximum deflection in each spring needed to stop the motion of the ingot. Neglect the mass of the springs.

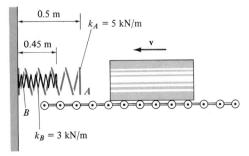

Prob. 14–25

14–26. The collar has a mass of 20 kg and is supported on the smooth rod. The attached springs are undeformed when $d = 0.5$ m. Determine the speed of the collar after the applied force $F = 100$ N causes it to be displaced so that $d = 0.3$ m. When $d = 0.5$ m the collar is at rest.

Probs. 14–26/14–27

14–27. The collar has a mass of 20 kg and is supported on the smooth rod. The attached springs are both compressed 0.4 m when $d = 0.5$ m. Determine the speed of the collar after the applied force $F = 100$ N causes it to be displaced so that $d = 0.3$ m. When $d = 0.5$ m the collar is at rest.

*14–28. A rocket of mass m is fired vertically from the surface of the earth, i.e., at $r = r_1$. Assuming that no mass is lost as it travels upward, determine the work it must do against gravity to reach a distance r_2. The force of gravity is $F = GM_em/r^2$ (Eq. 13–1), where M_e is the mass of the earth and r the distance between the rocket and the center of the earth.

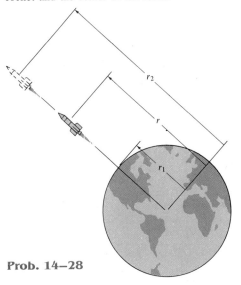

Prob. 14–28

14–29. The 5-lb cylinder is dropped from A with an initial speed of $v_A = 10$ ft/s onto the platform. Determine the maximum displacement of the platform caused by the collision. The spring has an unstretched length of 1.75 ft and is originally kept in compression by the 1-ft long cables attached to the platform. Neglect the mass of the platform and spring and any energy lost during the collision.

Prob. 14–29

14.4 Power and Efficiency

Power

Power is defined as the amount of work performed per unit of time. Hence, the *power* generated by a machine or engine that performs an amount of work dU within the time interval dt is

$$P = \frac{dU}{dt} \qquad (14\text{–}9)$$

Provided the work dU is expressed by $dU = \mathbf{F} \cdot d\mathbf{r}$, then it is also possible to write

$$P = \frac{dU}{dt} = \frac{\mathbf{F} \cdot d\mathbf{r}}{dt}$$

or

$$P = \mathbf{F} \cdot \mathbf{v} \qquad (14\text{–}10)$$

Hence, power is a *scalar,* where in the formulation $\mathbf{v}$ represents the velocity of the point which is acted upon by the force $\mathbf{F}.$

The basic units of power used in the SI and FPS systems are the watt (W) and horsepower (hp), respectively. These units are defined as

$$1 \text{ W} = 1 \text{ J/s} = 1 \text{ N} \cdot \text{m/s}$$

$$1 \text{ hp} = 550 \text{ ft} \cdot \text{lb/s}$$

For conversion between the two systems of units, 1 hp = 746 W.

The term "power" provides a useful basis for determining the type of motor or machine which is required to do a certain amount of work in a given time. For example, two pumps may each be able to empty a reservoir if given enough time; however, the pump having the larger power will complete the job sooner.

Efficiency

The *mechanical efficiency* of a machine is defined as the ratio of the output of useful power produced by the machine to the input of power supplied to the machine. Hence,

$$\varepsilon = \frac{\text{power output}}{\text{power input}} \qquad (14\text{--}11)$$

If energy applied to the machine occurs during the *same time interval* at which it is removed, then the efficiency may also be expressed in terms of the ratio of output energy to input energy; i.e.

$$\varepsilon = \frac{\text{energy output}}{\text{energy input}} \qquad (14\text{--}12)$$

Since machines consist of a series of moving parts, frictional forces will always be developed within the machine, and as a result, extra energy or power is needed to overcome these forces. Consequently, *the efficiency of a machine is always less than 1*.

PROCEDURE FOR ANALYSIS

When computing the power supplied to a body, one must first determine the unbalanced external force **F** acting on the body which causes the motion. This force is usually developed by a machine or engine placed either within or external to the body. If the body is accelerating, it may be necessary to draw its free-body diagram and apply the equation of motion ($\Sigma \mathbf{F} = m\mathbf{a}$) to determine **F**. Once **F** and the velocity **v** of the point where **F** is applied have been computed, the power is determined by multiplying the force magnitude by the component of velocity acting in the direction of **F**, i.e., $P = \mathbf{F} \cdot \mathbf{v} = Fv \cos \theta$.

In some problems the power may also be computed by calculating the work done per unit of time ($P_{avg} = \Delta U / \Delta t$, or $P = dU/dt$). Depending upon the problem, this work can be done either by the external or internal force of a machine or engine, by the weight of the body, or by an elastic spring force acting on the body.

Example 14–7

The motor M of the hoist shown in Fig. 14–15a operates with an efficiency of 0.85. Determine the power that must be supplied to the motor to lift the crate C having a weight of 75 lb. Point P on the cable is being drawn in with an acceleration of 4 ft/s^2, and at the instant shown its speed is 2 ft/s. Neglect the mass of the pulley and cable.

(a)

SOLUTION

In order to compute the power output of the motor, it is first necessary to determine the tension in the cable since this force is developed by the motor.

From the free-body diagram, Fig. 14–15b, we have

$$+\uparrow \Sigma F_y = ma_y; \qquad 2T - 75 = -\frac{75}{32.2}a_C \qquad (1)$$

(b)

Fig. 14–15

The acceleration of the crate can be obtained by using kinematics to relate the motion of the crate to the known motion of point P, Fig. 14–15a. Hence, by the methods of Sec. 12.8, the coordinates s_C and s_P in Fig. 14–15a can be related to a constant portion of cable length l which is changing in the vertical and horizontal directions. We have $2s_C + s_P = l$. Taking the second time derivative of this equation yields

$$2a_C = -a_P \qquad (2)$$

Since $a_P = +4$ ft/s^2, then $a_C = (-4 \text{ ft/s}^2)/2 = -2$ ft/s^2. Substituting this result into Eq. (1) and *retaining* the negative sign since the acceleration in *both* Eqs. (1) and (2) is considered positive downward, we have

$$2T - 75 = -\frac{75}{32.2}(-2)$$

$$T = 39.8 \text{ lb}$$

The power output, measured in units of horsepower, required to draw the cable in at a rate of 2 ft/s is therefore

$$P = \mathbf{T} \cdot \mathbf{v} = (39.8 \text{ lb})(2 \text{ ft/s})(1 \text{ hp}/550 \text{ lb} \cdot \text{ft/s})$$

$$= 0.145 \text{ hp}$$

This *power output* requires that the motor provide a *power input* of

$$\text{power input} = \frac{1}{\varepsilon}(\text{power output})$$

$$= \frac{1}{0.85}(0.145) = 0.170 \text{ hp} \qquad \qquad Ans.$$

Since the velocity of the crate is constantly changing, notice that this power requirement is *instantaneous*.

Example 14–8

The car shown in Fig. 14–16a has a mass of 2 Mg and an engine running efficiency of $\varepsilon = 0.63$. As it moves forward, the wind creates a drag resistance on the car of $F_D = 1.2v^2$ N, where v is the velocity in m/s. If the car is traveling at a speed of 50 m/s, determine the maximum power supplied by the engine.

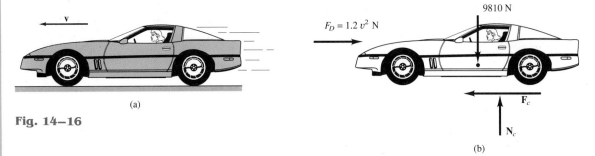

(a)

Fig. 14–16

(b)

SOLUTION

As shown on the free-body diagram, Fig. 14–16b, the normal force N_C and frictional force $\mathbf{F}_C$ represent the *resultant forces* of all four wheels. In particular, the unbalanced frictional force drives or pushes the car *forward*. This effect is, of course, created by the rotating motion of the rear wheels on the pavement and is developed by the power of the engine.

Applying the equation of motion in the x direction, we have

$$\xleftarrow{+}\Sigma F_x = ma_x; \qquad F_C - 1.2v^2 = 2000\ \frac{dv}{dt}$$

Since the car is traveling with *constant velocity*, $dv/dt = 0$. Hence, with $v = 50$ m/s,

$$F_C = 1.2(50)^2 = 3000\ \text{N}$$

The power output of the car is manifested by the driving (frictional) force $\mathbf{F}_C$. Thus

$$P = \mathbf{F}_C \cdot \mathbf{v} = (3000\ \text{N})(50\ \text{m/s}) = 150\ \text{kW}$$

The power supplied by the engine (power input) is therefore

$$\text{power input} = \frac{1}{\varepsilon}\ (\text{power output}) = \frac{1}{0.63}\ (150) = 238\ \text{kW} \quad \textit{Ans.}$$

PROBLEMS

14–30. A boy having a weight of 130 lb can run with constant speed up a staircase that is 20 ft high in 8 s. Determine the average power he produces.

14–31. A train having a weight of $4(10^6)$ lb is traveling at 20 ft/s. If it is brought to a stop, determine how many days a 75-W light bulb must burn to expend the same amount of energy.

*14–32. Determine the power input for a motor necessary to lift 300 lb at a constant rate of 5 ft/s. The efficiency of the motor is $\varepsilon = 0.65$.

14–33. The jeep has a weight of 2500 lb and an engine which transmits a power of 100 hp to *all* the wheels. Assuming the wheels do not slip on the ground, determine the angle θ of the largest incline the jeep can climb at a constant speed $v = 30$ ft/s.

$v = 30$ ft/s

Prob. 14–33

14–34. A motor for a mine hoist lifts a 5000-lb load from a speed of 2 ft/s to a speed of 10 ft/s in $t = 5$ s with constant acceleration. Determine the power developed by the motor when $t = 5$ s.

14–35. The locomotive of a train exerts a constant pull of 400 kN on the cars which have a total mass of 2 Gg. The cars have a total frictional resistance of 6.5 kN and begin traveling at 2 m/s up a slope of 1°. Determine the speed of the cars after they travel 2 km. What is the power output of the locomotive when it has reached this point?

*14–36. The motor M is used to hoist the loaded 500-kg elevator upward with a constant velocity $v_E = 8$ m/s. If the motor draws 60 kW of electrical power, determine the motor's efficiency. Neglect the mass of the pulleys and cable.

Probs. 14–36/14–37

14–37. The 500-kg elevator starts from rest and travels upward with a constant acceleration $a_c = 2$ m/s². Determine the power output of the motor M when $t = 3$ s. Neglect the mass of the pulleys and cable.

14–38. The 10-lb collar starts from rest at A and is lifted by applying a constant horizontal force of $F = 25$ lb to the cord. If the rod is smooth, determine the power developed by the force at the instant $\theta = 60°$.

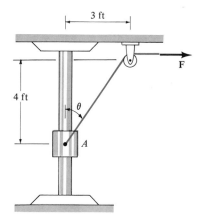

Prob. 14–38

14–39. A car has a mass m and accelerates along a horizontal straight road from rest such that the power is always a constant amount P. Determine how far it must travel to reach a speed of v.

*14–40. The 50-lb load is hoisted by the pulley system and motor M. If the crate starts from rest and by constant acceleration attains a speed of 15 ft/s after rising $s = 6$ ft, determine the power that must be supplied to the motor when $s = 6$ ft. The motor has an efficiency of $\varepsilon = 0.76$. Neglect the mass of the pulleys and cable.

Prob. 14–40

14–41. The crate has a mass of 150 kg and rests on a surface for which the coefficients of static and kinetic friction are $\mu_s = 0.3$ and $\mu_k = 0.2$, respectively. If the motor M supplies a cable force of $F = (8t^2 + 20)$ N, where t is in seconds, determine the power output developed by the motor when $t = 5$ s.

Prob. 14–41

14–42. The rocket has a mass of 3 Mg and is fired from rest. If the engine provides a constant vertical thrust $T = 150$ kN, determine the power output of the engine as a function of time. Neglect the change in weight, loss of fuel mass, and drag due to air resistance, i.e., $F_D = 0$.

Probs. 14–42/14–43

14–43. Solve Prob. 14–42 if the drag due to air resistance is $F_D = (0.06v^2)$ kN, where v is in m/s.

*14–44. The 150-lb crate rests on the rough surface for which the coefficients of static and kinetic friction are $\mu_s = 0.3$ and $\mu_k = 0.25$, respectively. A towing force $F = (50 + 0.3s^2)$ lb, where s is the horizontal displacement in ft, acts on the crate in the direction shown. If the crate is originally at rest, determine the power developed by the force when the crate has moved $s = 3$ ft.

Prob. 14–44

14-45. The block has a weight of 80 lb and rests on the floor for which the coefficient of kinetic friction is $\mu_k = 0.4$. If the motor draws in the cable at a constant rate of 6 ft/s, determine the power output of the motor at the instant $\theta = 30°$. Neglect the mass of the cable and pulleys.

3 ft

3 ft

Prob. 14–45

14.5 Conservative Forces and Potential Energy

Conservative Force

A particularly simple type of force acting on a particle is one that depends *only* on the particle's position and is independent of the particle's velocity and acceleration. Furthermore, if the work done by this force in moving the particle from one point to another is *independent of the path* followed by the particle, this force is called a *conservative force*. The weight of a particle and the force of an elastic spring are two examples of conservative forces often encountered in mechanics.

Weight. The work done by the weight of a particle is *independent of the path*; rather, it depends only on the particle's *vertical displacement*. If this displacement is Δy (positive upward), then from Eq. 14–3,

$$U = -W(\Delta y)$$

Elastic Spring. The work done by a spring force *acting on a particle* is *independent of the path* of the particle, but depends only on the extension or compression s of the spring. If the spring is elongated or compressed from position s_1 to a further position s_2, then from Eq. 14–4,

$$U = -(\tfrac{1}{2}ks_2^2 - \tfrac{1}{2}ks_1^2)$$

Friction. In contrast to a conservative force, consider the force of friction exerted *on a moving object* by a fixed surface. The work done by the frictional force *depends upon the path*—the longer the path, the greater the work. Consequently, *frictional forces are nonconservative*. The work is dissipated from the body in the form of heat.

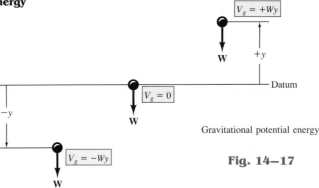

Gravitational potential energy

Fig. 14–17

Potential Energy

Energy may be defined as the capacity for doing work. When energy comes from the *motion* of the particle, it is referred to as *kinetic energy*. When it comes from the *position* of the particle, measured from a fixed datum or reference plane, it is called *potential energy*. Thus, potential energy is a measure of the amount of work a force will do when it moves from a given position to the datum. In mechanics, the potential energy due to both gravity (weight) and an elastic spring is important.

 Gravitational Potential Energy. If a particle is located a distance y *above* an arbitrarily selected datum, as shown in Fig. 14–17, the particle's weight **W** has positive *gravitational potential energy*, V_g, since **W** has the capacity of doing positive work when the particle is moved back down to the datum. Likewise, if the particle is located a distance y *below* the datum, V_g is negative since the weight does negative work when the particle is moved back up to the datum. At the datum $V_g = 0$.

 In general, if y is *positive upward,* the gravitational potential energy of the particle of weight W is thus*

$$\boxed{V_g = Wy} \tag{14–13}$$

 Elastic Potential Energy. When an elastic spring is elongated or compressed a distance s from its unstretched position, the elastic potential energy V_e due to the spring's configuration can be expressed as

$$\boxed{V_e = +\tfrac{1}{2}ks^2} \tag{14–14}$$

Here V_e is *always positive* since, in the deformed position, the force of the spring has the *capacity* for always doing positive work on the particle when the spring is returned to its unstretched position, Fig. 14–18.

---

*Here the weight is assumed to be *constant*. This assumption is suitable for small differences in elevation Δy. If the elevation change is significant, however, a variation of weight with elevation must be taken into account (see Prob. 14–59).

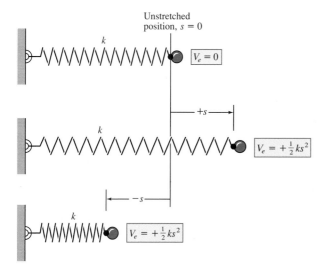

Elastic potential energy

Fig. 14–18

Potential Function

In the general case, if a particle is subjected to both gravitational and elastic forces, the particle's potential energy can be expressed as a *potential function*, which is the algebraic sum

$$V = V_g + V_e \qquad (14\text{–}15)$$

Measurement of V depends upon the location of the particle with respect to a selected datum in accordance with Eqs. 14–13 and 14–14. If the particle is located at an arbitrary point (x, y, z) in space, this potential function is then $V = V(x, y, z)$. The work done by a conservative force in moving the particle from point (x_1, y_1, z_1) to point (x_2, y_2, z_2) is measured by the *difference* of this function, i.e.,

$$U_{1-2} = V_1 - V_2 \qquad (14\text{–}16)$$

For example, the potential function for a particle of weight W suspended from a spring can be expressed in terms of its position, s, measured from a datum located at the unstretched length of the spring, Fig. 14–19. We have

$$V = V_g + V_e$$
$$= -Ws + \tfrac{1}{2}ks^2$$

If the particle moves from s_1 to a lower position s_2, then applying Eq. 14–16 it can be seen that the work of $\mathbf{W}$ and $\mathbf{F}_s$ is

$$U_{1-2} = V_1 - V_2 = (-Ws_1 + \tfrac{1}{2}ks_1^2) - (-Ws_2 + \tfrac{1}{2}ks_2^2)$$
$$= W(s_2 - s_1) - (\tfrac{1}{2}ks_2^2 - \tfrac{1}{2}ks_1^2)$$

Fig. 14–19

When the displacement path is infinitesimal, i.e., from point (x, y, z) to $(x + dx, y + dy, z + dz)$, Eq. 14–16 becomes

$$dU = V(x, y, z) - V(x + dx, y + dy, z + dz)$$

$$= -dV(x, y, z) \tag{14–17}$$

Provided both the force and displacement are defined using rectangular coordinates, then the work can also be expressed as

$$dU = \mathbf{F} \cdot d\mathbf{r} = (F_x\mathbf{i} + F_y\mathbf{j} + F_z\mathbf{k}) \cdot (dx\mathbf{i} + dy\mathbf{j} + dz\mathbf{k})$$

$$= F_x dx + F_y dy + F_z dz$$

Substituting this result into Eq. 14–17 and expressing the differential $dV(x, y, z)$ in terms of partial derivatives yields

$$F_x dx + F_y dy + F_z dz = -\left(\frac{\partial V}{\partial x}dx + \frac{\partial V}{\partial y}dy + \frac{\partial V}{\partial z}dz\right)$$

Since changes in x, y, and z are all independent of one another, this equation is satisfied provided

$$F_x = -\frac{\partial V}{\partial x}, \qquad F_y = -\frac{\partial V}{\partial y}, \qquad F_z = -\frac{\partial V}{\partial z} \tag{14–18}$$

Thus,

$$\mathbf{F} = -\frac{\partial V}{\partial x}\mathbf{i} - \frac{\partial V}{\partial y}\mathbf{j} - \frac{\partial V}{\partial z}\mathbf{k}$$

$$= -\left(\frac{\partial}{\partial x}\mathbf{i} + \frac{\partial}{\partial y}\mathbf{j} + \frac{\partial}{\partial z}\mathbf{k}\right)V$$

or

$$\boxed{\mathbf{F} = -\nabla V} \tag{14–19}$$

where ∇ (del) represents the vector operator $\nabla = (\partial/\partial x)\mathbf{i} + (\partial/\partial y)\mathbf{j} + (\partial/\partial z)\mathbf{k}$.

Equation 14–19 relates a force $\mathbf{F}$ to its potential function V and thereby provides a mathematical criterion for proving that $\mathbf{F}$ is conservative. For example, the gravitational potential function for a weight located a distance y above a datum is $V_g = Wy$, Eq. 14–13. To prove that $\mathbf{W}$ is conservative, it is necessary to show that it satisfies Eq. 14–19 (or Eq. 14–18), in which case

$$F_y = -\frac{\partial V}{\partial y}; \qquad\qquad F = -\frac{\partial}{\partial y}(Wy) = -W$$

The negative sign indicates that $\mathbf{W}$ acts downward, opposite to positive y, which is upward.

14.6 Conservation of Energy

When a particle is acted upon by a system of *both* conservative and nonconservative forces, the portion of the work done by the *conservative forces* can be written in terms of the difference in their potential energies using Eq.

14–16, i.e., $(\Sigma U_{1-2})_{\text{cons.}} = V_1 - V_2$. As a result, the principle of work and energy can be written as

$$T_1 + V_1 + (\Sigma U_{1-2})_{\text{noncons.}} = T_2 + V_2 \qquad (14\text{–}20)$$

Here $(\Sigma U_{1-2})_{\text{noncons.}}$ represents the work of the nonconservative forces acting on the particle. If *only conservative forces* are applied to the body, this term is zero and then we have

$$\boxed{T_1 + V_1 = T_2 + V_2} \qquad (14\text{–}21)$$

This equation is referred to as the *conservation of mechanical energy* or simply the *conservation of energy*. It states that during the motion the sum of the particle's kinetic and potential energies remains *constant*. For this to occur, kinetic energy must be transformed into potential energy, and vice versa. For example, if a ball of weight **W** is dropped from a height h above the ground (datum), Fig. 14–20, the potential energy of the ball is maximum before it is dropped, at which time its kinetic energy is zero. The total mechanical energy of the ball in its initial position is thus

$$E = T_1 + V_1 = 0 + Wh = Wh$$

Potential Energy (max)
Kinetic Energy (zero)

Potential Energy and
Kinetic Energy

Datum

Potential Energy (zero)
Kinetic Energy (max)

Fig. 14–20

When the ball has fallen a distance $h/2$, its speed can be determined by using $v^2 = v_0^2 + 2a_c(y - y_0)$, which yields $v = \sqrt{2g(h/2)} = \sqrt{gh}$. The energy of the ball at the mid-height position is therefore

$$E = V_2 + T_2 = W\frac{h}{2} + \frac{1}{2}\frac{W}{g}(\sqrt{gh})^2 = Wh$$

Just before the ball strikes the ground, its potential energy is zero and its speed is $v = \sqrt{2gh}$. Here, again, the total energy of the ball is

$$E = V_3 + T_3 = 0 + \frac{1}{2}\frac{W}{g}(\sqrt{2gh})^2 = Wh$$

Note that when the ball comes in contact with the ground, it deforms somewhat, and provided the ground is hard enough, the ball will rebound off the surface, reaching a new height h', which will be less than the height h from which it was first released. Neglecting air friction, the difference in height accounts for an energy loss, $E_l = W(h - h')$, occurring at the moment of collision. Portions of this loss produce noise, localized deformation of the ball and ground, and heat.

System of Particles

If a system of particles is *subjected only to conservative forces,* then an equation similar to Eq. 14–21 can be written for the particles. Applying the ideas of the preceding discussion, Eq. 14–8 $(\Sigma T_1 + \Sigma U_{1-2} = \Sigma T_2)$ becomes

$$\boxed{\Sigma T_1 + \Sigma V_1 = \Sigma T_2 + \Sigma V_2} \qquad (14-22)$$

Here, the sum of the system's initial kinetic and potential energy is equal to the sum of the system's final kinetic and potential energy. In other words, the system's kinetic energy and potential energy, caused by *both* the internal and external forces, remain constant; i.e., $\Sigma T + \Sigma V = \text{const.}$

PROCEDURE FOR ANALYSIS

The conservation of energy equation is used to solve problems involving *velocity, displacement,* and *conservative force systems.* It is generally *easier to apply* than the principle of work and energy. This is because the energy equation just requires specifying the particle's kinetic and potential energy at only *two points* along the path, rather than determining the work when the particle moves through a *displacement.* For application it is suggested that the following procedure be used.

Potential Energy. Draw two diagrams showing the particle located at its initial and final points along the path. If the particle is subjected to a vertical displacement, establish the fixed horizontal datum from which to measure the particle's gravitational potential energy V_g. Although this position can be selected arbitrarily, it is best to locate the datum either at the initial or final point of the path, since at the datum $V_g = 0$. Data pertaining to the elevation y of the particle from the datum and the extension or compression s of any connecting springs can be determined from the geometry associated with the two diagrams. Recall that the potential energy $V = V_g + V_e$. Here $V_g = Wy$, where y is positive upward from the datum, and $V_e = \frac{1}{2}ks^2$, which is always positive.

Conservation of Energy. Apply the equation $T_1 + V_1 = T_2 + V_2$. When computing the kinetic energy, $T = \frac{1}{2}mv^2$, the particle's speed v must be measured from an inertial reference frame.

It is important to remember that only problems involving conservative force systems may be solved by using the conservation of energy theorem. As stated previously, friction or other drag-resistant forces, which depend upon velocity or acceleration, are nonconservative. A portion of the work done by such forces is transformed into thermal energy used to heat up the surfaces of contact, and consequently this energy dissipates into the surroundings and may not be recovered. Therefore, problems involving frictional forces can either be solved by using the principle of work and energy written in the form of Eq. 14–20, if it applies, or the equation of motion.

The following example problems numerically illustrate application of the procedure described above.

Example 14–9

The boy and his bicycle shown in Fig. 14–21a have a total weight of 125 lb and a center of mass at G. If he is coasting, i.e., not pedaling, with a speed of 10 ft/s at the top of the hill A, determine the normal force exerted on both wheels of the bicycle when he arrives at B, where the radius of curvature of the road is $\rho = 50$ ft. Neglect friction.

Fig. 14–21

(a) (b)

SOLUTION

Since the normal force does *no work,* it must be obtained using the equation of motion, $\Sigma F_n = m(v^2/\rho)$. We can, however, determine the bicycle's speed at B using the conservation of energy equation. Why?

Potential Energy. Figure 14–21b shows the bicycle at points A and B. For convenience, the potential-energy datum has been established through the center of mass when the bicycle is located at B.

Conservation of Energy

$$\{T_A\} + \{V_A\} = \{T_B\} + \{V_B\}$$

$$\left\{\frac{1}{2}\left(\frac{125}{32.2}\right)(10)^2\right\} + \{(125)(30)\} = \left\{\frac{1}{2}\left(\frac{125}{32.2}\right)(v_B)^2\right\} + \{0\}$$

$$v_B = 45.1 \text{ ft/s}$$

(c)

Equation of Motion. Using the data tabulated on the free-body diagram when the bicycle is at B, Fig. 14–21c, we have

$$+\uparrow \Sigma F_n = ma_n; \qquad N_B - 125 = \frac{125}{32.2}\frac{(45.1)^2}{50}$$

$$N_B = 283 \text{ lb} \qquad\qquad Ans.$$

Example 14–10

The ram R shown in Fig. 14–22a has a mass of 100 kg and is released from rest 0.75 m from the top of a spring A that has a stiffness $k_A = $ 12 kN/m. If a second spring B, having a stiffness $k_B = 15$ kN/m, is "nested" in A, determine the maximum deflection of A needed to stop the downward motion of the ram. The unstretched length of each spring is indicated in the figure. Neglect the mass of the springs.

SOLUTION

Potential Energy. We will *assume* that the ram compresses *both* springs at the instant it comes to rest. The datum is located through the center of gravity of the ram at its initial position, Fig. 14–22b. When the kinetic energy is reduced to zero ($v_2 = 0$), A is compressed a distance s_A so that B compresses $s_B = s_A - 0.1$ m.

Conservation of Energy

$$T_1 + V_1 = T_3 + V_3$$

$$\{0\} + \{0\} = \{0\} + \{\tfrac{1}{2}k_A s_A^2 + \tfrac{1}{2}k_B(s_A - 0.1)^2 - Wh\}$$

$$\{0\} + \{0\} = \{0\} + \left\{\frac{1}{2}(12000)s_A^2 + \frac{1}{2}(15000)(s_A - 0.1)^2 - 981(0.75 + s_A)\right\}$$

Rearranging the terms,

$$13500s_A^2 - 2481s_A - 660.75 = 0$$

Using the quadratic formula, and solving for the positive root,* we have

$$s_A = 0.331 \text{ m} \qquad\qquad Ans.$$

Since $s_B = 0.331 - 0.1 = 0.231$ m, which is positive, indeed the assumption that *both* springs are compressed by the ram is correct.

*The second root, $s_A = -0.148$ m, does not represent the physical situation. Since positive s is measured downward, the negative sign indicates that spring A would have to be "extended" by an amount of 0.148 m to stop the ram.

(a)

(b)

Fig. 14–22

Example 14–11

A smooth 2-kg collar C, shown in Fig. 14–23a, fits loosely on the vertical shaft. If the spring is unstretched when the collar is in the dashed position A, determine the speed at which the collar is moving when $y = 1$ m, if (a) it is released from rest at A, and (b) it is released at A with an *upward* velocity $v_A = 2$ m/s.

(a)

SOLUTION

Part (a)

Potential Energy. For convenience, the datum is established through AB, Fig. 14–23b. When the collar is at C, the gravitational potential energy is $-(mg)y$, since the collar is *below* the datum, and the elastic potential energy is $\frac{1}{2}ks_{CB}^2$. Here $s_{CB} = 0.5$ m, which represents the *stretch* in the spring as computed in the figure.

Conservation of Energy

$$\{T_A\} + \{V_A\} = \{T_C\} + \{V_C\}$$

$$\{0\} + \{0\} = \{\tfrac{1}{2}mv_C^2\} + \{\tfrac{1}{2}ks_{CB}^2 - mgy\}$$

$$\{0\} + \{0\} = \left\{\frac{1}{2}(2)v_C^2\right\} + \left\{\frac{1}{2}(3)(0.5)^2 - 2(9.81)(1)\right\}$$

$$v_C = 4.39 \text{ m/s} \downarrow \qquad\qquad Ans.$$

This problem can also be solved by using the equation of motion or the principle of work and energy. Note that in *both* of these methods the variation of the magnitude and direction of the spring force must be taken into account (see Example 13–4). Here, however, the above method of solution is clearly advantageous since the calculations depend *only* on data calculated at the initial and final points of the path.

(b)

Fig. 14–23

Part (b)

Conservation of Energy. If $v_A = 2$ m/s, using the data in Fig. 14–23b, we have

$$\{T_A\} + \{V_A\} = \{T_C\} + \{V_C\}$$

$$\{\tfrac{1}{2}mv_A^2\} + \{0\} = \{\tfrac{1}{2}mv_C^2\} + \{\tfrac{1}{2}ks_{CB}^2 - mgy\}$$

$$\left\{\frac{1}{2}(2)(2)^2\right\} + \{0\} = \left\{\frac{1}{2}(2)v_C^2\right\} + \left\{\frac{1}{2}(3)(0.5)^2 - 2(9.81)(1)\right\}$$

$$v_C = 4.82 \text{ m/s} \downarrow \qquad\qquad Ans.$$

Note that the kinetic energy of the collar depends only on the *magnitude* of velocity, and therefore it is immaterial if the collar is moving up or down at 2 m/s when released at A.

PROBLEMS

14–46. Solve Prob. 14–8 using the conservation of energy equation.

14–47. Solve Prob. 14–10 using the conservation of energy equation.

*14–48. Solve Prob. 14–20 using the conservation of energy equation.

14–49. Solve Prob. 14–14 using the conservation of energy equation.

14–50. Solve Prob. 14–15 using the conservation of energy equation.

14–51. Solve Prob. 14–25 using the conservation of energy equation.

*14–52. The 20-lb collar is constrained to move on the smooth rod. It is attached to the three springs which are unstretched when $s = 0$. If the collar is displaced $s = 0.5$ ft and released from rest, determine its speed when $s = 0$.

Prob. 14–52

14–53. The collar has a weight of 8 lb. If it is pushed down so as to compress the spring 2 ft and then released from rest $(h = 0)$, determine its speed when it is displaced $h = 4.5$ ft. The spring is not attached to the collar. Neglect friction.

Prob. 14–53

14–54. The ball has a weight of 3 lb and is released from rest when $\theta = 0°$. Determine its speed just after the cord strikes the peg at B. Also, what is the speed of the ball and the tension in the cord when the ball reaches its lowest point C as shown?

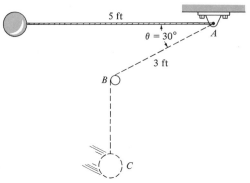

Prob. 14–54

14–55. Each of the two elastic rubberbands of the slingshot has an unstretched length of 200 mm. If it is pulled back to the position shown and released from rest, determine the speed of the 25-g pellet just after the rubberbands become unstretched. Neglect the mass of the rubberbands and the change in elevation of the pellet while it is constrained by the rubberbands. Each rubberband has a stiffness of $k = 50$ N/m.

Prob. 14–55

*14–56. The ball has a weight of 15 lb and is fixed to a rod having a negligible mass. If it is released from rest when $\theta = 0°$, determine the angle θ at which the compressive force in the rod becomes zero.

Probs. 14–56/14–57

14–57. The ball has a weight of 15 lb and is fixed to a rod having a negligible mass. If it is given a speed of 2 ft/s when $\theta = 0°$, determine the angle θ when the compressive force in the rod becomes 4 lb.

14–58. The cylinder has a mass of 20 kg and is released from rest when $h = 0$. Determine its speed when $h = 3$ m. The springs each have an unstretched length of 2 m.

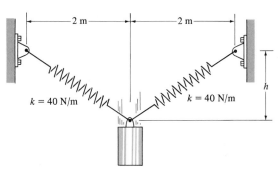

Prob. 14–58

14–59. If the mass of the earth is M_e, show that the gravitational potential energy of a body of mass m located a distance r from the center of the earth is $V_g = -GM_e m/r$. Recall that the gravitational force acting between the earth and the body is $F = G(M_e m/r^2)$, Eq. 13–1. For the calculation, locate the datum at $r \rightarrow \infty$. Also, prove that $\mathbf{F}$ is a conservative force.

*14–60. A 60-kg satellite is traveling in free flight along an elliptical orbit such that at A, where $r_A = 20$ Mm, it has a speed $v_A = 40$ Mm/h. What is the speed of the satellite when it reaches point B, where $r_B = 80$ Mm? *Hint:* See Prob. 14–59, where $M_e = 5.976(10^{24})$ kg and $G = 66.73(10^{-12})$ m³/(kg · s²).

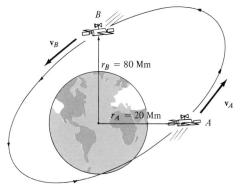

Prob. 14–60

14–61. The assembly consists of two blocks A and B which have a mass of 20 kg and 30 kg, respectively. Determine the speed of each block when B descends 1.5 m. The blocks are released from rest. Neglect the mass of the pulleys and cords.

Prob. 14–61

14–62. Four inelastic cables C are attached to a plate P and hold the 1-ft-long spring 0.25 ft in compression when *no weight* is on the plate. There is also an undeformed spring nested within this compressed spring. If the block, having a weight of 10 lb, is moving downward at $v = 4$ ft/s, when it is 2 ft above the plate, determine the maximum compression in each spring after it strikes the plate. Neglect the mass of the plate and spring and any energy lost in the collision.

2 ft

P C

0.75 ft

0.5 ft

$k = 30$ lb/in. $k' = 50$ lb/in.   **Probs. 14–62/14–63**

14–63. Four inelastic cables C are attached to a plate P and hold the 1-ft-long spring 0.25 ft in compression when *no weight* is on the plate. There is also an undeformed spring nested within this compressed spring. Determine the speed v of the 10-lb block when it is 2 ft above the plate, so that after it strikes the plate, it compresses the spring having a stiffness of 50 lb/in. an amount of 0.20 ft. Neglect the mass of the plate and spring and any energy lost in the collision.

*14–64.** The ride at an amusement park consists of a gondola which is lifted to a height of 120 ft at A. If it is released from rest and falls along the parabolic track, determine its speed at the instant $y = 20$ ft. Also determine the normal reaction of the tracks on the gondola at this instant. The gondola and passenger have a total weight of 500 lb. Neglect the effects of friction, the mass of the wheels and the size of the gondola.

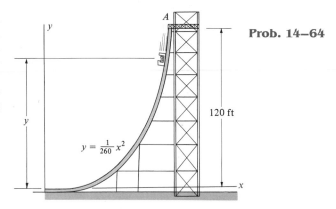

Prob. 14–64

120 ft

$y = \frac{1}{260} x^2$

14–65. The spring has a stiffness $k = 3$ lb/ft and an unstretched length of 2 ft. If it is attached to the 5-lb smooth collar and the collar is released from rest at A, determine the speed of the collar just before it strikes the end of the rod at B. Neglect the size of the collar.

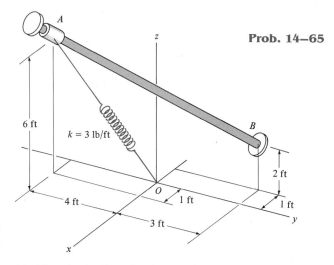

Prob. 14–65

6 ft

$k = 3$ lb/ft

4 ft

O 1 ft

3 ft

2 ft

1 ft

14–66. The double-spring bumper is used to stop the 1500-lb steel billet in the rolling mill. Determine the maximum deflection of the plate A caused by the billet if it strikes the plate with a speed of 8 ft/s. Neglect the mass of the rollers, the springs, and the plates A and B, and the energy lost during the collision.

$v = 8$ ft/s

$k_2 = 4500$ lb/ft

A B

$k_1 = 3000$ lb/ft   **Prob. 14–66**

15

Kinetics of a Particle: Impulse and Momentum

In this chapter we will integrate the equation of motion with respect to time and thereby obtain the principle of impulse and momentum. This principle is useful for solving problems involving force, velocity, and time. It will be shown that impulse and momentum principles also provide the necessary means for analyzing problems of impact, steady fluid flow, and systems which gain or lose mass.

15.1 Principle of Linear Impulse and Momentum

The equation of motion for a particle of mass m can be written as

$$\Sigma \mathbf{F} = m\mathbf{a} = m\frac{d\mathbf{v}}{dt} \qquad (15\text{--}1)$$

where $\mathbf{a}$ and $\mathbf{v}$ are both measured from an inertial frame of reference. Rearranging the terms and integrating between the limits $\mathbf{v} = \mathbf{v}_1$ at $t = t_1$ and $\mathbf{v} = \mathbf{v}_2$ at $t = t_2$, we have

$$\Sigma \int_{t_1}^{t_2} \mathbf{F} \, dt = m \int_{\mathbf{v}_1}^{\mathbf{v}_2} d\mathbf{v}$$

or

$$\Sigma \int_{t_1}^{t_2} \mathbf{F} \, dt = m\mathbf{v}_2 - m\mathbf{v}_1 \qquad (15\text{--}2)$$

This equation is referred to as the *principle of linear impulse and momentum*. From the derivation it can be seen that it is simply a time integration of the equation of motion. It provides a *direct means* of obtaining the particle's final velocity $\mathbf{v}_2$ after a specified time period when the particle's initial velocity is known and the forces acting on the particle are either constant or can be expressed as functions of time. Notice that if $\mathbf{v}_2$ is determined using the equation of motion, a two-step process is necessary; i.e., apply $\Sigma\mathbf{F} = m\mathbf{a}$ to obtain $\mathbf{a}$, then integrate $\mathbf{a} = d\mathbf{v}/dt$ to obtain $\mathbf{v}_2$.

Linear Impulse

The integral $\mathbf{I} = \int \mathbf{F} \, dt$ in Eq. 15–2 is defined as the *linear impulse*. This term is a vector quantity which measures the effect of a force during the time the force acts. The impulse vector acts in the same direction as the force, and its magnitude has units of force–time, e.g., N · s or lb · s. If the force is expressed as a function of time, the impulse may be determined by direct evaluation of the integral. In particular, if $\mathbf{F}$ acts in a *constant direction* during the time period t_1 to t_2, the magnitude of the impulse $\mathbf{I} = \int_{t_1}^{t_2} \mathbf{F} \, dt$ can be represented experimentally by the shaded area under the curve of force versus time, Fig. 15–1. However, if the force is constant in magnitude and direction, the resulting impulse becomes $\mathbf{I} = \int_{t_1}^{t_2} \mathbf{F}_c \, dt = \mathbf{F}_c(t_2 - t_1)$, which represents the shaded rectangular area shown in Fig. 15–2.

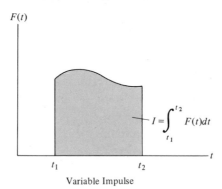

$$I = \int_{t_1}^{t_2} F(t)dt$$

Variable Impulse

Fig. 15–1

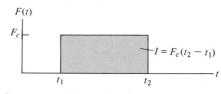

$$I = F_c(t_2 - t_1)$$

Fig. 15–2   Constant Impulse

Linear Momentum

Each of the two vectors of the form $\mathbf{L} = m\mathbf{v}$ in Eq. 15–2 is defined as the *linear momentum* of the particle. Since m is a scalar, the linear-momentum vector has the same direction as $\mathbf{v}$, and its magnitude mv has units of mass–velocity, e.g., kg · m/s, slug · ft/s.*

Principle of Linear Impulse and Momentum

For problem solving, Eq. 15–2 will be rewritten in the form

$$m\mathbf{v}_1 + \Sigma \int_{t_1}^{t_2} \mathbf{F} \, dt = m\mathbf{v}_2 \qquad (15\text{–}3)$$

which states that the initial momentum of the particle at t_1 plus the vector sum of all the impulses applied to the particle during the time interval t_1 to t_2 is equivalent to the final momentum of the particle at t_2. These three terms are illustrated graphically on the *impulse and momentum diagrams* shown in Fig. 15–3. Each of these diagrams graphically accounts for all the vectors in the equation $m\mathbf{v}_1 + \Sigma \int_{t_1}^{t_2} \mathbf{F} \, dt = m\mathbf{v}_2$. The two *momentum diagrams* are simply outlined shapes of the particle which indicate the direction and magnitude of the particle's initial and final momenta, $m\mathbf{v}_1$ and $m\mathbf{v}_2$, respectively, Fig. 15–3. Similar to the free-body diagram, the *impulse diagram* is an outlined shape of the particle showing all the impulses that act on the particle when it is located at some intermediate point along its path. In general, whenever the magnitude or direction of a force *varies,* the impulse of the force is determined by integration and represented on the impulse diagram as $\mathbf{I} = \int_{t_1}^{t_2} \mathbf{F} \, dt$. If the force is *constant* for the time interval $(t_2 - t_1)$, the impulse applied to the particle is $\mathbf{I} = \mathbf{F}_c(t_2 - t_1)$, acting in the same direction as $\mathbf{F}_c$.

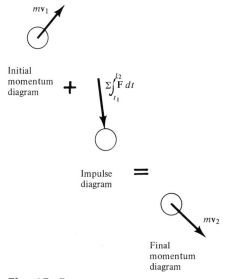

Fig. 15–3

Scalar Equations

If each of the vectors in Eq. 15–3 or on Fig. 15–3 is resolved into its x, y, and z components, we can write symbolically the following three scalar equations:

$$m(v_x)_1 + \Sigma \int_{t_1}^{t_2} F_x \, dt = m(v_x)_2$$

$$m(v_y)_1 + \Sigma \int_{t_1}^{t_2} F_y \, dt = m(v_y)_2 \qquad (15\text{–}4)$$

$$m(v_z)_1 + \Sigma \int_{t_1}^{t_2} F_z \, dt = m(v_z)_2$$

These equations represent the principle of linear impulse and momentum for the particle in the x, y, and z directions, respectively.

*Although the units for impulse and momentum are defined differently, show that Eq. 15–2 is dimensionally homogeneous.

PROCEDURE FOR ANALYSIS

The principle of linear impulse and momentum is used to solve problems involving *force, time,* and *velocity,* since these terms are involved in the formulation. For application it is suggested that the following procedure be used.

Free-Body Diagram.   Establish the x, y, z inertial frame of reference and draw the particle's free-body diagram in order to account for all the forces that produce impulses on the particle. The direction and sense of the particle's initial and final velocities should also be established. If any of these are unknown, assume that the sense of its components is in the direction of the positive inertial coordinate(s).

As an alternative procedure draw the impulse and momentum diagrams for the particle as discussed in reference to Fig. 15–3.*

Principle of Impulse and Momentum.   Apply the principle of linear impulse and momentum, $m\mathbf{v}_1 + \Sigma \int_{t_1}^{t_2} \mathbf{F} \, dt = m\mathbf{v}_2$. If motion occurs in the x-y plane, the two scalar component equations can be formulated by either resolving the vector components of $\mathbf{F}$ from the free-body diagram, or by using the data on the impulse and momentum diagrams.

If the problem involves the dependent motion of several particles, use the method outlined in Sec. 12.8 to relate their velocities. Make sure the positive coordinate directions used for writing these kinematic equations are the *same* as those used for writing the equations of impulse and momentum.

The following examples numerically illustrate application of this procedure.

*This procedure will be followed when developing the proofs and theory in the text.

Example 15–1

The 100-kg crate shown in Fig. 15–4a is originally at rest on the smooth horizontal surface. If a force of 200 N, acting at an angle of 45°, is applied to the crate for 10 s, determine the final velocity of the crate and the normal force which the surface exerts on the crate during the time interval.

(a)

SOLUTION

This problem can be solved using the principle of impulse and momentum since it involves force, velocity, and time.

Free-Body Diagram. This diagram is shown in Fig. 15–4b. Here it has been assumed that during the motion the crate remains on the surface and after 10 s the crate moves to the right with a velocity $\mathbf{v}_2$. Since all the forces acting on the crate are *constant*, the respective impulses are simply the product of the force magnitude and 10 s ($\mathbf{I} = \mathbf{F}_c(t_2 - t_1)$).

Principle of Impulse and Momentum. Resolving the vectors in Fig. 15–4b along the x,y axes, and applying Eqs. 15–4, yields

$(\xrightarrow{+})$

$$m(v_x)_1 + \Sigma \int_{t_1}^{t_2} F_x \, dt = m(v_x)_2$$

$$0 + 200(10) \cos 45° = 100v_2$$

$$v_2 = 14.1 \quad \text{m/s} \rightarrow \qquad Ans.$$

$(+\uparrow)$

$$m(v_y)_1 + \Sigma \int_{t_1}^{t_2} F_y \, dt = m(v_y)_2$$

$$0 + N_C(10) - 981(10) + 200(10) \sin 45° = 0$$

$$N_C = 840 \text{ N} \qquad Ans.$$

(b)

Fig. 15–4

Since no motion occurs in the y direction, direct application of $\Sigma F_y = 0$ gives the same result for N_C.

Note the alternative procedure of drawing the crate's impulse and momentum diagrams, Fig. 15–4c, prior to applying Eqs. 15–4.

Example 15-2

The crate shown in Fig. 15–5a has a weight of 50 lb and is acted upon by a force having a variable magnitude $P = (20t)$ lb, where t is in seconds. Compute the crate's velocity 2 s after **P** has been applied. The crate has an initial velocity $v_1 = 3$ ft/s down the plane, and the coefficient of kinetic friction between the crate and the plane is $\mu_k = 0.3$.

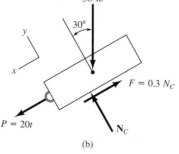

Fig. 15–5

(a)   (b)

SOLUTION

Free-Body Diagram. Fig. 15–5b. Since the magnitude of force $P = 20t$ *varies* with time, the impulse it creates must be determined by *integrating* over the 2-s time interval. The weight, normal force, and frictional force (which acts opposite to the direction of motion) are all *constant,* so that the impulse created by each of these forces is simply the magnitude of the force times 2 s.

Principle of Impulse and Momentum. Applying Eqs. 15–4 in the x direction, we have

$(+\swarrow)$ $\qquad\qquad m(v_x)_1 + \Sigma \int_{t_1}^{t_2} F_x\, dt = m(v_x)_2$

$$\frac{50}{32.2}(3) + \int_0^2 20t\, dt - 0.3N_C(2) + (50)(2) \sin 30° = \frac{50}{32.2}v_2$$

$$4.66 + 40 - 0.6N_C + 50 = 1.55v_2$$

The equation of equilibrium can be applied in the y direction. Why?

$+\nwarrow\Sigma F_y = 0;$ $\qquad\qquad N_C - 50 \cos 30° = 0$

Solving,

$$N_C = 43.3 \text{ lb} \qquad v_2 = 44.2 \text{ ft/s } \swarrow \qquad\qquad \textit{Ans.}$$

This problem has also been solved using the equation of motion in Example 13–3. The two methods of solution should be compared. Since *force, velocity,* and *time* are involved in the problem, application of the principle of impulse and momentum eliminates the need for using kinematics ($a = dv/dt$) and thereby yields an easier method for solution.

Example 15–3

Blocks A and B shown in Fig. 15–6a have a mass of 3 kg and 5 kg, respectively. If the system is released from rest, determine the velocity of block B in 6 s. Neglect the mass of the pulleys and cord.

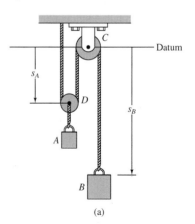

Datum

SOLUTION

Free-Body Diagram. Fig. 15–6b. Since the impulses of the blocks' weights are constant, the impulses of the cord tensions will also be constant. Furthermore, since the mass of pulley D is neglected, the cord tension $T_A = 2T_B$, Fig. 15–6c.

(a)

Principle of Impulse and Momentum
   Block A:

$(+\downarrow)$
$$m(v_A)_1 + \Sigma \int_{t_1}^{t_2} F_y \, dt = m(v_A)_2$$

$$0 - 2T_B(6) + 3(9.81)(6) = 3(v_A)_2 \qquad (1)$$

   Block B:

$(+\downarrow)$
$$m(v_B)_1 + \Sigma \int_{t_1}^{t_2} F_y \, dt = m(v_B)_2$$

$$0 + 5(9.81)(6) - T_B(6) = 5(v_B)_2 \qquad (2)$$

(b)

Kinematics. Since the blocks are subjected to dependent motion, the velocity of A may be related to that of B by using the kinematic analysis discussed in Sec. 12.8. A horizontal datum is established through the fixed point at C, Fig. 15–6a, and the changing positions of the blocks, s_A and s_B, are related to the constant total length l of the vertical segments of the cord by the equation

$$2s_A + s_B = l$$

Taking the time derivative yields

$$2v_A = -v_B \qquad (3)$$

As indicated by the negative sign, when B moves downward A moves upward.* Substituting this result into Eq. (1) and solving Eqs. (1) and (2) yields

$$(v_B)_2 = 35.8 \text{ m/s} \downarrow \qquad \textit{Ans.}$$
$$T_B = 19.2 \text{ N}$$

(c)

Fig. 15–6

*Note that the *positive* (downward) directions for $(\mathbf{v}_A)_2$ and $(\mathbf{v}_B)_2$ are *consistent* in Fig. 15–6a and b and in Eqs. (1) to (3). Why is this important?

PROBLEMS

15–1. A ball has a mass of 30 kg and is thrown upwards with a speed of 15 m/s. Determine how long it takes for it to stop. Also, how high does it rise before stopping? Use the principle of impulse and momentum for the solution.

15–2. A 2-lb ball is thrown in the direction shown with an initial speed of $v_A = 18$ ft/s. Determine the time needed for it to reach its highest point B and the speed at which it is traveling at B. Use the principle of impulse and momentum for the solution.

Prob. 15–2

15–3. A 40-g golf ball is struck in 3 ms by a driver such that it is given a velocity of $v_1 = 35$ m/s directed 30° from the horizontal. Determine the average force exerted on the ball and the momentum of the ball in $t = 1$ s after it leaves the ground.

*15–4. The 5-kg block is moving downward at $v_1 = 2$ m/s when it is 8 m from the sandy surface. Determine the impulse of the sand on the block necessary to stop its motion. Neglect the distance the block dents into the sand and assume the block does not rebound. Take the weight of the block to be nonimpulsive during the impact.

Prob. 15–4

15–5. A tankcar has a mass of 20 Mg and is freely rolling to the right with a speed of 0.75 m/s. If it strikes the barrier, determine the horizontal impulse needed to stop the car if the spring in the bumper B has a stiffness (a) $k \rightarrow \infty$ (bumper is rigid), and (b) $k = 15$ kN/m.

Prob. 15–5

15–6. A man kicks the 200-g ball such that it leaves the ground at an angle of 30° with the horizontal and strikes the ground at the same elevation a distance of 15 m away. Determine the impulse of his foot F on the ball. Neglect the impulse caused by the ball's weight while it is being kicked.

Prob. 15–6

15–7. If it takes 35 s for the 50-Mg towboat to increase its speed uniformly to 25 km/h, starting from rest, determine the force developed on the rope between the towboat and the barge. The propeller provides the propulsion force **F** which gives the towboat forward motion, whereas the barge moves freely. Also, determine F acting on the towboat. The barge has a mass of 75 Mg.

Prob. 15–7

*15–8. The tennis ball has a horizontal speed of 15 m/s when it is struck by the racket. If it then travels away at an angle of 25° from the horizontal and reaches a maximum altitude of 10 m, measured from the height of the racket, determine the magnitude of the net impulse of the racket on the ball. The ball has a mass of 180 g. Neglect the weight of the ball during the time the racket strikes the ball.

Prob. 15–8

15–9. As indicated by the derivation, the principle of impulse and momentum is valid for observers in *any* inertial reference frame. Show that this is so, by considering the 10-kg block which rests on the smooth surface and is subjected to a horizontal force of 6 N. If observer *A* is in a *fixed* frame *x*, determine the final speed of the block in 4 s if it has an initial speed of 5 m/s measured from the fixed frame. Compare the result with that obtained by an observer *B*, attached to the *x'* axis and moving at a constant velocity of 2 m/s relative to *A*.

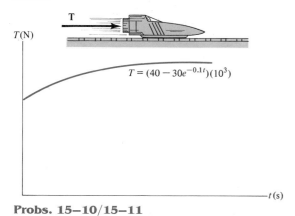

Prob. 15–9

15–10. The rocket sled has a mass of 3 Mg and starts from rest when $t = 0$. If the engines provide a horizontal thrust *T* which varies as shown in the graph, determine the sled's velocity in $t = 4$ s. Neglect air resistance, friction, and the loss of fuel during the motion.

$$T = (40 - 30e^{-0.1t})(10^3)$$

Probs. 15–10/15–11

15–11. The rocket sled has a mass of 3 Mg and starts from rest when $t = 0$. If the engines provide a horizontal thrust *T*, which varies as shown in the graph, determine the time needed for the sled to obtain a speed of 18 m/s. Neglect air resistance, friction, and the loss of fuel during the motion.

*15–12. The uniform beam has a weight of 5000 lb. Determine the average tension in each of the two cables *AB* and *AC* if the beam is given an upward speed of 8 ft/s in 1.5 s starting from rest. Neglect the mass of the cables.

Prob. 15–12

15–13. Block *A* weighs 10 lb and block *B* weighs 3 lb. If *B* is moving downward with a velocity of $(v_B)_1 = 3$ ft/s at $t = 0$, determine the velocity of *A* when $t = 1$ s. Assume that the horizontal plane is smooth. Neglect the mass of the pulleys and cord.

Probs. 15–13/15–14

15–14. Solve Prob. 15–13 if the coefficient of kinetic friction between the horizontal plane and block *A* is $\mu_A = 0.15$.

15–15. The block B has a mass of 70 kg. The winch delivers a horizontal towing force F to its cable at A which varies as shown in the graph. Determine the speed of the block when $t = 18$ s. Originally the block is moving upwards at $v_1 = 3$ m/s.

Prob. 15–15

*15–16.** The force acting on the 50-lb crate has a magnitude of $F = (2.4t^2 + 15t)$ lb, where t is in seconds. If the crate starts from rest, determine its speed when $t = 2$ s. Also determine the distance the crate moves in 2 s. Neglect friction.

Probs. 15–16/15–17

15–17. The force acting on the 50-lb crate has a magnitude of $F = (2.4t^2)$ lb, where t is in seconds. If the crate starts from rest, determine its speed when $t = 5$ s. The coefficients of static and kinetic friction between the crate and floor are $\mu_s = 0.3$ and $\mu_k = 0.2$, respectively.

15–18. The motor pulls on the cable at A with a force $F = (30 + t^2)$ lb, where t is in seconds. If the 34-lb crate is originally at rest at $t = 0$, determine its speed in $t = 4$ s. Neglect the mass of the cable and pulleys. *Hint:* First find the time needed to begin lifting the crate.

Prob. 15–18

15–19. The 50-kg block B is hoisted using the cable and motor arrangement shown. If the block is moving upwards at $v_1 = 2$ m/s when $t = 0$, and the motor develops a tension in the cord of $T = (500 + 120\sqrt{t})$ N, where t is in seconds, determine the velocity of the block when $t = 2$ s. Neglect the mass of the pulleys and cable.

Prob. 15–19

15.2 Principle of Linear Impulse and Momentum for a System of Particles

The principle of linear impulse and momentum for a system of particles moving relative to an inertial reference, Fig. 15–7, may be obtained from the equation of motion $\Sigma\mathbf{F}_i = \Sigma m_i \mathbf{a}_i$, Eq. 13–5, which may be rewritten as

$$\Sigma\mathbf{F}_i = \Sigma m_i \frac{d\mathbf{v}_i}{dt} \tag{15-5}$$

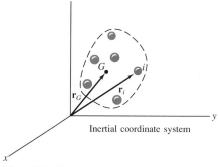

Inertial coordinate system

Fig. 15–7

The term on the left side represents only the sum of all the *external forces* acting on the system of particles. Recall that the internal forces $\mathbf{f}_i$ between particles do not appear with this summation, since by Newton's third law they occur in equal but opposite collinear pairs and therefore cancel out. Multiplying both sides of Eq. 15–5 by dt and integrating between the limits $t = t_1$, $\mathbf{v}_i = (\mathbf{v}_i)_1$ and $t = t_2$, $\mathbf{v}_i = (\mathbf{v}_i)_2$ yields

$$\Sigma m_i (\mathbf{v}_i)_1 + \Sigma \int_{t_1}^{t_2} \mathbf{F}_i \, dt = \Sigma m_i (\mathbf{v}_i)_2 \tag{15-6}$$

This equation states that the initial linear momenta of the system added vectorially to the impulses of all the *external forces* acting on the system during the time period t_1 to t_2 are equal to the system's final linear momenta.

By definition, the location of the mass center G of the system is determined from $m\mathbf{r}_G = \Sigma m_i \mathbf{r}_i$, where $m = \Sigma m_i$ is the total mass of all the particles, and $\mathbf{r}_G$ and $\mathbf{r}_i$ are defined in Fig. 15–7. Taking the time derivatives, we have

$$m\mathbf{v}_G = \Sigma m_i \mathbf{v}_i$$

which states that the total linear momentum of the system of particles is equivalent to the linear momentum of a "fictitious" aggregate particle of mass $m = \Sigma m_i$ moving with the velocity of the mass center G of the system. Substituting into Eq. 15–6 yields

$$m(\mathbf{v}_G)_1 + \Sigma \int_{t_1}^{t_2} \mathbf{F} \, dt = m(\mathbf{v}_G)_2 \tag{15-7}$$

This equation states that the initial linear momentum of the aggregate particle plus (vectorially) the external impulses acting on the system of particles during the time interval t_1 to t_2 is equal to the aggregate particle's final linear momentum. Since in reality all particles must have finite size to possess mass, the above equation justifies application of the principle of linear impulse and momentum to a rigid body represented as a single particle.

15.3 Conservation of Linear Momentum for a System of Particles

When the sum of the *external impulses* acting on a system of particles is *zero*, Eq. 15–6 reduces to a simplified form, namely,

$$\Sigma m_i(\mathbf{v}_i)_1 = \Sigma m_i(\mathbf{v}_i)_2 \qquad (15\text{--}8)$$

This equation is referred to as the *conservation of linear momentum*. It states that the vector sum of the linear momenta for a system of particles remains constant throughout the time period t_1 to t_2. Since $m\mathbf{v}_G = \Sigma m_i\mathbf{v}_i$, we can also write

$$(\mathbf{v}_G)_1 = (\mathbf{v}_G)_2 \qquad (15\text{--}9)$$

which indicates that the velocity $\mathbf{v}_G$ of the mass center for the system of particles does not change when no external impulses are applied to the system.

The conservation of linear momentum is often applied when particles collide or interact. For application, a careful study of the free-body diagram for the *entire* system of particles should be made. By doing this, one will be able to identify the forces as creating either external or internal impulses and thereby determine in what direction(s) linear momentum is conserved. As stated earlier, for the system the *internal impulses* will always cancel out since they occur in equal but opposite collinear pairs. If the time period over which the motion is studied is *very short,* some of the external impulses may also be neglected or considered approximately equal to zero. The forces causing these negligible impulses are called *nonimpulsive forces.* By comparison, forces which are very large, act for a very short period of time, and yet produce a significant change in momentum are called *impulsive forces.* They, of course, cannot be neglected in the impulse-momentum analysis.

Impulsive forces normally occur due to an explosion or the striking of one body against another, whereas nonimpulsive forces may include the weight of a body, the force imparted by a slightly deformed spring having a relatively small stiffness, or for that matter, any force that is very small compared to other larger (impulsive) forces. When making this distinction between impulsive and nonimpulsive forces, it is important to realize that it only applies during a *specific time period.* To illustrate, consider the effect of striking a baseball with a bat. During the *very short* time of interaction, the force of the bat on the ball (or particle) is impulsive since it drastically changes the ball's momentum. By comparison the ball's weight will have a negligible effect on the change in momentum, and therefore it is nonimpulsive. Consequently, it can be neglected from an impulse-momentum analysis during this time period. It should be pointed out, however, that if an impulse-momentum analysis is considered during the much longer time of flight after the ball-bat interaction, then the impulse of the ball's weight is important since it, along with air resistance, causes the change in the momentum of the ball.

PROCEDURE FOR ANALYSIS

Generally, the principle of linear impulse and momentum or the conservation of linear momentum is applied to a *system of particles* in order to determine the final velocities of the particles *just after* the time period considered. By applying these equations to the entire system, the internal impulses acting within the system, which may be unknown, are *eliminated* from the analysis. For application it is suggested that the following procedure be used.

Free-Body Diagram. Establish the *x*, *y*, *z* inertial frame of reference and draw the free-body diagram for each particle of the system in order to identify the internal and external forces. The conservation of linear momentum applies to the system in a given direction when *no external impulsive forces* act on the system in that direction. Also, establish the direction and sense of the particles' initial and final velocities. If the sense is unknown, assume it is along a positive inertial coordinate.

As an alternative procedure, draw the impulse and momentum diagrams for each particle of the system just before, during, and just after the impulsive forces are applied. Then investigate the impulse diagram in order to clearly distinguish the external impulsive and nonimpulsive forces from the system's internal impulses.

Momentum Equations. Apply the principle of linear impulse and momentum or the conservation of linear momentum in the appropriate directions. If the particles are subjected to dependent motion using cables and pulleys, kinematics as discussed in Sec. 12.8 can be used to relate the velocities.

If it is necessary to determine the *internal impulsive force* acting on only one particle of a system, the particle must be *isolated* (free-body diagram) and the principle of linear impulse and momentum must be applied *to the particle*. After the impulse $\int F \, dt$ is calculated, then, provided the time Δt for which the impulse acts is known, the *average impulsive force* F_{avg} can be determined from $F_{avg} = \int F \, dt / \Delta t$.

The following examples numerically illustrate application of this procedure.

Example 15–4

The 15-Mg boxcar A is coasting freely at 1.5 m/s on the horizontal track when it encounters a tank car B having a mass of 12 Mg and coasting at 0.75 m/s toward it as shown in Fig. 15–8a. If the cars meet and couple together, determine (a) the speed of both cars just after the coupling, and (b) the average force between them if the coupling takes place in 0.8 s.

(a)

SOLUTION

Part (a)

Free-Body Diagram.* As shown in Fig. 15–8b, we have considered *both* cars as a single system. By inspection, momentum is conserved in the x direction since the coupling force **F** is *internal* to the system and will therefore cancel out. It is assumed both cars, when coupled, move at v_2 in the positive x direction.

(b)

(c)

Fig. 15–8

Conservation of Linear Momentum

$$(\xrightarrow{+})\qquad m_A(v_A)_1 + m_B(v_B)_1 = (m_A + m_B)v_2$$
$$15\ 000(1.5) - 12\ 000(0.75) = 27\ 000v_2$$
$$v_2 = 0.5\ \text{m/s} \rightarrow \qquad \textit{Ans.}$$

Part (b). The average (impulsive) coupling force, $\mathbf{F}_{avg}$, can be determined by applying the principle of linear momentum to *either one* of the cars.

Free-Body Diagram. As shown in Fig. 15–8c, by isolating the boxcar the coupling force is *external* to the car.

Principle of Impulse and Momentum. Since $\int F\,dt = F_{avg}\,\Delta t = F_{avg}(0.8)$, we have

$$(\xrightarrow{+})\qquad m_A(v_A)_1 + \Sigma \int F\,dt = m_A v_2$$
$$15\ 000(1.5) - F_{avg}(0.8) = 15\ 000(0.5)$$
$$F_{avg} = 18.8\ \text{kN} \qquad \textit{Ans.}$$

Solution was possible since the boxcar's final velocity was obtained in Part (a). Try solving for F_{avg} by applying the principle of impulse and momentum to the tank car.

*Only horizontal forces are shown on the free-body diagram.

Example 15–5

The 1200-lb cannon shown in Fig. 15–9a fires an 8-lb projectile with a muzzle velocity of 1500 ft/s. If firing takes place in 0.03 s, determine (a) the recoil velocity of the cannon just after firing, and (b) the average impulsive force acting on the projectile. The cannon support is fixed to the ground and the horizontal recoil of the cannon is absorbed by two springs.

Fig. 15–9

SOLUTION

Part (a)

Free-Body Diagram.* As shown in Fig. 15–9b, we have considered the projectile and cannon as a single system, since then the impulsive forces, **F**, between the cannon and projectile are *internal* to the system and will therefore cancel from the analysis. Furthermore, during the time $\Delta t = 0.03$ s, the two recoil springs which are attached to the support each exert a *nonimpulsive force* $\mathbf{F}_s$ on the cannon. This is because Δt is very short, so that during this time the cannon only moves through a very small distance† s. Consequently, $F_s = ks \approx 0$. Hence it may be concluded that momentum for the system is conserved in the *horizontal direction*. Here we will assume that the cannon moves to the left, while the projectile moves to the right after firing.

Conservation of Linear Momentum.

$$(\xrightarrow{+}) \qquad m_c(v_c)_1 + m_p(v_p)_1 = -m_c(v_c)_2 + m_p(v_p)_2$$

$$0 + 0 = -\frac{1200}{32.2}(v_c)_2 + \frac{8}{32.2}(1500)$$

$$(v_c)_2 = 10 \text{ ft/s} \leftarrow \qquad\qquad Ans.$$

Part (b). The average impulsive force exerted by the cannon on the projectile can be determined by applying the principle of linear impulse and momentum to the projectile (or to the cannon).

Principle of Impulse and Momentum. Using the data on the free-body diagram, Fig. 15–9c, noting that $\int F\, dt = F_{avg}\,\Delta t = F_{avg}(0.03)$, we have

$$(\xrightarrow{+}) \qquad m(v_p)_1 + \Sigma \int F\, dt = m(v_p)_2$$

$$0 + F_{avg}(0.03) = \frac{8}{32.2}(1500)$$

$$F_{avg} = 12.4(10^3) \text{ lb} = 12.4 \text{ kip} \qquad Ans.$$

*Only horizontal forces are shown on the free-body diagram.

†If the cannon is firmly fixed to its support (no springs), the reactive force of the support on the cannon must be considered as an external impulse to the system, since the support would allow no movement of the cannon.

Example 15–6

The 350-Mg tugboat T shown in Fig. 15–10a is used to pull the 50-Mg barge B with a rope R. If the tugboat is coasting freely with an initial velocity of $(v_T)_1 = 3$ m/s while the rope is *slack*, determine the velocity of the tugboat *directly after* the rope becomes taut. Assume the rope does not stretch. Neglect the frictional effects of the water.

(a)

Fig. 15–10

(b)

SOLUTION

Free-Body Diagram.* As shown in Fig. 15–10b, we have considered the entire system (tugboat and barge). Hence, the impulsive force created by the rope and the barge is *internal* to the system, and therefore momentum of the system is conserved during the instant of towing.

Conservation of Momentum. Noting that $(v_B)_2 = (v_T)_2$, we have

$$(\xleftrightarrow{+}) \qquad m_T(v_T)_1 + m_B(v_B)_1 = m_T(v_T)_2 + m_B(v_B)_2$$
$$350(10^3)(3) + 0 = 350(10^3)(v_T)_2 + 50(10^3)(v_T)_2$$

Solving,

$$(v_T)_2 = 2.63 \text{ m/s} \leftarrow \qquad\qquad \textit{Ans.}$$

This value represents the tugboat's velocity *just after* the towing impulse. Use this result and show that the towing impulse is 131 kN · s.

The alternative procedure of drawing the system's impulse and momentum diagrams is shown in Fig. 15–10c. The conservation of momentum is formulated as above by writing the vector terms directly from the diagrams.

*Only horizontal forces are shown on the free-body diagram.

(c)

Example 15–7

A rigid pile P shown in Fig. 15–11a has a mass of 800 kg and is driven into the ground using a hammer H that has a mass of 300 kg. The hammer falls from rest from a height $y_0 = 0.5$ m and strikes the top of the pile. Determine the impulse which the hammer imparts on the pile if the pile is surrounded entirely by loose sand so that after striking the hammer does *not* rebound off the pile.

(a)

SOLUTION

Conservation of Energy. The velocity at which the hammer strikes the pile can be determined using the conservation of energy equation applied to the hammer. With the datum at the top of the pile, Fig. 15–11a, we have

$$T_0 + V_0 = T_1 + V_1$$

$$\tfrac{1}{2}m_H(v_H)_0^2 + W_H y_0 = \tfrac{1}{2}m_H(v_H)_1^2 + W_H y_1$$

$$0 + 300(9.81)(0.5) = \frac{1}{2}(300)(v_H)_1^2 + 0$$

$$(v_H)_1 = 3.13 \text{ m/s}$$

Free-Body Diagram. From the physical aspects of the problem, the free-body diagram of the hammer and pile, Fig. 15–11b, indicates that during the *short time* occurring just before to just after the *collision* the weights of the hammer and pile and the resistance force $\mathbf{F}_s$ of the sand are all *nonimpulsive*. Furthermore, the impulsive force $\mathbf{R}$ is internal to the system and therefore cancels. Consequently, momentum is conserved in the vertical direction.

(b)

Conservation of Momentum. Just after collision we will assume $(v_H)_2 = (v_P)_2 = v_2$.

$$(+\downarrow) \qquad m_H(v_H)_1 + m_P(v_P)_1 = m_H v_2 + m_P v_2$$

$$300(3.13) + 0 = 300v_2 + 800v_2$$

$$v_2 = 0.854 \text{ m/s}$$

Principle of Impulse and Momentum. The impulse which the hammer imparts to the pile can now be determined since $\mathbf{v}_2$ is known. From the free-body diagram for the hammer, Fig. 15–11c, we have

(c)

Fig. 15–11

$$(+\downarrow) \qquad m_H(v_H)_1 + \Sigma \int_{t_1}^{t_2} F_y \, dt = m_H v_2$$

$$300(3.13) - \int R \, dt = 300(0.854)$$

$$\int R \, dt = 683 \text{ N} \cdot \text{s} \qquad\qquad \textit{Ans.}$$

Try finding the impulse by applying the principle of impulse and momentum to the pile.

Example 15–8

A boy having a mass of 40 kg stands on the back of a 15-kg toboggan which is originally at rest, Fig. 15–12a. If he walks to the front B and stops, determine the distance the toboggan moves. Neglect friction between the bottom of the toboggan and the ground (ice).

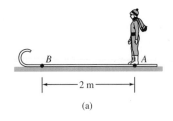

(a)

SOLUTION I

Free-Body Diagram.* The unknown frictional force of the boy's shoes on the bottom of the toboggan can be *excluded* from the analysis if the toboggan and boy on it are considered as a single system. In this way the frictional force $\mathbf{F}$ becomes internal and the conservation of momentum applies. Since both the initial and final momenta of the system are zero (because the initial and final velocities are zero), the system's momentum must also be zero when the boy is at some intermediate point between A and B, Fig. 15–12b.

Conservation of Momentum

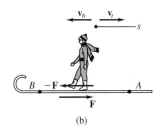

(b)

$(\overset{+}{\rightarrow})$ $\qquad\qquad\qquad -m_b v_b + m_t v_t = 0$ $\qquad\qquad$ (1)

Here the two unknowns v_b and v_t represent the velocities of the boy and the toboggan measured from a *fixed inertial reference, s*, on the ground. The *positions* of point A on the toboggan and the boy must be determined by integration. Assuming the initial position of point A to be at the origin, then at the intermediate position we have

$$-m_b s_b + m_t s_t = 0$$

If in the final position point A on the toboggan is located $s_t = +s$ to the right (in the same direction as $\mathbf{v}_t$), Fig. 15–12c, then when the boy is at B this point has the position $s_b = (2 - s)$ m to the left (in the same direction as $\mathbf{v}_b$). Hence,

$$-m_b(2 - s) + m_t s = 0 \qquad\qquad (2)$$

$$s = \frac{2m_b}{m_b + m_t} = \frac{2(40)}{40 + 15} = 1.45 \text{ m} \qquad\qquad \textit{Ans.}$$

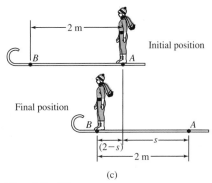

(c)

Fig. 15–12

SOLUTION II

The problem may also be solved by considering the relative motion of the boy with respect to the toboggan, $\mathbf{v}_{b/t}$. This velocity is related to the velocities of the boy and toboggan by the equation $\mathbf{v}_b = \mathbf{v}_t + \mathbf{v}_{b/t}$, Eq. 12–34. Since positive motion is assumed to be to the right in Eq. (1), $\mathbf{v}_b$ and $\mathbf{v}_{b/t}$ are negative, because the boy's motion is to the left. Hence, in scalar form, $-v_b = v_t - v_{b/t}$, and Eq. (1) then becomes $m_b(v_t - v_{b/t}) + m_t v_t = 0$. Integrating gives

$$m_b(s_t - s_{b/t}) + m_t s_t = 0$$

Assuming point A on the toboggan moves a distance s to the right, Fig. 15–12c, realizing $s_{b/t} = 2$ m, we have $m_b(s - 2) + m_t s = 0$, which is the same as Eq. (2).

*Only horizontal forces are shown on the free-body diagram.

PROBLEMS

*15–20. The car A has a weight of 4500 lb and is traveling to the right at 3 ft/s. Meanwhile a 3000-lb car B is traveling at 6 ft/s to the left. If the cars crash head-on and become entangled, determine their common velocity just after the collision. Assume that the brakes are not applied during collision.

$v_A = 3$ ft/s $v_B = 6$ ft/s

Prob. 15–20

15–21. A railroad car having a mass of 15 Mg is coasting at 1.5 m/s on a horizontal track. At the same time another car having a mass of 12 Mg is coasting at 0.75 m/s in the opposite direction. If the cars meet and couple together, determine the speed of both cars just after the coupling. Compute the difference between the total kinetic energy before and after coupling has occurred, and explain qualitatively what happened to this energy.

15–22. The man M weighs 150 lb and jumps onto the boat B which has a weight of 200 lb. If he has a horizontal component of velocity *relative to the boat* of 3 ft/s, just before he enters the boat, and the boat is traveling at $v_B = 2$ ft/s away from the pier when he makes the jump, determine the resulting velocity of the man and boat.

M

$v_B = 2$ ft/s

B

Prob. 15–22

15–23. A 20-g bullet is fired horizontally into the 300-g block which rests on the smooth surface. After the bullet becomes embedded into the block, the block moves to the right 0.3 m before momentarily coming to rest. Determine the speed $(v_B)_1$ of the bullet. The spring has a stiffness of $k = 200$ N/m and is originally unstretched.

$(v_B)_1$ $k = 200$ N/m

Prob. 15–23

*15–24. The boy and girl are each standing on ice skates and holding on to the ends of the rope. If they begin to pull in the rope, such that their relative speed of approach is $v_{\text{rel}} = 2$ m/s, determine their velocities at the instant they meet. If the rope is 5 m long, how long will it take before they embrace? The boy has a mass of 60 kg, and the girl has a mass of 50 kg.

5 m

Prob. 15–24

15–25. Blocks A and B have masses of 40 kg and 60 kg, respectively. They are placed on a smooth surface and the spring connected between them is stretched 2 m. If they are released from rest, determine the speeds of both blocks the instant the spring becomes unstretched.

$k = 180$ N/m

A B

Prob. 15–25

15–26. The block has a mass of 50 kg and rests on the surface of the cart having a mass of 75 kg. If the spring which is attached to the cart and not the block is compressed 0.2 m and the system is released from rest, determine the speed of the block after the spring becomes undeformed. Neglect the mass of the cart's wheels and spring in the calculation. Also neglect friction. Take $k = 300$ N/m.

B C

Probs. 15–26/15–27

15–27. The block has a mass of 50 kg and rests on the surface of the cart having a mass of 75 kg. If the spring which is attached to the cart and not the block is compressed 0.2 m and the system is released from rest, determine the speed of the block with respect to the cart after the spring becomes undeformed. Neglect the mass of the wheels and spring in the calculation. Also neglect friction. Take $k = 300$ N/m.

*15–28. A boy having a weight of 75 lb stands at the end of the 50-lb canoe. If the canoe is not tied to the pier P and the boy runs at 4 ft/s relative to the canoe, determine how far d the canoe has moved away from the pier at the instant he leaps off at A. Neglect water resistance. Originally d = 0.

Prob. 15–28

15–29. The boy jumps off the flat car at A with a velocity of v' = 4 ft/s relative to the car as shown. If he lands on the second flat car B, determine the final speed of both cars after the motion. Each car has a weight of 80 lb. The boy's weight is 60 lb. Both cars are originally at rest. Neglect the mass of the car's wheels.

Prob. 15–29

15–30. Two men A and B, each having a weight of 160 lb, stand on the 200-lb cart. Each can run with a speed of 3 ft/s measured relative to the cart. Determine the final speed of the cart if (a) A runs and jumps off, then B runs and jumps off the same end, and (b) both run at the same time and jump off at the same time. Neglect the mass of the wheels. The jumps are made horizontally.

Prob. 15–30

15–31. A boy A having a weight of 80 lb and a girl B having a weight of 65 lb stand motionless at the ends of the toboggan, which has a weight of 20 lb. If they exchange positions, A going to B and then B going to A's original position, determine the final position of the toboggan just after the motion. Neglect friction.

Probs. 15–31/15–32

*15–32. Solve Prob. 15–26 if B first moves to A, then A moves to B's original position.

15–33. The spring-loaded gun rests on the smooth surface and has a mass of 5 kg. It fires a ball having a mass of 1 kg with a velocity of v' = 6 m/s relative to the gun in the direction shown. If the gun is originally at rest, determine the horizontal distance d the ball is from the gun at the instant the ball strikes the ground at B. Neglect the height of the gun.

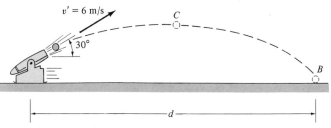

Probs. 15–33/15–34

15–34. The spring-loaded gun rests on the smooth surface and has a mass of 5 kg. It fires a ball having a mass of 1 kg with a velocity of v' = 6 m/s relative to the gun in the direction shown. If the gun is originally at rest, determine the horizontal distance the ball is from the gun at the instant the ball reaches its highest elevation C. Neglect the height of the gun.

15–35. The 10-lb projectile is fired from ground level with an initial velocity of $v_A = 80$ ft/s in the direction shown. When it reaches its highest point it explodes into two 5-lb fragments. If one fragment travels vertically upwards at 12 ft/s, determine the distance between the fragments after they strike the ground. Neglect the size of the gun.

Probs. 15–35/15–36

*15–36. The 10-lb projectile is fired from ground level with an initial velocity of 80 ft/s in the direction shown. When it reaches its highest point it explodes into two 5-lb fragments. If one fragment is seen to travel vertically upwards, and after they fall they are 150 ft apart, determine the speed of each fragment just after the explosion. Neglect the size of the gun.

15–37. The crate has a mass m and rests on the barge which has a mass M. The coefficient of kinetic friction between the crate and barge is μ. The rope is pulled (or jerked) such that it snaps and as a result the initial impulse gives the barge a sudden velocity $\mathbf{v}_0$. Determine the time the crate slides on the barge before coming to rest on the barge. Also, what is the final velocity of the barge? The crate and barge have zero velocity just before the impulse is applied.

Prob. 15–37

15–38. Using the data in Prob. 15–37, determine the distance the crate slides on the barge before coming to rest.

15–39. The block A has a mass of 5 kg and is placed on the smooth triangular block B having a mass of 30 kg. If the system is released from rest, determine the distance B moves from point O when A reaches the bottom. Neglect the size of block A and friction.

Probs. 15–39/15–40

*15–40. Solve Prob. 15–39 if the coefficient of kinetic friction between A and B is $\mu_k = 0.3$. Neglect friction between block B and the horizontal plane.

15.4 Impact

Central impact
(a)

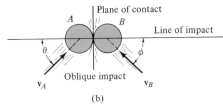

Oblique impact
(b)

Fig. 15–13

Impact occurs when two bodies collide with each other during a very *short* interval of time, causing relatively large (impulsive) forces to be exerted between the bodies. The striking of a hammer and nail, or a golf club and ball, are common examples of impact loadings.

In general, there are two types of impact. *Central impact* occurs when the direction of motion of the mass centers of the two colliding particles is along a line passing through the mass centers of the particles. This line is called the *line of impact,* Fig. 15–13*a*. When the motion of one or both of the particles is at an angle with the line of impact, Fig. 15–13*b*, the impact is said to be *oblique impact.*

Central Impact

To illustrate the method for analyzing the mechanics of impact, consider the case involving the central impact of two *smooth* particles *A* and *B* shown in Fig. 15–14.

1. The particles have the initial momenta shown in Fig. 15–14*a*. Provided $(v_A)_1 > (v_B)_1$, collision will eventually occur.
2. During the collision the material of the particles must be thought of as *deformable* or nonrigid. The particles will undergo a *period of deformation* such that they exert an equal but opposite deformation impulse $\int \mathbf{P} \, dt$ on each other, Fig. 15–14*b*.
3. Only at the instant of *maximum deformation* will both particles move with a common velocity $\mathbf{v}$, Fig. 15–14*c*.
4. Afterward a *period of restitution* occurs, in which case the material from which the particles are made will either return to its original shape or remain permanently deformed. The equal but opposite *restitution impulse* $\int \mathbf{R} \, dt$ pushes the particles apart from one another, Fig. 15–14*d*. In reality, the physical properties of any two bodies are such that the deformation impulse is *always greater* than that of restitution, i.e., $\int P \, dt > \int R \, dt$.
5. Just after separation the particles will have the final momenta shown in Fig. 15–14*e*, where $(v_B)_2 > (v_A)_2$.

In most problems the initial velocities of the particles will be *known* and it will be necessary to determine their final velocities $(v_A)_2$ and $(v_B)_2$. In this regard, *momentum* for the *system of particles* is *conserved* since during collision the internal impulses of deformation and restitution *cancel*. Hence, referring to Fig. 15–14*a* and *e*,

$$(\overset{+}{\rightarrow}) \qquad m_A(v_A)_1 + m_B(v_B)_1 = m_A(v_A)_2 + m_B(v_B)_2 \qquad (15\text{–}10)$$

In order to obtain a second equation, necessary to solve for $(v_A)_2$ and $(v_B)_2$, we must apply the principle of impulse and momentum to *each particle*. For example, during the deformation phase for particle *A*, Fig. 15–14*a, b,* and *c,* we have

$$(\overset{+}{\rightarrow}) \qquad m_A(v_A)_1 - \int P\,dt = m_A v$$

Before impact

(a)

For the restitution phase, Fig. 15–14c, d, and e,

$$(\overset{+}{\rightarrow}) \qquad m_A v - \int R\,dt = m_A(v_A)_2$$

The ratio of the restitution impulse to the deformation impulse is called the *coefficient of restitution, e.* From the above equations, this value for particle A is

Deformation Impulse

(b)

$$e = \frac{\displaystyle\int R\,dt}{\displaystyle\int P\,dt} = \frac{v - (v_A)_2}{(v_A)_1 - v}$$

In a similar manner, we can establish e by considering particle B, Fig. 15–14. This yields

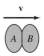

Maximum deformation

(c)

$$e = \frac{\displaystyle\int R\,dt}{\displaystyle\int P\,dt} = \frac{(v_B)_2 - v}{v - (v_B)_1}$$

If the unknown v is eliminated from the above two equations, the coefficient of restitution can be expressed in terms of the particles' initial and final velocities as

Restitution Impulse

(d)

$$(\overset{+}{\rightarrow}) \qquad \boxed{e = \frac{(v_B)_2 - (v_A)_2}{(v_A)_1 - (v_B)_1}} \qquad (15\text{–}11)$$

Provided a value for e is specified, Eqs. 15–10 and 15–11 may be solved simultaneously to obtain $(v_A)_2$ and $(v_B)_2$. In doing so, however, it is important to carefully establish a sign convention for defining the positive directions for both $\mathbf{v}_B$ and $\mathbf{v}_A$, and then use it *consistently* when writing *both* equations. As noted from the application above, and indicated symbolically by the arrow in parentheses, we have defined the positive direction to the right when referring to the motions of both A and B. Consequently, if a negative value results from the solution of either $(v_A)_2$ or $(v_B)_2$, it indicates motion is to the left.

After impact

(e)

Fig. 15–14

Coefficient of Restitution. With reference to Fig. 15–14a and e, it is seen that Eq. 15–11 states that the coefficient of restitution is equal to the ratio of the relative velocity of the particles' separation *just after impact,* $(v_B)_2 - (v_A)_2$, to the relative velocity of the particles' approach *just before impact,* $(v_A)_1 - (v_B)_1$. By measuring these relative velocities experimentally, it has been found that e varies appreciably with impact velocity as well as with the size and shape of the colliding bodies. For these reasons the coefficient of restitution is reliable only when used under conditions which closely approximate those which were known to exist when measurements of it were made. In general e has a value between zero and one, and one should be aware of the physical meaning of these two limits.

Elastic Impact (e = 1). If the collision between the two particles is *perfectly elastic,* the deformation impulse ($\int \mathbf{P}\, dt$) is equal and opposite to the restitution impulse ($\int \mathbf{R}\, dt$). Although in reality this can never be achieved, $e = 1$ for an elastic collision.

Plastic Impact (e = 0). The impact is said to be *inelastic or plastic* when $e = 0$. In this case there is no restitution impulse given to the particles ($\int \mathbf{R}\, dt = \mathbf{0}$), so that after collision both particles couple or stick *together* and move with a common velocity.

From the above derivation it should be evident that the principle of work and energy cannot be used for the analysis of this problem since it is not possible to know how the *internal forces* of deformation and restitution vary or move during the collision. By knowing the particle's velocities before and after collision, however, the energy loss during collision can be calculated on the basis of the difference in the particle's kinetic energy. This energy loss, $\Sigma U_{1-2} = \Sigma T_2 - \Sigma T_1$, occurs because some of the initial kinetic energy of the particle is transformed into thermal energy as well as creating sound and localized deformation of the material when the collision occurs. In particular, if the impact is *perfectly elastic,* no energy would be lost in the collision (see Prob. 15–43); whereas if the collision is *plastic,* the energy lost during collision would be a maximum.

PROCEDURE FOR ANALYSIS (CENTRAL IMPACT)

In most cases the *final velocities* of two smooth particles are to be determined *just after* they are subjected to direct central impact. Provided the coefficient of restitution, the mass of each particle, and each particle's initial velocity *just before* impact are known, the solution to the problem can be obtained using the following two equations:

1. The conservation of momentum applies to the system of particles, $\Sigma m v_1 = \Sigma m v_2$.
2. The coefficient of restitution, $e = ((v_B)_2 - (v_A)_2)/((v_A)_1 - (v_B)_1)$, relates the relative velocities of the particles along the line of impact, just before and just after collision.

When applying these two equations, the sense of an unknown velocity can be assumed. If the solution yields a negative magnitude, the velocity acts in the opposite sense.

Oblique Impact

When oblique impact occurs between two smooth particles, the particles move away from each other with velocities having unknown directions as well as unknown magnitudes. Provided the initial velocities are known, four unknowns are present in the problem. As shown in Fig. 15–15a, these unknowns may be represented as $(v_A)_2$, $(v_B)_2$, θ_2, and ϕ_2.

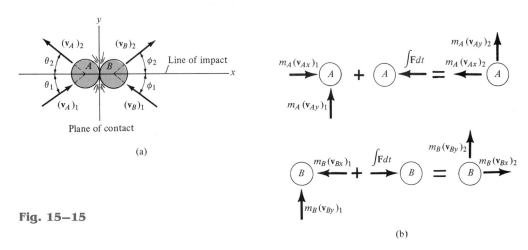

Fig. 15–15

PROCEDURE FOR ANALYSIS (OBLIQUE IMPACT)

If the y axis is established within the plane of contact and the x axis along the line of impact, the impulsive forces of deformation and restitution act *only in the x direction*, Fig. 15–15b. Resolving the velocity or momentum vectors into components along the x and y axes, Fig. 15–15b, it is possible to write four independent scalar equations in order to determine $(v_{Ax})_2$, $(v_{Ay})_2$, $(v_{Bx})_2$, and $(v_{By})_2$.

1. Momentum of the system is conserved *along the line of impact*, x axis, so that $\Sigma m(v_x)_1 = \Sigma m(v_x)_2$.
2. The coefficient of restitution, $e = ((v_{Bx})_2 - (v_{Ax})_2)/((v_{Ax})_1 - (v_{Bx})_1)$, relates the relative-velocity *components* of the particles *along the line of impact* (x axis).
3. Momentum of particle A is conserved along the y axis, perpendicular to the line of impact, since no impulse acts on the particle in this direction.
4. Momentum of particle B is conserved along the y axis, perpendicular to the line of impact, since no impulse acts on the particle in this direction.

Application of these four equations is illustrated numerically in Example 15–11.

Example 15–9

The bag A, having a weight of 6 lb, is released from rest at the position $\theta = 0°$, as shown in Fig. 15–16a. It strikes an 18-lb box B when $\theta = 90°$. If the coefficient of restitution between the bag and box is $e = 0.5$, determine the velocities of the bag and box just after impact.

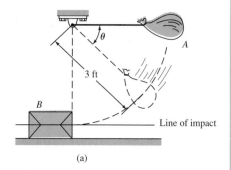

(a)

SOLUTION

This problem involves central impact. Why? Before analyzing the mechanics of the impact, however, it is first necessary to obtain the velocity of the bag *just before* it strikes the box.

Conservation of Energy. With the datum at $\theta = 0°$, Fig. 15–16b, we have

$$T_0 + V_0 = T_1 + V_1$$

$$0 + 0 = \frac{1}{2}\left(\frac{6}{32.2}\right)(v_A)_1^2 - 6(3)$$

$$(v_A)_1 = 13.9 \text{ ft/s}$$

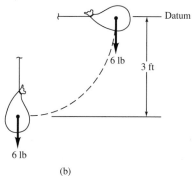

(b)

Conservation of Momentum. After impact we will assume A and B travel to the left. Applying the conservation of momentum to the system, we have

$(\xleftarrow{+})$
$$m_B(v_B)_1 + m_A(v_A)_1 = m_B(v_B)_2 + m_A(v_A)_2$$

$$0 + \frac{6}{32.2}(13.9) = \frac{18}{32.2}(v_B)_2 + \frac{6}{32.2}(v_A)_2$$

$$(v_A)_2 = 13.9 - 3(v_B)_2 \qquad (1)$$

Coefficient of Restitution. Realizing that for separation to occur after collision $(v_B)_2 > (v_A)_2$, Fig. 15–16c, we have

$(\xleftarrow{+})$
$$e = \frac{(v_B)_2 - (v_A)_2}{(v_A)_1 - (v_B)_1}$$

$$0.5 = \frac{(v_B)_2 - (v_A)_2}{13.9 - 0}$$

$$(v_A)_2 = (v_B)_2 - 6.95 \qquad (2)$$

(c)

Solving Eqs. (1) and (2) simultaneously yields

$$(v_A)_2 = -1.74 \text{ ft/s} = 1.74 \text{ ft/s} \rightarrow \quad \text{and} \quad (v_B)_2 = 5.21 \text{ ft/s} \leftarrow \ \textit{Ans.}$$

Fig. 15–16

Example 15–10

The ball B shown in Fig. 15–17a has a mass of 1.5 kg and is suspended from the ceiling by a 1-m-long elastic cord. If the cord is *stretched* downward 0.25 m and the ball is released from rest, determine how far the cord stretches after the ball rebounds from the ceiling. The stiffness of the cord is $k = 800$ N/m and the coefficient of restitution between the ball and ceiling is $e = 0.8$. The ball makes a central impact with the ceiling.

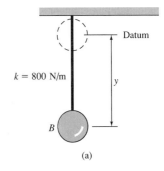

(a)

SOLUTION

It is necessary to first obtain the initial velocity of the ball *just before* it strikes the ceiling.

Conservation of Energy. With the datum located as shown in Fig. 15–17a, realizing that initially $y = y_0 = (1 + 0.25)$ m $= 1.25$ m, we have

$$T_0 + V_0 = T_1 + V_1$$

$$\tfrac{1}{2}m(v_B)_0^2 - W_B y_0 + \tfrac{1}{2}ks^2 = \tfrac{1}{2}m(v_B)_1^2 + 0$$

$$0 - 1.5(9.81)(1.25) + \frac{1}{2}(800)(0.25)^2 = \frac{1}{2}(1.5)(v_B)_1^2$$

$$(v_B)_1 = 2.97 \text{ m/s} \uparrow$$

(b)

Fig. 15–17

The interaction of the ball with the ceiling will now be considered using the principles of impact.* Note that the y axis, Fig. 15–17b, represents the line of impact for the ball. Since an unknown portion of the mass of the ceiling is involved in the impact, the conservation of momentum for the ball–ceiling system will not be written. The "velocity" of this portion of ceiling is zero since it remains at rest *both* before and after impact.

Coefficient of Restitution. Since $(\mathbf{v}_B)_2$ is downward, we have

$$(+\uparrow) \quad e = \frac{(v_B)_2 - (v_A)_2}{(v_A)_1 - (v_B)_1}; \quad 0.8 = \frac{-(v_B)_2 - 0}{0 - 2.97}$$

$$(v_B)_2 = 2.37 \text{ m/s} \downarrow$$

Conservation of Energy. The maximum stretch s in the cord may be determined by again applying the conservation of energy equation to the ball just after collision. Assuming that $y = y_3 = (1 + s_3)$ m, Fig. 15–17a, then

$$T_2 + V_2 = T_3 + V_3$$

$$\tfrac{1}{2}m(v_B)_2^2 + 0 = \tfrac{1}{2}m(v_B)_3^2 - W_B y_3 + \tfrac{1}{2}ks_3^2$$

$$\frac{1}{2}(1.5)(2.37)^2 = 0 - 9.81(1.5)(1 + s_3) + \frac{1}{2}(800)s_3^2$$

$$400s_3^2 - 14.72s_3 - 18.94 = 0$$

Solving this quadratic equation for the positive root yields

$$s_3 = 0.237 \text{ m} = 237 \text{ mm} \qquad \textit{Ans.}$$

*The weight of the ball is considered a nonimpulsive force.

Example 15–11

Two smooth disks A and B, having a mass of 1 and 2 kg, respectively, collide with initial velocities as shown in Fig. 15–18a. If the coefficient of restitution for the disks is $e = 0.75$, determine the x and y components of the final velocity of each disk after collision. Neglect friction.

(a)

SOLUTION

The problem involves *oblique impact*. Why? In order to seek a solution, we have established the x and y axes along the line of impact and the plane of contact, respectively, Fig. 15–18a.

Resolving each of the initial velocities into x and y components, we have

$$(v_{Ax})_1 = 3 \cos 30° = 2.60 \text{ m/s}, \qquad (v_{Ay})_1 = 3 \sin 30° = 1.50 \text{ m/s}$$

$$(v_{Bx})_1 = -1 \cos 45° = -0.707 \text{ m/s}, \quad (v_{By})_1 = -1 \sin 45° = -0.707 \text{ m/s}$$

Since the impact occurs only in the x direction (line of impact), Fig. 15–18b, the conservation of momentum for *both* disks can be applied in this direction. Why?

(b)

Conservation of "x" Momentum. In reference to the momentum diagrams, we have

$$(\overset{+}{\rightarrow}) \qquad m_A(v_{Ax})_1 + m_B(v_{Bx})_1 = m_A(v_{Ax})_2 + m_B(v_{Bx})_2$$

$$1(2.60) + 2(-0.707) = 1(v_{Ax})_2 + 2(v_{Bx})_2$$

$$(v_{Ax})_2 + 2(v_{Bx})_2 = 1.18 \tag{1}$$

Coefficient of Restitution (x). Both disks are *assumed* to have components of velocity in the $+x$ direction after collision, Fig. 15–18b.

$$(\overset{+}{\rightarrow}) \quad e = \frac{(v_{Bx})_2 - (v_{Ax})_2}{(v_{Ax})_1 - (v_{Bx})_1}; \quad 0.75 = \frac{(v_{Bx})_2 - (v_{Ax})_2}{2.60 - (-0.707)}$$

$$(v_{Bx})_2 - (v_{Ax})_2 = 2.48 \tag{2}$$

Solving Eqs. (1) and (2) for $(v_{Ax})_2$ and $(v_{Bx})_2$ yields

$$(v_{Ax})_2 = -1.26 \text{ m/s} = 1.26 \text{ m/s} \leftarrow \qquad (v_{Bx})_2 = 1.22 \text{ m/s} \rightarrow \textit{Ans.}$$

Conservation of "y" Momentum. The momentum of *each disk* is conserved in the y direction (plane of contact), since the disks are smooth and therefore *no impact* occurs in this direction. From Fig. 15–18b,

$$(+\uparrow) \qquad m_A(v_{Ay})_1 = m_A(v_{Ay})_2 \qquad (v_{Ay})_2 = 1.50 \text{ m/s} \uparrow \qquad \textit{Ans.}$$

$$(+\uparrow) \; m_B(v_{By})_1 = m_B(v_{By})_2 \quad (v_{By})_2 = -0.707 \text{ m/s} = 0.707 \text{ m/s} \downarrow \quad \textit{Ans.}$$

Show that when the velocity components are summed, one obtains the results shown in Fig. 15–18c.

Fig. 15–18

(c)

PROBLEMS

15–41. Disks A and B have a mass of 2 kg and 4 kg, respectively. If they have the velocities shown, and $e = 0.4$, determine their velocities just after direct central impact.

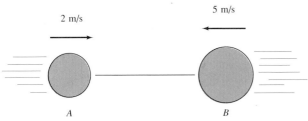

Prob. 15–41

15–42. The cue ball A is given an initial velocity $(v_A)_1 = 5$ m/s. If it makes a direct collision with ball B ($e = 0.8$), determine the velocity of B and the angle θ just after it rebounds from the cushion at C ($e' = 0.6$). Each ball has a mass of 0.4 kg. Neglect the size of each ball.

Prob. 15–42

15–43. If two disks A and B have the same mass and are subjected to direct central impact such that the collision is perfectly elastic ($e = 1$), prove that the kinetic energy before collision equals the kinetic energy after collision. The surface upon which they slide is smooth.

*15–44. Block A has a mass of 3 kg and is sliding on a rough horizontal surface with a velocity $(v_A)_1 = 2$ m/s when it makes a direct collision with block B, which has a mass of 2 kg and is originally at rest. If the collision is perfectly elastic ($e = 1$), determine the velocity of each block just after the collision and determine the distance between the blocks when they stop sliding. The coefficient of kinetic friction between the blocks and the plane is $\mu_k = 0.3$.

Probs. 15–44/15–45

15–45. Block A has a mass of 3 kg and is sliding on a rough horizontal surface with a velocity $(v_A)_1 = 2$ m/s when it makes a direct collision with block B, which has a mass of 2 kg and is originally at rest. If $e = 0.6$, determine the velocity of each block just after the collision and determine the distance between the blocks when they stop sliding. The coefficient of kinetic friction between the blocks and the plane is $\mu_k = 0.3$.

15–46. Determine the horizontal velocity $\mathbf{v}_A$ at which the girl must throw the ball so that it bounces once on the smooth surface and then lands into the cup at C. Take $e = 0.6$ and neglect the size of the cup.

Prob. 15–46

15–47. The 5-lb box B is dropped from rest 5 ft from the top of the 10-lb plate P, which is supported by the spring having a stiffness of $k = 30$ lb/ft. If $e = 0.6$ between the box and plate, determine the maximum compression imparted to the spring. Neglect the mass of the spring.

Prob. 15–47

*15–48. The three balls each have the same mass m. If A has a speed v just before a direct collision with B, determine the speed of C just after collision. The coefficient of restitution between each ball is e. Neglect the size of each ball.

Prob. 15–48

15–49. The 20-lb suitcase A is released from rest at C. After it slides down the smooth ramp it strikes the 10-lb suitcase B, which is originally at rest. If the coefficient of restitution between the suitcases is $e = 0.3$ and the coefficient of kinetic friction between the floor DE and each suitcase is $\mu_k = 0.4$, determine (a) the velocity of A just before impact, (b) the velocities of A and B just after impact, and (c) the distance B slides before coming to rest.

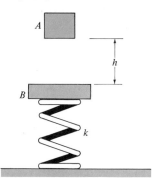

Prob. 15–49

15–50. Block A, having a mass m, is released from rest, falls a distance h and strikes the plate B having a mass $2m$. If the coefficient of restitution between A and B is e, determine the velocity of the plate just after collision. The spring has a stiffness k.

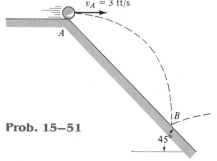

Prob. 15–50

15–51. The 0.2-lb ball bearing travels over the edge A with a velocity of $v_A = 3$ ft/s. Determine the speed at which it rebounds from the smooth inclined plane at B. Take $e = 0.8$.

Prob. 15–51

*15–52. The ball is dropped from rest and falls a distance of 4 ft before striking the smooth plane at A. If $e = 0.8$, determine the distance d to where it again strikes the surface at B.

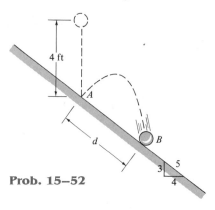

Prob. 15–52

15–53. The slider block B is confined to move within the smooth slot. It is connected to two springs, each of which has a stiffness of $k = 30$ N/m. They are originally stretched 0.5 m when $s = 0$ as shown. Determine the maximum distance, s_{max}, block B moves after it is hit by another block A which is originally traveling at $(v_A)_1 = 8$ m/s. Take $e = 0.4$ and the mass of each block to be $m = 1.5$ kg.

Probs. 15–53/15–54

15–54. In Prob. 15–53 determine the average net force exerted between the two blocks A and B during impact, if the impact occurs in 0.005 s.

15–55. A ball has a mass m and is dropped onto a surface from a height h. If the coefficient of restitution is e between the ball and the surface, determine the time needed for the ball to stop bouncing.

*15–56. A ball of negligible size and mass m is given a velocity of $\mathbf{v}_0$ on the center of the cart which has a mass M and is originally at rest. If the coefficient of restitution between the ball and walls A and B is e, determine the velocity of the ball and the cart just after the ball strikes A. Also, determine the total time needed for the ball to strike A, rebound, then strike B, and rebound and then return to the center of the cart.

Prob. 15–56

15–57. Two smooth disks A and B each have a mass of 0.5 kg. If both disks are moving with the velocities shown when they collide, determine their final velocities just after collision. The coefficient of restitution is $e = 0.75$.

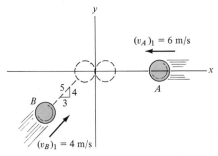

Prob. 15–57

15–58. Ball A strikes ball B with an initial velocity $(v_A)_1$ as shown. If both balls have the same mass and the collision is perfectly elastic, determine the angle θ after collision. Ball B is originally at rest. Neglect the size of each ball.

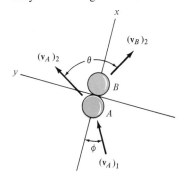

Prob. 15–58

15–59. Two balls A and B each have a mass of 2 kg and the initial velocities shown just before they collide. If the coefficient of restitution is $e = 0.5$, determine their speeds just after impact. Neglect the size of each ball.

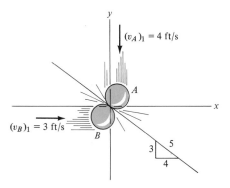

Prob. 15–59

*15–60. Two coins A and B, each having the same mass, slide on a smooth surface with the motion shown. Determine the speed of each coin after collision if they move off along the dashed paths. *Hint:* Since the line of impact has not been defined, apply the conservation of momentum along the x and y axes, respectively.

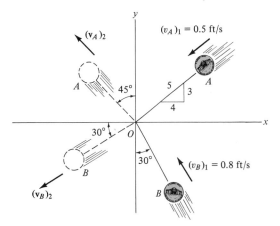

Prob. 15–60

193

15–61. If disk A is sliding forward along the tangent to disk B and strikes B with a velocity $\mathbf{v}$, determine the velocity of B after the collision and compute the loss of kinetic energy during the collision. Neglect friction. The coefficient of restitution is e, and each disk has the same size and mass m.

Prob. 15–61

15–62. The sphere of mass m falls and strikes the triangular block with a vertical velocity $\mathbf{v}$. If the block rests on a smooth surface and has a mass $3m$, determine its velocity just after the collision. The coefficient of restitution is e.

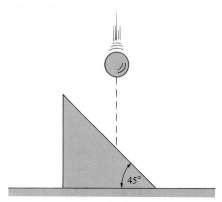

Prob. 15–62

15.5 Angular Momentum

Fig. 15–19

The *angular momentum* of a particle about point O is defined as the "moment" of the particle's linear momentum about O. Since this concept is analogous to finding the moment of a force about a point, the angular momentum, $\mathbf{H}_O$, is sometimes referred to as the *moment of momentum*.

Scalar Formulation

If a particle is moving along a curve lying in the x-y plane, Fig. 15–19, the angular momentum at any instant can be computed about point O (actually the z axis) by using a scalar formulation. The *magnitude* of $\mathbf{H}_O$ is

$$\boxed{(H_O)_z = (d)(mv)} \qquad (15\text{–}12)$$

Here d is the moment arm or perpendicular distance from O to the line of action of $m\mathbf{v}$. Common units for this magnitude are kg $\cdot$ m^2/s or slug $\cdot$ ft^2/s. The *direction* of $\mathbf{H}_O$ is defined by the right-hand rule. As shown in Fig. 15–19, the curl of the fingers of the right hand indicates the sense of rotation of $m\mathbf{v}$ about O, so that in this case the thumb (or $\mathbf{H}_O$) is directed perpendicular to the x-y plane along the $+z$ axis.

Vector Formulation

If the particle is moving along a space curve, Fig. 15–20, the vector cross product can be used to determine the *angular momentum* about O. In this case

$$\mathbf{H}_O = \mathbf{r} \times m\mathbf{v} \qquad (15\text{--}13)$$

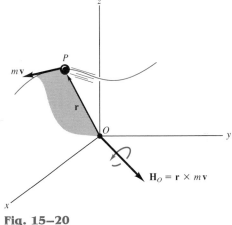

Fig. 15–20

Here $\mathbf{r}$ denotes a position vector drawn from the moment point O to the particle P. As shown in the figure, $\mathbf{H}_O$ is *perpendicular* to the shaded plane containing $\mathbf{r}$ and $m\mathbf{v}$.

In order to evaluate the cross product, $\mathbf{r}$ and $m\mathbf{v}$ should be expressed in terms of their Cartesian components, so that the angular momentum is determined by evaluating the determinant:

$$\mathbf{H}_O = \begin{vmatrix} \mathbf{i} & \mathbf{j} & \mathbf{k} \\ r_x & r_y & r_z \\ mv_x & mv_y & mv_z \end{vmatrix} \qquad (15\text{--}14)$$

15.6 Relation Between Moment of a Force and Angular Momentum

The moments about point O of all the forces acting on the particle may be related to the particle's angular momentum by using the equation of motion. If the mass of the particle is constant, we may write

$$\Sigma\mathbf{F} = m\dot{\mathbf{v}}$$

Performing a cross-product multiplication of each side of this equation by the position vector to obtain the moments of the forces about point O, we have

$$\Sigma\mathbf{M}_O = \mathbf{r} \times \Sigma\mathbf{F} = \mathbf{r} \times m\dot{\mathbf{v}}$$

From Appendix C, the derivative of $\mathbf{r} \times m\mathbf{v}$ can be written as

$$\dot{\mathbf{H}}_O = \frac{d}{dt}(\mathbf{r} \times m\mathbf{v}) = \dot{\mathbf{r}} \times m\mathbf{v} + \mathbf{r} \times m\dot{\mathbf{v}}$$

The first term on the right side, $\dot{\mathbf{r}} \times m\mathbf{v} = m(\dot{\mathbf{r}} \times \dot{\mathbf{r}}) = \mathbf{0}$, since the cross product of a vector with itself is zero. Hence, the above equation becomes

$$\Sigma\mathbf{M}_O = \dot{\mathbf{H}}_O \qquad (15\text{--}15)$$

This equation states that *the resultant moment about point O of all the forces acting on the particle is equal to the time rate of change of the particle's angular momentum about point O.* This result is similar to Eq. 15–1, i.e.,

$$\boxed{\Sigma \mathbf{F} = \dot{\mathbf{L}}} \qquad (15\text{--}16)$$

Here $\mathbf{L} = m\mathbf{v},$ so that *the resultant force acting on the particle is equal to the time rate of change of the particle's linear momentum.*

From the derivations, it is seen that Eqs. 15–15 and 15–16 are actually another way of stating Newton's second law of motion. In other sections of this book it will be shown that these equations have many practical applications when extended and applied to the solution of problems involving either a system of particles or rigid bodies.

System of Particles

An equation having the same form as Eq. 15–15 may be derived for the system of n particles shown in Fig. 15–21. The forces acting on the arbitrary ith particle of the system consist of a resultant *external force* $\mathbf{F}_i$ and a resultant *internal force* $\mathbf{f}_i = \displaystyle\sum_{j=1(j\neq i)}^{n} \mathbf{f}_{ij}.$ Expressing the moments of these forces about point O, using the form of Eq. 15–15, we have

$$(\mathbf{r}_i \times \mathbf{F}_i) + (\mathbf{r}_i \times \mathbf{f}_i) = (\dot{\mathbf{H}}_i)_O$$

Here $\mathbf{r}_i$ represents the position vector drawn from the origin O of an inertial frame of reference to the ith particle, and $(\mathbf{H}_i)_O$ is the angular momentum of the ith particle about O. Similar equations can be written for each of the other particles of the system. When the results are summed vectorially, the result is

$$\Sigma(\mathbf{r}_i \times \mathbf{F}_i) + \Sigma(\mathbf{r}_i \times \mathbf{f}_i) = \Sigma(\dot{\mathbf{H}}_i)_O$$

The second term is zero since the internal forces occur in equal but opposite pairs, and since each pair of forces has the same line of action, the moment of each pair of forces about point O is therefore zero. Hence, dropping the index notation, the above equation can be written in a simplified form as

$$\Sigma \mathbf{M}_O = \dot{\mathbf{H}}_O \qquad (15\text{--}17)$$

which states that *the sum of the moments about point O of all the external forces acting on a system of particles is equal to the time rate of change of the total angular momentum of the system of particles about point O.* Although point O has been chosen here as the origin of coordinates, it actually can represent any point *fixed* in the inertial frame of reference.

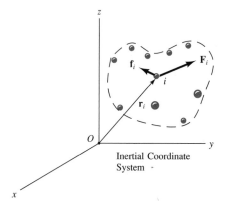

Inertial Coordinate System

Fig. 15–21

Example 15–12

The box shown in Fig. 15–22a has a mass m and is traveling down the smooth circular ramp such that when it is at the angle θ it has a speed v. Compute its angular momentum about point O at this instant and the rate of increase in its speed.

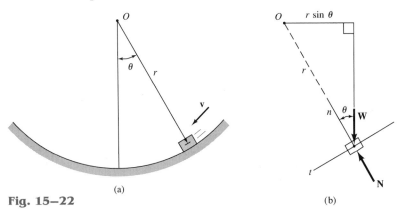

Fig. 15–22

(a)

(b)

SOLUTION

Since **v** is tangent to the path, applying Eq. 15–12 the angular momentum is

$$H_O = rmv \,\downarrow \qquad\qquad Ans.$$

The rate of increase in its speed (dv/dt) can be found by applying Eq. 15–15. From the free-body diagram of the block, Fig. 15–22b, it is seen that only the weight contributes a moment about point O. We have

$$\jmath + \Sigma M_O = \dot{H}_O; \qquad mg(r \sin \theta) = \frac{d}{dt}(rmv)$$

Since r and m are constant,

$$mgr \sin \theta = rm\frac{dv}{dt}$$

$$\frac{dv}{dt} = g \sin \theta \qquad\qquad Ans.$$

This same result can, of course, be obtained from the equation of motion applied in the tangential direction, Fig. 15–22b, i.e.,

$$+\swarrow \Sigma F_t = ma_t; \qquad mg \sin \theta = m\!\left(\frac{dv}{dt}\right)$$

$$\frac{dv}{dt} = g \sin \theta \qquad\qquad Ans.$$

197

15.7 Angular Impulse and Momentum Principles

Principle of Angular Impulse and Momentum

If Eq. 15–15 is rewritten in the form $\Sigma\mathbf{M}_O\,dt = d\mathbf{H}_O$ and integrated, we have, assuming that at time $t = t_1$, $\mathbf{H}_O = (\mathbf{H}_O)_1$ and at time $t = t_2$, $\mathbf{H}_O = (\mathbf{H}_O)_2$,

$$\Sigma \int_{t_1}^{t_2} \mathbf{M}_O\,dt = (\mathbf{H}_O)_2 - (\mathbf{H}_O)_1$$

or

$$(\mathbf{H}_O)_1 + \Sigma \int_{t_1}^{t_2} \mathbf{M}_O\,dt = (\mathbf{H}_O)_2 \qquad (15\text{–}18)$$

This equation is referred to as the *principle of angular impulse and momentum*. The initial and final angular momenta $(\mathbf{H}_O)_1$ and $(\mathbf{H}_O)_2$ are defined as the moment of the linear momentum of the particle $(\mathbf{H}_O = \mathbf{r} \times m\mathbf{v})$ at the instants t_1 and t_2, respectively. The second term on the left side, $\Sigma \int_{t_1}^{t_2} \mathbf{M}_O\,dt$, is called the *angular impulse*. It is computed on the basis of integrating, with respect to time, the moments of all the forces acting on the particle over the time interval t_1 to t_2. Since the moment of a force about point O is defined as $\mathbf{M}_O = \mathbf{r} \times \mathbf{F}$, the angular impulse may be expressed in vector form as

$$\text{angular impulse} = \int_{t_1}^{t_2} \mathbf{M}_O\,dt = \int_{t_1}^{t_2} (\mathbf{r} \times \mathbf{F})\,dt \qquad (15\text{–}19)$$

Here $\mathbf{r}$ is a position vector which extends from point O to any point on the line of action of $\mathbf{F}$.

In a similar manner, using Eq. 15–17, the principle of angular impulse and momentum for a system of particles may be written as

$$\Sigma(\mathbf{H}_O)_1 + \Sigma \int_{t_1}^{t_2} \mathbf{M}_O\,dt = \Sigma(\mathbf{H}_O)_2 \qquad (15\text{–}20)$$

Here the first and third terms represent the angular momenta of the system of particles $(\Sigma\mathbf{H}_O = \Sigma(\mathbf{r}_i \times m\mathbf{v}_i))$ at the instants t_1 and t_2. The second term is the vector sum of the angular impulses given to all the particles during the time period t_1 to t_2. Recall that these impulses are created only by the moments of the external forces acting on the system where, for the *i*th particle, $\mathbf{M}_O = \mathbf{r}_i \times \mathbf{F}_i$.

Vector Formulation. Using impulse and momentum principles, it is therefore possible to write two vector equations which define the particle motion; namely, Eqs. 15–3 and 15–18, restated as

$$m\mathbf{v}_1 + \Sigma \int_{t_1}^{t_2} \mathbf{F}\, dt = m\mathbf{v}_2$$

$$(\mathbf{H}_O)_1 + \Sigma \int_{t_1}^{t_2} \mathbf{M}_O\, dt = (\mathbf{H}_O)_2$$

(15–21)

Scalar Formulation. In general, the above equations may be expressed in x, y, z component form, yielding a total of six independent scalar equations. If the particle is confined to move in the x-y plane, three independent scalar equations may be written to express the motion, namely,

$$m(v_x)_1 + \Sigma \int_{t_1}^{t_2} F_x\, dt = m(v_x)_2$$

$$m(v_y)_1 + \Sigma \int_{t_1}^{t_2} F_y\, dt = m(v_y)_2$$

$$(H_O)_1 + \Sigma \int_{t_1}^{t_2} M_O\, dt = (H_O)_2$$

(15–22)

The first two of these equations represent the principle of linear impulse and momentum in the x and y directions, and the third equation represents the principle of angular impulse and momentum about the z axis.

Conservation of Angular Momentum

When the angular impulses acting on a particle are all zero during the time t_1 to t_2, Eq. 15–18 reduces to the following simplified form,

$$(\mathbf{H}_O)_1 = (\mathbf{H}_O)_2$$

(15–23)

This equation is known as the *conservation of angular momentum*. It states that from t_1 to t_2 the particle's angular momentum remains constant. Obviously, if no external impulse is applied to the particle, both linear and angular momentum will be conserved. In some cases, however, the particle's angular momentum will be conserved and linear momentum may not. An example of this occurs when the particle is subjected *only* to a *central force* (see Sec. 13.7). As shown in Fig. 15–23, the impulsive central force $\mathbf{F}$ is always directed toward point O as the particle moves along the path. Hence, the angular impulse (moment) created by $\mathbf{F}$ about the z axis passing through point O is always zero, and therefore angular momentum of the particle is conserved about this axis.

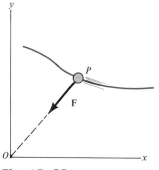

Fig. 15–23

Using Eq. 15–20, we can also write the conservation of angular momentum for a system of particles, namely,

$$\Sigma(\mathbf{H}_O)_1 = \Sigma(\mathbf{H}_O)_2 \tag{15–24}$$

In this case the summation must include the angular momenta of all n particles in the system.

PROCEDURE FOR ANALYSIS

When applying the principles of angular impulse and momentum, or the conservation of angular momentum, it is suggested that the following procedure be used.

Free-Body Diagram. Draw the particle's free-body diagram in order to determine any axis about which angular momentum may be conserved. For this to occur, the moments of the forces (or impulses) about the axis must be zero throughout the time period t_1 to t_2. Apart from the free-body diagram, the direction and sense of the particle's initial and final velocities should also be established.

An alternative procedure to that mentioned above requires drawing the impulse and momentum diagrams for the particle.

Momentum Equations. Apply the principle of angular impulse and momentum, $(\mathbf{H}_O)_1 + \Sigma \int_{t_1}^{t_2} \mathbf{M}_O \, dt = (\mathbf{H}_O)_2$, or if appropriate, the conservation of angular momentum, $(\mathbf{H}_O)_1 = (\mathbf{H}_O)_2$.

If other equations are needed for the problem solution, when appropriate, use the principle of linear impulse and momentum, the principle of work and energy, the equations of motion, or kinematics.

The following examples numerically illustrate application of the above procedure.

Example 15–13

The 5-kg block of negligible size rests on the smooth horizontal plane, Fig. 15–24a. It is attached at A to a slender rod of negligible mass. The rod is attached to a ball-and-socket joint at B. If a moment $M = (3t)$ N · m, where t is in seconds, is applied to the rod and a horizontal force $P = 10$ N is applied to the block as shown, determine the speed of the block in 4 s starting from rest.

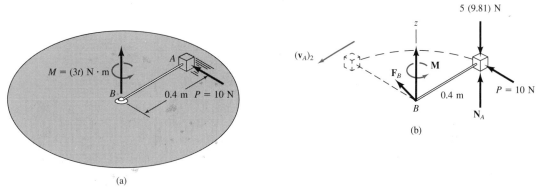

Fig. 15–24   (a)

SOLUTION

Free-Body Diagram. If we consider the system of both the rod and block, Fig. 15–24b, then the resultant force reaction $\mathbf{F}_B$ can be eliminated from the analysis by applying the principle of angular impulse and momentum about the z axis. Why? If this is done, the angular impulses created by the weight and normal reaction $\mathbf{N}_A$ are also eliminated, since they act parallel to the z axis and therefore create zero moment about this axis.

Principle of Angular Impulse and Momentum

$$(H_z)_1 + \Sigma \int_{t_1}^{t_2} M_z \, dt = (H_z)_2$$

$$(H_z)_1 + \int_{t_1}^{t_2} M \, dt + r_{BA}P(\Delta t) = (H_z)_2$$

$$0 + \int_0^4 3t \, dt + (0.4)(10)(4) = 5(v_A)_2(0.4)$$

$$24 + 16 = 2(v_A)_2$$

$$(v_A)_2 = 20 \text{ m/s} \qquad\qquad Ans.$$

Example 15–14

The ball B, shown in Fig. 15–25a, has a weight of 0.8 lb and is attached to a cord which passes through a hole at A in a smooth table. When the ball is $r_1 = 1.75$ ft from the hole, it is rotating around in a circle such that its speed is $v_1 = 4$ ft/s. If by applying a force $\mathbf{F}$ the cord is pulled downward through the hole with a constant speed $v_c = 6$ ft/s, determine (a) the speed of the ball at the instant it is $r_2 = 0.6$ ft from the hole, and (b) the amount of work done by the force $\mathbf{F}$ in shortening the radial distance r. Neglect the size of the ball.

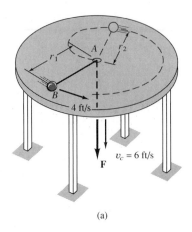

(a)

SOLUTION

Part (a)

Free-Body Diagram. As the ball moves from radial position r_1 to r_2, Fig. 15–25b, the three forces acting on it are all unknown; however, since the cord tension $\mathbf{F}$ passes through the z axis and $\mathbf{W}$ and $\mathbf{N}_B$ are parallel to it, the moments, or angular impulses created by the forces, are all *zero* about this axis. Hence, the conservation of angular momentum applies about the z axis.

Conservation of Angular Momentum. As indicated in Fig. 15–25b, the ball's velocity $\mathbf{v}_2$ is resolved into two components. The radial component, 6 ft/s, is known; however, it produces zero angular momentum about the z axis. Thus,

$$\mathbf{H}_1 = \mathbf{H}_2$$

$$r_1 m_B v_1 = r_2 m_B v_2'$$

$$1.75\left(\frac{0.8}{32.2}\right)4 = 0.6\left(\frac{0.8}{32.2}\right)v_2'$$

$$v_2' = 11.67 \text{ ft/s}$$

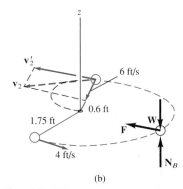

(b)

Fig. 15–25

The speed of the ball is thus

$$v_2 = \sqrt{(11.67)^2 + (6)^2}$$

$$= 13.1 \text{ ft/s} \qquad \textit{Ans.}$$

Part (b). The only force that does work on the ball is $\mathbf{F}$. (The normal force and weight do not move vertically.) The initial and final kinetic energies of the ball can be determined so that from the principle of work and energy we have

$$T_1 + \Sigma U_{1-2} = T_2$$

$$\frac{1}{2}\left(\frac{0.8}{32.2}\right)(4)^2 + U_F = \frac{1}{2}\left(\frac{0.8}{32.2}\right)(13.1)^2$$

$$U_F = 1.94 \text{ ft} \cdot \text{lb} \qquad \textit{Ans.}$$

Example 15–15

The 2-kg block shown in Fig. 15–26*a* rests on a smooth horizontal surface and is attached to an elastic cord that has a stiffness $k_c = 20$ N/m and is initially unstretched. If the block is given a velocity $(v_B)_1 = 1.5$ m/s, perpendicular to the cord, determine the rate at which the cord is being stretched and the speed of the block at the instant the cord is stretched 0.2 m.

(a)

SOLUTION

Free-Body Diagram. After the block has been launched, it slides along the dashed path shown in Fig. 15–26*b*. By inspection, angular momentum about point O (or the z axis) is *conserved*, since the moment of the central force $\mathbf{F}_e$ about O is always zero. Also, when the distance is 0.7 m, only the component of $(\mathbf{v}'_B)_2$ is effective in producing angular momentum of the block about O.

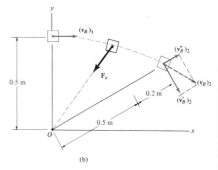

(b)

Conservation of Angular Momentum. The component $(\mathbf{v}'_B)_2$ can be obtained by applying the conservation of angular momentum about O (the z axis), i.e.,

$$(\mathbf{H}_O)_1 = (\mathbf{H}_O)_2$$

$\zeta +$
$$r_1 m_B (v_B)_1 = r_2 m_B (v'_B)_2$$
$$0.5(2)(1.5) = 0.7(2)(v'_B)_2$$
$$(v'_B)_2 = 1.07 \text{ m/s}$$

Fig. 15–26

Conservation of Energy. The speed of the block may be obtained by applying the conservation of energy equation at the point where the block was launched and at the point where the cord is stretched 0.2 m.

$$T_1 + V_1 = T_2 + V_2$$
$$\frac{1}{2}(2)(1.5)^2 + 0 = \frac{1}{2}(2)(v_B)_2^2 + \frac{1}{2}(20)(0.2)^2$$

Thus,
$$(v_B)_2 = 1.36 \text{ m/s} \qquad \textit{Ans.}$$

Having determined $(v_B)_2$ and its component $(v'_B)_2$, the rate of stretch of the cord $(v''_B)_2$ is determined from the Pythagorean theorem,

$$(v''_B)_2 = \sqrt{(v_B)_2^2 - (v'_B)_2^2}$$
$$= \sqrt{(1.36)^2 - (1.07)^2}$$
$$= 0.838 \text{ m/s} \qquad \textit{Ans.}$$

PROBLEMS

15–63. Determine the angular momentum $\mathbf{H}_O$ of each of the two particles about point O.

Probs. 15–63/15–64

*15–64. Determine the angular momentum $\mathbf{H}_P$ of each of the two particles about point P.

15–65. Determine the angular momentum of the 2-lb particle A about point O.

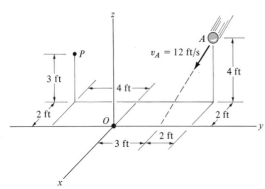

Probs. 15–65/15–66

15–66. Determine the angular momentum of the 2-lb particle A about point P.

15–67. A particle has a linear momentum $\mathbf{L} = \{-8\mathbf{i} + 14\mathbf{j} - 12\mathbf{k}\}$ kg · m/s. If it is located at point A (3 m, -4 m, 2 m), determine its angular momentum about point B (-1 m, -2 m, 8 m).

*15–68. The two spheres each have a mass of 3 kg and are rigidly attached to the rod of negligible mass. If a torque of $M = (6e^{0.2t})$ N · m, where t is in seconds, is applied to the rod as shown, determine the speed of each of the spheres in 2 s, starting from rest.

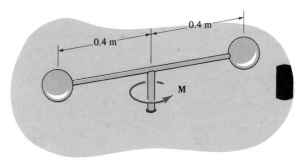

Prob. 15–68

15–69. The ball B has a mass of 10 kg and is attached to the end of a rod whose mass may be neglected. If the rod is subjected to a torque $M = (3t^2 + 5t + 2)$ N · m, where t is in seconds, determine the speed of the ball when $t = 2$ s. The ball has a speed $v = 2$ m/s when $t = 0$.

Prob. 15–69

15–70. A child having a mass of 50 kg holds her legs up as shown as she swings downward from rest at $\theta_1 = 30°$. Her center of mass is located at point G_1. When she is at the bottom position $\theta = 0°$, she *suddenly* lets her legs come down, shifting her center of mass to position G_2. Determine her speed in the upswing due to this sudden movement and the angle θ_2 to which she swings before coming to rest. Treat the child's body as a particle.

Prob. 15–70

15–71. A 4-lb ball B is traveling around in a circle of radius $r_1 = 3$ ft with a speed of $(v_B)_1 = 6$ ft/s. If the attached cord is pulled down through the hole with a constant speed of $v_r = 2$ ft/s, determine the ball's speed at the instant $r_2 = 2$ ft. How much work has to be done to pull down the cord? Neglect friction and the size of the ball.

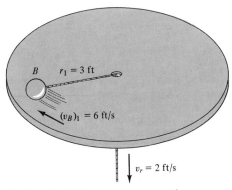

Prob. 15–71

*15–72. The elastic cord has an unstretched length $l_0 = 1.5$ ft and a stiffness $k = 12$ lb/ft. It is attached to a fixed point at A and a block at B, which has a weight of 2 lb. If the block is released from rest from the position shown, determine its speed when it reaches point C after it slides along the smooth guide. After leaving the guide, it is launched onto the smooth *horizontal* plane. Determine its speed at D at the instant the cord becomes unstretched. Also, calculate the angular momentum of the block about point A, at any instant after it passes point C.

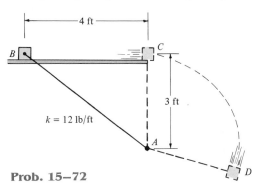

Prob. 15–72

15–73. An earth satellite of mass 700 kg is launched into a free-flight trajectory about the earth with an initial speed $v_A = 10$ km/s when the distance from the center of the earth is $r_A = 15$ Mm. If the launch angle at this position is $\phi_A = 70°$, determine the speed v_B of the satellite and its closest distance r_B from the center of the earth. The earth has a mass $M_e = 5.976(10^{24})$ kg. *Hint:* Under these conditions, the satellite is subjected only to the earth's gravitational force, $F = GM_e m_s/r^2$, Eq. 13–1. For part of the solution, use the conservation of energy (see Prob. 14–59.)

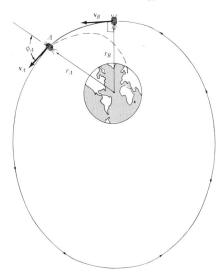

Prob. 15–73

205

15-74. The 800-lb roller coaster car starts from rest on the track having the shape of a cylindrical helix. If the helix descends 8 ft for every one revolution, determine the speed of the car in $t = 4$ s. Also, how far has the car descended in this time? Neglect friction.

Prob. 15–74

* 15.8 Steady Fluid Streams

Knowledge of the forces developed by steadily moving fluid streams is of importance in the design and analysis of turbines, pumps, blades, and fans. To illustrate how the principle of impulse and momentum may be used to determine these forces, consider the diversion of a steady stream of fluid (liquid or gas) by a fixed pipe, Fig. 15–27a. The fluid enters the pipe with a velocity $\mathbf{v}_A$ and exits with a velocity $\mathbf{v}_B$. The impulse and momentum diagrams for the fluid stream are shown in Fig. 15–27b. The force $\Sigma\mathbf{F}$, shown on the impulse diagram, represents the resultant of all the external forces acting on the fluid stream. It is this loading which gives the fluid stream an impulse whereby the original momentum of the fluid is changed in both its magnitude and direction. Since the flow is steady, $\Sigma\mathbf{F}$ will be *constant* during the time interval dt. During this time the fluid stream is in motion, and as a result a small amount of fluid, having a mass dm, is about to enter the pipe with a velocity $\mathbf{v}_A$ at time t. If this element of mass and the mass of fluid in the pipe is considered as a "closed system," then at time $t + dt$ a corresponding element of mass dm must leave the pipe with a velocity $\mathbf{v}_B$. The fluid stream *within* the pipe section has a mass m and an *average velocity* $\mathbf{v}$ which is constant during the time interval dt. Applying the principle of linear impulse and momentum to the fluid stream, we have

$$dm\, \mathbf{v}_A + m\mathbf{v} + \Sigma\mathbf{F}\, dt = dm\, \mathbf{v}_B + m\mathbf{v}$$

(a)

Fig. 15–27a

(b)

(c)

Fig. 15—27(b, c)

Force Resultant

Solving for the resultant force yields

$$\Sigma \mathbf{F} = \frac{dm}{dt}(\mathbf{v}_B - \mathbf{v}_A) \qquad (15\text{--}25)$$

Provided the motion of the fluid can be represented in the *x-y* plane, it is usually convenient to express this vector equation in the form of two scalar component equations, i.e.,

$$\Sigma F_x = \frac{dm}{dt}(v_{Bx} - v_{Ax})$$
$$\Sigma F_y = \frac{dm}{dt}(v_{By} - v_{Ay}) \qquad (15\text{--}26)$$

The term dm/dt is called the *mass flow* and indicates the constant amount of fluid which flows either into or out of the pipe per unit of time. If the cross-sectional areas and densities of the fluid at the entrance A and exit B are A_A, ρ_A and A_B, ρ_B, respectively, Fig. 15–27c, then *continuity of mass* requires that $dm = \rho\, dV = \rho_A(ds_A\, A_A) = \rho_B(ds_B\, A_B)$. Hence, during the time dt, since $v_A = ds_A/dt$ and $v_B = ds_B/dt$, we have

$$\frac{dm}{dt} = \rho_A v_A A_A = \rho_B v_B A_B = \rho_A Q_A = \rho_B Q_B \qquad (15\text{--}27)$$

Here $Q = vA$ is the volumetric *flow rate*, which measures the volume of fluid flowing per unit of time.

Fig. 15–27a

Moment Resultant

In some cases it is necessary to obtain the support reactions on the fluid-carrying device. If Eq. 15–25 does not provide enough information to do this, the principle of angular impulse and momentum must be used. The formulation of this principle applied to fluid streams can be obtained from Eq. 15–17, $\Sigma \mathbf{M}_O = \dot{\mathbf{H}}_O$, which states that the moment of all the external forces acting on the system about point O is equal to the time rate of change of angular momentum about O. In the case of the pipe shown in Fig. 15–27a, the flow is steady in the x-y plane; hence we have

$$(\curvearrowright+) \qquad \Sigma M_O = \frac{dm}{dt}(d_{OB}\, v_B - d_{OA}\, v_A) \qquad (15\text{–}28)$$

where the moment arms d_{OB} and d_{OA} are directed from O to the *center* of the openings at A and B.

PROCEDURE FOR ANALYSIS

The following procedure provides a method for solving problems involving steady flow.

Kinematic Diagram. In problems where the device is *moving,* a *kinematic diagram* may be helpful for determining the entrance and exit velocities of the fluid flowing onto the device, since a *relative-motion analysis* of velocity will be involved. The *kinematic diagram* in this case is simply a graphical representation of the velocities showing the vector addition of the relative-motion components. Once the velocity of the fluid flowing onto the device is determined, the mass flow is calculated using Eq. 15–27.

Free-Body Diagram. Draw a free-body diagram of the device which is directing the fluid in order to establish the forces $\Sigma \mathbf{F}$ acting on it. These external forces will include the support reactions, the weight of the device and the fluid contained within it, and the static pressure forces of the fluid at the entrance and exit sections of the device.*

Equations of Steady Flow. Apply the equations of steady flow, Eqs. 15–26 and 15–28, using the appropriate components of velocity and force shown on the kinematic and free-body diagrams.

The following examples numerically illustrate application of this procedure.

*In the SI system pressure is measured using the *pascal* (Pa), where 1 Pa = 1 N/m².

Example 15–16

Determine the reaction components which the pipe joint at A exerts on the elbow in Fig. 15–28a, if water flowing through the pipe is subjected to a static pressure of 100 kPa at A. The discharge at B is $Q_B = 0.2$ m³/s. Water has a density $\rho_w = 1000$ kg/m³, and the water-filled elbow has a mass of 20 kg and center of mass at G.

(a)

SOLUTION

The absolute velocity of flow at A and B and the mass flow rate can be obtained from Eq. 15–27. Since the density of water is constant, $Q_B = Q_A = Q$. Hence,

$$\frac{dm}{dt} = \rho_w Q = (1000 \text{ kg/m}^3)(0.2 \text{ m}^3/\text{s}) = 200 \text{ kg/s}$$

$$v_B = \frac{Q}{A_B} = \frac{0.2 \text{ m}^3/\text{s}}{\pi(0.05 \text{ m})^2} = 25.46 \text{ m/s} \downarrow$$

$$v_A = \frac{Q}{A_A} = \frac{0.2 \text{ m}^3/\text{s}}{\pi(0.1 \text{ m})^2} = 6.37 \text{ m/s} \rightarrow$$

Free-Body Diagram. As shown on the free-body diagram, Fig. 15–28b, the *fixed* connection at A exerts a resultant couple $\mathbf{M}_O$ and force components $\mathbf{F}_x$ and $\mathbf{F}_y$ on the elbow. Due to the static pressure of water in the pipe, the pressure force acting on the fluid at A is $F_A = p_A A_A$. Since 1 kPa = 1000 N/m²,

$$F_A = p_A A_A; \quad F_A = (100(10^3) \text{ N/m}^2)[\pi(0.1 \text{ m})^2] = 3141.6 \text{ N}$$

There is no static pressure acting at B, since the water is discharged at atmospheric pressure; i.e., the pressure measured by a gauge at B is equal to zero, $p_B = 0$.

(b)

Fig. 15–28

Equations of Steady Flow

$$\xrightarrow{+} \Sigma F_x = \frac{dm}{dt}(v_{Bx} - v_{Ax}); \quad -F_x + 3141.6 = 200(0 - 6.37)$$

$$F_x = 4.42 \text{ kN} \qquad \qquad \textit{Ans.}$$

$$+\uparrow \Sigma F_y = \frac{dm}{dt}(v_{By} - v_{Ay}); \quad -F_y - 20(9.81) = 200(-25.46 - 0)$$

$$F_y = 4.90 \text{ kN} \qquad \qquad \textit{Ans.}$$

If moments are summed about point O, Fig. 15–28b, $\mathbf{F}_x$, $\mathbf{F}_y$, and the static pressure $\mathbf{F}_A$ are eliminated, as well as the moment of momentum of the water entering at A, Fig. 15–28a. Hence,

$$\zeta+ \Sigma M_O = \frac{dm}{dt}(d_{OB}v_B - d_{OA}v_A)$$

$$M_O + 20(9.81)(0.125) = 200[(0.3)(25.46) - 0]$$

$$M_O = 1.50 \text{ kN} \cdot \text{m} \qquad \qquad \textit{Ans.}$$

Example 15–17

A 2-in.-diameter water jet having a velocity of 25 ft/s impinges upon a single moving blade, Fig. 15–29a. If the blade is moving at 5 ft/s away from the jet, determine the horizontal and vertical components of force which the blade is exerting on the water. What power does the water generate on the blade? Water has a specific weight of $\gamma_w = 62.4$ lb/ft³.

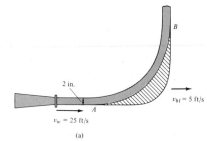

(a)

SOLUTION

Kinematic Diagram. From a fixed coordinate system, Fig. 15–29b, the rate at which water enters the blade is

$$\mathbf{v}_A = \{20\mathbf{i}\} \text{ ft/s}$$

The *relative-flow velocity* of the water onto the blade is $\mathbf{v}_{w/bl} = \mathbf{v}_w - \mathbf{v}_{bl} = 25\mathbf{i} - 5\mathbf{i} = \{20\mathbf{i}\}$ ft/s. Since the blade is moving with a velocity of $\mathbf{v}_{bl} = \{5\mathbf{i}\}$ ft/s, the velocity of flow at B measured from x, y is the vector sum, shown in Fig. 15–29b. Here,

$$\mathbf{v}_B = \mathbf{v}_{bl} + \mathbf{v}_{w/bl}$$
$$= \{5\mathbf{i} + 20\mathbf{j}\} \text{ ft/s}$$

(b)

The mass flow of water *onto* the blade undergoes a momentum change. Thus,

$$\frac{dm}{dt} = \rho_w(v_{w/bl})A_A = \frac{62.4}{32.2}(20)\left[\pi\left(\frac{1}{12}\right)^2\right] = 0.846 \text{ slug/s}$$

Free-Body Diagram. The free-body diagram of a section of water acting on the blade is shown in Fig. 15–29c. The weight of the water will be neglected in the calculation, since this force is small compared to the reactive components $\mathbf{F}_x$ and $\mathbf{F}_y$.

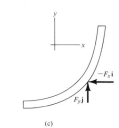

(c)

Fig. 15–29

Equations of Steady Flow

$$\Sigma\mathbf{F} = \frac{dm}{dt}(\mathbf{v}_B - \mathbf{v}_A)$$

$$-F_x\mathbf{i} + F_y\mathbf{j} = 0.846(5\mathbf{i} + 20\mathbf{j} - 25\mathbf{i})$$

Equating the respective **i** and **j** components gives

$$F_x = 0.846(20) = 16.9 \text{ lb} \leftarrow \qquad\qquad Ans.$$
$$F_y = 0.846(20) = 16.9 \text{ lb} \uparrow \qquad\qquad Ans.$$

The water exerts equal but opposite forces on the blade.

Since the water force which causes the blade to move forward horizontally with a velocity of 5 ft/s is $F_x = 16.9$ lb, then from Eq. 14–10 the power is

$$P = \mathbf{F} \cdot \mathbf{v}; \qquad\qquad P = \frac{16.9(5)}{550} = 0.154 \text{ hp} \qquad\qquad Ans.$$

*15.9 Propulsion with Variable Mass

In Sec. 15.8 the case in which a *constant* amount of mass dm entered and left a "*closed system*" was considered. There are, however, two other important cases involving mass flow, which are represented by a system which is either gaining or losing mass. In this section these cases will be discussed separately.

A System That Loses Mass

Consider a device which at an instant of time has a mass m and a forward velocity **v**, Fig. 15–30a. At this same instant the device is expelling an amount of mass m_e such that the mass flow velocity $\mathbf{v}_e$ is *constant* when the measurement is made a small distance away from the device. For the analysis, consider the "*closed system*" at an instant to include *both the mass m of the device and the expelled mass* m_e, as shown by the dashed line in the figure. The impulse and momentum diagrams for the system are shown in Fig. 15–30b. During the time interval dt, the velocity of the device is increased from **v** to **v** + d**v.** This increase in forward velocity, however, does not change the velocity $\mathbf{v}_e$ of the expelled mass, since this mass moves at a constant speed once it has been ejected. To increase the velocity of the device during the time dt, an amount of mass dm_e has been ejected and thereby gained in the exhaust. The impulses are created by $\Sigma \mathbf{F}_s$, which represents the resultant of all the external forces which *act on the system* in the direction of motion. This force resultant *does not include* the force which causes the device to move forward, since this force (called a *thrust*) is *internal to the system;* that is, the thrust acts with equal magnitude but opposite direction on the mass m of the device and the expelled exhaust mass m_e.* Applying the principle of impulse and momentum to the system, in reference to Fig. 15–30b, we have

(a)

$$(\overset{+}{\rightarrow}) \quad mv - m_e v_e + \Sigma F_s\, dt = (m - dm_e)(v + dv) - (m_e + dm_e)v_e$$

or

$$\Sigma F_s\, dt = -v\, dm_e + m\, dv - dm_e\, dv - v_e\, dm_e$$

Without loss of accuracy, the third term on the right side of this equation may be neglected since it is a "second-order" differential. Dividing by dt gives

(b)

Fig. 15–30

*$\Sigma \mathbf{F}_s$ represents the external resultant force *acting on the system,* which is different from $\Sigma \mathbf{F}$, the resultant force acting only on the device.

Fig. 15–31

$$\Sigma F_s = m\frac{dv}{dt} - (v + v_e)\frac{dm_e}{dt}$$

Noting that the relative velocity of the device as seen by an observer moving with the particles of the ejected mass is $v_{D/e} = (v + v_e)$, the final result can be written as

$$\Sigma F_s = m\frac{dv}{dt} - v_{D/e}\frac{dm_e}{dt} \qquad (15\text{–}29)$$

Here the term dm_e/dt represents the rate at which mass is being ejected.

To illustrate an application of Eq. 15–29, consider the rocket shown in Fig. 15–31, which has a weight **W** and is moving upward against an atmospheric drag force **F_D**. The system to be considered consists of the mass of the rocket and the mass of ejected gas m_e. Applying Eq. 15–29 to this system gives

$$(+\uparrow) \qquad -F_D - W = \frac{W}{g}\frac{dv}{dt} - v_{D/e}\frac{dm_e}{dt}$$

The last term of this equation represents the *thrust* **T** which the engine exhaust exerts on the rocket, Fig. 15–31. Recognizing that $dv/dt = a$, we may therefore write

$$(+\uparrow) \qquad T - F_D - W = \frac{W}{g}a$$

If a free-body diagram of the rocket is drawn, it becomes obvious that this equation represents an application of $\Sigma \mathbf{F} = m\mathbf{a}$ for the rocket.

A System That Gains Mass

A device such as a scoop or a shovel may gain mass as it moves forward. Consider, for example, the device shown in Fig. 15–32a, which at an instant of time has a mass m and is moving forward with a velocity **v.** At the same instant, the device is collecting a particle stream of mass m_i. The flow velocity **v_i** of this injected mass is constant and independent of the velocity **v.** It is required that $v > v_i$. The system to be considered at this instant includes both the mass of the device and the mass of the injected fluid, as shown by the dashed line in the figure. The impulse and momentum diagrams for this system are shown in Fig. 15–32b. With an increase in mass dm_i gained by the device, there is an assumed increase in velocity $d\mathbf{v}$ during the time interval dt. This increase is caused by the impulse created by ΣF_s, the resultant of all the external forces *acting on the system* in the direction of motion. The force summation does not include the retarding force of the injected mass acting on the device. Why? Applying the principle of impulse and momentum to the system, we have

$$(\overset{+}{\rightarrow}) \qquad mv + m_i v_i + \Sigma F_s\, dt = (m + dm_i)(v + dv) + (m_i - dm_i)v_i$$

(a)

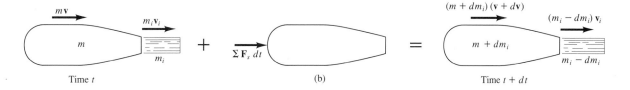

Time t (b)   Time $t + dt$

Fig. 15–32   (c)

Using the same procedure as in the previous case, we may write this equation as

$$\Sigma F_s = m\frac{dv}{dt} + (v - v_i)\frac{dm_i}{dt}$$

Since the relative velocity of the device as seen by an observer moving with the particles of the injected mass is $v_{D/i} = (v - v_i)$, the final result can be written as

$$\boxed{\Sigma F_s = m\frac{dv}{dt} + v_{D/i}\frac{dm_i}{dt}} \qquad (15\text{–}30)$$

where dm_i/dt is the rate of mass injected into the device. The last term in this equation represents the magnitude of force **R,** which the injected mass *exerts on the device*. Since $dv/dt = a$, Eq. 15–30 becomes

$$\Sigma F_s - R = ma$$

This is the application of $\Sigma\mathbf{F} = m\mathbf{a}$, Fig. 15–32c.

As in the case of steady flow, problems which are solved using Eqs. 15–29 and 15–30 should be accompanied by the necessary free-body diagram. With this diagram one can then determine ΣF_s *for the system* and isolate the force exerted on the device by the particle stream.

Example 15–18

The initial combined mass of a rocket and its fuel is m_0. A total mass m_f of fuel is consumed at a constant rate of $dm_e/dt = c$ and expelled at a constant speed of u relative to the rocket. Determine the velocity of the rocket at the instant the fuel runs out. Neglect the change in the rocket's weight with altitude and the drag resistance of the air. The rocket is fired vertically from rest.

System

SOLUTION

Since the rocket is losing mass as it moves upward, Eq. 15–29 can be used for the solution. The only *external force* acting on the *system* consisting of the rocket and a portion of the expelled mass is the weight $\mathbf{W}$, Fig. 15–33. Hence,

$$+\uparrow \; \Sigma F_s = m\frac{dv}{dt} - v_{D/e}\frac{dm_e}{dt}; \qquad -W = m\frac{dv}{dt} - uc \qquad (1)$$

At any given instant t during the flight, the mass of the rocket can be expressed as $m = m_0 - (dm_e/dt)t = m_0 - ct$. Since $W = mg$, Eq. (1) becomes

$$-(m_0 - ct)g = (m_0 - ct)\frac{dv}{dt} - uc$$

Fig. 15–33

Rearranging the terms and integrating, realizing that $v = 0$ at $t = 0$, we have

$$\int_0^v dv = \int_0^t \left(\frac{uc}{m_0 - ct} - g\right) dt$$

$$v = -u\ln(m_0 - ct) - gt\Big|_0^t$$

$$v = u\ln\left(\frac{m_0}{m_0 - ct}\right) - gt \qquad (2)$$

This equation gives the speed of the rocket at any instant of time. The time t' needed to consume all the fuel is

$$m_f = (dm_e/dt)t' = ct'$$

Hence,

$$t' = m_f/c$$

Substituting into Eq. (2) yields

$$v = u\ln\left(\frac{m_0}{m_0 - m_f}\right) - \frac{gm_f}{c} \qquad \qquad Ans.$$

Example 15–19

A chain of length l, Fig. 15–34a, has a mass m. Determine the magnitude of force $\mathbf{F}$ required to (a) raise the chain with a constant speed v_c, starting from rest when $y = 0$; and (b) lower the chain with a constant speed v_c, starting from rest when $y = l$.

SOLUTION

Part (a). As the chain is raised, all the suspended links are given a sudden impulse downward by each added link which is lifted off the ground. Thus, the *suspended portion* of the chain may be considered as a device which is *gaining mass*. A free-body diagram of a portion of the chain which is located at an arbitrary height y above the ground is shown in Fig. 15–34b. The system to be considered is the length of chain y which is suspended by $\mathbf{F}$ at any instant, including the next link which is about to be added but is still at rest. The forces acting on this system *exclude* the internal force $\mathbf{P}$ which the added link exerts on the suspended portion of the chain. Hence, $\Sigma F_s = F - (m/l)gy$.

It now becomes necessary to find the rate at which mass is being added to the system. The velocity $\mathbf{v}_c$ of the chain at a given instant is equivalent to $\mathbf{v}_{D/i}$. Why? Since v_c is constant, $dv_c/dt = 0$ and $dy/dt = v_c$. Integrating, using the initial condition that $y = 0$ at $t = 0$, gives $y = v_c t$. Thus, the mass of the system at any instant is $m = (m/l)y = (m/l)v_c t$, and therefore the *rate* at which mass is *added* to the suspended chain is

$$\frac{dm_i}{dt} = (m/l)v_c$$

(a)

(b)

(c)

Fig. 15–34

Applying Eq. 15–30 to the system, using this data, we have

$$+\uparrow \ \Sigma F_s = m\frac{dv_c}{dt} + v_{D/i}\frac{dm_i}{dt}$$

$$F - (m/l)gy = 0 + v_c(m/l)v_c$$

Hence,
$$F = (m/l)(gy + v_c^2) \qquad \textit{Ans.}$$

Part (b). When the chain is being lowered, the links which are expelled (given zero velocity) *do not* impart an impulse to the *remaining* suspended links. Why? Thus, the system in Part (a) cannot be considered. Instead, the equation of motion will be used to obtain the solution. At time t the portion of chain still off the floor is y. The free-body diagram for a suspended portion of the chain is shown in Fig. 15–34c. Thus,

$$+\uparrow \Sigma F = ma; \qquad F - (m/l)gy = 0$$

$$F = (m/l)gy \qquad \textit{Ans.}$$

PROBLEMS

15–75. Sand is deposited from a chute onto a conveyor belt which is moving at 0.5 m/s. If the sand is assumed to fall vertically on the belt at A at the rate of 4 kg/s, determine the required belt tension F_B to the right of A. The belt is free to move over the conveyor rollers and its tension to the left of A is $F_C = 400$ N.

Prob. 15–75

*15–76. The fan draws air through a vent with a speed of 12 ft/s. If the cross-sectional area of the vent is 2 ft^2, determine the thrust on the blade. The specific weight of the air is $\gamma_a = 0.076$ lb/ft^3.

Prob. 15–76

15–77. A jet of water has a cross-sectional area of 5 in^2. If it strikes the fixed blade with a speed of 60 ft/s, determine the magnitude of force which the water exerts on the blade. $\gamma_w = 62.4$ lb/ft^3.

Prob. 15–77

15–78. The nozzle discharges water at a constant flow rate of 2 ft^3/s. The cross-sectional area of the nozzle at A is 4 in^2 and at B the cross-sectional area is 12 in^2. If the static pressure due to the water at B is 2 lb/in^2, determine the magnitude of force which must be applied at section B to hold the nozzle in place. $\gamma_w = 62.4$ lb/ft^3.

Prob. 15–78

15–79. The static pressure of water at C is 40 lb/in^2. If water flows out of the pipe at A and B with velocities $v_A = 12$ ft/s and $v_B = 25$ ft/s, determine the horizontal and vertical components of force exerted on the elbow at C necessary to hold the pipe assembly in equilibrium. Neglect the weight of water within the pipe and the weight of the pipe. The pipe has a diameter of 0.75 in. at C, and at A and B the diameter is 0.5 in. $\gamma_w = 62.4$ lb/ft^3.

Prob. 15–79

*15–80. The boy at A aims the water hose in the direction shown. If the water is discharged at 30° from the horizontal, and the cross-sectional area of the water stream is approximately 2 in², determine the force the boy at B must exert on the shield in order to hold it perpendicular to the water. $\gamma_w = 62.4$ lb/ft³.

Prob. 15–80

15–81. The vane on a turbine is moving at 80 ft/s when a jet of water having a velocity of $v_A = 150$ ft/s strikes it. If the cross-sectional area of the jet is 1.5 in², and it is diverted as shown, determine the power developed by the water on the blade. $\gamma_w = 62.4$ lb/ft³.

Prob. 15–81

15–82. The hose on the fire boat ejects water in a stream which reaches a height of 6 m as shown. If the water is ejected at 0.08 m³/s, determine the horizontal thrust T provided by the engines of the boat needed to hold the boat steady. $\rho_w = 1$ Mg/m³.

Prob. 15–82

15–83. A rocket burns 2000 lb of fuel in 2 min. If the velocity of the gas is 6500 ft/s with respect to the rocket, determine the thrust of the rocket engine.

*15–84. A plow located on the front of a locomotive scoops up snow at the rate of 10 ft³/s and stores it in the train. If the locomotive is traveling at a constant speed of 10 mi/h, determine the resistance to motion caused by the shoveling. The specific weight of snow is $\gamma_s = 6$ lb/ft³.

15–85. A boat weighs 400 lb and is traveling forward on a river with a constant velocity of 45 ft/s. The river is flowing in the opposite direction at 10 ft/s. A scoop is placed in the water in the direction of motion of the boat. If the scoop collects 100 lb of water in the boat in 2 minutes, determine the horizontal thrust **T** which is required to propel the boat forward. Neglect the drag effect of the water on the boat.

15–86. The second stage of the two-stage rocket weighs 2000 lb (empty) and is launched from the first stage with a velocity of 3000 mi/h. The fuel in the second stage weighs 1000 lb. If it is consumed at the rate of 50 lb/s and ejected with a relative velocity of 8000 ft/s, determine the acceleration of the second stage at the instant the engine is fired. What is the rocket's acceleration after all the fuel is consumed? Neglect the effect of gravitation.

15–87. A rocket has an empty weight of 500 lb and carries 300 lb of fuel. If the fuel is burned at the rate of 15 lb/s and ejected with a relative velocity of 4400 ft/s, determine the maximum speed attained by the rocket starting from rest. Neglect the effect of gravitation on the rocket.

*15–88. The missile weighs 40 kip. The constant thrust provided by the turbojet engine is $T = 15$ kip. Additional thrust is provided by *two* rocket boosters B. There is 600 lb of propellant in each booster which is burned at a constant rate of 150 lb/s, with a relative exhaust velocity of 3000 ft/s. If the mass of the propellant lost by the turbojet engine can be neglected, determine the velocity of the missile after the 4 s burn time of the boosters. The initial velocity of the missile is 300 mi/h.

Prob. 15–88

15–89. The 12-Mg jet airplane has a constant speed of 950 km/h when it is flying along a horizontal straight-line path. Air enters the intake scoops S at the rate of 50 m³/s. If the engine burns fuel at the rate of 0.4 kg/s and the gas (air and fuel) is exhausted relative to the plane with a speed of 450 m/s, determine the resultant drag force exerted on the plane by air resistance. Assume that air has a constant density of 1.22 kg/m³. *Hint:* Since mass both enters and exits the plane, Eqs. 15–29 and 15–30 must be combined to yield

$$\Sigma F_s = m\frac{dv}{dt} - v_{D/e}\frac{dm_e}{dt} + v_{D/i}\frac{dm_i}{dt}$$

$v = 950$ km/h

S

Prob. 15–89

15–90. The jet is traveling at a speed of 500 mi/h, 30° with the horizontal. If the fuel is being spent at 3 lb/s, and the engine takes in air at 400 lb/s, whereas the exhaust gas (air and fuel) has a relative speed of 32,800 ft/s, determine the acceleration of the plane at this instant. The drag resistance of the air is $F_D = (0.7v^2)$ lb, where the speed is measured in ft/s. The jet has a weight of 15,000 lb. *Hint:* See Prob. 15–89.

500 mi/h

30°

Prob. 15–90

15–91. The rocket car has a mass of 3 Mg (empty) and carries 150 kg of fuel. If the fuel is consumed at a constant rate of 40 kg/s and ejected from the car with a relative velocity of 250 m/s, determine the maximum speed attained by the car starting from rest. The drag resistance due to the atmosphere is $F_D = (0.6v^2)$ N, where v is the speed measured in m/s.

Prob. 15–91

*15–92. The chain has a total length $L < d$ and a mass per unit length of m'. If a portion h of the chain is suspended over the table and released, determine the velocity of its end A as a function of its position y. Neglect friction.

h

y

A

d

Prob. 15–92

15–93. If the chain is lowered at a constant velocity $v = 4$ ft/s, determine the normal reaction exerted on the floor as a function of time. The chain has a weight of 5 lb/ft and a total length of 20 ft.

$v = 4$ ft/s

20 ft

Prob. 15–93

15–94. The car has a mass m_0 and is used to tow the smooth chain having a total length l and a mass per unit of length m'. If the chain is originally piled up, determine the tractive force F that must be supplied by the rear wheels of the car to maintain a constant speed v while the chain is being drawn out. Assume the chain is smooth.

Prob. 15–94

15–95. The chain has a total length of 10 ft and a weight of 2 lb/ft. Determine the magnitude of force **F** as a function of time which must be applied to the end A of the chain to suspend it if the platform upon which it rests is descending at a constant rate of 3 ft/s. Neglect the size of the chain. Initially $y = 0$ at $t = 0$.

Prob. 15–95

***15–96.** The rocket has an initial mass m_0, including the fuel. For practical reasons desired for the passengers, it is required that it maintain a constant upward acceleration a_0. If the fuel is expelled from the rocket at a relative speed $v_{e/r}$, determine the rate at which the fuel should be consumed to maintain the motion. Neglect air resistance, and assume that the gravitational acceleration is constant.

Prob. 15–96

Review 1:
Kinematics and Kinetics of a Particle

The various topics and problems presented in Chapters 12 through 15 have all been *categorized* in order to provide a *clear focus* for learning the various problem-solving principles involved. In engineering practice, however, it is most important to be able to *identify* an appropriate method for the solution of a particular problem. In this regard, one must fully understand the limitations and use of the equations of dynamics, and be able to recognize which equations and principles to use for the problem's solution. For these reasons, we will now summarize the equations and principles of particle dynamics and provide the opportunity for applying them to a variety of problems.

Kinematics

Problems in kinematics require a study only of the geometry of motion, and do not account for the forces causing the motion. When the equations of kinematics are applied, one should clearly establish a fixed origin and select an appropriate coordinate system used to define the position of the particle. Once the positive direction of each coordinate axis is established, then the directions of the components of position, velocity, and acceleration can be determined from the algebraic sign of their numerical quantities.

Rectilinear Motion

Variable Acceleration. If a mathematical (or graphical) relationship is established between *any two* of the *four* variables s, v, a, and t, then a *third* variable can be determined by solving one of the following equations which relates all three variables.

$$v = \frac{ds}{dt} \qquad a = \frac{dv}{dt} \qquad a\, ds = v\, dv$$

Constant Acceleration. Be *absolutely certain* that the acceleration is constant when using the following equations:

$$s = s_0 + v_0 t + \tfrac{1}{2} a_c t^2 \qquad v = v_0 + a_c t \qquad v^2 = v_0{}^2 + 2a_c(s - s_0)$$

Curvilinear Motion

x, y, z Coordinates. These coordinates are often used when the motion can be resolved into horizontal and vertical components. Also they are useful for studying projectile motion since the acceleration of the projectile is *always* downward.

$$v_x = \dot{x} \qquad a_x = \dot{v}_x$$
$$v_y = \dot{y} \qquad a_y = \dot{v}_y$$
$$v_z = \dot{z} \qquad a_z = \dot{v}_z$$

n, t, b Coordinates. These coordinates are particularly advantageous for studying the particle's *acceleration* along a known path. This is because the *t* and *n* components of **a** represent the separate changes in the magnitude and direction of the velocity, and these components can be readily formulated.

$$v = \dot{s}$$

$$a_t = \dot{v} = v \frac{dv}{ds}$$

$$a_n = \frac{v^2}{\rho}$$

where $\rho = \left| \dfrac{[1 + (dy/dx)^2]^{3/2}}{d^2y/dx^2} \right|$ when the path $y = f(x)$ is given.

r, θ, z Coordinates. These coordinates are used when data regarding the angular motion of the radial coordinate *r* is given to describe the particle's motion. Also, some paths of motion can conveniently be described using these coordinates.

$$v_r = \dot{r} \qquad a_r = \ddot{r} - r\dot{\theta}^2$$
$$v_\theta = r\dot{\theta} \qquad a_\theta = r\ddot{\theta} + 2\dot{r}\dot{\theta}$$
$$v_z = \dot{z} \qquad a_z = \ddot{z}$$

Relative Motion. If the origin of a translating coordinate system is established at particle *A*, then for particle *B*

$$\mathbf{r}_B = \mathbf{r}_A + \mathbf{r}_{B/A}$$
$$\mathbf{v}_B = \mathbf{v}_A + \mathbf{v}_{B/A}$$
$$\mathbf{a}_B = \mathbf{a}_A + \mathbf{a}_{B/A}$$

Kinetics

Problems in kinetics involve the analysis of forces which cause the motion. When applying the equations of kinetics, it is absolutely necessary that measurements of the motion be made from an *inertial coordinate system,* i.e., one that does not rotate and is either fixed or translates with constant velocity. If a problem requires *simultaneous solution* of the equations of kinetics and kinematics, then the coordinate systems selected for writing each of the equations should define the *positive directions* of the axes in the *same* manner.

Equations of Motion. These equations are used to solve for the particle's acceleration or the forces causing the motion. If they are used to determine a particle's position, velocity, or time of motion, then kinematics will also have to be considered in the solution. Before applying the equations of motion, *always draw a free-body diagram* to identify all the forces acting on the particle. Also, establish the direction of the particle's acceleration or its components. (A kinetic diagram may accompany the solution in order to graphically account for the *m***a** vector.)

$$\Sigma F_x = ma_x \qquad \Sigma F_n = ma_n \qquad \Sigma F_r = ma_r$$
$$\Sigma F_y = ma_y \qquad \Sigma F_t = ma_t \qquad \Sigma F_\theta = ma_\theta$$
$$\Sigma F_z = ma_z \qquad \Sigma F_b = 0 \qquad \Sigma F_z = ma_z$$

Work and Energy. The equation of work and energy represents an integrated form of the tangential equation of motion, $\Sigma F_t = ma_t$, combined with kinematics ($a_t\, ds = v\, dv$). *It is used to solve problems involving force, velocity, and displacement.* Before applying this equation, *always draw a free-body diagram* in order to identify the forces which do work on the particle.

$$T_1 + \Sigma U_{1-2} = T_2$$

where
$$T = \tfrac{1}{2}mv^2 \quad \text{(kinetic energy)}$$

$$U_F = \int_{s_1}^{s_2} F \cos\theta\, ds \quad \text{(variable force)}$$

$$U_{F_c} = F_c \cos\theta\, (s_2 - s_1) \quad \text{(constant force)}$$
$$U_W = W\, \Delta y \quad \text{(weight)}$$
$$U_s = -(\tfrac{1}{2}ks_2^2 - \tfrac{1}{2}ks_1^2) \quad \text{(elastic spring)}$$

If the forces acting on the particle are *conservative forces,* i.e., those that *do not* cause a dissipation of energy such as friction, then apply the conservation of energy equation. This equation is easier to use than the equation of work and energy since it applies only at *two points* on the path and *does not* require calculation of the work done by a force as the particle moves along the path.

$$T_1 + V_1 = T_2 + V_2$$

where
$$V_g = Wy \quad \text{(gravitational potential energy)}$$
$$V_e = \tfrac{1}{2}ks^2 \quad \text{(elastic potential energy)}$$

If the *power* developed by a force is to be calculated, use

$$P = \frac{dU}{dt} = \mathbf{F} \cdot \mathbf{v}$$

where **v** is the velocity of a particle acted upon by the force **F**.

Impulse and Momentum. The equation of *linear impulse and momentum* is an integrated form of the equation of motion, $\Sigma \mathbf{F} = m\mathbf{a}$, combined with kinematics ($\mathbf{a} = d\mathbf{v}/dt$). *It is used to solve problems involving force, velocity, and time*. Before applying this equation, one should *always draw the free-body diagram,* in order to identify all the forces that cause impulses on the particle. From the diagram the impulsive and nonimpulsive forces should be identified. Recall that the nonimpulsive forces can be neglected in the analysis. Also, establish the direction of the particle's velocity just before and just after the impulses are applied. As an alternative procedure, the impulse and momentum diagrams may accompany the solution in order to graphically account for the terms in the equation.

$$m\mathbf{v}_1 + \Sigma \int_{t_1}^{t_2} \mathbf{F}\, dt = m\mathbf{v}_2$$

If several particles are involved in the problem, consider applying the *conservation of momentum* to the system in order to eliminate the internal impulses from the analysis. This can be done in a specified direction, provided no external impulses act on the particles in that direction.

$$\Sigma m\mathbf{v}_1 = \Sigma m\mathbf{v}_2$$

If the problem involves impact and the coefficient of restitution e is given, then apply the following equation.

$$e = \frac{(v_B)_2 - (v_A)_2}{(v_A)_1 - (v_B)_1} \qquad \text{(along line of impact)}$$

The *principle of angular impulse and momentum* and the *conservation of angular momentum* may be applied about an axis in order to *eliminate* some of the unknown impulses acting on the particle during the time period when its motion is studied. Investigation of the particle's free-body diagram (or the impulse diagram) will aid in choosing the axis for application.

$$(\mathbf{H}_O)_1 + \Sigma \int_{t_1}^{t_2} \mathbf{M}_O\, dt = (\mathbf{H}_O)_2$$

$$(\mathbf{H}_O)_1 = (\mathbf{H}_O)_2$$

The following problems involve application of the above concepts. They are presented in *random order* so as to provide practice in identifying the various types of problems and developing the skills necessary for their solution.

PROBLEMS

R1–1. A projectile, initially at the origin, moves along a straight-line path through a fluid medium such that its velocity is defined as $v = 1800(1 - e^{-0.3t})$ mm/s, where t is measured in seconds. Determine the displacement of the projectile during the first 3 s.

R1–2. An electric train car, having a mass of 25 Mg, travels up a 10° incline with a constant speed of 80 km/h. Determine the power required to overcome the force of gravity.

R1–3. A sports car can accelerate at 6 m/s² and decelerate at 8 m/s². If the maximum speed it can attain is 60 m/s, determine the shortest time it takes to travel 900 m starting from rest and then stopping when $s = 900$ m.

*R1–4. The 2-kg collar C is free to slide along the smooth shaft AB. Determine the acceleration of collar C if (a) the shaft is fixed from moving, (b) collar A, which is fixed to shaft AB, moves to the left at a constant velocity of 3 m/s along the horizontal guide, and (c) collar A is subjected to an acceleration of 2 m/s² to the left. In all cases, the shaft moves in the vertical plane.

Prob. R1–4

R1–5. The speed of a train during the first minute of its motion has been recorded as follows:

$t(s)$	0	20	40	60
v(m/s)	0	16	21	24

Plot the v-t graph, approximating the curve as straight line segments between the given points. Determine the total distance traveled.

R1–6. A crate has a weight of 1500 lb. If it is pulled along the ground at a constant speed for a distance of 20 ft, and the towing cable makes an angle of 15° with the horizontal, determine the tension in the cable and the work done by the towing force. The coefficient of kinetic friction between the ground and the crate is $\mu_k = 0.55$.

R1–7. If a 150-lb crate is released from rest at A, determine its speed after it slides 30 ft down the plane. The coefficient of kinetic friction between the crate and plane is $\mu_k = 0.3$.

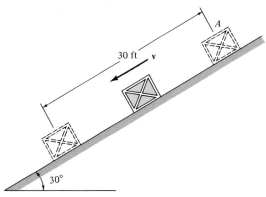

Prob. R1–7

*R1–8. A girl having a weight of 80 lb is skating around in a circle of radius $r_A = 20$ ft with a speed of $(v_A)_1 = 5$ ft/s while holding on to the end of a rope. If her partner starts to pull the rope inward with a constant speed of $v_r = 2$ ft/s, determine the girl's speed at the instant $r_B = 10$ ft. How much work is done by her partner after pulling in the rope? Neglect friction and assume the girl remains in a rigid position.

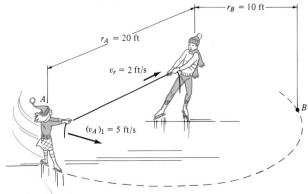

Prob. R1–8

R1–9. Determine the tension developed in the two cords and the acceleration of each block. Neglect the mass of the pulleys and cords. *Hint:* Since the system consists of *two* cords, relate the motion of block A to C, and of block B to C. Then, by elimination, relate the motion of A to B.

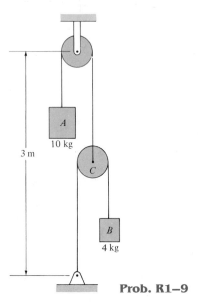

Prob. R1–9

R1–10. An automobile is traveling with a *constant speed* along a horizontal circular curve that has a radius of $\rho = 750$ ft. If the magnitude of acceleration is $a = 8$ ft/s², determine the speed at which the automobile is traveling.

R1–11. Disk A weighs 2 lb and is sliding on a smooth horizontal plane with a velocity of 3 ft/s. Disk B weighs 11 lb and is initially at rest. If after the impact A has a velocity of 1 ft/s directed along the positive x axis, determine the velocity of B after impact. How much kinetic energy is lost in the collision?

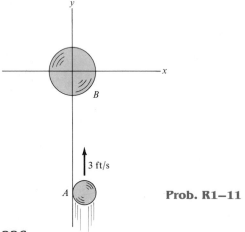

Prob. R1–11

*R1–12. At a given instant the 10-lb block A is moving downward with a speed of 6 ft/s. Determine its speed 2 s later. Block B has a weight of 4 lb, and the coefficient of kinetic friction between it and the horizontal plane is $\mu_k = 0.2$. Neglect the mass of the pulleys and cord.

Prob. R1–12

R1–13. A 20-kg block is originally at rest on a horizontal surface for which the coefficient of both static and kinetic friction is $\mu = 0.6$. If a horizontal force F is applied such that it varies with time as shown, determine the speed of the block in 10 s. *Hint:* First determine the time needed to overcome friction and start the block moving.

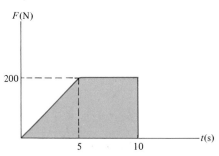

Prob. R1–13

R1–14. Two smooth billiard balls A and B have an equal mass of $m = 200$ g. If A strikes B with a velocity of $(v_A)_1 = 2$ m/s as shown, determine their final velocities just after collision. Ball B is originally at rest and the coefficient of restitution is $e = 0.75$.

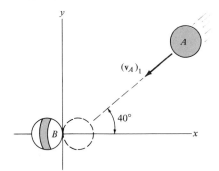

Prob. R1–14

R1–15. The 100-kg crate is subjected to the action of two forces, $F_1 = 800$ N and $F_2 = 1.5$ kN, as shown. If it is originally at rest, determine the distance it slides in order to attain a speed of 6 m/s. The coefficient of kinetic friction between the crate and the surface is $\mu_k = 0.2$.

Prob. R1–15

*R1–16. If a horizontal force of $P = 10$ lb is applied to block A, determine the acceleration of block B. Neglect friction. *Hint:* Show that $a_B = a_A \tan 15°$.

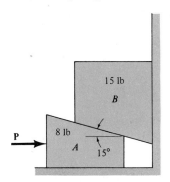

Prob. R1–16

R1–17. The spool, which has a mass of 4 kg, slides along the rotating rod. At the instant shown, the angular rate of rotation of the rod is $\dot{\theta} = 6$ rad/s and the rotation is increasing at $\ddot{\theta} = 2$ rad/s². At this same instant, the spool has a velocity of 3 m/s and an acceleration of 1 m/s², both measured relative to the rod and directed away from the center O when $r = 0.5$ m. Determine the radial frictional force and the normal force, both exerted by the rod on the spool at this instant.

Prob. R1–17

R1–18. A freight train starts from rest at $t = 0$ and travels with a constant acceleration of 0.5 ft/s². After a time t' it maintains a constant speed so that when $t = 160$ s the train has traveled 2000 ft. Determine the time t' and draw the v-t graph for the motion.

R1–19. The block has a mass of $m = 0.5$ kg and moves within the smooth vertical slot. If it starts from rest when the *attached* spring is in the unstretched position at A, determine the *constant* vertical force F which must be applied to the cord so that the block attains a speed of $v_B = 2.5$ m/s when it reaches B; $s_B = 0.15$ m. Neglect the mass of the cord and pulley. *Hint:* The work of $\mathbf{F}$ can be determined by finding the difference Δl in cord lengths AC and BC and using $U_F = F \, \Delta l$.

Prob. R1–19

227

*R1–20. An ivory ball having a mass of 200 g is released from rest at a height of 400 mm above a very large fixed metal surface. If the ball rebounds to a height of 325 mm above the surface, determine the coefficient of restitution between the ball and the surface.

R1–21. If a particle has an initial velocity of $v_0 = 12$ ft/s to the right, determine its position when $t = 10$ s, if $a = 2$ ft/s² to the left. Originally $s_0 = 0$.

R1–22. A 3-lb block, initially at rest at point A, slides along the smooth parabolic surface. Determine the normal force acting on the block when it reaches point B. Neglect the size of the block.

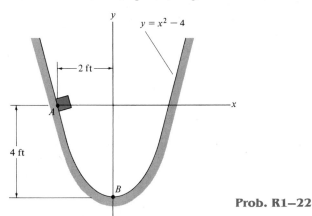

Prob. R1–22

R1–23. The crate, having a weight of 50 lb, is hoisted by the pulley system and motor M. If the crate starts from rest and, by constant acceleration, attains a speed of 12 ft/s after rising $s = 10$ ft, determine the power that must be supplied to the motor at the instant $s = 10$ ft. The motor has an efficiency of $\varepsilon = 0.74$.

Prob. R1–23

*R1–24. A rifle has a mass of 2.5 kg. If it is loosely gripped and a 1.5-g bullet is fired from it with a muzzle velocity of 1400 m/s, determine the recoil velocity of the rifle just after firing.

R1–25. Two planes, A and B, are flying at the same altitude. If their velocities are $v_A = 450$ mi/h and $v_B = 400$ mi/h such that the angle between their straight-line courses is $\theta = 75°$, determine the velocity of plane B with respect to plane A.

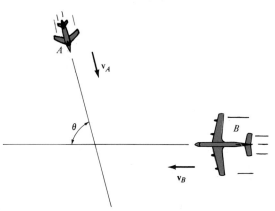

Prob. R1–25

R1–26. The 90-lb force is required to drag the 200-lb block 60 ft up the *rough* inclined plane at constant velocity. If the force is removed when the block reaches point B, and the block is then released from rest, determine the block's velocity when it slides back down the plane and reaches point A.

Prob. R1–26

228

R1–27. The boy of mass 40 kg is sliding down the spiral slide at a constant speed such that his position, measured from the top of the chute, has components $r = 2$ m, $\theta = (0.7t)$ rad, and $z = (-0.5t)$ m, where t is in seconds. Determine the components of force $\mathbf{F}_r$, $\mathbf{F}_\theta$, and $\mathbf{F}_z$ which he exerts on the chute at the instant $t = 2$ s.

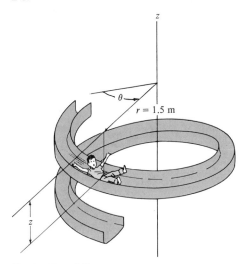

Prob. R1–27

*R1–28. A spring having a stiffness of 5 kN/m is compressed 400 mm. The stored energy in the spring is used to drive a machine which requires 80 W of power. Determine how long the spring can supply energy at the required rate.

R1–29. A tugboat T having a mass of 19 Mg is tied to a barge B having a mass of 75 Mg. If the rope is "elastic" such that it has a stiffness of $k = 600$ kN/m, determine the maximum stretch in the rope during the initial towing. Originally both the tugboat and barge are moving in the same direction with speeds of $(v_T)_1 = 15$ km/h and $(v_B)_1 = 10$ km/h, respectively. Neglect the resistance of the water. When taut, the rope becomes horizontal.

Prob. R1–29

R1–30. The conveyor belt delivers each 15-lb crate to the ramp at A such that the crate's velocity is $v_A = 4$ ft/s, directed down *along* the ramp. If the coefficient of kinetic friction between each crate and the ramp is $\mu_k = 0.3$, determine the speed at which each crate slides off the ramp at B. Assume that no tipping occurs.

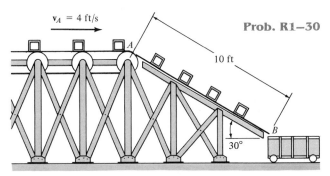

Prob. R1–30

R1–31. The slotted arm AB drives the pin C through the spiral groove described by the equation $r = 1.5\theta$, where θ is in radians and r is in feet. If the arm starts from rest when $\theta = 60°$ and is driven at an angular rate of $\dot{\theta} = 4t$ rad/s, determine the radial and transverse components of velocity and acceleration of the pin when $t = 1$ s.

Prob. R1–31

*R1–32. A skier starts from rest at A (30 ft, 0) and descends the smooth slope, which may be approximated by a parabola. If she has a weight of 120 lb, determine the normal force she exerts on the ground at the instant she arrives at point B.

Prob. R1–32

R1–33. The catapulting mechanism is used to propel the 15-kg slider A to the right along the smooth track. The propelling action is obtained by drawing the pulley attached to rod BC rapidly to the left by means of a piston P. If the piston applies a constant force of $F = 8$ kN to rod BC such that it moves it 0.2 m, determine the speed attained by the slider if it was originally at rest. Neglect the mass of the pulleys, cable, piston, and rod BC.

Prob. R1–33

R1–34. A railroad car having a mass of 15 Mg is coasting at 2 m/s on a horizontal track. At the same time another car having a mass of 10 Mg is coasting at 0.75 m/s in the opposite direction. If the cars meet and couple together, determine the speed of both cars just after the coupling. Compute the difference between the total kinetic energy before and after coupling has occurred, and explain qualitatively what happened to this energy.

R1–35. A boy at A throws a ball 45° from the horizontal such that it strikes the slope at B. Determine the speed at which the ball is thrown and the time of flight.

Prob. R1–35

*■R1–36. Packages having a mass of 2.5 kg ride on the surface of the conveyor belt. If the belt starts from rest and with constant acceleration increases to a speed of 0.75 m/s in 2 s, determine the maximum angle of tilt, θ, so that none of the packages slip on the inclined surface AB of the belt. The coefficient of static friction between the belt and each package is $\mu_s = 0.3$. At what angle ϕ do the packages first begin to slip off the surface of the belt if the belt is moving at a constant speed of 0.75 m/s?

Prob. R1–36

R1–37. To test the manufactured properties of 2-lb steel balls, each ball is released from rest as shown and strikes a 45° inclined surface. If the coefficient of restitution is $e = 0.8$, determine the distance s to where the ball strikes the horizontal plane at A. At what speed does the ball strike point A?

Prob. R1–37

R1–38. A car has a weight of 2500 lb and is traveling forward horizontally at 60 ft/s. If the coefficient of kinetic friction between the tires and the pavement is $\mu_k = 0.6$, determine the time needed to stop the car. The brakes are applied to all four wheels.

R1–39. Packages having a mass of 6 kg slide down a smooth chute and land horizontally with a speed of 3 m/s on the surface of a conveyor belt. If the coefficient of kinetic friction between the belt and a package is $\mu_k = 0.2$, determine the time needed to bring the package to rest on the belt if the belt is moving in the same direction as the package with a speed of $v = 1$ m/s.

Prob. R1–39

*R1–40. Block B rests on a smooth surface. If the coefficient of static friction between A and B is $\mu_s = 0.4$, determine the acceleration of each block if (a) $F = 6$ lb, and (b) $F = 50$ lb.

Prob. R1–40

R1–41. The roller-coaster car has a mass of 800 kg, including its passenger, and moves from the top of the hill A with a speed of $v_A = 3$ m/s. Determine the minimum height h of the hill crest so that the car travels around both inside loops without leaving the track. Neglect friction and the size of the car. What is the normal reaction on the car when the car is at B and when it is at C?

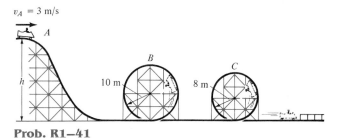

Prob. R1–41

R1–42. A spiral transition curve is used on railroads to connect a straight portion of the track with a curved portion. If the spiral is defined by the equation $y = (10^{-6})x^3$, where x and y are in feet, determine the magnitude of the acceleration of a train engine moving with a constant speed of 40 ft/s when it is at point $x = 600$ ft.

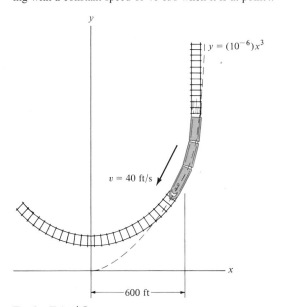

Prob. R1–42

R1–43. Determine the velocity of each block 2 s after the blocks are released from rest. Neglect the mass of the pulleys and cord.

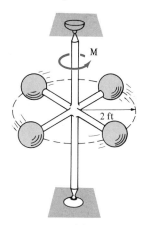

Prob. R1–43

*R1–44. The four 5-lb spheres are rigidly attached to the cross-bar frame. The frame has a negligible weight. If a couple moment $M = (5t)$ lb · ft, where t is in seconds, is applied as shown, determine the speed of each of the spheres in 4 s starting from rest.

Prob. R1–44

Review 1: Kinematics and Kinetics of a Particle

R1–45. The small 2-lb collar starting from rest at A slides down along the smooth wire. During the motion, the collar is acted upon by a force $\mathbf{F} = \{10\mathbf{i} + 6y\mathbf{j} + 2z\mathbf{k}\}$ lb, where x, y, z are in feet. Determine the collar's speed when it strikes the wall at B.

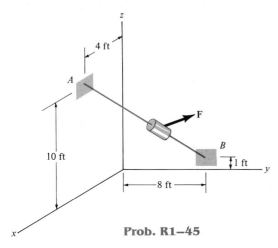

Prob. R1–45

R1–46. In each case, if the end of the cable at A is pulled down with a speed of 2 m/s, determine the speed at which block B rises.

Prob. R1–46

R1–47. A particle is moving along a circular path of 2-m radius such that its position as a function of time is given by $\theta = (5t^2)$ rad, where t is in seconds. Determine the magnitude of the particle's acceleration when $\theta = 30°$. The particle starts from rest when $\theta = 0°$.

*R1–48. Packages having a weight of 5 lb ride on the surface of the conveyor belt. If the belt travels at a constant speed of 2 ft/s and the coefficient of static friction between the belt and each package is $\mu_s = 0.4$, determine the angle θ at which the packages first begin to slip off the surface of the belt.

Prob. R1–48

R1–49. Four inelastic cables C are attached to a plate P and hold the spring 6 in. in compression when *no force* acts on the plate. If a block B, having a weight of 5 lb, is placed on the plate and the plate is pushed down 8 in. and released from rest, determine how high the block rises from the point where it was released. Neglect the mass of the plate.

Prob. R1–49

R1–50. The gantry structure is used to test the response of an airplane during a crash. The plane having a mass of 8 Mg is hoisted back until $\theta = 60°$, and then the pull-back cable AC is released when the plane is at rest. Determine the speed of the plane just before crashing into the ground, $\theta = 15°$. What is the maximum tension developed in the supporting cable during the motion? Neglect the effects of lift caused by the wings.

Prob. R1–50

232

16

Planar Kinematics of a Rigid Body

In this chapter kinematics or the study of the geometry of the planar motion for a rigid body will be discussed. This study is important for the design of gears, cams, and mechanisms used for machine operations. Furthermore, once the kinematics of a rigid body is thoroughly understood, it will be possible to apply the equations of motion, which relate the forces on the body to the body's motion.

A rigid body can be subjected to three types of planar motion, namely, translation, rotation about a fixed axis, and general plane motion. We will first define each of these motions, then discuss the analysis of each separately. In all cases it will be shown that rigid-body planar motion is completely specified provided the motions of any two points on the body are known. The more complex study of three-dimensional kinematics of a rigid body is presented in Chapter 20.

16.1 Rigid-Body Motion

When each of the particles of a rigid body moves along a path which is equidistant from a fixed plane, the body is said to undergo *planar motion*. As previously stated, there are three types of planar motion; in order of increasing complexity, they are:

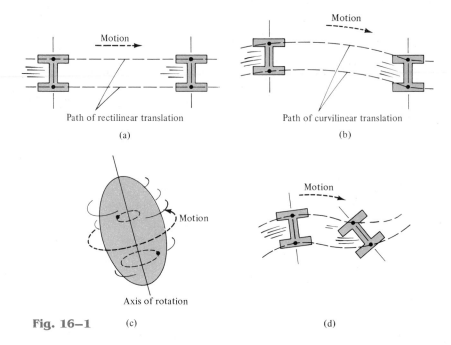

Fig. 16–1

1. *Translation*. This type of motion occurs if any line segment on the body remains parallel to its original direction during the motion. When the paths of motion for all the particles of the body are along equidistant straight lines, Fig. 16–1a, the motion is called *rectilinear translation*. However, if the paths of motion are along curved lines which are equidistant, Fig. 16–1b, the motion is called *curvilinear translation*.

2. *Rotation about a fixed axis*. When a rigid body rotates about a fixed axis, all the particles of the body, except those which lie on the axis of rotation, move along circular paths, Fig. 16–1c.

3. *General plane motion*. When a body is subjected to general plane motion, it undergoes a combination of translation *and* rotation, Fig. 16–1d. The translation occurs within a reference plane, and the rotation occurs about an axis perpendicular to the reference plane.

The above planar motions are exemplified by the moving parts of the crank mechanism shown in Fig. 16–2.

Fig. 16–2

16.2 Translation

Consider a rigid body which is subjected to either rectilinear or curvilinear translation in the *x-y* plane, Fig. 16–3.

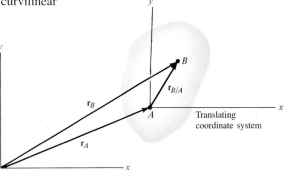

Fig. 16–3

Position

The locations of points *A* and *B* in the body are defined from the fixed *x, y* reference frame by using *position vectors* $\mathbf{r}_A$ and $\mathbf{r}_B$. The translating *x', y'* coordinate system is *fixed in the body* and has its origin located at *A*, hereafter referred to as the *base point*. The position of *B* with respect to *A* is denoted by the *relative-position vector* $\mathbf{r}_{B/A}$ ("**r** of *B* with respect to *A*"). By vector addition,

$$\mathbf{r}_B = \mathbf{r}_A + \mathbf{r}_{B/A}$$

Velocity

A relationship between the instantaneous velocities of *A* and *B* is obtained by taking the time derivative of the position equation, which yields $\mathbf{v}_B = \mathbf{v}_A + d\mathbf{r}_{B/A}/dt$. Here $\mathbf{v}_A$ and $\mathbf{v}_B$ denote *absolute velocities* since these vectors are measured from the *x, y* axes. The term $d\mathbf{r}_{B/A}/dt = \mathbf{0}$, since the *magnitude* of $\mathbf{r}_{B/A}$ is constant by definition of a rigid body, and because the body is translating the *direction* of $\mathbf{r}_{B/A}$ is constant. Therefore,

$$\mathbf{v}_B = \mathbf{v}_A$$

Acceleration

Taking the time derivative of the velocity equation yields a similar relationship between the instantaneous accelerations of *A* and *B*,

$$\mathbf{a}_B = \mathbf{a}_A$$

The above two equations indicate that *all points in a rigid body subjected to either curvilinear or rectilinear translation move with the same velocity and acceleration*. As a result, the kinematics of particle motion, discussed in Chapter 12, may also be applied to specify the kinematics of the particles located in a translating rigid body.

16.3 Rotation About a Fixed Axis

(a)

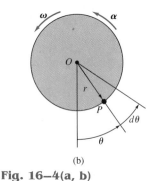

(b)

Fig. 16–4(a, b)

When a body is rotating about a fixed axis, any point P located in the body travels along a *circular path*. This motion depends upon the angular motion of the body about the axis. For this reason, we will first study the angular-motion properties of the body before analyzing the circular motion of P.

Angular Motion

Since a point is without dimension, it has no angular motion. Only lines or bodies undergo angular motion. To study its effects, we will consider the body shown in Fig. 16–4a and the angular motion of a radial line r located within the shaded plane.

Angular Position. At the instant shown, the *angular position* of r is defined by the angle θ, measured between a *fixed* reference line and r. Here r extends perpendicular to the axis of rotation from point O to a point P in the body.

Angular Displacement. The change in the angular position, often measured as a differential $d\boldsymbol{\theta}$, is called the *angular displacement*.* This vector has a *magnitude* of $d\theta$, which is measured in degrees, radians, or revolutions, where 1 rev = 2π rad. Since motion is about a *fixed axis,* the direction of $d\boldsymbol{\theta}$ is *always* along the axis. Specifically, the *direction* of $d\boldsymbol{\theta}$ is determined by the right-hand rule; that is, the fingers of the right hand are curled with the sense of rotation, so that in this case the thumb, or $d\boldsymbol{\theta}$, points upward, Fig. 16–4a. In two dimensions, as shown on the top view of the shaded plane, Fig. 16–4b, both θ and $d\boldsymbol{\theta}$ are directed counterclockwise and so the thumb points outward.

Angular Velocity. The time rate of change in the angular position is called the *angular velocity* $\boldsymbol{\omega}$ (omega). Since $d\theta$ occurs during an instant of time dt, then,

$(\curvearrowleft+)$

$$\omega = \frac{d\theta}{dt} \qquad (16\text{–}1)$$

This vector has a *magnitude* which is often measured in rad/s. It is expressed here in scalar form, since its *direction* is always along the axis of rotation, i.e., in the same direction as $d\boldsymbol{\theta}$, Fig. 16-4a. When indicating the angular motion in the shaded plane, Fig. 16–4b, we can refer to the sense of rotation as clockwise or counterclockwise. Here we have *arbitrarily* chosen counterclockwise rotations as *positive,* and indicated this by the curl shown in parentheses next to Eq. 16–1. Realize, however, that the directional sense of $\boldsymbol{\omega}$ is actually outward.

---

*It is shown in Sec. 20.1 that finite rotations or finite angular displacements are *not* vector quantities, although differential rotations $d\boldsymbol{\theta}$ are vectors.

Angular Acceleration. The *angular acceleration* $\boldsymbol{\alpha}$ (alpha) measures the time rate of change of the angular velocity. Hence, the *magnitude* of this vector may be written as

$(\zeta+)$

$$\alpha = \frac{d\omega}{dt}$$

$(16\text{--}2)$

Or taking the second time derivative of Eq. 16–1, it is possible to express α as

$(\zeta+)$

$$\alpha = \frac{d^2\theta}{dt^2}$$

$(16\text{--}3)$

The line of action of $\boldsymbol{\alpha}$ is the same as that for $\boldsymbol{\omega}$, Fig. 16–4a; however, its sense of *direction* depends upon whether $\boldsymbol{\omega}$ is increasing or decreasing with time. In particular, if $\boldsymbol{\omega}$ is decreasing $\boldsymbol{\alpha}$ is called an *angular deceleration* and therefore has a sense of direction which is opposite to $\boldsymbol{\omega}$.

By eliminating dt from Eqs. 16–1 and 16–2, it is possible to obtain a differential relation between the angular acceleration, angular velocity, and angular displacement, namely,

$(\zeta+)$

$$\alpha\, d\theta = \omega\, d\omega$$

$(16\text{--}4)$

The similarity between the differential relations for angular motion and those developed for rectilinear motion of a particle ($v = ds/dt$, $a = dv/dt$, and $a\, ds = v\, dv$) should be apparent.

Constant Angular Acceleration. If the angular acceleration of the body is constant, $\boldsymbol{\alpha} = \boldsymbol{\alpha}_c$, Eqs. 16–1, 16–2, and 16–4, when integrated, yield a set of formulas which relate the body's angular velocity, its angular position, and time. These equations are similar to Eqs. 12–4 to 12–6 used for rectilinear motion. The results are

$(\zeta+)$ $\omega = \omega_0 + \alpha_c t$ $(16\text{--}5)$

$(\zeta+)$ $\theta = \theta_0 + \omega_0 t + \frac{1}{2}\alpha_c t^2$ $(16\text{--}6)$

$(\zeta+)$ $\omega^2 = \omega_0^2 + 2\alpha_c(\theta - \theta_0)$ $(16\text{--}7)$

Constant Angular Acceleration

Here θ_0 and ω_0 are the initial values of the body's angular position and angular velocity, respectively.

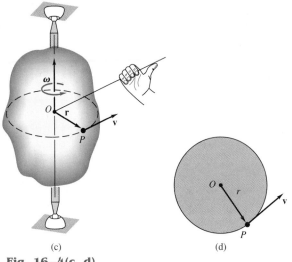

(c)

(d)

Fig. 16–4(c, d)

Motion of Point P

As the rigid body in Fig. 16–4c rotates, point P travels along a *circular path* of radius r and with center at point O. This path is contained in the shaded plane shown in top view, Fig. 16–4d.

Position. The position of P is defined by the position vector $\mathbf{r}$, which extends from O to P.

Velocity. The velocity of P has a magnitude which can be determined from the circular motion of P using its polar coordinate components $v_r = \dot{r}$ and $v_\theta = r\dot{\theta}$, Eqs. 12–25. Since r is constant, the radial component $v_r = \dot{r} = 0$, so that $v = v_\theta = r\dot{\theta}$. Because $\omega = \dot{\theta}$, Eq. 16–1, the result is

$$v = \omega r \qquad (16–8)$$

As shown in Figs. 16–4c and 16–4d, the *direction* of $\mathbf{v}$ is *tangent* to the circular path.

Both the magnitude and direction of $\mathbf{v}$ can be accounted for by using the cross product of $\boldsymbol{\omega}$ and $\mathbf{r}$ (see Appendix C). We have

$$\mathbf{v} = \boldsymbol{\omega} \times \mathbf{r} \qquad (16–9)$$

The order of the vectors in this formulation is important, since the cross product is not commutative, i.e., $\boldsymbol{\omega} \times \mathbf{r} \neq \mathbf{r} \times \boldsymbol{\omega}$. In this regard, notice in Fig. 16–4c how the correct direction of $\mathbf{v}$ is established by the right-hand rule. The fingers of the right hand are curled from $\boldsymbol{\omega}$ to $\mathbf{r}$ ($\boldsymbol{\omega}$ "cross" $\mathbf{r}$). The thumb indicates the correct direction of $\mathbf{v}$, that is, tangent to the path in the direction of motion.

Acceleration. For convenience, the acceleration of P will be expressed in terms of its normal and tangential components.* Using $a_t = dv/dt$ and $a_n = v^2/\rho$, noting that $\rho = r$, $v = \omega r$, and $\alpha = d\omega/dt$, we have

$$a_t = \alpha r \qquad (16\text{--}10)$$

$$a_n = \omega^2 r \qquad (16\text{--}11)$$

The *tangential component of acceleration*, Fig. 16–4e and f, represents the time rate of change in the velocity's magnitude. If the speed of P is increasing, then $\mathbf{a}_t$ acts in the same direction as $\mathbf{v}$; if the speed is decreasing, $\mathbf{a}_t$ acts in the opposite direction of $\mathbf{v}$, and finally, if the speed is constant, $\mathbf{a}_t$ is zero.

The *normal component of acceleration* represents the time rate of change in the velocity's direction. The *direction* of $\mathbf{a}_n$ is always towards O, the center of the circular path, Fig. 16–4e and f.

Like the velocity, the acceleration of point P may be expressed in terms of the vector cross product. Taking the time derivative of Eq. 16–9 yields

$$\mathbf{a} = \frac{d\mathbf{v}}{dt} = \frac{d\boldsymbol{\omega}}{dt} \times \mathbf{r} + \boldsymbol{\omega} \times \frac{d\mathbf{r}}{dt}$$

Recalling that $\boldsymbol{\alpha} = d\boldsymbol{\omega}/dt$, and using Eq. 16–9 ($d\mathbf{r}/dt = \mathbf{v} = \boldsymbol{\omega} \times \mathbf{r}$), we have

$$\mathbf{a} = \boldsymbol{\alpha} \times \mathbf{r} + \boldsymbol{\omega} \times (\boldsymbol{\omega} \times \mathbf{r}) \qquad (16\text{--}12)$$

By definition of the cross product, the first term on the right has a magnitude $a_t = \alpha r$, and by the right-hand rule, $\boldsymbol{\alpha} \times \mathbf{r}$ is in the direction of $\mathbf{a}_t$, Fig. 16–4e. Likewise, the second term has a magnitude $a_n = \omega^2 r$, and applying the right-hand rule twice, first $\boldsymbol{\omega} \times \mathbf{r}$ then $\boldsymbol{\omega} \times (\boldsymbol{\omega} \times \mathbf{r})$, it can be seen that this result is in the same direction as $\mathbf{a}_n$, Fig. 16–4e. Noting that this is also the *same* direction as $-\mathbf{r}$, we can express $\mathbf{a}_n$ in a much simpler form as $\mathbf{a}_n = -\omega^2 \mathbf{r}$. Hence, Equation 16–12 becomes

$$\begin{aligned} \mathbf{a} &= \mathbf{a}_t + \mathbf{a}_n \\ &= \boldsymbol{\alpha} \times \mathbf{r} - \omega^2 \mathbf{r} \end{aligned} \qquad (16\text{--}13)$$

Since $\mathbf{a}_t$ and $\mathbf{a}_n$ are perpendicular, Fig. 16–4e and f, if needed the magnitude of acceleration can be determined from the Pythagorean theorem; namely, $a = \sqrt{a_n^2 + a_t^2}$.

(e)

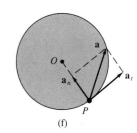

(f)

Fig. 16–4(e, f)

---

*Polar coordinates can also be used. Since $a_r = \ddot{r} - r\dot{\theta}^2$ and $a_\theta = r\ddot{\theta} + 2\dot{r}\dot{\theta}$, substituting $\dot{r} = \ddot{r} = 0$, $\dot{\theta} = \omega$, $\ddot{\theta} = \alpha$, we obtain Eqs. 16–10 and 16–11.

PROCEDURE FOR ANALYSIS

In order to determine the velocity and acceleration of a point located in a rigid body that is rotating about a fixed axis, it is first necessary to know the body's angular velocity and angular acceleration. If Eqs. 16–1 to 16–7 are used to obtain ω and α, it is important that a *positive direction* along the axis of rotation be established. By doing so, the sense of θ, ω, and α can be determined from the algebraic signs of their numerical quantities. In the examples that follow, the positive sense will be indicated by a curl along-side each kinematic equation as it is applied.

Angular Motion. If $\boldsymbol{\alpha}$ or $\boldsymbol{\omega}$ are unknown, then the relationships between the angular motions are defined by the differential equations:

$$\omega = \frac{d\theta}{dt} \qquad \alpha = \frac{d\omega}{dt} \qquad \alpha\,d\theta = \omega\,d\omega$$

If one is *absolutely certain* that the body's angular acceleration is *constant*, the following equations can be used:

$$\omega = \omega_0 + \alpha_c t$$
$$\theta = \theta_0 + \omega_0 t + \tfrac{1}{2}\alpha_c t^2$$
$$\omega^2 = \omega_0^2 + 2\alpha_c(\theta - \theta_0)$$

Motion of P. When the motion of a point P in the body is to be determined, it is suggested that a *kinematic diagram* accompany the problem solution. This diagram is simply a graphical representation showing the motion of the point.

In most cases the velocity and the two components of acceleration can be determined from the scalar equations

$$v = \omega r$$
$$a_t = \alpha r$$
$$a_n = \omega^2 r$$

However, if the geometry of the problem is difficult to visualize, the following vector equations should be used:

$$\mathbf{v} = \boldsymbol{\omega} \times \mathbf{r}$$
$$\mathbf{a}_t = \boldsymbol{\alpha} \times \mathbf{r}$$
$$\mathbf{a}_n = -\omega^2 \mathbf{r}$$

For application, $\mathbf{r}$, $\boldsymbol{\omega}$, and $\boldsymbol{\alpha}$ are expressed in terms of their $\mathbf{i}$, $\mathbf{j}$, $\mathbf{k}$ components and, if necessary, the cross products are computed by using a determinant expansion (see Eq. C–12).

The following examples numerically illustrate application of these equations.

Example 16–1

A cord is wrapped around a wheel which is initially at rest as shown in Fig. 16–5. If a force is applied to the cord and gives it an acceleration $a = (4t)$m/s^2, where t is in seconds, determine as a function of time (a) the angular velocity of the wheel, and (b) the angular position of line OP in radians.

SOLUTION

Part (a). The wheel is subjected to rotation about a fixed axis passing through point O. Thus, point P on the wheel has motion about a circular path, and therefore the acceleration of this point has both tangential and normal components. In particular, $(a_P)_t = (4t)$ m/s^2, since the cord is connected to the wheel and *tangent* to it at P. Hence, the angular acceleration of the wheel is

Fig. 16–5

$(\curvearrowright+)$
$$(a_P)_t = \alpha r$$
$$4t = \alpha(0.2)$$
$$\alpha = 20t \text{ rad/s}^2$$

The wheel's angular velocity ω can be determined by using $\alpha = d\omega/dt$, since this equation relates α, t, and ω. Integrating, with the initial condition that $\omega = 0$ at $t = 0$, yields

$(\curvearrowright+)$
$$\alpha = \frac{d\omega}{dt} = 20t$$
$$\int_0^\omega d\omega = \int_0^t 20t \, dt$$
$$\omega = 10t^2 \text{ rad/s} \downarrow \qquad\qquad Ans.$$

Why is it not possible to use Eq. 16–5 ($\omega = \omega_0 + \alpha_c t$) to obtain this result?

Part (b). The angular position θ of the radial line OP can be computed using $\omega = d\theta/dt$, since this equation relates θ, ω, and t. Using the initial condition $\theta = 0$ at $t = 0$, we have

$(\curvearrowright+)$
$$\frac{d\theta}{dt} = \omega = 10t^2$$
$$\int_0^\theta d\theta = \int_0^t 10t^2 \, dt$$
$$\theta = 3.33 \, t^3 \text{ rad} \qquad\qquad Ans.$$

Example 16–2

Disk A, shown in Fig. 16–6a, starts from rest and rotates with a constant angular acceleration of $\alpha_A = 2$ rad/s². If disk A is in contact with disk B and no slipping occurs between the disks, determine the angular velocity and angular acceleration of B just after A turns 10 revolutions.

SOLUTION

The angular velocity of A can be determined using the equations of *constant* angular acceleration since $\alpha_A = 2$ rad/s². There are 2π radians to one revolution, so that

$$\theta_A = 10 \text{ rev} \left(\frac{2\pi \text{ rad}}{1 \text{ rev}}\right) = 62.83 \text{ rad}$$

Thus,

$$(\nearrow+) \qquad \omega^2 = \omega_0^2 + 2\alpha_c(\theta - \theta_0)$$
$$\omega_A^2 = 0 + 2(2)(62.83 - 0)$$
$$\omega_A = 15.9 \text{ rad/s} \downarrow$$

As shown in Fig. 16–6b, the speed of the contacting point P on the rim of A is

$$(+\downarrow) \qquad v_P = \omega_A r_A = (15.9)(2) = 31.8 \text{ ft/s} \downarrow$$

The velocity is always tangent to the path of motion, so that the speed of point P' on B is the *same* as the speed of P on A. The angular velocity of B is therefore

$$(\downarrow+) \qquad \omega_B = \frac{v_{P'}}{r_B} = \frac{31.8}{1.5} = 21.2 \text{ rad/s} \uparrow \qquad Ans.$$

As shown in Fig. 16–6c, the *tangential components* of acceleration of both disks are equal, since the disks are in contact with one another. Hence,

$$(a_P)_t = (a_{P'})_t$$
$$\alpha_A r_A = \alpha_B r_B$$

$$\alpha_B = \alpha_A \left(\frac{r_A}{r_B}\right) = 2\left(\frac{2}{1.5}\right) = 2.67 \text{ rad/s}^2 \uparrow \qquad Ans.$$

Notice that the normal components of acceleration $(a_P)_n$ and $(a_{P'})_n$ act in *opposite directions,* since the paths of motion for both points are *different.* Furthermore, $(a_P)_n \neq (a_{P'})_n$, since the *magnitudes* of these components depend upon the radius and angular velocity of each disk, i.e., $(a_P)_n = \omega_A^2 r_A$ and $(a_{P'})_n = \omega_B^2 r_B$.

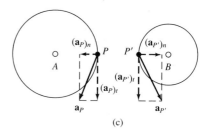

Fig. 16–6

PROBLEMS

16–1. A disk having a radius of 0.5 ft rotates with an initial angular velocity of 2 rad/s and has a constant angular acceleration of 1 rad/s^2. Determine the magnitude of the velocity and acceleration of a point on the rim of the disk when $t = 2$ s.

16–2. A wheel has its angular speed increased uniformly from 20 rad/s when $t = 0$ to 40 rad/s when $t = 35$ s. If the diameter of the wheel is 1 ft, determine the magnitudes of the normal and tangential components of acceleration of a point on the rim of the wheel when $t = 10$ s, and the total distance the point travels during the 35-s time period.

16–3. If the angular velocity of the drum is increased uniformly from 3 rev/min when $t = 0$ to 10 rev/min when $t = 4$ s, determine the magnitudes of the velocity and acceleration of point A on the drum when $t = 4$ s.

Prob. 16–3

*16–4. The angular position of a disk is defined by $\theta = (t + 4t^2)$ rad, where t is in minutes. Determine the number of revolutions, the angular velocity, and angular acceleration of the disk when $t = 90$ s.

16–5. The disk is originally rotating at $\omega_1 = 8$ rad/s. If it is subjected to a constant angular acceleration of $\alpha = 6$ rad/s^2, determine the magnitudes of the velocity and the n and t components of acceleration of point A at the instant $t = 3$ s.

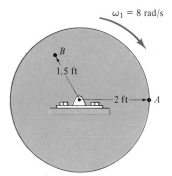

Probs. 16–5/16–6

16–6. The disk is originally rotating at $\omega_1 = 8$ rad/s. If it is subjected to an angular acceleration of $\alpha = 6$ rad/s^2, determine the magnitudes of the velocity and the n and t components of acceleration of point B just after the wheel undergoes 2 revolutions.

16–7. The gear A on the drive shaft of the outboard motor has a radius $r_A = 0.5$ in. and the meshed pinion gear B on the propeller shaft has a radius $r_B = 1.2$ in. Determine the angular velocity of the propeller in $t = 1.5$ s, if the drive shaft rotates with an angular acceleration $\alpha = (400t^3)$ rad/s^2, where t is in seconds. The propeller is originally at rest and the motor frame does not move.

Probs. 16–7/16–8

*16–8. For the outboard motor in Prob. 16–7, determine the magnitudes of the velocity and acceleration of a point P located on the tip of the propeller at the instant $t = 0.75$ s.

16–9. A textile mill uses the belt-and-pulley arrangement shown to transmit power. If an electric motor turns pulley A at 5 rad/s, compute the angular velocities of pulleys B and C. The hub at D is rigidly *connected* to C and turns with it.

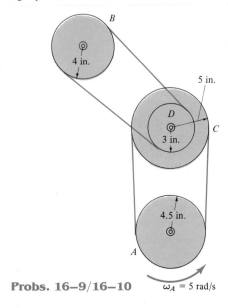

Probs. 16–9/16–10 $\omega_A = 5$ rad/s

16–10. A textile mill uses the belt-and-pulley arrangement shown to transmit power. When $t = 0$ an electric motor is turning pulley A with an angular velocity of $\omega_A = 5$ rad/s. If this pulley is subjected to a constant angular acceleration of 2 rad/s^2, determine the angular velocity of pulley B after B turns 6 revolutions. The hub at D is rigidly *connected* to pulley C and turns with it.

16–11. If disk A has an initial angular velocity of $\omega_1 = 6$ rad/s and a constant angular acceleration of $\alpha_A = 3$ rad/s^2, determine the magnitudes of the velocity and acceleration of block B in 2 s.

Probs. 16–11/16–12

*16–12. Solve Prob. 16–11 assuming that a motor gives the disk A an angular acceleration of $\alpha_A = (0.6t^2 + 0.75)$ rad/s^2 and initial angular velocity of $\omega_1 = 6$ rad/s, where t is in seconds.

16–13. A belt is wrapped around the inner hubs of pulleys A and B as shown. Each hub is fixed to its respective pulley and turns with it. If A has a constant angular acceleration $\alpha_A = 6$ rad/s^2, determine the velocity of block C in 3 s if it starts from rest. How high does it rise in 3 s?

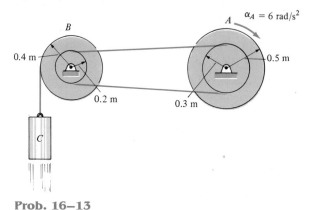

Prob. 16–13

16–14. A motor gives gear A an angular acceleration of $\alpha_A = (0.25\theta^3 + 0.5)$ rad/s^2, where θ is in radians. If this gear is initially turning at $(\omega_A)_1 = 20$ rad/s, determine the angular velocity of gear B after gear A undergoes an angular displacement of 10 rev.

Prob. 16–14

16–15. Determine the distance the load W is lifted in $t = 5$ s using the hoist. The shaft of the motor M turns with an angular velocity $\omega = 100(4 + t)$ rad/s, where t is in seconds.

Prob. 16–15

*16–16. Due to an increase in power, the motor M rotates the shaft A with an angular acceleration $\alpha = (0.06\theta^2)$ rad/s^2, where θ is in radians. If the shaft is initially turning at $\omega_1 = 50$ rad/s, determine the angular velocity of gear B after the shaft undergoes an angular displacement $\Delta\theta = 10$ rev.

Prob. 16–16

16–17. The angular acceleration of a wheel having a diameter of 2 ft is defined by $\alpha = (3 + 0.1\theta)$ rad/s^2, where θ is in radians. Determine the magnitudes of the velocity and acceleration of a point P on its rim when $\theta = 2$ rad. How much time is needed to reach this angular position? Originally, $\theta = 0$, and $\omega = 0$ when $t = 0$.

Prob. 16–17

16–18. Just after the fan is turned on, the motor gives the blade an angular acceleration $\alpha = (20e^{-0.6t})$ rad/s^2, where t is in seconds. Determine the speed of the tip P of one of the blades when $t = 3$ s. How many revolutions has the blade turned in 3 s? When $t = 0$ the blade is at rest.

Prob. 16–18

16–19. The rope of diameter d is wrapped around the tapered drum which has the dimensions shown. If the drum is rotating at a constant rate of ω, determine the upward acceleration of the block. Neglect the small horizontal displacement of the block.

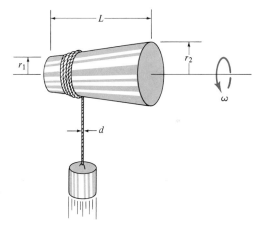

Prob. 16–19

245

*16.4 Absolute General Plane Motion Analysis

A body subjected to *general plane motion* undergoes a *simultaneous* translation and rotation. If the body is represented by a thin slab, the slab translates in the plane and rotates about an axis perpendicular to the plane. The analysis of this motion can be specified by knowing *both* the angular rotation of a line fixed in the body and the rectilinear motion of a point in the body. One way to define these motions is to use a position coordinate s to specify the location of the point along its path and an angular position coordinate θ to specify the orientation of the line. The two coordinates are then related using the geometry of the problem. By *direct application* of the time-differential equations $v = ds/dt$, $a = dv/dt$, $\omega = d\theta/dt$, and $\alpha = d\omega/dt$, the *motion* of the point and the *angular motion* of the line can then be related. In some cases, this procedure may also be used to relate the motions of one body to those of a connected body.

PROCEDURE FOR ANALYSIS

The following procedure provides a method for relating the velocity and acceleration of a point P undergoing rectilinear motion to the angular velocity and angular acceleration of a line contained in a body.

Position Coordinate Equation. Locate point P using a position coordinate s, which is measured from a *fixed origin* and is *directed along the straight-line path of motion* of point P. Also measure from a fixed reference line the angular position θ of a line lying in the body. From the dimensions of the body, relate s to θ, $s = f(\theta)$, using geometry and/or trigonometry.

Time Derivatives. Take the first derivative of $s = f(\theta)$ with respect to time to get a relationship between v and ω. Take the second time derivative to get a relationship between a and α.

This procedure is illustrated in the following example problems.

Example 16–3

At a given instant, the cylinder of radius r, shown in Fig. 16–7, has an angular velocity $\boldsymbol{\omega}$ and angular acceleration $\boldsymbol{\alpha}$. Determine the velocity and acceleration of its center G if it rolls without slipping.

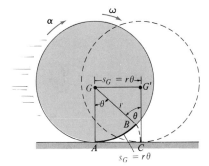

Fig. 16–7

SOLUTION

Position-Coordinate Equation. By inspection, point G moves *horizontally* to the right from G to G' as the cylinder rolls, Fig. 16–7. Consequently its new location G' will be specified by the *horizontal* position coordinate s_G, which is measured from the original position (G) of the cylinder's center. Notice also, that as the cylinder rolls (without slipping), points on its surface contact the ground such that the arc length AB of contact must be equal to the distance s_G. Consequently, motion from the colored to the dashed position requires the radial line GB to rotate θ to the position $G'C$. Since the arc $AB = r\theta$, then G travels a distance

$$s_G = r\theta$$

Time Derivatives. Taking successive time derivatives of this equation, realizing that r is constant, $\omega = d\theta/dt$, and $\alpha = d\omega/dt$, gives the necessary relationships

$$s_G = r\theta$$
$$v_G = r\omega \qquad\qquad \textit{Ans.}$$
$$a_G = r\alpha \qquad\qquad \textit{Ans.}$$

Example 16–4

The end of rod R shown in Fig. 16–8 maintains contact with the cam by means of a spring. If the cam rotates about an axis through point O with an angular acceleration α and angular velocity ω, compute the velocity and acceleration of R when the cam is in the arbitrary position θ.

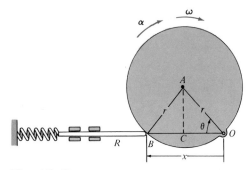

Fig. 16–8

SOLUTION

Position Coordinate Equation. Coordinates x and θ are chosen for the analysis in order to express the motion of the cam, which is defined by the angular motion of line OA, i.e., $\omega = d\theta/dt$ and the rectilinear motion of the rod (or horizontal component of the motion of point B), i.e., $v = dx/dt$. These coordinates are measured from the fixed point O and may be related to each other using trigonometry. Since $OC = CB = r \cos \theta$, Fig. 16–8, then

$$x = 2r \cos \theta$$

Time Derivatives. Using the chain rule of calculus, we have

$$\frac{dx}{dt} = -2r \sin \theta \frac{d\theta}{dt}$$

$$v = -2r\, \omega \sin \theta \qquad\qquad Ans.$$

$$\frac{dv}{dt} = -2r\left(\frac{d\omega}{dt}\right) \sin \theta - 2r\omega\left(\cos \theta \frac{d\theta}{dt}\right)$$

$$a = -2r\, \alpha \sin \theta - 2r\omega^2 \cos \theta \qquad\qquad Ans.$$

PROBLEMS

*16–20. The *Scotch yoke* is used to convert the constant circular motion of crank *OA* into translating motion of rod *BC*. If *OA* is rotating with a constant angular velocity $\omega = 5$ rad/s, determine the velocity and acceleration of *BC* for any angle θ of the crank.

Probs. 16–20/16–21

16–21. The *Scotch yoke* is used to convert the constant circular motion of crank *OA* into translating motion of rod *BC*. When $\theta = 30°$, *OA* is rotating with an angular velocity $\omega = 5$ rad/s and angular acceleration $\alpha = 2$ rad/s^2. Determine the velocity and acceleration of *BC* at this instant.

16–22. At the instant shown, $\theta = 60°$, and rod *AB* is subjected to a deceleration of 16 m/s^2 when the velocity is 10 m/s. Determine the angular velocity and angular acceleration of link *CD* at this instant.

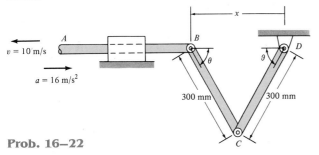

Prob. 16–22

16–23. Compute the angular velocity of rod *CD* at the instant $\theta = 30°$. Rod *AB* moves to the left at a constant rate $v_{AB} = 5$ m/s.

Probs. 16–23/16–24

*16–24. Compute the angular acceleration of rod *CD* at the instant $\theta = 30°$. Rod *AB* has zero velocity, i.e., $v_{AB} = 0$, and an acceleration of $a_{AB} = 2$ rad/s^2 to the right when $\theta = 30°$.

16–25. Determine the velocity and acceleration of the plate *C* as a function of θ, if the circular cam is rotating about the fixed point *O* with a constant angular velocity $\omega = 4$ rad/s.

Probs. 16–25/16–26

16–26. Determine the velocity and acceleration of the plate *C* at the instant $\theta = 30°$, if at this instant the circular cam is rotating about the fixed point *O* with an angular velocity $\omega = 4$ rad/s and an angular acceleration $\alpha = 2$ rad/s^2.

16–27. The bar remains in contact with the floor and with point *A*. If point *B* moves to the right with a constant velocity of v_B, determine the angular velocity and angular acceleration of the bar as a function of *x*.

Prob. 16–27

249

*16–28. The pins at A and B are confined to move in the vertical and horizontal tracks. If the slotted arm is causing A to move downward at $\mathbf{v}_A$, determine the velocity $\mathbf{v}_B$ at the instant shown.

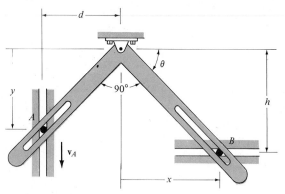

Prob. 16–28

16–29. The end A of the bar is moving downward along the slotted guide with a constant velocity $\mathbf{v}_A$. Determine the angular velocity ω and angular acceleration α of the bar as a function of its position y.

Prob. 16–29

16–30. The block moves to the left with a constant velocity $\mathbf{v}_0$. Determine the angular velocity and angular acceleration of the bar as a function of x.

Prob. 16–30

16–31. The bar is confined to move along the vertical and inclined planes. If the velocity of the roller at A is $v_A = 6$ ft/s when $\theta = 45°$, determine the bar's angular velocity and the velocity of roller B at this instant.

Probs. 16–31/16–32

*16–32. The bar is confined to move along the vertical and inclined planes. If the velocity of the roller at A is $v_A = 6$ ft/s, when $\theta = 45°$, determine the bar's angular acceleration and the acceleration of roller B at this instant.

16–33. The end A of the bar is moving to the left with a constant velocity $\mathbf{v}_A$. Determine the angular velocity ω and angular acceleration α of the bar as a function of its position x.

Probs. 16–33/16–34

16–34. The end A of the bar is moving to the left with an acceleration $\mathbf{a}_A$ and at the instant shown it has a velocity $\mathbf{v}_A$. Determine the angular velocity ω and angular acceleration α of the bar as a function of its position x.

16–35. Wheel *A* rolls without slipping over the surface of the *fixed* cylinder *B*. Determine the angular velocity of *A* if its center *C* has a speed $v_C = 5$ m/s. How many revolutions will *A* have made about its center just after link *DC* completes one revolution?

Prob. 16–35

16–37. The slotted yoke is pinned at *A* while end *B* is used to move the ram *R* horizontally. If the disk rotates with a constant angular velocity ω, determine the velocity and acceleration of the ram. The crank pin *C* is fixed to the disk and turns with it.

Prob. 16–37

*16–36. Bar *AB* rotates uniformly about the fixed pin *A* with a constant angular velocity ω. Determine the velocity and acceleration of block *C*, at the instant $\theta = 60°$.

Prob. 16–36

16.5 Relative-Motion Analysis: Velocity

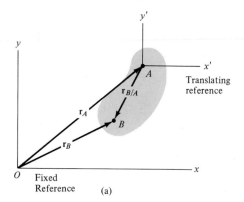

Fixed
Reference   (a)

It was pointed out in Sec. 16.1 that general plane motion of a rigid body consists of a *combination* of translation and rotation. Oftentimes it is convenient to view these "component" motions *separately* using a *relative-motion analysis.* Two sets of coordinate axes will be used to do this. The *x, y* coordinate system is fixed and will be used to measure the *absolute* positions, velocities, and accelerations of two points *A* and *B* in the body, Fig. 16–9a. The origin of the *x', y'* coordinate system will be attached to the selected "base point" *A,* which generally has a *known* motion. The axes of the *x', y'* coordinate system are *not fixed to the body;* rather they will be allowed to *translate* with respect to the fixed frame.

Position

The position vector $\mathbf{r}_A$ in Fig. 16–9a specifies the location of the "base point" *A,* whereas the relative-position vector $\mathbf{r}_{B/A}$ locates point *B* in the body with respect to *A.* By vector addition, the *position* of *B* can be determined from the equation

$$\mathbf{r}_B = \mathbf{r}_A + \mathbf{r}_{B/A}$$

Time *t*   Time *t + dt*

General plane
motion

(b)

Displacement

During an instant of time *dt,* points *A* and *B* undergo displacements $d\mathbf{r}_A$ and $d\mathbf{r}_B$ as shown in Fig. 16–9b. As stated above, here we will consider the motion by its component parts. For example, the *entire body* first *translates* by an amount $d\mathbf{r}_A$ so that *A,* the base point, moves to its *final position* and *B* moves to *B',* Fig. 16–9c. The body is then rotated by an amount $d\theta$ about *A,* so that *B'* undergoes a *relative displacement* $d\mathbf{r}_{B/A}$ and thus moves to its final position *B,* Fig. 16–9d. Note that since the body is *rigid,* $\mathbf{r}_{B/A}$ has a fixed magnitude and thus $d\mathbf{r}_{B/A}$ accounts only for a change in the *direction* of $\mathbf{r}_{B/A}$. Due to the rotation about *A* the magnitude of the relative displacement is $dr_{B/A} = r_{B/A}d\theta$ and the displacement of *B* in Fig. 16–9b is thus

$$d\mathbf{r}_B = d\mathbf{r}_A + d\mathbf{r}_{B/A}$$

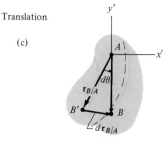

Translation

(c)

Rotation about the
base point *A*

(d)

Velocity

To determine the relationship between the velocities of points *A* and *B,* it is necessary to take the time derivative of the position equation, or simply divide the displacement equation by *dt.* This yields

$$\frac{d\mathbf{r}_B}{dt} = \frac{d\mathbf{r}_A}{dt} + \frac{d\mathbf{r}_{B/A}}{dt}$$

Fig. 16–9(a–d)

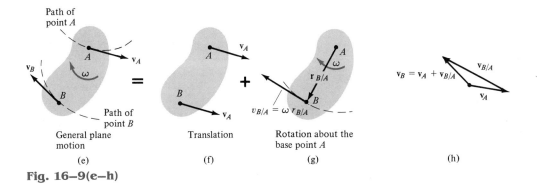

Path of point A

v_B

A ω v_A

B

Path of point B

General plane motion

(e)

$=$

A v_A

B

v_A

Translation

(f)

$+$

A ω

$r_{B/A}$

B

$v_{B/A} = \omega\, r_{B/A}$

Rotation about the base point A

(g)

$v_B = v_A + v_{B/A}$

$v_{B/A}$

v_A

(h)

Fig. 16–9(e–h)

The terms $d\mathbf{r}_B/dt = \mathbf{v}_B$ and $d\mathbf{r}_A/dt = \mathbf{v}_A$ are measured from the fixed x, y axes and represent the *absolute velocities* of points A and B, respectively. The magnitude of the third term is $d(r_{B/A}\theta)/dt = r_{B/A}\dot{\theta} = r_{B/A}\omega$, where ω is the angular velocity of the body at the instant considered. Note that this rotation is *absolute* since it is the *same* when measured from either the fixed x, y or translating x', y' axes. We will denote this third term as the *relative velocity* $\mathbf{v}_{B/A}$, since it represents the velocity of B with respect to A as measured by an observer fixed to the translating x', y' axes. Since the body is rigid, it is important to realize that this observer sees point B move along a *circular arc* that has a radius of curvature $r_{B/A}$. In other words, *the body appears to move as if it were rotating with an angular velocity* $\boldsymbol{\omega}$ *about the z' axis passing through A*. Consequently, $\mathbf{v}_{B/A}$ has a magnitude of $v_{B/A} = \omega r_{B/A}$ and a *direction* which is perpendicular to $r_{B/A}$. We therefore have

$$\boxed{\mathbf{v}_B = \mathbf{v}_A + \mathbf{v}_{B/A}} \qquad (16\text{--}14)$$

where

$\mathbf{v}_B$ = velocity of point B

$\mathbf{v}_A$ = velocity of the base point A

$\mathbf{v}_{B/A}$ = relative velocity of "B with respect to A."

  Since the relative motion is *circular*, the *magnitude* of $\mathbf{v}_{B/A}$ is $v_{B/A} = \omega r_{B/A}$ and the *direction* is perpendicular to $\mathbf{r}_{B/A}$.

Each of the three terms in Eq. 16–14 is represented graphically on the *kinematic diagrams* in Fig. 16–9e, f, and g. Here it is seen that at a given instant the velocity of B, Fig. 16–9e, is determined by considering the entire body to translate with a velocity of $\mathbf{v}_A$, Fig. 16–9f, and then rotate about the base point A with an angular velocity $\boldsymbol{\omega}$, Fig. 16–9g. Vector addition of these two effects, applied to B, yields $\mathbf{v}_B$, as shown in Fig. 16–9h.

  Since the relative velocity $\mathbf{v}_{B/A}$ represents the effect of *circular motion*, observed from translating axes having their origin at the base point A, this term can be expressed by the cross product $\mathbf{v}_{B/A} = \boldsymbol{\omega} \times \mathbf{r}_{B/A}$, Eq. 16–9.

Hence, for application, we can also write Eq. 16–14 as

$$\boxed{\mathbf{v}_B = \mathbf{v}_A + \boldsymbol{\omega} \times \mathbf{r}_{B/A}}$$ (16–15)

where

$$\mathbf{v}_B = \text{velocity of } B$$
$$\mathbf{v}_A = \text{velocity of the base point } A$$
$$\boldsymbol{\omega} = \text{angular velocity of the body}$$
$$\mathbf{r}_{B/A} = \text{relative-position vector drawn from } A \text{ to } B$$

To apply this equation, all of these terms must be represented in Cartesian vector form.

The velocity equation 16–14 or 16–15 may be used in a practical manner to study the motion of a rigid body which is either pin-connected to or in contact with other moving bodies. To obtain the necessary data when applying this equation, points A and B should generally be selected at joints which are pin-connected, or at points in contact with adjacent bodies which have a *known motion*. For example, both points A and B on link AB, Fig. 16–10a, have circular paths of motion. Hence, at the instant shown, the magnitudes of their velocities are determined by the angular motion of the wheel and link BC so that $v_A = \omega_A r$ and $v_B = \omega_{BC} l$. The *directions* of $\mathbf{v}_A$ and $\mathbf{v}_B$ are always *tangent* to their paths of motion, Fig. 16–10b. In the case of the wheel in Fig. 16–11, which rolls without slipping, point A can be selected at the ground. Here A momentarily has zero velocity since the ground does not move. Furthermore, the center of the wheel, B, moves along a horizontal path so that $\mathbf{v}_B$ is horizontal.

(a)

Fig. 16–11

$v_A = \omega_A r$

$v_B = \omega_{BC} l$

Fig. 16–10

(b)

PROCEDURE FOR ANALYSIS

Equation 16–14 relates the velocities of any two points A and B located on the *same* rigid body. It can be applied either by using Cartesian vector analysis, in the form of Eq. 16–15, or by writing the x and y scalar component equations directly. For application, it is suggested that the following procedure be used.

Kinematic Diagram. Establish the directions of the fixed x, y coordinates, and draw a kinematic diagram of the body. Indicate on it the velocities $\mathbf{v}_A$, $\mathbf{v}_B$ of points A and B, the angular velocity $\boldsymbol{\omega}$, and the relative position vector $\mathbf{r}_{B/A}$. Normally the base point A is selected as a point having a known velocity. Identify the *two* unknowns on the diagram. If the magnitudes of $\mathbf{v}_A$, $\mathbf{v}_B$, or $\boldsymbol{\omega}$ are unknown, the sense of direction of these vectors may be assumed.

If the velocity equation is to be applied *without* using a Cartesian vector analysis, then the magnitude and direction of the relative velocity $\mathbf{v}_{B/A}$ must be established. This can be done by drawing a kinematic diagram such as shown in Fig. 16–9g. Since the body is considered to be ''pinned'' momentarily at the base point A, the *magnitude* of $\mathbf{v}_{B/A}$ is $v_{B/A} = \omega r_{B/A}$. The *direction* is established from the diagram, such that $\mathbf{v}_{B/A}$ acts perpendicular to $\mathbf{r}_{B/A}$ in accordance with the rotational motion $\boldsymbol{\omega}$ of the body.*

Velocity Equation. To apply Eq. 16–15, express the vectors in Cartesian vector form and substitute them into the equation. Evaluate the cross product and then equate the respective $\mathbf{i}$ and $\mathbf{j}$ components to obtain two scalar equations. To obtain these scalar equations *directly*, write Eq. 16–14 in symbolic form, $\mathbf{v}_B = \mathbf{v}_A + \mathbf{v}_{B/A}$, and underneath each of the terms represent the vectors by their magnitudes and directions. To do this, use the data tabulated on the kinematic diagrams. The scalar equations are determined from the x and y components of these vectors. Solve these equations for the two unknowns. If the solution yields a *negative* answer for an *unknown*, it indicates the sense of direction of the vector is opposite to that shown on the kinematic diagram.

The following example problems numerically illustrate both methods of application.

*The notation $\mathbf{v}_B = \mathbf{v}_A + \mathbf{v}_{B/A\text{(pin)}}$ may be helpful in recalling that A is ''pinned.''

Example 16–5

The link shown in Fig. 16–12a is guided by two blocks at A and B, which move in the fixed slots. If the velocity of A is 2 m/s downward, determine the velocity of B at the instant $\theta = 45°$.

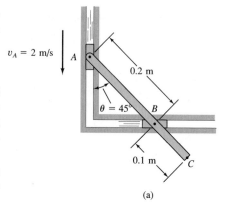

(a)

SOLUTION (VECTOR ANALYSIS)

Kinematic Diagram. Since points A and B are restricted to move along the fixed slots and $\mathbf{v}_A$ is directed downward, the velocity $\mathbf{v}_B$ must be directed horizontally to the right, Fig. 16–12b. This motion causes the link to rotate counterclockwise; that is, by the right-hand rule the angular velocity $\boldsymbol{\omega}$ is directed outward, perpendicular to the page (the plane of motion). Knowing the magnitude and direction of $\mathbf{v}_A$ and the lines of action of $\mathbf{v}_B$ and $\boldsymbol{\omega}$, it is possible to apply the velocity equation $\mathbf{v}_B = \mathbf{v}_A + \boldsymbol{\omega} \times \mathbf{r}_{B/A}$ to points A and B in order to solve for the two unknown magnitudes v_B and ω. Since $\mathbf{r}_{B/A}$ is needed, it is also shown in Fig. 16–12b.

Velocity Equation. Expressing each of the vectors in Fig. 16–12b in terms of their $\mathbf{i}$, $\mathbf{j}$, and $\mathbf{k}$ components, we have

$$\mathbf{v}_A = \{-2\mathbf{j}\} \text{ m/s}, \qquad \mathbf{v}_B = v_B\mathbf{i}, \qquad \boldsymbol{\omega} = \omega\mathbf{k}$$
$$\mathbf{r}_{B/A} = \{0.2 \sin 45°\mathbf{i} - 0.2 \cos 45°\mathbf{j}\} \text{ m}$$

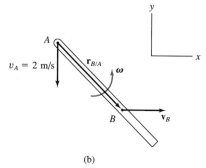

(b)

Fig. 16–12

Applying Eq. 16–15 to A, the base point, and B, we have

$$\mathbf{v}_B = \mathbf{v}_A + \boldsymbol{\omega} \times \mathbf{r}_{B/A}$$
$$v_B\mathbf{i} = -2\mathbf{j} + [\omega\mathbf{k} \times (0.2 \sin 45°\mathbf{i} - 0.2 \cos 45°\mathbf{j})]$$

or

$$v_B\mathbf{i} = -2\mathbf{j} + 0.2\omega \sin 45°\mathbf{j} + 0.2\omega \cos 45°\mathbf{i}$$

Equating the $\mathbf{i}$ and $\mathbf{j}$ components gives

$$v_B = 0.2\omega \cos 45°, \qquad 0 = -2 + 0.2\omega \sin 45°$$

Thus,

$$\omega = 14.1 \text{ rad/s} \text{↰}$$
$$v_B = 2 \text{ m/s} \rightarrow \qquad\qquad\qquad Ans.$$

Since both results are *positive*, the *directions* of $\mathbf{v}_B$ and $\boldsymbol{\omega}$ are indeed *correct* as shown in Fig. 16–12b. It should be emphasized that these results are *valid only* at the instant $\theta = 45°$. A recalculation for $\theta = 44°$ yields $v_B = 2.07$ m/s and $\omega = 14.4$ rad/s; whereas when $\theta = 46°$, $v_B = 1.93$ m/s and $\omega = 13.9$ rad/s, etc.

Having determined $\boldsymbol{\omega}$, as an exercise apply Eq. 16–15 to points A and C and show that $v_C = 3.16$ m/s, directed at $\theta = 18.4°$ up from the horizontal.

Example 16–6

The cylinder shown in Fig. 16–13a rolls freely on the surface of a conveyor belt which is moving at 2 ft/s. Assuming that no slipping occurs between the cylinder and the belt, determine the velocity of point A. The cylinder has a clockwise angular velocity $\omega = 15$ rad/s at the instant shown.

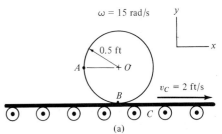

(a)

SOLUTION I (VECTOR ANALYSIS)

Kinematic Diagram. Since no slipping occurs, point B on the cylinder has the same velocity as the conveyor, Fig. 16–13b. Knowing the angular velocity of the cylinder, we will apply the velocity equation to B, the base point, and A to determine the x and y components of $\mathbf{v}_A$, i.e., $\mathbf{v}_A = \mathbf{v}_B + \boldsymbol{\omega} \times \mathbf{r}_{A/B}$.

Velocity Equation

$$\mathbf{v}_A = \mathbf{v}_B + \boldsymbol{\omega} \times \mathbf{r}_{A/B}$$
$$(v_A)_x\mathbf{i} + (v_A)_y\mathbf{j} = 2\mathbf{i} + (-15\mathbf{k}) \times (-0.5\mathbf{i} + 0.5\mathbf{j})$$
$$(v_A)_x\mathbf{i} + (v_A)_y\mathbf{j} = 2\mathbf{i} + 7.50\mathbf{j} + 7.50\mathbf{i}$$

(b)

so that

$$(v_A)_x = 2 + 7.50 = 9.50 \text{ ft/s} \qquad (1)$$
$$(v_A)_y = 7.50 \text{ ft/s} \qquad (2)$$

Thus,

$$v_A = \sqrt{(9.50)^2 + (7.50)^2} = 12.1 \text{ ft/s} \qquad \textit{Ans.}$$

$$\theta = \tan^{-1}\frac{7.50}{9.50} = 38.3° \qquad \textit{Ans.}$$

SOLUTION II (SCALAR COMPONENTS)

As an alternative procedure, the scalar components of $\mathbf{v}_A = \mathbf{v}_B + \mathbf{v}_{A/B}$ can be obtained directly. From the kinematic diagram showing the relative "circular" motion $\mathbf{v}_{A/B}$, Fig. 16–13c, we have

(c)

Fig. 16–13

$$v_{A/B} = \omega r_{A/B} = (15 \text{ rad/s})\left(\frac{0.5 \text{ ft}}{\cos 45°}\right) = 10.6 \text{ ft/s} \quad \angle^{45°}$$

Thus,

$$v_A = v_B + v_{A/B}$$
$$\left[\begin{matrix}(v_A)_x \\ \rightarrow\end{matrix}\right] + \left[\begin{matrix}(v_A)_y \\ \uparrow\end{matrix}\right] = \left[\begin{matrix}2 \text{ ft/s} \\ \rightarrow\end{matrix}\right] + \left[\begin{matrix}10.6 \text{ ft/s} \\ \angle^{45°}\end{matrix}\right]$$

Equating the x and y components gives the same results as before, namely,

$$(\xrightarrow{+}) \qquad (v_A)_x = 2 + 10.6 \cos 45° = 9.50 \text{ ft/s}$$
$$(+\uparrow) \qquad (v_A)_y = 0 + 10.6 \sin 45° = 7.50 \text{ ft/s}$$

257

Example 16–7

The collar C in Fig. 16–14a is moving downward with a velocity of 2 m/s. Determine the angular velocity of CB at this instant.

SOLUTION I (VECTOR ANALYSIS)
Kinematic Diagram. The downward motion of C causes B to move to the right. Also, CB rotates counterclockwise. It is subjected to general plane motion, Fig. 16–14b.

Velocity Equation

$$\mathbf{v}_B = \mathbf{v}_C + \boldsymbol{\omega}_{CB} \times \mathbf{r}_{B/C}$$
$$v_B\mathbf{i} = -2\mathbf{j} + \omega_{CB}\mathbf{k} \times (0.2\mathbf{i} - 0.2\mathbf{j})$$
$$v_B\mathbf{i} = -2\mathbf{j} + 0.2\omega_{CB}\mathbf{j} + 0.2\omega_{CB}\mathbf{i}$$

Equating the $\mathbf{i}$ and $\mathbf{j}$ components yields

$$v_B = 0.2\omega_{CB} \qquad\qquad (1)$$
$$0 = -2 + 0.2\omega_{CB} \qquad\qquad (2)$$

So that

$$\omega_{CB} = 10 \text{ rad/s} \; \uparrow \qquad\qquad \textit{Ans.}$$
$$v_B = 2 \text{ m/s} \rightarrow$$

SOLUTION II (SCALAR COMPONENTS)
The scalar component equations of $\mathbf{v}_B = \mathbf{v}_C + \mathbf{v}_{B/C}$ can be obtained directly. The kinematic diagram in Fig. 16–14c shows the relative "circular" motion $\mathbf{v}_{B/C}$. We have

$$\mathbf{v}_B = \mathbf{v}_C + \mathbf{v}_{B/C}$$
$$\begin{bmatrix} v_B \\ \rightarrow \end{bmatrix} = \begin{bmatrix} 2 \text{ m/s} \\ \downarrow \end{bmatrix} + \begin{bmatrix} \omega_{CB}(0.2 \sqrt{2} \text{ m}) \\ \nearrow 45° \end{bmatrix}$$

Resolving these vectors in the x and y directions yields

$(\overset{+}{\rightarrow})$ $\qquad\qquad v_B = 0 + \omega_{CB}(0.2 \sqrt{2} \cos 45°)$
$(+\uparrow)$ $\qquad\qquad 0 = -2 + \omega_{CB}(0.2 \sqrt{2} \sin 45°)$

which is the same as Eqs. (1) and (2).

(a)

(b)

(c)

Fig. 16–14

Example 16–8

The bar AB of the linkage shown in Fig. 16–15a has a clockwise angular velocity of 30 rad/s when $\theta = 60°$. Compute the angular velocities of members BC and DC at this instant.

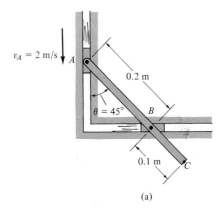

(a)

SOLUTION (VECTOR ANALYSIS)
Kinematic Diagram. The velocities of points B and C are defined by the rotation of links AB and DC about their fixed axes. The relative-position vectors and the angular velocity of each link are shown on the kinematic diagram in Fig. 16–15b. Here we will apply a Cartesian vector analysis to each link.

Velocity Equation. Expressing each vector in Cartesian vector form, we have

$$\boldsymbol{\omega}_{AB} = \{-30\mathbf{k}\}\ \text{rad/s}, \qquad \boldsymbol{\omega}_{BC} = \omega_{BC}\mathbf{k}, \qquad \boldsymbol{\omega}_{DC} = \omega_{DC}\mathbf{k}$$

$$\mathbf{r}_{B/A} = \{0.2 \cos 60°\mathbf{i} + 0.2 \sin 60°\mathbf{j}\}\ \text{m}, \qquad \mathbf{r}_{C/B} = \{0.2\mathbf{i}\}\ \text{m},$$

$$\mathbf{r}_{C/D} = \{-0.1\mathbf{j}\}\ \text{m}$$

For link AB (rotation about a fixed axis):

$$\begin{aligned}
\mathbf{v}_B &= \boldsymbol{\omega}_{AB} \times \mathbf{r}_{B/A} \\
&= (-30\mathbf{k}) \times (0.2 \cos 60°\mathbf{i} + 0.2 \sin 60°\mathbf{j}) \\
&= \{5.20\mathbf{i} - 3.0\mathbf{j}\}\ \text{m/s}
\end{aligned}$$

Fig. 16–15

(b)

For link BC (general plane motion):

$$\begin{aligned}
\mathbf{v}_C &= \mathbf{v}_B + \boldsymbol{\omega}_{BC} \times \mathbf{r}_{C/B} \\
&= 5.20\mathbf{i} - 3.0\mathbf{j} + (\omega_{BC}\mathbf{k}) \times (0.2\mathbf{i}) \\
&= 5.20\mathbf{i} + (0.2\omega_{BC} - 3.0)\mathbf{j}
\end{aligned}$$

For link DC (rotation about a fixed axis):

$$\mathbf{v}_C = \boldsymbol{\omega}_{DC} \times \mathbf{r}_{C/D}$$

$$5.20\mathbf{i} + (0.2\omega_{BC} - 3.0)\mathbf{j} = (\omega_{DC}\mathbf{k}) \times (-0.1\mathbf{j})$$

$$5.20\mathbf{i} + (0.2\omega_{BC} - 3.0)\mathbf{j} = 0.1\omega_{DC}\mathbf{i}$$

Equating the respective $\mathbf{i}$ and $\mathbf{j}$ components and solving, we have

$$\omega_{DC} = 52\ \text{rad/s} \curvearrowleft \qquad\qquad\textit{Ans.}$$

$$\omega_{BC} = 15\ \text{rad/s} \curvearrowleft \qquad\qquad\textit{Ans.}$$

Note that by inspection, Fig. 16–15a, $v_B = (0.2)(30) = 6$ m/s, $\searrow 30°$ and $\mathbf{v}_C$ is directed to the right. As an exercise, obtain ω_{BC} by applying $\mathbf{v}_C = \mathbf{v}_B + \mathbf{v}_{C/B}$ using scalar components.

PROBLEMS

16–38. The crankshaft AB is rotating at 500 rad/s about a fixed axis passing through A. Determine the speed of the piston P at the instant it is in the position shown.

Prob. 16–38

16–39. A bowling ball is cast on the "alley" with a backspin of $\omega = 10$ rad/s while its center O has a forward velocity of $v_O = 8$ m/s. Determine the velocity of the contact point A touching the alley.

Prob. 16–39

*16–40. The cylinder B rolls on the *fixed cylinder A* without slipping. If the connected bar CD is rotating with an angular velocity of $\omega_{CD} = 5$ rad/s, determine the angular velocity of cylinder B. Point C is a fixed point.

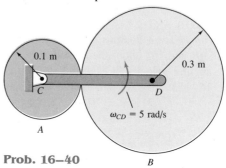

Prob. 16–40

16–41. At the instant shown the boomerang has an angular velocity of $\omega = 4$ rad/s, and its mass center G has a velocity of $v_G = 6$ in./s. Determine the velocity of point A at this instant.

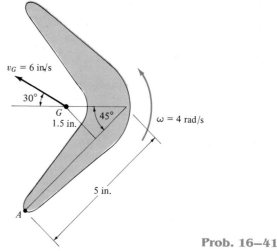

Prob. 16–41

16–42. The wheel is rotating with an angular velocity $\omega = 8$ rad/s. Determine the velocity of the collar A at this instant.

Prob. 16–42

16–43. If the angular velocity of link *AB* is $\omega_{AB} = 3$ rad/s, determine the velocity of the block at *C* and the angular velocity of the connecting link *CB* at the instant shown.

Prob. 16–43

*16–44. Part of an automatic transmission consists of a ring gear *R*, three equal planet gears *P*, the sun gear *S*, and the planet carrier *C*, which is shaded. If the sun gear is *fixed* and the ring gear is rotating at $\omega_R = 5$ rad/s, determine the angular velocity of the *planet carrier*. Note that *C* is pin-connected to the center of each of the planet gears.

Prob. 16–44

16–45. When the crank on the Chinese windlass is turning, the rope on shaft *A* unwinds while that on shaft *B* winds up. Determine the speed at which the block lowers if the crank is turning with an angular velocity $\omega = 4$ rad/s. What is the angular velocity of the pulley at *C*? The rope segments on each side of the pulley are both parallel and vertical, and the rope does not slip on the pulley.

Prob. 16–45

16–46. If the end of the cord is pulled downward with a speed $v_C = 120$ mm/s, determine the angular velocities of pulleys *A* and *B* and the speed of block *D*. Assume that the cord does not slip on the pulleys.

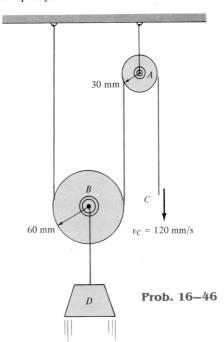

Prob. 16–46

261

16–47. If rod AB is rotating with an angular velocity $\omega_{AB} = 12$ rad/s, determine the angular velocity of rods BC and CD at the instant shown.

Prob. 16–47

16–50. If disk A has an angular velocity $\omega_A = 4$ rad/s, determine the angular velocity of disk D at the instant shown.

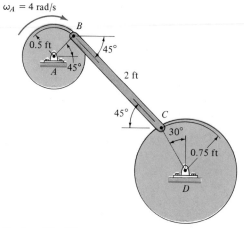

Prob. 16–50

*16–48. If the slider block A is moving downward at $v_A = 4$ m/s, determine the velocity of blocks B and C at the instant shown.

Probs. 16–48/16–49

16–49. If the slider block A is moving downward at $v_A = 4$ m/s, determine the velocity of point E at the instant shown.

16–51. If the link AB is rotating about the pin at A with an angular velocity of $\omega_{AB} = 5$ rad/s, determine the velocity of blocks C and E at the instant shown.

Prob. 16–51

16.6 Instantaneous Center of Zero Velocity

The velocity of any point B located in a rigid body can be obtained in a simple way if one chooses the base point A to be a point that has *zero velocity* at the instant considered. In this case, $\mathbf{v}_A = \mathbf{0}$, and therefore the velocity equation, $\mathbf{v}_B = \mathbf{v}_A + \boldsymbol{\omega} \times \mathbf{r}_{B/A}$, becomes $\mathbf{v}_B = \boldsymbol{\omega} \times \mathbf{r}_{B/A}$. For a body having general plane motion, point A so chosen is called the *instantaneous center of zero velocity* (*IC*) and it lies on the *instantaneous axis of zero velocity* (*IA*), Fig. 16–17a. The *IA* is always perpendicular to the plane used to represent the motion, and the intersection of the *IA* with this plane defines the location of the *IC*. Since point A is coincident with the *IC*, $\mathbf{v}_B = \boldsymbol{\omega} \times \mathbf{r}_{B/IC}$ and so point B moves momentarily about the *IC* in a *circular path*; in other words, the body appears to rotate about the *IA*. If the relative-position vector $\mathbf{r}_{B/IC}$ is established from the *IC* to point B as shown, then the *magnitude* of $\mathbf{v}_B$ is simply $v_B = \omega r_{B/IC}$, where ω is the angular velocity of the body. Due to the circular motion, the *direction* of $\mathbf{v}_B$ must always be *perpendicular* to $\mathbf{r}_{B/IC}$. For example, consider a wheel which rolls *without slipping*, Fig. 16–16a. In this case the point of *contact* with the ground has *zero velocity*. Hence this point represents the *IC* for the wheel, Fig. 16–16b. If it is imagined that the wheel is momentarily pinned at this point, the velocities of points A, B, O, and so on, can be found using $v = \omega r$. Here the radial distances $r_{A/IC}$, $r_{B/IC}$, and $r_{O/IC}$, shown in Fig. 16–16b, must be determined from the geometry of the wheel.

When a body is subjected to general plane motion, the point determined as the instantaneous center of zero velocity for the body can only be used for an *instant of time*. Since the body changes its position from one instant to the next, then for each position of the body a unique instantaneous center must be determined. The locus of points which define the *IC* during the body's motion is called a *centrode*, Fig. 16–17a. Thus, each point on the centrode acts as the *IC* for the body only for an instant of time.

Although the *IC* may be conveniently used to determine the velocity of any point in a body, it generally *does not have zero acceleration* and therefore it *should not* be used for finding the accelerations of points in a body.

(a)

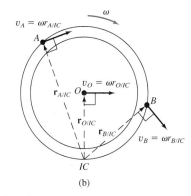

(b)

Fig. 16–16

Location of the IC

If the location of the *IC* is unknown, it may be determined by using the fact that the *relative-position vector* extending from the *IC* to a point is *always perpendicular* to the *velocity* of the point. Several possibilities exist:

1. *Given the velocity of a point in the body and the angular velocity of the body*. In this case if $\mathbf{v}_B$ and $\boldsymbol{\omega}$ are known, the *IC* is located along the line drawn perpendicular to $\mathbf{v}_B$ at B, such that the distance from B to the *IC* is $r_{B/IC} = v_B/\omega$, Fig. 16–17a. Note that the *IC* lies on that side of B which causes rotation about the *IC* which is consistent with the direction of motion caused by $\boldsymbol{\omega}$ and $\mathbf{v}_B$.

(a)

Fig. 16–17a

(b)

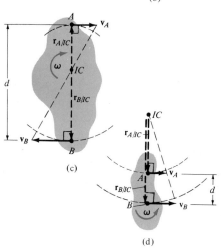

(c)

(d)

Fig. 16–17(b–d)

2. *Given the lines of action of two nonparallel velocities.* Consider the body in Fig. 16–17b, where the lines of action of the velocities $\mathbf{v}_A$ and $\mathbf{v}_B$ are known. From each of these lines construct at points A and B perpendicular line segments which then define the lines of action of $\mathbf{r}_{A/IC}$ and $\mathbf{r}_{B/IC}$, respectively. Extending these perpendiculars to their *point of intersection* as shown locates the IC at the instant considered. The magnitudes of $\mathbf{r}_{A/IC}$ and $\mathbf{r}_{B/IC}$ are generally determined from the geometry of the body and trigonometry. Furthermore, if the magnitude and sense of $\mathbf{v}_A$ are known, then the angular velocity of the body is determined from $\omega = v_A/r_{A/IC}$. Once computed, $\boldsymbol{\omega}$ can then be used to determine $v_B = \omega r_{B/IC}$.

3. *Given the magnitude and direction of two parallel velocities.* When the velocities of points A and B are parallel and have known magnitudes v_A and v_B, then the location of the IC is determined by proportional triangles. Examples are shown in Fig. 16–17c and d. In both cases $v_A = \omega r_{A/IC}$ and $v_B = \omega r_{B/IC}$ so that, $r_{A/IC} = r_{B/IC}(v_A/v_B)$. If d is the known distance between points A and B, then in Fig. 16–17c, $r_{A/IC} = d - r_{B/IC}$ and in Fig. 16–17d, $r_{A/IC} = r_{B/IC} - d$. Substituting each of these equations into the previous equation yields a solution for $r_{B/IC}$. As a special case, note that if the body is *translating*, $\mathbf{v}_A = \mathbf{v}_B$ and the IC would be located at infinity, in which case $r_{A/IC} = r_{B/IC} \rightarrow \infty$. This being the case, $\omega = v_A/\infty = v_B/\infty = 0$, as expected.

PROCEDURE FOR ANALYSIS

The velocity of a point in a body which is subjected to general plane motion can be determined with reference to its instantaneous center of zero velocity, provided the location of the IC is first established. This is done by using one of the methods described above. As shown on the kinematic diagram in Fig. 16–18, the body is imagined as "extended and pinned" at the IC such that it rotates about this pin with its instantaneous angular velocity $\boldsymbol{\omega}$. The *magnitude* of velocity for the arbitrary points A, B, and C in the body can then be determined by using the equation $v = \omega r$, where r is the radial line drawn from the IC to the point. The line of action of each velocity vector is *perpendicular* to its associated radial line, and the velocity has a *direction* which tends to move the point in a manner consistent with the angular rotation of the radial line, Fig. 16–18.

The following examples illustrate the method of determining the IC and computing the velocities of points in a rigid body subjected to general plane motion.

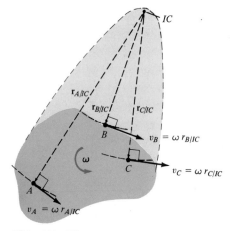

Fig. 16–18

Example 16–9

Determine the location of the instantaneous center of zero velocity for (a) the crankshaft *BC* shown in Fig. 16–19*a*; and (b) the link *CB* shown in Fig. 16–19*b*.

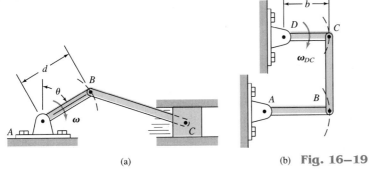

(a)   (b) **Fig. 16–19**

SOLUTION

Part (a). As shown in Fig. 16–19*a*, link *AB* rotates about the fixed pin *A*. Thus, point *B* has a speed of $v_B = \omega d$, caused by the clockwise rotation of *AB*. Also, $\mathbf{v}_B$ is perpendicular to *AB*, so that it acts at an angle θ from the horizontal as shown in Fig. 16–19*c*. The motion of point *B* causes the piston to move forward *horizontally* with a velocity $\mathbf{v}_C$. Consequently, point *C* on the crankshaft moves horizontally with this same velocity. When lines are drawn perpendicular to $\mathbf{v}_B$ and $\mathbf{v}_C$, Fig. 16–19*c*, they intersect at the *IC*. The magnitudes of $\mathbf{r}_{B/IC}$ and $\mathbf{r}_{C/IC}$ are determined strictly from the geometry of construction.

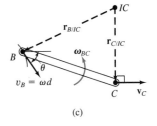

(c)

Part (b). Points *B* and *C* follow circular paths of motion since rods *AB* and *DC* are each subjected to rotation about a fixed axis, Fig. 16–19*b*. In particular, $v_C = \omega_{DC}b$. Since the velocity is always tangent to the path, at the instant considered, $\mathbf{v}_C$ on rod *DC* and $\mathbf{v}_B$ on rod *AB* are both directed vertically downward, along the axis of link *CB*, Fig. 16–19*d*. Furthermore, since *CB* is *rigid*, no relative displacement occurs between points *B* and *C*, so that $\mathbf{v}_B = \mathbf{v}_C$. Radial lines drawn perpendicular to these two velocities form parallel lines which intersect at "infinity;" i.e., $r_{C/IC} \rightarrow \infty$ and $r_{B/IC} \rightarrow \infty$. Thus, $\omega_{CB} = v_C/r_{C/IC} = (\omega_{DC}b)/\infty = 0$. As a result, rod *CB* momentarily *translates* with a speed of $v_C = \omega_{DC}b$. An instant later, however, *CB* will move to a new position, causing the instantaneous center to move to some finite location.

(d)

Example 16–10

Block D shown in Fig. 16–20a moves with a speed of 3 m/s. Determine the angular velocities of links BD and AB, and the velocity of point B at the instant shown.

(a)   **Fig. 16–20a**

SOLUTION

Since D moves to the right at 3 m/s, it causes arm AB to rotate about point A in a clockwise direction. Hence, $\mathbf{v}_B$ is directed perpendicular to AB as shown in Fig. 16–20b. The instantaneous center of zero velocity for BD is located at the intersection of the line segments drawn perpendicular to $\mathbf{v}_B$ and $\mathbf{v}_D$, Fig. 16–20b. From the geometry,

$$r_{B/IC} = 0.4 \tan 45° \text{ m} = 0.4 \text{ m}$$

$$r_{D/IC} = \frac{0.4 \text{ m}}{\cos 45°} = 0.566 \text{ m}$$

(b)

Since the magnitude of $\mathbf{v}_D$ is known, the angular velocity of the link is

$$\omega_{BD} = \frac{v_D}{r_{D/IC}} = \frac{3 \text{ m/s}}{0.566 \text{ m}} = 5.30 \text{ rad/s} \; \text{↰} \qquad Ans.$$

The velocity of B is therefore

$$v_B = \omega_{BD}(r_{B/IC}) = 5.30 \text{ rad/s}(0.4 \text{ m}) = 2.12 \text{ m/s} \quad \text{↙}45° \qquad Ans.$$

Link AB is subjected to rotation about a fixed axis passing through A, Fig. 16–20c, and since v_B is known, the angular velocity of AB is

$$\omega_{AB} = \frac{v_B}{r_{B/A}} = \frac{2.12 \text{ m/s}}{0.4 \text{ m}} = 5.30 \text{ rad/s} \; \text{↓} \qquad Ans.$$

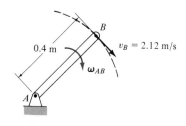

(c)   **Fig. 16–20(b, c)**

Example 16—11

The cylinder shown in Fig. 16–21a rolls without slipping between the two moving plates E and D. Determine the angular velocity of the cylinder and the velocity of its center C at the instant shown.

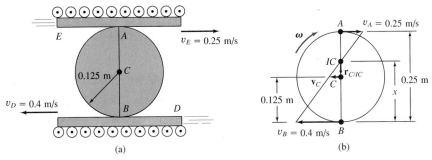

Fig. 16–21

SOLUTION

Since no slipping occurs, the contact points A and B on the cylinder have the same velocity as the plates E and D. Furthermore, the velocities $\mathbf{v}_A$ and $\mathbf{v}_B$ are *parallel*, so that by the proportionality of right triangles the IC is located at a point on line AB, Fig. 16–21b. Assuming this point to be a distance x from B, we have

$$v_B = \omega x; \qquad\qquad 0.4 = \omega x$$
$$v_A = \omega(0.25 - x); \qquad 0.25 = \omega(0.25 - x)$$

Dividing one of these equations into the other eliminates ω and yields

$$0.4(0.25 - x) = 0.25x$$

$$x = \frac{0.1}{0.65} = 0.154 \text{ m}$$

Hence, the angular velocity is

$$\omega = \frac{v_B}{x} = \frac{0.4}{0.154} = 2.60 \text{ rad/s } \downarrow \qquad\qquad Ans.$$

The velocity of point C is therefore

$$v_C = \omega r_{C/IC} = 2.60(0.154 - 0.125)$$
$$= 0.0754 \text{ m/s} \leftarrow \qquad\qquad Ans.$$

PROBLEMS

*16–52. Solve Prob. 16–40 using the method of instantaneous center of zero velocity.

16–53. Solve Prob. 16–38 using the method of instantaneous center of zero velocity.

16–54. Solve Prob. 16–42 using the method of instantaneous center of zero velocity.

16–55. Solve Prob. 16–43 using the method of instantaneous center of zero velocity.

*16–56. Solve Prob. 16–44 using the method of instantaneous center of zero velocity.

16–57. Solve Prob. 16–47 using the method of instantaneous center of zero velocity.

16–58. In each case show graphically how to locate the instantaneous center of zero velocity of link AB. Assume the geometry is known.

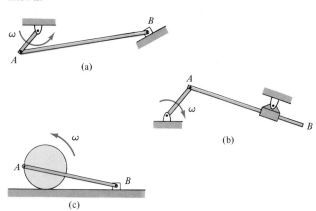

(a)

(b)

(c)

Prob. 16–58

16–59. The conveyor belt is moving to the right at $v = 8$ ft/s, and at the same instant the cylinder is rolling counterclockwise at $\omega = 2$ rad/s without slipping. Determine the velocities of the cylinder's center C and point B at this instant.

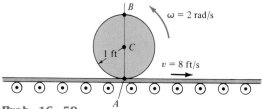

Prob. 16–59

*16–60. At the instant shown, the disk is rotating at $\omega = 4$ rad/s. If the end of the cord wrapped around the disk is fixed at D, determine the velocities of points A, B, and C.

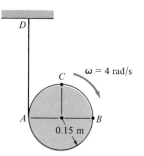

Prob. 16–60

16–61. As the car travels forward at 80 ft/s on a wet road, due to slipping the rear wheels have an angular velocity $\omega = 100$ rad/s. Determine the speeds of points A, B, and C caused by the motion.

Prob. 16–61

16–62. Determine the velocity of block C at the instant shown, if link AB is rotating at 8 rad/s.

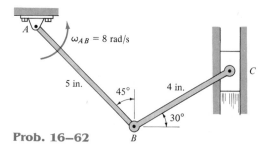

Prob. 16–62

16–63. The disk of radius r is confined to roll without slipping at A and B. If the plates have the velocities shown, determine the angular velocity of the disk.

Prob. 16–63

16–66. The planet gear A is pin-connected to the end of the link BC. If the link rotates about the fixed point B at 4 rad/s, determine the angular velocity of the ring gear R. The sun gear D is fixed from rotating.

Probs. 16–66/16–67

16–67. Solve Prob. 16–66 if the sun gear D is rotating clockwise at $\omega_D = 5$ rad/s while link BC rotates counterclockwise at $\omega_{BC} = 4$ rad/s.

*16–64. If link AB is rotating at $\omega_{AB} = 6$ rad/s, determine the angular velocities of links BC and CD at the instant shown.

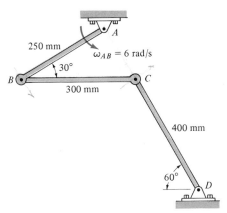

Prob. 16–64

*16–68. If the hub gear H and ring gear R have angular velocities $\omega_H = 5$ rad/s and $\omega_R = 20$ rad/s, respectively, determine the angular velocity ω_S of the spur gear S and the angular velocity of its attached arm OA.

Probs. 16–68/16–69

16–65. Due to slipping points A and B on the rim of the disk have the velocities shown. Determine the velocity of the center point C and point D at this instant.

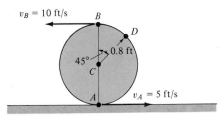

Prob. 16–65

16–69. If the hub gear H has an angular velocity $\omega_H = 5$ rad/s, determine the angular velocity of the ring gear R so that the arm OA attached to the spur gear S remains stationary ($\omega_{OA} = 0$). What is the angular velocity of the spur gear?

269

16-70. If bar AB has an angular velocity $\omega_{AB} = 4$ rad/s, determine the velocity of the slider block C at the instant shown.

Prob. 16-70

16-71. Mechanical toy animals often use a walking mechanism as shown idealized in the figure. If the driving crank AB is propelled by a spring motor such that $\omega_{AB} = 5$ rad/s, determine the velocity of the rear foot E at the instant shown.

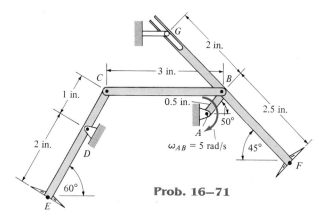

Prob. 16-71

*16-72. If the slider block A is moving to the right at $v_A = 8$ ft/s, determine the velocity of blocks B and C at the instant shown.

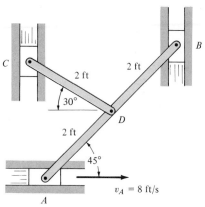

Prob. 16-72

16.7 Relative-Motion Analysis: Acceleration

An equation that relates the accelerations of two points in a rigid body subjected to general plane motion may be determined by differentiating the velocity equation $\mathbf{v}_B = \mathbf{v}_A + \mathbf{v}_{B/A}$ with respect to time. Thus,

$$\frac{d\mathbf{v}_B}{dt} = \frac{d\mathbf{v}_A}{dt} + \frac{d\mathbf{v}_{B/A}}{dt}$$

The terms $d\mathbf{v}_B/dt = \mathbf{a}_B$ and $d\mathbf{v}_A/dt = \mathbf{a}_A$ are measured from a set of *fixed* x, y axes and represent the *absolute accelerations* of points B and A. The last term represents the acceleration of B with respect to A as measured by an observer fixed to the translating x', y' axes having their origin at the base point A. In Sec. 16.5, it was shown that to this observer point B appears to move along a *circular arc* that has a radius of curvature $r_{B/A}$. In other words, *the body appears to move as if it were rotating about the z' axis passing through point A.* Consequently, $\mathbf{a}_{B/A}$ can be expressed in terms of its tangential and normal components of motion; i.e., $\mathbf{a}_{B/A} = (\mathbf{a}_{B/A})_t + (\mathbf{a}_{B/A})_n$, where $(a_{B/A})_t = \alpha r_{B/A}$ and $(a_{B/A})_n = \omega^2 r_{B/A}$. Hence, the above equation can be written in the form

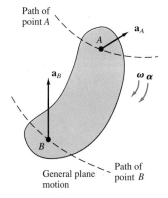

Path of point A

General plane motion

Path of point B

(a)

$$\mathbf{a}_B = \mathbf{a}_A + (\mathbf{a}_{B/A})_t + (\mathbf{a}_{B/A})_n \qquad (16\text{–}16)$$

where

$\mathbf{a}_B$ = acceleration of point B

$\mathbf{a}_A$ = acceleration of point A

$(\mathbf{a}_{B/A})_t$ = relative tangential acceleration component of "B with respect to A." Since the relative motion is circular, the *magnitude* of $(\mathbf{a}_{B/A})_t$ is $(a_{B/A})_t = \alpha r_{B/A}$ and the *direction* is perpendicular to $\mathbf{r}_{B/A}$

$(\mathbf{a}_{B/A})_n$ = relative normal acceleration component of "B with respect to A." Since the relative motion is circular, the *magnitude* of $(\mathbf{a}_{B/A})_n$ is $(a_{B/A})_n = \omega^2 r_{B/A}$ and the *direction* is always from B towards A

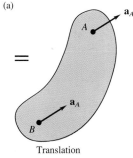

Translation

(b)

Each of the four terms in Eq. 16–16 is represented graphically on the *kinematic diagrams* shown in Fig. 16–22. Here it is seen that at a given instant the acceleration of B, Fig. 16–22a, is determined by considering the body to translate with an acceleration $\mathbf{a}_A$, Fig. 16–22b, and simultaneously rotate about the base point A with an instantaneous angular velocity $\boldsymbol{\omega}$ and angular acceleration $\boldsymbol{\alpha}$, Fig. 16–22c. Vector addition of these two effects, applied to B, yields $\mathbf{a}_B$, as shown in Fig. 16–22d. It should be noted from Fig. 16–22a that points A and B move along *curved paths*, and as a result the accelerations of these points have *both tangential and normal components*. (Recall that the acceleration of a point is *tangent to the path only* when the path is *rectilinear* or when it is an inflection point.)

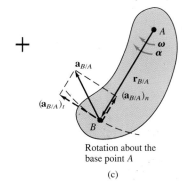

Rotation about the base point A

(c)

Since the relative-acceleration components represent the effect of *circular motion* observed from translating axes having their origin at the base point A, these terms can be expressed as $(\mathbf{a}_{B/A})_t = \boldsymbol{\alpha} \times \mathbf{r}_{B/A}$ and $(\mathbf{a}_{B/A})_n = -\omega^2 \mathbf{r}_{B/A}$, Eq. 16–13. Hence, Eq. 16–16 becomes

$$\mathbf{a}_B = \mathbf{a}_A + \boldsymbol{\alpha} \times \mathbf{r}_{B/A} - \omega^2 \mathbf{r}_{B/A} \qquad (16\text{–}17)$$

where

$\mathbf{a}_B$ = acceleration of point B

$\mathbf{a}_A$ = acceleration of the base point A

$\boldsymbol{\alpha}$ = angular acceleration of the body

$\boldsymbol{\omega}$ = angular velocity of the body

$\mathbf{r}_{B/A}$ = relative-position vector drawn from A to B

To apply this equation, all of these terms must be represented in Cartesian vector form.

$\mathbf{a}_B = \mathbf{a}_A + (\mathbf{a}_{B/A})_t + (\mathbf{a}_{B/A})_n$

(d)

Fig. 16–22

(a)

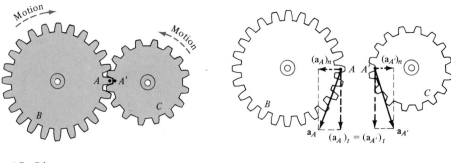

Fig. 16–23

(b)

If Eq. 16–16 or 16–17 is applied in a practical manner to study the acceleration of a rigid body which is pin-connected to two other bodies, it should be realized that points which are *coincident at the pin* move with the *same acceleration,* since the path of motion over which they travel is the *same.* For example, point B lying on either rod AB or BC of the crank mechanism shown in Fig. 16–23a has the same acceleration, since the rods are pin-connected at B. Here the motion of B is along a *curved path,* so that $\mathbf{a}_B$ is calculated on the basis of its tangential and normal components $(a_B)_t = \alpha r_{B/A}$ and $(a_B)_n = \omega^2 r_{B/A}$, which are defined by the angular motion of AB. At the other end of rod BC, however, point C moves along a *rectilinear path,* which is defined by the piston. Hence, in this case, the acceleration $\mathbf{a}_C$ is directed along the path, Fig. 16–23b.

If two bodies contact one another *without slipping,* and the *points in contact* move along *different paths,* the *tangential components* of acceleration of the points will be the *same;* however, the *normal components* will *not* be the same. For example, consider the two meshed gears in Fig. 16–24a. Point A is located on gear B and a coincident point A' is located on gear C. Due to the rotational motion, $(\mathbf{a}_A)_t = (\mathbf{a}_{A'})_t$; however, since both points follow different curved paths, $(\mathbf{a}_A)_n \neq (\mathbf{a}_{A'})_n$ and therefore $\mathbf{a}_A \neq \mathbf{a}_{A'}$, Fig. 16–24b.

Fig. 16–24   (a)   (b)

PROCEDURE FOR ANALYSIS

Equation 16–16 relates the accelerations of any two points A and B located on the *same* rigid body. Like the velocity equation, Eq. 16–14, it can be applied either by using a Cartesian vector analysis, in the form of Eq. 16–17, or by writing the x and y scalar component equations directly. For application, it is suggested that the following procedure be used.

Velocity Analysis. If the angular velocity $\boldsymbol{\omega}$ of the body is unknown, determine it by using a velocity analysis as discussed in Sec. 16.5 or 16.6. Determine also the velocities $\mathbf{v}_A$ and $\mathbf{v}_B$ of points A and B if these points move along curved paths.

Kinematic Diagrams. Establish the directions of the fixed x, y coordinates, and draw the kinematic diagram of the body. Indicate on it the accelerations of points A and B, $\mathbf{a}_A$ and $\mathbf{a}_B$, the angular velocity $\boldsymbol{\omega}$, the angular acceleration $\boldsymbol{\alpha}$, and the relative position vector $\mathbf{r}_{B/A}$. Normally the base point A is selected as a point having a known acceleration. In particular, if points A and B move along *curved paths,* their accelerations should be expressed in terms of their tangential and normal components, i.e., $\mathbf{a}_A = (\mathbf{a}_A)_t + (\mathbf{a}_A)_n$ and $\mathbf{a}_B = (\mathbf{a}_B)_t + (\mathbf{a}_B)_n$. Here $(a_A)_n = (v_A)^2/\rho_A$ and $(a_B)_n = (v_B)^2/\rho_B$, where ρ_A and ρ_B define the radii of curvature of the paths of points A and B, respectively. Identify the two unknowns on the diagram. If the magnitudes of $(\mathbf{a}_A)_t$, $(\mathbf{a}_B)_t$, or $\boldsymbol{\alpha}$ are unknown, the sense of direction of these vectors may be assumed.

If the acceleration equation is applied *without* using a Cartesian vector analysis, then the magnitude and direction of the relative acceleration components $(\mathbf{a}_{B/A})_t$ and $(\mathbf{a}_{B/A})_n$ must be established. This can be done by drawing a kinematic diagram such as shown in Fig. 16–22c. Since the body is considered to be momentarily "pinned" at the base point A, these components have *magnitudes* of $(a_{B/A})_t = \alpha r_{B/A}$ and $(a_{B/A})_n = \omega^2 r_{B/A}$. Their *directions* are established from the diagram such that $(\mathbf{a}_{B/A})_t$ acts perpendicular to $\mathbf{r}_{B/A}$, in accordance with the rotational motion $\boldsymbol{\alpha}$ of the body, and $(\mathbf{a}_{B/A})_n$ is directed from B toward A.*

Acceleration Equation. To apply Eq. 16–17, express the vectors in Cartesian vector form and substitute them into the equation. Evaluate the cross product and then equate the respective $\mathbf{i}$ and $\mathbf{j}$ components to obtain two scalar equations. To obtain these scalar equations *directly* write Eq. 16–16 in symbolic form, $\mathbf{a}_B = \mathbf{a}_A + (\mathbf{a}_{B/A})_t + (\mathbf{a}_{B/A})_n$, and underneath each of the terms represent the vectors by their magnitudes and directions. To do this, use the data tabulated on the kinematic diagrams. The scalar equations are determined from the x and y components of these vectors. Solve these equations for the two unknowns. If the solution yields a *negative* answer for an *unknown,* it indicates the sense of direction of the vector is opposite to that shown on the kinematic diagram.

The following example problems numerically illustrate both methods of application.

---

*Perhaps the notation $\mathbf{a}_B = \mathbf{a}_A + (\mathbf{a}_{B/A(\text{pin})})_t + (\mathbf{a}_{B/A(\text{pin})})_n$ may be helpful in recalling that A is pinned.

Example 16–12

The rod AB shown in Fig. 16–25a is confined to move along the inclined planes at A and B. If point A has an acceleration of 3 m/s² and a velocity of 2 m/s, both directed down the plane at the instant the rod becomes horizontal, determine the angular acceleration of the rod at this instant.

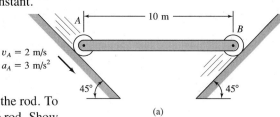

(a)

SOLUTION I (VECTOR ANALYSIS)

We will apply the acceleration equation to points A and B on the rod. To do so it is first necessary to determine the angular velocity of the rod. Show that it is $\omega = 0.283$ rad/s ↺ using either the velocity equation or the method of instantaneous centers.

Kinematic Diagram. The x, y axes are established in Fig. 16–25b. Since points A and B both move along straight-line paths, they have *no* components of acceleration in the normal direction. Why? There are two unknowns in Fig. 16–25b, namely, a_B and α.

(b)

Acceleration Equation. Expressing each of the vectors in Cartesian vector form, we have

$\mathbf{r}_{B/A} = \{10\mathbf{i}\}$ m $\qquad \boldsymbol{\omega} = \{0.283\mathbf{k}\}$ rad/s $\qquad \boldsymbol{\alpha} = \alpha\mathbf{k}$

$\mathbf{a}_A = \{3 \cos 45°\mathbf{i} - 3 \sin 45°\mathbf{j}\}$ m/s² $\qquad \mathbf{a}_B = a_B \cos 45°\mathbf{i} + a_B \sin 45°\mathbf{j}$

Applying Eq. 16–17 to points A and B on the rod, we have

$$\mathbf{a}_B = \mathbf{a}_A + \boldsymbol{\alpha} \times \mathbf{r}_{B/A} - \omega^2 \mathbf{r}_{B/A}$$

$a_B \cos 45°\mathbf{i} + a_B \sin 45°\mathbf{j}$

$\qquad = 3 \cos 45°\mathbf{i} - 3 \sin 45°\mathbf{j} + (\alpha\mathbf{k}) \times (10\mathbf{i}) - (0.283)^2(10\mathbf{i})$

Carrying out the cross product and equating the **i** and **j** components yields

$$a_B \cos 45° = 3 \cos 45° - (0.283)^2(10) \qquad (1)$$

$$a_B \sin 45° = -3 \sin 45° + \alpha(10) \qquad (2)$$

Solving, we have

$$a_B = 1.87 \text{ m/s}^2 \quad \angle^{45°}$$

$$\alpha = 0.344 \text{ rad/s}^2 \text{ ↺} \qquad\qquad Ans.$$

(c)

Fig. 16–25

SOLUTION II (SCALAR COMPONENTS)

As an alternative procedure, the scalar component equations (1) and (2) can be obtained directly. From the kinematic diagram, showing the relative acceleration components $(\mathbf{a}_{B/A})_t$ and $(\mathbf{a}_{B/A})_n$, Fig. 16–25c, we have

$$\mathbf{a}_B = \mathbf{a}_A + (\mathbf{a}_{B/A})_t + (\mathbf{a}_{B/A})_n$$

$$\begin{bmatrix} a_B \\ {\scriptstyle\angle 45°} \end{bmatrix} = \begin{bmatrix} 3 \text{ m/s}^2 \\ {\scriptstyle\searrow 45°} \end{bmatrix} + \begin{bmatrix} \alpha(10 \text{ m}) \\ \uparrow \end{bmatrix} + \begin{bmatrix} (0.283 \text{ rad/s})^2(10 \text{ m}) \\ \leftarrow \end{bmatrix}$$

Equating the x and y components yields Eqs. (1) and (2) and the solution proceeds as before.

Example 16–13

At a given instant, the cylinder of radius r, shown in Fig. 16–26a, has an angular velocity $\boldsymbol{\omega}$ and angular acceleration $\boldsymbol{\alpha}$. Determine the velocity and acceleration of its center G if it rolls without slipping.

(a)

SOLUTION

As the cylinder rolls, point G moves along a straight line, and point A, located on the rim of the cylinder, moves along a curved path called a *cycloid*,* Fig. 16–26b. We will apply the velocity and acceleration equations to these points.

Velocity Analysis. Since no slipping occurs, at the instant A contacts the ground, $\mathbf{v}_A = \mathbf{0}$. Thus, from the kinematic diagram in Fig. 16–26c we have

$$\mathbf{v}_G = \mathbf{v}_A + \boldsymbol{\omega} \times \mathbf{r}_{G/A}$$
$$v_G\mathbf{i} = \mathbf{0} + (-\omega\mathbf{k}) \times (r\mathbf{j})$$
$$v_G = \omega r \qquad (1) \quad \textit{Ans.}$$

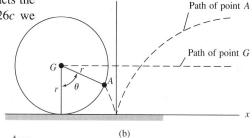
(b)

This same result can also be obtained directly by noting that point A represents the instantaneous center of zero velocity.

Kinematic Diagram. The acceleration of point G is horizontal since it moves along a straight-line path. *Just before* point A touches the ground, its velocity is directed downward along the y axis, Fig. 16–26b, and just after contact, its velocity is directed upward.† For this reason, point A begins to accelerate upward when it leaves the ground at A, Fig. 16–26d. No motion occurs in the x direction during the instant of ground contact. The magnitudes of $\mathbf{a}_A$ and $\mathbf{a}_G$ are unknown.

(c)

Acceleration Equation

$$\mathbf{a}_G = \mathbf{a}_A + \boldsymbol{\alpha} \times \mathbf{r}_{G/A} - \omega^2 \mathbf{r}_{G/A}$$
$$a_G\mathbf{i} = a_A\mathbf{j} + (-\alpha\mathbf{k}) \times (r\mathbf{j}) - \omega^2(r\mathbf{j})$$

Evaluating the cross product and equating the $\mathbf{i}$ and $\mathbf{j}$ components yields

$$a_G = \alpha r \qquad (2) \quad \textit{Ans.}$$
$$a_A = \omega^2 r \qquad (3)$$

Fig. 16–26 (d)

These important results, that $v_G = \omega r$ and $a_G = \alpha r$, were also obtained in Example 16–3. They apply to any circular object, such as a ball, pulley, disk, etc., that rolls *without* slipping. Also, the fact that $a_A = \omega^2 r$ indicates that the instantaneous center of zero velocity, point A, is not a point of zero acceleration.

---

*Although it is not necessary here, one can show that the equation of the path (cycloid) can be written in terms of θ as $x = r(\theta - \sin\theta)$ and $y = r(1 - \cos\theta)$.
†The slope of the path, dy/dx, is along the y axis at $\theta = 90°$.

Example 16–14

The spool shown in Fig. 16–27a unravels from the cord, such that at the instant shown it has an angular velocity of 3 rad/s and an angular acceleration of 4 rad/s². Determine the acceleration of point B.

SOLUTION I (VECTOR ANALYSIS)

The spool "appears" to be rolling downward without slipping at point A. Therefore, we can use the results of Example 16–13 to determine the acceleration of point G, i.e.,

$$a_G = \alpha r = 4(0.5) = 2 \text{ ft/s}^2$$

We will apply the acceleration equation between points G and B.

Kinematic Diagram. Point B moves along a *curved path* having an unknown radius of curvature.* Its acceleration is represented by its unknown x and y components as shown in Fig. 16–27b.

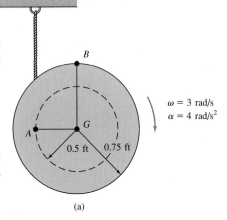

(a)

Acceleration Equation

$$\mathbf{a}_B = \mathbf{a}_G + \boldsymbol{\alpha} \times \mathbf{r}_{B/G} - \omega^2 \mathbf{r}_{B/G}$$

$$(a_B)_x\mathbf{i} + (a_B)_y\mathbf{j} = -2\mathbf{j} + (-4\mathbf{k}) \times (0.75\mathbf{j}) - (3)^2(0.75\mathbf{j})$$

Therefore, the component equations are

$$(a_B)_x = 4(0.75) = 3 \text{ ft/s}^2 \qquad (1)$$
$$(a_B)_y = -2 - 6.75 = -8.75 \text{ ft/s}^2 = 8.75 \text{ ft/s}^2 \qquad (2)$$

The magnitude and direction of $\mathbf{a}_B$ are therefore

$$a_B = \sqrt{(3)^2 + (8.75)^2} = 9.25 \text{ ft/s}^2 \qquad \textit{Ans.}$$

$$\theta = \tan^{-1}\frac{8.75}{3} = 71.1° \qquad \textit{Ans.}$$

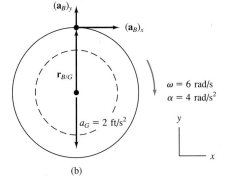

(b)

SOLUTION II (SCALAR COMPONENTS)

This problem may be solved by writing the scalar component equations directly. The kinematic diagram in Fig. 16–27c shows the relative acceleration components $(\mathbf{a}_{B/G})_t$ and $(\mathbf{a}_{B/G})_n$. Thus,

$$\mathbf{a}_B = \mathbf{a}_G + (\mathbf{a}_{B/G})_t + (\mathbf{a}_{B/G})_n$$

$$\begin{bmatrix} (a_B)_x \\ \rightarrow \end{bmatrix} + \begin{bmatrix} (a_B)_y \\ \uparrow \end{bmatrix}$$

$$= \begin{bmatrix} 2 \text{ ft/s}^2 \\ \downarrow \end{bmatrix} + \begin{bmatrix} 4 \text{ rad/s}^2(0.75 \text{ ft}) \\ \rightarrow \end{bmatrix} + \begin{bmatrix} (3 \text{ rad/s})^2(0.75 \text{ ft}) \\ \downarrow \end{bmatrix}$$

The x and y components yield Eqs. (1) and (2) shown above.

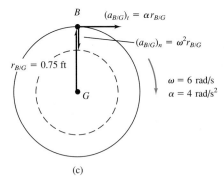

(c)

Fig. 16–27

*Realize that the path's radius of curvature ρ is *not* equal to the radius of the cylinder since the cylinder is *not* rotating about point G. Furthermore, ρ is *not* defined as the distance from A (IC) to B, since the location of the IC depends only on the velocity of a point and *not* the geometry of its path.

Example 16–15

The collar C in Fig. 16–28a is moving downward with an acceleration of 1 m/s^2. At the instant shown, it has a speed of 2 m/s which gives links CB and AB an angular velocity $\omega_{AB} = \omega_{CB} = 10$ rad/s. (See Example 16–7.) Determine the angular accelerations of CB and AB at this instant.

$a_C = 1$ m/s^2
$v_C = 2$ m/s

SOLUTION (VECTOR ANALYSIS)

Kinematic Diagrams. The kinematic diagrams of *both* links AB and CB are shown in Fig. 16–28b.

Acceleration Equations. Expressing each of the vectors in Cartesian vector form, we have

$$\mathbf{r}_{B/A} = \{-0.2\mathbf{j}\} \text{ m} \qquad \mathbf{r}_{B/C} = \{0.2\mathbf{i} - 0.2\mathbf{j}\} \text{ m}$$
$$\boldsymbol{\omega}_{AB} = \{10\mathbf{k}\} \text{ rad/s} \qquad \boldsymbol{\omega}_{CB} = \{10\mathbf{k}\} \text{ rad/s}$$
$$\boldsymbol{\alpha}_{AB} = \alpha_{AB}\mathbf{k} \qquad \boldsymbol{\alpha}_{CB} = \alpha_{CB}\mathbf{k}$$
$$\mathbf{a}_C = \{-1\mathbf{j}\} \text{ m/s}^2$$

Link AB is rotating about a fixed axis. Applying Eq. 16–13 to point B, we have

$$\mathbf{a}_B = \boldsymbol{\alpha}_{AB} \times \mathbf{r}_{B/A} - \omega_{AB}^2 \mathbf{r}_{B/A}$$
$$\mathbf{a}_B = (\alpha_{AB}\mathbf{k}) \times (-0.2\mathbf{j}) - (10)^2(-0.2\mathbf{j})$$
$$\mathbf{a}_B = 0.2\alpha_{AB}\mathbf{i} + 20\mathbf{j}$$

Link BC is subjected to general plane motion. Applying Eq. 16–17 to points B and C, using the above value for $\mathbf{a}_B$, yields

$$\mathbf{a}_B = \mathbf{a}_C + \boldsymbol{\alpha}_{CB} \times \mathbf{r}_{B/C} - \omega_{CB}^2 \mathbf{r}_{B/C}$$
$$0.2\alpha_{AB}\mathbf{i} + 20\mathbf{j} = -1\mathbf{j} + (\alpha_{CB}\mathbf{k}) \times (0.2\mathbf{i} - 0.2\mathbf{j}) - (10)^2(0.2\mathbf{i} - 0.2\mathbf{j})$$
$$0.2\alpha_{AB}\mathbf{i} + 20\mathbf{j} = -1\mathbf{j} + 0.2\alpha_{CB}\mathbf{j} + 0.2\alpha_{CB}\mathbf{i} - 20\mathbf{i} + 20\mathbf{j}$$

Fig. 16–28

Equating the respective $\mathbf{i}$ and $\mathbf{j}$ components, we obtain

$$0.2\alpha_{AB} = 0.2\alpha_{CB} - 20 \qquad (1)$$
$$20 = -1 + 0.2\alpha_{CB} + 20 \qquad (2)$$

Hence,

$$\alpha_{CB} = 5 \text{ rad/s}^2$$
$$\alpha_{AB} = -95 \text{ rad/s}^2$$

Thus,

$$\boldsymbol{\alpha}_{CB} = \{5\mathbf{k}\} \text{ rad/s}^2 \qquad \qquad Ans.$$
$$\boldsymbol{\alpha}_{AB} = \{-95\mathbf{k}\} \text{ rad/s}^2 \qquad \qquad Ans.$$

Example 16–16

The crankshaft AB of an engine turns with a clockwise angular accelera-tion of 20 rad/s^2, Fig. 16–29a. Determine the acceleration of the piston at the instant AB is in the position shown. At this instant $\omega_{AB} = 10$ rad/s and $\omega_{BC} = 2.43$ rad/s.

SOLUTION (VECTOR ANALYSIS)
Kinematic Diagrams. The kinematic diagrams for both AB and BC are shown in Fig. 16–29b.

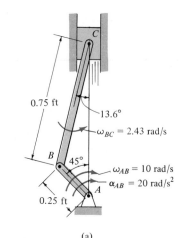

(a)

Acceleration Equations. Expressing each of the vectors in Cartesian vec-tor form yields

$$\mathbf{r}_{B/A} = \{-0.25 \sin 45°\mathbf{i} + 0.25 \cos 45°\mathbf{j}\} \text{ ft} = \{-0.177\mathbf{i} + 0.177\mathbf{j}\} \text{ ft}$$
$$\mathbf{r}_{C/B} = \{0.75 \sin 13.6°\mathbf{i} + 0.75 \cos 13.6°\mathbf{j}\} \text{ ft} = \{0.176\mathbf{i} + 0.729\mathbf{j}\} \text{ ft}$$
$$\boldsymbol{\omega}_{AB} = \{-10\mathbf{k}\} \text{ rad/s} \qquad \boldsymbol{\alpha}_{AB} = \{-20\mathbf{k}\} \text{ rad/s}^2$$
$$\boldsymbol{\omega}_{BC} = \{2.43\mathbf{k}\} \text{ rad/s} \qquad \boldsymbol{\alpha}_{BC} = \alpha_{BC}\mathbf{k}$$
$$\mathbf{a}_C = a_C\mathbf{j}$$

The connecting rod AB is subjected to rotation about a fixed axis. Applying Eq. 16–13 to point B, we have

$$\mathbf{a}_B = \boldsymbol{\alpha}_{AB} \times \mathbf{r}_{B/A} - \omega_{AB}^2\mathbf{r}_{B/A}$$
$$= (-20\mathbf{k}) \times (-0.177\mathbf{i} + 0.177\mathbf{j}) - (10)^2(-0.177\mathbf{i} + 0.177\mathbf{j})$$
$$= \{21.24\mathbf{i} - 14.16\mathbf{j}\} \text{ ft/s}^2$$

The crankshaft is subjected to general plane motion. Applying Eq. 16–17 to points B and C yields

$$\mathbf{a}_C = \mathbf{a}_B + \boldsymbol{\alpha}_{BC} \times \mathbf{r}_{C/B} - \omega_{BC}^2\mathbf{r}_{C/B}$$
$$a_C\mathbf{j} = 21.24\mathbf{i} - 14.16\mathbf{j} + (\alpha_{BC}\mathbf{k}) \times (0.176\mathbf{i} + 0.729\mathbf{j})$$
$$\qquad\qquad\qquad - (2.43)^2(0.176\mathbf{i} + 0.729\mathbf{j})$$
$$a_C\mathbf{j} = 21.24\mathbf{i} - 14.16\mathbf{j} + 0.176\alpha_{BC}\mathbf{j} - 0.729\alpha_{BC}\mathbf{i} - 1.04\mathbf{i} - 4.30\mathbf{j}$$

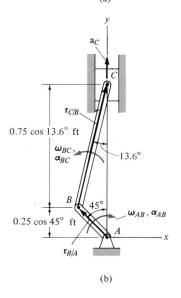

(b)

Equating the respective $\mathbf{i}$ and $\mathbf{j}$ components, we have

$$0 = 20.20 - 0.729\alpha_{BC}$$
$$a_C = 0.176\alpha_{BC} - 18.46$$

Fig. 16–29

Solving yields

$$\alpha_{BC} = 27.7 \text{ rad/s}^2 \text{ } \curvearrowleft$$
$$a_C = -13.6 \text{ ft/s}^2 \qquad\qquad\qquad \textit{Ans.}$$

The negative sign indicates that the piston is decelerating, i.e., $\mathbf{a}_C = \{-13.6\mathbf{j}\}$ ft/s^2. This causes the speed of the piston to decrease until the crankshaft AB becomes vertical, at which time the piston is momentarily at rest.

PROBLEMS

16–73. At a given instant the top A of the bar has the velocity and acceleration shown. Determine the acceleration of the bottom B of the bar, and the bar's angular acceleration at this instant.

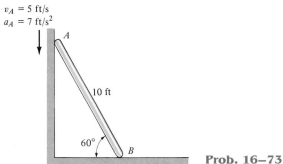

$v_A = 5$ ft/s
$a_A = 7$ ft/s^2

A

10 ft

$60°$

B

Prob. 16–73

16–74. The disk is moving to the left such that is has an angular acceleration $\alpha = 8$ rad/s^2 and angular velocity $\omega = 3$ rad/s at the instant shown. If it does not slip at A, determine the acceleration of point B.

$\omega = 3$ rad/s
$\alpha = 8$ rad/s^2

D

$30°$ $45°$
C
B 0.5 m

A

Probs. 16–74/16–75

16–75. The disk is moving to the left such that is has an angular acceleration $\alpha = 8$ rad/s^2 and angular velocity $\omega = 3$ rad/s at the instant shown. If it does not slip at A, determine the acceleration of point D.

*16–76. A cord is wrapped around the inner spool of the gear. If it is pulled with a constant velocity $\mathbf{v}$, determine the velocity and acceleration of points A and B. The gear rolls on the fixed gear rack.

B

$2r$

A G r

$\mathbf{v}$

Prob. 16–76

16–77. The hoop is cast on the rough surface such that it has an angular velocity of $\omega = 4$ rad/s and an angular deceleration of $\alpha = 5$ rad/s^2. Also, its center has a velocity of $v_O = 5$ m/s and a deceleration of $a_O = 2$ m/s^2. Determine the acceleration of point A at this instant.

$\omega = 4$ rad/s

A

$\alpha = 5$ rad/s^2

$a_O = 2$ m/s^2 O

0.3 m $45°$ $v_O = 5$ m/s

B

Probs. 16–77/16–78

16–78. The hoop is cast on the rough surface such that it has an angular velocity of $\omega = 4$ rad/s and an angular deceleration of $\alpha = 5$ rad/s^2. Also, its center has a velocity of $v_O = 5$ m/s and a deceleration of $a_O = 2$ m/s^2. Determine the acceleration of point B at this instant.

16–79. The disk rotates with an angular velocity $\omega = 5$ rad/s and an angular acceleration $\alpha = 6$ rad/s^2. Determine the angular accelerations of links AB and BC at this instant.

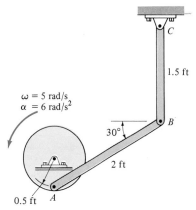

C

1.5 ft

$\omega = 5$ rad/s
$\alpha = 6$ rad/s^2

$30°$ B

2 ft

0.5 ft A

Prob. 16–79

*16–80. At a given instant the wheel is rotating with the angular velocity and angular acceleration shown. Determine the acceleration of block B at this instant.

Prob. 16–80

16–81. If rod AB has the angular motion at the instant shown, determine the acceleration of block C at this instant.

Prob. 16–81

16–82. At a given instant, link AB has an angular velocity $\omega_{AB} = 4$ rad/s and an angular acceleration $\alpha_{AB} = 12$ rad/s^2. Determine the angular velocity and angular acceleration of links BC and CD at this instant.

Prob. 16–82

16–83. At a given instant gears A and B have the angular motions shown. Determine the angular acceleration of gear C and the acceleration of its center point D at this instant. Note that the inner hub of gear C is in mesh with gear A and its outer rim is in mesh with gear B.

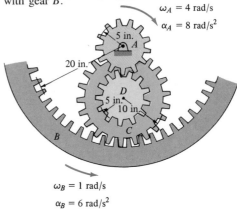

Prob. 16–83

*16–84. The wheel rolls without slipping such that at the instant shown it has an angular velocity ω and angular acceleration α. Determine the angular velocity and angular acceleration of the rod at this instant.

Prob. 16–84

16–85. The ends of the bar AB are confined to move along the paths shown. At a given instant, A has a velocity of $v_A = 4$ ft/s and an acceleration of $a_A = 7$ ft/s^2. Determine the angular velocity and angular acceleration of AB at this instant.

Prob. 16–85

16.8 Relative-Motion Analysis Using Rotating Axes

In the previous sections, the relative-motion analysis for velocity and acceleration was described using a translating coordinate system. This type of analysis is useful for determining the motion of points in the *same* rigid body, or the motion of points located in several pin-connected rigid bodies. In some problems, however, rigid bodies (mechanisms) are constructed such that *sliding* will occur at their connections. The kinematic analysis for such cases is best performed if the motion is analyzed using a coordinate system which both *translates* and *rotates*. Furthermore, this frame of reference is useful for analyzing the motions of two points on a mechanism which are *not* located in the *same* rigid body and for specifying the kinematics of particle motion when one of the particles is moving along a rotating path.

In the following analysis two equations are developed which relate the velocity and acceleration of a point to the origin of a moving frame of reference subjected to both a translation and a rotation in the plane.* Due to the generality in the derivation which follows, these two points may represent either two particles moving independently of one another or two points located in the same (or different) rigid bodies.

Position

Consider the two points A and B shown in Fig. 16–30a. Their location is specified by the position vectors $\mathbf{r}_A$ and $\mathbf{r}_B$, which are measured from the fixed X, Y, Z coordinate system. As shown in the figure, the "base point" A represents the origin of the x, y, z coordinate system, which is assumed to be both translating and rotating with respect to the X, Y, Z system. The position of B with respect to A is specified by the relative-position vector $\mathbf{r}_{B/A}$. The components of this vector may be expressed either in terms of unit vectors along the X, Y axes, i.e., $\mathbf{I}$ and $\mathbf{J}$, or by unit vectors along the x, y axes, i.e., $\mathbf{i}$ and $\mathbf{j}$. Although the magnitude of $\mathbf{r}_{B/A}$ is the same when measured in both coordinate systems, the direction of this vector will be measured differently if the x, y axes are not parallel to the X, Y axes. For the proof which follows, $\mathbf{r}_{B/A}$ will be measured relative to the moving x, y frame of reference. Thus, if B has coordinates (x_B, y_B), Fig. 16–30a, then

$$\mathbf{r}_{B/A} = x_B\mathbf{i} + y_B\mathbf{j}$$

Using vector addition, the three position vectors in Fig. 16–30a may be related by the equation

$$\boxed{\mathbf{r}_B = \mathbf{r}_A + \mathbf{r}_{B/A}} \qquad (16\text{–}18)$$

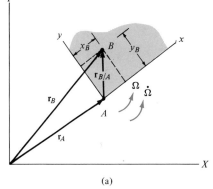

Fig. 16–30a

*The more general, three-dimensional motion of the points is developed in Sec. 20.4.

At the instant considered, point A has a velocity $\mathbf{v}_A$ and an acceleration $\mathbf{a}_A$, while the angular velocity and angular acceleration of the x, y axes are $\boldsymbol{\Omega}$ (omega) and $\dot{\boldsymbol{\Omega}} = d\boldsymbol{\Omega}/dt$, respectively. All these vectors are measured from the X, Y, Z frame of reference, although they may be expressed either in terms of $\mathbf{I}$, $\mathbf{J}$, $\mathbf{K}$ or $\mathbf{i}$, $\mathbf{j}$, $\mathbf{k}$ components. Since planar motion is specified, then by the right-hand rule $\boldsymbol{\Omega}$ and $\dot{\boldsymbol{\Omega}}$ are always directed *perpendicular* to the reference plane of motion, whereas $\mathbf{v}_A$ and $\mathbf{a}_A$ lie in this plane.

Velocity

The velocity of point B is determined by taking the time derivative of Eq. 16–18, which yields

$$\mathbf{v}_B = \mathbf{v}_A + \frac{d\mathbf{r}_{B/A}}{dt} \tag{16–19}$$

The last term in this equation is evaluated as follows:

$$\frac{d\mathbf{r}_{B/A}}{dt} = \frac{d}{dt}(x_B\mathbf{i} + y_B\mathbf{j})$$

$$= \frac{dx_B}{dt}\mathbf{i} + x_B\frac{d\mathbf{i}}{dt} + \frac{dy_B}{dt}\mathbf{j} + y_B\frac{d\mathbf{j}}{dt}$$

$$= \left(\frac{dx_B}{dt}\mathbf{i} + \frac{dy_B}{dt}\mathbf{j}\right) + \left(x_B\frac{d\mathbf{i}}{dt} + y_B\frac{d\mathbf{j}}{dt}\right) \tag{16–20}$$

The two terms in the first set of parentheses represent the components of velocity of point B as measured by an observer attached to the moving co-ordinate system. These terms will be denoted by vector $(\mathbf{v}_{B/A})_{\mathrm{rel}}$. In the second set of parentheses the instantaneous time rate of change of the unit vectors $\mathbf{i}$ and $\mathbf{j}$ is measured by an observer located in the fixed X, Y, Z coordinate system. These changes, $d\mathbf{i}$ and $d\mathbf{j}$, are due *only* to the instantaneous *rotation* $d\theta$ of the x, y, z axes, Fig. 16–30b. As shown, the *magnitudes* of both $d\mathbf{i}$ and $d\mathbf{j}$ equal 1 $(d\theta)$, since $i = j = 1$. The *direction* of $d\mathbf{i}$ is defined by $+\mathbf{j}$, since $d\mathbf{i}$ is tangent to the path described by the tip of $\mathbf{i}$ in the limit as $\Delta t \to dt$. Likewise, $d\mathbf{j}$ acts in the $-\mathbf{i}$ direction Fig. 16–30b. Hence,

$$\frac{d\mathbf{i}}{dt} = \frac{d\theta}{dt}(\mathbf{j}) = \Omega\mathbf{j} \qquad \frac{d\mathbf{j}}{dt} = \frac{d\theta}{dt}(-\mathbf{i}) = -\Omega\mathbf{i}$$

Viewing the axes in three dimensions, Fig. 16–30c, and noting that $\boldsymbol{\Omega} = \Omega\mathbf{k}$, we can express the above derivatives in terms of the cross product as

$$\frac{d\mathbf{i}}{dt} = \boldsymbol{\Omega} \times \mathbf{i} \qquad \frac{d\mathbf{j}}{dt} = \boldsymbol{\Omega} \times \mathbf{j} \tag{16–21}$$

Substituting these results into Eq. 16–20 and using the distributive property of the vector cross product, we obtain

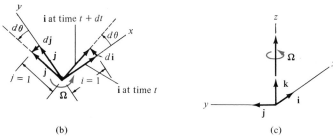

(b)

(c)

Fig. 16–30(b, c)

$$\frac{d\mathbf{r}_{B/A}}{dt} = \mathbf{v}_{B/A} + \mathbf{\Omega} \times (x_B\mathbf{i} + y_B\mathbf{j}) = (\mathbf{v}_{B/A})_{\text{rel}} + \mathbf{\Omega} \times \mathbf{r}_{B/A} \quad (16\text{–}22)$$

Hence, Eq. 16–19 becomes

$$\boxed{\mathbf{v}_B = \mathbf{v}_A + \mathbf{\Omega} \times \mathbf{r}_{B/A} + (\mathbf{v}_{B/A})_{\text{rel}}} \quad (16\text{–}23)$$

where

$\mathbf{v}_B$ = velocity of B, measured from the X, Y, Z reference

$\mathbf{v}_A$ = velocity of the origin A of the x, y, z reference, measured from the X, Y, Z reference

$(\mathbf{v}_{B/A})_{\text{rel}}$ = relative velocity of "B with respect to A," as measured by an observer attached to the *rotating* x, y, z reference

$\mathbf{\Omega}$ = angular velocity of the x, y, z reference, measured from the X, Y, Z reference

$\mathbf{r}_{B/A}$ = relative position of "B with respect to A"

Comparing Eq. 16–23 with Eq. 16–15 ($\mathbf{v}_B = \mathbf{v}_A + \mathbf{\Omega} \times \mathbf{r}_{B/A}$), which is valid for a translating frame of reference, it can be seen that the only difference between the equations is represented by the term $(\mathbf{v}_{B/A})_{\text{rel}}$.

When applying Eq. 16–23 it is often useful to understand what each of the terms represents. In order of appearance, they are as follows:

$\mathbf{v}_B$ {absolute velocity of B

(equals)

$\mathbf{v}_A$ {absolute velocity of origin of x, y, z frame

(plus) } motion of x, y, z frame

$\mathbf{\Omega} \times \mathbf{r}_{B/A}$ {angular velocity effect caused by rotation of x, y, z frame }

(plus)

$(\mathbf{v}_{B/A})_{\text{rel}}$ {relative velocity of B with respect to A } motion of B within x, y, z frame

283

Acceleration

The acceleration of point B, observed from the X, Y, Z coordinate system, may be expressed in terms of its motion measured with respect to the rotating or moving system of coordinates by taking the time derivative of Eq. 16–23, i.e.,

$$\frac{d\mathbf{v}_B}{dt} = \frac{d\mathbf{v}_A}{dt} + \frac{d\mathbf{\Omega}}{dt} \times \mathbf{r}_{B/A} + \mathbf{\Omega} \times \frac{d\mathbf{r}_{B/A}}{dt} + \frac{d(\mathbf{v}_{B/A})_{\text{rel}}}{dt}$$

$$\mathbf{a}_B = \mathbf{a}_A + \dot{\mathbf{\Omega}} \times \mathbf{r}_{B/A} + \mathbf{\Omega} \times \frac{d\mathbf{r}_{B/A}}{dt} + \frac{d(\mathbf{v}_{B/A})_{\text{rel}}}{dt} \qquad (16\text{–}24)$$

Here $\dot{\mathbf{\Omega}} = d\mathbf{\Omega}/dt$ is the angular acceleration of the x, y, z coordinate system. For planar motion $\mathbf{\Omega}$ is always perpendicular to the plane of motion, and therefore $\dot{\mathbf{\Omega}}$ measures *only the change in magnitude* of $\mathbf{\Omega}$. The derivative $d\mathbf{r}_{B/A}/dt$ in Eq. 16–24 is defined by Eq. 16–22, so that

$$\mathbf{\Omega} \times \frac{d\mathbf{r}_{B/A}}{dt} = \mathbf{\Omega} \times (\mathbf{v}_{B/A})_{\text{rel}} + \mathbf{\Omega} \times (\mathbf{\Omega} \times \mathbf{r}_{B/A}) \qquad (16\text{–}25)$$

Computing the time derivative of $(\mathbf{v}_{B/A})_{\text{rel}} = (v_{B/A})_x\mathbf{i} + (v_{B/A})_y\mathbf{j}$, we have

$$\frac{d(\mathbf{v}_{B/A})_{\text{rel}}}{dt} = \left[\frac{d(v_{B/A})_x}{dt}\mathbf{i} + \frac{d(v_{B/A})_y}{dt}\mathbf{j}\right] + \left[(v_{B/A})_x\frac{d\mathbf{i}}{dt} + (v_{B/A})_y\frac{d\mathbf{j}}{dt}\right]$$

The two terms in the first set of brackets represent the components of acceleration of point B as measured by an observer attached to the moving coordinate system. These terms will be denoted by vector $(\mathbf{a}_{B/A})_{\text{rel}}$. The terms in the second set of brackets can be simplified using Eqs. 16–21. Hence,

$$\frac{d(\mathbf{v}_{B/A})_{\text{rel}}}{dt} = (\mathbf{a}_{B/A})_{\text{rel}} + \mathbf{\Omega} \times (\mathbf{v}_{B/A})_{\text{rel}}$$

Substituting this equation and Eq. 16–25 into Eq. 16–24 and rearranging terms yields

$$\boxed{\mathbf{a}_B = \mathbf{a}_A + \dot{\mathbf{\Omega}} \times \mathbf{r}_{B/A} + \mathbf{\Omega} \times (\mathbf{\Omega} \times \mathbf{r}_{B/A}) + 2\mathbf{\Omega} \times (\mathbf{v}_{B/A})_{\text{rel}} + (\mathbf{a}_{B/A})_{\text{rel}}}$$

$$(16\text{–}26)$$

where

$\mathbf{a}_B$ = acceleration of B, measured from the X, Y, Z reference

$\mathbf{a}_A$ = acceleration of the origin A of the x, y, z reference, measured from the X, Y, Z reference

$(\mathbf{a}_{B/A})_{\text{rel}}, (\mathbf{v}_{B/A})_{\text{rel}}$ = relative acceleration and relative velocity of "B with respect to A," as measured by an observer attached to the *rotating* x, y, z reference

$\dot{\mathbf{\Omega}}, \mathbf{\Omega}$ = angular acceleration and angular velocity of the x, y, z reference, measured from the X, Y, Z reference

$\mathbf{r}_{B/A}$ = relative position of "B with respect to A"

If the motions of points A and B are along *curved paths,* it is often convenient to express the accelerations $\mathbf{a}_B$, $\mathbf{a}_A$, and $(\mathbf{a}_{B/A})_{\text{rel}}$ in Eq. 16–26 in terms of their normal and tangential components. If Eq. 16–26 is compared with Eq. 16–17, written in the form $\mathbf{a}_B = \mathbf{a}_A + \dot{\boldsymbol{\Omega}} \times \mathbf{r}_{B/A} + \boldsymbol{\Omega} \times (\boldsymbol{\Omega} \times \mathbf{r}_{B/A})$, which is valid for a translating frame of reference, it can be seen that the difference between the equations is represented by the terms $2\boldsymbol{\Omega} \times (\mathbf{v}_{B/A})_{\text{rel}}$ and $(\mathbf{a}_{B/A})_{\text{rel}}$. In particular, $2\boldsymbol{\Omega} \times (\mathbf{v}_{B/A})_{\text{rel}}$ is called the *Coriolis acceleration,* named after the French engineer G. C. Coriolis (1792–1843), who was the first to determine it. This term represents the difference in the acceleration of B as measured from non-rotating and rotating x, y, z axes. As indicated by the vector cross product, the Coriolis acceleration will *always* be perpendicular to both $\boldsymbol{\Omega}$ and $(\mathbf{v}_{B/A})_{\text{rel}}$. It is an important component of the acceleration which must be considered whenever rotating reference frames are used. This often occurs, for example, when studying the accelerations and forces which act on rockets, long-range projectiles, or other bodies having motions which are largely affected by the rotation of the earth.

The following interpretation of the terms in Eq. 16–26 may be useful when applying this equation to the solution of problems.

$\mathbf{a}_B$ $\quad\quad$ {absolute acceleration of B

$\quad\quad\quad\quad\quad\quad$ (equals)

$\mathbf{a}_A$ $\quad\quad$ {absolute acceleration of origin of x, y, z frame

$\quad\quad\quad\quad\quad\quad$ (plus)

$\dot{\boldsymbol{\Omega}} \times \mathbf{r}_{B/A}$ $\quad$ {angular acceleration effect caused by rotation of x, y, z frame $\quad$ } motion of x, y, z frame

$\quad\quad\quad\quad\quad\quad$ (plus)

$\boldsymbol{\Omega} \times (\boldsymbol{\Omega} \times \mathbf{r}_{B/A})$ {angular velocity effect caused by rotation of x, y, z frame

$\quad\quad\quad\quad\quad\quad$ (plus)

$2\boldsymbol{\Omega} \times (\mathbf{v}_{B/A})_{\text{rel}}$ {combined effect of B moving relative to x, y, z coordinates and rotation of x, y, z frame $\quad$ } interacting motion

$\quad\quad\quad\quad\quad\quad$ (plus)

$(\mathbf{a}_{B/A})_{\text{rel}}$ {relative acceleration of B with respect to A $\quad$ } motion of B within x, y, z frame

PROCEDURE FOR ANALYSIS

The following procedure provides a method for applying Eqs. 16–23 and 16–26 to the solution of problems involving the planar motion of particles or rigid bodies.

Coordinate Axes. Choose an appropriate location for the origin and proper orientation of the axes for both the X, Y, Z and moving x, y, z reference frames. Most often solutions are easily obtained if at the instant considered: (1) the origins are coincident, (2) the axes are collinear, and/or (3) the axes are parallel. The moving frame should be selected fixed to the body or device where the relative motion occurs.

Kinematic Equations. After defining the origin A of the moving reference and specifying the moving point B, Eqs. 16–23 and 16–26 should be written in symbolic form

$$\mathbf{v}_B = \mathbf{v}_A + \mathbf{\Omega} \times \mathbf{r}_{B/A} + (\mathbf{v}_{B/A})_{\text{rel}}$$

$$\mathbf{a}_B = \mathbf{a}_A + \dot{\mathbf{\Omega}} \times \mathbf{r}_{B/A} + \mathbf{\Omega} \times (\mathbf{\Omega} \times \mathbf{r}_{B/A}) + 2\mathbf{\Omega} \times (\mathbf{v}_{B/A})_{\text{rel}} + (\mathbf{a}_{B/A})_{\text{rel}}$$

Each of the vectors in these equations should be defined from the problem data and expressed in Cartesian vector form. This essentially requires a determination of (1) the motion of the moving reference, i.e., $\mathbf{v}_A$, $\mathbf{a}_A$, $\mathbf{\Omega}$, and $\dot{\mathbf{\Omega}}$; and (2) the motion of B measured with respect to the moving reference, i.e., $\mathbf{r}_{B/A}$, $(\mathbf{v}_{B/A})_{\text{rel}}$, and $(\mathbf{a}_{B/A})_{\text{rel}}$. The components of all these vectors may be selected along either the X, Y, Z axes or the x, y, z axes. The choice is arbitrary provided a consistent set of unit vectors is used.

Finally, substitute the data into the kinematic equations and perform the vector operations.

The following examples numerically illustrate this procedure.

Example 16-17

At the instant $\theta = 60°$, the rod in Fig. 16–31 has an angular velocity of 3 rad/s and an angular acceleration of 2 rad/s². At this same instant, the collar C is traveling outward along the rod such that when $x = 0.2$ m the velocity is 2 m/s and the acceleration is 3 m/s², both measured relative to the rod. Determine the Coriolis acceleration and the velocity and acceleration of the collar at this instant.

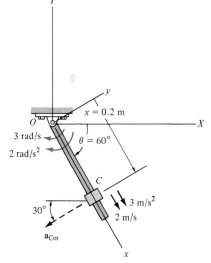

Fig. 16–31

SOLUTION

Coordinate Axes. The origin of both coordinate systems is located at point O, Fig. 16–31. Since motion of the collar is reported relative to the rod, the moving x, y, z frame of reference is *attached* to the rod.

Kinematic Equations

$$\mathbf{v}_C = \mathbf{v}_O + \mathbf{\Omega} \times \mathbf{r}_{C/O} + (\mathbf{v}_{C/O})_{\text{rel}} \qquad (1)$$

$$\mathbf{a}_C = \mathbf{a}_O + \dot{\mathbf{\Omega}} \times \mathbf{r}_{C/O} + \mathbf{\Omega} \times (\mathbf{\Omega} \times \mathbf{r}_{C/O}) + 2\mathbf{\Omega} \times (\mathbf{v}_{C/O})_{\text{rel}} + (\mathbf{a}_{C/O})_{\text{rel}} \qquad (2)$$

It will be simpler to express the data in terms of **i**, **j**, **k** component vectors rather than **I**, **J**, **K** components. Hence,

Motion of moving reference	*Motion of C with respect to moving reference*
$\mathbf{v}_O = \mathbf{0}$	$\mathbf{r}_{C/O} = \{0.2\mathbf{i}\}$ m
$\mathbf{a}_O = \mathbf{0}$	$(\mathbf{v}_{C/O})_{\text{rel}} = \{2\mathbf{i}\}$ m/s
$\mathbf{\Omega} = \{-3\mathbf{k}\}$ rad/s	$(\mathbf{a}_{C/O})_{\text{rel}} = \{3\mathbf{i}\}$ m/s²
$\dot{\mathbf{\Omega}} = \{-2\mathbf{k}\}$ rad/s²	

From Eq. (2) the Coriolis acceleration is defined as

$$\mathbf{a}_{\text{Cor}} = 2\mathbf{\Omega} \times (\mathbf{v}_{C/O})_{\text{rel}} = 2(-3\mathbf{k}) \times (2\mathbf{i}) = \{-12\mathbf{j}\} \text{ m/s}^2 \qquad ***Ans.***$$

This vector is shown dashed in Fig. 16–31. If desired, it may be resolved into **I**, **J** components acting along the X and Y axes, respectively.

The velocity and acceleration of the collar are determined by substituting the data into Eqs. (1) and (2) and evaluating the cross products, which yields

$$\mathbf{v}_C = \mathbf{v}_O + \mathbf{\Omega} \times \mathbf{r}_{C/O} + (\mathbf{v}_{C/O})_{\text{rel}}$$
$$= \mathbf{0} + (-3\mathbf{k}) \times (0.2\mathbf{i}) + 2\mathbf{i}$$
$$= \{2\mathbf{i} - 0.6\mathbf{j}\} \text{ m/s} \qquad ***Ans.***$$

$$\mathbf{a}_C = \mathbf{a}_O + \dot{\mathbf{\Omega}} \times \mathbf{r}_{C/O} + \mathbf{\Omega} \times (\mathbf{\Omega} \times \mathbf{r}_{C/O}) + 2\mathbf{\Omega} \times (\mathbf{v}_{C/O})_{\text{rel}} + (\mathbf{a}_{C/O})_{\text{rel}}$$
$$= \mathbf{0} + (-2\mathbf{k}) \times (0.2\mathbf{i}) + (-3\mathbf{k}) \times [(-3\mathbf{k}) \times (0.2\mathbf{i})] + 2(-3\mathbf{k}) \times (2\mathbf{i}) + 3\mathbf{i}$$
$$= \mathbf{0} - 0.4\mathbf{j} - 1.80\mathbf{i} - 12\mathbf{j} + 3\mathbf{i}$$
$$= \{1.20\mathbf{i} - 12.4\mathbf{j}\} \text{ m/s}^2 \qquad ***Ans.***$$

Example 16–18

The rod AB, shown in Fig. 16–32, rotates clockwise such that it has an angular velocity $\omega_{AB} = 2$ rad/s and angular acceleration $\alpha_{AB} = 4$ rad/s² when $\theta = 45°$. Determine the angular motion of rod DE at this instant. The collar at C is pin-connected to AB and slides over rod DE.

Fig. 16–32

SOLUTION

Coordinate Axes. The origins of both the fixed and moving frames of reference are located at D, Fig. 16–32. Furthermore, the x, y, z reference is attached to and rotates with rod DE.

Kinematic Equations

$$\mathbf{v}_C = \mathbf{v}_D + \mathbf{\Omega} \times \mathbf{r}_{C/D} + (\mathbf{v}_{C/D})_{\text{rel}} \qquad (1)$$

$$\mathbf{a}_C = \mathbf{a}_D + \dot{\mathbf{\Omega}} \times \mathbf{r}_{C/D} + \mathbf{\Omega} \times (\mathbf{\Omega} \times \mathbf{r}_{C/D}) + 2\mathbf{\Omega} \times (\mathbf{v}_{C/D})_{\text{rel}} + (\mathbf{a}_{A/B})_{\text{rel}} \qquad (2)$$

All vectors will be expressed in terms of $\mathbf{i}$, $\mathbf{j}$, $\mathbf{k}$ components.

Motion *of moving reference*	*Motion of C with respect* *to moving reference*
$\mathbf{v}_D = \mathbf{0}$	$\mathbf{r}_{C/D} = \{0.4\mathbf{i}\}$ m
$\mathbf{a}_D = \mathbf{0}$	$(\mathbf{v}_{C/D})_{\text{rel}} = (v_{C/D})_{\text{rel}}\mathbf{i}$
$\mathbf{\Omega} = -\omega_{DE}\mathbf{k}$	$(\mathbf{a}_{C/D})_{\text{rel}} = (a_{C/D})_{\text{rel}}\mathbf{i}$
$\dot{\mathbf{\Omega}} = -\alpha_{DE}\mathbf{k}$	

Motion of C: Since the collar moves along a *circular path,* its velocity and acceleration can be determined using Eqs. 16–9 and 16–13.

$$\mathbf{v}_C = \boldsymbol{\omega}_{AB} \times \mathbf{r}_{C/A} = (-2\mathbf{k}) \times (0.4\mathbf{i} + 0.4\mathbf{j}) = \{0.8\mathbf{i} - 0.8\mathbf{j}\} \text{ m/s}$$

$$\mathbf{a}_C = \boldsymbol{\alpha}_{AB} \times \mathbf{r}_{C/A} - \omega_{AB}^2 \mathbf{r}_{C/A} = (-4\mathbf{k}) \times (0.4\mathbf{i} + 0.4\mathbf{j}) - (2)^2(0.4\mathbf{i} + 0.4\mathbf{j}) = \{-3.2\mathbf{j}\} \text{ m/s}^2$$

Substituting the data into Eqs. (1) and (2), we have

$$\mathbf{v}_C = \mathbf{v}_D + \mathbf{\Omega} \times \mathbf{r}_{C/D} + (\mathbf{v}_{C/D})_{\text{rel}}$$

$$0.8\mathbf{i} - 0.8\mathbf{j} = \mathbf{0} + (-\omega_{DE}\mathbf{k}) \times (0.4\mathbf{i}) + (v_{C/D})_{\text{rel}}\mathbf{i}$$

$$0.8\mathbf{i} - 0.8\mathbf{j} = \mathbf{0} - 0.4\omega_{DE}\mathbf{j} + (v_{C/D})_{\text{rel}}\mathbf{i}$$

$$(v_{C/D})_{\text{rel}} = 0.8 \text{ m/s}, \quad \omega_{DE} = 2 \text{ rad/s} \qquad \textit{Ans.}$$

$$\mathbf{a}_C = \mathbf{a}_D + \dot{\mathbf{\Omega}} \times \mathbf{r}_{C/D} + \mathbf{\Omega} \times (\mathbf{\Omega} \times \mathbf{r}_{C/D}) + 2\mathbf{\Omega} \times (\mathbf{v}_{C/D})_{\text{rel}} + (\mathbf{a}_{C/D})_{\text{rel}}$$

$$-3.2\mathbf{j} = \mathbf{0} + (-\alpha_{DE}\mathbf{k}) \times (0.4\mathbf{i}) + (-2\mathbf{k}) \times [(-2\mathbf{k}) \times (0.4\mathbf{i})] + 2(-2\mathbf{k}) \times (0.8\mathbf{i}) + (a_{C/D})_{\text{rel}}\mathbf{i}$$

$$-3.2\mathbf{j} = -0.4\alpha_{DE}\mathbf{j} + 0.4\alpha_{DE}\mathbf{i} - 1.6\mathbf{i} - 1.6\mathbf{j} - 3.2\mathbf{j} + (a_{C/D})_{\text{rel}}\mathbf{i}$$

$$(a_{C/D})_{\text{rel}} = 1.6 \text{ m/s}^2 \qquad \alpha_{DE} = 0 \qquad \textit{Ans.}$$

Example 16–19

Two planes are flying at the same elevation and have the motions shown in Fig. 16–33. Determine the velocity and acceleration of A as measured by the pilot of B.

SOLUTION

Coordinate Axes. Since the relative motion of A with respect to B is being sought, the x, y, z axes are attached to plane B, Fig. 16–33. At the *instant* considered, the origin B coincides with the origin of the fixed X, Y, Z frame.

Kinematic Equations

$$\mathbf{v}_A = \mathbf{v}_B + \boldsymbol{\Omega} \times \mathbf{r}_{A/B} + (\mathbf{v}_{A/B})_{\text{rel}} \tag{1}$$

$$\mathbf{a}_A = \mathbf{a}_B + \dot{\boldsymbol{\Omega}} \times \mathbf{r}_{A/B} + \boldsymbol{\Omega} \times (\boldsymbol{\Omega} \times \mathbf{r}_{A/B}) + 2\boldsymbol{\Omega} \times (\mathbf{v}_{A/B})_{\text{rel}} + (\mathbf{a}_{A/B})_{\text{rel}} \tag{2}$$

Motion of moving reference:

$$\mathbf{v}_B = \{600\mathbf{j}\} \text{ km/h}$$

$$\mathbf{a}_B = (\mathbf{a}_B)_n + (\mathbf{a}_B)_t = \{900\mathbf{i} - 100\mathbf{j}\} \text{ km/h}^2$$

$$(a_B)_n = \frac{v_B^2}{\rho} = \frac{(600)^2}{400} = 900 \text{ km/h}^2$$

$$\boldsymbol{\Omega} = \{-1.5\mathbf{k}\} \text{ rad/h} \qquad \Omega = \frac{v_B}{\rho} = \frac{600 \text{ km/h}}{400 \text{ km}} = 1.5 \text{ rad/h} \curvearrowright$$

$$\dot{\boldsymbol{\Omega}} = \{0.25\mathbf{k}\} \text{ rad/h}^2 \qquad \dot{\Omega} = \frac{(a_B)_t}{\rho} = \frac{100 \text{ km/h}^2}{400 \text{ km}} = 0.25 \text{ rad/h}^2 \curvearrowleft$$

Fig. 16–33

Motion of A with respect to moving reference:

$$\mathbf{r}_{A/B} = \{-4\mathbf{i}\} \text{ km} \qquad (\mathbf{v}_{A/B})_{\text{rel}} = ? \qquad (\mathbf{a}_{A/B})_{\text{rel}} = ?$$

Substituting this data into Eqs. (1) and (2), realizing that $\mathbf{v}_A = \{700\mathbf{j}\}$ km/h and $\mathbf{a}_A = \{50\mathbf{j}\}$ km/h^2, we have

$$\mathbf{v}_A = \mathbf{v}_B + \boldsymbol{\Omega} \times \mathbf{r}_{A/B} + (\mathbf{v}_{A/B})_{\text{rel}}$$

$$700\mathbf{j} = 600\mathbf{j} + (-1.5\mathbf{k}) \times (-4\mathbf{i}) + (\mathbf{v}_{A/B})_{\text{rel}}$$

$$(\mathbf{v}_{A/B})_{\text{rel}} = \{94\mathbf{j}\} \text{ km/h} \qquad\qquad\qquad Ans.$$

$$\mathbf{a}_A = \mathbf{a}_B + \dot{\boldsymbol{\Omega}} \times \mathbf{r}_{A/B} + \boldsymbol{\Omega} \times (\boldsymbol{\Omega} \times \mathbf{r}_{A/B}) + 2\boldsymbol{\Omega} \times (\mathbf{v}_{A/B})_{\text{rel}} + (\mathbf{a}_{A/B})_{\text{rel}}$$

$$50\mathbf{j} = (900\mathbf{i} - 100\mathbf{j}) + (0.25\mathbf{k}) \times (-4\mathbf{i})$$

$$+ (-1.5\mathbf{k}) \times [(-1.5\mathbf{k}) \times (-4\mathbf{i})] + 2(-1.5\mathbf{k}) \times (94\mathbf{j}) + (\mathbf{a}_{A/B})_{\text{rel}}$$

$$(\mathbf{a}_{A/B})_{\text{rel}} = \{-1190\mathbf{i} + 151\mathbf{j}\} \text{ km/h}^2 \qquad\qquad\qquad Ans.$$

PROBLEMS

16–86. A ball B of negligible size rolls through the tube such that at the instant shown it has a velocity of 5 ft/s and an acceleration of 3 ft/s^2, measured relative to the tube. If the tube has an angular velocity of $\omega = 3$ rad/s and an angular acceleration of $\alpha = 5$ rad/s^2 at this same instant, determine the velocity and acceleration of the ball.

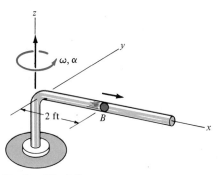

Prob. 16–86

16–87. The man stands on the platform at O and runs out toward the edge such that when he is at A, $y = 5$ ft, his mass center has a velocity of 2 ft/s and an acceleration of 3 ft/s^2, both measured with respect to the platform and directed along the y axis. If the platform has the angular motions shown, determine the velocity and acceleration of his mass center at this instant.

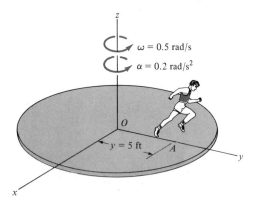

Prob. 16–87

*16–88. Block B moves along the slot in the platform with a constant speed of 2 ft/s, measured relative to the platform in the direction shown. If the platform is rotating at a constant rate of $\omega = 5$ rad/s, determine the velocity and acceleration of the block at the instant $\theta = 90°$.

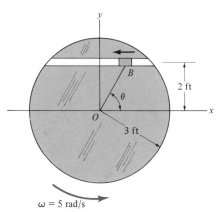

Probs. 16–88/16–89/16–90

16–89. Solve Prob. 16–88 if when $\theta = 90°$ the block has a velocity of 2 ft/s and an acceleration of 4 ft/s^2, measured relative to the platform in the direction shown, and at this same instant the platform has an angular velocity of $\omega = 5$ rad/s and an angular acceleration of $\alpha = 8$ rad/s^2.

16–90. Block B moves along the slot in the platform with a constant speed of 2 ft/s, measured relative to the platform in the direction shown. If the platform is rotating at a constant rate of $\omega = 5$ rad/s, determine the velocity and acceleration of the block at the instant $\theta = 60°$.

16–91. Rod AB rotates counterclockwise with a constant angular velocity of $\omega = 3$ rad/s. Determine the velocity of point C located on the double collar when $\theta = 30°$. The collar consists of two pin-connected slider blocks which are constrained to move along the circular path and the rod AB.

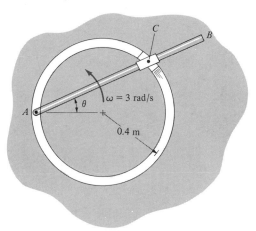

Probs. 16–91/16–92

*16–92. Rod AB rotates counterclockwise with a constant angular velocity of $\omega = 3$ rad/s. Determine the velocity and acceleration of point C located on the double collar when $\theta = 45°$. The collar consists of two pin-connected slider blocks which are constrained to move along the circular path and the rod AB.

16–93. A ride in an amusement park consists of a rotating arm AB having a constant angular velocity $\omega_{AB} = 2$ rad/s about point A and a car mounted at the end of the arm which has a constant angular velocity of $\omega' = \{-0.5\mathbf{k}\}$ rad/s, measured relative to the arm. At the instant shown, determine the velocity and acceleration of the passenger at C.

Probs. 16–93/16–94

16–94. Solve Prob. 16–93 if at the instant shown the arm has an angular acceleration of $\alpha_{AB} = 1$ rad/s² when $\omega_{AB} = 2$ rad/s and the car has a relative angular acceleration of $\alpha' = \{-0.6\mathbf{k}\}$ rad/s² when $\omega' = \{-0.5\mathbf{k}\}$ rad/s.

16–95. At the instant shown, the robotic arm AB is rotating counterclockwise at $\omega = 5$ rad/s and has an angular acceleration of $\alpha = 2$ rad/s². Simultaneously, the grip BC is rotating counterclockwise at $\omega' = 6$ rad/s and $\alpha' = 2$ rad/s², both measured with respect to a *fixed* reference. Determine the velocity and acceleration of the object held at the grip, C.

Prob. 16–95

*16–96. The "quick return" mechanism consists of a crank AB, slider block B, and slotted link CD. If the crank has the angular motions shown, determine the angular velocity and angular acceleration of the slotted link at this instant.

Prob. 16–96

291

16–97. At a given instant rod *AB* has the angular motions shown. Determine the angular velocity and angular acceleration of rod *CD* at this instant. There is a collar at *D*.

Prob. 16–97

16–98. At the instant $\theta = 60°$ rod *AB* has an angular velocity $\omega_{AB} = 3$ rad/s and an angular acceleration $\alpha_{AB} = 5$ rad/s². Determine the angular velocity and angular acceleration of rod *CD* at this instant. The collar at *C* is pin-connected to *CD* and slides over *AB*.

Prob. 16–98

16–99. The disk rotates with the angular motion shown. Determine the angular velocity and angular acceleration of the slotted link *AC* at this instant. The peg at *B* is fixed to the disk.

Prob. 16–99

*16–100. The two-link mechanism serves to amplify angular motion. Link *AB* has a pin at *B* which is confined to move within the slot of link *CD*. If at the instant shown *AB* (input) has an angular velocity $\omega_{AB} = 3$ rad/s and an angular acceleration $\alpha_{AB} = 2$ rad/s², determine the angular velocity and angular acceleration of *CD* (output) at this instant.

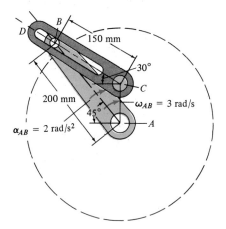

Prob. 16–100

17

Planar Kinetics of a Rigid Body: Force and Acceleration

In Chapter 16 the planar kinematics of rigid-body motion was presented in order of increasing complexity, that is, translation, rotation about a fixed axis, and general plane motion. The study of rigid-body kinetics in this chapter will be presented in somewhat the same order. The chapter begins by introducing a property of a body called the moment of inertia. Afterwards, a derivation of the equations of general plane motion for a symmetric rigid body is given. These equations are then applied to specific problems of rigid-body translation, rotation about a fixed axis, and finally general plane motion. A kinetic study of these motions is referred to as the kinetics of planar motions or simply *planar kinetics*. The more complex study of three-dimensional rigid-body kinetics, which includes planar motion of unsymmetrical rigid bodies, is presented in Chapter 21.

17.1 Moment of Inertia

Since a body has a definite size and shape, an applied nonconcurrent force system may cause the body to both translate and rotate. The translational aspects of the motion were studied in Chapter 13 and are governed by the equation $\mathbf{F} = m\mathbf{a}$. It will be shown in Sec. 17.2 that the rotational aspects, caused by the moment $\mathbf{M}$, are governed by an equation of the form $\mathbf{M} = I\boldsymbol{\alpha}$. The symbol I in this equation is termed the moment of inertia. By comparison, the *moment of inertia* is a measure of the resistance of a body to *angular acceleration* ($\mathbf{M} = I\boldsymbol{\alpha}$) in the same way that *mass* is a measure of the body's resistance to *acceleration* ($\mathbf{F} = m\mathbf{a}$).

We define the *moment of inertia* as the integral of the "second moment" about an axis of all the differential-size elements of mass dm which compose the body.* For example, consider the rigid body shown in Fig. 17–1. The body's moment of inertia about the z axis is

Fig. 17–1

$$I = \int_m r^2 \, dm \qquad (17\text{–}1)$$

Here the "moment arm" r is the perpendicular distance from the axis to the arbitrary element dm. Since the formulation involves r, the value of I is different for each axis about which it is computed. For example, if the axis coincides with the longitudinal axis for a slender rod, I will be small, since r is small for each element of the rod. If the axis is perpendicular to the rod, I will be large, because the rod will have a larger mass distributed farther from the axis. In the study of planar kinetics, the axis which is generally chosen for analysis passes through the body's mass center G and is always perpendicular to the plane of motion. The moment of inertia computed about this axis will be denoted as I_G. Note that because r is squared in Eq. 17–1 the mass moment of inertia is always a positive quantity. Common units used for its measurement are kg · m^2 or slug · ft^2.

---

* Another property of the body, which measures the symmetry of the body's mass with respect to a coordinate system, is the product of inertia. This property applies to the three-dimensional motion of a body and will be discussed in Chapter 21.

PROCEDURE FOR ANALYSIS

In this treatment, we will only consider symmetric bodies having surfaces which are generated by revolving a curve about an axis. An example of such a body of which the volume of revolution is generated about the z axis is shown in Fig. 17–2. If the body consists of material having a variable density, $\rho = \rho\,(x, y, z)$, the elemental mass dm of the body may be expressed in terms of its density and volume as $dm = \rho\,dV$. Substituting dm in Eq. 17–1, the body's moment of inertia is then computed using *volume elements* for integration, i.e.,

$$I = \int_V r^2 \rho\,dV \qquad (17\text{--}2)$$

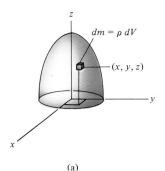

(a)

In the special case of ρ being a *constant*, this term may be factored out of the integral and the integration is then purely a function of geometry,

$$I = \rho \int_V r^2\,dV \qquad (17\text{--}3)$$

When the elemental volume chosen for integration has infinitesimal dimensions in all three directions, e.g., $dV = dx\,dy\,dz$, Fig. 17–2a, the moment of inertia of the body must be computed using "triple integration." The integration process can, however, be simplified to a *single integration* provided the chosen elemental volume has a differential size or thickness in only *one direction*. Shell or disk elements are often used for this purpose.

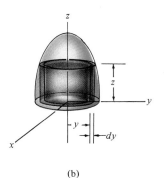

(b)

Shell Element. If a *shell element* having a height z, radius $r = y$, and thickness dy is chosen for integration, Fig. 17–2b, then the volume is $dV = (2\pi y)(z)\,dy$. This element may be used in Eq. 17–2 or 17–3 for computing the moment of inertia I_z of the body about the z axis, since the *entire element*, due to its "thinness," lies at the *same* perpendicular distance $r = y$ from the z axis (see Example 17–1).

Disk Element. If a disk element having a radius $r = y$ and a thickness dz is chosen for integration, Fig. 17–2c, then the volume is $dV = (\pi y^2)\,dz$. In this case, however, the element is *finite* in the radial direction, and consequently parts of it *do not* all lie at the *same radial distance r* from the z axis. As a result, Eq. 17–2 or 17–3 *cannot* be used to determine I_z directly. Instead, to perform the integration using this element, it is first necessary to determine the moment of inertia *of the element* about the z axis and then integrate this result (see Example 17–2).

(c)

Fig. 17–2

Example 17–1

Determine the moment of inertia of the right circular cylinder shown in Fig. 17–3a about the z axis. The density ρ of the material is constant.

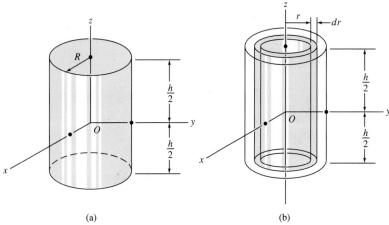

Fig. 17–3

(a)   **(b)**

SOLUTION

Shell Element. This problem may be solved using the *shell element* in Fig. 17–3b and single integration. The volume of the element is $dV = (2\pi r)(h)\, dr$, so that the mass is $dm = \rho\, dV = \rho(2\pi hr\, dr)$. Since the *entire element* lies at the same distance r from the z axis, the moment of inertia *of the element* is

$$dI_z = r^2\, dm = \rho 2\pi hr^3\, dr$$

Integrating over the entire region of the cylinder yields

$$I_z = \int_m r^2\, dm = \rho 2\pi h \int_0^R r^3\, dr = \frac{\rho\pi}{2} R^4 h$$

The mass of the cylinder is

$$m = \int_m dm = \rho 2\pi h \int_0^R r\, dr = \rho\pi h R^2$$

so that

$$I_z = \frac{1}{2} mR^2 \qquad\qquad Ans.$$

Example 17–2

A solid is formed by revolving the shaded area shown in Fig. 17–4a about the y axis. If the density of the material is 5 slug/ft^3, determine the moment of inertia about the y axis.

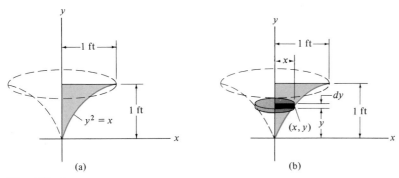

Fig. 17–4

SOLUTION

Disk Element. The moment of inertia will be computed using a *disk element,* as shown in Fig. 17–4b. Here the element intersects the curve at the arbitrary point (x, y) and has a mass

$$dm = \rho \, dV = \rho(\pi x^2) \, dy$$

Although all portions of the element are *not* located at the same distance from the y axis, it is still possible to determine the moment of inertia dI_y *of the element* about the y axis. In the preceding example it was shown that the moment of inertia of a cylinder about its longitudinal axis is $I = \frac{1}{2}mR^2$, where m and R are the mass and radius of the cylinder. Since the height of the cylinder is not involved in this formula, the moment of inertia of the disk element in Fig. 17–4b is

$$dI_y = \tfrac{1}{2}(dm)x^2 = \tfrac{1}{2}[\rho(\pi x^2) \, dy]x^2$$

Substituting $x = y^2$, $\rho = 5$ slug/ft^3, and integrating with respect to y, from $y = 0$ to $y = 1$ ft, yields the moment of inertia for the entire solid.

$$I_y = \frac{\pi(5)}{2} \int_0^1 x^4 \, dy = \frac{\pi(5)}{2} \int_0^1 y^8 \, dy = 0.873 \text{ slug} \cdot \text{ft}^2 \qquad Ans.$$

Parallel-Axis Theorem

If the moment of inertia of the body about an axis passing through the body's mass center is known, then the body's moment of inertia may be determined about any other *parallel axis* by using the *parallel-axis theorem*. This theorem can be derived by considering the body shown in Fig. 17–5. The z' axis passes through the body's mass center G, whereas the corresponding *parallel axis z* lies at a constant distance d away. Selecting the differential element of mass dm, which is located at point (x', y'), and using the Pythagorean theorem, $r^2 = (d + x')^2 + y'^2$, we can express the moment of inertia of the body about the z axis as

$$I = \int_m r^2 \, dm = \int_m [(d + x')^2 + y'^2] \, dm$$

$$= \int_m (x'^2 + y'^2) \, dm + 2d \int_m x' \, dm + d^2 \int_m dm$$

Since $r'^2 = x'^2 + y'^2$, the first integral represents I_G. The second integral equals *zero*, since the z' axis passes through the body's mass center, i.e., $\int x' \, dm = \bar{x} \int dm = 0$ since $\bar{x} = 0$. Finally, the third integral represents the total mass m of the body. Hence, the moment of inertia about the z axis can be written as

$$\boxed{I = I_G + md^2} \qquad (17\text{–}4)$$

where

I_G = moment of inertia about the z' axis passing through the mass center G
m = mass of the body
d = perpendicular distance between the parallel axes

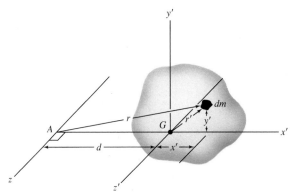

Fig. 17–5

Radius of Gyration

Occasionally, the moment of inertia of a body about a specified axis is reported in handbooks using the *radius of gyration, k*. When this length k and the body's mass m are known, the body's moment of inertia is determined from the equation

$$I = mk^2 \quad \text{or} \quad k = \sqrt{\frac{I}{m}} \qquad (17\text{–}5)$$

Note the *similarity* between the definition of k in this formula and r in the equation $dI = r^2\, dm$, which defines the moment of inertia of an elemental mass dm of the body about an axis.

Composite Bodies

If a body is constructed of a number of simple shapes such as disks, spheres, and rods, the moment of inertia of the body about any axis z can be determined by adding algebraically the moments of inertia of all the composite shapes computed about the z axis. Algebraic addition is necessary since a composite part must be considered as a negative quantity if it has already been counted as part of another part—for example a "hole" subtracted from a solid plate. The parallel-axis theorem is needed for the calculations if the center of mass of each composite part does not lie on the z axis. For the calculation then $I = \Sigma(I_G + md^2)$. Here I_G for each of the composite parts is either known or can be computed by integration.*

*See the table given on the inside back cover of this book.

Example 17–3

If the plate shown in Fig. 17–6a has a density of 8000 kg/m³ and a thickness of 10 mm, compute its moment of inertia about an axis directed perpendicular to the page and passing through point O.

(a)   (b)

Fig. 17–6

SOLUTION

The plate consists of two composite parts, the 250-mm-radius disk *minus* a 125-mm-radius disk, Fig. 17–6b. The moment of inertia about O can be determined by computing the moment of inertia of each of these parts about O and then *algebraically* adding the results. The computations are performed by using the parallel-axis theorem in conjunction with the data listed in the table on the inside back cover.

Disk. The moment of inertia of a thin disk about an axis perpendicular to the plane of the disk is $I_G = \frac{1}{2}mr^2$. The mass center of the disk is located at a distance of 0.25 m from point O. Thus,

$$m_d = \rho_d V_d = 8000 \text{ kg/m}^3[\pi(0.25 \text{ m})^2(0.01 \text{ m})] = 15.71 \text{ kg}$$
$$(I_O)_d = \tfrac{1}{2}m_d r_d^2 + m_d d^2$$
$$= \frac{1}{2}(15.71 \text{ kg})(0.25 \text{ m})^2 + (15.71 \text{ kg})(0.25 \text{ m})^2$$
$$= 1.473 \text{ kg} \cdot \text{m}^2$$

Hole. For the 125-mm-radius disk (hole), we have

$$m_h = \rho_h V_h = 8000 \text{ kg/m}^3[\pi(0.125 \text{ m})^2(0.01 \text{ m})] = 3.93 \text{ kg}$$
$$(I_O)_h = \tfrac{1}{2}m_h r_h^2 + m_h d^2$$
$$= \frac{1}{2}(3.93 \text{ kg})(0.125 \text{ m})^2 + (3.93 \text{ kg})(0.25 \text{ m})^2$$
$$= 0.276 \text{ kg} \cdot \text{m}^2$$

The moment of inertia of the plate about point O is therefore

$$I_O = (I_O)_d - (I_O)_h$$
$$= 1.473 - 0.276$$
$$= 1.20 \text{ kg} \cdot \text{m}^2 \qquad\qquad Ans.$$

Example 17–4

The pendulum consists of two thin rods, each having a weight of 10 lb and suspended from point O as shown in Fig. 17–7. Compute the pendulum's moment of inertia about an axis passing through (a) the pin at O, and (b) the mass center G of the pendulum.

SOLUTION

Part (a). Using the table on the inside back cover, the moment of inertia of rod OA about an axis perpendicular to the page and passing through the end point O of the rod is $I_O = \frac{1}{3}ml^2$. Hence,

$$(I_O)_{OA} = \frac{1}{3}ml^2 = \frac{1}{3}\left(\frac{10}{32.2}\right)(2)^2 = 0.414 \text{ slug} \cdot \text{ft}^2$$

Fig. 17–7

The same value is obtained using $I_G = \frac{1}{12}ml^2$ and the parallel-axis theorem

$$(I_O)_{OA} = \frac{1}{12}ml^2 + md^2 = \frac{1}{12}\left(\frac{10}{32.2}\right)(2)^2 + \left(\frac{10}{32.2}\right)(1)^2$$

$$= 0.414 \text{ slug} \cdot \text{ft}^2$$

For rod BC we have

$$(I_O)_{BC} = \frac{1}{12}ml^2 + md^2 = \frac{1}{12}\left(\frac{10}{32.2}\right)(2)^2 + \left(\frac{10}{32.2}\right)(2)^2$$

$$= 1.346 \text{ slug} \cdot \text{ft}^2$$

The moment of inertia of the pendulum is therefore

$$I_O = 0.414 + 1.346 = 1.76 \text{ slug} \cdot \text{ft}^2 \qquad \textit{Ans.}$$

Part (b). The mass center G will be located relative to the pin at O. Assuming this distance to be $\bar{y}$, Fig. 17–7, and using the formula for determining the mass center, we have

$$\bar{y} = \frac{\Sigma \tilde{y}m}{\Sigma m} = \frac{1\left(\dfrac{10}{32.2}\right) + 2\left(\dfrac{10}{32.2}\right)}{\left(\dfrac{10}{32.2}\right) + \left(\dfrac{10}{32.2}\right)} = 1.50 \text{ ft}$$

The moment of inertia I_G may be computed in the same manner as I_O, which requires successive applications of the parallel-axis theorem to transfer the moments of inertia of rods OA and BC to G. A more direct solution, however, involves using the result for I_O, i.e.,

$$I_O = I_G + md^2; \qquad 1.76 = I_G + \left(\frac{20}{32.2}\right)(1.50)^2$$

$$I_G = 0.362 \text{ slug} \cdot \text{ft}^2 \qquad \textit{Ans.}$$

PROBLEMS

17–1. Determine the moment of inertia I_y for the slender rod. The rod's density ρ and cross-sectional area A are constant. Express the result in terms of the rod's total mass m.

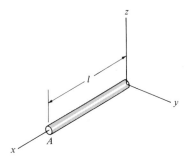

Prob. 17–1

17–2. Determine the moment of inertia of the thin ring about the z axis. The ring has a mass m.

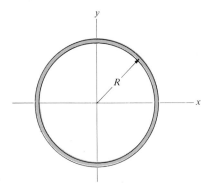

Prob. 17–2

17–3. The right circular cone is formed by revolving the shaded area around the x axis. Determine the moment of inertia I_x and express the result in terms of the total mass m of the cone. The cone has a constant density ρ.

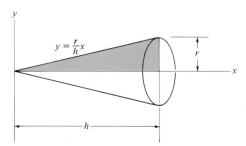

$$y = \frac{r}{h}x$$

Prob. 17–3

*17–4. The paraboloid is formed by revolving the shaded area around the x axis. Determine the radius of gyration k_x. The material has a constant density ρ.

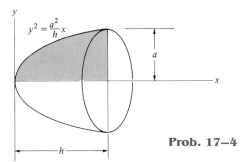

$$y^2 = \frac{a^2}{h}x$$

Prob. 17–4

17–5. A semiellipsoid is formed by rotating the shaded area about the x axis. Determine the moment of inertia of this solid with respect to the x axis and express the result in terms of the mass m of the solid. The material has a constant density ρ.

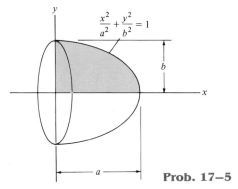

$$\frac{x^2}{a^2} + \frac{y^2}{b^2} = 1$$

Prob. 17–5

17–6. The sphere is formed by revolving the shaded area around the x axis. Determine the moment of inertia I_x and express the result in terms of the total mass m of the sphere. The sphere has a constant density ρ.

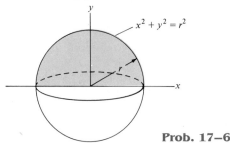

$$x^2 + y^2 = r^2$$

Prob. 17–6

17–7. The solid is formed by revolving the shaded area around the *x* axis. Determine the radius of gyration k_x. The specific weight of the material is $\gamma = 380 \text{ lb/ft}^3$.

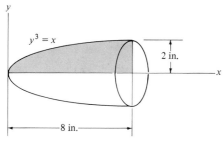

Prob. 17–7

*17–8. Determine the moment of inertia of the homogeneous triangular prism with respect to the *y* axis. Express the result in terms of the mass *m* of the prism. *Hint:* For integration, use thin plate elements parallel to the *x-y* plane and having a thickness of *dz*.

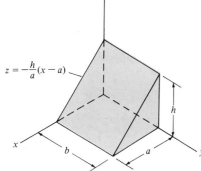

Prob. 17–8

17–9. The concrete solid is formed by rotating the shaded area about the *y* axis. Determine the moment of inertia I_y. The specific weight of concrete is $\gamma = 150 \text{ lb/ft}^3$.

Prob. 17–9

17–10. The slender rods have a weight of 3 lb/ft. Determine the moment of inertia of the assembly about an axis perpendicular to the page and passing through point *A*.

Prob. 17–10

17–11. The wheel consists of a thin ring having a mass of 10 kg and *four* spokes made from slender rods and each having a mass of 2 kg. Determine the wheel's moment of inertia about an axis perpendicular to the page and passing through point *A*.

Prob. 17–11

*17–12. Determine the moment of inertia of the solid steel assembly about the *xx* axis. Steel has a specific weight of $\gamma = 490 \text{ lb/ft}^3$.

Prob. 17–12

17–13. Determine the moment of inertia I_z of the frustum of the cone which has a conical depression. The material has a density of 200 kg/m³.

Prob. 17–13

17–14. Determine the moment of inertia of the wheel about an axis which is perpendicular to the page and passes through the center of mass G. The material has a specific weight of $\gamma = 90$ lb/ft³.

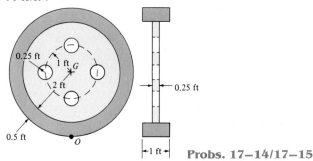

Probs. 17–14/17–15

17–15. Determine the moment of inertia of the wheel about an axis which is perpendicular to the page and passes through point O. The material has a specific weight of $\gamma = 90$ lb/ft³.

***17–16.** The pendulum consists of a disk having a mass of 6 kg and slender rods AB and DC which have a mass of 2 kg/m. Determine the length L of DC so that the center of mass is at the bearing O. What is the moment of inertia of the assembly about an axis perpendicular to the page and passing through point O?

Prob. 17–16

17–17. Determine the moment of inertia of the wire triangle about an axis perpendicular to the page and passing through point O. Also, locate the mass center G and determine the moment of inertia about an axis perpendicular to the page and passing through point G. The wire has a mass of 0.3 kg/m.

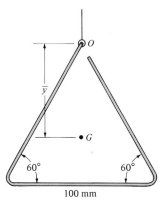

Prob. 17–17

17–18. The pendulum consists of two slender rods AB and OC which have a mass of 3 kg/m. The thin plate has a mass of 12 kg/m². Determine the location $\bar{y}$ of the center of mass G of the pendulum, then calculate the moment of inertia of the pendulum about an axis perpendicular to the page and passing through G.

Prob. 17–18

17.2 Planar Kinetic Equations of Motion

In the following analysis we will limit our study of planar kinetics to rigid bodies which, along with their loadings, are considered to be *symmetrical* with respect to a fixed reference plane.* The plane motion of such a rigid body may be analyzed in a fixed reference plane because the path of motion of each particle of the body is a plane curve parallel to the reference plane. Thus, for the analysis, the motion of the body may be represented by a *slab* having the same mass and exterior outline as the body. All the forces (and couples) acting on the body may then be projected onto the plane of this slab. An example of an arbitrary slab (body) of this type is shown in Fig. 17–8a. Here the *inertial frame of reference x, y, z* has its origin *coincident* with the arbitrary point P in the body. By definition, *these axes do not rotate and are either fixed or translate with constant velocity.*

Equation of Translational Motion

The external forces shown on the body in Fig. 17–8a symbolically represent the effect of gravitational, electrical, magnetic, or contact forces between adjacent bodies. Since this force system has been considered previously in Sec. 13.3, for the analysis of a system of particles, the results may be used here, in which case the particles are contained within the boundary of the body (or slab). Hence, if the equation of motion is applied to each of the particles of the body, and the results added vectorially, it may be concluded that

$$\Sigma \mathbf{F} = m\mathbf{a}_G$$

This equation is referred to as the *equation of motion* for the mass center of a rigid body. It states that *the sum of all the external forces acting on the body is equal to the body's mass times the acceleration of the mass center G.*

For motion of the body (or slab) in the x-y plane, the equation of motion for G may be written in the form of two independent scalar equations, namely,

$$\Sigma F_x = m(a_G)_x$$
$$\Sigma F_y = m(a_G)_y$$

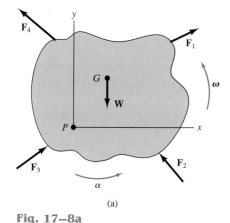

Fig. 17–8a

*By doing this, the rotational equation of motion reduces to a rather simplified form. The more general case of body shape and loading is considered in Chapter 21.

Particle free-body diagram

(b)

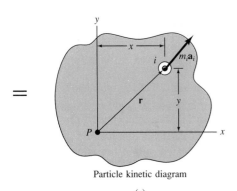

Particle kinetic diagram

(c)

Fig. 17–8(b, c)

Equation of Rotational Motion

We will now determine the effects caused by the moments of the external force system computed about an axis perpendicular to the plane of motion (the z axis) and passing through point P. As shown on the free-body diagram of the ith particle, Fig. 17–8b, $\mathbf{F}_i$ represents the *resultant external force* acting on the particle, and $\mathbf{f}_i$ is the *resultant of the internal forces* caused by interactions with adjacent particles. If the particle has a mass m_i and at the instant considered its acceleration is $\mathbf{a}_i$, then the kinetic diagram is constructed as shown in Fig. 17–8c. If moments of the forces acting on the particle are summed about point P, then we require

$$\mathbf{r} \times \mathbf{F}_i + \mathbf{r} \times \mathbf{f}_i = \mathbf{r} \times m_i \mathbf{a}_i$$

or

$$(\mathbf{M}_P)_i = \mathbf{r} \times m_i \mathbf{a}_i$$

We will now write this equation in terms of the acceleration $\mathbf{a}_P$ of point P. If the body has an angular acceleration $\boldsymbol{\alpha}$ and angular velocity $\boldsymbol{\omega}$, Fig. 17–8d, then using Eq. 16–17 we have

$$(\mathbf{M}_P)_i = m_i \mathbf{r} \times (\mathbf{a}_P - \omega^2 \mathbf{r} + \boldsymbol{\alpha} \times \mathbf{r})$$
$$= m_i[\mathbf{r} \times \mathbf{a}_P - \omega^2(\mathbf{r} \times \mathbf{r}) + \mathbf{r} \times (\boldsymbol{\alpha} \times \mathbf{r})]$$

The second term on the right side is zero since $\mathbf{r} \times \mathbf{r} = \mathbf{0}$. Expressing the vectors with Cartesian components and carrying out the cross-product operations yields

$$(M_P)_i \, \mathbf{k} = m_i\{(x\mathbf{i} + y\mathbf{j}) \times [(a_P)_x \mathbf{i} + (a_P)_y \mathbf{j}]$$
$$+ (x\mathbf{i} + y\mathbf{j}) \times [\alpha \mathbf{k} \times (x\mathbf{i} + y\mathbf{j})]\}$$
$$(M_P)_i \mathbf{k} = m_i[-y(a_P)_x + x(a_P)_y + \alpha x^2 + \alpha y^2]\mathbf{k}$$
$$(M_P)_i = m_i[-y(a_P)_x + x(a_P)_y + \alpha r^2]$$

Letting $m_i \rightarrow dm$, and integrating with respect to the entire mass m of the body, we obtain the resultant moment equation

$$\Sigma M_P = -\left(\int_m y \, dm\right)(a_P)_x + \left(\int_m x \, dm\right)(a_P)_y + \left(\int_m r^2 \, dm\right)\alpha$$

Here ΣM_P represents only the moment of the *external forces* acting on the body about point P. The resultant moment of the internal forces is zero, since for the entire body they occur in equal and opposite collinear pairs and thus the moment of each pair of forces about point P cancels. The integrals in the first and second terms on the right are used to locate the body's center of mass G with respect to P, since $\bar{y}m = \int y \, dm$ and $\bar{x}m = \int x \, dm$, Fig. 17–8d. Also, the last integral represents the body's moment of inertia computed about the z axis, i.e., $I_P = \int r^2 \, dm$. Thus,

$$\Sigma M_P = -\bar{y}m(a_P)_x + \bar{x}m(a_P)_y + I_P\alpha \qquad (17\text{–}6)$$

It is possible to reduce this equation to a simpler form if point P coincides with the mass center G for the body.* If this is the case, then $\bar{x} = \bar{y} = 0$, and therefore

$$\boxed{\Sigma M_G = I_G \alpha} \qquad (17\text{-}7)$$

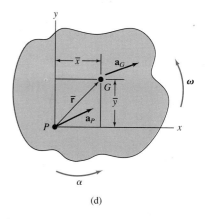

(d)

This rotational equation of motion states that the sum of the moments of all the external forces computed about the body's mass center G is equal to the product of the moment of inertia of the body about an axis passing through G and the body's angular acceleration.

Equation 17–6 can also be rewritten in terms of the x and y components of $\mathbf{a}_G$ and the body's moment of inertia I_G. If point G is located at point $(\bar{x}, \bar{y})$, Fig. 17–8d, then by the parallel-axis theorem, $I_P = I_G + m(\bar{x}^2 + \bar{y}^2)$. Substituting into Eq. 17–6 and rearranging terms, we get

$$\Sigma M_P = \bar{y}m[-(a_P)_x + \bar{y}\alpha] + \bar{x}m[(a_P)_y + \bar{x}\alpha] + I_G\alpha \qquad (17\text{-}8)$$

From the kinematic diagram of Fig. 17–8d, $\mathbf{a}_P$ can be expressed in terms of $\mathbf{a}_G$ as

$$\mathbf{a}_G = \mathbf{a}_P - \omega^2\bar{\mathbf{r}} + \boldsymbol{\alpha} \times \bar{\mathbf{r}}$$

$$(a_G)_x\mathbf{i} + (a_G)_y\mathbf{j} = (a_P)_x\mathbf{i} + (a_P)_y\mathbf{j} - \omega^2(\bar{x}\mathbf{i} + \bar{y}\mathbf{j}) + \alpha\mathbf{k} \times (\bar{x}\mathbf{i} + \bar{y}\mathbf{j})$$

Carrying out the cross product and equating the respective $\mathbf{i}$ and $\mathbf{j}$ components yields the two scalar equations.

$$(a_G)_x = (a_P)_x - \bar{x}\omega^2 - \bar{y}\alpha$$

$$(a_G)_y = (a_P)_y - \bar{y}\omega^2 + \bar{x}\alpha$$

From these equations, $[-(a_P)_x + \bar{y}\alpha] = [-(a_G)_x - \bar{x}\omega^2]$ and $[(a_P)_y + \bar{x}\alpha] = [(a_G)_y + \bar{y}\omega^2]$. Substituting these results into Eq. 17–8 and simplifying gives

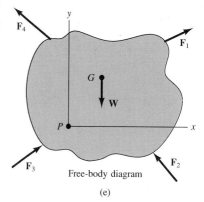

F₄, F₁, G, W, P, F₃, F₂

Free-body diagram

(e)

$$\Sigma M_P = -\bar{y}m(a_G)_x + \bar{x}m(a_G)_y + I_G\alpha \qquad (17\text{-}9)$$

This important result indicates that when moments of the external forces acting on the free-body diagram are summed about point P, Fig. 17–8e, they are equivalent to the sum of the "kinetic moments" of the components of $m\mathbf{a}_G$ about P plus the "kinetic moment" of $I_G\boldsymbol{\alpha}$, Fig. 17–8f. In other words, when the "kinetic moments," $\Sigma(\mathcal{M}_k)_P$, are computed, Fig. 17–8f, the vectors $m(\mathbf{a}_G)_x$ and $m(\mathbf{a}_G)_y$ are treated as sliding vectors, that is, they can act at any point along their line of action. In a similar manner, $I_G\boldsymbol{\alpha}$ can be treated as a free vector and can therefore act at any point. It is important to keep in mind that $m\mathbf{a}_G$ and $I_G\boldsymbol{\alpha}$ are not the same as a force or a couple. Instead, they are caused by the effects of forces and moments acting on the body. With this in mind we can therefore write Eq. 17–9 in a more general form as

$=$

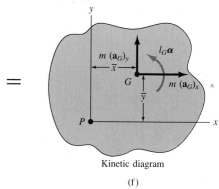

Kinetic diagram

(f)

Fig. 17–8(d–f)

$$\boxed{\Sigma M_P = \Sigma(\mathcal{M}_k)_P} \qquad (17\text{-}10)$$

---

*It also reduces to this same simple form $\Sigma M_P = I_P\alpha$ if point P is a *fixed point* (see Eq. 17–16) or the acceleration of point P is directed along the line PG.

General Application of the Equations of Motion

To summarize the above analysis, *three* scalar equations may be written to describe the general plane motion of a symmetrical rigid body.

$$\Sigma F_x = m(a_G)_x$$
$$\Sigma F_y = m(a_G)_y \qquad\qquad (17\text{–}11)$$
$$\Sigma M_G = I_G\alpha \quad \text{or} \quad \Sigma M_P = \Sigma(\mathcal{M}_k)_P$$

When applying these equations, one should *always* draw a free-body diagram, Fig. 17–8e, in order to account for the terms involved in ΣF_x, ΣF_y, ΣM_G or ΣM_P. In some problems it may also be helpful to draw the *kinetic diagram* for the body. This diagram graphically accounts for the terms $m(\mathbf{a}_G)_x$, $m(\mathbf{a}_G)_y$, and $I_G\alpha$, and it is especially convenient when used to determine the components of $m\mathbf{a}_G$ and the moment terms in $\Sigma(\mathcal{M}_k)_P$.*

17.3 Equations of Motion: Translation

When a rigid body undergoes a *translation*, Fig. 17–9a, all the particles of the body have the *same acceleration*, so that $\mathbf{a}_G = \mathbf{a}$. Furthermore, $\boldsymbol{\alpha} = \mathbf{0}$, in which case the rotational equation of motion applied at point G reduces to a simplified form, $\Sigma M_G = 0$. Application of this and the translational equations of motion will now be discussed for each of the two types of translation presented in Chapter 16.

Rectilinear Translation

When a body is subjected to *rectilinear translation,* all the particles of the body (slab) travel along parallel straight-line paths. The free-body and kinetic diagrams are shown in Fig. 17–9b. Since $I_G\boldsymbol{\alpha} = \mathbf{0}$, only $m\mathbf{a}_G$ is shown on the kinetic diagram. Hence, the equations of motion which apply in this case become

$$\Sigma F_x = m(a_G)_x$$
$$\Sigma F_y = m(a_G)_y \qquad\qquad (17\text{–}12)$$
$$\Sigma M_G = 0$$

The last equation requires that the sum of the moments of all the external forces computed about the body's center of mass be equal to zero. It is possible, of course, to sum moments about other points on or off the body, in which case the moment of $m\mathbf{a}_G$ must be taken into account. For example, if

---

*For this reason, the kinetic diagram will be used in the solution of an example problem whenever $\Sigma M_P = \Sigma(\mathcal{M}_k)_P$ is applied.

Fig. 17–9

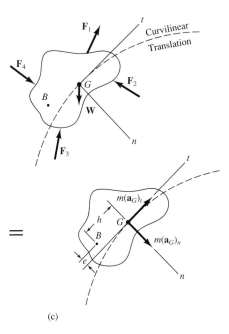

point A is chosen, which lies at a perpendicular distance d from the line of action of $m\mathbf{a}_G$, the following moment equation applies:

$$\zeta + \Sigma M_A = \Sigma(\mathcal{M}_k)_A; \qquad \Sigma M_A = (ma_G)d$$

Here the moment of the external forces about A (ΣM_A, free-body diagram) equals the moment of $m\mathbf{a}_G$ about A ($\Sigma(\mathcal{M}_k)_A$, kinetic diagram).

Curvilinear Translation

When a rigid body is subjected to *curvilinear translation,* all the particles of the body travel along *parallel curved paths.* For analysis, it is often convenient to use an inertial coordinate system having axes which are oriented in the normal and tangential directions of the path of motion, Fig. 17–9c. The three scalar equations of motion are then

$$\begin{array}{|c|}
\hline
\Sigma F_n = m(a_G)_n \\
\Sigma F_t = m(a_G)_t \\
\Sigma M_G = 0 \\
\hline
\end{array} \qquad (17\text{--}13)$$

where $(a_G)_t$ and $(a_G)_n$ represent, respectively, the magnitudes of the tangential and normal components of acceleration of point G.

If the moment equation $\Sigma M_G = 0$ is replaced by a moment summation about the arbitrary point B, Fig. 17–9c, it is necessary to account for the moments, $\Sigma(\mathcal{M}_k)_B$, of the two components $m(\mathbf{a}_G)_n$ and $m(\mathbf{a}_G)_t$ about this point. From the kinetic diagram, h and e represent the perpendicular distances (or "moment arms") from B to the lines of action of the components. If positive moments are assumed to be clockwise, the required moment equation becomes

$$\gamma + \Sigma M_B = \Sigma(\mathcal{M}_k)_B; \qquad \Sigma M_B = h[m(a_G)_n] - e[m(a_G)_t]$$

PROCEDURE FOR ANALYSIS

The following procedure provides a method for solving kinetic problems involving rigid-body *translation*.

Free-Body Diagram. Establish the x, y or n, t inertial coordinate system and draw the free-body diagram in order to account for all the external forces and moments that act on the body. The direction and sense of the acceleration of the body's mass center $\mathbf{a}_G$ should also be established. If the sense of its components *cannot* be determined, assume they are in the direction of the positive inertial coordinate axes. Identify the unknowns in the problem. If it is decided that the rotational equation of motion $\Sigma M_P = \Sigma(\mathcal{M}_k)_P$ is to be used in the solution, then it may be beneficial to draw the kinetic diagram, since it graphically accounts for the components $m(\mathbf{a}_G)_x$, $m(\mathbf{a}_G)_y$ or $m(\mathbf{a}_G)_t$, $m(\mathbf{a}_G)_n$ and is therefore convenient for "visualizing" the terms needed in the moment sum $\Sigma(\mathcal{M}_k)_P$.

Equations of Motion. Apply the three equations of motion, Eqs. 17–12 or Eqs. 17–13. To simplify the analysis, the moment equation $\Sigma M_G = 0$ can be replaced by the more general equation $\Sigma M_P = \Sigma(\mathcal{M}_k)_P$, where point P is usually located at the intersection of the lines of action of as many unknown forces as possible.

If the body is in contact with a *rough surface* and slipping occurs, use the frictional equation $F = \mu_k N$ to relate the normal force $\mathbf{N}$ to its associated frictional force $\mathbf{F}$.

Kinematics. Use kinematics if the velocity and position of the body are to be determined. For *rectilinear translation* with *variable acceleration* use

$$a_G = \frac{dv_G}{dt} \qquad a_G \, ds_G = v_G \, dv_G \qquad v_G = \frac{ds_G}{dt}$$

For *rectilinear translation* with *constant acceleration,* use

$$v_G = (v_G)_0 + a_G t$$
$$v_G^2 = (v_G)_0^2 + 2a_G[s_G - (s_G)_0]$$
$$s_G = (s_G)_0 + (v_G)_0 t + \tfrac{1}{2}a_G t^2$$

For *curvilinear translation,* use

$$(a_G)_n = \frac{v_G^2}{\rho} = \omega^2 \rho$$

$$(a_G)_t = \frac{dv_G}{dt} \qquad (a_G)_t \, ds_G = v_G \, dv_G \qquad (a_G)_t = \alpha\rho$$

The following examples numerically illustrate application of this procedure.

Example 17–5

The car shown in Fig. 17–10a has a mass of 2 Mg and a center of mass at G. Determine the car's acceleration if the "driving" wheels in the back are always slipping, whereas the front wheels freely rotate. Neglect the mass of the wheels. The coefficient of kinetic friction between the wheels and the road is $\mu_k = 0.25$.

(a)

SOLUTION I

Free-Body Diagram. As shown in Fig. 17–10b, the rear-wheel frictional force $\mathbf{F}_B$ pushes the car forward, and since *slipping occurs* the magnitude of this force is related to the magnitude of its associated normal force $\mathbf{N}_B$ by $F_B = 0.25N_B$. The frictional forces acting on the *front wheels* are *zero*, since these wheels have negligible mass.* There are three unknowns in the problem, namely N_A, N_B, and a_G. Here we will sum moments about the mass center. The car (point G) is assumed to accelerate to the left, i.e., in the negative x direction, Fig. 17–10b.

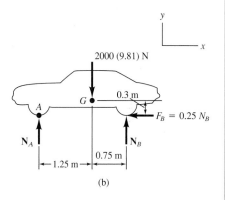

(b)

Equations of Motion

$$\xrightarrow{+} \Sigma F_x = m(a_G)_x; \qquad -0.25N_B = -2000a_G \qquad (1)$$

$$+\uparrow \Sigma F_y = m(a_G)_y; \qquad N_A + N_B - 2000(9.81) = 0 \qquad (2)$$

$$\zeta + \Sigma M_G = 0; \qquad -N_A(1.25) - 0.25N_B(0.3) + N_B(0.75) = 0 \qquad (3)$$

Solving,

$$a_G = 1.59 \text{ m/s}^2 \leftarrow \qquad\qquad Ans.$$
$$N_A = 6.88 \text{ kN}$$
$$N_B = 12.7 \text{ kN}$$

SOLUTION II

Free-Body and Kinetic Diagrams. If the "moment" equation is applied about point A, then the unknown N_A will be eliminated from the equation. To "visualize" the moment of $m\mathbf{a}_G$ about A we will include the kinetic diagram as part of the analysis, Fig. 17–10c.

$$=$$

(c)

Equation of Motion. We require

$$\zeta + \Sigma M_A = \Sigma(\mathcal{M}_k)_A; \qquad N_B(2) - 2000(9.81)(1.25) = 2000a_G(0.3)$$

Solving this and Eq. (1) for a_G leads to a simpler solution than that obtained from Eqs. (1) to (3).

Fig. 17–10

---

*If the mass of the front wheels were to be included in the analysis, the frictional force acting at A would be *directed to the right* to create the necessary counterclockwise rotation of the wheels. The problem solution for this case would be more involved since a general-plane-motion analysis of the wheels would have to be considered (see Sec. 17.5).

Example 17–6

The motorcycle shown in Fig. 17–11a has a mass of 125 kg and a center of mass at G_1, while the rider has a mass of 75 kg and a center of mass at G_2. If the coefficient of static friction between the wheels and the pavement is $\mu_s = 0.8$, determine if it is possible for the rider to do a "wheely," i.e., lift the front wheel off the ground. What acceleration is necessary to do this? Neglect the mass of the wheels and assume that the front wheel is free to roll.

0.3 m

0.6 m

B ◄0.4 m►◄0.5 m►◄0.7 m►A

(a)

SOLUTION

Free-Body and Kinetic Diagrams. In this problem we will consider both the motorcycle and the rider as the "system" to be analyzed. It is possible first to determine the location of the center of mass for this "system" by using the equations $\bar{x} = \Sigma \tilde{x}m/\Sigma m$ and $\bar{y} = \Sigma \tilde{y}m/\Sigma m$. Here, however, we will consider the separate weight and mass of each of its *component parts* as shown on the free-body and kinetic diagrams, Fig. 17–11b. Both parts move with the *same* acceleration and we have assumed that the front wheel is about to leave the ground, so that the normal reaction $N_A \approx 0$. In order for this to happen it is required that the friction force $F_B \leq 0.8 N_B$, otherwise slipping will occur. The three unknowns in the problem are N_B, F_B, and a_G.

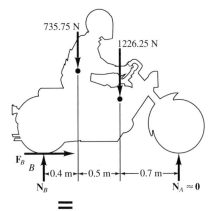

735.75 N

1226.25 N

F_B B

◄0.4 m►◄0.5 m►◄0.7 m►

N_B $N_A \approx 0$

=

Equations of Motion

$$\xrightarrow{+}\Sigma F_x = m(a_G)_x; \qquad F_B = (75 + 125)a_G \qquad (1)$$
$$+\uparrow \Sigma F_y = m(a_G)_y; \quad N_B - 735.75 - 1226.25 = 0 \qquad (2)$$
$$\zeta + \Sigma M_B = \Sigma(\mathcal{M}_k)_B; \quad (735.75)(0.4) + (1226.25)(0.9) = (75a_G)(0.9) + (125a_G)(0.6)$$

Solving,

$$a_G = 9.81 \text{ m/s}^2 \rightarrow \qquad \textit{Ans.}$$
$$N_B = 1.96 \text{ kN}$$
$$F_B = 1.96 \text{ kN}$$

75 a_G

0.3 m

125 a_G

0.6 m

B

(b)

Fig. 17–11

The maximum frictional force that can be developed at B is

$$(F_B)_{max} = \mu_s N_B = 0.8(1.96) = 1.57 \text{ kN}$$

Since this force is less than that required ($F_B = 1.96$ kN), it is *not possible* to lift the front wheel off the ground.

Note: If the maximum acceleration of the motorcycle is to be obtained, then $F_B = 0.8 N_B$ and $N_A \neq 0$. The problem would have to be solved in a manner similar to that of Example 17–5. Try it, and show that $(a_G)_{max} = 6.76 \text{ m/s}^2$.

Example 17–7

A uniform 50-kg crate rests on a horizontal surface for which the coefficient of kinetic friction is $\mu_k = 0.2$. Determine the crate's acceleration if a force of $P = 600$ N is applied to the crate as shown in Fig. 17–12a.

(a)   **Fig. 17–12**

SOLUTION

Free-Body Diagram. The force **P** can either cause the crate to slide or to tip over. As shown in Fig. 17–12b, it is assumed that the crate slides, so that $F = \mu_k N_C = 0.2N_C$. Also, the resultant normal force $\mathbf{N}_C$ acts at O, a distance x (where $0 < x \leq 0.5$ m) from the crate's center line.* The three unknowns are N_C, x, and a_G.

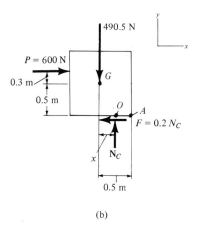

(b)

Equations of Motion

$$\xrightarrow{+}\Sigma F_x = m(a_G)_x; \qquad 600 - 0.2N_C = 50a_G \qquad (1)$$

$$+\uparrow \Sigma F_y = m(a_G)_y; \qquad N_C - 490.5 = 0 \qquad (2)$$

$$\zeta+\Sigma M_G = 0; \qquad -600(0.3) + N_C(x) - 0.2N_C(0.5) = 0 \qquad (3)$$

Solving, we obtain

$$N_C = 490 \text{ N}$$

$$x = 0.467 \text{ m}$$

$$a_G = 10.0 \text{ m/s}^2 \rightarrow \qquad\qquad\qquad Ans.$$

Since $x = 0.467$ m < 0.5 m, indeed the crate slides as originally assumed. If the solution had given a value of $x > 0.5$ m, the problem would have to be reworked with the assumption that tipping occurred. If this were the case, $\mathbf{N}_C$ would act at the *corner point A* and $F \leq 0.2N_C$.

*The line of action of $\mathbf{N}_C$ does not necessarily pass through the mass center G ($x = 0$), since $\mathbf{N}_C$ must counteract the tendency for tipping caused by **P.** See Sec. 8.1 of *Engineering Mechanics: Statics.*

Example 17–8

The 100-kg beam *BD* shown in Fig. 17–13a is supported by two rods having negligible mass. Determine the force created in each rod if at the instant $\theta = 30°$ the rods are both rotating with an angular velocity of $\omega = 6$ rad/s.

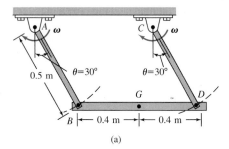

0.5 m $\theta = 30°$ $\theta = 30°$

G   *D*

B ⊢— 0.4 m —⊣— 0.4 m —⊣

(a)   **Fig. 17–13**

SOLUTION

Free-Body Diagram.   The beam moves with *curvilinear translation* since points *B* and *D* and the center of mass *G* move along circular paths, each path having the same radius of 0.5 m. Using normal and tangential coordinates, the free-body diagram for the beam is shown in Fig. 17–13b. Because of the *translation, G* has the *same* motion as the pin at *B*, which is connected to both the rod and the beam. By studying the angular motion of rod *AB*, Fig. 17–13c, note that the tangential component of acceleration acts downward to the left due to the clockwise direction of $\boldsymbol{\alpha}$. Furthermore, the normal component of acceleration is *always* directed toward the center of curvature (toward point *A* for rod *AB*). Since the angular velocity of *AB* is 6 rad/s, then

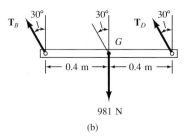

T_B 30°   30° T_D 30°

G

⊢— 0.4 m —⊣— 0.4 m —⊣

981 N

(b)

$$(a_G)_n = \omega^2 r = (6)^2(0.5) = 18 \text{ m/s}^2$$

The three unknowns are T_B, T_D, and $(a_G)_t$. The kinetic diagram has not been shown, since the directions of $(\mathbf{a}_G)_n$ and $(\mathbf{a}_G)_t$ have been established, and we will sum moments about *G*.

A α ω

0.5 m $\mathbf{a}_n$

B

$\mathbf{a}_t$

(c)

Equations of Motion

$$+\nwarrow \Sigma F_n = m(a_G)_n; \quad T_B + T_D - 981 \cos 30° = 100(18) \qquad (1)$$

$$+\nearrow \Sigma F_t = m(a_G)_t; \qquad 981 \sin 30° = 100(a_G)_t \qquad (2)$$

$$\zeta + \Sigma M_G = 0; \quad -(T_B \cos 30°)0.4 + (T_D \cos 30°)0.4 = 0 \qquad (3)$$

Simultaneous solution of these three equations gives

$$T_B = T_D = 1.33 \text{ kN} \quad \nwarrow_{30°} \qquad\qquad Ans.$$

$$(a_G)_t = 4.90 \text{ m/s}^2$$

PROBLEMS

17–19. The desk has a weight of 75 lb and a center of gravity at *G*. Determine its initial acceleration if a man pushes on it at *C* with a force of 60 lb. Also, find the vertical reactions on each of the two legs at *A* and at *B*. The coefficient of kinetic friction at *A* and *B* is $\mu_k = 0.2$.

Prob. 17–19

*17–20. The handcart has a mass of 200 kg and center of mass at *G*. Determine the normal reactions at each of the wheels *A* and *B* if a force of *P* = 50 N is applied to the handle. Neglect the mass and rolling resistance of the wheels.

Probs. 17–20/17–21

17–21. The handcart has a mass of 200 kg and center of mass at *G*. Determine the magnitude of the largest force *P* that can be applied to the handle so that the wheels at *A* or *B* continue to maintain contact with the ground. Neglect the mass and rolling resistance of the wheels.

17–22. The smooth 180-lb pipe has a length of 20 ft and a negligible diameter. It is carried on a truck as shown. Determine the maximum acceleration which the truck can have without causing the normal reaction at *A* to be zero. Also determine the horizontal and vertical components of force which the truck exerts on the pipe at *B*.

Prob. 17–22

17–23. The smooth 180-lb pipe has a length of 20 ft and a negligible diameter. It is carried on a truck as shown. If the truck accelerates at $a = 5$ ft/s², determine the normal reaction at *A* and the horizontal and vertical components of force which the truck exerts on the pipe at *B*.

*17–24. The jet aircraft has a total mass of 22 Mg and a center of mass at *G*. Initially at take-off the engines provide a thrust of $2T = 4$ kN and $T' = 1.5$ kN. Determine the acceleration of the plane and the normal reactions on the nose wheel and each of the *two* wing wheels located at *B*. Neglect the mass of the wheels and, due to low velocity, neglect any lift caused by the wings.

Prob. 17–24

17–25. The dresser has a weight of 80 lb and is pushed along the floor. If the coefficient of static friction at *A* and *B* is $\mu_s = 0.3$ and the coefficient of kinetic friction is $\mu_k = 0.2$, determine the smallest horizontal force **P** needed to cause motion. If this force is increased slightly, determine the acceleration of the dresser. Also, what are the normal reactions at *A* and *B* when it begins to move?

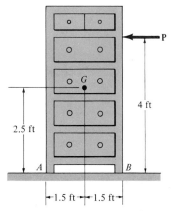

Prob. 17–25

315

17–26. The dresser has a weight of 80 lb and is pushed along the floor. If the coefficient of static friction at A and B is $\mu_s = 0.3$ and the coefficient of kinetic friction is $\mu_k = 0.2$, determine the maximum force horizontal **P** that can be applied without causing the dresser to tip over.

17–27. The truck carries the 800-lb crate which has a center of gravity at G_c. Determine the largest acceleration **a** of the truck so that the crate will not slip or tip on the truck bed. The coefficient of static friction between the crate and the truck is $\mu_s = 0.6$.

Probs. 17–27/17–28

*17–28. The truck has a weight of 8000 lb and center of gravity at G_t. It carries the 800-lb crate which has a center of gravity at G_c. Determine the normal reaction at *each* of its four tires if it accelerates at $a = 0.5$ ft/s². Also, what is the frictional force acting between the crate and the truck, and between each of the rear tires and the road? Assume that power is delivered only to the rear tires. The front tires are free to roll. Neglect the mass of the tires. The crate does not slip on the truck.

17–29. The sports car has been designed so that the 260-kg engine E and 165-kg transmission T have been placed over the front and rear wheels, respectively. Their mass centers are located at G_E and G_T. The remaining 512-kg mass of the body and frame is located at G_b. If power is supplied to the rear wheels only, determine the shortest time it takes for the car to reach a speed of 80 km/h, starting from rest. The front wheels are free rolling, and the coefficient of kinetic friction between the wheels and the road is $\mu_k = 0.4$. Neglect the mass of the wheels and driver.

Prob. 17–29

17–30. The pipe has a mass of 800 kg and is being towed behind the truck. If the acceleration of the truck is $a_t = 0.5$ m/s², determine the angle θ and the tension in the cable. The coefficient of kinetic friction between the pipe and the ground is $\mu_k = 0.1$.

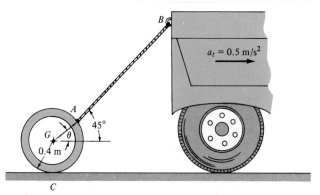

Prob. 17–30

17–31. The bicycle and rider have a mass of 80 kg with center of mass located at G. If the coefficient of kinetic friction at the rear tire is $\mu_B = 0.8$, determine the normal reactions at the tires A and B, and the deceleration of the rider, when the rear wheel locks for braking. What is the normal reaction at the rear wheel when the bicycle is traveling at constant velocity and the brakes are not applied? Neglect the mass of the wheels.

Probs. 17–31/17–32

*17–32. The bicycle and rider have a mass of 80 kg with center of mass located at G. Determine the minimum coefficient of kinetic friction between the road and the wheels so that the rear wheel B starts to lift off the ground when the rider applies the brakes to the front wheel. Neglect the mass of the wheels.

17–33. The van has a weight of 4500 lb and center of gravity at G_v. It carries a fixed 800-lb load which has a center of gravity at G_l. If the van is traveling at 40 ft/s, determine the shortest distance it skids before stopping at the moment the driver applies the brakes to *all* the wheels and causes each wheel to slip. The coefficient of kinetic friction between the wheels and the pavement is $\mu_k = 0.3$. Compare this distance with that of the van being empty. Neglect the mass of the wheels.

Probs. 17–33/17–34

17–34. The van has a weight of 4500 lb and center of gravity at G_v. It carries a fixed 800-lb load which has a center of gravity at G_l. If power is applied only to the rear wheels B, causing them to slip on the pavement, determine the acceleration of the van. The coefficient of kinetic friction between the wheels and the pavement is $\mu_k = 0.3$. Neglect the mass of the wheels. The front wheels are free to roll.

17–35. The portable crane has a mass of 500 kg and mass center at G. It is used to lift the 50-kg crate when the boom is in the horizontal position shown. Determine the largest initial angular acceleration of the boom, so that the front wheels E do not lose contact with the ground. Neglect the mass of the boom and hydraulic cylinder AC. The wheels at H are fixed from rolling.

Prob. 17–35

*17–36. The two 3-lb rods EF and HI are fixed (welded) to the link AC at E. Determine the internal axial force $\mathbf{E}_x$, shear force $\mathbf{E}_y$, and moment $\mathbf{M}_E$, which the bar AC exerts on FE at E if at the instant $\theta = 30°$ link AB has an angular velocity $\omega = 5$ rad/s and an angular acceleration $\alpha = 8$ rad/s^2 as shown.

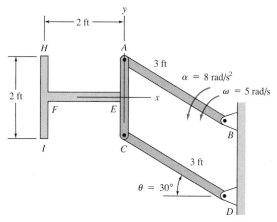

Prob. 17–36

17–37. The 1.6-Mg car shown has been "raked" by increasing the height $h = 0.2$ m of its center of mass. This was done by raising the springs on the rear axle. If the coefficient of kinetic friction between the rear wheels and the ground is $\mu_k = 0.3$, show that the car can accelerate slightly faster than its counterpart for which $h = 0$. Neglect the mass of the wheels and driver and assume the front wheels at B are free to roll while the rear wheels slip.

Prob. 17–37

317

17–38. The dragster has a mass of 1500 kg and a center of mass at G. If the coefficient of kinetic friction between the rear wheels and the pavement is $\mu_k = 0.6$, determine if it is possible for the driver to lift the front wheels, A, off the ground while the rear wheels are slipping. If so, what acceleration is necessary to do this? Neglect the mass of the wheels and assume that the front wheels are free to roll.

Probs. 17–38/17–39

17–39. The dragster has a mass of 1500 kg and a center of mass at G. If no slipping occurs, determine the friction force $\mathbf{F}_B$ which must be applied to each of the rear wheels B in order to develop an acceleration of $a = 6$ m/s². What are the normal reactions of each wheel on the ground? Neglect the mass of the wheels and assume that the front wheels are free to roll.

*17–40. The uniform bar BC has a weight of 40 lb and is pin-connected to the two links which have negligible mass. Determine the moment $\mathbf{M}$ that must be applied to link DC so that DC has the angular motion shown at the instant $\theta = 30°$. Also, what is the force developed in member AB?

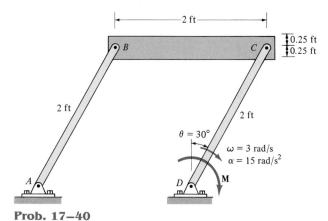

Prob. 17–40

17–41. The crate C has a weight of 150 lb and rests on the floor of a truck elevator for which the coefficient of static friction is $\mu_s = 0.4$. Determine the largest initial angular acceleration α, starting from rest, which the parallel links AB and DE can have without causing the crate to slip. No tipping occurs.

Prob. 17–41

17–42. The car having a mass of 1.40 Mg and mass center at G_c pulls a loaded trailer having a mass of 0.800 Mg and mass center at G_t. Determine the normal reactions at each of the car and trailer wheels if the driver applies the car's brakes and causes the rear wheels C car to skid. The coefficient of kinetic friction is $\mu_C = 0.4$. Assume the hitch at A is a pin or ball-and-socket joint. The wheels at B and D are free to roll. Neglect their mass and the mass of the driver.

Prob. 17–42

17.4 Equations of Motion: Rotation About a Fixed Axis

Consider the rigid body (or slab) shown in Fig. 17–14a, which is constrained to rotate in the vertical plane about a fixed axis perpendicular to the page and passing through the pin at O. The angular velocity and angular acceleration are caused by the external force system acting on the body. Because the body's center of mass G moves in a *circular path*, the acceleration of this point is represented by its tangential and normal components. The *tangential component of acceleration* has a *magnitude* of $(a_G)_t = \alpha r_G$ and must act in a *direction* which is *consistent* with the angular acceleration $\boldsymbol{\alpha}$. The *magnitude* of the *normal component of acceleration* is $(a_G)_n = \omega^2 r_G$. This component is *always directed* from point G to O regardless of the direction of $\boldsymbol{\omega}$.

The free-body and kinetic diagrams for the body are shown in Fig. 17–14b. The weight of the body, $W = mg$, and the pin reaction $\mathbf{F}_O$ are included on the free-body diagram since they represent external forces acting on the body. The two components $m(\mathbf{a}_G)_t$ and $m(\mathbf{a}_G)_n$, shown on the kinetic diagram, are associated with the tangential and normal acceleration components of the body's mass center. These vectors act in the same *direction* as the acceleration components and have *magnitudes* of $m(a_G)_t$ and $m(a_G)_n$. The $I_G\boldsymbol{\alpha}$ vector acts in the same *direction* as $\boldsymbol{\alpha}$ and has a *magnitude* of $I_G\alpha$, where I_G is the body's moment of inertia calculated about an axis which is perpendicular to the page and passing through G. From the derivation given in Sec. 17.2, the equations of motion which apply to the body may be written in the form

$$\Sigma F_n = m(a_G)_n = m\omega^2 r_G$$
$$\Sigma F_t = m(a_G)_t = m\alpha r_G \qquad (17\text{–}14)$$
$$\Sigma M_G = I_G\alpha$$

The moment equation may be replaced by a moment summation about any arbitrary point P on or off the body provided one accounts for the moments $\Sigma(\mathcal{M}_k)_P$ produced by $I_G\boldsymbol{\alpha}$, $m(\mathbf{a}_G)_t$, and $m(\mathbf{a}_G)_n$ about the point. In many problems it is convenient to sum moments about the pin at O in order to eliminate the *unknown* force $\mathbf{F}_O$. From the kinetic diagram, Fig. 17–14b, this requires

$$\zeta + \Sigma M_O = \Sigma(\mathcal{M}_k)_O; \qquad \Sigma M_O = r_G m(a_G)_t + I_G\alpha \qquad (17\text{–}15)$$

Note that the moment of $m(\mathbf{a}_G)_n$ is not included in the summation since the line of action of this vector passes through O. Substituting $(a_G)_t = r_G\alpha$, we may rewrite the above equation as $\zeta + \Sigma M_O = (I_G + mr_G^2)\alpha$. From the parallel-axis theorem, $I_O = I_G + md^2$, the term in parentheses represents the *moment of inertia of the body about the fixed axis of rotation passing through* O. Therefore, we can write the three equations of motion for the body as*

(a)

(b)

Fig. 17–14

---

*The result $\Sigma M_O = I_O\alpha$ can also be obtained *directly* from Eq. 17–6 by selecting point P to coincide with O, realizing that $(a_P)_x = (a_P)_y = 0$.

$$\Sigma F_n = m(a_G)_n = m\omega^2 r_G$$
$$\Sigma F_t = m(a_G)_t = m\alpha r_G \qquad\qquad (17\text{--}16)$$
$$\Sigma M_O = I_O \alpha$$

For applications, one should remember that "$I_O\alpha$" accounts for the "moment" of *both* $m(\mathbf{a}_G)_t$ *and* $I_G\boldsymbol{\alpha}$ about point O, Fig. 17–14*b*. In other words, $\Sigma M_O = \Sigma(\mathcal{M}_k)_O = I_O\alpha$, as indicated by Eqs. 17–15 and 17–16.

PROCEDURE FOR ANALYSIS

The following procedure provides a method for solving kinetic problems which involve the rotation of a body about a fixed axis.

Free-Body Diagram. Establish the inertial x, y or n, t coordinate system, and specify the direction and sense of the accelerations $(\mathbf{a}_G)_n$ and $(\mathbf{a}_G)_t$, and the angular acceleration $\boldsymbol{\alpha}$ of the body. If the sense of these vectors *cannot* be established, assume they are in the direction of the positive coordinate axes. Draw the free-body diagram to account for all the external forces that act on the body and compute the moment of inertia I_G or I_O. Identify the unknowns in the problem. If it is decided that the rotational equation of motion $\Sigma M_P = \Sigma(\mathcal{M}_k)_P$ is to be used, i.e., moments will *not* be summed about point G or O, then consider drawing the kinetic diagram in order to help "visualize" the "moment" developed by the components $m(\mathbf{a}_G)_n$, $m(\mathbf{a}_G)_t$, and $I_G\boldsymbol{\alpha}$ when writing the terms for the moment sum $\Sigma(\mathcal{M}_k)_P$.

Equations of Motion. Apply the three equations of motion, Eqs. 17–14 or 17–16.

Kinematics. Use kinematics if a complete solution cannot be obtained strictly from the equations of motion. In this regard, if the *angular acceleration is variable*, use

$$\alpha = \frac{d\omega}{dt}, \qquad \alpha\, d\theta = \omega\, d\omega \qquad \omega = \frac{d\theta}{dt}$$

If the *angular acceleration is constant*, use

$$\omega = \omega_0 + \alpha_c t$$
$$\theta = \theta_0 + \omega_0 t + \tfrac{1}{2}\alpha_c t^2$$
$$\omega^2 = \omega_0^2 + 2\alpha_c(\theta - \theta_0)$$

The following examples numerically illustrate application of this procedure.

Example 17–9

The 30-kg disk shown in Fig. 17–15a is pin-supported at its center. If it starts from rest, determine the number of revolutions it must make to attain an angular velocity of 20 rad/s. Also, what are the reactions at the pin? The disk is acted upon by a constant force $F = 10$ N, which is applied to a cord wrapped around its periphery, and a constant couple moment $M = 5$ N · m. Neglect the mass of the cord in the calculation.

(a)

SOLUTION

Free-Body Diagram. Fig. 17–15b. Note that the mass center is not subjected to an acceleration; however, the disk has a clockwise angular acceleration.

The moment of inertia of the disk is

$$I_O = \tfrac{1}{2}mr^2 = \frac{1}{2}(30)(0.2)^2 = 0.6 \text{ kg} \cdot \text{m}^2$$

The three unknowns are O_x, O_y, and α.

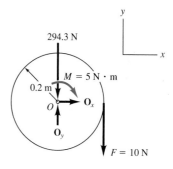

(b)

Fig. 17–15

Equations of Motion

$$\xrightarrow{+} \Sigma F_x = m(a_G)_x; \qquad O_x = 0 \qquad\qquad Ans.$$
$$+ \uparrow \Sigma F_y = m(a_G)_y; \qquad O_y - 294.3 - 10 = 0$$
$$O_y = 304 \text{ N} \qquad\qquad Ans.$$
$$\big\downarrow\!+ \Sigma M_O = I_O\alpha; \qquad -10(0.2) - 5 = -0.6\alpha \qquad \alpha = 11.7 \text{ rad/s}^2 \,\big\downarrow$$

Kinematics. Since α is constant and rotates clockwise as assumed, the number of radians the disk must turn to obtain a clockwise angular velocity of 20 rad/s is

$$\big\downarrow\!+ \qquad\qquad \omega^2 = \omega_0^2 + 2\alpha_c(\theta - \theta_0)$$
$$(-20)^2 = 0 + 2(-11.7)(\theta - 0)$$
$$\theta = -17.1 \text{ rad} = 17.1 \text{ rad} \,\big\downarrow$$

Hence,

$$\theta = 17.1 \text{ rad} \left(\frac{1 \text{ rev}}{2\pi \text{ rad}} \right) = 2.73 \text{ rev} \,\big\downarrow \qquad\qquad Ans.$$

Example 17–10

The drum shown in Fig. 17–16a has a mass of 60 kg and a radius of gyration $k_O = 0.25$ m. A cord of negligible mass is wrapped around the periphery of the drum and attached to a crate having a mass of 20 kg. If the crate is released, determine the drum's angular acceleration.

(a)

SOLUTION I

Free-Body Diagram. Here we will consider the drum and crate separately, Fig. 17–16b. Assuming the crate accelerates *downward* at **a**, it creates a *counterclockwise* angular acceleration **α** of the drum.

The moment of inertia of the drum is

$$I_O = mk_O^2 = (60)(0.25)^2 = 3.75 \text{ kg} \cdot \text{m}^2$$

There are five unknowns, namely O_x, O_y, T, a and α.

Equations of Motion. Applying the translational equations of motion $\Sigma F_x = m(a_G)_x$ and $\Sigma F_y = m(a_G)_y$ to the drum is of no consequence to the solution, since these equations involve the unknowns O_x and O_y. Thus, for the drum and crate, respectively

$$\zeta + \Sigma M_O = I_O \alpha; \qquad\qquad T(0.4) = 3.75\alpha \qquad\qquad (1)$$
$$+ \uparrow \Sigma F_y = m(a_G)_y; \qquad -20(9.81) + T = -20a \qquad (2)$$

Kinematics. Since the point of contact A between the cord and drum has a tangential component of acceleration **a,** Fig. 17–16a, then

$$\zeta + a = \alpha r; \qquad\qquad a = \alpha(0.4) \qquad\qquad (3)$$

Solving the above equations

$$T = 106 \text{ N}$$
$$a = 4.52 \text{ m/s}^2$$
$$\alpha = 11.3 \text{ rad/s}^2 \; \zeta \qquad\qquad Ans.$$

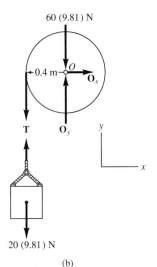

(b)

SOLUTION II

Free-Body and Kinetic Diagrams. The cable tension T can be eliminated from the analysis by considering the drum and crate as a single system, Fig. 17–16c. The kinetic diagram is shown since moments will be summed about point O.

Equations of Motion. Using Eq. (3) and applying the moment equation about point O to eliminate the unknowns O_x and O_y, we have

$$\zeta + \Sigma M_O = \Sigma(\mathcal{M}_k)_O; \qquad 20(9.81)(0.4) = 3.75\alpha + [20(0.4\alpha)]0.4$$
$$\alpha = 11.3 \text{ rad/s}^2 \; \zeta \qquad\qquad Ans.$$

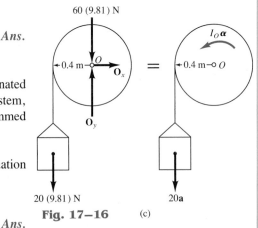

Fig. 17–16   (c)

Example 17–11

The unbalanced 50-lb flywheel shown in Fig. 17–17a has a radius of gyration of $k_G = 0.6$ ft about an axis passing through its mass center G. If it has a clockwise angular velocity of 8 rad/s at the instant shown, determine the horizontal and vertical components of reaction at the pin O.

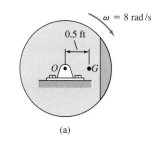

(a)

SOLUTION

Free-Body and Kinetic Diagrams. Since G moves in a *circular path*, $(a_G)_y = \alpha r_G$ and $(a_G)_x = \omega^2 r_G$. By inspection, the weight creates a clockwise angular acceleration $\boldsymbol{\alpha}$, so $I_G\alpha$ acts in a clockwise sense and $m(\mathbf{a}_G)_y$ acts downward, in accordance with the direction of $\boldsymbol{\alpha}$, Fig. 17–17b. Furthermore, $m(\mathbf{a}_G)_x$ acts to the left, toward the center of curvature O.

The moment of inertia of the flywheel about its mass center is determined from the radius of gyration and the mass; i.e., $I_G = mk_G^2 = (50/32.2)(0.6)^2 = 0.559$ slug · ft².

The three unknowns are O_x, O_y, and α.

Equations of Motion

$$\xleftarrow{+}\Sigma F_x = m\omega^2 r_G; \qquad O_x = \left(\frac{50}{32.2}\right)(8)^2(0.5) \tag{1}$$

$$+\uparrow \Sigma F_y = m\alpha r_G; \quad O_y - 50 = -\left(\frac{50}{32.2}\right)(\alpha)(0.5) \tag{2}$$

$$\zeta+\Sigma M_G = I_G\alpha; \qquad O_y(0.5) = 0.559\alpha \tag{3}$$

Solving,

$$\alpha = 26.4 \text{ rad/s}^2, \quad O_x = 49.7 \text{ lb}, \quad O_y = 29.5 \text{ lb} \qquad \textit{Ans.}$$

(b)

Fig. 17–17

Moments can also be summed about point O in order to eliminate $\mathbf{O}_x$ and $\mathbf{O}_y$ and thereby obtain a *direct solution* for $\boldsymbol{\alpha}$, Fig. 17–17b. This can be done in one of *two* ways, i.e., by using either $\Sigma M_O = \Sigma(\mathcal{M}_k)_O$ or $\Sigma M_O = I_O\alpha$. If the first of these equations is applied, then,

$$\zeta+\Sigma M_O = \Sigma(\mathcal{M}_k)_O; \quad 50(0.5) = 0.559\alpha + \left[\left(\frac{50}{32.2}\right)\alpha(0.5)\right](0.5)$$

$$50(0.5) = 0.947\alpha \tag{4}$$

If $\Sigma M_O = I_O\alpha$ is applied, then by the parallel-axis theorem the moment of inertia of the flywheel about O is

$$I_O = I_G + mr_G^2 = 0.559 + \left(\frac{50}{32.2}\right)(0.5)^2 = 0.947 \text{ slug · ft}^2$$

Hence, from the free-body diagram, Fig. 17–17b, we require

$$\zeta+\Sigma M_O = I_O\alpha; \qquad 50(0.5) = 0.947\alpha$$

which is the same as Eq. (4). Solving for α and substituting into Eq. (2) yields the answer for O_y obtained previously.

Example 17–12

The slender rod shown in Fig. 17–18a has a mass m and length l and is released from rest when $\theta = 0°$. Determine the horizontal and vertical components of force which the pin at O exerts on the rod at the instant $\theta = 90°$.

(a)

SOLUTION

Free-Body Diagram. The free-body diagram for the rod is shown when the rod is in the general position θ, Fig. 17–18b. For convenience, the force components at A are shown acting in the n and t directions. Note that α acts clockwise.

The moment of inertia of the rod about point A is

$$I_A = \tfrac{1}{3}ml^2$$

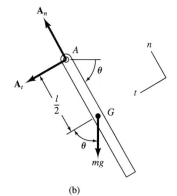

(b)

Fig. 17–18

Equations of Motion. Moments will be summed about A in order to eliminate the reactive forces there.*

$$+\nwarrow \Sigma F_n = m\omega^2 r_G; \qquad A_n - mg \sin \theta = m\omega^2(l/2) \qquad (1)$$
$$+\swarrow \Sigma F_t = m\alpha r_G; \qquad A_t + mg \cos \theta = m\alpha(l/2) \qquad (2)$$
$$\curvearrowright + \Sigma M_A = I_A \alpha; \qquad mg \cos \theta(l/2) = (\tfrac{1}{3}ml^2)\alpha \qquad (3)$$

Kinematics. For a given angle θ there are four unknowns in the above three equations: A_n, A_t, ω, and α. As shown by Eq. (3) α is *not constant*; rather, it depends upon the position θ of the rod. Using kinematics, α and ω can be related to θ by the equation

$$\curvearrowright + \qquad\qquad \omega \, d\omega = \alpha \, d\theta \qquad (4)$$

In order to solve for ω at $\theta = 90°$, eliminate α from Eqs. (3) and (4), which yields

$$\omega \, d\omega = (1.5 \, g/l) \cos \theta \, d\theta$$

Since $\omega = 0$ at $\theta = 0°$, we have

$$\int_0^\omega \omega \, d\omega = (1.5 \, g/l) \int_{0°}^{90°} \cos \theta \, d\theta$$

$$\omega^2 = 3 \, g/l$$

Substituting this value into Eq. (1) with $\theta = 90°$ and solving Eqs. (1) to (3) yields

$$\alpha = 0, \qquad A_t = 0, \qquad A_n = 2.5 \, mg \qquad\qquad \textit{Ans.}$$

---

*If $\Sigma M_A = \Sigma(\mathcal{M}_k)_A$ is used, one must account for the moments of $I_G \alpha$ and $m(\mathbf{a}_G)_t$ about A. Refer to Eq. 17–15.

PROBLEMS

17–43. The 10-lb rod is pin-connected to its support and has an angular velocity $\omega = 4$ rad/s when it is in the horizontal position shown. Determine its angular acceleration and the horizontal and vertical components of reaction which the pin exerts on the rod at this instant.

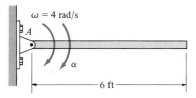

$\omega = 4$ rad/s

A

α

6 ft

Prob. 17–43

*17–44. The 10-kg wheel has a radius of gyration $k_A = 0.2$ m. If the wheel is subjected to a moment of $M = 50$ N $\cdot$ m, determine the magnitudes of the velocity and the normal and tangential components of acceleration of point P on its rim after it has rotated 4 revolutions, starting from rest. Also, compute the reactions which the fixed pin A exerts on the wheel during the motion.

P

0.3 m

M

A

Prob. 17–44

17–45. The fan blade has a mass of 2 kg and a moment of inertia $I_O = 0.18$ kg $\cdot$ m^2 about an axis passing through its center O. If it is subjected to a moment of $M = 3(1 - e^{-0.2t})$ N $\cdot$ m, where t is in seconds, determine its angular velocity in $t = 4$ s starting from rest.

O

M

Prob. 17–45

17–46. The cord is wrapped around the inner core of the spool. If a 5-lb weight B is suspended from the cord and released from rest, determine the spool's angular velocity in $t = 3$ s. Neglect the mass of the cord. The spool has a weight of 180 lb and the radius of gyration about the axle A is $k_A = 1.25$ ft. Solve the problem in two ways. First by considering the "system" consisting of the block and spool, and then by considering the block and spool separately.

2.75 ft

A

1.5 ft

s

B

Prob. 17–46

17–47. The 80-kg disk is supported by a pin at A. If it is released from rest from the position shown, determine the initial horizontal and vertical components of reaction at the pin.

A

1.5 m

Probs. 17–47/17–48

*17–48. The 80-kg disk is supported by a pin at A. If it is rotating clockwise at $\omega = 0.5$ rad/s when it is in the position shown, determine the horizontal and vertical components of reaction at the pin at this instant.

17–49. The two blocks A and B have a mass of m_A and m_B, respectively, where $m_B > m_A$. If the pulley can be treated as a disk of mass M, determine the acceleration of block A. Neglect the mass of the cord and any slipping on the pulley.

Prob. 17–49

17–50. The spool is supported on small rollers at A and B. Determine the angular acceleration of the spool and the normal forces at A and B if a vertical force of $P = 80$ N is applied to the cable. The spool has a mass of 60 kg and a centroidal radius of gyration $k_O = 0.65$ m. For the calculation neglect the mass of the cable and the mass of the rollers at A and B.

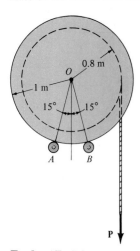

Prob. 17–50

17–51. The drum has a weight of 80 lb and a radius of gyration of $k_O = 0.4$ ft. If the cable, which is wrapped around the drum, is subjected to a vertical force of $P = 15$ lb, determine the time needed to increase the drum's angular velocity from $\omega_1 = 5$ rad/s to $\omega_2 = 25$ rad/s. Neglect the mass of the cable.

Probs. 17–51/17–52

*17–52. The drum has a weight of 80 lb and a radius of gyration of $k_O = 0.4$ ft. If the cable, which is wrapped around the drum, is subjected to a vertical force of $P = (3t^2 + 5)$ lb, where t is in seconds, determine the angular velocity of the drum in $t = 2$ s if at $t = 0$, $\omega_1 = 5$ rad/s. Neglect the mass of the cable.

17–53. The 20-kg roll of paper has a radius of gyration $k_A = 90$ mm about an axis passing through point A. It is pin-supported at both ends by two brackets AB. If the roll rests against a wall for which the coefficient of kinetic friction is $\mu_C = 0.2$ and a vertical force of 30 N is applied to the end of the paper, determine the angular acceleration of the roll as the paper unrolls.

Prob. 17–53

17–54. The pendulum consists of a uniform 5-kg plate and a 2-kg slender rod. Compute the horizontal and vertical components of reaction that the pin O exerts on the rod at the instant $\theta = 30°$, at which time its angular velocity is $\omega = 3$ rad/s.

Prob. 17–54

17–55. The kinetic diagram representing the general rotational motion of a rigid body about a fixed axis at O is shown in the figure. Show that $I_G\alpha$ may be eliminated by moving the vectors $m(\mathbf{a}_G)_t$ and $m(\mathbf{a}_G)_n$ to point P, located a distance $r_{GP} = k_G^2/r_{OG}$ from the center of mass G of the body. Here k_G represents the radius of gyration of the body about G. The point P is called the *center of percussion* of the body.

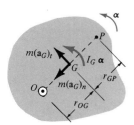

Prob. 17–55

*__17–56.__ Determine the position r_P of the center of percussion P of the 10-lb slender bar. (See Prob. 17–55.) What is the horizontal force $\mathbf{A}_x$ at the pin when the bar is struck at P with a force of $F = 20$ lb?

Prob. 17–56

17–57. The two blocks A and B having a weight of 10 lb and 5 lb, respectively, are attached to the ends of a cord which passes over a 3-lb pulley which has a radius of gyration of $k_O = 0.5$ ft. If the blocks are released from rest, determine their speed just after A descends 3 ft. The cord does not slip on the pulley. Neglect the mass of the cord. Solve the problem in two ways. First by considering the "system" consisting of both the blocks and the pulley, and then by considering the blocks and pulley separately.

Prob. 17–57

17–58. The 10-lb bar is pinned at its center O and connected to a torsional spring. The spring has a stiffness $k = 5$ lb · ft/rad, so that the torque $\mathbf{M}$ developed is $M = (5\theta)$ lb · ft, where θ is in radians. If the bar is released from rest when it is vertical at $\theta = 90°$, determine its angular velocity at the instant $\theta = 0°$.

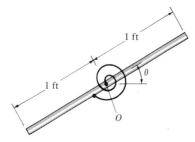

Prob. 17–58

327

17–59. The armature (slender rod) AB has a mass of 0.2 kg and can pivot about the pin at A. Movement is controlled by the electromagnet E, which exerts a horizontal attractive force on the armature at B of $F_B = (0.2(10^{-3})l^{-2})$ N, where l in meters is the gap between the armature and the magnet at any instant. If the armature lies in the horizontal plane, and is originally at rest, determine the speed of the contact at B the instant $l = 0.01$ m. Originally $l = 0.02$ m.

Prob. 17–59

*17–60. Disk A has a weight of 5 lb and disk B has a weight of 10 lb. If no slipping occurs between them, determine the couple **M** which must be applied to disk A to give it an angular acceleration of 4 rad/s^2.

Prob. 17–60

17–61. Determine the angular acceleration of the 25-kg diving board and the horizontal and vertical components of reaction at the pin A the instant the man jumps off. Assume that the board is thin, uniform, and rigid, and that at the instant the man jumps off the spring is compressed a maximum amount of 200 mm, $\omega = \mathbf{0}$, and the board is horizontal.

Prob. 17–61

17–62. The assembly consists of the 30-lb blocks A and B and a pulley which has a weight of 20 lb and a radius of gyration of $k_C = 0.5$ ft. Assuming the cord does not slip over the surface of the pulley, determine the force in the horizontal and vertical segments of the cord. Neglect both the mass of the cord and friction between the block at A and the horizontal plane.

Probs. 17–62/17–63

17–63. The assembly consists of the 30-lb blocks A and B and a pulley which has a weight of 20 lb and a radius of gyration of $k_C = 0.5$ ft. Assuming the cord does not slip over the surface of the pulley, determine the force in the horizontal and vertical segments of the cord. Neglect the mass of the cord and take the coefficient of kinetic friction between the block at A and the horizontal plane to be $\mu_k = 0.3$.

*17–64. The cylinder has a radius r and mass m and rests in the trough for which the coefficient of kinetic friction at A and B is μ. If a horizontal force **P** is applied to the cylinder, determine the cylinder's angular acceleration if it begins to spin.

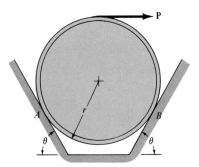

Prob. 17–64

17–66. The slender rod of length L and mass m is released from rest when $\theta = 0°$. Determine as a function of θ the normal and frictional forces which are exerted on the ledge at A as it falls downward. At what angle θ does it begin to slip if the coefficient of static friction at A is μ?

Prob. 17–66

17–65. The rod has a length L and mass m. If it is released from rest when $\theta \approx 0°$, determine its angular velocity as a function of θ. Also, express the horizontal and vertical components of reaction at the pin O as a function of θ.

Prob. 17–65

■17–67. A 40-kg boy sits on top of the large wheel which has a mass of 400 kg and a radius of gyration of $k_G = 5.5$ m. If the boy essentially starts from rest at $\theta = 0°$, and the wheel begins to freely rotate, determine the angle at which the boy begins to slip. The coefficient of static friction between the wheel and the boy is $\mu_s = 0.5$. Neglect the size of the boy in the calculation.

Prob. 17–67

17.5 Equations of Motion: General Plane Motion

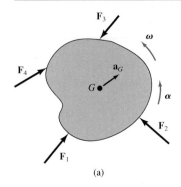

(a)

The rigid body (or slab) shown in Fig. 17–19a is subjected to general plane motion caused by the externally applied force system. The free-body and kinetic diagrams for the body are shown in Fig. 17–19b. The vector $m\mathbf{a}_G$ (shown dashed) has the same *direction* as the acceleration of the body's mass center, and $I_G\alpha$ acts in the same *direction* as the angular acceleration. If an x and y inertial coordinate system is chosen as shown, the three equations of motion may be written as

$$
\begin{aligned}
\Sigma F_x &= m(a_G)_x \\
\Sigma F_y &= m(a_G)_y \\
\Sigma M_G &= I_G\alpha
\end{aligned}
\qquad (17\text{–}17)
$$

In some problems it may be convenient to sum moments about some point P other than G. This is usually done in order to eliminate unknown forces from the moment summation. When used in this more general sense, the three equations of motion become

$$
\begin{aligned}
\Sigma F_x &= m(a_G)_x \\
\Sigma F_y &= m(a_G)_y \\
\Sigma M_P &= \Sigma(\mathcal{M}_k)_P
\end{aligned}
\qquad (17\text{–}18)
$$

where $\Sigma(\mathcal{M}_k)_P$ represents the moment sum of $I_G\alpha$ and $m\mathbf{a}_G$ (or its components) about P as determined by the data on the kinetic diagram.

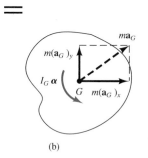

(b)

Fig. 17–19

PROCEDURE FOR ANALYSIS

The following procedure provides a method for solving kinetic problems involving general plane motion of a rigid body.

Free-Body Diagram. Establish the x, y inertial coordinate system and draw the free-body diagram for the body. Specify the direction and sense of the acceleration of the mass center, $\mathbf{a}_G$, and the angular acceleration $\boldsymbol{\alpha}$ of the body. Also, compute the moment of inertia I_G. Identify the unknowns in the problem. If it is decided that the rotational equation of motion $\Sigma M_P = \Sigma(\mathcal{M}_k)_P$ is to be used, then consider drawing the kinetic diagram in order to help "visualize" the "moments" developed by the components $m(\mathbf{a}_G)_x$, $m(\mathbf{a}_G)_y$, and $I_G\boldsymbol{\alpha}$ when writing the terms in the moment sum $\Sigma(\mathcal{M}_k)_P$.

Equations of Motion. Apply the three equations of motion, Eqs. 17–17 or 17–18.

Kinematics. Use kinematics if a complete solution cannot be obtained strictly from the equations of motion. In particular, if the body's motion is *constrained* due to its supports, additional equations may be obtained by using $\mathbf{a}_B = \mathbf{a}_A + \mathbf{a}_{B/A}$, which relate the accelerations of any two points A and B on the body (see Example 17–16).

Frictional Rolling Problems

There is a class of planar kinetics problems which deserves special mention. These problems involve wheels, cylinders, or bodies of similar shape, which roll on a *rough* plane surface. Because of the applied loadings, it may not be known if the body *rolls without slipping*, or if it *slides as it rolls*. For example, consider the homogeneous disk shown in Fig. 17–20a, which has a mass m and is subjected to a known horizontal force $\mathbf{P}$. Following the procedure outlined above, the free-body diagram is shown in Fig. 17–20b. Since $\mathbf{a}_G$ is directed to the right and $\boldsymbol{\alpha}$ is clockwise, we have

(a)

(b)

Fig. 17–20

$$\xrightarrow{+}\Sigma F_x = m(a_G)_x; \qquad P - F = ma_G \qquad (17\text{–}19)$$

$$+\uparrow\Sigma F_y = m(a_G)_y; \qquad N - mg = 0 \qquad (17\text{–}20)$$

$$\zeta+\Sigma M_G = I_G\alpha; \qquad Fr = I_G\alpha \qquad (17\text{–}21)$$

A fourth equation is needed since these *three equations* contain *four unknowns: F, N, α, and a_G.*

No Slipping. If the frictional force $\mathbf{F}$ is great enough to allow the disk to roll *without slipping*, then a_G may be related to α by the *kinematic equation**

$$(\zeta+) \qquad\qquad a_G = \alpha r \qquad (17\text{–}22)$$

The four unknowns are determined by *solving simultaneously* Eqs. 17–19 to 17–22. When the solution is obtained, the assumption of no slipping must be *checked.* Recall that no slipping occurs provided $F \leq \mu_s N$, where μ_s is the coefficient of static friction. If the inequality is satisfied, the problem is solved. However, if $F > \mu_s N$, the problem must be *reworked,* since then the disk slips as it rolls.

Slipping. In the case of slipping, α and a_G are *independent of one another* so that Eq. 17–22 does not apply. Instead, the magnitude of the frictional force is related to the magnitude of the normal force using the coefficient of kinetic friction μ_k, i.e.,

$$F = \mu_k N \qquad (17\text{–}23)$$

Here $\mathbf{F}$ acts to the left on the disk to prevent the slipping motion to the right, Fig. 17–20b. In this case Eqs. 17–19 to 17–21 and 17–23 are used for the solution. Examples 17–14 and 17–15 illustrate these concepts numerically.

---

*See Example 16–3 or 16-13.

Example 17–13

The spool in Fig. 17–21*a* has a mass of 8 kg and a radius of gyration of $k_G = 0.35$ m. If cords of negligible mass are wrapped around its inner hub and outer rim as shown, determine the spool's angular velocity 3 s after it is released from rest.

100 N

SOLUTION

Free-Body Diagram. Fig. 17–21*b*.   The 100-N force causes $\mathbf{a}_G$ to act upward. Also, $\boldsymbol{\alpha}$ acts clockwise since the spool winds around the cord at A.

There are three unknowns, namely T, a_G, and α. The moment of inertia of the spool about its mass center is

$$I_G = mk_G^2 = 8(0.35)^2 = 0.980 \text{ kg} \cdot \text{m}^2$$

(a)

Equations of Motion

$$+\uparrow \Sigma F_y = m(a_G)_y; \qquad T + 100 - 78.48 = 8a_G \qquad (1)$$

$$\circlearrowright + \Sigma M_G = I_G\alpha; \qquad 100(0.2) - T(0.5) = 0.980\alpha \qquad (2)$$

100 N

(b)

Fig. 17–21

Kinematics.   A complete solution is obtained if kinematics is used to relate a_G to α. In this case the spool "rolls without slipping" on the cord at A. Hence, we can use the results of Example 16–3 or 16–13, so that

$$(\circlearrowright+)a_G = r\alpha; \qquad a_G = 0.5\alpha \qquad (3)$$

Solving Eqs. (1) to (3), we have

$$\alpha = 10.3 \text{ rad/s}^2$$
$$a_G = 5.16 \text{ m/s}^2$$
$$T = 19.8 \text{ N}$$

Since α is constant, the angular velocity in 3 s is

$(\circlearrowright+)$
$$\omega = \omega_0 + \alpha_c t$$
$$= 0 + 10.3(3)$$
$$= 30.9 \text{ rad/s} \downarrow \qquad Ans.$$

Example 17–14

The 50-lb wheel shown in Fig. 17–22a has a radius of gyration $k_G = 0.70$ ft. If a 35-lb · ft couple moment is applied to the wheel, determine the acceleration of its mass center G. The coefficients of static and kinetic friction between the wheel and the plane at A are $\mu_s = 0.3$ and $\mu_k = 0.25$, respectively.

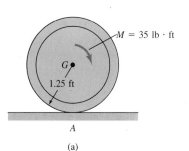

(a)

SOLUTION

Free-Body Diagram. By inspection of Fig. 17–22b, it is seen that the couple moment causes the wheel to have a clockwise angular acceleration of $\boldsymbol{\alpha}$. As a result, the acceleration of the mass center, $\mathbf{a}_G$, is directed to the right. The moment of inertia is

$$I_G = mk_G^2 = \frac{50}{32.2}(0.70)^2 = 0.761 \text{ slug} \cdot \text{ft}^2$$

The unknowns are N_A, F_A, a_G, and α.

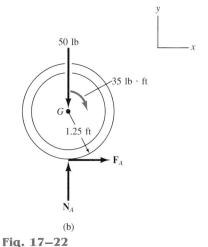

(b)   **Fig. 17–22**

Equations of Motion

$$\xrightarrow{+}\Sigma F_x = m(a_G)_x; \qquad F_A = \frac{50}{32.2}a_G \qquad (1)$$

$$+\uparrow \Sigma F_y = m(a_G)_y; \qquad N_A - 50 = 0 \qquad (2)$$

$$\zeta+\Sigma M_G = I_G\alpha; \qquad 35 - 1.25(F_A) = 0.761\alpha \qquad (3)$$

Kinematics (No Slipping). If this assumption is made, then

$$(\zeta+) \qquad\qquad a_G = 1.25\alpha \qquad (4)$$

Solving Eqs. (1) to (4),

$$N_A = 50.0 \text{ lb} \qquad F_A = 21.3 \text{ lb}$$
$$\alpha = 11.0 \text{ rad/s}^2 \qquad a_G = 13.7 \text{ ft/s}^2$$

The original assumption of no slipping requires $F_A \leq \mu_s N_A$. However, since 21.3 lb $>$ 0.3(50 lb) = 15 lb, the wheel slips as it rolls.

(Slipping). This requires $F_A = \mu_k N_A$, or

$$F_A = 0.25N_A \qquad (5)$$

Solving Eqs. (1) to (3) and (5) yields

$$N_A = 50.0 \text{ lb} \qquad F_A = 12.5 \text{ lb}$$
$$\alpha = 25.5 \text{ rad/s}^2$$
$$a_G = 8.05 \text{ ft/s}^2 \rightarrow \qquad\qquad Ans.$$

Example 17–15

The uniform slender pole shown in Fig. 17–23a has a mass of 100 kg and a moment of inertia $I_G = 75$ kg · m². If the coefficients of static and kinetic friction between the end of the pole and the surface are $\mu_s = 0.3$ and $\mu_k = 0.25$, respectively, determine the pole's angular acceleration at the instant the 400-N horizontal force is applied. The pole is originally at rest.

SOLUTION

Free-Body Diagram. Figure 17–23b. The path of motion of the mass center G will be along an unknown (dashed) curved path having a radius of curvature ρ which is initially parallel to the y axis. There is no normal or y component of acceleration since the pole is originally at rest, i.e., $\mathbf{v}_G = \mathbf{0}$, so that $(a_G)_y = v_G^2/\rho = 0$. We will assume the mass center accelerates to the right, and the pole has a clockwise angular acceleration of $\boldsymbol{\alpha}$. The unknowns are N_A, F_A, a_G, and α.

(a)

(b)

Fig. 17–23

Equations of Motion

$$\xrightarrow{+}\Sigma F_x = m(a_G)_x; \qquad 400 - F_A = 100a_G \qquad (1)$$

$$+\uparrow \Sigma F_y = m(a_G)_y; \qquad N_A - 981 = 0 \qquad (2)$$

$$\zeta+\Sigma M_G = I_G\alpha; \qquad F_A(1.5) - 400(1) = 75\alpha \qquad (3)$$

A fourth equation is needed for a complete solution.

Kinematics (No Slipping). In this case point A acts as a "pivot" so that

$$\zeta+a_G = \alpha r_{AG}; \qquad a_G = 1.5\alpha \qquad (4)$$

Solving Eqs. (1) to (4) yields

$$N_A = 981 \text{ N} \qquad F_A = 300 \text{ N}$$
$$a_G = 1 \text{ m/s}^2 \qquad \alpha = 0.667 \text{ rad/s}^2$$

Testing the original assumption of no slipping requires $F_A \le \mu_s N_A$. However, $300 > 0.3(981) = 294$ N. (Slips at A.)

(Slipping). For this case Eq. (4) does *not* apply. Instead, the frictional equation $F_A = \mu_k N_A$ is used, i.e.,

$$F_A = 0.25N_A \qquad (5)$$

Solving Eqs. (1) to (3) and (5) simultaneously yields

$$N_A = 981 \text{ N} \qquad F_A = 245 \text{ N} \qquad a_G = 1.55 \text{ m/s}^2$$
$$\alpha = -0.428 \text{ rad/s}^2 = 0.428 \text{ rad/s}^2 \, \zeta \qquad \qquad \textit{Ans.}$$

Example 17–16

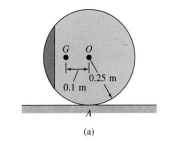

(a)

The 30-kg wheel shown in Fig. 17–24a has a mass center at G and a radius of gyration $k_G = 0.15$ m. If the wheel is originally at rest and released from the position shown, determine its angular acceleration. No slipping occurs.

SOLUTION

Free-Body and Kinetic Diagrams. The two unknowns $\mathbf{F}_A$ and $\mathbf{N}_A$ shown on the free-body diagram, Fig. 17–24b, can be eliminated from the analysis by summing moments about point A. The kinetic diagram accompanies the solution in order to illustrate application of $\Sigma(\mathcal{M}_k)_A$. Since the *path of motion* of G is *unknown*, the two components $m(\mathbf{a}_G)_x$ and $m(\mathbf{a}_G)_y$ must be shown on the kinetic diagram, Fig. 17–24b.

The moment of inertia is

$$I_G = mk_G^2 = 30(0.15)^2 = 0.675 \text{ kg} \cdot \text{m}^2$$

There are five unknowns, namely, N_A, F_A, $(a_G)_x$, $(a_G)_y$, and α.

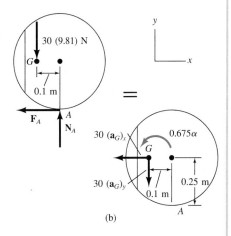

(b)

Equation of Motion. Applying the rotational equation of motion about point A, to eliminate N_A, and F_A, we have

$$\zeta + \Sigma M_A = \Sigma(\mathcal{M}_k)_A;$$
$$30(9.81)(0.1) = 0.675\alpha + 30(a_G)_x(0.25) + 30(a_G)_y(0.1) \quad (1)$$

There are three unknowns in this equation: $(a_G)_x$, $(a_G)_y$, and α.

Kinematics. Using kinematics, $(a_G)_x$, $(a_G)_y$, and α can be related. Since no slipping occurs, $a_O = \alpha r = \alpha(0.25)$, directed to the left, Fig. 17–24c. Also, $\boldsymbol{\omega} = \mathbf{0}$ since the wheel is originally at rest. Using Eq. 16–16, with point O as the base point, Fig. 17–24c and d, we have

$$\mathbf{a}_G = \mathbf{a}_O + (\mathbf{a}_{G/O})_t + (\mathbf{a}_{G/O})_n$$

$$\begin{bmatrix} (a_G)_x \\ \leftarrow \end{bmatrix} + \begin{bmatrix} (a_G)_y \\ \downarrow \end{bmatrix} = \begin{bmatrix} \alpha(0.25) \\ \leftarrow \end{bmatrix} + \begin{bmatrix} \alpha(0.1) \\ \downarrow \end{bmatrix} + \begin{bmatrix} 0 \end{bmatrix}$$

Absolute motion

(c)

Equating the respective horizontal and vertical components,

$$(a_G)_x = \alpha(0.25) \quad (2)$$
$$(a_G)_y = \alpha(0.1) \quad (3)$$

Solving Eqs. (1) to (3) yields

$$\alpha = 10.3 \text{ rad/s}^2 \, \zeta \qquad Ans.$$
$$(a_G)_x = 2.58 \text{ m/s}^2$$
$$(a_G)_y = 1.03 \text{ m/s}^2$$

As an exercise, show that $F_A = 77.4$ N and $N_A = 263$ N.

Fig. 17–24

Relative motion

(d)

335

PROBLEMS

*17–68. If the disk in Fig. 17–20*a* *rolls without slipping*, show that when moments are summed about the instantaneous center of zero velocity, *IC*, it is possible to use the moment equation $\Sigma M_{IC} = I_{IC}\alpha$, where I_{IC} represents the moment of inertia of the disk calculated about the instantaneous axis of zero velocity.

17–69. The uniform 50-lb board is suspended from cords at *C* and *D*. If these cords are subjected to constant forces of 30 lb and 45 lb, respectively, determine the acceleration of the board's center and its angular acceleration. Assume the board is a thin plate. Neglect the mass of the pulleys at *E* and *F*.

Probs. 17–69/17–70

17–70. Solve Prob. 17–69 if blocks having a weight of $W_A = 30$ lb and $W_B = 45$ lb are suspended from the cords at *A* and *B*.

17–71. The wheel has a mass of 80 kg and a radius of gyration $k_G = 0.25$ m. If it is subjected to a couple moment of $M = 50$ N · m, determine its angular acceleration. The coefficients of static and kinetic friction between the ground and the wheel are $\mu_s = 0.2$ and $\mu_k = 0.15$, respectively.

Prob. 17–71

*17–72. The spool and wire wrapped around its core have a mass of 20 kg and a centroidal radius of gyration $k_G = 250$ mm. If the coefficient of kinetic friction at the ground is $\mu_B = 0.1$, determine the angular acceleration of the spool when the 30-N · m couple is applied.

Prob. 17–72

17–73. The spool has a mass of 100 kg and a radius of gyration of $k_G = 0.3$ m. If the coefficients of static and kinetic friction at *A* are $\mu_s = 0.2$ and $\mu_k = 0.15$, respectively, determine the angular acceleration of the spool if $P = 50$ N.

Probs. 17–73/17–74/17–75

17–74. Solve Prob. 17–73 if the cord and force $P = 50$ N are directed vertically upwards.

17–75. The spool has a mass of 50 kg and a radius of gyration of $k_G = 0.3$ m. If the coefficients of static and kinetic friction at *A* are $\mu_s = 0.2$ and $\mu_k = 0.15$, respectively, determine the angular acceleration of the spool if $P = 600$ N.

*17–76. The spool has a mass of 500 kg and a radius of gyration $k_G = 0.380$ m. It rests on the inclined surface for which the coefficient of kinetic friction is $\mu_A = 0.15$. If the spool is released from rest, determine the initial tension in the cord and the angular acceleration of the spool.

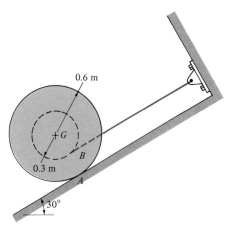

Prob. 17–76

17–77. The 20-kg punching bag has a radius of gyration about its center of mass G of $k_G = 0.4$ m. If it is subjected to a horizontal force of $F = 30$ N, determine the initial angular acceleration of the bag and the tension in the supporting cable AB.

Prob. 17–77

17–78. The 15-lb circular plate is suspended from a pin at A. If the pin is connected to a track which is given an acceleration of $a_A = 3$ ft/s², determine the horizontal and vertical components of reaction at A and the acceleration of the mass center G. The plate is originally at rest.

Prob. 17–78

17–79. The assembly consists of an 8-kg disk and a 10-kg bar which is pin-connected to the disk. If the system is released from rest, determine the initial angular acceleration of the disk. The coefficients of static and kinetic friction between the disk and the inclined plane are $\mu_s = 0.6$ and $\mu_k = 0.4$, respectively. Neglect friction at B.

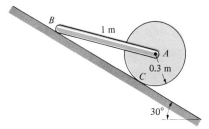

Probs. 17–79/17–80

*17–80. Solve Prob. 17–79 if the bar is removed.

17–81. By pressing down with the finger at B, a thin ring having a mass m is given an initial velocity v_1 and a backspin ω_1 when the finger is released. If the coefficient of kinetic friction between the table and the ring is μ, determine the distance the ring travels forward before backspinning stops.

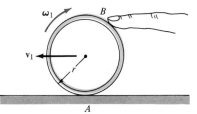

Prob. 17–81

17–82. A biomechanical model of the head and neck of a restrained passenger in an automobile is shown in the figure. Measurements indicate that the head plus half the neck, assumed pinned at A, have a mass of 0.28 slug and a moment of inertia about an axis through the center of gravity of $I_G = 0.024$ slug · ft². If an automobile and the passenger's torso are subjected to a horizontal deceleration $a_A = 15g = 483$ ft/s², when the head is in the position shown, determine the horizontal and vertical components of force which the neck exerts on the head at A at this instant. The restraining moment $\mathbf{M}_A$ is required to hold the head in the *nonrotating* equilibrium position shown, prior to the deceleration. Assume this moment remains constant during the deceleration.

Prob. 17–82

17–83. The uniform beam has a weight W. If it is originally at rest while being supported at A and B by cables, determine the tension in cable A if cable B suddenly fails. Assume the beam is a slender rod.

Prob. 17–83

*17–84. A cord is wrapped around each of the two disks. If disks A and B each have a mass of 10 kg and they are released from rest, determine the angular acceleration of each disk and the tension in cord C.

Prob. 17–84

Prob. 17–85

17–85. The uniform bar of mass m and length L is balanced in the vertical position when the horizontal force $\mathbf{P}$ is applied to the roller at A. Determine the bar's initial angular acceleration and the acceleration of its top point B.

17–86. The semicircular disk has a mass of 50 kg and is released from rest from the position shown. The coefficients of static and kinetic friction between the disk and the beam are $\mu_s = 0.5$ and $\mu_k = 0.3$, respectively. Determine the initial reactions at the pin A and roller B used to support the beam. Neglect the mass of the beam for the calculation.

Probs. 17–86/17–87

17–87. The semicircular disk has a mass of 50 kg and is released from rest from the position shown. The coefficients of static and kinetic friction between the disk and the beam are $\mu_s = 0.2$ and $\mu_k = 0.1$, respectively. Determine the initial reactions at the pin A and roller B used to support the beam. Neglect the mass of the beam for the calculation.

*17–88. The 2-kg slender bar is supported by cord BC and then released from rest at A. Determine the initial angular acceleration of the bar and the tension in the cord.

Prob. 17–88

17–89. The two pin-connected bars each have a weight of 10 lb/ft. If a moment of $M = 60$ lb · ft is applied to bar AB, determine the initial vertical reaction at C and the horizontal and vertical components of reaction at B. Neglect the size of the roller at C. The bars are initially at rest.

Prob. 17–89

17–90. The 10-lb disk and 4-lb block are released from rest. Determine the velocity of the block when $t = 3$ s. The coefficients of static and kinetic friction at A are $\mu_s = 0.2$ and $\mu_k = 0.15$, respectively. Neglect the mass of the cord and pulleys.

Prob. 17–90

17–91. The disk of mass m and radius r rolls without slipping on the circular path. Determine the normal force which the path exerts on the disk and the disk's angular acceleration if at the instant shown the disk has an angular velocity of ω.

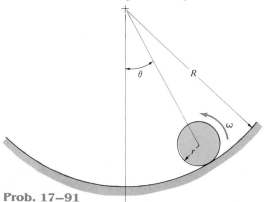

Prob. 17–91

*17–92. The semicircular disk has a mass of 10 kg. If it is rotating at $\omega = 4$ rad/s at the instant $\theta = 60°$, determine the normal and frictional forces it exerts on the ground at this instant. Assume the disk does not slip as it rolls.

Probs. 17–92/17–93

17–93. The semicircular disk having a mass of 10 kg is rotating at $\omega = 4$ rad/s at the instant $\theta = 60°$. If the coefficient of static friction at A is $\mu_s = 0.5$, determine if the disk slips at this instant.

17–94. The wheel has a mass of 50 kg and a radius of gyration $k_G = 0.4$ m. If it rolls without slipping down the inclined plank, determine the forces which the supports at A and B exert on the plank at the instant the wheel is located at the midpoint of the plank. The plank is uniform and has a mass of 20 kg. Assume that horizontal and vertical reactions occur at A, and a normal reaction occurs at B.

Prob. 17–94

■17–95. The cylinder has a mass m and is released from rest at $\theta = 0°$. If it rolls down the surface of the fixed semicylinder, determine the angle θ at which the cylinder just begins to slip on the surface. The coefficient of static friction between the cylinder and track is $\mu_s = 0.3$.

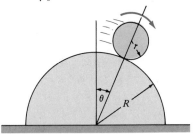

Prob. 17–95

*17–96. The 15-lb disk rests on a 5-lb plate. A cord is wrapped around the periphery of the disk and attached to the wall at B. If a torque $M = 40$ lb · ft is applied to the disk, determine the angular acceleration of the disk and the time needed for the end C of the plate to travel 3 ft and strike the wall. The coefficient of kinetic friction between the plate and disk is $\mu_k = 0.2$, and between the plate and floor $\mu_k' = 0.1$. Neglect the mass of the cord.

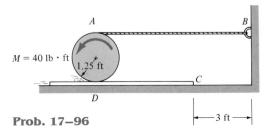

Prob. 17–96

17–97. The slender rod of length L and mass m is released from rest when $\theta = 0°$. If the ground is smooth, determine the angular acceleration of the rod just before $\theta = 90°$. Also, what is the reaction of the ground on the rod at this instant?

Prob. 17–97

17–98. The slender bar has a mass m and rests on the smooth floor A and against the smooth wall B. Determine its *initial* angular acceleration when it is released from rest from the position θ and allowed to slide downward.

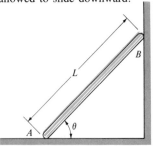

Prob. 17–98

18

Planar Kinetics of a Rigid Body: Work and Energy

It was shown in Chapter 14 that problems which involve force, velocity, and displacement can conveniently be solved by using the principle of work and energy, or if the force system is "conservative," the conservation of energy theorem. In this chapter we will apply work and energy methods to solve problems involving the planar motion of a rigid body. The more general discussion as applied to the three-dimensional motion of a rigid body is presented in Chapter 21.

Before discussing the principle of work and energy for a body, however, the methods for obtaining the body's kinetic energy when it is subjected to translation, rotation about a fixed axis, or general plane motion will be developed.

18.1 Kinetic Energy

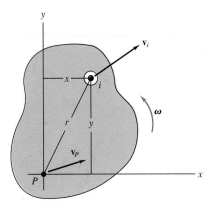

Fig. 18–1

Consider the rigid body shown in Fig. 18–1, which is represented here by a *slab* moving in the inertial *x-y* reference plane. An arbitrary *i*th particle of the body, having a mass dm, is located at r from the arbitrary point P. If at the *instant* shown the particle has a velocity $\mathbf{v}_i$, then the particle's kinetic energy is $T_i = \frac{1}{2}dm\, v_i^2$. The kinetic energy of the entire body is determined by writing similar expressions for each particle of the body and integrating the results, i.e.,

$$T = \frac{1}{2}\int_m dm\, v_i^2$$

This equation may be written in another manner by expressing $\mathbf{v}_i$ in terms of the velocity of point P, i.e., $\mathbf{v}_P$. If the body has an angular velocity $\boldsymbol{\omega}$, then from Fig. 18–1 we have

$$\mathbf{v}_i = \mathbf{v}_P + \mathbf{v}_{i/P}$$
$$= (v_P)_x\mathbf{i} + (v_P)_y\mathbf{j} + \omega\mathbf{k} \times (x\mathbf{i} + y\mathbf{j})$$
$$= ((v_P)_x - \omega y)\mathbf{i} + ((v_P)_y + \omega x)\mathbf{j}$$

The square of the magnitude of $\mathbf{v}_i$ is thus

$$\mathbf{v}_i \cdot \mathbf{v}_i = v_i^2 = [(v_P)_x - \omega y]^2 + [(v_P)_y + \omega x]^2$$
$$= (v_P^2)_x - 2(v_P)_x\omega y + \omega^2 y^2 + (v_P^2)_y + 2(v_P)_y\omega x + \omega^2 x^2$$
$$= v_P^2 - 2(v_P)_x\omega y + 2(v_P)_y\omega x + \omega^2 r^2$$

Substituting into the equation of kinetic energy yields

$$T = \frac{1}{2}\left(\int_m dm\right)v_P^2 - (v_P)_x\omega\left(\int_m y\, dm\right) + (v_P)_y\omega\left(\int_m x\, dm\right) + \frac{1}{2}\omega^2\left(\int_m r^2\, dm\right)$$

The first integral on the right represents the entire mass m of the body. Since $\bar{x}m = \int x\, dm$ and $\bar{y}m = \int y\, dm$, the second and third integrals locate the body's center of mass G with respect to P. The last integral represents the body's moment of inertia I_P, computed about the z axis passing through point P. Thus,

$$T = \frac{1}{2}mv_P^2 - (v_P)_x\, \omega\bar{y}\, m + (v_P)_y\, \omega\bar{x}\, m + \frac{1}{2}I_P\, \omega^2 \qquad (18\text{–}1)$$

This equation reduces to a simpler form if point P coincides with the mass center G for the body, in which case $\bar{x} = \bar{y} = 0$, and therefore

$$T = \frac{1}{2}mv_G^2 + \frac{1}{2}I_G\omega^2 \qquad (18\text{–}2)$$

Here I_G is the mass moment of inertia of the body about an axis which is perpendicular to the plane of motion and passes through the mass center. Both terms on the right side are always *positive,* since the velocities are squared. Furthermore, it may be verified that these terms have units of length times force, common units being m · N or ft · lb. Recall, however, that in the SI system the unit of energy is the joule (J), where 1 J = 1 m · N.

Translation

When a rigid body of mass m is subjected to either rectilinear or curvilinear *translation*, the kinetic energy due to rotation is zero, since $\boldsymbol{\omega} = \mathbf{0}$. From Eq. 18–2, the kinetic energy of the body is therefore

$$T = \tfrac{1}{2}mv_G^2$$

where v_G is the magnitude of the translational velocity at the instant considered, Fig. 18–2.

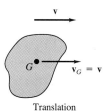

Translation

Fig. 18–2

Rotation About a Fixed Axis

When a rigid body is *rotating about a fixed axis* passing through point O, Fig. 18–3, the body has both *translational* and *rotational* kinetic energy as defined by Eq. 18–2, i.e.,

$$T = \tfrac{1}{2}mv_G^2 + \tfrac{1}{2}I_G\omega^2 \qquad (18–3)$$

The body's kinetic energy may be formulated in another manner by noting that $v_G = r_{G/O}\omega$, in which case $T = \tfrac{1}{2}(I_G + mr_{G/O}^2)\omega^2$. By the parallel-axis theorem, the terms inside the parentheses represent the moment of inertia I_O of the body about an axis perpendicular to the plane of motion and passing through point O. Hence,*

$$T = \tfrac{1}{2}I_O\omega^2 \qquad (18–4)$$

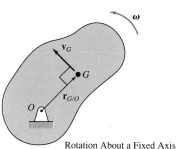

Rotation About a Fixed Axis

Fig. 18–3

From the derivation, this equation may be substituted for Eq. 18–3, since it accounts for *both* the translational kinetic energy of the body's mass center and the rotational kinetic energy of the body computed about the mass center.

General Plane Motion

When a rigid body is subjected to general plane motion, Fig. 18–4, it has an angular velocity $\boldsymbol{\omega}$ and its mass center has a velocity $\mathbf{v}_G$. Hence, the kinetic energy is defined by Eq. 18–2, i.e.,

$$T = \tfrac{1}{2}mv_G^2 + \tfrac{1}{2}I_G\omega^2 \qquad (18–5)$$

Here it is seen that the total kinetic energy of the body consists of the *scalar* sum of the body's *translational* kinetic energy, $\tfrac{1}{2}mv_G^2$, and *rotational* kinetic energy about its mass center, $\tfrac{1}{2}I_G\omega^2$.

Because energy is a scalar quantity, the total kinetic energy for a system of *connected* rigid bodies is the sum of the kinetic energies of all the moving parts. Depending upon the type of motion, the kinetic energy of *each body* is found by applying Eq. 18–2 or the alternative forms mentioned above.

General plane motion

Fig. 18–4

---

*The similarity between this derivation and that of $\Sigma M_O = I_O\alpha$, Eq. 17–16, should be noted. Also note that the same result can be obtained from Eq. 18–1 by selecting point P at O, realizing that $\mathbf{v}_O = \mathbf{0}$.

Example 18–1

The system of three elements shown in Fig. 18–5a consists of a 6-kg block B, a 10-kg disk D, and a 12-kg cylinder C. A continuous cord of negligible mass is wrapped around the cylinder, passes over the disk, and is then attached to the block. If the block is moving downward with a speed of 0.8 m/s and the cylinder rolls without slipping, determine the total kinetic energy of the system at this instant.

(a)

$v_E = 0.8$ m/s

ω_C

$r_{E/IC} = 0.2$ m

$r_{G/IC} = 0.1$ m

IC

Fig. 18–5   (b)

SOLUTION

By inspection, the block is translating, the disk rotates about a fixed axis, and the cylinder has general plane motion. Hence, in order to compute the kinetic energy of the disk and cylinder, it is first necessary to determine ω_D, ω_C, and v_G, Fig. 18–5a. From the *kinematics* of the disk,

$$v_B = r_D \omega_D; \qquad 0.8 \text{ m/s} = (0.1 \text{ m})\omega_D; \qquad \omega_D = 8 \text{ rad/s}$$

Since the cylinder rolls without slipping, the instantaneous center of zero velocity is at the point of contact with the ground, Fig. 18–5b, hence,

$$v_E = r_{E/IC}\omega_C; \qquad 0.8 \text{ m/s} = (0.2 \text{ m})\omega_C; \qquad \omega_C = 4 \text{ rad/s}$$
$$v_G = r_{G/IC}\omega_C; \qquad v_G = (0.1 \text{ m})(4 \text{ rad/s}) = 0.4 \text{ m/s}$$

The kinetic energy of the block is

$$T_B = \tfrac{1}{2}m_B v_B^2; \quad T_B = \frac{1}{2}(6 \text{ kg})(0.8 \text{ m/s})^2 = 1.92 \text{ J}$$

The kinetic energy of the disk is

$$T_D = \tfrac{1}{2}I_D \omega_D^2; \quad T_D = \tfrac{1}{2}(\tfrac{1}{2}m_D r_D^2)\omega_D^2$$
$$= \frac{1}{2}\left[\frac{1}{2}(10 \text{ kg})(0.1 \text{ m})^2\right](8 \text{ rad/s})^2 = 1.60 \text{ J}$$

Finally, the kinetic energy of the cylinder is

$$T_C = \tfrac{1}{2}mv_G^2 + \tfrac{1}{2}I_G\omega_C^2; \quad T_C = \tfrac{1}{2}mv_G^2 + \tfrac{1}{2}(\tfrac{1}{2}m_C r_C^2)\omega_C^2$$
$$= \frac{1}{2}(12 \text{ kg})(0.4 \text{ m/s})^2 + \frac{1}{2}\left[\frac{1}{2}(12 \text{ kg})(0.1 \text{ m})^2\right](4 \text{ rad/s})^2 = 1.44 \text{ J}$$

The total kinetic energy of the system is therefore

$$T = T_B + T_D + T_C$$
$$= 1.92 \text{ J} + 1.60 \text{ J} + 1.44 \text{ J} = 4.96 \text{ J} \qquad \textit{Ans.}$$

18.2 The Work of a Force

Several types of forces are often encountered in planar kinetics problems involving a rigid body. The work of each of these forces has been presented in Sec. 14.1 and is listed below as a summary.

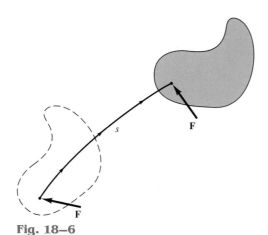

Fig. 18–6

Work of a Variable Force

If an external force **F** acts on a rigid body the work done by the force when it moves along the path s, Fig. 18–6, is defined as

$$U = \int_s F \cos \theta \, ds \qquad (18\text{–}6)$$

Here θ is the angle between the ''tails'' of the force vector and the differential displacement. In general, the integration must account for the variation of the force's direction and magnitude.

345

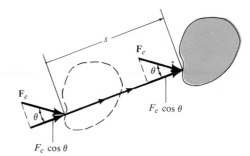

Fig. 18–7

Work of a Constant Force

If an external force $\mathbf{F}_c$ acts on a rigid body, Fig. 18–7, and maintains a constant magnitude F_c and constant direction θ, while the body undergoes a translation s, Eq. 18–6 can be integrated so that the work becomes

$$U_{F_c} = (F_c \cos \theta)s \qquad (18\text{–}7)$$

Here $F_c \cos \theta$ represents the magnitude of the component of force in the direction of displacement.

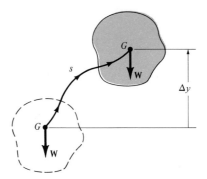

Fig. 18–8

Work of a Weight

The weight of a body does work only when the body's center of mass G undergoes a *vertical displacement* Δy. If this displacement is *upward,* Fig. 18–8, the work is negative, since the weight and displacement are in the opposite directions.

$$U_W = -W\,\Delta y \qquad (18\text{–}8)$$

Likewise, if the displacement is downward $(-\Delta y)$ the work becomes positive. Here the elevation change is considered to be small so that $\mathbf{W}$, which is caused by gravitation, is constant.*

---

*If the difference in elevation is large, use the results of Prob. 14–28.

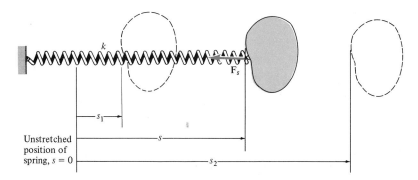

Fig. 18–9

Work of a Spring Force

If a linear elastic spring is attached to a body, the spring force $F_s = ks$ *acting on the body* does work when the spring either stretches or compresses from s_1 to a *further* position s_2. In both cases the work will be *negative* since the *displacement of the body* is always in the opposite direction to the force, Fig. 18–9. The work done is

$$U_s = -(\tfrac{1}{2}ks_2^2 - \tfrac{1}{2}ks_1^2) \qquad (18\text{–}9)$$

where $|s_2| > |s_1|$.

Forces That Do No Work

There are some external forces that do no work when the body is displaced. These forces can act either at *fixed points* on the body or they can have a direction *perpendicular to their displacement*. Examples include the reactions at a pin support about which a body rotates, the normal reaction acting on a body that moves along a fixed surface, and the weight of a body when the center of gravity of the body moves in a *horizontal plane*. A rolling resistance force $\mathbf{F}_r$ acting on a body as it *rolls without slipping* over a rough surface also does no work, Fig. 18–10.* This is because, during any *instant of time dt*, $\mathbf{F}_r$ acts at a point on the body which has *zero velocity* (instantaneous center, *IC*). In other words, for any differential rolling movement ds_{IC} of the body's *IC*, the work is $dU = F_r \, ds_{IC} = F_r(v_{IC}dt) = 0$, since $v_{IC} = 0$.

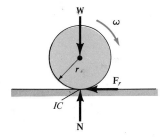

Fig. 18–10

*The work done by the frictional force *when the body slips* has been discussed in Sec. 14.3.

18.3 The Work of a Couple

(a)

Translation
(b)

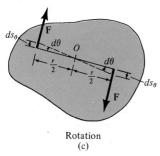

Rotation
(c)

Fig. 18–11

Recall that a *couple* consists of a pair of noncollinear forces which have equal magnitudes and opposite directions. When a body subjected to a couple undergoes general plane motion, the two forces do work *only* when the body undergoes a *rotation*. To show this, consider the body in Fig. 18–11a, which is subjected to a couple having a magnitude of $M = Fr$. Any general differential displacement of the body can be considered as a separate translation and rotation. When the body *translates* such that the *component of displacement* along the line of action of the forces is ds_t, Fig. 18–11b, clearly the "positive" work of one force *cancels* the "negative" work of the other. Consider now a differential rotation $d\theta$ of the body about an axis which is perpendicular to the plane of the couple and intersects the plane at point O, Fig. 18–11c. (For the derivation any other point in the plane may also be considered.) As shown, each force undergoes a displacement $ds_\theta = (r/2)\,d\theta$ in the direction of the force; hence, the total work done is

$$dU_M = F\left(\frac{r}{2}d\theta\right) + F\left(\frac{r}{2}d\theta\right) = (Fr)\,d\theta$$

$$= M\,d\theta$$

Here the line of action of $d\theta$ is parallel to the line of action of M. This is *always the case for general plane motion*, since **M** and $d\boldsymbol{\theta}$ are perpendicular to the plane of motion. Furthermore, the resultant work is *positive* when **M** and $d\boldsymbol{\theta}$ are in the *same direction* and *negative* if these vectors are in *opposite directions*.

When the body rotates in the plane through a finite angle θ measured in radians, from θ_1 to θ_2, the work of a couple is

$$U_M = \int_{\theta_1}^{\theta_2} M\,d\theta \qquad (18\text{–}10)$$

If the couple moment **M** has a *constant magnitude*, then

$$U_M = M(\theta_2 - \theta_1) \qquad (18\text{–}11)$$

Here the work is *positive* provided **M** and $(\boldsymbol{\theta}_2 - \boldsymbol{\theta}_1)$ are in the same direction.

18.4 Principle of Work and Energy

In Sec. 14.2 the principle of work and energy was developed for a particle. By applying this principle to each of the particles of a rigid body and adding the results algebraically, since energy is a scalar, the principle of work and energy for a rigid body may be developed. In this regard, the body's initial and final kinetic energies have been defined by the formulations in Sec. 18.1, and the work done by the *external* forces and couples is defined by formulations in Secs. 18.2 and 18.3. Note that the work of the body's *internal forces* does not have to be considered since the body is rigid. These forces occur in equal but opposite collinear pairs, so that when the body moves, the work of one force cancels that of its counterpart. Furthermore, since the body is rigid, *no relative movement* between these forces occurs, so that no internal work is done. Thus the principle of work and energy for a rigid body may be written as

$$\boxed{T_1 + \Sigma U_{1-2} = T_2} \qquad (18\text{--}12)$$

This equation states that the body's initial translational *and* rotational kinetic energy plus the work done by all the external forces and couples acting on the body as the body moves from its initial to its final position is equal to the body's final translational *and* rotational kinetic energy.

When several rigid bodies are pin-connected, connected by inextensible cables, or in mesh with one another, this equation may be applied to the entire system of connected bodies. In all these cases the internal forces, which hold the various members together, do no work and hence are eliminated from the analysis.

PROCEDURE FOR ANALYSIS

The principle of work and energy is used to solve kinetics problems that involve *velocity, force,* and *displacement,* since these terms are involved in the formulation. For application, it is suggested that the following procedure be used.

Kinetic Energy (Kinematic Diagrams). Determine the kinetic-energy terms T_1 and T_2 by applying the equation $T = \frac{1}{2}mv_G^2 + \frac{1}{2}I_G\omega^2$ or an appropriate form of this equation developed in Sec. 18.1. In this regard, *kinematic diagrams* for velocity may be useful for determining v_G and ω, or for establishing a *relationship* between v_G and ω.*

Work (Free-Body Diagram). Draw a free-body diagram of the body when it is located at an intermediate point along the path, in order to account for all the forces and couples which do work on the body as it moves along the path. The work of each force and couple can be computed using the appropriate formulations outlined in Secs. 18.2 and 18.3. Since *algebraic addition* of the work terms is required, it is important that the proper sign of each term be specified. Specifically, work is *positive* when the force (couple) is in the *same direction* as its displacement (rotation); otherwise, it is negative.

Principle of Work and Energy. Apply the principle of work and energy, $T_1 + \Sigma U_{1-2} = T_2$. Since this is a scalar equation, it can be used to solve for only one unknown when it is applied to a single rigid body. This is in contrast to the three scalar equations of motion, Eqs. 17–11, which may be written for the same body.

The following example problems numerically illustrate application of this procedure.

---

*A brief review of Secs. 16.5 to 16.7 may prove helpful in solving problems, since computations for kinetic energy require a kinematic analysis of velocity.

Example 18–2

The 30-kg disk shown in Fig. 18–12a is pin-supported at its center. Determine the number of revolutions it must make to attain an angular velocity of 20 rad/s starting from rest. It is acted upon by a constant force $F = 10$ N, which is applied to a cord wrapped around its periphery, and a constant couple moment $M = 5$ N · m. Neglect the mass of the cord in the calculation.

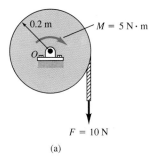

$M = 5$ N · m

0.2 m

$F = 10$ N

(a)

Fig. 18–12

SOLUTION

Kinetic Energy. Since the disk rotates about a fixed axis, the kinetic energy can be computed using $T = \frac{1}{2}I_O\omega^2$, where the moment of inertia is $I_O = \frac{1}{2}mr^2$. Initially, the disk is at rest, so that

$$T_1 = 0$$

$$T_2 = \tfrac{1}{2}I_O\omega_2^2 = \frac{1}{2}\left[\frac{1}{2}(30)(0.2)^2\right](20)^2 = 120 \text{ J}$$

Work (Free-Body Diagram). As shown in Fig. 18–12b, the pin reactions O_x and O_y and the weight (294.3 N) do no work, since they are not displaced. The *couple moment,* having a constant magnitude, does positive work $U_M = M\theta$ as the disk *rotates* through a clockwise angle of θ rad, and the *constant force* **F** does positive work $U_{F_c} = Fs$ as the cord *moves* downward $s = \theta r = \theta(0.2)$ m.

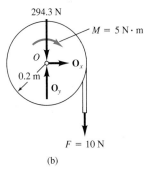

294.3 N

$M = 5$ N · m

O O_x

0.2 m

O_y

$F = 10$ N

(b)

Principle of Work and Energy

$$\{T_1\} + \{\Sigma U_{1-2}\} = \{T_2\}$$
$$\{T_1\} + \{M\theta + Fs\} = \{T_2\}$$
$$\{0\} + \{5\theta + (10)\theta(0.2)\} = \{120\}$$

$$\theta = 17.1 \text{ rad} = 17.1 \text{ rad}\left(\frac{1 \text{ rev}}{2\pi \text{ rad}}\right) = 2.73 \text{ rev} \qquad \textit{Ans.}$$

This problem has also been solved in Example 17–9. Compare the two methods of solution and note that since force, velocity, and displacement θ are involved, a work-energy approach yields a more direct solution.

Example 18–3

The uniform 5-kg bar shown in Fig. 18–13a is acted upon by the 30-N force which always acts perpendicular to the bar as shown. If the bar has an initial clockwise angular velocity $\omega_1 = 10$ rad/s when $\theta = 0°$, determine its angular velocity at the instant $\theta = 90°$.

30 N

θ

0.6 m

A

(a)

Fig. 18–13

SOLUTION

Kinetic Energy (Kinematic Diagrams). Two kinematic diagrams of the bar when $\theta = 0°$ (position 1) and $\theta = 90°$ (position 2) are shown in Fig. 18–13b. The initial kinetic energy may be computed with reference to either the fixed point of rotation A or the center of mass G. If A is considered, then

$$T_1 = \tfrac{1}{2}I_A\omega_1^2 = \frac{1}{2}\left[\frac{1}{3}(5)(0.6)^2\right](10)^2 = 30 \text{ J}$$

If point G is considered, then

$$T_1 = \tfrac{1}{2}m(v_G)_1^2 + \tfrac{1}{2}I_G\omega_1^2 = \frac{1}{2}(5)(3)^2 + \frac{1}{2}\left[\frac{1}{12}(5)(0.6)^2\right](10)^2 = 30 \text{ J}$$

In the final position

$$T_2 = \frac{1}{2}\left[\frac{1}{3}(5)(0.6)^2\right]\omega_2^2 = 0.3\omega_2^2$$

Work (Free-Body Diagram). Fig. 18–13c. The reactions A_x and A_y do no work, since these forces do not move. The 49.05-N weight, centered at G, moves downward through a vertical distance $\Delta y = 0.3$ m. The 30-N force moves tangent to its path of length $\tfrac{1}{2}\pi(0.6)$ m.

Principle of Work and Energy

$$\{T_1\} + \{\Sigma U_{1-2}\} = \{T_2\}$$

$$\{30\} + \{49.05(0.3) + 30[\tfrac{1}{2}\pi(0.6)]\} = \{0.3\omega_2^2\}$$

$$\omega_2 = 15.6 \text{ rad/s} \,\downdownarrows \qquad\qquad Ans.$$

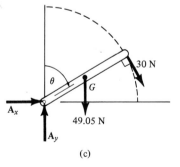

(1)

G $(v_G)_1 = 3$ m/s

0.3 m $\omega_1 = 10$ rad/s

A ω_2 G (2)

$(v_G)_2$

(b)

30 N

θ G

A_x 49.05 N

A_y

(c)

Example 18–4

The wheel shown in Fig. 18–14a weighs 40 lb and has a radius of gyration $k_G = 0.6$ ft about its mass center G. If it is subjected to a clockwise couple moment of 15 lb · ft and rolls from rest without slipping, determine its angular velocity after its center G moves 0.5 ft. The spring has a stiffness $k = 10$ lb/ft and is initially unstretched when the couple is applied.

Fig. 18–14

(a)

SOLUTION

Kinetic Energy (Kinematic Diagram). Since the wheel is initially at rest,

$$T_1 = 0$$

The kinematic diagram of the wheel when it is in the final position is shown in Fig. 18–14b. The velocity of the mass center $(\mathbf{v}_G)_2$ can be related to the angular velocity $\boldsymbol{\omega}_2$ by using the instantaneous center of zero velocity (*IC*), i.e., $(v_G)_2 = 0.8\omega_2$. Hence, the final kinetic energy is

$$T_2 = \tfrac{1}{2}m(v_G)_2^2 + \tfrac{1}{2}I_G(\omega_2)^2$$

$$= \frac{1}{2}\left(\frac{40}{32.2}\right)(0.8\omega_2)^2 + \frac{1}{2}\left[\frac{40}{32.2}(0.6)^2\right](\omega_2)^2$$

$$= 0.621(\omega_2)^2$$

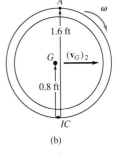

(b)

Work (Free-Body Diagram). As shown in Fig. 18–14c, only the spring force $\mathbf{F}_s$ and the couple moment do work. The normal force does not move along its line of action and the frictional force does *no work*, since the wheel does not slip as it rolls.

The work of $\mathbf{F}_s$ may be computed using $U_s = -\tfrac{1}{2}ks^2$. (Why is the work negative?) Since the wheel does not slip when the center G moves 0.5 ft, the wheel rotates $\theta = s_G/r_{G/IC} = 0.5/0.8 = 0.625$ rad, Fig. 18–14b. Hence, the spring stretches $s_A = \theta\, r_{A/IC} = 0.625(1.6) = 1$ ft.

(c)

Principle of Work and Energy

$$\{T_1\} + \{\Sigma U_{1-2}\} = \{T_2\}$$

$$\{T_1\} + \{M\theta - \tfrac{1}{2}ks^2\} = \{T_2\}$$

$$\{0\} + \{15(0.625) - \frac{1}{2}(10)(1)^2\} = \{0.621(\omega_2)^2\}$$

$$\omega_2 = 2.65 \text{ rad/s } \downarrow \qquad \textit{Ans.}$$

Example 18–5

The 10-kg rod shown in Fig. 18–15a is constrained so that its ends move along the grooved slots. The rod is initially at rest when $\theta = 0°$. If the slider block at B is acted upon by a horizontal force $P = 50$ N, determine the angular velocity of the rod at the instant $\theta = 45°$. Neglect the mass of blocks A and B. (Why can the principle of work and energy be used to solve this problem?)

(a)

SOLUTION

Kinetic Energy (Kinematic Diagrams). Two kinematic diagrams of the rod, when it is in the initial position 1 and final position 2, are shown in Fig. 18–15b. When the rod is in position 1, $T_1 = 0$ since $(v_G)_1 = \omega_1 = 0$. In position 2 the angular velocity is ω_2 and the velocity of the mass center is $(v_G)_2$. Hence, the kinetic energy is

$$T_2 = \tfrac{1}{2}m(v_G)_2^2 + \tfrac{1}{2}I_G(\omega_2)^2$$
$$= \frac{1}{2}(10)(v_G)_2^2 + \frac{1}{2}\left[\frac{1}{12}(10)(0.8)^2\right](\omega_2)^2$$
$$= 5(v_G)_2^2 + 0.267(\omega_2)^2 \tag{1}$$

The two unknowns $(v_G)_2$ and ω_2 may be related via the instantaneous center of zero velocity for the rod, Fig. 18–15b. It is seen that as A moves downward with a velocity $(v_A)_2$, B moves horizontally to the left with a velocity $(v_B)_2$. Knowing these directions, the IC may be determined as shown in the figure. Hence,

$$(v_G)_2 = r_{G/IC}\omega_2 = (0.4 \tan 45°)\omega_2$$
$$= 0.4\omega_2$$

Substituting into Eq. (1), we have

$$T_2 = 5(0.4\omega_2)^2 + 0.267(\omega_2)^2 = 1.067(\omega_2)^2$$

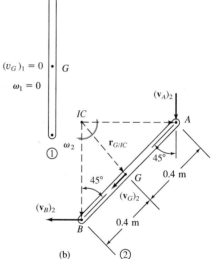

(b)

Work (Free-Body Diagram). Fig. 18–15c. The normal forces N_A and N_B do no work as the rod is displaced. Why? The 98.1-N weight is displaced a vertical distance of $\Delta y = (0.4 - 0.4 \cos 45°)$ m; whereas the 50-N force moves a horizontal distance of $s = (0.8 \sin 45°)$ m.

Principle of Work and Energy
$$\{T_1\} + \{\Sigma U_{1-2}\} = \{T_2\}$$
$$\{T_1\} + \{W\Delta y + Ps\} = \{T_2\}$$
$$\{0\} + \{98.1(0.4 - 0.4 \cos 45°) + 50(0.8 \sin 45°)\} = \{1.067(\omega_2)^2\}$$

Solving for ω_2 gives

$$\omega_2 = 6.11 \text{ rad/s} \downarrow \qquad \textit{Ans.}$$

(c)

Fig. 18–15

PROBLEMS

18–1. Solve Prob. 17–44 using the principle of work and energy.

18–2. Solve Prob. 17–57 using the principle of work and energy.

18–3. Solve Prob. 17–58 using the principle of work and energy.

*18–4. At a given instant the body of mass m has an angular velocity $\boldsymbol{\omega}$ and its mass center has a velocity $\mathbf{v}_G$. Show that its kinetic energy can be represented as $T = \frac{1}{2}I_{IC}\omega^2$, where I_{IC} is the moment of inertia of the body computed about the instantaneous axis of zero velocity, located at a distance $r_{G/IC}$ from the mass center as shown.

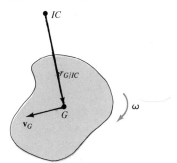

Prob. 18–4

18–5. The wheel is made from a 5-kg thin ring and two 2-kg slender rods. If the torsional spring attached to the wheel's center has a stiffness of $k = 2$ N · m/rad, so that the torque on the center of the wheel is $M = (2\theta)$ N · m, where θ is in radians, determine the maximum angular velocity of the wheel if it is rotated two revolutions and released from rest.

Prob. 18–5

18–6. The 4-kg slender rod is subjected to the force and couple moment. When it is in the position shown it has an angular velocity of $\omega_1 = 6$ rad/s. Determine its angular velocity at the instant it has rotated downward 90°. The force is always applied perpendicular to the axis of the rod. Motion occurs in the vertical plane.

Prob. 18–6

18–7. The soap-box car has a weight of 110 lb, including the passenger but *excluding* its four wheels. Each wheel has a weight of 5 lb and a radius of gyration $k = 0.3$ ft, computed about an axis passing through the wheel's axle. Determine the car's speed after it has traveled 100 ft starting from rest. The wheels roll without slipping. Neglect air resistance.

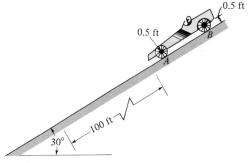

Prob. 18–7

*18–8. The pendulum consists of two slender rods each having a density of 4 kg/m. If it is acted upon by a moment of $M = 50$ N · m and released from the position shown, determine its angular velocity when it has rotated (a) 90° and (b) 180°. Motion occurs in the vertical plane.

Prob. 18–8

18–9. The revolving door consists of four doors which are attached to an axle AB. Each door can be assumed to be a 50-lb thin plate. Friction at the axle contributes a resultant moment of 2 lb · ft which resists the rotation of the doors. If a man passes through one door by always pushing with a force $P = 15$ lb perpendicular to the plane of the door as shown, determine the door's angular velocity after it has rotated 90°. The doors are originally at rest.

Prob. 18–9

18–10. The linkage consists of two 8-lb rods AB and CD and a 10-lb bar AD. When $\theta = 0°$, rod AB is rotating with an angular velocity $\omega_{AB} = 2$ rad/s. If rod CD is subjected to a couple moment $M = 15$ lb · ft and bar AD is subjected to a horizontal force $P = 20$ lb as shown, determine ω_{AB} at the instant $\theta = 90°$.

Prob. 18–10

18–11. The 20-kg disk is originally at rest and the spring holds it in equilibrium. A couple moment of $M = 30$ N · m is then applied to the disk as shown. Determine its angular velocity at the instant its mass center G has moved $s = 0.8$ m down along the inclined plane. The disk rolls without slipping.

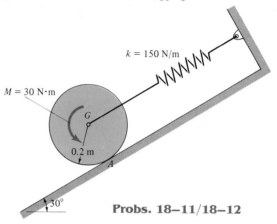

Probs. 18–11/18–12

*18–12. The 20-kg disk is originally at rest and the spring holds it in equilibrium. A couple moment of $M = 30$ N · m is then applied to the disk as shown. Determine how far s the center of mass of the disk travels down along the incline, measured from the equilibrium position, before it stops. The disk rolls without slipping.

18–13. The drum has a mass of 50 kg and a radius of gyration about the pin at O of $k_O = 0.23$ m. Starting from rest, the suspended 15-kg block B is allowed to fall 3 m without applying the brake ACD. Determine the speed of the block at this instant. If the coefficient of kinetic friction at the brake pad C is $\mu_k = 0.5$, determine the force P that must be applied at the brake handle which will then stop the block after it descends *another* 3 m. Neglect the thickness of the handle.

Probs. 18–13/18–14

18–14. The drum has a mass of 50 kg and a radius of gyration about the pin at O of $k_O = 0.23$ m. If the 15-kg block is moving downward at 3 m/s, and a force of $P = 100$ N is applied to the brake arm, determine how far the block descends from the instant the brake is applied until it stops. Neglect the thickness of the handle. The coefficient of kinetic friction at the brake pad is $\mu_k = 0.5$.

18–15. The spool has a weight of 500 lb and a radius of gyration of $k_G = 1.75$ ft. A horizontal force of $P = 15$ lb is applied to a cable wrapped around its inner core. If the spool is originally at rest, determine its angular velocity after the mass center G has moved $s = 6$ ft. The spool rolls without slipping. Neglect the mass of the cable.

Prob. 18–15

*18–16. The wheel has a mass of 100 kg and a radius of gyration $k_O = 0.40$ m. A motor supplies a torque $M = (40\theta + 900)$ N · m, where θ is in radians, about the drive shaft at O. Determine the speed of the loading car, which has a mass of 300 kg, after it travels $s = 4$ m. Initially the car is at rest when $s = 0$ and $\theta = 0$. Neglect the mass of the attached cable and the mass of the car's wheels.

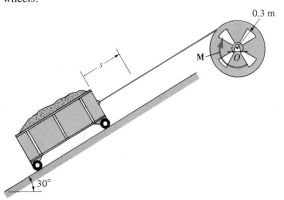

Prob. 18–16

18–17. The spool has a mass of 60 kg and a radius of gyration $k_G = 0.3$ m. If it is released from rest, determine how far it descends down the smooth plane before it attains an angular velocity of $\omega = 6$ rad/s. Neglect friction and the mass of the cord which is wound around the central core.

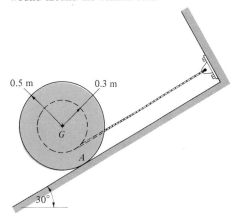

Probs. 18–17/18–18

18–18. Solve Prob. 18–17 if the plane is rough, such that the coefficient of kinetic friction at A is $\mu_A = 0.2$.

18–19. The drum is pinned at its center O and is subjected to a moment of $M = 500$ N · m. If it has a mass of 80 kg and radius of gyration of $k_O = 0.58$ m, determine the speed of the 130-kg car when it has moved up the plane 10 m. The car starts from rest. Neglect the mass of both the pulley at A and the car's wheels.

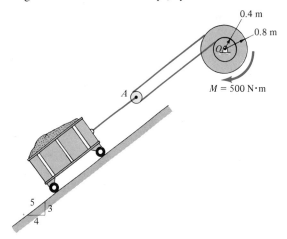

Prob. 18–19

357

*18–20. The assembly consists of two 15-lb slender rods and a 20-lb disk. If the spring is unstretched when $\theta = 45°$ and the assembly is released from rest at this position, determine the angular velocity of rod AB at the instant $\theta = 0°$. The disk rolls without slipping.

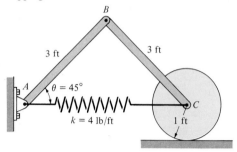

Prob. 18–20

18–21. The 50-lb crate C is lifted using the hoist shown. A motor delivers 50 lb · ft of torque to the pinion gear D, which is in mesh with the larger gear A, and the attached drum over which the cable winds. If A has a weight of 120 lb and a radius of gyration $k_A = 0.6$ ft, determine the velocity of the crate when it has risen 5 ft starting from rest. Neglect the mass of the pinion gear and the cable and pulley at B.

Probs. 18–20/18–21

18–22. Solve Prob. 18–21 assuming that pulley B is a disk having a weight of 20 lb.

18–23. The uniform slender bar has a mass m and a length L. It is subjected to a uniform distributed load w_0 which is always directed perpendicular to the axis of the bar. If it is released from rest from the position shown, determine its angular velocity at the instant it has rotated 90°. Solve the problem for rotation in (a) the horizontal plane, and (b) the vertical plane.

Probs. 18–23/18–24

*18–24. Solve Prob. 18–23 if the distributed load is triangular, which varies from zero at the pin O to w_0 at its other end.

18–25. The two 2-kg gears A and B are attached to the ends of a 3-kg slender bar. The gears roll within the fixed ring gear C, which lies in the horizontal plane. If a 10-N · m torque is applied to the center of the bar as shown, determine the number of revolutions the bar must rotate starting from rest in order for it to have an angular velocity of $\omega_{AB} = 20$ rad/s. For the calculation, assume the gears can be approximated by thin disks. What is the result if the gears lie in the vertical plane?

Prob. 18–25

18–26. The two bars are released from rest at the position θ. Determine their angular velocities at the instant they become horizontal. Neglect the mass of the roller at C. Each bar has a mass m and length L.

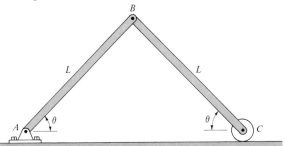

Prob. 18–26

18–27. A ball of mass m and radius r is cast onto the horizontal surface such that it rolls without slipping. Determine the required speed v_G of its mass center G so that it rolls completely around the loop of radius $R + r$ without leaving the track.

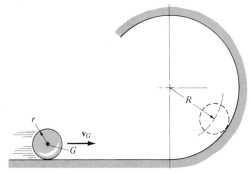

Prob. 18–27

18.5 Conservation of Energy

When a force system acting on a rigid body consists only of *conservative forces*, the conservation of energy theorem may be used to solve a problem which otherwise would be solved using the principle of work and energy. This theorem is often easier to apply since the work of a conservative force is *independent of the path* and only depends upon the initial and final positions of the body. It was shown in Sec. 14.5 that the work of a conservative force may be expressed as the difference in the body's potential energy measured from an arbitrarily selected reference or datum.

Gravitational Potential Energy

Since the total weight of a body can be considered concentrated at its center of gravity, the *gravitational potential energy* of the body is determined by knowing the height of the body's center of gravity above or below a horizontal datum. Measuring y_G as *positive upward*, the gravitational potential energy of the body is thus

$$V_g = W\, y_G \qquad (18\text{–}13)$$

Here the potential energy is *positive* when y_G is positive, since the weight has the ability to do *positive work* when the body is moved back to the datum, Fig. 18–16. Likewise, if the body is located *below* the datum ($-y_G$), the gravitational potential energy is *negative*, since the weight does *negative work* when the body is moved back to the datum.

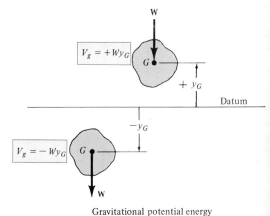

Gravitational potential energy

Fig. 18–16

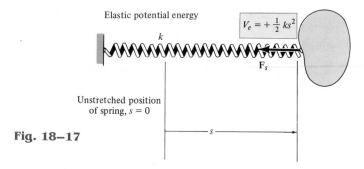

Fig. 18–17

Elastic Potential Energy

The force developed by an elastic spring is also a conservative force. The *elastic potential energy* which a spring imparts to an attached body when the spring is elongated or compressed from an initial undeformed position ($s = 0$) to a final position s, Fig. 18–17, is

$$V_e = +\tfrac{1}{2}ks^2 \qquad (18\text{--}14)$$

In the deformed position, the spring force acting *on the body* always has the capacity for doing positive work when the spring is returned back to its original undeformed position (see Sec. 14.5).

Conservation of Energy

In general, if a body is subjected to both gravitational and elastic forces, the total *potential energy* is expressed as a potential function V represented as the algebraic sum

$$V = V_g + V_e \qquad (18\text{--}15)$$

Here measurement of V depends upon the location of the body with respect to a selected datum in accordance with Eqs. 18–13 and 18–14.

Realizing that the work of conservative forces can be written as a difference in their potential energies, i.e., $(\Sigma U_{1-2})_{\text{cons}} = V_1 - V_2$, Eq. 14–16, we can rewrite the principle of work and energy for a rigid body as

$$T_1 + V_1 + (\Sigma U_{1-2})_{\text{noncons}} = T_2 + V_2 \qquad (18\text{--}16)$$

Here $(\Sigma U_{1-2})_{\text{noncons}}$ represents the work of the nonconservative forces, such as friction, acting on the body. If this term is zero, then

$$T_1 + V_1 = T_2 + V_2 \qquad (18\text{--}17)$$

This is the equation of the conservation of mechanical energy for the body. It states that the *sum* of the potential and kinetic energies of the body remains *constant* when the body moves from one position to another. It also applies to a system of smooth, pin-connected rigid bodies, bodies connected by inextensible cords, and bodies in mesh with other bodies. In all these cases the forces acting at the points of contact are *eliminated* from the analysis, since they occur in equal and opposite collinear pairs and each pair of forces moves through an equal distance when the system undergoes a displacement.

PROCEDURE FOR ANALYSIS

The conservation of energy equation is used to solve problems involving *velocity, displacement,* and *conservative force systems.* For application it is suggested that the following procedure be used.

Potential Energy. Draw two diagrams showing the body located at its initial and final positions along the path. If the center of mass of the body is subjected to a *vertical displacement,* determine where to establish the fixed horizontal datum from which to measure the body's gravitational potential energy V_g. Although the location of the datum is arbitrary, it is advantageous to place it through G when the body is either at its initial or final position, since at the datum $V_g = 0$. Data pertaining to the elevation y of the body's mass center from the datum and the extension or compression of any connecting springs can be determined from the geometry associated with the two diagrams. Recall that the potential energy $V = V_g + V_e$. Here $V_g = Wy_G$, where y_G is positive upward from the datum, and $V_e = \frac{1}{2}ks^2$, which is always positive.

Kinetic Energy. The kinetic-energy terms T_1 and T_2 are determined from $T = \frac{1}{2}mv_G^2 + \frac{1}{2}I_G\omega^2$ or an appropriate form of this equation as developed in Sec. 18.1. In this regard, kinematic diagrams for velocity may be useful for determining v_G and ω, or for establishing a *relationship* between these quantities.

Conservation of Energy. Apply the conservation of energy equation $T_1 + V_1 = T_2 + V_2$.

It is important to remember that *only problems involving conservative force systems may be solved by using this equation.* As stated in Sec. 14.5, friction or other drag-resistant forces, which depend upon velocity or acceleration, are nonconservative. The work of such forces is transformed into thermal energy used to heat up the surfaces of contact, and consequently this energy is dissipated into the surroundings and may not be recovered. Therefore, problems involving frictional forces can either be solved by using the principle of work and energy written in the form of Eq. 18–16, if it applies, or the equations of motion.

The following example problems numerically illustrate application of the above procedure.

Example 18–6

The 10-kg rod AB shown in Fig. 18–18a is confined so that its ends move in the horizontal and vertical slots. The spring has a stiffness of $k = 800$ N/m and is unstretched when $\theta = 0°$. Determine the angular velocity of AB when $\theta = 0°$, if AB is released from rest when $\theta = 30°$. Neglect the mass of the slider blocks.

$k = 800$ N/m

(a)

SOLUTION

Potential Energy. The two diagrams of the rod, when it is located at its initial and final positions, are shown in Fig. 18–18b. The datum, used to measure the gravitational potential energy, is placed in line with the rod when $\theta = 0°$.

When the rod is in position 1, the center of gravity G is located *below the datum* so that the gravitational potential energy is *negative*. Furthermore, (positive) elastic potential energy is stored in the spring, since it is stretched a distance of $s_1 = (0.4 \sin 30°)$ m. Thus,

$$V_1 = -Wy_1 + \tfrac{1}{2}ks_1^2$$

$$= -98.1(0.2 \sin 30°) + \frac{1}{2}(800)(0.4 \sin 30°)^2 = 6.19 \text{ J}$$

When the rod is in position 2, the potential energy of the rod is zero, since the spring is unstretched, $s_2 = 0$, and the center of gravity G is located at the datum. Thus,

$$V_2 = 0$$

Kinetic Energy. The rod is released from rest from position 1, thus $(\mathbf{v}_G)_1 = \mathbf{0}$ and $\boldsymbol{\omega}_1 = \mathbf{0}$, and

$$T_1 = 0$$

In position 2 the angular velocity is $\boldsymbol{\omega}_2$ and the rod's mass center has a velocity of $(\mathbf{v}_G)_2$. Using *kinematics*, $(\mathbf{v}_G)_2$ can be related to $\boldsymbol{\omega}_2$ as shown in Fig. 18–18c. At the instant considered, the instantaneous center of zero velocity (*IC*) for the rod is at point A; hence, $(v_G)_2 = (r_{G/IC})\omega_2 = (0.2)\omega_2$. Thus,

$$T_2 = \tfrac{1}{2}m(v_G)_2^2 + \tfrac{1}{2}I_G(\omega_2)^2$$

$$= \frac{1}{2}(10)(0.2\omega_2)^2 + \frac{1}{2}\left[\frac{1}{12}(10)(0.4)^2\right](\omega_2)^2 = 0.267\omega_2^2$$

Conservation of Energy

$$\{T_1\} + \{V_1\} = \{T_2\} + \{V_2\}$$

$$\{0\} + \{6.19\} = \{0.267\omega_2^2\} + \{0\}$$

$$\omega_2 = 4.82 \text{ rad/s} \uparrow$$

Ans.

$y_1 = (0.2 \sin 30°)$ m

Datum

$s_1 = (0.4 \sin 30°)$ m

98.1 N

①

$s_2 = 0$

98.1 N

②

(b)

$(v_G)_2$

ω_2

G

$r_{G/IC}$

IC

0.2 m

B

(c)

Fig. 18–18

Example 18–7

The disk shown in Fig. 18–19a has a weight of 30 lb and is attached to a spring which has a stiffness $k = 2$ lb/ft and an unstretched length of 1 ft. If the disk is released from rest in the position shown and rolls without slipping, determine its angular velocity at the instant it is displaced 3 ft.

SOLUTION

Potential Energy. Two diagrams of the disk, when it is located in its initial and final positions, are shown in Fig. 18–19b. A gravitational datum is not needed here since the weight is not displaced vertically. From the problem geometry the spring is stretched $s_1 = (\sqrt{3^2 + 4^2} - 1) = 4$ ft and $s_2 = (4 - 1) = 3$ ft in the initial and final positions, respectively. Hence,

$$V_1 = \tfrac{1}{2}ks_1^2 = \frac{1}{2}(2)(4)^2 = 16 \text{ J}$$

$$V_2 = \tfrac{1}{2}ks_2^2 = \frac{1}{2}(2)(3)^2 = 9 \text{ J}$$

Kinetic Energy. The disk is released from rest so that $(\mathbf{v}_G)_1 = \mathbf{0}, \boldsymbol{\omega}_1 = \mathbf{0}$, and

$$T_1 = 0$$

Since the disk rolls without slipping, $(\mathbf{v}_G)_2$ can be related to $\boldsymbol{\omega}_2$ from the instantaneous center of zero velocity, Fig. 18–19c. Hence, $(v_G)_2 = 0.75\omega_2$. Thus,

$$\begin{aligned}
T_2 &= \tfrac{1}{2}m(v_G)_2^2 + \tfrac{1}{2}I_G(\omega_2)^2 \\
&= \frac{1}{2}\left(\frac{30}{32.2}\right)(0.75\omega_2)^2 + \frac{1}{2}\left[\frac{1}{2}\left(\frac{30}{32.2}\right)(0.75)^2\right](\omega_2)^2 \\
&= 0.393(\omega_2)^2
\end{aligned}$$

Conservation of Energy

$$\{T_1\} + \{V_1\} = \{T_2\} + \{V_2\}$$
$$\{0\} + \{16\} = \{0.393(\omega_2)^2\} + \{9\}$$
$$\omega_2 = 4.22 \text{ rad/s} \text{ ↾} \qquad Ans.$$

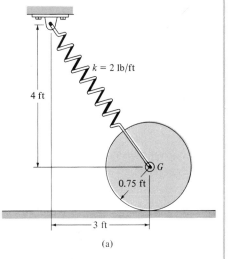

$k = 2$ lb/ft

4 ft

0.75 ft

3 ft

(a)

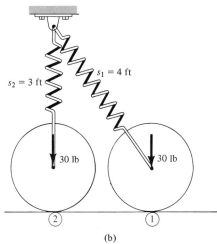

$s_2 = 3$ ft

$s_1 = 4$ ft

30 lb

30 lb

(2)

(1)

(b)

ω_2

$(v_G)_2$

0.75 ft

IC

Fig. 18–19

Example 18–8

The 10-kg homogeneous disk shown in Fig. 18–20a is attached to a uniform 5-kg rod AB. If the assembly is released from rest when $\theta = 60°$, determine the angular velocity of the rod when $\theta = 0°$. Assume that the disk rolls without slipping. Neglect friction along the guide and the mass of the collar at B.

(a)

SOLUTION

Potential Energy. Two diagrams for the rod and disk, when they are located at their initial and final positions, are shown in Fig. 18–20b. For convenience the datum, which is horizontally fixed, passes through point A.

When the system is in position 1, the rod's weight has positive potential energy. Thus,

$$V_1 = W_R y_1 = 49.05(0.3 \sin 60°) = 12.74 \text{ J}$$

When the system is in position 2, both the weight of the rod and the weight of the disk have zero potential energy. Why? Thus,

$$V_2 = 0$$

Kinetic Energy. Since the entire system is at rest in the initial position,

$$T_1 = 0$$

(b)

In the final position the rod has an angular velocity $(\omega_R)_2$ and its mass center has a velocity $(v_G)_2$, Fig. 18–20c. Since the rod is *fully extended* in this position, the disk is momentarily at rest, so $(\omega_D)_2 = \mathbf{0}$ and $(v_A)_2 = \mathbf{0}$. For the rod $(v_G)_2$ can be related to $(\omega_R)_2$ from the instantaneous center of zero velocity, which is located at point A, Fig. 18–20c. Hence, $(v_G)_2 = r_{G/IC}(\omega_R)_2$ or $(v_G)_2 = 0.3(\omega_R)_2$. Thus,

$$T_2 = \tfrac{1}{2}m_R(v_G)_2^2 + \tfrac{1}{2}I_G(\omega_R)_2^2 + \tfrac{1}{2}m_D(v_A)_2^2 + \tfrac{1}{2}I_A(\omega_D)_2^2$$

$$= \frac{1}{2}(5)(0.3(\omega_R)_2)^2 + \frac{1}{2}\left[\frac{1}{12}(5)(0.6)^2\right](\omega_R)_2^2 + 0 + 0$$

$$= 0.3(\omega_R)_2^2$$

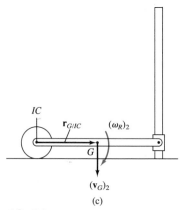

(c)

Conservation of Energy

$$\{T_1\} + \{V_1\} = \{T_2\} + \{V_2\}$$

$$\{0\} + \{12.74\} = \{0.3(\omega_R)_2^2\} + \{0\}$$

$$(\omega_R)_2 = 6.52 \text{ rad/s} \downarrow \qquad \textit{Ans.}$$

Fig. 18–20

PROBLEMS

*18–28. Solve Prob. 18–7 using the conservation of energy equation.

18–29. Solve Prob. 18–17 using the conservation of energy equation.

18–30. Solve Prob. 18–26 using the conservation of energy equation.

18–31. Solve Prob. 18–27 using the conservation of energy equation.

*18–32. Solve Prob. 18–20 using the conservation of energy equation.

18–33. The spool has a mass of 50 kg and a radius of gyration $k_O = 0.280$ m. If the 20-kg block A is released from rest, determine its velocity just after it descends 2 m. Neglect the mass of the cable.

Prob. 18–33

18–34. An automobile tire has a mass of 7 kg and radius of gyration $k_G = 0.3$ m. If it is released from rest at A on the incline, determine its angular velocity when it reaches the horizontal plane. The tire rolls without slipping.

Prob. 18–34

18–35. The jeep and passenger have a total weight of 1600 lb, not including the four wheels. Each wheel has a weight of 20 lb and a radius of gyration $k_G = 0.4$ ft about an axis passing through its center. If the jeep starts from rest and coasts down the hill in neutral gear, determine the distance it must travel to attain a speed of 15 ft/s. The tires do not slip on the ground.

Prob. 18–35

*18–36. When the slender 10-kg bar AB is horizontal it is at rest and the spring is unstretched. Determine the bar's angular velocity when it is released and has rotated downwards 90°.

Prob. 18–36

18–37. The 50-lb wheel has a radius of gyration about its mass center G of $k_G = 0.7$ ft. If it rolls without slipping, determine its angular velocity when it has rotated clockwise 90° from the position shown. The spring AB has a stiffness of $k = 1.20$ lb/ft and an unstretched length of 0.5 ft. The wheel is released from rest.

Prob. 18–37

18–38. The 80-lb cylinder is attached to the 10-lb slender rod which is pinned from point A. At the instant $\theta = 30°$ it has an angular velocity of $\omega_1 = 1$ rad/s as shown. Determine the largest angle θ to which the rod swings before it momentarily stops.

Prob. 18–38

18–39. The 80-lb cylinder is attached to the 10-lb slender rod which is pinned from point A. At the instant $\theta = 30°$ it has an angular velocity of $\omega_1 = 4$ rad/s as shown. Determine the angular velocity of the rod at the instant $\theta = 90°$.

*18–40. The 25-lb slender rod AB is attached to a spring BC which has an unstretched length of 4 ft. If the rod is released from rest when $\theta = 30°$, determine its angular velocity at the instant $\theta = 90°$.

Prob. 18–40

18–41. The assembly consists of two 8-lb bars which are pin-connected to the two 10-lb disks. If the bars are released from rest when $\theta = 60°$, determine their angular velocities at the instant $\theta = 0°$. Assume the disks roll without slipping.

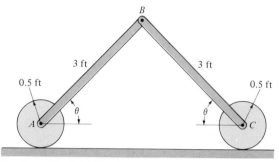

Probs. 18–41/18–42

18–42. The assembly consists of two 8-lb bars which are pin-connected to the two 10-lb disks. If the bars are released from rest when $\theta = 60°$, determine their angular velocities at the instant $\theta = 30°$. Assume the disks roll without slipping.

18–43. The garage door AB has a weight of 80 lb and can be approximated as a thin plate. During operation, a small roller at B is guided along the fixed horizontal track, while the two arms AC located on each side of the door are used to guide the bottom of the door along a circular arc having a center at D and a radius of 4 ft. If the end A of the door is given a horizontal velocity of 3 ft/s to the left and the door has a clockwise angular velocity $\omega_{AB} = 2$ rad/s when $\theta = 0°$ and ADC is vertical, determine the velocity of A at the instant the door becomes fully open, i.e., $\theta = 90°$. Each of the two springs has a stiffness of 50 lb/ft and an unstretched length of 1 ft. Neglect the mass of the arms ADC.

Prob. 18–43

*18–44. The disk is rotating about pin A in the vertical plane with an angular velocity $\omega_1 = 2$ rad/s when $\theta = 0°$. If it has a weight of 15 lb, determine its angular velocity at the instant $\theta = 90°$. Also, compute the horizontal and vertical components of reaction at A at this instant.

Prob. 18–44

18–45. The uniform window shade AB has a total weight of 0.4 lb. When it is released, it winds around the spring-loaded core O. Motion is caused by a spring within the core, which is coiled so that it exerts a torque $M = (0.3(10^{-3})\theta)$ lb · ft, where θ is in radians, on the core. If the shade is released from rest, determine the angular velocity of the core at the instant the shade is completely rolled up, i.e., after 12 revolutions. When this occurs, the spring becomes uncoiled and the radius of gyration of the shade about the axle at O is $k_O = 0.9$ in. *Note:* The elastic potential energy of a torsional spring is $V_e = \frac{1}{2}k\theta^2$, where $M = k\theta$. Here $k = 0.3(10^{-3})$ lb · ft/rad.

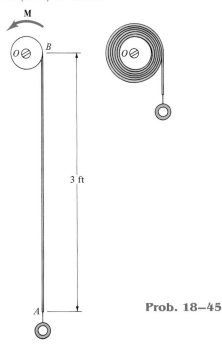

Prob. 18–45

18–46. The spool at A has a mass of 12 kg and a radius of gyration $k_C = 225$ mm. Pulley B has a mass of 7 kg and may be considered as a disk. If the system is released from rest and the cord does not slip, determine the angular velocity of the pulley after the spool descends 0.5 m. Neglect the mass of the cord.

Prob. 18–46

18–47. The slender rod has a mass of 10 kg and is released from rest when it is vertical. Determine its angular velocity at the instant it becomes horizontal. The attached block at D has a mass of 4 kg. The mass of the cable and pulley at C can be neglected.

Prob. 18–47

367

*18–48. The gear has a mass m and may be treated as a uniform disk. If it is released from rest at $\theta = 0°$, and rolls along the circular gear rack, determine the angular velocity of the radial line AB at the instant $\theta = 90°$.

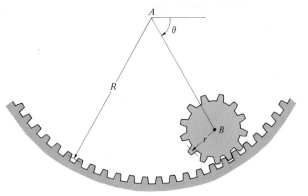

Prob. 18–48

18–49. The assembly consists of a 5-lb slender rod AC to which is pin-connected a 12-lb disk and spring BD. If the rod is brought into the horizontal position $\theta = 0°$, and the disk is given a counterclockwise rotation of 3 rad/s when the rod is released from rest, determine the angular velocity of the rod at the instant $\theta = 90°$. The spring has an unstretched length of 1 ft.

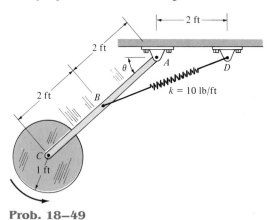

Prob. 18–49

18–50. The garage door CD has a mass of 15 kg and can be treated as a thin plate. If it is released from rest from the horizontal position shown, determine its angular velocity at the instant the two side bars ABC become horizontal. Each of the two side springs has a stiffness of $k = 30$ N/m and is originally unstretched. Neglect the mass of bar ABC.

Prob. 18–50

18–51. The slender 15-kg bar is initially at rest and standing in the vertical position when the bottom end A is displaced slightly to the right. If the horizontal surface is smooth, determine the speed at which A strikes the corner D. The bar is constrained to move in the vertical plane. Neglect the mass of the cord BC.

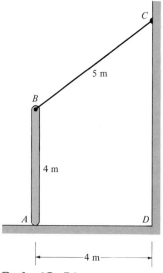

Prob. 18–51

19

Planar Kinetics of a Rigid Body: Impulse and Momentum

In Chapter 15 it was shown that problems which involve force, velocity, and time can conveniently be solved by using the principle of linear or angular impulse and momentum. In this chapter we will extend these concepts somewhat in order to determine the impulse and momentum relationships that apply to a rigid body. The three planar motions which will be considered are translation, rotation about a fixed axis, and general plane motion. The more general discussion of momentum principles applied to the three-dimensional motion of a rigid body is presented in Chapter 21.

19.1 Linear and Angular Momentum

In this section we will develop formulations for the linear and angular momentum of a rigid body which is symmetric with respect to an inertial x-y reference plane. As a result, we will represent the body as a slab having the same mass as the body and moving in the reference plane.

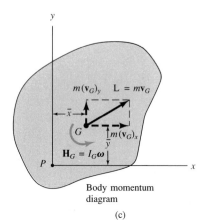

Fig. 19—1

Linear Momentum

The linear momentum of a rigid body is determined by summing vectorially the linear momenta of all the particles of the body, i.e., $\mathbf{L} = \Sigma m_i \mathbf{v}_i$. We can simplify this expression by noting that $\Sigma m_i \mathbf{v}_i = m\mathbf{v}_G$ (see Sec. 15.2) so that

$$\mathbf{L} = m\mathbf{v}_G \qquad (19\text{–}1)$$

This equation states that the body's linear momentum ʀs a vector quantity having a *magnitude* mv_G, which is commonly measured in units of kg · m/s or slug · ft/s, and a *direction* defined by $\mathbf{v}_G$, the velocity of the mass center.

Angular Momentum

Consider the body in Fig. 19–1a, which is subjected to general plane motion. At the instant shown, the arbitrary point P has a velocity $\mathbf{v}_P$ and the body has an angular velocity $\boldsymbol{\omega}$. If the velocity of the *i*th particle of the body is to be determined, Fig. 19–1a, then

$$\mathbf{v}_i = \mathbf{v}_P + \mathbf{v}_{i/P}$$

The angular momentum of particle i about point P is equal to the "moment" of the particle's linear momentum about P, Fig. 19–1a. Thus,

$$(\mathbf{H}_P)_i = \mathbf{r} \times m_i\, \mathbf{v}_i$$

Expressing $\mathbf{v}_i$ in terms of $\mathbf{v}_P$ and using Cartesian vectors, we have

$$(H_P)_i\, \mathbf{k} = m_i(x\mathbf{i} + y\mathbf{j}) \times [(v_P)_x\mathbf{i} + (v_P)_y\mathbf{j} + \omega\mathbf{k} \times (x\mathbf{i} + y\mathbf{j})]$$
$$(H_P)_i = -m_i\, y(v_P)_x + m_i\, x(v_P)_y + m_i\, \omega r^2$$

Letting $m_i \to dm$ and integrating over the entire mass m of the body, we obtain

$$H_P = -\left(\int_m y\, dm\right)(v_P)_x + \left(\int_m x\, dm\right)(v_P)_y + \left(\int_m r^2\, dm\right)\omega$$

Here H_P represents the angular momentum of the body about an axis (the z axis) perpendicular to the plane of motion and passing through point P. Since $\bar{x}m = \int x\, dm$ and $\bar{y}m = \int y\, dm$, the integrals for the first and second terms on the right are used to locate the body's center of mass G with respect to P. Also, the last integral represents the body's moment of inertia computed about the z axis, i.e., $I_P = \int r^2\, dm$. Thus,

$$H_P = -\bar{y}m(v_P)_x + \bar{x}m(v_P)_y + I_P\omega \qquad (19\text{–}2)$$

This equation reduces to a simpler form if point P coincides with the mass center G for the body,* in which case $\bar{x} = \bar{y} = 0$. Hence, the angular momentum of the body, computed about an axis passing through point G, is therefore

---

*It *also* reduces to the same simple form, $H_P = I_P\omega$, if point P is a *fixed point* (see Eq. 19–9) or the velocity of point P is directed along the line PG.

$$H_G = I_G\omega \qquad (19\text{--}3)$$

This equation states that the angular momentum of the body computed about point G is equal to the product of the moment of inertia of the body about an axis passing through G and the body's angular velocity. It is important to realize that $\mathbf{H}_G$ is a vector quantity having a *magnitude* $I_G\omega$, which is commonly measured in units of kg · m²/s or slug · ft²/s, and a *direction* defined by $\boldsymbol{\omega}$, which is always perpendicular to the slab.

Equation 19–2 can also be rewritten in terms of the x and y components of the velocity of the body's mass center, $(v_G)_x$ and $(v_G)_y$, and the body's moment of inertia I_G. If point G is located at coordinates $(\bar{x}, \bar{y})$, Fig. 19–1b, then by the parallel-axis theorem, $I_P = I_G + m(\bar{x}^2 + \bar{y}^2)$. Substituting into Eq. 19–2 and rearranging terms, we have

$$H_P = \bar{y}m[-(v_P)_x + \bar{y}\omega] + \bar{x}m[(v_P)_y + \bar{x}\omega] + I_G\omega \qquad (19\text{--}4)$$

From the kinematic diagram of Fig. 19–1b, $\mathbf{v}_G$ can be expressed in terms of $\mathbf{v}_P$ as

$$\mathbf{v}_G = \mathbf{v}_P + \boldsymbol{\omega} \times \bar{\mathbf{r}}$$
$$(v_G)_x\mathbf{i} + (v_G)_y\mathbf{j} = (v_P)_x\mathbf{i} + (v_P)_y\mathbf{j} + \omega\mathbf{k} \times (\bar{x}\mathbf{i} + \bar{y}\mathbf{j})$$

Carrying out the cross product and equating the respective $\mathbf{i}$ and $\mathbf{j}$ components yields the two scalar equations

$$(v_G)_x = (v_P)_x - \bar{y}\omega$$
$$(v_G)_y = (v_P)_y + \bar{x}\omega$$

Substituting these results into Eq. 19–4 yields

$$H_P = -\bar{y}m(v_G)_x + \bar{x}(v_G)_y + I_G\omega \qquad (19\text{--}5)$$

With reference to Fig. 19–1c, *this result indicates that when the angular momentum of the body is computed about point P it is equivalent to the moment of the linear momentum* $m\mathbf{v}_G$ *or its components* $m(\mathbf{v}_G)_x$ *and* $m(\mathbf{v}_G)_y$ *about P plus the angular momentum* $I_G\boldsymbol{\omega}$. Note that since $\boldsymbol{\omega}$ is a free vector, $\mathbf{H}_G$ *can act at any point on the body* (or slab) provided it preserves its same magnitude and direction. Furthermore, since angular momentum is equal to the "moment" of the linear momentum, the *line of action of* $\mathbf{L}$ *must pass through the body's mass center G* in order to preserve the correct magnitude of $\mathbf{H}_P$ when "moments" are computed about P, Fig. 19–1c. As a result of these ideas, we will now consider three types of motion.

Translation

When a rigid body of mass m is subjected to either rectilinear or curvilinear *translation*, Fig. 19–2a, its mass center has a velocity of $\mathbf{v}_G = \mathbf{v}$ and $\boldsymbol{\omega} = \mathbf{0}$. Hence, the linear momentum and the angular momentum, computed about G, become

$$\boxed{\begin{aligned} L &= mv_G \\ H_G &= 0 \end{aligned}}$$

(19–6)

If the angular momentum is computed about any other point A on or off the body, Fig. 19–2a, the "moment" of the linear momentum $\mathbf{L}$ must be computed about the point. Since d is the "moment arm" as shown in the figure, then in accordance with Eq. 19–5, $H_A = (d)(mv_G)$.

Rotation About a Fixed Axis

When a rigid body is *rotating about a fixed axis* passing through point O, Fig. 19–2b, the linear momentum and the angular momentum computed about G are

$$\boxed{\begin{aligned} L &= mv_G \\ H_G &= I_G\omega \end{aligned}}$$

(19–7)

It is sometimes convenient to compute the angular momentum of the body about point O. In this case it is necessary to account for the "moments" of *both* $\mathbf{L}$ and $\mathbf{H}_G$ about O. Noting that $\mathbf{L}$ (or $\mathbf{v}_G$) is always *perpendicular to* $\mathbf{r}_{G/O}$, we have

$$H_O = I_G\omega + r_{G/O}(mv_G) \tag{19–8}$$

This equation may be *simplified* by first substituting $v_G = r_{G/O}\omega$, in which case $H_O = (I_G + mr_{G/O}^2)\omega$, and, by the parallel-axis theorem, noting that the terms inside the parentheses represent the moment of inertia I_O of the body about an axis perpendicular to the plane of motion and passing through point O. Hence,*

$$\boxed{H_O = I_O\omega}$$

(19–9)

For the computation, then, either Eq. 19–8 or 19–9 can be used.

---

*The similarity between this derivation and that of Eq. 17–16 ($\Sigma M_O = I_O\alpha$) and Eq. 18–4 ($T = \frac{1}{2}I_O\omega^2$) should be noted. Also note that the same result can be obtained from Eq. 19–2 by selecting point P at O, realizing that $(v_O)_x = (v_O)_y = 0$.

General Plane Motion

When a rigid body is subjected to general plane motion, Fig. 19–2c, the linear momentum and the angular momentum computed about G become

$$\boxed{\begin{aligned} L &= mv_G \\ H_G &= I_G\omega \end{aligned}}$$

(19–10)

If the angular momentum is computed about a point A located either on or off the body, Fig. 19–2c, it is necessary to compute the moments of *both* $\mathbf{L}$ and $\mathbf{H}_G$ about this point. In this case,

$$H_A = I_G\omega + (d)(mv_G)$$

Here d is the moment arm, as shown in the figure.

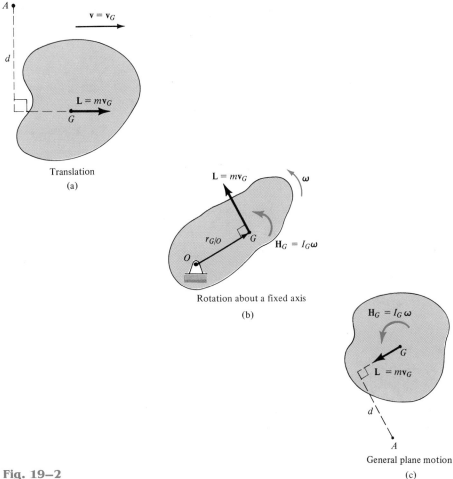

Translation
(a)

Rotation about a fixed axis
(b)

General plane motion
(c)

Fig. 19–2

19.2 Principle of Impulse and Momentum

Like the case for particle motion, the principle of impulse and momentum for a rigid body is developed by *combining* the equation of motion with kinematics. The resulting equation will allow a *direct solution to problems involving force, velocity, and time.*

Principle of Linear Impulse and Momentum

The equation of translational motion for a rigid body can be written as $\Sigma \mathbf{F} = m\mathbf{a}_G = m \, (d\mathbf{v}_G/dt)$. Since the mass of the body is constant,

$$\Sigma \mathbf{F} = \frac{d}{dt}(m\mathbf{v}_G)$$

Multiplying both sides by dt and integrating from $t = t_1$, $\mathbf{v}_G = (\mathbf{v}_G)_1$ to $t = t_2$, $\mathbf{v}_G = (\mathbf{v}_G)_2$ yields

$$\Sigma \int_{t_1}^{t_2} \mathbf{F} \, dt = m(\mathbf{v}_G)_2 - m(\mathbf{v}_G)_1 \qquad (19\text{--}11)$$

This equation is referred to as the *principle of linear impulse and momentum.* It states that the sum of all the impulses created by the *external force system* which acts on the body during the time interval t_1 to t_2 is equal to the change in the linear momentum of the body during the time interval.

Principle of Angular Impulse and Momentum

If the body has *general plane motion* we can write $\Sigma M_G = I_G \alpha = I_G(d\omega/dt)$. Since the moment of inertia is constant,

$$\Sigma M_G = \frac{d}{dt}(I_G \omega)$$

Multiplying both sides by dt and integrating from $t = t_1$, $\omega = \omega_1$ to $t = t_2$, $\omega = \omega_2$ gives

$$\Sigma \int_{t_1}^{t_2} M_G \, dt = I_G \omega_2 - I_G \omega_1 \qquad (19\text{--}12)$$

In a similar manner, for *rotation about a fixed axis* passing through point O, Eq. 17–16 ($\Sigma M_O = I_O \alpha$) when integrated becomes

$$\Sigma \int_{t_1}^{t_2} M_O \, dt = I_O \omega_2 - I_O \omega_1 \qquad (19\text{--}13)$$

Equations 19–12 and 19–13 are referred to as the *principle of angular impulse and momentum.* Both equations state that the sum of the angular impulses acting on the body during the time interval t_1 to t_2 is equal to the change in the body's angular momentum during this time interval. In particular, the angular impulse considered is determined by integrating the moments about point G or O of all the external forces and couples applied to the body.

To summarize the preceding concepts, if motion is occurring in the x-y plane, using impulse and momentum principles the following *three scalar equations* may be written which describe the *planar motion* of the body:

$$m(v_{Gx})_1 + \Sigma \int_{t_1}^{t_2} F_x \, dt = m(v_{Gx})_2$$

$$m(v_{Gy})_1 + \Sigma \int_{t_1}^{t_2} F_y \, dt = m(v_{Gy})_2 \qquad (19\text{–}14)$$

$$I_G \omega_1 + \Sigma \int_{t_1}^{t_2} M_G \, dt = I_G \omega_2$$

The first two of these equations represent the principle of linear impulse and momentum in the x-y plane, Eq. 19–11, and the third equation represents the principle of angular impulse and momentum about the z axis, Eq. 19–12.

The terms in Eqs. 19–14 can be graphically accounted for by drawing a set of impulse and momentum diagrams for the body, Fig. 19–3. Note that the linear momenta $m\mathbf{v}_G$ are applied at the body's mass center, Fig. 19–3a and c; whereas the angular momenta $I_G\boldsymbol{\omega}$ are free vectors, and therefore, like a couple moment, they may be applied at any point on the body. When the impulse diagram is constructed, Fig. 19–3b, vectors $\mathbf{F}$ and $\mathbf{M}$ which vary with time are indicated by the integrals. However, if $\mathbf{F}$ and $\mathbf{M}$ are *constant* from t_1 to t_2, integration of the impulses yields $\mathbf{F}(t_2 - t_1)$ and $\mathbf{M}(t_2 - t_1)$, respectively. Such is the case for the body's weight $\mathbf{W}$, Fig. 19–3b.

Equations 19–14 may also be applied to an entire system of connected bodies rather than to each body separately. Doing this eliminates the need to include reactive impulses which occur at the connections since they are *internal* to the system. The resultant equations may be written in symbolic form as

$$\left(\Sigma \frac{\text{syst. linear}}{\text{momentum}} \right)_{x1} + \left(\Sigma \frac{\text{syst. linear}}{\text{impulse}} \right)_{x(1-2)} = \left(\Sigma \frac{\text{syst. linear}}{\text{momentum}} \right)_{x2}$$

$$\left(\Sigma \frac{\text{syst. linear}}{\text{momentum}} \right)_{y1} + \left(\Sigma \frac{\text{syst. linear}}{\text{impulse}} \right)_{y(1-2)} = \left(\Sigma \frac{\text{syst. linear}}{\text{momentum}} \right)_{y2}$$

$$\left(\Sigma \frac{\text{syst. angular}}{\text{momentum}} \right)_{O1} + \left(\Sigma \frac{\text{syst. angular}}{\text{impulse}} \right)_{O(1-2)} = \left(\Sigma \frac{\text{syst. angular}}{\text{momentum}} \right)_{O2}$$

$$(19\text{–}15)$$

As indicated, the system's angular momentum and angular impulse must be computed with respect to the *same fixed reference point O* for all the bodies of the system.

Initial momentum diagram

(a)

+

Impulse diagram

(b)

=

Final momentum diagram

(c)

Fig. 19–3

PROCEDURE FOR ANALYSIS

Impulse and momentum principles are used to solve kinetics problems that involve *velocity, force,* and *time* since these terms are involved in the formulation. For application it is suggested that the following procedure be used.

Free-Body Diagram. Establish the x, y, z inertial frame of reference and draw the free-body diagram in order to account for all the forces that produce impulses on the body. The direction and sense of the initial and final velocity of the body's mass center, $\mathbf{v}_G$, and the body's angular velocity $\boldsymbol{\omega}$ should also be established. Also compute the moment of inertia I_G or I_O.

As an alternative procedure draw the impulse and momentum diagrams for the body or system of bodies. Each of these diagrams represents an outlined shape of the body which graphically accounts for the data required for each of the three terms in Eqs. 19–14 or 19–15, Fig. 19–3. Note that these diagrams are particularly helpful in order to visualize the "moment" terms used in the principle of angular impulse and momentum, when it has been decided that application of this equation is to be about a point other than the body's mass center G or a fixed point O.

Principle of Impulse and Momentum. Apply the three scalar equations 19–14 (or 19–15). In cases where the body is rotating about a fixed axis, Eq. 19–13 may be substituted for the third of Eqs. 19–14.

Kinematics. If more than three equations are needed for a complete solution, it may be possible to relate the velocity of the body's mass center to the body's angular velocity using *kinematics.* If the motion appears to be complicated, kinematic (velocity) diagrams may be helpful in obtaining the necessary relation.

In general, a method to be used for the solution of a particular type of problem should be decided upon *before* attempting to solve the problem. As stated above, the *principle of impulse and momentum* is most suitable for solving problems which involve *velocity, force,* and *time*. For some problems, however, a combination of the equation of motion and its two integrated forms, the principle of work and energy and the principle of impulse and momentum, will yield the most direct solution to the problem.

Example 19–1

The disk shown in Fig. 19–4a weighs 20 lb and is pin-supported at its center. If it is acted upon by a constant couple moment of 4 lb · ft and a force of 10 lb which is applied to a cord wrapped around its periphery, determine the angular velocity of the disk two seconds after starting from rest. What are the force components of reaction at pin A?

(a)

Fig. 19–4

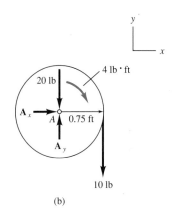

(b)

SOLUTION

Free-Body Diagram Fig. 19–4b. The disk's mass center does not move; however, the loading causes the disk to rotate clockwise.

The moment of inertia of the disk about its fixed axis of rotation is

$$I_A = \tfrac{1}{2}mr^2 = \frac{1}{2}\left(\frac{20}{32.2}\right)(0.75)^2 = 0.175 \text{ slug} \cdot \text{ft}^2$$

Principle of Impulse and Momentum

$(\xrightarrow{+})$ $$m(v_{Ax})_1 + \Sigma \int_{t_1}^{t_2} F_x \, dt = m(v_{Ax})_2$$

$$0 + A_x(2) = 0$$

$(+\uparrow)$ $$m(v_{Ay})_1 + \Sigma \int_{t_1}^{t_2} F_y \, dt = m(v_{Ay})_2$$

$$0 + A_y(2) - 20(2) - 10(2) = 0$$

$(\curvearrowright +)$ $$I_A\omega_1 + \Sigma \int_{t_1}^{t_2} M_A \, dt = I_A\omega_2$$

$$0 + 4(2) + [10(2)](0.75) = 0.175\omega_2$$

Solving these equations yields

$$A_x = 0 \qquad\qquad\qquad Ans.$$
$$A_y = 30 \text{ lb} \qquad\qquad Ans.$$
$$\omega_2 = 132 \text{ rad/s} \;\downarrow \qquad Ans.$$

Example 19–2

The block shown in Fig. 19–5a has a mass of 6 kg. It is attached to a cord which is wrapped around the periphery of a 20-kg disk that has a moment of inertia $I_A = 0.40$ kg · m². If the block is initially moving downward with a speed of 2 m/s, determine its speed in 3 s. Neglect the mass of the cord in the calculation.

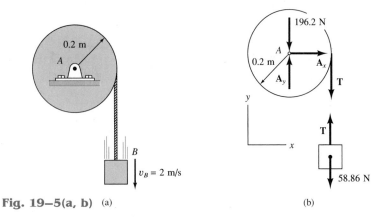

Fig. 19–5(a, b) (a)   (b)

SOLUTION I
Free-Body Diagram. The free-body diagrams of the block and disk are shown in Fig. 19–5b. All the forces are *constant* since the weight of the block causes the motion. The downward motion of the block, $\mathbf{v}_B$, causes $\boldsymbol{\omega}$ of the disk to be clockwise.

Principle of Impulse and Momentum. We can eliminate $\mathbf{A}_x$ and $\mathbf{A}_y$ from the analysis by applying the principle of angular momentum about point A. Hence

Disk
$(\curvearrowright+)$
$$I_A\omega_1 + \Sigma \int M_A\, dt = I_A\omega_2$$
$$0.40(\omega_1) + T(3)(0.2) = 0.4\omega_2$$

Block
$(+\uparrow)$
$$m_B(v_B)_1 + \Sigma \int F_y\, dt = m_B(v_B)_2$$
$$-6(2) + T(3) - 58.86(3) = -6(v_B)_2$$

Kinematics. Since $\omega = v_B/r$, then $\omega_1 = 2/0.2 = 10$ rad/s and $\omega_2 = (v_B)_2/0.2 = 5(v_B)_2$. Substituting and solving the equations simultaneously for $(v_B)_2$ yields

$$(v_B)_2 = 13.0 \text{ m/s} \downarrow \qquad\qquad Ans.$$

SOLUTION II

Impulse and Momentum Diagrams. We can obtain $(v_B)_2$ *directly* by considering the *system* consisting of the block, the cord, and the disk. The impulse and momentum diagrams have been drawn to clarify application of the principle of angular impulse and momentum about point A, Fig. 19–5c.

Principle of Angular Impulse and Momentum. Realizing that $\omega_1 = 10$ rad/s and $\omega_2 = 5(v_B)_2$, we have

$$\left(\sum \begin{array}{c}\text{syst. angular} \\ \text{momentum}\end{array} \right)_{A1} + \left(\sum \begin{array}{c}\text{syst. angular} \\ \text{impulse}\end{array} \right)_{A(1-2)} = \left(\sum \begin{array}{c}\text{syst. angular} \\ \text{momentum}\end{array} \right)_{A2}$$

$$(\curvearrowright+) \quad 6(2)(0.2) + 0.4(10) + 58.86(3)(0.2) = 6(v_B)_2(0.2) + 0.4(5(v_B)_2)$$

$$(v_B)_2 = 13.0 \text{ m/s} \downarrow \qquad\qquad Ans.$$

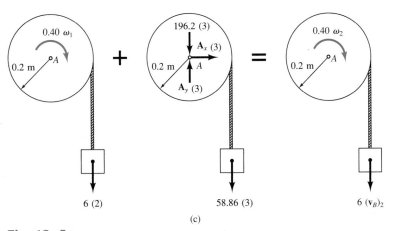

Fig. 19–5c

Example 19–3

The 100-kg spool shown in Fig. 19–6a has a radius of gyration $k_G = 0.35$ m. A cable is wrapped around the central hub of the spool and a variable horizontal force having a magnitude of $P = (t + 10)$ N is applied, where t is measured in seconds. If the spool is initially at rest, determine its angular velocity in 5 s. Assume that the spool rolls without slipping at A.

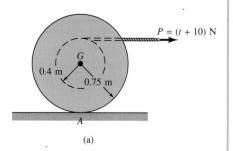

(a)

SOLUTION

Free-Body Diagram. By inspection of the free-body diagram, Fig. 19–6b, the *variable* force **P** will cause the friction force F_A to be variable, and thus the impulses created by both **P** and F_A must be determined by integration. The loading causes the mass center to have a velocity v_G to the right, and the spool has a clockwise angular velocity ω, Fig. 19–6c.

The moment of inertia of the spool about its mass center is

$$I_G = mk_G^2 = 100(0.35)^2 = 12.25 \text{ kg} \cdot \text{m}^2$$

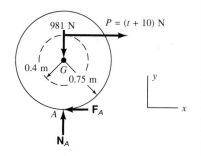

Principle of Impulse and Momentum

$(\xrightarrow{+})$
$$m(v_G)_1 + \Sigma \int F_x \, dt = m(v_G)_2$$

$$0 + \int_0^5 (t + 10) \, dt - \int F_A \, dt = 100(v_G)_2 \qquad (1)$$

$(\curvearrowright +)$
$$I_G\omega_1 + \Sigma \int M_G \, dt = I_G\omega_2$$

$$0 + \left[\int_0^5 (t + 10) \, dt\right](0.4) + \left(\int F_A \, dt\right)(0.75) = 12.25 \, \omega_2 \qquad (2)$$

(c)

Fig. 19–6

Kinematics. Since the spool does not slip, the instantaneous center of zero velocity is at point A, Fig. 19–6c. Hence, the velocity of G can be expressed in terms of the spool's angular velocity as $(v_G)_2 = 0.75\omega_2$. Substituting this, evaluating the integral for the impulse caused by **P** (i.e., 62.5 N · s), and eliminating the unknown impulse $\int F_A \, dt$ between Eqs. (1) and (2), we obtain

$$\omega_2 = 1.05 \text{ rad/s} \quad \curvearrowright \qquad\qquad\qquad Ans.$$

Example 19–4

The Charpy impact test is used in materials testing to determine the energy absorption characteristics of a material during impact. The test is performed using the pendulum shown in Fig. 19–7a that has a mass m, mass center at G, and a radius of gyration k_G about G. Determine the distance r_P from the pin at A to the point P where the impact with the specimen S should occur so that the horizontal force at the pin is essentially zero during the impact. For the computation, assume the specimen absorbs all the pendulum's kinetic energy during the time it falls and thereby stops the pendulum from swinging at $\theta = 0°$.

(a)

SOLUTION

Free-Body Diagram. As shown on the free-body diagram, Fig. 19–7b, the conditions of the problem require the horizontal impulse at A to be zero. Just before impact, we assume the pendulum has a clockwise angular velocity $\boldsymbol{\omega}_1$ and the mass center of the pendulum is moving to the left at $(v_G)_1 = \bar{r}\omega_1$.

Principle of Impulse and Momentum. We will apply the principle of angular impulse and momentum about point A. Thus,

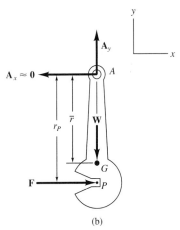

$(\curvearrowright+)$ $$I_A\omega_1 - \left(\int F\,dt\right)r_P = 0$$

$(\xrightarrow{+})$ $$-m(\bar{r}\omega_1) + \int F\,dt = 0$$

Eliminating the impulse $\int F\,dt$ and substituting $I_A = mk_G^2 + m\bar{r}^2$ yields

$$[mk_G^2 + m\bar{r}^2]\omega_1 - m(\bar{r}\omega_1)r_P = 0$$

Factoring out $m\omega_1$ and solving for r_P yields

$$r_P = \bar{r} + \frac{k_G^2}{\bar{r}} \qquad\qquad Ans.$$

(b)

Fig. 19–7

The point P, so defined, is called the *center of percussion*. By placing the striking point at P, the force developed at the pin will be minimized. Many sports rackets, clubs, etc. are designed so that the object being struck occurs at the center of percussion. As a consequence, no "sting" or little sensation occurs in the hand of the player. (Also see Probs. 17–55 and 19–7.)

PROBLEMS

19–1. Solve Prob. 17–45 using the principle of impulse and momentum.

19–2. Solve Prob. 17–46 using the principle of impulse and momentum.

19–3. Solve Prob. 17–51 using the principle of impulse and momentum.

*19–4. Solve Prob. 17–52 using the principle of impulse and momentum.

19–5. Show that if a slab is rotating about a fixed axis perpendicular to the slab and passing through its mass center G, the angular momentum is the same when computed about any other point P on the slab.

Prob. 19–5

19–6. At a given instant, the body has a linear momentum $\mathbf{L} = m\mathbf{v}_G$ and an angular momentum $\mathbf{H}_G = I_G\boldsymbol{\omega}$ computed about its mass center. Show that the angular momentum of the body computed about the instantaneous center of zero velocity IC can be expressed as $\mathbf{H}_{IC} = I_{IC}\boldsymbol{\omega}$, where I_{IC} represents the body's moment of inertia computed about the instantaneous axis of zero velocity. As shown, the IC is located at a distance of $r_{G/IC}$ away from the mass center G.

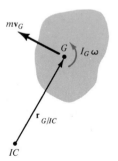

Prob. 19–6

19–7. The rigid body (slab) has a mass m and is rotating with an angular velocity $\boldsymbol{\omega}$ about an axis passing through the fixed point O. Show that the momenta of all the particles composing the body can be represented by a single vector having a magnitude mv_G and acting through point P, called the *center of percussion*, which lies at a distance $r_{P/G} = k_G^2/r_{G/O}$ from the mass center G. Here k_G is the radius of gyration of the body, computed about an axis perpendicular to the plane of motion and passing through G.

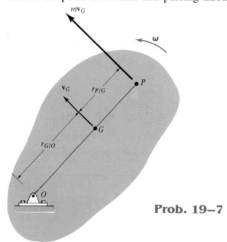

Prob. 19–7

*19–8. The spacecraft has a mass of 1200 kg and a moment of inertia $I_G = 900$ kg · m^2 about an axis passing through G and directed perpendicular to the page. If it is traveling forward with a speed $v_G = 800$ m/s and executes a turn by means of two jets, which provide a constant thrust of 400 N for 0.3 s, determine the spacecraft's angular velocity just after the jets are turned off.

Prob. 19–8

19–9. A flywheel has a mass of 60 kg and a radius of gyration of $k_G = 150$ mm about an axis of rotation passing through its mass center. If a motor supplies a clockwise torque having a magnitude of $M = (5t)$ N · m, where t is in seconds, to the flywheel, determine its angular velocity in $t = 3$ s. Initially the flywheel is rotating clockwise at $\omega_1 = 2$ rad/s.

19–10. The 10-lb rectangular plate is at rest on a smooth *horizontal* floor. If it is given the horizontal impulses shown, determine its angular velocity and the velocity of the mass center.

Prob. 19–10

19–11. The spool has a mass of 30 kg and a radius of gyration $k_O = 0.25$ m. Block A has a mass of 25 kg and block B has a mass of 10 kg. If they are released from rest, determine their speed in $t = 2$ s starting from rest. Neglect the mass of the ropes.

Prob. 19–11

*19–12. The spool has a weight of 30 lb and a radius of gyration $k_O = 0.45$ ft. A cord is wrapped around its inner hub and its end subjected to a horizontal force $P = 5$ lb. Determine the spool's angular velocity in 4 s starting from rest. Assume the spool rolls without slipping.

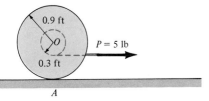

Probs. 19–12/19–13

19–13. Work Prob. 19–12 if the coefficient of static friction is $\mu_A = 0.3$.

19–14. The inner hub of the wheel rests on the inclined track. If it does not slip at A, determine its angular velocity $t = 2$ s after it is released from rest. The wheel has a weight of 30 lb and a radius of gyration of $k_G = 1.30$ ft.

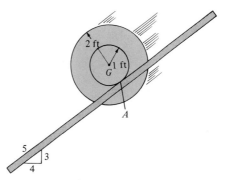

Prob. 19–14

19–15. The spool has a weight of 75 lb and a radius of gyration of $k_O = 1.20$ ft. If the block B weighs 60 lb, and a force of $P = 25$ lb is applied to the cord, determine the speed of the block in $t = 5$ s starting from rest. Neglect the mass of the cord.

Probs. 19–15/19–16

*19–16. The spool has a weight of 75 lb and a radius of gyration of $k_O = 1.20$ ft. If the block B weighs 60 lb, determine the horizontal force P which must be applied to the cord in order to give the block an upward speed of 10 ft/s in $t = 4$ s, starting from rest. Neglect the mass of the cord.

19–17. Spool B is at rest and spool A is rotating at 6 rad/s when the slack in the cord connecting them is taken up. Determine the angular velocity of each spool immediately after the cord is jerked tight by the spinning of spool A. A and B have weights and radii of gyration $W_A = 30$ lb, $k_A = 0.8$ ft and $W_B = 15$ lb, $k_B = 0.6$ ft, respectively.

Prob. 19–17

19–18. A torque $M = (0.05e^{0.5t})$ N · m, where t is in seconds, is applied to gear A. If gear A is originally rotating at 5 rad/s, determine its angular velocity when $t = 2$ s. The two smaller *equivalent* gears B and C are pinned at their centers. The mass and centroidal radii of gyration of the gears about axes perpendicular to the page and passing through their centers are given in the figure.

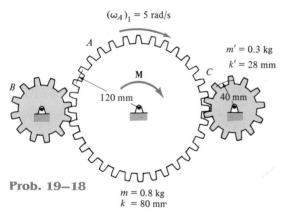

Prob. 19–18

$m = 0.8$ kg
$k = 80$ mm

19–19. The two gears A and B have weights and radii of gyration of $W_A = 15$ lb, $k_A = 0.5$ ft, and $W_B = 10$ lb, $k_B = 0.35$ ft, respectively. If a couple moment having a magnitude of $M = 2(1 - e^{-0.5t})$ lb · ft is applied to gear B, determine the angular velocity of each gear in $t = 5$ s, starting from rest.

Prob. 19–19

*19–20. If the hoop has a weight W and radius r and is thrown onto a *rough surface* with a velocity $\mathbf{v}_G$ parallel to the surface, determine the amount of backspin, ω, it must be given so that it stops spinning at the same instant that its forward velocity is zero. It is not necessary to know the coefficient of kinetic friction at A for the calculation.

Prob. 19–20

19–21. The ball of mass m and radius r rolls along an inclined plane for which the coefficient of static friction is μ. If the ball is released from rest, determine the maximum angle θ for the incline so that it rolls without slipping at A.

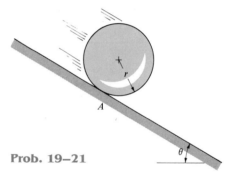

Prob. 19–21

19–22. The slender rod of mass m and length L is released from rest when it is in the vertical position. If it falls and strikes the soft ledge at A without rebounding, determine the impulse which the ledge exerts on the rod.

Prob. 19–22

19–23. The square plate has a mass m and is suspended at its corner by a cord. If it receives a horizontal impulse I at corner B, determine the location y of the point P about which the plate appears to rotate during the impact.

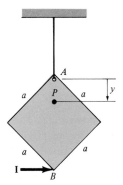

Prob. 19–23

*19–24. The inner hub of the wheel rests on the horizontal track. If it does not slip at A, determine the speed of the 10-lb block when $t = 2$ s after it is released from rest. The wheel has a weight of 30 lb and a radius of gyration $k_G = 1.30$ ft. Neglect the mass of the pulley and cord.

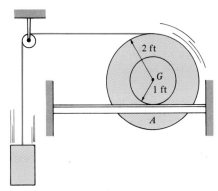

Prob. 19–24

19–25. The rod of length L and mass m lies on a smooth horizontal surface and is subjected to a force P at its end A as shown. Determine the location d of the point about which the rod begins to turn, i.e., the point that has zero velocity.

Prob. 19–25

19–26. The 10-lb cylinder rests on the 20-lb dolly. If the system is released from rest, determine the angular velocity of the cylinder in $t = 2$ s. The cylinder does not slip on the dolly. Neglect the mass of the wheels on the dolly.

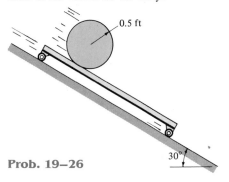

Prob. 19–26

19–27. A cord is wrapped around the rim of each 10-lb disk. If disk B is released from rest, determine the angular velocity of disk A in 2 s. Neglect the mass of the cord.

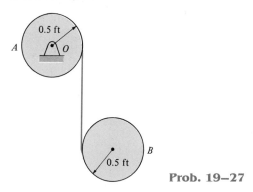

Prob. 19–27

385

19.3 Conservation of Momentum

Conservation of Linear Momentum

If the sum of all the *linear impulses* acting on a system of connected rigid bodies is *zero,* the linear momentum of the system is constant or conserved. Consequently, the first two of Eqs. 19–15 reduce to the form

$$\left(\sum \begin{array}{c} \text{syst. linear} \\ \text{momentum} \end{array} \right)_1 = \left(\sum \begin{array}{c} \text{syst. linear} \\ \text{momentum} \end{array} \right)_2 \qquad (19\text{–}16)$$

This equation is referred to as the *conservation of linear momentum.*

Without inducing appreciable errors in the computations, it may be possible to apply Eq. 19–16 in a specified direction for which the linear impulses are small or *nonimpulsive.* Specifically, nonimpulsive forces occur when small forces act over very short periods of time. For example, the impulse created by the force of a tennis racket hitting a ball during a very short time interval Δt is large, whereas the impulse of the weight of the ball during this time is small by comparison and may therefore be neglected in the motion analysis of the ball during Δt.

Conservation of Angular Momentum

The angular momentum of a system of connected rigid bodies is conserved about the system's center of mass G, or a fixed point O, when the sum of all the angular impulses created by the external forces acting on the system is zero or appreciably small (nonimpulsive) when computed about these points. The third of Eqs. 19–15 then becomes

$$\left(\sum \begin{array}{c} \text{syst. angular} \\ \text{momentum} \end{array} \right)_{O1} = \left(\sum \begin{array}{c} \text{syst. angular} \\ \text{momentum} \end{array} \right)_{O2} \qquad (19\text{–}17)$$

This equation is referred to as the *conservation of angular momentum.* In the case of a single rigid body, Eq. 19–17 applied to point G becomes $(I_G\omega)_1 = (I_G\omega)_2$. To illustrate an application of this equation, consider a swimmer who executes a somersault after jumping off a diving board. By tucking his arms and legs in close to his chest, he *decreases* his body's moment of inertia and thus *increases* his angular velocity ($I_G\omega$ must be constant). If he straightens out just before entering the water, his body's moment of inertia is *increased* and his angular velocity *decreases.* Since the weight of his body creates a linear impulse during the time of motion, this example also illustrates that the angular momentum of a body is conserved and yet the linear momentum is *not.* Such cases occur whenever the external forces creating the linear impulse pass through either the center of mass of the body or a fixed axis of rotation.

PROCEDURE FOR ANALYSIS

Provided the initial velocity of the body is known, the conservation of linear or angular momentum is used to determine the final velocity of a body *just after* the time period considered. Furthermore, by applying these equations to a *system* of bodies, the internal impulses acting within the system, which may be unknown, are eliminated from the analysis, since they occur in equal but opposite collinear pairs. For application it is suggested that the following procedure be used.

Free-Body Diagram. Establish the x, y inertial frame of reference and draw the free-body diagram for the body or system of bodies during the time of impact. From this diagram classify each of the applied forces as being either ''impulsive'' or ''nonimpulsive.'' In general, for short times, ''nonimpulsive forces'' consist of the weight of a body, the force of a slightly deformed spring, or any force that is *known to be small* when compared to an impulsive force. It will also be possible to tell if the conservation of linear or angular momentum can be applied. Specifically, the *conservation of linear momentum* applies in a given direction when *no* external impulsive forces act on the body or system in that direction; whereas the *conservation of angular momentum* applies about a fixed point O or at the mass center G of a body or system of bodies when all the external impulsive forces acting on the body or system create zero moment (or zero angular impulse) about O or G.

As an alternative procedure, consider drawing the impulse and momentum diagrams for the body or system of bodies. These diagrams are particularly helpful in order to visualize the ''moment'' terms used in the conservation of angular momentum equation, when it has been decided that angular momenta are to be computed about a point other than the body's mass center G.

Conservation of Momentum. Apply the conservation of linear or angular momentum in the appropriate directions.

Kinematics. Use *kinematics* if further equations are necessary for the solution of a problem. If the motion appears to be complicated, kinematic (velocity) diagrams may be helpful in obtaining the necessary kinematic relations.

If it is necessary to determine an *internal impulsive force* acting on only one body of a system, the body must be *isolated* (free-body diagram) and the principle of linear or angular impulse and momentum must be applied *to the body*. After the impulse $\int F \, dt$ is calculated, then, provided the time Δt for which the impulse acts is known, the *average impulsive force* F_{avg} can be determined from $F_{avg} = \int F \, dt / \Delta t$.

The following examples numerically illustrate application of this procedure.

Example 19–5

The 10-kg wheel shown in Fig. 19–8a has a moment of inertia $I_G = 0.156 \text{ kg} \cdot \text{m}^2$. Assuming that the wheel does not slip or rebound, determine the minimum velocity $\mathbf{v}_G$ it must have to just roll over the obstruction at A.

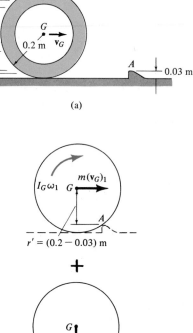

(a)

SOLUTION

Impulse and Momentum Diagrams. Since no slipping or rebounding occurs, the wheel essentially *pivots* about point A during contact. This condition is shown in Fig. 19–8b, which indicates, respectively, the momentum of the wheel *just before impact,* the impulses given to the wheel *during impact,* and the momentum of the wheel *just after impact.* Only two impulses (forces) act on the wheel. By comparison, the impulse at A is much greater than that caused by the weight, and since the time of impact is very short, the weight can be considered nonimpulsive. The impulsive force $\mathbf{F}$ at A has both an unknown magnitude and an unknown direction θ. To eliminate this force from the analysis, note that angular momentum about A is essentially *conserved* since $(98.1\Delta t)d \approx 0$.

Conservation of Angular Momentum. With reference to Fig. 19–8b,

$$(\circlearrowleft +) \qquad (\mathbf{H}_A)_1 = (\mathbf{H}_A)_2$$

$$r'm(v_G)_1 + I_G\omega_1 = rm(v_G)_2 + I_G\omega_2$$

$$(0.2 - 0.03)(10)(v_G)_1 + (0.156)(\omega_1) = (0.2)(10)(v_G)_2 + (0.156)(\omega_2)$$

Kinematics. Since no slipping occurs, $\omega = v_G/r = v_G/0.2 = 5v_G$. Substituting this into the above equation and simplifying yields

$$(v_G)_2 = 0.892(v_G)_1 \qquad (1)$$

*Conservation of Energy.** In order to roll over the obstruction, the wheel must pass the dashed position 3 shown in Fig. 19–8c. Hence, if $(v_G)_2$ (or $(v_G)_1$) is to be a minimum, it is necessary that the kinetic energy of the wheel at position 2 be equal to the potential energy at position 3. Constructing the datum through the center of gravity, as shown in the figure, and applying the conservation of energy equation, we have

$$\{T_2\} + \{V_2\} = \{T_3\} + \{V_3\}$$

$$\left\{ \frac{1}{2}(10)(v_G)_2^2 + \frac{1}{2}(0.156)(\omega_2)^2 \right\} + \{0\} = \{0\} + \{(98.1)(0.03)\}$$

Substituting $\omega_2 = 5(v_G)_2$ and Eq. (1) into this equation, and simplifying, yields

$$3.98(v_G)_1^2 + 1.55(v_G)_1^2 = 2.94$$

Thus,

$$(v_G)_1 = 0.729 \text{ m/s} \rightarrow \qquad Ans.$$

(b)

Fig. 19–8

(c)

*This principle *does not* apply during *impact,* since energy is *lost* during the collision; however, just after impact it can be used.

Example 19–6

The 5-kg slender rod shown in Fig. 19–9a is pinned at O and is initially at rest. If a 4-g bullet is fired into the rod with a velocity of 400 m/s, as shown in the figure, determine the angular velocity of the rod just after the bullet becomes embedded in it.

SOLUTION

Impulse and Momentum Diagrams. The impulse which the bullet exerts on the rod can be eliminated from the analysis and the angular velocity of the rod just after impact can be determined by considering the bullet and rod as a single system. To clarify the principles involved, the impulse and momentum diagrams are shown in Fig. 19–9b. The momentum diagrams are drawn *just before and just after impact*. During impact, the bullet and rod exchange equal but opposite *internal impulses* at A. As shown on the impulse diagram, the impulses that are external to the system are due to the reactions at O and the weights of the bullet and rod. Since the time of impact, Δt, is very short, the rod moves only a slight amount and so the "moments" of these impulses about point O are essentially zero, and therefore angular momentum is conserved about this point.

(a)

Fig. 19–9

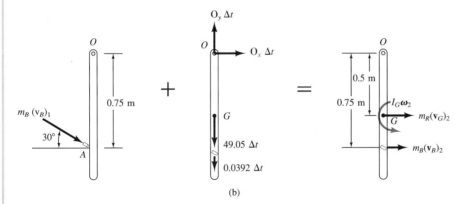

(b)

Conservation of Angular Momentum. From Fig. 19–9b, we have

$(\downarrow+)$ $\qquad \Sigma(H_O)_1 = \Sigma(H_O)_2$

$m_B(v_B)_1 \cos 30°(0.75 \text{ m}) = m_B(v_B)_2(0.75 \text{ m}) + m_R(v_G)_2(0.5 \text{ m}) + I_G\omega_2$

$(0.004)(400 \cos 30°)(0.75) = (0.004)(v_B)_2(0.75) + (5)(v_G)_2(0.5) + \left[\dfrac{1}{12}(5)(1)^2\right]\omega_2$

or

$$1.039 = 0.003(v_B)_2 + 2.50(v_G)_2 + 0.417\omega_2 \qquad (1)$$

Kinematics. Since the rod is pinned at O, from Fig. 19–9c we have

$$(v_G)_2 = 0.5\omega_2 \qquad (v_B)_2 = 0.75\omega_2$$

Substituting into Eq. (1) and solving yields

$$\omega_2 = 0.622 \text{ rad/s} \uparrow \qquad\qquad Ans.$$

(c)

19.4 Eccentric Impact

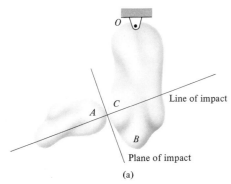

Line of impact

Plane of impact

(a)

Fig. 19—10a

Impact occurs when two bodies collide with each other during a very *short* interval of time. During the collision very large equal but opposite impulsive forces are created between the two bodies at their point of contact. The concepts involving central and oblique impact of particles have been presented in Sec. 15.4. We will expand this treatment and discuss the eccentric impact of two bodies. *Eccentric impact* occurs when the line connecting the *mass centers of* the two bodies *does not* coincide with the line of impact.* This type of impact often occurs when one or both of the bodies is constrained to rotate about a fixed axis. Consider, for example, the collision between the two bodies A and B, shown in Fig. 19–10a, which collide at point C. Body B rotates about an axis passing through point O, whereas body A both rotates and translates. It is assumed that just before collision B is rotating counterclockwise with an angular velocity $(\omega_B)_1$, and the velocity of the contact point C located on A is $(\mathbf{u}_A)_1$. Kinematic diagrams for both bodies just before collision are shown in Fig. 19–10b. Provided the bodies are smooth at point C, the *impulsive forces* they exert on each other *are directed along the line of impact*. Hence, the component of velocity of point C on body B, which is directed along the line of impact, is $(v_B)_1 = (\omega_B)_1 r$, Fig. 19–10b. Likewise, on body A the component of velocity $(\mathbf{u}_A)_1$ along the line of impact is $(\mathbf{v}_A)_1$. In order for a collision to occur, $(v_A)_1 > (v_B)_1$.

During the impact an equal but opposite impulsive force $\mathbf{P}$ is exerted between the bodies which *deforms* their shapes at the point of contact. The resulting impulse is shown on the impulse diagrams for both bodies, Fig. 19–10c. Note that the impulsive force created at point C, on the rotating body B, creates impulsive pin reactions at the supporting pin O. On these diagrams it is assumed the impact creates forces which are much larger than the weights of the bodies. Hence, the impulses created by the weights are not shown since they are negligible compared to $\int \mathbf{P}\, dt$ and the reactive impulses at O. When the deformation of point C is a maximum, C on both the bodies moves with a common velocity $\mathbf{v}$ along the line of impact, Fig. 19–10d. A period of *restitution* then occurs in which the bodies tend to regain their original shapes. The restitution phase creates an equal but opposite impulsive force $\mathbf{R}$ acting between the bodies as shown on the impulse diagram, Fig. 19–10e. After restitution the bodies move apart such that point C on body B has a velocity $(\mathbf{v}_B)_2$ and point C on body A has a velocity $(\mathbf{u}_A)_2$, Fig. 19–10f, where $(v_B)_2 > (v_A)_2$.

In general, a problem involving the impact of two bodies requires determining the *two unknowns* $(v_A)_2$ and $(v_B)_2$, assuming $(v_A)_1$ and $(v_B)_1$ are known (or can be determined using kinematics, energy methods, the equations of motion, etc.) To solve this problem two equations must be written. The *first equation* generally involves application of *the conservation of angular momentum to the two bodies*. In the case of bodies A and B, we can state that

---

*When these lines coincide, central impact occurs and the problem can be analyzed as discussed in Sec. 15–4.

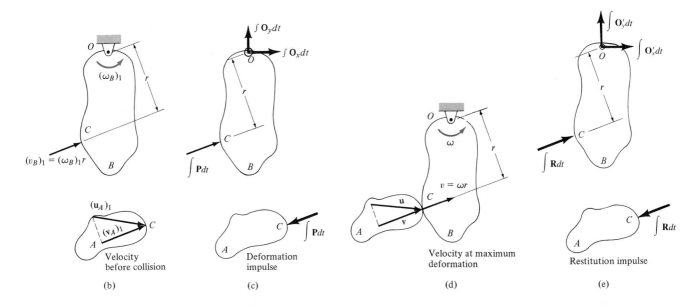

(b) Velocity before collision

(c) Deformation impulse

(d) Velocity at maximum deformation

(e) Restitution impulse

angular momentum is conserved about point O since the impulses at C are internal to the system and the impulses at O create zero moment (or zero angular impulse) about point O. The *second equation* is obtained using the definition of the *coefficient of restitution, e,* which is a ratio of the restitution impulse to the deformation impulse. To establish a useful form of this equation we must first apply the principle of angular impulse and momentum about point O to bodies B and A separately. Combining the results, we then obtain the necessary equation. Proceeding in this manner, the principle of impulse and momentum applied to body B from the time just before the collision to the instant of maximum deformation, Fig. 19–10b, c, and d, becomes

$$(\downarrow+) \qquad I_O(\omega_B)_1 + r \int P \, dt = I_O\omega \qquad (19\text{–}18)$$

Here I_O is the moment of inertia of body B about point O. Similarly, applying the principle of angular impulse and momentum from the instant of maximum deformation to the time just after the impact, Fig. 19–10d, e and f, yields

$$(\downarrow+) \qquad I_O\omega + r \int R \, dt = I_O(\omega_B)_2 \qquad (19\text{–}19)$$

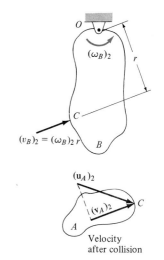

Velocity after collision

Fig. 19–10(b–f)   (f)

Solving Eqs. 19–18 and 19–19 for $\int P \, dt$ and $\int R \, dt$, respectively, and formulating e, we have

$$e = \frac{\int R \, dt}{\int P \, dt} = \frac{r(\omega_B)_2 - r\omega}{r\omega - r(\omega_B)_1} = \frac{(v_B)_2 - v}{v - (v_B)_1}$$

391

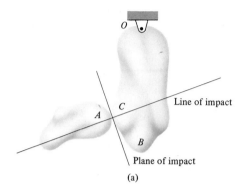

Fig. 19–10a

In the same manner, we may write an equation which relates the magnitudes of velocity $(v_A)_1$ and $(v_A)_2$ of body A. The result is

$$e = \frac{(v_A)_2 - v}{v - (v_A)_1}$$

Combining the above equations by eliminating the common velocity v yields the desired result, i.e.,

$(\xrightarrow{+})$
$$e = \frac{(v_B)_2 - (v_A)_2}{(v_A)_1 - (v_B)_1}$$
(19–20)

This equation is identical to Eq. 15–11, which was derived for the central impact occurring between two particles. Equation 19–20, however, states that the coefficient of restitution, measured at the points of contact (C) between two colliding bodies, is equal to the ratio of the relative velocity of *separation* of the points *just after impact* to the relative velocity at which the points *approach* one another *just* before impact. In deriving Eq. 19–20, we assumed that the points of contact for both bodies move to the right *both* before and after impact. If motion of any one of the contacting points occurs to the left, the velocity of this point is considered a negative quantity in Eq. 19–20.

As stated previously, when Eq. 19–20 is used in conjunction with the conservation of angular momentum for the bodies, it provides a useful means of obtaining the velocities of two colliding bodies just after collision.

Example 19–7

The 10–lb slender rod is suspended from the pin at A, Fig. 19–11a. If a 2-lb ball B is thrown at the rod and strikes its center with a horizontal velocity of 30 ft/s, determine the angular velocity of the rod just after impact. The coefficient of restitution is $e = 0.4$.

Fig. 19–11

(a)

SOLUTION

Conservation of Angular Momentum. Consider the ball and rod as a system, Fig. 19–11b. Angular momentum is conserved about point A since the impulsive force between the rod and ball is *internal*. Also, the *weights* of the ball and rod are *nonimpulsive*. Noting the directions of the velocities of the ball and rod juster after impact as shown on the kinematic diagram, Fig. 19–11c, we require

(b)

$$(\downarrow+) \qquad (H_A)_1 = (H_A)_2$$

$$m_B(v_B)_1(1.5) = m_B(v_B)_2(1.5) + m_R(v_G)_2(1.5) + I_G\omega_2$$

$$\left(\frac{2}{32.2}\right)(30)(1.5) = \left(\frac{2}{32.2}\right)(v_B)_2(1.5) + \left(\frac{10}{32.2}\right)(v_G)_2(1.5) + \left(\frac{1}{12}\frac{10}{32.2}(3)^2\right)\omega_2$$

Since $(v_G)_2 = 1.5\omega_2$ then

$$2.795 = 0.09317(v_B)_2 + 0.9317\omega_2 \qquad (1)$$

Coefficient of Restitution. With reference to Fig. 19–11c, we have

$$e = \frac{(v_G)_2 - (v_B)_2}{(v_B)_1 - (v_G)_1}; \qquad 0.4 = \frac{(1.5\omega_2) - (v_B)_2}{30 - 0}$$

$$12.0 = 1.5\omega_2 - (v_B)_2$$

(c)

Solving,

$$(v_B)_2 = -6.52 \text{ ft/s} = 6.52 \text{ ft/s} \leftarrow$$

$$\omega = 3.65 \text{ rad/s} \uparrow \qquad\qquad Ans.$$

PROBLEMS

*19–28. The turntable T of a record player has a mass of 0.75 kg and a radius of gyration $k_z = 125$ mm. It is *turning freely* at $\omega_T = 2$ rad/s when a 50-g record (thin disk) falls on it. Determine the final angular velocity of the turntable just after the record stops slipping on the turntable.

Prob. 19–28

19–29. A thin disk of mass m has an angular velocity ω_1 while rotating on a smooth surface. Determine its new angular velocity just after the hook at its edge strikes the peg P and the disk starts to rotate about P without rebounding.

Prob. 19–29

19–30. The 2-kg rod ACB supports the two 4-kg disks at its ends. If both disks are given a clockwise angular velocity of $(\omega_A)_1 = (\omega_B)_1 = 5$ rad/s while the rod is held stationary and then released, determine the angular velocity of the rod after both disks have stopped spinning relative to the rod due to frictional resistance at the pins A and B. Motion is in the *horizontal plane*. Neglect friction at pin C.

Probs. 19–30/19–31

19–31. Solve Prob. 19–30 if disk A is given an angular velocity of $(\omega_A)_1 = 5$ rad/s clockwise, and disk B has an angular velocity of $(\omega_B)_1 = 3$ rad/s counterclockwise.

*19–32. Determine the height h at which the disk of mass m must be struck with a horizontal force $\mathbf{F}$ so that no frictional force develops between it and the ground at A.

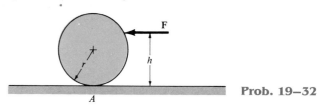

Prob. 19–32

19–33. Each of the two slender rods and the disk have the same mass m. Also, the length of each rod is equal to the diameter d of the disk. If the assembly is rotating with an angular velocity of ω_1 when the rods are directed outwards, determine the angular velocity of the assembly if by internal means the rods are brought to an upright vertical position.

Prob. 19–33

19–34. A horizontal circular platform has a weight of 300 lb and a radius of gyration $k_z = 8$ ft about the z axis passing through its center O. The platform is free to rotate about the z axis and is initially at rest. A man having a weight of 150 lb throws a 15-lb block off the edge of the platform with a horizontal velocity of 5 ft/s, *measured relative to the platform*. Compute the angular velocity of the platform if the block is thrown (a) tangent to the platform, along the $+t$ axis, and (b) outward along a radial line, or $+n$ axis. Neglect the size of the man.

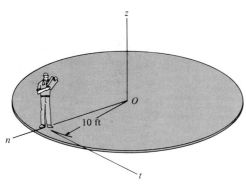

Prob. 19–34

19–35. The disk has a mass m and radius r. If it strikes a rough step having a height $\frac{1}{8}r$ as shown, determine the smallest angular velocity ω_1 the disk can have and not rebound off the step when it strikes it.

Prob. 19–35

*19–36. The rod of mass m and length L is released from rest without rotating. When it falls a distance L, the end A strikes the hook S, which provides a permanent connection. Determine the angular velocity ω of the rod after it has rotated 90°. Treat the rod's weight during impact as a nonimpulsive force.

Prob. 19–36

19–37. The solid ball has a mass m and rolls without slipping along a horizontal plane with an angular velocity of ω_1. Provided it does not slip or rebound, determine its angular velocity as it just starts to roll up the inclined plane.

Prob. 19–37

19–38. The plank has a weight of 30 lb, center of gravity at G, and it rests on the two sawhorses at A and B. If the end D is raised 2 ft above the top of the sawhorses and is released from rest, determine how high end C will rise from the top of the sawhorses after the plank falls so that it rotates clockwise about A, strikes and pivots on the sawhorse at B, and rotates clockwise off the sawhorse at A.

Prob. 19–38

19–39. The solid ball of mass m is dropped with a velocity $\mathbf{v}_1$ onto the edge of the rough step. If it rebounds horizontally off the step with a velocity $\mathbf{v}_2$, determine the angle θ at which contact occurs. Assume no slipping occurs when the ball strikes the step. The coefficient of restitution is e.

Prob. 19–39

*19–40. The rod has a length L and mass m. A smooth collar having a negligible size and one fourth the mass of the rod is placed on the rod at its midpoint. If the rod is freely rotating at $\boldsymbol{\omega}$ about its end and the collar is released, determine the rod's angular velocity at the instant the collar flies off the rod. Also, what is the speed of the collar as it leaves the rod?

Prob. 19–40

19–41. The pendulum consists of a 10-lb sphere and 4-lb rod. If it is released from rest when $\theta_1 = 90°$, determine the angle θ_2 of rebound after the sphere strikes the floor. Take $e = 0.8$.

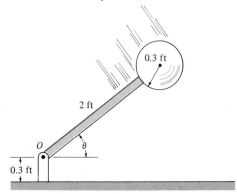

Prob. 19–41

19–42. The pendulum consists of a 10-lb solid ball and 4-lb rod. If it is released from rest when $\theta_1 = 0°$, determine the angle θ_2 of rebound after the ball strikes the wall and the pendulum swings up to the point of momentary rest. Take $e = 0.6$.

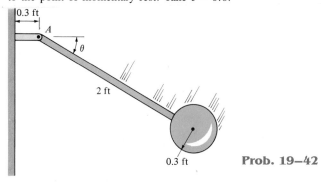

Prob. 19–42

19–43. A 7-g bullet having a velocity of 800 m/s is fired into the edge of the 5-kg disk as shown. Determine the angular velocity of the disk just after the bullet becomes embedded in it. Also, calculate how far θ the disk will swing until it stops. The disk is originally at rest.

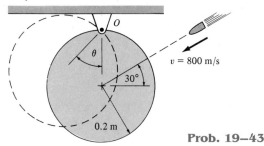

Prob. 19–43

*19–44. The 6-lb slender rod AB is released from rest when it is in the *horizontal position* so that it begins to rotate clockwise. A 1-lb ball is thrown at the rod with a velocity $v = 50$ ft/s. The ball strikes the rod at C at the instant the rod is in the vertical position as shown. Determine the angular velocity of the rod just after the impact. Take $e = 0.7$.

Prob. 19–44

Review 2: Planar Kinematics and Kinetics of a Rigid Body

Having presented the various topics in planar kinematics and kinetics in Chapters 16 through 19, we will now summarize these principles and provide an opportunity for applying them to the solution of various types of problems.

Kinematics

Here we are interested in studying the geometry of motion, without concern for the forces which cause the motion. Before solving a planar kinematics problem, it is *first* necessary to *classify the motion* as being either rectilinear or curvilinear translation, rotation about a fixed axis, or general plane motion. In particular, problems involving general plane motion can be solved either with reference to a fixed axis (absolute motion analysis) or using a translating or rotating frame of reference (relative motion analysis). The choice generally depends upon the type of constraints and the problem geometry. In all cases, application of the necessary equations may be clarified by drawing a kinematic diagram. Remember that the *velocity* of a point is always *tangent* to its path of motion, and the *acceleration* of a point can have *components* in the *n-t* directions when the path is *curved*.

Translation. When the body moves with rectilinear or curvilinear translation, *all* the points on the body have the *same motion.*

$$\mathbf{v}_B = \mathbf{v}_A, \qquad \mathbf{a}_B = \mathbf{a}_A$$

Rotation About a Fixed Axis.

Angular Motion

Variable Angular Acceleration.   Provided a mathematical relationship is given between *any two* of the *four* variables θ, ω, α, and t, then a *third* variable can be determined by solving one of the following equations which relate all three variables.

$$\omega = \frac{d\theta}{dt} \qquad \alpha = \frac{d\omega}{dt} \qquad \alpha \, d\theta = \omega \, d\omega$$

Constant Angular Acceleration.   The following equations apply when it is *absolutely certain* that the angular acceleration is constant.

$$\theta = \theta_0 + \omega_0 t + \tfrac{1}{2}\alpha_c t^2 \qquad \omega = \omega_0 + \alpha t \qquad \omega^2 = \omega_0^2 + 2\alpha_c(\theta - \theta_0)$$

Motion of Point P

$$v = \omega r \qquad\qquad \mathbf{v} = \boldsymbol{\omega} \times \mathbf{r}$$

$$a_t = \alpha r \quad a_n = \omega^2 r \qquad \mathbf{a} = \boldsymbol{\alpha} \times \mathbf{r} + \boldsymbol{\omega} \times (\boldsymbol{\omega} \times \mathbf{r})$$

General Plane Motion—Relative Motion Analysis.

Recall that when *translating axes* are placed at the "base point" *A*, the *relative motion* of point *B* with respect to *A* is simply *circular motion* of *B* about *A*. The following equations apply to two points *A* and *B* located on the *same* rigid body.

$$\mathbf{v}_B = \mathbf{v}_A + \mathbf{v}_{B/A}$$

$$\mathbf{a}_B = \mathbf{a}_A + \mathbf{a}_{B/A}$$

Rotating and translating axes are often used to analyze the motion of rigid bodies which are connected together by collars or slider blocks.

$$\mathbf{v}_B = \mathbf{v}_A + \boldsymbol{\Omega} \times \mathbf{r}_{B/A} + (\mathbf{v}_{B/A})_{\text{rel}}$$

$$\mathbf{a}_B = \mathbf{a}_A + \dot{\boldsymbol{\Omega}} \times \mathbf{r}_{B/A} + \boldsymbol{\Omega} \times (\boldsymbol{\Omega} \times \mathbf{r}_{B/A}) + 2\boldsymbol{\Omega} \times (\mathbf{v}_{B/A})_{\text{rel}} + (\mathbf{a}_{B/A})_{\text{rel}}$$

Kinetics

To analyze the forces which cause the motion we must use the principles of kinetics. When applying the necessary equations, it is important to first establish the inertial coordinate system and define the positive directions of the axes. The *directions* should be the *same* as those selected when writing any equations of kinematics provided *simultaneous solution* of equations becomes necessary.

Equations of Motion.   These equations are used to determine accelerated motions or forces causing the motion. If used to determine position, velocity, or time of motion, then kinematics will have to be considered for part of the solution. Before applying the equations of motion, *always draw a free-body diagram* in order to identify all the forces acting on the body. Also, establish the directions of the acceleration of the mass center and the angular acceleration of the body. (A kinetic diagram may also be drawn in order to graphically represent $m\mathbf{a}_G$ and $I_G\boldsymbol{\alpha}$. This diagram is particularly convenient for resolving $m\mathbf{a}_G$ into components and for identifying the terms in the moment sum $\Sigma(\mathcal{M}_k)_P$.)

The three equations of motion are

$$\Sigma F_x = m(a_G)_x$$
$$\Sigma F_y = m(a_G)_y$$
$$\Sigma M_G = I_G\alpha \quad \text{or} \quad \Sigma M_P = \Sigma(\mathcal{M}_k)_P$$

In particular, if the body is rotating about a fixed axis, moments may also be summed about point O on the axis, in which case

$$\Sigma M_O = \Sigma(\mathcal{M}_k)_O = I_O\alpha$$

Work and Energy. *The equation of work and energy is used to solve problems involving force, velocity, and displacement.* Before applying this equation, *always draw a free-body diagram* of the body in order to identify the forces which do work. Recall that the kinetic energy of the body consists of that due to translational motion of the mass center, $\mathbf{v}_G$, *and* rotational motion of the body, $\boldsymbol{\omega}$.

$$T_1 + \Sigma U_{1-2} = T_2$$

where

$$T = \tfrac{1}{2}mv_G^2 + \tfrac{1}{2}I_G\omega^2$$

$$U_F = \int F\cos\theta\, ds \quad \text{(variable force)}$$

$$U_{F_c} = F_c\cos\theta\,(s_2 - s_1) \quad \text{(constant force)}$$
$$U_W = W\,\Delta y \quad \text{(weight)}$$
$$U_s = -(\tfrac{1}{2}ks_2^2 - \tfrac{1}{2}ks_1^2) \quad \text{(spring)}$$
$$U_M = M\theta \quad \text{(constant couple)}$$

If the forces acting on the body are conservative forces, then apply the conservation of energy equation. This equation is easier to use than the equation of work and energy, since it applies only at *two points* on the path and *does not* require calculation of the work done by a force as the body moves along the path.

$$T_1 + V_1 = T_2 + V_2$$

where

$$V_g = Wy \quad \text{(gravitational potential energy)}$$
$$V_e = \tfrac{1}{2}ks^2 \quad \text{(elastic potential energy)}$$

Impulse and Momentum. The principles of linear and angular impulse and momentum are used to solve problems involving force, velocity, and time. Before applying the equations, *draw a free-body diagram* in order to identify all the forces which cause linear and angular impulses on the body. Also, establish the directions of the velocity of the mass center and the angular velocity of the body just before and just after the impulses are applied. (As an alternative procedure, the impulse and momentum diagrams may accompany the solution in order to graphically account for the terms in the equations. These diagrams are particularly advantageous when computing the angular impulses and angular momenta about a point other than the body's mass center.)

$$m(\mathbf{v}_G)_1 + \Sigma \int \mathbf{F}\, dt = m(\mathbf{v}_G)_2$$

$$(\mathbf{H}_G)_1 + \Sigma \int \mathbf{M}_G\, dt = (\mathbf{H}_G)_2$$

or

$$(\mathbf{H}_O)_1 + \Sigma \int \mathbf{M}_O\, dt = (\mathbf{H}_O)_2$$

Conservation of Momentum. If nonimpulsive forces or no impulsive forces act on the body in a particular direction, or the motions of several bodies are involved in the problem, then consider applying the conservation of linear or angular momentum for the solution. Investigation of the free-body diagram (or the impulse diagram) will aid in determining the directions for which the impulsive forces are zero, or axes for which the impulsive forces cause zero angular momentum. When this occurs, then

$$m(\mathbf{v}_G)_1 = m(\mathbf{v}_G)_2$$
$$(\mathbf{H}_O)_1 = (\mathbf{H}_O)_2$$

The problems that follow involve application of all the above concepts. They are presented in *random order* so that practice may be gained at identifying the various types of problems and developing the skills necessary for their solution.

PROBLEMS

R2–1. The 40-lb cylinder rests on an inclined surface having a coefficient of kinetic friction $\mu_k = 0.1$. A cord is wrapped around the outer surface of the cylinder. If the cylinder is released from rest, determine the velocity of the center of the cylinder one second after it is released. Neglect the mass of the cord.

Prob. R2–1

R2–2. The double pendulum consists of two rods. Rod AB has a constant angular velocity of 3 rad/s, and rod BC has a constant angular velocity of 2 rad/s, both measured counterclockwise. Determine the velocity and acceleration of point C at the instant shown.

Prob. R2–2

R2–3. A chain that has a negligible mass is draped over a sprocket which has a mass of 2 kg and a radius of gyration of $k_O = 50$ mm. If the 4-kg block A is released from rest in the position shown, $s = 1$ m, determine the angular velocity which the chain imparts to the sprocket when $s = 2$ m.

Prob. R2–3

*R2–4. A tape having a thickness s wraps around the wheel which is turning at a constant rate ω. Assuming the unwrapped portion of tape remains horizontal, determine the acceleration of point P on the tape when the radius is r. *Hint:* Since $v_P = \omega r$, take the time derivative and note that $dr/dt = \omega(s/2\pi)$.

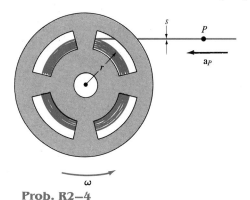

Prob. R2–4

401

R2–5. The 2-lb bottle rests on the checkout conveyor at a grocery store. If the coefficient of static friction is $\mu_s = 0.2$, determine the largest acceleration which the conveyor can have without causing the bottle to slip or tip. The center of mass is at G.

8 in.

a

1.5 in.

Prob. R2–5

R2–6. The 20-lb solid ball is cast on the floor such that it has a backspin of $\omega = 15$ rad/s and its center has an initial horizontal velocity of $v_G = 20$ ft/s. If the coefficient of kinetic friction between the floor and the ball is $\mu_A = 0.3$, determine the distance it travels before it stops spinning.

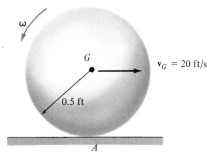

ω

G

$v_G = 20$ ft/s

0.5 ft

A

Probs. R2–6/R2–7

R2–7. Determine the backspin ω which should be given to the 20-lb ball so that when its center is given an initial horizontal velocity of $v_G = 20$ ft/s it stops spinning and translating at the same instant. The coefficient of kinetic friction is $\mu_A = 0.3$.

*R2–8. Gear A is pinned at B and rotates along the periphery of the circular gear rack R. If A has a weight of 4 lb and a radius of gyration of $k_B = 0.5$ ft, determine the angular momentum of gear A about point C when $\omega_{CB} = 30$ rad/s and (a) $\omega_R = 0$, (b) $\omega_R = 20$ rad/s.

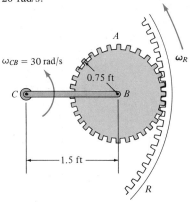

$\omega_{CB} = 30$ rad/s

A

0.75 ft

ω_R

C

B

1.5 ft

R

Prob. R2–8

R2–9. A large roll of paper having a mass of 20 kg and radius $r = 150$ mm is resting over the edge of a table. If the roll is disturbed slightly from its equilibrium position, determine the angle θ at which it begins to leave the table edge A as it falls. The centroidal radius of gyration of the roll is $k_G = 75$ mm.

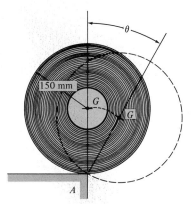

θ

150 mm

G

G

A

Prob. R2–9

R2–10. Knowing that the angular velocity of link CB is $\omega_{CB} = 6$ rad/s, determine the velocity of the collar A at the instant shown. Use Eq. 16–14.

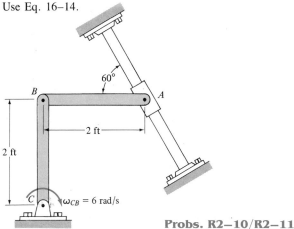

Probs. R2–10/R2–11

R2–11. Solve Prob. R2–10 using the method of instantaneous center of zero velocity.

*R2–12. A 20-kg roll of paper, originally at rest, is pin-supported at its ends to bracket AB. The roll rests against a wall for which the coefficient of kinetic friction at C is $\mu_C = 0.3$. If a force of 40 N is applied uniformly to the end of the sheet, determine the initial angular acceleration of the roll and the tension in the bracket as the paper unwraps. For the calculation, treat the roll as a cylinder.

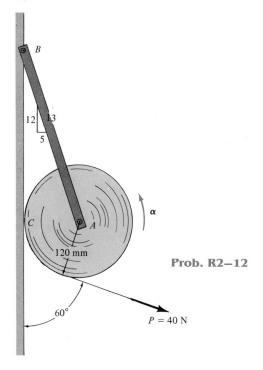

Prob. R2–12

R2–13. The spool and wire wrapped around its core have a mass of 20 kg and a centroidal radius of gyration $k_G = 250$ mm. If the coefficient of kinetic friction at the ground is $\mu_B = 0.1$, determine the angular acceleration of the spool when the 30-N · m couple moment is applied.

Prob. R2–13

R2–14. The spool has a weight of 30 lb and a radius of gyration of $k_O = 1.40$ ft. If a force of 40 lb is applied to the supporting cord at A as shown, determine the angular velocity of the spool in $t = 3$ s starting from rest. Neglect the mass of the pulley and cord.

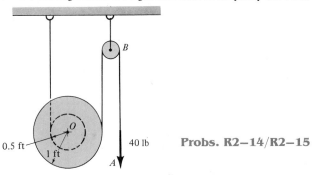

Probs. R2–14/R2–15

R2–15. Solve Prob. R2–14 if a 40-lb block is suspended from the cord at A, rather than applying the 40-lb force.

*R2–16. If the support at B is suddenly removed, determine the initial horizontal and vertical components of reaction which the pin A exerts on the rod ACB. Segments AC and CB each have a weight of 10 lb.

Prob. R2–16

R2–17. Locate the center of mass G of the baseball bat by determining $\bar{y}$. Then, compute the moment of inertia of the bat about an axis which passes through G and is perpendicular to the plane of the page. For the calculation, consider the bat to be composed of a truncated cone and cylinder. Neglect the size of the lip at A. The density of wood is $\rho_w = 750$ kg/m^3.

Prob. R2–17

R2–18. Link CD has an angular velocity of 3 rad/s at the instant shown. Determine the angular velocity of link AB at this instant.

Prob. R2–18

R2–19. The assembly consists of a uniform rod AB having a mass of 0.2 kg and two solid spheres C and D having a mass of 0.4 kg and 0.6 kg, respectively. Determine $\bar{x}$, which locates the center of mass G, and then calculate the moment of inertia about an axis perpendicular to the page and passing through G.

Prob. R2–19

*R2–20. The cart and its contents have a mass of 40 kg and a mass center at G, excluding the wheels. Each of the two wheels has a mass of 2 kg and a radius of gyration of $k_O = 0.120$ m. If the cart is released from rest from the position shown, determine its speed after it travels 4 m down the incline. The coefficient of kinetic friction is $\mu_A = 0.3$ between the incline and A. The wheels roll without slipping at B.

Prob. R2–20

R2–21. The links AB and BC each have a mass of 6 kg. An elastic cable is attached to A and C and has an unstretched length of 0.2 m. If the links are extended out into the horizontal position, then released from rest ($\theta \approx 0°$) so that they move out of equilibrium, compute the velocity of C at the instant $\theta = 45°$. Neglect the mass of the roller.

Prob. R2–21

R2–22. If the ball has a weight of 15 lb and is thrown onto a *rough surface* with a velocity of 6 ft/s parallel to the surface, determine the amount of backspin, ω, it must be given so that it stops spinning at the same instant that its forward velocity is zero. It is not necessary to know the coefficient of kinetic friction at A for the calculation.

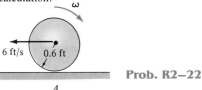

Prob. R2–22

R2–23. The wheel has a mass of 20 kg and a radius of gyration of $k_A = 0.13$ m. The attached spring is unstretched and is restricted from stretching more than 300 mm due to the chain. If the wheel is subjected to a torque of $M = 40$ N · m, determine its angular velocity just before the chain becomes taut. Assume no slipping at B. Neglect the mass of the chain.

Prob. R2–23

*R2–24. The tub of the mixer has a weight of 70 lb and a radius of gyration about its center of gravity of $k_G = 1.3$ ft. If a constant torque of 60 lb · ft is applied to the dumping wheel, determine the angular velocity of the tub when it has turned 90°. Originally the tub is at rest when $\theta = 0°$.

Prob. R2–24

R2–25. A flywheel has its angular speed increased uniformly from 15 rad/s to 60 rad/s in 120 s. If the diameter of the wheel is 2 ft, determine the normal and tangential components of acceleration of a point on the rim of the wheel when $t = 120$ s, and the total distance the point travels during the time period.

R2–26. At the instant $\theta = 45°$ link CD has an angular velocity of $\omega_{CD} = 4$ rad/s and an angular acceleration of $\alpha_{CD} = 2$ rad/s^2. Determine the angular velocity and angular acceleration of member AB.

Prob. R2–26

R2–27. The 3-kg ball rests on a horizontal surface and is being pushed forward by moving the piston P with a force of 2 N. If the ball rolls without slipping at C and the coefficient of kinetic friction between the piston and the ball at D is $\mu_D = 0.20$, compute the velocity of the ball's center of mass G after the piston moves 50 mm. Neglect the mass of the piston and assume that the guides at A and B are smooth. The ball is originally at rest.

Prob. R2–27

*R2–28. Compute the velocity of rod R for any angle θ of the cam C if the cam rotates with a constant angular velocity ω. The pin connection at O does not cause an interference with the motion of A on C.

Prob. R2–28

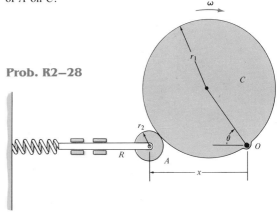

R2–29. The lawn roller has a mass of 80 kg and a radius of gyration of $k_G = 0.175$ m. If it is pushed forward with a force of 200 N when the handle is at 45°, determine its angular acceleration. The coefficients of static and kinetic friction between the ground and the roller are $\mu_s = 0.12$ and $\mu_k = 0.10$, respectively.

Prob. R2–29

R2–30. A 4-kg disk A is mounted on arm BC, which has a negligible mass. If a constant torque of $M = (5e^{0.5t})$ N · m, where t is in seconds, is applied to the arm C, determine the angular velocity of BC in 2 s starting from rest. Solve the problem assuming that (a) the disk is set in a smooth bearing at B so that it rotates with curvilinear translation and (b) the disk is fixed to the shaft BC.

Prob. R2–30

R2–31. The 10-kg rod AB is pin-connected at A and subjected to a couple moment of $M = 15$ N · m. If the rod is released from rest when the spring is unstretched, at $\theta = 30°$, determine the rod's angular velocity at the instant $\theta = 60°$. As the rod rotates, the spring always remains horizontal, because of the roller support at C.

Prob. R2–31

***R2–32.** The assembly weighs 10 lb and has a radius of gyration $k_G = 0.6$ ft about its center of mass G. The kinetic energy of the assembly is 31 ft-lb when it is in the position shown. If it is rolling counterclockwise on the surface without slipping, determine its linear momentum at this instant.

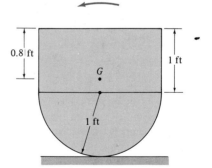

Prob. R2–32

R2–33. The pendulum consists of a 30-lb sphere and a 10-lb slender rod. Compute the reaction at the pin O just after the cord AB is cut.

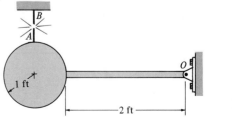

Prob. R2–33

R2–34. The center of the pulley is being lifted vertically with an acceleration of 4 m/s² at the instant it has a velocity of 2 m/s. If the cable does not slip on the pulley's surface, determine the accelerations of the cylinder B and point C on the pulley.

Prob. R2–34

R2–35. The 50-kg cylinder has an angular velocity of 30 rad/s when it is brought into contact with the horizontal surface at C. If the coefficient of kinetic friction is $\mu_k = 0.2$, determine how long it takes for the cylinder to stop spinning. What force is developed at the pin A during this time? The axis of the cylinder is connected to *two* symmetrical links. (Only AB is shown.) For the computation, neglect the weight of the links.

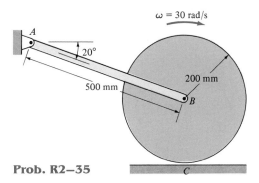

Prob. R2–35

*R2–36. The board rests on the surface of two drums. At the instant shown, it has an acceleration of 0.5 m/s² to the right, while at the same instant points on the outer rim of each drum have an acceleration with a magnitude of 3 m/s². If the board does not slip on the drums, determine its speed due to the motion.

Prob. R2–36

R2–37. The uniform slender rod AB has a mass of 4 kg and a length of 2 m. If it is suspended horizontally by a spring at A and a cord at B, determine the angular acceleration of the rod and the acceleration of the rod's mass center at the instant the cord at B is cut. *Hint:* The stiffness of the spring is not needed for the calculation.

Prob. R2–37

R2–38. The tire has a mass of 9 kg and a radius of gyration $k_O = 225$ mm. If it is released from rest and rolls down the plane without slipping, determine the speed of its center O when $t = 3$ s.

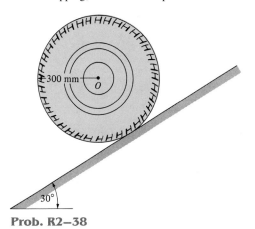

Prob. R2–38

Review 2: Planar Kinematics and Kinetics of a Rigid Body

R2–39. The double pulley consists of two wheels which are attached to one another and turn at the same rate. The pulley has a mass of 15 kg and a radius of gyration $k_O = 110$ mm. If the block at A has a mass of 40 kg, determine the speed of the block in 3 s after a constant force of 2 kN is applied to the rope wrapped around the inner hub of the pulley. The block is originally at rest. Neglect the mass of the rope.

Prob. R2–39

*R2–40. The spool has a weight of 150 lb and a radius of gyration of $k_O = 2.25$ ft. If a cord is wrapped around its inner core and the end is pulled with a horizontal force of $P = 40$ lb, determine the angular velocity of the spool after the center O has moved 10 ft to the right. The spool starts from rest and does not slip as it rolls. Neglect the mass of the cord.

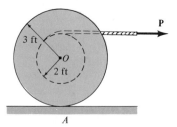

Probs. R2–40/R2–41

R2–41. Determine the magnitude of the horizontal force **P** that must be applied to the cord which is wrapped around the inner core of the spool, so that the spool attains an angular velocity of $\omega = 5$ rad/s when the center O has moved 10 ft to the right. Neglect the mass of the cord. The spool has a weight of 150 lb and a radius of gyration of $k_O = 2.25$ ft. It starts from rest and does not slip as it rolls.

R2–42. At the instant shown, two forces act on the 30-lb slender rod which is pinned at O. Determine the magnitude of force **F** and the initial angular acceleration of the rod so that the horizontal reaction which the *pin exerts on the rod* is 5 lb directed to the right.

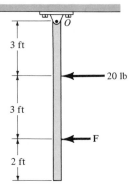

Prob. R2–42

R2–43. The disk A maintains a constant clockwise angular velocity of 30 rad/s. If the 20-kg disk B is initially at rest when it is brought into contact with A, determine the time required for B to attain the same angular velocity as A. The coefficient of kinetic friction between the two disks is $\mu_k = 0.3$. Neglect the mass of bar BC.

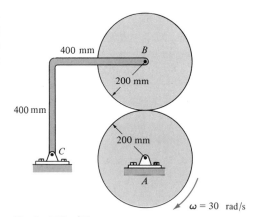

Prob. R2–43

408

*R2–44. The door has a weight of 200 lb and a center of gravity at G. Determine how far the door moves in 2 s, starting from rest, if a man pushes on it at C with a horizontal force of 30 lb. Also, find the vertical reactions at the rollers A and B.

Prob. R2–44

R2–45. The ring has a mass of 10 kg, a center of mass at G, and a radius of gyration $k_G = 135$ mm. If its angular velocity is $\omega_1 = 2$ rad/s when it is in the position shown, compute its angular acceleration at this instant and the normal force of the ring on the ground. Assume that slipping does not occur.

Prob. R2–45

R2–46. The uniform 150-lb stone (rectangular block) is being turned over on its side by pulling the vertical cable *slowly* upward until the stone begins to tip. If it then falls freely ($\mathbf{T} = \mathbf{0}$) from an essentially balanced at rest position, determine the speed at which the corner A strikes the pad at B. The stone does not slip at its corner C as it falls.

Prob. R2–46

R2–47. The cylinder having a mass of 5 kg is initially at rest when it is placed into contact with the wall B and the rotor at A. If the rotor always maintains a constant clockwise angular velocity $\omega = 6$ rad/s, determine the initial angular acceleration of the cylinder. The coefficient of kinetic friction at the contacting surfaces B and C is $\mu_k = 0.2$.

Prob. R2–47

*R2–48. A concrete solid is formed by rotating the shaded area about the y axis. Determine the moment of inertia I_y. The density of material is $\rho = 150$ lb/ft³.

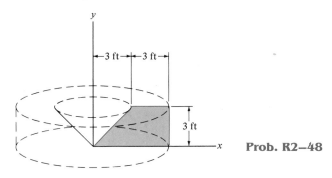

Prob. R2–48

R2–49. A uniform rod having a weight of 10 lb is pin-supported at A from a roller which rides on a horizontal track. If the rod is originally at rest, and a horizontal force of $F = 15$ lb is applied to the roller, determine the acceleration of the roller. Neglect the mass of the roller and its size d in the computations.

R2–50. Solve Prob. R2–49 assuming that the roller at A is replaced by a slider block having a negligible mass. The coefficient of kinetic friction between the block and the track is $\mu_k = 0.2$. Neglect the dimension d and the size of the block in the computations.

Probs. R2–49/R2–50

20

Three-Dimensional Kinematics of a Rigid Body

Three types of planar rigid-body motion have been presented in Chapter 16: translation, rotation about a fixed axis, and general plane motion. In this chapter the three-dimensional motion of a rigid body, which consists of rotation about a fixed point and general motion, will be discussed. This is followed by a general study of the motion of particles and rigid bodies using coordinate systems which both translate and rotate. The analysis of these motions is more complex than planar-motion analysis since the angular acceleration of the body will measure a change in both the *magnitude* and *direction* of the body's angular velocity. In order to simplify the motion's three-dimensional aspects, throughout the chapter we will make complete use of vector analysis.*

*20.1 Rotation About a Fixed Point

When a rigid body rotates about a fixed point, the distance r from the point to a particle P located in the body is the *same* for *any position* of the body. Thus, the path of motion for the particle lies on the *surface of a sphere* having a radius r and centered at the fixed point. Since motion along this path occurs only from a series of rotations made during a finite time interval, we will first develop a familiarity with some of the properties of rotational displacements.

*A brief review of vector analysis is given in Appendix C.

Euler's Theorem

This theorem states that two "component" rotations about different axes passing through a point are equivalent to a single resultant rotation about an axis passing through the point. If more than two rotations are applied, they can be combined into pairs, and each pair can be further reduced to combine into one rotation.

Finite Rotations

If the component rotations used in Euler's theorem are *finite,* it is important that the *order* in which they are applied be maintained. This is because finite rotations do *not* obey the law of vector addition, and hence they cannot be classified as vector quantities. To show this, consider the two finite rotations $\theta_1 + \theta_2$ applied to the block in Fig. 20–1a. Each rotation has a magnitude of 90° and a direction defined by the right-hand rule, as indicated by the arrow. The resultant orientation of the block is shown at the right. When these two rotations are applied in the order $\theta_2 + \theta_1$, as shown in Fig. 20–1b, the resultant position of the block is *not* the same as it is in Fig. 20–1a. Consequently, *finite rotations* do not obey the commutative law of addition ($\theta_1 + \theta_2 \neq \theta_2 + \theta_1$), and therefore *they cannot be classified as vectors.* If smaller, yet finite, rotations had been used to illustrate this point, e.g., 10° instead of 90°, the *resultant* orientation of the block after each combination of rotations would also be different; however, in this case, only by a small amount.

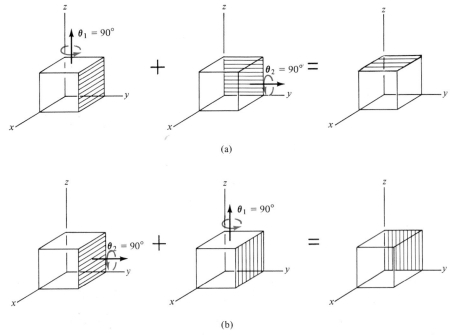

Fig. 20–1

412

Infinitesimal Rotations

When defining the angular motions of a body subjected to three-dimensional motion, only rotations which are *infinitesimally small* will be considered. *Such rotations may be classified as vectors, since they can be added vectorially in any manner.* For example, let us for purposes of simplicity consider the rigid body itself to be a sphere which is allowed to rotate about its central fixed point O, Fig. 20–2a. If we impose two infinitesimal rotations $d\boldsymbol{\theta}_1 + d\boldsymbol{\theta}_2$ on the body, it is seen that point P moves along the path $d\boldsymbol{\theta}_1 \times \mathbf{r} + d\boldsymbol{\theta}_2 \times \mathbf{r}$ and ends up at P'. Had the two successive rotations occurred in the order $d\boldsymbol{\theta}_2 + d\boldsymbol{\theta}_1$, then the resultant displacements of P would have been $d\boldsymbol{\theta}_2 \times \mathbf{r} + d\boldsymbol{\theta}_1 \times \mathbf{r}$. Since the vector cross product obeys the distributive law, by comparison $(d\boldsymbol{\theta}_1 + d\boldsymbol{\theta}_2) \times \mathbf{r} = (d\boldsymbol{\theta}_2 + d\boldsymbol{\theta}_1) \times \mathbf{r}$. Hence infinitesimal rotations $d\boldsymbol{\theta}$ are vectors, since these quantities have both a magnitude and direction for which the order of (vector) addition is not important, i.e., $d\boldsymbol{\theta}_1 + d\boldsymbol{\theta}_2 = d\boldsymbol{\theta}_2 + d\boldsymbol{\theta}_1$. Furthermore, as shown in Fig. 20–2a, the two "component" rotations $d\boldsymbol{\theta}_1$ and $d\boldsymbol{\theta}_2$ are equivalent to a single resultant rotation $d\boldsymbol{\theta} = d\boldsymbol{\theta}_1 + d\boldsymbol{\theta}_2$, a consequence of Euler's theorem.

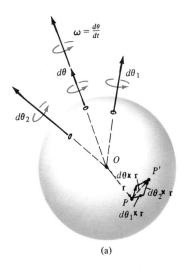

(a)

Angular Velocity

If the body is subjected to an angular rotation $d\boldsymbol{\theta}$ about a fixed point, the angular velocity of the body is defined by the time derivative,

$$\boldsymbol{\omega} = \dot{\boldsymbol{\theta}} \qquad (20\text{–}1)$$

The line specifying the direction of $\boldsymbol{\omega}$, which is collinear with $d\boldsymbol{\theta}$, is referred to as the *instantaneous axis of rotation*, Fig. 20–2b. In general, this axis changes direction during each instant of time. Since $\boldsymbol{\omega}$ is a vector quantity, it follows from vector addition that if the body is subjected to two component angular motions, $\boldsymbol{\omega}_1 = \dot{\boldsymbol{\theta}}_1$ and $\boldsymbol{\omega}_2 = \dot{\boldsymbol{\theta}}_2$, the resultant angular velocity is $\boldsymbol{\omega} = \boldsymbol{\omega}_1 + \boldsymbol{\omega}_2$.

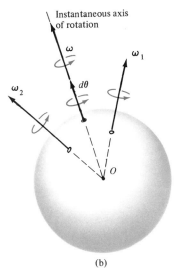

(b)

Fig. 20–2

Angular Acceleration

The body's angular acceleration is determined from the time derivative of the angular velocity, i.e.,

$$\boldsymbol{\alpha} = \dot{\boldsymbol{\omega}} \qquad (20\text{–}2)$$

For motion about a fixed point, $\boldsymbol{\alpha}$ must account for a change in *both* the magnitude and direction of $\boldsymbol{\omega}$, so that, in general, $\boldsymbol{\alpha}$ is not directed along the instantaneous axis of rotation, Fig. 20–3.

As the direction of the instantaneous axis of rotation (or the line of action of $\boldsymbol{\omega}$) changes in space, the locus of points defined by the axis generates a fixed *space cone*. If the change in this axis is viewed with respect to the rotating

Fig. 20–3

413

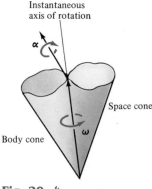

Instantaneous
axis of rotation

Space cone

Body cone

Fig. 20—4

body, the locus of the axis generates a *body cone,* Fig. 20–4. At any given instant, these cones are tangent along the instantaneous axis of rotation, and when the body is in motion, the body cone appears to roll either on the inside or the outside surface of the fixed space cone. Provided the paths defined by the open ends of the cones are described by the head of the $\boldsymbol{\omega}$ vector, $\boldsymbol{\alpha}$ must act tangent to these paths at any given instant, since the time rate of change of $\boldsymbol{\omega}$ is equal to $\boldsymbol{\alpha}$, Fig. 20–4.

Velocity

Once $\boldsymbol{\omega}$ is specified, the velocity of any point P on a body rotating about a fixed point can be determined using the same methods as for a body rotating about a fixed axis (Sec. 16.3). Hence, by the cross product,

$$\boxed{\mathbf{v} = \boldsymbol{\omega} \times \mathbf{r}} \tag{20–3}$$

Here $\mathbf{r}$ defines the position of P measured from the fixed point O, Fig. 20–3.

Acceleration

If $\boldsymbol{\omega}$ and $\boldsymbol{\alpha}$ are known at a given instant, the acceleration of any point P on the body can be obtained by time differentiation of Eq. 20–3, which yields

$$\boxed{\mathbf{a} = \boldsymbol{\alpha} \times \mathbf{r} + \boldsymbol{\omega} \times (\boldsymbol{\omega} \times \mathbf{r})} \tag{20–4}$$

The form of this equation is the same as that developed in Sec. 16.3, which defines the acceleration of a point located on a body subjected to rotation about a fixed axis.

*20.2 The Time Derivative of a Vector Measured from a Fixed and Translating-Rotating System

In many types of problems involving the motion of a body about a fixed point, the angular velocity $\boldsymbol{\omega}$ is specified in terms of its component angular motions. For example, the disk in Fig. 20–5 spins about the horizontal y axis at $\boldsymbol{\omega}_s$ while it rotates or precesses about the vertical z axis at $\boldsymbol{\omega}_p$. Therefore, its resultant angular velocity is $\boldsymbol{\omega} = \boldsymbol{\omega}_s + \boldsymbol{\omega}_p$. If the angular acceleration $\boldsymbol{\alpha}$ of such a body is to be determined, it is sometimes easier to compute the time derivative of $\boldsymbol{\omega}$, Eq. 20–2, by using a coordinate system which has a *rotation* defined by one or more of the components of $\boldsymbol{\omega}$.* For this reason, and for other uses later, an equation will presently be derived that relates the time

---

*In the case of the spinning disk, Fig. 20–5, the x, y, z axes may be given an angular velocity of $\boldsymbol{\omega}_p$.

derivative of any vector **A** defined from a translating-rotating reference to its derivative defined from a fixed reference.

Consider the x, y, z axes of the moving frame of reference to have an angular velocity $\boldsymbol{\Omega}$ which is measured from the fixed X, Y, Z axes, Fig. 20–6a. In the following discussion, it will be convenient to express vector **A** in terms of its **i**, **j**, **k** components, which define the directions of the moving axes. Hence,

$$\mathbf{A} = A_x\mathbf{i} + A_y\mathbf{j} + A_z\mathbf{k}$$

In general, the time derivative of **A** must account for the change in both the vector's magnitude and direction. However, if this derivative is taken *with respect to the moving frame of reference*, only a change in the magnitudes of the components of **A** must be accounted for, since the directions of the components do not change with respect to the moving reference. Hence,

$$(\dot{\mathbf{A}})_{xyz} = \dot{A}_x\mathbf{i} + \dot{A}_y\mathbf{j} + \dot{A}_z\mathbf{k} \qquad (20\text{–}5)$$

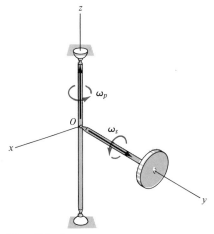

Fig. 20–5

When the time derivative of **A** is taken *with respect to the fixed frame of reference*, the *directions* of **i**, **j**, and **k** change only on account of the rotation $\boldsymbol{\Omega}$ of the axes and not their translation. Hence, in general,

$$(\dot{\mathbf{A}})_{XYZ} = \dot{A}_x\mathbf{i} + \dot{A}_y\mathbf{j} + \dot{A}_z\mathbf{k} + A_x\dot{\mathbf{i}} + A_y\dot{\mathbf{j}} + A_z\dot{\mathbf{k}}$$

The time derivatives of the unit vectors will now be considered. For example, $\dot{\mathbf{i}} = d\mathbf{i}/dt$ represents only a change in the direction of **i** with respect to time, since **i** has a fixed magnitude of 1 unit. As shown in Fig. 20–6b, the change, $d\mathbf{i}$, is *tangent to the path* described by the head of **i** as **i** moves due to the rotation $\boldsymbol{\Omega}$. Accounting for both the magnitude and direction of $d\mathbf{i}$, we can therefore define $\dot{\mathbf{i}}$ using the cross product, $\dot{\mathbf{i}} = \boldsymbol{\Omega} \times \mathbf{i}$. In general,

$$\dot{\mathbf{i}} = \boldsymbol{\Omega} \times \mathbf{i} \qquad \dot{\mathbf{j}} = \boldsymbol{\Omega} \times \mathbf{j} \qquad \dot{\mathbf{k}} = \boldsymbol{\Omega} \times \mathbf{k}$$

These formulations were also developed in Sec. 16.8, regarding planar motion of the axes. Substituting the results into the above equation and using Eq. 20–5 yields

$$\boxed{(\dot{\mathbf{A}})_{XYZ} = (\dot{\mathbf{A}})_{xyz} + \boldsymbol{\Omega} \times \mathbf{A}} \qquad (20\text{–}6)$$

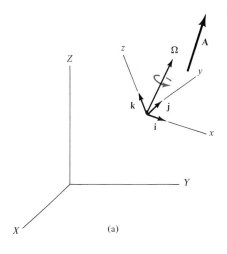

(a)

This result is rather important and it will be used throughout Sec. 20.4 and Chapter 21. In words, it states that *the time derivative of **A** as observed from the fixed X, Y, Z frame of reference is equal to the time rate of change of **A** as observed from the x, y, z translating-rotating frame of reference, Eq. 20–5, plus $\boldsymbol{\Omega} \times \mathbf{A}$, the change of **A** caused by the rotation of the x, y, z frame.*

The following two example problems numerically illustrate the use of this equation for obtaining the angular acceleration of a body rotating about a fixed point.

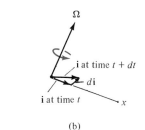

(b)

Fig. 20–6

Example 20–1

The disk shown in Fig. 20–7 is spinning about its horizontal axis with a constant angular velocity $\omega_s = 3$ rad/s, while the horizontal platform upon which the disk is mounted is rotating about the vertical axis at a constant rate $\omega_p = 1$ rad/s. Determine the angular acceleration of the disk and the velocity and acceleration of point A on the disk when it is in the position shown.

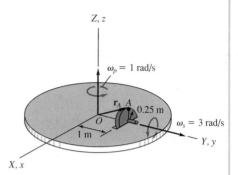

SOLUTION

Point O represents a fixed point of rotation for the disk if one considers a hypothetical extension of the disk to this point. To determine the velocity and acceleration of point A, it is first necessary to determine the resultant angular velocity $\boldsymbol{\omega}$ and angular acceleration $\boldsymbol{\alpha}$ of the disk, since these vectors are used in Eqs. 20–3 and 20–4.

Fig. 20–7

Angular Velocity. The angular velocity is simply the vector addition of the two component motions. Thus,

$$\boldsymbol{\omega} = \boldsymbol{\omega}_s + \boldsymbol{\omega}_p = \{3\mathbf{j} + 1\mathbf{k}\} \text{ rad/s}$$

Angular Acceleration. Since the magnitude of $\boldsymbol{\omega}$ is constant, only a change in its direction, as seen from a fixed reference, creates the angular acceleration $\boldsymbol{\alpha}$ of the disk. One way to obtain $\boldsymbol{\alpha}$ is to compute the time derivative of *each of the two components* of $\boldsymbol{\omega}$ using Eq. 20–6. At the instant shown in Fig. 20–7, imagine the fixed X, Y, Z and a rotating x, y, z frame to be in coincidence. If the rotating x, y, z frame is chosen to have an angular velocity of $\boldsymbol{\Omega} = \boldsymbol{\omega}_p = \{1\mathbf{k}\}$ rad/s, then $\boldsymbol{\omega}_s$ will *always* be directed along the y (not Y) axis, and the time rate of change of $\boldsymbol{\omega}_s$ as seen from x, y, z is zero; i.e., $(\dot{\boldsymbol{\omega}}_s)_{xyz} = \mathbf{0}$ (the magnitude of $\boldsymbol{\omega}_s$ is constant). Consequently,

$$(\dot{\boldsymbol{\omega}}_s)_{XYZ} = (\dot{\boldsymbol{\omega}}_s)_{xyz} + \boldsymbol{\omega}_p \times \boldsymbol{\omega}_s = \mathbf{0} + (1\mathbf{k}) \times (3\mathbf{j}) = \{-3\mathbf{i}\} \text{ rad/s}^2$$

By the same choice of axes rotation, $\boldsymbol{\Omega} = \boldsymbol{\omega}_p$, or even with $\boldsymbol{\Omega} = \mathbf{0}$, the time derivative $(\dot{\boldsymbol{\omega}}_p)_{xyz} = \mathbf{0}$, since $\boldsymbol{\omega}_p$ is *always* directed along the z (or Z) axis and has a constant magnitude. Hence,

$$(\dot{\boldsymbol{\omega}}_p)_{XYZ} = (\dot{\boldsymbol{\omega}}_p)_{xyz} + \boldsymbol{\omega}_p \times \boldsymbol{\omega}_p = \mathbf{0} + \mathbf{0} = \mathbf{0}$$

The angular acceleration is therefore

$$\boldsymbol{\alpha} = (\dot{\boldsymbol{\omega}})_{XYZ} = (\dot{\boldsymbol{\omega}}_s)_{XYZ} + (\dot{\boldsymbol{\omega}}_p)_{XYZ} = \{-3\mathbf{i}\} \text{ rad/s}^2 \qquad \textit{Ans.}$$

Velocity and Acceleration. Since $\boldsymbol{\omega}$ and $\boldsymbol{\alpha}$ have been determined, the velocity and acceleration of point A can be computed using Eqs. 20–3 and 20–4. Realizing that $\mathbf{r}_A = \{1\mathbf{j} + 0.25\mathbf{k}\}$ m, Fig. 20–7, we have

$$\mathbf{v}_A = \boldsymbol{\omega} \times \mathbf{r}_A = (3\mathbf{j} + 1\mathbf{k}) \times (1\mathbf{j} + 0.25\mathbf{k}) = \{-0.25\mathbf{i}\} \text{ m/s} \qquad \textit{Ans.}$$

$$\mathbf{a}_A = \boldsymbol{\alpha} \times \mathbf{r}_A + \boldsymbol{\omega} \times (\boldsymbol{\omega} \times \mathbf{r}_A)$$
$$= (-3\mathbf{i}) \times (1\mathbf{j} + 0.25\mathbf{k}) + (3\mathbf{j} + 1\mathbf{k}) \times [(3\mathbf{j} + 1\mathbf{k}) \times (1\mathbf{j} + 0.25\mathbf{k})]$$
$$= \{0.5\mathbf{j} - 2.25\mathbf{k}\} \text{ m/s}^2 \qquad \textit{Ans.}$$

Example 20–2

At the instant $\theta = 60°$, the gyrotop in Fig. 20–8 has three components of angular motion directed as shown and having magnitudes which are defined as follows:

spin: $\omega_s = 10$ rad/s, increasing at the rate of 6 rad/s^2
nutation: $\omega_n = 3$ rad/s, increasing at the rate of 2 rad/s^2
precession: $\omega_p = 5$ rad/s, increasing at the rate of 4 rad/s^2

Determine the angular velocity and angular acceleration of the top.

SOLUTION

Angular Velocity. The top is rotating about the fixed point O. If the fixed and rotating frames are considered to be coincident at the instant shown, then the angular velocity can be expressed in terms of i, j, k components, appropriate to the x, y, z frame, i.e.,

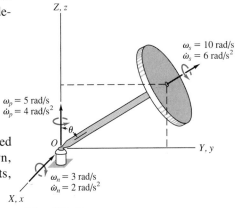

Fig. 20–8

$$\boldsymbol{\omega} = -\omega_n \mathbf{i} + \omega_s \sin \theta \mathbf{j} + (\omega_p + \omega_s \cos \theta)\mathbf{k}$$
$$= -3\mathbf{i} + 10 \sin 60°\mathbf{j} + (5 + 10 \cos 60°)\mathbf{k}$$
$$= \{-3\mathbf{i} + 8.66\mathbf{j} + 10\mathbf{k}\} \text{ rad/s} \qquad Ans.$$

Angular Acceleration. As in the solution of Example 20–1, the angular acceleration $\boldsymbol{\alpha}$ will be determined by investigating separately the time rate of change of *each of the angular velocity components* as observed from the fixed reference X, Y, Z. *This is done by using Eq. 20–6 and choosing an Ω for the x, y, z reference so that the component of $\boldsymbol{\omega}$ which is being considered is viewed as having a constant direction when observed from x, y, z.*

Careful examination of the motion of the top reveals that $\boldsymbol{\omega}_s$ has a *constant direction* relative to x, y, z if these axes rotate at $\Omega = \omega_n + \omega_p$. Since $\boldsymbol{\omega}_n$ always lies in the fixed X-Y plane, this vector has a *constant direction* if the motion is viewed from axes x, y, z having a rotation of $\Omega = \omega_p$ (not $\Omega = \omega_s + \omega_p$). Finally, the component $\boldsymbol{\omega}_p$ is *always directed* along the Z axis so that here it is not necessary to think of x, y, z as rotating, i.e., $\Omega = 0$. Expressing the data in terms of the i, j, k components, we therefore have

For ω_s, $\Omega = \omega_n + \omega_p$;

$$(\dot{\boldsymbol{\omega}}_s)_{XYZ} = (\dot{\boldsymbol{\omega}}_s)_{xyz} + (\omega_n + \omega_p) \times \omega_s$$
$$= (6 \sin 60°\mathbf{j} + 6 \cos 60°\mathbf{k}) + (-3\mathbf{i} + 5\mathbf{k}) \times (10 \sin 60°\mathbf{j} + 10 \cos 60°\mathbf{k})$$
$$= \{-43.30\mathbf{i} + 20.20\mathbf{j} - 22.98\mathbf{k}\} \text{ rad/s}^2$$

For ω_n, $\Omega = \omega_p$;

$$(\dot{\boldsymbol{\omega}}_n)_{XYZ} = (\dot{\boldsymbol{\omega}}_n)_{xyz} + \omega_p \times \omega_n = -2\mathbf{i} + (5\mathbf{k}) \times (-3\mathbf{i}) \doteq \{-2\mathbf{i} - 15\mathbf{j}\} \text{ rad/s}^2$$

For ω_p, $\Omega = 0$;

$$(\dot{\boldsymbol{\omega}}_p)_{XYZ} = (\dot{\boldsymbol{\omega}}_p)_{xyz} + \mathbf{0} \times \omega_p = \{4\mathbf{k}\} \text{ rad/s}^2$$

Thus, the angular acceleration of the top is

$$\boldsymbol{\alpha} = (\dot{\boldsymbol{\omega}}_s)_{XYZ} + (\dot{\boldsymbol{\omega}}_n)_{XYZ} + (\dot{\boldsymbol{\omega}}_p)_{XYZ}$$
$$= \{-45.3\mathbf{i} + 5.20\mathbf{j} - 19.0\mathbf{k}\} \text{ rad/s}^2 \qquad Ans.$$

*20.3 General Motion

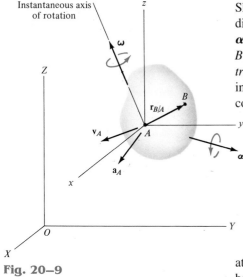

Instantaneous axis of rotation

Fig. 20–9

Shown in Fig. 20–9 is a rigid body subjected to general motion in three dimensions for which the angular velocity is $\boldsymbol{\omega}$ and the angular acceleration is $\boldsymbol{\alpha}$. If point A has a known motion of $\mathbf{v}_A$ and $\mathbf{a}_A$, the motion of any other point B may be determined by using a relative-motion analysis. In this section a *translating coordinate system* will be used to define the relative motion, and in the next section a reference that is both rotating and translating will be considered.

If the origin of the translating coordinate system x, y, z ($\boldsymbol{\Omega} = \mathbf{0}$) is located at the "base point" A, then, at the instant shown, the motion of the body may be regarded as the sum of an instantaneous translation of the body having a motion of $\mathbf{v}_A$ and $\mathbf{a}_A$ and a rotation of the body about an instantaneous axis passing through the base point. Since the body is rigid, the motion of point B measured by an observer located at A is the same as *motion of the body about a fixed point*. This relative motion occurs about the instantaneous axis of rotation and is defined by $\mathbf{v}_{B/A} = \boldsymbol{\omega} \times \mathbf{r}_{B/A}$, Eq. 20–3, and $\mathbf{a}_{B/A} = \boldsymbol{\alpha} \times \mathbf{r}_{B/A} + \boldsymbol{\omega} \times (\boldsymbol{\omega} \times \mathbf{r}_{B/A})$, Eq. 20–4. For translating axes the relative motions are related to absolute motions by $\mathbf{v}_B = \mathbf{v}_A + \mathbf{v}_{B/A}$ and $\mathbf{a}_B = \mathbf{a}_A + \mathbf{a}_{B/A}$, Eqs. 16–14 and 16–16, so that the absolute velocity and acceleration of point B can be determined from the equations

$$\mathbf{v}_B = \mathbf{v}_A + \boldsymbol{\omega} \times \mathbf{r}_{B/A} \tag{20–7}$$

and

$$\mathbf{a}_B = \mathbf{a}_A + \boldsymbol{\alpha} \times \mathbf{r}_{B/A} + \boldsymbol{\omega} \times (\boldsymbol{\omega} \times \mathbf{r}_{B/A}) \tag{20–8}$$

These two equations are identical to those describing the general plane motion of a rigid body, Eqs. 16–15 and 16–17. However, difficulty in application arises for three-dimensional motion, because $\boldsymbol{\alpha}$ measures the change in *both* the magnitude and direction of $\boldsymbol{\omega}$. (Recall that for general plane motion $\boldsymbol{\alpha}$ and $\boldsymbol{\omega}$ are always parallel or perpendicular to the plane of motion, and therefore $\boldsymbol{\alpha}$ measures only a change in the magnitude of $\boldsymbol{\omega}$.) In some problems the constraints or connections of a body will require that the directions of the angular motions or displacement paths of points on the body be defined. As illustrated in the following example, this information is useful for obtaining some of the terms in the above equations.

Example 20–3

One end of the rigid bar CD shown in Fig. 20–10a slides along the grooved wall slot AB, and the other end slides along the vertical member EF. If the collar at C is moving toward B at a speed of 3 m/s, determine the velocity of the collar at D and the angular velocity of the bar at the instant shown. The bar is connected to the collars at its end points by ball-and-socket joints.

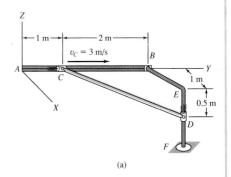

(a)

SOLUTION

Bar CD is subjected to general motion. Why? The velocity of point D on the bar may be related to the velocity of point C by the equation

$$\mathbf{v}_D = \mathbf{v}_C + \boldsymbol{\omega} \times \mathbf{r}_{D/C}$$

The fixed and translating frames of reference are assumed to coincide at the instant considered, Fig. 20–10b. We have

$$\mathbf{v}_D = -v_D\mathbf{k} \qquad \mathbf{v}_C = \{3\mathbf{j}\} \text{ m/s}$$
$$\mathbf{r}_{D/C} = \{1\mathbf{i} + 2\mathbf{j} - 0.5\mathbf{k}\} \text{ m} \qquad \boldsymbol{\omega} = \omega_x\mathbf{i} + \omega_y\mathbf{j} + \omega_z\mathbf{k}$$

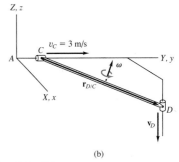

(b)

Fig. 20–10

Substituting these quantities into the above equation gives

$$-v_D\mathbf{k} = 3\mathbf{j} + \begin{vmatrix} \mathbf{i} & \mathbf{j} & \mathbf{k} \\ \omega_x & \omega_y & \omega_z \\ 1 & 2 & -0.5 \end{vmatrix}$$

Expanding and equating the respective $\mathbf{i}, \mathbf{j}, \mathbf{k}$ components yields

$$-0.5\omega_y - 2\omega_z = 0 \qquad (1)$$
$$0.5\omega_x + 1\omega_z + 3 = 0 \qquad (2)$$
$$2\omega_x - 1\omega_y + v_D = 0 \qquad (3)$$

These equations contain four unknowns.* A fourth equation can be written if the direction of $\boldsymbol{\omega}$ is specified. In particular, any component of $\boldsymbol{\omega}$ acting along the bar's axis has no effect on moving the collars. This is because the bar is *free to rotate* about its axis. Therefore, if $\boldsymbol{\omega}$ is specified as acting *perpendicular* to the axis of the bar, then $\boldsymbol{\omega}$ must have a unique magnitude to satisfy the above equations. Perpendicularity is guaranteed provided the dot product of $\boldsymbol{\omega}$ and $\mathbf{r}_{D/C}$ is zero (see Eq. C–14). Hence,

$$\boldsymbol{\omega} \cdot \mathbf{r}_{D/C} = (\omega_x\mathbf{i} + \omega_y\mathbf{j} + \omega_z\mathbf{k}) \cdot (1\mathbf{i} + 2\mathbf{j} - 0.5\mathbf{k}) = 0$$
$$1\omega_x + 2\omega_y - 0.5\omega_z = 0 \qquad (4)$$

Solving Eqs. (1) to (4) simultaneously yields

$$\omega_x = -4.86 \text{ rad/s} \quad \omega_y = 2.29 \text{ rad/s} \quad \omega_z = -0.571 \text{ rad/s} \quad \textit{Ans.}$$
$$v_D = 12.0 \text{ m/s} \downarrow \qquad \textit{Ans.}$$

*Although this is the case the magnitude of $\mathbf{v}_D$ can be obtained. For example, solve Eqs. (1) and (2) for ω_y and ω_x in terms of ω_z and substitute into Eq. (3). It will be noted that ω_z will *cancel out*, which will allow a solution for v_D.

PROBLEMS

20–1. The motion of the top is such that at the instant shown it is rotating about the z axis at $\omega_1 = 0.6$ rad/s, while it is spinning at $\omega_2 = 8$ rad/s. Determine the angular velocity and angular acceleration of the top at this instant. Express the result as a Cartesian vector.

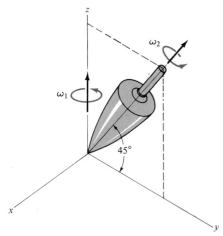

Prob. 20–1

20–2. The ladder of the fire truck rotates about the z axis with an angular velocity $\omega_1 = 0.15$ rad/s, which is increasing at 0.8 rad/s^2. At the same instant it is rotating upward at a constant rate $\omega_2 = 0.6$ rad/s. Compute the velocity and acceleration of point A located at the top of the ladder at this instant.

Prob. 20–2

20–3. Gear A is *fixed* while gear B is free to rotate on the shaft S. If the shaft is turning about the z axis at $\omega_z = 5$ rad/s, which is increasing at 2 rad/s^2, determine the velocity and acceleration of point P at the instant shown. The face of gear B lies in a vertical plane.

Prob. 20–3

*20–4. At the instant shown, the radar dish is rotating about the z axis at $\omega_1 = 3$ rad/s, which is increasing at 4 rad/s^2. Also, at this instant the angle of tilt $\phi = 30°$, and $\dot{\phi} = 2$ rad/s, $\ddot{\phi} = 6$ rad/s^2. Determine the angular velocity and angular acceleration of the dish at this instant.

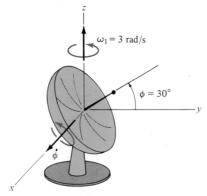

Prob. 20–4

20–5. The drill pipe P turns at a constant angular rate $\omega_P = 4$ rad/s. Determine the angular velocity and angular acceleration of the conical rock bit, which rolls without slipping. Also, what are the velocity and acceleration of point A? Establish x outward, y to the right, and z upwards.

Prob. 20–5

20–6. The tower crane is rotating about the z axis with a constant rate $\omega_1 = 0.25$ rad/s, while the boom OA is rotating downward with a constant rate $\omega_2 = 0.4$ rad/s. Compute the velocity and acceleration of point A located at the top of the boom at the instant shown.

20–7. At the instant shown, the tower crane is rotating about the z axis with an angular velocity $\omega_1 = 0.25$ rad/s, which is increasing at 0.6 rad/s^2. The boom OA is rotating downward with an angular velocity $\omega_2 = 0.4$ rad/s, which is increasing at 0.8 rad/s^2. Compute the velocity and acceleration of point A located at the top of the boom at this instant.

*20–8. The truncated cone rotates about the z axis at a constant rate $\omega_z = 0.4$ rad/s without slipping on the horizontal plane. Determine the velocity and acceleration of point A on the cone.

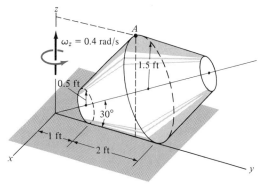

Probs. 20–8/20–9

20–9. The truncated cone rotates about the z axis at $\omega_z = 0.4$ rad/s without slipping on the horizontal plane. If at this same instant ω_z is increasing at $\dot{\omega}_z = 0.5$ rad/s^2, determine the velocity and acceleration of point A on the cone.

20–10. The cone A rolls without slipping such that at the instant shown, $\omega = 4$ rad/s and $\alpha = 3$ rad/s^2. Determine the velocity and acceleration of point A at this instant.

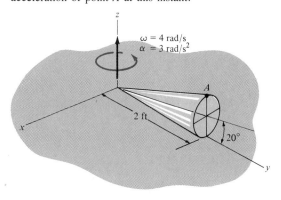

Probs. 20–6/20–7   Prob. 20–10

20–11. The differential of an automobile allows the two rear wheels to rotate at different speeds when the automobile travels along a curve. For operation, the rear axles are attached to the wheels at one end and have beveled gears A and B on their other ends. The differential case D is placed over the left axle, but can rotate about C, independent of the axle. The case supports a pinion gear E on a shaft, which meshes with gears A and B. Finally, a ring gear G is *fixed* to the differential case so that the case rotates with the ring gear when the latter is driven by the drive pinion H. This gear, like the differential case, is free to rotate about the left wheel axle. If the drive pinion is turning at $\omega_H = 100$ rad/s and the pinion gear E is spinning about its shaft at $\omega_E = 30$ rad/s, compute the angular velocity, ω_A and ω_B, of each axle.

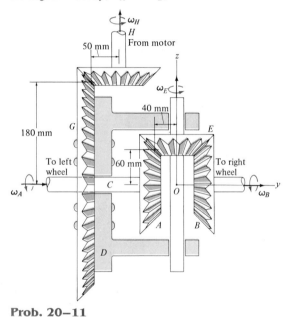

Prob. 20–11

*20–12. Gear B is connected to the rotating shaft, while the plate gear A is fixed. If the shaft is turning at a constant rate of $\omega_z = 10$ rad/s about the z axis, determine the magnitudes of the angular velocity and the angular acceleration of gear B. Also, determine the magnitudes of the velocity and acceleration of point P.

Prob. 20–12

20–13. If the rod is attached to smooth collars A and B at its end points by ball-and-socket joints, determine the speed of B if A is moving downward at a constant speed of 8 ft/s. Also, determine the angular velocity of the rod if it is directed perpendicular to the axis of the rod.

Prob. 20–13

20–14. The rod *AB* is attached to collars at its ends by ball-and-socket joints. If collar *A* has a speed $v_A = 20$ ft/s at the instant shown, determine the speed of collar *B*.

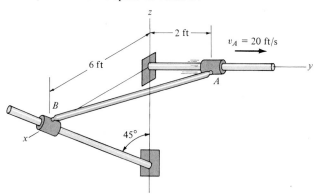

Prob. 20–14

20–15. The rod is attached to smooth collars *A* and *B* at its ends using ball-and-socket joints. Determine the speed of *B* if *A* is moving at $v_A = 6$ ft/s. Also, determine the angular velocity of the rod if it is directed perpendicular to the axis of the rod.

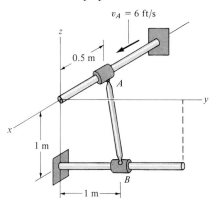

Prob. 20–15

*20–16. Rod *AB* is attached to collars at its ends by using ball-and-socket joints. If collar *A* moves along the fixed rod with a velocity of $v_A = 8$ ft/s, determine the angular velocity of the rod and the velocity of collar *B* at the instant shown. Assume that the rod's angular velocity is directed perpendicular to the axis of the rod.

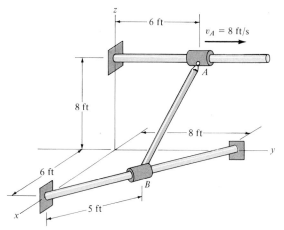

Probs. 20–16/20–17

20–17. Rod *AB* is attached to collars at its ends by using ball-and-socket joints. If collar *A* moves along the fixed rod with a velocity of $v_A = 8$ ft/s and has an acceleration of $a_A = 4$ ft/s^2 at the instant shown, determine the angular acceleration of the rod and the acceleration of collar *B* at this instant. Assume that the rod's angular velocity and angular acceleration are directed perpendicular to the axis of the rod.

20–18. Rod *AB* is attached to a disk and a collar by ball-and-socket joints. If the disk is rotating at a constant angular velocity of 2 rad/s, determine the velocity and acceleration of the collar at *A* at the instant shown. Assume that the angular velocity of the rod is directed perpendicular to the axis of the rod.

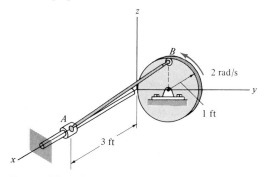

Prob. 20–18

423

20–19. The triangular plate ABC is supported at A by a ball-and-socket and at point C by the x-z plane. The side AB lies in the x-y plane. At the instant $\theta = 60°$, $\dot\theta = 2$ rad/s and point C has the coordinates shown. Determine the angular velocity of the plate and the velocity of point C at this instant.

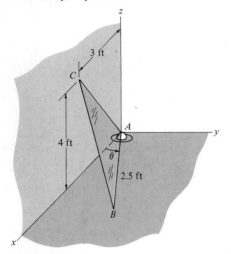

Prob. 20–19

*20–20. The rod ABC is supported at A by a ball-and-socket and at point C by the y-z plane. The segment AB lies in the x-y plane. At the instant $\theta = 60°$, $\dot\theta = 2$ rad/s, and point C has the coordinates shown. Determine the angular velocity of the rod and the velocity of point C at this instant.

Prob. 20–20

20–21. The rod AB is attached to collars at its ends by ball-and-socket joints. If collar A has a velocity of $v_A = 5$ ft/s, determine the angular velocity of the rod and the velocity of collar B at the instant shown. Assume the angular velocity of the rod is directed perpendicular to the rod.

Probs. 20–21/20–22

20–22. The rod AB is attached to the collars at its ends by ball-and-socket joints. If collar A has an acceleration $\mathbf{a}_A = \{6\mathbf{i}\}$ ft/s², and a velocity $\mathbf{v}_A = \{5\mathbf{i}\}$ ft/s, determine the angular acceleration of the rod and the acceleration of collar B at the instant shown. Assume the angular velocity and angular acceleration of the rod are both directed perpendicular to the rod.

20–23. Rod AB is attached to the rotating arm using ball-and-socket joints. If AC is rotating with a constant angular velocity of 8 rad/s about the pin at C, determine the angular velocity of link BD at the instant shown.

Probs. 20–23/20–24

*20–24. Rod *AB* is attached to the rotating arm using ball-and-socket joints. If *AC* is rotating about point *C* with an angular velocity of 8 rad/s and has an angular acceleration of 6 rad/s² at the instant shown, determine the angular velocity and angular acceleration of link *BD* at this instant.

20–25. Disk *A* is rotating at a constant angular velocity of 10 rad/s. If rod *BC* is joined to the disk and a collar by ball-and-socket joints, determine the velocity of collar *B* at the instant shown, and the rod's angular velocity if ω_{BC} is directed perpendicular to the axis of the rod.

Probs. 20–25/20–26

20–26. If the disk in Prob. 20–25 has an angular acceleration $\alpha = 5$ rad/s², determine the acceleration of the collar *B* at the instant shown, when $\omega = 10$ rad/s. Also, what is the rod's angular acceleration α_{BC} if it is directed perpendicular to the axis of the rod?

20–27. Solve Prob. 20–25 if the connection at *B* consists of a pin as shown in the figure below, rather than a ball-and-socket joint. *Hint:* The constraint allows rotation of the rod both along bar *DE* (**j** direction) and along the axis of the pin (**n** direction). Since there is no rotational component in the **u** direction, i.e., perpendicular to **n** and **j** where **u** = **j** × **n,** an additional equation for solution can be obtained from $\boldsymbol{\omega} \cdot \mathbf{u} = 0$. The vector **n** is in the same direction as $\mathbf{r}_{B/C} \times \mathbf{r}_{D/C}$.

Prob. 20–27

*20–28. The rod assembly is supported at *B* by a ball-and-socket joint and at *A* by a clevis connection. If the collar at *B* moves in the *x*-*z* plane with a speed $v_B = 5$ ft/s, determine the velocity of points *A* and *C* on the rod assembly. *Hint:* See Prob. 20–27.

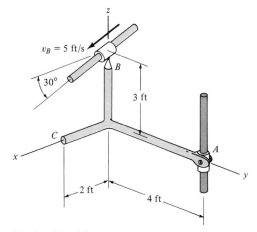

Prob. 20–28

*20.4 Relative-Motion Analysis Using Translating and Rotating Axes

The most general way to analyze the three-dimensional motion of a rigid body requires the use of a system of x, y, z axes which both translate and rotate relative to a second frame X, Y, Z. This analysis also provides a means for determining the motions of two points located on separate members of a mechanism, and for determining the relative motion of one particle with respect to another when one or both particles are moving along *rotating paths*. In this section two equations will be developed which relate the velocities and accelerations of two points A and B, of which one point moves relative to a frame of reference subjected to both translation and rotation. Because of the generality in the derivation, A and B may represent either two particles moving independently of one another or two points located in the same or different rigid bodies.

As shown in Fig. 20–11, the locations of points A and B are specified relative to the X, Y, Z frame of reference by position vectors $\mathbf{r}_A$ and $\mathbf{r}_B$. The base point A represents the origin of the x, y, z coordinate system, which is translating and rotating with respect to X, Y, Z. At the instant considered, the velocity and acceleration of point A are $\mathbf{v}_A$ and $\mathbf{a}_A$, respectively, and the angular velocity and angular acceleration of the x, y, z axes are $\mathbf{\Omega}$ and $\dot{\mathbf{\Omega}} = d\mathbf{\Omega}/dt$, respectively. All these vectors are *measured* with respect to the X, Y, Z frame of reference, although they may be expressed in Cartesian component form along either set of axes.

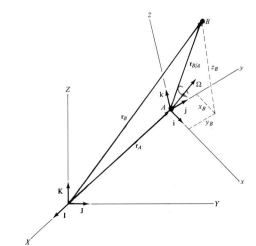

Fig. 20–11

Position

If the position of "B with respect to A" is specified by the *relative-position vector* $\mathbf{r}_{B/A}$, Fig. 20–11, then, by vector addition,

$$\mathbf{r}_B = \mathbf{r}_A + \mathbf{r}_{B/A} \qquad (20\text{–}9)$$

where

$$\mathbf{r}_B = \text{position of } B$$
$$\mathbf{r}_A = \text{position of the origin } A$$
$$\mathbf{r}_{B/A} = \text{relative position of ``}B \text{ with respect to } A\text{''}$$

Velocity

The velocity of point B measured from X, Y, Z is determined by taking the time derivative of Eq. 20–9, which yields

$$(\dot{\mathbf{r}}_B)_{XYZ} = (\dot{\mathbf{r}}_A)_{XYZ} + (\dot{\mathbf{r}}_{B/A})_{XYZ}$$

The first two terms represent $\mathbf{v}_B$ and $\mathbf{v}_A$. The last term is evaluated by applying Eq. 20–6, since $\mathbf{r}_{B/A}$ is measured between two points in a rotating reference. Hence,

$$(\dot{\mathbf{r}}_{B/A})_{XYZ} = (\dot{\mathbf{r}}_{B/A})_{xyz} + \boldsymbol{\Omega} \times \mathbf{r}_{B/A} = (\mathbf{v}_{B/A})_{\text{rel}} + \boldsymbol{\Omega} \times \mathbf{r}_{B/A} \quad (20\text{–}10)$$

Here $(\mathbf{v}_{B/A})_{\text{rel}}$ is the relative velocity of B with respect to A measured from x, y, z. Thus,

$$\mathbf{v}_B = \mathbf{v}_A + \boldsymbol{\Omega} \times \mathbf{r}_{B/A} + (\mathbf{v}_{B/A})_{\text{rel}} \qquad (20\text{–}11)$$

where

$$\mathbf{v}_B = \text{velocity of } B$$
$$\mathbf{v}_A = \text{velocity of the origin } A \text{ of the } x, y, z \text{ frame of reference}$$
$$(\mathbf{v}_{B/A})_{\text{rel}} = \text{relative velocity of ``}B \text{ with respect to } A\text{'' as measured by an}$$
$$\text{observer attached to the rotating } x, y, z \text{ frame of reference}$$
$$\boldsymbol{\Omega} = \text{angular velocity of the } x, y, z \text{ frame of reference}$$
$$\mathbf{r}_{B/A} = \text{relative position of ``}B \text{ with respect to } A\text{''}$$

Acceleration

The acceleration of point B measured from X, Y, Z is determined by taking the time derivative of Eq. 20–11, which yields

$$(\dot{\mathbf{v}}_B)_{XYZ} = (\dot{\mathbf{v}}_A)_{XYZ} + (\dot{\boldsymbol{\Omega}})_{XYZ} \times \mathbf{r}_{B/A} + \boldsymbol{\Omega} \times (\dot{\mathbf{r}}_{B/A})_{XYZ} + (\dot{\mathbf{v}}_{B/A})_{XYZ}$$

The time derivatives defined in the first and second terms represent $\mathbf{a}_B$ and $\mathbf{a}_A$ respectively. The fourth term is evaluated using Eq. 20–10, and the last term is evaluated by applying Eq. 20–6, which yields

$$(\dot{\mathbf{v}}_{B/A})_{XYZ} = (\dot{\mathbf{v}}_{B/A})_{xyz} + \boldsymbol{\Omega} \times (\mathbf{v}_{B/A})_{rel}$$
$$= (\mathbf{a}_{B/A})_{rel} + \boldsymbol{\Omega} \times (\mathbf{v}_{B/A})_{rel}$$

Here $(\mathbf{a}_{B/A})_{rel}$ is the relative acceleration of B with respect to A measured from x, y, z. Substituting this result and Eq. 20–10 into the above equation and simplifying, we have

$$\boxed{\mathbf{a}_B = \mathbf{a}_A + \dot{\boldsymbol{\Omega}} \times \mathbf{r}_{B/A} + \boldsymbol{\Omega} \times (\boldsymbol{\Omega} \times \mathbf{r}_{B/A}) + 2\boldsymbol{\Omega} \times (\mathbf{v}_{B/A})_{rel} + (\mathbf{a}_{B/A})_{rel}}$$

$$(20\text{–}12)$$

where

$\mathbf{a}_B$ = acceleration of B

$\mathbf{a}_A$ = acceleration of the origin A of the x, y, z frame of reference

$(\mathbf{a}_{B/A})_{rel}, (\mathbf{v}_{B/A})_{rel}$ = relative acceleration and relative velocity of "B with respect to A" as measured by an observer attached to the rotating x, y, z frame of reference

$\dot{\boldsymbol{\Omega}}, \boldsymbol{\Omega}$ = angular acceleration and angular velocity of the x, y, z frame of reference

$\mathbf{r}_{B/A}$ = relative position of "B with respect to A"

Equations 20–11 and 20–12 are identical to those used in Sec. 16.8 for analyzing relative plane motion.* In that case, however, application was simplified since $\boldsymbol{\Omega}$ and $\dot{\boldsymbol{\Omega}}$ have a *constant direction* which is always perpendicular to the plane of motion. For three-dimensional motion, $\dot{\boldsymbol{\Omega}}$ must be computed by using Eq. 20–6, since it depends upon the change in both the magnitude and direction of $\boldsymbol{\Omega}$. Furthermore, in some problems, calculation of $\mathbf{v}_A$, $\mathbf{a}_A$ and $(\mathbf{v}_{B/A})_{rel}$, $(\mathbf{a}_{B/A})_{rel}$ must also be performed by using Eq. 20–6, since these quantities depend upon the angular rate at which $\mathbf{r}_A$ and $\mathbf{r}_{B/A}$ are "swinging" as measured from their respective frames of reference.

---

*Refer to Sec. 16.8 for an interpretation of the terms.

PROCEDURE FOR ANALYSIS

The following procedure provides a method for applying Eqs. 20–11 and 20–12 to solve problems involving the three-dimensional motion of particles or rigid bodies.

Coordinate Axes. Define the location and orientation of the X, Y, Z and x, y, z coordinate axes. Most often solutions are easily obtained if at the instant considered: (1) the origins are *coincident,* (2) the axes are collinear, and/or (3) the axes are parallel. Since several components of angular velocity may be involved in a problem, the calculations will be reduced if the x, y, z axes are selected such that only one component of angular velocity is observed in this frame ($\mathbf{\Omega}_{\text{rel}}$) and the frame rotates with $\mathbf{\Omega}$ defined by the other components of angular velocity.

Kinematic Equations. After the origin of the moving reference, A, is defined and the moving point B is specified, Eqs. 20–11 and 20–12 should be written in symbolic form as

$$\mathbf{v}_B = \mathbf{v}_A + \mathbf{\Omega} \times \mathbf{r}_{B/A} + (\mathbf{v}_{B/A})_{\text{rel}}$$

$$\mathbf{a}_B = \mathbf{a}_A + \dot{\mathbf{\Omega}} \times \mathbf{r}_{B/A} + \mathbf{\Omega} \times (\mathbf{\Omega} \times \mathbf{r}_{B/A}) + 2\mathbf{\Omega} \times (\mathbf{v}_{B/A})_{\text{rel}} + (\mathbf{a}_{B/A})_{\text{rel}}$$

Vectors $\mathbf{\Omega}$, $\mathbf{r}_A$, and $\mathbf{r}_{B/A}$ should be defined from the problem data and represented in Cartesian form. *Motion of the moving reference ($\mathbf{v}_A$, $\mathbf{a}_A$, and $\dot{\mathbf{\Omega}}$)* is determined by applying Eq. 20–6 to compute the time derivatives of $\mathbf{r}_A$ and $\mathbf{\Omega}$. In a similar manner, if $\mathbf{r}_{B/A}$ has an angular motion $\mathbf{\Omega}_{\text{rel}}$ when observed from the moving reference, then the *motion of B with respect to the moving reference,* $(\mathbf{v}_{B/A})_{\text{rel}}$ and $(\mathbf{a}_{B/A})_{\text{rel}}$, must be determined by applying Eq. 20–6 to compute the time derivatives of $\mathbf{r}_{B/A}$. In all cases, the kinematic quantities which are involved in computing the time derivatives of $\mathbf{r}_A$, $\mathbf{r}_{B/A}$, and $\mathbf{\Omega}$ should be treated as *variables.* After the final forms of $\mathbf{v}_A$, $\mathbf{a}_A$, $(\mathbf{v}_{B/A})_{\text{rel}}$, $(\mathbf{a}_{B/A})_{\text{rel}}$ and $\mathbf{\Omega}$ are obtained, numerical problem data may be substituted and the kinematic terms evaluated. The components of all these vectors may be selected either along the X, Y, Z axes or along x, y, z. The choice is arbitrary, provided a consistent set of unit vectors is used.

Finally, substitute the data into the kinematic equations and perform the vector operations.

The following two examples numerically illustrate this procedure.

Example 20–4

A motor M and attached rod AB have the angular motion shown in Fig. 20–12. A collar C on the rod is located 0.25 m from A, and is moving downward along the rod with a velocity of 3 m/s and an acceleration of 2 m/s^2. Determine the velocity and acceleration of C at this instant.

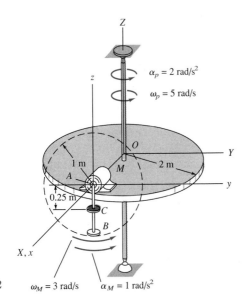

Fig. 20–12

$\alpha_p = 2$ rad/s^2

$\omega_p = 5$ rad/s

$\omega_M = 3$ rad/s $\alpha_M = 1$ rad/s^2

SOLUTION

Coordinate Axes. The origin of the fixed X, Y, Z reference is chosen at the center of the platform, and the origin of the moving x, y, z frame at point A, Fig. 20–12. Since the collar is subjected to two components of angular motion, $\boldsymbol{\omega}_p$ and $\boldsymbol{\omega}_M$, it is viewed as having an angular velocity of $\boldsymbol{\Omega}_{\text{rel}} = \boldsymbol{\omega}_M$ in x, y, z and the x, y, z axes are attached to the platform so that $\boldsymbol{\Omega} = \boldsymbol{\omega}_p$.

Kinematic Equations. Equations 20–11 and 20–12, applied to points C and A, become

$$\mathbf{v}_C = \mathbf{v}_A + \boldsymbol{\Omega} \times \mathbf{r}_{C/A} + (\mathbf{v}_{C/A})_{\text{rel}} \tag{1}$$

$$\mathbf{a}_C = \mathbf{a}_A + \dot{\boldsymbol{\Omega}} \times \mathbf{r}_{C/A} + \boldsymbol{\Omega} \times (\boldsymbol{\Omega} \times \mathbf{r}_{C/A}) + 2\boldsymbol{\Omega} \times (\mathbf{v}_{C/A})_{\text{rel}} + (\mathbf{a}_{C/A})_{\text{rel}} \tag{2}$$

Motion of Moving Reference

$$\boldsymbol{\Omega} = \boldsymbol{\omega}_p = 5\mathbf{k}$$

$$\dot{\boldsymbol{\Omega}} = (\dot{\boldsymbol{\omega}}_p)_{xyz} + \boldsymbol{\omega}_p \times \boldsymbol{\omega}_p = 2\mathbf{k} + 5\mathbf{k} \times 5\mathbf{k} = 2\mathbf{k}$$

$$\mathbf{r}_A = 2\mathbf{i}$$

$$\left.\begin{aligned}
\mathbf{v}_A &= \dot{\mathbf{r}}_A = (\dot{\mathbf{r}}_A)_{xyz} + \boldsymbol{\omega}_p \times \mathbf{r}_A = \mathbf{0} + 5\mathbf{k} \times 2\mathbf{i} = 10\mathbf{j} \\
\mathbf{a}_A &= \ddot{\mathbf{r}}_A = [(\ddot{\mathbf{r}}_A)_{xyz} + \boldsymbol{\omega}_p \times (\dot{\mathbf{r}}_A)_{xyz}] + \dot{\boldsymbol{\omega}}_p \times \mathbf{r}_A + \boldsymbol{\omega}_p \times \dot{\mathbf{r}}_A \\
&= \mathbf{0} + \mathbf{0} + 2\mathbf{k} \times 2\mathbf{i} + 5\mathbf{k} \times 10\mathbf{j} = -50\mathbf{i} + 4\mathbf{j}
\end{aligned}\right\}$$

Since point A moves in a circular path lying in the X-Y plane we can also apply Eqs. 16–15 and 16–17 to obtain these same results.

Motion of C with Respect to Moving Reference

$$\boldsymbol{\Omega}_{\text{rel}} = \boldsymbol{\omega}_M = 3\mathbf{i} \qquad \dot{\boldsymbol{\Omega}}_{\text{rel}} = \dot{\boldsymbol{\omega}}_M = 1\mathbf{i} \big\}$$

Note that $\boldsymbol{\omega}_M$ does not change direction relative to x, y, z.

$$\mathbf{r}_{C/A} = -0.25\mathbf{k}$$

$$\begin{aligned}
(\mathbf{v}_{C/A})_{\text{rel}} &= \dot{\mathbf{r}}_{C/A} = (\dot{\mathbf{r}}_{C/A})_{xyz} + \boldsymbol{\omega}_M \times \mathbf{r}_{C/A} \\
&= -3\mathbf{k} + [3\mathbf{i} \times (-0.25\mathbf{k})] = 0.75\mathbf{j} - 3\mathbf{k}
\end{aligned}$$

$$\begin{aligned}
(\mathbf{a}_{C/A})_{\text{rel}} &= \ddot{\mathbf{r}}_{C/A} = [(\ddot{\mathbf{r}}_{C/A})_{xyz} + \boldsymbol{\omega}_M \times (\dot{\mathbf{r}}_{C/A})_{xyz}] + \dot{\boldsymbol{\omega}}_M \times \mathbf{r}_{C/A} + \boldsymbol{\omega}_M \times \dot{\mathbf{r}}_{C/A} \\
&= [-2\mathbf{k} + 3\mathbf{i} \times (-3\mathbf{k})] + [(1\mathbf{i}) \times (-0.25\mathbf{k})] + [(3\mathbf{i}) \times (0.75\mathbf{j} - 3\mathbf{k})] \\
&= 18.25\mathbf{j} + 0.25\mathbf{k}
\end{aligned}$$

Try to obtain these same results by applying Eqs. 20–11 and 20–12 between the x, y, z axes and coincident (rotating) x', y', z' axes having $\boldsymbol{\Omega} = 3\mathbf{i}$, $\dot{\boldsymbol{\Omega}} = 1\mathbf{i}$.

Substituting the data into Eqs. (1) and (2) yields

$$\begin{aligned}
\mathbf{v}_C &= \mathbf{v}_A + \boldsymbol{\Omega} \times \mathbf{r}_{C/A} + (\mathbf{v}_{C/A})_{\text{rel}} \\
&= 10\mathbf{j} + [5\mathbf{k} \times (-0.25\mathbf{k})] + (0.75\mathbf{j} - 3\mathbf{k}) \\
&= \{10.8\mathbf{j} - 3\mathbf{k}\} \text{ m/s} \qquad\qquad\qquad\qquad \textit{Ans.}
\end{aligned}$$

$$\begin{aligned}
\mathbf{a}_C &= \mathbf{a}_A + \dot{\boldsymbol{\Omega}} \times \mathbf{r}_{C/A} + \boldsymbol{\Omega} \times (\boldsymbol{\Omega} \times \mathbf{r}_{C/A}) + 2\boldsymbol{\Omega} \times (\mathbf{v}_{C/A})_{\text{rel}} + (\mathbf{a}_{C/A})_{\text{rel}} \\
&= (-50\mathbf{i} + 4\mathbf{j}) + [2\mathbf{k} \times (-0.25\mathbf{k})] + 5\mathbf{k} \times [5\mathbf{k} \times (-0.25\mathbf{k})] \\
&\quad + 2[5\mathbf{k} \times (0.75\mathbf{j} - 3\mathbf{k})] + (18.25\mathbf{j} + 0.25\mathbf{k}) \\
&= \{-57.5\mathbf{i} + 22.2\mathbf{j} + 0.25\mathbf{k}\} \text{ m/s}^2 \qquad\qquad \textit{Ans.}
\end{aligned}$$

Example 20–5

The pendulum shown in Fig. 20–13 consists of two rods. AB is pin-supported at A and swings only in the Y-Z plane, whereas a bearing at B allows the attached rod BD to spin about rod AB. At a given instant, the rods have the angular motions shown. If a collar C, located 0.2 m from B, has a velocity of 3 m/s and an acceleration of 2 m/s^2 along the rod, determine the velocity and acceleration of the collar at this instant.

$\dot{\omega}_1 = 1.5$ rad/s^2

$\omega_1 = 4$ rad/s

0.5 m

$\omega_2 = 5$ rad/s

3 m/s

2 m/s^2

0.2 m

$\dot{\omega}_2 = 6$ rad/s^2

Fig. 20–13

SOLUTION I $(\Omega = \omega_1, \Omega_{\text{rel}} = \omega_2)$

Coordinate Axes. Since the rod is rotating about the fixed point A, the origin of the X, Y, Z frame will be chosen at A. Motion of the collar is conveniently observed from B, so the origin of the x, y, z frame is located at this point.

Kinematic Equations

$$\mathbf{v}_C = \mathbf{v}_B + \Omega \times \mathbf{r}_{C/B} + (\mathbf{v}_{C/B})_{\text{rel}} \qquad (1)$$

$$\mathbf{a}_C = \mathbf{a}_B + \dot{\Omega} \times \mathbf{r}_{C/B} + \Omega \times (\Omega \times \mathbf{r}_{C/B}) + 2\Omega \times (\mathbf{v}_{C/B})_{\text{rel}} + (\mathbf{a}_{C/B})_{\text{rel}} \qquad (2)$$

Motion of Moving Reference

$$\Omega = \omega_1 = 4\mathbf{i}, \quad \dot{\Omega} = (\dot{\omega}_1)_{xyz} + \omega_1 \times \omega_1 = 1.5\mathbf{i} + 0 = 1.5\mathbf{i}$$

$$\mathbf{r}_B = -0.5\mathbf{k}$$

$$\left.\begin{aligned}
\mathbf{v}_B &= \dot{\mathbf{r}}_B = (\dot{\mathbf{r}}_B)_{xyz} + \omega_1 \times \mathbf{r}_B = 0 + 4\mathbf{i} \times (-0.5\mathbf{k}) = 2\mathbf{j} \\
\mathbf{a}_B &= \ddot{\mathbf{r}}_B = [(\ddot{\mathbf{r}}_B)_{xyz} + \omega_1 \times (\dot{\mathbf{r}}_B)_{xyz}] + \dot{\omega}_1 \times \mathbf{r}_B + \omega_1 \times \dot{\mathbf{r}}_B \\
&= [0 + 0] + [1.5\mathbf{i} \times (-0.5\mathbf{k})] + 4\mathbf{i} \times 2\mathbf{j} = 0.75\mathbf{j} + 8\mathbf{k}
\end{aligned}\right\}$$

This is the same as applying Eqs. 16–15 and 16–17 since B has circular motion in the X-Y plane.

Motion of C with Respect to Moving Reference

$$\left.\begin{aligned}
&\Omega_{\text{rel}} = \omega_2 = 5\mathbf{k}, \quad \dot{\Omega}_{\text{rel}} = \dot{\omega}_2 = -6\mathbf{k}\right\} \qquad \text{Note that } \omega_2 \text{ does not change direction relative to } x, y, z. \\
&\mathbf{r}_{C/B} = 0.2\mathbf{j}
\end{aligned}$$

$$\left.\begin{aligned}
(\mathbf{v}_{C/B})_{\text{rel}} &= \dot{\mathbf{r}}_{C/B} = (\dot{\mathbf{r}}_{C/B})_{xyz} + \omega_2 \times \mathbf{r}_{C/B} = 3\mathbf{j} + 5\mathbf{k} \times 0.2\mathbf{j} = -1\mathbf{i} + 3\mathbf{j} \\
(\mathbf{a}_{C/B})_{\text{rel}} &= \ddot{\mathbf{r}}_{C/B} = [(\ddot{\mathbf{r}}_{C/B})_{xyz} + \omega_2 \times (\dot{\mathbf{r}}_{C/B})_{xyz}] + \dot{\omega}_2 \times \mathbf{r}_{C/B} + \omega_2 \times \dot{\mathbf{r}}_{C/B} \\
&\quad (2\mathbf{j} + 5\mathbf{k} \times 3\mathbf{j}) + (-6\mathbf{k} \times 0.2\mathbf{j}) + [5\mathbf{k} \times (-1\mathbf{i} + 3\mathbf{j})] \\
&= -28.8\mathbf{i} - 3\mathbf{j}
\end{aligned}\right\}$$

The same results can be obtained by applying Eqs. 20–11 and 20–12 between the x, y, z axes and coincident (rotating) x', y', z' axes having $\Omega = 5\mathbf{k}$, $\dot{\Omega} = -6\mathbf{k}$.

Substituting the data into Eqs. (1) and (2) yields

$$\mathbf{v}_C = \mathbf{v}_B + \Omega \times \mathbf{r}_{C/B} + (\mathbf{v}_{C/B})_{\text{rel}} = 2\mathbf{j} + 4\mathbf{i} \times 0.2\mathbf{j} + (-1\mathbf{i} + 3\mathbf{j})$$

$$= \{-1\mathbf{i} + 5\mathbf{j} + 0.8\mathbf{k}\} \text{ m/s} \qquad \textit{Ans.}$$

$$\mathbf{a}_C = \mathbf{a}_B + \dot{\Omega} \times \mathbf{r}_{C/B} + \Omega \times (\Omega \times \mathbf{r}_{C/B}) + 2\Omega \times (\mathbf{v}_{C/B})_{\text{rel}} + (\mathbf{a}_{C/B})_{\text{rel}}$$

$$= (0.75\mathbf{j} + 8\mathbf{k}) + (1.5\mathbf{i} \times 0.2\mathbf{j}) + [4\mathbf{i} \times (4\mathbf{i} \times 0.2\mathbf{j})]$$

$$+ 2[4\mathbf{i} \times (-1\mathbf{i} + 3\mathbf{j})] + (-28.8\mathbf{i} - 3\mathbf{j})$$

$$= \{-28.8\mathbf{i} - 5.45\mathbf{j} + 32.3\mathbf{k}\} \text{ m/s}^2 \qquad \textit{Ans.}$$

SOLUTION II $(\mathbf{\Omega} = \boldsymbol{\omega}_1 + \boldsymbol{\omega}_2, \mathbf{\Omega}_{\text{rel}} = \mathbf{0})$

Motion of Moving Reference

$$\mathbf{\Omega} = \boldsymbol{\omega}_1 + \boldsymbol{\omega}_2 = 4\mathbf{i} + 5\mathbf{k}$$

From the constraints of the problem $\boldsymbol{\omega}_1$ does not change direction; however, the direction of $\boldsymbol{\omega}_2$ is changed by $\boldsymbol{\omega}_1$. Thus, a simple way of obtaining $\dot{\mathbf{\Omega}}$ is to consider x', y', z' axes coincident with the X, Y, Z axes at A, such that the primed axes have an angular velocity $\boldsymbol{\omega}_1$. Then

$$\dot{\mathbf{\Omega}} = \dot{\boldsymbol{\omega}}_1 + \dot{\boldsymbol{\omega}}_2 = [(\dot{\boldsymbol{\omega}}_1)_{x'y'z'} + \boldsymbol{\omega}_1 \times \boldsymbol{\omega}_1] + [(\dot{\boldsymbol{\omega}}_2)_{x'y'z'} + \boldsymbol{\omega}_1 \times \boldsymbol{\omega}_2]$$
$$= (1.5\mathbf{i} + \mathbf{0}) + (-6\mathbf{k} + 4\mathbf{i} \times 5\mathbf{k}) = 1.5\mathbf{i} - 20\mathbf{j} - 6\mathbf{k}$$

Also,
$$\mathbf{r}_B = -0.5\mathbf{k}$$

Here again only $\boldsymbol{\omega}_1$ changes the direction of $\mathbf{r}_B$ so that the time derivatives of $\mathbf{r}_B$ can be computed from the primed axes used above, which are rotating at $\boldsymbol{\omega}_1$. Hence,

$$\mathbf{v}_B = \dot{\mathbf{r}}_B = (\dot{\mathbf{r}}_B)_{x'y'z'} + \boldsymbol{\omega}_1 \times \mathbf{r}_B$$
$$= \mathbf{0} + 4\mathbf{i} \times (-0.5\mathbf{k}) = 2\mathbf{j}$$
$$\mathbf{a}_B = \ddot{\mathbf{r}}_B = [(\ddot{\mathbf{r}}_B)_{x'y'z'} + \boldsymbol{\omega}_1 \times (\dot{\mathbf{r}}_B)_{x'y'z'}] + \dot{\boldsymbol{\omega}}_1 \times \mathbf{r}_B + \boldsymbol{\omega}_1 \times \dot{\mathbf{r}}_B$$
$$= [\mathbf{0} + \mathbf{0}] + [1.5\mathbf{i} \times (-0.5\mathbf{k})] + 4\mathbf{i} \times 2\mathbf{j} = 0.75\mathbf{j} + 8\mathbf{k}$$

Motion of C with Respect to Moving Reference

$$\mathbf{\Omega}_{\text{rel}} = \mathbf{0}$$
$$\dot{\mathbf{\Omega}}_{\text{rel}} = \mathbf{0}$$
$$\mathbf{r}_{C/B} = 0.2\mathbf{j}$$
$$(\mathbf{v}_{C/B})_{\text{rel}} = \dot{\mathbf{r}}_{C/B} = (\dot{\mathbf{r}}_{C/B})_{xyz} + \mathbf{\Omega}_{\text{rel}} \times \mathbf{r}_{C/B}$$
$$= 3\mathbf{j} + \mathbf{0} = 3\mathbf{j}$$
$$(\mathbf{a}_{C/B})_{\text{rel}} = \ddot{\mathbf{r}}_{C/B} = [(\ddot{\mathbf{r}}_{C/B})_{xyz} + \mathbf{\Omega}_{\text{rel}} \times (\dot{\mathbf{r}}_{C/B})_{xyz}] + \dot{\mathbf{\Omega}}_{\text{rel}} \times \mathbf{r}_{C/B} + \mathbf{\Omega}_{\text{rel}} \times \dot{\mathbf{r}}_{C/B}$$
$$= [2\mathbf{j} + \mathbf{0}] + \mathbf{0} + \mathbf{0} = 2\mathbf{j}$$

Substituting the data into Eqs. (1) and (2) yields

$$\mathbf{v}_C = \mathbf{v}_B + \mathbf{\Omega} \times \mathbf{r}_{C/B} + (\mathbf{v}_{C/B})_{\text{rel}}$$
$$= 2\mathbf{j} + [(4\mathbf{i} + 5\mathbf{k}) \times (0.2\mathbf{j})] + 3\mathbf{j}$$
$$= \{-1\mathbf{i} + 5\mathbf{j} + 0.8\mathbf{k}\} \text{ m/s} \hspace{2cm} \textit{Ans.}$$

$$\mathbf{a}_C = \mathbf{a}_B + \dot{\mathbf{\Omega}} \times \mathbf{r}_{C/B} + \mathbf{\Omega} \times (\mathbf{\Omega} \times \mathbf{r}_{C/B}) + 2\mathbf{\Omega} \times (\mathbf{v}_{C/B})_{\text{rel}} + (\mathbf{a}_{C/B})_{\text{rel}}$$
$$= (0.75\mathbf{j} + 8\mathbf{k}) + [(1.5\mathbf{i} - 20\mathbf{j} - 6\mathbf{k}) \times (0.2\mathbf{j})]$$
$$\qquad + (4\mathbf{i} + 5\mathbf{k}) \times [(4\mathbf{i} + 5\mathbf{k}) \times 0.2\mathbf{j}] + 2[(4\mathbf{i} + 5\mathbf{k}) \times 3\mathbf{j}] + 2\mathbf{j}$$
$$= \{-28.8\mathbf{i} - 5.45\mathbf{j} + 32.3\mathbf{k}\} \text{ m/s}^2 \hspace{1.5cm} \textit{Ans.}$$

PROBLEMS

20–29. Solve Example 20–5 by fixing x, y, z axes to rod BD so that $\Omega = \omega_1 + \omega_2$. In this case the collar appears only to move radially outward along BD, hence $\Omega_{C/B} = 0$.

20–30. Solve Example 20–5 such that the x, y, z axes move with curvilinear translation, $\Omega = 0$, in which case the collar appears to have both an angular velocity $\Omega_{C/B} = \omega_1 + \omega_2$ and radial motion.

20–31. At a given instant, the antenna has an angular motion $\omega_1 = 3$ rad/s and $\dot{\omega}_1 = 2$ rad/s² about the z axis. At this same instant $\theta = 30°$, the angular motion about the x axis is $\omega_2 = 1.5$ rad/s, and $\dot{\omega}_2 = 4$ rad/s². Determine the velocity and acceleration of the signal horn A at this instant. The distance from O to A is $d = 3$ ft.

Prob. 20–31

*20–32. At a given instant, rod BD is rotating about the x axis with an angular velocity $\omega_{BD} = 2$ rad/s and an angular acceleration $\alpha_{BD} = 5$ rad/s². Also, $\theta = 45°$ and link AC is rotating downward such that $\dot{\theta} = 4$ rad/s and $\ddot{\theta} = 2$ rad/s². Determine the velocity and acceleration of point A on the link at this instant.

Prob. 20–32

20–33. At a given instant, the crane is moving along the track with a velocity $v_{CD} = 8$ m/s and acceleration of 9 m/s². Simultaneously, it has the angular motions shown. If the trolley T is moving outwards along the boom AB with a relative speed of 3 ft/s and relative acceleration of 5 ft/s², determine the velocity and acceleration of the trolley.

Prob. 20–33

20–34. At a given instant, the rod has the angular motions shown, while the collar C is moving down *relative* to the rod with a velocity of 6 ft/s and an acceleration of 2 ft/s². Determine the collar's velocity and acceleration at this instant.

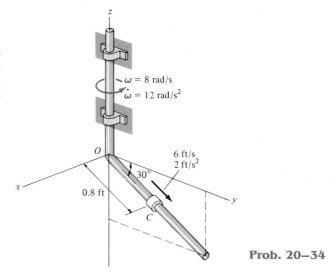

Prob. 20–34

20–35. The boom AB of the crane is rotating about the z axis with an angular velocity $\omega_z = 0.75$ rad/s and an angular acceleration of $\dot{\omega}_z = 2$ rad/s^2. At the same instant, $\theta = 60°$ and the boom is rotating upward at a constant rate $\dot{\theta} = 0.5$ rad/s. Determine the velocity and acceleration of the tip B of the boom at this instant.

Prob. 20–35

20–37. At the instant shown, the tractor is traveling forward at a constant speed of $v_O = 2$ ft/s, and the boom ABC is rotating about the z axis with an angular velocity of $\omega_1 = 0.8$ rad/s and an angular acceleration $\dot{\omega}_1 = 1.30$ rad/s^2. At this same instant the boom is rotating with $\omega_2 = 3$ rad/s when $\dot{\omega}_2 = 2$ rad/s^2, both measured relative to the tractor frame. Determine the velocity and acceleration of point P on the bucket at this instant.

Prob. 20–37

***20–36.** At the instant shown, the arm AB is rotating about the fixed pin A with an angular velocity $\omega_1 = 4$ rad/s and angular acceleration $\dot{\omega}_1 = 3$ rad/s^2. At this same instant, rod BD is rotating relative to rod AB with an angular velocity $\omega_2 = 5$ rad/s which is increasing at $\dot{\omega}_2 = 7$ rad/s^2. Also, the collar C is moving along rod BD with a velocity of 3 m/s and an acceleration of 2 m/s^2, both measured relative to the rod. Determine the velocity and acceleration of the collar at this instant.

Prob. 20–36

20–38. At the instant shown, the arm AB is rotating about the fixed bearing with an angular velocity $\omega_1 = 2$ rad/s and angular acceleration $\dot{\omega}_1 = 6$ rad/s^2. At the same instant, rod BD is rotating relative to rod AB at $\omega_2 = 7$ rad/s, which is increasing at $\dot{\omega}_2 = 1$ rad/s^2. Also, the collar C is moving along rod BD with a velocity $\dot{r} = 2$ ft/s and a *deceleration* $\ddot{r} = -0.5$ ft/s^2, both measured relative to the rod. Determine the velocity and acceleration of the collar at this instant.

Prob. 20–38

20–39. At the given instant, the rod is spinning about the z axis with an angular velocity $\omega_1 = 3$ rad/s and angular acceleration $\dot{\omega}_1 = 4$ rad/s². At this same instant, the disk is spinning with $\omega_2 = 2$ rad/s when $\dot{\omega}_2 = 1$ rad/s², both measured *relative* to the rod. Determine the velocity and acceleration of point P on the disk at this instant.

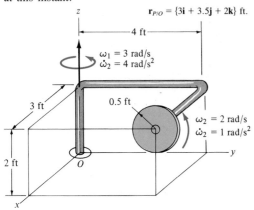

$\mathbf{r}_{P/O} = \{3\mathbf{i} + 3.5\mathbf{j} + 2\mathbf{k}\}$ ft.

Prob. 20–39

***20–40.** At the instant shown the portal crane is moving forward with a velocity $v_C = 6$ ft/s and an acceleration $a_C = 2$ ft/s². Simultaneously, it has the relative angular motions shown at $\theta = 60°$. Determine the velocity and acceleration of point A at the top of boom AB.

Prob. 20–40

20–41. At the instant shown, the rod AB is rotating about the z axis with an angular velocity $\omega_1 = 4$ rad/s and an angular acceleration $\dot{\omega}_1 = 3$ rad/s². At this same instant, the circular rod has an angular motion relative to the rod as shown in the figure. If the collar C is moving down around the circular rod with a speed of 3 in./s, which is increasing at 8 in./s², both measured relative to the rod, determine the collar's velocity and acceleration at this instant.

Prob. 20–41

20–42. At the instant shown, the arm OA of the conveyor belt is rotating about the z axis with a constant angular velocity $\omega_1 = 6$ rad/s, while at the same instant the arm is rotating upward at a constant rate $\omega_2 = 4$ rad/s. If the conveyor is running at a constant rate $\dot{r} = 5$ ft/s, determine the velocity and acceleration of the package P at the instant shown. Neglect the size of the package.

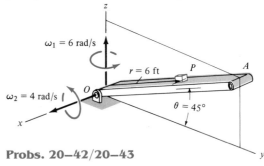

Probs. 20–42/20–43

20–43. At the instant shown, the arm OA of the conveyor belt is rotating about the z axis with a constant angular velocity $\omega_1 = 6$ rad/s, while at the same instant the arm is rotating upward at a constant rate $\omega_2 = 4$ rad/s. If the conveyor is running at a rate $\dot{r} = 5$ ft/s, which is increasing at $\ddot{r} = 8$ ft/s², determine the velocity and acceleration of the package P at the instant shown. Neglect the size of the package.

*20–44. At the instant shown, the industrial manipulator is rotating about the z axis at $\omega_1 = 5$ rad/s, and about joint B at $\omega_2 = 2$ rad/s. Determine the velocity and acceleration of the grip A at this instant when $\phi = 30°$, $\theta = 45°$, and $r = 1.6$ m.

20–45. At the instant shown, the industrial manipulator is rotating about the z axis at $\omega_1 = 5$ rad/s, which is increasing at the rate $\dot{\omega}_1 = 7$ rad/s^2. Simultaneously, the arm BA is rotating about joint B at $\omega_2 = 2$ rad/s, which is increasing at the rate $\dot{\omega}_2 = 6$ rad/s^2, and the arm BA is telescoping outward (elongating) at $\dot{r} = 0.5$ m/s, which is increasing at $\ddot{r} = 1.3$ m/s^2. Determine the velocity and acceleration of the grip A at this instant when $\phi = 30°$, $\theta = 45°$, and $r = 1.6$ m.

Probs. 20–44/20–45

21

Three-Dimensional Kinetics of a Rigid Body

In general, as a body moves through space it has a simultaneous translation and rotation at a given instant. The kinetic aspects of the *translation* have been discussed in Chapter 17, where it was shown that a system of external forces acting on the body may be related to the acceleration of the body's mass center by the equation $\Sigma\mathbf{F} = m\mathbf{a}_G$. In this chapter emphasis is placed primarily upon the *rotational* aspects of rigid-body motion, since motion of the body's mass center, defined by $\Sigma\mathbf{F} = m\mathbf{a}_G$, is treated in the same manner as particle motion.

The rotational equations of motion relate the body's components of angular motion to the moment components created by the external forces about some point located either on or off the body. To apply these equations, it is first necessary to formulate the moments and products of inertia of the body and to compute the body's angular momentum. Afterward, the principles of impulse and momentum and work and energy will be at our disposal for solving problems in three dimensions. The rotational equations of motion are developed in Sec. 21.4. Of special interest are problems involving the motion of an unsymmetrical body about a fixed axis; motion of a gyroscope, Sec. 21.5; and torque-free motion, Sec. 21.6.

*21.1 Moments and Products of Inertia

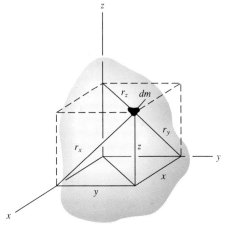

Fig. 21–1

When studying the planar kinetics of a body, it was necessary to introduce the moment of inertia I_G, which was computed about an axis perpendicular to the plane of motion and passing through the mass center G. For the kinetic analysis of three-dimensional motion it will sometimes be necessary to calculate six inertial quantities. These terms, called the moments and products of inertia, describe in a particular way the distribution of mass for a body relative to a given coordinate system that has a specified orientation and point of origin.

Moment of Inertia

Consider the rigid body shown in Fig. 21–1. The *moment of inertia* for a differential element dm of the body about any one of the three coordinate axes is defined as the product of the mass of the element and the square of the shortest distance from the axis to the element. For example, as noted in the figure, $r_x = \sqrt{y^2 + z^2}$, so that the mass moment of inertia of dm about the x axis is

$$dI_{xx} = r_x^2 \, dm = (y^2 + z^2) \, dm$$

The moment of inertia I_{xx} for the body is determined by integrating this expression over the entire mass of the body. Hence, for each of the axes, we may write

$$
\begin{aligned}
I_{xx} &= \int_m r_x^2 \, dm = \int_m (y^2 + z^2) \, dm \\[2mm]
I_{yy} &= \int_m r_y^2 \, dm = \int_m (x^2 + z^2) \, dm \\[2mm]
I_{zz} &= \int_m r_z^2 \, dm = \int_m (x^2 + y^2) \, dm
\end{aligned}
\tag{21–1}
$$

Here it is seen that the moment of inertia is *always a positive quantity,* since it is the summation of the product of the mass $dm,$ which is always positive, and distances squared.

Product of Inertia

The *product of inertia* for a differential element dm is defined with respect to a set of *two orthogonal planes* as the product of the mass of the element and the perpendicular (or shortest) distances from the planes to the element. For example, with respect to the x-z and y-z planes, the product of inertia dI_{xy} for the element dm shown in Fig. 21–1 is

$$dI_{xy} = xy \, dm$$

Note also that $dI_{yx} = dI_{xy}.$ By integrating over the entire mass, the product of inertia of the body for each combination of planes may be expressed as

$$I_{xy} = I_{yx} = \int_m xy \; dm$$

$$I_{yz} = I_{zy} = \int_m yz \; dm \qquad\qquad (21-2)$$

$$I_{xz} = I_{zx} = \int_m xz \; dm$$

Unlike the moment of inertia, which is always positive, the product of inertia may be positive, negative, or zero. The result depends upon the signs of the two defining coordinates, which vary independently from one another. In particular, if either one or both of the orthogonal planes are *planes of symmetry* for the mass, the *product of inertia* with respect to these planes will be *zero*. In such cases, elements of mass will occur in *pairs*, located on each side of the plane of symmetry. On one side of the plane the product of inertia for the element will be positive, while on the other side the product of inertia for the corresponding element will be negative, the sum therefore yielding zero. Examples of this are shown in Fig. 21–2. In the first case, Fig. 21–2*a*, the *y-z* plane is a plane of symmetry, and hence for point O, $I_{xz} = I_{xy} = 0$. Computation for I_{yz} will yield a *positive* result, since all elements of mass are located using only positive y and z coordinates. For the cylinder, with the coordinate axes located as shown in Fig. 21–2*b*, the *x-z* and *y-z* planes are both planes of symmetry. Thus, for point O, $I_{zx} = I_{yz} = I_{xy} = 0$.

 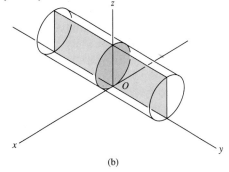

(a)   (b)

Fig. 21–2

Parallel-Axis and Parallel-Plane Theorems

The techniques of integration which are used to determine the moment of inertia of a body were described in Sec. 17.1. Also discussed were methods to determine the moment of inertia of a composite body, i.e., a body that is composed of simpler segments, as tabulated on the inside back cover of the book. In both of these cases the *parallel-axis theorem* is often used for the calculations. This theorem, which was developed in Sec. 17.1, is used to transfer the moment of inertia of a body from an axis passing through its mass center G to a parallel axis passing through some other point. In this regard, if

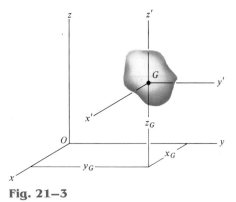

Fig. 21–3

G has coordinates x_G, y_G, z_G defined from the x, y, z axes, Fig. 21–3, then the parallel-axis equations used to calculate the moments of inertia about the x, y, z axes are

$$I_{xx} = (I_{x'x'})_G + m(y_G^2 + z_G^2)$$
$$I_{yy} = (I_{y'y'})_G + m(x_G^2 + z_G^2)$$
$$I_{zz} = (I_{z'z'})_G + m(x_G^2 + y_G^2)$$

(21–3)

The products of inertia of a body or a composite are computed in the same manner as the body's moments of inertia. Here, however, the *parallel-plane theorem* is important. This theorem is used to transfer the products of inertia of the body from a set of three orthogonal planes passing through the body's mass center to a corresponding set of three parallel planes passing through some other point O. Defining the perpendicular distances between the planes as x_G, y_G, and z_G, Fig. 21–3, the parallel-plane equations can be written as

$$I_{xy} = (I_{x'y'})_G + mx_G y_G$$
$$I_{yz} = (I_{y'z'})_G + my_G z_G$$
$$I_{zx} = (I_{z'x'})_G + mz_G x_G$$

(21–4)

The derivation of these formulas is similar to that given for the parallel-axis equation, Sec. 17.1.

Inertia Tensor

The inertial properties of a body are completely characterized by nine terms, six of which are independent of one another. This set of terms is defined using Eqs. 21–1 and 21–2 and can be written as

$$\begin{pmatrix} I_{xx} & -I_{xy} & -I_{xz} \\ -I_{yx} & I_{yy} & -I_{yz} \\ -I_{zx} & -I_{zy} & I_{zz} \end{pmatrix}$$

This array is called an *inertia tensor*. It has a unique set of values for a body when it is computed for each location of the origin O and orientation of the coordinate axes.

For point O we can specify a unique axes inclination for which the products of inertia for the body are zero when computed with respect to these axes. When this is done, the inertia tensor is said to be "diagonalized" and may be written in the simplified form

$$\begin{pmatrix} I_x & 0 & 0 \\ 0 & I_y & 0 \\ 0 & 0 & I_z \end{pmatrix}$$

Here $I_x = I_{xx}$, $I_y = I_{yy}$, and $I_z = I_{zz}$ are termed the *principal moments of inertia* for the body, which are computed from the *principal axes of inertia*. Of these three principal moments of inertia, one will be a maximum and another a minimum of the body's moment of inertia.

Mathematical determination of the directions of principal axes of inertia will not be discussed here (see Prob. 21–18). There are many cases, however, in which the principal axes may be determined by inspection. From the previous discussion it was noted that if the coordinate axes are oriented such that *two* of the three orthogonal planes containing the axes are planes of *symmetry* for the body, then all the products of inertia for the body are zero with respect to the coordinate planes, and hence the coordinate axes are principal axes of inertia. For example, the x, y, z axes shown in Fig. 21–2b represent the principal axes of inertia for the cylinder at point O.

Moment of Inertia About an Arbitrary Axis

Consider the body shown in Fig. 21–4, where the nine elements of the inertia tensor have been computed for the x, y, z axes having an origin at O. Here we wish to compute the moment of inertia of the body about the Oa axis, for which the direction is defined by the unit vector $\mathbf{u}_a$. By definition $I_{aa} = \int b^2 \, dm$, where b is the *perpendicular distance* from dm to Oa. If the position of dm is located using $\mathbf{r}$, then $b = r \sin \theta$, which represents the *magnitude* of the cross product $\mathbf{u}_a \times \mathbf{r}$. Hence, the moment of inertia can be expressed as

$$I_{aa} = \int_m |(\mathbf{u}_a \times \mathbf{r})|^2 \, dm = \int_m (\mathbf{u}_a \times \mathbf{r}) \cdot (\mathbf{u}_a \times \mathbf{r}) \, dm$$

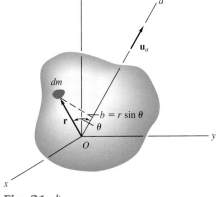

Fig. 21–4

Provided $\mathbf{u}_a = u_x\mathbf{i} + u_y\mathbf{j} + u_z\mathbf{k}$ and $\mathbf{r} = x\mathbf{i} + y\mathbf{j} + z\mathbf{k}$, so that $\mathbf{u}_a \times \mathbf{r} = (u_yz - u_zy)\mathbf{i} + (u_zx - u_xz)\mathbf{j} + (u_xy - u_yx)\mathbf{k}$, then, after substituting and performing the dot-product operation, we can write the moment of inertia as

$$I_{aa} = \int_m [(u_yz - u_zy)^2 + (u_zx - u_xz)^2 + (u_xy - u_yx)^2] \, dm$$

$$= u_x^2 \int_m (y^2 + z^2) \, dm + u_y^2 \int_m (z^2 + x^2) \, dm + u_z^2 \int_m (x^2 + y^2) \, dm$$

$$- 2u_xu_y \int_m xy \, dm - 2u_yu_z \int_m yz \, dm - 2u_zu_x \int_m zx \, dm$$

Recognizing the integrals to be the moments and products of inertia of the body, Eqs. 21–1 and 21–2, we have

$$\boxed{I_{aa} = I_{xx}u_x^2 + I_{yy}u_y^2 + I_{zz}u_z^2 - 2I_{xy}u_xu_y - 2I_{yz}u_yu_z - 2I_{zx}u_zu_x} \quad (21\text{–}5)$$

Thus, if the inertia tensor is specified for the x, y, z axes, the moment of inertia of the body about the inclined Oa axis can be computed by using Eq. 21–5. For the calculation the direction cosines u_x, u_y, u_z of the axes must be determined. These terms specify the cosines of the coordinate direction angles α, β, γ made between the Oa axis and the x, y, z axes, respectively (see Appendix C).

Example 21–1

Determine the moment of inertia of the bent rod shown in Fig. 21–5a about the Aa axis. The mass of each of the three segments is shown in the figure.

SOLUTION

The moment of inertia I_{Aa} can be computed by using Eq. 21–5. It is first necessary, however, to determine the moments and products of inertia of the rod about the x, y, z axes. This is done using the formula for the moment of inertia of a slender rod, $I = \frac{1}{12}ml^2$, and the parallel-axis and parallel-plane theorems, Eqs. 21–3 and 21–4. Dividing the rod into three parts and locating the mass center of each segment, Fig. 21–5b, we have

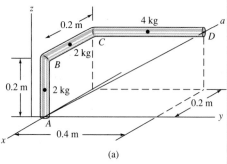

$$I_{xx} = \left[\frac{1}{12}(2)(0.2)^2 + 2(0.1)^2\right] + [0 + 2(0.2)^2]$$

$$+ \left[\frac{1}{12}(4)(0.4)^2 + 4((0.2)^2 + (0.2)^2)\right] = 0.480 \text{ kg} \cdot \text{m}^2$$

$$I_{yy} = \left[\frac{1}{12}(2)(0.2)^2 + 2(0.1)^2\right] + \left[\frac{1}{12}(2)(0.2)^2 + 2((-0.1)^2 + (0.2)^2)\right]$$

$$+ [0 + 4((-0.2)^2 + (0.2)^2)] = 0.453 \text{ kg} \cdot \text{m}^2$$

$$I_{zz} = [0 + 0] + \left[\frac{1}{12}(2)(0.2)^2 + 2(0.1)^2\right]$$

$$+ \left[\frac{1}{12}(4)(0.4)^2 + 4((-0.2)^2 + (0.2)^2)\right] = 0.400 \text{ kg} \cdot \text{m}^2$$

$$I_{xy} = [0 + 0] + [0 + 0] + [0 + 4(-0.2)(0.2)] = -0.160 \text{ kg} \cdot \text{m}^2$$

$$I_{yz} = [0 + 0] + [0 + 0] + [0 + 4(0.2)(0.2)] = 0.160 \text{ kg} \cdot \text{m}^2$$

$$I_{zx} = [0 + 0] + [0 + 2(0.2)(-0.1)] + [0 + 4(0.2)(-0.2)]$$
$$= -0.200 \text{ kg} \cdot \text{m}^2$$

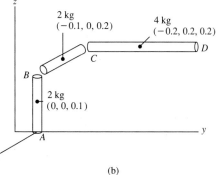

Fig. 21–5

The Aa axis is defined by the unit vector

$$\mathbf{u}_{Aa} = \frac{\mathbf{r}_D}{r_D} = \frac{-0.2\mathbf{i} + 0.4\mathbf{j} + 0.2\mathbf{k}}{\sqrt{(-0.2)^2 + (0.4)^2 + (0.2)^2}} = -0.408\mathbf{i} + 0.816\mathbf{j} + 0.408\mathbf{k}$$

Thus,

$$u_x = -0.408, \qquad u_y = 0.816, \qquad u_z = 0.408$$

Substituting the computed data into Eq. 21–5 yields

$$I_{Aa} = I_{xx}u_x^2 + I_{yy}u_y^2 + I_{zz}u_z^2 - 2I_{xy}u_xu_y - 2I_{yz}u_yu_z - 2I_{zx}u_zu_x$$

$$= 0.480(-0.408)^2 + (0.453)(0.816)^2 + 0.400(0.408)^2 - 2(-0.160)(-0.408)(0.816)$$
$$- 2(0.160)(0.816)(0.408) - 2(-0.200)(0.408)(-0.408)$$

$$= 0.168 \text{ kg} \cdot \text{m}^2 \qquad\qquad\qquad \textit{Ans.}$$

PROBLEMS

21–1. Show that the sum of the moments of inertia of a body, $I_{xx} + I_{yy} + I_{zz}$, is independent of the orientation of the x, y, z axes and thus depends only on the location of the origin.

21–2. Determine the moments of inertia I_x and I_y of the paraboloid of revolution. The mass of the paraboloid is 20 slug.

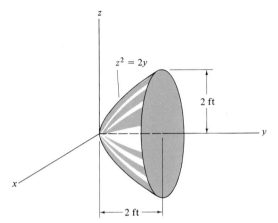

Prob. 21–2

21–3. Determine the moment of inertia I_y and the product of inertia I_{xy} of the body formed by revolving the shaded area about the line $x = 5$ ft. Express the result in terms of the density of the material, ρ.

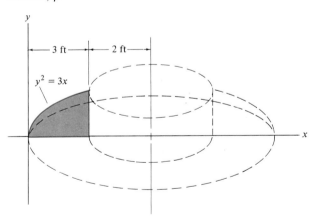

Prob. 21–3

*21–4. Determine the radii of gyration k_x and k_y for the solid formed by revolving the shaded area about the y axis. The density of the material is ρ.

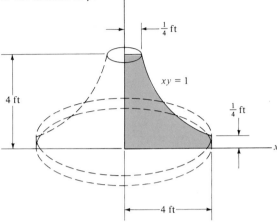

Prob. 21–4

21–5. Determine the product of inertia I_{yz} for the homogeneous prism. The density of the material is ρ. Express the result in terms of the total mass m of the prism.

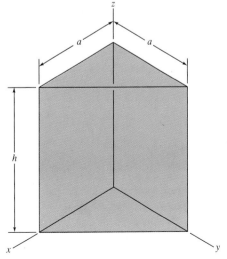

Prob. 21–5

445

21–6. Determine by direct integration the product of inertia I_{yz} for the homogeneous tetrahedron. The mass density of the material is ρ. Express the result in terms of the total mass m of the solid.

Prob. 21–6

21–7. Determine the moment of inertia of the cone about the z' axis. The weight of the cone is 15 lb, the *height* is $h = 1.5$ ft, and the radius is $r = 0.5$ ft.

Prob. 21–7

*21–8. Determine the moment of inertia I_{xx} and the product of inertia I_{xy} for the bent rod. The rod has a density of 2 kg/m.

Prob. 21–8

21–9. The bent rod has a weight of 1.5 lb/ft. Locate the center of gravity $G(\bar{x}, \bar{y})$ and determine the principal moments of inertia $I_{x'}$, $I_{y'}$, and $I_{z'}$ of the rod with respect to the x', y', z' axes.

Prob. 21–9

21–10. Rod AB has a weight of 6 lb and each sphere has a weight of 8 lb. Determine the moment of inertia of the assembly about the x axis.

Prob. 21–10

21–11. Determine the moment of inertia of the composite body about the aa axis. The cylinder weighs 20 lb, and each hemisphere weighs 10 lb.

Prob. 21–11

*21–12. Compute the moment of inertia of the rod-and-disk assembly about the xx axis. The disks each have a weight of 12 lb. The two rods each have a weight of 4 lb and their ends extend to the rim of the disks.

Prob. 21–12

21–13. Determine the moment of inertia of the 4-kg circular plate about the axis of rod OA.

Prob. 21–13

21–14. Compute the moment of inertia of the rod-and-thin-ring assembly about the z axis. The rods and ring have a mass of 2 kg/m.

Prob. 21–14

21–15. The assembly consists of two square plates A and B which have a mass of 3 kg each and a rectangular plate C which has a mass of 4.5 kg. Determine the moments of inertia with respect to the principal x, y, z axes.

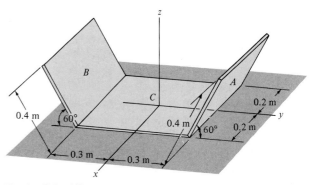

Prob. 21–15

*21–16. Determine the moments of inertia for the homogeneous cylinder of mass m with respect to the x', y', z' coordinate axes.

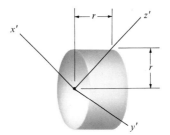

Prob. 21–16

21–17. Determine the elements of the inertia tensor for the cube with respect to the xyz coordinate system. The mass of the cube is m.

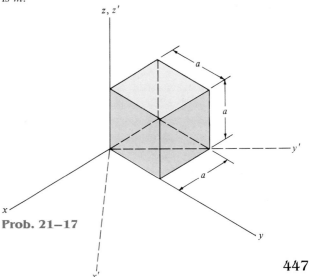

Prob. 21–17

447

*21.2 Angular Momentum

In this section we will develop the necessary equations used to determine the angular momentum of a rigid body about an arbitrary point. This formulation will provide the necessary means for developing both the principle of impulse and momentum and the equations of rotational motion for a rigid body.

Consider the rigid body in Fig. 21–6, which has a total mass m and center of mass located at G. The X, Y, Z coordinate system represents an inertial frame of reference, and hence, its axes are fixed or translate with a constant velocity. The angular momentum as measured from this reference will be computed relative to the arbitrary point A. The position vectors $\mathbf{r}_A$ and $\boldsymbol{\rho}_A$ are drawn from the origin of coordinates to point A and from A to the ith particle of the body. If the particle's mass is m_i, the angular momentum about point A is

$$(\mathbf{H}_A)_i = \boldsymbol{\rho}_A \times m_i \mathbf{v}_i$$

where $\mathbf{v}_i$ represents the particle's velocity as measured from the X, Y, Z coordinate system. If the body has an angular velocity $\boldsymbol{\omega}$ at the instant considered, $\mathbf{v}_i$ may be related to the velocity of A by applying Eq. 20–7, i.e.,

$$\mathbf{v}_i = \mathbf{v}_A + \boldsymbol{\omega} \times \boldsymbol{\rho}_A$$

Thus,
$$(\mathbf{H}_A)_i = \boldsymbol{\rho}_A \times m_i(\mathbf{v}_A + \boldsymbol{\omega} \times \boldsymbol{\rho}_A)$$
$$= (\boldsymbol{\rho}_A \, m_i) \times \mathbf{v}_A + \boldsymbol{\rho}_A \times (\boldsymbol{\omega} \times \boldsymbol{\rho}_A) \, m_i$$

For the entire body, summing all the particles of the body requires an integration, i.e., $m_i \rightarrow dm$ so that

$$\mathbf{H}_A = \left(\int_m \boldsymbol{\rho}_A \, dm \right) \times \mathbf{v}_A + \int_m \boldsymbol{\rho}_A \times (\boldsymbol{\omega} \times \boldsymbol{\rho}_A) \, dm \qquad (21\text{–}6)$$

Fig. 21–6

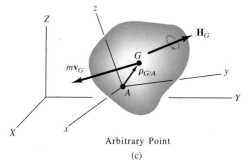

Fig. 21–7

Fixed Point O

If A becomes a *fixed point* O in the body, Fig. 21–7a, then $\mathbf{v}_A = \mathbf{0}$ and Eq. 21–6 reduces to

$$\mathbf{H}_O = \int_m \boldsymbol{\rho}_O \times (\boldsymbol{\omega} \times \boldsymbol{\rho}_O)\ dm \qquad (21\text{–}7)$$

Center of Mass G

If A is located at the *center of mass* G of the body, Fig. 21–7b, then $\int_m \boldsymbol{\rho}_A\ dm = \mathbf{0}$ and

$$\mathbf{H}_G = \int_m \boldsymbol{\rho}_G \times (\boldsymbol{\omega} \times \boldsymbol{\rho}_G)\ dm \qquad (21\text{–}8)$$

Arbitrary Point A

In general, A may be some point other than O or G, Fig. 21–7c, in which case Eq. 21–6 may nevertheless be simplified to the following form (see Prob. 21–19).

$$\mathbf{H}_A = \boldsymbol{\rho}_{G/A} \times m\mathbf{v}_G + \mathbf{H}_G \qquad (21\text{–}9)$$

Here it can be seen that the angular momentum consists of two parts—the moment of the linear momentum $m\mathbf{v}_G$ of the body about point A added (vectorially) to the angular momentum $\mathbf{H}_G$. Equation 21–9 may also be used for computing the angular momentum of the body about a fixed point O; the results, of course, will be the same as those computed using the more convenient Eq. 21–7.

Rectangular Components of $\mathbf{H}$

To make practical use of Eqs. 21–7 to 21–9, the angular momentum must be expressed in terms of its scalar components. For this purpose, it is convenient to choose a second set of x, y, z axes having an arbitrary orientation relative to the X, Y, Z axes, Fig. 21–7. For a general explanation, note that Eqs. 21–7 to 21–9 all contain the form

$$\mathbf{H} = \int_m \boldsymbol{\rho} \times (\boldsymbol{\omega} \times \boldsymbol{\rho}) \, dm$$

Expressing $\mathbf{H}$, $\boldsymbol{\rho}$, and $\boldsymbol{\omega}$ in terms of x, y, and z components, we have

$$H_x\mathbf{i} + H_y\mathbf{j} + H_z\mathbf{k} = \int_m (x\mathbf{i} + y\mathbf{j} + z\mathbf{k}) \times [(\omega_x\mathbf{i} + \omega_y\mathbf{j} + \omega_z\mathbf{k})$$
$$\times \ (x\mathbf{i} + y\mathbf{j} + z\mathbf{k})] \, dm$$

Expanding the cross products and combining terms yields

$$H_x\mathbf{i} + H_y\mathbf{j} + H_z\mathbf{k} = \left[\omega_x \int_m (y^2 + z^2) \, dm - \omega_y \int_m xy \, dm - \omega_z \int_m xz \, dm\right]\mathbf{i}$$
$$+ \left[-\omega_x \int_m xy \, dm + \omega_y \int_m (x^2 + z^2) \, dm - \omega_z \int_m yz \, dm\right]\mathbf{j}$$
$$+ \left[-\omega_x \int_m zx \, dm - \omega_y \int_m yz \, dm + \omega_z \int_m (x^2 + y^2) \, dm\right]\mathbf{k}$$

Equating the respective $\mathbf{i}$, $\mathbf{j}$, $\mathbf{k}$ components and recognizing that the integrals represent the moments and products of inertia, we obtain

$$\boxed{\begin{aligned} H_x &= I_{xx}\omega_x - I_{xy}\omega_y - I_{xz}\omega_z \\ H_y &= -I_{yx}\omega_x + I_{yy}\omega_y - I_{yz}\omega_z \\ H_z &= -I_{zx}\omega_x - I_{zy}\omega_y + I_{zz}\omega_z \end{aligned}} \tag{21–10}$$

These three equations represent the scalar form of the $\mathbf{i}$, $\mathbf{j}$, and $\mathbf{k}$ components of $\mathbf{H}_O$ or $\mathbf{H}_G$ (given in vector form by Eqs. 21–7 and 21–8). The angular momentum of the body about the arbitrary point A, other than the fixed point O or the center of mass G, may also be expressed in scalar form. Here it is necessary to use Eq. 21–9 and to represent $\boldsymbol{\rho}_{G/A}$ and $\mathbf{v}_G$ as Cartesian vectors, carry out the cross-product operation, and substitute the components, Eqs. 21–10, for $\mathbf{H}_G$.

Equations 21–10 may be simplified further if the x, y, z coordinate axes are oriented such that they become *principal axes of inertia* for the body at the point. When these axes are used, computation of the products of inertia $I_{xy} = I_{yz} = I_{zx} = 0$, and if the principal moments of inertia about the x, y, z axes are represented as $I_x = I_{xx}$, $I_y = I_{yy}$, and $I_z = I_{zz}$, the three components of angular momentum become

$$\boxed{H_x = I_x\omega_x \qquad H_y = I_y\omega_y \qquad H_z = I_z\omega_z} \tag{21–11}$$

Principle of Impulse and Momentum

Now that the means for computing the angular momentum for a body have been presented, the *principle of impulse and momentum*, as discussed in Sec. 19.2, may be used to solve kinetics problems which involve *force, velocity, and time*. For this case, the following two vector equations are available:

$$m(\mathbf{v}_G)_1 + \Sigma \int_{t_1}^{t_2} \mathbf{F}\, dt = m(\mathbf{v}_G)_2 \qquad (21\text{--}12)$$

$$\mathbf{H}_1 + \Sigma \int_{t_1}^{t_2} \mathbf{M}_O\, dt = \mathbf{H}_2 \qquad (21\text{--}13)$$

In three dimensions each vector term can be represented by three scalar components, and therefore a total of *six scalar equations* can be written. Three equations relate the linear impulse and momentum in the x, y, z directions, and three equations relate the body's angular impulse and momentum about the x, y, z axes. Before applying Eqs. 21–12 and 21–13 to the solution of problems, the material in Secs. 19.2 and 19.3 should be reviewed.

*21.3 Kinetic Energy

In order to apply the principle of work and energy to the solution of problems involving general rigid-body motion, it is first necessary to formulate expressions for the kinetic energy of the body. In this regard, consider the rigid body shown in Fig. 21–8, which has a total mass m and center of mass located at G. The kinetic energy of the ith particle of the body having a mass m_i and velocity $\mathbf{v}_i$, measured relative to the inertial X, Y, Z frame of reference, is

$$T_i = \tfrac{1}{2} m_i\, v_i^2 = \tfrac{1}{2} m_i\, (\mathbf{v}_i \cdot \mathbf{v}_i)$$

Provided the velocity of an arbitrary point A in the body is known, $\mathbf{v}_i$ may be related to $\mathbf{v}_A$ by the equation $\mathbf{v}_i = \mathbf{v}_A + \boldsymbol{\omega} \times \boldsymbol{\rho}_A$, where $\boldsymbol{\omega}$ is the instantaneous angular velocity of the body, measured from the X, Y, Z coordinate system, and $\boldsymbol{\rho}_A$ is a position vector drawn from A to i. Using this expression for $\mathbf{v}_i$, the kinetic energy for the particle may be written as

$$T_i = \tfrac{1}{2} m_i\, (\mathbf{v}_A + \boldsymbol{\omega} \times \boldsymbol{\rho}_A) \cdot (\mathbf{v}_A + \boldsymbol{\omega} \times \boldsymbol{\rho}_A)$$
$$= \tfrac{1}{2}(\mathbf{v}_A \cdot \mathbf{v}_A)\, m_i + \mathbf{v}_A \cdot (\boldsymbol{\omega} \times \boldsymbol{\rho}_A)\, m_i + \tfrac{1}{2}(\boldsymbol{\omega} \times \boldsymbol{\rho}_A) \cdot (\boldsymbol{\omega} \times \boldsymbol{\rho}_A)\, m_i$$

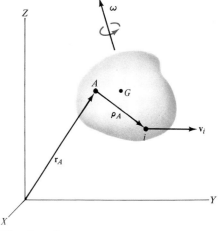

Fig. 21–8

The kinetic energy for the entire body is obtained by summing the kinetic energies of all the particles of the body. This requires an integration, since $m_i \to dm$, so that

$$T = \tfrac{1}{2} m(\mathbf{v}_A \cdot \mathbf{v}_A) + \mathbf{v}_A \cdot \left(\boldsymbol{\omega} \times \int_m \boldsymbol{\rho}_A\, dm \right) + \tfrac{1}{2} \int_m (\boldsymbol{\omega} \times \boldsymbol{\rho}_A) \cdot (\boldsymbol{\omega} \times \boldsymbol{\rho}_A)\, dm$$

The last term on the right may be rewritten using the vector identity $\mathbf{a} \times \mathbf{b} \cdot \mathbf{c} = \mathbf{a} \cdot \mathbf{b} \times \mathbf{c}$, where $\mathbf{a} = \boldsymbol{\omega}$, $\mathbf{b} = \boldsymbol{\rho}_A$, and $\mathbf{c} = \boldsymbol{\omega} \times \boldsymbol{\rho}_A$. The final result is therefore,

451

$$T = \tfrac{1}{2}m(\mathbf{v}_A \cdot \mathbf{v}_A) + \mathbf{v}_A \cdot \left(\boldsymbol{\omega} \times \int_m \boldsymbol{\rho}_A \, dm\right) + \tfrac{1}{2}\boldsymbol{\omega} \cdot \int_m \boldsymbol{\rho}_A \times (\boldsymbol{\omega} \times \boldsymbol{\rho}_A) \, dm$$

$$(21\text{–}14)$$

This equation is rarely used because of the computations involving the integrals. Simplification occurs, however, if the reference point A is either a fixed point O or the center of mass G.

Fixed Point O

If A is a *fixed point* O in the body, Fig. 21–7a, then $\mathbf{v}_A = \mathbf{0}$, and using Eq. 21–7, we can express Eq. 21–14 as

$$T = \tfrac{1}{2}\boldsymbol{\omega} \cdot \mathbf{H}_O$$

If the x, y, z axes represent the principal axes of inertia for the body, then $\boldsymbol{\omega} = \omega_x\mathbf{i} + \omega_y\mathbf{j} + \omega_z\mathbf{k}$ and $\mathbf{H}_O = I_x\omega_x\mathbf{i} + I_y\omega_y\mathbf{j} + I_z\omega_z\mathbf{k}$. Substituting into the above equation and performing the dot product operations yields

$$T = \tfrac{1}{2}I_x\omega_x^2 + \tfrac{1}{2}I_y\omega_y^2 + \tfrac{1}{2}I_z\omega_z^2 \qquad (21\text{–}15)$$

Center of Mass G

If A is located at the *center of mass* G of the body, Fig. 21–7b, then $\int_m \boldsymbol{\rho}_A \, dm = \mathbf{0}$ and, using Eq. 21–8, we can write Eq. 21–14 as

$$T = \tfrac{1}{2}mv_G^2 + \tfrac{1}{2}\boldsymbol{\omega} \cdot \mathbf{H}_G$$

In a manner similar to that for a fixed point, the last term on the right side may be represented in scalar form, in which case

$$T = \tfrac{1}{2}mv_G^2 + \tfrac{1}{2}I_x\omega_x^2 + \tfrac{1}{2}I_y\omega_y^2 + \tfrac{1}{2}I_z\omega_z^2 \qquad (21\text{–}16)$$

Here it is seen that the kinetic energy consists of two parts; namely, the translational kinetic energy of the mass center, $\tfrac{1}{2}mv_G^2$, and the body's rotational kinetic energy.

Principle of Work and Energy

Using one of the above expressions for computing the kinetic energy of a body, the *principle of work and energy* may be applied to solve kinetics problems which involve *force, velocity, and displacement*. For this case only one scalar equation can be written for each body, i.e.,

$$T_1 + \Sigma U_{1-2} = T_2 \qquad (21\text{–}17)$$

Before applying this equation to the solution of problems the material in Chapter 18 should be reviewed.

Example 21–2

The rod in Fig. 21–9a has a weight of 1.5 lb/ft. Determine its angular velocity just after the end A falls onto the hook at E. The hook provides a permanent connection for the rod due to the spring-lock mechanism S. Just before striking the hook the rod is falling downward with a speed $(v_G)_1 = 10$ ft/s.

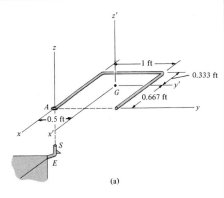

(a)

SOLUTION

The principle of impulse and momentum will be used since impact occurs.

Impulse and Momentum Diagrams. Fig. 21–9b. During the short time Δt, the impulsive force $\mathbf{F}$ acting at A changes the momentum of the rod. (The impulse created by the rod's weight $\mathbf{W}$ during this time is small compared to $\int \mathbf{F} \, dt$, so that it is neglected, i.e., the weight is a nonimpulsive force.) Hence, the angular momentum of the rod is *conserved* about point A since the moment of $\int \mathbf{F} \, dt$ about A is zero.

Conservation of Angular Momentum. Equation 21–9 must be used for computing the angular momentum of the rod, since A does not become a *fixed point* until *after* the impulsive interaction with the hook. Thus, with reference to Fig. 21–9b, $(\mathbf{H}_A)_1 = (\mathbf{H}_A)_2$, or

$$\mathbf{r}_{G/A} \times m(\mathbf{v}_G)_1 = \mathbf{r}_{G/A} \times m(\mathbf{v}_G)_2 + (\mathbf{H}_G)_2 \qquad (1)$$

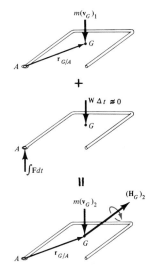

(b)

Fig. 21–9

From Fig. 21–9a, $\mathbf{r}_{G/A} = \{-0.667\mathbf{i} + 0.5\mathbf{j}\}$ ft. Furthermore, the primed axes are principal axes of inertia for the rod because $I_{x'y'} = I_{x'z'} = I_{z'y'} = 0$. Hence, from Eqs. 21–11, $(\mathbf{H}_G)_2 = I_{x'}\omega_x\mathbf{i} + I_{y'}\omega_y\mathbf{j} + I_{z'}\omega_z\mathbf{k}$. The principal moments of inertia are $I_{x'} = I_{y'} = 0.0155$ slug $\cdot$ ft^2, $I_{z'} = 0.0401$ slug $\cdot$ ft^2 (see Prob. 21–9). Substituting into Eq. (1), we have

$$(-0.667\mathbf{i} + 0.5\mathbf{j}) \times \left[\left(\frac{4.5}{32.2}\right)(-10\mathbf{k})\right] = (-0.667\mathbf{i} + 0.5\mathbf{j}) \times \left[\left(\frac{4.5}{32.2}\right)(-v_G)_2\mathbf{k}\right]$$

$$+ \ 0.0155\omega_x\mathbf{i} + 0.0155\omega_y\mathbf{j} + 0.0401\omega_z\mathbf{k}$$

Expanding and equating the respective $\mathbf{i}$, $\mathbf{j}$, and $\mathbf{k}$ components yields

$$-0.699 = -0.0699v_G + 0.0155\omega_x \qquad (2)$$
$$-0.932 = -0.0932v_G + 0.0155\omega_y \qquad (3)$$
$$0 = 0.0401\omega_z \qquad (4)$$

Kinematics. There are four unknowns in the above equations; however, another equation may be obtained by relating $\boldsymbol{\omega}$ to $(\mathbf{v}_G)_2$ using *kinematics*. Since $\omega_z = 0$ (Eq. 4) and after impact the rod rotates about the fixed point A, Eq. 20–3 may be applied, in which case $\mathbf{v}_G = \boldsymbol{\omega} \times \mathbf{r}_{G/A}$, or

$$-(v_G)_2\mathbf{k} = (\omega_x\mathbf{i} + \omega_y\mathbf{j}) \times (-0.667\mathbf{i} + 0.5\mathbf{j})$$
$$-(v_G)_2 = 0.5\omega_x + 0.667\omega_y \qquad (5)$$

Solving Eqs. (2) to (5) simultaneously yields

$$(\mathbf{v}_G)_2 = \{-8.62\mathbf{k}\} \text{ ft/s} \qquad \boldsymbol{\omega} = \{-6.21\mathbf{i} - 8.29\mathbf{j}\} \text{ rad/s} \qquad \textit{Ans.}$$

Example 21–3

A 5-N · m torque is applied to the vertical shaft CD shown in Fig. 21–10a, which allows the 10-kg gear A to turn freely about CE. Assuming that gear A starts from rest, determine the angular velocity of CD after it has turned two revolutions. Neglect the mass of shaft CD and axle CE and assume that gear A is approximated by a thin disk. Gear B is fixed.

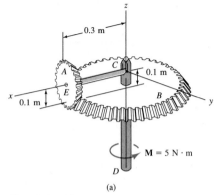

(a)

Fig. 21–10a

SOLUTION
The principle of work and energy may be used for the solution. Why?

Work. If shaft CD, the axle CE, and gear A are considered as a system of connected bodies, only the applied torque $\mathbf{M}$ does work. For two revolutions of CD, this work is $\Sigma U_{1-2} = (5 \text{ N} \cdot \text{m})(4\pi \text{ rad}) = 62.83 \text{ J}$.

Kinetic Energy. Since the gear is initially at rest, its initial kinetic energy is zero. A kinematic diagram for the gear is shown in Fig. 21–10b. If the angular velocity of CD is taken as $\boldsymbol{\omega}_{CD}$, then the angular velocity of gear A is $\boldsymbol{\omega}_A = \boldsymbol{\omega}_{CD} + \boldsymbol{\omega}_{CE}$. The gear may be imagined as a portion of a massless extended body which is rotating about the *fixed point* C. The instantaneous axis of rotation for this body is along line CH, because both points C and H on the body (gear) have zero velocity and must therefore lie on this axis. This requires that the components $\boldsymbol{\omega}_{CD}$ and $\boldsymbol{\omega}_{CE}$ be related by the equation $\omega_{CD}/0.1 \text{ m} = \omega_{CE}/0.3 \text{ m}$ or $\omega_{CE} = 3\omega_{CD}$. Thus,

$$\boldsymbol{\omega}_A = -\omega_{CE}\mathbf{i} + \omega_{CD}\mathbf{k} = -3\omega_{CD}\mathbf{i} + \omega_{CD}\mathbf{k} \qquad (1)$$

The x, y, z axes in Fig. 21–10a represent *principal axes of inertia* at C for the gear. Since point C is a fixed point of rotation, Eq. 21–15 may be applied to determine the kinetic energy, i.e.,

$$T = \tfrac{1}{2}I_x\omega_x^2 + \tfrac{1}{2}I_y\omega_y^2 + \tfrac{1}{2}I_z\omega_z^2 \qquad (2)$$

Using the parallel-axis theorem, the moments of inertia of the gear about point C are as follows:

(b)

Fig. 21–10b

$$I_x = \frac{1}{2}(10)(0.1)^2 = 0.05 \text{ kg} \cdot \text{m}^2$$

$$I_y = I_z = \frac{1}{4}(10)(0.1)^2 + 10(0.3)^2 = 0.925 \text{ kg} \cdot \text{m}^2$$

Since $\omega_x = -3\omega_{CD}$, $\omega_y = 0$, $\omega_z = \omega_{CD}$, Eq. (2) becomes

$$T_A = \frac{1}{2}(0.05)(-3\omega_{CD})^2 + 0 + \frac{1}{2}(0.925)(\omega_{CD})^2 = 0.6875\omega_{CD}^2$$

Principle of Work and Energy. Applying the principle of work and energy, we obtain

$$T_1 + \Sigma U_{1-2} = T_2 \qquad 0 + 62.83 = 0.6875\omega_{CD}^2$$

$$\omega_{CD} = 9.56 \text{ rad/s} \qquad\qquad Ans.$$

PROBLEMS

21–18. If a body contains *no planes of symmetry*, the principal moments of inertia can be computed mathematically. To show how this is done, consider the rigid body which is spinning with an angular velocity ω directed along one of its principal axes of inertia. If the principal moment of inertia about this axis is I, the angular momentum can be expressed as $\mathbf{H} = I\omega = I\omega_x\mathbf{i} + I\omega_y\mathbf{j} + I\omega_z\mathbf{k}$. The components of $\mathbf{H}$ may also be expressed by Eqs. 21–10, where the inertia tensor is assumed to be known. Equate the $\mathbf{i}$, $\mathbf{j}$, and $\mathbf{k}$ components of both expressions for $\mathbf{H}$ and consider ω_x, ω_y, and ω_z to be unknown. The solution of these three equations is obtained provided the determinant of the coefficients is zero. Show that this determinant, when expanded, yields the cubic equation

$$I^3 - (I_{xx} + I_{yy} + I_{zz})I^2 + (I_{xx}I_{yy} + I_{yy}I_{zz} + I_{zz}I_{xx}$$
$$- I_{xy}^2 - I_{yz}^2 - I_{zx}^2)I - (I_{xx}I_{yy}I_{zz} - 2I_{xy}I_{yz}I_{zx}$$
$$- I_{xx}I_{yz}^2 - I_{yy}I_{zx}^2 - I_{zz}I_{xy}^2) = 0$$

The three positive roots of I, obtained from the solution of this equation, represent the principal moments of inertia I_x, I_y, and I_z.

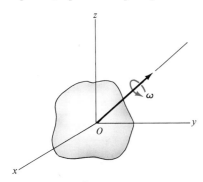

Prob. 21–18

21–19. Show that if the angular momentum of a body is computed with respect to an arbitrary point A, then $\mathbf{H}_A$ can be expressed by Eq. 21–9. This requires substituting $\boldsymbol{\rho}_A = \boldsymbol{\rho}_G + \boldsymbol{\rho}_{G/A}$ into Eq. 21–6 and expanding, noting that $\int \boldsymbol{\rho}_G \, dm = \mathbf{0}$ by definition of the mass center and $\mathbf{v}_G = \mathbf{v}_A + \boldsymbol{\omega} \times \boldsymbol{\rho}_{G/A}$.

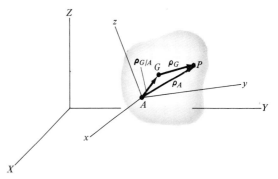

Prob. 21–19

*21–20.** Determine the kinetic energy of the 7-kg disk and 1.5-kg rod when the assembly is rotating about the z axis at $\omega = 5$ rad/s.

Probs. 21–22/21–23

21–21. Determine the magnitude of the angular momentum of the 7-kg disk and 1.5-kg rod about point C. The shaft BA is rotating at $\omega = 5$ rad/s.

455

21–22. Rod *AB* has a weight of 6 lb and is attached to two smooth collars at its end points by ball-and-socket joints. If collar *A* is moving downward at a speed of 8 ft/s, determine the kinetic energy of the rod at the instant shown. Assume that at this instant the angular velocity of the rod is directed perpendicular to the rod's axis. *Note:* The motion has been specified in Prob. 20–13.

$v_A = 8$ m/s

3 ft

7 ft

2 ft

B

6 ft

Probs. 21–20/21–21

21–23. Compute the magnitude of the angular momentum of rod *AB* in Prob. 21–22 about the mass center *G* at the instant shown. *Note:* The motion has been specified in Prob. 20–13.

*21–24. The 2-kg gear *A* rolls on the fixed plate gear *B*. Determine the angular velocity of rod *AO* about the *z* axis after it rotates one revolution about the *z* axis starting from rest. The rod is acted upon by the constant moment of *M* = 5 N · m. Neglect the mass of rod *OB*. Assume that gear *A* is a uniform disk having a radius of 100 mm.

$M = 5$ N · m

100 mm

300 mm

O

A

B

100 mm

C

Prob. 21–24

21–25. The 4-lb rod *AB* is attached to the disk and collar using ball-and-socket joints. If the disk has a constant angular velocity of 2 rad/s, compute the kinetic energy of the rod when it is in the position shown. Assume the angular velocity of the rod is directed perpendicular to the axis of the rod. *Hint:* The motion has been specified in Prob. 20–18.

B

2 rad/s

y

A

1 ft

3 ft

x

Probs. 21–25/21–26

21–26. Compute the angular momentum of rod *AB* in Prob. 21–25 at the instant shown. *Hint:* The motion has been specified in Prob. 20–18.

21–27. The rod assembly has a mass of 2.5 kg/m and is rotating with a constant angular velocity of $\boldsymbol{\omega} = \{2\mathbf{k}\}$ rad/s when the looped end at *C* encounters a hook at *S*, which provides a permanent connection. Determine the angular velocity of the assembly immediately after impact.

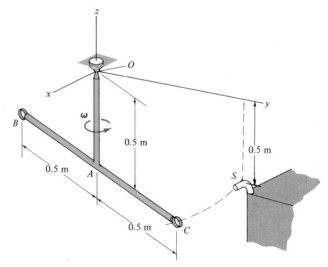

O

x

y

B

ω

0.5 m

0.5 m

A

0.5 m

0.5 m

S

0.5 m

C

Prob. 21–27

*21–28. The 25-lb thin plate is suspended from a ball-and-socket joint at O. A 0.2-lb bullet is fired with a velocity of $\mathbf{v} = \{-300\mathbf{i} - 250\mathbf{j} + 300\mathbf{k}\}$ ft/s into the plate and becomes embedded in the plate at point A. Determine the angular momentum and the instantaneous axis of rotation of the plate just after impact.

Prob. 21–28

21–29. The 15-lb plate is subjected to a force $F = 8$ lb which is always directed perpendicular to the face of the plate. If the plate is originally at rest, determine its angular velocity after it has rotated one revolution (360°). The plate is supported by ball-and-socket joints at A and B.

Prob. 21–29

21–30. The 4-lb rod AB is attached to the 1-lb collar at A and a 2-lb link BC using ball-and-socket joints. If the rod is released from rest in the position shown, determine the angular velocity of the link after it has rotated 180°.

Prob. 21–30

21–31. Rod AB has a weight of 6 lb and is attached to two smooth collars at its ends by ball-and-socket joints. If collar A is moving downward with a speed of 8 ft/s when $z = 3$ ft, determine the speed of collar A at the instant $z = 0$. The spring has an unstretched length of 2 ft. Neglect the mass of the collars. Assume the angular velocity of rod AB is perpendicular to its axis. See Prob. 20–11.

Prob. 21–31

*21–32. The 5-kg thin plate is suspended at O using a ball-and-socket joint. It is rotating with a constant angular velocity $\boldsymbol{\omega} = \{2\mathbf{k}\}$ rad/s when the corner A strikes the hook at S, which provides a permanent connection. Determine the angular velocity of the plate immediately after impact.

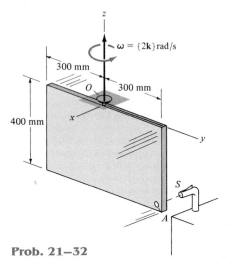

Prob. 21–32

21–33. The 15-kg rectangular plate is free to rotate about the y axis because of the bearing supports at A and B. When the plate is balanced in the vertical plane, a 3-g bullet is fired into it, perpendicular to its surface, with a velocity $\mathbf{v} = \{-2000\mathbf{i}\}$ m/s. Compute the angular velocity of the plate at the instant it has rotated $180°$. If the bullet strikes corner D with the same velocity $\mathbf{v}$, instead of at C, does the angular velocity remain the same? Why or why not?

Prob. 21–33

21–34. The circular disk has a weight of 15 lb and is mounted on the shaft AB at an angle of $45°$ with the horizontal. Determine the angular velocity of the shaft when $t = 2$ s if a torque $M = (4e^{0.1t})$ lb · ft, where t is in seconds, is applied to the shaft. The shaft is originally spinning at $\omega_1 = 8$ rad/s when the torque is applied.

Prob. 21–34

21–35. Each of the two disks has a weight of 10 lb. The axle AB weighs 3 lb. If the assembly is rotating about the z axis at $\omega_z = 6$ rad/s, determine its angular momentum about the z axis and its kinetic energy. The disks roll without slipping.

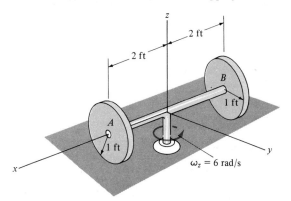

Prob. 21–35

*21–36. The rod assembly is supported at G by a ball-and-socket joint. Each rod has a mass of 0.5 kg/m. If the assembly is originally at rest and an impulse of $\mathbf{I} = \{-8\mathbf{k}\}$ N · s is applied at D, determine the angular velocity of the assembly just after the impact.

Prob. 21–36

21–37. The assembly consists of a 4-kg rod *AB* which is connected to link *OA* and the collar at *B* by ball-and-socket joints. When $\theta = 0°$, the system is at rest, the spring is unstretched, and a 7-N · m couple moment is applied to the link at *O*. Determine the angular velocity of the link at the instant $\theta = 90°$. Neglect the mass of the link.

21–38. The space capsule has a mass of 3.5 Mg and the radii of gyration are $k_x = k_z = 0.8$ m and $k_y = 0.5$ m. If it is traveling with a velocity $\mathbf{v}_G = \{600\mathbf{j}\}$ m/s, compute its angular velocity just after it is struck by a meteoroid having a mass of 0.60 kg and a velocity of $\mathbf{v}_m = \{-200\mathbf{i} - 400\mathbf{j} + 200\mathbf{k}\}$ m/s. Assume that the meteoroid embeds itself into the capsule at point *A* and that the capsule initially has no angular velocity.

Prob. 21–37

Prob. 21–38

*21.4 Equations of Motion

Having become familiar with the techniques used to describe both the inertial properties and the angular momentum of a body, we can now write the equations which describe the motion of the body in their most useful forms.

Equations of Translational Motion

The *translational motion* of a body is defined in terms of the acceleration of the body's mass center, which is measured from an inertial *X*, *Y*, *Z* reference. The equation of translational motion for the body can be written in vector form as

$$\Sigma \mathbf{F} = m\mathbf{a}_G \qquad (21\text{–}18)$$

or by the three scalar equations

$$\begin{aligned} \Sigma F_x &= m(a_G)_x \\ \Sigma F_y &= m(a_G)_y \\ \Sigma F_z &= m(a_G)_z \end{aligned} \qquad (21\text{-}19)$$

Here, $\Sigma \mathbf{F} = \Sigma F_x\mathbf{i} + \Sigma F_y\mathbf{j} + \Sigma F_z\mathbf{k}$ represents the sum of all the external forces acting on the body.

Equations of Rotational Motion

In Sec. 15.6, we developed Eq. 15–17, namely,

$$\Sigma\mathbf{M}_O = \dot{\mathbf{H}}_O \tag{21–20}$$

which states that the sum of the moments about a fixed point O of all the external forces acting on a system of particles (contained in a rigid body) is equal to the time rate of change of the total angular momentum of the body about point O. When moments of the external forces acting on the particles are summed about the system's *mass center* G, one again obtains the same simple form of Eq. 21–20, relating the moment summation $\Sigma\mathbf{M}_G$ to the angular momentum $\mathbf{H}_G$. To show this, consider the system of particles in Fig. 21–11, where x, y, z represents an inertial frame of reference and the x', y', z' axes, with origin at G, *translate* with respect to this frame. In general, G is *accelerating*, so by definition the translating frame is *not* an inertial reference. The angular momentum of the ith particle with respect to this frame is, however,

$$(\mathbf{H}_i)_G = \mathbf{r}_{i/G} \times m_i\mathbf{v}_{i/G}$$

where $\mathbf{r}_{i/G}$ and $\mathbf{v}_{i/G}$ represent the relative position and relative velocity of the ith particle with respect to G. Taking the time derivative gives

$$(\dot{\mathbf{H}}_i)_G = \dot{\mathbf{r}}_{i/G} \times m_i\mathbf{v}_{i/G} + \mathbf{r}_{i/G} \times m_i\dot{\mathbf{v}}_{i/G}$$

By definition, $\mathbf{v}_{i/G} = \dot{\mathbf{r}}_{i/G}$. Thus, the first term on the right side is zero since the cross product of equal vectors is zero. Also, $\mathbf{a}_{i/G} = \dot{\mathbf{v}}_{i/G}$, so that

$$(\dot{\mathbf{H}}_i)_G = \mathbf{r}_{i/G} \times m_i\mathbf{a}_{i/G}$$

Similar expressions can be written for the other particles of the system. When the results are summed for all the particles, we get

$$\dot{\mathbf{H}}_G = \Sigma(\mathbf{r}_{i/G} \times m_i\mathbf{a}_{i/G})$$

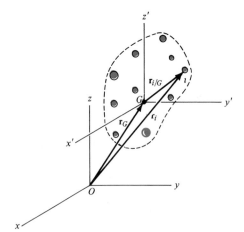

Fig. 21–11

Here $\dot{\mathbf{H}}_G$ is the time rate of change of the total angular momentum of the body computed relative to point G.

The relative acceleration for the ith particle is defined by the equation $\mathbf{a}_{i/G} = \mathbf{a}_i - \mathbf{a}_G$, where $\mathbf{a}_i$ and $\mathbf{a}_G$ represent, respectively, the accelerations of the ith particle and point G measured with respect to the *inertial frame of reference*. Substituting and expanding, using the distributive property of the vector cross product, yields

$$\dot{\mathbf{H}}_G = \Sigma(\mathbf{r}_{i/G} \times m_i\mathbf{a}_i) - (\Sigma m_i\mathbf{r}_{i/G}) \times \mathbf{a}_G$$

By definition of the mass center, the sum $(\Sigma m_i\mathbf{r}_{i/G}) = (\Sigma m_i)\bar{\mathbf{r}}$ is equal to zero, since the position vector $\bar{\mathbf{r}}$ relative to G is zero. Hence, the last term in the above equation is zero. Using the equation of motion, the product $m_i\mathbf{a}_i$ may be replaced by the resultant *external force* $\mathbf{F}_i$ acting on the ith particle. Denoting $\Sigma\mathbf{M}_G = \Sigma\mathbf{r}_{i/G} \times \mathbf{F}_i$, the final result may be written as

$$\Sigma\mathbf{M}_G = \dot{\mathbf{H}}_G \qquad (21\text{-}21)$$

The rotational equation of motion for the body will now be developed from either Eq. 21–20 or 21–21. In this regard, the scalar components of the angular momentum $\mathbf{H}_O$ or $\mathbf{H}_G$ are defined by Eqs. 21–10 or, if principal axes of inertia are used either at point O or G, by Eqs. 21–11. If these components are computed about x, y, z axes that are *rotating* with an angular velocity $\boldsymbol{\Omega}$, which may be *different* from the body's angular velocity $\boldsymbol{\omega}$, then the time derivative $\dot{\mathbf{H}} = d\mathbf{H}/dt$, as used in Eqs. 21–20 and 21–21, must account for the rotation of the x, y, z axes as measured from the inertial X, Y, Z axes. Hence, the time derivative of $\mathbf{H}$ must be determined from Eq. 20–6, in which case Eqs. 21–20 and 21–21 become

$$\boxed{\begin{aligned} \Sigma\mathbf{M}_O &= (\dot{\mathbf{H}}_O)_{xyz} + \boldsymbol{\Omega} \times \mathbf{H}_O \\ \Sigma\mathbf{M}_G &= (\dot{\mathbf{H}}_G)_{xyz} + \boldsymbol{\Omega} \times \mathbf{H}_G \end{aligned}} \qquad (21\text{-}22)$$

Here $(\dot{\mathbf{H}})_{xyz}$ is the time rate of change of $\mathbf{H}$ measured from the x, y, z reference.

There are three ways in which one can define the motion of the x, y, z axes. Obviously, motion of this reference should be chosen to yield the simplest set of moment equations for the solution of a particular problem.

***x*, *y*, *z* Axes Having Motion $\boldsymbol{\Omega} = 0$.** If the body has general motion, the x, y, z axes may be chosen with origin at G, such that the axes only *translate* relative to the inertial X, Y, Z frame of reference. Doing this would certainly simplify Eq. 21–22, since $\boldsymbol{\Omega} = \mathbf{0}$. However, the body may have a rotation $\boldsymbol{\omega}$ about these axes, and therefore the moments and products of inertia of the body would have to be expressed as *functions of time*. In most cases this would be a difficult task, so that such a choice of axes has restricted value.

x, y, z Axes Having Motion $\boldsymbol{\Omega} = \boldsymbol{\omega}$. The x, y, z axes may be chosen such that they are *fixed in and move with the body*. The moments and products of inertia of the body relative to these axes will be *constant* during the motion. Since $\boldsymbol{\Omega} = \boldsymbol{\omega}$, Eqs. 21–22 become

$$\Sigma \mathbf{M}_O = (\dot{\mathbf{H}}_O)_{xyz} + \boldsymbol{\omega} \times \mathbf{H}_O$$
$$\Sigma \mathbf{M}_G = (\dot{\mathbf{H}}_G)_{xyz} + \boldsymbol{\omega} \times \mathbf{H}_G$$

(21–23)

We may express each of these vector equations as three scalar equations using Eqs. 21–10. Neglecting the subscripts O and G yields

$$\Sigma M_x = I_{xx}\,\dot{\omega}_x - (I_{yy} - I_{zz})\omega_y\omega_z - I_{xy}(\dot{\omega}_y - \omega_z\omega_x) - I_{yz}(\omega_y^2 - \omega_z^2)$$
$$- I_{zx}(\dot{\omega}_z + \omega_x\omega_y)$$

$$\Sigma M_y = I_{yy}\,\dot{\omega}_y - (I_{zz} - I_{xx})\omega_z\omega_x - I_{yz}(\dot{\omega}_z - \omega_x\omega_y) - I_{zx}(\omega_z^2 - \omega_x^2)$$
$$- I_{xy}(\dot{\omega}_x + \omega_y\omega_z) \qquad (21–24)$$

$$\Sigma M_z = I_{zz}\,\dot{\omega}_z - (I_{xx} - I_{yy})\omega_x\omega_y - I_{zx}(\dot{\omega}_x - \omega_y\omega_z) - I_{xy}(\omega_x^2 - \omega_y^2)$$
$$- I_{yz}(\dot{\omega}_y + \omega_z\omega_x)$$

Notice that for a rigid body symmetric with respect to the xy reference plane, and undergoing general plane motion in this plane, $I_{xz} = I_{yz} = 0$, and $\omega_x = \omega_y = d\omega_x/dt = d\omega_y/dt = 0$. Equations 21–24 reduce to the form $\Sigma M_x = \Sigma M_y = 0$, and $\Sigma M_z = I_{zz}\alpha_z$ (where $\alpha_z = \dot{\omega}_z$), which is essentially the third of Eqs. 17–11 or 17–16 depending upon the choice of point G or O for summing moments.

If the x, y, and z axes are chosen as *principal axes of inertia*, the products of inertia are zero, $I_{xx} = I_x$, etc., and Eqs. 21–24 reduce to the form

$$\Sigma M_x = I_x\dot{\omega}_x - (I_y - I_z)\omega_y\omega_z$$
$$\Sigma M_y = I_y\dot{\omega}_y - (I_z - I_x)\omega_z\omega_x$$
$$\Sigma M_z = I_z\dot{\omega}_z - (I_x - I_y)\omega_x\omega_y$$

(21–25)

This set of equations is known historically as the *Euler equations of motion*, named after the Swiss mathematician Leonhard Euler (1707–1783). They *only* apply for moments summed either about point O or G.

When applying these equations it should be realized that $\dot{\omega}_x$, $\dot{\omega}_y$, and $\dot{\omega}_z$ represent the time derivatives of the magnitudes of the x, y, z components of $\boldsymbol{\omega}$. Since the x, y, z axes are rotating at $\boldsymbol{\Omega} = \boldsymbol{\omega}$, then, from Eq. 20–6, it may be noted that $(\dot{\boldsymbol{\omega}})_{XYZ} = (\dot{\boldsymbol{\omega}})_{xyz} + \boldsymbol{\omega} \times \boldsymbol{\omega}$. Since $\boldsymbol{\omega} \times \boldsymbol{\omega} = \mathbf{0}$, $(\dot{\boldsymbol{\omega}})_{XYZ} = (\dot{\boldsymbol{\omega}})_{xyz}$. This important result indicates that the required time derivative of $\boldsymbol{\omega}$ can be obtained either by first finding the components of $\boldsymbol{\omega}$ along the x, y, z axes and then taking the time derivative of the magnitudes of these components, i.e., $(\dot{\boldsymbol{\omega}})_{xyz}$, or by first finding the time derivative of $\boldsymbol{\omega}$ with respect to the X, Y, Z axes, i.e., $(\dot{\boldsymbol{\omega}})_{XYZ}$, and then determining the components $\dot{\omega}_x$, $\dot{\omega}_y$, and $\dot{\omega}_z$. In practice, it is generally easier to compute $\dot{\omega}_x$, $\dot{\omega}_y$, and $\dot{\omega}_z$ on the basis of finding $(\dot{\boldsymbol{\omega}})_{XYZ}$.

x, y, z Axes Having Motion $\boldsymbol{\Omega} \neq \boldsymbol{\omega}$. To simplify the calculations for the time derivative of $\boldsymbol{\omega}$, it is often convenient to choose the x, y, z axes having an angular velocity $\boldsymbol{\Omega}$ which is different from the angular velocity $\boldsymbol{\omega}$ of the body. This is particularly suitable for the analysis of spinning tops and gyroscopes which are *symmetrical* about their spinning axis.* When this is the case, the moments and products of inertia remain constant during the motion.

Equations 21–22 are applicable for such a set of chosen axes. Each of these two vector equations may be reduced to a set of three scalar equations which are derived in a manner similar to Eqs. 21–25,† i.e.,

$$
\begin{aligned}
\Sigma M_x &= I_x \dot{\omega}_x - I_y \Omega_z \omega_y + I_z \Omega_y \omega_z \\
\Sigma M_y &= I_y \dot{\omega}_y - I_z \Omega_x \omega_z + I_x \Omega_z \omega_x \\
\Sigma M_z &= I_z \dot{\omega}_z - I_x \Omega_y \omega_x + I_y \Omega_x \omega_y
\end{aligned}
\tag{21–26}
$$

Here Ω_x, Ω_y, Ω_z represent the x, y, z components of $\boldsymbol{\Omega}$, measured from the inertial frame of reference.

Any one of these sets of moment equations, Eqs. 21–24, 21–25, or 21–26, represents a series of three first-order nonlinear differential equations. These equations are "coupled," since the angular-velocity components are present in all the terms. Success in determining the solution for a particular problem therefore depends upon what is unknown in these equations. Difficulty certainly arises when one attempts to solve for the unknown components of $\boldsymbol{\omega}$, given the external moments as functions of time. Further complications can arise if the moment equations are coupled to the three scalar equations of translation, Eqs. 21–19. This can happen because of the existence of kinematic constraints which relate the rotation of the body to the translation of its mass center, as in the case of a hoop which rolls without slipping. Problems necessitating the simultaneous solution of differential equations generally require application of numerical methods with the aid of a computer. In many engineering problems, however, one is required to determine the applied moments acting on the body, given information about the motion of the body. Fortunately, many of these types of problems have direct solutions, so that there is no need to resort to computer techniques.

---

*A detailed discussion of such devices is given in Sec. 21.5.

†See Prob. 21–41.

PROCEDURE FOR ANALYSIS

The following procedure provides a method for solving problems involving the three-dimensional motion of a rigid body.

Free-Body Diagram. Draw a *free-body diagram* of the body at the instant considered and specify the x, y, z coordinate system. The origin of this reference must be located either at the body's mass center G, or at point O, considered fixed in an inertial reference frame and located either in the body or on a massless extension of the body. Depending upon the nature of the problem, a decision should be made as to what type of rotational motion $\boldsymbol{\Omega}$ this coordinate system should have, i.e., $\boldsymbol{\Omega} = \mathbf{0}$, $\boldsymbol{\Omega} = \boldsymbol{\omega}$, or $\boldsymbol{\Omega} \neq \boldsymbol{\omega}$. When deciding, one should keep in mind that the moment equations are simplified when the axes move in such a manner that they represent principal axes of inertia for the body at all times.

Kinematics. Compute the necessary moments and products of inertia and the components of the body's angular velocity and angular acceleration. In some cases a *kinematic diagram* might be helpful, since it provides a graphical aid for determining the acceleration of the body's mass center and for computing the angular acceleration. The angular velocity can usually be determined from the constraints of the body or from the given problem data. If this vector is changing direction, Eq. 20–6 $[(\dot{\boldsymbol{\omega}})_{XYZ} = (\dot{\boldsymbol{\omega}})_{xyz} + \boldsymbol{\Omega} \times \boldsymbol{\omega}]$ must be used to determine the components of angular acceleration. In particular, note that if $\boldsymbol{\Omega} = \boldsymbol{\omega}$ then $(\dot{\boldsymbol{\omega}})_{XYZ} = (\dot{\boldsymbol{\omega}})_{xyz}$.

Equations of Motion. Apply either the two vector equations 21–18 and 21–22, or the six scalar component equations appropriate for the x, y, z coordinate axes chosen for the problem.

The following example problems numerically illustrate application of this procedure.

Example 21–4

The gear shown in Fig. 21–12a has a mass of 10 kg and is mounted at an angle of 10° with a rotating shaft having negligible mass. If $I_z = 0.1$ kg · m², $I_x = I_y = 0.05$ kg · m², and the shaft is rotating with a constant angular velocity of $\omega_{AB} = 30$ rad/s, determine the reactions that the bearing supports A and B exert on the shaft at the instant shown.

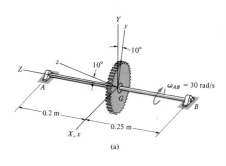

(a)

SOLUTION

Free-Body Diagram Fig. 21–12b. The origin of the x, y, z coordinate system is located at the gear's center of mass G, which is also a fixed point. The axes are fixed in and rotate with the gear; hence $\mathbf{\Omega} = \mathbf{\omega}$.

Kinematics. As shown in Fig. 21–12c, the angular velocity $\mathbf{\omega}$ of the gear is constant in magnitude and is always directed along the axis of the shaft AB. Since this vector is measured from the X, Y, Z inertial frame of reference, for any position of the x, y, z axes $\mathbf{\omega}$ has x, y, z components of

$$\omega_x = 0, \qquad \omega_y = -30 \sin 10°, \qquad \omega_z = 30 \cos 10°$$

(b)

Since $\mathbf{\omega}$ has a constant direction and magnitude when observed from X, Y, Z, $(\dot{\mathbf{\omega}})_{XYZ} = (\dot{\mathbf{\omega}})_{xyz} = \mathbf{0}$. Hence, $\dot{\omega}_x = \dot{\omega}_y = \dot{\omega}_z = 0$. Also, since G is a fixed point, $(a_G)_x = (a_G)_y = (a_G)_z = 0$.

Equations of Motion. Applying Eqs. 21–25 yields

$$\Sigma M_x = I_x \dot{\omega}_x - (I_y - I_z)\omega_y \omega_z$$

$$- (A_Y)(0.2) + (B_Y)(0.25) = 0 - (0.05 - 0.1)(-30 \sin 10°)(30 \cos 10°)$$

$$- 0.2A_Y + 0.25B_Y = -7.70 \qquad (1)$$

$$\Sigma M_y = I_y \dot{\omega}_y - (I_z - I_x)\omega_z \omega_x$$

$$A_X(0.2) \cos 10° - B_X(0.25) \cos 10° = 0 + 0$$

$$A_X = 1.25B_X \qquad (2)$$

$$\Sigma M_z = I_z \dot{\omega}_z - (I_x - I_y)\omega_x \omega_y$$

$$A_X(0.2) \sin 10° - B_X(0.25) \sin 10° = 0$$

$$A_X = 1.25B_X$$

(c)

Fig. 21–12

Applying Eqs. 21–19, we have

$\Sigma F_X = m(a_G)_X;$	$A_X + B_X = 0$	(3)
$\Sigma F_Y = m(a_G)_Y;$	$A_Y + B_Y - 98.1 = 0$	(4)
$\Sigma F_Z = m(a_G)_Z;$	$0 = 0$	

Solving Eqs. (1) to (4) simultaneously gives

$$A_X = B_X = 0, \; A_Y = 71.6 \text{ N}, \; B_Y = 26.4 \text{ N} \qquad \textit{Ans.}$$

Example 21–5

The airplane shown in Fig. 21–13a is in the process of making a steady *horizontal* turn at the rate of ω_p. During this motion, the airplane's propeller is spinning at the rate of ω_s. If the propeller has two blades, determine the moments which the propeller shaft exerts on the propeller when the blades are in the vertical position. For simplicity, assume the blades to be a uniform slender bar having a moment of inertia I about an axis perpendicular to the blades and passing through their center, and having zero moment of inertia about a longitudinal axis.

(a)

SOLUTION

Free-Body Diagram Fig. 21–13b.   The effect of the connecting shaft on the propeller is indicated by the resultants $\mathbf{F}_R$ and $\mathbf{M}_R$. (The propeller's weight is assumed to be negligible.) The x, y, z axes are taken to be *fixed to the propeller* so that $\mathbf{\Omega} = \boldsymbol{\omega} = \boldsymbol{\omega}_s + \boldsymbol{\omega}_p$. Furthermore, the axes so chosen represent the principal axes of inertia for the propeller.

Kinematics.   The moments of inertia I_x and I_y are equal ($I_x = I_y = I$) and $I_z = 0$. The angular velocity at the instant shown, as measured from a fixed X, Y, Z frame coincident with x, y, z, Fig. 21–13c, is $\boldsymbol{\omega} = \boldsymbol{\omega}_s + \boldsymbol{\omega}_p = \omega_s\mathbf{i} + \omega_p\mathbf{k}$, so that the components of $\boldsymbol{\omega}$ are

$$\omega_x = \omega_s, \qquad \omega_y = 0, \qquad \omega_z = \omega_p$$

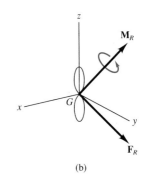

(b)

Since $\boldsymbol{\omega}$ is *changing direction*, the time derivative of $\boldsymbol{\omega}$ will be computed with respect to the fixed X, Y, Z axes and *then* $(\dot{\boldsymbol{\omega}})_{XYZ}$ will be resolved into components along the moving x, y, z axes. To simplify this calculation, a third coordinate system x', y', z' will be chosen which has an angular velocity $\mathbf{\Omega}' = \boldsymbol{\omega}_p$ and is coincident with the X, Y, Z axes at the instant shown. Thus

$$(\dot{\boldsymbol{\omega}})_{XYZ} = (\dot{\boldsymbol{\omega}})_{x'y'z'} + \mathbf{\Omega}' \times \boldsymbol{\omega} \qquad (1)$$
$$= (\dot{\boldsymbol{\omega}}_s)_{x'y'z'} + (\dot{\boldsymbol{\omega}}_p)_{x'y'z'} + \boldsymbol{\omega}_p \times (\boldsymbol{\omega}_s + \boldsymbol{\omega}_p)$$
$$= \mathbf{0} + \mathbf{0} + \boldsymbol{\omega}_p \times \boldsymbol{\omega}_s + \boldsymbol{\omega}_p \times \boldsymbol{\omega}_p$$
$$= \mathbf{0} + \mathbf{0} + \omega_p\mathbf{k} \times \omega_s\mathbf{i} + \mathbf{0} = \omega_p\omega_s\mathbf{j}$$

Since $\mathbf{\Omega} = \boldsymbol{\omega}$, $(\dot{\boldsymbol{\omega}})_{XYZ} = (\dot{\boldsymbol{\omega}})_{xyz}$ and therefore

$$\dot{\omega}_x = 0, \qquad \dot{\omega}_y = \omega_p\omega_s, \qquad \dot{\omega}_z = 0$$

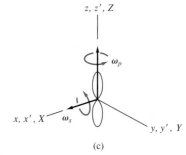

(c)

Fig. 21–13

Equations of Motion.   Using Eqs. 21–25, we have

$$\Sigma M_x = I_x\dot{\omega}_x - (I_y - I_z)\omega_y\omega_z = I(0) - (I - 0)(0)\omega_p$$
$$M_x = 0 \qquad\qquad Ans.$$
$$\Sigma M_y = I_y\dot{\omega}_y - (I_z - I_x)\omega_z\omega_x = I(\omega_p\omega_s) - (0 - I)\omega_p\omega_s$$
$$M_y = 2I\omega_p\omega_s \qquad\qquad Ans.$$
$$\Sigma M_z = I_z\dot{\omega}_z - (I_x - I_y)\omega_x\omega_y = 0(0) - (I - I)(\omega_s)(0)$$
$$M_z = 0 \qquad\qquad Ans.$$

PROBLEMS

21–39. Derive the scalar form of the rotational equation of motion along the *x* axis, when $\Omega \neq \omega$ and the moments and products of inertia of the body are *not constant* with respect to time.

***21–40.** Derive the scalar form of the rotational equation of motion along the *x* axis, when $\Omega \neq \omega$ and the moments and the products of inertia of the body are *constant* with respect to time.

21–41. Derive the Euler equations of motion for $\Omega \neq \omega$, i.e., Eqs. 21–26.

21–42. The 40-kg flywheel (disk) is mounted 20 mm off its true center at *G*. If the shaft is rotating at a constant speed $\omega = 8$ rad/s, determine the maximum reactions exerted on the journal bearings at *A* and *B*.

500 mm

$\omega = 8$ rad/s

20 mm

0.75 m

1.25 m

Prob. 21–42

21–43. The 4-lb bar rests along the smooth corners of an open box. At the instant shown the box has a velocity of $\mathbf{v} = \{5\mathbf{k}\}$ ft/s and an acceleration $\mathbf{a} = \{2\mathbf{k}\}$ ft/s². Determine the *x*, *y*, *z* components of force which the corners exert on the bar.

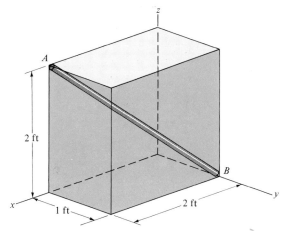

Probs. 21–43/21–44

***21–44.** The 4-lb bar rests along the smooth corners of an open box. At the instant shown the box has a velocity of $\mathbf{v} = \{3\mathbf{j}\}$ ft/s and an acceleration of $\mathbf{a} = \{-6\mathbf{j}\}$ ft/s². Determine the *x*, *y*, *z* components of force which the corners exert on the bar.

21–45. The 20-lb plate is mounted on the shaft *AB* so that the plane of the plate makes an angle of $\theta = 30°$ with the vertical. If the shaft is turning with an angular velocity of 25 rad/s, determine the vertical reactions at the bearing supports *A* and *B* when the plate is in the position shown.

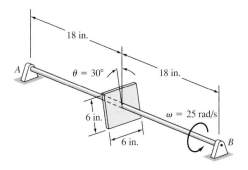

18 in.

A

$\theta = 30°$

18 in.

6 in.

$\omega = 25$ rad/s

6 in.

B

Prob. 21–45

21–46. The 25-lb disk *A* is *fixed* to rod *BCD*, which has negligible mass. Determine the torque **T** which must be applied to the vertical shaft so that the shaft has an angular acceleration of $\alpha = 6$ rad/s².

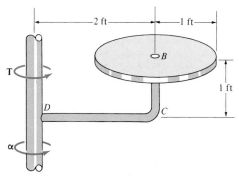

2 ft

1 ft

T

B

1 ft

D

C

α

Probs. 21–46/21–47

21–47. Solve Prob. 21–46, assuming rod *BCD* has a weight of 2 lb/ft.

467

*21–48. The rod assembly has a weight of 5 lb/ft. It is supported at *B* by a smooth journal bearing, which develops *x* and *y* force reactions, and at *A* by a smooth thrust bearing, which develops *x*, *y*, and *z* force reactions. If a 50-lb · ft torque is applied along rod *AB*, determine the components of reaction at the bearing at the instant the assembly has an angular velocity $\omega = 10$ rad/s.

Prob. 21–48

21–49. The 20-lb disk is mounted on the horizontal shaft *AB* such that its plane forms an angle of 10° with the vertical. If the shaft rotates with an angular velocity of 3 rad/s, determine the vertical reactions developed at the bearings when the disk is in the position shown.

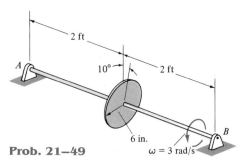

Prob. 21–49

21–50. The crankshaft is constructed from a rod which has a weight of 3 lb/ft. Determine the *x* and *z* components of reaction at bearings *A* and *B* at the instant the shaft is in the position shown.

Prob. 21–50

21–51. The rod assembly has a mass of 1.5 kg/m. A resultant frictional moment **M,** developed at the journal bearings *A* and *B*, causes the shaft to decelerate at 1 rad/s² when it has an angular velocity $\omega = 6$ rad/s. Determine the reactions at the bearings at the instant shown. What is the frictional moment **M** causing the deceleration? Establish the axes at *G* with +*y* upward, and +*z* directed along the shaft axis towards *A*.

Prob. 21–51

*21–52. The 5-kg rod *AB* is supported by a rotating arm. The support at *A* is a journal bearing, which develops reactions normal to the rod. The support at *B* is a thrust bearing, which develops reactions both normal to the rod and along the axis of the rod. Neglecting friction, determine the *x*, *y*, *z* components of reaction at these supports when the frame rotates with a constant angular velocity of ω = 10 rad/s.

Prob. 21–52

21–53. The rod assembly is supported by bearings at *A* and *B*. The mass of each rod is 1.5 kg/m. If the shaft *AB* is rotating with a speed ω = 5 rad/s, determine the reactions at the journal bearings when the assembly is in the position shown. Also, what is the shaft's angular acceleration?

Prob. 21–53

21–54. The rod assembly is supported by a ball-and-socket joint at *C* and a journal bearing support at *D*, which develops only *x* and *y* force reactions. The rods have a mass of 0.75 kg/m. Determine the angular acceleration of the rods and the components of reaction at the supports at the instant ω = 8 rad/s as shown.

Prob. 21–54

21–55. The cylinder has a mass of 30 kg and is mounted on an axle which is supported by bearings at *A* and *B*. If the axle is turning at a constant rate of 40 rad/s, determine the components of force acting at the bearings at the instant shown.

Prob. 21–55

*21–56. The uniform hatch door, having a mass of 15 kg and a mass center at *G*, is supported in the horizontal plane by bearings at *A* and *B*. If a vertical force *F* = 300 N is applied to the door as shown, determine the components of reaction at the bearings and the angular acceleration of the door. The bearing at *A* will resist a component of force in the *y* direction, whereas the bearing at *B* will not. For the calculation, assume the door to be a thin plate and neglect the size of the bearing. The door is originally at rest. Set *a* = 100 mm, *b* = 200 mm, *c* = 150 mm, and *d* = 30 mm.

Prob. 21–56

469

21–57. The man sits on a swivel chair which is rotating with a constant angular velocity of 3 rad/s. He holds the uniform 5-lb rod AB horizontal. He suddenly gives it an angular acceleration of 2 rad/s^2, measured relative to him, as shown. Determine the required force and moment components at the grip, A, necessary to do this. Establish axes at G with $+z$ upward, and $+y$ directed along the axis of the rod towards A.

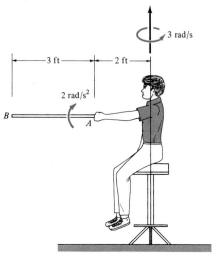

Prob. 21–57

21–58. The 4-kg slender rod AB is pinned at A and held at B by a cord. The axle CD is supported at its ends by ball-and-socket joints and is rotating with a constant angular velocity of 2 rad/s. Determine the tension developed in the cord and the magnitude of force developed at the pin A.

Probs. 21–58/21–59

21–59. Determine the tension in the cord and the magnitude of force at the pin A in Prob. 21–58 if the supporting axle has an angular acceleration $\alpha = 1.5$ rad/s^2 at the instant $\omega = 2$ rad/s. Assume that α acts in the same direction as ω shown in Prob. 21–58.

****21–60.** The bent uniform rod ACD has a weight of 5 lb/ft and is supported at A by a pin and at B by a cord. If the vertical shaft rotates with a constant angular velocity of $\omega = 20$ rad/s, determine the x, y, z components of force and moment developed at A and the tension in the cord.

Prob. 21–60

21–61. The rod AB supports the 10-lb sphere. If the rod is pinned at A to the vertical shaft which is rotating at a constant rate of $\omega = \{7\mathbf{k}\}$ rad/s, determine the angle θ of the rod during the motion. Neglect the mass of the rod in the calculation.

Probs. 21–61/21–62

21–62. The rod AB supports the 10 lb sphere. If the rod is pinned at A to the vertical shaft which is rotating with an angular acceleration of $\alpha = \{2\mathbf{k}\}$ rad/s^2 and at the instant shown the shaft has an angular velocity of $\omega = \{7\mathbf{k}\}$ rad/s, determine the angle θ of the rod at this instant. Neglect the mass of the rod in the calculation.

*21.5 Gyroscopic Motion

In this section the equations used for analyzing the motion of a body (or top) which is symmetrical with respect to an axis and moving about a fixed point lying on the axis will be developed. These equations will then be applied to study the motion of a particularly interesting device, the gyroscope.

The body's motion will be analyzed using *Euler angles* ϕ, θ, ψ (phi, theta, psi). To illustrate how these angles are formed, reference is made to the top shown in Fig. 21–14a. The top is attached to point O and has an orientation relative to the fixed X, Y, Z axes at some instant of time as shown in Fig. 21–14d. To define this final position, a second set of x, y, z axes will be needed. For purposes of discussion, assume that this reference is fixed in the top. Starting with the X, Y, Z and x, y, z axes in coincidence, Fig. 21–14a, the final position of the top is determined using the following three orderly steps:

1. Rotate the top about the Z (or z) axis through an angle ϕ ($0 \leqslant \phi < 2\pi$), Fig. 21–14b.
2. Rotate the top about the x axis through an angle θ ($0 \leqslant \theta \leqslant \pi$), Fig. 21–14c.
3. Rotate the top about the z axis through an angle ψ ($0 \leqslant \psi < 2\pi$) to obtain the final position, Fig. 20–14d.

When the top is in motion, in general each of the Euler angles is changing with time. Although the three finite rotations defined above are not vectors, the infinitesimal rotations $d\boldsymbol{\phi}$, $d\boldsymbol{\theta}$, and $d\boldsymbol{\psi}$ are vectors, and thus the angular velocity $\boldsymbol{\omega}$ of the top can be expressed in terms of the time derivatives of the Euler angles. The angular-velocity components $\dot{\phi}$, $\dot{\theta}$, and $\dot{\psi}$ are known as the

(a)

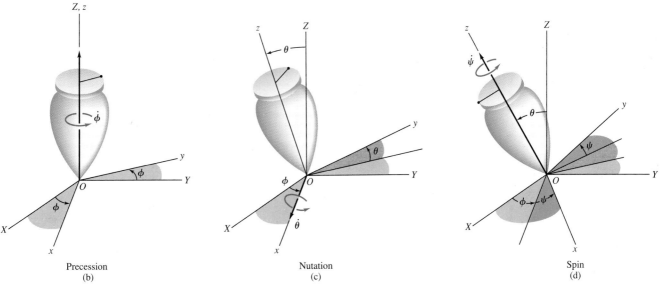

Precession	Nutation	Spin
(b)	(c)	(d)

Fig. 21–14

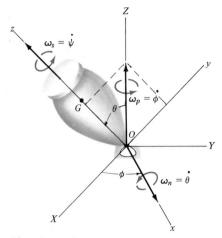

Fig. 21–15

precession, nutation, and *spin,* respectively. Their positive directions are shown in Fig. 21–14. It is seen that these vectors are not all perpendicular to one another; however, $\boldsymbol{\omega}$ of the top can still be expressed in terms of these three components.

In our case the body (top) is symmetric with respect to the z or spin axis. If we consider the top in Fig. 21–15, for which the x axis is oriented such that at the instant considered the spin angle $\psi = 0$ and *the x, y, z axes follow the motion of the body only in nutation and precession,* i.e., $\boldsymbol{\Omega} = \boldsymbol{\omega}_p + \boldsymbol{\omega}_n$, then the nutation and spin are always directed along the x and z axes, respectively. Hence, the angular velocity of the body is specified only in terms of the Euler angle θ, i.e.,

$$\boldsymbol{\omega} = \omega_x \mathbf{i} + \omega_y \mathbf{j} + \omega_z \mathbf{k}$$
$$= \dot{\theta}\mathbf{i} + (\dot{\phi} \sin \theta)\mathbf{j} + (\dot{\phi} \cos \theta + \dot{\psi})\mathbf{k} \qquad (21\text{–}27)$$

Since motion of the axes is not affected by the spin component,

$$\boldsymbol{\Omega} = \Omega_x \mathbf{i} + \Omega_y \mathbf{j} + \Omega_z \mathbf{k}$$
$$= \dot{\theta}\mathbf{i} + (\dot{\phi} \sin \theta)\mathbf{j} + (\dot{\phi} \cos \theta)\mathbf{k} \qquad (21\text{–}28)$$

The x, y, z axes in Fig. 21–15 represent *principal axes of inertia* of the body for *any* spin of the body about these axes. Hence, the moments of inertia are constant and will be represented as $I_{xx} = I_{yy} = I$ and $I_{zz} = I_z$. Since $\boldsymbol{\Omega} \neq \boldsymbol{\omega}$, the Euler Eqs. 21–26 are used to establish the rotational equations of motion. Substituting into these equations the respective angular-velocity components defined by Eqs. 21–27 and 21–28, their corresponding time derivatives, and the moment of inertia components yields

$$\Sigma M_x = I(\ddot{\theta} - \dot{\phi}^2 \sin \theta \cos \theta) + I_z \dot{\phi} \sin \theta (\dot{\phi} \cos \theta + \dot{\psi})$$
$$\Sigma M_y = I(\ddot{\phi} \sin \theta + 2\dot{\phi}\dot{\theta} \cos \theta) - I_z \dot{\theta}(\dot{\phi} \cos \theta + \dot{\psi}) \qquad (21\text{–}29)$$
$$\Sigma M_z = I_z(\ddot{\psi} + \ddot{\phi} \cos \theta - \dot{\phi}\dot{\theta} \sin \theta)$$

Each moment summation applies only at the fixed point O or the center of mass G of the body. Since the equations represent a coupled set of nonlinear second-order differential equations, in general a closed-form solution may not be obtained. Instead, the Euler angles ϕ, θ, and ψ may be obtained graphically as functions of time using numerical analysis and computer techniques.

A special case, however, does exist for which simplification of Eqs. 21–29 is possible. Commonly referred to as *steady precession,* it occurs when the nutation angle θ, precession $\dot{\phi}$, and spin $\dot{\psi}$ all remain *constant.* Equations 21–29 then reduce to the form

$$\boxed{\Sigma M_x = -I\dot{\phi}^2 \sin \theta \cos \theta + I_z \dot{\phi} \sin \theta (\dot{\phi} \cos \theta + \dot{\psi})} \qquad (21\text{–}30)$$

$$\Sigma M_y = 0$$
$$\Sigma M_z = 0$$

Equation 21–30 may be further simplified by noting that, from Eq. 21–27, $\omega_z = \dot{\phi} \cos \theta + \dot{\psi}$, so that

$$\Sigma M_x = -I\dot{\phi}^2 \sin \theta \cos \theta + I_z\dot{\phi}(\sin \theta)\omega_z$$

or

$$\boxed{\Sigma M_x = \dot{\phi} \sin \theta(I_z\omega_z - I\dot{\phi} \cos \theta)} \qquad (21\text{–}31)$$

It is interesting to note what effects the spin $\dot{\psi}$ has on the moment about the x axis. In this regard consider the spinning rotor shown in Fig. 21–16. Here $\theta = 90°$, in which case Eq. 21–30 reduces to the form

$$\Sigma M_x = I_z\dot{\phi}\dot{\psi}$$

or

$$\boxed{\Sigma M_x = I_z\Omega_y\omega_z} \qquad (21\text{–}32)$$

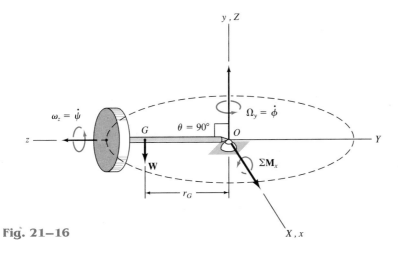

Fig. 21–16

From the figure it is seen that vectors $\Sigma\mathbf{M}_x$, $\boldsymbol{\Omega}_y$, and $\boldsymbol{\omega}_z$ all act along their respective *positive axes* and therefore they are mutually perpendicular. Instinctively, one would expect the rotor to fall down under the influence of gravity! However, this is not the case at all provided the product $I_z\Omega_y\omega_z$ is correctly chosen to counterbalance the moment $\Sigma M_x = Wr_G$ of the rotor's weight about O. This unusual phenomenon of rigid-body motion is often referred to as the *gyroscopic effect*.

473

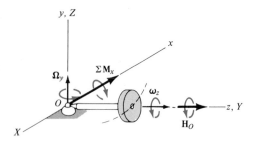

Fig. 21–17

Perhaps a more intriguing demonstration of the gyroscopic effect comes from studying the action of a *gyroscope,* frequently referred to as a *gyro.* A gyro is a rotor which spins at a very high rate about its axis of symmetry. This rate of spin is considerably greater than its precessional rate of rotation about the vertical axis. Hence, for all practical purposes, the angular momentum of the gyro can be assumed directed along its axis of spin. Thus, for the gyro rotor shown in Fig. 21–17, $\omega_z \gg \Omega_y$, and the magnitude of the angular momentum about point O, as computed by Eqs. 21–11, reduces to the form $H_O = I_z \omega_z$. Since both the magnitude and direction of $\mathbf{H}_O$ are constant as observed from x, y, z, direct application of Eq. 21–22 yields

$$\boxed{\Sigma \mathbf{M}_x = \mathbf{\Omega}_y \times \mathbf{H}_O} \tag{21-33}$$

Using the right-hand rule applied to the cross product, it is seen that $\mathbf{\Omega}_y$ always swings $\mathbf{H}_O$ (or $\boldsymbol{\omega}_z$) toward the sense of $\Sigma \mathbf{M}_x$. In effect, the *change in direction* of the gyro's angular momentum, $d\mathbf{H}_O$, causes an angular impulse about O, i.e., $d\mathbf{H}_O = \Sigma \mathbf{M}_x \, dt$, Eq. 21–20. Note that since $H_O = I_z \omega_z$ and ΣM_x, Ω_y, and H_O are mutually perpendicular, Eq. 21–33 reduces to Eq. 21–32.

When a gyro is mounted in gimbal rings, Fig. 21–18, it becomes *free* of external moments applied to its base. Thus, in theory, its angular momentum $\mathbf{H}$ will never precess but, instead, maintain its same fixed orientation along the axis of spin when the base is rotated. This type of gyroscope is called a *free gyro* and is useful as a gyrocompass when the spin axis of the gyro is directed north. In reality, the gimbal mechanism is never completely free of friction, so that such a device is useful only for the local navigation of ships and aircraft. The gyroscopic effect is also useful as a means of stabilizing both the rolling motion of ships at sea and the trajectories of missiles and projectiles. Furthermore, this effect is of significant importance in the design of shafts and bearings for rotors which are subjected to forced precessions.

Fig. 21–18

Example 21–6

The top shown in Fig. 21–19a has a mass of 0.5 kg and is precessing about the vertical axis at a constant angle of $\theta = 60°$. If it spins with an angular velocity $\omega_s = 100$ rad/s, determine the precessional velocity ω_p. Assume that the axial and transverse moments of inertia of the top are $4.5(10^{-4})$ kg · m^2 and $12.0(10^{-4})$ kg · m^2, respectively, measured with respect to the fixed point O.

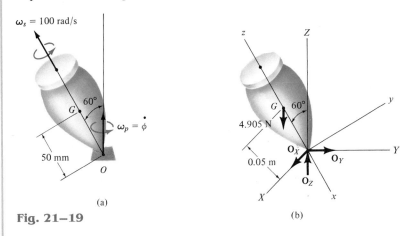

(a)

Fig. 21–19

(b)

SOLUTION

Equation 21–30 will be used for the solution since the motion is a *steady precession*. As shown on the free-body diagram, Fig. 21–19b, the coordinate axes are established in the usual manner, that is, with the positive z axis in the direction of spin, the positive Z axis in the direction of precession, and the positive x axis in the direction of the moment $\Sigma\mathbf{M}_x$ (refer to Fig. 21–15). Thus,

$$\Sigma M_x = -I\dot{\phi}^2 \sin\theta \cos\theta + I_z\dot{\phi} \sin\theta(\dot{\phi} \cos\theta + \dot{\psi})$$

$$4.905(0.05)\sin 60° = -12.0(10^{-4})\dot{\phi}^2 \sin 60° \cos 60°$$
$$+ 4.5(10^{-4})\dot{\phi} \sin 60°(\dot{\phi} \cos 60° + 100)$$

or

$$\dot{\phi}^2 - 120.0\dot{\phi} + 654.0 = 0 \qquad (1)$$

Solving this quadratic equation for the precession gives

$$\dot{\phi} = 114 \text{ rad/s} \qquad \text{(high precession)} \qquad \textit{Ans.}$$

and

$$\dot{\phi} = 5.72 \text{ rad/s} \qquad \text{(low precession)} \qquad \textit{Ans.}$$

In reality low precession of the top would generally be observed since high precession would require a large kinetic energy.

Example 21–7

The 1-kg disk shown in Fig. 21–20a is spinning about its axis with a constant angular velocity $\omega_D = 70$ rad/s. The block at B has a mass of 2 kg, and by adjusting its position s one can change the precession of the disk about its supporting pivot at O. Compute the position s which will enable the disk to have a constant precessional velocity $\omega_p = 0.5$ rad/s about the pivot. Neglect the weight of the shaft.

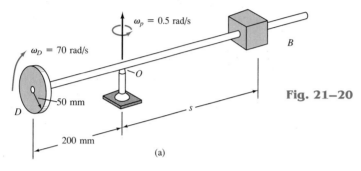

Fig. 21–20

(a)

SOLUTION

The free-body diagram of the disk is shown in Fig. 21–20b, where F represents the force reaction of the shaft on the disk. The origin for both the x, y, z and X, Y, Z coordinate systems is located at point O, which represents a *fixed point* for the disk. (Although point O does not lie on the disk, imagine a massless extension of the disk to this point.) In the conventional sense, the Z axis is chosen along the axis of precession, and the z axis is along the axis of spin, so that $\theta = 90°$. Since the precession is *steady*, Eq. 21–31 may be used for the solution. This equation reduces to

$$\Sigma M_x = \dot{\phi} I_z \omega_z$$

which is the same as Eq. 21–32. Substituting the required data gives

$$9.81(0.2) - F(0.2) = 0.5\left[\frac{1}{2}(1)(0.05)^2\right](-70)$$

$$F = 10.0 \text{ N}$$

As shown on the free-body diagram of the shaft and block B, Fig. 21–20c, summing moments about the x axis requires

$$(19.62)s = (10.0)(0.20)$$

$$= 0.102 \text{ m} = 102 \text{ mm} \qquad Ans.$$

(b)

(c)

*21.6 Torque-Free Motion

When the only external force acting on a body is caused by gravitation, the general motion of the body is referred to as *torque-free motion*. This type of motion is characteristic of planets, artificial satellites, and projectiles—provided the effects of air friction are neglected.

In order to describe the characteristics of this motion, the distribution of the body's mass will be assumed *axisymmetric*. The satellite shown in Fig. 21–21a is an example of such a body, where the z axis represents an axis of symmetry. The origin of the x, y, z coordinates is located at the mass center G, such that $I_{zz} = I_z$ and $I_{xx} = I_{yy} = I$ for the body. If it is assumed that gravitation is the only external force present, the summation of moments about the mass center is zero. From Eq. 21–21, this requires the angular momentum of the body to be constant, i.e.,

$$\mathbf{H}_G = \text{const}$$

At the instant considered, it will be assumed that the inertial frame of reference is oriented such that the positive Z axis is directed along $\mathbf{H}_G$ and the y axis lies in the plane formed by the z and Z axes, Fig. 21–21a. The Euler angle formed between Z and z is θ, and therefore, with this choice of axes the angular momentum may be expressed as

$$\mathbf{H}_G = H_G \sin \theta \mathbf{j} + H_G \cos \theta \mathbf{k}$$

Furthermore, using Eqs. 21–11, we have

$$\mathbf{H}_G = I\omega_x \mathbf{i} + I\omega_y \mathbf{j} + I_z\omega_z \mathbf{k}$$

where ω_x, ω_y, ω_z represent the x, y, z components of the body's angular velocity. Equating the respective **i, j,** and **k** components of the above two equations yields

$$\omega_x = 0, \qquad \omega_y = \frac{H_G \sin \theta}{I}, \qquad \omega_z = \frac{H_G \cos \theta}{I_z} \qquad (21\text{–}34)$$

or

$$\boxed{\boldsymbol{\omega} = \frac{H_G \sin \theta}{I}\mathbf{j} + \frac{H_G \cos \theta}{I_z}\mathbf{k}} \qquad (21\text{–}35)$$

In a similar manner, equating the respective **i, j,** and **k** components of Eq. 21–27 to those of Eq. 21–34, we obtain

$$\dot{\theta} = 0$$

$$\dot{\phi} \sin \theta = \frac{H_G \sin \theta}{I}$$

$$\dot{\phi} \cos \theta + \dot{\psi} = \frac{H_G \cos \theta}{I_z}$$

Fig. 21–21a

(a)

477

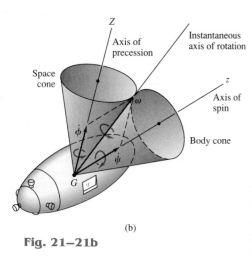

Fig. 21–21b

Solving, we get

$$
\boxed{
\begin{aligned}
\theta &= \text{const} \\[4pt]
\dot{\phi} &= \frac{H_G}{I} \\[6pt]
\dot{\psi} &= \frac{I - I_z}{II_z} H_G \cos\theta
\end{aligned}
}
\qquad (21\text{–}36)
$$

Thus, for torque-free motion of an axisymmetrical body, the angle θ formed between the angular-momentum vector and the spin of the body remains constant. Furthermore, the angular momentum $\mathbf{H}_G$, precession $\dot{\phi}$, and spin $\dot{\psi}$ for the body remain constant at all times during the motion. Eliminating H_G from the second and third of Eqs. 21–36 yields the following relationship between the spin and precession:

$$
\boxed{\dot{\psi} = \frac{I - I_z}{I_z}(\dot{\phi}) \cos\theta}
\qquad (21\text{–}37)
$$

As shown in Fig. 21–21b, the body precesses about the Z axis, which is fixed in direction, while it spins about the z axis. These two components of angular motion may be analyzed by using a simple cone model, introduced in Sec. 20.1. The *space cone* defining the precession is fixed from rotating, since the precession has a fixed direction, while the *body cone* rotates around the space cone's outer surface without slipping. On this basis, an attempt should be made to imagine the motion. The interior angle of each cone is chosen such that the resultant angular velocity of the body is directed along the line of contact of the two cones. This line of contact represents the instantaneous axis of rotation for the body cone, and hence the angular velocity of both the body cone and the body must be directed along this line. Since the spin is a function of the moments of inertia I and I_z of the body, Eq. 21–36, the cone model in Fig. 21–21b is satisfactory for describing the motion, provided $I > I_z$. Torque-free motion which meets these requirements is called *regular precession*. If $I < I_z$, the spin is negative and the precession positive. This motion is represented by the satellite motion shown in Fig. 21–22 ($I < I_z$). The cone model may again be used to represent the motion; however, to preserve the correct vector addition of spin and precession to obtain the angular velocity $\boldsymbol{\omega}$, the inside surface of the body cone must roll on the outside surface of the (fixed) space cone. This motion is referred to as *retrograde precession*.

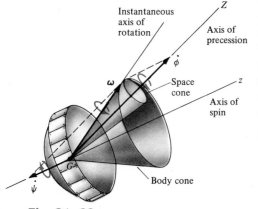

Fig. 21–22

Example 21—8

The motion of a football is observed using a slow-motion projector. From the film, the spin of the football is seen to be directed 30° from the horizontal, as shown in Fig. 21–23a. Also, the football is precessing about the vertical axis at a rate $\dot{\phi} = 3$ rad/s. If the ratio of the axial to transverse moments of inertia of the football is $\frac{1}{3}$, measured with respect to the center of mass, determine the magnitude of the football's spin and its angular velocity. Neglect the effect of air resistance.

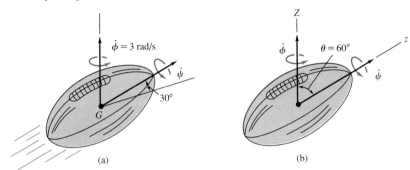

(a)   (b)

Fig. 21—23

SOLUTION

Since the weight of the football is the only force acting, the motion is torque-free. In the conventional sense, if the z axis is established along the axis of spin and the Z axis along the precession axis, as shown in Fig. 21–23b, then the angle $\theta = 60°$. Applying Eq. 21–37 gives

$$\dot{\psi} = \frac{I - I_z}{I_z}(\dot{\phi}) \cos \theta = \frac{I - \frac{1}{3}I}{\frac{1}{3}I}(3) \cos 60°$$

$$= 3 \text{ rad/s} \qquad\qquad Ans.$$

Using Eqs. 21–34, where $H_G = \dot{\phi}I$ (Eq. 21–36), we have

$$\omega_x = 0$$

$$\omega_y = \frac{H_G \sin \theta}{I} = \frac{3I \sin 60°}{I} = 2.60 \text{ rad/s}$$

$$\omega_z = \frac{H_G \cos \theta}{I_z} = \frac{3I \cos 60°}{\frac{1}{3}I} = 4.50 \text{ rad/s}$$

Thus,

$$\omega = \sqrt{(\omega_x)^2 + (\omega_y)^2 + (\omega_z)^2}$$

$$= \sqrt{(0)^2 + (2.60)^2 + (4.50)^2}$$

$$= 5.20 \text{ rad/s} \qquad\qquad Ans.$$

PROBLEMS

21–63. Show that the angular velocity of a body, in terms of Euler angles ϕ, θ, and ψ, may be expressed as $\boldsymbol{\omega} = (\dot{\phi}\sin\theta\sin\psi + \dot{\theta}\cos\cdot\psi)\mathbf{i} + (\dot{\phi}\sin\theta\cos\psi - \dot{\theta}\sin\psi)\mathbf{j} + (\dot{\phi}\cos\theta + \dot{\psi})\mathbf{k}$, where $\mathbf{i}$, $\mathbf{j}$, and $\mathbf{k}$ are directed along the x, y, z axes shown in Fig. 21–14d.

*21–64. A thin rod is initially coincident with the Z axis when it is given three rotations defined by the Euler angles $\theta = 45°$, $\phi = 30°$, and $\psi = 60°$. If these rotations are given in the order stated, determine the coordinate direction angles of the axis of the rod with respect to the X, Y, and Z axes. Is this direction the same for any order of the rotations? Why?

21–65. The propeller on a single-engine airplane has a mass of 15 kg and a centroidal radius of gyration of 0.3 m computed about the axis of spin. When viewed from the front of the airplane, the propeller is turning clockwise at 350 rad/s about the spin axis. If the airplane enters a vertical curve having a radius $\rho = 80$ m and is traveling at 200 km/h, determine the gyroscopic bending moment which the propeller exerts on the bearings of the engine when the airplane is in its lowest position.

21–66. An airplane descends at a steep angle and then levels off horizontally to land. If the propeller is turning clockwise when observed from the rear of the plane, determine the direction in which the plane tends to turn as caused by the gyroscopic effect.

21–67. The toy gyroscope consists of a rotor R which is attached to the frame of negligible mass. If the rotor is spinning about its axle at an angular speed $\omega_R = 150$ rad/s, determine the constant angular velocity ω_p at which the frame is precessing about the pivot point at O. OA moves in the horizontal plane. The rotor has a mass of 200 g and a radius of gyration $k_{OA} = 20$ mm about OA.

Prob. 21–67

*21–68. The top consists of a thin disk that has a weight of 8 lb and a radius of 0.3 ft. The rod has a negligible mass and a length of 0.5 ft. If the top is spinning with an angular velocity $\omega_s = 300$ rad/s, determine the steady-state precessional angular velocity ω_p of the rod when $\theta = 40°$.

Probs. 21–68/21–69

21–69. Solve Prob. 21–68 when $\theta = 90°$.

21–70. The rotor assembly on the engine of a jet airplane consists of the turbine, drive shaft, and compressor. The total mass is 700 kg, the radius of gyration about the shaft axis is $k_{AB} = 0.35$ m, and the mass center is at G. If the rotor has an angular velocity $\omega_{AB} = 1000$ rad/s, and the plane is pulling out of a vertical curve while traveling at 250 m/s, determine the components of reaction at the bearings A and B due to the gyroscopic effect.

Prob. 21–70

21–71. The top has a mass of 90 g, a center of mass at G, and a radius of gyration $k = 18$ mm about its axis of symmetry. About any transverse axis acting through point O the radius of gyration is $k_t = 35$ mm. If the top is pinned at O and the precession is $\omega_p = 0.5$ rad/s, determine the spin ω_s.

Prob. 21–71

21–73. The 30-lb wheel is rolling without slipping. If it has a radius of gyration $k_{AB} = 1.2$ ft about its axle AB, and the vertical drive shaft is turning at 8 rad/s, determine the normal reaction the wheel exerts on the ground at C.

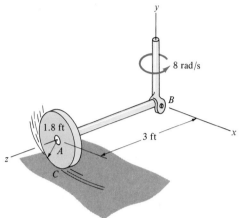

Prob. 21–73

***21–72.** The motor weighs 50 lb and has a radius of gyration of 0.2 ft about the z axis. The shaft of the motor is supported by bearings at A and B, and is turning at a constant rate of $\boldsymbol{\omega}_s = \{100\mathbf{k}\}$ rad/s, while the frame has an angular velocity of $\boldsymbol{\omega}_y = \{2\mathbf{j}\}$ rad/s. Determine the moment which the bearing forces at A and B exert on the shaft due to this motion.

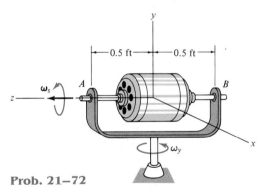

Prob. 21–72

21–74. The car is traveling at $v_C = 100$ km/h around the horizontal curve having a radius of curvature of 80 m. If each wheel has a mass of 16 kg, a radius of gyration $k_G = 300$ mm about its spinning axis, and a diameter of 400 mm, determine the difference between the normal forces of the rear wheels caused by the gyroscopic effect. The distance between the wheels is 1.30 m.

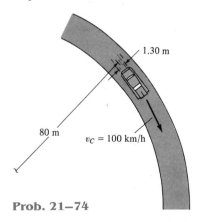

Prob. 21–74

481

21–75. The 10-lb sphere spins with an angular speed of $\omega_s =$ 1500 rpm about the axis of the horizontal rod. If the counterbalance block weighs 6 lb, determine the precession of rod DC. The cross frame has negligible mass and is supported by ball-and-socket joints at points C and D.

Prob. 21–75

21–77. The projectile shown is subjected to torque-free motion. The transverse and axial moments of inertia are I and I_z, respectively. If θ represents the angle between the precessional axis Z and the axis of symmetry z, and β is the angle between the angular velocity ω and the z axis, show that β and θ are related by the equation $\tan \theta = (I/I_z) \tan \beta$.

Prob. 21–77

21–78. The radius of gyration about an axis passing through the axis of symmetry of the 2.5-Mg satellite is $k_z = 2.3$ m, and about any transverse axis passing through the center of mass G, $k_t = 3.4$ m. If the satellite has a steady-state precession of two revolutions per hour about the Z axis, determine the rate of spin about the z axis.

Prob. 21–78

*21–76. The conical top has a mass of 0.8 kg, and the moments of inertia are $I_x = I_y = 3.5(10^{-3})$ kg·m² and $I_z = 0.8(10^{-3})$ kg·m². If it spins freely in the ball-and-socket joint at A with an angular velocity $\omega_s = 750$ rad/s, compute the precession of the top about the axis of the shaft AB.

Prob. 21–76

21–79. While the rocket is in free flight, it has a spin of 3 rad/s and precesses about an axis measured 10° from the axis of spin. If the ratio of the axial to transverse moments of inertia of the rocket is 1/15, computed about axes which pass through the mass center G, determine the angle which the resultant angular velocity makes with the spin axis. Construct the body and space cones used to describe the motion. Is the precession regular or retrograde?

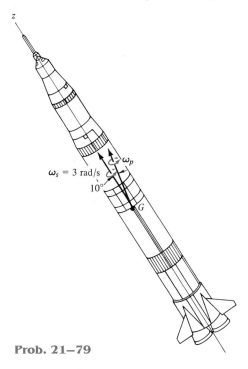

Prob. 21–79

*21–80. The 4-kg disk is thrown in the air with a spin $\omega_z = 6$ rad/s. If the angle θ is measured as 160°, determine the precession about the Z axis.

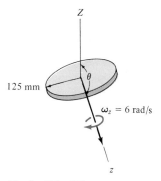

Prob. 21–80

21–81. The satellite has a mass of 1.8 Mg, and about axes passing through the mass center G the axial and transverse radii of gyration are $k_z = 0.8$ m and $k_t = 1.2$ m, respectively. If it is spinning at $\omega_s = 6$ rad/s when it is launched, determine its angular momentum. Precession occurs about the Z axis.

Prob. 21–81

22

Vibrations

A *vibration* is the periodic motion of a body or system of connected bodies displaced from a position of equilibrium. In general, there are two types of vibration, free and forced. *Free vibration* occurs when the motion is maintained by gravitational or elastic restoring forces, such as the swinging motion of a pendulum or the vibration of an elastic rod. *Forced vibration* is caused by an external periodic or intermittent force applied to the system. Both of these types of vibration may be either damped or undamped. *Undamped* vibrations can continue indefinitely because frictional effects are neglected in the analysis. Since in reality both internal and external frictional forces are present, the motion of all vibrating bodies is actually *damped*.

In this chapter we will study the characteristics of the above types of vibrating motion. The analysis will apply to those bodies which are constrained to move only in one direction. These single-degree-of-freedom systems require only one coordinate to specify completely the position of the system at any time. The analysis of multi-degree-of-freedom systems is based on this simplified case and is thoroughly treated in textbooks devoted to vibrational theory.

*22.1 Undamped Free Vibration

Equilibrium position

(a)

$W = mg$

$F = kx$

N_B

(b)

Fig. 22–1

The simplest type of vibrating motion is undamped free vibration, represented by the model shown in Fig. 22–1a. The block has a mass m and is attached to a spring having a stiffness k. Vibrating motion occurs when displacing the block a distance x from its equilibrium position and allowing the spring to restore it to its original position. As the spring pulls on the block, the block will attain a velocity such that it will proceed to move out of equilibrium when $x = 0$. Provided the supporting surface is smooth, oscillation will continue indefinitely.

The time-dependent path of motion of the block may be determined by applying the equation of motion to the block when it is in the displaced position x. The free-body diagram is shown in Fig. 22–1b. The elastic restoring force $F = kx$ is always directed toward the equilibrium position, whereas the acceleration **a** is assumed to act in the direction of *positive displacement*. Noting that $a = d^2x/dt^2 = \ddot{x}$, we have

$$\stackrel{+}{\rightarrow}\Sigma F_x = ma_x; \qquad\qquad -kx = m\ddot{x}$$

Here it is seen that the acceleration is proportional to the position. Motion described in this manner is called *simple harmonic motion*. Rearranging the terms into a "standard form" gives

$$\ddot{x} + p^2x = 0 \qquad\qquad (22\text{–}1)$$

The constant p is called the *circular frequency*, expressed in rad/s, and in this case

$$\boxed{p = \sqrt{\frac{k}{m}}} \qquad\qquad (22\text{–}2)$$

Equation 22–1 may also be obtained by considering the block to be suspended, as shown in Fig. 22–2a, and measuring the displacement y from the block's *equilibrium position*. The free-body diagram is shown in Fig. 22–2b. When the block is in equilibrium, the spring exerts an upward force of $F = W = mg$ on the block. Hence, when the block is displaced a distance y downward from this position, the magnitude of the spring force is $F = W + ky$.

k

Equilibrium position

y

$F = W + ky$

W

(a)

(b)

Fig. 22–2

Applying the equation of motion gives

$$+\downarrow \Sigma F_y = ma_y; \qquad -W - ky + W = m\ddot{y}$$

or

$$\ddot{y} + p^2 y = 0$$

which is the same form as Eq. 22–1, where p is defined by Eq. 22–2.

Equation 22–1 is a homogeneous, second-order, linear, differential equation with constant coefficients. It can be shown, using the methods of differential equations, that the general solution of this equation is

$$\boxed{x = A \sin pt + B \cos pt} \qquad (22\text{–}3)$$

where A and B represent two constants of integration. The block's velocity and acceleration are determined by taking successive time derivatives, which yields

$$v = \dot{x} = Ap \cos pt - Bp \sin pt \qquad (22\text{–}4)$$

$$a = \ddot{x} = -Ap^2 \sin pt - Bp^2 \cos pt \qquad (22\text{–}5)$$

When Eqs. 22–3 and 22–5 are substituted into Eq. 22–1, the differential equation is indeed satisfied, and therefore Eq. 22–3 represents the true solution to Eq. 22–1.

The integration constants A and B in Eq. 22–3 are generally determined from the initial conditions of the problem. For example, suppose that the block in Fig. 22–1a has been displaced a distance x_1 to the right from its equilibrium position and given an initial (positive) velocity $\mathbf{v}_1$ directed to the right. Substituting $x = x_1$ at $t = 0$ into Eq. 22–3 yields $B = x_1$. Since $v = v_1$ at $t = 0$, using Eq. 22–4 we obtain $A = v_1/p$. If these values are substituted into Eq. 22–3, the equation describing the motion becomes

$$x = \frac{v_1}{p} \sin pt + x_1 \cos pt \qquad (22\text{–}6)$$

Equation 22–3 may also be expressed in terms of simple sinusoidal motion. Let

$$A = C \cos \phi \qquad (22\text{–}7)$$

and

$$B = C \sin \phi \qquad (22\text{–}8)$$

where C and ϕ are new constants to be determined in place of A and B. Substituting into Eq. 22–3 yields

$$x = C \cos \phi \sin pt + C \sin \phi \cos pt$$

Since $\sin(\theta + \phi) = \sin \theta \cos \phi + \cos \theta \sin \phi$, then

$$\boxed{x = C \sin(pt + \phi)} \qquad (22\text{–}9)$$

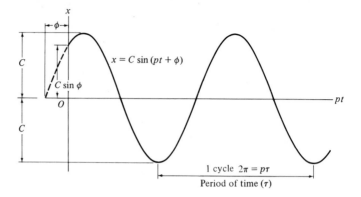

Fig. 22–3

If this equation is plotted on an x versus pt axis, the graph shown in Fig. 22–3 is obtained. The maximum displacement of the block from its equilibrium position is defined as the *amplitude* of vibration. From either the figure or Eq. 22–9 the amplitude is C. The angle ϕ is called the *phase angle* since it represents the amount by which the curve is displaced from the origin when $t = 0$. The constants C and ϕ are related to A and B by Eqs. 22–7 and 22–8. Squaring and adding these two equations, the amplitude becomes

$$C = \sqrt{A^2 + B^2} \tag{22–10}$$

If Eq. 22–8 is divided by Eq. 22–7, the phase angle is

$$\phi = \tan^{-1}\frac{B}{A} \tag{22–11}$$

Note that the sine curve, Eq. 22–9, completes one *cycle* in time $t = \tau$ (tau) when $p\tau = 2\pi$ or

$$\tau = \frac{2\pi}{p} \tag{22–12}$$

This length of time is called a *period,* Fig. 22–3. Using Eq. 22–2, the period may also be represented as

$$\tau = 2\pi\sqrt{\frac{m}{k}} \tag{22–13}$$

The *frequency f* is defined as the number of cycles completed per unit of time, which is the reciprocal of the period:

$$f = \frac{1}{\tau} = \frac{p}{2\pi} \tag{22–14}$$

or

$$f = \frac{1}{2\pi}\sqrt{\frac{k}{m}} \tag{22–15}$$

The frequency is expressed in cycles/s. This ratio of units is called a *hertz* (Hz), where 1 Hz = 1 cycle/s = 2π rad/s.

When a body or system of connected bodies is given an initial displacement from its equilibrium position and released, it will vibrate with a definite frequency known as the *natural frequency*. This type of vibration is called *free vibration*, provided no external forces except gravitational or elastic forces act on the body during the motion. Also, if the *amplitude* of vibration remains *constant*, the motion is said to be *undamped*. The undamped free vibration of a body having a single degree of freedom has the same characteristics as simple harmonic motion of the block and spring discussed above. Consequently, the body's motion is described by a differential equation of the *same form* as Eq. 22–1, i.e.,

$$\boxed{\ddot{x} + p^2 x = 0} \tag{22–16}$$

Hence, if the circular frequency p of the body is known, the period of vibration τ, natural frequency f, and other vibrating characteristics of the body can be established using the above equations.

PROCEDURE FOR ANALYSIS

As in the case of the block and spring, the circular frequency p of a rigid body or system of connected rigid bodies having a single degree of freedom can be determined using the following procedure:

Free-Body Diagram. Draw the free-body diagram of the body when the body is displaced by a *small amount* from its equilibrium position. Locate the body with respect to its equilibrium position by using an appropriate *inertial coordinate q*. The acceleration of the body's mass center $\mathbf{a}_G$ or the body's angular acceleration $\boldsymbol{\alpha}$ should have a sense which is in the positive direction of the position coordinate. If it is decided that the rotational equation of motion $\Sigma M_P = \Sigma(\mathcal{M}_k)_P$ is to be used, then it may be beneficial to also draw the kinetic diagram since it graphically accounts for the components $m(\mathbf{a}_G)_x$, $m(\mathbf{a}_G)_y$, and $I_G\boldsymbol{\alpha}$, and thereby makes it convenient for visualizing the terms needed in the moment sum $\Sigma(\mathcal{M}_k)_P$.

Equation of Motion. Apply the equation of motion to relate the elastic or gravitational restoring forces and couples acting on the body to the body's accelerated motion.

Kinematics. Using kinematics, express the body's accelerated motion in terms of the second time derivative of the position coordinate, $\ddot{q}$. Substitute this result into the equation of motion and determine p by rearranging the terms so that the resulting equation is of the form $\ddot{q} + p^2 q = 0$.

The following examples illustrate this procedure.

Example 22–1

Determine the period of vibration for the simple pendulum shown in Fig. 22–4a. The bob has a mass m and is attached to a cord of length l. Neglect the size of the bob.

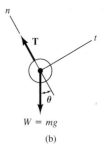

(a)   (b)

Fig. 22–4

SOLUTION

Free-Body Diagram. Motion of the system will be related to the position coordinate $(q =)\theta$, Fig. 22–4b. When the bob is displaced by an angle θ, the *restoring force* acting on the bob is created by the *weight component* $mg \sin \theta$. Furthermore, $\mathbf{a}_t$ acts in the direction of *increasing* s (or θ).

Equation of Motion. Applying the equation of motion in the *tangential direction,* since it involves the restoring force, yields

$$+ \nearrow \Sigma F_t = ma_t; \qquad -mg \sin \theta = ma_t \qquad (1)$$

Kinematics. $a_t = d^2s/dt^2 = \ddot{s}$. Furthermore, s may be related to θ by the equation $s = l\theta$, so that $a_t = l\ddot{\theta}$. Hence, Eq. (1) reduces to the form

$$\ddot{\theta} + \frac{g}{l} \sin \theta = 0 \qquad (2)$$

The solution of this equation involves the use of an elliptic integral. For *small displacements,* however, $\sin \theta \approx \theta$, in which case

$$\ddot{\theta} + \frac{g}{l}\theta = 0 \qquad (3)$$

Comparing this equation with Eq. 22–16 ($\ddot{x} + p^2x = 0$), which is the "standard form" for simple harmonic motion, it is seen that $p = \sqrt{g/l}$. From Eq. 22–12, the period of time required for the bob to make one complete swing is therefore

$$\tau = \frac{2\pi}{p} = 2\pi\sqrt{\frac{l}{g}} \qquad \qquad Ans.$$

This interesting result, originally discovered by Galileo Galilei through experiment, indicates that the period depends only on the length of the cord and not on the mass of the pendulum bob.

The solution of Eq. (3) is given by Eq. 22–3, where $p = \sqrt{g/l}$ and θ is substituted for x. Like the block and spring, the constants A and B in this problem may be determined if, for example, one knows the displacement and velocity of the bob at a given instant.

Example 22–2

The 10-kg rectangular plate shown in Fig. 22–5a is suspended at its center from a rod having a torsional stiffness $k = 1.5$ N · m/rad. Determine the natural period of vibration of the plate when it is given a small angular displacement θ in the plane of the plate.

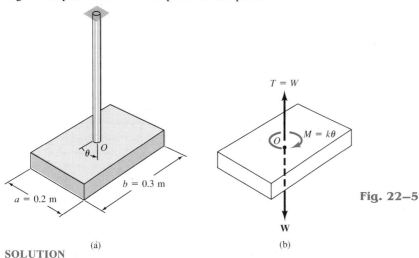

Fig. 22–5

(a)   (b)

SOLUTION

Free-Body Diagram. Fig. 22–5b. Since the plate is displaced in its own plane, the torsional *restoring* moment created by the rod is $M = k\theta$. This moment acts in the direction opposite to the angular displacement θ. The angular acceleration $\ddot{\theta}$ acts in the direction of *positive* θ.

Equation of Motion

$$\Sigma M_O = I_O \alpha; \qquad\qquad -k\theta = I_O \ddot{\theta}$$

or

$$\ddot{\theta} + \frac{k}{I_O} \theta = 0$$

Since this equation is in the "standard form," the circular frequency is $p = \sqrt{k/I_O}$.

The moment of inertia of the plate about an axis coincident with the rod is $I_O = \frac{1}{12}m(a^2 + b^2)$. Hence,

$$I_O = \frac{1}{12}(10)[(0.2)^2 + (0.3)^2] = 0.108 \text{ kg} \cdot \text{m}^2$$

The natural period of vibration is, therefore,

$$\tau = \frac{2\pi}{p} = 2\pi\sqrt{\frac{I_O}{k}} = 2\pi\sqrt{\frac{0.108}{1.5}} = 1.69 \text{ s} \qquad\qquad Ans.$$

Example 22–3

The bent rod shown in Fig. 22–6a has a negligible mass and supports a 5-kg collar at its end. Determine the natural period of vibration for the system.

(a)

SOLUTION

Free-Body and Kinetic Diagrams. Fig. 22–6b. Here the rod is displaced by a small amount θ from the equilibrium position. Since the spring is subjected to an initial compression of x_{st} for statical equilibrium, then when the displacement $x > x_{st}$ the spring exerts a force of $F_s = kx - kx_{st}$ on the rod. To obtain the "standard form," Eq. 22–16, $5\mathbf{a}_y$ acts *upward*, which is in accordance with positive θ displacement.

Equation of Motion. Moments will be summed about point B to eliminate the unknown reaction at this point. Since θ is small,

$$\zeta + \Sigma M_B = \Sigma(\mathcal{M}_k)_B; \quad kx(0.1) - kx_{st}(0.1) + 49.05(0.2) = -5a_y(0.2)$$

The second term on the left side, $-kx_{st}(0.1)$, represents the moment created by the spring force which is necessary to hold the collar in *equilibrium*, i.e., at $x = 0$. Since this moment is equal and opposite to the moment $49.05(0.2)$ created by the weight of the collar, these two terms cancel in the above equation, so that

$$kx(0.1) = -5a_y(0.2) \qquad (1)$$

Kinematics. The positions of the spring and the collar may be related to the angle θ, Fig. 22–6c. Since θ is small, $x = 0.1\theta$ and $y = 0.2\theta$. Therefore, $a_y = \ddot{y} = 0.2\ddot{\theta}$. Substituting into Eq. (1) yields

$$400(0.1\theta)0.1 = -5(0.2\ddot{\theta})0.2$$

Rewriting this equation in standard form gives

$$\ddot{\theta} + 20\theta = 0$$

Compared with $\ddot{x} + p^2 x = 0$ (Eq. 22–16), we have

$$p^2 = 20, \qquad p = 4.47 \text{ rad/s}$$

The natural period of vibration is therefore,

$$\tau = \frac{2\pi}{p} = \frac{2\pi}{4.47} = 1.40 \text{ s} \qquad Ans.$$

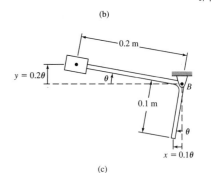

(b)

(c)

Fig. 22–6

Example 22–4

A 10-lb block is suspended from a cord that passes over a 15-lb disk, as shown in Fig. 22–7a. The spring has a stiffness $k = 200$ lb/ft. Determine the natural period of vibration for the system.

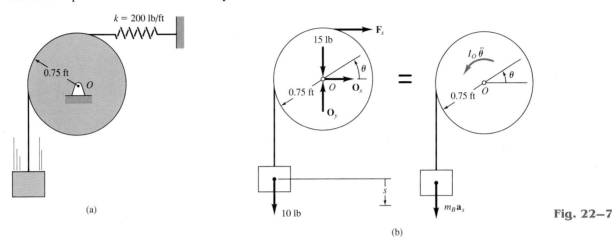

(a)

(b)

Fig. 22–7

SOLUTION

Free-Body and Kinetic Diagrams. Fig. 22–7b. The *system* consists of the disk, which undergoes a rotation defined by the angle θ, and the block, which translates by an amount s. The vector $I_O\ddot{\theta}$ acts in the direction of *positive* θ, and consequently $m_B\mathbf{a}_s$ acts downward in the direction of *positive* s.

Equation of Motion. Summing moments about point O to eliminate the reactions $\mathbf{O}_x$ and $\mathbf{O}_y$, realizing that $I_O = \frac{1}{2}mr^2$, yields

$\zeta+\Sigma M_O = \Sigma(\mathcal{M}_k)_O;$

$$10(0.75) - F_s(0.75) = \frac{1}{2}\left(\frac{15}{32.2}\right)(0.75)^2\ddot{\theta} + \left(\frac{10}{32.2}\right)a_s(0.75) \quad (1)$$

Kinematics. As shown on the kinematic diagram in Fig. 22–7c, a small positive displacement θ of the disk causes the block to lower by an amount $s = 0.75\theta$; hence, $a_s = \ddot{s} = 0.75\ddot{\theta}$. When $\theta = 0°$, the spring force required for *equilibrium* of the disk is 10 lb, acting to the right. For position θ, the spring force is $F_s = (200$ lb/ft$)(0.75\theta$ ft$) + 10$ lb. Substituting these results into Eq. (1) and simplifying yields

$$\ddot{\theta} + 368\theta = 0$$

Hence,

$$p^2 = 368 \qquad p = 19.2 \text{ rad/s}$$

Therefore, the natural period of vibration is

$$\tau = \frac{2\pi}{p} = \frac{2\pi}{19.2} = 0.328 \text{ s} \qquad Ans.$$

(c)

PROBLEMS

22–1. When a 20-lb weight is suspended from a spring, the spring is stretched a distance of 4 in. Determine the natural frequency and the period of vibration for a 10-lb weight attached to the same spring.

22–2. Determine the frequency of vibration for the block. The springs are originally compressed Δ.

Prob. 22–2

22–3. A spring has a stiffness of 600 N/m. If a 4-kg block is attached to the spring, pushed 50 mm above its equilibrium position, and released from rest, determine the equation which describes the block's motion. Assume that positive displacement is measured downward.

*22–4. When a 3-kg block is suspended from a spring, the spring is stretched a distance of 60 mm. Determine the natural frequency and the period of vibration for a 0.2-kg block attached to the same spring.

22–5. An 8-kg block is suspended from a spring having a stiffness $k = 80$ N/m. If the block is given an upward velocity of 0.4 m/s when it is 90 mm above its equilibrium position, determine the equation which describes the motion and the maximum upward displacement of the block measured from the equilibrium position. Assume that positive displacement is measured downward.

22–6. A 2-lb weight is suspended from a spring having a stiffness of $k = 2$ lb/in. If the weight is pushed 1 in. upward from its equilibrium position and then released from rest, determine the equation which describes the motion. What is the amplitude and the natural frequency of the vibration?

22–7. A 6-lb weight is suspended from a spring having a stiffness of $k = 3$ lb/in. If the weight is given an upward velocity of 20 ft/s when it is 2 in. above its equilibrium position, determine the equation which describes the motion and the maximum upward displacement of the weight, measured from the equilibrium position. Assume positive displacement is downward.

*■22–8. A spring is stretched 175 mm by an 8-kg block. If the block is displaced 100 mm downward from its equilibrium position and given a downward velocity of 1.50 m/s, determine the differential equation which describes the motion. Assume that positive displacement is measured downward. Use the Runge-Kutta method to determine the position of the block, measured from its unstretched position, when $t = 0.22$ s. Use a time increment of $\Delta t = 0.02$ s.

22–9. If the block in Prob. 22–8 is given an upward velocity of 4 m/s when it is displaced downward a distance of 60 mm from its equilibrium position, determine the equation which describes the motion. What is the amplitude of the motion? Assume that positive displacement is measured downward.

22–10. A pendulum has a 0.4-m-long cord and is given a tangential velocity of 0.2 m/s toward the vertical from a position $\theta = 0.3$ rad. Determine the equation which describes the angular motion.

22–11. Determine to the nearest degree the maximum angular displacement of the bob in Prob. 22–10 if it is initially displaced $\theta = 0.2$ rad from the vertical and given a tangential velocity of 0.4 m/s away from the vertical.

*22–12. The square plate has a mass m and is suspended at its corner by the pin O. Determine the period of oscillation if it is displaced a small amount and released.

Prob. 22–12

22–13. The body of arbitrary shape has a mass m, mass center at G, and a radius of gyration about G of k_G. If it is displaced by a slight amount θ from its equilibrium position and released, determine the period of vibration.

Prob. 22–13

22–14. The uniform beam is supported at its ends by two springs A and B, each having the same stiffness k. When nothing is supported on the beam, it has a period of vertical vibration of 0.83 s. If a 50-kg mass is placed at its center, the period of vertical vibration is 1.52 s. Compute the stiffness of each spring and the mass of the beam.

Prob. 22–14

22–15. The thin hoop of mass m is supported by a knife-edge. Determine the period of oscillation for small amplitudes of swing.

Prob. 22–15

*22–16. The semicircular disk weighs 20 lb. Determine the period of oscillation if it is displaced a small amount and released.

Prob. 22–16

22–17. While standing in an elevator, the man holds a pendulum which consists of an 18-in. cord and a 0.5-lb bob. If the elevator is descending with an acceleration $a = 4$ ft/s^2, determine the period of vibration for small amplitudes of swing.

Prob. 22–17

22–18. The disk has a weight of 10 lb and rolls without slipping on the horizontal surface as it oscillates about its equilibrium position. If the disk is displaced, by rolling it counterclockwise 0.4 rad, determine the equation which describes its oscillatory motion when it is released.

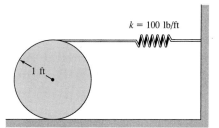

Prob. 22–18

22–19. The platform AB when empty has a mass of 400 kg, center of mass at G_1, and period of oscillation $\tau_1 = 2.38$ s. If a car having a mass of 1.2 Mg and center of mass at G_2 is placed on the platform, the period of oscillation becomes $\tau_2 = 3.16$ s. Determine the moment of inertia of the car about an axis passing through G_2.

Prob. 22–19

*22–20.** The uniform rod has a mass m and is supported by the pin O. If the rod is given a small displacement and released, determine the period of free vibration. The springs are unstretched when the rod is in the position shown.

Prob. 22–20

22–21. The pointer on a metronome supports a 0.4-lb slider A, which is positioned at a fixed distance from the pivot O of the pointer. When the pointer is displaced, a torsional spring at O exerts a restoring moment on the pointer having a magnitude $M = (1.2\theta)$ lb · ft, where θ represents the angle of displacement from the vertical, measured in radians. Determine the period of vibration when the pointer is displaced a small amount θ and released. Neglect the mass of the pointer.

Prob. 22–21

22–22. The 50-lb spool is attached to two springs. If the spool is displaced a small amount and released, determine the period of vibration. The radius of gyration of the spool is $k_G = 1.5$ ft. The spool rolls without slipping.

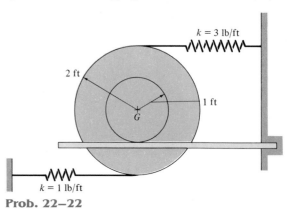

Prob. 22–22

22–23. The 20-lb rectangular plate has a natural period of vibration of $\tau = 0.3$ s, as it oscillates around the axis of rod AB. Determine the torsional stiffness k, measured in lb · ft/rad, of the rod. Neglect the mass of the rod.

Prob. 22–23

*22–24. The 6-lb weight is attached to the rods of negligible mass. Determine the frequency of vibration of the weight when it is displaced slightly and released.

$k = 5$ lb/ft

2 ft

3 ft

Prob. 22–24

22–25. The 25-lb weight is fixed to the end of the rod assembly. If both springs are unstretched when the assembly is in the position shown, determine the natural period of vibration for the weight when it is displaced slightly and released. Neglect the size of the block and the mass of the rods.

6 in. — 6 in.

$k = 2$ lb/in. $k = 2$ lb/in.

12 in.

Prob. 22–25

22–26. If the wire AB is subjected to a tension force of 20 lb, determine the equation which describes the motion when the 5-lb weight is displaced 2 in. horizontally and released from rest.

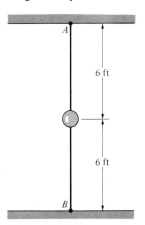

A

6 ft

6 ft

B

Prob. 22–26

22–27. The plate of mass m is supported by three symmetrically placed cords of length l as shown. If the plate is given a slight rotation about a vertical axis through its center and released, determine the period of oscillation.

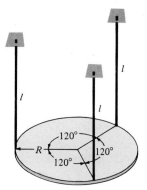

l

l

l

120°

R

120°

120°

Prob. 22–27

*22–28. The bar has a length l and mass m. It is supported at its ends by rollers of negligible mass. If it is given a small displacement and released, determine the frequency of vibration.

Prob. 22–28

22–29. The cylinder of radius r and mass m is displaced a small amount along the curved surface. If it rolls without slipping, determine the frequency of oscillation when it is released.

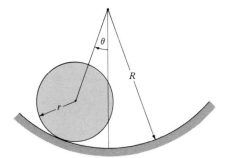

Prob. 22–29

22–30. Determine the frequency for small oscillations of the 10-lb sphere when the rod is displaced a slight distance and released. Neglect the size of the sphere and the mass of the rod. The spring has an unstretched length of 1 ft.

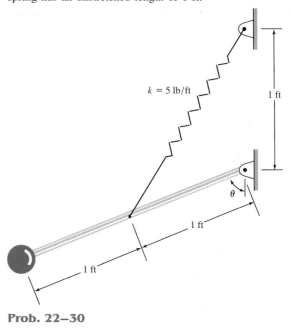

$k = 5$ lb/ft

Prob. 22–30

*22.2 Energy Methods

The simple harmonic motion of a body, discussed in Sec. 22.1, is due only to gravitational and elastic restoring forces acting on the body. Since these types of forces are *conservative*, it is possible to use the conservation of energy equation to obtain the body's natural frequency or period of vibration. To show how to do this, consider the block and spring in Fig. 22–8. When the block is displaced an arbitrary amount x from the equilibrium position, the kinetic energy is $T = \frac{1}{2}mv^2 = \frac{1}{2}m\dot{x}^2$ and the potential energy is $V = \frac{1}{2}kx^2$. By the conservation of energy equation, Eq. 14–21, it is necessary that

$$T + V = \text{const}$$
$$\tfrac{1}{2}m\dot{x}^2 + \tfrac{1}{2}kx^2 = \text{const} \qquad (22\text{–}17)$$

Equilibrium position

k

m

Fig. 22–8

The differential equation describing the *accelerated motion* of the block can be obtained by *differentiating* this equation with respect to time, i.e.,

$$m\dot{x}\ddot{x} + kx\dot{x} = 0$$

$$\dot{x}(m\ddot{x} + kx) = 0$$

Since the velocity $\dot{x}$ is not *always* zero in a vibrating system,

$$\ddot{x} + p^2 x = 0, \qquad p = \sqrt{k/m}$$

which is the same as Eq. 22–1.

If the energy equation is written for a *system of connected bodies,* the natural frequency or the equation of motion can also be determined by time differentiation. Here it is *not necessary* to dismember the system to account for reactive and connective forces which do no work. Also, by this method, the circular frequency p may be obtained *directly*. For example, consider the total mechanical energy of the block and spring in Fig. 22–8 when the block is at its *maximum displacement*. In this position the block is temporarily at rest, so that the kinetic energy is zero and the potential energy, stored in the spring, is a maximum. Therefore, Eq. 22–17 becomes $\frac{1}{2}kx_{max}^2 = $ const. At the instant the block passes the equilibrium position, the kinetic energy of the block is a maximum and the potential energy of the spring is zero. Hence, Eq. 22–17 becomes $\frac{1}{2}m(\dot{x})_{max}^2 = $ const. Since the vibrating motion of the block is *harmonic,* the *solution* for the displacement and velocity may be written in the form of Eq. 22–9 and its time derivative, i.e.,

$$x = C \sin (pt + \phi), \qquad \dot{x} = Cp \cos (pt + \phi)$$

so that
$$x_{max} = C, \qquad \dot{x}_{max} = Cp$$

Applying the conservation of energy equation ($T + V = $ const) yields

$$V_{max} = T_{max}; \qquad \frac{1}{2}kx_{max}^2 = \frac{1}{2}m\dot{x}_{max}^2 = \text{const}$$

or
$$kC^2 = mC^2p^2$$

Solving for p yields
$$p = \sqrt{\frac{k}{m}}$$

which is identical to Eq. 22–2.

PROCEDURE FOR ANALYSIS

The following procedure provides a method for determining the circular frequency p of a body or system of connected bodies using the conservation of energy equation.

Energy Equation. Draw the body when it is displaced by a *small amount* from its equilibrium position and define the location of the body from its equilibrium position by an appropriate position coordinate q. Formulate the equation of energy for the body, $T + V = $ const, in terms of the position coordinate.* Recall that, in general, the kinetic energy must account for both the body's translational and rotational motion, $T = \frac{1}{2}mv_G^2 + \frac{1}{2}I_G\omega^2$, Eq. 18–2, and that the potential energy is the sum of the gravitational and elastic potential energies of the body, $V = V_g + V_e$, Eq. 18–15. In particular, V_g should be measured from a datum for which $q = 0$ (equilibrium position).

Time Derivative. Take the time derivative of the energy equation and factor out the common terms. The resultant differential equation represents the equation of motion for the system. The value of p is obtained after rearranging the terms in the standard form $\ddot{q} + p^2q = 0$.

The following examples illustrate this procedure.

*It is suggested that the material in Sec. 18.5 be reviewed.

Example 22–5

The thin hoop shown in Fig. 22–9a is supported by a peg at O. Determine the period of oscillation for small amplitudes of swing. The hoop has a mass m.

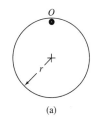

(a)

SOLUTION

Energy Equation. A diagram of the hoop when it is displaced a small amount $(q =)\theta$ from the equilibrium position is shown in Fig. 22–9b. Using the table on the inside back cover of the book and the parallel-axis theorem to determine I_O, we can express the kinetic energy as

$$T = \tfrac{1}{2}I_O\omega^2 = \tfrac{1}{2}[mr^2 + mr^2]\dot{\theta}^2 = mr^2\dot{\theta}^2$$

(b)

Fig. 22–9

If a horizontal datum is placed through the center of gravity of the hoop when $\theta = 0$, then the center of gravity moves upward $r(1 - \cos \theta)$ in the displaced position. For *small angles,* $\cos \theta$ may be replaced by the first two terms of its series expansion, $\cos \theta = 1 - \theta^2/2 + \cdots$. Therefore, the potential energy is

$$V = mgr\left[1 - \left(1 - \frac{\theta^2}{2}\right)\right] = mgr\frac{\theta^2}{2}$$

The total energy in the system is

$$T + V = mr^2\dot{\theta}^2 + mgr\frac{\theta^2}{2}$$

Time Derivative

$$mr^2 2\dot{\theta}\ddot{\theta} + mgr\theta\dot{\theta} = 0$$
$$mr\dot{\theta}(2r\ddot{\theta} + g\theta) = 0$$

Since $\dot{\theta}$ is not always equal to zero, from the terms in parentheses

$$\ddot{\theta} + \frac{g}{2r}\theta = 0$$

Hence,

$$p = \sqrt{\frac{g}{2r}}$$

so that

$$\tau = \frac{2\pi}{p} = 2\pi\sqrt{\frac{2r}{g}} \qquad\qquad\qquad Ans.$$

Example 22–6

A 10-kg block is suspended from a cord wrapped around a 5-kg disk, as shown in Fig. 22–10a. If the spring has a stiffness $k = 200$ N/m, determine the natural period of vibration for the system.

(a)

SOLUTION

Energy Equation. A diagram of the block and disk when they are displaced by respective amounts s and θ from the equilibrium position is shown in Fig. 22–10b. Since $s = 0.15\theta$, the kinetic energy of the system is

$$T = \tfrac{1}{2}m_b v_b^2 + \tfrac{1}{2}I_O \omega_d^2$$

$$= \frac{1}{2}(10)(0.15\dot{\theta})^2 + \frac{1}{2}\left[\frac{1}{2}(5)(0.15)^2\right](\dot{\theta})^2$$

$$= 0.141(\dot{\theta})^2$$

Establishing the datum at the equilibrium position of the block and realizing that the spring stretches s_{st} for equilibrium, we can write the potential energy as

$$V = \tfrac{1}{2}k(s_{st} + s)^2 - Ws$$

$$= \frac{1}{2}(200)(s_{st} + 0.15\theta)^2 - 98.1(0.15\theta)$$

The total energy for the system is, therefore,

$$T + V = 0.141(\dot{\theta})^2 + 100(s_{st} + 0.15\theta)^2 - 14.72\theta$$

(b)

Fig. 22–10

Time Derivative

$$0.282(\dot{\theta})\ddot{\theta} + 200(s_{st} + 0.15\theta)0.15\dot{\theta} - 14.72\dot{\theta} = 0$$

Since $s_{st} = 98.1/200 = 0.4905$ m, the above equation reduces to the standard form

$$\ddot{\theta} + 16\theta = 0$$

so that

$$p = \sqrt{16} = 4 \text{ rad/s}$$

Thus,

$$\tau = \frac{2\pi}{p} = \frac{2\pi}{4} = 1.57 \text{ s} \qquad\qquad Ans.$$

PROBLEMS

22–31. Solve Prob. 22–12 using energy methods.

*22–32. Solve Prob. 22–13 using energy methods.

22–33. Solve Prob. 22–15 using energy methods.

22–34. Solve Prob. 22–16 using energy methods.

22–35. The machine has a mass m and is uniformly supported by four springs each having a stiffness k. Determine the period of vertical vibration.

Prob. 22–35

*22–36. Determine the differential equation of motion of the block of mass m when it is displaced slightly and released. Motion occurs in the vertical plane. The springs are originally unstretched and each is attached to the block.

Prob. 22–36

22–37. If the disk has a mass of 8 kg, determine the natural frequency of vibration. The springs are originally unstretched.

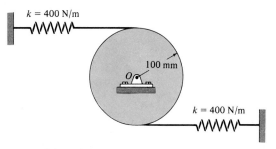

$k = 400 \text{ N/m}$

100 mm

O

$k = 400 \text{ N/m}$

Prob. 22–37

22–38. The 7-kg disk is pin-connected at its midpoint. Determine the period of vibration of the disk if the springs have sufficient tension in them to prevent the cord from slipping on the disk as it oscillates. *Hint:* Assume that the initial stretch in each spring is δ_O. This term will cancel out after taking the time derivative of the energy equation.

100 mm

O

$k = 600 \text{ N/m}$ $k = 600 \text{ N/m}$

Prob. 22–38

22–39. Determine the period of vibration of the pendulum. Consider the two rods to be slender, each having a weight $W = 8$ lb/ft.

2 ft

1 ft — 1 ft

Prob. 22–39

*22–40. Determine the natural frequency of vibration of the 20-lb disk. Assume the disk does not slip on the inclined surface.

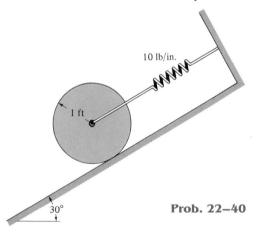

10 lb/in.

1 ft

30°

Prob. 22–40

22–41. Determine the period of vibration of the 10-lb semicircular disk.

0.5 ft

Prob. 22–41

22–42. The 5-lb sphere is attached to a rod of negligible mass. Determine the natural frequency of vibration of the sphere. Neglect the size of the sphere.

1 ft — 0.5 ft

$k = 10$ lb/ft

Prob. 22–42

22–43. Determine the period of vibration of the 3-kg sphere. Neglect the mass of the rod and the size of the sphere.

$k = 500$ N/m

O

300 mm — 300 mm

Prob. 22–43

*22–44. The slender rod has a weight $W = 4$ lb/ft. If it is supported in the horizontal plane by a ball-and-socket joint at A and a cable at B, determine the natural frequency of vibration when the end B is given a small horizontal displacement and then released.

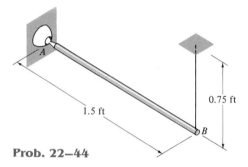

A

1.5 ft

0.75 ft

B

Prob. 22–44

22–45. The bar has a mass of 8 kg and is suspended from two springs such that when it is in equilibrium, the springs make an angle of 45° with the horizontal as shown. Determine the period of vibration if the bar is pulled down a short distance and released. Each spring has a stiffness of $k = 40$ N/m.

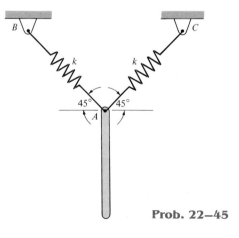

B C

k k

45° 45°

A

Prob. 22–45

*22.3 Undamped Forced Vibration

Undamped forced vibration is considered to be one of the most important types of vibrating motion in engineering work. The principles which describe the nature of this motion may be applied to the analysis of forces which cause vibration in various types of machines and structures.

Periodic Force

The block and spring shown in Fig. 22–11a provide a convenient "model" which represents the vibrational characteristics of a system subjected to a periodic force $F = F_O \sin \omega t$. This force has a maximum magnitude of F_O and a *forcing frequency* ω. The free-body diagram for the block when it is displaced a distance x is shown in Fig. 22–11b. Applying the equation of motion yields

$$\xrightarrow{+}\Sigma F_x = ma_x; \qquad\qquad F_O \sin \omega t - kx = m\ddot{x}$$

or

$$\ddot{x} + \frac{k}{m}x = \frac{F_O}{m} \sin \omega t \qquad\qquad (22\text{–}18)$$

This equation is referred to as a nonhomogeneous second-order differential equation. The general solution consists of a complementary solution, x_c, *plus* a particular solution, x_p.

The *complementary solution* is determined by setting the term on the right side of Eq. 22–18 equal to zero and solving the resulting homogeneous equation, which is equivalent to Eq. 22–1. The solution is defined by Eq. 22–3, i.e.,

$$x_c = A \sin pt + B \cos pt \qquad\qquad (22\text{–}19)$$

where p is the circular frequency, $p = \sqrt{k/m}$, Eq. 22–2.

Fig. 22–11 (a) (b)

Since the motion is periodic, the *particular solution* of Eq. 22–18 may be determined by assuming a solution of the form

$$x_p = C \sin \omega t \qquad (22\text{–}20)$$

where C is a constant. Taking the second time derivative and substituting into Eq. 22–18 yields

$$-C\omega^2 \sin \omega t + \frac{k}{m}(C \sin \omega t) = \frac{F_O}{m} \sin \omega t$$

(a)

Factoring out $\sin \omega t$ and solving for C gives

$$C = \frac{F_O/m}{\dfrac{k}{m} - \omega^2} = \frac{F_O/k}{1 - \left(\dfrac{\omega}{p}\right)^2} \qquad (22\text{–}21)$$

Substituting into Eq. 22–20, we obtain the particular solution

$$\boxed{x_p = \frac{F_O/k}{1 - \left(\dfrac{\omega}{p}\right)^2} \sin \omega t} \qquad (22\text{–}22)$$

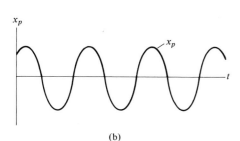

(b)

The *general solution* is therefore

$$x = x_c + x_p = A \sin pt + B \cos pt + \frac{F_O/k}{1 - \left(\dfrac{\omega}{p}\right)^2} \sin \omega t \qquad (22\text{–}23)$$

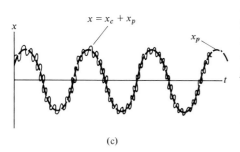

(c)

Here x describes two types of vibrating motion of the block. The *complementary solution* x_c defines the *free vibration*, which depends upon the circular frequency $p = \sqrt{k/m}$ and the constants A and B, Fig. 22–12a. Specific values for A and B are obtained by evaluating Eq. 22–23 at a given instant when the displacement and velocity are known. The *particular solution* x_p describes the *forced vibration* of the block caused by the applied force $F = F_O \sin \omega t$. Fig. 22–12b. The resultant vibration x is shown in Fig. 22–12c. Since all vibrating systems are subject to *friction*, the free vibration, x_c, will in time dampen out. For this reason the free vibration is referred to as *transient,* and the forced vibration is called *steady state,* since it is the only vibration that remains, Fig. 22–12d.

From Eq. 22–21 it is seen that the *amplitude* of forced vibration depends upon the *frequency ratio* ω/p. If the *magnification factor* MF is defined as the ratio of the amplitude of steady-state vibration, $(x_p)_{max}$, to the static deflection F_O/k, which is caused by the amplitude of the periodic force F_O, then, from Eq. 22–22,

$$\text{MF} = \frac{(x_p)_{max}}{F_O/k} = \frac{1}{1 - \left(\dfrac{\omega}{p}\right)^2} \qquad (22\text{–}24)$$

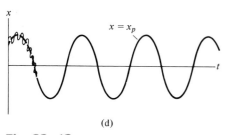

(d)

Fig. 22–12

This equation is graphed in Fig. 22–13, where it is seen that for $\omega \approx 0$, the MF ≈ 1. In this case, because of the very low frequency $\omega \ll p$, the vibration of the block will be in phase with the applied force **F.** If the force or displacement is applied with a frequency close to the natural frequency of the system, i.e., $\omega/p \approx 1$, the amplitude of vibration of the block becomes extremely large. This condition is called *resonance,* and in practice, resonating vibrations can cause tremendous stress and rapid failure of parts. When the cyclic force $F_O \sin \omega t$ is applied at high frequencies ($\omega > p$), the value of the MF becomes negative, indicating that the motion of the block is out of phase with the force. Under these conditions, as the block is displaced to the right, the force acts to the left, and vice versa. For extremely high frequencies ($\omega \gg p$) the block remains almost stationary, and hence the MF is approximately zero.

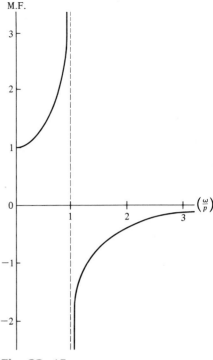

Fig. 22–13

Periodic Support Displacement

Forced vibrations can also arise from the periodic excitation of the support of a system. The model shown in Fig. 22–14a represents the periodic vibration of a block which is caused by harmonic movement $\delta = \delta_O \sin \omega t$ of the support. The free-body diagram for the block in this case is shown in Fig. 22–14b. The coordinate x is measured from the point of zero displacement of the support, i.e., when the radial line OA coincides with OB, Fig. 22–14a. Therefore, general displacement of the spring is $(x - \delta_O \sin \omega t)$. Applying the equation of motion yields

$$\overset{+}{\rightarrow} \Sigma F_x = ma_x; \qquad -k(x - \delta_O \sin \omega t) = m\ddot{x}$$

or

$$\ddot{x} + \frac{k}{m}x = \frac{k\delta_O}{m} \sin \omega t \qquad (22\text{–}25)$$

By comparison, this equation is identical to the form of Eq. 22–18, *provided* F_O is *replaced* by $k\delta_O$. If this substitution is made into the solutions defined by Eqs. 22–21 to 22–23, the results are appropriate for describing the motion of the block when subjected to the support displacement $\delta = \delta_O \sin \omega t$.

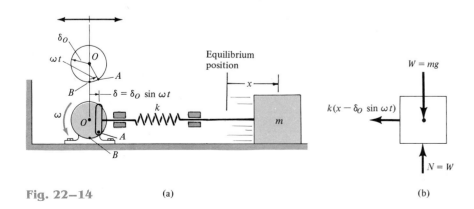

Fig. 22–14

(a)

(b)

Example 22–7

The instrument shown in Fig. 22–15 is rigidly attached to a platform P, which in turn is supported by *four* springs, each having a stiffness $k = 800$ N/m. Initially the platform is at rest when the floor is subjected to a displacement $\delta = 10 \sin (8t)$ mm, where t is in seconds. If the instrument is constrained to move vertically, and the total mass of the instrument and platform is 20 kg, determine the vertical displacement y of the platform, measured from the equilibrium position, as a function of time. What floor vibration is required to cause resonance?

Fig. 22–15

SOLUTION

Since the induced vibration is caused by the displacement of the supports, the motion is described by Eq. 22–23, with F_O replaced by $k\delta_O$, i.e.,

$$y = A \sin pt + B \cos pt + \frac{\delta_O}{1 - \left(\dfrac{\omega}{p}\right)^2} \sin \omega t \qquad (1)$$

Here $\delta = \delta_O \sin \omega t = 10 \sin (8t)$ mm, so that

$$\delta_O = 10 \text{ mm}, \qquad \omega = 8 \text{ rad/s}$$

$$p = \sqrt{\frac{k}{m}} = \sqrt{\frac{4(800)}{20}} = 12.6 \text{ rad/s}$$

From Eq. 22–22, with $k\delta_O$ replacing F_O, the amplitude of vibration caused by the floor displacement is

$$(y_p)_{\max} = \frac{\delta_O}{1 - \left(\dfrac{\omega}{p}\right)^2} = \frac{10}{1 - \left(\dfrac{8}{12.6}\right)^2} = 16.7 \text{ mm} \qquad (2)$$

Hence, Eq. (1) and its time derivative become

$$y = A \sin (12.6t) + B \cos (12.6t) + 16.7 \sin (8t)$$
$$\dot{y} = A(12.6) \cos (12.6t) - B(12.6) \sin (12.6t) + 133.3 \cos (8t)$$

The constants A and B are evaluated from these equations. Since $y = \dot{y} = 0$ at $t = 0$, then

$$0 = 0 + B + 0; \qquad\qquad B = 0$$
$$0 = A(12.6) - 0 + 133.3; \qquad A = -10.6$$

The vibrating motion is therefore described by the equation

$$y = -10.6 \sin (12.6t) + 16.7 \sin (8t) \qquad\qquad Ans.$$

Resonance will occur when the amplitude of vibration caused by the floor displacement approaches infinity. From Eq. (2), this requires

$$\omega = p = 12.6 \text{ rad/s} \qquad\qquad Ans.$$

*22.4 Viscous Damped Free Vibration

The vibration analysis considered thus far has not included the effects of friction or damping in the system, and as a result, the solutions obtained are only in close agreement with the actual motion. Since all vibrations die out in time, the presence of damping forces should be included in the analysis.

In many cases damping is attributed to the resistance created by the substance, such as water, oil, or air, in which the system vibrates. Provided the body moves slowly through this substance, the resistance to motion is directly proportional to the body's speed. The type of force developed under these conditions is called a *viscous damping force*. The magnitude of this force may be expressed by an equation of the form

$$F = c\dot{x} \qquad (22\text{-}26)$$

where the constant c is called the *coefficient of viscous damping* and has units of N · s/m or lb · s/ft.

The vibrating motion of a body or system having viscous damping may be characterized by the block and spring shown in Fig. 22–16a. The effect of damping is provided by the *dashpot* connected to the block on the right side. Damping occurs when the piston P moves to the right or left within the enclosed cylinder. The cylinder contains a fluid, and the motion of the piston is retarded since the fluid must flow around or through a small hole in the piston. The dashpot is assumed to have a coefficient of viscous damping c.

If the block is displaced a distance x from its equilibrium position, the resulting free-body diagram is shown in Fig. 22–16b. Both the spring force kx and the damping force $c\dot{x}$ oppose the forward motion of the block, so that applying the equation of motion yields

$$\xrightarrow{+}\Sigma F_x = ma_x; \qquad -kx - c\dot{x} = m\ddot{x}$$

or

$$m\ddot{x} + c\dot{x} + kx = 0 \qquad (22\text{-}27)$$

This linear, second-order, homogeneous, differential equation has solutions of the form

$$x = e^{\lambda t}$$

where e is the base of the natural logarithm and λ is a constant. The value of λ (lambda) may be obtained by substituting this solution into Eq. 22–27, which yields

$$m\lambda^2 e^{\lambda t} + c\lambda e^{\lambda t} + k e^{\lambda t} = 0$$

or

$$e^{\lambda t}(m\lambda^2 + c\lambda + k) = 0$$

Since $e^{\lambda t}$ is always positive, a solution is possible provided

$$m\lambda^2 + c\lambda + k = 0$$

Equilibrium position

(a)

(b)

Fig. 22–16

Hence, by the quadratic formula, the two values of λ are

$$\lambda_1 = -\frac{c}{2m} + \sqrt{\left(\frac{c}{2m}\right)^2 - \frac{k}{m}}$$

$$\lambda_2 = -\frac{c}{2m} - \sqrt{\left(\frac{c}{2m}\right)^2 - \frac{k}{m}}$$

(22–28)

The general solution of Eq. 22–27 is therefore a linear combination of exponentials which involves both of these roots. There are three possible combinations of λ_1 and λ_2 which must be considered for the general solution. Before discussing these combinations, however, it is first necessary to consider the definition of the *critical damping coefficient* c_c as the value of c which makes the radical in Eqs. 22–28 equal to zero; i.e.,

$$\left(\frac{c_c}{2m}\right)^2 - \frac{k}{m} = 0$$

or

$$c_c = 2m\sqrt{\frac{k}{m}} = 2mp$$

(22–29)

Here the value of p is the circular frequency $p = \sqrt{k/m}$, Eq. 22–2.

Overdamped System

When $c > c_c$, the roots λ_1 and λ_2 are both real. The general solution of Eq. 22–27 may then be written as

$$x = Ae^{\lambda_1 t} + Be^{\lambda_2 t}$$

(22–30)

Motion corresponding to this solution is *nonvibrating*. The effect of damping is so strong that when the block is displaced and released, it simply creeps back to its original position without oscillating. The system is said to be *overdamped*.

Critically Damped System

If $c = c_c$, then $\lambda_1 = \lambda_2 = -c_c/2m = -p$. This situation is known as *critical damping*, since it represents a condition where c has the smallest value necessary to cause the system to be nonvibrating. Using the methods of differential equations, it may be shown that the solution to Eq. 22–27 for critical damping is

$$x = (A + Bt)e^{-pt}$$

(22–31)

Underdamped System

Most often $c < c_c$, in which case the system is referred to as *underdamped*. In this case the roots λ_1 and λ_2 are complex numbers and it may be shown that the general solution of Eq. 22–27 can be written as

$$x = D[e^{-(c/2m)t} \sin (p_d t + \phi)] \qquad (22\text{–}32)$$

where D and ϕ are constants generally determined from the initial conditions of the problem. The constant p_d is called the *damped natural frequency* of the system. It has a value of

$$p_d = \sqrt{\frac{k}{m} - \left(\frac{c}{2m}\right)^2} = p\sqrt{1 - \left(\frac{c}{c_c}\right)^2} \qquad (22\text{–}33)$$

where the ratio c/c_c is called the *damping factor*.

The graph of Eq. 22–32 is shown in Fig. 22–17. The initial limit of motion, D, diminishes with each cycle of vibration, since motion is confined within the bounds of the exponential curve. Using the damped natural frequency p_d, the period of damped vibration may be written as

$$\tau_d = \frac{2\pi}{p_d} \qquad (22\text{–}34)$$

Since $p_d < p$, Eq. 22–33, the period of damped vibration, τ_d, will be greater than that of free vibration, $\tau = 2\pi/p$.

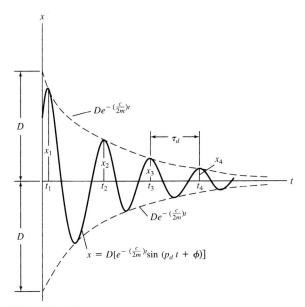

Fig. 22–17

*22.5 Viscous Damped Forced Vibration

The most general case of single-degree-of-freedom vibrating motion occurs when the system includes the effects of forced motion and induced damping. The analysis of this particular type of vibration is of practical value when applied to systems having significant damping characteristics.

If a dashpot is attached to the block and spring shown in Fig. 22–11a, the differential equation which describes the motion becomes

$$m\ddot{x} + c\dot{x} + kx = F_O \sin \omega t \qquad (22\text{–}35)$$

A similar equation may be written for a block and spring having a periodic support displacement, Fig. 22–14a, which includes the effects of damping. In that case, however, F_O is replaced by $k\delta_O$. Since Eq. 22–35 is nonhomogeneous, the general solution is the sum of a complementary solution, x_c, and a particular solution, x_p. The complementary solution is determined by setting the right side of Eq. 22–35 equal to zero and solving the homogeneous equation, which is equivalent to Eq. 22–27. The solution is therefore given by Eq. 22–30, 22–31, or 22–32, depending upon the values of λ_1 and λ_2. Because all systems contain friction, however, this solution will dampen out with time. Only the particular solution, which describes the *steady-state vibration* of the system, will remain. Since the applied forcing function is harmonic, the steady-state motion will also be harmonic. Consequently, the particular solution will be of the form

$$x_p = A' \sin \omega t + B' \cos \omega t \qquad (22\text{–}36)$$

The constants A' and B' are determined by taking the necessary time derivatives and substituting them into Eq. 22–35, which after simplification yields

$$[-A'm\omega^2 - cB'\omega + kA'] \sin \omega t + [-B'm\omega^2 + cA'\omega + kB'] \cos \omega t$$
$$= F_O \sin \omega t$$

Since this equation holds for all time, the constant coefficients of sin ωt and cos ωt may be equated; i.e.,

M.F.

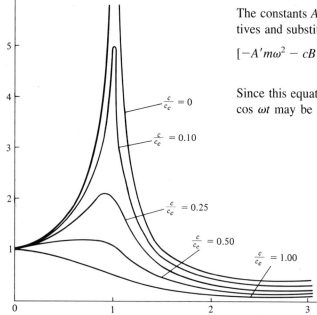

$$\frac{c}{c_c} = 0$$

$$\frac{c}{c_c} = 0.10$$

$$\frac{c}{c_c} = 0.25$$

$$\frac{c}{c_c} = 0.50$$

$$\frac{c}{c_c} = 1.00$$

$\left(\frac{\omega}{p}\right)$

Fig. 22–18

$$-A'm\omega^2 - cB'\omega + kA' = F_O$$
$$-B'm\omega^2 + cA'\omega + kB' = 0$$

Solving for A' and B', realizing that $p^2 = k/m$, yields

$$A' = \frac{\left(\dfrac{F_O}{m}\right)(p^2 - \omega^2)}{(p^2 - \omega^2)^2 + \left(\dfrac{c\omega}{m}\right)^2}$$

$$\tag{22-37}$$

$$B' = \frac{-F_O\dfrac{c\omega}{m^2}}{(p^2 - \omega^2)^2 + \left(\dfrac{c\omega}{m}\right)^2}$$

It is also possible to express Eq. 22–36 in a form similar to Eq. 22–9,

$$x_p = C' \sin(\omega t - \phi') \tag{22-38}$$

in which case the constants C' and ϕ' are

$$C' = \frac{F_O/k}{\sqrt{\left[1 - \left(\dfrac{\omega}{p}\right)^2\right]^2 + \left(2\dfrac{c}{c_c}\dfrac{\omega}{p}\right)^2}}$$

$$\tag{22-39}$$

$$\phi' = \tan^{-1}\left(\frac{c\omega/k}{1 - \left(\dfrac{\omega}{p}\right)^2}\right)$$

The angle ϕ' represents the phase difference between the applied force and the resulting steady-state vibration of the damped system.

The *magnification factor* MF has been defined in Sec. 22.3 as the ratio of the amplitude of deflection caused by the forced vibration to the deflection caused by a static force $\mathbf{F}_O$. From Eq. 22–38, the forced vibration has an amplitude of C'; thus,

$$\text{MF} = \frac{C'}{F_O/k} = \frac{1}{\sqrt{\left[1 - \left(\dfrac{\omega}{p}\right)^2\right]^2 + \left(2\dfrac{c}{c_c}\dfrac{\omega}{p}\right)^2}} \tag{22-40}$$

The MF is plotted in Fig. 22–18 versus the frequency ratio ω/p for various values of the damping factor c/c_c. It can be seen from this graph that the magnification of the amplitude increases as the damping factor decreases. Resonance obviously occurs only when the damping factor is zero and the frequency ratio equals 1.

Example 22–8

The 30-kg electric motor shown in Fig. 22–19 is supported by *four* springs, each spring having a stiffness of 200 N/m. If the rotor R is unbalanced such that its effect is equivalent to a 4-kg mass located 60 mm from the axis of rotation, determine the amplitude of vibration when the rotor is turning at $\omega = 10$ rad/s. The damping factor is $c/c_c = 0.15$.

Fig. 22–19

SOLUTION

The periodic force which causes the motor to vibrate is the centrifugal force due to the unbalanced rotor. This force has a constant magnitude of

$$F_O = ma_n = mr\omega^2 = 4 \text{ kg}(0.06 \text{ m})(10 \text{ rad/s})^2 = 24 \text{ N}$$

Oscillation in the vertical direction may be expressed in the periodic form $F = F_O \sin \omega t$, where $\omega = 10$ rad/s. Thus,

$$F = 24 \sin 10t$$

The stiffness of the entire system of four springs is $k = 4(200) = 800$ N/m. Therefore, the circular frequency of vibration is

$$p = \sqrt{\frac{k}{m}} = \sqrt{\frac{800}{30}} = 5.16 \text{ rad/s}$$

Since the damping factor is known, the steady-state amplitude may be determined from the first of Eqs. 22–39, i.e.,

$$C' = \frac{F_O/k}{\sqrt{\left[1 - \left(\dfrac{\omega}{p}\right)^2\right]^2 + \left(2\dfrac{c}{c_c}\dfrac{\omega}{p}\right)^2}}$$

$$= \frac{24/800}{\sqrt{\left[1 - \left(\dfrac{10}{5.16}\right)^2\right]^2 + \left[2(0.15)\dfrac{10}{5.16}\right]^2}}$$

$$= 0.0107 \text{ m} = 10.7 \text{ mm} \qquad\qquad\qquad Ans.$$

*22.6 Electrical Circuit Analogues

The characteristics of a vibrating mechanical system may be represented by an electric circuit. Consider the circuit shown in Fig. 22–20a, which consists of an inductor L, a resistor R, and a capacitor C. When a voltage $E(t)$ is applied, it causes a current of magnitude i to flow through the circuit. As the current flows past the inductor the voltage drop is $L(di/dt)$, when it flows across the resistor the drop is Ri, and when it arrives at the capacitor the drop is $(1/C) \int i \, dt$. Since current cannot flow past a capacitor, it is only possible to measure the charge q acting on the capacitor. The charge may, however, be related to the current by the equation $i = dq/dt$. Thus, the voltage drops which occur across the inductor, resistor, and capacitor may be written as $L \, d^2q/dt^2$, $R \, dq/dt$, and q/C, respectively. According to Kirchhoff's voltage law, the applied voltage balances the sum of the voltage drops around the circuit. Therefore,

$$L\frac{d^2q}{dt^2} + R\frac{dq}{dt} + \frac{1}{C}q = E(t) \qquad (22\text{–}41)$$

(a)

(b)

Fig. 22–20

Consider now the model of a single-degree-of-freedom mechanical system, Fig. 22–20b, which is subjected to both a general forcing function $F(t)$ and damping. The equation of motion for this system was established in the previous section and can be written as

$$m\frac{d^2x}{dt^2} + c\frac{dx}{dt} + kx = F(t) \qquad (22\text{–}42)$$

By comparison, it is seen that Eqs. 22–41 and 22–42 have the same form, hence mathematically the problem of analyzing an electric circuit is the same as that of analyzing a vibrating mechanical system. The analogues between the two equations are given in Table 22–1.

This analogy has important application to experimental work, for it is much easier to simulate the vibration of a complex mechanical system using an electric circuit, which can be constructed on an analogue computer, than to make an equivalent mechanical spring and dashpot model.

Table 22–1 Electrical-Mechanical Analogues

Electrical		Mechanical	
Electric charge	q	Displacement	x
Electric current	i	Velocity	dx/dt
Voltage	$E(t)$	Applied force	$F(t)$
Inductance	L	Mass	m
Resistance	R	Viscous damping coefficient	c
Reciprocal of capacitance	$1/C$	Spring stiffness	k

PROBLEMS

22–46. If the block is subjected to the impressed force $F = F_O \cos \omega t$, show that the differential equation of motion is $\ddot{y} + (k/m)y = (F_O/m) \cos \omega t$, where y is measured from the equilibrium position of the block. What is the general solution of this equation?

$F = F_O \cos \omega t$

Prob. 22–46

22–47. The 20-lb block is attached to a spring having a stiffness of 20 lb/ft. A force of $F = (6 \cos 2t)$ lb, where t is in seconds, is applied to the block. Determine the maximum speed of the block after frictional forces cause the free vibrations to dampen out.

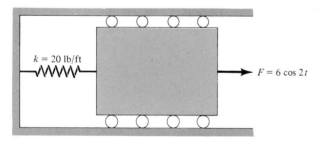

$k = 20$ lb/ft

$F = 6 \cos 2t$

Prob. 22–47

*22–48. A 5-lb weight is suspended from a vertical spring having a stiffness of 50 lb/ft. An impressed force of $F = (0.25 \sin 8t)$ lb, where t is in seconds, is acting on the weight. Determine the equation of motion of the weight when it is pulled down 3 in. from the equilibrium position and released from rest.

22–49. A 5-kg block is suspended from a spring having a stiffness of 300 N/m. If the block is acted upon by a vertical force $F = (7 \sin 8t)$ N, where t is in seconds, determine the equation which describes the motion of the block when it is pulled down 100 mm from the equilibrium position and released from rest at $t = 0$. Assume that positive displacement is measured downward.

$k = 300$ N/m

$F = 7 \sin 8t$ **Prob. 22–49**

22–50. A 7-kg block is suspended from a spring that has a stiffness of $k = 350$ N/m. The block is drawn downward 70 mm from the equilibrium position and released from rest at $t = 0$. If the support moves with an impressed displacement $\delta = (20 \sin 4t)$ mm, where t is in seconds, determine the equation which describes the vertical motion of the block. Assume that positive displacement is measured downward.

22–51. The electric motor turns an eccentric flywheel which is equivalent to an unbalanced 0.25-lb weight located 10 in. from the axis of rotation. If the static deflection of the beam is 1 in. because of the weight of the motor, determine the angular velocity of the flywheel at which resonance will occur. The motor weighs 150 lb. Neglect the mass of the beam.

ω

Probs. 22–51/22–52/22–53

*22–52. What will be the amplitude of steady-state vibration of the motor in Prob. 22–51 if the angular velocity of the flywheel is 20 rad/s?

22–53. Determine the angular velocity of the motor in Prob. 22–51 to produce an amplitude of vibration of 0.25 in.

22–54. The fan has a mass of 25 kg and is fixed to the end of a horizontal beam that has a negligible mass. The fan blade is mounted eccentrically on the shaft such that it is equivalent to an unbalanced 3.5-kg mass located 100 mm from the axis of rotation. If the static deflection of the beam is 50 mm as a result of the weight of the fan, determine the angular speed of the fan at which resonance will occur. *Hint:* See the first part of Example 22–8.

Probs. 22–54/22–55/22–56

22–55. What is the amplitude of steady-state vibration of the fan in Prob. 22–54 when the angular speed of the fan is 10 rad/s? *Hint:* See the first part of Example 22–8.

*22–56. What will be the amplitude of steady-state vibration of the fan in Prob. 22–54 if the angular speed of the fan is 18 rad/s? *Hint:* See the first part of Example 22–8.

22–57. The engine is mounted on a foundation block which is spring-supported. Describe the steady-state vibration of the system if the block and engine have a total weight $W = 1500$ lb and the engine, when running, creates an impressed force $F = (50 \sin 2t)$ lb, where t is in seconds. Assume that the system vibrates only in the vertical direction, with the positive displacement measured downward, and that the total stiffness of the springs can be represented as $k = 2000$ lb/ft.

Probs. 22–57/22–58

22–58. Determine the rotational speed ω of the engine in Prob. 22–57 which will cause resonance.

22–59. The 80-lb block is attached to a spring, the end of which is subjected to a periodic support displacement $\delta_A = (0.5 \sin 8t)$ ft, where t is in seconds. Determine the amplitude of the steady-state motion.

Prob. 22–59

*22–60. The instrument is centered uniformly on a platform P, which in turn is supported by *four* springs, each spring having a stiffness $k = 130$ N/m. If the floor is subjected to a vibration $\omega = 7$ Hz, having a vertical displacement amplitude $\delta_O = 0.17$ ft, determine the vertical displacement amplitude of the platform and instrument. The instrument and the platform have a total weight of 18 lb.

Prob. 22–60

517

22–61. The electric motor has a mass of 50 kg and is supported by *four springs,* each spring having a stiffness of 100 N/m. If the motor turns a disk D which is mounted eccentrically, 20 mm from the disk's center, determine the angular rotation ω at which resonance occurs. Assume that the motor only vibrates in the vertical direction.

Prob. 22–61

22–62. The small block at A has a mass of 4 kg and is mounted on the bent rod having negligible mass. If the rotor at B causes a harmonic movement $\delta_B = (0.1 \cos 15t)$ m, where t is in seconds, determine the amplitude of vibration of the block.

Prob. 22–62

22–63. The 450-kg trailer is pulled with a constant speed over the surface of a bumpy road, which may be approximated by a cosine curve having an amplitude of 50 mm and wave length of 4 m. If the two springs s which support the trailer each have a stiffness of 800 N/m, determine the speed v which will cause the greatest vibration (resonance) of the trailer. Neglect the weight of the wheels.

Probs. 22–63/22–64

*****22–64.** Determine the amplitude of vibration of the trailer in Prob. 22–63 if the speed $v = 15$ km/h.

22–65. A block having a weight of 7 lb is suspended from a spring having a stiffness $k = 75$ lb/ft. The support to which the spring is attached is given a simple harmonic motion which may be expressed by $\delta = (0.15 \sin 2t)$ ft, where t is in seconds. If the damping factor is $c/c_c = 0.8$, determine the phase angle ϕ of forced vibration.

22–66. Determine the magnification factor of the block, spring, and dashpot combination in Prob. 22–65.

22–67. A block having a mass of 0.5 slug is suspended from a spring having a stiffness of $k = 0.6$ lb/in. If a dashpot provides a damping force of 0.2 lb on the block when the speed of the block is $v = 1$ ft/s, determine the period of free vibration.

*****22–68.** The 4-kg circular disk is attached to three springs, each spring having a stiffness of $k = 180$ N/m. If the disk is immersed in a fluid and given a downward velocity of 0.3 m/s at the equilibrium position, determine the equation which describes the motion. Assume that positive displacement is measured downward, and that fluid resistance acting on the disk furnishes a damping force having a magnitude of $F = (60|v|)$ N, where v is in m/s.

120°   **Prob. 22–65**

22–69. A 5-lb weight is suspended from a spring having a stiffness of $k = 6$ lb/in. The support to which the spring is attached is given simple harmonic motion which may be expressed by $\delta = (1.5 \sin 2t)$ in. where t is in seconds. If the damping coefficient is $c = 0.08$ lb·s/in., determine the phase angle ϕ of forced vibration.

22–70. Determine the magnification ratio of the spring and dashpot combination in Prob. 22–69.

22–71. The 200-lb electric motor is fastened to the midpoint of the simply supported beam. It is found that the beam deflects 2 in. when the motor is not running. The rotor turns an eccentric flywheel which is equivalent to an unbalanced weight of 1 lb located 5 in. from the axis of rotation. If the rotor is turning at 100 rpm, determine the amplitude of steady-state vibration. The damping factor is $c/c_c = 0.20$. Neglect the mass of the beam.

Prob. 22–71

*22–72. A spring has a stiffness of $k = 2$ lb/in. If a 5-lb weight is attached to the spring and is given an upward velocity of 2 ft/s from the equilibrium position, determine the position of the weight as a function of time. Assume that motion takes place in a medium which furnishes a retarding force F (pounds) having a magnitude numerically equal to four times the speed v of the weight, where v is in ft/s.

22–73. A block having a mass of 0.8 kg is suspended from a spring having a stiffness of 120 N/m. If a dashpot provides a damping force of 2.5 N when the speed of the block is 0.2 m/s, determine the period of free vibration.

Prob. 22–74

22–74. The 20-kg block is subjected to the action of the harmonic force $F = (90 \cos 6t)$ N, where t is in seconds. Write the equation which describes the steady-state motion.

22–75. The bar has a weight of 6 lb. If the stiffness of the spring is $k = 8$ lb/ft and the dashpot has a damping coefficient $c = 60$ lb·s/ft, determine the differential equation which describes the motion in terms of the angle θ of the bar's rotation.

Prob. 22–75

*22–76. The bell-crank mechanism consists of a bent rod, having a negligible mass, and an attached 5-lb weight. Determine the critical damping coefficient c_c and the damping natural frequency for small vibrations about the equilibrium position. Neglect the size of the weight.

Prob. 22–76

22–77. The damping factor, c/c_c, may be determined experimentally by measuring the successive amplitudes of vibrating motion of a system. If two of these maximum displacements can be approximated by x_1 and x_2, as shown in Fig. 22–17, show that the ratio $\ln x_1/x_2 = 2\pi(c/c_c)/\sqrt{1 - (c/c_c)^2}$. The quantity $\ln x_1/x_2$ is called the *logarithmic decrement*.

22–78. The block shown in Fig. 22–16 has a mass of 20 kg and the spring has a stiffness $k = 600$ N/m. When the block is displaced and released, two successive amplitudes are measured as $x_1 = 150$ mm and $x_2 = 87$ mm. Determine the coefficient of viscous damping, c. *Hint:* See Prob. 22–77.

22–79. The barrel of a cannon has a mass of 700 kg, and after firing it recoils a distance of 0.64 m. If it returns to its original position by means of a single recuperator having a damping coefficient of 2 kN · s/m, determine the required stiffness of each of the two springs fixed to the base and attached to the barrel so that the barrel recuperates without vibration.

*22–80. A block having a mass of 7 kg is suspended from a spring that has a stiffness $k = 600$ N/m. If it is given an upward velocity of 0.6 m/s from its equilibrium position at $t = 0$, determine the position of the block as a function of time. Assume that positive displacement of the block is downward, and that motion takes place in a medium which furnishes a damping force $F = (50|v|)$ N, where v is measured in m/s.

22–81. Determine the differential equation of motion for the damped vibratory system shown. What type of motion occurs?

$k = 100$ N/m

25 kg

$c = 200$ N · s/m $c = 200$ N · s/m   **Prob. 22–81**

22–82. Determine the mechanical analogue for the electrical circuit. What differential equations describe the mechanical and electrical systems?

C

L

$E = E_O \sin \omega t$ **Prob. 22–82**

22–83. Draw the electrical circuit that is equivalent to the mechanical system shown. What is the differential equation which describes the motion of the current in the circuit?

k

m

c

Prob. 22–83

*22–84. Draw the electrical circuit that is equivalent to the mechanical system shown. What is the differential equation which describes the motion of the current in the circuit?

k c k

m

Prob. 22–84

22–85. Draw the electric circuit that is equivalent to the mechanical system shown. Determine the differential equation which describes the motion of the current in the circuit.

k

m

c

$F = F_O \cos \omega t$

Prob. 22–85

A

Mathematical Expressions

Quadratic Formula

If $ax^2 + bx + c = 0$, then $x = \dfrac{-b \pm \sqrt{b^2 - 4ac}}{2a}$

Hyperbolic Functions

$\sinh x = \dfrac{e^x - e^{-x}}{2}$, $\cosh x = \dfrac{e^x + e^{-x}}{2}$, $\tanh x = \dfrac{\sinh x}{\cosh x}$

Trigonometric Identities

$\sin^2 \theta + \cos^2 \theta = 1$

$\sin (\theta \pm \phi) = \sin \theta \cos \phi \pm \cos \theta \sin \phi$

$\sin 2\theta = 2 \sin \theta \cos \theta$

$\cos (\theta \pm \phi) = \cos \theta \cos \phi \mp \sin \theta \sin \phi$

$\cos 2\theta = \cos^2 \theta - \sin^2 \theta$

$\cos \theta = \pm \sqrt{\dfrac{1 + \cos 2\theta}{2}}$, $\sin \theta = \pm \sqrt{\dfrac{1 - \cos 2\theta}{2}}$

$\tan \theta = \dfrac{\sin \theta}{\cos \theta}$

$1 + \tan^2 \theta = \sec^2 \theta$

$1 + \operatorname{ctn}^2 \theta = \csc^2 \theta$

Power–Series Expansions

$\sin x = x - \dfrac{x^3}{3!} + \dfrac{x^5}{5!} - \dfrac{x^7}{7!} + \cdots$

$\cos x = 1 - \dfrac{x^2}{2!} + \dfrac{x^4}{4!} - \dfrac{x^6}{6!} + \cdots$

$\sinh x = x + \dfrac{x^3}{3!} + \dfrac{x^5}{5!} + \cdots$

$\cosh x = 1 + \dfrac{x^2}{2!} + \dfrac{x^4}{4!} + \cdots$

Derivatives

$$\frac{d}{dx}(u^n) = nu^{n-1}\frac{du}{dx}$$

$$\frac{d}{dx}(uv) = u\frac{dv}{dx} + v\frac{du}{dx}$$

$$\frac{d}{dx}\left(\frac{u}{v}\right) = \frac{v\frac{du}{dx} - u\frac{dv}{dx}}{v^2}$$

$$\frac{d}{dx}(\sin u) = \cos u\frac{du}{dx}$$

$$\frac{d}{dx}(\cos u) = -\sin u\frac{du}{dx}$$

$$\frac{d}{dx}(\tan u) = \sec^2 u\frac{du}{dx}$$

$$\frac{d}{dx}(\cot u) = -\csc^2 u\frac{du}{dx}$$

$$\frac{d}{dx}(\sec u) = \tan u \sec u\frac{du}{dx}$$

$$\frac{d}{dx}(\csc u) = -\csc u \cot u\frac{du}{dx}$$

$$\frac{d}{dx}(\sinh u) = \cosh u\frac{du}{dx}$$

$$\frac{d}{dx}(\cosh u) = \sinh u\frac{du}{dx}$$

Integrals

$$\int x^n \, dx = \frac{x^{n+1}}{n+1} + C, \ n \neq -1$$

$$\int \frac{dx}{a + bx} = \frac{1}{b} \ln (a + bx) + C$$

$$\int \frac{dx}{a + bx^2} = \frac{1}{2\sqrt{-ba}} \ln \left[\frac{\sqrt{a} + 2\sqrt{-b}}{\sqrt{a} - x\sqrt{-b}}\right] + C, \ a > 0, \ b < 0$$

$$\int \frac{x \, dx}{a + bx^2} = \frac{1}{2b} \ln (bx^2 + a) + C$$

$$\int \frac{x^2 \, dx}{a + bx^2} = \frac{x}{b} - \frac{a}{b\sqrt{ab}} \tan^{-1} \frac{x\sqrt{ab}}{a} + C$$

$$\int \frac{dx}{a^2 - x^2} = \frac{1}{2a} \ln \left[\frac{a + x}{a - x}\right] + C, \ a^2 > x^2$$

$$\int \sqrt{a + bx} \, dx = \frac{2}{3b} \sqrt{(a + bx)^3} + C$$

$$\int x\sqrt{a + bx} \, dx = \frac{-2(2a - 3bx)\sqrt{(a + bx)^3}}{15b^2} + C$$

$$\int x^2\sqrt{a + bx} \, dx = \frac{2(8a^2 - 12abx + 15b^2x^2)\sqrt{(a + bx)^3}}{105b^3} + C$$

$$\int \sqrt{a^2 - x^2} \, dx = \frac{1}{2}\left[x\sqrt{a^2 - x^2} + a^2 \sin^{-1} \frac{x}{a}\right] + C, \ a > 0$$

$$\int x\sqrt{a^2 - x^2} \, dx = -\frac{1}{3}\sqrt{(a^2 - x^2)^3} + C$$

$$\int x^2\sqrt{a^2 - x^2} \, dx = -\frac{x}{4}\sqrt{(a^2 - x^2)^3} + \frac{a^2}{8}\left(x \sqrt{a^2 - x^2} + a^2 \sin^{-1} \frac{x}{a}\right) + C, \ a > 0$$

$$\int \sqrt{x^2 \pm a^2}\, dx = \frac{1}{2}\left[x\sqrt{x^2 + a^2} \pm a^2 \ \ln\,(x + \sqrt{x^2 \pm a^2})\right] + C$$

$$\int x\sqrt{x^2 \pm a^2}\, dx = \frac{1}{3}\sqrt{(x^2 \pm a^2)^3} + C$$

$$\int x^2\sqrt{x^2 \pm a^2}\, dx = \frac{x}{4}\sqrt{(x^2 \pm a^2)^3} \pm \frac{a^2}{8}\, x\sqrt{x^2 \pm a^2} - \frac{a^4}{8}\, \ln\,(x + \sqrt{x^2 \pm a^2}) + C$$

$$\int \frac{dx}{\sqrt{a + bx}} = \frac{2\sqrt{a + bx}}{b} + C$$

$$\int \frac{x\,dx}{\sqrt{x^2 \pm a^2}} = \sqrt{x^2 \pm a^2} + C$$

$$\int \frac{dx}{\sqrt{a + bx + cx^2}} = \frac{1}{\sqrt{c}}\, \ln\left[\sqrt{a + bx + cx^2} + x\sqrt{c} + \frac{b}{2\sqrt{c}}\right] + C,\ c > 0$$

$$= \frac{1}{\sqrt{-c}}\, \sin^{-1}\left(\frac{-2cx - b}{\sqrt{b^2 - 4ac}}\right) + C,\ c < 0$$

$$\int \sin x\, dx = -\cos x + C$$

$$\int \cos x\, dx = \sin x + C$$

$$\int x \cos(ax)\, dx = \frac{1}{a^2}\, \cos\,(ax) + \frac{x}{a}\, \sin\,(ax) + C$$

$$\int x^2 \cos(ax)\, dx = \frac{2x}{a^2}\, \cos\,(ax) + \frac{a^2x^2 - 2}{a^3}\, \sin\,(ax) + C$$

$$\int e^{ax}\, dx = \frac{1}{a}e^{ax} + C$$

$$\int x\, e^{ax}\, dx = \frac{e^{ax}}{a^2}(ax - 1) + C$$

$$\int \sinh x\, dx = \cosh x + C$$

$$\int \cosh x\, dx = \sinh x + C$$

B

Numerical and Computer Analysis

Occasionally the application of the laws of mechanics will lead to a system of equations for which a closed-form solution is difficult or impossible to obtain. When confronted with this situation, engineers will often use a numerical method which in most cases can be programmed on a microcomputer or "programmable" pocket calculator. Here we will briefly present a computer program for solving a set of linear algebraic equations and three numerical methods which can be used to solve an algebraic or transcendental equation, evaluate a definite integral, and solve an ordinary differential equation. Application of each method will be explained by example, and an associated computer program written in Microsoft BASIC, which is designed to run on most personal computers, is provided.* A text on numerical analysis should be consulted for further discussion regarding a check of the accuracy of each method and the inherent errors that can develop.

Linear Algebraic Equations

Application of the equations of static equilibrium or the equations of motion sometimes requires solving a set of linear algebraic equations. The computer program listed in Fig. B–1 can be used for this purpose. It is based on the

---

* Similar types of programs can be written or purchased for programmable pocket calculators.

```
 1 PRINT"Linear system of equations":PRINT
 2 DIM A(10,11)
 3 INPUT"Input number of equations : ",N
 4 PRINT
 5 PRINT"A  coefficients"
 6 FOR I = 1 TO N
 7 FOR J = 1 TO N
 8 PRINT "A(";I;",";J;
 9 INPUT")=",A(I,J)
10 NEXT J
11 NEXT I
12 PRINT
13 PRINT"B  coefficients"
14 FOR I = 1 TO N
15 PRINT "B(";I;
16 INPUT")=",A(I,N+1)
17 NEXT I
18 GOSUB 25
19 PRINT

20 PRINT"Unknowns"
21 FOR I = 1 TO N
22 PRINT "X(";I;")=";A(I,N+1)
23 NEXT I
24 END
25 REM Subroutine Guassian
26 FOR M=1 TO N
27 NP=M
28 BG=ABS(A(M,M))
29 FOR I = M TO N
30 IF ABS(A(I,M))<=BG THEN 33
31 BG=ABS(A(I,M))
32 NP=I
33 NEXT I
34 IF NP=M THEN 40
35 FOR I = M TO N+1
36 TE=A(M,I)
37 A(M,I)=A(NP,I)
38 A(NP,I)=TE

39 NEXT I
40 FOR I = M+1 TO N
41 FC=A(I,M)/A(M,M)
42 FOR J = M+1 TO N+1
43 A(I,J)=A(I,J)-FC*A(M,J)
44 NEXT J
45 NEXT I
46 NEXT M
47 A(N,N+1)=A(N,N+1)/A(N,N)
48 FOR I = N-1 TO 1 STEP -1
49 SM=0
50 FOR J=I+1 TO N
51 SM=SM+A(I,J)*A(J,N+1)
52 NEXT J
53 A(I,N+1)=(A(I,N+1)-SM)/A(I,I)
54 NEXT I
55 RETURN
```

Fig. B-1

method of a Gauss elimination and can solve at most ten equations with ten unknowns. To do so, the equations should first be written in the following general format:

$$A_{11}x_1 + A_{12}x_2 + \cdots + A_{1n}x_n = B_1$$
$$A_{21}x_1 + A_{22}x_2 + \cdots + A_{2n}x_n = B_2$$

.

.

.

$$A_{n1}x_1 + A_{n2}x_2 + \cdots + A_{nn}x_n = B_n$$

The "A" and "B" coefficients are "called" for when running the program. The output presents the unknowns $x_1, \ldots, x_n$.

Example B-1

Solve the two equations

$$3x_1 + x_2 = 4$$
$$2x_1 - x_2 = 10$$

SOLUTION

When the program begins to run, it first calls for the number of equations (2), then the A coefficients in the following sequence: $A_{11} = 3$, $A_{12} = 1$, $A_{21} = 2$, $A_{22} = -1$ and finally the B coefficients $B_1 = 4$, $B_2 = 10$. The output appears as:

	Unknowns	
	$X(1) = 2.8$	*Ans.*
	$X(2) = -4.4$	*Ans.*

Simpson's Rule

This numerical method can be used to determine the area under a curve given as a graph or as an explicit function $y = f(x)$. Likewise it can be used to compute the value of a definite integral which involves the function $y = f(x)$. To do so, the area must be subdivided into an *even number* of strips or intervals having a width h. The curve between three consecutive ordinates is approximated by a parabola, and the entire area or definite integral is then determined from the formula

$$\int_{x_0}^{x_n} f(x)\, dx \simeq \frac{h}{3}[y_0 + 4(y_1 + y_3 + \cdots + y_{n-1})$$
$$+ 2(y_2 + y_4 + \cdots + y_{n-2}) + y_n] \quad \text{(B--1)}$$

The computer program for this equation is given in Fig. B–2. For its use, one must first enter the function (line 6), then the upper and lower limits of the integral (lines 7 and 8), and finally the number of intervals (line 9). The value of the integral is then given as the output.

```
1 PRINT"Simpson's rule":PRINT
2 PRINT" To execute this program :":PRINT
3 PRINT"   1- Modify right-hand side of the equation given below,
4 PRINT"       then press RETURN key"
5 PRINT"   2- Type  RUN 6":PRINT:EDIT 6
6 DEF FNF(X)=LOG(X)
7 PRINT:INPUT" Enter Lower Limit = ",A
8 INPUT" Enter Upper Limit = ",B
9 INPUT" Enter Number (even) of Intervals = ",N%
10 H=(B-A)/N%:AR=FNF(A):X=A+H
11 FOR J%=2 TO N%
12 K=2*(2-J%+2*INT(J%/2))
13 AR=AR+K*FNF(X)
14 X=X+H:NEXT J%
15 AR=H*(AR+FNF(B))/3
16 PRINT" Integral = ",AR
17 END
```

Fig. B–2

Example B–2

Evaluate the definite integral $\int_2^5 \ln x \, dx$

SOLUTION

The interval $x_0 = 2$ to $x_6 = 5$ will be divided into six equal parts ($n = 6$), each having a width $h = (5 - 2)/6 = 0.5$. We then compute $y = f(x) = \ln x$ at each point of subdivision.

n	x_n	y_n
0	2	0.693
1	2.5	0.916
2	3	1.099
3	3.5	1.253
4	4	1.386
5	4.5	1.504
6	5	1.609

Thus, Eq. B–1 becomes

$$\int_2^5 \ln x \, dx \simeq \frac{0.5}{3}[0.693 + 4(0.916 + 1.253 + 1.504)$$
$$+ 2(1.099 + 1.386) + 1.609]$$
$$\simeq 3.66 \qquad\qquad Ans.$$

This answer is equivalent to the exact answer to three significant figures. Obviously, accuracy to a larger number of significant figures can be improved by selecting a smaller interval h (or larger n).

Using the computer program, we enter the function $\ln x$, line 6 in Fig. B–2, the upper and lower limits 2 and 5, and the number of intervals $n = 6$. The output appears as

$$\text{Integral} = 3.66082 \qquad\qquad Ans.$$

The Secant Method

This numerical method is used to find the real roots of an algebraic or transcendental equation $f(x) = 0$. The method derives its name from the fact that the formula used is established from the slope of the secant line to the graph $y = f(x)$. This slope is $(f(x_n) - f(x_{n-1}))/(x_n - x_{n-1})$ and the secant formula is

$$x_{n+1} = x_n - f(x_n) \left(\frac{x_n - x_{n-1}}{f(x_n) - f(x_{n-1})} \right) \qquad (B-2)$$

For application it is necessary to provide two initial guesses, x_0 and x_1, and thereby evaluate x_2 from Eq. B–2 ($n = 1$). One then proceeds to reapply Eq. B–2 with x_1 and the calculated value of x_2 and obtain x_3 ($n = 2$), etc., until the value $x_{n+1} \simeq x_n$. One can see this will occur if x_n is approaching the root of the function $f(x) = 0$ since the correction term on the right of Eq. B–2 will tend toward zero. In particular, the larger the slope the smaller the correction to x_n and the faster the root will be found. Likewise, if the slope is very small in the neighborhood of the root, the method leads to large corrections for x_n and convergence to the root is slow and may even lead to a failure to find it. In such cases other numerical techniques must be used for solution.

A computer program based on Eq. B–2 is listed in Fig. B–3. The user must supply the function (line 7) along with two initial guesses x_0 and x_1 used to approximate the solution. The output specifies the value of the root. If it cannot be determined it is so stated.

```
1 PRINT"Secant method":PRINT
2 PRINT" To execute this program :":PRINT
3 PRINT"    1) Modify right hand side of the equation given below,"
4 PRINT"       then press RETURN key."
5 PRINT"    2) Type  RUN 7"
6 PRINT:EDIT 7
7 DEF FNF(X)=.5*SIN(X)-2*COS(X)+1.3
8 INPUT"Enter point #1 =",X
9 INPUT"Enter point #2 =",X1
10 IF X=X1 THEN 14
11 EP=.00001:TL=2E-20
12 FP=(FNF(X1)-FNF(X))/(X1-X)
13 IF ABS(FP)>TL THEN 15
14 PRINT"Root can not be found.":END
15 DX=FNF(X1)/FP
16 IF ABS(DX)>EP THEN 19
17 PRINT "Root = ";X1;"        Function evaluated at this root = ";FNF(X1)
18 END
19 X=X1:X1=X1-DX
20 GOTO 12
```

Fig. B–3

Example B–3

Determine the root of the equation

$$f(x) = 0.5 \sin x - 2 \cos x + 1.30 = 0$$

SOLUTION

Guesses of the initial roots will be $x_0 = 45°$ and $x_1 = 30°$. Applying Eq. B–2,

$$x_2 = 30° - (-0.1821)\frac{(30° - 45°)}{(-0.1821 - 0.2393)} = 36.48°$$

Using this value in Eq. B–2, along with $x_1 = 30°$, we have

$$x_3 = 36.48° - (-0.0108)\frac{36.48° - 30°}{(-0.0108 + 0.1821)} = 36.89°$$

Repeating the process with this value and $x_2 = 36.48°$ yields

$$x_4 = 36.89° - (0.0005)\left(\frac{36.89° - 36.48°}{(0.0005 + 0.0108)}\right) = 36.87°$$

Thus $x = 36.9°$ is appropriate to three significant figures.

If the problem is solved using the computer program, we enter the function, line 7 in Fig. B–3. The first and second guesses must be in radians. Choosing these to be 0.8 rad, and 0.5 rad, the result appears as

Root = 0.6435022.
Function evaluated at this root = 1.66893E–06.

This result converted from radians to degrees is therefore

$$x = 36.9° \qquad\qquad Ans.$$

Runge-Kutta Method

This numerical method is used to solve an ordinary differential equation. It consists of applying a set of formulas which are used to find specific values of y for corresponding incremental values h in x. The formulas given in general form are as follows:

First-Order Equation. To integrate $\dot{x} = f(t, x)$ step-by-step use

$$x_{i+1} = x_i + \frac{1}{6}(k_1 + 2k_2 + 2k_3 + k_4) \tag{B-3}$$

where

$$
\begin{aligned}
k_1 &= hf(t_i, x_i) \\
k_2 &= hf\left(t_i + \frac{h}{2}, x_i + \frac{k_1}{2}\right) \\
k_3 &= hf\left(t_i + \frac{h}{2}, x_i + \frac{k_2}{2}\right) \\
k_4 &= hf(t_i + h, x_i + k_3)
\end{aligned}
\tag{B-4}
$$

Second-Order Equation. To integrate $\ddot{x} = f(t, x, \dot{x})$ use

$$x_{i+1} = x_i + h\left[\dot{x}_i + \frac{1}{6}(k_1 + k_2 + k_3)\right]$$

$$\dot{x}_{i+1} = \dot{x}_i + \frac{1}{6}(k_1 + 2k_2 + 2k_3 + k_4) \tag{B-5}$$

where

$$
\begin{aligned}
k_1 &= hf(t_i, x_i, \dot{x}_i) \\
k_2 &= hf\left(t_i + \frac{h}{2}, x_i + \frac{h}{2}\dot{x}_i, \dot{x}_i + \frac{k_1}{2}\right) \\
k_3 &= hf\left(t_i + \frac{h}{2}, x_i + \frac{h}{2}\dot{x}_i + \frac{h}{4}k_1, \dot{x}_i + \frac{k_2}{2}\right) \\
k_4 &= hf\left(t_i + h, x_i + h\dot{x}_i + \frac{h}{2}k_2, \dot{x}_i + k_3\right)
\end{aligned}
\tag{B-6}
$$

To apply these equations, one starts with initial values $t_i = t_0$, $x_i = x_0$ and $\dot{x}_i = \dot{x}_0$ (for the second-order equation). Choosing an increment h for t_0 the four constants k are computed and these results are substituted into Eq. B–3 or B–5 in order to compute $x_{i+1} = x_1$, $\dot{x}_{i+1} = x_1$, corresponding to $t_{i+1} = t_1 = t_0 + h$. Repeating this process using t_1, x_1, $\dot{x}_1$ and h, values for x_2, $\dot{x}_2$ and $t_2 = t_1 + h$ are then computed, etc.

Computer programs which solve first and second-order differential equations by this method are listed in Figs. B–4 and B–5, respectively. In order to use these programs, the operator specifies the function $\dot{x} = f(t, x)$ or $\ddot{x} = f(t, x, \dot{x})$ (line 7), the initial values t_0, x_0, $\dot{x}_0$ (for second-order equation), the final time t_n, and the step size h. The output gives the values of t, x, and $\dot{x}$ for each time increment until t_n is reached.

```
1 PRINT"Runge-Kutta Method for 1-st order Differential Equation":PRINT
2 PRINT" To execute this program :";PRINT
3 PRINT"    1) Modify right hand side of the equation given below,"
4 PRINT"       then Press RETURN key"
5 PRINT"    2) Type  RUN 7"
6 PRINT:EDIT 7
7 DEF FNF(T,X)=5*T+X
8 CLS:PRINT" Initial Conditions":PRINT
9 INPUT"Input  t   = ",T
10 INPUT"       x   = ",X
11 INPUT"Final  t   = ",T1
12 INPUT"step size  = ",H:PRINT
13 PRINT"        t                x"
14 IF T>=T1+H THEN 23
15 PRINT USING"#####.#####";T;X
16 K1=H*FNF(T,X)
17 K2=H*FNF(T+.5*H,X+.5*K1)
18 K3=H*FNF(T+.5*H,X+.5*K2)
19 K4=H*FNF(T+H,X+K3)
20 T=T+H
21 X=X+(K1+K2+K2+K3+K3+K4)/6
22 GOTO 14
23 END
```

Fig. B–4

```
1 PRINT"Runge-Kutta Method for 2-nd order Differential Equation":PRINT
2 PRINT" To execute this program :":PRINT
3 PRINT"    1) Modify right hand side of the equation given below,"
4 PRINT"       then Press RETURN key"
5 PRINT"    2) Type  RUN 7"
6 PRINT:EDIT 7
7 DEF FNF(T,X,XD)=
8 INPUT"Input  t   = ",T
9 INPUT"       x   = ",X
10 INPUT"    dx/dt = ",XD
11 INPUT"Final  t   = ",T1
12 INPUT"step size  = ",H:PRINT
13 PRINT"        t           x           dx/dt"
14 IF T>=T1+H THEN 24
15 PRINT USING"#####.#####";T;X;XD
16 K1=H*FNF(T,X,XD)
17 K2=H*FNF(T+.5*H,X+.5*H*XD,XD+.5*K1)
18 K3=H*FNF(T+.5*H,X+(.5*H)*(XD+.5*K1),XD+.5*K2)
19 K4=H*FNF(T+H,X+H*XD+.5*H*K2,XD+K3)
20 T=T+H
21 X=X+H*XD+H*(K1+K2+K3)/6
22 XD=XD+(K1+K2+K2+K3+K3+K4)/6
23 GOTO 14
24 END
```

Fig. B–5

Example B–4

Solve the differential equation $\dot{x} = 5t + x$. Obtain the results for two steps using time increments of $h = 0.02$ s. At $t_0 = 0$, $x_0 = 0$.

SOLUTION

This is a first-order equation, so Eqs. B–3 and B–4 apply. Thus, for $t_0 = 0$, $x_0 = 0$, $h = 0.02$, we have

$$k_1 = 0.02[0 + 0] = 0$$

$$k_2 = 0.02[5(0.1) + 0] = 0.001$$

$$k_3 = 0.02[5(0.01) + 0.0005] = 0.00101$$

$$k_4 = 0.02[5(0.02) + 0.00101] = 0.00202$$

$$x_1 = 0 + \frac{1}{6}[0 + 2(0.001) + 2(0.00101) + 0.00202] = 0.00101$$

Using the values $t_1 = 0 + 0.02 = 0.02$ and $x_1 = 0.00101$ with $h = 0.02$, the value for x_2 is now computed from Eqs. B–3 and B–4.

$$k_1 = 0.02[5(0.02) + 0.00101] = 0.00202$$

$$k_2 = 0.02[5(0.03) + 0.00202] = 0.00304$$

$$k_3 = 0.02[5(0.03) + 0.00253] = 0.00305$$

$$k_4 = 0.02[5(0.04) + 0.00406] = 0.00408$$

$$x_2 = 0.001 + \frac{1}{6}[0.00202 + 2(0.00304) + 2(0.00305) + 0.00408]$$

$$x_2 = 0.00405 \qquad \qquad Ans.$$

To solve this problem using the computer program in Fig. B–4, the function is first entered on line 7, then the data $t_0 = 0$, $x_0 = 0$, $t_n = 0.04$, and $h = 0.02$ is specified. The results appear as

t	x	
0.00000	0.00000	
0.02000	0.00101	
0.04000	0.00405	*Ans.*

C

Vector Analysis

The following discussion provides a brief review of the vector analysis used in statics. A more detailed treatment of these topics is given in *Engineering Mechanics: Statics*.

Vector

A vector, **A,** is a quantity which has a magnitude and direction, and adds according to the parallelogram law. As shown in Fig. C–1, **A = B + C,** where **A** is the *resultant vector* and **B** and **C** are *component vectors*.

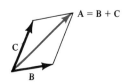

Fig. C–1

Unit Vector

A unit vector, $\mathbf{u}_A$, has a magnitude of one "dimensionless" unit and acts in the same direction as **A.** It is determined by dividing **A** by its magnitude A, i.e.,

$$\mathbf{u}_A = \frac{\mathbf{A}}{A} \qquad \text{(C–1)}$$

Fig. C–2

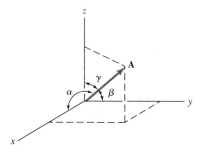

Fig. C–3

Cartesian Vector Notation

The directions of the positive x, y, z axes are defined by the Cartesian unit vectors, $\mathbf{i}$, $\mathbf{j}$, $\mathbf{k}$, respectively.

As shown in Fig. C–2, vector $\mathbf{A}$ is formulated by the addition of its x, y, z components as

$$\mathbf{A} = A_x\mathbf{i} + A_y\mathbf{j} + A_z\mathbf{k} \qquad (C-2)$$

The *magnitude* of $\mathbf{A}$ is determined from

$$A = \sqrt{A_x^2 + A_y^2 + A_z^2} \qquad (C-3)$$

The *direction* of $\mathbf{A}$ is defined in terms of its *coordinate direction angles*, α, β, γ, measured from the *tail* of $\mathbf{A}$ to the *positive x, y, z* axes, Fig. C–3. These angles are determined from the *direction cosines* which represent the $\mathbf{i}$, $\mathbf{j}$, $\mathbf{k}$ components of the unit vector $\mathbf{u}_A$; i.e., from Eqs. C–1 and C–2,

$$\mathbf{u}_A = \frac{A_x}{A}\mathbf{i} + \frac{A_y}{A}\mathbf{j} + \frac{A_z}{A}\mathbf{k} \qquad (C-4)$$

so that the direction cosines are

$$\cos\alpha = \frac{A_x}{A}, \qquad \cos\beta = \frac{A_y}{A}, \qquad \cos\gamma = \frac{A_z}{A} \qquad (C-5)$$

Hence $\mathbf{u}_A = \cos\alpha\,\mathbf{i} + \cos\beta\,\mathbf{j} + \cos\gamma\,\mathbf{k}$, and using Eq. C–3, it is seen that

$$\cos^2\alpha + \cos^2\beta + \cos^2\gamma = 1 \qquad (C-6)$$

The Cross Product

The cross product of two vectors $\mathbf{A}$ and $\mathbf{B}$, which yields the resultant vector $\mathbf{C}$, is written as

$$\mathbf{C} = \mathbf{A} \times \mathbf{B} \qquad (C-7)$$

and reads $\mathbf{C}$ equals $\mathbf{A}$ "cross" $\mathbf{B}$. The *magnitude* of $\mathbf{C}$ is

$$C = AB \sin\theta \qquad (C-8)$$

where θ is the angle made between the *tails* of $\mathbf{A}$ and $\mathbf{B}$ ($0° \le \theta \le 180°$). The *direction* of $\mathbf{C}$ is determined by the right-hand rule, whereby the fingers of the right hand are curled *from* $\mathbf{A}$ *to* $\mathbf{B}$ and the thumb points in the direction of $\mathbf{C}$, Fig. C–4. This vector is perpendicular to the plane containing vectors $\mathbf{A}$ and $\mathbf{B}$.

The vector cross product is *not* commutative, i.e., $\mathbf{A} \times \mathbf{B} \ne \mathbf{B} \times \mathbf{A}$. Rather,

$$\mathbf{A} \times \mathbf{B} = -\mathbf{B} \times \mathbf{A}$$

Fig. C–4

The distributive law is valid, i.e.,

$$\mathbf{A} \times (\mathbf{B} + \mathbf{D}) = \mathbf{A} \times \mathbf{B} + \mathbf{A} \times \mathbf{D} \qquad (C\text{--}10)$$

And, the cross product may be multiplied by a scalar m in any manner, i.e.,

$$m(\mathbf{A} \times \mathbf{B}) = (m\mathbf{A}) \times \mathbf{B} = \mathbf{A} \times (m\mathbf{B}) = (\mathbf{A} \times \mathbf{B})m \qquad (C\text{--}11)$$

If $\mathbf{A}$ and $\mathbf{B}$ are expressed in Cartesian component form, then the cross product, Eq. C–7, may be evaluated by expanding the determinant

$$\mathbf{C} = \mathbf{A} \times \mathbf{B} = \begin{vmatrix} \mathbf{i} & \mathbf{j} & \mathbf{k} \\ A_x & A_y & A_z \\ B_x & B_y & B_z \end{vmatrix} \qquad (C\text{--}12)$$

which yields

$$\mathbf{C} = (A_y B_z - A_z B_y)\mathbf{i} - (A_x B_z - A_z B_x)\mathbf{j} + (A_x B_y - A_y B_x)\mathbf{k}$$

Recall that the cross product is used in statics to define the moment of a force $\mathbf{F}$ about point O, in which case

$$\mathbf{M}_O = \mathbf{r} \times \mathbf{F} \qquad (C\text{--}13)$$

where $\mathbf{r}$ is a position vector directed from point O to *any point* on the line of action of $\mathbf{F}$.

The Dot Product

The dot product of two vectors $\mathbf{A}$ and $\mathbf{B}$, which yields a scalar, is defined as

$$\mathbf{A} \cdot \mathbf{B} = AB \cos \theta \qquad (C\text{--}14)$$

and reads $\mathbf{A}$ "dot" $\mathbf{B}$. The angle θ is formed between the *tails* of $\mathbf{A}$ and $\mathbf{B}$ $(0° \leqslant \theta \leqslant 180°)$.

The dot product is commutative, i.e.,

$$\mathbf{A} \cdot \mathbf{B} = \mathbf{B} \cdot \mathbf{A} \qquad (C\text{--}15)$$

The distributive law is valid, i.e.,

$$\mathbf{A} \cdot (\mathbf{B} + \mathbf{D}) = \mathbf{A} \cdot \mathbf{B} + \mathbf{A} \cdot \mathbf{D} \qquad (C\text{--}16)$$

And, scalar multiplication can be performed in any manner, i.e.,

$$m(\mathbf{A} \cdot \mathbf{B}) = (m\mathbf{A}) \cdot \mathbf{B} = \mathbf{A} \cdot (m\mathbf{B}) = (\mathbf{A} \cdot \mathbf{B})m \qquad (C\text{--}17)$$

If $\mathbf{A}$ and $\mathbf{B}$ are expressed in Cartesian component form, then the dot product, Eq. C–14, can be determined from

$$\mathbf{A} \cdot \mathbf{B} = A_x B_x + A_y B_y + A_z B_z \qquad (C\text{--}18)$$

The dot product may be used to determine the *angle θ formed between two vectors*. From Eq. C–14,

$$\theta = \cos^{-1}\left(\frac{\mathbf{A} \cdot \mathbf{B}}{AB}\right) \tag{C–19}$$

It is also possible to find the *component of a vector in a given direction* using the dot product. For example, the magnitude of the component (or projection) of vector $\mathbf{A}$ in the direction of $\mathbf{B}$, Fig. C–5, is defined by $A \cos \theta$. From Eq. C–14, this magnitude is

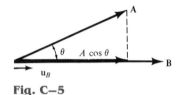

Fig. C–5

$$A \cos \theta = \mathbf{A} \cdot \frac{\mathbf{B}}{B} = \mathbf{A} \cdot \mathbf{u}_B \tag{C–20}$$

where $\mathbf{u}_B$ represents a unit vector acting in the direction of $\mathbf{B}$, Fig. C–5.

Differentiation and Integration of Vector Functions

The rules for differentiation and integration of the sums and products of scalar functions apply as well to vector functions. Consider, for example, the two vector functions $\mathbf{A}(s)$ and $\mathbf{B}(s)$. Provided these functions are smooth and continuous for all s, then

$$\frac{d}{ds}(\mathbf{A} + \mathbf{B}) = \frac{d\mathbf{A}}{ds} + \frac{d\mathbf{B}}{ds} \tag{C–21}$$

$$\int (\mathbf{A} + \mathbf{B}) \, ds = \int \mathbf{A} \, ds + \int \mathbf{B} \, ds \tag{C–22}$$

For the cross product,

$$\frac{d}{ds}(\mathbf{A} \times \mathbf{B}) = \left(\frac{d\mathbf{A}}{ds} \times \mathbf{B}\right) + \left(\mathbf{A} \times \frac{d\mathbf{B}}{ds}\right) \tag{C–23}$$

Similarly, for the dot product,

$$\frac{d}{ds}(\mathbf{A} \cdot \mathbf{B}) = \frac{d\mathbf{A}}{ds} \cdot \mathbf{B} + \mathbf{A} \cdot \frac{d\mathbf{B}}{ds} \tag{C–24}$$

Answers

CHAPTER 12

12–1. 12.5 m/s, 2.55 s.

12–2. -1.96 ft/s^2, 28.6 s.

12–3. $v_{avg} = 0.222$ m/s, $v_{sp} = 2.22$ m/s.

12–5. $h_B = 26.1$ m, $h_{max} = 28.0$ m.

12–6. $d = 13.0$ ft, $v_{avg} = -1.00$ ft/s, $v_{sp} = 2.60$ ft/s.

12–7. 54.0 m.

12–9. 5.12 m.

12–10. 0.371 ft/s^2.

12–11. 517 ft, 616 ft.

12–13. $s = 12.5$ m at $t = 2$ s.

12–14. $v = 322$ m/s, $t = 19.3$ s.

12–15. $s = -18.0$ ft, $s_{Tot} = 46.0$ ft.

12–17. $s = -27.0$ ft, $s_{Tot} = 69.0$ ft.

12–18. 11.9 m, 0.250 m/s.

12–19. 16.9 ft.

12–21. 10.3 s, 4.11 km.

12–22. (a) 6.25 m/s. (b) 7.23 m/s. (c) 1.42 m.

12–23. 6.53 m, 3.27 s.

12–25. 31.6 m/s, 16.5 m.

12–26. $\Delta s = 152$ ft, $s_{A \, Tot} = 41.0$ ft, $s_{B \, Tot} = 200$ ft.

12–27. 11.2 km/s.

12–29. $0 \leqslant t < 30$, $a = 0.4$, $s = 0.2t^2$,
$30 < t \leqslant 50$, $a = 0$, $s = 12t - 180$.

12–30. $0 \leqslant t \leqslant 5$, $v = 3t^2$, $5 \leqslant t \leqslant 10$,
$v = t^2 + 20t - 50$, $s = 917$ m.

12–31. $0 \leqslant t < 4$, $a = 1.25$, $s = 0.625t^2$,
$4 < t < 10$, $a = 0$, $s = 5t - 10$,
$10 < t \leqslant 15$, $a = -1$, $s = 15t - 0.5t^2 - 60$, $s = 52.5$ m.

12–33. $0 \leqslant t < 30$, $v = 0.8t$, $a = 0.8$, $30 < t \leqslant 40$, $v = 24$, $a = 0$.

12–34. $v_{max} = 11.25$ ft/s, $0 \leqslant t \leqslant 15$, $s = 0.375t^2$,
$15 \leqslant t \leqslant 51.9$, $s = 11.25t - 84.375$.

12–35. $0 \leqslant t \leqslant 30$, $v = 0.00333t^3$, $s = 0.000833t^4$, $30 \leqslant t \leqslant 60$,
$v = 15t - 360$, $s = 7.5t^2 - 360t + 4725$.

12–37. $0 \leqslant t \leqslant 10$, $s = 0.133t^3$, $10 \leqslant t \leqslant 30$, $s = 0.5t^2 + 30t - 217$,
$v_{avg} = 37.8$ ft/s, $s_{Tot} = 1133$ ft.

12–38. $0 \leqslant t < 4$, $a = 3.5$, $4 < t < 5$, $a = 0$, $5 < t \leqslant 8$, $a = 4$,
$a_{max} = 4$ m/s^2.

12–39. $0 \leqslant t \leqslant 4$, $s = 1.75t^2$, $4 \leqslant t \leqslant 5$, $s = 14t - 28$, $5 \leqslant t \leqslant 8$,
$s = 2t^2 - 6t + 22$, $s_{Tot} = 102$ m.

12–41. $v_{40} = 12.7$ ft/s, $v_{90} = 22.8$ ft/s, $v_{200} = 36.1$ ft/s.

12–42. $v_{max} = 55.0$ ft/s, $t' = 14.6$ s.

12–43. $0 \leqslant s < 100$, $a = 0.16$ s, $100 < s \leqslant 200$, $a = 0.01s + 3$,
$a_{100} = 16.0$ m/s^2, $a_{150} = 4.50$ m/s^2.

12–45. $0 \leqslant s \leqslant 300$, $v = 4.90 \, s^{1/2}$, $300 \leqslant s \leqslant 450$,
$v = (-0.04s^2 + 48s - 3600)^{1/2}$, $t = 5.77$ s.

12–46. $0 \leqslant t < 13$, $v = 0.75 \, t$, $13 \leqslant t \leqslant 160$, $v = 9.77$, $t' = 13.0$ s.

12–47. $s_{Tot} = 9$ km, $\Delta s = 3.61$ km, $\searrow\,56.3°$, $v_{avg} = 2.61$ m/s,
$v_{sp} = 6.52$ m/s.

12–49. $a = 13.5$ m/s^2, $\alpha = 63.7°$, $\beta = 82.4°$, $\gamma = 27.6°$.

12–50. (18.0 ft, 6.20 ft, 4.00 ft).

12–51. 3.05 m/s, 7.48 m.

12–53. 9.68 m/s, 16.8 m/s^2.

12–54. (a) $s_A = 1.40$ m, $s_B = 3$ m, (b) $\mathbf{r}_A = \{1.38\mathbf{i} + 0.195\mathbf{j}\}$ m,
$\mathbf{r}_B = \{-2.82\mathbf{i} + 0.873\mathbf{j}\}$ m, (c) $\Delta r = 4.26$ m.

12–55. $v = 168$ mm/s, $\alpha = 91.0°$, $\beta = 87.7°$, $\gamma = 2.49°$,
$a = 198$ mm/s^2, $\alpha = 114°$, $\beta = 110°$, $\gamma = 32.0°$.

Answers

12–57. $(9.0 \text{ m}, 0)$, $v_A = 27.4$ m/s, $v_B = 10.1$ m/s, $t = 2.0$ s.

12–58. $(4 \text{ ft}, 3 \text{ ft})$, $v_A = 4.47$ ft/s, $v_B = 8.06$ ft/s.

12–59. $y = \left(\dfrac{x}{4}\right)^{1/4} [1 + 8x^{1/2}]$, $v = 162$ in./s, $a = 215$ in./s.

12–61. 2.69 ft/s, 0.02 ft/s².

12–62. 4.22 m/s, 0.130 m/s².

12–63. $v_x = 0.398$ ft/s, $v_y = 3.98$ ft/s.

12–65. 8.34 m, 0.541 m/s².

12–66. 19.8 m/s, $\measuredangle$ 36.2°.

12–67. 19.4 m/s, 4.54 s.

12–69. 18.2 m/s, 12.7 m.

12–70. 0.400 s, 2.42 ft.

12–71. 166 ft.

12–73. 36.7 ft/s, 11.5 ft.

12–74. 0.242 s, 0.714 m.

12–75. $x = 13.3$ ft, $y = -7.10$ ft.

12–77. $\theta_C = 75.3°$, $\theta_D = 14.7°$, $\Delta t = 1.45$ s.

12–78. 45°, 10.2 m.

12–79. 8.83 m.

12–81. 20.8 ft/s.

12–82. $v_B = 21.6$ ft/s < 23 ft/s, (he can catch the ball).

12–83. $\Delta t = \dfrac{2v_0}{g}\left[\dfrac{\sin(\theta_1 - \theta_2)}{\cos\theta_2 + \cos\theta_1}\right]$.

12–85. 35.0 ft/s², 2.64 $(10)^3$ ft.

12–86. $v_x = \dfrac{v_0}{\sqrt{1 + b^2}}$, $v_y = \dfrac{bv_0}{\sqrt{1 + b^2}}$, $a_n = \dfrac{2cv_0^2}{[1 + b^2]^{3/2}}$.

12–87. $(0, -9 \text{ ft})$, 450 ft/s².

12–89. 256 ft/s, 5.25 (10^3) ft.

12–90. 12.6 ft/s².

12–91. 21.6 ft/s², 40.5 ft.

12–93. 0.921 m/s².

12–94. 97.2 ft/s, 42.6 ft/s².

12–95. 115 ft/s, 58.4 ft/s².

12–97. 1.96 m/s, 0.930 m/s².

12–98. $a_n = 8.00$ ft/s², $a_t = 24$ ft/s².

12–99. 1.68 s.

12–101. 2.32 m/s².

12–102. $v_r = 0$, $v_\theta = 100$ ft/s, $a_r = -25.0$ ft/s², $a_\theta = -3.2$ ft/s².

12–103. $\dot\theta = 0.075$ rad/s, $a = 2.25$ ft/s².

12–105. 24.1 ft/s, 8.17 ft/s².

12–106. 3.03 m/s, 12.6 m/s².

12–107. $v_r = \dfrac{-20}{\sqrt{1 + \theta^2}}$, $v_\theta = \dfrac{20\,\theta}{\sqrt{1 + \theta^2}}$,

$v_r = -14.1$ ft/s, $v_\theta = 14.1$ ft/s.

12–109. $v_r = -0.230$ ft/s, $v_\theta = 1.35$ ft/s, $a_r = -1.47$ ft/s²,

$a_\theta = -0.706$ ft/s².

12–110. $v_r = 0.425$ ft/s, $v_\theta = 0.425$ ft/s,

$a_r = 0.425$ ft/s², $a_\theta = 0.595$ ft/s².

12–111. $v_r = 0$, $v_\theta = 4.8$ ft/s, $v_z = -0.664$ ft/s, $a_r = -2.88$ ft/s²,

$a_\theta = 0$, $a_z = -0.365$ ft/s².

12–113. $v_r = 32$ ft/s, $v_\theta = 50.3$ ft/s, $a_r = -161$ ft/s², $a_\theta = 319$ ft/s².

12–114. $v_r = -93.6$ ft/s, $v_\theta = 27.0$ ft/s, $a_r = -68.7$ ft/s²,

$a_\theta = -41.3$ ft/s².

12–115. $v_r = -14.1$ m/s, $v_\theta = 14.1$ m/s, $a_r = -1.41$ m/s²,

$a_\theta = -1.41$ m/s².

12–117. $v_r = a\dot\theta$, $v_\theta = a\theta\dot\theta$, $a_r = -a\theta\dot\theta^2$, $a_\theta = 2a\dot\theta^2$.

12–118. $v_r = ake^{k\theta}\dot\theta$, $v_\theta = ae^{k\theta}\dot\theta$, $a_r = ae^{k\theta}(\dot\theta)^2[k^2 - 1]$,

$a_\theta = 2ake^{k\theta}\dot\theta^2$.

12–119. $v_r = 6$ ft/s, $v_\theta = 18.3$ ft/s, $a_r = -67.1$ ft/s², $a_\theta = 66.3$ ft/s².

12–121. 640 m/s, 657 m/s².

12–122. 155 m/s, 67.1 m/s².

12–123. $v_r = 0.9$ ft/s, $v_\theta = 2.46$ ft/s,

$a_r = -12.1$ ft/s², $a_\theta = 5.40$ ft/s².

12–125. $v_r = -16.9$ m/s, $v_\theta = 4.87$ m/s,

$a_r = -89.4$ m/s², $a_\theta = -53.7$ m/s².

12–126. $\dot{\mathbf{a}} = (\dddot{r} - 3\dot r\dot\theta^2 - 3r\dot\theta\ddot\theta)\mathbf{u}_r + (3\ddot r\dot\theta + r\dddot\theta + 3\dot r\ddot\theta - r\dot\theta^3)\mathbf{u}_\theta + (\dddot z)\mathbf{u}_z$.

12–127. $v_r = 0.242$ m/s, $v_\theta = 0.943$ m/s,

$a_r = -2.33$ m/s², $a_\theta = 1.74$ m/s².

12–129. $\mathbf{u}_b = -0.349\mathbf{i} - 0.0127\mathbf{j} - 0.937\mathbf{k}$

12–130. 1.33 ft ↑.

12–131. 12.0 ft/s →.

12–133. 8.0 ft/s ↑.

12–134. 58 ft/s ↑.

12–135. 0.667 m/s ↑.

12–137. 1.82 s, 5.29 m/s ↑.

12–138. 1.5 ft/s ↑.

12–139. 2 ft/s ↑.

12–141. $v_B = (-3(y^2 + 16)^{1/2}/y)$ m/s

12–142. 5 m/s, 7 m/s².

12–143. 2.4 ft/s →, 3.85 ft/s² →.

12–145. 0.513 ft/s².

12–146. 525 mi/hr ⬂ 19.1°.

12–147. 23.4 mi.

12–149. 21.7 ft/s 18.0° ⬋, 36.9 s.

12–150. 70.9 ft/s 7.11° ⬈, 22.6 s.

12–151. $v_{A/B} = 6.69$ m/s ⬈ 53.3°, $a_{A/B} = 1.52$ m/s² 41.9° ⬈.

12–153. 15.1°.

12–154. 98.4 ft/s 67.6° ⬈, 19.8 ft/s² $\measuredangle$ 57.4°.

12–155. 5.75 m/s, 17.8 m/s, ⬂ 76.2°, 9.81 m/s² ↓.

CHAPTER 13

13–1. 296 N, 8.47 m/s².

13–2. 199 $(10)^{18}$ N.

13–3. 35.3 (10^{21}) N.

13–5. 0.343 m/s², 0.436 m/s².

13–6. $T_B = T_A = 30$ lb, $a_A = 6.44$ ft/s², $a_B = 4.83$ ft/s².

13–7. 7.23 m/s.

13–9. $T = \dfrac{mg}{2}\sin 2\theta$.

13–10. $T = mg\cos\theta[\sin\theta - \mu\cos\theta]$.

13–11. 64.9 ft/s.

13–13. 286 lb.

13–14. 206 lb.

13–15. $0 \le t < 10$, $F = 652$ lb, $10 < t \le 30$, $F = 109$ lb.

13–17. 64.4 ft.

13–18. 13.1 lb.

13–19. 3.62 m/s.

13–21. 0.297 m/s².

13–22. 0.0595 m/s².

13–23. 5.30 ft., 1.82 s.
13–25. $a_A = 9.66$ ft/s^2, $a_B = 15.0$ ft/s^2.
13–26. 8.91 m/s.
13–27. 247 mm.
13–29. 1.75 m/s^2.
13–30. $v = \left[\dfrac{mg}{k} - \left(\dfrac{mg}{k} - v_0^2\right)e^{-\frac{2kh}{m}}\right]^{1/2}$, $v = \sqrt{\dfrac{mg}{k}}$.
13–31. $v = \sqrt{\dfrac{mg}{k}}\left(\dfrac{e^{2t\sqrt{\frac{kg}{m}}} - 1}{e^{2t\sqrt{\frac{kg}{m}}} + 1}\right)$, $v_t = \sqrt{\dfrac{mg}{k}}$.
13–33. 39.5 mm.
13–34. (a) $a = \dfrac{P}{m} - 3\mu g$, (b) $a = \dfrac{P}{2m} - 2\mu g$.
13–35. $a_B = 0$, $a_C = 4.11$ m/s^2, $a_D = 0.162$ m/s^2.
13–37. $P = 2mg\left(\dfrac{\mu \cos \theta + \sin \theta}{\cos \theta - \mu \sin \theta}\right)$.
13–38. 78.7°.
13–39. 1.06 s.
13–41. $F = m(a_B + g)\sqrt{4y^2 + d^2}/(4y)$.
13–42. 188 mm.
13–43. $s = \dfrac{evLl}{v_0^2\, wm}$.
13–45. $v_{B/AC} = a_0 (\sin \theta)t$, $s = \frac{1}{2}a_0 (\sin \theta)\, t^2$.
13–46. 1.99 (10^{30}) kg.
13–47. 80.2 ft/s.
13–49. $N = 277$ lb, $F = 13.4$ lb.
13–50. 15.6 ft/s.
13–53. 962 N.
13–54. $T = 3mg \sin \theta$.
13–55. $a_t = 16.1$ ft^2/s, $T = 46.9$ lb.
13–57. 2.10 m/s.
13–58. 2.09 m/s.
13–59. 4.90 lb.
13–61. $T = 361$ N, $a_t = 3.36$ m/s^2.
13–62. $F = 1.63$ kN, $N_s = 720$ N.
13–63. $F = 1.23$ kN, $N_s = 924$ N.
13–65. 7.31 lb, 6.00 ft/s^2.
13–66. 1.95 m/s^2, 647 N.
13–67. 0.969 m/s.
13–69. 29.4 ft/s^2, 69.0 lb.
13–70. $v = \sqrt{9.81 \cos \theta - 5.81}$, $N = 58.9 \cos \theta - 23.2$, $\theta = 0°$.
13–71. $F_r = -92.5$ N, $F_\theta = 317$ N, $F_z = 35.6$ N.
13–73. $F_r = -1.54$ kN, $F_\theta = 0$, $F_z = 2.45$ kN.
13–74. $F_r = -100$ N, $F_\theta = 0$, $F_z = 2.45$ kN.
13–75. $p = 12.6$ lb, $N = 4.22$ lb.
13–77. 0.300 lb.
13–78. $\theta = \tan^{-1}\left(\dfrac{4r_c}{g}\,\dot{\theta}_0^2\right)$.
13–79. 12.0 m/s^2 5.71° ⬂.
13–81. $N_C = 3.03$ N, $F = 1.81$ N.
13–82. $N_C = 25.6$ N, $F_{OA} = 0$.
13–83. $N_C = 20.7$ N, $F_{OA} = 0$.
13–85. -0.0108 lb.

13–86. 0.163 lb.
13–87. $F_r = -1.35$ kN, $F_\theta = -0.300$ kN, $F_z = 2.94$ kN.
13–89. $F = 5.07$ kN, $N = 2.74$ kN.
13–90. 113 lb.
13–91. 26.7 lb.
13–93. $r = 42.2$ Mm, $v = 3.07$ km/s.
13–94. (a) $r = 194$ (10^3) mi, (b) 392 (10^3) mi,
(c) 194 (10^3) mi $< r <$ 392 (10^3) mi, (d) $r >$ 392 (10^3) mi.
13–95. $\Delta v = \sqrt{\dfrac{GM_e}{r_0}}(\sqrt{2} - \sqrt{e + 1})$.
13–97. $\Delta v = 2.57$ km/s.
13–98. 9.60 km/s, 6.50 Mm.
13–99. $v_p = 7.82$ km/s, $T = 1.97$ h.
13–101. $v_B = 1.20$ km/s, $\Delta v = 2.17$ km/s.
13–102. $v_p = 55.3$ (10^3) ft/s, $r = 484$ (10^6) mi.
13–105. 11.5 Mm/h, 27.3 Mm.
13–106. $\Delta v_a = 466$ m/s (decrease), $\Delta v_p = 2.27$ km/s (decrease).

CHAPTER 14

14–1. $U_w = 4.12$ kJ, $U_{NF} = -2.44$ kJ.
14–2. 3.80 kN.
14–3. 2.72 in.
14–5. 16.0 ft/s, 7.98 ft.
14–6. 1.05 ft.
14–7. 4.52 m/s.
14–9. 0.365 ft/s.
14–10. 8.79 ft/s.
14–11. 1.13 kN.
14–13. 642 lb/ft, 18.0 ft/s.
14–14. $v_B = 7.22$ ft/s, $N_B = 27.1$ lb, $v_C = 17.0$ ft/s, $N_C = 133$ lb, $v_D = 18.2$ ft/s.
14–15. 179 mm.
14–17. $U = [\frac{1}{2}k(y - d) + mg](y - d)$.
14–18. 0.313 ft.
14–19. 30.5 m/s, 1.74 kN.
14–21. 6.89 m/s.
14–22. 3.41 m.
14–23. For A, 6.08 m/s, for B, 4.08 m/s.
14–25. $s_A = 0.255$ m, $s_B = 0.205$ m.
14–26. 2.36 m/s.
14–27. 2.34 m/s.
14–29. 0.0735 ft.
14–30. 0.591 hp.
14–31. 5.20 days.
14–33. 47.2°.
14–34. 95.4 hp.
14–35. 4.12 MW, 10.3 m/s.
14–37. 35.4 kW.
14–38. 0.229 hp.
14–39. $s = \dfrac{mv^3}{3P}$.
14–41. 1.13 kW.
14–42. $P = (6.03t)$ MW.
14–43. $P = \left(\dfrac{6.72(e^{1.79t} - 1)}{(e^{1.79t} + 1)}\right)$ MW.

14–45. 0.211 hp.

14–46. 3.82 m/s.

14–47. 8.79 ft/s.

14–49. $v_B = 7.22$ ft/s, $N_B = 27.1$ lb, $v_C = 17.0$ ft/s, $N_C = 133$ lb, $v_D = 18.2$ ft/s.

14–50. 179 mm.

14–51. $s_A = 0.255$ m, $s_B = 0.205$ m.

14–53. 13.9 ft/s.

14–54. $v_B = 12.7$ ft/s, $v_C = 15.0$ ft/s, $T = 13.5$ lb.

14–55. 2.86 m/s.

14–57. 39.7°.

14–58. 6.97 m/s.

14–61. $v_A = 1.54$ m/s ↑ , $v_B = 4.62$ m/s ↓ .

14–62. 0.195 ft.

14–63. 20.4 ft/s.

14–65. 27.2 ft/s.

14–66. 1.29 ft.

CHAPTER 15

15–1. 1.53 s, 11.5 m.

15–2. 0.280 s, 15.6 ft/s.

15–3. 467 N, 1.25 kg · m/s ⦨ 14.2°.

15–5. (a) 15 kN · s, (b) 15 kN · s.

15–6. 2.61 N · s ⦨ 30°.

15–7. $T = 14.9$ kN, $F = 24.8$ kN.

15–9. For A, 7.40 m/s, for B, 5.40 m/s.

15–10. 20.4 m/s.

15–11. 3.64 s.

15–13. 10.5 ft/s →.

15–14. 6.00 ft/s →.

15–15. 21.8 m/s.

15–17. 40.2 ft/s.

15–18. 10.1 ft/s.

15–19. 6.90 m/s.

15–21. 0.5 m/s →, 16.9 kJ.

15–22. 3.29 ft/s.

15–23. 120 m/s.

15–25. $v_A = 3.29$ m/s →, $v_B = 2.19$ m/s ←.

15–26. $v_B = 0.379$ m/s →.

15–27. $v_{B/C} = 0.632$ m/s →.

15–29. $v_A = 1.58$ m/s ←, $v_B = 0.904$ ft/s ←.

15–30. (a) 2.26 ft/s, (b) 1.85 ft/s.

15–31. 0.364 ft ←.

15–33. 3.18 m.

15–34. 1.65 m.

15–35. 145 ft.

15–37. $t = v_0/\mu g \left(1 + \dfrac{m}{M}\right)$, $v = \dfrac{Mv_0}{m + M}$.

15–38. $x_{C/b} = \left| v_0^2/2\mu g \left(1 + \dfrac{m}{M}\right) \right|$.

15–39. 71.4 mm →.

15–41. $(v_B)_2 = 1.73$ m/s ←, $(v_A)_2 = 4.53$ m/s ←.

15–42. 3.24 m/s, 43.9°.

15–45. $(v_A)_2 = 0.720$ m/s, $(v_B)_2 = 1.92$ m/s →, $\Delta d = 0.538$ m.

15–46. 8.42 ft/s.

15–47. 1.31 ft.

15–49. $(v_A)_1 = 19.7$ ft/s, $(v_A)_2 = 11.1$ ft/s, $(v_B)_2 = 17.0$ ft/s, $d = 11.3$ ft.

15–50. $(v_B)_2 = \frac{1}{3}(\sqrt{2gh})(1 + e)$.

15–51. 6.59 ft/s.

15–53. 1.53 m.

15–54. 1.68 kN.

15–55. $t_{\text{Tot}} = \sqrt{\dfrac{2h}{g}} \left(\dfrac{1 + e}{1 - e}\right)$.

15–57. $v_A = 1.35$ m/s →, $v_B = 5.89$ m/s 32.9° ⦥.

15–58. 90°.

15–59. $v_A = 2.46$ ft/s, $v_B = 3.09$ ft/s.

15–61. $(v_B)_2 = \sqrt{3}v(1 + e)/4$ ⦨ 30, $\Delta T = \frac{3}{16}mv^2(e^2 - 1)$.

15–62. $v_B = \frac{1}{4}(e + 1)v$.

15–63. $(H_A)_O = 72$ kg · m²/s ↓, $(H_B)_O = 59.5$ kg · m²/s ↓.

15–65. $\mathbf{H}_A = \{-1.83\mathbf{i} - 0.913\mathbf{k}\}$ slug · ft²/s.

15–66. $\mathbf{H}_B = \{-5.17\mathbf{i} + 0.304\mathbf{j} - 2.74\mathbf{k}\}$ slug · ft²/s.

15–67. $\mathbf{H}_B = \{108\mathbf{i} + 96\mathbf{j} + 40\mathbf{k}\}$ kg · m²/s.

15–69. 3.47 m/s.

15–70. $v_2 = 2.53$ m/s, $\theta = 27.0°$.

15–71. $v_2 = 9.22$ ft/s 1, $\Sigma U_{1-2} = 3.04$ ft · lb.

15–73. $v_B = 10.2$ km/s, $r_B = 13.8$ Mm.

15–74. $v = 20.2$ ft/s, $h = 6.36$ ft.

15–75. 402 N.

15–77. 242 lb.

15–78. 303 lb.

15–79. $F_x = 19.5$ lb, $F_y = 1.96$ lb.

15–81. 17.5 hp.

15–82. 844 N.

15–83. $3.36(10^3)$ lb.

15–85. 1.42 lb.

15–86. $a_{\text{init}} = 133$ ft/s², $a_{\text{finial}} = 200$ ft/s².

15–87. $2.07 (10^3)$ ft/s.

15–89. 11.5 kN.

15–90. 37.5 ft/s².

15–91. 12.2 m/s.

15–93. $R = (20t + 2.48)$ lb.

15–94. $F = m'v^2$.

15–95. $F = (6t + 0.0559)$ lb.

REVIEW PROBLEMS 1

R1–1. 1.84 m.

R1–2. 946 kW.

R1–3. 23.8 s.

R1–5. 980 m.

R1–6. 744 lb, 14.4 ft · kip.

R1–7. 21.5 ft/s.

R1–9. $a_A = 0.755$ m/s², $a_B = 1.51$ m/s², $T_A = 90.6$ N, $T_B = 45.3$ N.

R1–10. 77.5 ft/s.

R1–11. 0.575 ft/s 71.6° ⦥, 0.192 ft · lb.

R1–13. 24.8 m/s.

R1–14. $(v_A)_2 = 1.30$ m/s ⦨ 8.47°, $(v_B)_2 = 1.34$ m/s ←.

R1–15. 0.933 m.
R1–17. $F_s = 68$ N, $N_s = 153$ N.
R1–18. 27.3 s.
R1–19. 38.5 N.
R1–21. 20 ft.
R1–22. 51.0 lb.
R1–23. 1.80 hp.
R1–25. $v_{B/A} = 675$ mph, $_{40.1°}$ ⤢.
R1–26. 38.5 ft/s.
R1–27. $F_r = -39.2$ N, $F_\theta = 0$, $F_z = 392$ N.
R1–29. 0.221 m.
R1–30. 13.1 ft/s.
R1–31. $v_r = 6$ ft/s, $v_\theta = 18.3$ ft/s, $a_r = -67.1$ ft/s^2, $a_\theta = 66.3$ ft/s^2.
R1–33. 14.6 m/s.
R1–34. 0.9 m/s, 22.7 kJ.
R1–35. 2.09 s, 50.9 ft/s.
R1–37. 1.90 ft, 16.9 ft/s.
R1–38. 3.11 s.
R1–39. 1.02 s.
R1–41. 24.5 m, $N_B = 0$, $N_C = 9.81$ kN.
R1–42. 1.81 ft/s^2.
R1–43. $v_B = 55.2$ ft/s, $v_A = 27.6$ ft/s.
R1–45. 47.8 ft/s.
R1–46. 0.5 m/s ↑ .
R1–47. 29.0 m/s^2.
R1–49. 2.22 ft.
R1–50. 13.5 m/s, 147 kN.

CHAPTER 16

16–1. 2 ft/s, 8.02 ft/s^2.
16–2. $a_t = 0.286$ ft/s^2, $a_n = 331$ ft/s^2, $s = 525$ ft.
16–3. 0.838 m/s, 0.890 m/s^2.
16–5. $v = 52$ ft/s, $a_t = 12$ ft/s^2, $a_n = 1352$ ft/s^2.
16–6. $v_B = 22$ ft/s, $a_t = 9$ ft/s^2, $a_n = 322$ ft/s^2.
16–7. 211 rad/s.
16–9. $\omega_C = 4.50$ rad/s, $\omega_B = 3.38$ rad/s.
16–10. 10.6 rad/s.
16–11. 1.80 m/s ↓ , 0.450 m/s^2 ↓ .
16–13. 10.8 m/s ↑ , 16.2 m.
16–14. 465 rad/s.
16–15. 2.89 m.
16–17. 3.52 ft/s, 12.8 ft/s^2, 1.15 s.
16–18. 48.7 ft/s, 8.54 rev.
16–19. $a = \left(\dfrac{\omega^2}{2\pi}\right)\left(\dfrac{r_2 - r_1}{L}\right)d$.
16–21. 650 mm/s, -1615 mm/s^2.
16–22. -19.2 rad/s, -183 rad/s^2.
16–23. 4.17 rad/s.
16–25. $v = 0.480 \cos\theta$ m/s, $a = -1.92 \sin\theta$ m/s^2.
16–26. 0.416 m/s, -0.752 m/s^2.
16–27. $\dot\theta = -v_B h/(h^2 + x^2)$, $\alpha = 2v_B^2 hx/(h^2 + x^2)^2$.
16–29. $\omega = -rv_A/y\sqrt{y^2 - r^2}$, $\alpha = rv_A^2(2y^2 - r^2)/y^2(y^2 - r^2)^{3/2}$.
16–30. $\dot\theta = \dfrac{v_0}{a}\sin^2\theta$, $\ddot\theta = \left(\dfrac{v_0}{a}\right)^2 \sin 2\theta \sin^2\theta$.

16–31. 1.08 rad/s, 4.39 ft/s.
16–33. $\omega = -\dfrac{v_A \tan\theta}{x}$, $\alpha = \dfrac{v_A^2 \tan\theta}{x^2}(2 + \tan^2\theta)$.
16–34. $\omega = -\dfrac{v_A \tan\theta}{x}$, $\alpha = \dfrac{\tan\theta}{x^2}[v_A^2(2 + \tan^2\theta) - a_A x]$.
16–35. 33.3 rad/s, 1 rev.
16–37. $v = lr\omega[r + d\cos\theta/(d + r\cos\theta)^2]$,
$a = lr\omega^2 \sin\theta\,[2r^2 - d^2 + dr\cos\theta/(d + r\cos\theta)^3]$.
16–38. 50 m/s ↓ .
16–39. 9.20 m/s → .
16–41. 9.33 in/s ⤡ $_{16.5°}$.
16–42. 2.40 m/s → .
16–43. 2.45 rad/s ↑ , 2.20 ft/s ← .
16–45. 4 rad/s ↑ , 100 mm/s ↓ .
16–46. $\omega_A = 4$ rad/s ↓ , $\omega_B = 1$ rad/s ↑ , $v_D = 60$ mm/s ↑ .
16–47. $\omega_{CD} = 10.0$ rad/s ↑ , $\omega_{BC} = 1.33$ rad/s ↓ .
16–49. 4.00 m/s ⤡ $_{52.7°}$.
16–50. 10.3 rad/s ↓ .
16–51. $v_C = 2.89$ ft/s ← , $v_E = 9.33$ ft/s ↓ .
16–53. 50 m/s ↓ .
16–54. 2.40 m/s → .
16–55. 2.45 rad/s ↑ , 2.20 ft/s ← .
16–57. $\omega_{CD} = 10.0$ rad/s ↑ , $\omega_{BC} = 1.33$ rad/s ↓ .
16–59. $v_C = 6$ ft/s → , $v_B = 4$ ft/s → .
16–61. $v_A = 60$ ft/s → , $v_C = 220$ ft/s ← , $v_B = 161$ ft/s ⤢ $_{60.3°}$.
16–62. 77.3 in./s ↑ .
16–63. 1.5 v/r.
16–65. $v_C = 2.50$ ft/s ← , $v_D = 9.43$ ft/s ⤣ $_{55.8°}$.
16–66. 6 rad/s ↑ .
16–67. 8.50 rad/s ↑ .
16–69. $\omega_s = 15$ rad/s ↑ , $\omega_R = 3$ rad/s ↑ .
16–70. 1.04 m/s → .
16–71. 4.42 in./s $_{30°}$ ⤢ .
16–73. 0.332 rad/s^2 ↓ , 7.88 ft/s^2 ← .
16–74. 2.25 m/s^2, ⤡ $_{32.6°}$.
16–75. 10.0 m/s^2, $_{2.02°}$ ⤤ .
16–77. 5.94 m/s^2, $_{53.9°}$ ⤤ .
16–78. 6.21 m/s^2, $_{45.6°}$ ⤢ .
16–79. $\alpha_{AB} = 4.81$ rad/s^2 ↓ , $\alpha_{BC} = 5.21$ rad/s^2 ↑ .
16–81. 2.41 m/s^2 ↓ .
16–82. $\omega_{BC} = 5.58$ rad/s ↓ , $\omega_{CD} = 5.44$ rad/s ↑ , $\alpha_{BC} = 54.4$ rad/s^2, $\alpha_{CD} = 103$ rad/s^2.
16–83. 10.7 rad/s^2 ↑ , 14.1 in./s^2 ∠ $_{18.4°}$.
16–85. 4.73 rad/s ↑ , 131 rad/s^2 ↓ .
16–86. $\mathbf{v}_B = \{5\mathbf{i} + 6\mathbf{j}\}$ ft/s, $\mathbf{a}_B = \{-15\mathbf{i} + 40\mathbf{j}\}$ ft/s^2.
16–87. $\mathbf{v}_A = \{-2.5\mathbf{i} + 2\mathbf{j}\}$ ft/s, $\mathbf{a}_A = \{-3\mathbf{i} + 1.75\mathbf{j}\}$ ft/s^2.
16–89. $\mathbf{v}_B = \{-12\mathbf{i}\}$ ft/s, $\mathbf{a}_B = \{-20\mathbf{i} - 70\mathbf{j}\}$ ft/s^2.
16–90. $\mathbf{v}_B = \{-12\mathbf{i} + 5.77\mathbf{j}\}$ ft/s, $\mathbf{a}_B = \{-28.9\mathbf{i} - 70\mathbf{j}\}$ ft/s^2.
16–91. 2.40 m/s.
16–93. $\mathbf{v}_C = \{-7.00\mathbf{i} + 17.3\mathbf{j}\}$ ft/s, $\mathbf{a}_C = \{-34.6\mathbf{i} - 15.5\mathbf{j}\}$ ft/s^2.
16–94. $\mathbf{v}_C = \{-7.00\mathbf{i} + 17.3\mathbf{j}\}$ ft/s, $\mathbf{a}_C = \{-38.8\mathbf{i} - 6.84\mathbf{j}\}$ ft/s^2.
16–95. $\mathbf{v}_C = \{-0.944\mathbf{i} + 2.02\mathbf{j}\}$ m/s, $\mathbf{a}_C = \{-11.2\mathbf{i} - 4.15\mathbf{j}\}$ m/s^2.
16–97. 10.0 rad/s ↓ , 24.0 rad/s^2 ↓ .
16–98. 9.00 rad/s ↓ , 249 rad/s^2 ↓ .
16–99. $\omega_{AC} = 0$, $\alpha_{AC} = 14.4$ rad/s^2 ↓ .

Answers

CHAPTER 17

17-1. $I_y = \frac{1}{3}ml^2$.

17-2. $I_z = mR^2$.

17-3. $I_x = \frac{3}{10}mr^2$.

17-5. $I_x = \frac{2}{5}mb^2$.

17-6. $I_x = \frac{2}{5}mr^2$.

17-7. 1.20 in.

17-9. 2.25 slug $\cdot$ ft^2.

17-10. 2.17 slug $\cdot$ ft^2.

17-11. 7.67 kg $\cdot$ m^2.

17-13. 1.53 kg $\cdot$ m^2.

17-14. 118 slug $\cdot$ ft^2.

17-15. 293 slug $\cdot$ ft^2.

17-17. 0.450 (10^{-3}) kg $\cdot$ m^2, $\bar{y} = 57.7$ mm, 0.150 (10^{-3}) kg $\cdot$ m^2.

17-18. 0.888 m, 5.61 kg $\cdot$ m^2.

17-19. $a_G = 13.3$ ft/s^2, $N_A = 44.0$ lb, $N_B = 61.0$ lb.

17-21. 2.00 kN.

17-22. $a_{max} = 77.3$ ft/s$^2 \rightarrow$, $B_x = 432$ lb, $B_y = 180$ lb.

17-23. $N_A = 120$ lb, $B_x = 73.9$ lb, $B_y = 69.7$ lb.

17-25. $P = 24$ lb, $a_G = 3.22$ ft/s^2, $N_A = 65.3$ lb, $N_B = 14.8$ lb.

17-26. 53.3 lb.

17-27. 16.1 ft/s^2.

17-29. 11.2 s.

17-30. 18.6°.

17-31. (a) $a_B = 2.26$ m/s^2, $N_B = 226$ N, $N_A = 559$ N, (b) $N_B = 454$ N.

17-33. (a) 82.8 ft, (b) 82.8 ft.

17-34. 5.59 ft/s^2.

17-35. 20.0 rad/s^2.

17-37. $a_G|_{h=0.2\ m} = 1.41$ m/s^2, $a_G|_{h=0} = 1.38$ m/s^2.

17-38. impossible, $a_G = 5.89$ m/s$^2 < 39.2$ m/s^2.

17-39. $F_B = 4.50$ kN, $N_A = 1.78$ kN, $N_B = 5.58$ kN.

17-41. 9.67 rad/s^2.

17-42. $N_B = 4.70$ kN, $N_C = 2.31$ kN, $N_D = 3.78$ kN.

17-43. $A_x = 14.9$ lb, $A_y = 2.50$ lb, $\alpha = 8.05$ rad/s^2 $\downarrow$.

17-45. 20.8 rad/s.

17-46. 2.48 rad/s $\downarrow$.

17-47. $A_x = 0$, $A_y = 262$ N.

17-49. $a = (m_B - m_A)g/(\frac{1}{2}M + m_B + m_A)$.

17-50. $N_A = N_B = 346$ N, $\alpha = 2.52$ rad/s^2.

17-51. 1.06 s.

17-53. 7.28 rad/s^2.

17-54. $O_x = 54.8$ N, $O_y = 41.2$ N.

17-57. 7.69 ft/s.

17-58. 10.9 rad/s.

17-59. 0.548 m/s.

17-61. $A_x = 0$, $A_y = 289$ N, $\alpha = 23.1$ rad/s^2.

17-62. $T_A = 13.3$ lb, $T_B = 16.7$ lb.

17-63. $T_A = 18.3$ lb, $T_B = 20.7$ lb.

17-65. $\omega = \sqrt{(3g/L)}\,[1 - \cos\theta]$, $O_x = mg\sin\theta\,[2.25\cos\theta - 1.5]$, $O_y = mg(1 - 1.5\cos\theta + 1.5\cos^2\theta - 0.75\sin^2\theta)$.

17-66. $F = 2.5\ mg\sin\theta$, $N = 0.25\ mg\cos\theta$, $\theta = \tan^{-1}(0.1\ \mu)$.

17-67. 29.8°

17-69. 16.1 ft/s^2, 5.80 rad/s^2.

17-70. 0.860 rad/s^2, 5.92 ft/s^2.

17-71. 4.10 rad/s^2.

17-73. 1.30 rad/s^2.

17-74. 0.5 rad/s^2.

17-75. 31.2 rad/s^2.

17-77. 5.62 rad/s^2, 196 N.

17-78. $A_x = 0.466$ lb, $A_y = 15$ lb, $a_G = 1.00$ ft/s^2.

17-79. 13.4 rad/s^2.

17-81. $s = \dfrac{\omega_1 r}{\mu g}(v_1 - \frac{1}{2}\omega_1 r)$.

17-82. $A_x = 42.2$ lb, $A_y = 22.0$ lb.

17-83. $T_A = \frac{4}{7}W$.

17-85. $\alpha = \dfrac{6P}{mL}\downarrow$, $a_B = \dfrac{2P}{m}\rightarrow$.

17-86. $A_x = 139$ N, $A_y = 252$ N, $B_y = 180$ N.

17-87. $A_x = 34.3$ N, $A_y = 200$ N, $B_y = 143$ N.

17-89. $B_x = 16.7$ lb, $B_y = 18.8$ lb, $N_C = 31.2$ lb.

17-91. $\alpha = \dfrac{2g}{3r}\sin\theta$, $N = m\left[g\cos\theta + \dfrac{r^2\omega^2}{R-r}\right]$.

17-93. Does not slip, $F = 20.1$ N $< \mu N$.

17-94. $N_B = 297$ N $\nwarrow_{30°}$, $A_x = 1.63$ N, $A_y = 344$ N.

17-95. 33.5°.

17-97. $\alpha = \dfrac{3g}{2l}$, $N = \frac{1}{4}mg$.

17-98. $\alpha = \dfrac{1.5g}{l}\cos\theta$.

CHAPTER 18

18-1. $v = 23.8$ m/s, $a_n = 1.88\ (10^3)$ m/s^2, $a_t = 37.5$ m/s^2, $A_x = 0$, $A_y = 98.1$ N.

18-2. 7.69 ft/s.

18-3. 10.9 rad/s.

18-5. 14.1 rad/s.

18-6. 8.25 rad/s.

18-7. 55.2 ft/s.

18-9. 2.45 rad/s.

18-10. 5.74 rad/s.

18-11. 11.0 rad/s.

18-13. 2.58 m/s, 141 N.

18-14. 9.75 m.

18-15. 1.32 rad/s.

18-17. 0.661 m.

18-18. 0.859 m.

18-19. 2.58 m/s.

18-21. 10.4 ft/s.

18-22. 8.37 ft/s.

18-23. (a) $\omega = \sqrt{\dfrac{3\pi w_0}{2m}}$, (b) $\omega = \sqrt{\dfrac{3\pi w_0}{2m} + \dfrac{3g}{L}}$.

18-25. 0.891 rev.

18-26. $\omega_{BC} = \omega_{AB} = \sqrt{\dfrac{3g}{L}}\sin\theta$.

18-27. $v_G = 3\sqrt{\dfrac{3gR}{7}}$.

18-29. 0.661 m.

18–30. $\omega_{BC} = \omega_{AB} = \sqrt{\dfrac{3g}{L} \sin \theta}$.

18–31. $v_G = 3\sqrt{\dfrac{3gR}{7}}$.

18–33. 2.58 m/s.

18–34. 19.8 rad/s.

18–35. 7.02 ft.

18–37. 1.80 rad/s.

18–38. 39.3°.

18–39. 2.57 rad/s.

18–41. $\omega_{AB} = \omega_{BC} = 5.28$ rad/s.

18–42. $\omega_{AB} = \omega_{BC} = 2.21$ rad/s.

18–43. 10.5 ft/s.

18–45. 85.1 rad/s.

18–46. 11.6 rad/s.

18–47. 5.48 rad/s.

18–49. 5.04 rad/s.

18–50. 2.49 rad/s.

18–51. 4.00 m/s.

CHAPTER 19

19–1. 20.8 rad/s.

19–2. 2.48 rad/s.

19–3. 1.06 s.

19–9. 18.7 rad/s.

19–10. 119 rad/s, 64.5 ft/s $\measuredangle$ 22.5°.

19–11. $v_A = 8.83$ m/s $\downarrow$, $v_B = 5.30$ m/s $\uparrow$.

19–13. 12.7 rad/s.

19–14. 14.4 rad/s.

19–15. 4.26 ft/s.

19–17. $\omega_A = 1.70$ rad/s, $\omega_B = 5.10$ rad/s.

19–18. 23.4 rad/s.

19–19. $\omega_A = 47.4$ rad/s, $\omega_B = 75.6$ rad/s.

19–21. $\theta = \tan^{-1}(3.5 \, \mu)$.

19–22. $I = m\sqrt{\dfrac{gL}{6}}$.

19–23. $y = \dfrac{\sqrt{2}}{3}a$.

19–25. $d = \frac{2}{3}L$.

19–26. $\omega = 0$.

19–27. 51.5 rad/s.

19–29. $\omega_2 = \frac{1}{3}\omega_1$.

19–30. 0.0906 rad/s.

19–31. 0.0181 rad/s.

19–33. $\omega' = \frac{11}{3}\omega_1$.

19–34. (a) $\omega = 0.025$ rad/s, (b) $\omega = 0$.

19–35. $\omega_1 < 1.02\sqrt{\dfrac{g}{r}}$.

19–37. $\omega_2 = (\frac{2}{7} + \frac{5}{7}\cos\theta)\,\omega_1$.

19–38. 0.500 ft.

19–39. $\theta = \tan^{-1}\left(\sqrt{\dfrac{7e}{5}}\right)$.

19–41. 39.8°.

19–42. 39.8°.

19–43. 3.23 rad/s, 32.8°.

REVIEW PROBLEMS 2

R2–1. 20.4 ft/s.

R2–2. $v_C = 25.3$ in./s, $\measuredangle$ 63.4°, $a_C = 73.8$ in./s^2, 32.5° $\searrow$.

R2–3. 41.8 rad/s.

R2–5. 6.04 ft/s^2.

R2–6. 5.75 ft.

R2–7. 100 rad/s.

R2–9. 52.0°.

R2–10. 24 ft/s.

R2–11. 24 ft/s.

R2–13. 8.89 rad/s^2.

R2–14. 65.6 rad/s.

R2–15. 27.8 rad/s.

R2–17. 0.393 m, 0.0885 kg · m^2.

R2–18. $\omega_{AB} = 6$ rad/s $\downarrow$.

R2–19. 216 mm, 0.0264 kg · m^2.

R2–21. 4.72 m/s.

R2–22. 25 rad/s.

R2–23. 8.59 rad/s.

R2–25. $a_t = 0.375$ ft/s^2, $a_n = 3600$ ft/s^2, $s = 4500$ ft.

R2–26. 1.89 rad/s, 15.6 rad/s^2.

R2–27. 195 mm/s.

R2–29. 5.0 rad/s^2.

R2–30. (a) 68.7 rad/s, (b) 68.8 rad/s.

R2–31. 4.60 rad/s.

R2–33. 6.14 lb.

R2–34. 8 m/s^2, $\uparrow$, 50.6 m/s^2, $\measuredangle$ 9.09°.

R2–35. 1.64 s, 48.7 N.

R2–37. 4.90 m/s^2, 14.7 m/s^2.

R2–38. 9.42 m/s.

R2–39. 24.1 m/s.

R2–41. 49.1 lb.

R2–42. 12.1 rad/s^2, 30.0 lb.

R2–43. 0.714 s.

R2–45. 12.5 rad/s^2, 88.8 N.

R2–46. 11.9 ft/s.

R2–47. 14.2 rad/s^2.

R2–49. 193 ft/s^2.

R2–50. 167 ft/s^2.

CHAPTER 20

20–1. $\omega = \{5.66\mathbf{j} + 6.26\mathbf{k}\}$ rad/s, $\alpha = \{-3.39\mathbf{i}\}$ rad/s^2.

20–2. $\mathbf{v}_A = \{-5.20\mathbf{i} - 12.0\mathbf{j} + 20.8\mathbf{k}\}$ ft/s,
$\mathbf{a}_A = \{-24.1\mathbf{i} - 13.3\mathbf{j} - 7.20\mathbf{k}\}$ ft/s^2.

20–3. $\mathbf{v}_P = \{-1.6\mathbf{i}\}$ m/s, $\mathbf{a}_P = \{-0.32\mathbf{i} - 12.0\mathbf{j} - 8.00\mathbf{k}\}$ m/s^2.

20–5. $\omega = \{-4\mathbf{j}\}$ rad/s, $\alpha = \{16\mathbf{i}\}$ rad/s^2, $\mathbf{v}_A = \{-0.283\mathbf{i}\}$ m/s,
$\mathbf{a}_A = \{-1.13\mathbf{j} - 1.13\mathbf{k}\}$ m/s^2.

20–6. $\mathbf{v}_A = \{-8.66\mathbf{i} + 8.00\mathbf{j} - 13.9\mathbf{k}\}$ ft/s,
$\mathbf{a}_A = \{-4.00\mathbf{i} - 7.71\mathbf{j} - 3.20\mathbf{k}\}$ ft/s^2.

20–7. $\mathbf{v}_A = \{-8.66\mathbf{i} + 8.00\mathbf{j} - 13.9\mathbf{k}\}$ ft/s,
$\mathbf{a}_A = \{-24.8\mathbf{i} + 8.29\mathbf{j} - 30.9\mathbf{k}\}$ ft/s^2.

20–9. $\mathbf{v}_A = \{1.80\mathbf{i}\}$ ft/s, $\mathbf{a}_A = \{-0.750\mathbf{i} - 0.720\mathbf{j} - 0.831\mathbf{k}\}$ ft/s^2.

20–10. $\mathbf{v}_A = \{-14.1\mathbf{i}\}$ ft/s, $\mathbf{a}_A = \{-4.60\mathbf{i} - 56.5\mathbf{j} - 88.0\mathbf{k}\}$ ft/s^2.

20–11. $\omega_A = 47.8$ rad/s, $\omega_B = 7.78$ rad/s.

20–13. $v_B = 12.0$ ft/s, $\omega = \{0.980\mathbf{i} - 1.06\mathbf{j} - 1.47\mathbf{k}\}$ rad/s.

Answers

20–14. 9.43 ft/s.

20–15. $v_B = 3.00$ ft/s, $\boldsymbol{\omega} = \{1.33\mathbf{i} + 2.67\mathbf{j} + 3.33\mathbf{k}\}$ rad/s.

20–17. $\boldsymbol{\alpha} = \{0.413\mathbf{i} + 0.622\mathbf{j} - 0.412(10^{-3})\mathbf{k}\}$ rad/s^2,
$\mathbf{a}_B = \{-5.98\mathbf{i} + 7.98\mathbf{j}\}$ ft/s^2.

20–18. $\mathbf{v}_A = \{0.667\mathbf{i}\}$ ft/s, $\mathbf{a}_A = \{-0.148\mathbf{i}\}$ ft/s^2.

20–19. $\boldsymbol{\omega} = \{1.50\mathbf{i} + 2.60\mathbf{j} + 2.00\mathbf{k}\}$ rad/s, $\mathbf{v}_C = \{10.4\mathbf{i} - 7.79\mathbf{k}\}$ ft/s.

20–21. $\boldsymbol{\omega} = \{-0.375\mathbf{i} + 0.5\mathbf{j} + 0.625\mathbf{k}\}$ rad/s, $\mathbf{v}_B = \{-3.75\mathbf{j}\}$ ft/s.

20–22. $\boldsymbol{\alpha} = \{-1.43\mathbf{i} + 0.600\mathbf{j} + 1.34\mathbf{k}\}$ rad/s^2, $\mathbf{a}_B = \{-14.3\mathbf{j}\}$ ft/s^2.

20–23. $\boldsymbol{\omega}_{BD} = \{-2\mathbf{i}\}$ rad/s.

20–25. $\mathbf{v}_B = \{-0.333\mathbf{j}\}$, $\boldsymbol{\omega} = \{0.204\mathbf{i} - 0.612\mathbf{j} + 1.36\mathbf{k}\}$ rad/s.

20–26. $\boldsymbol{\alpha}_{BC} = \{1.24\mathbf{i} - 0.306\mathbf{j} + 1.44\mathbf{k}\}$ rad/s^2, $\mathbf{a}_B = \{7.98\mathbf{j}\}$ m/s^2.

20–27. $\boldsymbol{\omega} = \{0.769\mathbf{i} - 2.31\mathbf{j} + 0.513\mathbf{k}\}$ rad/s, $\mathbf{v}_B = \{-0.333\mathbf{j}\}$ m/s.

20–30. $\mathbf{v}_C = \{3\mathbf{i} + 6\mathbf{j} - 3\mathbf{k}\}$ m/s,
$\mathbf{a}_C = \{-13\mathbf{i} + 28.5\mathbf{j} - 10.2\mathbf{k}\}$ m/s^2.

20–31. $\mathbf{v}_A = \{-7.79\mathbf{i} - 2.25\mathbf{j} + 3.90\mathbf{k}\}$ ft/s,
$\mathbf{a}_A = \{8.30\mathbf{i} - 35.2\mathbf{j} + 7.02\mathbf{k}\}$ ft/s^2.

20–33. $\mathbf{v}_T = \{-9.50\mathbf{i} + 3\mathbf{j} + 1.20\mathbf{k}\}$ m/s,
$\mathbf{a}_T = \{-14.4\mathbf{i} - 4.23\mathbf{j} + 4.20\mathbf{k}\}$ m/s^2.

20–34. $\mathbf{v}_C = \{-5.54\mathbf{i} + 5.20\mathbf{j} - 3.00\mathbf{k}\}$ ft/s,
$\mathbf{a}_C = \{-91.5\mathbf{i} - 42.6\mathbf{j} - 1.00\mathbf{k}\}$ ft/s^2.

20–35. $\mathbf{v}_B = \{-17.3\mathbf{i} + 18.8\mathbf{j} + 10\mathbf{k}\}$ ft/s,
$\mathbf{a}_B = \{-19.1\mathbf{i} + 24.0\mathbf{j} - 8.66\mathbf{k}\}$ ft/s^2.

20–37. $\mathbf{v}_P = \{-9.80\mathbf{i} + 14.4\mathbf{j} + 48.0\mathbf{k}\}$ ft/s,
$\mathbf{a}_P = \{-160\mathbf{i} + 5.16\mathbf{j} - 13.0\mathbf{k}\}$ ft/s^2.

20–38. $\mathbf{v}_C = \{-1.73\mathbf{i} - 5.77\mathbf{j} + 7.06\mathbf{k}\}$ ft/s,
$\mathbf{a}_C = \{9.88\mathbf{i} - 72.8\mathbf{j} + 0.365\mathbf{k}\}$ ft/s^2.

20–39. $\mathbf{v}_P = \{-10.5\mathbf{i} + 9.00\mathbf{j} - 1.00\mathbf{k}\}$ ft/s,
$\mathbf{a}_P = \{-41.0\mathbf{i} - 17.5\mathbf{j} - 0.500\mathbf{k}\}$ ft/s^2.

20–41. $\mathbf{v}_C = \{-20.0\mathbf{i} + 24.0\mathbf{j} - 3.00\mathbf{k}\}$ in./s,
$\mathbf{a}_C = \{-145\mathbf{i} - 24.0\mathbf{j} + 8.00\mathbf{k}\}$ in./s^2.

20–42. $\mathbf{v}_P = \{-25.5\mathbf{i} - 13.4\mathbf{j} + 20.5\mathbf{k}\}$ ft/s,
$\mathbf{a}_P = \{161\mathbf{i} - 249\mathbf{j} - 39.6\mathbf{k}\}$ ft/s^2.

20–43. $\mathbf{v}_P = \{-25.5\mathbf{i} - 13.4\mathbf{j} + 20.5\mathbf{k}\}$ ft/s,
$\mathbf{a}_P = \{161\mathbf{i} - 243\mathbf{j} - 33.9\mathbf{k}\}$ ft/s^2.

20–45. $\mathbf{v}_A = \{-8.66\mathbf{i} + 2.62\mathbf{j} + 1.91\mathbf{k}\}$ m/s,
$\mathbf{a}_A = \{-38.3\mathbf{i} - 38.7\mathbf{j} + 11.8\mathbf{k}\}$ m/s^2.

CHAPTER 21

21–2. $I_y = 26.7$ slug-ft^2, $I_x = 53.3$ slug-ft^2.

21–3. $I_y = 4.48\,(10^3)\,\rho$, $I_{xy} = 636\,\rho$.

21–5. $I_{yz} = \dfrac{mah}{6}$.

21–6. $I_{yz} = \dfrac{mah}{5}$.

21–7. 0.0961 slug-ft^2.

21–9. $\bar{x} = -0.667$ ft, $\bar{y} = 0.500$ ft, $I_x{}' = 0.0272$ slug-ft^2,
$I_y{}' = 0.0155$ slug-ft^2, $I_z{}' = 0.0427$ slug-ft^2.

21–10. 0.888 slug-ft^2.

21–11. 1.13 slug-ft^2.

21–13. 0.0945 kg $\cdot$ m^2.

21–14. 0.429 kg $\cdot$ m^2.

21–15. $I_x = 1.36$ kg $\cdot$ m^2, $I_y = 0.380$ kg $\cdot$ m^2, $I_z = 1.26$ kg $\cdot$ m^2.

21–17. $I_x = I_y = I_z = \dfrac{2ma^2}{3}$, $I_{xy} = \dfrac{-a^2m}{4}$, $I_{yz} = \dfrac{a^2m}{4}$, $I_{xz} = -\dfrac{a^2m}{4}$.

21–21. 0.727 kg $\cdot$ m^2/s.

21–22. 6.46 ft $\cdot$ lb.

21–23. 1.57 slug $\cdot$ ft^2/s.

21–25. 0.0920 ft $\cdot$ lb.

21–26. $\mathbf{H} = \{0.0207\mathbf{i} - 0.00690\mathbf{j} + 0.0690\mathbf{k}\}$ ft $\cdot$ lb $\cdot$ s.

21–27. $\boldsymbol{\omega} = \{-0.444\mathbf{j} + 0.444\mathbf{k}\}$ rad/s.

21–29. 58.4 rad/s.

21–30. 19.7 rad/s.

21–31. 18.2 ft/s.

21–33. 21.4 rad/s.

21–34. 87.2 rad/s.

21–35. $\mathbf{H} = \{16.6\mathbf{k}\}$ slug-ft^2/s, $T = 72.1$ ft $\cdot$ lb.

21–37. 20.2 rad/s.

21–38. $\boldsymbol{\omega} = \{0.0536\mathbf{i} + 0.0536\mathbf{k}\}$ rad/s.

21–39. $\Sigma M_x = \dfrac{d}{dt}(I_x\omega_x - I_{xy}\omega_y - I_{xz}\omega_z) - \Omega_z(I_y\omega_y - I_{yz}\omega_z - I_{yx}\omega_x) +$
$\Omega_y(I_z\omega_z - I_{zx}\omega_x - I_{zy}\omega_y)$.

21–41. $\Sigma M_x = I_x\dot{\omega}_x - I_y\Omega_z\omega_y + I_z\Omega_y\omega_z$.

21–42. $F_A = 277$ N, $F_B = 166$ N.

21–43. $A_x = -2.12$ lb, $A_y = 1.06$ lb, $B_x = 2.12$ lb, $B_y = -1.06$ lb,
$B_z = 4.25$ lb.

21–45. $F_A = 8.83$ lb, $F_B = 11.2$ lb.

21–46. 21.0 lb $\cdot$ ft.

21–47. 23.4 lb $\cdot$ ft.

21–49. $F_B = 5.01$ lb, $F_A = 4.99$ lb.

21–50. $A_x = -8.97$ lb, $B_x = -9.66$ lb, $A_z = -6.07$ lb,
$B_z = 15.2$ lb.

21–51. $M = 0.0320$ N $\cdot$ m, $A_x = -1.08$ N, $A_y = 13.2$ N,
$B_x = 1.08$ N, $B_y = 13.3$ N.

21–53. $\dot{\omega}_y = 25.9$ rad/s^2, $B_x = 9.59$ N, $B_z = 12.3$ N, $A_x = -10.8$ N,
$A_z = 12.3$ N.

21–54. $\dot{\omega}_z = 200$ rad/s^2, $D_x = -37.5$ N, $D_y = -12.9$ N,
$C_x = -37.5$ N, $C_y = -11.1$ N, $C_z = 36.8$ N.

21–55. $A_x = B_x = 0$, $A_y = -1.09$ kN, $B_y = 1.38$ kN.

21–57. $A_x = 0$, $A_y = 4.89$ lb, $A_z = 5.47$ lb, $M_x = -8.43$ lb $\cdot$ ft,
$M_y = M_z = 0$.

21–58. $T = 23.3$ N, $F_A = 51.7$ N.

21–59. $T = 23.3$ N, $F_A = 51.8$ N.

21–61. 70.8°.

21–62. 70.8°.

21–65. 328 N $\cdot$ m.

21–66. To the right when viewed from above.

21–67. 4.91 rad/s.

21–69. 1.19 rad/s.

21–70. $A_y = -B_y = 13.7$ kN.

21–71. 3.63 (10^3) rad/s.

21–73. 77.7 lb.

21–74. 107 N.

21–75. 0.820 rad/s.

21–78. 2.33 rev/h.

21–79. 0.673° regular precession.

21–81. 12.5 Mg $\cdot$ m^2/s.

CHAPTER 22

22-1. $\tau = 0.452$ s, $f = 2.21$ Hz.

22-2. $f = \dfrac{1}{\pi}\sqrt{\dfrac{k}{m}}$.

22-3. $x = -0.05\cos(12.2t)$ m.

22-5. $x = (-0.126\sin(3.16t) - 0.09\cos(3.16t))$ m, $C = 0.155$ m.

22-6. $y = (0.0833\cos 19.7t)$ ft, $C = 1$ in., $f = 3.13$ Hz.

22-7. $y = (-1.44\sin 13.9t - 0.167\cos 13.9t)$ ft, $C = 1.45$ ft.

22-9. $x = (-0.534\sin(7.49\,t) + 0.06\cos(7.49t))$ m, $C = 0.538$ m.

22-10. $\theta = (-0.101\sin(4.95\,t) + 0.3\cos(4.95t))$ rad.

22-11. $16.3°$.

22-13. $\tau = 2\pi\sqrt{\dfrac{k_G{}^2 + d^2}{gd}}$.

22-14. 21.2 kg, 609 N/m.

22-15. $\tau = 2\pi\sqrt{\dfrac{2r}{g}}$.

22-17. 1.45 s.

22-18. $\theta = 0.4\cos 20.7t$.

22-19. 2.50 Mg $\cdot$ m^2.

22-21. 0.167 s.

22-22. 2.67 s.

22-23. 90.8 lb $\cdot$ ft/rad.

22-25. 0.910 s.

22-26. $x = 0.167\cos 6.55t$.

22-27. $\tau = 2\pi\sqrt{\dfrac{l}{2g}}$.

22-29. $f = \dfrac{1}{2\pi}\sqrt{\dfrac{2}{3}\left(\dfrac{g}{R - r}\right)}$.

22-30. $f = 0.597$ Hz.

22-31. $\tau = 6.10\sqrt{\dfrac{a}{g}}$.

22-33. $\tau = 2\pi\sqrt{\dfrac{2r}{g}}$.

22-34. 1.18 s.

22-35. $\tau = \pi\sqrt{\dfrac{m}{k}}$.

22-37. 2.25 Hz.

22-38. 0.339 s.

22-39. 1.52 s.

22-41. 0.970 s.

22-42. 0.851 Hz.

22-43. 0.487 s.

22-45. 2.81 s.

22-46. $y = A\sin pt + B\cos pt + \dfrac{F_0/m}{(k/m - \omega^2)}\cos \omega t$.

22-47. 0.685 ft/s.

22-49. $y = (361\sin 7.75t + 100\cos 7.75t - 350\sin 8t)$ mm.

22-50. $y = (-16.6\sin 7.07t + 70\cos 7.07t + 29.4\sin 4t)$ mm.

22-51. 19.7 rad/s.

22-53. 19.0 rad/s.

22-57. $x_P = (0.0276\sin 2t)$ ft.

22-58. 6.55 rad/s.

22-59. 0.625 in.

22-61. 2.83 rad/s.

22-62. 5.95 mm.

22-63. 1.20 m/s.

22-65. $9.89°$.

22-66. 0.997.

22-67. 1.66 s.

22-69. $1.54°$.

22-70. 1.01.

22-71. 0.0269 in.

22-73. 0.666 s.

22-74. $x = 0.119\cos(6t - 83.9°)$ m.

22-75. $1.55\ddot{\theta} + 540\dot{\theta} + 200\theta = 0$.

22-78. 18.9 N $\cdot$ s/m.

22-79. 714 N/m.

22-81. Overdamped, $\ddot{y} + 16\dot{y} + 4y = 0$.

22-82. $m\ddot{x} + kx = F_0\sin\omega t$, $L\ddot{q} + \dfrac{1}{c}q = E_0\sin\omega t$.

22-83. $L\ddot{q} + R\dot{q} + \dfrac{1}{c}q = 0$.

22-85. $L\ddot{q} + R\dot{q} + \dfrac{1}{c}q = E_0\cos\omega t$.

Index

Geometric Properties of Line and Area Elements

Centroid Location	Centroid Location	Area Moment of Inertia

Circular arc segment

Circular sector area

$$I_x = \tfrac{1}{4}r^4(\theta - \tfrac{1}{2}\sin 2\theta)$$
$$I_y = \tfrac{1}{4}r^4(\theta + \tfrac{1}{2}\sin 2\theta)$$

Quarter and semicircular arcs

Quarter circular area

$$I_x = \tfrac{1}{16}\pi r^4$$
$$I_y = \tfrac{1}{16}\pi r^4$$

Trapezoidal area

Semicircular area

$$I_x = \tfrac{1}{8}\pi r^4$$
$$I_y = \tfrac{1}{8}\pi r^4$$

Semiparabolic area

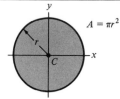

Circular area

$$I_x = \tfrac{1}{4}\pi r^4$$
$$I_y = \tfrac{1}{4}\pi r^4$$

Exparabolic area

Rectangular area

$$I_x = \tfrac{1}{12}bh^3$$
$$I_y = \tfrac{1}{12}hb^3$$

Parabolic area

Triangular area

$$I_x = \tfrac{1}{36}bh^3$$

Center of Gravity and Mass Moment of Inertia of Homogeneous Solids

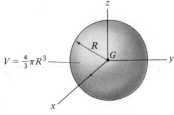

Sphere

$$V = \tfrac{4}{3}\pi R^3$$

$$I_{xx} = I_{yy} = I_{zz} = \tfrac{2}{5}mR^2$$

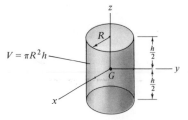

Cylinder

$$V = \pi R^2 h$$

$$I_{xx} = I_{yy} = \tfrac{1}{12}m(3R^2 + h^2) \qquad I_{zz} = \tfrac{1}{2}mR^2$$

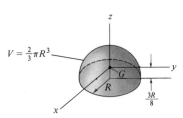

Hemisphere

$$V = \tfrac{2}{3}\pi R^3$$

$$I_{xx} = I_{yy} = 0.259mR^2 \qquad I_{zz} = \tfrac{2}{5}mR^2$$

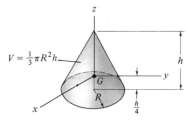

Cone

$$V = \tfrac{1}{3}\pi R^2 h$$

$$I_{xx} = I_{yy} = \tfrac{3}{80}m(4R^2 + h^2) \qquad I_{zz} = \tfrac{3}{10}mR^2$$

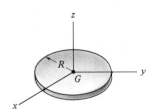

Thin circular disk

$$I_{xx} = I_{yy} = \tfrac{1}{4}mR^2 \qquad I_{zz} = \tfrac{1}{2}mR^2$$

Thin plate

$$I_{xx} = \tfrac{1}{12}mb^2 \qquad I_{yy} = \tfrac{1}{12}ma^2 \qquad I_{zz} = \tfrac{1}{12}m(a^2 + b^2)$$

Thin ring

$$I_{xx} = I_{yy} = \tfrac{1}{2}mR^2 \qquad I_{zz} = mR^2$$

Slender rod

$$I_{xx} = I_{yy} = \tfrac{1}{12}ml^2 \qquad I_{x'x'} = I_{y'y'} = \tfrac{1}{3}ml^2 \qquad I_{zz} = 0$$